P9-AFF-676

Biotechnology

Second Edition

Volume 10

Special Processes

Biotechnology

Second Edition

Fundamentals

Volume 1
Biological Fundamentals

Volume 2
Genetic Fundamentals and
Genetic Engineering

Volume 3
Bioprocessing

Volume 4
Measuring, Modelling and Control

Products

Volume 5a
Recombinant Proteins, Monoclonal
Antibodies and Therapeutic Genes

Volume 5b
Genomics and Bioinformatics

Volume 6
Products of Primary Metabolism

Volume 7
Products of Secondary Metabolism

Volumes 8a and b
Biotransformations I and II

Special Topics

Volume 9
Enzymes, Biomass, Food and Feed

Volume 10
Special Processes

Volumes 11a–c
Environmental Processes I–III

Volume 12
Legal, Economic and
Ethical Dimensions

A Multi-Volume Comprehensive Treatise

Biotechnology

Second, Completely Revised Edition

Edited by
H.-J. Rehm and G. Reed
in cooperation with
A. Pühler and P. Stadler

Volume 10

Special Processes

Edited by
H.-J. Rehm

Weinheim · New York · Chichester · Brisbane · Singapore · Toronto

Series Editors:
Prof. Dr. H.-J. Rehm
Institut für Mikrobiologie
Universität Münster
Correnstraße 3
D-48149 Münster
FRG

Dr. G. Reed
1029 N. Jackson St. #501-A
Milwaukee, WI 53202-3226
USA

Prof. Dr. A. Pühler
Biologie VI (Genetik)
Universität Bielefeld
P.O. Box 100131
D-33501 Bielefeld
FRG

Prof. Dr. P. J. W. Stadler
Artemis Pharmaceuticals
Geschäftsführung
Pharmazentrum Köln
Neurather Ring
D-51063 Köln
FRG

Volume Editor:
Prof. Dr. H.-J. Rehm
Institut für Mikrobiologie
Universität Münster
Correnstraße 3
D-48149 Münster
FRG

Library of Congress Card No.: applied for

British Library Cataloguing-in-Publication Data:
A catalogue record for this book is available from the British Library

Die Deutsche Bibliothek – CIP-Cataloguing-in-Publication Data:

A catalogue record for this book
is available from Die Deutsche Bibliothek
ISBN 3-527-28320-X

Printed on acid-free paper.

Composition and Printing: Zechner, Datenservice und Druck, D-67346 Speyer.
Bookbinding: J. Schäffer, D-67269 Grünstadt.
Printed in the Federal Republic of Germany

Preface

In recognition of the enormous advances in biotechnology in recent years, we are pleased to present this Second Edition of "Biotechnology" relatively soon after the introduction of the First Edition of this multi-volume comprehensive treatise. Since this series was extremely well accepted by the scientific community, we have maintained the overall goal of creating a number of volumes, each devoted to a certain topic, which provide scientists in academia, industry, and public institutions with a well-balanced and comprehensive overview of this growing field. We have fully revised the Second Edition and expanded it from ten to twelve volumes in order to take all recent developments into account.

These twelve volumes are organized into three sections. The first four volumes consider the fundamentals of biotechnology from biological, biochemical, molecular biological, and chemical engineering perspectives. The next four volumes are devoted to products of industrial relevance. Special attention is given here to products derived from genetically engineered microorganisms and mammalian cells. The last four volumes are dedicated to the description of special topics.

The new "Biotechnology" is a reference work, a comprehensive description of the state-of-the-art, and a guide to the original literature. It is specifically directed to microbiologists, biochemists, molecular biologists, bioengineers, chemical engineers, and food and pharmaceutical chemists working in industry, at universities or at public institutions.

A carefully selected and distinguished Scientific Advisory Board stands behind the series. Its members come from key institutions representing scientific input from about twenty countries.

The volume editors and the authors of the individual chapters have been chosen for their recognized expertise and their contributions to the various fields of biotechnology. Their willingness to impart this knowledge to their colleagues forms the basis of "Biotechnology" and is gratefully acknowledged. Moreover, this work could not have been brought to fruition without the foresight and the constant and diligent support of the publisher. We are grateful to VCH for publishing "Biotechnology" with their customary excellence. Special thanks are due to Dr. Hans-Joachim Kraus and Karin Dembowsky, without whose constant efforts the series could not be published. Finally, the editors wish to thank the members of the Scientific Advisory Board for their encouragement, their helpful suggestions, and their constructive criticism.

H.-J. Rehm
G. Reed
A. Pühler
P. Stadler

Scientific Advisory Board

Contents

Contributors

Prof. Dr. Garabed Antranikian
Technische Universität Hamburg-Harburg
Arbeitsbereich 2-09
Denickestraße 15
D-21071 Hamburg
Germany
Chapter 4

Dr. Claudia Bardouille
SIGMA-ALDRICH Chemie GmbH
Eschenstraße 5
D-82024 Taufkirchen
Germany
Chapter 2

Dr. Constanzo Bertoldo
Technische Universität Hamburg-Harburg
Arbeitsbereich 2-09
Denickestraße 15
D-21071 Hamburg
Germany
Chapter 4

Prof. Dr. Dr. Ralf G. Berger
Institut für Lebensmittelchemie
Universität Hannover
Wunstorfer Straße 14
D-30453 Hannover
Germany
Chapter 13

Dr. Helmut Brandl
Institut für Naturwissenschaften
Universität Zürich
Winterthurer Strasse 190
CH-8057 Zürich
Switzerland
Chapter 8

Dr. Hubert Cramer
Lehrstuhl für Biotechnologie
Universität Würzburg
Biozentrum
Am Hubland
D-97074 Würzburg
Germany
Chapter 19

Dr. René Fakoussa
Institut für Mikrobiologie und Biotechnologie
Universität Bonn
Meckenheimer Allee 168
D-53115 Bonn
Germany
Chapter 7

Dr. Klaus Frobel
Medizinische Chemie
Pharmaforschungszentrum
Bayer AG
D-42096 Wuppertal
Germany
Chapter 3

Dr. Günter Fuhr
Lehrstuhl für Membranphysiologie
Institut für Biologie
Humboldt-Universität zu Berlin
D-10115 Berlin
Germany
Chapter 19

Dr. Christian Hasse
Klinik für Allgemeinchirurgie
Universität Marburg
D-53033 Marburg
Germany
Chapter 19

Prof. Dr. Geoffrey M. Gadd
Department of Biological Sciences
University of Dundee
Dundee DD1 4HN
Scotland
Chapter 9

Dr. Martin Hofrichter
Department of Applied Chemistry and Microbiology
University of Helsinki, Viikki Biocenter
P.O. Box 56
Fin-00014 Helsinki
Finland
Chapter 7

Prof. Rajesh S. Gokhale
National Institute of Immunology
Aruna Asaf Ali Marg
New Delhi 110067
India
Chapter 12

Dr. Udo Hölker
Botanisches Institut
Universität Bonn
Kirschallee 1
D-53115 Bonn
Germany
Chapter 7

Dr. Wolfgang Groß
Institut für Pflanzenphysiologie und Mikrobiologie
Freie Universität Berlin
Königin-Luise-Straße 12–16a
D-14195 Berlin
Germany
Chapter 5

Dr. Anette Jork
Lehrstuhl für Biotechnologie
Universität Würzburg
Biozentrum
Am Hubland
D-97074 Würzburg
Germany
Chapter 19

Dr. Ralf Grote
Technische Universität Hamburg-Harburg
Arbeitsbereich 2-09
Denickestraße 15
D-21071 Hamburg
Germany
Chapter 4

Prof. Dr. Jürgen Klein
Heideweg 10
D-45529 Hattingen
Germany
Chapter 7

Prof. Dr. Ulrike Lindequist
ImaB Institut für Marine Biotechnologie e.V.
Walter-Rathenau-Straße 49a
D-17489 Greifswald
Germany
Chapter 15

Dr. Susanne Metzger
Medizinische Chemie
Pharmaforschungszentrum
Bayer AG
D-42096 Wuppertal
Germany
Chapter 3

Dr. Frank Niepold
BBA für Land- und Forstwirtschaft
Institut für Pflanzenschutz in Ackerbau und Grünland
Messeweg 11/12
D-38104 Braunschweig
Germany
Chapter 16

Dr. Otto Pulz
IGV – Institut für Getreideverarbeitung
Arthur-Schemmert-Allee 40–41
D-14558 Potsdam-Rehbrücke
Germany
Chapter 5

Prof. Dr. Hans-Jürgen Rehm
Institut für Mikrobiologie
Universität Münster
Corrensstraße 3
D-48149 Münster
Germany
Chapter 20

Dr. Matthias Rothmund
Klinik für Allgemeinchirurgie
Universität Marburg
D-53033 Marburg
Germany
Chapter 19

Dr. Klaus Rudolph
Institut für Pflanzenpathologie und Pflanzenschutz
Universität Göttingen
Grisebachstraße 6
D-37077 Göttingen
Germany
Chapter 16

Dr. Wolfgang Sand
Institut für Allgemeine Botanik und Botanischer Garten
Abteilung Mikrobiologie
Universität Hamburg
Ohnhorststraße 18
D-22609 Hamburg
Germany
Chapter 10

Dr. Karl Scheibenbogen
IGV – Institut für Getreideverarbeitung
Arthur-Schemmert-Allee 40–41
D-14558 Potsdam-Rehbrücke
Germany
Chapter 5

Dr. Helmut Schmiers
Institut für Energieverfahrenstechnik und Chemieingenieurwesen
Reiche Zeche
TU Bergakademie Freiberg
D-09599 Freiberg
Germany
Chapter 7

Prof. Dr. Heide Schnabl
Institut für Landwirtschaftliche Botanik
Universität Bonn
Karlrobert-Kreiten-Straße 13
D-53115 Bonn
Germany
Chapter 17

Dr. Jens Schrader
Karl-Winnacker-Institut
DECHEMA e. V.
Theodor-Heuss-Allee 25
D-60486 Frankfurt/Main
Germany
Chapter 13

Dr. Thomas Schweder
Institut für Marine Biotechnologie e.V.
Walter-Rathenau-Str. 49a
D-17489 Greifswald
Germany
Chapter 15

Dr. Dieter Sell
DECHEMA e.V.
Theodor-Heuss-Allee 25
D-60486 Frankfurt/Main
Germany
Chapter 1

Prof. Dr. Sakayu Shimizu
Division of Applied Life Sciences
Graduate School of Agriculture
Kyoto University
Kitashirakawa-Oiwakecho
Sakyo-ku, Kyoto 606-8502
Japan
Chapter 11

Dr. Christoph Sinder
DMT – Gesellschaft für Forschung und Prüfung mbH
Am Technologiepark 1
D-45307 Essen
Germany
Chapter 7

Prof. Dr. Alexander Steinbüchel
Institut für Mikrobiologie
Universität Münster
Corrensstraße 3
D-48149 Münster
Germany
Chapters 7 and 14

Dr. Anna Suurnäkki
Tietotie 2
Espoo
P.O. Box 1500
Fin-02044 VTT
Finland
Chapter 18

Dr. Maija Tenkanen
Tietotie 2
Espoo
P.O. Box 1500
Fin-02044 VTT
Finland
Chapter 18

Dr. Frank Thürmer
Lehrstuhl für Biotechnologie
Universität Würzburg
Biozentrum
Am Hubland
D-97074 Würzburg
Germany
Chapter 19

Prof. Dr. Arno Tiedtke
Institut für Allgemeine Zoologie und Genetik
Universität Münster
Schloßplatz 5
D-48149 Münster
Germany
Chapter 6

Dr. Dipika Tuteja
National Institute of Immunology
Aruna Asaf Ali Marg
New Delhi 110067
India
Chapter 12

Prof. Dr. Liisa Viikari
Tietotie 2
Espoo
P.O. Box 1500
Fin-02044 VTT
Finland
Chapter 18

Dr. Heiko Zimmermann
Lehrstuhl für Membranphysiologie
Institut für Biologie
Humboldt-Universität zu Berlin
D-10115 Berlin
Germany
Chapter 19

Prof. Dr. Ulrich Zimmermann
Lehrstuhl für Biotechnologie
Universität Würzburg
Am Hubland
Biozentrum
D-97074 Würzburg
Germany
Chapter 19

Introduction

HANS-JÜRGEN REHM
Münster, Germany

This volume presents various biotechnological processes beyond the subject matter of the preceding volumes. The enormous progress in biotechnology during the past few years made it impossible to describe all fields of biotechnology in the other volumes. This volume concentrates on those special areas of microbiology primarily treated in Volume 6b (1988) and also in other volumes of the 1st Edition of *Biotechnology*.

Chapter 1 focuses on the production of energy by bioelectrical fuel cells – currently a field of enormous interest.

Chapters 2 and 3 explore the conservation of human and animal cells. Here the pertinent problems are of special interest to scientists working in medical areas of biotechnology. The biotechnological screening methods elucidated here represent the most up-to-date state of research.

Chapters 4–6 concentrate on microorganisms with a wide potential for new applications which have not been exhausted yet. For example, biotechnology using protozoa promises to develop into a highly important area in future research. Extremophilic bacteria, cyanobacteria, and microalgae have shown great potential for biotechnological applications.

Chapters 7–10 treat the enormous progress in the field of inorganic biotechnology.

Chapters 11–14 delve into the biotechnological production of vitamins, macrolides, special aromatic substances, and microbially degradable biopolyesters. These treatments fulfill both: complementation of earlier contributions to this topic in Volumes 7 and 8b and presentation of even more recent advances which have not been incorporated there.

Chapters 15–20 explore special processes within biotechnology. They incorporate new developments since the publication of the 1st Edition of *Biotechnology*, especially in the areas of marine biotechnology, biotechnology in space, biotechnology in the paper industry, but also in the new field of implanting endocrine cells by immobilization. In the final chapter I briefly describe other fields of biotechnological research which may prove to be of importance for future industrial applications of biotechnology.

As it is impossible to present all the burgeoning fields of importance in biotechnology such as biodiversity, nanotechnology, biotechnological analysis (e.g., DNA chip technology), novel foods, separate supplements to the *Biotechnology* series will be devoted to these and other topics after the completion of the 2nd Edition.

The fields of bionics will come up to transfer natural processes into industrial application. The so-called "lotus effect" (adapting the hydrophobic structure of *Lotus* leaves for industrial exploitation) is a classical example of bionics, which has become the domain of a new field of scientific research closely related to biotechnology with its own specialized scientific society.

The authors of this volume have worked hard to make their extensive knowledge and expertise accessible to the readers. I hope that this volume will stimulate reader interest in these newly emerging areas of biotechnology.

I wish to extend my warmest thanks to my friend and colleague Dr. GERALD REED for his invaluable editorship and to Ms. KARIN DEMBOWSKY at WILEY-VCH for her consistently helpful and constructive support in ushering this volume into print.

Münster, February 2001 Hans-Jürgen Rehm

Energy by Microbial Processes

1 Bioelectrochemical Fuel Cells

DIETER SELL
Frankfurt/Main, Germany

1 Introduction

Today only the direct incineration or gasification of biomass is of practical importance for the use of biological resources for industrial energy generation. In a broader sense the incineration of non-regenerative, fossil sources of energy – coal, petroleum, and natural gas – also represents an energetic use of biological metabolism (OSTEROTH, 1989).

A further possibility will be illustrated in this chapter: the application of fuel cell devices which are powered by biological redox reactions, performed by intact microorganisms or isolated enzymes.

Typically, a fuel cell works by the oxidation of a reduced fuel at an anode with concomitant transfer of electrons, via a circuit, to a suitable electron acceptor molecule, e.g., oxygen, at the cathode. The power output of the fuel cell depends on the number and rate of transfer of electrons and on the potential difference between anode and cathode. In theory, fuel cells are the most effective devices for the conversion of chemical energy to electrical energy because they avoid the limitations of the Carnot cycle (BOCKRIS and REDDY, 1970).

Microorganisms are able to utilize a vast range of organic and inorganic substances, which can be regarded as potential fuels for combustion in a fuel cell. The direct combustion of these substances cannot be performed due to lack of suitable electrocatalysis. Microorganisms dispose of a large set of enzymes for the highly efficient and controlled oxidation of different substrates with concomitant electron transfer by way of a sequence of reactions to a terminal electron acceptor (oxygen for aerobic catabolism). The key enzymes for these processes are the oxidoreductases, which catalyze the electron transfer for the redox reactions involved, a property that is highly desirable for good fuel cell performance.

2 History of Bioelectrochemical Fuel Cells

The concept of "bioelectrochemical fuel cells" was extremely popular in the early 1960s when it was taken up by NASA (National Aeronautics and Space Administration, USA). NASA's interest in studying the production of electricity by means of biochemical reactions stemmed from the association with the problems of waste management in the closed system of a space shuttle during longer space flights: human waste should be reprocessed in order to attain a closed or nearly closed ecology in the spacecraft (LEWIS, 1966). Therefore, algae and bacteria were among the first organisms used in this phase of experiments with bioelectrochemical fuel cells. Even in current publications this concept of electricity generation from waste materials is proposed for outer space applications (PARK and ZEIKUS, 2000).

In 1963 bioelectrochemical fuel cells were already commercially available and they were promoted as power sources for radios, powered radio beacons, signal lights and other apparatus at sea (Anonymous, Financial Times, 1963a, b). However, these fuel cells were not a commercial success and they soon disappeared from the market. With the successful development of technical alternatives, e.g., photovoltaic technologies for the energy supply on space flights and later on for many different fields of application, interest in bioelectrochemical systems declined for a while until there was a revival of interest in the late 1970s/early 1980s, stimulated by the bottleneck in world oil supplies and the growing importance of biotechnology as a field with broad industrial applications.

The generation of electric energy was, of course, just one aspect of the research activities engaged in by groups all over the world (VIDELA and ARIVA, 1975; HIGGINS and HILL, 1979; TANAKA et al., 1983). However, it was soon realized that bioelectrochemical fuel cells were also promising for a different application in the area of measurement and control due to the fact that the voltage or current generated by such a fuel cell is directly dependent on the activity of the biological component, be it an isolated enzyme or an intact microorganism.

Work on bioelectrochemical fuel cells, therefore, represents the first systematic approach to investigating the interaction of biological systems with electrodes and for this reason can be regarded as the forerunner of the development of electrochemical biosensors, bioelectrochemical syntheses, and modern bioelectronics.

The first experiments in the field of bioelectrochemical fuel cells were already performed in 1911. It was POTTER who first postulated a relationship between electrode potentials and microbial activity (POTTER, 1911). POTTER measured the potential difference between two electrodes, one of which was inserted into a bacterial culture and the other into a sterile culture medium. He actually made what can be considered to be the first biochemical fuel cell battery by assembling six cells, each of which consisted of a yeast–glucose half-cell and a glucose half-cell without microorganisms. Similarly, in 1931 COHEN studied the potential differences arising between various cultures and sterile media; he also built a bacterial battery which produced a small current for a short period of time. Such an experimental set-up is shown in Fig. 1. COHEN found that the potential of a vigorously growing bacterial culture amounted to 0.5–1 V over the control medium. The greatest deficiency of the microbial half-cell was that its current output was generally very low (10^{-5}–10^{-6} A).

3 Global Dimensions of Bioenergetics

From an energetic standpoint, the existence of the whole terrestrial biosphere, including man, depends solely on extraterrestrial solar radiation. The decisive factor is that it is not solar energy itself that is the driving force for all life processes, but the conversion of electromagnetic radiation into electrochemical, and ultimately chemical, energy that takes place in living organisms (RENGER, 1983).

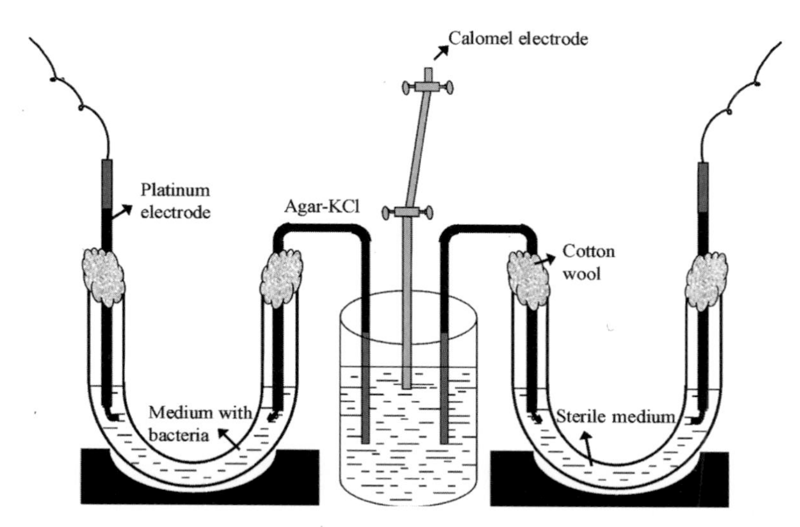

Fig. 1. Early bioelectrochemical fuel cell developed by COHEN (1931). The half-cell on the left side contained a glucose medium with a growing bacterial culture; the half cell on the right side just contained a sterile glucose medium. Both half-cells were connected via an agar-KCl bridge (slightly modified according to COHEN, 1931).

Photoautotrophic organisms are capable of this conversion. They can use sunlight in the visible range (400–800 nm) to synthesize biomass, a process referred to as photosynthesis (ZIEGLER, 1983).

The crucial phase of this fundamental bioenergetic reaction took place during evolution in the cyanophytic stage. Cyanobacteria developed a complex molecular system enabling water to degrade into molecular oxygen by means of sunlight on the one hand, and into hydrogen bonded to suitable carriers by chemical means on the other hand.

One consequence was that the hitherto anaerobic atmosphere of the earth became aerobic.

This was advantageous for the group of chemoorganotrophic organisms which cannot use sunlight directly as a source of energy; instead they metabolize exogenic nutrients from which they derive energy to sustain their vital functions.

Oxygen provided these organisms with a reactant which facilitated a substantially more effective exploitation of substrates.

Thus, for instance, under anaerobic conditions glucose can only be degraded to ethanol and carbon dioxide by yeasts, producing 197 kJ mol^{-1}. Under aerobic conditions it can be completely degraded into carbon dioxide and water, supplying the organism with 2,874 kJ mol^{-1} (LEHNINGER, 1977).

Only the availability of oxygen, permitting an increased energy yield, enabled heterotrophic organisms to develop highly organized mobile forms of life with a high energy turnover.

This bioenergetic principle was strongly favored throughout the further evolution processes: With all forms of life above the evolution stage of cyanobacteria and with the overwhelming majority of bacteria, the central system for the realization of fundamental bioenergetic reactions that has prevailed is the oxidation of carrier-bonded hydrogen using oxygen as the oxidant with the concomitant formation of water. Sunlight causes water to be broken down by photosynthesis by which carrier-bonded hydrogen and oxygen are produced. This process is reversed by heterotrophic processes consuming oxygen, and water is produced again (Fig. 2).

It will be shown that the universality of this metabolic concept is very advantageous for

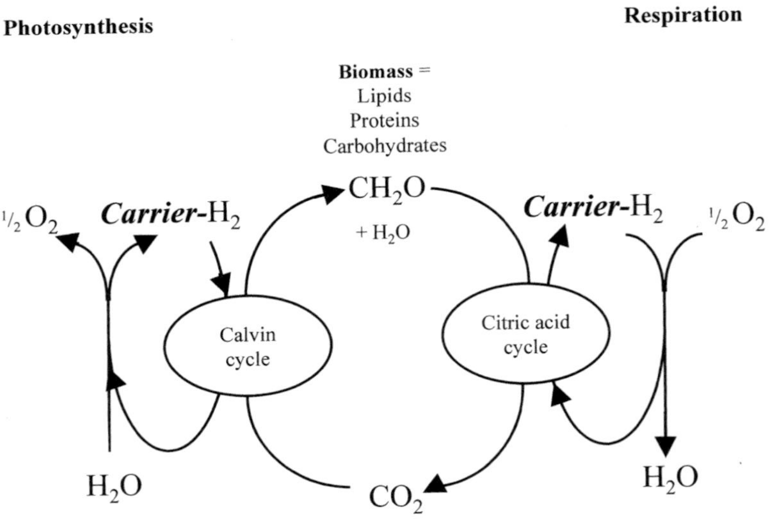

Fig. 2. The system of carrier-bonded hydrogen as fundamental bioenergetic principle: Photosynthesis as a source and heterotrophic processes as sink for this universal energy carrier which can be used directly for intracellular energy production in the catabolic metabolism or can be used for anabolic processes.

the development of bioelectrochemical fuel cells. Some forms of carrier-bonded hydrogen formed during metabolic processes are suitable forms of fuel for combustion in a bioelectrochemical fuel cell or can easily be transformed into a combustional form. Due to this fact, potentially a broad spectrum of microorganisms, heterotrophic as well as (photo-) autotrophic, can be used as biocatalysts for energy production by means of a fuel cell.

With an estimated annual conversion of 200–300 Gt (gigaton = 10^9 tons) carbon, photosynthesis and its heterotrophic conversion are the quantitatively most important chemical processes on the earth's surface (KRIEB, 1981).

Considered on a global scale, when biomass builds up by photosynthesis the degree of effectiveness is not high. Approximately 0.12% of solar radiation energy on the whole earth is converted into chemically bonded energy in the form of biomass. Nevertheless approximately 170 Gt per year of biomass accumulate from the photosynthesis of all land and water plants. The energy content of this biomass amounts to about 3,000 EJ (Exa Joule) = $3 \cdot 10^{21}$ J, corresponding to tenfold of mankind's annual energy needs (GRATHWOHL, 1983).

The orders of magnitude of these figures explain why diverse investigations are being conducted to determine whether and how bioenergetic reactions can be relied on as a basis for industrial energy production.

4 A Brief Introduction to Fuel Cells

A fuel cell is a device for the direct energy conversion of chemical energy to electrical energy. It requires an anode, a cathode, a supporting electrolyte medium to connect the two electrodes, and an external circuit to utilize the energy. Reactants must be supplied to both electrodes as a source for the electron transfer reactions; catalysts must be present to provide a rapid rate of reaction at each electrode.

To understand the principle of bioelectrochemical fuel cells a look at the simplest and best known of all technical fuel cells will help: the hydrogen–oxygen fuel cell, which today can already be found as a prototype in cars, buses, etc.

4.1 The Hydrogen–Oxygen Fuel Cell

The hydrogen–oxygen fuel cell is the most highly developed fuel cell, since hydrogen is the most reactive fuel known. The principle of operation of such a fuel cell is illustrated in Fig. 3. This cell operates with hydrogen gas as the fuel and oxygen gas as the oxidant. The fuel cell consists of two electrodes, an anode at which oxidation occurs and a cathode at which reduction occurs. The electrodes are in contact with an electrolyte. Often an ion-exchange membrane separates the anode and cathode compartments. In operation, the fuel, hydrogen gas, passes over the surface of the anode and is electrochemically oxidized to hydrogen ions, which enter the electrolyte and migrate towards the cathode. Oxygen gas passes over the surface of the cathode and is reduced, combining with the hydrogen ions from the elec-

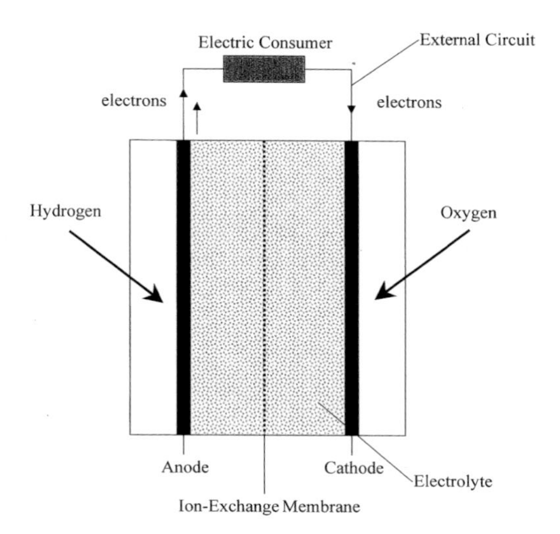

Fig. 3. Principle of a hydrogen–oxygen fuel cell.

trolyte to form water. The net result of the operation of the fuel cell is the combination of hydrogen and oxygen to form water with electrons flowing through the external circuit. This flow of electrons can be utilized to power a machine, a light bulb, or anything that runs on electricity. Other fuels for technical fuel cells are methane, natural gas, carbon monoxide, methanol, etc. More details about the state of the art in the field of fuel cell research are summarized by KORDESCH and SIMADER (1996).

The electricity output of the hydrogen–oxygen fuel cell is determined by the electrode reactions, the fuel oxidation at the anode and the oxidant reduction at the cathode. The reactions in a hydrogen–oxygen fuel cell (with an acid electrolyte) are the following:

Cathode: $O^2 + 4H^+ + 4e^- \rightarrow 2H_2O$ $\quad E^0 = 1.23\ \text{V}$ (1)

Anode: $H_2 \rightarrow 2H^+ + 2e^-$ $\quad E^0 = 0\ \text{V}$ (2)

Total: $2H_2 + O_2 \rightarrow 2H_2O$ $\quad \Delta E^0 = 1.23\ \text{V}$ (3)

E^0 represents the standard half-cell potential, assuming reversible operation.

The theoretical amount of useful work, W_u, that can be obtained depends on the net release of Gibbs' free energy, ΔG, as the reaction proceeds from reactants to products. Both values are equal and are related to the net reversible potential difference ΔE^0 for the anodic and cathodic half-cell reactions according to:

$$-\Delta G = W_u = n\ F\ \Delta E^0 \tag{4}$$

In this equation n is the number of electrons transferred per mole of fuel and F the Faraday constant (96,500 coulombs per mole).

With regard to the first law of thermodynamics, the free energy is related to the change in enthalpy, ΔH, and the change in entropy, ΔS, at constant temperatures, T:

$$\Delta G = \Delta H - T\ \Delta S \tag{5}$$

From this, the efficiency of a fuel cell can be calculated as follows:

$$\text{Efficiency} = \frac{\Delta G}{\ } = 1 - \frac{T\ \Delta S}{\ } \tag{6}$$

The theoretical efficiency of fuel cell systems can be as high as 85–90%; this is considerably higher than the 40–60% theoretical efficiencies for the mechanical-thermal energy transfer in a combustion engine. The reason is that the fuel cell reactions are not coupled with an increase in temperature, so that the $T\ \Delta S$ term amounts to only ~10% of ΔH (BOCKRIS and REDDY, 1970).

5 Principle of Bioelectrochemical Fuel Cells

In a bioelectrochemical fuel cell the electrode reactions are similar to those in the hydrogen–oxygen fuel cell, except for the fact that the fuel is not hydrogen gas but a form of carrier-bonded hydrogen, produced by physiological redox reactions. If we again look at Fig. 2 and recall the metabolic concept of carrier-bonded hydrogen as fuel for the intracellular generation of energy, we can draw interesting parallels between the reaction in a hydrogen–oxygen fuel cell, a bioelectrochemical fuel cell, and even the catabolic reactions in a living organism.

The overall reaction of a hydrogen–oxygen fuel cell and a bioelectrochemical fuel cell [with the coenzyme NAD (nicotine adenine dinucleotide) as the assumed hydrogen carrier] are:

Hydrogen–oxygen fuel cell:

$H_2 + {}^1/_2\,O_2 = H_2O$ $\quad E^0 = 1.23\ \text{V}$, $\Delta G = 237\ \text{kJ mol}^{-1}$ (7)

Bioelectrochemical fuel cell:

$\textit{Carrier}\text{-}H_2 + {}^1/_2\,O_2 = H_2O + \textit{Carrier}$

$E^0 = 1.14\ \text{V}$, $\Delta G = 220\ \text{kJ mol}^{-1}$ (8)

It is remarkable that the combustion of hydrogen bonded to NAD supplies only slightly less energy than the combustion of hydrogen gas. Another interesting fact is that the overall reaction for the processes in a bioelectrochemical fuel cell is the same as for the metabolic combustion of $NADH_2$ (the hydrogen-loaded form of NAD) via the respiratory chain.

This parallel led to the conclusion that a fuel cell is nothing but a functional technical construction analogous to the energy-supplying apparatus that works within every living cell. The following quotation from the early days of bioelectrochemical fuel cell research summarizes this as follows: "Perhaps the most highly refined fuel cell system today is a living organism, a mechanism that catalytically (enzymes) burns (oxidizes) food (fuel) in an electrolyte (cytoplasm or blood) to produce energy, some of which is electric" (LEWIS, 1966).

The amounts of intracellular electricity may be described by the following example from our own metabolism: the entire current in all the mitochondria of a resting, adult human being is 80–100 A. With a given potential difference of 1.14 V ($NADH_2$–O_2) this results in power of approximately 100 W. This demonstrates the fact that the possibility of converting chemical energy from biomass (e.g., energy crops) or even carbohydrate wastes into electric energy by means of living organisms in fuel cell devices has a sound foundation and opens up a perspective for a regenerative energy source based on biotechnology.

5.1 Classification of Bioelectrochemical Fuel Cells

The concept of a bioelectrochemical fuel cell can be introduced by reference to the analogy to the hydrogen–oxygen fuel cell in Fig. 3. Energy for life processes is derived from the combustion of electron-rich substances or food materials, which are oxidized stepwise by a cascade of redox reactions in the energy metabolism. There are different ways how these redox reactions can be exploited for energy production within a fuel cell.

5.1.1 Indirect Bioelectrochemical Fuel Cells

High-energy, electron-rich substances, such as carbohydrates, lipids, and proteins, are usually not electroactive themselves (they cannot be directly oxidized at an electrode), but some intermediates formed during biological oxidation may often be active at an electrode. Furthermore, some final products of fermentation processes, such as ethanol, ammonia, hydrogen sulfide (which in a broader sense can be regarded as different forms of carrier-bonded hydrogen), or even hydrogen, can serve as fuels in bioelectrochemical fuel cells. In the literature this concept is referred to as an "*indirect (or product) bioelectrochemical fuel cell*" because there is no direct interaction between organism and electrode. The vessel in which the electroactive product is formed by fermentation may even be located away from the bioelectrochemical fuel cell. Thus, certain substrates which are not electroactive (e.g., most carbohydrates) can be transformed into electroactive substances biochemically. It is significant that living organisms are able to make use of vastly more complex organic molecules than a technical fuel cell can handle and to do this under mild conditions of temperature, pressure, and concentration of reactants. Accordingly, microorganisms with their enzyme systems are catalysts which convert a great variety of organic substrates into useful fuels for oxidation in fuel cells.

Early examples of indirect bioelectrochemical fuel cells were described by BERK and CANFIELD (1964). In this research work, photosynthetic bacteria and algae were used as biocatalysts; hydrogen was produced as one of the fermentation products under illumination. This hydrogen was then combusted at the fuel cell anode, where it was oxidized.

Another example described by BRAKE et al. (1963) is the urea fuel cell. This cell is also one of the indirect type in which the urease enzyme generated ammonia from the substrate urea. The urea and the enzyme were in contact with the anode in the fuel cell described and the ammonia that was produced was then electrochemically oxidized at the anode of the fuel cell.

These fuel cells were able to generate currents in the range of microamperes up to a few milliamperes.

5.1.2 Direct Bioelectrochemical Fuel Cells

It has also been demonstrated that the redox reactions of a living cell's energy metabolism can directly contribute to the reaction at the anode of a fuel cell: here, redox enzymes with their cofactors are responsible for an electron flow towards the anode. This kind of interaction between physiological energy metabolism and the electrode of a fuel cell led to the classification as a "direct bioelectrochemical fuel cell". In such a direct cell the biocatalyst has to be close to the electrode and may not be located away from it as with indirect bioelectrochemical fuel cells. ROHRBACK et al. (1962) introduced a direct bioelectrochemical fuel cell with glucose oxidase as the biocatalyst and glucose as the substrate. In contact with the fuel cell's anode, it was postulated that the enzyme directly supplied electrons to the anode while oxidizing glucose to gluconolactone. WINGARD et al. (1982) discussed different research approaches to direct bioelectrochemical fuel cells using either whole microorganisms or isolated enzymes. These authors also compared different models for the interaction of biocatalysts and the electrode surface.

The currents which could be achieved with direct bioelectrochemical fuel cells were in the range of only a few microamperes.

The principle of the anodic reactions in a direct and an indirect bioelectrochemical fuel cell is illustrated in Fig. 4.

Of course, the diversion of electrons from the processes of the physiological energy metabolism in a direct bioelectrochemical fuel cell only occurs to a small extent. Due to steric hindrance most redox enzymes (and whole microorganisms in any case) lack direct elec-

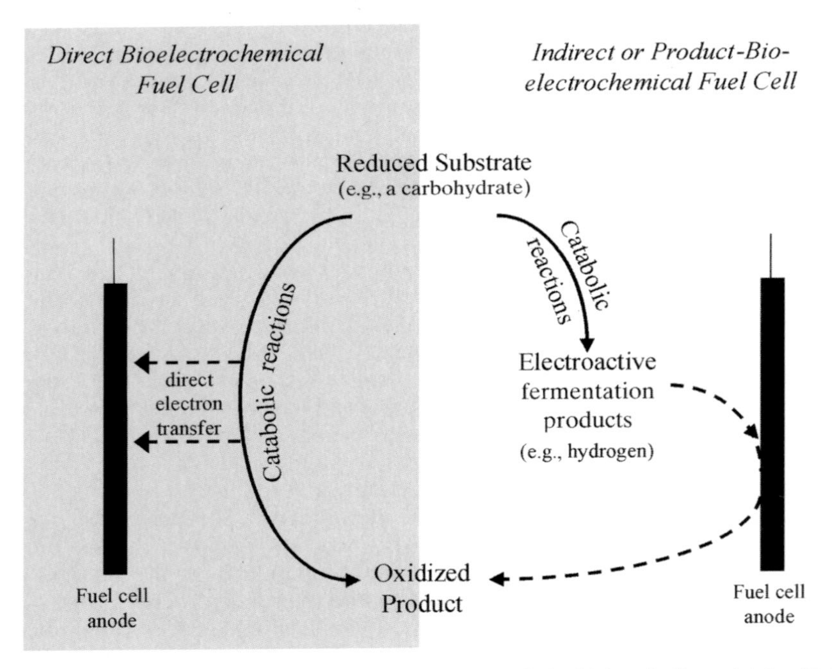

Fig. 4. Principle of the anodic reactions in a direct (left side) or indirect (right side) bioelectrochemical fuel cell.

tron transfer with conductive supports. This is the reason why direct bioelectrochemical fuel cells based on enzyme reactions suffered from low electricity output in the early stages of their development. This is a severe limitation to the current available from this kind of bioelectrochemical fuel cell. The physiological barrier to the direct transfer of electrons from intact microorganisms is the outer cell membrane and the cell wall. If electrons from physiological redox reactions are used for energy production in the fuel cell the organisms themselves cannot derive energy from them. From the standpoint of the microorganisms these barriers prevent them from losing metabolic energy – from the standpoint of fuel cell efficiency these barriers are a serious drawback. Direct bioelectrochemical fuel cells based on intact microorganisms, therefore, also suffered from poor currents and low charge outputs.

5.1.3 Application of Redox Mediators

Substantial progress with direct bioelectrochemical fuel cells was achieved by applying redox mediators, which increased the current available by at least one order of magnitude.

Redox mediators can be used to increase the electron transfer from microorganisms or enzymes to electrodes. Redox mediators are molecules that are able to divert electrons from the metabolic electron transfer systems of organisms or enzymes and convey them to the anode of the fuel cell. Although it was quite early known that certain substances are able to divert electrons from energy metabolism, the systematic application of these mediator substances in bioelectrochemical fuel cells did not begin until the late 1970s/early 1980s.

A suitable mediator for a bioelectrochemical fuel cell should fulfill the following requirements, it should be:

- water-soluble,
- readily reducible by the biocatalyst,
- readily oxidizable at the anode of the fuel cell,
- non-toxic to the biocatalyst,
- not degradable by the biocatalyst,
- a substance with a standard potential which is as negative as possible so that the potential difference between anode and cathode is at its maximum.

One postulated mode of action of redox mediators is illustrated in Fig. 5. In the microbial metabolism carrier-bonded hydrogen in the form of $NADH_2$ is delivered from metabolic processes and is oxidized by NADH dehydrogenase, the first enzyme complex of the membrane-bound respiratory chain. The respiratory chain is a highly developed electron transport system which shuttles the electrons from the hydrogen carriers $NADH_2$ or $FADH_2$ in a sequential and orderly fashion to oxygen as the terminal electron acceptor. Here, the cytochrome oxidase complex catalyzes the reduction of oxygen. The respiratory chain is the central cellular system for generating an organism's metabolic energy in the form of ATP (adenosine triphosphate). (For further details see MITCHELL, 1979 and NICHOLLS, 1982.)

As shown in Fig. 5, a redox mediator is postulated to interact with a component of the respiratory chain, here the coenzyme Q, which is a quinonoid substance.

Fig. 6 shows the structures of two redox mediators, 2-hydroxy-1,4-naphtoquinone (A) and thionine (C). The figure also shows the structure of the quinonoid substance menadione (B), which is a component of the respiratory chain in many bacteria (it substitutes for coenzyme Q which is commonly present in higher organisms and has a similar structure except for a mononuclear quinonoid component). These structural similarities between menadione, the physiological electron transport molecule, and 2-hydroxy-1,4-naphtoquinone (HNQ), the artificial electron-transport molecule utilized for better performance of bioelectrochemical fuel cells, support the theory of interaction between redox mediators and the respiratory chain. The similarities of thionine (C) with the isoalloxazine group (D) of FAD (flavin adenine dinucleotide) or FMN (flavin mononucleotide) also back up this theory, but indicate that the NADH-dehydrogenase of the respiratory chain could be the target for this redox mediator. BENNETTO et al. (1987) report that thionine also reacts rapidly

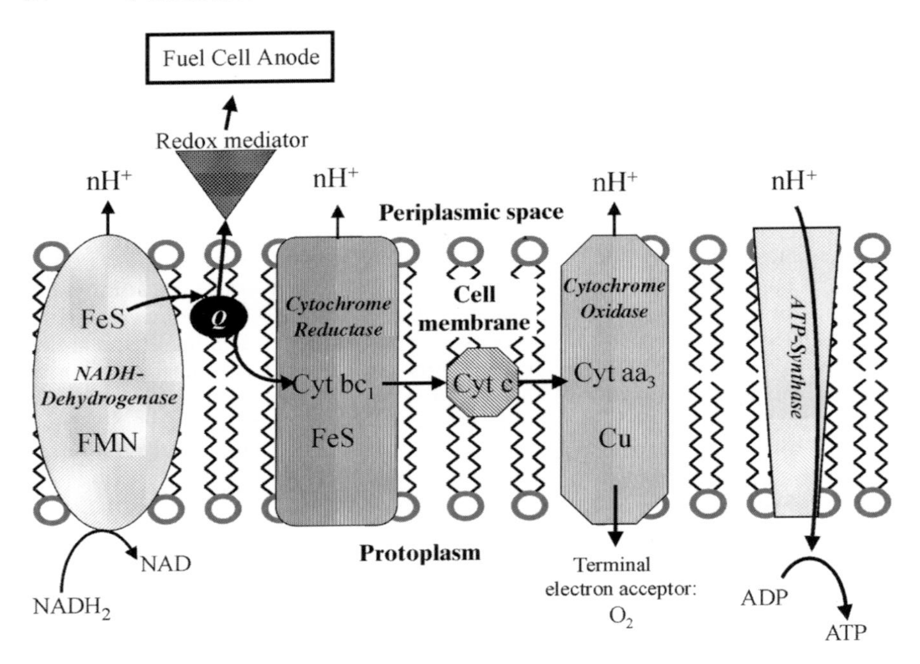

Fig. 5. Redox mediators: Interaction with one component of the respiratory chain as one postulated mode of action.

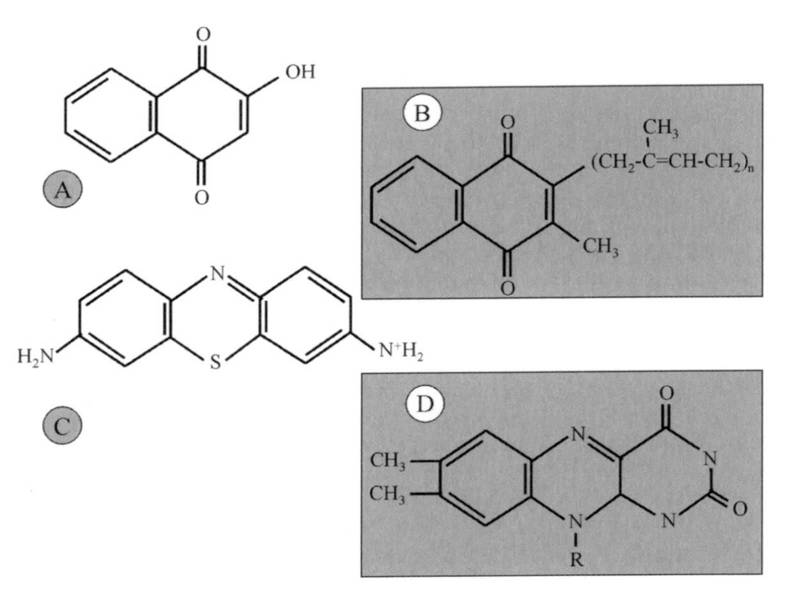

Fig. 6. Structure of the redox mediators 2-hydroxy-1,4-naphtoquinone (A) and thionine (C) and of the physiological hydrogen carriers menadione (B) and the isoalloxazine group of FAD or FMN (D).

with free $NADH_2$ in buffer solutions. In contrast, HNQ is not reduced at all by free $NADH_2$. This suggests that $NADH_2$ alone cannot reduce HNQ within a microorganism and that electrons must therefore come from a different intermediate. Interestingly, HNQ is reduced by $NADH_2$ in the presence of microbial cell-membrane fractions.

The list of substances that can be used as redox mediators is very long, extending from inorganic complexes, e.g., potassium ferrocyanide, to organic redox dyes, such as toluidine blue, methylene blue, thionine, or various quinones (e.g., 2-hydroxy-1,4-naphtoquinone). It can be assumed that different redox mediators also address different targets in the microbial metabolism and that a given enzyme only reacts with certain redox mediators. Extensive information on various redox mediators can be found in the literature (e.g., FULTZ and DURST, 1982).

5.2 Cathode Systems in Bioelectrochemical Fuel Cells

The electrochemical reaction at the cathode completes the electric circuit of a fuel cell system. The electrons, which are generated in the anode reaction, reduce an oxidant, which is present in the cathodic compartment. As stated before, the amount of useful work that can be obtained from a fuel cell is determined by the potential difference between the anodic and the cathodic half-cell reactions. Whereas the potential of the anodic reaction should be as negative as possible, the potential of the cathodic reaction should be as positive as possible. The majority of bioelectrochemical fuel cells described in the literature were equipped with an oxygen cathode. Only a few authors have made use of different oxidants for cathodic reactions, such as potassium ferricyanide (ALLEN and BENNETTO, 1993) or even tried to utilize "biocathodes", where different microorganisms (*Desulfovibrio* sp. or algae) were applied to reduce the oxidants nitrate, sulfate, or carbonate anions (ROHRBACK et al., 1962). The oxygen cathodes used in bioelectrochemical fuel cells sometimes caused problems because the oxygen (but also other oxidants) directly oxidized the reduced form of redox mediators or other reduced species which were generated for the anodic reactions by the biocatalysts. This "short circuit" within the electrolyte lowered the electricity output. For this reason, the anodic and cathodic compartments had to be separated by an ion-exchange membrane, which prevented the direct contact of reductants and oxidants.

To some extent this problem could be overcome by introducing an oxygen gas diffusion cathode (SELL et al., 1989). The advantage of this type of electrode is that no bubbling of oxygen into the cathodic compartment is necessary because the oxygen is taken directly from the air via the outer plane and the pore structure of this kind of electrode.

5.3 Examples of Different Approaches to Bioelectrochemical Fuel Cells

SUZUKI and KARUBE (1983) described different approaches to indirect bioelectrochemical fuel cells; one of their examples was a fuel cell which was supplied with hydrogen generated by a culture of *Clostridium butyricum* immobilized in a packed bed bioreactor that was fed with wastewater from an alcohol factory. The fuel cell was operated for 20 d and produced a continuous current from 13 to 15 mA. Their fuel cell can be regarded as a "classical" hydrogen–oxygen fuel cell which utilized hydrogen that was produced fermentatively. The achievable current is determined by the amount of the available hydrogen which can be determined by the dimension of the packed bed bioreactor.

The application of redox mediators results in a substantial increase in the current and electricity yields obtainable from direct bioelectrochemical fuel cells. ALLEN and BENNETTO (1993) report that in fuel cells with *Proteus vulgaris* as the biocatalyst and thionine as the redox mediator about 50% of the glucose substrate was oxidatively converted into electricity. In the case of sucrose, coulombic yields approached 100% of the theoretical maximum. The electric charge obtained from the oxidation of the substrates was calculated by the integration of current vs. time plots. The coulombic efficiency was defined as the percentage charge obtained compared to the theoretical charge obtainable from the complete oxidation of the substrate.

The application of redox mediators also led to an increase in achievable currents from bioelectrochemical fuel cells. This can be proved by the following comparison: A fuel cell system operated without a mediator reaches maximum currents in the microampere range. An example of such a system was described by YAHIRO et al. (1963). These authors used different enzymes, one being glucose oxidase, as biocatalysts. With glucose as the substrate the described fuel cell showed a current of approximately 500 μA, corresponding to a current density, which takes into account the fuel cell and electrode dimensions, of 16 $\mu A\ cm^{-2}$.

A fuel cell with methanol dehydrogenase as the biocatalyst and methanol as the substrate, mediated either by phenazine methosulfate or phenazine ethosulfate, was described by TURNER et al. (1982). This fuel cell provided an increased maximum current of 4 mA corresponding to a current density of 0.8 $mA\ cm^{-2}$. The current is about ten times higher than in the fuel cell without a redox mediator; the current density was increased by a factor of 50.

A different approach to improving the fuel cell's current output was chosen by KREYSA et al. (1990, 1993). These authors used a packed bed electrode as the fuel cell's anode. The surface of this packed bed anode was approximately 200 cm^2 compared with only 12 cm^2 with a plate anode. The maximum current of the fuel cell with *Escherichia coli* as the biocatalyst and glucose as the substrate was 1 mA with the plate anode, corresponding to a current density of 0.17 $mA\ cm^{-2}$ (KREYSA et al., 1990) and 36 mA with the packed bed anode, corresponding to a current density of 5 $mA\ cm^{-2}$. 2-Hydroxy-1,4-naphtoquinone was used as the redox mediator; *Wolinella succinogenes* and formic acid were applied as an alternative biocatalyst/substrate combination. With both types of anode, the plate anode and the packed bed anode, the fuel cell voltage decreased when the load in the external circuit was lowered, whereas the current output was rose. This behavior is typical of all fuel cells, not only of bioelectrochemical fuel cells, and is due to polarization effects within the cell. Polarization effects in fuel cells may result from slow mass transfer of substances towards or away from the electrode. Furthermore they may be caused by the slow transport of ions through the electrolyte or slow intermediate steps in electrode reactions. By the application of a packed bed anode the polarization effects could be significantly decreased due to an increased charge transfer rate as a consequence of the shorter distance between biocatalyst and electrode surface.

Another approach to overcome the polarization effects and to enhance the electron flow from physiological redox reaction sites to an electrode is the concept of functionalizing the electrode surface; in this area different approaches can be found:

- Redox mediators are assembled as monolayers on the electrode surface to stimulate the electrical contact between redox enzymes and electrode.
- Redox enzymes themselves are immobilized at the electrode surface.
- Cofactors or other redox-active biomaterials are assembled on electrode supports.

Bioelectrochemical fuel cells with such modified anodes have been described (WINGARD et al., 1982; PARK et al., 1997; WILLNER et al., 1998), but have not yet made a substantial breakthrough with regard to current output and fuel cell performance.

The general principle of a bioelectrochemical fuel cell can be summarized from all the different approaches described (see Fig. 7).

The biocatalyst (a microorganism in this example) is in the center of the anodic compartment. The metabolic reactions of this biocatalyst are the driving force for the generation of substances which are subsequently oxidized at the anode of the fuel cell. These substances may be electroactive fermentation products in an "indirect" bioelectrochemical fuel cell or a direct electron flow from the biocatalyst in a "direct" bioelectrochemical fuel cell which can be enhanced by a redox mediator.

A suitable cathode (an oxygen electrode in this example) completes the electric circuit. The biocatalyst's reactions induce the flow of electrons in the external circuit; the electrons

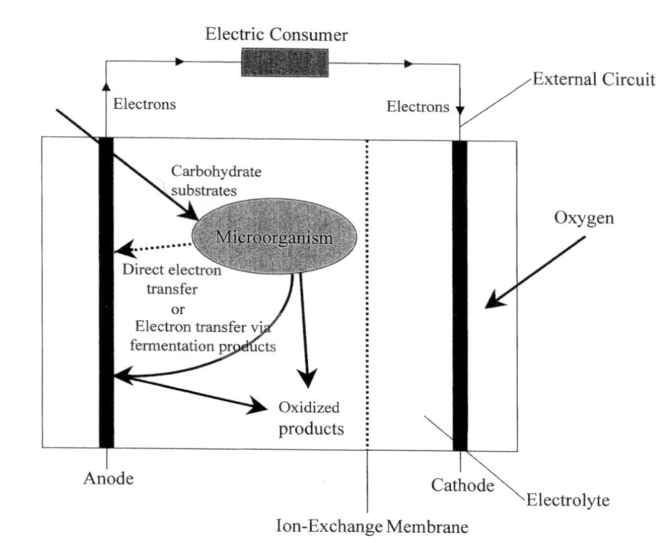

Fig. 7. General principle of a bioelectrochemical fuel cell. A biocatalyst (a microorganism in this example) is in the center of the anodic compartment. Alternatively a direct or an indirect transfer of electrons to the electrode can be realized. An oxygen electrode is used as the cathode.

originate from the substrate, which is metabolized. In order to develop and optimize bioelectrochemical fuel cells a great number of biocatalysts and substrates have been investigated:

In the literature, microorganisms were used in most of the research approaches described concerning the application of bioelectrochemical fuel cells for the generation of electric energy. Only a minority of researchers have made use of cell-free extracts or isolated enzymes because long-term experiments, especially in the early phase of research activities, could only be performed with intact organisms, which some authors characterized as self-reproducing catalysts for fuel cell experiments.

Some early attempts to develop bioelectrochemical fuel cells aimed at utilizing different kinds of organic waste. This approach still seems to be attractive. In Tab. 1 some representative research papers are listed, and the conversion of waste substrates is a dominant group within this list. Experimental approaches with model substrates, such as glucose, sucrose, etc., should be regarded as part of the basic research work for the exploration of the general fundamentals of bioelectrochemical fuel cells.

The different approaches in Tab. 1 reflect the variety of different fuel cell concepts which can be implemented by applying different biocatalysts. The application of photoautotrophic organisms could open up perspectives for the development of biotechnological photovoltaic cells in the future.

Another interesting aspect which has already been applied is the development of biosensors on the basis of bioelectrochemical fuel cells.

6 The Role of Bioelectrochemical Fuel Cell Research in the Development of Biosensors and Bioelectronics

The effective coupling of the electron flow of metabolic redox systems to a suitable elec-

Tab. 1. Representative Approaches to Bioelectrochemical Fuel Cells from the Literature

Substrate	Biocatalyst	Reference
Organic waste	algae, fecal bacteria	REYNOLDS and KONIKOFF (1963)
Humus, sugar waste, wastewater	microbial consortia	HABERMANN and POMMER (1991)
Urea	*Bacillus pasteurii*	BRAKE et al. (1963)
Complex organic medium	various bacteria	COHEN (1931)
Glucose	*Saccharomyces cerevisiae*	VIDELA and ARIVA (1975)
Glucose	*Saccharomyces cerevisiae*	HALME and ZHANG (1995)
Dextrose	yeast	KONIKOFF et al. (1963)
Sucrose	*Proteus vulgaris*	BENNETTO et al. (1985b)
Lactate	*Shewanella putrifaciens*	KIM (1998)
Maltose, glucose, succinate	various bacteria	ROLLER et al. (1984)
Lactose wastes	*Escherichia coli*	ROLLER et al. (1983)
Hydrogen (from fermentation processes)	*Clostridium butyricum*	BRAKE et al. (1963)
Hydrogen (from fermentation processes)	*Enterobacter aerogenes*	TANISHO et al. (1989)
Cellulosic wastes	*Bacillus* sp.	AKIBA et al. (1987)
C_1-Compounds	methanol- and formate-dehydrogenase	YUE and LOWTHER (1986)
Amino acids, ethanol	amino acid oxidase, alcohol dehydrogenase	YAHIRO et al. (1963)
Ethanol	alcohol dehydrogenase	DAVIS et al. (1983)
Various	$NAD(P)^+$-dependent dehydrogenases	WILLNER et al. (1998)
Light	marine algae and *Rhodospirillum rubrum*	BERK and CANFIELD (1964)
Light	*Anabaena variabilis*	TANAKA et al. (1988)
Light	*Synechococcus* sp.	YAGISHITA et al. (1993)
Light	*Anabaena variabilis*	YAGISHITA et al. (1996)

trode, a principle which was first developed for and first applied in bioelectrochemical fuel cells, forms the basis for commercially attractive biosensor systems for industrial and clinical monitors.

It is possible to adapt a bioelectrochemical fuel cell with only a few modifications for sensor use, since the electron flow resulting from the catabolic processes of the biocatalyst is readily available for measurement by amperometric or other methods. Under appropriate conditions the current output of a bioelectrochemical fuel cell is dependent on the number of microorganisms inside the fuel cell and, moreover, it is dependent on the concentration of the added substrate.

One early fuel cell sensor was described by MELE et al. (1979). These authors applied *Pseudomonas ovalis* in a bioelectrochemical fuel cell to monitor the catabolic activity and glucose concentration in a fermentation system of this strain. They found that there is a good correlation between the metabolic reactions and the fuel cell's voltage and current output.

A sensor for the sensitive detection of primary alcohols based on a bioelectrochemical fuel cell containing the alcohol dehydrogenase enzyme was introduced by DAVIS et al. (1983). These authors used N,N,N′,N′-tetramethyl-4-phenylene diamine (TMPD) as the redox mediator and could detect methanol and butanol in the range of $2–10 \cdot 10^{-9}$ mol L^{-1}. The addition of secondary and tertiary alcohols did not result in a current or voltage signal of the fuel cell.

BENNETTO et al. (1985a) demonstrated the dependence of coulombic output and maximum currents on the amounts of glucose fed to the bioelectrochemical fuel cell with *Proteus vulgaris* and thionine as the redox mediator. Regarding the good correlation of the coulombic output to the quantity of glucose added, the authors suggested the development of whole-cell biosensors on the basis of bioelectrochemical fuel cells.

MIYABYASHI et al. (1987) aimed to determine the number of microorganisms in a fermentation vessel. A flow-through dual fuel cell system was used to establish the potential generated by microbial populations. With such fuel cell equipment it was possible to determine the cell mass of *Saccharomyces cerevisiae* in the region between 0.25 to 13 g L^{-1} wet cells. No redox mediator was used in these experiments.

A selection of different sensors based on the fuel cell principle is shown in Tab. 2.

Instead of a fuel cell configuration, a potential may be applied to the biological half-cell by means of a defined power source or a potentiostat. In combination with a standard reference electrode such as a silver/silver chloride electrode, and a suitable counter-electrode such a system forms a poised potential cell which allows much more sensitive and accurate amperometric measurements than a fuel cell sensor can perform. The development of biosensors with electrochemical transducers very much focused on this sensitive type of sensor, whereas sensors of the fuel cell type have not played an important part in recent years.

The main design modifications made in the area of electrochemical biosensors in the past were miniaturization, immobilization of the biocatalysts, "tailoring" of the electrode surface to make it compatible for a rapid electron exchange with the biocatalyst, and the application of new materials, such as conductive polymers. These efforts have already been successful, although developments in this field are still ongoing. The interaction between the biocatalyst applied for biosensory purposes and an analyte can be electrochemically transduced into a detectable signal. Amperometric and potentiometric electrodes, fiber optics, and thermistors are generally the preferred transducers here, the former being by far the most important. For further information on biosensors and the state of the art in this field the reader is referred to SCHELLER et al. (1997).

Developments in the field of electrochemical biosensors may also prove fruitful for the further development of bioelectrochemical fuel cells and bioelectrochemical fuel cell sen-

Tab. 2. Bioelectrochemical Fuel Cells with Application as Biosensors for Different Parameters

Biocatalyst	Measurement Parameter	Reference
Clostridium butyricum	biological oxygen demand	KARUBE et al. (1977)
Pseudomonas ovalis	glucose	MELE et al. (1979)
Clostridium butyricum	formic acid	MATSUNAGA et al. (1980)
Methanol dehydrogenase	methanol	PLOTKIN et al. (1981)
Escherichia coli	glucose	HANAZATO and SHIONO (1983)
Escherichia coli	lactose	ROLLER et al. (1983)
Alcohol dehydrogenase	primary alcohols	DAVIS et al. (1983)
Anabaena variabilis	CO_2	BENNETTO et al. (1984)
Lactobacillus fermenti	vitamin B_1	SCHELLER et al. (1985)
Proteus vulgaris	glucose	BENNETTO et al. (1985a)
Hansenula anomala	lactate	SCHELLER et al. (1985)
Methylomonas methylovora	ethanol, methanol	BENNETTO et al. (1987)
Saccharomyces cerevisiae	cell number	MIYABYASHI et al. (1987)
Pseudomonas sp.	methanol, ethanol	BIRCH et al. (1987)
Various	cell number	PATCHETT et al. (1988)
Synechococcus sp.	herbicides	RAWSON et al. (1989)
Escherichia coli	toxicity testing	KREYSA et al. (1990)
Sewage sludge	metabolic activity during wastewater treatment	HOLTMANN and SELL (2000)

sors. Fuel cell sensors offer advantages for application in areas where cheap, robust sensor systems are required, e.g., in the environmental sector. Sonex, a company located in Berlin, Germany, is currently developing a toxicity test based on a bioelectrochemical fuel cell for application in wastewater treatment or the testing of soil and groundwater samples.

Research activities in the author's laboratory aim at developing a system for on-line monitoring of the metabolic activity of the microbial consortium in sewage sludge during wastewater treatment. A flow-through fuel cell has been constructed which is supplied with a steady stream of waste from a laboratory wastewater treatment plant. The fuel cell measurements are taken without a redox mediator.

The fuel cell applied is depicted in Fig. 8. Experiments with different organic freights which were supplied to the wastewater reactor showed that the different microbial activities of the sewage sludge could sufficiently be monitored with the fuel cell sensor (HOLTMANN and SELL, 2000).

7 Conclusions

Bioelectrochemical fuel cells have not reached a technological breakthrough yet. Today, only one bioelectrochemical fuel cell is commercially available; it can be ordered on the internet under: *http://134.225.167.114/NCBE/MATERIALS/fuelcell.html*. This fuel cell is offered for educational purposes and can be obtained at a price of 40 GBP. A drawing of this fuel cell, which is typical of the setup of many bioelectrochemical fuel cells, is given in Fig. 9. The cell is constructed from perspex and contains an anode and a cathode, on the upper cell side there are different inlets for a dosage of electrolyte, biocatalysts, or other ingredients.

The application of such a fuel cell for educational purposes in schools can deal with the following items: comparison of the rates of substrate utilization, comparison of the rates of mediator reduction and oxygen uptake, prediction of current and voltage, etc.

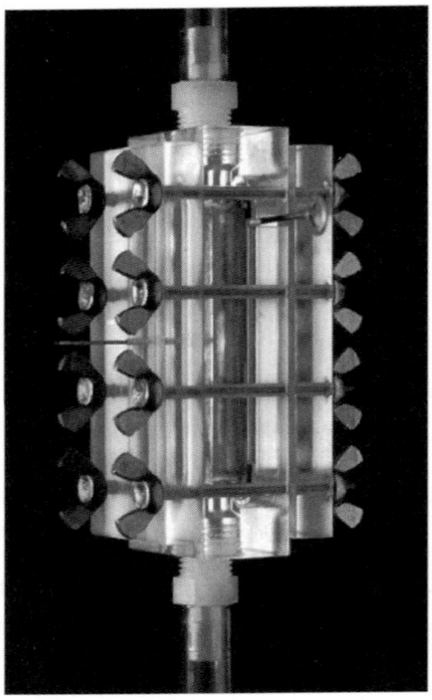

Fig. 8. Photo of a flow-through bioelectrochemical fuel cell for sensory applications (HOLTMANN and SELL, 2000).

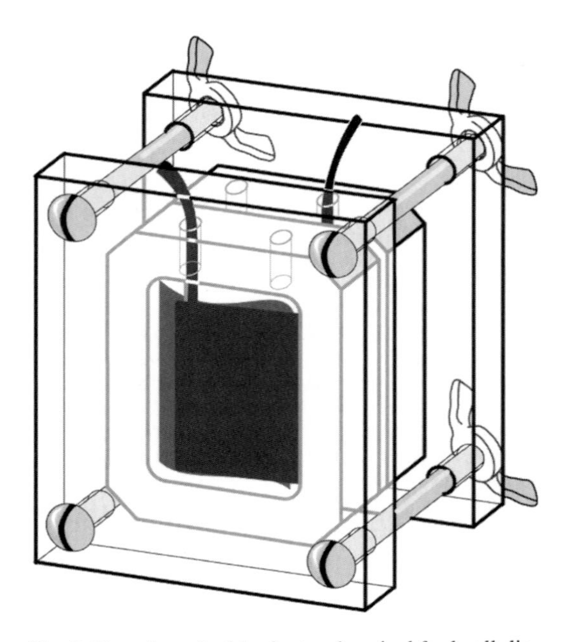

Fig. 9. Drawing of a bioelectrochemical fuel cell distributed for experiments in classroom teaching.

Research on bioelectrochemical fuel cells can now look back on close to 100 years of history. The concept of such fuel cells still opens up a vision of electricity generation from organic waste material or even from sunlight under the mild conditions of a biotechnological process. Due to the work of many groups of researchers all over the world during the last decades, the concept of bioelectrochemical fuel cells has become a demonstrated scientific reality. However, considerable work on bioelectrochemical fundamentals and electron transfer mechanisms remains to be done before practical use in the area of electricity production may also become a reality. There still seems to be a long way to go. Nonetheless, there may be potential applications for low-power direct-current bioelectrochemical fuel cells, such as maintaining telecommunications, surveying instruments, or meteorological measurement equipment in remote areas.

Further applications can be expected in the area of measurement and control, where bioelectrochemical fuel cell sensors can be used as monitors for different parameters.

8 References

AKIBA, T., BENNETTO, H. P., STIRLING, J. L., TANAKA, K. (1987), Electricity production from alcalophilic organisms, *Biotechnol. Lett.* **9**, 611–616.

ALLEN, M. A., BENNETTO, P. (1993), Microbial fuel-cells, *Appl. Biochem. Biotechnol.* **39/40**, 27–40.

Anonymous (1963a), Bacteria operate the radio, *The Financial Times*, Friday November 22, 13.

Anonymous (1963b), Bacterial Fuel Cells, *The Financial Times*, Friday September 13, 11.

BENNETTO, H. P., TANAKA, K. MATSUDA, K. (1984), Bio-fuel cell containing algae, in: *Charge and Field Effects in Biosystems* (ALLEN, M. J., USHERWOOD, P. N. R., Eds.), pp. 515–522. Tunbridge Wells: Abacus Press.

BENNETTO, H. P., DELANEY, G. M., MASON, J. R., ROLLER, S. D., STIRLING, J. L., THURSTON, C. F. (1985a), Electron and energy transduction from microbes, *Industrial Biotechnology*, October/November, 85–88.

BENNETTO, H. P., DELANEY, G. M., MASON, J. R., ROLLER, S. D., STIRLING, J. L., THURSTON, C. F. (1985b), The sucrose fuel cell: efficient biomass conversion using a microbial catalyst, *Biotechnol. Lett.* **7**, 699–704.

BENNETTO, H. P., BOX, J., DELANEY, G. M., MASON, J. M., ROLLER, S. D. et al. (1987), Redox-mediated electrochemistry of whole microorganisms: from fuel cells to biosensors, in: *Biosensors* (TURNER, A. P. F., KARUBE, I., WILSON, G. S., Eds.), pp. 291–314. Oxford: Oxford University Press.

BERK, R.S., CANFIELD, J. H. (1964), Bioelectrochemical Energy Conversion, *Appl. Microbiol.* **12**, 10–12.

BIRCH, S. W., TURNER, A. P. F., ASHBY, R. E. (1987), An inexpensive on-line alcohol sensor for fermentation monitoring and control, *Proc. Biochem.* **22**, 37–42.

BOCKRIS, J. O. M., REDDY, A. K. N. (1970), *Modern Electrochemistry* Vol. 2, pp. 1361–1366. New York: Plenum Press.

BRAKE, J., MOMYER, W., CAVALLO, J., SILVERMAN, H. (1963), Biochemical fuel cells, part 2, Proc. 17th Annual Power Sources Conference, 56–59. Fort Monmouth, NJ: US Army, Power Source Division.

COHEN, B. (1931), The bacterial culture as an electrical half-cell, *J. Bacteriol.* **21**, 18.

DAVIS, G., HILL, H. A. O., ASTON, W. J., HIGGINS, I. J., TURNER, A. P. F. (1983), Bioelectrochemical fuel cell and sensor based on a quinoprotein, alcohol dehydrogenase, *Enzyme Microb. Technol.* **5**, 383–388.

FULTZ, M. L., DURST, R. A. (1982), Mediator compounds for the electrochemical study of biological redox systems: a compilation, *Analyt. Chim. Acta* **140**, 1–18.

GRATHWOHL, M. (1983), Sekundärenergie aus Sonnenenergie, in: *Energieversorgung*, pp. 272–276. Berlin: Walter de Gruyter.

HABERMANN, W., POMMER, E. H. (1991), Biological fuel cells with sulphide storage capacity, *Appl. Microbiol. Biotechnol.* **35**, 128–133.

HALME, A., ZHANG, X.-C. (1995), Modelling of a microbial fuel cell process, *Biotechnol Lett.* **17**, No. 8, 809–814.

HANAZATO, Y., SHIONO, S. (1983), Bioelectrode using two hydrogen ion sensitive field effect transistors and a platinum wire pseudo reference electrode, in: *Chemical Sensors* (SEIYAMA, T., FUEKI, K., SHIOKAWA, J., SUZUKI, S., Eds.), pp. 513–518. Chem. Symp. Series, Vol. 17, Kodansha/Elsevier, Tokyo.

HIGGINS, I. J., HILL, H. A. O. (1979), Microbial generation and interconversion of energy sources, in: *Microbial Technology, Current State, Future Prospects* (BULL, A. T., ELLWOOD, D. C., RATLEDGE, C., Eds.), pp. 359–377. Cambridge: Cambridge University Press.

HOLTMANN, D., SELL, D. (2000), Einsatz eines Sensors zur Online-Bestimmung der mikrobiellen Aktivität, *UWSF – Z. Umweltchem. Ökotox.* **12**, 62–63.

Karube, I., Matsunaga, T., Suzuki, S. (1977), A new microbial electrode for BOD estimation, *J. Solid Phase Biochem.* **2**, 97–104.

Kim, B. H., Park, D. H., Shin, P. K., Chang, I. S., Kim, H. J. (1998), Biofuel cell without electron transfer mediator, *Eur. Patent Appl.* EP 0 827 229 A2.

Konikoff, J. J., Reynolds, I. W., Harris, E. S. (1963), Electrical energy from biological systems, *Aerospace Med.* **34**, 1129–1133.

Kordesch, K., Simader, G. (1996), *Fuel Cells and their Applications*. Weinheim: VCH.

Kreysa, G., Sell, D., Krämer, P. (1990), Bioelectrochemical fuel cells, *Ber. Bunsenges. Phys. Chem.* **94**, 1042–1045.

Kreysa, G., Schenck, K., Sell, D., Vuorilehto, K. (1993), Bioelectrochemical hydrogen production, *Int. J. of Hydrogen Energy* **19**, 673–676.

Krieb, K.-H. (1981), Alternative Energiequellen, in: *Energieversorgung* (Gemper, B., Ed.), pp. 38–43. München: Verlag Vahlen.

Lehninger, A. L. (1977), *Biochemie*, pp. 369–398. Weinheim, New York: Verlag Chemie.

Lewis, K. (1966), Symposium of bioelectrochemistry of microorganisms, IV. Biochemical fuel cells, *Bacteriol. Rev.* **30**, 101–113.

Matsunaga, T., Karube, I., Suzuki, S. (1980), A specific microbial sensor for formic acid, *Eur. J. Appl. Microbiol. Biotechnol.* **10**, 235–243.

Mele, M. F. L., de Cardos, M. J., Videla, H. A. (1979), A biofuel cell as a bioelectrochemical sensor of glucose oxidation, *Anales Asoc. Quim. Argentina* **67**, 125–138.

Mitchell, P. (1979), Keilin's respiratory chain concept and its chemiosmotic consequences, *Science* **206**, 1148–1159.

Miyabayashi, A., Danielsson, B., Mattiasson, B. (1987), Development of a flow-cell system with dual fuel cell electrodes for continuous monitoring of microbial populations, *Biotechnol. Tech.* **1**, 219–224.

Nicholls, D. G. (1982), *Bioenergetics: An Introduction to the Chemiosmotic Theory*. London/New York: Academic Press.

Osteroth, D. (1989), *Von der Kohle zur Biomasse*. Berlin: Springer-Verlag.

Park, D. H., Kim, B. H., Moore, B., Hill, H. A. O., Song, M. K., Rhee, H. W. (1997), Electrode reaction of *Desulfovibrio desulfuricans* modified with organic conductive compounds, *Biotechnol. Tech.* **11**, 145–148.

Park, H. D., Zeikus, J. G. (2000), Electricity generation in microbial fuel cells using neutral red as an electrophore, *Appl. Environ. Microbiol.* Apr. 2000, 1292.

Patchett, R. A., Kelly, A. F., Kroll, R. G. (1988), Use of a microbial fuel cell for the rapid enumeration of bacteria, *Appl. Microbiol. Biotechnol.* **28**, 26–31.

Plotkin, E. V., Higgins, I. J., Hill, H. A. O. (1981), Methanol dehydrogenase bioelectrochemical cell and alcohol detector, *Biotechnol. Lett.* **3**, 187–192.

Potter, M. C. (1911), Electrochemical effects accompanying the decomposition of organic compounds, *Proc. Univ. Durham Philos. Soc.* **4**, 260–274.

Rawson, D. M., Willmer, A. J., Turner, A. P. F. (1989), Whole-cell biosensors for environmental monitoring, *Biosensors* **4**, 299–311.

Renger, G. (1983), Energy conservation, in: *Biophysics* (Hoppe, W., Lohmann, W., Markl, H., Ziegler, H., Eds.). Berlin: Springer-Verlag.

Reynolds, I. W., Konikoff, J. J. (1963), A preliminary report on two bioelectrogenic systems, *Dev. Ind. Microbiol.* **4**, 59–69.

Rohrback, G. H., Scott, W. R., Canfield, J. H. (1962), Biochemical fuel cells, Proc. 16th annual power sources conference, 18–21. Fort Monmouth, NJ: US Army Power Source Division.

Roller, S. D., Bennetto, H. P., Delaney, G. M., Mason, J. R., Stirling, J. L., Thurston, C. F. (1983), Bio-fuel cell for the utilisation of lactose wastes, *Proc. 1st World Conf. Commercial Applications and Implications of Biotechnology*, pp. 655–663. London: Online Publications.

Roller, S. D., Bennetto, H. P., Delaney, J. R., Stirling J. F., Thurston, C. F. (1984), Electron-transfer coupling in microbial fuel cells: 1. Comparison of redox mediator reduction rates and respiration rates of bacteria, *J. Chem. Tech. Biotechnol.* **34B**, 3–12.

Scheller, F. W., Schubert, F., Renneberg, R., Mueller, H.-G., Jaenchen, M., Weise, H. (1985), Biosensors: Trends and commercialization, *Biosensors* **1**, 135–160.

Scheller, F. W., Schubert, F., Fedrowitz, J. (1997) *Frontiers in Biosensorics* Vols. I and II. Basel: Birkhäuser Verlag.

Sell, D., Krämer, P., Kreysa, G. (1989), Use of an oxygen gas diffusion cathode and a three-dimensional packed bed anode in a bioelectrochemical fuel cell, *Appl. Microbiol. Biotechnol.* **31**, 211–213.

Suzuki, S., Karube, I. (1983), Energy production with immobilized cells, *Appl. Biochem. Bioeng.* **4**, 281–310.

Tanaka, K., Vega, C. A., Tamamushi, R. (1983), Mediating effects of ferric chelate compounds as coupled mediators in microbial fuel cells, *Bioelectrochem. Bioenerg.* **11**, 289–297.

Tanaka, K., Kashiwagi, N., Ogawa, T. (1988), Effects of light on the electrical output of bioelectrochemical fuel cells containing *Anabaena variabilis* M-2, *J. Chem. Tech. Biotechnol.* **42**, 235–240.

Tanisho, S., Kamiya, N., Wakao, N. (1989), Microbial fuel cell using *Enterobacter aerogenes*, *Bioelectrochem. Bioenerg.* **21**, 25–32.

TURNER, A. P. F., ASTON, W. J., HIGGINS, I. J., DAVIS, G., HILL, H. A. O. (1982), Applied aspects of bioelectrochemistry: Fuel cells, sensors and bioorganic synthesis, *Biotechnol. Bioeng. Symp.* **12**, 401–412.

VIDELA, H. A., ARIVA, A. J. (1975), The response of a bioelectrochemical cell with *Saccharomyces cerevisiae* metabolizing glucose under various fermentation conditions, *Biotechnol. Bioeng.* **17**, 1529–1543.

WILLNER, I., ARAD, G., KATZ, E. (1998), A biofuel cell based on pyrroloquinoline quinone and microperoxidase-11 monolayer-functionalized electrodes, *Bioelectrochem. Bioenerget.* **44**, 209–214.

WINGARD, L. B., SHAW, C. H., CASTNER, J. F. (1982), Bioelectrochemical fuel cells, *Enzyme Microb. Technol.* **4**, 137–142.

YAGISHITA, T., HORIGOME, T., TANAKA, K. (1993), Effects of light, CO_2 and inhibitors on the current output of biofuel cells containing the photosynthetic organism *Synechococcus* sp., *J. Chem. Tech. Biotechnol.* **56**, 393–399.

YAGISHITA, T., SAWAYAMA, S., TSUKAHARA, K., OGI, T. (1996), Photosynthetic bio-fuel cells using immobilized cyanobacterium *Anabaena variabilis* M-3 in: *Progr. Biotechnol.* **11**, 563–569.

YAHIRO, A. T., LEE, S. M., KIMBLE, D. O. (1963), Bioelectrochemistry, I. Enzyme utilizing Bio-Fuel-Cell studies, *Biochim. Biophys. Acta* **88**, 375–383.

YUE, P. L., LOWTHER, K. (1986), Enzymatic oxidation of C_1 compounds in a biochemical fuel cell, *Chem. Eng. J.* **33**, B69–B77.

ZIEGLER, H. (1983), in: *Lehrbuch der Botanik* (STRASBURGER, E., Ed.), pp. 214–225. Stuttgart, New York: G. Fischer.

Special Microorganisms of Biotechnological Interest

2 Maintenance of Cell Cultures – with Special Emphasis on Eukaryotic Cells

CLAUDIA BARDOUILLE

Taufkirchen, Germany

1 Introduction

The application and use of cell cultures has seen dramatic changes in the last 120 years starting with Roux performing experiments with embryonic chick cells in 1885, the establishment of the human HeLa cell line in 1948 to the large-scale production of biopharmaceuticals in cell lines (MIZRAHI and LAZAR, 1988; LUBINIECKI, 1998). Today, thousands of animal cell lines originating from different species and tissues displaying specific characteristics are available. Several of these lines are used in large-scale manufacturing processes producing vaccines, therapeutic proteins, cytokines, and monoclonal antibodies (reviewed in BARDOUILLE, 1999). The use of continuous cell lines instead of animals or primary cells has been accepted due to benefits such as reproducibility and standardization as well as ethical reasons. However, quality control and advanced techniques for authentication of cell lines are needed to ensure safety and purity of products.

Various industrial proteins, antibodies, and human pharmaceuticals are also targets for plant production systems as reviewed by HOOD and JILKA (1999). Genetic engineering techniques have widened the scope and opened a range of new applications for both animal and plant cell cultures.

The correct handling of cell cultures is of major importance to meet quality and regulatory requirements and to produce reliable and safe cell stocks. This should not only be the case for production purposes, but for any work connected with cell cultures. Similar to chemicals with defined quality, standardization in the work with cell cultures ensures reproducibility and comparability. This chapter outlines methods for standardizing procedures for the maintenance of cell cultures including authentication, safety and regulatory issues. A look at trends in this area will describe potential future developments.

2 Maintenance

Research into external requirements of cell cultures such as media or tissue culture vessels and a better understanding of internal characteristics like cell cycle and growth phases has led to improvements in cell culture techniques. Correct maintenance of cell lines begins with choosing suitable raw or source materials, which can be distinguished into two main categories. First, the cell lines themselves have to be regarded as one category. The second category includes all components cells come into contact with such as basic media, serum, hormones, plastic ware, and attachment substrates. Routine cell culture techniques and cell banking principles are vital to ensure consistent quality and reproducibility. This includes appropriate cryopreservation and storage facilities.

2.1 Terminology

Terms and definitions in the cell culture field are still subject to discussion and partly to controversy (SCHAEFFER, 1989; HAYFLICK, 1990). Here, the following categories will be used:

Primary cell culture

A culture that has been derived from tissues, organs, or cells directly from the organism without passage. Generally, a mixture of cell types can be found representing the original tissue material.

Finite cell line

Derived from normal tissue or cells having a finite *in vitro* life span determined as population doublings (e.g., cell lines MRC-5, WI-38). The number of population doublings depends on the age of the donor, with fetal material undergoing 50–70 doublings before senescence occurs on the one hand and a very limited life span for adult-derived material on the other hand. Finite cell lines often display tissue-specific characteristics such as differentiation markers, but these are subject to change with *in vitro* age.

Continuous cell line (CCL)

Cell line with unlimited generational potential (indefinite passage, e.g., CHO, BHK-21, HeLa). Cells show a higher growth rate and cloning efficiency with a reduced requirement for serum and growth factors. Tumorigenicity is increased and contact-inhibition is often lost. CCLs display a reduced genetic stability and chromosomal abnormalities and often loose tissue-specific functions or characteristics.

More detailed information regarding terminology can be found in SCHAEFFER (1990). It is necessary to give clear definitions describing cell cultures as authors might use specific terms such as cell strain or cell line differently.

2.2 Source Materials

Cell lines should be obtained from a reputable source where quality control and authentication methods are applied routinely. For applications such as vaccine production cell banks are available from culture collections, which have been manufactured and documented to standards being acknowledged by the World Health Organisation (WHO). Guidelines by the Food and Drug Administration (FDA) point out that the history and general characteristics of cell lines should be documented as well as specific procedures that were used to generate the cells (e.g., fusion, selection, cloning; CBER, 1999). A detailed description including species, strain, age, gender, tissue, and physiological condition of donor should be available whether if cell cultures are used for the production of pharmaceuticals or basic research.

Medium and all other substrates or additives used during cell culture are a potential source of introducing impurities or contamination. These should therefore be obtained from a validated and certified supplier (HESSE and WAGNER, 2000). Tissue-culture tested material should be preferred and is available for most reagents, hormones, and other additives.

Special care should be taken choosing serum, a main constituent in complete medium for cell lines, which in most cases will be of fetal calf origin. Serum is undefined and varies from batch to batch making tests such as cloning efficiency, plating efficiency, and growth promotion necessary for selection. In case of the production of native or recombinant proteins or presence of specific markers, expression should be verified with each serum batch as well. Tab. 1 lists advantages and disadvantages using serum in cell culture. Various aspects of using serum for the manufacture of pharmaceuticals especially with regard to viral safety have been discussed at a recent meeting regarding biological standardization (BROWN et al., 1999).

The growing concern with biosafety, especially the outbreak of bovine spongiform encephalopathy (BSE), is met by sourcing bovine serum from appropriate geographical regions certified to be free of a range of adventitious agents (ASHER, 1999). These are currently countries such as New Zealand, Australia, and the USA, however, certificates of analysis should be obtained from suppliers with details of origin, quality, and treatment of serum.

Several serum-free or protein-free media formulations have been developed for a limited range of cell lines, which are mainly used for large-scale biopharmaceutical production such as CHO, BHK-21, and hybridoma lines. Addition of specific hormones or growth factors, often of recombinant origin, can still be necessary making alternatives to animal sera costly and impractical. However, several suppliers offer ready-to-use media and the development in this area is continuing with serum-free media suitable for more than one cell line and improved adaptation processes.

2.3 General Cell Culture Practice and Banking Principles

Cell cultures must be maintained, cryopreserved, and stored correctly to ensure availability of an authenticated and constant supply. Here, only general principles will be presented. Detailed descriptions of laboratory layout and equipment, procedures, and protocols for animal cell cultures have been published by FRESHNEY (2000) and DOYLE and GRIFFITHS (1998) and are also available on CD-ROM as multimedia guide including video sequences (FRESHNEY, 1999). Protocols for maintaining plant cell cultures have been de-

Tab. 1. Advantages and Disadvantages Using Serum

Advantages	Disadvantages
Complex mixture of growth factors, hormones, and other components supports growth of wide range of eukaryotic cell types	chemically not defined, quantity and quality of constituents varies
Binds and neutralizes toxins	extensive testing necessary before use
Contains protease inhibitors	lack of reproducibility due to batch variations
Facilitates attachment and spreading of cells	standardization of experimental and production protocols difficult
Increases buffering capacity	risk of contamination
Carrier proteins such as transferrin and albumin make essential components bioavailable	contains growth and metabolism inhibitors
Protects cells in agitated bioreactors from mechanical damage and shear forces	influence on upstream and downstream processing due to high protein content in serum
	availability and costs fluctuating

scribed by PAYNE et al. (1993) and DIXON and GONZALES (1995). Appropriate training in these techniques is essential and is offered by various institutes or organizations including cell culture collections.

Theoretically, cell lines could be maintained in culture indefinitely, however, this is not advisable. Genetic mutations accumulating with continuous passages, the increased risk of microbial or cellular contamination and simple practicable reasons necessitate the storage of well-defined stocks. Other cell cultures have only a limited life span and, therefore, cannot be cultured for prolonged periods of time. Of particular importance is the availability of authenticated and validated cells stocks for production purposes.

Cell banking principles and the seed stock concept have been reviewed by several authors (see HAY, 1988; WIEBE and MAY, 1990; FACKLAM and GEYER, 1991) and have been recommended by the FDA (see CBER, 1999). Here, the main issues will be pointed out that any laboratory or production facility should be aiming at. Cryopreservation procedures will be described in Sect. 3.

Cryopreserved stocks should be prepared as soon as possible after receiving or developing a new cell line. This should be performed following the procedure described in Fig. 1. The Master Cell Bank (MCB) is the stock material that is used for the production of all Working Cell Banks (WCB). As the name indicates, only the WCB is used for production or experimental processes. After the MCB has been authenticated and passed all tests (as, e.g., required by regulatory authorities for production purposes), a WCB should be produced from this stock. Once the WCB is depleted, a new bank is produced from a MCB vial. This procedure allows the manufacture of WCBs that have undergone a comparable number of population doublings. Operations for the production of both banks should be documented well, including all steps and material involved such as media, subculture, and cell counts.

It is advisable to store banks in the vapor phase of liquid nitrogen for long-term preservation, which is also recommended from suppliers of plastic vials used in most cases. MCB and WCB should be stored in separate nitrogen storage vessels avoiding loss of all material

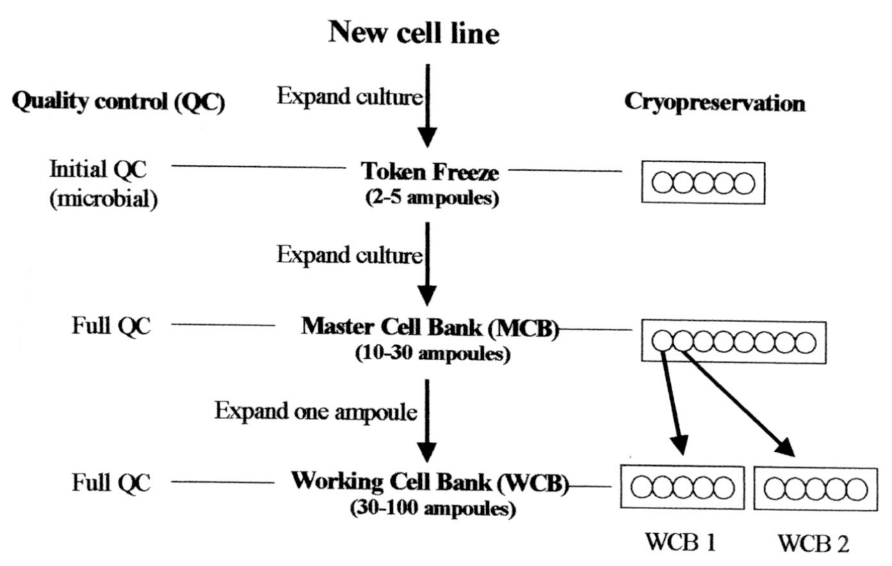

Fig. 1. Cell banking procedure.

in case of accidents. An aliquot can also be deposited with cell bank repositories (see Sect. 6) offering safe storage in continuously monitored nitrogen vessels. Cell cultures shown to be free of contamination should be stored separately from material that is either not tested, known to be infected or pathogenic organisms. Reports on viruses such as hepatitis B and vesicular stomatitis virus retaining infectivity in the liquid nitrogen phase indicate the risk connected to potential cross-contamination (BYERS, 1999; TEDDER et al., 1995).

3 Cryopreservation

The preservation of cell cultures requires different and more complex methods than the preparation of bacteria stocks or other microorganisms. Cell cultures can be preserved by storage at ultra-low temperatures, so-called cryopreservation. Cell stocks that have been prepared, frozen, and stored properly can be kept for years as chemical and biochemical reactions causing alterations are minimized at temperatures below $-130\,°C$. These cells can then be used as consistent and reproducible seed stock for production purposes.

The cryopreservation process is a key element in the preparation of cell stocks. It has been extensively covered in a recent publication by AVIS and WAGNER (1999) including large-scale cryopreservation of cells and microorganisms and a particular stress on requirements for industrial applications.

During cryopreservation damage of cells can occur through

(1) osmotic stress (concentrated solutes),
(2) intracellular ice crystal formation,
(3) mechanical stress (e.g., external ice crystals).

The formation of intracellular ice is generally thought to be the most damaging factor during this process.

In view of these factors, several points have to be considered to freeze cells successfully and obtain good viability after thawing. These are

- suitable cryoprotective agent,
- slow cooling rate,
- storage below −130 °C,
- rapid thawing.

In order to reduce cellular damage, cryoprotectants are added to the cell suspension before freezing. Although the mode of action is not fully understood, it has been shown that these reagents protect cells from osmotic stress. Most commonly used cryoprotectants are dimethyl sulfoxide (DMSO) or glycerol for animal cells (SIEGEL, 1999). The choice of cryoprotectant is often empirical, however, most animal cell lines can be frozen in DMSO, while glycerol is often used for cryopreservation of insect cells. Certain contra-indications exist with cell lines that are subject to functional changes upon contact with specific cryoprotectants. This is, e.g., the case for cell lines that differentiate in the presence of DMSO, such as the hematopoietic precursor line HL-60. Plant cells are often frozen using a mixture of cryoprotectants, e.g., DMSO, glycerol, and sucrose (BENSON, 1994).

In addition to the various cryoprotectants, fetal calf serum at concentrations between 20–90% is added for freezing most animal cell lines. However, just as the development of serum- and protein-free media is progressing, advances in the use of serum- or protein-free solutions for cryopreservation are taking place. So far, success is limited to protocols designed for preserving cells such as CHO and BHK-21 (see, e.g., HESSE et al., 1999).

Suitable freezing and thawing rates are also needed to minimize cellular injury mainly by ice crystals. To avoid intracellular ice formation cells have to loose water, i.e., shrink. This is achieved by slow freezing rates of −1 °C to −3 °C min^{-1} allowing cells to maintain an osmotic equilibrium leading to shrinkage and only little or no intracellular ice before freezing. Cells that have undergone this process will be functional on thawing. However, contrary to the freezing step, the thawing procedure has to be rapid limiting time for ice recrystallization. In addition, cells are generally started in a protein-enriched medium with up to 20% fetal calf serum to provide high concentrations of nutrients, hormones, and other essential factors. This allows quick recovery after exposure of cells to mechanical and osmotic stress.

As already mentioned in the previous section, storage of frozen cells should be below −130 °C, preferably in the vapor phase of liquid nitrogen (−140 °C to −180 °C) or in the liquid phase at −196 °C.

A range of methods and equipment are used for the actual freezing process (SIEGEL, 1999). However, the most reproducible method resulting in up to 100% viability for the majority of cell lines is the programmable controlled-rate freezer. It allows uniform cooling for animal and plant cells and provides freezing records for documentation purposes. Freezing rates can be varied to optimize programs for different cell lines, which is of special importance if cell cultures from different species such as mammalian, insect, and fish are handled.

Various elements affect cell survival, which can be broadly separated as factors before, during and after cryopreservation. Cells should be prepared for freezing in their exponential growth phase displaying a viability of more than 90%. Cell size and concentration, method of cooling in freezing devices, and finally storage temperature influence the final outcome. Detailed protocols and practical implications on cryopreservation and resuscitation of cell cultures can be found in FRESHNEY (1999, 2000). A cryopreservation manual covering animal and plant cells as well as microorganisms is also available from NALGE Nunc International Corp. (SIMIONE, 1998).

Although the majority of cell lines can be cryopreserved successfully, individual lines may need extensive process optimization to achieve good viability after thawing. However, viability is not the only criterion. Further characterization and quality control tests are necessary ensuring pheno- and genotype are comparable before and after freezing and detecting any contamination that may have been introduced during the procedure. Tests needed here depend on regulatory-specific requirements and are connected to the cell line source and application. An overview of the most common quality and authentication assays is given in Sect. 4.

4 Quality Control

Aim of a properly devised quality control system for cell cultures is the identification of cell material and the demonstration of stable and pure cell stocks. The quality control of cell lines encompasses all stages from initial receipt to final working cell bank. In this section some of the main tests for quality control and authentication will be discussed. However, it should be pointed out that the test regime could differ depending on the use of the material and country-specific differences for approval. It is, therefore, recommended to obtain information on regulatory requirements from relevant authorities before the onset of any project.

Quality control can be divided into two parts, the microbial tests for adventitious agents and the authentication assays identifying a cell line (Fig. 2). The cell banking process described in Sect. 2.3 will allow the control of cell line deviation as much as it is practicable and feasible in view of setting up cell stock. Good laboratory and manufacturing practice during all processes and vigilance is also essential to avoid contamination. Some issues have to be stressed as these have great implications for any laboratory.

(1) Cell cultures should be obtained from well-documented sources, where quality is regularly controlled (e.g., culture collections). The history of a cell line together with its origin, method for establishment, number of population doublings, medium, and additives should be documented (FACKLAM and GEYER, 1991).
(2) Quality and identity should be tested at the earliest stage and at regular intervals thereafter.
(3) Untested cell cultures have to be isolated. Physical isolation in a dedicated quarantine laboratory should be preferred. If only limited facilities are available, chronological isolation can be performed by working with tested and contamination-free material first, followed by non-tested, and finally contaminated cultures.
(4) It is strongly recommended not to use any antibiotics in routine culture as these
   - mask contamination and delay detection,
   - lead to development of resistant microorganisms,
   - may affect cell growth and functions,
   - invalidate sterility tests, and

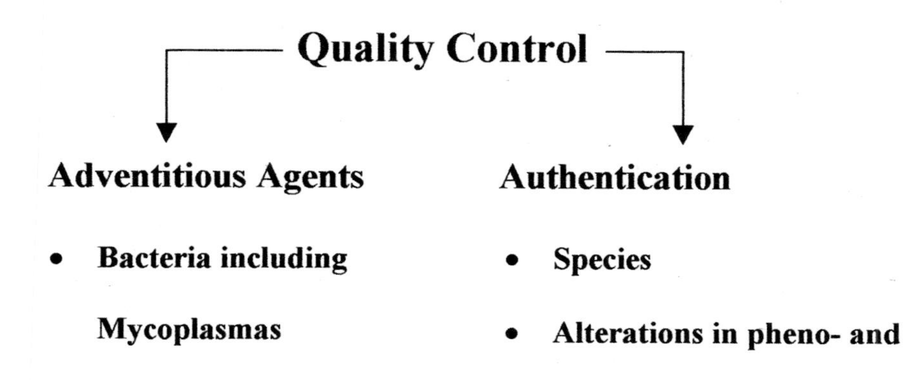

Fig. 2. Quality control of cell lines.

- are not applied in biopharmaceutical production.

The number of cell lines handled in a laboratory varies, but due to the numerous applications and varied cell lines available the risk of cellular cross-contamination is increasing. Identifying the species is also part of a risk assessment (see Sect. 5). It allows suitable safety measurements to be taken working with cell lines carrying potentially greater risks (e.g., of human or primate origin).

4.1 Adventitious Agents

Microbial quality control is concerned with the testing of cell lines, but also with all other reagents and media components that come into contact with the culture. Suppliers of cell culture media, serum, and reagents will perform routine tests being specified in the accompanying Certificate of Analysis.

Likely sources for microbial contamination are poor laboratory conditions and inadequately trained personnel, however, the most common source are untested cell lines.

4.1.1 Bacteria (Excluding Mycoplasmas) and Fungi

Gross contamination with bacteria or fungi can be detected by microscopic examination in most cases if antibiotic-free medium is used. Some microorganisms grow slowly and may be mistaken as cell debris. Test systems for detecting bacteria and fungi will generally involve inoculation of fluid broths or agar (FACKLAM and GEYER, 1991; HAY, 1988).

4.1.2 Mycoplasmas

Mycoplasmas are the smallest self-replicating organisms and lack a cell wall, which distinguishes them from all other bacteria. This major characteristic is reflected in their respective taxonomy term, the class *Mollicutes* (Latin for "soft skin"). An extensive review on molecular biology and pathogenicity has been published by RAZIN et al. (1998).

Mycoplasmas are not as easily detected as bacteria and can have various effects on cell cultures (DOYLE and GRIFFITHS, 1998). An infection can induce

- changes in cell growth rate,
- morphological alterations,
- chromosomal aberrations,
- changes in amino acid and nucleic acid metabolism,
- membrane alterations and cell transformation.

The problem is very often underestimated, but experience within culture collections has shown that up to 30% of new cell lines received were positive for mycoplasmas (FLECKENSTEIN et al., 1994). The actual number of infected cultures might even be higher as mostly prescreened cell lines are deposited with culture collections.

4.2 Authentication

The number and diversity of cell lines used in laboratories and production facilities worldwide is still rising. Assays ensuring correct identity and excluding intraspecies and interspecies cross-contamination are often neglected in the research and development environment, however, production programs have to include appropriate quality control methods for authentication.

Cross-contamination cases are a real threat having drastic consequences for resources spent and invalidated or at least questioned an enormous amount of research in the past. The classic example, contamination of cell lines with HeLa (NELSON-REES et al., 1981), is well known and several cell lines have been identified to either contain a subpopulation of HeLa cells or actually consist only of HeLa cells. A late example of misidentification has been reported by DIRKS et al. (1999a), where the popular human endothelial cell line ECV304 was shown to be a subclone of the human bladder line T24. MARKOVIC and MARKOVIC (1998) have published a more recent review on cell cross-contamination. Authentication methods, especially DNA fingerprinting assays are a valuable tool for identifying cell lines and cross-contamination and will be discussed in Sect. 4.2.3.

4.2.1 Isoenzyme Analysis

Speciation of cell lines can be performed by analyzing various isoenzymes such as glucose-6-phosphate dehydrogenase and lactate dehydrogenase. The different electrophoretic mobility results in a specific pattern characteristic for each species and is visible on the gel used for separation. Disadvantages of this method are that intraspecies contamination cannot be detected and standardized patterns for more exotic species (e.g., fish, insects) are not available.

4.2.2 Karyotyping

The karyotype analysis can be used for confirming species and detecting gross chromosomal abnormalities or rearrangements. It is most useful for normal diploid cells as karyotypes are expected to be stable here. The application for continuous cell lines is problematic, as variable polyploidy is often present in the original material (PETRICCIANI and HORAUD, 1998). The time-consuming assay requires experienced operators and not all species are characterized well enough using karyology. However, cytogenetic analysis remains an acknowledged method for determination of identity and stability of cell lines (WHO, 1998).

4.2.3 DNA Fingerprinting

The most powerful tool for identifying genetic stability and cross-contamination is the DNA fingerprint produced by classic Southern blot techniques or PCR methods. Characteristic DNA profiles can be obtained using multilocus probes such as those developed by JEFFREYS et al. (1985) for cell lines from animal and plant species. A description of this method can be found in DOYLE and GRIFFITHS (1998). DIRKS et al. (1999b) reported the application of sequential fingerprint techniques using different single- and multiple-locus variable numbers of tandem repeat (VNTR) DNA markers. This has led to the establishment of a searchable electronic database enabling rapid authentication. DNA fingerprinting methods allow detection of interspecies and intraspecies cross-contamination distinguishing, e.g., closely related lines from the same mouse strain (STACEY et al., 1992).

Other assays for validation of cell lines or cell stocks may be required and will vary depending on process and product. These may include tests for tumorigenicity and extended genotypic characterization of vectors or inserted genes. Expression rates have to be determined if a product is synthesized by the cell culture.

Guidelines on the testing portfolio required for the manufacture of biological products have been published (WHO, 1998; Human Medicines Evaluation Unit, 1997). Industry and regulatory bodies aim to improve the process for development and registration of biopharmaceuticals in Europe, USA, and Japan. Harmonization of different national regulations and guidelines is prepared by the International Conference on Harmonisation of Technical Requirements for the Registration of Pharmaceuticals for Human use (ICH). This harmonization process also involves safety issues, which will be discussed in the following section.

5 Safety and Regulatory Considerations

The acceptance of cell cultures as substrate for the production of biologicals has been through impressive changes. In the past, primary cultures from healthy animals were considered to be the only safe source for the manufacture of vaccines or pharmaceuticals. Soon after human diploid cells became acceptable for these purposes, continuous cell lines were introduced and are now widespread especially as genetically engineered cells for the production of biologicals (reviewed by HAYFLICK, 1989 and LÖWER, 1995).

The main hazard connected to cell lines is not the cell but the possibility of contaminating adventitious agents such as viruses or mycoplasmas. Therefore, a thorough risk assessment is essential considering issues such as

- origin of cells (e.g., primate, human),
- history (e.g., genetic modifications),
- level of characterization and authentication,
- risk of carrying endogenous human pathogens,
- expression of potentially hazardous proteins,
- volume handled,
- production facilities and procedures.

A comprehensive publication on all aspects of safety in cell culture can be found in STACEY et al. (1998).

High risk is linked to cells from human and primate origin and here especially with cells derived from peripheral blood, lymphoid cells, and neuronal tissues as these may harbor known or unknown human pathogens. The Advisory Committee on Dangerous Pathogens published laboratory safety guidelines including risk assessment in the UK (SHEELEY, 1998).

Apart from the cell cultures, non-cell associated risks exist. These are substances derived from animal origin (e.g., trypsin), carcinogens and irritants like trypan blue and DMSO, additives to the media (e.g., cisplatin for drug resistant cells), liquid nitrogen, and compressed gasses. Protection of work and operator has to be achieved by dedicated cell culture laboratories. Most cell lines can be handled in containment level-2 facilities but the individual risk assessment may necessitate a higher level.

6 Culture Collections

The extensive use of cell cultures in biotechnology and other biological disciplines and the ever increasing number of cell lines has led to the establishment of centers capable of supplying microbe-free and authenticated cell stocks. Culture collections fulfill a range of important functions that can be divided into

(1) collection, maintenance, and distribution of cell cultures,
(2) safe deposit,
(3) patent deposit,
(4) other services.

A centralized collection facilitates availability of cell cultures for present or potential future interest. Authenticated and consistent cell stocks with a high level of characterization are supplied and thus ensure reproducibility needed for experimental and industrial purposes. Fully authenticated cell banks like the World Health Organisation (WHO) Vero stock for the production of vaccines are held at the ECACC and other culture collections (WHO, 1989).

Most culture collections offer services for safe and patent deposits. Safe deposits of cell cultures at two geographically separate locations are strongly recommended to avoid loss of valuable material. Culture collections possess dedicated, up-to-date storage facilities, which are maintained and monitored with utmost care. A deposit in a recognized International Depositary Authority (IDA) is necessary if an invention using biological material is registered as patent. Culture collections can act as IDA under the "Budapest Treaty on the International Recognition of the Deposit of Microorganisms for the Purposes of Patent Procedure". Patents are thus regarded as valid by all participating countries and their patent offices. Tab. 2 features culture collections in Europe holding animal and plant cell cultures. An overview of all collections offering plasmids, bacteria, viruses, and other organisms is published by the European Culture Collections' Organisation (ECCO, 1999).

The specific expertise and knowledge within culture collections allow other services to be offered such as training, cell banking, quality control, and large-scale culture systems.

Tab. 2. Collections Holding Cell Cultures

Country	Acronym	Animal Cells	Plant Cells	Tel/Fax[a] Correspondent	e-mail	Internet Site	IDA
Bulgaria	NBIMCC	+	–	Tel +359 2 720 865 Assoc. Prof. Dr. Todor Donev	nbimcc@olb.net		+
France	CNCM	+	–	Tel +33 1 45688251 Fax +33 1 45688236 Yvanne Cerisier			+
Germany	DSMZ	+	+	Tel +49 531 2616 312 / 311 Fax +49 531 2616 418 Dr. Barbara Lehnberg Dr. Manfred Kracht	ble@dsmz.de mkr@dsmz.de	http://www.dsmz.de	+
Italy	ICLC	+	–	Tel +39 010 5737474 Fax +39 010 5737295 Dr. Barbara Parodi	iclc@ist.unige.it	http://www.biotech.ist.unige. it/interlab/iclc/	+
Japan	Riken Gene Bank	+	+	3-1-1 Koyadai, Tsukuba Science City 305 Iboraki, Japan		http://rtcso.riken.go.jp	+
Russia	RCCC	+	+	Tel +7 812 2474296 Fax +7 812 2470341 Dr. George Pinaev	pinaev@link.cytspb. rssi.ru		–
Turkey	HÜKÜK	+	–	Tel +90 312 2873600 Fax +90 312 2873606 Dr. S. Ismet Gürhan	sapens-d@tr-net.net.tr	http://www.ahis.gov.tr	–
United Kingdom	ECACC	+	–	Tel +44 1980 612512 Fax +44 1980 611315 Dr. David Lewis	ecacc@camr.org.uk	http://www.camr.org.uk /ecacc.htm	+
United States of America	ATCC	+	–	PO Box 1549, Manassas, Virginia, 20108 1549 USA		http://www.atcc.org	+

[a] For full address of European collections, see ECCO brochure (1999).

7 Trends and Future Developments

Animal cell technology has firmly established itself in the field of biotechnology. Next to the well-established lines used for decades for the production of vaccines and pharmaceuticals, new cell cultures and, probably even more important, new applications are entering the arena.

Genetic engineering allows high production levels of native or novel therapeutics. The use of non-mammalian cell types with a lower risk of harboring human pathogens is investigated and already used for research and manufacture. The field of tissue engineering strongly depends on animal cells and the development of new techniques. Organotypical systems, as, e.g., artificial liver, bone, or skin are constructed using three-dimensional matrices. Encapsulation of cells followed by transplantation into patients could provide treatment or cure for various illnesses (BARDOUILLE, 1999). Since new methods for the derivation of embryonic stem (ES) cells and progenitor cells, their expansion *ex vivo* and modification have been successfully established, multiple cell types might be available in the future (HEATH, 2000). These could be used for tissue and organ repair as well as tumor modeling investigating the use of vaccines or gene therapy.

Cytotoxicity assays benefit from the range of cell cultures available and the introduction of automated systems as, e.g., in cell-based high throughput screening (HTS) of potential pharmaceuticals. Medical devices used for implantation can be screened using continuous cell lines (e.g., osteoblasts) and primary cultures (CENNI, 1999). These applications have a direct impact on the number of animals needed for screening by either reducing or replacing animal tests depending on the system and regulatory framework. HTS is also an example of the exponential increase in data collection and the field of bioinformatics connected to it. The collection, organization and provision of data will be a major activity within animal biotechnology as already seen in the human genome sequencing project and the new field of proteomics. With the use of chip-based technology, treatment could be tailored to individual patients improving efficiency.

Other developments that will accelerate the entry of animal cells into biotechnology are formulations for serum- and protein-free media and optimized bioreactors for large-scale cultivation.

Although cell cultures have become ethically accepted especially with a view to reducing or replacing animal tests, new problems have to be faced. These can be found in areas such as genetic engineering and cloning, but the future will undoubtedly question other innovations. The progress will highlight advantages and disadvantages, which have to be met with fair and open communication.

8 References

Advisory Committee on Dangerous Pathogens (1996), Microbiological risk assessment: An interim report, The Stationery Office, London.

ASHER, D. M. (1999), Bovine sera used in the manufacture of biologicals: Current concerns and policies of the U.S. Food and Drug administration regarding the transmissible spongiform encephalopathies **99**, 41–44. Basel: Karger.

AVIS, K. E., WAGNER, C. M. (Eds.) (1999), *Cryopreservation: Applications in Pharmaceuticals and Biotechnology*. Denver, CO: Interpharm Press.

BARDOUILLE, C. (1999), Animal cells used in manufacturing, in: *Encyclopedia of Bioprocess Technology: Fermentation, Biocatalysis and Bioseparation* (FLICKINGER, M. C., DREW, S. W., Eds.), pp. 170–179. New York: John Wiley & Sons.

BENSON, E. E. (1994), Cryopreservation, in: *Plant Cell Culture – A Practical Approach* (DIXON, R. A., GONZALES, R. A., Eds.), pp. 147–167. Oxford: IRL Press.

BROWN, F., CARTWRIGHT, T., HORAUD, F., SPIESER, J.-M. (Eds.) (1999), Animal sera, animal sera derivatives and substitutes used in the manufacture of pharmaceuticals: viral safety and regulatory aspects, *Dev. Biol. Stand.* **99**.

BYERS, K. B. (1999), Risks associated with liquid nitrogen cryogenic storage systems, *J. Am. Biol. Safety Ass.* **3**, 143–146.

CBER (1999), Guidance for Industry: Content and Format of Chemistry, Manufacturing and Controls; Information and Establishment Description Information for a Vaccine or Related Product, Center for Biologics Evaluation and Research,

U.S. Department of Health and Human Services, Food and Drug Administration (*http://www.fda.gov/cber/guidelines.htm*).

CENNI, E., CIAPETTI, G., GRANCHI, D., ARCIOLA, C. R., SAVARINO, L. et al. (1999), Established cell lines and primary cultures in testing medical devices *in vitro*, *Toxicology in vitro* **13**, 801–810.

DIRKS, W. G., MACLEOD, R. A. F., DREXLER, H. G. (1999a), ECV304 (endothelial) is really T24 (bladder carcinoma): Cell line cross-contamination at source, *In vitro Cell. Dev. Biol.* **35**, 558–559.

DIRKS, W., MACLEOD, R. A. F., JÄGER, K., MILCH, H., DREXLER, H. G. (1999b), First searchable database for DNA profiles of human cell lines: sequential use of fingerprint techniques for authentication, *Cell. Mol. Biol.* **45**, 841–853.

DIXON, R. A., GONZALES, R. A. (Eds.) (1995), *Plant Cell Culture – A Practical Approach*, 2nd Edn. Oxford: IRL Press.

DOYLE, A., GRIFFITHS, J. B. (Eds.) (1998), *Cell and Tissue Culture: Laboratory Procedures in Biotechnology*. Chichester: John Wiley & Sons.

ECCO (1999), *European Culture Collections: Microbial Diversity in Safe Hands*, 3rd Edn. Espoo: VTT.

FACKLAM, T. J., GEYER, S. (1991), The preparation and validation of stock cultures of mammalian cells, *Bioprocess Technol.* **13**, 54–85.

FLECKENSTEIN, E., UPHOFF, C. C., DREXLER, H. G. (1994), Effective treatment of mycoplasma contamination in cell lines with enrofloxazin (Baytril), *Leukemia* **8**, 1424–1434.

FRESHNEY, R. I. (1999), *Freshney's Culture of Animal Cells: A Multimedia Guide*. New York: Wiley-Liss.

FRESHNEY, R. I. (2000), *Culture of Animal Cells: A Manual of Basic Technique*, 4th Edn. New York: Wiley-Liss.

HAY, R. J. (1988), The seed stock concept and quality control for cell lines, *Anal. Biochem.* **171**, 225–237.

HAYFLICK, L. (1989), History of cell substrates used for human biologicals, *Dev. Biol. Stand.* **70**, 11–26.

HAYFLICK, L. (1990), In the interest of clearer communication, *In vitro Cell. Dev. Biol.* **26**, 1–6.

HEATH, C. A. (2000), Cells for tissue engineering, *TIBTECH* **18**, 17–19.

HESSE, F., WAGNER, R. (2000), Developments and improvements in the manufacturing of human therapeutics with mammalian cell cultures, *TIBTECH* **18**, 173–180.

HESSE, F., SCHARFENBERG, K., WAGNER, R. (1999), Cryopreservation under protein-free medium conditions: a reliable way to safe cell banking, in: *Animal Cell Technology: Products from Cells, Cells as Products* (BERNARD, A., GRIFFITHS, B., NOÉ, W., WURM, F., Eds.), pp. 501–503. Dordrecht: Kluwer Academic Publishers.

HOOD, E. E., JILKA, J. M. (1999), Plant-based production of xenogenic proteins, *Curr. Opin. Biotechnol.* **10**, 382–386.

Human Medicines Evaluation Unit (1997): ICH Topic Q 5 D – Quality of biotechnological products: Derivation and characterisation of cell substrates used for the production of biotechnological/biological products, European Agency for the Evaluation of Medicinal Products, ICH Technical co-ordination, London (*http://www.eudra.org/emea. html*).

JEFFREYS, A. J., WILSON, V., THEIN, S. L. (1985), Individual specific "fingerprints" of human DNA, *Nature* **316**, 76–79.

LÖWER, J. (1995), Acceptability of continuous cell lines for the production of biologicals, *Cytotechnology* **18**, 15–20.

LUBINIECKI, A. S. (1998), Historical reflections on cell culture engineering, *Cytotechnology* **28**, 139–145.

MARKOVIC, O., MARKOVIC, N. (1998), Cell cross-contamination in cell cultures: the silent and neglected danger, *In vitro Cell. Dev. Biol.* **34**, 1–8.

MIZRAHI, A., LAZAR, A. (1988), Media for cultivation of animal cells: an overview, *Cytotechnology* **1**, 199–214.

NELSON-REES, W. A., DANIELS, D. W., FLANDERMEYER, R. R. (1981), Cross-contamination of cells in cell culture, *Science* **212**, 446–452.

PAYNE, G., BRINGI, V., PRINCE, C., SHULER, M. (1993), *Plant Cell and Tissue Culture in Liquid Systems*. New York: Wiley-Liss.

PETRICCIANI, J. C., HORAUD, F. N. (1998), Karyology and tumorigenicity testing requirements: Past, present and future, *Dev. Biol. Stand.* **93**, 5–13.

RAZIN, S., YOGEV, D., NAOT, Y. (1998), Molecular biology and pathogenicity of mycoplasmas, *Microbiol. Mol. Biol. Rev.* **62**, 1094–1156.

SCHAEFFER, W. I. (1989), In the interest of clear communication, *In vitro Cell. Dev. Biol.* **25**, 389–390.

SCHAEFFER, W. I. (1990), Terminology associated with cells, tissue and organ culture, molecular biology and molecular genetics, *In vitro Cell. Dev. Biol.* **26**, 97–101.

SHEELY, H. (1998), Risk assessment, in: *Safety in Cell and Tissue Culture* (STACY, G., DOYLE, A., HAMBLETON, P., Eds.), pp. 173–188. Dordrecht: Kluwer Academic Publishers.

SIEGEL, W. H. (1999), Cryopreservation of mammalian cell cultures, in: *Cryopreservation: Applications in Pharmaceuticals and Biotechnology* (AVIS, K. E., WAGNER, C. M., Eds.), pp. 371–406. Denver, CO: Interpharm Press.

SIMIONE, F. P. (1998), *Cryopreservation Manual, American Type Culture Collection in cooperation with NALGE Nunc International Corp*.

STACEY, G., BOLTON, B., DOYLE, A., GRIFFITHS, B. (1992), DNA fingerprinting – a valuable new technique for the characterisation of cell lines, *Cytotechnology* **9**, 211–216.

STACEY, G., DOYLE, A., HAMBLETON, P. (Eds.) (1998), *Safety in Cell and Tissue Culture*. Dordrecht: Kluwer Academic Publishers.

TEDDER, R. S., ZUCKERMAN, M. A., GOLDSTONE, A. H., HAWKINS, A. E., FIELDING, A. et al. (1995), Hepatitis B transmission from a contaminated cryopreservation tank, *Lancet* **346**, 137–140.

WHO (1989), WHO cell banks of continuous cell lines for the production of biologicals, *Technical Report Series* 756, World Health Organisation, Geneva.

WHO Expert Committee on Biological Standardisation and Executive Board (1998), Requirements for the use of animal cells as *in vitro* substrates for the production of biologicals, *Technical Report Series* No. 50, World Health Organisation, Geneva.

WIEBE, M. E., MAY, L. H. (1990) Cell banking, *Bioprocess Technol.* **10**, 147–160.

3 New Methods of Screening in Biotechnology

KLAUS FROBEL
SUSANNE METZGER
Wuppertal, Germany

1 Introduction

Drug screening involves a sequential approach to selecting the most appropriate chemical entity – the development or clinical candidate – for drug development, which itself is characterized by a complex process of clinical and toxicological studies to determine the safety and efficacy of a new drug before it is launched to the market. The identification of a promising development candidate is a multidisciplinary effort which usually takes several years of intensive research. One essential part is the selection of the most adequate biological assay which must reflect the pathological mechanism of the particular target disease. As also clearly indicated by modern pathogenetics, diseases do not usually have single causes. Several mechanistic alternatives can, therefore, be chosen for *in vitro* testing, translating into a complex battery of complementary biological assays. If the selected biological assay allows for an efficient process for picking the most bioactive compound by qualitative and quantitative means – the drug screening process – the particular assay qualifies for screening.

A second task is the identification of a small organic molecule – the "lead structure" – which can serve as the starting molecule for further optimization by synthetic modification. Criteria for optimization are biological activity, and also pharmacokinetic parameters such as uptake and metabolism and others. The very early stages of drug screening are, therefore, determined by two processes: lead structure identification and lead structure optimization. Both, assays and compounds submitted to the screening process have to correlate to the particular stage of drug finding (Fig. 1). As to library formats and building techniques there are, therefore, different approaches in building "targeted" and "general" compound libraries. In this context, libraries are understood as collections or pools of real or virtual (*in silico*) small organic molecules.

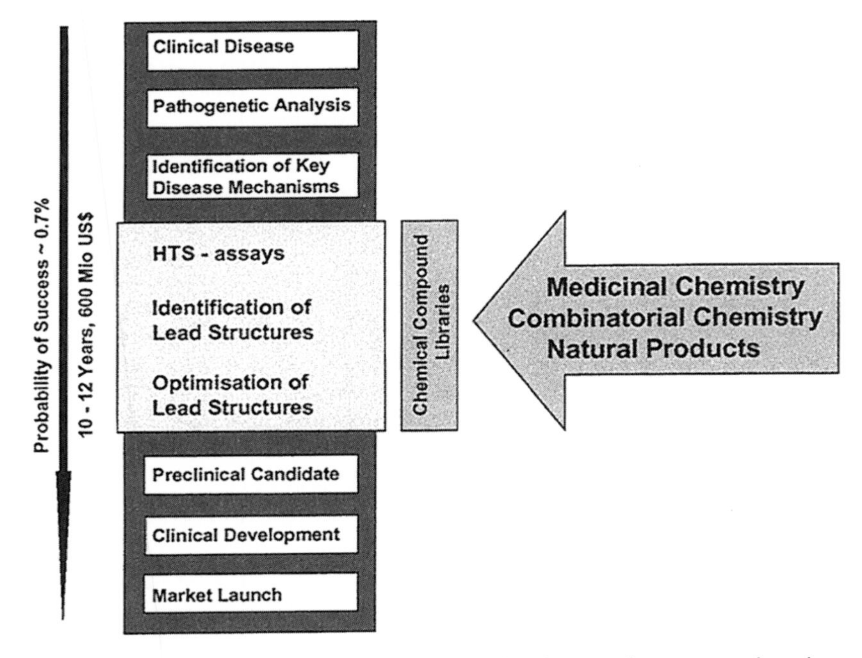

Fig. 1. The most modern technologies of functional genomics, gene engineering, computational design, and organic synthetic chemistry have to join in order to meet the task of identifying an innovative drug candidate.

Although regulatory requirements and specific technical details are different, what has been said above goes beyond pharmaceutical applications and, therefore, also applies to other life science indications.

2 Screening of Bioactive Compounds

2.1 Characteristics of High-Throughput Screening (HTS)

2.1.1 Targets and the Impact of Genomics

Targets for drug discovery, either related to human diseases or to human pathogenic organisms, are usually enzymes, receptors, structural proteins, transcription factors, or even complete complex pathways, such as signal transduction. The currently most important ones are kinases, hormone receptors, G-protein coupled receptors (GPCRs), ion channels, and immune system modulating factors (Fox et al., 2000). It is estimated (Drews, 2000), that genomics in combination with bioinformatics (Lyall, 1998) will identify up to 5,000 to 10,000 potential new targets for drug discovery. This includes a significant number of "orphan" targets (Williams, 2000) which need "de-orphaning" before qualifying for the standard HTS drug discovery process. One way of de-orphaning is to use HTS as an instrument to identify a bioactive compound which can clarify the physiological role of the target protein. In this case, the identified bioactive ligand serves as a tool compound for the set-up of respective HTS assays. Therefore, it may or may not fulfil the criteria of a lead-structure for a small molecule drug by itself, thus including proteins, peptides, and regulatory factors (Lenz et al., 2000).

2.1.2 Increasing Throughput: Automation and Miniaturization

Integrated HTS technologies have become the primary tool for drug discovery, comprising powerful sample logistics, liquid handling, efficient methods of screening, and potent data management systems. Typical characteristics of HTS are the following:

- increasing numbers of test points to be measured with up to 100,000 per day in UHTS (ultra high-throughput screening),
- automated processing of assay protocols, favoring simple and fast procedures (mix and measure),
- single-time-point measurement instead of kinetic studies, as well as decreasing numbers of references and controls to save measurement time,
- more sensitive detection technologies with fast read times to reduce consumption of reagents and chemical library compounds,
- dense sample arrays to reduce logistics workflows and measurement times (384- and 1,536-well microtiter plates are currently becoming the standard format for HTS),
- development of miniaturized and paralleled technologies.

Beyond miniaturization of sample volumes it is required to reduce workloads also with regard to assay development time, sample preparation efforts, reagent and sample transfer, and assessment and validation of results. The above-mentioned features of HTS represent a strong driving force for the development of increasingly sensitive detection methods, new and reliable miniaturized robotics with a high degree of flexibility for fast changes in process flow, and manufacturing of readily available reagents, kits and custom assay development.

2.2 Biochemical Assays

2.2.1 Readout Methods

2.2.1.1 Radiolabeled Readouts

Radioimmunoassays based on isotopically labeled (^{3}H, ^{125}I, ^{32}P, ^{14}C) ligands represent a valuable method for quantifying receptor binding. A huge variety of radioimmunoassays is commercially available and detailed laboratory protocols for these assays are described (VOGEL and VOGEL, 1997). Due to their broad applicability in biochemical, cell-based, and tissue-based assays they can be considered a valuable instrument for lower-throughput investigations as required in the follow up test cascades during early drug development. However, the use of radiolabeled ligands and filtration assays to measure high-affinity binding interactions for cell–surface receptors has largely been replaced by homogeneous methods (SUNDBERG, 2000), a trend which is expected to continue (FOX et al., 2000).

Scintillation proximity assays (SPA) offer a homogeneous approach widely applied to detect binding of radiolabeled ligands to protein targets. SPA measures the energy transfer from radiolabeled tracers to a solid phase carrying the immobilized target proteins. Tracers emit low energy radiation which is dissipated into the assay medium in the unbound state, but activates fluorescence emission of the solid phase when bound to the target molecule, i.e., in close proximity to the support. The solid phase can either be suspended microspheres (BOSWORTH and TOWERS, 1989; PICARDO and HUGHES, 1997) or the microtiter plate itself (FlashPlate™ Technology, BROWN et al., 1997). Downsides are limited stability of the radioactive reagent and relatively long read times by scintillation counters (HERTZBERG and POPE, 2000).

2.2.1.2 Fluorescent Readouts

Fluorescence detection methods are orders of magnitudes more sensitive than colorimetric or UV absorbance readouts, but suffer from side effects such as autofluorescence of samples and plates, inner filter effects, and light scattering. This holds especially true for readouts based on fluorescence intensity (FI) measurements. Several approaches to overcome these shortcomings have been developed and adapted to high-throughput formats (POPE et al., 1999; WEBER, 1997).

Fluorescence labeling: Optimized fluorescence labels covering the whole range of wavelengths can be easily incorporated into proteins, polynucleotides, and tracers via coupling to a variety of functional groups by means of standardized methods. Furthermore, dyes have been developed for specific reversible detection and quantification of single-stranded or double-stranded DNA (HAUGLAND, 1999; KESSLER, 1992).

Besides that, antibodies can be labeled by means of fused enzymes which convert fluorigenic substrates, thus giving rise to a catalytically amplified signal (KESSLER, 1992).

Fluorescence polarization (FP), and equivalently fluorescence anisotropy, is a measure for changes in effective volume of a molecule. The polarization of a molecule excited with polarized light decays with time as the molecule tumbles in solution. Residual polarization of emitted light is proportional to the rotational relaxation time of the molecule, which in turn depends on its molecular size (POPE et al., 1999). In the case of receptor binding studies, binding of a tracer to the protein changes its molecular size and the measured tracer polarization is a weighted average of bound and free tracer polarization, thus providing a real-time measurement of the fraction of free and bound molecules (CHECOVICH et al., 1995; NASIR and JOLLEY, 1999).

Fluorescence polarization high-throughput readers for microtiter plate formats are available since a few years (PERSIDIS, 1998). An increasing number of fluorescence polarization assays for drug discovery have been described (SPORTSMAN et al., 1997) and are commercially available. Long lifetime labels have been developed to extend the applicability of this technique to protein–protein or protein–DNA (or RNA) interactions, of which rotational correlation times were otherwise too high to be measured within the lifetime of conventional fluorescence labels (TERPETSCHING et al., 1997). Applications of FP for HTS assays are listed in Tab. 1.

Fluorescence resonance energy transfer (FRET) is a measure for the efficiency of radiationless energy transfer between a donor and an acceptor fluorophore. It depends on the distance between two dyes (proportional to r^{-6}), their spectral overlap, and their spatial orientation relative to each other (POPE et al., 1999). It, therefore, offers a method for detecting changes in the distance between two labeled molecules or parts of a molecule, a fact

Tab. 1. Applications of Fluorescence Polarization in Screening Assays

Principle	Application Principle	Reference
Binding assays	ligand receptor binding protein–protein interaction	many applications are reviewed by NASIR and JOLLEY (1999) and JAMESON and SAWYER (1995)
	protein–nucleic acid interaction	HILL and ROYER (1997)
	DNA–DNA hybridization	MURAKAMI et al. (1991)
Detection of degenerative enzymes	protease activity	NASIR and JOLLEY (1999), JOLLEY (1996)
	nuclease activity	RNAse assay: HEYDUK (1996)
	general protease detection kits	SCHADE et al. (1996), BOLGER and CHECOVICH (1994)
Combination of enzyme activity and binding	phosphatase and kinase assays	direct binding and competition assay: SEETHALA and MENZEL (1998), PARKER et al. (2000)
	cAMP assay	see HERTZBERG and POPE (2000) for references
β-Galactosidase	well described, stable, and robust readout increased sensitivity due to availability of bioluminescent and chemiluminescent readout	BRONSTEIN et al. (1994)
β-Lactamase	very sensitive readout with a cell-permeable substrate, but the substrate is not yet commercially available	ZLOKARNIK (2000) ZLOKARNIK et al. (1998)
Aequorin	sensitive, stable, and free of intracellular background; flash-type emission requires sophisticated detection system requires Ca^{2+} (mainly used in second messenger assays)	KENDALL and BADMINTON (1998) TANAHASHI et al. (1990)
Secreted placental alkaline phosphatase (SEAP)	extracellular readout and, therefore, non-invasive readout (bioluminescent and chemiluminescent) extremely sensitive, but may suffer from endogeneous activity	CULLEN (2000)

which can be used for a variety of screening assays, both biochemical and intracellular (MERE et al., 1999; SELVIN, 1995; WU and BRAND, 1994). Energy transfer efficiency for a given donor–acceptor pair is described by the so-called Förster radius, which can be interpreted as the distance at which 50% of the maximum energy transfer occurs. In most applications, the donor and acceptor dyes are different, so that FRET can be detected ratiometrically, measuring simultaneously the appearance of acceptor fluorescence and simultaneous quenching of donor fluorescence. Ratiometric measurements offer an additional internal calibration parameter with respect to substrate and product concentrations. Bead-based variants measuring FRET type fluorescence transfer to immobilized receptors have been developed (HERTZBERG and POPE, 2000).

Homogeneous time-resolved-fluorescence detection (HTRF) represents a development for discriminating bio-specific signals over background contributions in fluorescence detection and, therefore, have particular merits compared to miniaturization. Dyes such as rare earth cryptates have excited state lifetimes in the millisecond range. Time-gated reading is used to record the signal at a relatively long time interval after excitation. At this point of time background signals with lifetimes in the low nanosecond range have already decayed (HEMMILÄ, 1997). Using these long-lifetime fluorophores in combination with a suitable acceptor molecule such as the fluorescent protein allophycocyanin, energy transfer experiments can also be performed in a time resolved mode (KOLB, et al., 1997; MATHIS, 1995), resulting in very long Förster distances up to 90 Å (POPE et al., DDT 1999). Such time-resolved energy transfer experiments (TRET) are a powerful tool for measuring protein–protein interactions even in multiple-protein complexes.

2.2.1.3 Microvolume and Single Molecule Detection

Detection methods such as microvolume techniques and single molecule detection (SMD) allow for miniaturization of assay formats and increase in sensitivity at the same time. The technologies described below are compatible to automated HTS or are expected to be compatible in the near future (HERTZBERG and POPE, 2000). These detection methods make use of measuring in very small, confocal sample volumes, thereby reducing background contribution and offering access to molecular parameters which were unexplored for biophysical detection with conventional methods.

- Fluorimetric microvolume assay technology (FMAT) is a method which detects binding of fluorescent ligands to microbeads or cells carrying the receptors of interest. Measurement is performed in very small confocal sample volumes on a single-bead (or single cell) level by statistical analysis. Therefore, background contribution to the signal can be eliminated (MIRAGLIA et al., 1999).
- Two photon excitation (SOINI et al., 2000; HÄNNINEN et al., 2000) is a more sophisticated variation which also measures single beads in microvolumes. Simultaneous absorption of two photons by the fluorescence dyes allows measurements with extremely low background.
- Fluorescence correlation spectroscopy (FCS) measures fluorescence on a single molecule level, detecting concentration fluctuations within very small sample volumes. Statistical analysis then gives access to translational diffusion coefficients of labeled molecules or, in the case of enzymatic processes, to reaction kinetics. Since diffusion coefficients depend on molecular mass, FCS can be applied for detecting receptor–ligand binding events (MOORE et al., 1999; MAITI et al., 1997; RIGLER, 1995).
- A fast and sensitive variant of FCS called confocal fluorescence coincidence analysis detects fluctuations that occur coincidentally for two fluorophores and can thereby monitor association and dissociation between labeled molecules or molecular fragments (WINKLER et al., 1999).
- Fluorescence intensity distribution analysis (FIDA) is a methodology based on measuring the distribution of fluorescence brightness (mean photon count rate) in a confocal volume per time. Changes in brightness occur upon binding events and have the advantage of being independent of molecular mass or size in contrast to macroscopic methods described above (KASK et al., 1999).
- Fluorescence lifetime assay repertoire (FLARe) detects the lifetime of a fluorophore's excited state which depends on the molecular environment and which is, therefore, sensitive to the interaction between tracer and target molecule which again is independent of mass or size (FRENCH et al., 1999).

2.2.2 Ligand Binding Assays

Ligand binding assays, often also referred to as immunoassays, are based on the recognition of a specific ligand by the target protein, in most cases a receptor. Besides binding of low molecular weight ligands such as hormones, interaction of peptides, DNA or RNA and protein ligands with target proteins are also evaluated, either in heterogeneous or homogeneous set-ups (BOCK, 2000).

Readouts for ligand binding assays are often designed in a way that a high-affinity ligand of the target protein is furnished with a label, forming a so called tracer. Detection itself is then usually done as a competitive binding assay, where the sample to be screened competes with the tracer for the binding site of the target protein. Using highly specific tracers, purification of the target protein can be avoided and crude cell extracts or membrane preparations can be directly used as protein source.

In contrast, direct binding assays using labeled proteins are feasible, if ligands are immobilized on a suitable support. However, the reduction to praxis is often difficult since it requires uniform immobilization chemistry which is difficult to achieve for diverse sample libraries. An exception are peptide libraries which are readily synthesized on a variety of solid supports (WALTER et al., 2000). Furthermore, one has to be aware of the fact that immobilization steps can impact on ligand binding kinetics and may, therefore, result in steric hindrance and reduced accessibility of the immobilized protein. General design with emphasis on technical implementation and validation of receptor binding assays have been described by SCHEIRER (1997).

2.2.3 Enzyme Assays

Meeting the needs of HTS set-ups, enzyme inhibition assays, which search for either competitive or allosteric inhibitors, are performed as single-time point measurements in the linear range of a reaction time course. Quantification of product generation is preferred over measuring remaining unreacted substrate and conditions are balanced in a way that strong binders for only one of multiple inhibition sites are likely to be found (SCHEIRER, 1997). General practical approaches concerning design, set-up, and evaluation of enzymatic assays are described, e.g., by TIPTON (1993).

A huge variety of substrates for enzymatic assays are based on *fluorescence intensity* (FI) detection (HAUGLAND, 1999). Hydrolytic enzyme substrates consist of functionalized fluorescence dyes to which the cleavable moiety is directly attached via ether or ester linkages. Release of the fluorophore usually shifts the emission maximum considerably, allowing to follow the reaction rate directly in solution without any need for separation steps (KNIGHT, 1995). Fluorigenic, low-background enzyme substrates are available for a broad spectrum of wavelengths, offering the option for multicolor experiments. Substrates for oxidizing enzymes have long been available, since oxidation of many fluorescent dyes destroys its fluorescence. By designing substrates which become fluorescent upon oxidation, improved readouts with low intrinsic background have been created. Reduced derivatives of common dyes, such as dihydrofluoresceins, are readily oxidized by peroxidases in the presence of hydrogen peroxide, offering direct and indirect readouts for peroxidases and for hydrogen peroxide producing enzymes, including oxidases (by monitoring hydrogen peroxide production) and many metabolic enzymes (oxidizing the product metabolites by their specific oxidases and monitoring hydrogen peroxide production).

An elegant method to assay ATP utilizes *bioluminescent* readouts based on extremely sensitive ATP dependent luciferases (LUNDIN, 2000). Multiple enzymatic steps can be coupled to a final bioluminescent ATP readout. Standardized assay kits are becoming available, e.g., a general inorganic pyrophosphate readout based on ATP sulfurylase with detection limits of pyrophosphate in the picomolar range and a broad linear range. This homogeneous, robust, and sensitive readout offers access to a wide variety of ATP and PPi consuming or releasing enzymes, e.g., NDP kinases (KARAMOHAMED and NYREN, 1999) or DNA polymerase.

As discussed above, *fluorescence polarization* is a method to detect changes in molecular weight and, therefore, represents another valuable detection method for degradative enzymes, given that the change of mass upon substrate cleavage is big enough. This strategy has successfully been implemented for assaying protease (BOLGER and CHECOVICH, 1994; SCHADE et al., 1996) and nuclease activity (HEYDUK et al., 1996) . Combining enzymatic reactions with a subsequent binding step using substrate- or product-specific antibodies (see Tab. 1), the application range of FP as been further extended.

Internally quenched, *FRET*-based endopeptidase substrates are obtained by placing a donor–acceptor pair of dyes at two different sites of a peptide substrate. Upon enzymatic cleavage, the two dyes are spatially separated from each other, resulting in free, unquenched fluorescence of the donor. Acceptor molecules do not have to be fluorescent themselves, efficient absorption of the donor fluorescence is sufficient. The same principle is applicable to oligonucleotides as well, offering detection methods for applications such as assaying nucleases, polymerases, helicases and more (MERE et al., 1999).

2.3 Functional Assays

Cell-based assays do have the intrinsic advantage of being functional assays which do not rely on pure binding events. Cell permeability, resistance to fast metabolism or elimination, and non-toxicity of the bioactive compound are assessed in parallel. However, potential leads may be missed due to a lack of one of these features, properties which usually can be overcome in the process of lead optimization.

Assays can be performed in a variety of organisms such as bacteria, yeast, and mammalian cells, and can be assigned to three categories which are reporter gene assays, second messenger assays, and two-hybrid techniques.

2.3.1 Reporter Gene Assays

Reporter gene assays quantify the expression level of a specific reporter gene product which is used as marker for activation of the targeted pathway (WALLACE and GOLDMAN, 1997). A marker protein should be conveniently quantified, non-toxic and not present in the unmodified cell. Reporter gene product quantification is usually done by measuring catalytic turnover of exogenously added substrates after cell lysis. Due to their low intrinsic background and large dynamic ranges, bioluminescent and chemiluminescent readouts have brought about a substantial gain in sensitivity as compared to the conventional chloramphenicol transferase (CAT) reporter gene system (KRICKA et al., 2000; BRONSTEIN et al., 1994; SCHEIRER, 1997). In contrast, fluorescence readouts give much better sensitivity, but have a higher background signal due to non-specific fluorescence and light scattering. Development of cell permeable substrates for several reporter gene assays have provided access to real-time and kinetic measurements. An increasing set of reporter genes for a variety of applications is becoming available.

Non-disruptive assays have been developed by using reporter enzymes with cell permeable substrates, secreted reporter enzymes or intrinsically fluorescent reporter proteins (Tab. 2), offering the possibility for real-time measurements.

2.3.2 Second Messenger Assays

A variety of fluorescent ion indicators and voltage-sensitive probes based on fluorescence intensity and FRET mechanisms have been developed for detecting second messenger signals, such as calcium or cAMP levels, in response to receptor stimulation or ion channel activation (GONZÁLEZ et al., 1999; DENYER et al., 1998). The calcium dependent photoprotein aequorin offers an intracellularly available, sensitive probe for Ca^{2+} flux (KENDALL and BADMINTON, 1998), avoiding the need for loading cells with fluorescent probes.

In contrast to reporter gene assays, cytotoxic compounds do not generate false positive hits in second messenger assays, thereby generating favorably low hit rates. Imaging systems have been developed, allowing detection of calcium flux and voltage gated ion channels with very high sensitivity (DENYER et al., 1998; SULLIVAN, 1999).

Tab. 2. Reporter Genes of Cell-Based HTS Assays

Reporter Gene System	Characteristics	Reference
Green fluorescent protein (GFP)	non-invasive approach without requirement for a substrate; relatively low sensitivity and slow improved mutants (wavelengths, brightness) have been engineered	MISTELLI and SPECTOR (1997) CHALFIE et al. (1994)
Firefly luciferase	highly efficient and well described reporter gene system cell-permeable substrates are meanwhile available	GELMINI et al. (2000) BRONSTEIN et al. (1994)
Bacterial luciferase	similar to firefly luciferase, but mutants with a broad spectrum of temperature stability are available for a variety of applications	GELMINI et al. (2000) BRONSTEIN et al. (1994)
β-Galactosidase	well described, stable, and robust readout increased sensitivity due to availability of bioluminescent and chemiluminescent readout	BRONSTEIN et al. (1994)
β-Lactamase	very sensitive readout with cell-permeable substrate, but the substrate is not yet commercially available	ZLOKARNIK, 2000 ZLOKARNIK et al., 1998
Aequorin	sensitive, stable, and free of intracellular background; flash-type emission requires sophisticated detection system requires Ca^{2+} (mainly used in second messenger assays	KENDALL and BADMINTON (1998) TANAHASHI et al. (1990)
Secreted placental alkaline phosphatase (SEAP)	extracellular readout and therefore non-invasive readout (bioluminescent and chemiluminescent) extremely sensitive, but may suffer from endogeneous activity	CULLEN (2000)

2.3.3 Two-Hybrid Assays for Measuring Protein–Protein Interactions

Two-hybrid assays provide a powerful tool to analyze protein–protein-interactions in living cells (FIELDS and SONG, 1989; CAGNEY et al., 2000), regarding that the complexity of interactions between multiple proteins involved in certain pathways is difficult to establish in biochemical assays (HUANG and SCHREIBER, 1997). The two interacting target proteins are fused to two separated domains of specific transcription factors. Upon contact between the target proteins, the activity of the transcription factor is restored and can be detected via a reporter gene assay such as galactosidase (see above). Due to its low cost, its robustness, and easy genetic manipulation, yeast is the by far most frequently used organism for two-hybrid assay development (MUNDER and HINNEN, 1999). Gain in sensitivity has been achieved using a reversed two-hybrid screening system (VIDAL and ENDOH, 1999) in which the interaction of two target proteins results in the activation of a toxic or lethal marker. Viability of such negative mutants is recovered only when protein–protein interaction is disrupted by a compound, resulting in a very low intrinsic background and good sensitivity with detection limits in the submicromolar range. Systems for prokaryotic (DOVE et al., 1997) and mammalian cells (BUCHERT et al., 1997) have also been described.

2.4 Miniaturization of Technology

2.4.1 Miniaturized Detection Systems

Serial readers such as scintillation counters, luminometers, and fluorescence plate readers, have the advantage of being robust, widely spread, and creating a conveniently small scope of data. They have been successfully adapted to miniaturized MTP assay formats (BURBAUM, 1998). In contrast, imaging systems which allow parallel detection of multiple assay points in one detection step had to be developed for HTS applications (RAMM, 1999; PERSIDIS, 1998). They are especially advantageous when measuring slow readouts. Detectors are based on charge-coupled device (CCD) cameras and are available for scintillation counting, fluorescence, and luminescence detection (HERTZBERG and POPE, 2000). Numbers of sample points per detection step are only restricted by the resolution of the imager so that imaging is the method of choice for free-format arrays. They furthermore allow repeated parallel and time-resolved measurements. A technical challenge for the optical system results from the requirement of uniform illumination and detection over the whole plate area, a problem which is tackled by a sophisticated calibration software (RAMM, 1999). In addition, instruments for cell-based applications detecting fast calcium mobilization by fluorescence intensity are on the market. (DENYER et al., 1998; SULLIVAN et al., 1999) and on FRET (GONZÁLES et al., 1999)

2.4.2 Miniaturized Liquid Handling

Reliable liquid handling systems – automated pipetting devices or bulk reagent delivery systems – are crucial to HTS. Although still being miniaturized, conventional syringe pipettors are rarely used for applications in the sub-microliter range. Systems which can cope with very small volumes have been described (ROGERS, 1997; PERSIDIS, 1998) and have been reviewed recently by ROSE (1999). All systems described there have been shown to handle biochemical solutions as well as cell suspensions, making them suitable for all kinds of HTS assays.

- Conventional syringe pipettors have been developed to handle volumes below 1 µL and are often preferred because of their robust operation and existing long-term experience. Pipetting is fast, allowing pipetting of 1,536-well plates within 20 s, and also quite reproducible, resulting in standard deviations of 10% or less. Multi-channel dispensers have been developed for increased throughput.
- Pin transfer systems cover the low nanoliter to low microliter range and offer fast, cost-effective, and robust operation. Efficient sample transfer and mixing can be achieved without dead volume. However, standard deviations of transferred volumes can reach up to 30%, thus limiting the range of applications for this system to date.
- Inkjet liquid delivery systems can be subdivided into selenoid and piezoelectric techniques which both can handle extremely low volumes down to a few picoliters. These systems offer a very broad volume range with low standard deviations, allowing for the recording of dose–response curves without intermediate dilution steps. Contact free liquid delivery further reduces the danger of sample carry-over to other wells.

Pin transfer and inkjet technologies are rather young technologies and have just started to disseminate into HTS robotics.

2.4.3 Flow Technologies

In flow injection analysis, assay reagents are continuously added, giving a constant background signal. Sample compounds are injected sequentially, giving rise to bio-specific signals only when active compounds are present. Chip technology represents a miniaturized format of this approach, where fluid channels and reservoirs are etched on a silicon chip and samples are injected by means of a microcapillary.

Extremely reduced amounts of target protein or cells and test compounds are required for this technique and standard deviations are low due to the controlled constant environment. Quantitative and kinetic measurements are feasible, and efforts to increase throughput are underway. A variety of applications, including enzyme and immunoassays, has been reviewed by BOUSSE et al. (2000).

Flow cytometry is a technology which allows rapid quantification of fluorescence in living cells and can be applied for analysis of marker protein expression levels or second messengers whenever a suitable fluorescence label is available. Furthermore, cell viability can be detected, for example to analyze antimicrobial drug–cell interactions (DAVEY and KELL, 1996).

In contrast, a microphysiometer consists of a cytosensor which is incorporated in a flow chamber containing immobilized cells. The sensor is capable of monitoring physiological parameters, such as a pH change, within the environment of the cells. For example, acidification rates upon receptor activation of the immobilized cells (MCCONNELL et al., 1992) are detected in a stopped-flow mode. This approach gives access to a label-free analysis of orphan receptors under physiological conditions (ALAJOKI et al., 1997).

3 Chemical Compound Libraries

Traditionally, organic synthesis of small compounds is targeted to one specific end product or a small group of compounds, and the challenge for the organic chemist is to use all his skill and intuition to find the most efficient and intelligent synthetic route. The combinatorial approach makes use of these skills in a somewhat different way. Combinatorial chemistry is targeted to the preparation of all possible combinations of a group of chemical building blocks, with the aim of identifying the most active of the resulting molecules. Although the term "Combinatorial Chemistry" still has a different meaning to different people, ranging from spatially separated automated parallel synthesis leading to libraries of distinct compounds to manual mix and split techniques, which yield pooled compound mixtures, there is a common attempt to systematically expand the diversity of chemical structure applied to the biological screening process. Thus, in a modern medicinal chemistry laboratory environment, a comprehensive set of various methods is on hand, providing whatever techniques seem appropriate for the efficient solution of a particular synthetic problem. This also includes data processing and material logistics required for the handling of large numbers of chemical compounds in an integrated HTS workflow. Although there are powerful techniques which allow for the construction of libraries on solid support (JACOBS et al., 1998), most commonly the chemical compound libraries are submitted to biological screening in solution.

3.1 Combinatorial Libraries

3.1.1 The Combinatorial Chemistry Paradigm

In contrast to the classical synthetic strategy, which generally follows a linear route, combinatorial chemistry achieves parallel synthesis by using identical reactions for a multiple set of chemical building blocks permuted in a systematic fashion. As an example: Three reaction steps (A, B, C) are used with three different building blocks (1, 2, 3), yielding a library of 27 structurally different combinations. In the first reaction A, the three combinations a1, a2, a3 are obtained; the second step B gives the 9 ab-, and the step C the 27 abc-permutations. The number of steps is additive (three reaction types A, B, C were performed), the number of final chemical compounds grows exponentially (3^3). In analogy, 10 building blocks would yield 1,000, 100 building blocks already 1,000,000 different chemical compounds in a three-step reaction sequence. This methodology proves powerful to produce a large number of chemical compounds with a comparatively low input on chemical reaction steps to perform, but it requires special techniques to control the reactions and to achieve a spatially separated format, if one aims for distinct compounds. It is a major challenge to identify building sequences and the respective chemical methodology which allow for at least three permutation sites to gain benefit from the combinatorial versus the linear approach (THOMPSON and ELLMAN, 1996; BALKENHOL et al., 1996; JUNG, 1999; SCHREIBER, 2000). Therefore, a library is regarded a combinatorial one, if its creation is built upon a permutation synthesis sequence independent of the technologies used, thus including split and mix approaches as well as high-throughput parallel synthesis set-ups.

3.1.2 Combinatorial Library Formats

Depending on the specific task, one has to decide between the synthesis of a pool library, receiving a *mixture of compounds*, or the synthesis of a library consisting of a compilation of *distinct compounds*. A pool library always requires deconvolution steps after biological testing: most commonly a follow-up sequence comprises resynthesis of sub-libraries of reduced size and re-testing – an iterative process which finally allows for the correlation of biological activity with one particular chemical structure (WILLIARD and TARTAR, 1997). Since the structural characterization of individuals in complex compound mixtures is limited, one has to rely on a strict process control. A missing compound due to different reaction yield could entirely mislead the biological screening result. The value of compositional analysis and fingerprinting of large libraries is mostly too unclear to guarantee compositional quality of (large) compound libraries (FITCH et al., 1998). Although pool libraries were the very early examples of solid phase combinatorial libraries using different "split and mix" or "split and combine" procedures developed independently by FURKA (1991), LAM et al. (1991), and HOUGHTEN (1991), they have to date proved to be most successful with oligopeptides and oligonucleotides. The split and mix approach permits the synthesis of a multitude of individual compounds being attached to polymer beads in one batch thus minimizing the synthetic effort per compound. Repetitive recombining and splitting of the target drug intermediates loaded resins allows for multistep synthetic sequences.

For solid-phase combinatorial libraries there are further alternatives for deconvolution: using highly loaded resins a "pick & analyse" approach becomes feasible. Further options are the use of coding technologies. Chemical coding implies a parallel coding synthesis sequence, which must not interfere with the target molecule synthesis (orthogonal chemical synthesis). Extended reaction times and special decoding techniques limit the practical use of these technologies. More convenient is radio frequency encoding using radiotags. Small reaction containers are individually marked by a radiotag allowing for individual tracking and automated final sorting of the reaction containers at the very end of the synthesis sequence. The repetitive sequence of pooling reaction containers in one synthetic step followed by an automated resorting for the subsequent synthetic transformation reduces the synthetic workload. The final sorting process allows for individual product characterization and a high quality of chemical compound libraries (HINZEN, 2000; JACOBS and NI, 1998).

3.1.3 Solution-Phase Techniques

Although it seems unlikely to perform combinatorial chemistry in solution following a multistep synthesis sequence, a number of technologies have emerged which yield high-quality chemical compounds (KIBBEY, 1997). Some multicomponent reactions can be carried out efficiently in solution. Among those, the ones in which the final step in product formation is irreversible proceed with good yields (BANNWARTH, 2000).

Limitations to solution chemistry are uncompleted product formation and lack of intermediate product purification tools. In a short synthesis sequence this is often of minor relevance, and the final product matches the required quality standards. Use of special extraction techniques such as solid-phase extraction or "fluorous work-up" (BANNWARTH,

2000) have provided high performance routines to produce large numbers of compounds in parallel fashion. Polymer-assisted solution phase chemistry (PASP) is a reagent quenching approach, which removes unreacted starting materials or impurities by covalent binding to a reactive polymer. Thus, reactions can be driven to completion by addition of excess starting material, which eventually is removed by the polymer scavenger and subsequent filtration. Although coming close to the traditional synthesis workflow, significant progress has been made in automated preparative chromatography, providing medium-size, high-quality distinct compound libraries (OBRECHT and VILLALGORDO, 1998).

3.1.4 Solid-Phase Techniques

Elaborate work-up procedures can be avoided, if the target molecule or its respective precursors are covalently linked to a polymer bead. This simplifies the work-up procedure to a mere filtration combined with the advantage, that large excess of reagents can be used driving chemical reactions to completion. Higher product yields can, therefore, be expected as compared to the same reaction in solution with equimolar amounts of reagents. The price one has to pay for this advantage is that the synthetic sequence is extended by two additional reaction steps: coupling to the bead at the very beginning and the decoupling step at the very end, since by far most screening assays are performed in solution. Solid support synthesis requires several components which have to match: the polymer, suitable for organic solvent reactions and sufficient loading capacity; the linker, which covalently hooks the target molecule without interference during the synthetic sequence and is still "labile" enough to be eventually cleaved by a specific reaction, and a synthetic sequence adapted towards synthetic transformations on polymer-bound organic molecules (DÖRWALD, 2000; OBRECHT and VILLALGORDO, 1998). Consequently, library design and synthesis requires significant input into methodological work (BUNIN, 1998).

3.2 Targeted Screening Libraries

Since there is not one single route to create screening compound libraries, the method of choice depends very much upon the chemical matter which has to be addressed. Any *a priori* information as to what structure type may lead to an improvement in biological activity, guides the design and selection of the specific library building technology (GORDON, 1998).

3.2.1 Lead Structure Optimization

The standard work flow of the drug finding process (see Fig. 1) delivers lead structures from the primary screening run. Several "learning cycles" have to follow, which establish a structure–activity relationship (SAR) and guide the optimization process. Synthetic methodologies come into place to allow for fast synthesis of medium to small size compound libraries, which are targeted towards specific pharmacophoric properties. Molecular design methods can contribute significantly favoring structure templates and substitution patterns. They may use quantitative SAR methodology (KUBINYI, 1993; HOPFINGER et al., 2000) just based upon the small molecule ligand, genetic algorithms or – in a very favorable situation – protein structures derived form NMR and/or X-ray experiments.

3.2.2 Molecular Mechanism of Bioactivity

Particularly with the use of enzymes as screening targets, the molecular mechanism of action can serve as a guideline for the design of targeted libraries. A generic structure motif, which is, e.g., a transition state analog, can therefore serve as a central core or scaffold which is decorated with diverse substituents. The P1, P2 and P1′, P2′ positions often direct the enzyme specificity of the small organic molecule ligand within one class of enzymes, which is a crucial prerequisite for a favorable therapeutic profile (WHITTAKER, 1998).

3.2.3 Covalent Marker Libraries

Chemical libraries can be designed as "photoprobes" for the identification of target drugs (DORMÁN and PRESTWICH, 2000) to determine the affinity and selectivity of drug–target interactions. All members of this chemical compound library carry photoreactive groups. Those compounds, which are able to bind to the target protein, either form a new covalent linkage (photoaffinity labeling and photoimmobilization) or cleaving a specific bond (photodeprotection) initiated by irradiation of the photophore at a defined wavelength (DORMÁN, 2000).

3.3 General Screening Libraries

In contrast to the targeted library design, general screening libraries have to be designed independent of the targeted biological screening assay. In a modern, automated, and highly specialized HTS unit, assays can change on a weekly basis. As a consequence, the adaptation of sample logistics to each assay individually would be a much slower and cost intensive process, if one considers a typical general library size of more than 500,000 distinct compounds. Statistical design tools come into place, which increase the hit rate and the quality of the identified lead structures, thus reducing attrition rates on secondary screening levels and during lead structure optimization cycles.

3.3.1 Library Diversity Design

The most important design requirement is to create a library which presents the chemical space of structural diversity best. Hence, a combinatorial synthetic general compound library is often complemented by other sources of chemical matter such as natural products. Although diversity is a common term used, it is not yet possible to give a clear definition. Synthetic chemists are tempted to see chemical diversity only from the structural point of view. It has to be emphasized though, that structurally very similar compounds often give rise to completely different biological responses. A minor structural change, e.g., one that makes the amine part essential for the pharmacophoric interaction more or less basic, can reduce or increase the biological effect by several orders of magnitude. It translates to success or failure in biological screening. The same is true for the scaffold (three-dimensional configuration) of the organic molecule. If the change of stereochemistry modulates the spacial positioning or dynamics (floppiness) of pharmacophoric groups, affinity towards the biological target protein may be lost or gained.

Thus, for a statistical design of large chemical compound libraries, it is more convenient to address diversity by looking at missing similarity (DOMINIK, 2000). Since combinatorial libraries always are clustered around templates, it is important to compose a diverse general library out of a large set of different structural clusters, each being decorated with a "pharmacophoric" set of substituents in a combinatorial fashion. To make the small organic molecule interact with the target protein in a specific manner, the ligand has to carry organic functional groups, which allow binding to the protein (Fig. 2).

As a statistical first line "back of the envelope" 2D-design tool, the Tanimoto coefficient serves well to quantify molecular similarity (FLOWER, 1998; XUE et al., 2000). Each compound receives a chemical barcode as a two-dimensional substructure descriptor, which allows for a convenient comparison of a pair of structures (Fig. 3). The mean Tanimoto coefficient of >0.8 within one combinatorial sub-library indicates structural similarity, a mean coefficient of <0.4 between different structure clusters indicates significant dissimilarity. The standard approach implies the *in silico* creation of a virtual library followed by filtering with statistical design tools, before preparative work is started (FINN, 1996).

3.3.2 Design Tools for Library Tuning

Beyond diversity, there are further design tools which support the probability of success in a randomized screening approach. It is of utmost relevance to identify lead structures of high quality, i.e., a chemical structure which al-

Pharmacophoric properties of a drug-like compound are determined by:

- ➔ Number
- ➔ Type
- ➔ Spatial arrangement

of chemical groups which can interact with the biological target (e.g. receptor, enzyme)

Typical examples for chemical groups with pharmacophoric properties are:

• H-bond acceptor groups	(e.g. >=O, OH, 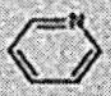)
• H-bond donor groups	(e.g. COOH, OH, NH_3^+)
• Coulomb interaction	(e.g. $R\text{-}N^+H_3$, COO^-)
• Dipolar interaction	(e.g. F, CF_3, NO_2)
• Non-polar interaction	(e.g. aromatic rings, aliphatic chains)

Fig. 2. In order to gain enough free energy of interaction with a protein target molecule, a drug-like small chemical compound has to be decorated with appropriate substituents.

Comparison between a pair of bit strings (barcodes)
[absence (0) or presence (1) of structural keys]

$$\text{Tanimoto coefficient: Tan Coeff (a, b)} = \frac{N_{ab}}{N_a + N_b - N_{ab}}$$

N_{ab} = number of bits which equal 1 in both bit strings
N_a = number of bits which equal 1 in bit string a (structure a)
N_b = number of bits which equal 1 in bit string b (structure b)

values between 0,0 and 1,0
1,0: a pair with identical 2D structures
0,0: a pair wthout common structure fragments

Fig. 3. As a common indicator for structural diversity, the lack of similarity is used, quantified by the Tanimoto coefficient. The chemical structures are described by "barcodes" (bit strings) derived from substructure (structural keys) analysis. A Tan coeff >0.85 can serve as indicator for similar screening results ("similar property principle").

ready shows a maximum of drug-like features thus shortening the lead structure optimization process. Cell permeability is one of the most important criteria. As state of the art, the "rule of five" is applied: Poor absorption or cell permeation is more likely when there are more than 5 H bond donors, 10 H bond acceptors, the molecular weight is >500, and the calculated log$p$ is >5 (LIPINSKI et al., 1997). Early prediction methods for pharmacokinetics open new possibilities for optimization of a drug discovery library before doing any experimental screening (DARVAS et al., 2000a) addressing ADME (adsorption, distribution, metabolism, excretion).

Chemical compound stability is an important issue to address coping with large compound repositories. Beyond statistical spot checks, there are systematic tools extrapolating thermal decomposition by the Arrhenius equation (DARVAS et al., 2000b).

4 Outlook

Modern HTS and combinatorial synthesis technologies have changed the drug identification process dramatically. Functional genomics has and will further dominate the target identification and screening assay development process in the future. Interlocking bio-informatics and chemo-informatics, pharmacophore-informatics will gain significant relevance as a new discipline, using pattern recognition and sophisticated *in silico* techniques to further guide the screening library design on the basis of the huge increase of multi-million data created by high-throughput screening technologies per day.

5 References

ALAJOKI, M. L., BAXTER, G. T., BEMISS, W. R., BLAU, D., BOUSSE, L. J. et al. (1997), High-performance microphysiometry in drug discovery, in: *High Throughput Screening: The Discovery of Bioactive Substances* (DEVLIN, J. P., Ed.), pp. 427–442. New York: Marcel Dekker.

BALKENHOL, F., VON DEM BUSSCHE-HÜNNENFELD, C., LANSKY, A., ZECHEL, C. (1996), Combinatorial synthesis of small organic molecules, *Angew. Chem.* (Int. Edn. Engl.) **35**, 2288–2337

BANNWARTH, W. (2000), Combinatorial chemistry in solution, in: *Combinatorial Chemistry, a Practical Approach* 1st Edn. (BANNWARTH, W., FELDER, E., Eds.), pp. 5–45. Weinheim: Wiley-VCH.

BOCK, J. L. (2000), The new era of automated immunoassay, *Am. J. Clin. Pathol.* **113**, 528–546.

BOLGER, R., CHECOVICH, W. (1994), A new protease activity assay using fluorescence polarization, *Biotechniques* **17**, 585–589.

BOSWORTH, N., TOWERS, P. (1989), Scintillation proximity assay, *Nature* **341**, 167–168.

BOUSSE, L., COHEN, C., NIKIFOROV, T., CHOW, A., KOPF-SILL, A. R. et al. (2000), Electrokinetically controlled microfluidic analysis systems, *Ann. Rev. Biophys. Biomol. Struct.* **29**, 155–181.

BRONSTEIN, I., FORTIN, J., STANLEY, P. E., STEWART, G. S. A. B., KRICKA, L. J. (1994), Chemoluminescent and bioluminescent reporter gene assays, *Anal. Biochem.* **219**, 169–181.

BROWN, B. A., CAIN, M., BROADBENT, J., TOMPKINS, S., HENRICH, G. et al. (1997), FlashPlate™ technology, in: *High Throughput Screening: The Discovery of Bioactive Substances* (DEVLIN, J. P., Ed.), pp. 317–328. New York: Marcel Dekker.

BUCHERT, M., SCHNEIDER, S., ADAMS, M. T., HETFI, H. P., MOELLING, K., HOVENS, C. M. (1997), Useful vectors for the two-hybrid system in mammalian cells, *Biotechniques* **23**, 396–402.

BUNIN, B. A. (1998), *The Combinatorial Index*. San Diego, CA: Academic Press.

BURBAUM, J. J. (1998), Miniaturization technologies in HTS: How fast, how small, how soon? *Drug Discovery Today* **3**, 313–322.

CAGNEY, G., UETZ, P., FILEDS, S. (2000), High-throughput screening for protein–protein interactions using two-hybrid assays, *Methods Enzymol.* **328**, 3–16.

CHALFIE, M., EUSKIRCHEN, Y., TU, G., WARD, W. W., PRASHER, D. C. (1994), GFP as a marker for gene expression, *Science* **263**, 802–805.

CHECOVICH, W. J., BOLGER, R. E., BURKE, T. (1995), Fluorescence polarization – a new tool for cell and molecular biology, *Nature* **375**, 254–256.

CULLEN, B. R. (2000), Utility of the secreted placental alkaline phosphatase reporter enzyme, *Methods Enzymol* **326**, 159–164.

DARVAS, F., DORMÁN, G., PAPP, Á. (2000a), Diversity measures for enhancing ADME admissibility of combinatorial libraries, *J. Chem. Inf. Comput. Sci.* **40**, 314–322.

DARVAS, F., KARANCSI, T., SLÉGEL, P., DORMÁN, G. (2000b), High-throughput stability estimation for combinatorial libraries and repositories, *Gen. Eng. News* **20**, 30–31.

DAVEY, H. M., KELL, D. B. (1996), Flow cytometry and cell sorting of heterogeneous microbial populations: The importance single cell analyses, *Microbiol. Rev.* **60**, 641–696.

DENYER, J., WORLEY, J., COX, B., ALLENBY, G., BANKS, M. (1998), HTS approaches to voltage-gated ion channel drug discovery, *Drug Discovery Today* **3**, 323–332.

DÖRWALD, F. Z. (2000), *Organic Synthesis on Solid Phase*. Weinheim: Wiley-VCH.

DOMINIK, A. (2000), Computer-assisted library design, in: *Combinatorial Chemistry, a Practical Approach* 1st Edn. (BANNWARTH, W., FELDER, E.,

Eds.), pp. 277–327. Weinheim: Wiley-VCH.

DORMÁN, G. (2000), Photoaffinity labeling in biological signal transduction, in: *Topics in Current Chemistry* 1st Edn. Vol. 211 (WALDMANN, H., Ed.), pp. 169–225. Berlin: Springer-Verlag.

DORMÁN, G., PRESTWICH, G. D. (2000), Using photolabile ligands in drug discovery and development, *Trends Biotechnol.* **18**, 64–77.

DOVE, S. L., JOUNG, J. K., HOCHSCHILD, A. (1997), Activation of prokaryotic transcription through arbitrary protein–protein contacts, *Nature* **386**, 627–630.

DREWS, J. (2000), Drug discovery: a historical perspective, *Science* **287**, 1960–1964.

FIELDS, S., SONG, O. (1998), A novel genetic system to detect protein–protein interactions, *Nature* **340**, 125–137.

FINN, P. W. (1996), Computer-based screening of compound databases for the identification of novel leads, *Drug Design Today* **1**, 363–370.

FITCH, W. L., LOOK, G. C., DETRE, G. (1998), Analytical chemistry issues in combinatorial organic synthesis, in: *Combinatorial Chemistry and Molecular Diversity in Drug Discovery* 1st Edn. (GORDON, E. M., KERWIN, J. F., Jr. Eds), pp. 349–368. New York: Wiley-Liss.

FLOWER, D. R. (1998), On the properties of bit string-based measures of chemical similarity, *J. Chem. Inf. Comput. Sci.* **38**, 379–386.

FOX, S. J., YUND, M. A., FARR-JONES, S. (2000), Assay innovations vital to improving HTS, *Drug Disc. Dev.* 40–43.

FRENCH, T., BAILEY, B., STUMBO, D. P., MODLIN, D. N. (1999), A time resolved fluorimeter for high-throughput screening, *Proc. Soc. Opt. Eng.* **3603**, 272–280.

FURKA, A., SEBESTYEN, F., ASGEDOM, M., DIBO, G. (1991), General method for rapid synthesis of multicomponent peptide mixtures, *Int. J. Pept. Protein Res.* **37**, 487–493.

GELMINI, S., PINZANI, P., PAZZAGLI, M. (2000), Luciferase gene as reporter: comparison with the CAT gene and use in transfection and microinjection of mammalian cells, *Methods Enzymol.* **305**, 557–576.

GONZÁLEZ, J. E., OADES, K., LEYCHKIS, Y., HAROOTUNIAN, A., NEGULESCU, P. A. (1999), Cell based assays and instrumentation for screening ion-channel targets, *Drug Discovery Today* **4**, 431–439.

GORDON, E. M. (1998), Strategies in the design and synthesis of chemical libraries, in: *Combinatorial Chemistry and Molecular Diversity in Drug Discovery* 1st Edn. (GORDON, E. M., KERWIN, J. F. Jr., Eds.), pp. 17–38. New York: Wiley-Liss.

HÄNNINEN, P., SOINI, A., MELTOLA, N., SOINI, J., SOUKKA, J., SOINI, E. (2000), A new microvolume technique for bioaffinity assays using two-photon excitation, *Nature Biotechnol.* **18**, 548–550.

HAUGLAND, R. P. (1999), *Handbook of Fluorescent Probes and Research Chemicals* (CD-ROM), 7th Edn. Eugene: Molecular Probes, Inc.

HEMMILÄ, I. (1997), Time resolved fluorimetry: Advantages and potentials, in: *High Throughput Screening: The Discovery of Bioactive Substances* (DEVLIN, J. P., Ed.), pp. 361–376. New York: Marcel Dekker.

HERTZBERG, R. P., POPE, A. J. (2000), High-throughput screening: new technology for the 21st century, *Curr. Opin. Chem. Biol.* **4**, 445–451.

HEYDUK, T., MA, Y., TAND, H., EBRIGHT, R. H. (1996), Fluorescence anisotropy: Rapid, quantitative assay for protein–DNA and protein–protein interaction, *Methods Enzymol.* **274**, 492–503.

HILL, J. J., ROYER, C. A. (1997), Fluorescence approaches to study of protein–nucleic acid complexation, *Methods Enzymol.* **278**, 390–416.

HINZEN, B. (2000), Encoding strategies for combinatorial libraries, in: *Combinatorial Chemistry, a Practical Approach* 1st Edn. (BANNWARTH, W., FELDER, E., Eds.), pp. 239–242. Weinheim: Wiley-VCH.

HOPFINGER, A. J., DUCA, J. S. (2000), Extraction of pharmacophore information from high-throughput screens, *Curr. Opin. Biotechnol.* **11**, 97–103.

HOUGHTEN, R. A., PINILLA, C., BLONDELLE, S. E., APPEL, J. R., DOOLEY, C. T., CUERVO, J. H. (1991), Generation and use of synthetic peptide combinatorial libraries for basic research and drug discovery, *Nature* **354**, 84–86.

HUANG, J., SCHREIBER, S. L. (1997), A yeast genetic system for selecting small molecule inhibitors of protein–protein–interactions in nano-droplets, *Proc. Natl. Acad. Sci. USA* **94**, 13396–13401.

JACOBS, J. W., NI, Z.-J. (1998), Encoded combinatorial chemistry, in: *Combinatorial Chemistry and Molecular Diversity in Drug Discovery* 1st Edn. (GORDON, E. M., KERWIN, J. F. Jr., Eds), pp. 271–290. New York: Wiley-Liss.

JACOBS, J. W., PATEL, D. V., YUAN, Z., HOLMES, C. P., SCHULLEK, J. et al. (1998), Light-directed chemical synthesis of positionally encoded peptide arrays, in: *Combinatorial Chemistry and Molecular Diversity in Drug Discovery* 1st Edn. (GORDON, E. M., KERWIN, J. F. Jr., Eds), pp. 111–131. New York: Wiley-Liss.

JAMESON, D. M., SAWYER, W. H. (1995), Fluorescence anisotropy applied to biomolecular interactions, *Methods Enzymol.* **246**, 283–300.

JOLLEY, M. E. (1996), Fluorescence polarization assays for the detection of proteases and their inhibitors, *J. Biomol. Screening* **1**, 33–38.

JUNG, G. (Ed.) (1999), *Combinatorial Chemistry – Synthesis, Analysis, Screening*. Weinheim: Wiley-VCH

KARAMOHAMED, S., NYREN, P. (1999), Real-time detection and quantification of adenosine triphosphate sulfurylase activity by a bioluminometric approach, *Anal. Biochem.* **271**, 81–85.

KASK, P., PALO, K., ULLMANN, D., GALL, K. (1999), Fluorescence-intensity distribution analysis and its application in biomolecular detection technology, *Proc. Natl. Acad. Sci. USA* **96**, 13756–13761.

KENDALL, J. M., BADMINTON, M. N. (1998), *Aequorea victoria* bioluminescence moves into an exciting new era, *Trends Biotechnol.* **16**, 216–224.

KESSLER, C. (1992), General aspects of nonradioactive labeling and detection, in: *Nonradioactive Labeling and Detection of Biomolecules* (KESSLER, C., Ed.), pp. 1–23, Berlin: Springer-Verlag.

KIBBEY, E. C. (1997), Analytical tools for solution-phase synthesis, in: *A Practical Guide to Combinatorial Chemistry* 1st Edn. (CZARNIK, A. W., DEWITT, S. H., Eds.), pp. 249–277. Washington, DC: American Chemical Society.

KNIGHT, C. G. (1995), Fluorimetric assays of proteolytic enzymes, *Methods Enzymol.* **248**, 18–34.

KOLB, A. J., BURKE, J. W., MATHIS, G. (1997), Homogeneous, time-resolved fluorescence method for drug discovery, in: *High Throughput Screening: The Discovery of Bioactive Substances* (DEVLIN, J. P., Ed.), pp. 345–360. New York: Marcel Dekker.

KRICKA, L. J., VOYTA, J. C., BRONSTEIN, I. (2000), Chemiluminescent methods for detecting and quantitating enzyme activity, *Methods Enzymol.* **305**, 417–427.

KUBINYI, H. (Ed.) (1993), *3D-QSAR in Drug Design: Theory, Methods and Applications*. Leiden: ESCOM.

LAM, K. S., SALMON, S. E., HERSH, E. M., HRUBY, V. J., KAZMIERSKI, W. M., KNAPP, R. J. (1991), A new type of synthetic peptide library for identifying ligand-binding activity, *Nature* **354**, 82–84.

LENZ, G. R., NASH, H. M., JINDAL, S. (2000), Chemical ligands, genomics and drug discovery, *Drug Discovery Today* **5**, 145–156.

LIPINSKI, C. A., LOMBARDO, F., DOMINY, B. W., FEENEY, P. J. (1997), Experimental and computational approaches to estimate solubility and permeability in drug discovery and development settings, *Adv. Drug Delivery Rev.* **23**, 3–25.

LUNDIN, A. (2000), Use of firefly luciferase in ATP-related assays of biomass, enzymes, and metabolites, *Methods Enzymol.* **305**, 346–370.

LYALL, A. (1998), Bioinformatics: A revolution in drug discovery, *Scrip Reports*. Richmond: PJB Publications.

MAITI, S., HAUPTS, U., WEBB, W. W. (1997), Fluorescence correlation spectroscopy: diagnostics for sparse molecules, *Proc. Natl. Acad. Sci. USA* **94**, 11753–11757.

MATHIS, G. (1995), Probing molecular interactions with homogeneous techniques based on rare earth cryptates and fluorescence energy transfer, *Clin. Chem.* **41**, 1391–1397.

MCCONNELL, H. M., OWICKI, J. C., PARCE, J. W., MILLER, D. L., BAXTER, G. T. et al. (1992), The cytosensor microphysiometer: biological applications of silicon technology, *Science* **257**, 1906–1912.

MERE, L., BENNETT, T., COASSIN, P., ENGLAND, P., HAMMAN, B. et al. (1999), Miniaturized FRET assays and microfluidics: key components for ultra-high-throughput screening, *Drug Discovery Today* **4**, 363–369.

MIRAGLIA, S., SWARTZMAN, E. E., MELLENTIN-MICHELOTTI, J., EVANGELISTA, L., SMITH, C. et al. (1999), Homogeneous cell- and bead-based assays for high throughput screening using fluorometric microvolume assay technology, *J. Biomol. Screening* **4**, 193–204.

MISTELI, T., SPECTOR, D. L. (1997), Applications of the green fluorescent protein in cell biology and biotechnology, *Nature Biotechnol.* **15**, 961–964.

MOORE, K. J., TURCONI, S., ASHMAN, S., RUEDIGER, M., HAUPTS, U. et al. (1999), Single molecule detection technologies in miniaturized high throughput screening: Fluorescence correlation spectroscopy, *J. Biomol. Screening* **4**, 335–354.

MUNDER, T., HINNEN, A. (1999), Yeast cells as tools for target-oriented screening, *Appl. Microbiol. Biotechnol.* **52**, 311–320.

MURAKAMI, A., NAKAURA, M., NAKATSUJI, Y., NAGAHARA, S., TRAN-CONG, Q., MAKINO, K. (1991), Fluorescent-labeled oligonucleotide probes: Detection of hybrid formation in solution by fluorescence polarization spectroscopy, *Nucleic Acids Res.* **19**, 4097–4102.

NASIR, M. S., JOLLEY, M. E. (1999), Fluorescence polarization: An analytical tool for immunoassay and drug discovery, *Comb. Chem. High Throughput Screening* **4**, 177–190.

OBRECHT, D., VILLALGORDO, J. M. (1998), *Solid Supported Combinatorial and Parallel Synthesis of Small-Molecular-Weight Compound Libraries*. 1st Edn. Oxford: Elsevier Science.

PARKER, G. J., LAW, T. L., LENOCH, F. J., BOLGER, R. E. (2000), Development of high throughput screening assays using fluorescence polarization: Nuclear receptor-ligand–binding and kinase/phosphatase assays, *J. Biomol. Screening* **5**, 77–88.

PERSIDIS, A. (1998), High-throughput screening, *Nature Biotechnol.* **16**, 488–489.

PICARDO, M., HUGHES, K. T. (1997), Scintillation proximity assays, in: *High Throughput Screening: The Discovery of Bioactive Substances* (DEVLIN, J. P., Ed.), pp. 307–316. New York: Marcel Dekker.

POPE, A. J., HAUPTS, U. M., MOORE, K. J. (1999), Homogeneous fluorescence readouts for miniaturized high-throughput screening: theory and practice, *Drug Discovery Today* **4**, 350–362.

RAMM, P. (1999), Imaging systems in assay screening, *Drug Discovery Today* **4**, 401–410.

RIGLER, R. F. (1995), Fluorescence correlations, single molecule detection and large number screening. Applications in biotechnology, *J. Biotechnol.* **41**, 177–186.

ROGERS, M. V. (1997), High-throughput screening: Liquid handling systems for 384-well applications, *Drug Discovery Today* **2**, 395–396.

ROSE, L. D. (1999), Microdispensing technologies in drug discovery, *Drug Discovery Today* **4**, 411–419.

SCHADE, S. Z., JOLLEY, M. E, SARAUER, B. J, SIMONSON, L. G. (1996), BODIPY-Casein, a pH-independent protein substrate for protease assays using fluorescence polarization, *Anal. Biochem.* **243**, 1–7.

SCHEIRER, W. (1997), Bioassay design and implementation, in: *High Throughput Screening: The Discovery of Bioactive Substances* (DEVLIN, J. P., Ed.), pp. 401–412. New York: Marcel Dekker.

SCHREIBER, S. L. (2000), Target-oriented and diversity-oriented organic synthesis in drug discovery, *Science* **287**, 1964–1969.

SEETHALA, R., MENZEL, R. (1998), A fluorescence polarization competition immunoassay for tyrosine kinases, *Anal. Biochem.* **255**, 257–262.

SELVIN, P. R. (1995), Fluorescence resonance energy transfer, *Methods Enzymol.* **246**, 300–334.

SOINI, T. J., SOUKKA, J. M., MELTOLA, N. J., SOINI, A. E., SOINI, E., HÄNNINEN, P. E. (2000), Ultra sensitive bioaffinity assay for micro volumes, *Single Mol.* **1**, 203–206.

SPORTSMAN, R. J., LEE, S. K., DILLEY, H., BUKAR, R. (1997), Fluorescence polarization, in: *High Throughput Screening: The Discovery of Bioactive Substances* (DEVLIN, J. P., Ed.), pp. 389–400. New York: Marcel Dekker.

SULLIVAN, E., TUCKER, E. M., DALE, I. L. (1999), Measurement of $[Ca^{2+}]$ using the fluorometric imaging plate reader (FLIPR), *Methods Mol. Biol.* **114**, 125–133.

SUNDBERG, S. A. (2000), High-throughput and ultra-high-throughput screening: solution- and cell-based approaches, *Curr. Opin. Biotechnol.* **11**, 47–53.

TANAHASHI, H., ITO, T., INOUYE, S., TSUJI, F. I., SAKAKI, Y. (1990), Photoprotein aequorin: use as a reporter enzyme in studying gene expression in mammalian cells, *Gene* **96**, 249–255.

TERPETSCHING, E., SZMACINSKI, H., LAKOWICZ, J. R. (1997), Long lifetime metal-ligand complexes as probes in biophysics and clinical chemistry, *Methods Enzymol.* **278**, 295–321.

THOMPSON, L. A., ELLMAN, J. A. (1996), Synthesis and applications of small molecule libraries, *Chem. Rev.* **96**, 555–600.

TIPTON, K. F. (1992), Principles of enzyme assay and kinetic studies, in: *Enzyme Assays. A Practical Approach* (EISENTHAL, R., DANSON, M. J., Eds.), pp. 1–58. Oxford: IRL Press.

VIDAL, M., ENDOH, H. (1999), Prospects for drug screening using the reverse two-hybrid system, *Trends Biotechnol.* **17**, 374–381.

VOGEL, H., G., VOGEL, W. H. (1997), *Drug Discovery and Evaluation: Pharmacological Assays*. Heidelberg: Springer-Verlag.

WALLACE, R. W., GOLDMAN, M. E. (1997), Reporter gene assay applications, in: *High Throughput Screening: The Discovery of Bioactive Substances* (DEVLIN, J. P., Ed.), pp 279–306. New York: Marcel Dekker.

WALTER, G., BUSSOW, K., CAHILL, D., LUEKING, A., LEHRACH, H. (2000), Protein arrays for gene expression and molecular interaction screening, *Curr. Opin. Microbiol.* **3**, 298–302.

WEBER, G. (1997), Fluorescence in biophysics: accomplishments and deficiencies, *Methods Enzymol.* **278**, 1–14.

WHITTAKER, M. (1998), Discovery of protease inhibitors using targeted libraries, *Curr. Opin. Chem. Biol.* **2**, 386–396.

WILLIAMS, C. (2000), Biotechnology match making: Screening orphan ligands and receptors, *Curr. Opin. Biotechnol.* **11**, 42–46.

WILLIARD, X., TARTAR, A. (1997), Deconvolution tools in solution–phase synthesis, in: *A Practical Guide to Combinatorial Chemistry* 1st Edn. (CZARNIK, A. W., DEWITT, S. H., Eds.), pp. 249–277. Washington, DC: American Chemical Society.

WINKLER, T., KETTLING, U., KOLTERMANN, A., EIGEN, M. (1999), Confocal fluorescence coincidence analysis: An approach to ultra high-throughput screening, *Proc. Natl. Acad. Sci. USA* **96**, 1375–1378.

WU, P., BRAND, L. (1994), Resonance energy transfer: Methods and applications, *Anal. Biochem.* **218**, 1–13.

XUE, L., GODDEN, J. W., BAJORATH, J. (2000), Evaluation of descriptors and mini-fingerprints for the identification of molecules with similar activity, *J. Chem. Inf. Comput. Sci.* **40**, 1227–1234.

ZLOKARNIK, G. (2000), Fusions to beta-lactamase as a reporter for gene expression in live mammalian cells, *Methods Enzymol.* **326**, 221–244.

ZLOKARNIK, G., NEGULESCU, P. A., KNAPP, T. E., MERE, L., BURRES, N. et al. (1998), Quantitation of transcription and clonal selection of single living cells with beta-lactamase as reporter, *Science* **279**, 84–88.

4 Biocatalysis under Extreme Conditions

COSTANZO BERTOLDO
RALF GROTE
GARABED ANTRANIKIAN
Hamburg, Germany

1 Introduction

The industrial application of biocatalysts began in 1915 with the introduction of the first dehairing (pancreatic) enzyme by Dr. RÖHM. Since that time enzymes have found wider application in various industrial processes and production. The most important fields of enzyme application are nutrition, pharmaceuticals, diagnostics, detergents, textile and leather industries. There are more than 3,000 enzymes known to date that catalyze different biochemical reactions among the estimated 7,000. Interestingly, only 100 enzymes are being used industrially. The world market for industrial enzymes, which includes enzymes for research and diagnosis, is estimated to be around 1 billion US$. The products derived from these enzymes are estimated to represent a value of more than 100 billion US$. For various industrial applications there is a great demand for enzymes of high specificity and stability. Extreme environments provide a unique resource of microorganisms and novel biocatalysts. Microorganisms that live under extreme conditions are defined as extremophiles. Many parts of the world are considered extreme such as geothermal environments, polar regions, acid and alkaline springs, and cold pressurized depths of the oceans. As conditions become increasingly demanding, extreme environments become exclusively populated by microorganisms belonging to the bacterial and archaeal domains (prokaryotes). It is very likely that higher organisms are unable to survive under extreme conditions due to their cellular complexity and compartmentation. The realization that extreme environments harbor a different kind of prokaryote lineage has resulted in a complete reassessment of our concept of microbial evolution and has given con siderable impetus to extremophile research (LEUSCHNER and ANTRANIKIAN, 1995; LADENSTEIN and ANTRANIKIAN, 1998; NIEHAUS et al., 1999; HORIKOSHI and GRANT, 1988).

It is worth mentioning that modern biotechnology, which provides a whole new repertoire of methods and products, still tries to mimic nature, thus demanding continuous efforts in the isolation and characterization of novel microorganisms. In this chapter, we will focus on biocatalysts that are produced by microorganisms that are living under extreme conditions. In particular, we will focus on organisms that thrive at high temperatures and produce enzymes that are of biotechnological value.

2 Extreme Environments as a Resource for Novel Microorganisms and Enzymes

2.1 Microorganisms Growing at the Freezing Point of Water

Many parts of the world seldom or never reach temperatures above 5 °C. This is mainly due to the fact that almost 70% of the Earth is covered by deep oceans which can be considered as cold environments. Furthermore, the polar regions provide a permanently cold terrestrial habitat which is surrounded by an aquatic belt of melting ice. Numerous microorganisms, in particular bacteria, yeast, unicellular algae, and fungi have successfully colonized cold environments. These microorganisms, generally defined as "psychrophiles", are able to grow at temperatures close to 0 °C and have developed various adaptations in the form of finely tuned structural changes at various levels, e.g., of their enzymes, enabling them to compensate for the deleterious effect of low temperature. Most of the cold adapted microorganisms have been characterized from the Arctic and Antarctic seawaters; the latter are not only a cold habitat (around −1 °C) but also exert a high selective pressure on endemic microorganisms because the temperature of their environments is constant. Despite the harsh conditions the density of bacterial cells in the Antarctic oceans is as high as the density reported in temperate waters. The microorganisms which are able to grow at or close to the freezing point of water can be divided into two main groups: psychrophiles and psychrotolerants. Psychrophilic microorganisms are defined by an optimum temperature for growth at about 15 °C, a maximum growth temperature at about 20 °C, and a minimum temperature for growth at 0 °C or lower. In comparison, psychrotolerant microorganisms generally do not grow at 0 °C, and have optimum and maximum growth temperatures above 20 °C (MORITA, 1975). Psychrotolerant microorganisms growing at around 5 °C can also be found. In general, psychrophiles have significantly narrower growth temperature ranges and lower optimum/maximum growth temperatures compared to psychrotolerant microorganisms (Tab. 1a). Psychrophiles can be found in permanently cold environments such as the deep sea, glaciers, mountain regions, in soils, in fresh or saline waters associated with cold-blooded animals such as fish or crustaceans. The first psychrophilic bacterium was isolated by FORSTER in 1887 from preserved fish. Recently, a systematic investigation has been carried out in order to understand the rules governing their molecular adaptation to low temperatures. These fundamental aspects are closely associated with a strong biotechnological interest aiming at the exploitation of these microorganisms and their cell components such as membranes, polysaccharides, and enzymes. The specific activity of wild-type cold adapted enzymes and some of their recombinant forms produced by Antarctic and Arctic microorganisms have been determined for several enzymes, including alcohol dehydrogenase, α-amylase, aspartate transcarbamylase, metal protease, citrate synthase, subtilisin, triose phosphate isomerase, and xylanase. In general, cold adapted enzymes have higher specific activity at low and moderate temperatures than that of their mesophilic counterparts, and are inactivated easily by a slight increase in temperature. These properties can be extremely useful in various applications. The possible applications of cold active enzyme are as detergents for cold washing and food additives such as polyunsaturated flavor-modifying agents. In the dairy industry, β-galactosidase is applied to reduce the amount of lactose in milk, which is responsible for lactose intolerance in approximately two thirds of the world's population. The clarification of fruit juices is achieved by the addition of cold active pectinases. Further applications of cold active enzymes are found in the tenderizing of meat by proteases and improving the baking process by the addition of amylases, proteases, and xylanases. Cold active enzymes are also used in biosensors for environmental applications and in cleaning of contact lenses (RUSSELL and HAMAMOTO, 1998).

Some of the advantages of using psychrophiles and their enzymes in biotechnological application are the rapid termination of the process by moderate heat treatment, higher yields of thermosensitive components, modulation of the (stereo-)specificity of enzyme-catalyzed reactions, cost saving by elimination of expensive heating/cooling process steps, and finally the capacity for on-line monitoring under environmental conditions (GERDAY et al., 2000; FELLER et al., 1996).

2.2 Microbial Life at the Boiling Point of Water

Microorganisms that are adapted to grow optimally at high temperatures (60–108 °C) have been isolated from high-temperature terrestrial and marine habitats. The most common biotopes are volcanically and geothermal heated hydrothermal vent systems such as solfataric fields, neutral hot springs, and submarine hot vents. Submarine hydrothermal systems are situated in shallow and abyssal depth. They consist of hot fumaroles, springs, sediments, and deep-sea vents with temperatures up to 400 °C ("black smokers") (STETTER, 1996). Shallow marine hydrothermal systems are located at the beaches of Vulcano, Naples and Ischia (Italy), São Míguel (Azores), and Djibouti (Africa). Examples of deep-sea hydrothermal systems are the Guaymas Basin (depth 1,500 m), the East Pacific Rise (depth 2,500 m), both off the coast of Mexico, the Mid-Atlantic Ridge (depth 3,700 m), and the Okinawa Trough (depth 1,400 m) (GROTE et al., 1999; JEANTHON et al., 1998, 1999; GONZÁLES et al., 1998; CANGANELLA et al., 1998). Because of their ability to convert volcanic gases and sulfur compounds at high temperatures, hyperthermophilic communities living in such hydrothermal vents are expected to play an important role in marine ecological, geochemical and volcanic processes (HUBER et al., 1990). Shallow as well as deep-sea hydrother-

Tab. 1. Some Representatives of Microorganisms Living under Extreme Conditions

a Microorganisms Living at the Freezing Point of Water

Psychrophilic Microorganisms	Optimal Growth [°C]
Vibrio sp.	<20
Micrococcus cryophilus	<20
Arthrobacter glacialis	<20
Vibrio psychroerythreus	<20
Aquaspirillum articum	<20

b Microbial Life at the Boiling Point of Water

Microorganism	Optimal Growth [°C]
Moderate thermophiles (50–60 °C)	
Bacillus acidocaldarius	50
Bacillus stearothermophilus	55
Extreme thermophiles (60–80 °C)	
Thermus aquaticus	70
Thermoanaerobacter ethanolicus	65
Clostridium thermosulfurogenes	60
Fervidobacterium pennivorans	75
Hyperthermophiles (80–110 °C)	
Thermotoga maritima	90
Aquifex pyrophilus	85
Archeoglobus fulgidus	83
Methanopyrus kandleri	88
Sulfolobus sulfataricus	88
Thermococcus aggregans	88
Pyrobaculum islandicum	100
Pyrococcus furiosus	100
Pyrodictium occultum	105
Pyrolobus fumarii	106

c Microorganisms Growing at Extreme pH Values

Acidophilic Microorganisms	Optimal Growth [°C]	pH
Sarcina ventriculi	37	4.0
Thiobacillus ferrooxidans	37	2.5
Alicyclobacillus acidocaldarius	55	2.0–6.0
Picrophilus oshimae	60	0.7
Picrophilus torridus	60	0.7
Thermoplasma acidophilum	60	2.0
Sulfolobus acidocaldarius	75	2.5
Acidianus infernus	75	2.0

Alkaliphilic Microorganisms	[°C]	pH
Many cyanobacteria		6.0–8.0
Spirulina sp.		8.0–10.0
Chromatium sp.		8.5
Bacillus sp.		11.5

Tab. 1. Continued

Alkaliphilic Microorganisms	[°C]	pH
Anaerobranca gottschallkii	55	9.0
Thermococcus alcaliphilus	85	9.0
Thermococcus acidoaminivorans	85	9.0

d Survival with Salts

Halophilic Microorganisms	NaCl Concentration [M] Required for Growth		
	Minimum	Optimum	Maximum
Dunaliella spp.	0.3		5.0
Clostridium halophilum	0.15	0.6	6.0
Haloanaerobium praevalens	0.8	2.2	4.3
Halobacterium sp.		2.0	5.5
Halobacterium denitrificans	1.5	2.5	4.5
Haloferax vulcanii	1.0	1.5	3.0
Methanohalobium evestigatum			4.3

mal systems harbor members of various genera including *Pyrococcus*, *Pyrodictium*, *Igneococcus*, *Thermococcus*, *Methanococcus*, *Archaeoglobus*, and *Thermotoga*. So far, members of the genus *Methanopyrus* have been found only at greater depths, whereas *Aquifex* was isolated exclusively from shallow hydrothermal vents (STETTER, 1998). Recently, interesting biotopes of extreme- and hyperthermophiles were discovered in deep, geothermally heated oil reservoirs around 3,500 m below the bed of the North Sea and the permafrost soil of North Alaska (LIEN et al., 1998; STETTER et al., 1993).

Microorganisms capable of growing optimally at temperatures between 50 °C and 60 °C are designated as moderate thermophiles. Most of these microorganisms belong to many different taxonomic groups of eu- and prokaryotic microorganisms such as protozoa, fungi, algae, streptomycetes, and cyanobacteria, which comprise mainly mesophilic species. It can be assumed that moderate thermophiles, which are closely related phylogenetically to mesophilic organisms, may be secondarily adapted to life in hot environments. Extreme thermophiles, which grow optimally between 60 °C and 80 °C, are widely distributed among the genera *Bacillus*, *Clostridium*, *Thermoanaerobacter*, *Thermus*, *Fervidobacterium*, *Thermotoga*, and *Aquifex*. Most of the hyperthermophiles on the other hand, grow optimally between 80 °C and 108 °C (STETTER, 1996, 1998). Interestingly, and as shown in Fig. 1 the majority of the hyperthermophiles isolated to date belongs to the archaeal domain of life and no eukaryotic organism has been found that can grow at the boiling point of water. A 16S rDNA-based universal phylogenetic tree (WOESE et al., 1990; WOESE and FOX, 1977) shows a tripartite division of the living world consisting of the domains Bacteria, Archaea, and Eukarya (Fig. 1). The Archaea consist of two major kingdoms: the Crenarchaeota (some genera are *Sulfolobus*, *Picrophilus*, *Pyrodictium*, *Pyrolobus*, *Pyrobaculum*, and *Thermoproteus*) and the Euryarchaeota which include hyperthermophiles (some genera are *Thermococcus* and *Pyrococcus*), methanogenes (e.g., *Methanococcus*, *Methanobacterium*, and *Methanosarcina*), sulfate reducer (*Archaeoglobus*), and halophiles (including genera such as *Halobacterium* and *Halococcus*). Short phylogenetic branches indicate a rather slow clock of evolution. Deep branching points are evidence for early separation of the two groups. The separation of the Bacteria from the Eukarya–Archaea lineage is the deepest and earliest branching point known so far (STETTER, 1996). Hyperthermophiles are represented among all the deepest and shortest lineages, including the genera *Aquifex* and *Thermotoga* within the Bacteria and *Pyrodictium*, *Pyrobaculum*, *Thermoproteus*, *Desul-*

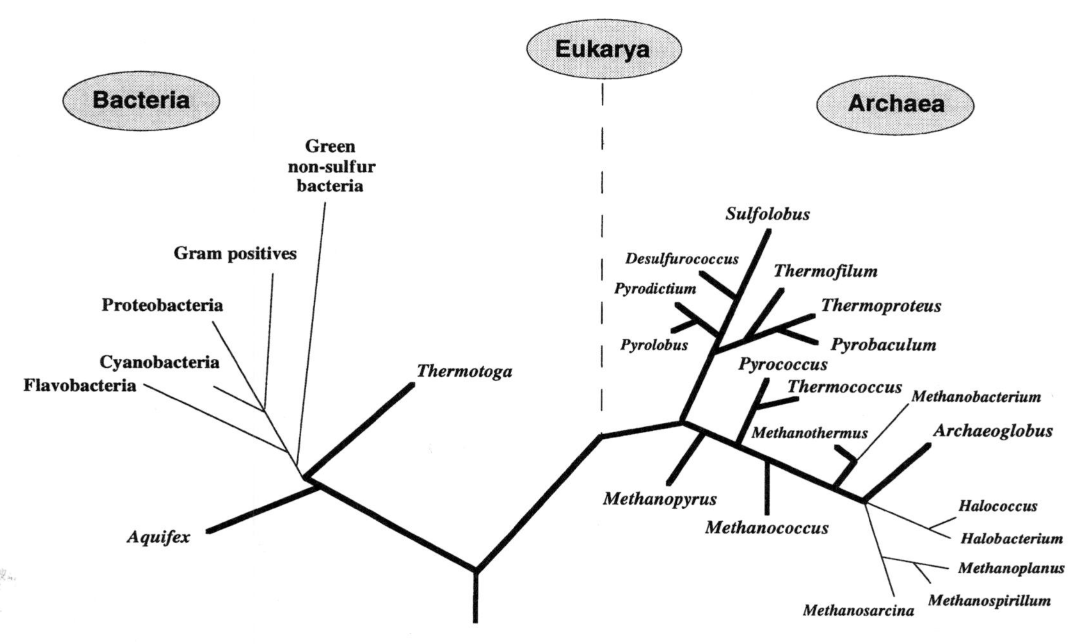

Fig. 1. Phylogenetic tree of organisms. Hyperthermophilic species are highlighted by bold branches. Branching order and branch lengths are based upon 16/18S rDNA sequence comparisons. The tree is a modified version of that from BLÖCHL et al. (1995).

furococcus, *Sulfolobus*, *Methanopyrus*, *Pyrococcus*, *Thermococcus*, *Methanococcus*, and *Archaeoglobus* within the Archaea (Fig. 1).

The relative abundance of Archaea and Bacteria in high-temperature environments was, until recently, mainly studied by cultivation-based techniques. Because of the frequent isolation of Archaea from these habitats, it was assumed that Archaea dominate the high-temperature biotopes (BAROSS and DEMING, 1995; STETTER et al., 1990). Recently, the application of molecular biological methods revealed a quite different picture. Slot-blot hybridizations of rRNA utilizing oligonucleotide probes targeting the 16S rRNA of Archaea and Bacteria revealed that Bacteria seem to be the major population of the microbial community along a thermal gradient at a shallow submarine hydrothermal vent near Milos Island (Greece) (SIEVERT et al., 2000). Bacteria made up at least 78% (mean 95%) of the prokaryotic rRNA. Along the steepest temperature gradient, the proportion of archaeal rRNA increased. Nevertheless, even in the hottest sediment layer archaeal rRNA made up only around 12% of the prokaryotic rRNA. These results suggest that Archaea may generally be of lower abundance in hot environments than could be assumed from cultivation-based experiments. However, the factors that allow Bacteria to dominate in high-temperature habitats, that were once believed to be the realm of Archaea, remain unknown. Most of these microorganisms that can be found in low-salinity and submarine environments are strict anaerobes. Terrestrial solfataric fields as they can be found in Italy or Iceland, harbor members of the genera *Pyrobaculum*, *Thermoproteus*, *Thermofilum*, *Desulfurococcus*, and *Methanothermus*. *Pyrobaculum islandicum* and *Thermoproteus tenax* are able to grow chemolithoautotrophically, gaining energy by anaerobic reduction of S^0 by H_2. In contrast to these strictly anaerobic microorganisms, *Pyrobaculum aerophilum* and *Aeropyrum pernix* are able to use oxygen as final electron acceptor. *Methanothermus fervidus*, on the other hand, is highly sensitive towards oxygen and can only survive in low-redox environments at temperatures between 65 °C and 97 °C. Some microorganisms from marine environments such as members of the genera *Archaeoglobus*, *Methanococcus*, and *Methanopyrus* are able to grow chemolithoautotrophically, gaining energy by the reduction of SO_4^{2-} by H_2 (*Archaeoglobus lithotrophicus* and *A. fulgidus*) or by the reduction of CO_2 by H_2 (*Methanococcus jannaschii*, *Methanopyrus kandleri*). Other members of the hyperthermophilic genera *Staphylothermus*, *Pyrococcus*, *Thermococcus*, and *Pyrodictium* are adapted to marine environments (NaCl concentration: about 30 g L^{-1}). Most of them gain energy by fermentation of polysaccharides, peptides, amino acids, and sugars (STETTER, 1996, 1999). Consequently, such thermophilic microorganisms have been found to be producers of interesting polymer degrading enzymes of industrial relevance (NIEHAUS et al., 1999).

Among the bacterial domain of life, members of the genera *Aquifex* and *Thermotoga* represent the deepest phylogenetic branches. Within this genus, *Thermotoga maritima* and *T. neapolitana* are the most thermophilic species with a maximal growth temperature of about 90 °C. The representatives of the order of Thermotogales also provide a resource of unique thermoactive enzymes. The enzymatic systems produced by extreme- and hyperthermophiles will be presented in this chapter. Some microorganisms living at high temperature are reported in Tab. 1b.

2.3 Microorganisms Growing at Extremes of pH

Solfataric fields are the most important biotopes of microorganisms that prefer to live under both thermophilic and acidic conditions. Solfataric soils consist of two different layers which can be easily distinguished by their characteristic colors: the upper, aerobic layer has an ochre color due to the presence of ferric iron. The layer below, which is anaerobic, appears rather blackish-blue owing to the presence of ferrous iron. According to the chemical parameters of the two layers, different kinds of microorganisms can be isolated from these habitats. Thermophilic acidophiles belonging to the genera *Sulfolobus*, *Acidianus*, *Thermoplasma*, and *Picrophilus*, with growth optima between 60 °C and 90 °C and pH 0.7 to 5.0 are commonly found in the aerobic upper layer whereas slightly acidophilic or neutrophilic anaerobes such as *Thermoproteus tenax* or *Methanothermus fervidus* can be isolated from the lower layer. Species of *Thermoplasma* (growth optima: pH 2 and 60 °C) have been found in hot springs, solfataras, and coal refuse piles (HORIKOSHI and GRANT, 1998). Their closest known phylogenetic relatives, also found in solfataras, are species of the genus *Picrophilus*, which are so far the most extreme acidophiles with growth close to pH 0. *Picrophilus oshimae* and *P. torridus* are both aerobic, heterotrophic Archaea that grow optimally at 60 °C and pH 0.7 and utilize various polymers such as starch and proteins as carbon source.

Members of the genus *Sulfolobus* are strict aerobes growing either autotrophically, heterotrophically, or facultative heterotrophically. During autotrophic growth, S^0, S^{2-}, and H_2 are oxidized to sulfuric acid or water as end products. *Sulfolobus metallicus* and *S. brierley* are able to grow by oxidation of sulfidic ores. A dense biofilm of these microorganisms is responsible for the microbial ore leaching process, in which heavy metal ions such as Fe^{2+}, Zn^{2+}, and Cu^{2+} are solubilized. During heterotrophic growth, a range of sugars and proteinaceous substrates is utilized. A spore-forming sulfur-oxidizing thermoacidophile, designated *Thiobacillus caldus* ($T_{max.}$ 55 °C), was isolated by HALLBERG and LINDSTRÖM (1994). Other hyperthermophilic acidophiles have been affiliated to the genera *Metallosphaera* (growth range: 50–80 °C, pH 1–4.5), *Acidianus* (growth range: 60–95 °C, pH 1.5–5), and *Stygioglobus* (growth range: 57–89 °C, pH 1–5.5) (STETTER, 1996).

On the other hand, the alkaliphiles that grow at high pH values are widely distributed throughout the world. They have been found in carbonate-rich springs and alkaline soils, where the pH can be around 10.0 or even higher, although the internal pH is maintained around 8.0. In such places, several species of cyanobacteria and *Bacillus* are normally abundant and provide organic matter for diverse groups of heterotrophs. Alkaliphiles require alkaline environments and sodium ions not only for growth but also for sporulation and germination. Sodium ion-dependent uptakes of nutrients have been reported in alkaliphiles. Many alkaliphiles require various nutrients for growth; few alkaliphilic *Bacillus* strains can grow in simple minimal media containing glycerol, glutamic acid, and citric acid (HORIKOSHI, 1998). In general, the cultivation temperature is in the range of 20–55 °C. Furthermore, many haloalkaliphiles isolated from alkaline hypersaline lakes can grow in alkaline media containing 20% NaCl. The soda lakes in the Rift Valley of Kenya and similar lakes found in a few other places on earth are highly alkaline with pH values between 11.0 and 12.0 and represent a typical habitat where alkaliphilic microorganisms can be isolated. Thermophilic anaerobic spore-forming alkaliphiles, thermoalkaliphilic clostridia, were isolated from sewage plants. Very recently, two thermoalkaliphilic bacteria, *Anaerobranca gottschallkii* and *A. horikoshii* have been isolated from Lake Bogoria in Kenya and from Yellowstone National Park, respectively. The new isolates represent a new line within the *Clostridium/Bacillus* subphylum (PROWE and ANTRANIKIAN, 2000). The two archaeal thermoalkaliphiles identified to date are *Thermococcus alcaliphilus* and *T. acidoaminivorans*, both growing at 85 °C and pH 9.0. The main industrial application of alkaliphilic enzymes is in the detergent industry, where they account for approximately 30% of the total worldwide enzyme production. Alkaline enzymes have been also used in the hide-dehairing process, where dehairing is carried out at pH values between 8.0 and 10.0 (HORIKOSHI, 1998) (Tab. 1c).

2.4 Halophilic Microorganisms

The halophiles comprise Bacteria and Archaea that grow optimally at NaCl concentrations above those of seawater (>0.6 M NaCl). In general, halophilic microorganisms are classified as moderate halophiles if they can grow at salt concentrations between 0.4 and 3.5 M NaCl and as extreme halophiles if they require NaCl concentrations above 2 M for growth (GRANT et al., 1998). Halophiles have been mainly isolated from saline lakes, such as the Great Salt Lake in Utah (salinity >2.6 M) and from evaporitic lagoons and coastal salterns with NaCl concentrations between 1 M and 2.6 M (GRANT et al., 1998). Saline soils are less well explored. Bulk salinity measurements of 1.7–3.4 M NaCl have been reported for saltern soils (VENTOSA and BIETO, 1995). Saline soils constitute less stable biotopes than hypersaline waters since they are subjected to periodic significant dilution during rainy periods. It can be assumed that microbial survival under these oscillating conditions would be even more difficult. There is no doubt, that almost all hypersaline habitats harbor significant populations of specifically adapted microorganisms. However, it remains unclear what substrates might be available for growth in these biotopes. Hypersaline lakes often contain up to 1 g L^{-1} of dissolved organic carbon. In many of these lakes, primary producers such as cyanobacteria, anoxygenic phototrophic bacteria, and algae may be the main source of organic compounds (GRANT et al., 1998).

It has been speculated that organic compatible solutes produced by many of the phototrophs as a means of counterbalancing the osmotic stress contribute significantly to the input of carbon sources. It is noteworthy that, despite the typically large surface-to-volume ratios, hypersaline environments are low in dissolved oxygen (<2 mg L^{-1}) and might be essentially anaerobic (for reviews, read GRANT et al., 1998; GRANT and TINDALL, 1986).

In a study of aerobic heterotrophs in marine salterns it has been shown that bacterial halophiles were predominant up to 2 M NaCl. Above this concentration archaeal halophiles become predominant, almost to the exclusion of bacteria. Halophilic primary producers mainly belong to the cyanobacteria and anoxygenic phototrophic sulfur bacteria. The former ones often thrive in eutrophic salterns forming large floating mats. The latter group, on the other hand, grows either in anaerobic sediments or in the water column where they are responsible for the characteristic red color of high-salinity habitats (RODRIGUEZ-VALERA et al., 1981). The range of heterotrophic Bacteria comprises proteobacteria, actinomycetes, and gram-positive rods and cocci. Fermentative anaerobes as well as sulfur oxidizers, sulfate reducers, and nitrate reducers are also present and give rise to the assumption that all kinds of metabolic features may be found in high-salinity environments (Tab. 1d). Halophilic Bacteria do not belong to one homogeneous group but rather fall into many bacterial taxa in which the capability to grow at high salt concentrations is a secondary adaptation.

The term "halobacteria" refers to the red-pigmented extremely halophilic Archaea, members of the family Halobacteriaceae, and the only family in the order Halobacteriales (VENTOSA and BIETO, 1995). Most halobacteria require 1.5 M NaCl in order to grow and to retain the structural integrity of the cell. Halobacteria can be distinguished from halophilic bacteria by their archaeal characteristic, in particular the presence of ether-linked lipids (ROSS et al., 1981). Most halobacteria are colored red or orange due to the presence of carotenoids, but some species are colorless, and those with gas vesicles form opaque, white or pink colonies. A purple hue may be seen in halobacteria that form the bacteriorhodopsin-containing purple membrane (GRANT et al., 1998). Halobacteria are the most halophilic organisms known so far and form the dominant microbial population when hypersaline waters approach saturation. Interestingly, the reddening caused by halobacterial blooms has an impact on the evaporation rates in saltern. It is known that the carotenoid pigments of halobacteria trap solar radiation, thus increasing the ambient temperature and evaporation rates (Tab. 1d).

The singular physiology of halophilic microorganisms that have to cope with 4 M ion concentration inside and outside the cells has theoretically evolved potentially interesting enzymes that might be capable of working under conditions of low water activity that could be imposed by other than salts, e.g., solvents. The reality is that, despite a range of potentially exploitable properties, halophiles have not yet had much of an impact on the commercial scene. However, there is still considerable interest, as evidenced by about 20% of all patent applications for extremophiles to date being concerned with halophiles in one form or other. Interestingly, halophilic and marine halotolerant bacteria produce and/or accumulate organic osmolytes (compatible solutes) for osmotic equilibrium. These metabolically compatible hygroscopic compounds not only protect living cells in a low-water environment but also exhibit an enzyme-stabilizing effect *in vitro* against a variety of stress factors such as heating, freezing, urea, and other denaturants. The ectoine-type osmolytes (2-methyl-1,4,5,6-tetrahydropyrimidine derivatives) represent the most abundant class of stabilizing solutes, typical for aerobic chemoheterotrophic halophilic and/or halotolerant bacteria (DA COSTA et al., 1998). The extrinsic stabilization effect of ectoines and other compatible solutes is most likely based on solvent-modulating properties of these compounds. The osmolytes already referred to have considerable potential as effective stabilizers of the hydration shell of proteins, and hence could be highly efficient stress protectants and stabilizers of biomolecules suitable for vaccines where refrigeration might be not available, or industrial enzymes functioning under extreme conditions. A number of enzymes were shown to be totally protected against heating and freeze–thaw cycles in the presence of a range of compatible solutes (GALINSKY, 1993). However, there was a significant variation in the degree of freeze and heat protection, dependent on compatible solute and enzyme under investigation. Recently, it has been shown that a number of hyperthermophilic microorganisms is also able to produce a variety of compatible solutes that were found to be effective in enzyme stabilization (DA COSTA et al., 1998).

2.5 Cultivation of Extremophilic Microorganisms

Extremophiles are receiving increasing interest because they provide a unique source of biocatalysts and cell components. However, until recently only low cell yields could be obtained making application studies very difficult. The main reason for this has to be ascribed to the difficulties related to produce and purify large quantities of biocatalysts and cell components. Moreover, extremophilic microorganisms require special equipment to reach and maintain the optimal cultivation temperatures and extreme pH. There are two different approaches to overcome this problems: recombinant DNA technique for increasing enzyme production in mesophilic hosts or innovative bioreactor design to improve biomass yield. Because the accumulation of toxic compounds is thought to be responsible for low biomass yields, dialysis fermentation with a number of extremophiles has been performed for effective removal of low-molecular-mass components from fermentation broth. Applying dialysis membrane reactors, a dramatic increase in cell yields was achieved (KRAHE et al., 1996). The cultivation of the hyperthermophilic archaeon *Pyrococcus furiosus* (growth at 90 °C), the thermoacidophile *Sulfolobus shibatae* (growth at 75 °C, pH 3.5), and the halophile *Marinococcus* M52 (growth at 35 °C, pH 7.5, and 10% NaCl) resulted in cell yields of 2.6 g L^{-1}, 114 g L^{-1} and 132 g L^{-1} (cell dry weight), respectively. For *P. furiosus* the optimum stirrer speed was 1,800 rpm and neither hydrogen nor the metabolic products were found responsible for the comparatively low cell yield. In the case of *S. shibatae*, which grows at low pH values, the choice of an appropriate membrane was crucial. Cuprophan membrane, which consists of regenerated cellulose and polyamide membrane was damaged after 2 d of operation probably due to enzyme action. A porous, non-transparent polyethersulfonic membrane was found to be stable. The fermentation processes can be scaled-up from 3 L over 30 L up to 300 L (Fig. 2). The pilot plant scale offers the possibility of transferring the fermentation performance into industrial standards. In recent experiments it was shown that even the results of the 1 L dialysis reactor can be reproduced in the 30 L reactor using external dialysis modules.

In addition to dialysis fermentation technique, also the application of a novel microfiltration (MF) bioreactor, based on a microfiltration hollow-fiber module located inside the traditional fermentation vessel, has been designed for improving both biomass yield and enzyme productivity. Using the cultivation of the thermoacidophilic archaeon *Sulfolobus solfataricus* as a model, a biomass of 35 g L^{-1} dry weight was obtained which was almost 20-fold higher than results obtained in conventional batch fermenters (SCHIRALDI et al., 1999).

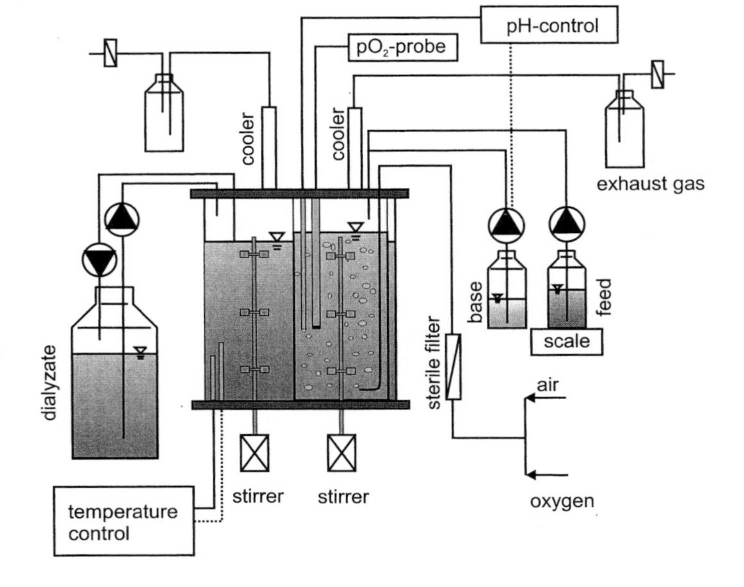

Fig. 2. Schematic presentation of a dialysis fermenter for suspended cells. Dimensions: length 300 mm, diameter 200 mm. Working volume of the culture chamber 1.2 L. Volume of the dialysis chamber 5 L. The figure was kindly provided by D. KÖSTER, C. FUCHS, and H. MÄRKL (Bioprocess and Biochemical Engineering, Technical University Hamburg-Harburg, Germany).

3 Biochemical Basis of Heat Stability

Due to the small cell size of hyperthermophiles, any heat protection by insulation against the hot environment is impossible. Therefore, all cell components have to be heat resistant, either intrinsically or by stabilization within the cells. The molecular basis of heat resistance is so far obscure and is still under investigation (LADENSTEIN and ANTRANIKIAN, 1998; STETTER, 1999; SCANDURRA et al., 1998; VETRIANI et al., 1998). As mentioned before, extreme- and hyperthermophiles belong to two different domains of life that are phylogenetically very divergent. Therefore, the molecular mechanisms of heat adaptation may be rather different depending on the phylogenetic position. Cell components such as lipids, nucleic acids, and proteins are usually very sensitive to heat. It has been reported that membrane lipids of the hyperthermophilic bacterium *Thermotoga maritima* are based on *n*-fatty acids, diabolic acid (15,16-dimethyl-30-triacontanedioic acid), and a novel glycerol ether lipid (15,16-dimethyl-30-glyceryloxytriacontanonic acid) (DE ROSA et al., 1989). The presence of an ether lipid may significantly increase stability of membranes against hydrolysis at the high growth temperature. In contrast, membranes of all archaea (including even mesophilic species) contain lipids derived from diphytanylglycerol diether or its dimer di(biphytanyl)diglycerol tetraether, which possess a remarkable resistance against hydrolysis at high temperatures and acidic pH (DE ROSA and GAMBACORTA, 1994; KATES, 1992).

Thermal resistance of the DNA double helix could be generally improved by increasing the GC content. However, GC analysis in extreme thermophiles and hyperthermophiles revealed that this is not the case. Interestingly, some genomes from hyperthermophiles such as *Acidianus infernus*, *Methanothermus sociabilis*, *Methanococcus igneus*, and *Pyrococcus furiosus* exhibit GC contents between 30 and 40 mol%, which are close to the lowest possible for maintenance of sufficient information. In principle, covalently linked (circular) double-stranded DNA of prokaryotes may be already much more stable than linear DNA (STETTER, 1998). All hyperthermophiles known so far possess reverse gyrase, a unique type I DNA topoisomerase that causes positive supertwists, and, therefore, may further stabilize the DNA molecule (FORTERRE et al., 1996). In contrast to bacteria, archaeal hyperthermophiles possess very basic histones that can be copurified with the DNA (GRAYLING et al., 1994; THOMM et al., 1982). *In vitro* addition of *Sulfolobus* histones to DNA has been demonstrated to increase the melting temperature dramatically (REDDY and SURYANARAYANA, 1988). The secondary structure of RNAs appears to be stabilized toward thermal destruction by an increased content of GC base pairs within the stem areas and post-transitionally by modifying bases and sugars (KAWAI et al., 1992; EDMONDS et al., 1991; WOESE et al., 1991). At the upper temperature border of growth of hyperthermophiles, the function of heat-shock proteins appears to become essential. At 108 °C about 80% of the soluble proteins of a crude extract from *Pyrodictium occultum* consist of a heat-inducible molecular chaperone designated thermosome (PHIPPS et al., 1991, 1993). The thermosome is related to the heat-shock protein of *Sulfolobus shibatae* and the eukaryotic TCP1 (TRENT et al., 1991; JEREZ, 1988). With the thermosome fully induced, cultures of *Pyrodictium occultum* are able to survive 1 h of autoclaving at 121 °C (PHIPPS et al., 1993). One can only speculate about the upper temperature border of life. Since the stability of biomolecules at temperatures above 100 °C decreases rapidly, the maximal growth temperature at which microbial life can exist may possibly be found between 110 °C and 150 °C. Within this temperature range heat-sensitive biomolecules could possibly be resynthesized at biologically feasible rates.

Several enzymes from hyperthermophiles have been purified and characterized (NIEHAUS et al., 1999). As a general rule, they show an extraordinary heat stability even *in vitro*. *Pyrococcus woesei* harbors an amylase that is active at 130 °C (KOCH et al., 1991). Within the bacterial domain *Thermotoga maritima* MSB8 possesses an extremely heat-resistant, toga-associated xylanase that is optimally active at 105 °C (WINTERHALTER and LIEBL, 1995). Even complex enzymes such as DNA-dependent RNA polymerases or glutamate dehydrogenases show a remarkable heat stability (SCHWERDTFEGER et al., 1999; ZILLIG et al., 1985). The principles of heat stabilization of thermoactive enzymes are so far not elucidated. The overall amino acid composition is very similar to homologous mesophilic enzymes. However, trends commonly associated with elevated thermostability in proteins include a relatively small solvent-exposed surface area (CHAN et al., 1995), increased packing density that reduces cavities in the hydrophobic core (LADENSTEIN and ANTRANIKIAN, 1998; BRITTON et al., 1995; ANDERSON et al., 1993; HURLEY et al., 1992), an increase in core hydrophobicity (SPASSOV et al., 1995), decreased length of surface loops (RUSSELL et al., 1994), and hydrogen bonds between polar residues (TANNER et al., 1996). Deeper insights will most likely be provided after comparison of three-dimensional structures obtained by X-ray analysis of crystallized enzymes as was shown in the case of glutamate dehydrogenase from *Thermotoga maritima* and *Pyrococcus furiosus* (KNAPP et al., 1997; YIP et al., 1995).

4 Screening Strategies for Novel Enzymes

Biotechnologically relevant enzymes can be detected in wild-type or recombinant host strains by classical or molecular biological methods. Hydrolytic enzymes are mostly screened by phenotypic detection of the corresponding activity using dyed substrates such as red-amylopectin, red-pullulan, or azo-casein. Positive strains – either wild-type or recombinant – can easily be distinguished from inactive ones by the formation of clearing zones around the colonies as was reported for the screening of a recombinant pullulanase from *Fervidobacterium pennivorans* Ven5 (BERTOLDO et al., 1999). In contrast to enzyme assays using crude extracts, this technique allows for the screening of a large number of strains in reasonable time. However, the described method can only be successfully applied when the enzyme activity is fully expressed in the strain of interest. Novel screening techniques on the DNA level have been shown to overcome this obstacle and supplement the classical screening methods. For example, genes encoding a desired enzyme can be detected and isolated by specific (non-)radioactive probes. By using this technique, a family B DNA polymerase from *Pyrobaculum islandicum* has been recently cloned, expressed, and characterized (KÄHLER and ANTRANIKIAN, 2000; NIEHAUS et al., 1997). Screening of gene libraries is another promising strategy for the detection of biotechnologically relevant genes and their corresponding enzymes. A genomic library consists of a pool of recombinant microorganisms (bacteria or yeast) or bacteriophages derived from total genomic DNA. The clones should carry as inserts a large number of (possibly overlapping) fragments covering the entire genome several times over. A cDNA cloning experiment is appropriate if it is known that the organism or cell type in question is transcribing the relevant gene at reasonably high levels. If it is known that a particular gene is well expressed, a cDNA cloning experiment is more likely to lead to a successful outcome than genomic DNA cloning, at least as far as the cloning of mammalian genes is concerned. For example, only 5% of the human genome are expected to code for proteins or RNAs corresponding to 70,000 genes. Therefore up to 95% of genomic DNA may encode "chunk" DNA such as satellite DNA, various types of interspersed elements, or pseudogenes (LETOVSKY et al., 1998). However, a cDNA clone will never give details about important signal structures such as promoters, enhancers, and transcription factor binding sites which are present upstream of a coding region and are determining the tissue and developmental specificity of gene expression because these elements are never transcribed and are, therefore, not part of the mRNA. Nowadays, kits for the construction of gene libraries and even certain types of ready-to-go libraries are commercially available but in any case, the screening still has to be carried out by the individual researcher. Another approach, which is also being used by commercial companies, is to collect environmental samples containing heterogeneous populations of uncultured microbes from diverse ecosystems. The genetic material is extracted from these organisms, eliminating the need to grow and maintain the organisms in cultures in the laboratory. Since even small samples yield sufficient DNA, the impact on sensitive environments is minimized. Gene expression libraries created from microbial DNA are subsequently used for the production of biomolecules. Diversa (San Diego, CA, USA) estimates that its gene expression libraries currently contain the complete genomes of over 1 million unique microorganisms, comprising a vast resource of genetic material that can be screened for valuable commercial products (MATHUR, personal communication). The source of the genes, however, remains obscure, since it is impossible to trace the nucleotide sequence back to the original (micro-)organism.

5 Starch Processing Enzymes

Starch from cultivated plants represents a ubiquitous and easily accessible source of energy. In plant cells or seeds, starch is usually deposited in the form of large granules in the cytoplasm. Starch is composed exclusively of α-glucose units that are linked by α-1,4- or α-1,6-glycosidic bonds. The two high-molecular-weight components of starch are amylose (15–25%), a linear polymer consisting of α-1,4-linked glucopyranose residues, and amylopectin (75–85%), a branched polymer containing, in addition to α-1,4-glycosidic linkages, α-1,6-linked branch points occurring every 17–26 glucose units. α-Amylose chains, which are not soluble in water but form hydrated micelles, are polydisperse and their molecular weights vary from hundreds to thousands. The molecular weight of amylopectin may be as high as 100 million and in solution such a polymer has colloidal or micellar forms.

Because of the complex structure of starch, cells require an appropriate combination of hydrolyzing enzymes for its depolymerization to oligosaccharides and smaller sugars, such as glucose and maltose. They can be simply classified into two groups: endo-acting or endo-hydrolases and exo-acting enzymes or exo-hydrolases (Fig. 3). Endo-acting enzymes, such as α-amylase (α-1,4-glucan-4-glucanohydrolase; EC 3.2.1.1), hydrolyze linkages in the interior of the starch polymer in a random fashion which leads to the formation of linear and branched oligosaccharides. The sugar reducing groups are liberated in the α-anomeric configuration. Most of the starch hydrolyzing enzymes belong to the α-amylase family which contains a characteristic catalytic $(\beta/\alpha)_8$-barrel domain. α-Amylases belong to two families, 13 and 57 of the glycosyl hydrolase families. Family 13 has approximately 150 members from eukarya and bacteria. Family 57, on the other hand, has only three members from Bacteria and Archaea (NIEHAUS et al., 1999). Throughout the α-amylase family, only eight amino acid residues are invariant, seven at the active site and a glycine in a short turn. On the structural level, there are to date no X-ray structures of amylolytic enzymes derived from hyperthermophiles.

Exo-acting starch hydrolases include β-amylase, glucoamylase, and α-glucosidase. These enzymes attack the substrate from the non-reducing end, producing small and well-defined oligosaccharides. β-Amylase (EC 3.2.1.2), also referred to as α-1,4-D-glucan maltohydrolase or saccharogen amylase, hydrolyzes α-1,4-glucosidic linkages to remove successive maltose units from the non-reducing ends of the starch chains, producing β-maltose by an inversion of the anomeric configuration of the maltose. β-Amylase belongs to family 14 of the glycosyl hydrolases, having 11 members from eukarya and bacteria.

Glucoamylase (EC 3.2.1.3) hydrolyzes terminal α-1,4-linked-D-glucose residues successively from non-reducing ends of the chains, releasing β-D-glucose. Glucoamylase, which is typically a fungal enzyme, has several names: α-1,4-D-glucan hydrolase, amyloglucosidase, and γ-amylase. Most forms of the enzyme can hydrolyze α-1,6-D-glucosidic bonds when the next bond in sequence is α-1,4. However, *in vitro* this enzyme also hydrolyzes α-1,6- and α-1,3-D-glucosidic bonds in other polysaccharides with high molecular weights.

α-Glucosidase (EC 3.2.1.20), or α-D-glucoside glucohydrolase, attacks the α-1,4-linkages of oligosaccharides that are produced by the

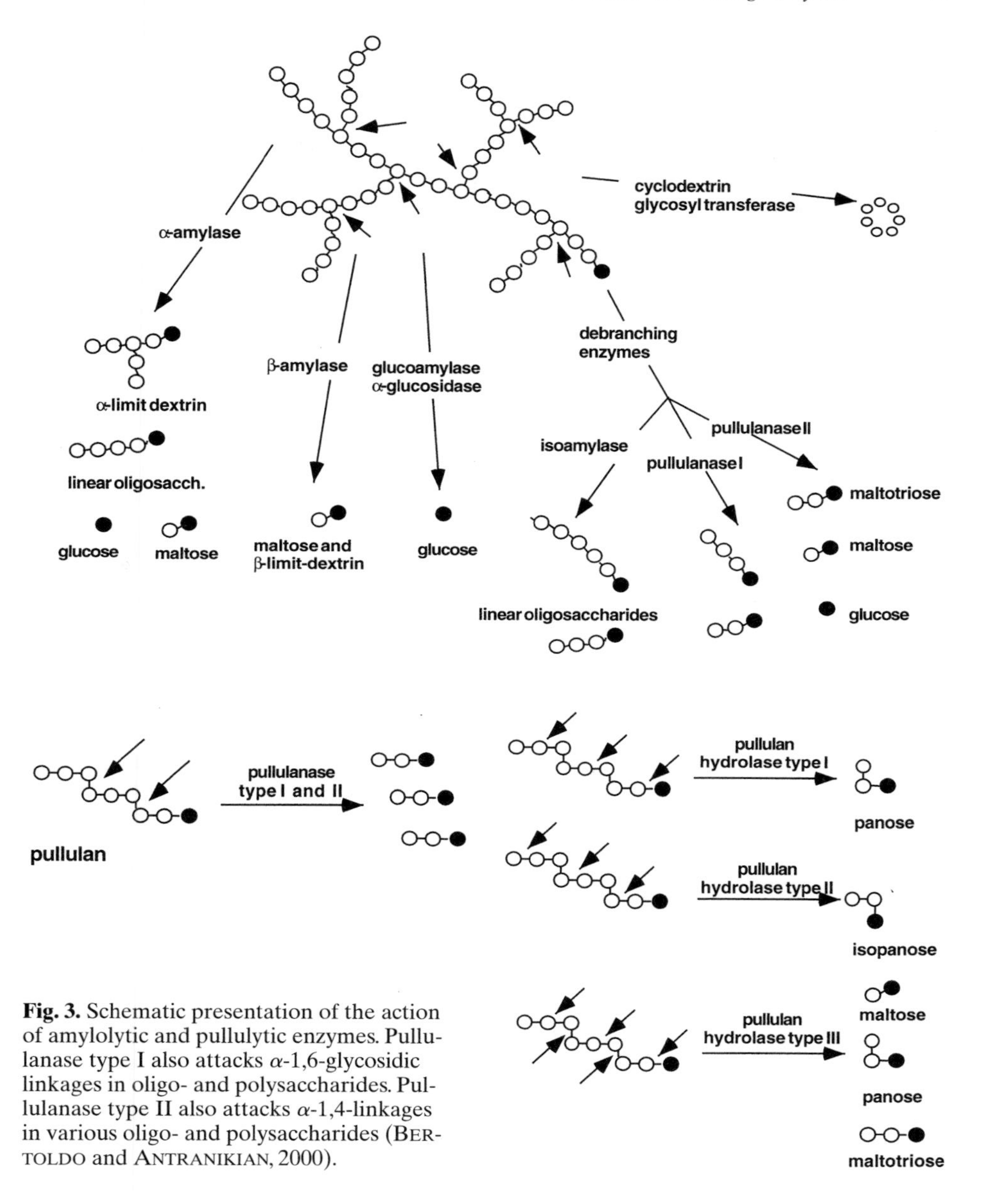

Fig. 3. Schematic presentation of the action of amylolytic and pullulytic enzymes. Pullulanase type I also attacks α-1,6-glycosidic linkages in oligo- and polysaccharides. Pullulanase type II also attacks α-1,4-linkages in various oligo- and polysaccharides (BERTOLDO and ANTRANIKIAN, 2000).

action of other amylolytic enzymes. Unlike glucoamylase, α-glucosidase liberates glucose with an α-anomeric configuration. α-Glucosidases are members of the family 15 and the very diverse family 31 of the glycosyl hydrolases.

Isoamylase (EC 3.2.1.68), or glycogen-6-glucanohydrolase, is a debranching enzyme specific for α-1,6-linkages in polysaccharides such as amylopectin, glycogen, and β-limit dextrin, but it is unable to hydrolyze the α-1,6-linkages in pullulan or branched oligosaccharide; therefore, it has limited action on α-limit dextrin. Enzymes capable of hydrolyzing α-1,6-glycosidic bonds in pullulan are defined as pullulanases (Fig. 3). Pullulan is a linear α-glucan consisting of maltotriose units joined by α-1,6-glycosidic linkages. It is produced by *Aureobasidium pullulans* with a chain length of 480 maltotriose units. On the basis of substrate specificity and product formation, pullulanases have been classified into two groups: pullulanase type I and pullulanase type II. Pullulanase type I (EC 3.2.1.41) specifically hydrolyzes the α-1,6-linkages in pullulan as well as in branched oligosaccharides (debranching enzyme), and its degradation products are maltotriose and linear oligosaccharides, respectively. Pullulanase type I is unable to attack α-1,4-linkages in α-glucans and belongs to the family 13 of the glycosyl hydrolases. Pullulanase type II, or amylopullulanase, attacks α-1,6-glycosidic linkages in pullulan and α-1,4-linkages in branched and linear oligosaccharides (Fig. 3). The enzyme has a multiple specificity and is able to fully convert polysaccharides (e.g., amylopectin) to small sugars (Dp1, Dp2, Dp3; Dp = degree of polymerization) in the absence of other enzymes, such as α-amylase or β-amylase.

In contrast to the previously described pullulanases, pullulan hydrolases type I and type II are unable to hydrolyze α-1,6-glycosidic linkages in pullulan or in branched substrates. They can attack α-1,4-glycosidic linkages in pullulan leading to the formation of panose or isopanose. Pullulan hydrolase type I or neopullulanase (EC 3.2.1.135) hydrolyzes pullulan to panose (α-6-D-glucosylmaltose). Pullulan hydrolase type II or isopullulanase (EC 3.2.1.57) hydrolyzes pullulan to isopanose (α-6-maltosylglucose) (Fig. 3).

Finally, cyclodextrin glycotransferase (CGTase; EC 2.4.1.19) or α-1,4-D-glucan-α-4-D-(α-1,4-D-glucano)-transferase, is an enzyme that is generally found in bacteria and was recently discovered in archaea. This enzyme produces a series of non-reducing cyclic dextrins from starch, amylose, and other polysaccharides. α-, β-, and γ-cyclodextrins are rings formed by 6, 7, and 8 glucose units that are linked by α-1,4-bonds, respectively (Fig. 3).

5.1 Thermoactive Amylolytic Enzymes

5.1.1 Heat Stable Amylases and Glucoamylases

Extremely thermostable α-amylases have been characterized from the hyperthermophilic Archaea *Pyrococcus furiosus* (KOCH et al., 1990), *Pyrococcus woesei* (FRILLINGOS et al., 2000; KOCH et al., 1991), and *Thermococcus profundus* (LEE et al., 1996; CHUNG et al., 1995). The optimal temperatures for the activity of these enzymes are 100 °C, 100 °C, and 80 °C, respectively. Thermoactive amylolytic enzymes have been also detected in hyperthermophilic Archaea of the genera *Sulfolobus*, *Thermophilum*, *Desulfurococcus*, and *Staphylothermus* (LEE et al., 1996; BRAGGER et al., 1989). Molecular cloning of the corresponding genes and their expression in heterologous hosts circumvent the problem of insufficient expression in the natural host. The gene encoding an extracellular α-amylase from *P. furiosus* has been recently cloned and the recombinant enzyme has been expressed in *Bacillus subtilis* and *E. coli* (DONG et al., 1997a; JØRGENSEN et al., 1997). This is the first report on the expression of an archaeal gene derived from an extremophile in a *Bacillus* strain. The high thermostability of the pyrococcal extracellular α-amylase (thermal activity even at 130 °C) in the absence of metal ions, together with its unique product pattern and substrate specificity, makes this enzyme an interesting candidate for industrial application. In addition, an intracellular α-amylase gene from *P. furiosus* has been cloned and sequenced (LADERMANN et al., 1993). It was interesting to note that the four highly conserved regions usually identified in α-amylases are not found in this enzyme. α-Amylases with lower thermostability and thermoactivity have been isolated from the Archaea *Thermococcus profundus* (KWAK et al., 1998), *Pyrococcus* sp. KOD1 (TACHIBANA et al., 1996), and the bacteria *Thermotoga maritima* (LIEBL et al., 1997) and *Dictioglomus thermophilum* (FUKUSUMI et al., 1988). The genes encoding these enzymes were successfully expressed in *E. coli*. Similar to the amylase from *Bacillus licheniformis*, which is commonly used in liquefaction, the enzyme from *T. maritima* requires Ca^{2+} for activity. Further investigations have shown that the extreme hyperthermophilic archaeon *Pyrodictium abyssi* can grow on various polysaccharides and also secretes a heat stable amylase (unpublished results).

Unlike α-amylase, the production of glucoamylase seems to be very rare in extremely thermophilic and hyperthermophilic bacteria and archaea. Among the thermophilic anaerobic bacteria, glucoamylases have been purified and characterized from *Clostridium thermohydrosulfuricum* 39E (HYUN and ZEIKUS, 1985), *Clostridium thermosaccharolyticum* (SPECKA et al., 1991), and *Thermoanaerobacterium thermosaccharolyticum* DSM 571 (GANGHOFNER et al., 1998). Recently, it has been shown that the thermoacidophilic archaea *Thermoplasma acidophilum*, *Picrophilus torridus*, and *Picrophilus oshimae* produce heat and acid stable glucoamylases. The purified archaeal glucoamylases are optimally active at pH 2 and 90 °C. Catalytic activity is still detectable at pH 0.5 and 100 °C. This represents the first report on the production of glucoamylases in thermophilic archaea (unpublished results).

5.1.2 α-Glucosidases

α-Glucosidases are present in thermophilic archaea and bacteria. An intracellular α-glucosidase has been purified from *P. furiosus*. The enzyme exhibits optimal activity at pH 5.0 to 6.0 over a temperature range of 105 °C to 115 °C; the half-life at 98 °C is 48 h. An extracellular α-glucosidase from the thermophilic archaeon *Thermococcus* strain AN1 (PILLER et al., 1996) was purified and its molecular characteristics determined. The monomeric enzyme (60 kDa) is optimally active at 98 °C. The purified enzyme has a half-life around 35 min, which is increased to around 215 min in the presence of 1% (w/v) dithiothreitol and 1% (w/v) BSA. The substrate preference of the enzyme is: *p*-nitrophenyl-α-D-glucoside > nigerose > panose > palatinose > isomaltose > maltose and turanose. No activity was found with starch, pullulan, amylose, maltotriose, maltotetraose, isomaltotriose, cellobiose, and β-gentiobiose. The enzyme is also active at 130 °C. The gene encoding α-glucosidase from *Thermococcus hydrothermalis* (GALICHET and BELARBI, 1999) was cloned by complementation of a *Saccharomyces cerevisiae* deficiency maltase-deficient mutant strain. The cDNA clone isolated encodes an open reading frame corresponding to a protein of 242 amino acids. The protein shows 42% identity to a *Pyrococcus horikoshii* unknown ORF but no similarities were obtained with other polysaccharidase sequences.

5.1.3 Thermoactive Pullulanases and CGTases

Thermostable and thermoactive pullulanases from extremophilic microorganisms have been detected in *Thermococcus celer*, *Desulfurococcus mucosus*, *Staphylothermus marinus*, and *Thermococcus aggregans* (CANGANELLA et al., 1994). Temperature optima between 90 °C and 105 °C, as well as remarkable thermostability even in the absence of substrate and calcium ions, have been observed. Most thermoactive pullulanases identified to date belong to the type II group, which attack α-1,4- and α-1,6-glycosidic linkages. They have been purified from *P. furiosus*, *Thermococcus litoralis* (BROWN and KELLY, 1993), *T. hydrothermalis* (GANTALET and DUCHIRON, 1998), and *Pyrococcus* strain ES4 (SCHULIGER et al., 1993). Pullulanases type II from *P. furiosus* and *P. woesei* have been expressed in *E. coli* (DONG et al., 1997b; RÜDIGER et al., 1995). The unfolding and refolding of the pullulanase from *P. woesei* has been investigated using guanidinium chloride as denaturant. The monomeric enzyme (90 kDa) was found to be very resistant to chemical denaturation and the transition midpoint for guanidinium chloride-induced unfolding was determined to be 4.86 ± 0.29 M for intrinsic fluorescence and 4.90 ± 0.31 M for far-UV CD changes. The unfolding process was reversible. Reactivation of the completely denatured enzyme (in 7.8 M guanidinium chloride) was obtained upon removal of the denaturant by stepwise dilution, 100% reactivation was observed when refolding was carried out via a guanidinium chloride concentration of 4 M in the first dilution step. Particular attention has been paid to the role of Ca^{2+}, which activates and stabilizes this archaeal pullulanase against thermal inactivation. The enzyme binds two Ca^{2+} ions with a K_d of 0.080 ± 0.010 μM and a Hill coefficient H of 1.00 ± 0.10. This cation significantly enhances the stability of the pullulanase against guanidinium chloride-induced unfolding. The refolding of the pullulanase, on the other hand, was not affected by Ca^{2+} (SCHWERDTFEGER et al., 1999). Very recently, the genes encoding the pullulanases from *T. hydrothermalis* (ERRA-PUJADA, 1999), *Desulfurococcus mucosus* (DUFFNER et al., 2000), and *T. aggregans* (NIEHAUS et al., 2000) have been isolated and expressed in mesophilic hosts. Since the latter enzyme attacks α-1,4- as well as α-1,6-glycosidic linkages in pullulan, it has been classified as pullulan-hydrolase type III. Pullulan is converted to maltotriose, maltose, panose, and glucose (NIEHAUS et al., 2000). The aerobic thermophilic bacterium *Thermus caldophilus* GK-24 produces a thermostable pullulanase of type I when grown on starch (KIM et al., 1996). This enzyme debranches amylopectin by attacking specifically α-1,6-glycosidic linkages. The pullulanase is optimally active at 75 °C and pH 5.5, is thermostable up to 90 °C, and does not require Ca^{2+} for either activity or stability. The first debranching enzyme (pullulanase type I) from an anaerobic thermophile was identified in the bacterium *Fervidobacterium pennivorans* Ven5 which was cloned and expressed in *E. coli* (KOCH et al., 1997; BERTOLDO et al., 1999). In contrast to pullulanase type II from *P. woesei* (specific to both α-1,6- and α-1,4-glycosidic linkages) the enzyme from *F. pennivorans* Ven5 attacks exclusively the α-1,6-glycosidic linkages in polysaccharides. This thermostable debranching enzyme leads to formation of long chain linear polysaccharides from amylopectin.

Thermostable cyclodextrin glycosyltransferases (CGTases) are produced by *Thermoanaerobacter* species, *Thermoanaerobacterium thermosulfurigenes*, and *Anaerobranca gottschallkii* (PROWE et al., 1996; PEDERSEN et al., 1995; WIND et al., 1995). Recently, a CGTase, with optimal temperature at 100 °C, was purified from a newly isolated archaeon, *Thermococcus* sp. This is the first report on the presence of a thermostable CGTase in a hyperthermophilic archaeon (TACHIBANA et al., 1999). The enzyme from this strain has been cloned and sequenced. The gene of 2,217 nucleotides encodes a protein with a MW of 83 kDa. The ability of extreme thermophiles and hyperthermophiles to produce heat-stable glycosyl hydrolases is summarized in Tab. 2.

The finding of extremely thermophilic bacteria and archaea capable of producing novel thermostable starch-hydrolyzing enzymes is a valuable contribution to the starch-processing industry. By using robust starch modifying enzymes from thermophiles, innovative and environmentally friendly processes can be developed, aimed at the formation of products of high added value for the food industry. New and enhanced functionality can be obtained by changing the structural properties of starch. In order to prevent retrogradation, starch-modifying enzymes can be used at higher temperatures. The use of the extremely thermostable amylolytic enzymes can lead to valuable products, which include innovative starch-based materials with gelatin-like characteristics and defined linear dextrins that can be used as fat substitutes, texturizers, aroma stabilizers, and prebiotics (CRABB and MITCHINSON, 1997; COWAN, 1996). CGTases are used for the production of cyclodextrins that can be used as a gelling, thickening or stabilizing agent in jelly desserts, dressing, confectionery, dairy and meat products. Due to the ability of cyclodextrins to form inclusion complexes with a variety of organic molecules, cyclodextrins improve the solubility of hydrophobic compounds in aqueous solution. This is of interest for the pharmaceutical and cosmetic industries. Cyclodextrin production is a multistage process in which starch is first liquefied by a heat-stable amylase and in the second step a less thermostable CGTase from *Bacillus* sp. is used. The application of heat-stable CGTase in jet cooking, where temperatures up to 105 °C are achieved, will allow liquefaction and cyclization to take place in one step.

Tab. 2. Starch Hydrolyzing Enzymes from Extreme Thermophilic and Hyperthermophilic Archaea and Bacteria

Enzymes	Organism[a]	Enzyme Properties			
		Optimal Temperature [°C]	Optimal pH	MW [kDa]	Remarks
α-Amylase	*Desulfurococcus mucosus* (85)	85	5.5	74	purified/cloned
	Pyrococcus furiosus (100)	100	6.5–7.5	129	purified/cloned/intracellular
		100	7.0	68	purified/cloned/extracellular
	Pyrococcus sp. KOD1	90	6.5	49.5	purified/cloned/extracellular
	Pyrococcus woesei (100)	100	5.5	68	purified/extracellular
	Pyrodictium abyssi (98)	100	5.0	–	crude extract[b]
	Staphylothermus marinus (90)	100	5.0	–	crude extract
	Sulfolobus solfataricus (88)	–	–	240	extracellular
	Thermococcus celer (85)	90	5.5		crude extract
	Thermococcus profundus DT5432 (80)	80	5.5	42	purified/cloned/"Amy S"
	Thermococcus profundus (80)	80	4.0–5.0	42	purified/"Amy L"
	Thermococcus aggregans (85)	95	6.5	–	cloned
	Dyctyoglomus thermophilum Rt46B.1 (73)	90	5.5	75	purified/cloned/cytoplasmic fraction
	Thermotoga maritima MSB8 (90)	85–90	7.0	61	purified/cloned/lipoprotein
Pullulanase type I	*Fervidobacterium pennavorans* Ven5 (75)	80	6	190 (93)	purified/cloned
	Thermotoga maritima MSB8 (90)	90	6.0	93 (subunit)	cloned/type I[b]
	Thermus caldophilus GK24 (75)	75	5.5	65	purified/cell-associated
Pullulanase type II	*Desulfurococcus mucosus* (88)	100	5.0	74	purified/cloned
	Pyrococcus woesei (100)	100	6.0	90	purified/cloned/cell-associated
	Pyrodictium abyssi (98)	100	9.0	–	crude extract
	Thermococcus celer (85)	90	5.5	–	crude extract

Tab. 2. Continued

Enzymes	Organism[a]	Enzyme Properties			
		Optimal Temperature [°C]	Optimal pH	MW [kDa]	Remarks
	Thermococcus litoralis (90)	98	5.5	119	purified/extracell./ glycoprotein
	Thermococcus hydrothermalis (80)	95	5.5	128	purified/extracell./ glycoprotein
	Thermococcus aggregans (85)	100	6.5	83	purified/cloned
Glucoamylase	*Thermoplasma acidophilum* (60)	90	6.5	141	purified
	Picrophilus oshimae (60)	90	2.0	140	purified
	Picrophilus torridus (60)	90	2.0	133	purified
CGTase	*Thermococcus* sp. (75)	100	2.0	83	purified
	Thermoanaerobacterium thermosulfurigenes (60)	80	4.0–4.5	68	purified/cloned/crystallized
	Anaerobranca gottschallkii (55)	70	8.0	66	purified
α-Glucosidase	*Thermococcus* strain AN1 (80)	130	–	63	purified/extracell./ glycoprotein
	Thermococcus hydrothermalis (80)	–	–	–	cloned

[a] Values in brackets give the optimal growth temperature for each organism in °C.
[b] Unpublished results
– not determined

6 Cellulose-Hydrolyzing Enzymes

Cellulose commonly accounts for up to 40% of the plant biomass. It consists of glucose units linked by β-1,4-glycosidic bonds with a polymerization grade of up to 15,000 glucose units in a linear mode. Although cellulose has a high affinity to water, it is completely insoluble. Natural cellulose compounds are structurally heterogeneous and have both amorphous and highly ordered crystalline regions. The degree of crystallinity depends on the source of the cellulose and the higher crystalline regions are more resistant to enzymatic hydrolysis. Cellulose can be hydrolyzed into glucose by the synergistic action of at least three different enzymes: endoglucanase, exoglucanase (cellobiohydrolase), and β-glucosidase. Synonyms for cellulases (EC 3.2.1.4) are β-1,4-D-glucan glucano-hydrolases, endo-β-1,4-glucanases, or carboxymethyl cellulases. This enzyme is an endoglucanase, which hydrolyzes cellulose in a random manner as endo-hydrolase producing various oligosaccharides, cellobiose, and glucose. The enzyme catalyzes the hydrolysis of β-1,4-D-glycosidic linkages in cellulose but can also hydrolyze 1,4-linkages in β-D-glucans containing 1,3-linkages. Cellulases belong to the family 12 of the glycosyl hydrolases (HENRISSAT, 1991).

Exoglucanases, β-1,4-cellobiosidases, exo-cellobiohydrolases, or β-1,4-cellobiohydrolases (EC 3.2.1.91) hydrolyze β-1,4-D-glycosidic linkages in cellulose and cellotetraose, releasing cellobiose from the non-reducing end of the chain. They belong to family 6 of glycosyl hydrolases.

β-Glucosidases (EC 3.2.1.21), gentobiases, cellobiases, or amygdalases catalyze the hydrolysis of terminal, non-reducing β-D-glucose residues releasing β-D-glucose. These enzymes belong to family 3 of glycosyl hydrolases and have a wide specificity for β-D-glucosides. They are able to hydrolyze β-D-galactosides, α-L-arabinosides, β-D-xylosides, and β-D-fucosides.

6.1 Thermostable Cellulases

Thermostable cellulases active towards crystalline cellulose are of great biotechnological interest. Several cellulose degrading enzymes from various thermophilic organisms have been cloned, purified, and characterized. A thermostable cellulase from *Thermotoga maritima* MSB8 has been characterized (BRONNENMEIER et al., 1995). The enzyme is rather small with a molecular weight (MW) of 27 kDa and it is optimally active at 95 °C and between pH 6.0 and 7.0. Two thermostable endocellulases, CelA and CelB, were purified from *Thermotoga neapolitana*. CelA (MW of 29 kDa) is optimally active at pH 6 at 95 °C, while CelB (MW of 30 kDa) has a broader optimal pH range (pH 6 to 6.6) at 106 °C. The genes encoding these two endocellulases have been identified (BOK et al., 1998). Cellulase and hemicellulase genes have been found clustered together on the genome of the thermophilic anaerobic bacterium *Caldocellum saccharolyticum* which grows on cellulose and hemicellulose as sole carbon sources. The gene for one of the cellulases (*celA*) was isolated and was found to consist of 1,751 amino acids. This is the largest cellulase gene described to date (TEO et al., 1995).

A large cellulolytic enzyme (CelA) with the ability to hydrolyze microcrystalline cellulose was isolated from the extremely thermophilic bacterium *Anaerocellum thermophilum* (ZVERLOV et al., 1998). The enzyme has an apparent molecular mass of 230 kDa and exhibits significant activity towards Avicel and is most active towards soluble substrates such as CM-cellulose (CMC) and β-glucan. Maximal activity was observed at pH 5–6 and 85–95 °C. The thermostable exo-acting cellobiohydrolase from *Thermotoga maritima* MSB8 has a MW of 29 kDa and is optimally active at 95 °C at pH 6.0–7.5 with a half-life of 2 h at 95 °C. The enzyme hydrolyzes Avicel, CM-Cellulose, and β-glucan forming cellobiose and cellotriose (BRONNENMEIER et al., 1995). A thermostable cellobiase is produced by *Thermotoga* sp. FjSS3-B1 (RUTTERSMITH and DANIEL, 1991). The enzyme is highly thermostable and shows maximal activity at 115 °C at pH 6.8–7.8. The thermostability of this enzyme is salt dependent. This cellobiase is active against amorphous cellulose and CM-cellulose.

Recently, a thermostable endoglucanase, which is capable of degrading β-1,4-bonds of β-glucans and cellulose, has been identified in the archaeon *Pyrococcus furiosus*. The gene encoding this enzyme has been cloned and sequenced in *E. coli* and has significant amino acid sequence similarities with endoglucanases from glucosyl hydrolases family 12. The purified recombinant endoglucanase hydrolyzes β-1,4- but not β-1,3-glycosidic linkages and has the highest specific activity with cellopentaose and cellohexaose as substrates (BAUER et al., 1999). In contrast to this, several β-glucosidases have been detected in archaea. In fact, archaeal β-glucosidases have been found in *Sulfolobus solfataricus* MT4, *S. acidocaldarius*, *S. shibatae* (GROGAN, 1991), and *P. furiosus* (KENGEN et al., 1993). The enzyme from the latter microorganism (homotetramer, 56 kDa/subunit) is very stable and shows optimal activity at 102 °C to 105 °C with a half-life of 3.5 d at 100 °C and 13 h at 110 °C (VOORHORST et al., 1995; KENGEN et al., 1993). The β-glucosidase from *S. solfataricus* MT4 has been purified and characterized (NUCCI et al., 1993). The enzyme is a homotetramer (56 kDa per subunit) and very resistant to various denaturants with activity up to 85 °C (PISANI et al., 1990). The gene for this β-glucosidase has been cloned and overexpressed in *E. coli* (CUBELLIS et al., 1990; MORACCI et al., 1993; PRISCO et al., 1994) (Tab. 3).

Cellulose hydrolyzing enzymes are widespread in fungi and Bacteria. Less thermoactive cellulases have already found various biotechnological applications. The most effective enzyme of commercial interest is the cellulase produced by *Trichoderma* sp. (TEERI et al., 1998). Cellulases were also obtained from strains of *Aspergillus*, *Penicillium*, and Basidiomycetes (TOMME et al., 1995). Cellulolytic enzymes can be used in alcohol production to improve juice yields and effective color extraction of juices. The presence of cellulases in detergents causes color brightening, softening, and improves particulate soil removal. Cellulase (Denimax® Novo Nordisk) is also used for the "biostoning" of jeans instead of using stones. Other applications of cellulases include the pretreatment of cellulosic biomass and forage crops to improve nutritional quality and digestibility, enzymatic saccharification of agricultural and industrial wastes, and production of fine chemicals.

7 Xylan-Degrading Enzymes

Xylan is a heterogeneous molecule that constitutes the main polymeric compound of hemicellulose, a fraction of the plant cell wall, which is a major reservoir of fixed carbon in nature. The main chain of the heteropolymer is composed of xylose residues linked by β-1,4-glycosidic bonds. Approximately half of the xylose residues have substitution at O-2 or O-3 positions with acetyl-, arabinosyl- and glucuronosyl-groups. The complete degradation of xylan requires the action of several enzymes (for a detailed description see reviews of SUNNA and ANTRANIKIAN, 1997a and SUNNA et al., 1996a). The endo-β-1,4-xylanase (EC 3.2.1.8), or β-1,4-xylan xylanohydrolase, hydrolyzes β-1,4-xylosidic linkages in xylans, while β-1,4-xylosidase, β-xylosidase, β-1,4-xylan xylohydrolase, xylobiase, or exo-β-1,4-xylosidase (EC 3.2.1.37) hydrolyzes β-1,4-xylans and xylobiose by removing the successive xylose residues from the non-reducing termini. α-Arabinofuranosidase or arabinosidase (EC 3.2.1.55) hydrolyzes the terminal non-reducing α-L-arabinofuranoside residues in α-L-arabinosides. The enzyme also acts on α-L-arabinofuranosides, α-L-arabinans containing either (1,3)- or (1,5)-linkages. Glucuronoarabinoxylan endo-β-1,4-xylanase, feraxan endoxylanase or glucuronoarabinoxylan β-1,4-xylanohydrolase (EC 3.2.1.136) attacks β-1,4-xylosyl linkages in some glucuronoarabinoxylans. This enzyme also shows high activity towards feruloylated arabinoxylans from cereal plant cell walls. Acetyl xylan esterase (EC 3.1.1.6) removes acetyl groups from xylan.

7.1 Thermostable Xylanases

So far, only few extreme thermophilic microorganisms are able to grow on xylan and secrete thermoactive xylanolytic enzymes (Tab. 3). Members of the order Thermotogales and *Dictyoglomus thermophilum* Rt46B.1 have been described to produce xylanases that are active and stable at high temperatures (SUNNA and ANTRANIKIAN, 1997a; GIBBS et al., 1995). The most thermostable endoxylanases that have been described so far are those derived from *Thermotoga* sp. strain FjSS3-B.1 (SIMPSON et al., 1991), *Thermotoga maritima* (WINTERHALTER and LIEBL, 1995), *T. neapolitana* (BOK et al., 1994), and *T. thermarum* (SUNNA et al., 1996b). These enzymes, that are active between 80 and 105 °C, are mainly cell associated and most probably localized within the toga (SUNNA et al., 1996a; WINTERHALTER and LIEBL, 1995; RUTTERSMITH et al., 1992; SCHUMANN et al., 1991). Several genes encoding xylanases have already been cloned and sequenced. The gene from *T. maritima*, encoding a thermostable xylanase has been cloned and expressed in *E. coli*. Comparison between the *T. maritima* recombinant xylanase and the commercially available enzyme, Pulpenzyme™ (YANG and ERIKSSON, 1992) indicates that the thermostable xylanase could be of interest for application in the pulp and

Tab. 3. Production of Thermoactive Cellulases, Xylanases, and Chitinase by some Representatives of Extreme Thermophilic and Hyperthermophilic Archaea and Bacteria

Enzymes	Organism[a]	Enzyme Properties			
		Optimal Temperature [°C]	Optimal pH	MW [kDa]	Remarks
Endo-glucanase	*Thermotoga maritima* MSB8 (80)	95	6.0–7.5	27	purified/cloned/cellulase I
	Thermotoga neapolitana (80)	95	6.0	29	purified/cloned/CellA
		106	6.0–6.5	30	purified/cloned/CellB
Exoglucanase	*Thermotoga maritima* MSB8	95	6.0–7.5	29	purified/cellulase II
	Thermotoga sp. strain FjSS3-B.1 (80)	115	6.8–7.8	36	purified/cell-associated
	Anaerocellum thermophilum Rt46B.1 (100)	85	6.5	31	cloned
	Pyrococcus furiosus (100)	100	6.0	35.9	cloned
β-Glucosidase	*Pyrococcus furiosus* (100)	102–105	–	230/58	purified/cloned
	Sulfolobus solfataricus (88)	105	5.3	240/56	purified/cloned
	Thermotoga maritima MSB8	75	6.2	95 (47)	purified/cloned
	Thermotoga sp. strain FjSS3-B.1 (80)	80	7.0	100 (75)	purified/toga-associated
Endo-xylanases	*Pyrodictium abyssi* (98)	110	5.5	–	crude extract
	Dyctyoglomus thermophilum Rt46B.1 (73)	85	6.5	31	purified/cloned
	Thermotoga maritima MSB8 (80)	92	6.2	120	purified/toga-associated/XynA
		105	5.4	40	purified/toga-associated/XynB
	Thermotoga sp. strain FjSS3-B.1 (80)	105	5.3	31	purified/cloned/toga-associated
		85	6.3	40	purified/cloned
	Thermotoga neapolitana (80)	85	5.5	37	purified
		95	5.5–6.0	119	purified/cloned
	Thermotoga thermarum (77)	80	6.0	105/150	purified/toga-associated/ endoxylanase 1
		90–100	7.0	35	purified/endoxylanase 2
Chitinase	*Pyrococcus kodakaraensis* (95)	85	5.0	135	purified/cloned

[a] Values in the brackets give the optimal growth temperature for each organism in °C.
– not determined

paper industry (CHEN et al., 1997). A xylanase has been found in the archaeon *Thermococcus zilligii* strain AN1, which grows optimally at 75 °C. The enzyme has a molecular mass of 95 kDa and a unique N-terminal sequence. The pH optimum for activity is pH 6.0, and the half-life at 100 °C is 8 min. Another archaeal xylanase with a temperature optimum of 110 °C was found in the hyperthermophilic archaeon *Pyrodictium abyssi* (unpublished data).

Xylanases from bacteria and eukarya comprise families 10 and 11 of glycosyl hydrolases and have a wide range of potential biotechnological applications. They are already produced on an industrial scale and are used as feed additives for increasing poultry feed efficiency diets (ANNISON, 1992; CLASSEN, 1996), and in wheat flour for improving dough handling and the quality of baked products (MAAT et al., 1992). In recent years, the major interest in thermostable xylanases is found in enzyme-aided bleaching of paper (VIIKARI et al., 1994). More than 2 million tons of chlorine and chlorine derivatives are used annually in the United States for pulp bleaching. The chlorinated lignin derivatives generated by this process constitute a major environmental problem caused by the pulp and paper industry (MCDONOUGH, 1992). Recent investigations have demonstrated the feasibility of enzymatic treatments as alternatives to chlorine bleaching for the removal of residual lignin from pulp (VIIKARI et al., 1994). Treatment of craft pulp with xylanase leads to a release of xylan and residual lignin without undue loss of other pulp components. Xylanase treatment at elevated temperatures opens up the cell wall structure, thereby facilitating lignin removal in subsequent bleaching stages. Candidate xylanases for this important, potential market would have to satisfy several criteria:

(1) they must lack cellulolytic activity to avoid hydrolysis of the cellulose fibers,
(2) their molecular weight should be low enough to facilitate their diffusion in the pulp fibers,
(3) they must be stable and active at high temperature and at alkaline pH, and
(4) one must be able to obtain high yields of enzyme at very low cost.

All of the xylanases currently available from commercial suppliers can only partially fulfill the criteria. Xylanases from moderate thermophilic microorganisms are rapidly denatured at temperatures above 70 °C. Several of non-chlorine bleaching stages used in commercial operations are performed well above this temperature; consequently, pulp must be cooled before treatment with the available enzymes and reheated for subsequent processing steps (CHEN et al., 1997).

8 Chitin Degradation

Chitin is a linear β-1,4 homopolymer of N-acetyl-glucosamine residues and it is one of the most abundant natural biopolymers on earth. Particularly in the marine environment, chitin is produced in enormous amounts and its turnover is due to the action of chitinolytic enzymes. Chitin is the major structural component of most fungi and invertebrates (GOODAY, 1990, 1994), while for soil or marine bacteria chitin serves as a nutrient. Chitin degradation is known to proceed with the endo-acting chitin hydrolase (chitinase A; EC 3.2.1.14) and the chitin oligomer degrading exo-acting hydrolases (chitinase B) and N-acetyl-D-glycosaminidase (trivial name: chitobiase; EC 3.2.1.52).

Endo- and exo-chitinases comprise three glycosyl hydrolase families, i.e., family 18, 19, and 20. Chitinases, endo-β-N-acetyl-D-glucosaminidases (EC 3.2.1.96) and di-N-acetylchitobiases from eukarya, bacteria, and virus belong to family 18. The N-acetyl-D-glucosamine oligomeric product retains its C1 anomeric configuration. Family 19 contains only chitinases from eukarya and bacteria and in contrast to the family 18, the product has inverted anomeric configuration. Family 20 contains β-hexosaminidases and chitobiases. Chitobiase degrades only small N-acetyl-D-glucosamine oligomers (up to pentamers) and the released N-acetyl-D-glucosamine monomers retain their C1 anomeric configuration.

Chitin exhibits interesting properties that make it a valuable raw material for several applications (COHEN-KUPIEC and CHET, 1998;

KAS, 1997; MUZZARELLI, 1997; GEORGOPAPADAKOU and TKACZ, 1995; SPINDLER et al., 1990; CHANDY and SHARMA, 1990). It has been estimated that the annual worldwide formation rate and steady state amount of chitin is on the order of 10^{10} to 10^{11} tons per year. Therefore, application of thermostable chitin hydrolyzing enzymes (chitinases) is expected for effective utilization of this abundant biomass. Although a large number of chitin hydrolyzing enzymes has been isolated and their corresponding genes have been cloned and characterized, only few thermostable chitin hydrolyzing enzymes are known. These enzymes have been isolated from the thermophilic bacterium *Bacillus licheniformis* X-7u (TAKAYANAGI et al., 1991), *Bacillus* sp. BG-11 (BHARAT and HOONDAL, 1998), and *Streptomyces thermoviolaceus* OPC-520 (TSUJIBO et al., 1995).

The extreme thermophilic anaerobic archaeon *Thermococcus chitonophagus* has been reported to hydrolyze chitin (HUBER et al., 1995b). This is the first extremophilic archaeon, which produces chitinase(s) and N-acetylglucosaminidase(s), however, sequence and structural information for archaeal chitinases have not yet been reported. Very recently, the gene encoding a chitinase from a hyperthermophilic archaeon *Pyrococcus kodakaraensis* KOD1 was cloned, sequenced, and expressed in *E. coli*. The purified recombinant protein is optimally active at 85 °C and pH 5.0. The enzyme produces chitobiose as the major end product (Tab. 3).

9 Proteolytic Enzymes

Proteases are involved in the conversion of proteins to amino acids and peptides. They have been classified according to the nature of their catalytic site in the following groups: serine, cysteine, aspartic, or metalloproteases (Tab. 4).

A variety of heat-stable proteases has been identified in hyperthermophilic Archaea belonging to the genera *Desulfurococcus*, *Sulfolobus*, *Staphylothermus*, *Thermococcus*, *Pyrobaculum*, and *Pyrococcus*. It has been found that most proteases from extremophiles belong to the serine type, are stable at high temperatures even in the presence of high concentrations of detergents and denaturing agents. A heat-stable serine protease was isolated from cell-free supernatants of the hyperthermophilic archaeon *Desulfurococcus* strain $Tok_{12}S_1$ (COWAN et al., 1987). Recently, a cell associated serine protease was characterized from *Desulfurococcus* strain SY that showed a half-life of 4.3 h at 95 °C (HANZAWA et al., 1996). A globular serine protease from *Staphylothermus marinus* was found to be extremely thermostable. This enzyme, which is bound to the stalk of a filiform glycoprotein complex, named tetrabrachion, has a residual activity even at 135 °C after 10 min of incubation (MAYR et al., 1996). The properties of extracellular serine proteases from a number of *Thermococcus* species have been analyzed (KLINGEBERG et al., 1991). The extracellular enzyme from *T. stetteri* has a molecular mass of 68 kDa and is highly stable and resistant to chemical denaturation as illustrated by a half-life of 2.5 h at 100 °C and retention of 70% of its activity in the presence of 1% SDS (KLINGEBERG et al., 1995). A novel intracellular serine proteinase (pernilase) from the aerobic hyperthermophilic archaeon *Aeropyrum pernix* K1 was purified and characterized. At 90 °C, the pernilase has a broad pH profile and an optimum at pH 9.0 for peptide hydrolysis. Several proteases from hyperthermophiles have been cloned and sequenced but in general, their expression in mesophilic hosts is difficult. A gene encoding a subtilisin-like serine protease, named aereolysin has been cloned from *Pyrobaculum aerophilum*. The protein was modeled based on structures of subtilisin-type proteases (VÖLKL et al., 1995). Multiple proteolytic activities have been observed in *P. furiosus*. The cell envelope associated serine protease of *P. furiosus* called pyrolysin was found highly stable with a half-life of 20 min at 105 °C (EGGEN et al., 1990). The pyrolysin gene was cloned and sequenced and it was shown that this enzyme is a subtilisin-like serine protease (VOORHORST et al., 1996). A serine protease from *Aquifex pyrophilus* was cloned and weakly expressed in *E. coli* as active and processed forms. The activity of the enzyme was highest at 85 °C and pH 9. The half-life of the protein (6 h at 105 °C) makes it one of the most heat stable proteases known to date.

Tab. 4. Properties of Thermoactive Proteolytic Enzymes from Extreme Thermophilic and Hyperthermophilic Archaea and Bacteria

Enzymes	Organism[a]	Enzyme Properties			
		Optimal Temperature [°C]	Optimal pH	MW [kDa]	Remarks
Serine protease	*Desulfurococcus mucosus* (85)	95	7.5	52	purified
	Pyrococcus furiosus (100)	85	6.3	124 (19) 105/80	protease I/purified pyrolysin/purified/cloned
	Pyrobaculum aerophilum (95)	–	–	–	cloned
	Thermococcus aggregans (75)	90	7.0	–	crude extract
	Thermococcus celer (85)	95	7.5	–	crude extract
	Thermococcus litoralis (90)	95	9.5	–	crude extract
	Thermococcus stetteri (75)	85	8.5	68	purified/cloned
	Staphylothermus marinus (90)	–	9.0	140	stable up to 135 °C
	Sulfolobus solfataricus (88)	–	6.5–8	118 (52)	purified
	Aeropyrum pernix K1 (90)	90	9.0	50	purified
	Aquifex pyrophilus (90)	85	7.0–9.0	43	purified
	Fervidobacterium pennavorans (70)	80	10	130	purified/keratin hydrolysis
	Thermobacteroides proteolyticus (65)	85	9.0–9.5	–	crude extract
Thiol protease	*Pyrococcus* sp. KOD1 (95)	110	7	44	purified
Acidic protease	*Sulfolobus acidocaldarius* (70)	90	2.0	–	cloned
Aminopeptidase I	*Sulfolobus solfataricus* (88)	–	–	>450	crude extract
Aminopeptidase II		–	–	170	crude extract
Endopeptidase I, II, III		–	–	115, 32, 27	crude extract
Carboxypeptidase		–	–	160	crude extract

[a] Values in the brackets give the optimal growth temperature for each organism in °C.
– not determined

Proteases have also been characterized from the thermoacidophilic Archaea *Sulfolobus solfataricus* (BURLINI et al., 1992) and *S. acidocaldarius* (FUSEK et al., 1990; LIN and TANG, 1990). In addition to the serine proteases other types of enzymes have been identified in extremophiles: a thiol protease from *Pyrococcus* sp. KOD1 (FUJIWARA et al., 1996; MORIKAWA et al., 1994), a propylpeptidase (PEPase), and a new type protease from *P. furiosus* (HARWOOD et al., 1997; HALIO et al., 1996; ROBINSON et al., 1995; BLUMENTALS et al., 1992). An extracellular protease, which is designated aeropyrolysin, was purified from *Aeropyrum pernix* K1 (JCM 9820). The enzyme activity is completely inhibited by EDTA and EGTA, indicating that it is a metalloprotease. The enzyme is highly resistant to denaturing reagents and highly thermostable showing a half-life of 2.5 h at 120 °C and 1.2 h at 125 °C in the presence of 1 mM $CaCl_2$. These results indicate that this enzyme is one of the most thermostable extracellular metalloproteases reported to date. Thermostable serine proteases were also detected in a number of extreme thermophilic bacteria belonging to the genera *Thermotoga* and *Fervidobacterium* (unpublished results). The enzyme system from *Fervidobacterium pennivorans* is able to hydrolyze feather keratin forming amino acids and peptides. The enzyme is optimally active at 80 °C and pH 10.0 (FRIEDRICH and ANTRANIKIAN, 1996).

The amount of proteolytic enzymes produced worldwide on commercial scale is the largest compared to the other biotechnologically used enzymes. Serine alkaline proteases are used as additives to household detergents for laundering, where they have to resist denaturation by detergents and alkaline conditions. Proteases showing high keratinolytic and elastolytic activities are used for soaking in the leather industry. Proteases are also used as catalysts for peptide synthesis using their reverse reaction.

10 DNA Processing Enzymes

10.1 Polymerase Chain Reaction (PCR)

DNA polymerases (EC 2.7.7.7) are the key enzymes in the replication of cellular information present in all life forms. They catalyze, in the presence of Mg^{2+}-ions, the addition of a deoxyribonucleoside 5′-triphosphate onto the growing 3′-OH end of a primer strand, forming complementary base pairs to a second strand. More than 100 DNA polymerase genes have been cloned and sequenced from various organisms, including thermophilic bacteria and archaea. Several native and recombinant enzymes have been purified and characterized (PERLER et al., 1996). Thermostable DNA polymerases play a major role in a variety of molecular biological applications, e.g., DNA amplification, sequencing, or labeling.

One of the most important advances in molecular biology during the last 10 years is the development of polymerase chain reactions (PCR; ERLICH et al., 1988; SAIKI et al., 1988; MULLIS et al., 1986). The first described PCR procedure utilized the Klenow fragment of *E. coli* DNA polymerase I, which was heat labile and had to be added during each cycle following the denaturation and primer hybridization steps. Introduction of thermostable DNA polymerases in PCR facilitated the automation of the thermal cycling part of the procedure. The DNA polymerase I from the bacterium *Thermus aquaticus*, called *Taq* polymerase, was the first thermostable DNA polymerase characterized (CHIEN et al., 1976; KALEDIN et al., 1980) and applied in PCR.

Taq polymerase has a 5′-3′-exonuclease activity, but no detectable 3′-5′-exonuclease activity (LONGLEY et al., 1990). Due to the absence of a 3′-5′-exonuclease activity, this enzyme is unable to excise mismatches and as a result, the base insertion fidelity is low (LING et al., 1991; KEOHAVONG and THILLY, 1989; DUNNING et al., 1988; TINDALL and KUNKEL, 1988). The use of high fidelity DNA polymerases is essential for reducing the increase of amplification errors in PCR products that will be cloned, sequenced, and expressed. Several thermostable DNA polymerases with 3′-5′-exonuclease-dependent proofreading activity have been described and the error rates (number of misincorporated nucleotides per base synthesized) for these enzymes have been determined. A thermostable DNA polymerase from *Thermotoga maritima* (HUBER et al., 1986) was reported to have a 3′-5′-exonuclease activity (BOST et al., 1994). Archaeal proofreading polymerases such as *Pwo* pol (FREY and SUPPMANN, 1995) from *Pyrococcus woesei* (ZILLIG et al., 1987), *Pfu* pol (LUNDBERG et al., 1991) from *Pyrococcus furiosus* (FIALA and

STETTER, 1986), Deep Vent™ pol (PERLER et al., 1996) from *Pyrococcus* strain GB-D (JANNASCH et al., 1992), or Vent™ pol (CARIELLO et al., 1991; MATTILA et al., 1991) from *Thermococcus litoralis* (NEUNER et al., 1990) have an error rate that is up to 10-fold lower than that of *Taq* polymerase. The 9°N-7 DNA polymerase from *Thermococcus* sp. strain 9°N-7 has a 5-fold higher 3′-5′-exonuclease activity than *T. litoralis* DNA polymerase (SOUTHWORTH et al., 1996). However, *Taq* polymerase was not replaced by these DNA polymerases because of their low extension rates among other factors. DNA polymerases with higher fidelity are not necessarily suitable for amplification of long DNA fragments because of their potentially strong exonuclease activity (BARNES, 1994). The recombinant KOD1 DNA polymerase from *Pyrococcus* sp. strain KOD1 has been reported to show low error rates (similar values to those of *Pfu*), high processivity (persistence of sequential nucleotide polymerization), and high extension rates, resulting in an accurate amplification of target DNA sequences up to 6 kb (TAKAGI et al., 1997). In order to optimize the delicate competition of polymerase and exonuclease activity, the exo-motif 1 (BLANCO et al., 1991; MORRISON et al., 1991) of the 9°N-7 DNA polymerase was mutated in an attempt to reduce the level of exonuclease activity without totally eliminating it (SOUTHWORTH et al., 1996). Recently, a family B DNA polymerase from the hyperthermophilic archaeon *Pyrobaculum islandicum* has been cloned and sequenced. Under suitable assay conditions for PCR, the enzyme was able to amplify DNA fragments of up to 1,500 bp (KÄHLER and ANTRANIKIAN, 2000).

An additional problem in the performance of PCR is the generation of non-specific templates prior to thermal cycling. Several approaches have been made to prevent the elongation of polymerase before cycling temperatures are reached. Although using wax as a mechanical barrier between DNA and the enzyme, more sophisticated methods were invented like the inhibition of *Taq* polymerase by a neutralizing antibody at mesophilic temperatures (KELLOGG et al., 1994; SCALICE et al., 1994; SHARKEY et al., 1994) or heat mediated activation of the immobilized enzyme (NILSSON et al., 1997).

Recently, the PCR technique has been improved to allow low error synthesis of long amplificates (20–40 kb) by adding small amounts of thermostable, archaeal proofreading DNA polymerases, containing 3′-5′-exonuclease activity, to *Taq* or other non-proofreading DNA polymerases (BARNES, 1994; CHENG et al., 1994; COHEN, 1994). In this long PCR, the reaction conditions are optimized for long extension by adding different components such as gelatin, Triton X-100, or bovine serum albumin to stabilize the enzymes and mineral oil to prevent evaporation of water in the reaction mixture. In order to enhance specificity, glycerol (CHA et al., 1992) or formamide (SARKAR et al., 1990) are added.

10.2 High Temperature Reverse Transcription

The technique of DNA amplification has been extended to include RNA as the starting template by first converting RNA to cDNA employing either avian myeloblastosis virus reverse transcriptase (AMV-RT) or Moloney murine leukemia virus (M-MuLV) RT (FROHMAN et al., 1988; KAWASAKI et al., 1988; POWELL et al., 1987). The resultant first-strand complementary DNA (cDNA) can be used for generating cDNA libraries, quantifying the levels of gene expression, or determining unknown sequences of either the 3′- or 5′-ends of messenger RNA strands. The latter application is often referred to as RACE (rapid amplification of cDNA ends) "anchored" PCR (LOH et al., 1989) or "one sided" PCR (OHARA et al., 1989). A significant problem in using mesophilic viral reverse transcriptase (RT) is the occurrence of stable secondary RNA structures at low temperatures (KOTEWICZ et al., 1988). Many thermostable DNA polymerases, e.g., *Taq* polymerase (JONES and FOULKES, 1989; KALEDIN et al., 1980) and the DNA polymerases from *Thermus thermophilus* (AUER et al., 1995; MYERS and GELFAND, 1991; RÜTTIMANN et al., 1985) or *Thermus caldophilus* (PARK et al., 1993) can use RNA as a template in the presence of Mn^{2+} instead of Mg^{2+}. The DNA polymerase from *T. thermophilus* was reported to be 100-fold more efficient in a coupled RT-PCR than *Taq* polymerase

(Myers and Gelfand, 1991). Although we could not find any fidelity values, it is probable that the use of Mn^{2+}, like in the case of mesophilic DNA polymerases (Dong et al., 1993; El-Deiy et al., 1984), may increase the error rate and reduce the fidelity of thermostable DNA polymerases. The DNA polymerase from *Thermus filiformis* has been reported to use Mg^{2+} in RT-PCR, yielding products comparable to those synthesized by *T. thermophilus* DNA polymerase in the presence of Mn^{2+} (Perler et al., 1996).

10.3 DNA Sequencing

DNA sequencing by the Sanger method (Sanger et al., 1977) has undergone countless refinements in the last 20 years. A major step forward was the introduction of thermostable DNA polymerases leading in the cycle sequencing procedure. This method uses repeated cycles of temperature denaturation, annealing, and extension with dideoxy-termination to increase the amount of sequencing product by recycling the template DNA. Due to this "PCR-like" amplification of the sequencing products several problems could have been overcome. Caused by the cycle denaturation, only fmoles of template DNA are required, no separate primer annealing step is needed, and unwanted secondary structures within the template are resolved at high temperature elongation.

The first enzyme used for cycle sequencing was the thermostable DNA polymerase I from *Thermus aquaticus* (Gyllensten, 1989; Murray, 1989; Innis et al., 1988). As described by Longley et al. (1990) the enzyme displays 5′-3′-exonuclease activity which is undesirable because of the degradation of sequencing fragments. This enzymatic activity could be deleted by the construction of N-terminal truncated variants. One of them which lacks the N-terminal 289 amino acids was termed "Stoffel fragment" (Lawyer et al., 1993). Two other variants lacked the first 235 and 278 amino acids (Barnes, 1992, 1994). This deletion goes along with an improvement in the fidelity of polymerization (Barnes, 1992). One of the disadvantages over conventional sequencing with T7 polymerase was the inefficient incorporation of chain-terminating dideoxynucleotides by *Taq* polymerase into DNA (Innis et al., 1988). Mutagenesis analysis of the dNTP binding site revealed that only a single residue is critical for the selectivity. Therefore, Phe 667 Tyr exchange in *Taq* polymerase decreased the discrimination against ddNTP several thousand folds (Tabor and Richardson, 1995) resulting in longer reads (Reeve and Fuller, 1995) and improved signals (Fan et al., 1996).

Another drawback in sequencing efficiency is the ability of DNA polymerases to pyrophosphorolysis resulting in removal of dideoxynucleotides by pyrophosphate. This backward reaction has been suppressed by adding a thermostable pyrophosphatase from *Thermoplasma acidophilum* (Vander et al., 1997). Degradation of the inorganic pyrophosphate, therefore, ends up in a more efficient termination reaction.

Several other bacterial thermostable polymerases are described for the use in cycle sequencing (Tab. 5), namely *Bst* DNA polymerase from *Bacillus stearothermophilus* (Mead et al., 1991) or *Tfl* DNA polymerase from *Thermus flavus* (Rao and Saunders, 1992). The most thermostable DNA polymerases are derived from hyperthermophilic Archaea and are, therefore, highly desirable for the application in cycle reactions. Unlike the above-mentioned Pol I-like polymerases, these archaeal α-like DNA polymerases exhibit strong 3′-5′-exonuclease activity, which is detrimental for DNA sequencing. Due to this intrinsic proofreading activity, the incorporation of dideoxynucleotides is inhibited almost completely. By comparing the primary sequence of several α-type polymerases active sites responsible for 3′-5′-exonucleolytic activity were detected and altered by site specific mutagenesis ending up in polymerases suitable for cycle sequencing reactions (Perler et al., 1996; Southworth et al., 1996; Kong et al., 1993; Sears et al., 1992).

10.4 Ligase Chain Reaction

A variety of analytical methods is based on the use of thermostable ligases. Of considerable potential is the construction of sequencing primers by high temperature ligation of

Tab. 5. Applications of Thermostable DNA Polymerases and Ligases[a]

Polymerases	Organism	Application			
		PCR[b]	High Fidelity PCR[c]	Reverse Transcription	DNA Sequencing
Bacterial DNA polymerases					
Taq pol I	*Thermus aquaticus*	+	–	+ (weak; Mn^{2+})[d]	+[e]
Tth pol	*Thermus thermophilus*	+	–	+ (Mn^{2+})	+
Tfi pol	*Thermus filiformis*	+	n.i.	+ (Mg^{2+})	n.i.
Tfl pol	*Thermus flavus*	+	n.i.	–	+
Tca pol	*Thermus caldophilus* GK24	+	–	+ (weak; Mn^{2+})	n.i.
*Bst*I pol	*Bacillus stearothermophilus*	+	–	–	+
Tma pol	*Thermotoga maritima*	+	+	n.i.	+
Archaeal DNA polymerases					
Pwo pol	*Pyrococcus woesei*	+	+	–	–
Pfu pol	*Pyrococcus furiosus*	+	+	–	±[f]
DeepVent pol	*Pyrococcus* sp. GB-D	+	+	–	±[f]
KOD1 pol	*Pyrococcus* sp. KOD1	+	+	–	–
Vent pol	*Thermococcus litoralis*	+	+	–	±[f]
9°N-7 pol	*Thermococcus* sp. 9°N-7	+	+	–	±[f]

[a] Data derived from references cited in text or from commercial sources.
[b] PCR amplificates up to 40 kb could be obtained by applying a combination of a DNA polymerase with high extension rate (e.g., *Taq* pol or *Tth* pol) and a proofreading DNA polymerase with 3′-5′-exonuclease activity (e.g., *Pwo* pol or *Pfu* pol).
[c] Fidelity is enhanced by a proofreading DNA polymerase with 3′-5′-exonuclease activity.
[d] Efficiency of reverse transcription is more than 100-fold weaker than with *Tth* pol.
[e] *Taq* pol I has undergone several modifications to enhance sequencing procedure as there are elimination of 5′-3′-exonuclease activity either by N-terminal deletion or point mutation and reduced discrimination against ddNTP also by point mutation.
[f] Only applicable as exo^--mutation with deleted 3′-5′-exonuclease activity.
+ Suitable; – not suitable; n.i. no information available

hexameric primers (SZYBALSKI, 1990), the detection of trinucleotide repeats through repeat expansion detection (RED; SCHALLING et al., 1993), or DNA detection by circularization of oligonucleotides (NILSSON et al., 1994).

Tremendous improvements have been made in the field of heritable diseases. A powerful analytical method for detection of single base mutations in specific nucleotide sequences utilizes DNA ligases (LANDEGREN et al., 1988). Two oligonucleotides are hybridized to a DNA template, so that the 3′-end of the first one is adjacent to the 5′-end of the second one. In case that the two oligonucleotides are perfectly base paired, a DNA ligase can link them covalently (WU and WALLACE, 1989). A major drawback of this method is the detection of relative small amounts of ligated product and the high background due to unspecific ligation by the applied T4 DNA ligase. These problems have been overcome by the invention of the ligase chain reaction (LCR; BARANY, 1991). In a preceding step a thermostable ligase links two adjacent primers at a temperature above 60 °C. This product is amplified exponentially in the presence of a second set of complementary oligonucleotides when denaturation, annealing, and ligation is repeated several times comparable to the polymerase chain reaction. The specificity of the ligation reaction is dramatically enhanced by performing the reaction near the melting point of the primers. Remarkably, the first thermostable DNA ligase was described in 1984. It was derived from *Thermus thermophilus* HB8 (TAKAHASHI et al., 1984) and displayed a wide temperature range between 15 °C and 85 °C with an optimum at 70 °C. Over the years several additional thermostable DNA ligases were discovered. Bacterial enzymes were derived and cloned from *Thermus scotoductus* (JONSSON et al., 1994) and *Rhodothermus marinus* (THORBJARNARDOTTIR et al., 1995). Recent studies in the crude extract of 103 strains of the genera *Thermus*, *Bacillus*, *Rhodothermus*, and *Hydrogenobacter* revealed the presence of thermostable DNA ligases in 23 of the *Thermus* strains (HJÖRLEIFSDOTTIR et al., 1997). Up to now an archaeal DNA ligase has been described from *Desulfurolobus ambivalens* (KLETZIN, 1992). Unlike bacterial enzymes, this ligase is NAD^+-independent but ATP-dependent similar to the enzymes from bacteriophages, eukaryotes, and viruses.

11 Other Thermoactive Enzymes of Biotechnological Interest

In addition to polymer degrading and DNA modifying enzymes, other enzymes from extremophiles are expected to play a role in industrial processes involving reactions like transesterification, peptide-, oligosaccharide- and phospholipid synthesis.

11.1 Glucose Isomerases

Glucose isomerase (GI) or xylose isomerase (XI) (D-xylose ketol-isomerase; EC 5.3.1.5) catalyzes the reversible isomerization of D-glucose and D-xylose to D-fructose and D-xylulose, respectively. The enzyme has the largest market in the food industry because of its application in the production of high-fructose corn syrup (HFCS). HFCS, an equilibrium mixture of glucose and fructose, is 1.3 times sweeter than sucrose. Glucose isomerase is widely distributed in mesophilic microorganisms. Intensive research efforts are directed toward improving the suitability of GI for industrial application. In order to achieve fructose concentration of 55% the reaction must approach 110 °C. Improved thermostable GIs have been engineered from mesophilic enzymes (CRABB and MITCHINSON, 1997). The gene encoding a xylose isomerase (XylA) of *Thermus flavus* AT62 was cloned and the DNA sequence was determined. XylA (MW 185 kDa; 45 kDa/subunit) has an optimum temperature at 90 °C and pH 7.0; divalent cations such as Mn^{2+}, Co^{2+}, and Mg^{2+} are required for enzyme activity (PARK et al., 1997). *Thermoanaerobacterium* strain JW/SL-YS 489 forms a xylose isomerase (MW, 200 kDa; 50 kDa/subunit) which is optimally active at pH 6.4 (60 °C) or pH 6.8 (80 °C). Like other xylose isomerases (Xis), this enzyme requires Mn^{2+}, Co^{2+}, or Mg^{2+} for thermal stability (stable for 1 h at 82 °C in the absence of substrate). The gene encoding the XI of *Thermoanaerobacterium* strain JW/SL-YS 489 was cloned and expressed in *E. coli*. Comparison of the deduced amino acid sequence with sequences of other xylose isomerases showed that the enzyme has 98% homology with a xylose isomerase from a closely related bacterium, *Thermoanaerobacterium saccharolyticum* B6A-RI (LIU et al., 1996). A thermostable GI was purified and characterized from *Thermotoga maritima*. The enzyme is stable up to 100 °C, with a half-life of 10 min at 115 °C (BROWN et al., 1993). Interestingly, the GI from *Thermotoga neapolitana* displays a catalytic efficiency at 90 °C, which is 2–14 times higher than any other thermoactive GIs at temperatures between 60 °C and 90 °C (VIEILLE et al., 1995).

11.2 Alcohol Dehydrogenases

The secondary specific alcohol dehydrogenase (ADH), which catalyzes the oxidation of secondary alcohols and, less readily, the reverse reaction (the reduction of ketones) has a promising future in biotechnology (COWAN, 1992). Although ADHs are widely distributed among microorganisms, only few examples derived from hyperthermophilic microorganisms are currently known. Among the extreme thermophilic Bacteria, *Thermoanaerobacter ethanolicus* 39E was shown to produce an ADH, whose gene was cloned and expressed in *E. coli* (BURDETTE et al., 1997). Interestingly, a mutant has been found to posses an advantage over the wild-type enzyme by using the more stable cofactor NAD instead of NADP. In extreme thermophilic Archaea, ADHs have been studied from *Sulfolobus solfataricus* (PEARL et al., 1993; AMMENDOLA et al., 1992; RELLA et al., 1987) and from *Thermococcus stetteri* (MA et al., 1994). The enzyme from *S. solfataricus* requires NAD as cofactor and contains Zn ions. In contrast to ADHs from Bacteria and Eukarya, the enzyme from *T. stetteri* lacks metal ions. The enzyme catalyzes preferentially the oxidation of primary alcohols, using NADP as cofactor and it is very thermostable, showing half-lives of 15 min at 98 °C and 2 h at 85 °C. Compared to mesophilic enzymes, the ADH from *T. litoralis* represents a new type of alcohol-oxidizing enzyme system. The genes of ADHs from *S. solfataricus* and *Sulfolobus* sp. strain RC3 were expressed at high level in *E. coli* and the recombinant enzymes were purified and characterized (CANNIO et al., 1996).

11.3 Esterases

In the field of biotechnology, esterases are gaining increasing attention because of their application in organic biosyntheses. In aqueous solution, esterases catalyze the hydrolytic cleavage of esters to form the constituent acid and alcohol, whereas in organic solutions, transesterification reaction is promoted. Both the reactants and the products of transesterification are usually highly soluble in the organic phase and the reactants may even form the organic phase themselves. Solvent-stable esterases are formed by the extreme thermophilic Bacteria such as *Caldocellum saccharolyticum* (LUTHI et al., 1990) and the archaeon *Sulfolobus acidocaldarius* (SOBEK and GORISCH, 1988). Recently, the *Pyrococcus furiosus* esterase gene was cloned in *E. coli* and the functional properties were determined. The archaeal enzyme is the most thermostable (a half-life of 50 min at 126 °C) and thermoactive (temperature optimum 100 °C) esterase known to date (IKEDA and CLARK, 1998).

12 Conclusion

The steady increase in the number of newly isolated extremophilic microorganisms and the related discovery of their enzymes by academic and industrial institutions document the enormous potential of extremophiles for application in future biotechnological processes. Enzymes from extremophilic microorganisms offer versatile tools for sustainable developments in a variety of industrial application as they show important environmental benefits, e.g., they are biodegradable, specific, stable under extreme conditions, show improved use of raw materials and decreased amounts of waste products. Although major advances have been made in the last decade, our knowledge of the physiology, metabolism, enzymology and genetics of this fascinating group of extremophilic microorganisms and their related enzymes is still limited. In-depth information on the molecular properties of the enzymes and their genes, however, has to be obtained in order to analyze the structure and function of proteins being catalytically active around the boiling and freezing point of water and extreme of pHs. New techniques like genomics, proteomics, protein engineering, and gene shuffling will lead to the production of tailor-made enzymes that are highly specific for countless industrial applications. Due to the unusual properties of enzymes from extremophiles, they are expected to fill the gap between biological and chemical industrial processes.

Acknowledgements

Thanks are due to the Deutsche Bundesstiftung Umwelt and to the Fonds der Chemischen Industrie for financial support.

13 References

AMMENDOLA, S., RAIA, C. A., CARUSO, C., CAMARDELLA, L., AURIA, S. et al. (1992), Thermostable NAD dependent alcohol deydrogenase from *Sulfolobus solfataricus*: gene protein sequence determination and relationship to other alcohol dehydrogenases, *Biochemistry* **31**, 12514–12523.

ANDERSON, D. E., HURLEY, J. H., NICHOLSON, H., BAASE, W. A., MATTHEWS, B. W. (1993), Hydrophobic core repacking and aromatic : aromatic interaction in the thermostable mutant of T4 lysozyme Ser 117→Phe, *Protein Sci.* **2**, 1285–1290.

ANNISON, G. (1992), Commercial enzyme supplementation of wheat-based diets raises ileal glycanase activitites and improves apparent metabolisable energy starch and pentosan digestibilities in broiler chickens, *Anim. Feed Sci. Technol.* **38**, 105–121.

ANTRANIKIAN, G. (1992), Microbial degradation of starch, in: *Microbial Degradation of Natural Products* Vol. 2 (WINKELMANN, G., Ed.), pp. 28–50. Weinheim: VCH.

AUER, T., LANDRE, P. A., MYERS, T. W. (1995), Properties of the 5′-3′-exonuclease/ribonuclease H activity of *Thermus thermophilus* DNA polymerase, *Biochemistry* **34**, 4994–5002.

BARANY, F. (1991), Genetic disease detection and DNA amplification using cloned thermostable ligase, *Proc. Natl. Acad. Sci. USA* **88**, 189–193.

BARNES, W. M. (1992), The fidelity of Taq polymerase catalyzing PCR is improved by an N-terminal deletion, *Gene* **112**, 29–35.

BARNES, W. M. (1994), PCR amplification of up to 35-kb DNA fragments with high fidelity and high yield from lambda bacteriophage templates, *Proc. Natl. Acad. Sci. USA* **91**, 2216–2220.

BAROSS, J. A., DEMING, J. W. (1995), Growth at high temperatures: isolation and taxonomy, physiology, and ecology, in: *The Microbiology of Deep-Sea Hydrothermal Vents* (KARL, D. M., Ed.), pp. 169–217. Boca Raton, FL: CRC Press.

BAUER, M. W., DRISKILL, L. E., CALLEN, W., SNEAD, M. A., MATHUR, E. J., KELLY, R. M. (1999), An endoglucanase, EglA, from the hyperthermophilic archaeon *Pyrococcus furiosus* hydrolizes β-1,4 bonds in mixed-linkage (1→3), (1→4)-β-D-glucans and cellulose, *J. Bacteriol.* **181**, 284–290.

BERTOLDO, C., ANTRANIKIAN, G. (2000), Amylolytic enzymes from hyperthermophiles, *Methods Enzymol.* **330**, 269–283.

BERTOLDO, C., DUFFNER, F., JØRGENSEN, P. L., ANTRANIKIAN, G. (1999), Pullulanase type I from *Fervidobacterium pennavorans* Ven5: Cloning, sequencing, and expression of the gene and biochemical characterization of the recombinant enzyme, *Appl. Environ. Microbiol.* **65**, 2084–2091.

BHARAT, B., HOONDAL, G. S. (1998), Isolation, purification and properties of a thermostable chitinase from an alkalophilic *Bacillus* sp. BG-11, *Biotechnol. Lett.* **20**, 157–159.

BLANCO, L., BERNAD, A., BLASCO, M. A., SALAS, M. (1991), A general structure for DNA-dependent DNA polymerases, *Gene* **100**, 27–38.

BLÖCHL, E., BURGGRAF, S., FIALA, G., LAUERER, G., HUBER, G. et al. (1995), Isolation, taxonomy and phylogeny of hyperthermophilic microorganisms, *World J. Microbiol. Biotechnol.* **11**, 3–16.

BLUMENTALS, I. I., ROBINSON, A. S., KELLY, R. M. (1992), Characterization of sodiumdodecyl sulphate-resistant proteolytic activity in the hyperthermophilic archaebacterium *Pyrococcus furiosus*, *Appl. Environ. Microbiol.* **56**, 1992–1998.

BOK, J. D., GOERS, S. K., EVELEIGH, D. E. (1994), Cellulase and xylanase systems of *Thermotoga neapolitana*, *ACS Symp. Ser.* **566**, 54–65.

BOK, J. D., DINESH, A, YERNOOL, D. A., EVELEIGH, D. (1998), Purification, characterization and molecular analysis of thermostable cellulases Cel A and Cel B from *Thermotoga neapolitana*. *Appl. Environ. Microbiol.* **64**, 4774–4781.

BOST, D. A., STOFFEL, S., LANDRE, P., LAWYER, F. C., AKERS, J. et al. (1994), Enzymatic characterization of *Thermotoga maritima* DNA polymerase and a truncated form, Ultma DNA polymerase, *FASEB J.* **8**, A1395.

BRAGGER, J. M., DANIEL, R. M., COOLBEAR, T., MORGAN, H. W. (1989), Very stable enzyme from extremely thermophilic archaebacteria and eubacteria, *Appl. Microbiol. Biotechnol.* **31**, 556–561.

BRITTON, K. L., BAKER, P. J., BORGES, K. M. M., ENGEL, P. C., PASQUO, A. et al. (1995), Insights into thermal stability from a comparison of the glutamate dehydrogenases from *Pyrococcus furiosus* and *Thermococcus litoralis*, *Eur. J. Biochem.* **229**, 688–695.

BRONNENMEIER, K., KERN, A., LIEBL, W., STAUDENBAUER, W. L. (1995), Purification of *Thermotoga maritima* enzymes for the degradation of cellulose materials, *Appl. Environ. Microbiol.* **61**, 1399–1407.

BROWN, S. H., KELLY, R. M. (1993), Characterization of amylolytic enzymes, having both α-1,4 and α-1,6 hydrolytic activity, from the thermophilic archaea *Pyrococcus furiosus* and *Thermococcus litoralis*, *Appl. Environ. Microbiol.* **59**, 2614–2621.

BROWN, S. H., SJHOLM, C., KELLY, R. M. (1993), Purification and characterization of a higly thermostable glucose isomerase produced by the extremely thermophilic eubacterium *Thermotoga maritima*, *Biotechnol. Bioeng.* **41**, 878–886.

BURDETTE, D., SECUNDO, F., PHILLIPS, R., DONG, J., SCOTT, R., ZEIKUS, J. G. (1997), Biophysical and mutagenic analysis of *Thermoanaeobacter ethanolicus* 39E secondary alcohol deydrogenase and enzyme biochemical characterization, *Biochem. J.* **316**, 112–122.

BURLINI, N., MAGNANI, P., VILLA, A., MACCHI, F., TORTORA, P., GUERRITORE, A. (1992), A heat-stable serine proteinase from the extreme thermophilic archaebacterium *Sulfolobus solfataricus*, *Biochim. Biophys. Acta* **1122**, 283–292.

CANGANELLA, F., ANDRADE, C., ANTRANIKIAN, G. (1994), Characterization of amylolytic and pullulytic enzymes from thermophilic archaea and from a new *Fervidobacterium* species, *Appl. Microbiol. Biotechnol.* **42**, 239–245.

CANGANELLA, F., JONES, W. J., GAMBACORTA, A., ANTRANIKIAN, G. (1998), *Thermococcus guaymasensis* sp. nov. and *Thermococcus aggregans* sp. nov., two novel thermophilic archaea isolated from the Guaymas Basin hydrothermal vent site, *Int. J. Syst. Bacteriol.* **48**, 1181–1185.

CANNIO, R., FIORENTINO, G., CARPINELLI, P., ROSSI, M., BARTOLUCCI, S. (1996), Cloning and overexpression in *Escherichia coli* of the genes encoding NAD-dependent alcohol deydrogenase from two *Sulfolobus* species, *J. Bacteriol.* **178**, 301–305.

CARIELLO, N. F., SWENBERG, J. A., SKOPEK, T. R. (1991), Fidelity of *Thermococcus litoralis* DNA polymerase (Vent) in PCR determined by denaturating gradient gel electrophoresis, *Nucl. Acids Res.* **19**, 4193–4198.

CHA, R. S., ZARBEL, H., KEOHAVONG, P., THILLY, W. G. (1992), Mismatch amplification mutation assay (MAMA): Application to the c-H-*ras* gene, *PCR Methods Appl.* **2**, 14–20.

CHAN, M. K., MUKUND, S., KLETZIN, A., ADAMS, M. W., REES, D. C. (1995), Structure of a hyperthermophilic tungstopterin enzyme, aldehyde ferredoxin oxidoreductase, *Science* **267**, 1463–1469.

CHANDY, T., SHARMA, C. P. (1990), Chitosan as a biomaterial, *Biomater. Artif. Cells Artif. Organs* **31**, 1–24.

CHEN, C. C., ADOLPHSON, R., DEAN, F. D. J., ERIKSSON, K. E. L., ADAMAS, M. W. W., WESTPHELING, J. (1997), Release of lignin from kraft pulp by a hyperthermophilic xylanase from *Thermotoga maritima*, *Enzyme Microbiol. Technol.* **20**, 39–45.

CHENG, S., FOCKLER, C., BARNES, W. M., HIGUSHI, R. (1994), Effective amplification of long targets from cloned inserts and human genomic DNA, *Proc. Natl. Acad. Sci. USA* **91**, 5695–5699.

CHIEN, A., EDGAR, D. B., TRELA, J. M. (1976), Deoxyribonucleic acid polymerase from the extreme thermophile *Thermus aquaticus*, *J. Bacteriol.* **127**, 1550–1557.

CHUNG, Y. C., KOBAYASHI, T., KANAI, H., AKIBA, T., KUDO, T. (1995), Purification and properties of extracellular amylase from the hyperthermophilic archaeon *Thermococccus profundus* DT5432, *Appl. Environ. Microbiol.* **61**, 1502–1506.

CLASSEN, H. L. (1996), Cereal grain starch and exogenous enzymes in poultry diets, *Anim. Feed Sci. Technol.* **33**, 791–794.

COHEN, J. (1994), "Long PCR" leaps into larger DNA sequences, *Science* **263**, 1564–1565.

COHEN-KUPIEC, R., CHET, I. (1998), The molecular biology of chitin digestion, *Curr. Opin. Biotechnol.* **9**, 270–277.

COWAN, D. (1996), Industrial enzyme technology, *TIBTECH* **14**, 177–178.

COWAN, D. A., SMOLENSKI, K. A., DANIEL, R. M., MORGAN, H. W. (1987), An extremely thermostable extracellular proteinase from a strain of the archaebacterium *Desulfurococcus* growing at 88 °C, *Biochem. J.* **247**, 121–133.

CRABB, W. D., MITCHINSON, C. (1997), Enzymes involved in the processing of starch to sugars, *TIBTECH* **15**, 349–352.

CUBELLIS, M. V., ROZZO, C., MONTECUCCHI, P., ROSSI, M. (1990), Isolation and sequencing of a new β-galactosidase-encoding archaebacterial gene, *Gene* **94**, 89–94.

DA COSTA, M. S., SANTOS, H., GALINSKI, E. A. (1998), An overview of the role and diversity of compatible solutes in bacteria and archaea, *Adv. Biochem. Eng. Biotechnol.* **61**, 117–153.

DE ROSA, M., GAMBACORTA, A., HUBER, R., LANZOTTI, V., NICOLAUS, B. et al. (1989), Lipid structures in *Thermotoga maritima*, pp. 167–173.

DE ROSA, M., GAMBACORTA, A. (1994), Archaeal lipids, in: *Chemical Methods in Prokaryotic Systematics* (GOODFELLOW, M., O'DONNELL, A. G., Eds.), pp. 199–264. New York: John Wiley & Sons.

DONG, G., VIEILLE, C., SAVCHENKO, A., ZEIKUS, J. G. (1997a), Cloning, sequencing and expression of the gene encoding extracellular α-amylase from *Pyrococcus furiosus* and biochemical characterisation of the recombinant enzyme, *Appl. Environ. Microbiol.* **63**, 3569–3576.

DONG, G., VIEILLE, C., ZEIKUS, J. G. (1997b), Cloning, sequencing and expression of the gene amylopullulanase from *Pyrococcus furiosus* and biochemical characterisation of the recombinant enzyme, *Appl. Environ. Microbiol.* **63**, 3577–3584.

DONG, Q., COPELAND, W. C., WANG, T. S. (1993), Mutational studies of human DNA polymerase alpha. Identification of residues critical for deoxynucleotide binding and misinsertion fidelity of DNA synthesis, *J. Biol. Chem.* **268**, 24163–24174.

DUFFNER, F., BERTOLDO, C., ANDERSEN, J. T., WAGNER, K., ANTRANIKIAN, G. (2000), A new thermoactive pullulanase from *Desulfurococcus mucosus*: cloning, sequencing and characterization of the recombinant enzyme after expressio in *Bacillus subtilis*, *J. Bacteriol.* **182**, 6331–6338.

DUNNING, A. M., TALMUD, P., HUMPHRIES, S. E. (1988), Errors in the polymerase chain reaction, *Nucl. Acids Res.* **16**, 10393.

EDMONDS, C. G., CRAIN, P. F., GUPTA, R., HASHIZU-

ME, T., HOCART, C. H. et al. (1991), Posttranscriptional modification of tRNA in thermophilic archaea (archaebacteria), *J. Bacteriol.* **173**, 3138–3148.

EGGEN, H. I. L., GEERLING, A., WATTS, J., DE VOS, W. M. (1990), Characterization of pyrolysin, a hyperthermoactive serine protease from the archaeobacterium *Pyrococcus furiosus*, *FEMS Microbiol. Lett.* **71**, 17–20.

EL-DEIRY, W. S., DOWNEY, K. M., SO, A. G. (1984), Molecular mechanisms of manganese mutagenesis, *Proc. Natl. Acad. Sci. USA* **81**, 7378–7382.

ERLICH, H. A., GELFAND, D. H., SAIKI, R. K. (1988), Specific DNA amplification-product review, *Nature* **331**, 461–462.

FABER, K., FRANSSEN, M. C. R. (1993), Prospect for the increased application of biocatalysts in organic transformations, *Trends Biotechnol.* **11**, 461–470.

FAN, J., RANU, R. S., SMITH, C., RUAN, C., FULLER, C. W. (1996), DNA sequencing with alpha-33PI-labeled ddNTP terminators: a new approach to DNA sequencing with Thermo Sequenase DNA polymerase, *Biotechniques* **21**, 1132–1137.

FELLER, G., NARINX, E., ARPIGNY, J. L., AITTALEB, M., BAISE, E. et al. (1996), Enzyme from extremophilic microorganisms, *FEMS Microbiol. Rev.* **18**, 189–202.

FIALA, G., STETTER, K. O. (1986), *Pyrococcus furiosus* sp. nov. represents a novel genus of marine heterotrophic archaebacteria growing optimally at 100 °C, *Arch. Microbiol.* **145**, 56–61.

FORTERRE, P., BERGERAT, A., LOPEZ-GARCIA, P. (1996), The unique DNA topology and DNA topoisomerases of hyperthermophilic archaea, *FEMS Microbiol. Rev.* **18**, 237–248.

FREY, B., SUPPMANN, B. (1995), Demonstration of the Expand PCR System's greater fidelity and higher yields with a lacI-based fidelity assay, *Biochemica* **2**, 34–35.

FRIEDRICH, A., ANTRANIKIAN, G. (1996), Keratin degradation by *Fervidobacterium pennavorans*, a novel thermophilic anaerobic species of the order Thermotogales, *Appl. Environ. Microbiol.* **62**, 2875–2882.

FRILLINGOS, S., LINDEN, A., NIEHAUS, F., VARGAS, C., NIETO, J. J. et al. (2000), Drainas C cloning and expression of *alpha*-amylase from the hyperthermophilic archaeon *Pyrococcus woesei* in the moderately halophilic bacterium *Halomonas elongata*, *J. Appl. Microbiol.* **88**, 495–503.

FROHMAN, M. A., DUSH, M. K., MARTIN, G. R. (1988), Rapid production of full-length cDNAs from rare transcripts: Amplification using a single gene specific oligonucleotide primer, *Proc. Natl. Acad. Sci. USA* **85**, 8998–9002.

FUJIWARA, S., OKUYAMA, S., IMANAKA, T. (1996), The world of archaea: genome analysis, evolution and thermostable enzymes, *Gene* **179**, 165–170.

FUKUSUMI, S., KAMIZONO, A., HORINOUCHI, S., BEPPU, T. (1988), Cloning and nucleotide sequence of a heat stable amylase gene from an anaerobic thermophile, *Dictyoglomus thermophilum*, *Eur. J. Biochem.* **174**, 15–23.

FUSEK, M., LIN, X. L., TANG, J. (1990), Enzymic properties of thermopsin, *J. Biol. Chem.* **265**, 1496–1501.

GALICHET, A., BELARBI, A. (2000), Cloning of an α-glucosidase gene from *Thermococcus hydrothermalis* by functional complementation of a *Saccharomyces cerevisiae* mal11 mutant strain, *FEBS Lett.* **458**, 188–192.

GALINSKI, E. A. (1993), Compatible solutes of halophilic eubacteria: molecular principles, water-soluble interactions, stress protection, *Experientia* **49**, 487–496.

GANGHOFNER, D., KELLERMANN, J., STAUDENBAUER, W. L., BRONNENMEIER, K. (1998), Purification and properties of an amylopullulanase, a glucoamylase, and an alpha-glucosidase in the amylolytic enzyme system of *Thermoanaerobacterium thermosaccharolyticum*, *Biosci. Biotechnol. Biochem.* **62**, 302–308.

GANTELET, H., DUCHIRON, F. (1998), Purification and properties of a thermoactive and thermostable pullulanase from *Thermococcus hydrothermalis*, a hyperthermophilic archaeon isolated from a deep sea hydrothermal vent, *Appl. Microbiol. Biotechnol.* **49**, 770–777.

GEORGOPAPADAKOU, N. H., TKACZ, J. S. (1995), The fungal cell wall as a drug target, *Trends Microbiol.* **3**, 98–104.

GERDAY, C., AITTALEB, M., BENTAHIR, M., CHESSA, J. P., CLAVERIE, P. et al. (2000), Cold adapted enzymes: from fundamentals to biotechnology, *TIBTECH* **18**, 103–107.

GIBBS, M. D., REEVES, R. A., BERGQUIST, P. L. (1995), Cloning, sequencing and expression of a xylanase gene from the extreme thermophile *Dictyoglomus thermophilum* Rt46B.1 and activity of the enzyme on fiber-bound substrate, *Appl. Environ. Microbiol.* **61**, 4403–4408.

GONZÁLES, J. M., MASUCHI, Y., ROBB, F. T., AMMERMAN, J. W., MAEDER, D. L. et al. (1998), *Pyrococcus horikoshii* sp. nov., a hyperthermophilic archaeon isolated from a hydrothermal vent at the Okinawa Trough, *Extremophiles* **2**, 123–130.

GOODAY, G. W. (1990), Physiology of microbial degradation of chitin and chitosan, *Biodegradation* **1**, 177–190.

GOODAY, G. W. (1994), Physiology of microbial degradation of chitin and chitosan, in: *Biochemistry of Microbial Degradation* (RATLEDGE, C., Ed.), pp. 279–312. Dordrecht: Kluwer.

GRANT, W. D. (1992), Hypersaline environments, in: *Trends in Microbial Ecology* (GUERRERO, R., PE-

DRÓS ALIÒ, Eds.), pp. 2216–2219. Spanish Society for Microbiology.

GRANT, W. D., TINDALL, B. J. (1986), Microbes in extreme environment, in (HERBERT, R. A., CODD, C. O., Eds.), pp. 22–53. New York: Academic Press.

GRANT, W. D., GEMMELL, R. T., MCGENITY, T. J. (1998), Halophiles, in: *Halophiles* (HORIKOSHI, K., GRANT, W. D., Eds.), pp. 93–133. New York: Wiley-Liss.

GRAYLING, R. A., SANDMAN, K., REEVE, J. N. (1994), Archaeal DNA binding proteins and chromosome structure, in: *Molecular Biology of Archaea* (PFEIFER, F., PALM, P., SCHLEIFER, K. H., Eds.), pp. 82–90. Stuttgart, Jena, New York: Gustav Fischer Verlag.

GROGAN, D. W. (1991), Evidence that β-galactosidase of *Sulfolobus solfataricus* is only one of several activities of a thermostable β-D-glycosidase, *Appl. Environ. Microbiol.* **57**, 1644–1649.

GROTE, R., LI, L., TAMOAOKA, J., KATO, C., HORIKOSHI, K., ANTRANIKIAN, G. (1999), *Thermococcus siculi* sp. nov., a novel hyperthermophilic archaeon isolated from a deep-sea hydrothermal vent at the Mid-Okinawa Trough, *Extremophiles* **3**, 55–62.

GYLLENSTEN, U. B. (1989), PCR and DNA sequencing, *Biotechniques* **7**, 700–708.

HALBERG, K. B., LINDSTROM, E. B. (1994), Characterization of *Thiobacillus caldus* sp. nov., a moderately thermophilic acidophile, *Microbiology* **140**, 3451–3456.

HALIO, S. B., BLUMENTALS, I., SHORT, S. A., MERRIL, B. M., KELLY, R. (1996), Sequence, expression in *Escherichia coli* and analysis of the gene encoding a novel intracellular protease (PfpI) from the hyperthermophilic archaeon *Pyrococcus furiosus*, *J. Bacteriol.* **178**, 2605–2612.

HANZAWA, S., HOAKI, T., JANNASCH, H. W., MARUYAMA, T. (1996), An extremely thermostable serine protease from a hyperthermophilic archaeon *Desulfurococcus* strain SY, isolated from a deep-sea hydrothermal vent, *J. Mar. Biotechnol.* **4**, 121–126.

HARWOOD, V. J., DENSON, J. D., RINSON BIDLE, K. A., SCHREIER, H. J. (1997), Overexpression and characterization of a propyl endopeptidase from the hyperthermophilic archaeon *Pyrococcus furiosus*, *J. Bacteriol.* **179**, 3613–3618.

HENRISSAT, B. (1991), A classification of glycosyl hydrolases based on amino acid sequence similarity, *Biochem. J.* **280**, 309–316.

HJÖRLEIFSDOTTIR, S., RITTERBUSCH, W., PETURSDOTTIR, S. K., KRISTJANSSON, J. K. (1997), Thermostabilities of DNA ligases and DNA polymerases from four genera of thermophilic eubacteria, *Biotechnol. Lett.* **19**, 147–150.

HORIKOSHI, K. (1998), Alkaliphiles, in: *Extremophiles – Microbial Life in Extreme Environments* (HORIKOSHI, K., GRANT, W. D., Eds.), p. 155–179. New York: Wiley-Liss.

HORIKOSHI, K., GRANT, W. D. (Eds.) (1998), *Extremophiles – Microbial Life in Extreme Environments*. New York: Wiley-Liss.

HUBER, R., LANGWORTHY, T. A., KÖNIG, H., THOMM, M., WOESE, C. R. et al. (1986), *Thermotoga maritima* sp. nov. represents a new genius of unique extremely thermophilic eubacteria growing up to 90 °C, *Arch. Microbiol.* **144**, 324–333.

HUBER, R., KRISTJANSSON, J. K., STETTER, K. O. (1987), *Pyrobaculum* gen. nov., a new genus of neutrophilic, rod-shaped archaebacteria from continental solfataras growing optimally at 100 °C, *Arch. Microbiol.* **149**, 95–101.

HUBER, R., STOFFERS, P., CHEMINEE, J. L., RICHNOW, H. H., STETTER, K. O. (1990), Hyperthermophilic archaebacteria within the crater and open-sea plume of erupting MacDonald Seamount, *Nature* **345**, 179–182.

HUBER, R., BURGGRAF, S., MAYER, T., BARNS, S. M., ROSSNAGEL, P., STETTER, K. O. (1995a), Isolation of a hyperthermophilic archaeum predicted by *in situ* RNA analysis, *Nature* **376**, 57–58.

HUBER, R., STÖÖHR, J., HOHENHAUS, S., RACHEL, R., BURGGRAF, S. et al. (1995b), *Thermococcus chitonophagus* sp. nov., a novel chitin degrading, hyperthermophilic archeum from the deep-sea hydrothermal vent environment, *Arch. Microbiol.* **164**, 255–264.

HURLEY, J. H., BAASE, W. A., MATTHEWS, B. W. (1992), Design and structural analysis of alternative hydrophobic core packing arrangements in bacteriophage T4 lysozyme, *J. Mol. Biol.* **224**, 1143– 1159.

HYUN, H. H., ZEIKUS, J. G. (1985), Regulation and genetic enhancement of glucoamylase and pullulanase production in *Clostridium thermohydrosulfuricum*, *J. Bacteriol.* **164**, 1146–1152.

IKEDA, M., CLARK, S. C. (1998), Molecular cloning of extremely thermostable esterase gene from hyperthermophilic archeon *Pyrococcus furiosus* in *Escherichia coli*, *Biotechnol. Bioeng.* **57**, 624–629.

INNIS, M. A., MYAMBO, K. B., GELFAND, D. H., BROW, M. A. (1988), DNA sequencing with *Thermus aquaticus* DNA polymerase and direct sequencing of polymerase chain reaction-amplified DNA, *Proc. Natl. Acad. Sci. USA* **85**, 9436–9440.

JANNASCH, H. W., HUBER, R., BELKIN, S., STETTER, K. O. (1988), *Thermotoga neapolitana* sp. nov. of the extremely thermophilic, eubacterial genus *Thermotoga*, *Arch. Microbiol.* **150**, 103–104.

JANNASCH, H. W., WIRSEN, C. O., MOLYNEAUX, S. J., LANGWORTHY, T. A. (1992), Comparative physiological studies on hyperthermophilic archaea isolated from deep-sea hot vents with emphasis on *Pyrococcus* strain GB-D, *Appl. Environ. Microbiol.* **58**, 3472–3481.

JEANTHON, C., L'HARIDON, S., REYSENBACH, A. L., VERNET, M., MESSNER, P. et al. (1998), *Methanococcus infernus* sp. nov., a novel hyperthermophilic lithotrophic methanogen isolated from a deep-sea hydrothermal vent, *Int. J. Syst. Bacteriol.* **48**, 913–919.

JEANTHON, C., L'HARIDON, S., REYSENBACH, A. L., CORRE, E., VERNET, M. et al. (1999), *Methanococcus vulcanius* sp. nov., a novel hyperthermophilic methanogen isolated from East Pacific Rise, and identification of *Methanococcus* sp. DSM 4213T as *Methanococcus fervens* sp. nov., *Int. J. Syst. Bacteriol.* **49**, 583–589.

JEREZ, C. A. (1988), The heat shock response in meso- and thermoacidophilic bacteria, *FEMS Microbiol. Lett.* **56**, 254–261.

JONES, M. D., FOULKES, N. S. (1989), Reverse transcription of mRNA by *Thermus aquaticus* DNA polymerase, *Nucl. Acids Res.* **17**, 8387–8388.

JONSSON, Z. O., THORBJARNARDÍOTTIR, S. H., EGGERTSSON, G., PALSDOTTIR, A. (1994), Sequence of the DNA ligase-encoding gene from *Thermus scotoductus* and conserved motifs in DNA ligases, *Gene* **151**, 177–180.

JØRGENSEN, S., VORGIAS, C. E., ANTRANIKIAN, G. (1997), Cloning, sequencing and expression of an extracellular α-amylase from the hyperthermophilic archeon *Pyrococcus furiosus* in *Escherichia coli* and *Bacillus subtilis*, *J. Biol. Chem.* **272**, 16335–16342.

KÄHLER, M., ANTRANIKIAN, G. (2000), Cloning and characterization of a family B DNA polymerase from the hyperthermophilic crenarchaeon *Pyrobaculum islandicum*, *J. Bacteriol.* **182**, 655–663.

KALEDIN, A. S., SLIUSARENKO, A. G., GORODETSKII, S. I. (1980), Isolation and properties of DNA polymerase from extreme thermophylic bacteria *Thermus aquaticus* YT-1, *Biokhimiya* **45**, 644–651.

KAS, H. S. (1997), Chitosan: Properties, preparations and application to microparticulate systans, *J. Microencapsulation* **14**, 689–712.

KATES, M. (1992), Archaebacterial lipids: structure, biosynthesis and function, in: *The Archaebacteria: Biochemistry and Biotechnology* (DANSON, M. J., HOUGH, D. W., LUNT, G. G., Eds.), pp. 51–72. London, Chapel Hill: Portland Press.

KAWAI, G., HASHIZUME, T., YASUDA, M., MIYAZAWA, T., MCCLOSKEY, J. A., YOKOYAMA, S. (1992), Conformational rigidity of N^4-acetyl-2′-O-methylcytidine found in tRNA of extremely thermophilic archaebacteria (Archaea), *Nucleosides Nucleotides* **11**, 759–771.

KAWASAKI, E. S., CLARK, S. S., COYNE, M. Y., SMITH, S. D., CHAMPLIN, R. et al. (1988), Diagnosis of chronic myeloid and acute lymphocytic leukemias by detection of leukemia-specific mRNA sequences amplified *in vitro*, *Proc. Natl. Acad. Sci. USA* **85**, 5698–5702.

KELLOGG, D. E., RYBALKIN, I., CHEN, S., MUKHAMEDOVA, N., VLASIK, T. et al. (1994), TaqStart Antibody: "hot start" PCR facilitated by a neutralizing monoclonal antibody directed against Taq DNA polymerase, *Biotechniques* **16**, 1134–1137.

KENGEN, S. W. M., LUESINK, E. J., STAMS, A. J. M., ZEHNDER, A. J. B. (1993), Purification and characterization of an extremely thermostable β-glucosidase from the hyperthermophilic archaeon *Pyrococcus furiosus*, *Eur. J. Biochem.* **213**, 305–312.

KEOHAVONG, P., THILLY, W. G. (1989), Fidelity of PCR polymerases in DNA amplification, *Proc. Natl. Acad. Sci. USA* **86**,9253–9257.

KIM, C. H., NASHIRU, O., KO, J. H. (1996), Purification and biochemical characterisation of pullulanase type I from *Thermus caldophilus* GK-24, *FEMS Microbiol. Lett.* **138**, 147–152.

KLETZIN, A. (1992), Molecular characterisation of a DNA ligase gene of the extremely thermophilic archaeon *Desulfurolobus ambivalens* shows close phylogenetic relationship to eukaryotic ligases, *Nucleic Acids Res.* **20**, 5389–5396.

KLINGEBERG, M., HASHWA, F., ANTRANIKIAN, G. (1991), Properties of extremely thermostable proteases from anaerobic hyperthermophilic bacteria, *Appl. Microbiol. Biotechnol.* **34**, 715–719.

KLINGEBERG, M., GALUNSKY, B., SJOHOLM, C., KASCHE, V., ANTRANIKIAN, G. (1995), Purification and properties of highly thermostable, SDS resistant and stereospecific proteinase from the extreme thermophilic archaeon *Thermococcus stetteri*, *Appl. Environ. Microbiol.* **61**, 3098–3104.

KNAPP, S., DEVOS, W., RICE, D., LADENSTEIN, R. (1997), Crystal structure of glutamate dehydrogenase from hyperthermophilic eubacterium *Thermotoga maritima* at 3.0 Å resolution, *J. Mol. Biol.* **267**, 916–932.

KOCH, R. P., ZABLOWSKI, A., SPREINAT, K., ANTRANIKIAN, G. (1990), Extremely thermostable amylolytic enzyme from the archaeobacterium *Pyrococcus furiosus*, *FEMS Microbiol. Lett.* **71**, 21–26.

KOCH, R., SPREINAT, K., LEMKE, K., ANTRANIKIAN, G. (1991), Purification and properties of a hyperthermoactive α-amylase from the archaeobacterium *Pyrococcus woesei*, *Arch. Microbiol.* **155**, 572–578.

KOCH, R., CANGANELLA, F., HIPPE, H., JAHNKE, K. D., ANTRANIKIAN, G. (1997), Purification and properties of a thermostable pullulanase from a newly isolated thermophilic anaerobic bacterium *Fervidobacterium pennavorans* Ven5, *Appl. Environ. Microbiol.* **63**, 1088–1094.

KONG, H., KUCERA, R. B., JACK, W. E. (1993), Characterization of a DNA polymerase from the hyperthermophile archaea *Thermococcus litoralis*. Vent DNA polymerase, steady state kinetics, thermal stability, processivity, strand displacement, and exonuclease activities, *J. Biol. Chem.* **268**, 1965–1975.

KOTEWICZ, M. L., SAMPSON, C. M., D'ALLESSIO, J. M., GERARD, G. F. (1988), Isolation of cloned Moloney murine leukemia virus reverse transcriptase lacking ribonuclease H activity, *Nucleic Acids Res.* **16**, 265–277.

KRAHE, M., ANTRANIKIAN, G., MÄRKL, H. (1996), Fermentation of extremophilic microorganisms, *FEMS Microbiol. Rev.* **18**, 271–285.

KRISTJANSSON, J. K., HREGGVIDSSON, G. O. (1995), Ecology and habitats of extremophiles, *World J. Microbiol. Biotechnol.* **11**, 17–25.

KWAK, Y. S., AKEBA, T., KUDO, T. (1998), Purification and characterization from hyperthermophilic archaeon *Thermococcus profundus*, which hydrolyzes both α-1,4 and α-1,6 glucosidic linkages, *J. Ferment. Bioeng.* **86**, 363–367.

LADENSTEIN, R., ANTRANIKIAN, G. (1998), Proteins from hyperthermophiles: stability and enzymatic catalysis close to the boiling point of water, *Adv. Biochem. Eng./Biotechnol.* **61**, 37–85.

LADERMAN, K. A., ASADA, K., UEMORI, T., MUKAI, H., TAGUCHI, Y. et al. (1993), Alpha-amylase from the hyperthermophilic archaebacterium *Pyrococcus furiosus*. Cloning and sequencing of the gene and expression in *Escherichia coli*, *J. Biol. Chem.* **268**, 24402–24407.

LANDEGREN, U., KAISER, R., SANDERS, J., HOOD, L. (1988), A ligase-mediated gene detection technique, *Science* **241**, 1077–1080.

LAWYER, F. C., STOFFEL, S., SAIKI, R. K., CHANG, S. Y., LANDRE, P. A. et al. (1993), High-level expression, purification, and enzymatic characterization of full-length *Thermus aquaticus* DNA polymerase and a truncated form deficient in 5′- to 3′-exonuclease activity, *PCR Methods Appl.* **2**, 275–287.

LEE, J. T., KANAI, H., KOBAYASHI, T., AKIBA, T., KUDO, T. (1996), Cloning, nucleotide sequence and hyperexpression of α-amylase gene from an archaeon, *Thermococcus profundus*, *J. Ferment. Bioeng.* **82**, 432–438.

LETOVSKY, S. I., COTTINGHAM, R. W., PORTER, C. J., LI, P. W. D. (1998), GDB: the Human Genome Database, *Nucleic Acids Res.* **26**, 94–99.

LEUSCHNER, C., ANTRANIKIAN, G. (1995), Heat-stable enzymes from extremely thermophilic and hyperthermophilic microorganisms, *World J. Microbiol. Biotechnol.* **11**, 95–114.

LIEBL, W., STEMPLINGER, I., RUILE, P. (1997), Properties and gene structure of the *Thermotoga maritima* α-amylase AmyA, a putative lipoprotein of a hyperthermophilic bacterium, *J. Bacteriol.* **179**, 941–948.

LIEN, T., MADSEN, M., RAINEY, F. A., BIRKELAND, N. K. (1998), *Petrotoga mobilis* sp. nov., from a North Sea oil-production well, *Int. Syst. Bacteriol.* **48**, 1007–1013.

LIN, X. L., TANG, J. (1990), Purification, characterization and gene cloning of thermopsine, a thermostable acidic protease from *Sulfolobus acidocaldarius*, *J. Biol. Chem.* **265**, 1490–1495.

LING, L. L., KEOHAVONG, P., DIAS, C., THILLY, W. G. (1991), Optimization of the polymerase chain reaction with regard to fidelity: modified T7, Taq, and Vent DNA polymerases, *PCR Methods Appl.* **1**, 63–69.

LIU, S. Y., WIEGEL, J., GHERARDINI, F. C. (1996), Purification and cloning of a thermostable xylose (glucose) isomerase with an acidic pH optimum from *Thermoanaerobacterium* strain JW/SL-YS 489, *J. Bacteriol.* **178**, 5938–5945.

LOH, E. Y., ELLIOTT, J. F., CWIRLA, S., LANIER, L. L., DAVIS, M. M. (1989), Polymerase chain reaction with single-sided specifity: analysis of T cell receptor delta chain, *Science* **243**, 217–220.

LONGLEY, M. J., BENNETT, S. E., MOSBAUGH, D. W. (1990), Characterization of the 5′- to 3′-exonuclease associated with *Thermus aquaticus* DNA polymerase, *Nucleic Acids Res.* **18**, 7317–7322.

LUNDBERG, K. S., SHOEMAKER, D. D., ADAMS, M. W. W., SHORT, J. M., SORGE, J. A., MARTHUR, E. J. (1991), High-fidelity amplification using a thermostable polymerase isolated from *Pyrococcus furiosus*, *Gene* **108**, 1–6.

LUTHI, E., JASMAT, N. B., BERGQUIST, P. L. (1990), Overproduction of an acetylxylan esterase from the extreme thermophilic bacterium *Caldocellum saccharolyticum*, *Appl. Microbiol. Biotechnol.* **19**, 2677–2683.

MA, K., ROBB, F. T., ADAMS, M. W. W. (1994), Purification and characterization of NADP specific alcohol deydrogenase from the thermophilic archaeon *Thermococcus hydrothermalis*, *Appl. Environ. Microbiol.* **60**, 562–568.

MAAT, J., ROZA, M., VERBAKEL, J., STAM, H., SANTOS, H. et al. (1992), Xylanases and their application in bakery, in: *Progress in Biotechnology* Vol. 7 (VISSER, J. et al., Eds.), pp. 349–360. Amsterdam: Elsevier.

MATTILA, P., KORPELA, J., TENKANEN, T., PITKÄNEN, K. (1991), Fidelity of DNA synthesis by the *Thermococcus litoralis* DNA polymerase – an extremely heat stable enzyme with proofreading activity, *Nucleic Acids Res.* **19**, 4967–4973.

MAYR, J., LUPAS, A., KELLERMANN, J., ECKERSKORN, C., BAUMEISTER, W., PETERS, J. (1996), A hyperthermostable protease of the subtilisin family bound to the surface of the layer of the archaeon *Staphylothermus marinus*, *Curr. Biol.* **6**, 739–749.

MCDONOUGH, T. J. (1992), A survey of mechanical pulp bleaching in Canada: single-stage hydrosulfite lines are the rule, *Pulp Pap. Can.* **93**, 57.

MEAD, D. A., MCCLARY, J. A., LUCKEY, J. A., KOS-

TICHKA, A. J., WITNEY, F. R., SMITH, L. M. (1991), Bst DNA polymerase permits rapid sequence analysis from nanogram amounts of template, *Biotechniques* **11**, 76–78.

MORACCI, M., CIARAMELLA, M., NUCCI, R. R., PEARL, L. H., SANDERSON, I., TRINCONE, A., ROSSI, M. (1993), Thermostable β-glycosidase from *Sulfolobus solfataricus*, *Biocatalysis* **11**, 89–103.

MORIKAWA, M., IZAWA, Y., RASHID, N., HOAKI, T., IMANAKA, T. (1994), Purification and characterization of a thermostable Thiol protease from a newly isolated hyperthermophilic *Pyrococcus* sp., *Appl. Environ. Microbiol.* **60**, 4559–4566.

MORITA, R. Y. (1975), Psychrophilic bacteria, *Bacteriol. Rev.* **39**, 144–167.

MORRISON, A., BELL, J. B., KUNKEL, T. A., SUGINO, A. (1991), Eukaryotic DNA polymerase amino acid sequence required for 3′-5′-exonuclease activity, *Proc. Natl. Acad. Sci. USA* **88**, 9473–9477.

MULLIS, K., FALOONA, F., SAIKI, R., HORN, G., ERLICH, H. A. (1986), Specific enzymatic amplification of DNA *in vitro*: the polymerase chain reaction, *Cold Spring Harbor Symp. Quant. Biol.* **51**, 263–273.

MURRAY, V. (1989), Improved double-stranded DNA sequencing using the linear polymerase chain reaction, *Nucleic Acids Res.* **17**, 11.

MUZZARELLI, R. A. (1997), Human enzymatic activities related to the therapeutic administration of chitin derivatives, *Cell. Mol. Life Sci.* **53**, 131–140.

MYERS, T. W., GELFAND, D. H. (1991), Reverse transcription and DNA amplification by a *Thermus thermophilus* DNA polymerase, *Biochemistry* **30**, 7661–7665.

NEUNER, A., JANNASCH, H. W., BELKIN, S., STETTER, K. O. (1990), *Thermococcus litoralis* sp. nov., a new species of extremely thermophilic marine archaebacteria, *Arch. Microbiol.* **153**, 205–207.

NIEHAUS, F., FREY, B., ANTRANIKIAN, G. (1997), Cloning and characterisation of a thermostable α-DNA polymerase from the hyperthermophilic archaeon *Thermococcus* sp. TY, *Gene* **204**, 153–158.

NIEHAUS, F., BERTOLDO, C., KÄHLER, M., ANTRANIKIAN, G. (1999), Extremophiles as a source of novel enzymes for industrial application, *Appl. Microbiol. Biotechnol.* **51**, 711–729.

NIEHAUS, F., PETERS, A., GROUDIEVA, T., ANTRANIKIAN, G. (2000), Cloning, expression and biochemical characterisation of a unique thermostable pullulan-hydrolysing enzyme from the hyperthermophilic archaeon *Thermococcus aggregans*, *FEMS Microbiol. Lett.* **190**, 223–239.

NILSSON, M., MALMGREN, H., SAMIOTAKI, M., KWIATKOWSKI, M., CHOWDHARY, B. P., LANDEGREN, U. (1994), Padlock probes: circularizing oligonucleotides for localized DNA detection, *Science* **265**, 2085–2088.

NILSSON, J., BOSNES, M., LARSEN, F., NYGREN, P. A., UHLÉN, M., LUNDEBERG, J. (1997), Heat-mediated activation of affinity-immobilized Taq DNA polymerase, *Biotechniques* **22**, 744–751.

NUCCI, R., MORACCI, M., VACCARO, C., VESPA, N., ROSSI, M. (1993), Exo-glucosidase activity and substrate specificity of the β-glucosidase isolated from the extreme thermophilic *Sulfolobus solfataricus*, *Biotechnol. Appl. Biochem.* **17**, 239–250.

OHARA, O., DORIT, R. L., GILBERT, W. (1989), One-sided polymerase chain reaction: the amplification of cDNA, *Proc. Natl. Acad. Sci. USA* **86**, 5673–5677.

PARK, B. C., KOH, S., CHANG, C., SUH, S. W., LEE, D. S., BYUN, S. M. (1997), Cloning and expression of the gene for xylose isomerase from *Thermus flavus* AT62 in *Escherichia coli*, *Appl. Biochem. Biotechnol.* **62**, 15–27.

PARK, J. H., KIM, J. S., KWON, S. T., LEE, D. S. (1993), Purification and characterization of *Thermus caldophilus* GK24 DNA polymerase, *Eur. J. Biochem.* **214**, 135–140.

PEARL, L. H., HEMMINGS, A. M., SICA, F., MAZZARELLA, L., RAIA, C. A. et al. (1993), Crystallization and preliminary X-ray analysis of an NAD-dependent alcohol deydrogenase from the extreme thermophilic archaeobacterium *Sulfolobus solfataricus*, *J. Mol. Biol.* **229**, 561–563.

PEDERSEN, S., JENSEN, B. F., DIJKHUIZEN, L., JØRGENSEN, S. T., DIJKSTRA, B. W. (1995), A better enzyme for cyclodextrins, *Chemtech* **12**, 19–25.

PERLER, F. B., KUMAR, S., KONG, H. (1996), Thermostable DNA polymerases, *Adv. Protein Chem.* **48**, 377–435.

PHIPPS, B. M., HOFFMANN, A., STETTER, K. O., BAUMEISTER, W. (1991), A novel ATPase complex selectively accumulated upon heat shock is a major cellular component of thermophilic archaebacteria, *EMBO J.* **10**, 1711–1722.

PHIPPS, B. M., TYPKE, D., HEGERL, R., VOLKER, S., HOFFMANN, A. et al. (1993), Structure of a molecular chaperone from a thermophilic archaebacterium, *Nature* **361**, 475– 477.

PILLER, K., DANIEL, R. M., PETACH, H. H. (1996), Properties and stabilization of an extracellular *alpha*-glucosidase from the extremely thermophilic archaebacteria *Thermococcus* strain AN1: enzyme activity at 130°, *Biochim. Biophys. Acta* **1292**, 197–205.

PISANI, F. M., RELLA, R., RAIA, C. A., ROZZO, C., NUCCI, R. et al. (1990), Thermostable β-galactosidase from the archaebacterium *Sulfolobus solfataricus* – purification and properties, *Eur. J. Biochem.* **187**, 321–328.

POWELL, L. M., WALLIS, S. C., PEASE, R. J., EDWARDS, Y. H., KNOTT, T. J., SCOTT, J. (1987), A novel form of tissue-specific RNA processing produces apolipoprotein-B48 in intestine, *Cell* **50**, 831–840.

PRISCO, A., MORACCI, M., ROSSI, M., CIARAMELLA,

M. (1994), A gene encoding a putative membrane protein homologous to the major facilitator superfamily of transporters maps upstream of the β-glycosidase gene in the archaeon *Sulfolobus solfataricus*, *J. Bacteriol.* **177**, 1616–1619.

PROWE, S. G., ANTRANIKIAN, A. (2000), *Anaerobranca gottschallkii* sp. nov., a novel thermoalkaliphilic bacterium that grows anaerobically at high pH and temperature, *3rd Int. Congr. on Extremophiles 2000*, Hamburg, Germany, Book of Abstracts, p. 107.

PROWE, S., VAN DE VOSSENBERG, J., DRIESSEN, A., ANTRANIKIAN, G., KONINGS, W. (1996), Sodium-coupled energy transduction in the newly isolated thermoalkaliphic, strain LBS3, *J. Bacteriol.* **178**, 4099.

RAO, V. B., SAUNDERS, N. B. (1992), A rapid polymerase-chain-reaction-directed sequencing strategy using a thermostable DNA polymerase from *Thermus flavus*, *Gene* **113**, 17–23.

REDDY, T., SURYANARAYANA, T. (1988), Novel histone-like DNA-binding proteins in the nucleoid from the acidothermophilic archaebacterium *Sulfolobus acidocaldarius* that protect DNA against thermal denaturation, *Biochim. Biophys. Acta* **949**, 87–96.

REEVE, M. A., FULLER, C. W. (1995), A novel thermostable polymerase for DNA sequencing, *Nature* **376**, 796–797.

RELLA, R., RAIA, C. A., PENSA, M., PISANI, F. M., GAMBACORTA, A. et al. (1987), A novel archeabacterial NAD dependent alcohol deydrogenase: purification and properties, *J. Biochem.* **167**, 475–479.

ROBINSON, K. A., BARTLEY, D. A., ROBB, F. T., SCHREIER, H. J. (1995), A gene from the hyperthermophile *Pyrococcus furiosus* whose deduced product is homologous to members of the prolyl oligopeptidase family of proteases, *Gene* **152**, 103–106.

RODRIGUEZ-VALERA, F., RUIZ-BERRAQUERO, F., RAMOS-CORMENZANA, M. (1981), Characteristics of the heterotrophic bacterial populations in hypersaline environments of different salt concentrations, *Microb. Ecol.* **7**, 235–243.

ROSS, H. N. M., COLLINS, M. D., TINDALL, B. J., GRANT, W. D. (1981), A rapid method for detection of archaeabacterial lipids in halophilic bacteria, *J. Gen. Microbiol.* **123**, 75–80.

RÜDIGER, A., JØRGENSEN, P. L., ANTRANIKIAN, G. (1995), Isolation and characterization of a heat stable pullulanase from the hyperthermophilic archeon *Pyrococcus woesei* after cloning and expression of its gene in *Escherichia coli*, *Appl. Environ. Microbiol.* **61**, 567–575.

RUSSELL, N. J., HAMAMOTO, T. (1998), Psychrophiles, in: *Extremophiles – Microbial Life in Extreme Environments* (HORIKOSHI, K., GRANT, W. D., Eds.), pp. 25–47. New York: Wiley-Liss.

RUSSELL, R. J. M., HOUGH, D. W., DANSON, M. J., TAYLOR, G. L. (1994), The crystal structure of citrate synthase from the thermophilic Archaeon, *Thermoplasma acidophilum*. Structure. 2: 1157–1167. Hydrothermal vents, *Planet. Space Sci.* **43**, 115–122.

RUTTERSMITH, L. D., DANIEL, R. M. (1991), Thermostable cellobiohydrolase from the thermophilic eubacterium *Thermotoga* sp. strain FjSS3-B.1: Purification and properties, *Biochem. J.* **277**, 887–890.

RUTTERSMITH, L. D., DANIEL, R. M., SIMPSON, H. D. (1992), Cellulolytic and hemicellulolytic enzymes functional above 100 °C, *Ann. NY Acad. Sci.* **672**, 137–141.

RÜTTIMANN, C., COTORAS, M., ZALDIVAR, J., VICUÑA, R. (1985), DNA polymerases from the extremely thermophilic bacterium *Thermus thermophilus* HB, *Eur. J. Biochem.* **149**, 41–46.

SAIKI, R. K., GELFAND, D. H., STOFFE, S., SCHARF, S. J., HIGUCHI, R. et al. (1988), Primer-directed enzymatic amplification of DNA with thermostable DNA polymerase, *Science* **239**, 487–491.

SANGER, F., NICKLEN, S., COULSON, A. R. (1977), DNA sequencing with chain-terminating inhibitors, *Proc. Natl. Acad. Sci. USA* **74**, 5463–5467.

SARKAR, G., KAPELNER, S., SOMMER, S. S. (1990), Formamide can dramatically improve the specificity of PCR, *Nucleic Acids Res.* **18**, 7465.

SCALICE, E. R., SHARKEY, D. J., DAISS, J. L. (1994), Monoclonal antibodies prepared against the DNA polymerase from *Thermus aquaticus* are potent inhibitors of enzyme activity, *J. Immunol. Methods* **12**, 147–163.

SCANDURRA, R., CONSALVI, V., CHIARALUCE, R., POLITI, L., ENGEL, P. C. (1998), Protein thermostability in extremophiles, *Société Française de Biochimie et Biologie Moleculaire* **80**, 933–941.

SCHALLING, M., HUDSON, T. J., BUETOW, K. H., HOUSEMAN, D. E. (1993), Direct detection of novel expanded trinucleotides repeats in the human genome, *Nature Genet.* **4**, 135–139.

SCHIRALDI, C., MARULLI, F., DI LERNIA, I., MARTINO, A., DE ROSA, M. (1999), A microfiltration bioreactor to achieve high cell density in *Sulfolobus solfataricus* fermentation, *Extremophiles* **3**, 199–204.

SCHULIGER, J. W., BROWN, S. H., BAROSS, J. A., KELLY, R. M. (1993), Purification and characterisation of a novel amylolytic enzyme from ES4, a marine hyperthermophilic archaeum, *Mol. Mar. Biol. Biotechnol.* **2**, 76–87.

SCHUMANN, J., WRBA, A., JAENICKE, R., STETTER, K. O. (1991), Topographical and enzymatic characterization of amylase from the extreme thermophilic eubacterium *Thermotoga maritima*, *FEBS Lett.* **282**, 122–126.

SCHWERDTFEGER, R. M., CHIARALUCE, R., CONSALVI, V., SCANDURRA, R., ANTRANIKIAN, G. (1999), Stability, refolding and Ca^{2+} binding of pullulanase from the hyperthermophilic archaeon *Pyrococcus woesei*, *Eur. J. Biochem.* **264**, 479–487.

SEARS, L. E., MORAN, L. S., KISSINGER, C., CREASEY, T., PERRY, T. et al. (1992), CircumVent thermal cycle sequencing and alternative manual and automated DNA sequencing protocols using the highly thermostable VentR (exo-)DNA polymerase, *Biotechniques* **13**, 626–633.

SHARKEY, D. J., SCALICE, E. R., CHRISTY, K. G. Jr., ATWOOD, S. M., DAISS, J. L. (1994), Antibodies as thermolabile switches: high temperature triggering for the polymerase chain reaction, *Biotechnology* **12**, 506–509.

SIEVERT, S. M., ZIEBIS, W., KUEVER, J., SAHM, K. (2000), Relative abundance of Archaea and Bacteria along a thermal gradient of a shallow-water hydrothermal vent quantified by rRNA slot-slot hybridization, *Microbiology* **146**, 1287–1293.

SIMPSON, H. D., HAUFLER, U. R., DANIEL, R. M. (1991), An extremely thermostable xylanase from the thermophilic eubacterium *Thermotoga*, *Biochem. J.* **277**, 177–185.

SOBEK, H., GORISCH, H. (1988), Purification and characterization of a heat stable esterase from the thermoacidophilic archaeobacterium *Sulfolobus acidocaldarius*, *Appl. Environ. Microbiol.* **61**, 729–733

SOUTHWORTH, M. W., KONG, H., KUCERA, R. B., WARE, J., JANNASCH, H. W., PERLER, F. B. (1996), Cloning of thermostable DNA polymerases from hyperthermophilic marine Archaea with emphasis on *Thermococcus* sp. 9 degrees N-7 and mutations affecting 3′-5′-exonuclease activity, *Proc. Natl. Acad. Sci. USA* **93**, 5281–5285.

SPASSOV, V. Z., KARSHIKOFF, A. D., LADENSTEIN, R. (1995), The optimization of protein- solvent interactions: Thermostability and the role of hydrophobic and electrostatic interactions, *Protein Sci.* **4**, 1516–1527.

SPECKA, U., MAYER, F., ANTRANIKIAN, G. (1991), Purification and properties of a thermoactive glucoamylase from *Clostridium thermosaccharolyticum*, *Appl. Environ. Microbiol.* **57**, 2317–2323.

SPINDLER, K. D., SPINDLER-BARTH, M., LONDERSHAUSEN, M. (1990), Chitin metabolism: a target for drugs against parasites, *Parasitol. Res.* **76**, 283–288.

STETTER, K. O. (1996), Hyperthermophilic prokaryotes, *FEMS Microbiol. Rev.* **18**, 149–158.

STETTER, K. O. (1998), Hyperthermophiles: Isolation, classification and properties, in: *Extremophiles – Microbial Life in Extreme Environments* (HORIKOSHI, K., GRANT, W. D., Eds.), pp. 1–24. New York: Wiley-Liss.

STETTER, K. O. (1999), Extremophiles and their adaptation to hot environments, *FEBS Lett.* **452**, 22–25.

STETTER, K. O., KÖNIG, H., STACKEBRANDT, E. (1983), *Pyrodictium* gen. nov., a new genus of submarine disc-shaped sulfur reducing archaebacteria growing optimally at 105 °C, *Syst. Appl. Microbiol.* **4**, 535–551.

STETTER, K. O., FIALA, G., HUBER, G., HUBER, R., SEGERER, A. (1990), Hyperthermophilic microorganisms, *FEMS Microbiol. Rev.* **75**, 117–124.

STETTER, K. O., HUBER, R., BLÖCHL, E., KURR, M., EDEN, R. D. et al. (1993), Hyperthermophilic archaea are thriving in deep North Sea and Alaskan oil reservoirs, *Nature* **365**, 743–745.

SUNNA, A., ANTRANIKIAN, G. (1997), Growth and production of xylanolytic enzymes by the extreme thermophilic anaerobic bacterium *Thermotoga thermarum*, *Appl. Microbiol. Biotechnol.* **45**, 671–676.

SUNNA, A., MORACCI, M., ROSSI, M., ANTRANIKIAN, G. (1996a), Glycosyl hydrolases from hyperthermophiles, *Extremophiles* **1**, 2–13.

SUNNA, A., PULS, J., ANTRANIKIAN, G. (1996b), Purification and characterization of two thermostable endo-1,4-β-D-xylanases from *Thermotoga thermarum*, *Biotechnol. Appl. Biochem.* **24**, 177–185.

SZYBALSKI, W. (1990), Proposal for sequencing DNA using ligation of hexamers to generate sequential elongation primers (SPEL-6), *Gene* **90**, 177–178.

TABOR, S., RICHARDSON, C. C. (1995), A single residue in DNA polymerases of the *Escherichia coli* DNA polymerase I family is critical for distinguishing between deoxy- and dideoxyribonucleotides, *Proc. Natl. Acad. Sci. USA* **92**, 6339–6343.

TACHIBANA, Y., MENDEZ, L. M., FUJIWARA, S., TAKAGI, M., IMANAKA, T. (1996), Cloning and expression of the α-amylase gene from the hyperthermophilic archeon *Pyrococcus* sp. KOD1 and characterization of the enzyme, *J. Ferment. Bioeng.* **82**, 224–232.

TACHIBANA, Y., KURAMURA, A., SHIRASAKA, N., SUZUKI, Y., YAMAMOTO, T. et al. (1999), Purification and characterization of an extremely thermostable cyclomaltodextrin glucanotransferase from a newly isolated hyperthermophilic archaeon, a *Thermococcus* sp., *Appl. Environ. Microbiol.* **65**, 1991–1997.

TAKAGI, M., NISHIOKA, M., KAKIHARA, H., KITABAYASHI, M., INOUE, H. et al. (1997), Characterization of DNA polymerase from *Pyrococcus* sp. strain KOD1 and its application to PCR, *Appl. Environ. Microbiol.* **63**, 4504–4510.

TAKAHASHI, M., YAMAGUCHI, E., UCHIDA, T. (1984), Thermophilic DNA ligase. Purification and properties of the enzyme from *Thermus thermophilus* HB8, *J. Biol. Chem.* **259**, 10041–10047.

TAKAYANAGI, T., AJISAKA, K., TAKIGUCHI, Y., SHIMAHARA, K. (1991), Isolation and characterization of thermostable chitinases from *Bacillus licheniformis* X-7u, *Biochim. Biophys. Acta* **1078**, 404–410.

TANNER, J. J., HECHT, R. M., KRAUSE, K. L. (1996), Determinants of enzyme thermostability observed in the molecular structure of *Thermus aquaticus* D-glyceraldehyde-3-phosphate dehydrogenase at 25 Å resolution, *Biochemistry* **35**, 2597–2609.

TEERI, T. T., KOIVULA, A., LINDER, M., WOHLFAHRT, G., DIVNE, C., JONES, T. A. (1998), *Trichoderma reesei* cellobiohydrolases: why so efficient on crystalline cellulose? *Biochem. Soc. Trans.* **26**, 173–178.

TEO, V. S., SAUL, D. J., BERGQUIST, P. L. (1995), *Cel*A, another gene coding for a multidomain cellulase from the extreme thermophile *Caldocellum saccharolyticum*, *Appl. Microbiol. Biotechnol.* **43**, 291–296.

THOMM, M., STETTER, K. O., ZILLIG, W. (1982), Histone-like proteins in eu- and archaebacteria, *Zbl. Bakt. Hyg.*, I. Abt. Orig. C3, 128–139.

THORBJARNARDOTTIR, S. H., JONSSON, Z. O., ANDRESSON, O. S., KRISTJANSSON, J. K., EGGERTSSON, G., PALSDOTTIR, A. (1995), Cloning and sequence analysis of the DNA ligase-encoding gene of *Rhodothermus marinus*, and overproduction, purification and characterization of two thermophilic DNA ligases, *Gene* **161**, 1–6.

TINDALL, K. R., KUNKEL, T. A. (1988), Fidelity of DNA synthesis by the *Thermus aquaticus* DNA polymerase, *Biochemistry* **27**, 6008–6013.

TOMME, P., WARREN, R. A., GILKES, N. R. (1995), Cellulose hydrolysis by bacteria and fungi, *Adv. Microb. Physiol.* **37**, 1–81.

TRENT, J. D., NIMMESGERN, E., WALL, J. S., HARTL, U. F., HORWICH, A. L. (1991), A molecular chaperone from a thermophilic archaebacterium is related to the eukaryotic protein t-complex polypeptide-1, *Nature* **354**, 490–493.

TSUJIBO, H., ENDO, H., MIYAMOTO, K., INAMORI, Y. (1995), Expression in *Escherichia coli* of a gene encoding a thermostable chitinase from *Streptomyces thermoviolaceus* OPC-520, *Biosci. Biotechnol. Biochem.* **59**, 145–146.

VANDER, H. P. B., DAVIS, M. C., CUNNIFF, J. J., RUAN, C., MCARDLE, B. F. et al. (1997), Thermo Sequenase DNA polymerase and *T. acidophilum* pyrophosphatase: new thermostable enzymes for DNA sequencing, *Biotechniques* **22**, 758–762.

VENTOSA, A., BIETO, J. J. (1995), Biotechnological applications and potentialities of halophilic microorganisms, *World J. Microbiol. Technol.* **11**, 85–94.

VETRIANI, C., MAEDER, D. L., TOLLIDAY, N., YIP, K. S. P., STILLMAN, T. J. et al. (1998), Protein thermostability above 100 °C: A key role for ionic interactions, *Proc. Natl. Acad. Sci. USA* **95**, 12300–12305.

VIEILLE, C., HESS, J. M., KELLY, R. M., ZEIKUS, J. G. (1995), xylA cloning and sequencing and biochemical characterization of xylose isomerase from *Thermotoga neapolitana*, *Appl. Environ. Microbiol.* **61**, 1867–1875.

VIIKARI, L., KANTELINEN, A., SUNDQUIST, J,, LINKO, M. (1994), Xylanases in bleaching. From an idea to industry, *FEMS Microbiol. Lett.* **13**, 335–350.

VÖLKL, P., MARKIEWICZ, P., STETTER, K. O., MILLER, J. H. (1995), The sequence of a subtilisin-type protease (aerolysin) from the hyperthermophilic archaeon *Pyrobaculum aerophilum* reveals sites important to thermostability, *Protein Sci.* **3**, 1329–1340.

VOORHORST, W. G. B., EGGEN, R. I. L., LUESINK, E. J., DE VOS, W. M. (1995), Characterization of the *cel*B gene coding for β-glucosidase from the hyperthermophilic archaeon *Pyrococcus furiosus* and its expression and site-directed mutation in *Escherichia coli*, *J. Bacteriol.* **177**, 7105–7110.

VOORHORST, W. G. B., EGGEN, R. I. L., GEERLING, A. C. M., PLATTEEUW, C., SIEZEN, R. J., DE VOS, W. M. (1996), Isolation and characterization of the hyperthermostable serine protease, pyrolysin, and its gene from the hyperthermophilic archaeon *Pyrococcus furiosus*, *J. Biol. Chem.* **271**, 20426–20431.

WIND, R., LIEBL, W., BUITLAAR, R., PENNINGA, D., SPREINAT, A. et al. (1995), Cyclodextrin formation by the thermostable α-amylase of *Thermoanaerobacterium thermosulfurigenes* EM1 and reclassification of the enzyme as α-cyclodextrin glycosyltransferase, *Appl. Environ. Microbiol.* **61**, 1257–1265.

WINTERHALTER, C., LIEBL, W. (1995), Two extremely thermostable xylanases of the hyperthermophilic bacterium *Thermotoga maritima* MSB8, *Appl. Environ. Microbiol.* **61**, 1810–1815.

WOESE, C. R., FOX, G. E. (1977), Phylogenetic structure of the prokaryotic domain. The primary kingdoms, *Proc. Natl. Acad. Sci. USA.* **74**, 5088–5090.

WOESE, C. R., KANDLER, O., WHEELIS, M. L. (1990), Toward a natural system of organisms (1990): Proposal for the domains archaea, bacteria and eukarya, *Proc. Natl. Acad. Sci. USA* **87**, 4576–4579.

WOESE, C. R., ACHENBACH, L., ROUVIERE, P., MANDELCO, L. (1991), Archaeal phylogeny: reexamination of the phylogenetic position in light of certain composition-induced artifacts, *Syst. Appl. Microbiol.* **14**, 364–371.

WU, D. Y., WALLACE, R. B. (1989), Specificity of the nick-closing activity of bacteriophage T4 DNA ligase, *Gene* **76**, 245–254.

YANG, J. L., ERIKSSON, K. E. L. (1992), Use of hemicellulolytic enzymes as one stage in bleaching of kraft pulps, *Holzforsch.* **46**, 481–488.

YIP, K. S. P., STILLMAN, T. J., BRITTON, K. L., ARTYMIUK, P. J., BAKER, P. J. et al. (1995), The structure of *Pyrococcus furiosus* glutamate dehydrogenase reveals a key role for ion-pairs networks in maintaining enzyme stability at extreme temperatures, *Structure* **3**, 1147–1158.

ZILLIG, W., STETTER, K. O., SCHNABEL, R., THOMM, M. (1985), DNA-dependent RNA polymerases of the archaebacteria, in: *The Bacteria* Vol. III (WOESE, C. R., WOLFE, R. S., Eds.), pp. 499–524. Orlando, FL: Academic Press.

ZILLIG, W., HOLZ, I., KLENK, H. P., TRENT, J., WUNDERL, S. et al. (1987), *Pyrococcus woesei* sp. nov., an ultra-thermophilic marine archaebacterium, represents a novel order, *Thermococcales*, *Syst. Appl. Microbiol.* **9**, 62–70.

ZVERLOV, V., MAHR, S., RIEDEL, K., BRONNENMEIER, K. (1998), Properties and gene structure of a bifunctional cellulolytic enzyme (*Cel*A) from the extreme thermophile *Anaerocellum thermophilum* with separate glycosyl hydrolase family 9 and 48 catalytic domains, *Microbiology* **144**, 457–465.

5 Biotechnology with Cyanobacteria and Microalgae

OTTO PULZ
KARL SCHEIBENBOGEN
Potsdam-Rehbrücke, Germany
WOLFGANG GROß
Berlin, Germany

1 Introduction and Historical Background

The term algae describes a large and heterogeneous group of photosynthesizing organisms which strongly differ in morphology, physiology, and habitat. Apart from taxonomic criteria of systematization, one can distinguish between macro- and microalgae according to size and morphology. Microalgae, as individual organisms, have a size of only a few micrometers while the largest representatives of macroalgae have thalli up to 60 m in length, thus belonging to the giants under the plants. In general usage, the term microalgae includes the photosynthesizing prokaryotic cyanobacteria. Based on qualitative and quantitative composition of the photosynthetically active pigments, algae can also be grouped into the red algae, green algae, blue algae (cyanobacteria), and brown algae, thereby further subdividing the macro- and microalgae. Worldwide, macroalgae are found mainly in the light-flooded regions of marine coastal waters, where they are usually attached to rocks and produce biomass with surprising productivity (RICHMOND, 1986).

For centuries, certain algal species like the microalga *Spirulina* (*Arthrospira*) or the macroalga *Porphyra* (*Nori*) have been used directly as food. Since the beginning of this century, aqueous extracts from macroalgae as phycocolloids like agar or alginates have gained in economic importance in Europe and worldwide. At present 5.4 million tons of macroalgae are harvested and processed annually with a turnover of 4.9 billion US$ (CRITCHLEY and OHNO, 1998).

Phylogenetically algae, especially microalgae, belong to the oldest and ecologically most variable group of organisms of the earth.

Algae are

- the first oxygen producers,
- the most important extant CO_2 consumers,
- the basis of the food chain of the oceans,
- the most important primary producers.

The ecological amplitude of microalgae is far wider than that of macroalgae. They inhabit nearly all the ecosystems of the world, from the desert to the polar sea. They belong to those microorganisms which are often utilized by man inadvertently, e.g., in ploughed land and rice paddy fields, as nitrogen fixing soil algae or in ecophysiological integration with simple animals or fungi to colonize new habitats.

Over millions of years, these microorganisms have succeeded in developing the oxygen atmosphere of the earth from CO_2, sunlight, and some minerals. They are the main source of fossil fuels, mineral sediments, such as limestone, and a food source for humans and animals. Over these millions of years of evolution, microalgae have developed numerous physiological systems which allow for an adaptation to various, partly extreme ecological conditions and habitats, e.g.,

- temperature tolerance: enables growth either in cold, arctic regions, or in thermal springs and solfatara soils;
- light demand: high variability with regard to exploitation of the light spectrum due to the diversity of photosynthetically active pigments. Adaptability to different light intensities allows algal growth at extremely low light intensities, such as in water depths of 200 m, as well as on sites of high energetic light intensity such as strong ultraviolet radiation in high mountain ranges;
- pH and osmotolerance: some algae prefer extremely acid or alkaline habitats, ranging from pH 0.5 to 11 or require for growth certain minerals in high concentrations.

However, even today the thousands of species of microalgae continue to be a poorly tapped biotechnological resource (BENEMANN, 1990). Future developments of economic significance are apparent, e.g., agricultural and aquacultural exploitation, pharmacologically effective preparations from microalgae, products for cosmetic applications or dietetic products (BELAY, 1993; PICCARDI et al., 1999), and, with growing importance, technologies for environmental sanitation.

The health promoting effect of marine algae has been already described in Chinese references from 2700 BC, in this case for macroalgae. The addition of marine algae to food with the aim to overcome iodine deficiency, especially in mountain regions, is known from ancient times. In Asia, algal extracts prepared by boiling are used as beauty agents for skin and hair care.

The usage of microalgae as food is not widespread. Human consumption of cyanobacteria of the genus *Nostoc* is reported from Mongolia, China, Japan, and Peru and seems to be valid for other Asian countries like Myanmar. Legendary are the descriptions of the natural mass cultures of Mexican and African *Spirulina* and their use by local populations as health promoting food.

The mass cultivation in the sense of agriculture – like production or even biotechnological use of microalgae for different purposes – is fairly recent.

During World War II mass cultivation of diatoms was tested for industrial purposes because of their high lipid content. But it took 10–20 years before we can register a surprising enhancement of research activities with microalgae in the USA, Israel, Japan, The Netherlands, and Germany with the focus on diatoms and chlorococcalean green algae. These initiatives led to numerous pilot scale attempts and are reflected in the, even today, very popular book of BURLEW (1953). In the following decades numerous trials were started worldwide to exploit phototrophic microorganisms, especially as protein rich food and feed source but also for energetical, pharmaceutical, and environmental purposes. Examples of the extended literature are MUSAFAROW and TAUBAJEW (1974), SHELEF and SOEDER (1980), DILOW (1985), and RICHMOND (1986).

Already in the 1970s, their use in biotechnology had an upswing, due to the attempt to close the protein gap in world nutrition with microalgae. Acceptance problems of the direct use of green microalgal biomass for human nutrition and arising competition by other resources, such as soya, resulted in temporary degression. Over the last 5–10 years, biotechnology with microalgae has gained momentum not only for environmentally relevant implementations but also increasingly for the preparation of valuable substances (Tab. 1) (PULZ and SCHEIBENBOGEN, 1998).

Depending on the species, the specific strain, and the growth conditions, microalgae have a wide range of components which can be used in food and feed stuffs, and are becoming attractive for health food, pharmacy, and cosmetics. Since the protein gap in human nutrition, especially in developing countries, was filled more efficiently by traditional agricultural methods than with microalgae, the aims of microalgal cultivation changed to higher value products with better economy like astaxanthin or polyunsaturated fatty acids.

2 Biology of Biotechnologically Important Cyanobacteria and Microalgae

More than any other plant group, algae bring together taxonomists, physiologists, and biotechnologists. The identification and recognition of a specific algal strain as well as knowledge of its metabolism and physiology are often prerequisites for biotechnological screening and applications. The taxonomy tries to arrange the algal species in a natural, phylogenetic scheme which coincides very often with certain morphological, biochemical, or physiological properties. Thus, the biotechnologist may concentrate his search for a specific compound or characteristic on the most promising algal group. Knowing what is going on inside an algal cell, is often invaluable to predict effects of changing growth parameters or the production of natural products. Instead of time consuming and costly trials, purposeful optimization of culture techniques is possible. However, all three professions have, at least in part very different questions they want to answer in their research. The exchange of information will help to promote microalgal biotechnology and truly will have synergistic effects not only in applied phycology but also in basic research.

The term microalga is unrelated to algal taxonomy. It includes all microscopic algae,

Tab. 1. Market Estimations for Microalgal Products

	Product	US$ kg^{-1}	Market Size US$ $\cdot 10^6$
Biomass	health food	15–28	180–200
	functional food	25–52	growing
	feed additive	10–130	fast growing
	aquaculture	50–150	fast growing
	soil conditioners	>10	promising
Coloring substances	astaxanthin	>3,000	>50
	phycocyanin	>500	>10
	phycoerythrin	>10,000	> 2
Antioxidants	β-carotene	>750	>25
	superoxide dismutase	>1,000	promising
	tocopherol	30–40	stagnant
	AO-extracts	20–35	12–20
PUFA	ARA		20
	EPA		>500
	DHA		30
	PUFA extracts	30–80	10
Special products	toxins		1–3
	isotopes		<5

usually unicellular or filamentous, separating these from the macroscopic thallous macrophytes, such as kelp. Cyanobacteria are not part of the algae in a taxonomic sense because of their prokaryotic nature. However, with regard to biotechnology prokaryotic and eukaryotic phototrophs share a number of characteristics and, therefore, are conveniently treated as a whole. Eukaryotic algae represent an extremely heterogeneous group of plants with a long evolutionary history spanning up to 1.9 billion years. The first higher plants appeared only about 250 million years ago and are, therefore, newcomers. Bearing this in mind, it is quite clear that algae are not simply single-celled spinach, but may have very different physiology and genetics. The evolutionary distances calculated from ribosomal RNA sequences between the green alga *Chlamydomonas* and higher plants is 0.1 units, between *Chlamydomonas* and mammals is 0.21 units, and between *Chlamydomonas* and *Euglena* is 0.46 units. This demonstrates the enormous distance between the various algal groups.

The number of microalgal species is somewhere in the range of 200,000 (Norton et al., 1996). Assuming that only 1% of the strains exhibit sufficient productivity under laboratory conditions, there are still 2,000 species left from which about 50–100 have been tested for biotechnology. Considering this enormous resource of very different species (Fig. 1), it appears that genetic engineering of established strains of microalgae is certainly not the only way to introduce new strains into biotechnology. For example, it seems very unlikely to increase the overall productivity of a strain by genetic engineering, however, recent developments have already shown that genetically transformed algae have a very high potential in special applications.

2.1 Cyanobacteria

Cyanobacteria inhabit a wide variety of environments, including freshwater and salt water, soil and stones, neutral to alkaline hot springs, or polar regions. Layers of cyanobacteria often form so-called stromatolites which date back approximately 3.5 billion years. About 2,000 species in 150 genera are recognized but the taxonomy is extremely difficult due to the paucity of morphological characteristics. Thus, names of genera have changed frequently and a number of subspecies are

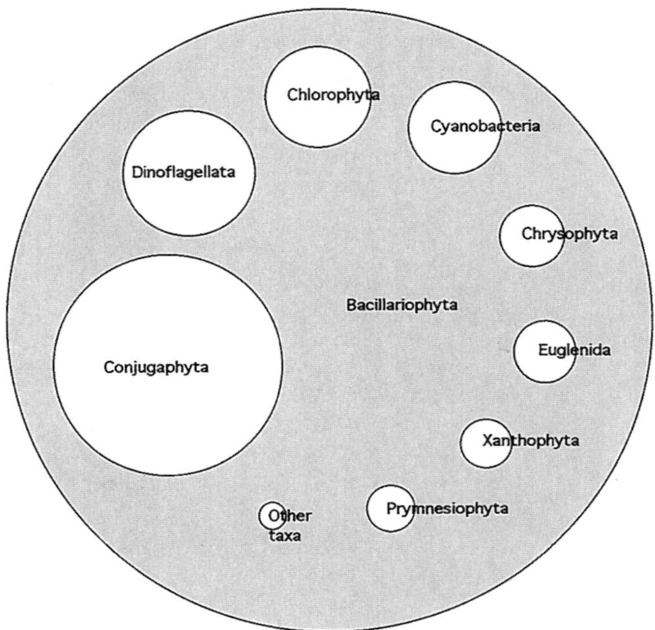

Fig. 1. Comparison of algal biodiversity. The size of the cicles represents the number of microalgal species in each group.

known. With the aid of genomic fingerprinting considerable changes in the cyanobacterial taxonomy can be expected in the future.

Only unicellular and filamentous cyanobacteria are known, the latter being dominant, either branched or unbranched. None of these forms are flagellated, but some species are able to glide, e.g., *Oscillatoria*. Many species are imbedded in a thick layer of extracellular polysaccharides. This mucilage may represent up to 28% of the total polysaccharides (e.g., *Anabaena flos-aquae*).

Most cyanobacteria are obligate phototrophs, only some species can grow photoorganotrophic or even heterotrophic. Reproduction occurs by fission. No true sexual reproduction is known. Cyanobacteria are gram-negative with a typical peptidoglucan cell wall. The chloroplast of plants is derived from cyanobacterial progenitors through endosymbiosis. Therefore, the photosynthetic apparatus is similar to higher plants, while other metabolic pathways are more similar to bacteria. The photosynthetic pigments are chlorophyll a, phycocyanins, and phycoerythrin in most species, only the prochlorophytes contain chlorophyll a and b and no phycobiliproteins. Cyanobacteria have a very efficient CO_2-concentrating mechanism, increasing the intracellular level about 1000-fold compared to the outside (TSUZUKI and MIYACHI, 1991). In addition, cells contain polyhedral structures, called carboxysomes, which contain the enzymes ribulose-1,5-bisphosphate carboxylase and carbonic anhydrase. In this way, cyanobacteria are able to thrive even in a low-CO_2 atmosphere. Many species are able to fix nitrogen with the help of nitrogenase usually housed in specialized cells, called heterocysts. Storage products are a highly branched α1,4-polyglucan (cyanophycean starch), polyphosphate granules, and cyanophycin, a polymer of asparagine and arginine.

Due to the prokaryotic nature of the cells, the most promising advances in genetic engineering of microalgae have been made in this group. In recent years transformation protocols have been developed for a number of strains and genes. In addition, with the sequencing of the complete genome of *Synechocystis* PCC 6803, a mile stone in cyanobacterial molecular biology has been reached.

For genetic engineering the genus *Synechococcus* has mainly been used. Uptake of vectors has been accomplished either by natural uptake or by electroporation. Promising results have been obtained for the expression of mosquito larvicides (CHUNGJATUPORNCHAI, 1990), salmon growth hormone, and superoxide dismutase as well as altered biochemical pathways for xanthophyll and lipid biosynthesis.

Several hundreds of unialgal or axenic strains are available from algal and bacterial culture collections.

2.2 Eukaryotic Algae (taxonomy according to MARGULIS et al., 1990)

2.2.1 Rhodophyta

Together with the green algae, the rhodophytes represent the oldest group of eukaryotic algae. Therefore, red algae form an evolutionary distinct branch with many characteristic properties (RAGAN and CHAPMAN, 1978). About 4,100 red algal species in 695 genera are recognized, the majority of which are marine macrophytes. Roughly 30 species can be considered microalgae, all belonging to the subclass Bangiophycidae. The red microalgae inhabit marine, freshwater as well as terrestic habitats. In addition to chlorophyll a, red algae contain high amounts of the biliproteins, phycoerythrin and phycocyanin. The main storage product is starch outside the plastid and floridoside, a glycerol galactoside. The high content of galactans in the cell wall makes red algae an interesting group for biotechnology.

Porphyridium cruentum is easily cultivated in artificial sea water medium without the requirement of vitamin B_{12} which is necessary for other red microalgae. This is one reason why up to now only *Porphyridium* species are used biotechnologically for the production of arachidonic acid, pigments (phycocyanin and phycoerythrin), and extracellular polysaccharides. It can be expected that genetic engineering of *Porphyridium* will be a useful tool in the near future. Another interesting rhodophyte is the acido- and thermophilic *Galdieria sulphuraria*, which not only grows best at pH 1–2 and 45 °C, but can also grow heterotrophically (GROSS and SCHNARRENBERGER, 1995). This is a unique exception among the rhodophytes.

Red microalgae can be obtained from various culture collections, e.g., UTEX or Algobank.

2.2.2 Glaucocystophyta

The Glaucocystophyta consist of only three species in three genera. All members are flagellated unicells, their plastids (cyanells) are surrounded by a peptidoglucan wall and resemble cyanobacteria. Starch is accumulated outside the phycobilisome-containing cyanells. All glaucocystophytes are found in freshwater habitats and are very rare. *Cyanophora paradoxa*, with a doubling time of 2 d, is probably the fastest growing species.

2.2.3 Chlorophyta

The systematics of green algae has undergone major changes in the last decades. Although there are still a number of questions regarding the taxonomic position of some orders, four classes within the Chlorophyta are well characterized:

Prasinophyceae

This small group of primitive flagellated unicells is found mainly in marine and brackish environments, but also in freshwater. About 120 species in 13 genera are recognized. The cells are small (10–15 μm) and covered with organic scales.

Some prasinophytes exhibit high growth rates and are used in outdoor cultures as food for, e.g., bivalve mollusk larvae (LAING and AYALA, 1990).

Chlorophyceae

The largest group within the green algae are the Chlorophyceae with about 2,500 species in 350 genera. Most species are unicellular or filamentous freshwater forms. The two major groups have very different cell walls, the Chlamydomonales have a proteinaceous wall while the Chlorococcales have cell walls comparable to higher plants. Especially species of the genera *Chlorella*, *Dunaliella*, *Chlamydomonas*, *Scenedesmus*, and *Nannochloris* are used commercially in aquaculture as well as for the production of valuable compounds.

Chlorella sorokiniana and *Chlamydomonas reinhardtii* have successfully been transformed by introducing genes via particle gun, glass beads, or electroporation into the cells.

Ulvophyceae

Although this group contains more than 1,100 species in 100 genera, only few species may be considered as microalgae (e.g., *Pseudendoclonium*, *Chlorocystis*). Typical ulvophytes are thallous forms found at the sea shores. So far, no microalga of this group has been used for biotechnological purposes.

Charophyceae

This group is phylogenetically closely related to higher plants and exhibits a physiology much more similar to the cormophytes than to any algal group. With the exception of the Charales, roughly all 45 species (5 genera) of the Charophyceae can be considered as microalgae inhabiting soil, snow fields, but also freshwater.

Members of the genus *Stichococcus* often dominate in outdoor cultures when the temperature is above 25 °C but they are not used biotechnologically, yet. *Klebsormidium* species also show very high growth rates in simple growth media.

Strains from all four groups are available from a number of culture collections (UTEX, SAG, CCAP). A large number of *Chlamydomonas* mutants are available from the Chlamydomonas Genetics Center (CGC).

2.2.4 Conjugaphyta

This group was formerly included within the Charophyceae but has now been placed into a separate phylum. About 12,000 species in 56 genera are recognized which are either unicellular (80%) or filamentous (20%). Nearly all species are found in freshwater, colonizing peat bogs, rivers, and streams. Vegetative cells lack a flagellum but some species are able to glide by the excretion of mucilage. Some filamentous forms have a remarkable drought tolerance, surviving decades of desiccation.

Protoplasts have been obtained with high yield from *Micrasterias* species, *Cosmarium turpinii* (BERLINER and WENC, 1976), and *Mesotaenium caldariorum* (BERKELMAN and LAGARIAS, 1990). Although, several species show high growth rates (e.g., *Mougeotia*, *Mesotaenium*, *Cosmarium*), their biotechnological applications are scarce. Potent growth promoters for unicellular species have been reported (LEFÈVRE and JAKOB, 1949) and may prove to be beneficial for mass cultures.

Strains can be obtained from an extensive collection at the SVCK in Hamburg.

2.2.5 Euglenida

The euglenoids represent a very divers and isolated group of flagellated unicells with roughly 800 species in about 43 genera. Phylogenetically the Euglenida share both plant- and animal-like characteristics and are related to the protozoan group Kinetoplastida. About 70% of the species are colorless and rely on soluble organic substrates or live phagotrophically. Their complex cell wall, called pellicle, consists of proteinaceous, interlocking strips beneath the plasma membrane. The pellicle is highly flexible in some species and allows them to perform crawling movements. The euglenoids are usually not planktonic, but inhabit almost all other freshwater habitats. Several species are eutrophic and common in wastewater or polluted areas. Physiological studies on *Euglena gracilis* have shown that this alga (and possibly all Euglenida) exhibit a number of unique pathways and metabolites (RAGAN and CHAPMAN, 1978). Their storage product is paramylum starch, a β1–3 polyglucan, outside the plastid, which is characteristic for this group. A number of bioactive compounds have been reported for euglenoids but have so far not been exploited for biotechnology. The accumulation and resistance towards heavy metals of some euglenoids may have potential biotechnological applications. Most Euglenida are difficult to grow under axenic conditions and may require complex growth media. An exception is *Euglena gracilis* which can be grown in mineral medium and has successfully been used in large scale mass cultures (SIEGELMAN and GUILLARD, 1971).

Stock cultures can be obtained from several culture collections, e.g., UTEX or SAG.

2.2.6 Prymnesiophyta

The Prymnesiophyta, or sometimes called Haptophyta, consist of about 500 species in 50 genera which are flagellated or coccoid unicells. The yellow-green to brownish color of the prymnesiophytes derives from xanthophylls, mainly fucoxanthin and is similar to the color of diatoms and xanthophytes (see below). The storage product is chrysolaminarin (a soluble β1,3-polyglucan). Prymnesiophytes are mostly found in the marine plankton, but also freshwater and terrestic habitats are known. Some species are phagotrophic. Especially marine species often form massive blooms (e.g., *Chrysochromulina*, *Emiliania*) with negative effects on the ecosystem.

A large group within the Prymnesiophyta are coccolithophorids possessing calcified structures (coccoliths) on the cell surface. The formation of coccoliths is an important geological process which has formed massive limestone deposits during the Cretaceous period. Massive blooms of the coccolithophorid *Emiliania huxleyi* can amount to almost 100,000 t of calcite (HOLLIGAN et al., 1983). Thus, coccolithophorids play an important role in the carbon transfer from the atmosphere into sediments. Considering the increasing concentration of CO_2 in the atmosphere, coccolithophorids might prove to be an effective agent for the reversion of this development.

Despite the fact that a number of species form blooms, prymnesiophytes are not easy to cultivate under laboratory conditions. Further studies are certainly needed to improve the culture conditions for this interesting algal group.

Many unusual natural products are found in prymnesiophytes, e.g., long chain polyunsaturated ketones and esters (C_{37}–C_{39}) as well as unique sterols. The production of dimethyl sulfide (DMS) has been observed in several species and may represent a significant part of the global sulfur cycle. Carbohydrates are also excreted by some species like the marine *Phaeocystis pouchetii*, which regularly forms extensive blooms giving the water a jelly-like appearance, due to the accumulation of mucilage. The excreted polysaccharides may amount to 16–64% of total cellular polysaccharides. Blooms of *Prymnesium* species are dreaded for their exotoxins which threaten fish farms or natural populations. Several species are also used as feed for shellfish larvae, especially oysters (LAING and AYALA, 1990).

The largest collections of strains are available from Algobank and UTEX.

2.2.7 Xanthophyta

The yellow-green algae consist of 600 species in more than 90 genera. They are mostly unicellular or filamentous and prefer freshwater habitats or soil. Their major storage products are chrysolaminarin and oil. The xanthophyte *Bumilleriopsis filiformis* has been cultured in mineral medium with a doubling time of 14 h (HESSE, 1974). Members of the genus *Tribonema* exhibit similar growth rates and form late winter blooms in still waters. Despite their apparently high potential, xanthophytes have no practical usage in biotechnology.

Strains can be obtained from several culture collections, e.g., UTEX or SAG.

2.2.8 Chrysophyta

Also called golden algae because of their high content of fucoxanthin, the Chrysophyta represent a group of about 1,000 species separated into 120 genera. The majority of the species are flagellated unicells living as freshwater plankton. The cells are either naked or covered with elaborate silica scales. Although most species can photosynthesize, mixotrophic growth is very common in this group. A number of species grow mainly phagotrophically. Their storage products are chrysolaminarin and oil in droplets. Although some species grow relatively fast in culture, the addition of vitamins, other organic nutrients, or soil extracts are necessary for most species for good growth. Despite the size of this group relatively few species are available in culture collection.

2.2.9 Raphidophyta

The Raphidophyta, closely related to the Chrysophyta, represent a very small group with 9 species in 4 genera. All species are unicellular, living in freshwater as well as marine habitats, but are generally not abundant. However, blooms of the marine species *Chattonella* and *Fibrocapsa* cause red tides. Raphidophytes usually require vitamins and other organic supplements for growth. The species *Olistodiscus luteus* and *Heterosigma* sp. have been used for physiological studies and show high growth rates. The first species may be of special interest for biotechnology because it is known to excrete mannitol (HELLEBUST, 1965). However, no biotechnological application has been developed for this algal class, yet. In culture collections strains of this group may be found also classified as xanthophytes or chrysophytes.

2.2.10 Eustigmatophyta

These coccoid unicells (12 species in 6 genera) are relatively rare and occur in freshwater and marine habitats. The genus *Nannochloropsis* contains some relatively fast growing species. None of the species have practical importance. Members of all genera of the eustigmatophytes are available from culture collections, e.g., SAG or UTEX.

2.2.11 Bacillariophyta

The Bacillariophyta or diatoms represent an enormously large group with more than 100,000 species in 250 genera. Because of the variability of the morphological characteristics as well as the similarity between species, the number of species may even be much larger. All species are unicellular and unflagellated in their vegetative stage. Only few species form pseudofilaments or colonial aggregates. The habitat is very divers and includes freshwater, marine as well as terrestic environments. Due to the high concentration of fucoxanthin, all diatoms have a brownish color. Their main reserve products are oil in droplets (e.g., *Phaeodactylum tricornutum*: 20–60% oil per d.wt.) and chrysolaminarin (see above). Characteristic for this group is the frustule, a silica cell covering consisting of two valves with elaborate fine structure. The Bacillariophyta are separated into three groups according to their overall morphology: Coscinodiscophyceae (centric diatoms), Fragilariophyceae (pennate diatoms without a raphe), and Bacillariophyceae (pennate diatoms exhibiting a raphe).

Centric as well as pennate diatoms are widely used in aquaculture as well as the production of oils. Especially the discovery of high concentrations of eicosapentaenoic acid (EPA) in diatoms has boosted their biotechnological use. Several diatoms are even a pest in outdoor cultures: The pennate diatom *Phaeodactylum tricornutum* is dominant in outdoor cultures at temperatures between 10–23 °C, even though it is rarely a significant part of the algal community in its natural habitat. Because of the high productivity and the accumulation of oils, bacillariophytes have long been the target for large scale oil production. First steps have even been done to genetically engineer diatoms (APT et al., 1996). Considering the enormous number of diatom species, this group continues to be a promising source for biotechnologically interesting organisms. The largest collection of marine and freshwater diatoms are provided by the CCMP and the Loras College, respectively.

2.2.12 Cryptophyta

This rather small group contains 60 freshwater and marine species in 20 genera. Most cryptophytes are photosynthetically active motile unicells which contain phycobiliproteins but no phycobilisomes like red algae or cyanobacteria. The cells are very small and especially marine species are often very fragile. In general, cryptophytes seem to dominate in cool, oligotrophic waters often forming blooms in winter and early spring. Some species produce toxic substances while others are used in aquaculture as feed for rotifers and clams. A small number of species is available in various culture collections.

2.2.13 Phaeophyta

The brown algae consist of about 900 species, almost exclusively macrophytes, in 250 genera. Only a few filamentous species in the group Ectocarpales can be considered microalgae. However, these are slow growing, marine species.

2.2.14 Dinoflagellata

The Dinoflagellata represent a large and very diverse group with about 4,000 species in 550 genera. Evolutionarily the Dinoflagellata are relatively isolated with some affiliation to the Ciliophora (e.g., *Paramecium, Tetrahymena*). Their storage products are oil and starch in the cytosol. About half of the species are colorless, the other half has often only limited photosynthetic capabilities and also has to rely on organic compounds. Nevertheless, some species contribute significantly to the primary production especially in coastal waters. The flagellated unicells inhabit freshwater and sea water. Dinoflagellates are the main cause of red tides with accumulation of toxins and often devastating effects for coastal fishery. On the other hand, *Gymnodinium* species are the preferred feed for fish larvae in aquaculture. However, only few dinoflagellates have been successfully grown under laboratory conditions, most species require complex growth media or natural waters supplemented with vitamins, etc. Growth conditions are also limited by the required light–dark cycle and the detrimental effect of agitation for many species (see also Sect. 3.2). Nevertheless, dinoflagellates exhibit a variety of unique sterols, fatty acids (docosahexaenoic acid, DHA), and pigments which have a high potential for biotechnology. Especially their high content of DHA has prompted the large scale cultivation of colorless species. Compared to the size of this group, only few species are available from culture collections (e.g., UTEX, CCMP, NEPCC) because many dinoflagellates are difficult to maintain in culture.

2.3 Maintenance

Strains can either be isolated from field samples or obtained from one of the numerous culture collections (Tab. 2). If axenic cultures are needed, the isolation from field samples is very often too time consuming and difficult. Stock cultures are usually maintained on agar slants placed near a north-facing window or in special culture rooms. Some strains will not grow on solidified media (e.g., many cryptophytes) and have to be kept in liquid culture, either agitated or not. Care has to be taken that cultures receive only low light intensities and constant temperatures. Cryopreservation offers a very effective way of long-term storage of algal cells as it has been shown for *Euglena*, *Porphyridium*, or *Chlamydomonas*. For this procedure, cells are slowly frozen usually in the presence of a protecting agent (e.g., dimethyl sulfoxide, methanol, or glycerol) and kept either at minus 70 °C or in liquid nitrogen (APT and BEHRENS, 1999).

Media for stock cultures are usually described in detail by the various culture collections. For mass culture, however, these media have to be modified in order to meet the demand for fast growth and high cell densities. Finding the optimal growth medium usually involves numerous trials with different compositions and concentrations of nutrients. Some basic considerations for the composition of media have been published by MCLACHLAN (1964) and UPITIS (1983).

3 Techniques and Technologies to Produce Biomass of Cyanobacteria and Microalgae

Microalgae have an extremely high productivity compared to higher plants. Some strains of high-temperature adapted green algae, e.g., *Chlorella pyrenoidosa*, are able to double their cell number every 2.5 h (OH-HAMA and MIYACHI, 1988). In contrast to higher plants, photo-

Tab. 2. List of Algal and Cyanobacterial Culture Collections

Acronym	Culture Collection
ACOI	The Culture Collection of Algae, University of Coimbra, Portugal
ALGOBANK	Microalgae Strain Bank, Université de Caen, France
ATCC	American Type Culture Collection, Rockville, Maryland, USA
CALU	Culture Collection of Algae, Laboratory of Microbiology, Biological Institute of St. Petersburg, State University, St. Petersburg-Stary Peterhof, Russia
CAUP	Culture Collection of Algae, Department of Cryptogamic Botany, Faculty of Science, Charles University, Prague, Czech Republic
CCALA	Culture Collection of Autotrophic Organisms, Institute of Botany, Academy of Sciences, Trebon, Czech Republic
CCMP	Provasoli – Guillard National Center for Culture of Marine Phytoplankton, Bigelow Laboratory for Ocean Sciences, West Boothbay Harbor, Maine, USA
CCAP	Culture Collection of Algae and Protozoa. c/o Freshwater Biological Association, Ambleside, Cumbria, UK and Scottish Marine Biological Association, Oban, Scotland, UK, a.k.a. SMBA
CGC	Chlamydomonas Genetics Center, Duke University, Durham, North Carolina, USA
CMARC CSIRO	Microalgae Research Centre, Hobart, Tasmania, Australia
DMMSU	Culture Collection of Algae, Department of Microbiology, Faculty of Biology, Moscow State University, Moscow, Russia
HPDP	Culture Collection of Algae, Institute of Hydrobiology, Academy of Sciences of Ukraine, Kiev, Ukraine
IAM	Culture Collection of the Institute of Applied Microbiology, University of Tokyo, Japan
IBASU	Culture Collection of Algae, Department of Cryptogamic Plants, N. G. Kholodny Institute of Botany, Academy of Sciences of Ukraine, Kiev, Ukraine
IPPAS	Culture Collection of Microalgae, K.A.Timiryazev Institute of Plant Physiology, Russian Academy of Sciences, Moscow, Russia
LABIK	Collection of Algal Cultures, Laboratory of Algology, V. L. Komarov Botanical Institute, Russian Academy of Sciences, St. Petersburg, Russia
LORAS	The Loras College Freshwater Diatom Culture Collection, Department of Biology, Loras College, Dubuque, Iowa 52004-0178 USA
MBI	Marine Biotechnology Institute, Kamaishi, Japan
NEPCC	North-East Pacific Culture Collection, Department of Botany, University of British Columbia, Vancouver, B.C., Canada
PACC	Plovdiv Algal Culture Collection, Laboratory of Algology, Plovdiv University "Paisiy Hilendarski", Plovdiv, Bulgaria
PCC	Pasteur Culture Collection of Cyanobacterial Strains, Institut Pasteur, Paris, France
PGC	Peterhof Genetic Collection of Strains of Green Algae, St. Petersburg State University, St. Petersburg-Stary Peterhof, Russia
SVCK	Sammlung von Conjugatenkulturen, Inst. Allgem. Botanik Hamburg, Germany
THALLIA	Thallia Pharmaceuticals SA, Ecole Centrale de Paris, Grande voie des vignes, Chatenay Malabry, France
UMACC	University of Malaya Algae Culture Collection, Kuala Lumpur, Malaysia
UTCC	University of Toronto Culture Collection of Algae and Cyanobacteria, Toronto, ON, Canada
UTEX	The Culture Collection of Algae at The University of Texas at Austin, Texas, USA

For a more detailed list of culture collections see NORTON et al. (1996).

synthetically inactive cells (such as roots or trunks) are absent, therefore, microalgae are a favored subject of research for the determination of complex photosynthetic processes, especially with regard to quantum yield. Although in well mixed microalgal cultures the antenna complex does not react to an increased light intensity to the extent which is used by phanerogams to convert efficiently high quantum yields to utilizable energy, this is compensated largely by the mobility and the constant supply of carbon dioxide, minerals, and water. Photosynthetic production of biomass is generally represented by the following equation:

$$\alpha CO_2 + \beta H_2O + \gamma NH_3 + \text{minerals} + nh\nu \rightarrow \alpha\{\text{CHON minerals}\} + O_2 \quad (1)$$

where $nh\nu$ is the number of energy quanta of a certain frequency which is required for the production of 1 mol O_2. Due to other energy consuming reactions such as dark respiration or maintenance metabolism, the yield of fixed carbon will be reduced with regard to the oxygen quantity produced. Therefore, α must not necessarily have the value 1. Based on the classical model of photosynthesis, 10 mol quanta produce sufficient ATP and NADPH to liberate 1 mol O_2. The actual requirement of quanta for photosynthesis, however, is controversial and given by the different authors between 6 and 16 quanta (RAVEN, 1988; PULZ and SCHEIBENBOGEN, 1998).

3.1 Cultivating Phototrophic Microorganisms

The basic ability of photoautotrophic microorganisms to utilize light quanta in the visual range as energy source for metabolism, is limited mainly by high population densities. In contrast to heterotrophic microorganisms, such as yeasts or bacteria, where the distribution of substrate by mixing is solved technologically, the supply of photons is dependent on the surface area. Light dissipation in the cell suspension is achieved by numerous quantum interactions with exponential decrease along the boundary normal (KIRK, 1994). This means light introduction into a given volume of algal cell suspension in connection with reducing layer thickness is the key question of photobioreactor design. There is nearly no end to the multiple technical approaches and systems for the cultivation of microalgae.

Not only light conditions are changing for microalgae which come into culture but other dramatic changes occur for each single cell in comparison with nature (Tab. 3). Therefore, it is most likely that no single type of photobioreactor configuration will be used in the future but a design diversity according to microalgal ecophysiologal requirements will develop.

The biological and technical potential of algal biotechnology has been described in detail in several reviews (BECKER, 1994; PULZ and SCHEIBENBOGEN, 1998; BOROWITZKA and BOROWITZKA, 1988). Most of the commercial plants are based on open pond technologies, for which detailed descriptions for their construction and productivity exist (SOEDER, 1986). These systems, however, seem to have reached their limits. The gap between the theoretical biological potential of microalgal biomass and the productivity actually achieved may be overcome by developing closed cultivation systems. It is rather difficult to compare the open pond technology with closed systems

Tab. 3. Environments for Microalgae in Nature and Photobioreactors

	Dimension	Nature	Photobioreactor
Cell density	cells m L^{-1}	10^3	10^8
Cell distance	μm	1,350	50
Cell distance	n cell diameter	250	10
Displacement	m s^{-1}	10^{-4}	0.3–1.5
Light	$\mu E \cdot m^{-2} s^{-1}$	10–300	10–1,800
Light fluctuations		diurnal	ultrashort
Shear stress		low	high
Mineral supply		limited	surplus
CO_2 supply		limited	surplus
O_2 concentration	mg L^{-1}	7–9	10–45
pH value		stable	variable
Temperature		stable	variable

and indoor photobioreactors because of the strong variations of the prevailing boundary conditions. It appears that open systems are predominating in mass cultivation because of cost considerations, and that photobioreactors receive increasing attention for the preparation of valuable substances and for special applications (BROUERS et al., 1989; RICHMOND, 1990). Fig. 2 shows a flow sheet of general algal production.

3.2 Lab-Scale Photobioreactors

On the basis of numerous studies for the preparation of valuable substances from phototrophic microorganisms, LEE (1986) and later PULZ (PULZ and SCHEIBENBOGEN, 1998) have summarized the development of lab-scale photobioreactors. As routinely used for heterotrophic organisms, shaker flasks represent the most common means to cultivate phototrophs. They are easy to sterilize and cause low shear forces within the cell suspen-

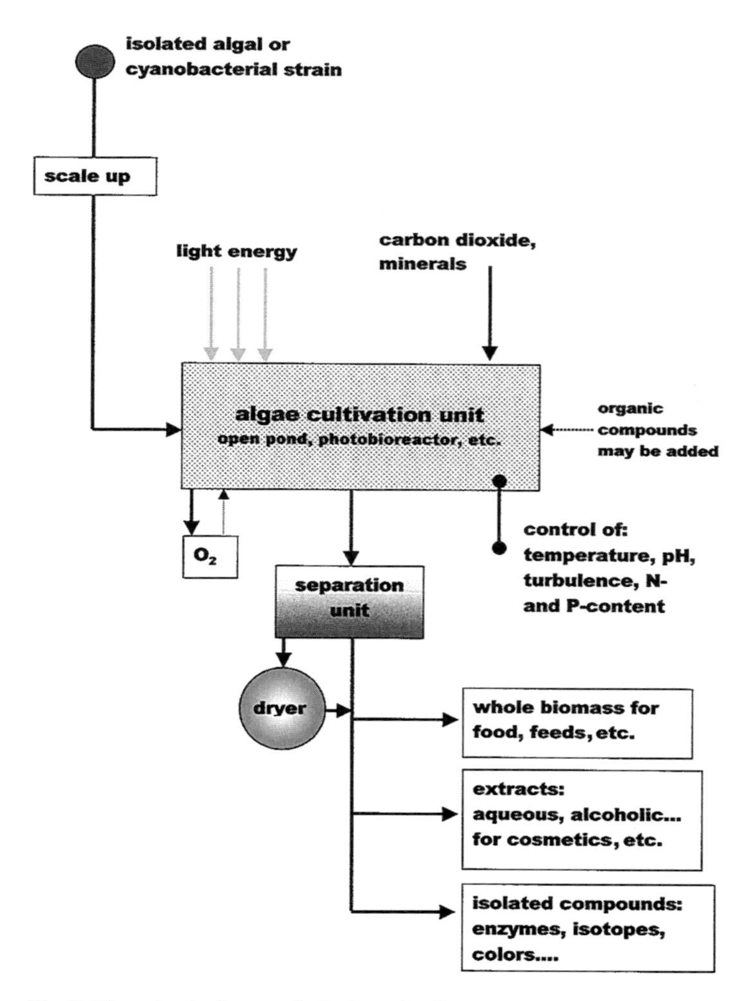

Fig. 2. Flow sheet of general algal production.

sion, thus applicable to grow most species in up to 1 L volume. Because light transmission and diffusion of CO_2 are poor under these conditions, growth rates of microalgae are rather moderate. Due to the high technical level of cylindrical reactors of classical biotechnology, glass containers are commonly applied for experimental cultivation on the laboratory scale. However, these cylinders, which are mostly designed as fully sterilizable stirrer reactors surrounded by lamps, are characterized by a high luminance at the surface in combination with a large dark volume. To increase the photic volume, several modifications have been introduced in sterilizable lab-scale photoreactors. All these systems include steel vessels with internal illumination units in tumbler heads or other strong agitation devices, which increases the growth rate of some species several times. However, the resulting turbulence which is needed to avoid photoinhibition may lead to inhibition of growth due to the shear forces created. Therefore, tubular cylinders of transparent material, double walled for easy temperature control and aerated with CO_2 enriched gas for mixing, seem to be the most convenient device to cultivate algae in small volumes.

To overcome the problem of insufficient photic volume within cell suspensions, fiber optics for uniform light dissipation have been suggested. Proposals refer to concentrically fitted radiating cylinders which are supplied by light from the exterior via fiber optics including illumination of immobilized *Rhodospirillum* for hydrogen production. Especially in Japan much work has been done to utilize light diffusing implementations to supply energy for suspended algae. The diameter of the radiation bodies ranges from 1 cm (MORI, 1986) to 1 mm (MATSUNAGA et al., 1991) and create surface to volume ratios higher than 700 m^{-1}. The radiation device occupies up to 60% of reaction space. Although much energy is lost during light transfer from source to the cell, high productivity confirms the excellent efficiency of this concept. By use of a Fresnel lens unit as sunlight collecting device, even natural radiation might be used for energy supply, however, the high technical expense is unlikely to meet the production value (HIRATA et al., 1996). Furthermore, the poor ability for scale-up may limit such systems for academic purposes. The use of heat durable fibers, which emit light evenly along their length, could be an interesting alternative for sterile production: A closely packed fiber unit turns conventional fermentation vessels into photobioreactors, thus reducing investment costs.

Immobilization of microalgal cells has been proposed for various reasons. The design of immobilized cell photobioreactors depends on the purpose for which the organisms will be used. Packed-bed and fluidized-bed reactors as well as air-lift systems are mentioned in the literature, but parallel plate types are preferred for effective utilization of sunlight. Although growth rates of immobilized phototrophs are generally lower than those of free cells, it is emphasized that the content of pigments and storage substances, as well as production of extracellular products, especially hydrogen, could be improved remarkably in most cases (ROBINSON et al., 1986; HALL et al., 1993).

Immobilization techniques range from hollow fiber systems, encapsulation by transparent hollow micro-spheres to self-immobilization in planar plate loop reactors or in ultrathin layer systems in which adhesive forces between two vertically arranged foils cause a suspension layer thickness of 0.2-1 mm. By the defined porosity of the fixing matrix, extracellular and intracellular products are trapped and enriched for downstream processing.

It should be pointed out that not all the cultivation systems are suited for all species. For example, the dinoflagellates which are supposed to bear a high potential for pharmaceutical substances, appear to be too fragile to be cultivated in traditional systems. Special culture techniques, such as dialysis culture systems have been proposed for shear force sensitive organisms and are still being investigated (BOROWITZKA, 1995).

3.3 Mass Cultivation Techniques

3.3.1 Open Cultivation Systems

Open cultivation systems comprise natural or artificial ponds, raceway ponds, and so-called inclined surface systems. They represent the classical processes of algal biomass production. All of them require large areas. If appropriate areas are available in regions of balanced and sunny climate (e.g., waste land in the vicinity of lakes) at reasonable costs, the investment efforts will be relatively low even at scaling up to 100 ha (BENEMANN et al., 1987). No investment costs for pond construction will occur in natural waters. The best known natural waters which are used for microalgal production are the lake Texcoco in Mexico and more for local purposes, the lake Chad in Africa. In both cases *Spirulina* (*Arthrospira*), one of the most important microalga, is harvested from natural habitats. In several alkaline crater lakes in central Myanmar a considerable *Spirulina* production has been developed in natural waters over the last 10 years (Fig. 3).

Natural and artificial ponds are generally used for the cultivation of fast growing, naturally occurring or extremophilic species. The cultivation of the extremely halophilic and high light-tolerant green alga *Dunaliella salina* for β-carotene production in Western Australia is a good example. In the shallow unlined ponds of more than 50 ha the algae are harvested continuously through a widely branched tubing system, nutrient-enriched sea water being added through pumps. Evaporation regulates the temperature of the suspension and maintains hypercritical salt concentrations so that the growth of other species is inhibited. Due to the lack of any stirring mechanisms, CO_2 and biomass are not evenly distributed, which is the cause of relatively low productivity not exceeding 1 g d.wt. m^{-2} d^{-1} (BOROWITZKA and BOROWITZKA, 1988). Nevertheless, the *Dunaliella* business is attractive enough that HENKEL bought both Australian production facilities to serve the health food market.

At optimal temperatures, obtained by utilizing the cooling water of power stations, *Spirulina* was also cultivated successfully in Europe. The yields amount to 24 t dry biomass per ha shallow lagoons per 6 months, which corresponds to a growth rate of 13 g d.wt. m^{-2} d^{-1} (DUJARDIN et al., 1992).

In pools of the dimension described above, the problem of stirring is of fundamental importance because large amounts of energy are required for the prevention of concentration gradients and algal sedimentation. Floating bodies with impellers pulled over the surface of the ponds (VENKATARAMAN and BECKER, 1985) or raceway ponds are energy-saving solutions (Fig. 4). The latter are characterized by parallel, loop-like channels, several kilometers in length; several paddle wheels combined with aerating units produce an economic motion of the algal suspension and a uniform nutrient supply. Thus, the production of biomass

Fig. 3. Natural lake in Myanmar used for *Spirulina* production.

Fig. 4. Paddle wheel driven raceway pond in Asia.

in a plant of several hectares requires relatively low operating expenses (BENEMANN et al., 1987). The walls can be constructed from plastic sheet covered earth wall liners or by concrete. Numerous variations are recorded according to local ideas and material supply. Detailed calculations on the design of raceway ponds have been published by BENEMANN (1990). In Asia (Japan, Taiwan) as well as in Ukraine, Belorussia, and Moldavia, metal walled circular ponds with rotating stirrer arms are still in use for biomass production and for cleaning of wastewater from sugar refineries. Thorough mixing of the suspension in a raceway pond with paddle wheels has no significant influence on the productivity when using suspension velocities of 1–30 cm s^{-1}. Therefore, in many production plants an intermittent raceway running regime is used, mainly to avoid settling of microalgae.

It is still not clear whether a synergistic relationship exists between light/dark cycles (flash light effects) and mechanisms of material transport connected with highly turbulent suspension flows. The energy input through paddle wheels as stirring organs of raceway ponds is rather expensive, especially in larger plants. In addition, the increased shear forces created by paddle wheels have negative effects on numerous algal species.

The growth of biomass in raceway ponds is also dependent on the prevailing regional climate. Mean growth rates for Chlorophyceae and suspension depths of approximately 20 cm, from 15–25 g d.wt. m^{-2} d^{-1} are recorded in tropical regions, and in California (VENKATARAMAN and BECKER, 1985). In Southern Europe up to 28 g d.wt. m^{-2} d^{-1} (PULZ, 1992), in Central Europe and Japan mean values of only 12–15 g d.wt. m^{-2} d^{-1} have been reported.

At an average radiation of the surface of the ponds of 16.7 MJ m^{-2} d^{-1} (PAR) a theoretical productivity of 130 g d.wt. m^{-2} d^{-1} can be calculated on the basis of a photosynthetic efficiency of sunlight of approximately 18%. The calculated average conversion of 1–2% in large plants implies that raceway ponds are a light limited cultivation system because of self-shading and poor material transport. In this regard, BENEMANN (1990) refers to mixotrophic algal growth during the treatment of high organic loads of sewage. The peak values of up to 100 g d.wt. m^{-2} d^{-1} biomass or above 40 g d.wt. m^{-2} d^{-1} as the annual average support the notion of light limitation. Nevertheless, the main mass producers of microalgal biomass like Dainippon Ink or Cyanotech are still using the raceway-pond technology successfully and Innogas is performing new trials with this technology even in Europe.

In the inclined surface plant with an active surface of 900 m^2, developed by SETLIK et al. (1970) in the Czech town Trebon, the algal suspension flows over terraces of a defined inclination so that a layer thickness below 1 cm and turbulent flow prevent any shaded volumes (Fig. 5). For example, at the IGV (Institut für Getreideverarbeitung – Institute for Cereal Processing, Potsdam-Rehbrücke, Germany) a peak productivity of 19 g d.wt. m^{-2} d^{-1} with *Scenedesmus* was obtained. PULZ (1992) reported a productivity of 24 g d.wt. m^{-2} d^{-1} for *Chlorella* which was achieved in a smaller plant of this type. This exceeds the productivity of raceway ponds by 100%. This superiority with regard to the suspension volume used is expressed even more clearly by the ratio of 0.95 g d.wt. L^{-1} d^{-1} and approximately 0.01 g d.wt. L^{-1} d^{-1} for raceway ponds. Thus, the additional constructive and energetic expenditures seem to have been justified. The fundamental influence of the illuminated surface with regard to the total volume of the suspension (S/V-ratio) is obvious. Inclined surface systems with a small dark volume of the collecting pond with thin suspension layers and a corresponding S/V-ratio of 20–100 m^{-1} (compared to 3–10 m^{-1} in raceway ponds) fulfill the requirements, however the investment and energy costs are increased drastically. Nevertheless, a larger plant (5,000 m^2) of this type was built in Rupite, Bulgaria, for the successful production of *Chlorella* with peak values of 37 g d.wt. L^{-1} d^{-1} during summer (FOURNADZHIEVA and PILLARSKY, 1993). Similar plants have been built in sunny regions of the Third World.

The Czech group intends to construct large scale facilities which are made of several 500 m long and 5–10 m wide meandering lanes with an inclination of 1.6% and layer thickness of about 6 mm. The productivity is expected to be very high because of low layer thickness, however, this reactor remains an open system.

3.3.2 Closed and Semi-Closed Outdoor Photobioreactors

Attempts to make the cultivation conditions independent of environment are accompanied by a sharp increase of the investment and operating costs. This was justified in earlier years only by the production of very high value products, e.g., radioactive labeled substances, performed both, in the USA and in the former Soviet Union. The future development of microalgal biotechnology will take place in the field of high value, algal specific substances which are produced under completely controlled and species specific conditions (GROBBELAAR et al., 1996; PICCARDI et al., 1999). This is only possible in closed systems.

Closed reactors have a number of advantages:

- low CO_2 losses,
- reduced risk of contamination,
- temperature regulation,
- controllable hydrodynamics,

Fig. 5. Inclined surface type photobioreactor.

- reproducible cultivation conditions,
- higher flexibility with regard to environmental influences,
- significantly smaller space requirements.

One of the main configuration advantages in closed reactors is probably the controllable design of layer thickness, light pass, and light supply. The most simple closed photobioreactor is a lab-scale vertically arranged glass tube supplied from the bottom with air (airlift). In aquaculture this principle is used in larger scale photobioreactors in configurations of up to several hundred liters containing plastic bags hanging in a support. Tank cultures may consist of smaller internal illuminated metallic vessels (up to 5 m^3) or fiberglass tanks. Both are used in aquaculture, *Chlorella* production, or in pharmaceutical production (OSINGA et al., 1999).

Tank/vessel cultures in non-illuminated fermenters with (facultative) heterotrophic algae gained industrial importance over the last years. While, e.g., in Japan or in the Czech Republic, *Chlorella* is grown heterotrophically already for years, recently the production of polyunsaturated fatty acids (PUFA) with heterotrophically grown microalgae started to be profitable (OSINGA et al., 1999).

3.3.3 Tubular Photobioreactors

Especially the group of MATERASSI and TREDICI (PICCARDI et al., 1999) in Italy as well as GUDIN and CHAUMONT (KRETSCHMER et al., 1995) in southern France have invested much effort into the development of outdoor reactors. Generally these tubular systems are arranged in a horizontal serpentine form and made of glass or plastic tubes. The culture suspension is re-circulated either by a pump or – more preferably – by air-lift technology. Temperature is controlled by floating or submerging the tubes on or in a pool of water, oxygen degassing is guaranteed by flexible tube elements. Using *Porphyridium cruentum* in a 100 m^2 culture unit (tube diameter 6 cm) a productivity of 20–25 g d.wt. m^{-2} d^{-1} (corresponding to approximately 0.4 g d.wt. L^{-1} d^{-1}) was achieved during 2 months of steady-state continuous culture. The company Thallia SA, formerly Heliosynthese SA, developed this type further and tried to launch commercial production (GUDIN and CHAUMONT, 1983). However, these plans failed and the company was apparently closed down.

A manifold tubular reactor was established in Israel, using polycarbonate tubes of reduced diameter (32 mm) for the cultivation of *Spirulina* and *Anabaena*. In Italy, too, several attempts were made to use closed tubular systems. Here, a maximum productivity of 25 g d.wt. m^{-2} d^{-1} (*Spirulina*) was achieved in a 10 m^3 serpentine bioreactor with intermittent culture circulation. Further improvements by constructing a two-plane tubular photobioreactor led to a mean daylight productivity of about 30 g d.wt. m^{-2} d^{-1} corresponding to 1.5 g d.wt. L^{-1} d^{-1}. Using strongly curved outdoor tubular reactors with high flow rates (0.97 m s^{-1}) generating Reynolds numbers above 4,000 within the tubes (26 mm inner diameter), the growth rate could be increased by about 17%, compared to straight tubes with 1.2 g d.wt. L^{-1} d^{-1}, i.e., 23 g d.wt. m^{-2} d^{-1} areal productivity. In Italy an average volumetric productivity for closed tubular reactors of 0.8 g d.wt. L^{-1} d^{-1} *Spirulina* per year was calculated (ZITELLI et al., 1996; PULZ and SCHEIBENBOGEN, 1998).

LEE et al. (1995) developed with their alpha-shaped 300 L tubular loop reactor a similar approach to increase turbulence and produced up to 72.5 g d.wt. m^{-2} d^{-1} *Chlorella pyrenoidosa* biomass (about 2.9 g d.wt. L^{-1} d^{-1}). DVORIN (DVORIN, 1992; KRETSCHMER et al., 1995) reported a plant of 130 m^3 size in Tadzhikistan (Sagdiana) which operated with a hydrodynamic regime generating Reynolds numbers over 8,000 for efficient mass transfer. Productivity of green algal biomass, however, was far below 0.1 g d.wt. L^{-1} d^{-1} (data for 1991). Another plant to produce *Chlorella* biomass commercially was established in Turkmenistan in 1978. With a 10 m^3 tubular device an annual capacity of 1 t, i.e., 1 g d.wt. L^{-1} d^{-1} and 0.7 g d.wt. L^{-1} d^{-1} *Spirulina* was achieved (SEITGELDIYEV, 1992; SEITGELDIYEV and AKYEV, 1995).

TREDICI and ZITELLI (1995) investigated the scale-up of photobioreactors to commercial size. They stated that incorrect evaluation of the efficiencies can lead to economic disaster.

In the case of PhotoBioreactors Ltd. (PBL, Spain) an inclined and near horizontal tubular system was built in 1990, using 200 km of polyethylene pipe (12 mm in diameter). However, incorrect management of the system led to poor growth and PBL was shut down before starting operations. In 1998/99 a tubular system, called "bio-fence", specially developed for aquaculture, was launched by the British company Applied Photosynthetic Ltd. with production volumes up to 1.2 m^3. Here, plastic tubes are arranged vertically to be exposed to light in a photostage with potentially significant higher productivity compared to the use of plastic bags (see above).

Horizontal arrangements were preferred in the construction of US American (Spektorova et al., 1997) and Italian developments (Microalgae spa). Again, these closed photobioreactors with industrial approach are developed for aquaculture and defined biomass production. The company Microalgae spa produces in two industrial modules consisting of tubes of 160 km length and a volume of 170 m^3 about 30 t dry matter per year. This represents a volumetric productivity of estimated 0.6 g d.wt. L^{-1} d^{-1} (Piccardi et al., 1999).

The helical tubular system commonly called Biocoil seems to be another promising alternative. The bioreactor consists of coiled polyethylene tubes (30–60 mm in diameter) around an open circular framework and was first realized by Addavita Ltd. in the UK. Algal suspension is re-circulated either by pumps or according to the air-lift principle. In Australia, 40–100 L laboratory reactors have been intensively studied and successfully scaled up to 1 m^3 outdoor pilot plants (Chrismadha and Borowitzka, 1994; Borowitzka, 1996). The productivity is comparable to other tubular systems. Several 5 m^3 reactors are planned to be established in the USA for aquaculture purposes (Borowitzka, 1996, personal communication).

Two very recent developments of tubular photobioreactors should be mentioned:

On Hawaii, Aquasearch Inc. issued a closed system for large scale microalgal production which is based on soft plastic tubes of approximately 50 cm diameter lying on the ground, partly filled with the algal suspension for growth. If the walls have been occupied inside by immobilized algae the whole plastic can be discarded and replaced. The AGM (Aquasearch growth module) has a capacity of 24 m^3 and is computer controlled (Piccardi et al., 1999).

After successful pilot plant tests using a 7 m^3 tubular photobioreactor over a period of almost 3 years, the German company Preussag initiated the construction of a production plant of 700 m^3 size. Using horizontal flow the glass tubes used were oriented vertically in order to utilize diffuse light (Fig. 6a and b). This configuration is a result of a long term investigation at the IGV (Pulz and Scheibenbogen,

Fig. 6. a One of 20 collecting containers which are part of the 700 m^3 closed type reactor. **b** A total of 500 km glass tubes have been installed within one algal production plant.

1998). The 700 m^3 total production volume is subdivided into 20 modules of 35 m^3. These are installed in a 10,000 m^2 glasshouse located near Wolfsburg in Germany. The production is estimated to be at least 130–150 t a^{-1}.

3.3.4 Plate Type Reactors

If light energy has to be available continuously to the cells, a lamination of the photobioreactor directed to the light source seems to be the best solution. This basic principle of a laminar concept has been practiced by plants for thousands of years. Compared with tubular systems, plate type geometry seems to have an identical configuration potential and high surface/volume ratio, but some advantages with respect to compactness (narrow U-turns, wall thickness). The most simple approach is found in the application of widely used polyethylene bags. With 50 L turbidostatic cultures of *Tetraselmis* yields are in the range of 20–30 g d.wt. m^{-2} d^{-1}. Illumination on both sides of the tanks (10 cm wide) and simple aeration of the cyanobacterium *Spirulina maxima* with CO_2-enriched air have resulted in a maximum increase of 1.17 g d.wt. L^{-1} d^{-1}. Reviews of early results can be found in PULZ and SCHEIBENBOGEN (1998) and PULZ (1992).

One of the first scientists who tested the horizontal meandrian channels was FALLOWFIELD (1991) followed by TREDICI and MATERASSI (1992) and PULZ (1992) who also favored the idea of arranging the rectangular channels vertically in the plate geometry. In contrast to the air-lift driven vertical alveolar panels (VAPs) of TREDICI and MATERASSI, PULZ used a closed system with horizontal flow and pumps (PBR) (Fig. 7). The flow rate must be high enough to maintain the turbulent flow in the channels (32·27 mm) and the distance to the temperature-regulated degassing container has to be kept short to avoid critical gradients (e.g., O_2, CO_2 concentrations, temperature). Parallel plates need only little spacing so that 42 plates (6 m^3 plant) will require only approximately 100 m^2 at a surface/volume ratio of 60 m^{-1}. Thus, such photobioreactor design meets the requirements for improved areal outputs:

(1) the compact arrangement distributes over-saturating solar energy evenly to the largest number of cells possible and
(2) the narrow light path and strong turbulence streaming move the algal cells in and out of the photic volume with the highest frequency possible (RICHMOND, 1996). Despite some light limitation caused by self-shading of the plates, even under the rather unfavorable climatic conditions of Germany, maximum biomass productivity (*Chlorella*) of 1.3 g d.wt. L^{-1} d^{-1} was obtained. This corresponds to an areal growth of 130 g d.wt. m^{-2} d^{-1}, which is considerably higher than reported for raceway ponds and conventional tubular reactors (PULZ et al., 1995). With vertical alveolar panels the S/V-ratio

Fig. 7. Vertically arranged meandrian channels forming the plate type photobioreactor series.

may be increased to 80 m^{-1}, the productivity related to the thin artificially illuminated plates of a surface of 0.3 m^2 amounting to 20–30 g d.wt. m^{-2} d^{-1} for *Spirulina* or 3 g d.wt. L^{-1} d^{-1} for *Tetraselmis* (TREDICI et al., 1994).

4 Cyanobacteria and Microalgae Exploitation

4.1 Biomass

Biomass of microalgae as powder, concentrated suspension, or in applied form is the primary product level in microalgal biotechnology. This biomass is obtained by microalgal growth, mechanical harvesting from growth media, and subsequent drying.

The production process consists of the following steps:

(1) Cultivation of microalgae in natural waters, ponds or photobioreactors or monitoring growth in natural waters.
(2) Harvesting microalgal cells or filaments from solution to get a slurry of 10–20% d.wt. by
- filtration,
- centrifugation or separation,
- flocculation, and
- floating.

(3) Dehydration of microalgal slurry by
- sun-drying,
- drum-drying,
- fluid-bed drying,
- spray drying.

If coarse structures occur, for example after sun- or drum-drying, milling should be applied.

The final product is usually a green powder, which is packed in bags and shipped to customers all over the world, ready to be used for food or feed purposes. In aquaculture the drying step is normally omitted because live feed of often very distinct microalgal taxa is necessary. While in invertebrate feeding or in aquaculture the biomass of different microalgae is the exclusive nutritional basis, in vertebrate feeding microalgae are a minor component in the diet, a so-called feed additive, or in human consumption a food supplement.

In all cases both, traditional experience in different parts of the world like Africa, Asia, and Mexico as well as numerous recent investigations confirm the health promoting effects of microalgal biomass in humans and animals. There seem to be no individual substances responsible for the beneficial effects. It appears that the more or less complex combination of bioactive compounds with synergistic effects and nearly optimal biological binding and balance stabilizes and protects animal cells and tissues.

The health promoting effects of microalgal biomass in the diet cannot be explained by nutritional aspects like protein supply or amino acid composition because the amounts supplied are too low. However, it was found in animals (fishes, birds, pigs, rabbits) and in *in vitro* experiments with human blood cells that the addition of less than one percent dry matter of microalgal biomass in the diet initiates the immune response of the organisms.

4.1.1 Human Diet

Even today, the consumption of microalgal biomass is restricted to very few taxa, e.g., *Chlorella*, *Scenedesmus*, *Dunaliella*, *Spirulina*, and with less importance *Nostoc*, *Aphanizomenon*, and *Porphyridium*. It can be expected that the exploitation of the biological diversity of microalgae will be hampered for a long time by food safety regulations for human consumption. Nevertheless the analysis of microalgal biomass of currently used taxa shows a very promising composition and biological value with special reference, e.g., to polyunsaturated fatty acids, vitamins, or minerals.

The bulk of the microalgal market is represented by *Chlorella* and *Spirulina* with approximately 2,000 t a^{-1} and 3,000 t a^{-1}, respectively. Both taxa are almost exclusively cultivated in open ponds, only a small amount of *Chlorella* is grown heterotrophically in tanks, mainly in Japan. Worldwide the market, both in developing and in industrial countries, is established predominantly as health food and not as protein source. During the past decades, the use of microalgal biomass was predominately in the health food market with more than 75% of the annual microalgal biomass production being used for the manufacture of powders, tablets, capsules, or pastilles. Countless combinations of microalgae or mixtures with other health foods can be found in the market. Of the numerous attempts to explain the health promoting effects of microalgal biomass, a general immune-modulating effect is most likely responsible (BELAY, 1993; OSINGA et al., 1999). Therefore, health foods are expected to be a stable market in the future. Currently most products launched to serve the health food market are supplied as powder and tablets. However, algal extracts in various product forms appear to create a second generation of microalgal products in this market:

- *Chlorella* health drinks (*Chlorella* growth factor),
- *Dunaliella* carotenoid enriched oily extracts (capsules) (BOROWITZKA, 1996; MASJUK, 1973),
- *Spirulina* liquid CO_2 extracts (antioxidant capsules).

Functional food or nutraceuticals produced with microalgal biomass are sensorially much more convenient and variable, thus, combining health benefits with attractiveness to consumers. The market of functional foods is believed to be the most dynamic sector in the food industry and could constitute 20–30% of the whole food market, within the next few years, growing rapidly. Food supplemented with microalgal biomass might have other positive influences, e.g., prebiotic effects or mineral fortification. We investigated prebiotic effects with *Spirulina* biomass both in pure and in functional food applications. The results show a positive effect on intestinal lactobacilli, which are regarded as beneficial, by still unknown components of *Spirulina*. *Spirulina* biomass as extracts or processed in pasta, biscuits, and other functional food products support the function of the digestive tract, e.g., healthy intestinal lactobacilli. All the *Spirulina* samples used stimulated the growth of all lactobacilli species tested. The addition of *Spirulina* biomass and of a derived aqueous extract led to at least a 10-fold increase in growth rate of the lactobacilli compared to control. The effect on *Lactobacillus acidophilus* is especially evident.

In Germany, food production and distribution companies have started serious activities to market functional food with microalgae and cyanobacteria. Examples are pasta, bread, yogurt, and soft drinks. Similar developments are observed, e.g., in Japan, USA, China, and Thailand.

4.1.2 Animal Feed

Survival, growth, development, productivity, and fertility of animals are a reflection of their health. Feed quality is the most important exogenous factor influencing animal health, especially in connection with intensive breeding conditions and the recent trend to avoid "chemicals", like antibiotics.

After decades of trials where animals were fed with high amounts of microalgae – up to 50% of the common feed – in order to exploit their protein content (RICHMOND, 1986), smaller doses were investigated especially in eastern Europe (MUSAFAROW and TAUBAJEW, 1974). There is evidence today, that very small amounts of microalgal biomass, almost exclusively of the genera *Chlorella*, *Scenedesmus*, and *Spirulina*, positively affect the physiology of animals. In particular, an unspecific immune response and a boosting of the immune system of the animals were observed including the following effects (BELAY, 1993):

- cholesterol reduction,
- protection against nephrotoxicity,
- anticancer effects,
- radiation protection,
- antiviral effects,
- immune-modulatory effects.

Even more important for animal production are other factors such as

- improved health status as indicated by viability, disease resistance, and improved ectoderm associated structures (skin, hair, feathers, nails);

- lower lethality in chicken;
- enhancement of feed utilization and growth for faster weight gain in piglets;
- higher egg production by increased laying performance in hens;
- better fertility with respect to reproduction efficiency (Piccardi et al., 1999).

Such economical effects led to a significant increase in the use of microalgal biomass as feed additives, especially in poultry production. Another very promising application for microalgal biomass or even extracts is the pet food market, where not only the health promoting effects, but also the external appearance of the pet (shiny hair, beautiful feathers) are of consumer importance. Studies on minks and rabbits support such effects for pets (Kretschmer et al., 1995).

4.1.3 Aquaculture

The global market of aquaculture products like fish and shellfish ranges between 40–50 billion US$ annually (New, 1999) with a strong growing trend (8% p.a.) especially in Asian-Pacific regions. Worldwide at least two trends emerge with respect to microalgal applications:

(1) The more and more sophisticated production of microalgal species meeting the feeding requirements of invertebrates or vertebrate larvae and
(2) the introduction of commonly produced microalgae into the fish feed to achieve similar positive effects as in animal feeding.

Usually the production of live microalgal biomass as starting feed for, e.g., larvae is performed locally. Very different, mostly technical inadequate equipment is used at a high cost level. Algal production is costly, often being the major cost item in aquacultural production. Cost estimates for microalgal production in the aquaculture area normally range between 50–150 US$, with peak values of 1,000 US$ per kg d.wt. (Spektorova et al., 1997).

As the basis of the food chain in nature, microalgae play a key role in aquaculture, especially mariculture, being the food source for larvae of many species of mollusks, crustaceans, and fishes. In addition, microalgae serve as food source for the zooplankton production (rotifers, copepods), which in turn is used as feed for rearing fish larvae (Lavens and Sorgeloos, 1996).

More than 40 species of microalgae are used in aquaculture worldwide depending on special requirements of local seafood production.

In the following some of the most important genera are listed:

Bacillariophyta:	*Skeletonema*, *Chaetoceros*, *Phaeodactylum*, *Nitzschia*, *Thalassiosira*
Prymnesiophyta:	*Isochrysis*, *Pavlova*
Prasinophyceae:	*Tetraselmis*
Chlorophyceae:	*Chlorella*, *Dunaliella*, *Scenedesmus*
Cyanobacteria:	*Spirulina* (*Arthrospira*)

Apart from feeding larvae and zooplankton, often with special microalgae, the addition of *Spirulina* and *Chlorella* to common fish feed compositions seems to be a promising market. Initially, the color enhancing effects of phycocyanin-containing *Spirulina* biomass or carotenoids from *Dunaliella* were exploited in ornamental fish. In recent years, questions of feed utilization and health status in the dense aquacultural fish populations became more important. Here, the addition of microalgae can, depending on concentrations, directly enhance the immune system of fish, as our own investigations on carp have shown.

Also, the addition of microalgal derived astaxanthin to feed formulations increases the coloring of the muscles of salmonids. This has a high biotechnological potential and culture techniques for *Haematococcus pluvialis* are quite well developed for this purpose (Piccardi et al., 1999; Kretschmer et al., 1995).

4.1.4 Biofertilizer

Historically macroalgae were used as soil fertilizer in coastal regions all over the world. The rational background for this interesting utilization of macroalgae or their extraction residues is the increase in water binding capacity and mineral composition of the soils (GUIRY and BLUNDEN, 1991). These properties are exploited today, using liquid fertilizers produced from macroalgae as initial coverage of, e.g., abandoned mining lands in order to avoid erosion and to initiate floral succession. This market segment amounts to approximately 5 million US$ p.a. The important role of microalgae in the soil ecosystem has often been neglected. The beneficial effects originate not only from the production of polymers for particle adherence and water storage in soils or nitrogen fixing but also from algae-derived bioactive compounds which influence higher plants (METTING, 1996; BOROWITZKA, 1995; ÖRDÖG et al., 1996). Soil microalgae should be regarded by microalgal biotechnologists as a promising area to find new species with unexpected properties. While N_2 fixation with microalgae (*Anabaena*, *Nostoc*) is important for rice production in tropical and subtropical agriculture, in more arid regions the surface solidification against erosion is also of interest. During the last decade, plant growth regulators both, from macro- and from microalgae gained increasing attention. Substances or extracts were found which promote germination, leaf or stem growth, or flowering.

A future trend seems to be the use of the biological activity of microalgal products against plant diseases caused by viruses or bacteria. It is likely that microalgae can be a source of a new class of biological plant protecting substances.

4.2 Valuable Substances from Microalgae

Today, microalgal biomass and extracts from biomass have their firm position in the market.

As shown above, in the health food market, with its main centers in the USA, Canada, Japan, and Europe, approximately 3,000 t *Spirulina* and 2,000 t *Chlorella* biomass are sold annually. There is an increasing demand for sophisticated products from microalgae, which are often closely related to the taxonomic position and physiology of microalgae. Especially the phylogenetically archaic cyanobacteria produce numerous substances which exhibit antioxidative effects, polyunsaturated fatty acids, heat induced proteins, or immunologically effective, virostatic compounds. Some of these substances are even excreted by the algae (BELAY, 1993; COHEN, 1999).

4.2.1 Polyunsaturated Fatty Acids

Only plants are able to synthesize polyunsaturated fatty acids (PUFA). Therefore, microalgae supply whole food chains with these vital components. Besides being a primary source of PUFA, these fatty acids from microalgae have further advantages over fish oils, such as lack of unpleasant odor, reduced risk of chemical contamination, and better purification potential. Therefore, microalgal PUFA have a very promising biotechnological market both, for food and feed, e.g., health-promoting purified PUFA are added to infant milk formulas. Hens are fed with special microalgae like heterotrophically grown *Schizochytrium* to produce "OMEGA eggs". Both applications turned out to be highly profitable.

The importance of microalgae as supplier of γ-linolenic acid was slightly weakened by the use of evening primrose oil. However, the preparation of eicosapentaenoic acid (EPA) and docosahexaenoic acid (DHA) from marine organisms with phototrophic capability such as the dinoflagellate *Crypthecodinium* for baby food or the health food market is an innovative approach (RADMER, 1996; APT and BEHRENS, 1999).

For lipid based cosmetics, like cremes or lotions, ethanolic or supercritical CO_2-extracts are gaining commercial importance, because they can provide both nourishing and protecting effects to the skin. For future developments in skin care other lipid classes from microalgae like glyco- and phospholipids should not be neglected (MULLER-FEUGA, 1997). The unicellular Rhodophyceae *Porphyridium cruentum*, having a well investigated autoecology, is regarded as another excellent future source of PUFA and related products (COHEN, 1999).

Other potential microalgae like *Phaeodactylum tricornutum*, *Isochrysis galbana*, *Monodus subterraneus*, or *Nannochloropsis* sp. are in part, well investigated, mainly as EPA producers (COHEN, 1999). The introduction of PUFA-rich components into human nutrition by functional food, aimed at prevention of cardiovascular diseases appears to have a future market potential.

4.2.2 Polysaccharides

Macroalgal polysaccharides like agar, alginates, or carrageenans are economically the most important products from algae. They are used in diverse fields of industry because of their rheological gelling or thickening properties. During the last years raw material shortages and pollution problems led to a fortification in R&D activities to use microalgae, transgene microalgae, protoplast fusion, or macroalgal cell cultures as biotechnological source (COHEN, 1999; GROBBELAAR and NEDDAL, 1996). Algal polysaccharides are also of pharmacological importance. The results of screening programs to test *in vitro* immunologically relevant effects of polysaccharides from microalgae have shown that certain highly sulfated polysaccharides can trigger either the cellular or the humoral stimulation of the human immune system (NAMIKOSHI, 1996). Effective polysaccharide fractions were found mainly in cyanobacteria; but also compounds from Rhodophyta and Chlorophyta have shown impressive efficiency. The results correlate with results from studies on animal feeding. Recent literature and the patent situation imply optimistic results (COHEN, 1999).

4.2.3 Antioxidants

During evolution over billions of years, microalgae as the phylogenetically oldest organisms have adapted uniquely to extreme habitats. Due to their phototrophic life, they are exposed to high oxygen and radical stresses. This has resulted in the development of numerous efficient protective systems against oxidative and radical stressors. The protective mechanisms are able to prevent the accumulation of free radicals and reactive oxygen species and thus to counteract cell damaging activities. In cultures of photosynthetically active microorganisms of high cell density, molecular oxygen is produced and an oxygen oversaturation is observed. In closed photobioreactors, even at less intensive photosynthetic conditions, O_2 concentrations can be as high as 50 mg L^{-1}. Such conditions will promote the endogenous detoxification process towards oxidative attack by an accumulation of highly effective antioxidative scavenger complexes, which are protecting the cell from damage by free radicals (superoxide anion or hydroxyl radical) (TUTOUR, 1990). For example, the antioxidative potential of *Spirulina platensis* can increase 2.3-fold during O_2 stress.

Of particular note is their high content in

- lipophilic scavengers, such as carotenoids, especially β-carotene and α-tocopherol;
- minerals and trace elements of antioxidative effect, such as zinc and selenium;
- enzymatic scavengers, such as catalase, superoxide dismutase, and peroxidase;
- polyphenols;
- vitamins, such as vitamin C and E.

The oxidative and radical protective potential of lipophilic and hydrophilic microalgal extracts can be characterized by different chemical test systems. The antioxidant radical scavenging activity of lipophilic extracts can be tested by the Rancimat test, electron spin resonance (ESR) spectroscopy, and photochemical luminescence (PCL) and compared with relevant reference antioxidants. Because the antioxidant components originate from a natural source, the application in cosmetics for preserving and protecting purposes is developing rapidly. In combination with other antioxidative or bioactive substances from microalgae, especially sun-protecting cosmetics represent an area of high demand. For functional food/nutraceuticals the radical scavenger capacity of microalgal products is of growing interest, especially in the beverage market segment and in pharmaceutical applications for therapy of oxidation associated diseases, like inflammations.

4.2.4 Colors and Coloring Food Products

Beside the chlorophylls of primary photosynthetic activity, microalgae contain a multitude of pigments which are associated with light incidence. The pigments improve the efficiency of light energy utilization (phycobiliproteins) of plants and protect them against solar radiation (carotenoids) and related effects.

Carotenoids from microalgae have a firm position in the market:

- β-carotene from *Dunaliella* in health food as vitamin A precursor,
- astaxanthin from *Haematococcus* in aquacultures for coloring muscles in fish,
- lutein, zeaxantin, canthaxantin for chicken skin coloration or for pharmaceutical purposes.

Unique to algae are the phycobiliproteins, such as phycocyanin and phycoerythrin. Some preparations are already in the stage of development for food and cosmetics. The development certainly will go beyond applications in diagnostics and photodynamic therapy and extend to cosmetics, nutrition, and pharmacy (ARAD and YARON, 1992).

4.2.5 Toxins and other Substances with Biological Activity

The most impressive demonstration of the ability of microalgae and cyanobacteria to produce highly effective bioactive compounds are toxins, which in algal blooms become dangerous to animals and humans, especially if such blooms occur in drinking water reservoirs. There are several freshwater algae, which can form toxic blooms especially the cyanobacteria *Microcystis*, *Anabaena*, and *Aphanizomenon*. Marine algal blooms become dangerous via the human consumption of shellfish. Three degrees of poisoning are distinguished here:

- PSP paralytic shellfish poisoning: caused by the water soluble neurotoxic substances saxitoxin, neosaxitoxin, and gonyautoxin in different derivatives – produced by the dinoflagellate *Alexandrium lusitanicum* – interrupting potential conduction in the neurons;
- DSP diarrhetic shellfish poisoning: induced by the polar polyoxo-substances okadaic acid and dynophysotoxin – produced mainly by *Dinophysis* species – resulting in strong diarrheic symptoms;
- ASP amnesic shellfish poisoning: caused by the amino acid domoic acid – produced mainly by the diatom *Nitzschia pungens* – which creates amnesic effects by acting as a glutamic acid antagonist.

The therapeutical value of all the toxins has not yet been investigated (LUCKAS, 1995; CARMICHAEL, 1992). Several, partly extensive, screening programs were performed in the USA, Australia, Germany, and France to find new substances with biological activity from microalgae and cyanobacteria. These studies and similar other investigations revealed the following effects:

- Cytotoxic activity which is important in anticancer drugs (GERWICK, 1994; SIRENKO et al., 1999).
- Antiviral activities were again found, mainly in cyanobacteria but also in apochlorotic diatoms and the conjugaphyte *Spirogyra*, where certain sulfolipids were active, e.g., against the herpes simplex virus (PATTERSON et al., 1994; MULLER-FEUGA, 1997).
- Antimicrobial activity was investigated to find new antibiotics. Although the success rate was about 1% (BOROWITZKA, 1995), there seem to be some promising substances from microalgae like the cyanobacterium *Scytonema*.
- Antifungal activity was found in different extracts of cyanobacteria.
- Antihelminthic effects are known from *Spirogyra* and *Oedogonium*.

4.2.6 Stable Isotopes in Microalgae

Because phototrophic microalgae can be cultivated under strictly controlled conditions, they are the ideal choice to incorporate stable isotopes from inorganic C-, H-, and N-sources. The various stable isotope labeled biochemicals cannot only be used for scientific purposes (molecular structure or physiological investigations) but also for clinical purposes like gastrointestinal or breath diagnosis tests (RADMER, 1996).

4.3 Ecological Applications

The protection and preservation of the natural basis of life is not only an ethical demand but also essential for sustainable economical and social developments. They initiate technological progress and employment. Persistent main objectives for algal biotechnology are the improvement of existing systems for wastewater treatment, the reduction of problematic emissions, the establishment of material circuits, and water recycling.

The consumption of inorganic nutrients by autotrophically growing microalgae may be used for the reduction of water or gas loads (Fig. 8). The key substances of water eutrophy, e.g., ammonia, nitrate, and phosphate as well as important industrial and agricultural waste gas (e.g., ammonia and carbon dioxide) are the main nutrients for algae. These algae can be exploited for many purposes, e.g., for feed, feed additives, or the extraction of valuable components if aspects of contamination are taken into account.

In this respect, a number of applications have been developed worldwide in the following fields:

- development of a process and pilot plant of disposal of inorganic loads, especially nitrate and phosphate from circulating process water of aquaculture by microalgae,
- heavy metal biosorption by viable microalgae, nonviable micro- and macroalgae and algal by-products,
- utilization of carbon dioxide from industrial exhaust gas,
- disposal of contaminants from agricultural wastewater,
- purification of wastewater of the biogas production,
- tertiary wastewater purification.

There are very innovative approaches regarding wastewater treatment and water recycling. Micro- and macroalgae, sometimes in combination with other microorganisms, are utilized to treat municipal, agricultural, food, industrial, organic as well as aqua- and mariculture wastewater/effluents. The level of practical application must be judged not only by the description of the relevant processes, but also by other important aspects, such as the exploitation of the resulting biomasses, technical solutions, and ecological feasibility.

4.3.1 Integrated Fish Production with Closed Material Circuit

The field of aquaculture is internationally highly developed to ensure a high quality pro-

Fig. 8. Pilot plant for the utilization of carbon dioxide waste from lime furnaces to produce algal biomass.

tein supply of the human population, although still scarce in central Europe. The FAO (Food and Agricultural Organization) predicted an increase in world fish demand. In fish farming only 30–50% of the nutrient supply can be converted into fish products. Therefore, an accumulation of nitrogen and phosphorus occurs. The term "integrated production" which efficiently links several areas of biological production with ecological engineering is a main aim for the future. Thus, the aspect of an environmentally inadequate production is reduced to a minimum by reintegrating by-products and environment affecting wastes as completely as possible. That means to create closed circuits and to utilize by-products in a value creating process. For example, such a system for agriculture could contain the following components: composting plant, algal cultivation plant, fish aquaculture plant, horticulture, compact heat-power station, and constructed wetlands. Such a combination of several aspects of ecological engineering leads to both, waste reduction and lower production costs. A number of beneficial effects due to the combined approach are likely to reduce the expenses for fish farming even further. In combination with a likewise integrated marketing strategy, fish from recirculation systems may be even competitive with imports of red perch or Victoria perch currently dominating the local mass market in Europe.

4.3.2 Heavy Metal Sorption and Accumulation

Non-viable micro- and macroalgae are able to adsorb heavy metal ions. Compared with other biosorbents, such as fungi, bacteria, yeasts, and live microalgae, they have the advantage of being easily available, cheap, and of having high heavy metal sorption capacities. Sorption tests have confirmed that significant heavy metal accumulations are feasible with low-cost sorbing agents. Under optimum conditions, non-viable macroalgae and by-products, which are cheap and easily available, show high sorption capacities and efficiencies of heavy metal cations, as well as favorable sorption kinetics.

The application of algal biosorbents is called for if conventional processes of heavy metal disposal at relatively low heavy metal concentrations are uneconomic. In addition, they may be added to conventional precipitation as a "safety filter" in order to respond to varying input concentrations. Mobile plants for the treatment of contaminated surface, ground, and wash waters are feasible.

4.3.3 Microalgae as a Photoautotrophic Component in Systems of Closed Material Cycle

To create closed equilibrated life support systems (CELSS) for space applications, different completely closed aquatic systems to study material cycle and population dynamics have been developed. Generally, microalgae are favored for conceptions of miniature ecosystems or bioregenerative life support systems (BLSS) because of their fast growth and consequently, their high physiological potential to fix CO_2 and to release O_2.

5 Genetic Engineering

Even more than as agricultural crops, microalgae are potentially important targets for genetic engineering. Although, their complexity on the cellular level is usually not less than that in higher plants, the lack of differentiation make microalgae a far simpler system for genetic manipulations. In addition, most vegetative stages of microalgae are haploid, abolishing the problems connected with allelic genes. New genes are introduced into the cells either by natural uptake (cyanobacteria), electroporation, biolistic methods, or via glass beads (KINDLE et al., 1990). However, the success of transformation is highly dependent on the correct gene cloning system, i.e., expression vector, promoter, resistance towards endogenous restriction enzymes (APT and BEHRENS, 1999). The aim of genetic engineering should not be the improvement of overall fitness of a strain but rather the production of valuable products and bioactive compounds. The adaptation of cells to new environments is usually too complex for genetic engineering and is more easily accomplished by natural selection and screening of new species.

6 Conclusion

Microalgal biotechnology – today still in its infancy – can be seen as a gateway to a multi-billion dollar industry. We have just started to tap the enormous biological resource of microalgal species growing in all ecological niches and their physiological potentials. One can expect that future trends in microalgal biotechnology will lead to a diversity of technical solutions of photobioreactors for cultivating microalgae. These will be adapted to the autecological demands of strains and to applications for biomass, valuable substances, and ecology. An exhaustive inventory of species in all regions accompanied with proper taxonomic handling and strain collection could be a basis for future success.

There are some indications to suppose that aquacultural and aquatic applications will be a profitable field for microalgae in the next 10 years. Both ecological applications in the sense of wastewater treatment and the agricultural use of microalgae for N_2 fixation or as soil conditioners have promising economic potentials.

Pharmaceutical applications seem to be of more future potential, nevertheless, some promising candidates exist, while microalgae in cosmetics, functional food, and animal feed could reach the level of mass products very soon.

7 References

APT, K. A., BEHRENS, P. W. (1999), Commercial developments in microalgal biotechnology, *J. Phycol.* **35**, 215–226.

APT, K. E., KROTH-PANCIC, P. G., GROSSMAN, A. R. (1996), Stable nuclear transformation of the diatom *Phaeodactylum tricornutum*, *Mol. Gen. Genet.* **252**, 572–579.

ARAD, S. A., YARON, A. (1992), Natural pigments from red algae, *Trends Food Sci. Technol.* **3**, 92–96.

BECKER, E. W. (1994), *Microalgae*. Cambridge: Cambridge University Press.

BELAY, A. (1993), Current knowledge on potential health benefits of *Spirulina*, *J. Appl. Phycol.* **5**, 235–240.

BENEMANN, J. R. (1990), *Microalgae Biotechnology: Products, Processes and Opportunities*. Washington, DC: OMEC Intern.

BENEMANN, J. R., TILLET, D. M., WEISMAN, J. C. (1987), Microalgae biotechnology, *Trends Biotechnol.* **5**, 47–52.

BERKELMAN, T., LAGARIAS, J. C. (1990), Calcium transport in the green alga *Mesotaenium caldariorum*, *Plant Physiol.* **93**, 748–757.

BERLINER, M. D., WENC, K. A. (1976), Protoplasts of *Cosmarium* as a potential protein source, *Appl. Environ. Microbiol.* **32**, 436–437.

BOROWITZKA, M. A. (1995), Microalgae as source of pharmaceuticals and other biologically active compounds, *J. Appl. Algol.* **7**, 3–15.

BOROWITZKA, M. A. (1996), Tubular photobioreactors, in: *Proc. 7th Int. Conf. Appl. Algol.*, p. 25. South Africa: Knysna.

BOROWITZKA, M. A., BOROWITZKA, L. J. (1988), *Microalgal Biotechnology*. Cambridge: Cambridge University Press.

BROUERS, M., DEJONG, H., SHI, D. J., HALL, D. O. (1989), Immobilized cells, in: *Algal and Cyanobacterial Biotechnology* (CRESSWELL, R. C., REES, T. A. V., SHAH, N., Eds.), pp. 272. New York: Longman Scientific & Technical.

BURLEW, J. S. (1953), *Algal Culture from Laboratory to Pilot Plant*. Washington, DC: Carnegie Inst.

CARMICHAEL, W. W. (1992), A review – Cyanobacteria secondary metabolites – the cyanotoxins, *J. Appl. Bacteriol.* **72**, 445–459.

CHRISMADHA, T., BOROWITZKA, M. A. (1994), Growth and lipid production of *Phaeodactylum*, in: *Algal Biotechnology in the Asia–Pacific Region* (PHANG, S. M., LEE, Y. K., BOROWITZKA, M. A., WHITTON, B. A., Eds.), pp. 122. Kuala Lumpur: Institute of Advanced Studies, University of Malaya.

CHUNGJATUPORNCHAI, W. (1990), Expression of the mosquitocidal-protein genes of *Bacillus thuringiensis* ssp. *israelensis* and the herbicide-resistance gene Bar in *Synechocystis* PCC6803, *Curr. Microbiol.* **21**, 283–288.

COHEN, Z. (1999), *Chemicals from Microalgae*. London: Taylor & Francis.

CRITCHLEY, A. T., OHNO, M. (1998), *Seaweed Resources of the World*. Yokosuka: JICA.

DILOW, C. (1985), *Mass Cultivation and Use of Microalgae*. Sofia: Bulgarian Academy of Sciences.

DUJARDIN, E., SIRONVAL, C., BOMBART, P., BROUERS, M. (1992), Controlled algae cultures at Liège, in: *Proc. 1st Eur. Workshop on Microalgal Biotechnology*, pp. 87–91. Potsdam-Rehbrücke, Germany.

DVORIN, S. A. (1992), A factory of commercial cultivation of microalgae, in: *Proc. 1st Eur. Workshop on Microalgal Biotechnology*, pp. 91–93. Potsdam-Rehbrücke, Germany.

FALLOWFIELD, H. J. (1991), Fluid mixing and photobioreactor apparatus, *U. K. Patent* 2.235.210A.

FOURNADZHIEVA, S., PILLARSKY, P. (1993), Mass culture and application of algae in Bulgaria, in: *Proc. 6th Int. Conf. Appl. Algol.*, pp. 20. Ceske Budjehovice, Czech Republic.

GERWICK, W. H. (1994), Screening cultured marine microalgae for anticancer-type activity, *J. Appl. Algol.* **6**, 143–149.

GROBBELAAR, J. U., NEDDAL, L. (1996), Growth of algae under turbulent conditions and intermitted light, in: *Proc. 7th Int. Conf. Appl. Algol.*, p. 40. Knysna, South Africa.

GROBBELAAR, J. U., KROON, B. M. A., WHITTON, B. A. (1996), Opportunities from micro- and macroalgae, *J. Appl. Phycol.* **8**, 261–464.

GROSS, W., SCHNARRENBERGER, C. (1995), Heterotrophic growth of two strains of the acidothermophilic red alga *Galdieria sulphuraria*, *Plant Cell Physiol.* **4**, 633–638.

GUDIN, C., CHAUMONT, D. (1983), Solar biotechnology study, in: *Workshop and EC Contractor's Meeting* (PALZ, W., PIRRWITZ, D., Eds), pp. 184. Capri: D. Reidel, Dordrecht.

GUIRY, M. D., BLUNDEN, G. (1991), *Seaweed Resources in Europe*. Chichester: John Wiley & Sons.

HALL, D. O., GARBISU, C., KANNAIYAN, S., LICHTL, R., MARKOV, S. et al. (1993), Photobioreactors with immobilized cyanobacteria for production of fuels and chemicals, in: *Proc. 6th Int. Conf. Appl. Algol.*, p. 5/1. Ceske Budjehovice, Czech Republic.

HELLEBUST, J. A. (1965), Excretion of some organic compounds by marine phytoplankton, *Limnol. Oceanogr.* **10**, 192–206.

HESSE, M. (1974), Growth and synchronization of the alga *Bumilleriopsis filiformis* (Xanthophyceae), *Planta* **120**, 135–146 (in German).

HIRATA, S., HAYASHITANI, M., TAYA, M., TONE, S. (1996), Carbon dioxide fixation in batch culture of *Chlorella* sp. using a photobioreactor with a sunlight collecting device, *J. Ferment. Bioeng.* **81**, 470–475.

HOLLIGAN, P. M., VIOLLIER, M., HARBOUR, D. S., CAMUS, P., CHAMPAGNE-PHILIPPE, M. (1983), Satellite and ship studies of coccolithophore production along a continental shelf edge, *Nature* **304**, 339–342.

KINDLE, K. L., RICHARDS, K. L., STERN, D. B. (1990), Engineering the chloroplast genome: Techniques and capabilities for chloroplast transformation in *Chlamydomonas reinhardtii*, *Proc. Natl. Acad. Sci. USA* **88**, 1721–1725.

KIRK, J. T. O. (1994), *Light and Photosynthesis in Aquatic Ecosystems*. Cambridge: Cambridge University Press.

KRETSCHMER, P., PULZ, O., GUDIN, C., SEMENENKO, V. (1995), Biotechnology of microalgae. *Proc. 2nd Eur. Workshop*, Potsdam-Rehbrücke: IGV Institute for Cereal Processing.

LAING, I., AYALA, F. (1990), Commercial mass culture techniques for producing microalgae, in: *Introduction to Applied Phycology* (AKATSUKA, I., Ed.), pp. 447–477. The Hague: SPB Acad. Publ.

LAVENS, P., SORGELOOS, P. (1996), Manual on the production and use of live food for aquaculture, pp. 7–42, FAO Fisheries Techn. *Paper* 361. Rome.

LEE, Y. K. (1986), Enclosed bioreactors for the mass cultivation of photosynthetic microorganisms: the future trend, *Trends Biotechnol.* **4**, 186–194.

LEE, Y. K., DING, S. Y., LOW, C. S., CHANG, Y. C., FORDAY, W. L., CHEW, P. C. (1995), Design and performance of an alpha-type tubular photobioreactor for mass culture of microalgae, *J. Appl. Phycol.* **7**, 47–52.

LEFÈVRE, M., JAKOB, H. (1949), Sur quelques propriétés des substances actives tirées des cultures d'algues d'eau douce, *C. R. Acad. Sci.* **229**, 234–236.

LUCKAS, B. (1995), Selective detection of algal toxins from shellfishes, *Chemie in unserer Zeit* **29**, 68–75 (in German).

MARGULIS, L., CORLISS, J. O., MELKONIAN, M., CHAPMAN, D. I. (1990), *Handbook of Protoctista*. Boston, MA: Jones and Bartlett Publ.

MASJUK, N. P. (1973), Morphology, taxonomy, ecology, geographical distribution and utilization of *Dunaliella*. Kiew: Naukowa Duma.

MATSUNAGA, T., TAKEYAMA, H., SUDO, H., OYAMA, N., ARIURA, S. et al. (1991), Glutamate production from carbon dioxide by marine *Synechococcus* sp., *Appl. Biochem. Biotechnol.* **28/29**, 157–167.

MCLACHLAN, J. (1964), Some considerations on the growth of marine algae in artificial medium, *Can. J. Microbiol.* **10**, 769–782.

METTING, F. B. (1996), Biodiversity and application of microalgae, *J. Ind. Microbiol.* **17**, 477–489.

MORI, K. (1986), Photoautotrophic solar bioreactor, *Biotech. Bioeng. Symp.* **15**, 331–345.

MULLER-FEUGA, A. (1997), *Marine Microalgae*. Plouzane: Editions Ifremer.

MUSAFAROW, A. M., TAUBAJEW, T. T. (1974), *Chlorella*. Tachkent: FAN.

NAMIKOSHI, M. (1996), Bioactive compounds produced by cyanobacteria, *J. Int. Microbiol. Biotechnol.* **17**, 373–384.

NEW, M. B. (1999), Global aquaculture: current trends and challenges for the 21st century, *World*

Aquaculture **3**, 8–14.

NORTON, T. A., MELKONIAN, M., ANDERSEN, R. A. (1996), Algal biodiversity, *Phycologia* **35**, 308–326.

OH-HAMA, T., MIYACHI, S. (1988), *Chlorella*, in: *Microalgal Biotechnology* (BOROWITZKA, M. A., BOROWITZKA, L. J., Eds), pp. 3–26. Cambridge: Cambridge University Press.

ÖRDÖG, V., SZIGETI, J., PULZ, O. (1996), *Proc. Conf. Progress in Plant Sciences from Plant Breeding to Growth Regulation*. Pannon University Mosonmagyarovar, Hungary.

OSINGA, R., TRAMPER, J., BURGESS, J. G., WIJFELLS, R. H. (1999), Marine bioprocess engineering, Proceedings, *Progress in Industrial Microbiology* **35**, 1–413. Amsterdam: Elsevier.

PATTERSON, G. M. L., LARSEN, L. K., MOORE, R. E. (1994), Bioactive natural products from blue-green algae, *J. Appl. Algol.* **6**, 151–157.

PICCARDI, R., MATERASSI, R., TREDICI, M. (1999), Algae and human affairs in the 21st century. *Abstracts* Int. Conf. Appl. Algol., Firenze: Universita degli Studi di Firenze.

PULZ, O. (1992), Cultivation techniques for microalgae in open and closed ponds, in: *Proc. 1st Eur. Workshop on Microalgal Biotechnology*, p. 61–67. Potsdam-Rehbrücke, Germany.

PULZ, O., SCHEIBENBOGEN, K. (1998), Photobioreactors: design and performance with respect to light energy input, in: *Advances in Biochemical Engineering Biotechnology* (SCHEPER, T., Ed.), pp. 123–152. Berlin, Heidelberg: Springer-Verlag.

PULZ, O., GERBSCH, N., BUCHHOLZ, R. (1995), Light energy supply in plate type and light diffusing optical fiber bioreactors, *J. Appl. Phycol.* **7**, 145–149.

RADMER, R. J. (1996), Algal diversity and commercial algal products, *Bioscience* **46**, 263–270.

RAGAN, M. A., CHAPMAN, D. J. (1978), *A Biochemical Phylogeny of the Protists*. New York: Academic Press.

RAVEN, J. (1988), Limits to growth, in: *Microalgal Biotechnology* (BOROWITZKA, M. A., BOROWITZKA, L. J., Eds.), pp. 331–335. Cambridge: Cambridge University Press.

RICHMOND, A. (1986), *Handbook of Microalgal Mass Culture*. Boca Raton, FL: CRC Press.

RICHMOND, A. (1990), Large scale microalgal culture and applications, in: *Progress in Physiological Research* Vol. 7 (ROUND, F. E., CHAPMAN, D. J., Eds.). Bristol: Biopress.

RICHMOND, A. (1996), Photobioreactor design, in: *Proc. 7th Int. Conf. Appl. Algol.*, p. 60. Knysna, South Africa.

ROBINSON, P. K., MAK, A. L., TREVAN, M. D. (1986), Immobilized algae: a review, *Process Biochem.* **21**, 122–134.

SEITGELDIYEV, N. (1992), Industrial photoreactors for producing *Chlorella* with solar energy, in: *Proc. 1st Eur. Workshop on Microalgal Biotechnology*, pp. 105–107. Potsdam-Rehbrücke, Germany.

SEITGELDIYEV, N., AKYEV, A. Y. (1995), A strategy for gas supply at large scale production, in: *Proc. 2nd Eur. Workshop on Microalgal Biotechnology*, pp. 25–27. Potsdam-Rehbrücke, Germany.

SETLIK, I., SUST, V., MALEK, I. (1970), Dual purpose open circulation units for large scale culture of algae, *Algol. Stud.* (Trebon) **1**, 111–124.

SHELEF, G., SOEDER, C. J. (1980), *Algae Biomass Production and Use.* Amsterdam: Elsevier.

SIEGELMAN, H. W., GUILLARD, R. R. L. (1971), Large-scale culture of algae, in: *Methods in Enzymology* Vol. 23 (COLOWICK, S. P., KAPLAN, N. O., Eds.), pp. 110–115. New York: Academic Press.

SIRENKO, L. A., KIRPENKO, Y. A., KIRPENKO, N. I. (1999), Influence of metabolites of certain algae on human and animal cell cultures, *Int. J. Algae* **1**, 122–126.

SOEDER, C. J. (1986), An historical outline of applied algology, in: *Handbook of Microalgal Mass Culture* (RICHMOND, A., Ed.), pp. 25. Boca Raton, FL: CRC Press.

SPEKTOROVA, L., CRESWELL, R. L., VAUGHAN, D. (1997), Closed tubular cultivators, *World Aquaculture* **6**, 39–43.

TREDICI, M. R., MATERASSI, R. (1992), From open ponds to alveolar panels: the Italian experience, *J. Appl. Phycol.* **4**, 221–231.

TREDICI, M., BIAGIOLINI, S., CHINI ZITELLI, G., MONTAINI, E., FAVILLI, F. et al. (1994), Fully-controllable photobioreactors, in: *Proc. 6th Eur. Congress on Biotechnology*, p. 1011. Florence, Italy.

TREDICI, M. R., CHINI ZITELLI, G. (1995), Scale-up of photoreactors to commercial size, in: *Proc. 2nd Eur. Workshop on Microalgal Biotechnology*, pp. 21–25. Potsdam-Rehbrücke, Germany.

TSUZUKI, M., MIYACHI, S. (1991), CO_2 syndrome in *Chlorella*, *Can. J. Bot.* **69**, 1003–1007.

TUTOUR, L. B. (1990), Antioxidative activities of algal extracts, *Phytochemistry* **29**, 3759–3763.

UPITIS, W. W. (1983), *Macro- and Microelements in Optimization of Microalgal Nutrition.* Riga: Sinatne (in Russian).

VENKATARAMAN, L. V., BECKER, E. W. (1985), *Biotechnology and Utilization of Algae.* Mangalore, India: Sharada Press.

ZITELLI, G., TOMASELLI, V., TREDICI, M. R. (1996), One-year cultivation of Athrospira, in: *Proc. 7th Int. Conf. Appl. Algol.*, p. 84. Knysna, South Africa.

6 Biotechnology with Protozoa

ARNO TIEDTKE
Münster, Germany

1 Introduction

A vast number of biotechnologically produced substances are derived from microorganisms of bacterial and fungal origin. They include enzymes, vitamins, biopolymers, antibiotics, and other compounds of commercial relevance. Highly productive strains of these microorganisms, that have been improved genetically over decades, are available. Bakers' yeast *Saccharomyces cerevisiae* and *Escherichia coli* are probably the best investigated microorganisms used in biotechnology and their genomes were among the first sequenced ones. Both are GRAS (generally recognized as safe) species and are used as the work horses in the production of heterologous proteins. However, in screenings for new compounds these well known species may not offer too many news. New organisms in biotechnology offer the possibility to detect novel products.

The role of protozoa in biotechnology has largely been ignored. We know that protozoa play important roles in treatment of communal and industrial wastewaters and some as parasites of man. Malaria tropica, caused by the apicomplexan *Plasmodium falciparum*, is killing more than two million people per year. *Trypanosoma* species are the causative agents of sleeping disease and Chagas disease. They form a minority when compared to the estimated number of about 65,000 protist species. The protists, comprising the protozoa and the protophyta (microalgae) are a highly diverse group of single-celled eukaryotes that have conquered a great variety of habitats from sea and lakes to soil and deserts and have adapted a corresponding variety of feeding and surviving strategies. Why then, have the protists been so poorly exploited in biotechnology? One answer is, that the heterotrophic protists, due to their specialized feeding strategies are difficult to culture. However, as will be shown later this difficulties can be overcome. But even those species that can be cultivated easily, like the ciliates of the genus *Tetrahymena*, amoebae, euglenoid flagellates, and dinoflagellates have only recently become used in biotechnology. Working with one of the best studied protists, *Tetrahymena thermophila*, T. CECH detected ribozymes and was awarded the Nobel Prize in Medicine and Physiology in 1989 for his studies on catalytic RNA (CECH, 1990). Another milestone is the detection of *Tetrahymena* telomerase by E. BLACKBURN (1992). This enzyme creates the ends of chromosomes, or telomeres of the developing new macronucleus after conjugation (the sexual form of reproduction in ciliated protists). *Tetrahymena* needs many telomeres, and therefore, has an abundance of the enzyme, that is normally rare in other cells. Protist biotechnology is at its beginning; it will flourish if more research is invested in this new field (KIY, 1998).

2 The Protists – an Overview

Protists are unicellular organisms whose ancestors were the first eukaryotes on earth. They arose from symbiotic association of various prokaryotes about two billion years ago. Studies on comparative sequencing of the small subunit ribosomal RNA show the tremendous genetic diversity of recent protists. Since they comprise species with photoautotrophic and heterotrophic modes of life they have been classified into microalgae (protophyta) and protozoa, an old-fashioned division. Protists seem to fall into three major groups as illustrated in Fig. 1 (BARDELE, 1997). The *early group* lacks mitochondria and is represented by the Microspora, *Giardia*, certain amoeboflagellates, trichomonads, and polymastigotes. It is a matter of debate whether the members of this group have lost their mitochondria or never possessed them. The detection of mitochondrial chaperones favors the first possibility. The *middle group* contains the Euglenozoa, which comprise the Kinetoplastida as well as heterotrophic and green euglenoids. This group furthermore contains the vast majority of the paraphyletic rhizopods and the slime molds (*Physarum* and *Dictyostelium*). At the top of the tree there is a sudden explosive radiation of the *crown group* with a lot of uncertainty with respect to the precise branching order. There is the mono-

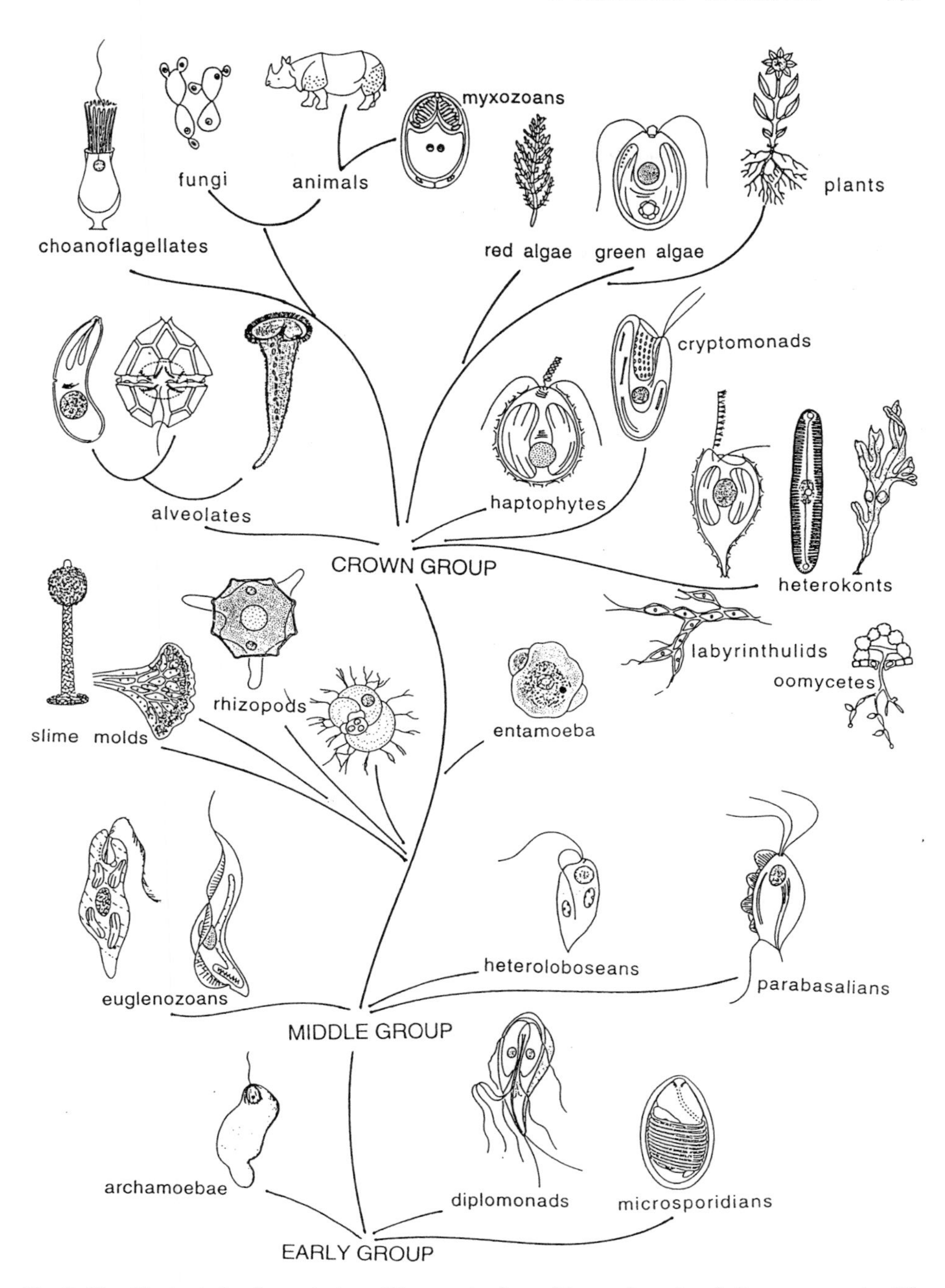

Fig. 1. The illustrated eukaryote tree. The precise branching pattern in all three groups is still largely unsettled (BARDELE, 1997).

phyletic group of the Alveolata comprising the Apicomplexa (the former Sporozoa containing the malaria parasite *Plasmodium*), the dinoflagellates (Dinozoa), and the ciliates (Ciliophora) containing the fresh water ciliates *Paramecium* and *Tetrahymena*. Another member of the crown group is the "Chromista", which comprise the cryptomonads, the haptophytes, and the heterokont algae with the chrysophytes, diatoms, and brown algae. All members of the "Chromista" possess tripartite tubular mastigonemes on their flagella. The net slime molds (*Labyrinthula*) and the plant pathogenic oomycetes (*Phytophthora*) are related to the heterokont algae. In conclusion: there is no monophyletic taxon "Protista" and protists thus must be defined as eukaryotes of a certain organizational level, namely that of a unicellular eukaryote.

3 Products Obtained from Protists

3.1 Bioassays

Three bioassays, all using the ciliate *Tetrahymena* (Fig. 2) as a test organism have been reported. PORRINI and TESTOLIN (1984) developed an assay to test the nutritional value of meat products with *Tetrahymena*, as this ciliate is known to have very similar demands on essential amino acids as man. Results with *Tetrahymena* were obtained quicker than with assays using rats or mice as test organisms. HUBER et al. (1990) presented a *Tetrahymena* growth inhibition assay to test for toxicity of environmental chemicals. The assay proved to give better results than the "activated sludge respiration test" recommended by the OECD (YOSHIOKA et al., 1986). PAULI and BERGER (1996) developed a chemosensory response assay to monitor the toxicity of chemical pollution of aquatic environments. The test is based on the avoidance reaction *Tetrahymena* exhibits when the cell encounters an unfavorable environment. In comparison to the standard aquatic toxicity tests using fish, *Daphnia*, and algae, the behavioral test is highly sensitive, more rapid, and less expensive (Fig. 3). Recently, a growth inhibition microtest using *Tetrahymena thermophila* as a test organism has been developed by PAULI and BERGER (2000). The assay is commercially distributed (PROTOTOXKIT F™ Ltd., Deinze, Belgium).

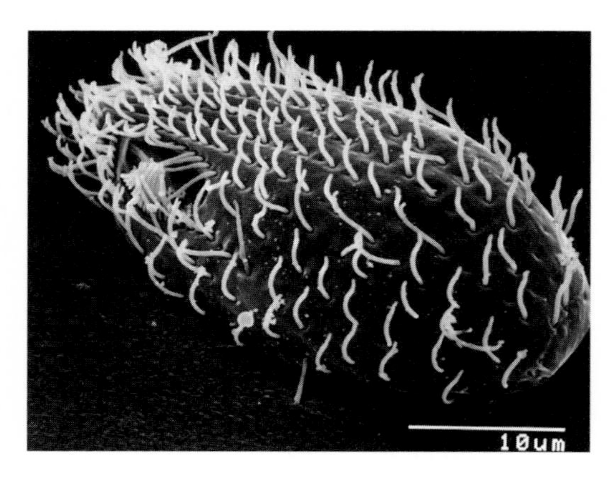

Fig. 2. Scanning electron microscopic image of the ciliated protist *Tetrahymena thermophila*. The cavity in the upper left region is the oral apparatus, the food gathering organelle of this about 50 μm long ciliated protist (KIY and TIEDTKE, 1993).

3.2 Small Bioactive Compounds

Heterotrophic protists are believed to have no secondary metabolism like plants. Therefore, the chance to detect new bioactive compounds may be less promising compared to photoautotrophic organisms. However, many heterotrophic protists have developed special defensive organelles against their predators. Many of these dense core vesicles are docked at the cell surface where they become discharged upon encounter with a predator. Blepharismin is a compound of the pigment granules of the ciliate *Blepharisma japonicum* and has been shown to inhibit growth of gram-positive bacteria, e.g., of methicillin-resistant strains of *Staphylococcus aureus* (PANT et al., 1997). The chemical structure of blepharismin is related to stentorin, a compound obtained from colored vesicles of the ciliated protist *Stentor coeruleus*. Stentorin and blepharismin (Figs. 4 and 5) may become used as phototherapeutics due to their chemical relation to

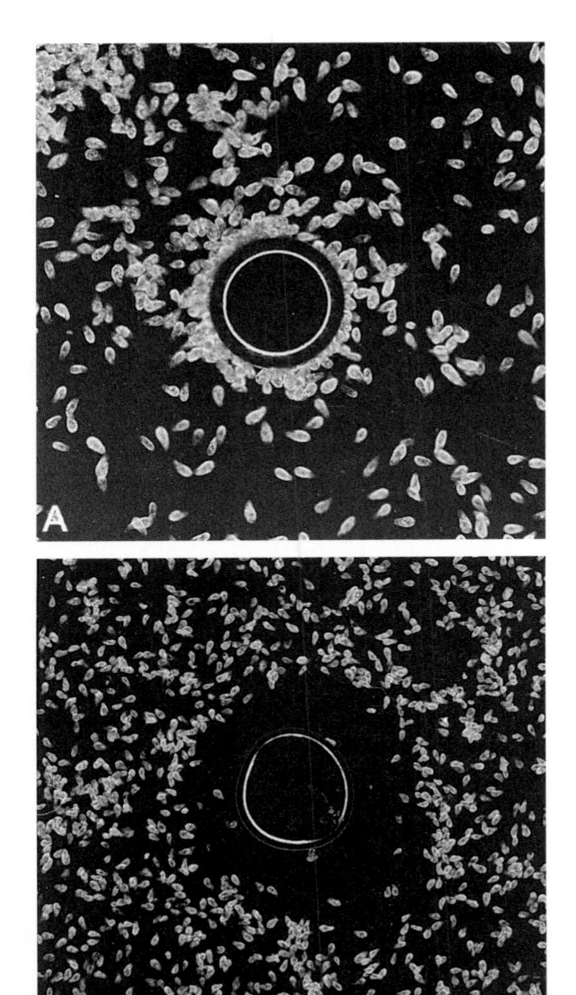

Fig. 3. Avoidance and escape behavior of the ciliated protist *Tetrahymena* in the absence (**A**) and presence of 10 μmol L^{-1} 2,4-dinitroanilins) **B** (courtesy of W. PAULI and S. BERGER, Freie Universität Berlin, Germany).

hypericin, a compound of highly specific anti-HIV activity (TAO et al., 1993; MAEDA et al., 1997). A cytotoxic compound – related to ergosterol – has been extracted from the colorless euglenoid flagellate *Astasia longa* (KAYA et al., 1995). The so-called astasin is novel as it contains a carbohydrate moiety. It inhibits growth of human lymphoma cells and kills 50% of the cells at a concentration of 10 μg mL^{-1}. So far there are only a few bioactive compounds of protist origin known and these may be not of immediate therapeutic value. However, they may serve as leading structures for the design of new drugs by combinatorial chemistry.

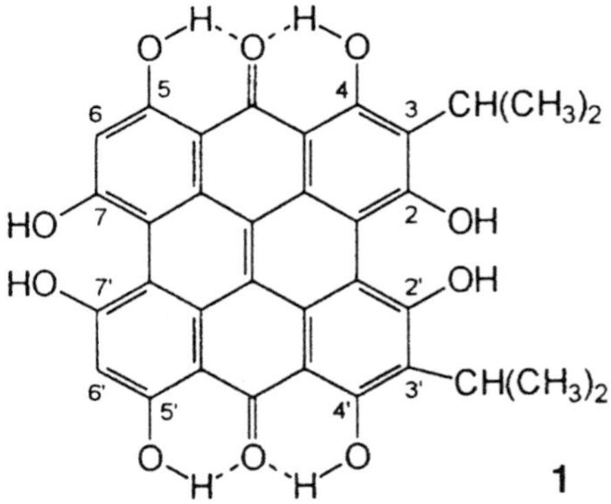

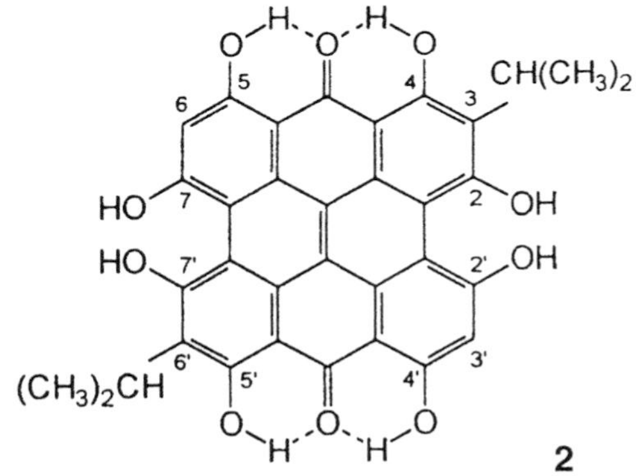

Fig. 4. Structures of stentorin extracted from the pigment granules of the ciliated protist *Stentor coeruleus* (TAO et al., 1994).

3.3 Long-Chain Polyunsaturated Fatty Acids

Polyunsaturated fatty acids (PUFAs) produced by photoautotrophic protists (micro-

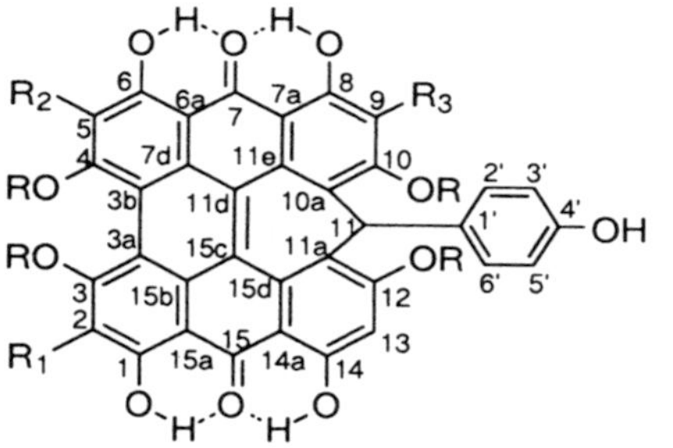

Fig. 5. Structures of blepharismin (BL) extracted from the pigment granules of the ciliate *Blepharisma japonicum* (MAEDA et al., 1997).
BL-1: $R_1=R_2=Et$, $R_3=H$; BL-2: $R_1=Et$, $R_2=i\text{-}Pr$, $R_3=H$; BL-3: $R_1=R_2=i\text{-}Pr$, $R_3=H$; BL-4: $R_1=Et$, $R_2=i\text{-}Pr$, $R_3=Me$ or $R_1=i\text{-}Pr$, $R_2=Et$, $R_3=Me$; BL-5: $R_1=R_2=i\text{-}Pr$, $R_3=Me$.

algae) are a primary source of these compounds in the entire food web up to man. Polyunsaturated *omega*-6-fatty acids, e.g., γ-linolenic acid (GLA, octadecatrienoic acid) and arachidonic acid (ARA, eicosatetraenoic acid) are essential precursors of the eicosanoids such as prostaglandins, leukotrienes, thromboxanes, and prostacyclins. A good GLA source is the oil extracted from seeds of primrose and blackcurrant, which contains about 15% of this fatty acid. However, an increasing demand for *omega*-3-PUFAs, especially DHA (docosahexaenoic acid) and EPA (eicosapentaenoic acid) as supplement of baby food and other foods, made marine protists attractive as producers of PUFAs. The oil extracted from the dinoflagellate *Cryptothecodinium cohnii* contains up to 40% DHA, but no EPA. Since the biomass of this protist contains up to 30% oil, DHA from this protist has become an important alternative to fish oil as the so far largest source of the long-chain polyunsaturated fatty acid DHA (BARCLAY et al., 1994). The bad smell of fish oil (absent in protist oil) and the possibility to produce the protist oil under controlled conditions and quality are in support of the single-cell oil. A good source for ARA and EPA is the green flagellated protist *Euglena gracilis*. When grown under low light conditions it increasingly accumulates these PUFAs (WARD, 1995). Another protist, the ciliate *Tetrahymena thermophila* contains up to 30% GLA in its membrane lipids and phospholipids. Since it does not store GLA-oil in reasonable amounts its commercial use as a GLA source is presently not lucrative.

3.4 Enzymes

Enzymes are universal tools in analytical and synthetic biochemistry, molecular biology, and numerous other fields of research and industrial production. The exploration of thermoresistant enzymes from thermoacidophilic archaea shows that the detection of new organisms offers the chance to detect new enzymes. The protists represent a highly diverse collection of single-celled organisms, that live and survive in a great variety of environments. Only a few have been cultured in amounts necessary to screen for new enzymes, and therefore it is the chosen few, that are used for enzyme production. *Tetrahymena* is probably the fastest growing protist and has been shown to yield high amounts of biomass and contains a number of enzymes worth to become explored. An enzyme that proved to degrade effectively the nerve gas agent SOMAN (O-1,2,2-trimethylpropylmethylphosphonofluoride), namely diisopropylfluorophosphatase (DFPase) has been obtained from *Tetrahymena thermophila* (LANDIS et al., 1986). The DFPase activity remained stable over a broad pH range of pH 4–10 and temperatures of 25–55 °C. Another enzyme, L-asparaginase, described by KYRIADIS et al. (1990) was found to inhibit proliferation of breast cancer cells. *Tetrahymena* is also known to secrete a great number of different acid hydrolases into the culture medium (MÜLLER, 1972). KIY and TIEDTKE (1991) and KIY et al. (1996) developed techniques for mass cultivation and easy harvesting of spent culture medium. They encaged *Tetrahymena thermophila* in Ca-alginate hollow spheres and later used a perfused bioreactor to harvest the spent culture medium (Fig. 6). Using continuous fermentation conditions 34,760 U of acid phosphatase and 9,944 U of β-hexosaminidase, e.g., were obtained per day from 2 L spent culture medium (KIY et al., 1996) (Fig. 7). The different glycosidases are of particular interest, because they are valuable

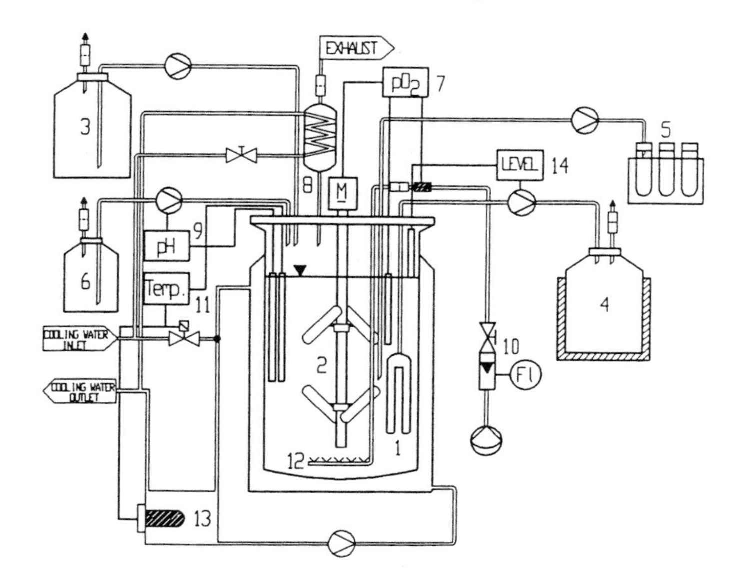

Fig. 6. Schematic diagram of the perfusion bioreactor used for high cell density cultures of *Tetrahymena*: 1 perfusion module, 2 paddle impeller, 3 fresh medium, 4 harvest, 5 sample port, 6 acid tank, 7 pO_2 controller, 8 reflux cooler, 9 pH controller, 10 mass flow controller, 11 temperature controller, 12 air sparger, 13 thermostat, 14 level controller. The culture medium was composed of 2% skimmed milk powder, 0.5 % yeast extract, 0.003% Sequestrene and 0.01% silicon oil to prevent foaming (KIY and TIEDTKE, 1992b).

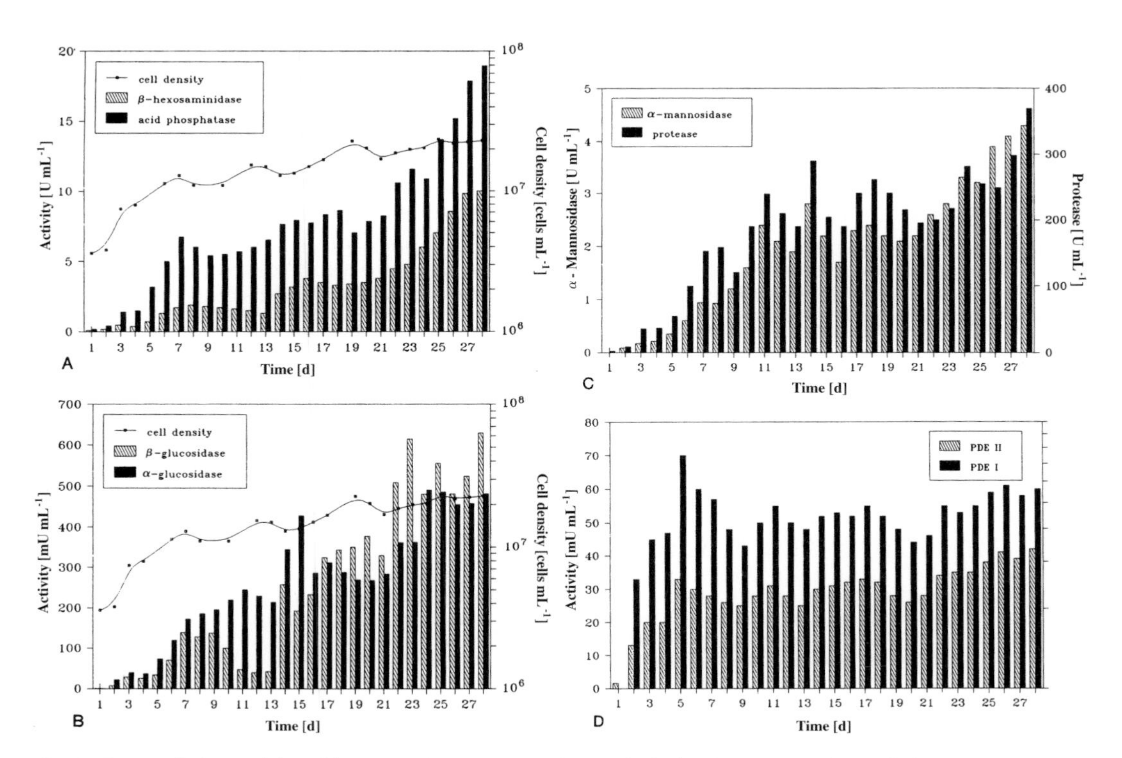

Fig. 7. Extracellular activity of lysosomal enzymes and growth during long-term high cell density cultivation of *Tetrahymena thermophila*. At d_2 continuous perfusion of the bioreactor was started at a rate of 0.69 d^{-1}. **A** acid phosphatase and β-hexosaminidase; **B** α- and β-glucosidase; **C** α-mannosidase and protease; **D** phosphodiesterase I and II (KIY et al., 1996).

tools for analysis and stereospecific synthesis of sugar moieties of glycans, glycoproteins, and glycolipids (KIY et al., 1995). Another acid hydrolase, phospholipase A_1 has been recently purified from spent culture medium of *T. thermophila* (GUBERMAN et al., 1999). PLA_1 splits specifically the ester bond of the fatty acid at *sn*-1 position of phospholipids. This enzyme is valuable for two reasons: first it is not available on the commercial market and secondly its availability completes the enzymatic tool kit for analysis and synthesis of phospholipids, since all other phospholipases are on the market. To improve the yield of PLA_1 a screening protocol for detection of mutants releasing higher activities of this phospholipase into the spent culture medium has been set up. Mutant strains releasing about twice as much PLA_1 activity have been obtained (HARTMANN et al., 2000). As *Tetrahymena* a related ciliate, *Paramecium tetraurelia* has been shown to release acid hydrolases, e.g., large amounts of a cysteine protease into the culture medium. Structure and function analysis of the purified protease revealed its relation to the cathepsin L family of cysteine proteases (VÖLKEL et al., 1996). A related cysteine protease has also been obtained from *Tetrahymena* (KARRER et al., 1993; SUZUKI et al., 1998). The propeptide of cathepsin L selectively inhibits the proteolytic activity of the mature enzyme.

This can be used to develop improved inhibitors of the cathepsin L family proteases that are believed to be involved in progression of arthritis and in tumor growth and metastasis.

4 Cultivation of Protists at Larger Scale

Many heterotrophic protists have developed special modes of feeding. They often require complex media or living food organisms. In the latter case the food organisms, e.g., nonpathogenic bacteria or photoautotrophic flagellates like *Dunaliella* or *Chlorella* species, are cultivated separately. Mass cultivation of these protists, if possible, is difficult and tedious. However, it has become possible to obtain mass cultures of even these protists due to improvements in cultivation techniques. KIY and MARQUARDT (1995) used spinner flasks, originally developed for cultivation of mammalian tissue cells, to grow the ciliates *Engelmanniella mobilis* and *Blepharisma americanum* in mass cultures.

Techniques for mass cultivation have been developed. SCHÖNEFELD et al. (1986) grew the ciliate *Paramecium tetraurelia* in a 250 L airlift fermenter on a 200 L scale. They obtained sufficient biomass for biochemical characterization of adenylate cyclase from cilia using a cheap axenic cultivation medium based on skim milk powder. Cheap culture media and techniques for continuous fermentation have been developed for *Tetrahymena*. This was done to explore the biotechnological potential of the ciliate *Tetrahymena thermophila*, which secretes high activities of several lysosomal hydrolases into the culture medium. Unprecedented short generation times of 1.4 h and cell densities of $2.2 \cdot 10^7$ cells mL^{-1} corresponding to dry weights of 54g L^{-1} were obtained as illustrated in Figs. 8 and 9 (KIY and TIEDTKE, 1992a, b). Progress in bioreactor technology and culture medium formulation allowed mass cultivation of other protist species. The amoebae *Acanthamoeba castellanii* and *Hartmannella vermiformis*, e.g., were successfully cultivated in a chemostat. At steady state conditions, when the growth rate equaled the dilution rate, generation times of 25 and 16 h and maximum cell densities of 2.5 and $3.6 \cdot 10^6$ cells mL^{-1}, respectively, were obtained (WEEKERS and VOGELS, 1994). The so far highest cell densities of 10^8 cells mL^{-1} were obtained when *Leishmania*, parasitic kinetoplastid protists causing leishmaniasis, e.g., Kala Azar (*L. donovani*) or Oriental Sore (*L. tropica*) were cultured on axenic media (ZHANG et al., 1995).

The dinoflagellate *Crypthecodinium cohnii* is cultivated at the so far largest scale seen among the heterotrophic protists. Martek Biosciences Corp. (Columbia, MD, USA) is growing this protist on a 150,000 L scale to produce a single-cell oil containing high amounts of the long chain polyunsaturated fatty acid docosahexaenoic acid (DHA). At the end of the batch fermentation a biomass of 40 g L^{-1} is obtained (CHEN, 1996).

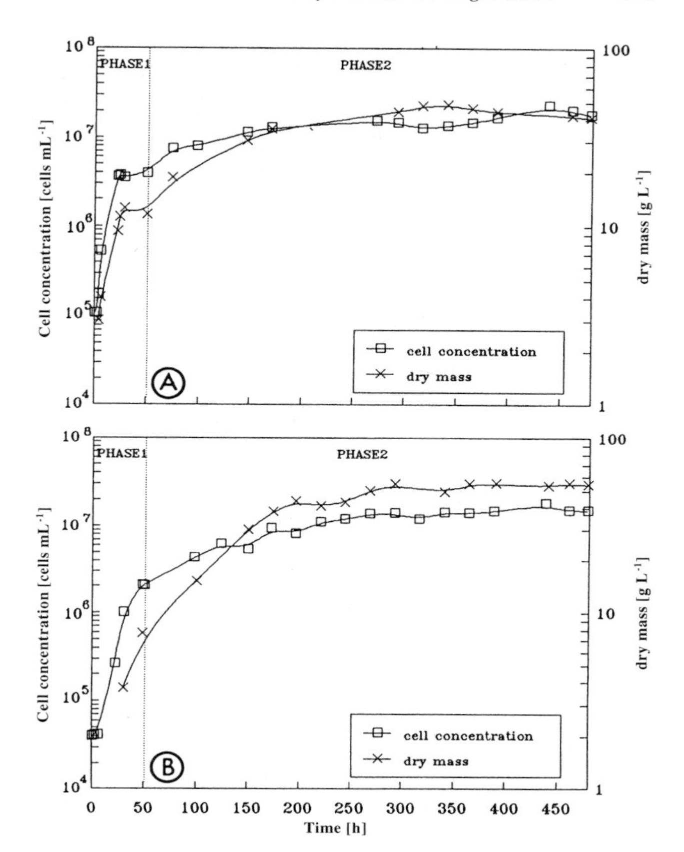

Fig. 8. Growth of *Tetrahymena* during long-term cultivation in a perfused system. In both fermentations continuous perfusion (perfusion rate per day, 1.0) was started at t_{50}. **A** Growth kinetics of *Tetrahymena thermophila* SB 281. **B** Growth kinetics of *Tetrahymena pyriformis* GL (KIY and TIEDTKE, 1992b).

Some of the photoautotrophic protists – often also referred to as microalgae – may be mentioned as well. The costs of inorganic cultivation media for photoautotrophic protists are considerably lower than for heterotrophic ones, but the autotrophs grow slower and reach lower cell concentrations. Many species are, therefore, cultivated in shallow open ponds instead of bioreactors.

Dunaliella is grown in salt water ponds. Is used for the production of β-carotene, a carotenoid used as natural food pigment and supplement of vitamins and health food. *Haematococcus pluvialis* is cultivated to produce another carotenoid, astaxanthin, giving shrimps and salmon their typical red-orange color. In salmon and shrimp farming, where the natural food is limited astaxanthin supplemented food is fed to the farmed animals to obtain the desired appetizing color. Cyanotech Corp. (Kailua-Kona, Hawaii, USA) produces astaxanthin (brand name: NatuRose) from *Haematococcus*, grown at large scale in open ponds. *Porphyridium cruentum*, also referred as a single-celled red algae, is a good source for production of PUFAs, especially arachidonic acid (ARA). This and other polyunsaturated fatty acids have a growing market as supplements of baby nutrition products. Babies especially need these essential fatty acids, when they are not supplied with mother's milk.

The green flagellate *Euglena gracilis* regularly lives as a photoautotroph. In the absence of light, however, it can switch to a heterotrophic way of life. Cultivated in rich organic nutrient media the cells produce and deposit the glucose polymer, β-1,3-glucan (paramylum), in higher amounts than under autotrophic condi-

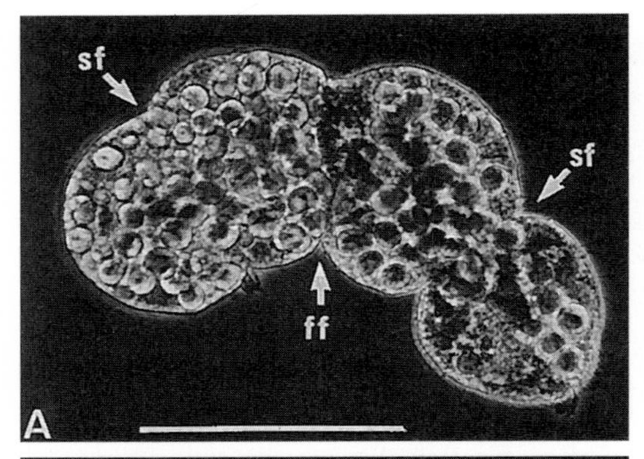

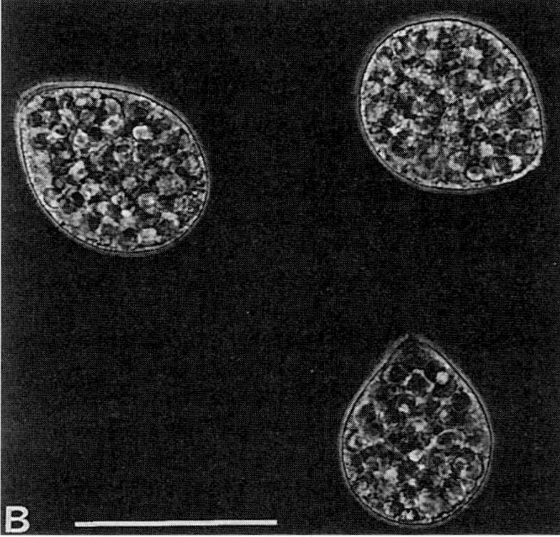

Fig. 9. Phase-contrast micrographs of *Tetrahymena thermophila* cells cultured in skimmed-milk medium. **A** Cell chain consisting of four cells: ff, first fission line; sf, second fission line; bar, 50 µm. **B** Early log phase cells containing many food vacuoles: bar, 50 µm (KIY and TIEDTKE, 1992a).

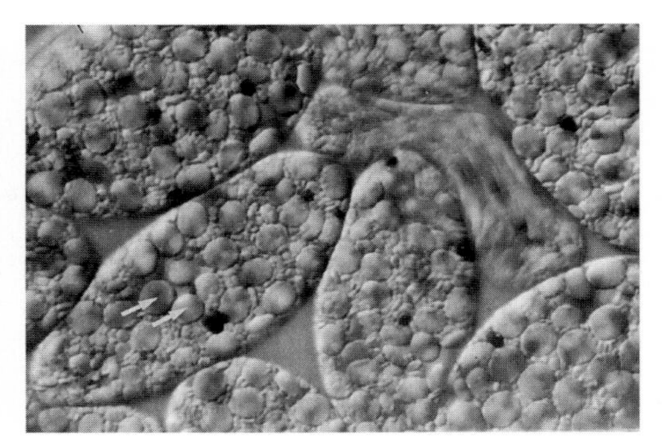

Fig. 10. Interference phase contrast micrograph of *Euglena gracilis* cultured in the dark. Many paramylum granules (arrows) are seen. The average length of a cell is 50 µm (courtesy of Drs. H. G. RUPPEL and D. BÄUMER, University of Bielefeld, Germany).

tions (Fig. 10). (Preliminary studies have shown that foils made of paramylum support wound healing by improving the regeneration of wound epithelia; SU et al., 1997; PORTERA et al., 1997.)

5 Transformation of Protists and Expression of Foreign Genes

Only few organisms are currently in use for expression of foreign eukaryotic genes. These comprise fast growing hosts like some species of bacteria and yeasts, and slowly growing baculo virus-infected insect cells and mammalian tissue cells. Whereas bacteria and yeast hosts express heterologous genes in high yields at low cost, mammalian cells have advantages when correctly modified and folded proteins are needed. Mammalian cell cultures, however, are expensive owing to strict growth requirements and slow growth rates and are less suitable for the large-scale production of recombinant proteins. Alternative systems, complementing the currently used hosts, are welcome, since no universal expression system for foreign genes exists at the present (GEISSE et al., 1996). The slime mold *Dictyostelium discoideum* represents an alternative protein expression system. It has simple nutrient requirements, grows fast, and stable cell lines are available. High copy number integrating and extrachromosomal plasmids exist for the production of recombinant proteins. N- or O-glycosylated or otherwise posttranslationally modified proteins – whose genes were directed into the secretory pathway – are recovered from spent culture fluids. Yields of 1 mg L^{-1} of both human anti-thrombin III and *Schistosoma japonicum* glutathione-S-transferase have been obtained from the culture fluids (EMSLIE et al., 1995). Several protocols for the transformation of both the somatic macronucleus and the germline nucleus have been developed for the ciliate *Tetrahymena thermophila*. Microinjection and electroporation have been successfully used as methods for transformation of both the macronuclei in vegetatively growing cells and the newly formed macronuclear anlagen in conjugating cells (TONDRAVI and YAO, 1986; BRUNK and NAVAS, 1988; ORIAS et al., 1988; GAERTIG and GOROVSKY, 1992). More recently the biolistic particle bombardment technique has heen applied successfully to transform (in conjugating cells) the micronuclei as well. Genetically stable transformants were obtained, when constructs containing the new sequence integrated into the non-coding up- and downstream sequences of the target gene were used. This protocol allowed the direct replacement of the target gene by the new gene obviously by homologous recombination (CASSIDY-HANLEY et al., 1997; HAI and GOROVSKY, 1997).

GAERTIG and coworkers (1999) developed a novel *Tetrahymena* expression system for foreign genes. They exploited a *Tetrahymena* strain carrying a mutation in either of the two β-tubulin genes. The mutation – lys350 is substituted by methionine – produces an increased sensitivity to the microtubule drug paclitaxel. When a vector containing a foreign gene flanked by the upstream and downstream sequences of the particular β-tubulin gene is targeted to this locus by homologous recombination, the resulting knocking out of the β-tubulin gene restores the resistance of the cell to paclitaxel. Transformed cells thus can be selected by their resistance to the drug and the foreign gene is expressed under the control of the β-tubulin promoter. The system has been used to express the gene of the major surface protein of the ciliated fish parasite *Ichtyophthirius multifiliis*, which causes "white spot" disease in economically important freshwater fish. The surface protein is apparently responsible for protective immunity and, therefore, a candidate for vaccine production. Unfortunately *Ichtyophthirius multifiliis* can only be grown in association with its host, making it difficult to obtain sufficient quantities of the surface protein. The expression of the *Ichtyophthirius multifiliis* surface protein in *Tetrahymena* – the latter can be cultured in mass – makes possible the large-scale production of the parasite's surface protein as a potential vaccine against the parasite. Since the surface protein of the parasite is not only expressed in *Tetrahymena* but also transported to the cell surface, it can be tested, whether the transformed *Tetrahymena* cells can be used as a living vaccine to immunize farmed catfish against *Ichtyophthirius multifiliis*.

Not only free-living but also parasitic protists of man like *Plasmodium* and *Leishmania* have been successfully transformed (HA et al., 1996; HYDE, 1996). In the case of *Leishmania*, cells transfected with the gene of the human tumor suppressor protein p53 produced a phosphorylated form of p53. The recombinant protein was shown to bind specifically to its enhancer sequence, demonstrating that it had been synthesized in biologically active conformation (ZHANG et al., 1995). Pathogenic protists like *Leishmania*, which are highly adapted to the human host metabolism, may become an attractive heterologous expression system for human proteins. This is emphasized by the fact that nowadays these kinetoplastids can be cultivated to high cell densities of up to 10^{11} cells L^{-1}. If non-infective strains of *Leishmania* perform as well, then this heterologous expression system may yield recombinant proteins in amounts so far obtained only with *E. coli* and yeasts.

6 Conclusion

Protists show an unparalleled diversity of forms and functions. This diversity has enabled the protists to colonize every conceivable habitat on earth. The tremendous diversity of protist forms and living modes might be paralleled by equally diverse biochemical pathways. The number of 65,000 protist species is probably estimated too low, because only a small number of protists of the marine biosphere as a global resource for biotechnology is presently known (COWAN, 1997). The protists comprise a so far untapped pool of new and potentially valuable compounds. The difficulties in mass cultivation of protists specialized in feeding, one of the reasons why protists so far have only rarely been used in biotechnology, will be overcome with improvement of cell culturing techniques suited for protists. The development of new techniques to use different protists for heterologous protein expression may allow the production of commercially interesting products from protists that withstand mass culturing. An integrated effort of scientists from different fields of general and applied research and technology is needed to improve the untapped biotechnological potential of protists.

Acknowledgement
The author thanks Prof. Dr. LEIF RASMUSSEN, Odense University, Odense, Denmark, for critical reading and many helpful comments on the manuscript.

7 References

BARCLAY, W. R., MEAGER, K. M., ABRIL, J. R. (1994), Heterotrophic production of long-chain omega-3 fatty acids utilizing algae and algae-like microorganisms, *J. Appl. Phycol.* **6**, 123–129.

BARDELE, C. F. (1997), On the symbiotic origin of protists, their diversity, and their pivotal role in teaching systematic biology, *Ital. J. Zool.* **64**, 107–113.

BLACKBURN, E. H. (1992), Telomerases, *Annu. Rev. Biochem.* **61**, 113–129.

BRUNK, C. F., NAVAS, P. (1988), Transformation of *Tetrahymena thermophila* by electroporation – parameters effecting cell survival, *Exp. Cell Res.* **161**, 525–532.

CASSIDY-HANLEY, D., BOWEN, J., LEE, J. H., COLE, E., VER-PLANK, L. A. et al. (1997), Germline and somatic transformation of mating *Tetrahymena thermophila* by particle bombardment, *Genetics* **146**, 135–147.

CECH, T. R. (1990), Selbstspleissen und enzymatische Aktivität einer intervenierenden Sequenz der RNA von *Tetrahymena* (Nobel-Vortrag), *Angew. Chem.* **102**, 745–755.

CHEN, F. (1996), High-cell density culture of microalgae in heterotrophic growth, *TIBTECH* **14**, 421–426.

COWAN, D. A. (1997), The marine biosphere: a global resource for biotechnology, *TIBTECH* **15**, 421–426.

EMSLIE, K. R., SLADE, M. B., WILLIAMS, K. L. (1995), From virus to vaccine: developments using the simple eukaryote, *Dictyostelium discoideum*, *Trends Microbiol.* **3**, 476–479.

GAERTIG, J., GOROVSKY, M. A. (1992), Efficient mass transformation of *Tetrahymena thermophila* by electroporation of conjugants, *Proc. Natl. Acad. Sci. USA* **89**, 9196–9200.

GAERTIG, J., GAO, Y., TISHGARTEN, T., CLARK, T. G., DICKERSON, H. W. (1999), Surface display of a parasite antigen in the ciliate *Tetrahymena thermophila*, *Nature Biotechnol.* **17**, 462–465.

GEISSE, S., GRAM, H., KLEUSER, B., KOCHER, H. P. (1996), Eukaryotic expression systems: a comparison, *Protein Expr. Purif.* **8**, 271–282.

GUBERMAN, A., HARTMANN, M., TIEDTKE, A., FLORIN-CHRISTENSEN, J., FLORIN-CHRISTENSEN, M. (1999), A method for the preparation of *Tetrahymena thermophila* phospholipase A_1 suitable for large-scale production, *J. Appl. Microbiol.* **86**, 226–230.

HA, D. S., SCHWARZ, J. K., TURCO, S. J., BEVERLY, S. M. (1996), Use of the green fluorescent protein as a marker in transfected *Leishmania*, *Mol. Biochem. Parasitol.* **77**, 57–64.

HAI, B., GOROVSKY, M. A. (1997), Germ line knockout heterokyons of an essential, α-tubulin gene enable high-frequency gene replacement and a test of gene transfer from somatic to germ-line nuclei in *Tetrahymena thermophila*, *Proc. Natl. Acad. Sci. USA* **94**, 1310–1315.

HARTMANN, M., GUBERMAN, A., FLORIN-CHRISTENSEN, M., TIEDTKE, A. (2000), Screening for and characterization of phospholipase A_1 hypersecretory mutants of *Tetrahymena thermophila*, *Appl. Microbiol. Biotechnol.* **54**, 390–396.

HUBER, H. C., HUBER, W., RITTER, U. (1990), Einfache *in vitro* Prüfsysteme zur Toxizitätsbestimmung von Umweltchemikalien; Mikrokulturen menschlicher Lymphozyten und monoxenische Ciliatenkulturen, *Zbl. Hyg.* **189**, 511–526.

HYDE, J. E. (1996), Transformation of malaria parasites: the barriers come down, *Trends Microbiol.* **4**, 43–45.

KARRER, K. M., PFEIFFER, S. L., DITOMAS, M. E. (1993), Two distinct gene subfamilies within the family of cysteine protease genes, *Proc. Natl. Acad. Sci. USA* **90**, 3063–3067.

KAYA, K., SANO, T., SHIRAISHI, F. (1995), Astasin, a novel cytotoxic carbohydrate-conjugated ergosterol from the colorless euglenoid *Astasia longa*,

Biochim. Biophys. Acta **1255**, 201–204.

KIY, T. (1998), Heterotrophic protists – a new challenge in biotechnology? *Protist* **149**, 17–21.

KIY, T., MARQUARDT, R. (1995), Method for mass culturing of ciliates, *Eur. Patent* EP-A-0 752 470.

KIY, T., TIEDTKE, A. (1991), Lysosomal enzymes produced by immobilized *Tetrahymena thermophila*, *Appl. Microbiol. Biotechnol.* **35**, 14–18.

KIY, T., TIEDTKE, A. (1992a), Mass cultivation of *Tetrahymena thermophila* yielding high cell densities and short generation times, *Appl. Microbiol. Biotechnol.* **37**, 576–579.

KIY, T., TIEDTKE, A. (1992b), Continuous high-cell-density fermentation of the ciliated protozoon *Tetrahymena* in a perfused bioreactor, *Appl. Microbiol. Biotechnol.* **38**, 141–146.

KIY, T., TIEDTKE, A. (1993), Protozoen in der Biotechnologie: Ciliaten auf neuen Wegen, *Bioengineering* **9**, 22–36.

KIY, T., HÖRSCH, B., MARQUARDT, R. (1995), Process for preparing glycosides using glycosidases from ciliates, *Eur. Patent* EP-A-0 725 144.

KIY, T., SCHEIDGEN-KLEYBOLDT, G., TIEDTKE, A. (1996), Production of lysosomal enzymes by continuous high-cell-density fermentation of the ciliated protozoon *Tetrahymena thermophila* in a perfused bioreactor, *Enzyme Microb. Technol.* **18**, 268–274.

KYRIAKIDIS, D. A., TSIRKA, S. A. E., TSAVDARIDIS, I. K., ILIADIS, S. N., KORTSARIS, A. H. (1990), Antiproliferative activity of L-asparaginase of *Tetrahymena pyriformis* on human breast cancer cell lines, *Mol. Cell. Biochem.* **96**, 137–142.

LANDIS, W. G., HALEY, M. V., JOHNSON, D. W. (1986), Kinetics of the DFPase activity in *Tetrahymena thermophila*, *J. Protozool.* **33**, 216–218.

MAEDA, M., NAOKI, H., MATSUOKA, T., KATO, Y., KOTSUKI, H. et al. (1997), Blepharismin 1–5, novel photoreceptor from the unicellular organism *Blepharisma japonicum*, *Tetrahedron Lett.* **38**, 7411–7414.

MÜLLER, M. (1972), Secretion of acid hydrolases and its intracellular source in *Tetrahymena pyriformis*, *J. Cell Biol.* **52**, 478–487.

ORIAS, E., LARSON, D., HU, Y.-F., YU, G.-L., KARTTUNEN, J. et al. (1988), Replacement of the macronuclear ribosomal RNA genes of a mutant of *Tetrahymena* using electroporation, *Gene* **17**, 295–301.

PANT, B., KATO, Y., KUMAGAI, T., MATSUOKA, T., SUGIYAMA, M. (1997), Blepharismin produced by a protozoan *Blepharisma* functions as an antibiotic effective against methicillin-resistant *Staphylococcus aureus*, *FEMS Microbiol. Lett.* **155**, 67–71.

PAULI, W., BERGER, S. (1996), Proc. Int. Workshop on a Protozoan Test Protocol with *Tetrahymena* in Aquatic Toxicity Testing, *Umweltbundesamt-Texte* **34/36**, Berlin, Germany.

PAULI, W., BERGER, S. (2000), New Toxkit microtest with the protozoan ciliate, *Tetrahymena*, in: *New Microtests for Routine Toxicity Screening and Biomonitoring* (PEROONE, G., JANSSEN, C., DECOEN, W., Eds.), pp. 169–178. New York, Boston, Dordrecht, London, Moscow: Kluwer Academic/Plenum Publishers.

PORRINI, M., TESTOLIN, G. (1984), Comparison of enzymatic, microbiological and biological assays for the determination of food protein quality, *Nutr. Rep. Int.* **30**, 1165–1172.

PORTERA, C. A., LOVE, E. J., MEMORE, L., ZHANG, L. Y., MULLER, A. et al. (1997), Effect of macrophage stimulation on collagen biosynthesis in the healing wound, *Am. Surg.* **63**, 125–130.

SCHÖNEFELD, U., ALFERMANN, A. W., SCHULTZ, J. E. (1986), Economic mass cultivation of *Paramecium tetraurelia* on a 200 liter scale, *J. Protozool.* **33**, 222–225.

SU, C. H., SUN, C. S., JUAN, S. W., HU, C. H., KE, W. T., SHEU, M. T. (1997), Fungal mycelia as the source of chitin and polysaccharides and their applications as skin substitutes, *Biomaterials* **18**, 1169–1174.

SUZUKI, K.-M., HAYASHI, N., HOSOYA, N., TAKAHASHI, T., KOSAKA, T., HOSOYA, H. (1998), Secretion of tetrain, a *Tetrahymena* cysteine protease, as a mature enzyme and its identification as a member of the cathepsin L subfamily, *Eur. J. Biochem.* **254**, 6–13.

TAO, N., ORLANDO, M., HYON, J. S. (1993), A new photoreceptor molecule from *Stentor coeruleus*, *J. Am. Chem. Soc.* **115**, 2526–2528.

TONDRAVI, M. M., YAO, M. C. (1986), Transformation of *Tetrahymena thermophila* by microinjection of ribosomal RNA genes, *Proc. Natl. Acad. Sci. USA* **83**, 4369–4373.

VÖLKEL, H., KURZ, U., LINDER, J., KLUMPP, S., GNAU, V. et al. (1996), Cathepsin-L is an intracellular and extracellular protease in *Paramecium tetraurelia*, *Eur. J. Biochem.* **238**, 198–206.

WARD, O. (1995), Microbial production of long-chain PUFAs, *INFORM* **6**, 683–688.

WEEKERS, P. H., VOGELS, G. D. (1994), Axenic cultivation of the free-living soil amoeba, *Acanthamoeba castellanii* and *Hartmanella vermiformis* in a chemostat, *J. Microbiol. Methods* **19**, 13–18.

YOSHIOKA, Y., NAGASE, H., OSE, Y., SATO, T. (1986), Evaluation of the test method "activated sludge, respiration inhibition test" proposed by the OECD, *Ecotoxicol. Environ. Saf.* **12**, 206–212.

ZHANG, W. W., CHAREST, H., MATLASHEWSKI, G. (1995), The expression of biologically active human p53 in *Leishmania* cells: a novel eukaryotic system to produce recombinant proteins, *Nucleic Acids Res.* **23**, 4073–4080.

“Inorganic” Biotechnology

7 Biotechnology of Coal

JÜRGEN KLEIN
Hattingen, Germany

RENE FAKOUSSA
UDO HÖLKER
Bonn, Germany

MARTIN HOFRICHTER
Helsinki, Finland

HELMUT SCHMIERS
Freiberg, Germany

CHRISTOPH SINDER
Essen, Germany

ALEXANDER STEINBÜCHEL
Münster, Germany

List of Abbreviations

ABCDE system	Microbial effector system consisting of **a**lkaline substances, (oxidative) **b**iocatalysts, **c**helators, **d**etergents and **e**sterases
ABTS	2,2′-Azinobis(3-ethylbenzthiazoline-6-sulfonate)
BMS	Benzylmethylsulfide
BTEX	Benzene, toluene, ethyl benzene, xylene
DBT	Dibenzothiophene
DMF	N,N-Dimethylformamide
ESR	Electron spin resonance
GC-MS	Gas chromatography-mass spectrometry
GPC	Gel permeation chromatography
HRP	Horseradish peroxidase
LiP	Lignin peroxidase
MnP	Manganese peroxidase
MW	Molecular weight
NADPH	Reduced nicotinamide adenine dinucleotide
NMR	Nuclear magnetic resonance
PAH	Polycyclic aromatic hydrocarbon
PHA	Polyhydroxyalkanoic acids
SBP	Soy bean peroxidase

1 Introduction

Coal is a chemically and physically heterogeneous, combustible, organic rock. It consists mainly of the elements carbon, hydrogen, and oxygen with lesser amounts of sulfur and nitrogen. Coal seams originate from peat deposited in swamps. The transformation of plant material into peat and coal is called the coalification process (STACH et al., 1982), which is divided into two stages: The biochemical or diagenetic stage and the geochemical or metamorphic stage. The diagenetic stage begins with the decomposition of the vegetable matter and ends with the formation of brown coal, generally called lignite. The geochemical stage begins with the transformation of brown coal/lignite into sub-bituminous coal and is finished with the formation of anthracite. Coal, therefore, consists, besides mineral inclusions, of a complex mixture of carbon compounds. Our present knowledge contributes to a complex picture – though of a statistic nature – of the chemical structure of coal. The coal molecules consist of multi-ring aromatic units linked by bridge bonds. These bridges are the weakest links in the coal structure so that any thermal and chemical, but also microbial, attack becomes effective in these points first. The mineral inclusions are the ash forming compounds which are distributed as discrete particles of mineral matter throughout the coal.

Coal is the most abundant fossil fuel in the world. It comprises about 75% of the total world resources of fossil fuels. Total coal resources of the world are about $9{,}600 \cdot 10^9$ t CE (CE = coal equivalent; 1 kg CE = $29.3 \cdot 10^3$ kJ = $27.8 \cdot 10^3$ BTU). Only 16% of the total or about $1{,}500 \cdot 10^9$ t CE can be extracted economically (Jahrbuch Bergbau, 2000).

A world total of $3.7 \cdot 10^9$ t of "hard" (bituminous) coal and $8.9 \cdot 10^8$ t of lignite (brown coal) was produced in 1998 (Anonymous, 2000).

The main customers for hard coal are the coke oven plants, for the production of metallurgical coke, and as for the lignite the power generating plants, both kinds of plants usually located in the immediate vicinity of the mining sites. 90% of the lignite produced around the world is used for generation of electricity and heat in power plants (including transmitted-heat generation plants). The world's largest producer of lignite is Germany, with 166 million tonnes, followed by the former USSR (94 million tonnes), the USA (81 million tonnes), and Poland (62 million tonnes).

Coals will remain the most important resource for energy supply, since they are available in sufficient quantities even in western industrial countries at present and for the next 300 years. Due to the different coal resources and yearly consumption, there are great differences in the coverage of lignite resources, of about 270 years for Germany and 1,100 years for North America (BRECHT et al., 1996).

New conversion technologies for coal are urgently needed to reduce environmental damage by this energy source. Thus the European Community developed a special research program supporting "Clean Coal Technologies" (European Commission, 1995).

One approach, still at the fundamental research stage, is the utilization of biotechnological processes to convert hard coal or lignite into a clean, cost-effective energy source or into value added products which can be used by further biotechnological or chemical syntheses. A microbial, enzymatic or enzyme-mimetic technology which operates at moderate temperatures and pressures would have great advantages compared to physical and chemical coal conversion technologies.

The idea that coal might be acted on by microorganisms is not new, at least 12 articles on the subject were published during the period 1908–1932. POTTER (1908) reported that certain bacteria were active in oxidizing amorphous lignite. Some years later, GALLE (1910) first isolated pure bacterial cultures from lignite samples. The German coal chemists FISCHER and FUCHS (1927a, b) published two articles about the ability of fungi to grow on lignite and hard coals. By chance, the authors observed that white and greenish mycelia were often formed on moist coal samples stored in the laboratory for other purposes. Studying this phenomenon more in detail, they discovered various molds (filamentous microfungi) colonizing lignite, coal briquettes, and even coke and hard coal. Microscopic studies revealed that these molds belong to the deuteromycetes (*Penicillium* spp., *Aspergillus* spp.) as well as to the yeast-like fungi (*Torula* spp.). The microflora of natural coal deposits was later systematically studied by LIESKE and HOFMANN (1928), who described numerous microbial species from mining areas. The authors also first thought about potential biotechnological applications of coal and coal-colonizing microbes, e.g., the use of brown coal as fertilizer in agriculture (LIESKE, 1929, 1931). The investigations came to a preliminary end in 1932 marked by the publication of the summarizing article "*Biology and Coal Research*" ("*Biologie und Kohleforschung*"; FISCHER, 1932).

Over the following decades however, the scientific interest in this field flagged, due to various difficulties, e.g., concerning the microbial adaptation and enrichment and transferring mini-scale results into a promising bench scale or pilot scale. It was FAKOUSSA in the beginning of the 1980s who investigated again the activities of microorganisms towards hard coal. He showed that bacteria could utilize and also solubilize part of the organic phase of hard coal (FAKOUSSA, 1981). One year later, COHEN and GABRIELE (1982) published an article in which they showed that wood-rot fungi can quantitatively solubilize the low-rank coal leonardite. The spectacular result, in which solid coal particles placed in contact with fungal mycelium were converted to black liquid droplets, attracted considerable attention and initiated a lot of research programs in the United States and in Germany to gain deeper insight into coal microbiology and to elucidate the microbial mechanisms of these effects.

Chapter 16 of Volume 6b of *Biotechnology* (1988), entitled "*Coal in Biotechnology*" was a summary of the worldwide research in the field of biotechnology of coal up to 1988. This chapter summarizes new results in this area obtained in the years 1988–2000, without repeating the formerly elucidated fundamental results, bottlenecks, and properties of coal as described in the preceding *Biotechnology* publication.

2 Biodegradation and Modification of Coal

On the first view, hard coal and lignite seem to be resistant to microbial colonization. So we cannot observe extensive microbial growth on coal pieces collected in open cast or underground mines under normal circumstances. But when the conditions for microbes are improved, e.g., by supplying water, mineral salts, or additional carbon and nitrogen sources, considerable fungal and bacterial growth on coal can be achieved. This section should show that – although coal is relatively resistant to microbial attack – there are microorganisms which are able to modify the structure of coal through different mechanisms. It focuses on the microbial conversion of lower rank coals (lignites) and the two main transformation principles: solubilization and depolymerization. In addition, a summarizing survey of the microbial modification of hard coal is given.

A historical overview showing the progress in the research field over the last 20 years is summarized in Tab. 1 (Fakoussa and Hofrichter, 1999).

2.1 Hard Coal

2.1.1 Bacteria

Certain bacteria are able to grow with hard coal as sole source of carbon and energy. In an extensive screening, comprising about 3,100 cultivation experiments, bacterial strains from suitable locations were enriched using ground hard coals as selective carbon source. Among others, forest fire regions (0.5 to 20 years old) were used as screening locations since the structure of char coal is related to that of hard coal. All in all, growth was only observed in 0.2% of all cases, which represents an exceptionally low rate (Fakoussa, 1981). Among the positive isolates were pleomorphic bacteria probably belonging to the mycobacteria or nocardia. The surface of their cell walls was extremely hydrophobic and thus, they formed cell aggregations on the surface of the medium liquid which could only be dispersed by using detergents or oily substances. The poor growth rates, the distance to the coal particles, and the fate of low-molecular weight substances in the culture liquid gave indication that these bacteria only grow on water-soluble coal substances, which are released through surface pores (Fakoussa, 1981, 1990).

During the same screening program, a particularly interesting bacterium was enriched and later identified as *Pseudomonas fluorescens*. This strain showed some remarkable properties:

- the bacterium released an efficient surfactant into the medium, which lowered the surface tension to about 25.5 mN m^{-2} (this value is even lower than that of a saturated SDS solution)
- the physicochemical properties of coal particles were altered during the cultivation (e.g., color, wettability, extractability)
- the culture supernatant changed its color to brownish indicating that coal substances were released. These substances had molecular weights in the range between 50 and 100 kDa. Infrared spectra and esterification experiments showed a high content of carboxylic and hydroxylic groups, which is consistent with an oxidative attack of the coal particles (Fakoussa, 1988).

It has been assumed that an extracellular biocatalyst turns the hard coal substances into a more hydrophilic status, i.e., they become partly soluble in water and then, can – at least in part – be taken up by the bacterium.

To sum up, the results of this screening program suggest that hard coals with a high content of volatile organic matter are more easily attacked by bacteria and it has been concluded that the bacterial attack on hard coal is – at this stage – not efficient enough for an application in commercial coal conversion processes.

2.1.2 Fungi

A few comprehensive screening programs were carried out to find fungal strains being capable of modifying the physicochemical

Tab. 1. Development and Main Advances in the Microbiology and Biotechnology of Coal Achieved over the last two Decades (modified after FAKOUSSA and HOFRICHTER, 1999)

Year	Step of Progress	Reference
1981	effects on hard coals by bacteria (*Pseudomonas* spp.), simultaneous biotenside secretion	FAKOUSSA (1981)
1982	solubilization of lignite to droplets on agar plates by action of wood-decaying basidiomycetous fungi	COHEN and GABRIELE (1982)
1986[a]	acceleration of solubilization by pretreatment of coal (fungi + bacteria)	SCOTT et al. (1986), GRETHLEIN (1990), and others
1987[a]	first solubilization mechanism elucidated: production of alkaline substances (fungi + bacteria)	QUIGLEY et al. (1987, 1988b, 1989a)
1988[a]	second mechanism elucidated: production of chelating agents (fungi)	COHEN et al. (1990), QUIGLEY et al. (1988b, 1989b)
1989	first product on market: solubilized lignite as fertilizer	Arctech, Virginia (USA)
1991[a]	evidence that chelators alone are not responsible for all effects	FAKOUSSA and WILLMANN (1991), FAKOUSSA (1994)
1994	decolorization and reduction of MW of soluble lignite derived humic acids proves catalytic, i.e., enzymatic attack (basidiomycetous fungi)	WILLMANN (1994), RALPH and CATCHESIDE (1994b), HOFRICHTER and FRITSCHE (1996, 1997a)
1991[a]	improved analysis by ^{13}C-solid state NMR, MW determination, e.g., ultrafiltration, gel permeation chromatography	WILLMANN and FAKOUSSA (1991), POLMAN and QUIGLEY (1991), RALPH and CATCHESIDE (1996), HOFRICHTER and FRITSCHE (1996, 1997a), HENNING et al. (1997), and others
1996	*in vitro* systems shown to preferentially polymerize humic acids without regulation of the fungus (laccase)	WILLMANN (1994), FROST (1996), and others
1997[a]	*in vitro* systems based on fungal Mn peroxidase shown to depolymerize humic acids and to attack coal particles including the matrix	HOFRICHTER and FRITSCHE (1997b), HOFRICHTER et al. (1999)
1997[a]	involvement of unspecific hydrolases (esterases) in the solubilization of brown coal by deuteromycetous fungi	HÖLKER et al. (1997b, 1999a)
1997[a]	first fine chemical produced successfully from heterogeneous humic acid mixtures by bacterial pure cultures: polyhydroxyalkanoates (PHA, "Bioplastic")	STEINBÜCHEL and FÜCHTENBUSCH (1997), FÜCHTENBUSCH and STEINBÜCHEL (1999)

[a] = and following years

structure of hard coal particles or derived products (e.g., asphaltenes, organic hard coal extracts). Already FAKOUSSA (1981, 1988, 1990) tested, in addition to bacteria, yeasts and filamentous fungi from suitable locations (e.g., former forest fire sites, hard coal samples) with respect to their ability to utilize hard coal as sole source of carbon and energy and to release brown colored substances from powdered hard coal. As the result, three filamentous fungi and two yeasts were enriched, which grew – although with very slow rate – with powdered hard coal as sole carbon source.

An extensive screening involving more than 750 strains of filamentous fungi was carried out to select strains which modify an untreated

German hard coal (Bublitz et al., 1994; Hofrichter et al., 1997a). Among the strains tested were representatives of different taxonomic groups of filamentous fungi: zygomycetes, ascomycetes and deuteromycetes as well as basidiomycetes. Only 6 of the 750 strains tested acted noticeably on the hard coal. Tight connections were developed between the fungal hyphae or rhizomorphs and the coal particles, which in turn were split into smaller pieces. Moreover, the wettability of the particles increased, which became visible by the attachment of fungal guttation droplets on the hydrophobic hard coal surface. Interestingly, all hard coal "eroding fungi" belonged to the litter-decomposing and wood-decaying basidiomycetes. The most active fungus, *Coprinus sclerotigenis*, was studied more in detail with respect to the formation of low-molecular mass products from powdered hard coal. 2-Hydroxybiphenyl, alkylated benzenes, polycyclic aromatic hydrocarbons (PAHs), and branched alkanes were extractable after fungal treatment. It was assumed that the non-oxidized compounds (alkylated benzenes and alkanes, PAHs) might be part of the mobile phase of hard coal and were liberated from the micropores through mechanic effects of the fungal hyphae (biodeterioration of coal). In contrast, the formation of 2-hydroxybiphenyl was probably brought about by an enzymatic, bond-cleaving process, because the same compound was also detected after the fungal treatment of hard coal-derived asphaltene powder lacking micropores and a mobile phase; however, it remained unclear which enzyme was responsible for this.

Powdered Polish hard coal and its chloroform extracts were exposed as cosubstrates to the basidiomycetous fungus *Piptoporus betulinus* belonging to the wood-decaying white-rot fungi (Osipowicz et al., 1994). The fungus acidified the medium drastically during its growth (from pH 6.0 to 1.0!), which was associated with the production of organic acids. *Piptoporus betulinus* acted on both the native coal and its organic extract. On the basis of spectroscopic data it was concluded that the biotransformation resulted in the partial dearomatization and depolymerization of the coal substances. Another white-rot fungus, *Panus tigrinus*, was found to convert an asphaltene – obtained through the hydrogenation of German hard coal – in a similar way (Hofrichter et al., 1997a). In the presence of wood shavings, the fungal treatment resulted in the decrease of predominant molecular masses of the asphaltene (from 0.4 to 0.3 kDa), and a novel low-molecular mass fraction (~ 0.1 kDa) consisting probably of monoaromatic compounds was formed. It was concluded that the ligninolytic enzyme system of *Panus tigrinus* attacked the asphaltene unspecifically while it degraded the lignin in the wood shavings.

The most efficient modification of hard coal observed so far has been reported for Spanish bituminous and sub-bituminous coals, which were converted under cometabolic conditions (Sabouraud maltose broth) to a tar-like mass by the action of a deuteromycetous fungus isolated from native hard coal samples (Monistrol and Laborda, 1994). Unfortunately, the strain lost its hard coal solubilizing ability over the following years, but the authors succeeded in isolating new molds (*Trichoderma* sp. M2, *Penicillium* sp. M4) with hard coal modifying and solubilizing activities, although the formation of tar-like products was not observed again (Laborda et al. 1997, 1999). The authors detected oxidative and hydrolytic enzyme activities in fungal supernatants which had been grown in the presence of hard coal, but the actual role of these enzymes (phenoloxidases, peroxidases, esterases) in hard coal transformation is still not clear.

In summary, hard coal is much more resistant towards fungal attack than lower rank coals which is a result of its higher hydrophobicity, the higher proportion of condensed aromatic rings and the lower oxygen content (Fakoussa, and Trüper, 1983). Nevertheless, there are some fungal organisms – both deuteromycetes and basidiomycetes – which are capable of modifying the physicochemical structure of hard coal and even liberating low-molecular mass compounds. The biochemical processes, however, underlying these phenomena are only poorly understood. Probably, a whole string of factors, e.g., mechanical effects of fungal hyphae, microbial surfactants, and extracellular enzymes (oxidative and/or hydrolytic ones), might be responsible for the structural changes.

2.2 Lignite (Brown Coal)

2.2.1 Mechanisms: Solubilization, Depolymerization, Utilization

One has to distinguish between two different principles of the structural modification of lignite: solubilization and depolymerization (HOFRICHTER et al., 1997b, c; CATCHESIDE and RALPH, 1997; FRITSCHE et al., 1999; KLEIN et al., 1999) (Fig. 1). The *solubilization* of lignite, which leads to the formation of black liquids (Fig. 2a), is a mainly non-enzymatic dissolving process preferably at higher pH values (pH 7–10) due to the microbial formation of alkaline substances and/or chelating agents and surfactants (QUIGLEY et al., 1989a, b); in addition, recent studies gave indication that certain hydrolytic enzymes may force the solubilization process (HÖLKER et al., 1999b). The term "liquefaction", although often used in earlier publications, should be avoided in connection with microbial activities towards coal, because it is already occupied by the "pure" chemical processes of coal conversion.

The *depolymerization* of lignite or derived macromolecules (coal humic acids) is an enzymatic process at lower pH (pH 3–6) that results in the cleavage of bonds inside the coal molecule and leads to the formation of yellowish, fulvic acid-like substances ("bleaching") with lower molecular masses. As an example, Fig. 2b shows the decolorization of an agar plate containing a high-molecular mass humic acid from lignite by a basidiomycetous fungus.

In general, coal solubilization and depolymerization are facilitated when oxidatively pretreated or naturally highly oxidized coals (e.g., weathered lignite, leonardite, HNO_3^- or $H_2O_2^-$ treated coals) are exposed to microbes (SCOTT et al., 1986; STRANDBERG and LEWIS 1987b; WARD et al., 1988; HOFRICHTER et al., 1997b; HENNING et al., 1997).

In addition to the structural modifications mentioned above, a number of microorganisms (bacteria, molds, yeasts) is able to grow on lignite *utilizing* parts of the mobile phase (HODEK, 1994), which comprises of a complex mixture of low-molecular mass aromatics and wax-like aliphatics, as sole carbon source (KUCHER and TUROVSKII, 1977; WARD, 1985; RALPH and CATCHESIDE, 1993; WILLMANN, 1994). Studies using low-molecular mass model substances gave indication that, e.g., phenols, benzoic acids, biphenyls, and bi-

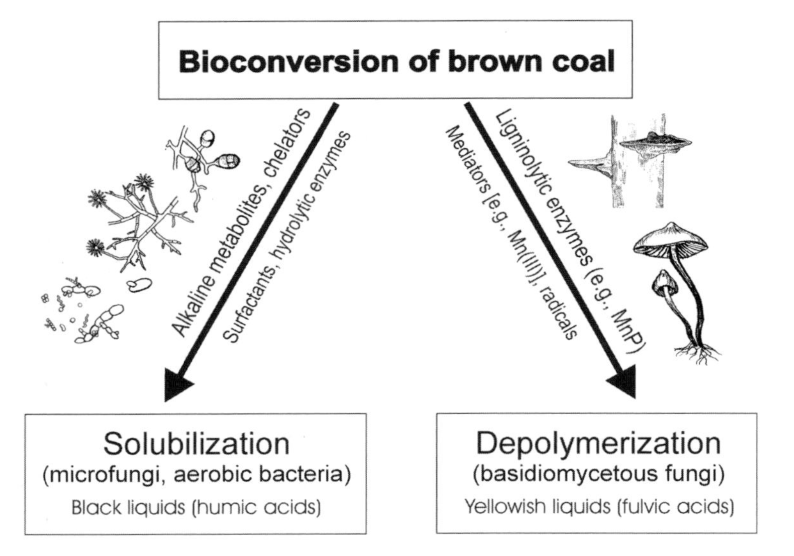

Fig. 1. The two main structural modifications of brown coal by microorganisms (modified after FRITSCHE et al., 1999).

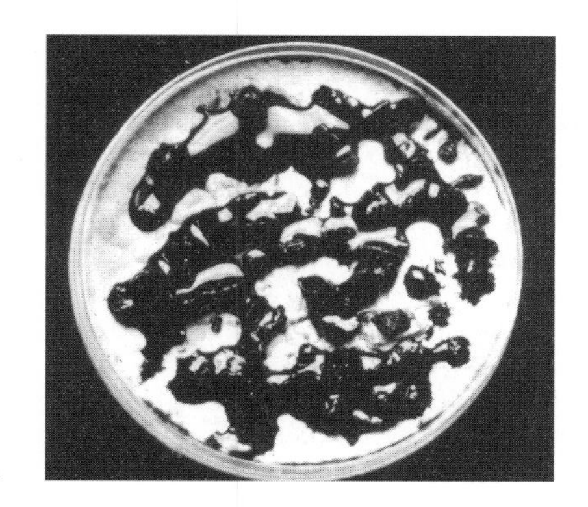

a

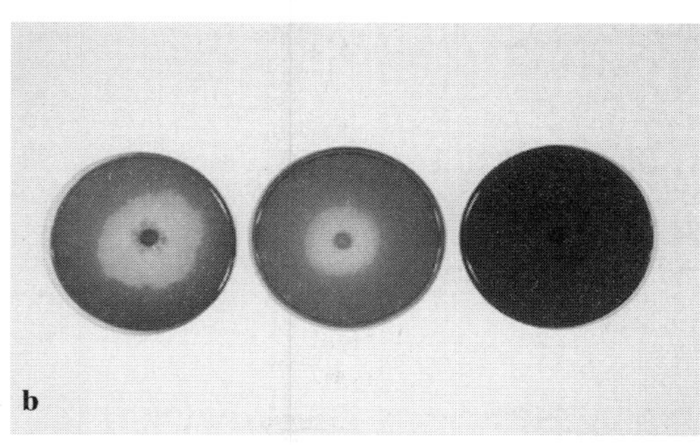

b

Fig. 2a Solubilization of brown coal by a filamentous microfungus. The action of alkaline substances, chelators, and hydrolytic enzymes secreted by *Trichoderma atroviride* results in the formation of black droplets. **b** Depolymerization of high-molecular weight coal substances by a ligninolytic fungus (*Nematoloma frowardii* b19). The agar plates contained coal humic acids extracted with NaOH from Rhenish brown coal (modified after HOFRICHTER and FRITSCHE, 1996). Control without fungus (right), 10-days old culture (middle), 20-days old culture (left).

phenylethers as well as various cycloalkanes, *n*-alkanes and *n*-alkanols are among these substances (RALPH and CATCHESIDE, 1994a; ENGESSER et al., 1994; SCHUMACHER and FAKOUSSA, 1999; FRITSCHE and HOFRICHTER, 2000). Residual cellulose and hemicelluloses that can be found in certain lignites (e.g., xylite = coal that preserved the wood structure) might be an additional carbon source for microorganisms (COHEN and GABRIELE, 1982). Usually, the microorganisms grow slowly on coal particles, but the growth is noticeably stimulated when naturally weathered or chemically pre-oxidized coals are used and additional N-, P-, and S-sources as well as mineral salts are added (WARD, 1985; FAKOUSSA, 1992). In some cases, the utilization of lignite might be accompanied by its solubilization. Thus, HÖLKER et al. (1999b) have reported that $^{14}CO_2$ was evolved from radioactively labeled coal (^{14}C-methoxylated German lignite) by a coal-solubilizing fungus.

FAKOUSSA (1991) has proposed the so-called ABC-system, that was later extended to the ABCDE-system (FAKOUSSA and HOFRICHTER, 1999), to describe all possible mechanisms which are involved in lignite bioconversion:

- **A**lkaline substances
- **B**iocatalysts (oxidative enzymes)
- **C**helators
- **D**etergents (surfactants, biotensides)
- **E**sterases

Fig. 3 depicts the targets in the coal structure of these effectors by the example of a brown coal model.

While the process of lignite solubilization is typical for microfungi, above all deuteromycetes (e.g., *Trichoderma* spp., *Fusarium* spp.), and certain bacteria (e.g., *Pseudomonas* spp.), the depolymerization of coal is evidently limited to the ligninolytic basidiomycetes (wood-decaying and litter-decomposing fungi, e.g., *Phanerochaete chrysosporium*, *Nematoloma frowardii*). Nevertheless, there is some overlap of these abilities and thus, some white-rot fungi (e.g., *Trametes* (*Coriolus*) *versicolor*, *Phanerochaete chrysosporium*), which are active bleachers of coal humic acid (RALPH and CATCHESIDE, 1994a; FROST, 1996), can also solubilize coal, but under other conditions and/or by using other mechanisms (COHEN and GABRIELE, 1982; TORZILLI and ISBISTER, 1994). Microorganisms utilizing part of coal as growth substrate can be found among many groups of aerobic microorganisms and it should also be mentioned that lignite derived products are used as fertilizers and soil conditioners in agriculture (YANG et al., 1985; FORTUN et al., 1986; GYORI, 1986; ALEKSANDROV et al., 1988). In Tab. 2, selected micro organisms are listed, which solubilize, depolymerize or utilize lignite.

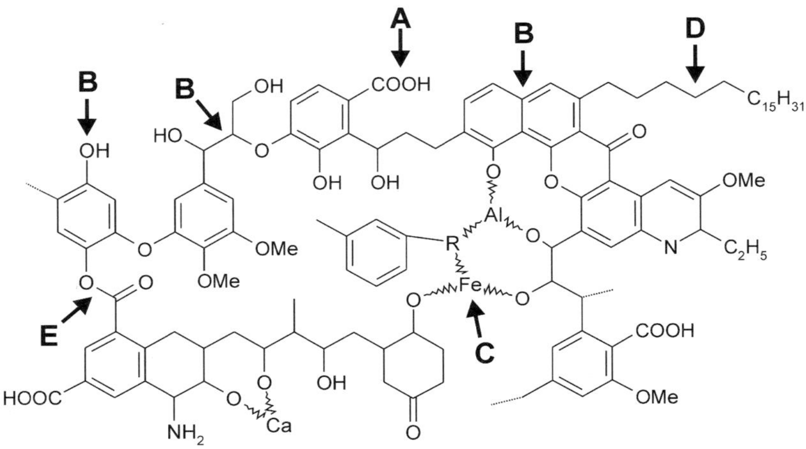

Fig. 3. The so-called ABCDE-mechanism of microbial attack of brown coal (FAKOUSSA, 1991; HOFRICHTER and FAKOUSSA, in press; the brown coal model was modifiedly adapted from HÜTTINGER and MICHENFELDER, 1987). Coal structures that are attacked by microbial agents are indicated by arrows. Alkaline substances (A), biocatalysts, i.e., oxidative enzymes (B), chelating agents (C), detergents (D), esterases and other hydrolytic enzymes (E).

2.2.2 Solubilization of Lignite

2.2.2.1 Non-Enzymatic Effectors

The alkaline solubilization of lignite was the first discovered mechanism of microbial coal conversion (QUIGLEY et al., 1987, 1988b) and later confirmed by several authors (MAKA et al., 1989; RUNNION and COMBIE, 1990a; TORZILLI and ISBISTER, 1994). It is attributed to the high content of carboxylic and phenolic functionalities (−COOH, −OH) in coal humic acids which can be deprotonated at higher pH (>8), resulting in the formation of water-soluble humates (typically dark-brown to black-colored). Microorganisms have two possibilities to adjust such alkaline conditions:

(1) They may produce and secrete respective metabolites, i.e., ammonia (NH_3) or biogenic amines ($R-CH_2-NH_2$), in response to high nitrogen levels in the medium. Thus, for example, the Sabouraud-medium used for many coal solubilization studies contains high amounts of peptone and nitrate, other solubilization media glutamate and nitrate (FAISON, 1991; WARD, 1993; MONISTROL and LABORDA, 1994; HÖLKER et al., 1995). The drastic influence of high nitrogen levels on the solubilization of lignite, regardless whether complex nitrogen sources or inorganic salts were used, was demonstrated in different studies for both fungi and bacteria (QUIGLEY et al., 1989a; TORZILLI and ISBISTER, 1994; HOFRICHTER et al. 1997b).

(2) Microorganisms may also alkalify their surrounding medium through the utilization of organic acids used as growth substrates, which results in the formation of free bases due to the remaining mono- and dibasic metals (Na^+, K^+, Ca^{2+}). This phenomenon is probably responsible for the enhanced coal solubilization by *Fusarium oxysporum* when sodium gluconate is used as carbon source instead of glucose (HÖLKER et al., 1995).

Leonardite solubilization was already early linked to the effect of chelating agents removing complexing metal ions that form coordinative bridging elements (e.g., Ca^{2+}, Fe^{3+}, Al^{3+}; QUIGLEY et al., 1988a). COHEN et al. (1990)

Tab. 2. Selected Microorganisms that have Solubilizing or Depolymerizing Activities towards Brown Coal or Utilize it as a Growth Substrate (Hofrichter and Fakoussa, in press)

Organisms	Effects on Brown Coal and Derived Products	References
Bacteria		
Actinomycetes		
Streptomyces spp.	solubilization	Gupta et al. (1988)
S. setonii	solubilization	Strandberg and Lewis (1987a)
Eubacteria		
Bacillus sp.	solubilization	Quigley et al. (1989a)
B. licheniformis	solubilization	Polman et al. (1994a)
Pseudomonas cepacia	solubilization (depolymerization)	Crawford and Gupta (1991)
Basidiomycetous fungi		
Wood-decaying white-rot fungi		
Clitocybula dusenii	depolymerization	Ziegenhagen and Hofrichter (1998)
Nematoloma frowardii	depolymerization	Hofrichter and Fritsche (1996)
Phanerochaete chrysosporium	solubilization, depolymerization	Torzilli and Isbister (1994), Ralph and Catcheside (1994b)
Trametes (*Coriolus*) *versicolor*	solubilization, utilization, depolymerization	Cohen and Gabriele (1982), Frost (1996), Fakoussa and Frost (1999)
Litter-decomposing fungi		
Agrocybe praecox	depolymerization	Steffen et al. (1999)
Stropharia rugosoannulata	depolymerization	Hofrichter and Fritsche (1996)
Isolates RBS 1k[a], RBS 1b[a]	depolymerization	Willmann (1994), Willmann and Fakoussa (1997a, b)
Wood-decaying brown-rot fungi		
Poria monticola	solubilization, utilization	Cohen and Gabriele (1982)
Deuteromycetous and ascomycetous fungi		
Alternaria sp.	solubilization	Hofrichter et al. (1997b)
Aspergillus terreus	utilization	Ward (1985)
Fusarium oxysporum	solubilization	Hölker et al. (1995)
Neurospora crassa	solubilization	Patel et al. (1996)
Paecilomyces spp.	solubilization, utilization	Ward (1985), Scott et al. (1986)
Penicillium citrinum	solubilization, utilization	Polman et al. (1994b)
Trichoderma atroviride	solubilization	Hölker et al. (1997b, 1999a)
Yeasts-like fungi		
Candida bombicola	solubilization	Breckenridge and Polman (1994)
Candida sp.	utilization	Ward (1985)
Candida tropicalis	utilization	Kucher and Turovskii (1977)
Zygomycetous fungi		
Cunninghamella sp.	solubilization	Ward and Sanders (1989),
Mucor lausannesis	utilization	Ward (1985, 1993)

[a] Enriched and isolated from brown coal from an open cast mining region.

could even demonstrate that the biosolubilization of leonardite by a strain of *Trametes* (*Coriolus*) *versicolor* was only brought about by ammonium oxalate [$C_2O_4(NH_4)_2$], which is excreted by this fungus in high amounts. By testing different chelator molecules (oxalate, citrate, tartrate, dihydroxybenzoic acids, etc.), it was later found that oxalate is indeed one of the most effective agents for solubilizing special kinds of lignite (leonardite, pre-oxidized coals), probably due to its small molecular size (FAKOUSSA, 1994). On the other hand, it was shown that not highly oxidized coals (e.g., certain German lignites) were only solubilized to a limited extent by an oxalate treatment. It was concluded that metal removing via chelation cannot have a great influence on the solubilization of lignites with low ash and corresponding carboxylic groups contents (FAKOUSSA, 1994).

The use of chemo- and biosurfactants (= detergents) was a simple attempt to achieve a partial solubilization of coal or to extract specific compounds out of it. POLMAN and coworkers used three chemosurfactants (Tween 80, Triton-X-100, SDS) and two biological surfactants (from the bacterium *Bacillus licheniformis* and the yeast *Candida bombicola*) to "dissolve" different coals or coal fractions (POLMAN et al., 1994a; BRECKENRIDGE and POLMAN, 1994). Some of the surfactants indeed brought about the extraction of different low- and high-molecular weight components, probably by interacting with aliphatic coal moieties (paraffins; see Fig. 3). Under culture conditions in the laboratory, however, the significance of surfactants for the solubilization of coal is rather low compared with that of alkaline and chelating agents, since their natural concentrations are mostly below the physicochemically effective levels (FAKOUSSA, 1992).

2.2.2.2 Involvement of Hydrolases in Lignite Solubilization

Solubilization of lignite has also been attributed to hydrolytic enzymes (e.g., esterases) which may cleave hydrolyzable bonds inside the coal molecule. First indication for a coal-solubilizing esterase activity was found in the white-rot fungus *Trametes* (*Coriolus*) *versicolor* during the growth in the presence of leonardite (CAMPBELL et al., 1988). However, this activity was relatively low compared to that of chelating agents (COHEN et al., 1990). CRAWFORD and GUPTA (1991) described a non-oxidative enzyme that was proposed to be involved in the depolymerization of humic acids obtained from weathered Vermont lignite by several gram-positive and gram-negative bacteria. During their action on soluble coal polymers, the bacteria produced similar enzymes which progressively converted the principal broad polymer peak of alkali-solubilized coal (about 174 kDa) into a much sharper peak of about 113 kDa. In some cases, products showing even lower MWs also appeared upon prolonged incubation of the reaction mixture. Part of the coal-solubilizing activity was inactivated by boiling the culture supernatant. WILLMANN and FAKOUSSA (1992) have described the induction of an extracellular esterase activity of a newly isolated fungal strain from an open cast mining area by lignite particles and the separated bitumen fraction of this coal.

More detailed results were obtained for the coal-solubilizing activity of the deuteromycete *Trichoderma atroviride* (HÖLKER et al., 1997a, b; HÖLKER, 1998), and later also for other *Trichoderma* spp. (LABORDA et al., 1997, 1999). *Trichoderma atroviride* was found to secrete a heat-sensitive, partly inducible agent that facilitated the solubilization of Rhenish lignite and had hydrolytic properties (HÖLKER et al., 1999a). There are indications that this mold produces an unusual esterase in the presence of lignite, which cleaves ester bonds that are sterically blocked and not accessible to "normal" esterases (HÖLKER et al., 1999b). In addition, *T. atroviride* produces effective non-enzymatic agents (alkaline substances, chelators) that effect dissolving of coal humic substances. The mechanism of coal solubilization by *T. atroviride* is thus attributed to a "mixed action" of non-enzymatic agents and hydrolytic enzymes. The coal solubilizing activity of this fungus is used for a coal fermentation process in a new kind of bioreactor (HÖLKER, 2000). With regard to the involvement of esterases in coal solubilization/liquefaction it seems to be important that esterases are not known to act by the mediation of a messenger molecule ("mediator"). This leads to steric obstacles, as the whole esterase molecule cannot invade the tight macromolecular network of coal. Therefore, more research is needed to prove their actual role in detail. From the point of view of coal chemistry, it will have to be clarified how many ester bonds lignite actually contains and how these bonds contribute to the stability of the whole coal molecule (KLEIN et al., 1999).

Recent analytical search for possible targets for extracellular hydrolytic enzymes in lignite structure revealed that 1–2% of the total carbon in lignite (Bergheim, lithotype A) can be attributed to carboxylic esters. Saponification with 2.5% BaOH at 130 °C for 12 h decreased the number of ester bonds in lignite by 90% increasing the solubility of lignite at pH 9 tenfold (GROßE, 2000). Even the insoluble matrix can be partially hydrolyzed by strong bases (GROßE, 2000). Chemical analysis of the biodegradation products showed that up to 50% of ester bonds in lignite were not detectable following a fungal attack by *T. atroviride* (GROßE, unpublished data). The study performed by LISSNER and THAU (1953) shows that ester groups are mainly in the aliphatic portion of lignite (bitumen). This portion of ester groups can be characterized by determining the so called "saponification number" which can be assessed by saponification of the bitumen and titration of the carboxyl acids with chlorine acids. The presence of ester groups was confirmed by IR spectra of Australian lignite by SUPALUKNARI and coworkers (1988). They found a relationship between the amounts of ester groups and the H/C-ratio in lignite and concluded that lignite with a higher H/C-ratio contained a higher concentration of ester groups than lignites with a lower H/C-ratio. The presence of ester groups in humic substances of soil, which can be seen as a precursor of lignite, was documented by GRASSET and AMBLES (1998). The authors could show that the main portion of humic substances was soluble in an acid milieu after saponification. The identification of the saponification products by gas chromatography revealed that ester groups were localized mainly in the aliphatic fractions of the soil.

2.2.3 Depolymerization of Lignite by Oxidative Enzymes

Lignin and other plant macromolecules were the main parent materials in the formation of lignite. Via two phases – first a biochemical stage and second a geochemical stage – the remaining plant material (debris) was transformed into peat and further to coal (STACH et al., 1982). Though coalification was accompanied by characteristic changes in the physicochemical properties and structure of the plant material, typical structures of the original lignin have been preserved in coal, which is not least evident from the other designation of coal: lignite (DURIE et al., 1960; HAYATSU et al., 1979; HATCHER, 1990). Therefore, it seems to be worth investigating the potential of ligninolytic microorganisms to convert lignite.

The complex lignin polymer can only be degraded by unspecific extracellular mechanisms, and there are only a few groups of microorganisms in nature that are able to do this. Most active lignin degraders can be found among the basidiomycetes (higher fungi) causing white-rot of wood (KIRK and FARRELL, 1987; BLANCHETTE, 1991; HATAKKA, 1994). The ligninolytic enzyme system consists of peroxidases (manganese peroxidase, lignin peroxidase, other peroxidases), phenol oxidases (laccases), supporting enzymes (e.g., H_2O_2-generating oxidases), and various low-molecular weight agents (organic acids, e.g., oxalate, malate, fatty acids/lipids, Mn-ions). Certain effects of these biocatalytic systems on brown coal and derived coal humic acids have been investigated in the last 10 years and are described and discussed as follows.

2.2.3.1 Peroxidases

Lignin peroxidase (LiP; EC 1.11.1.14) was the first discovered ligninolytic peroxidase. It was found in the basidiomycetous fungus *Phanerochaete chrysosporium*, which causes a strong white-rot of wood (TIEN and KIRK, 1983). Later, the enzyme was also found in other basidiomycetes (HATAKKA, 1994). LiP is a glycoprotein containing iron protoporphyrin IX (heme) as a prosthetic group and requires H_2O_2 for catalytic activity (TIEN and KIRK, 1984). The enzyme is expressed in multiple forms and has a catalytic cycle similar to that of horseradish peroxidase (DUNFORD, 1991; WARIISHI et al., 1991). LiP has a broad substrate specificity for aromatic compounds and oxidizes both non-phenolic and phenolic rings. The latter are oxidized by one-electron abstraction to phenoxy radicals undergoing disproportionation to quinones (ODIER et al., 1988), whereas the former are oxidized to aryl cation radicals (SCHOEMAKER et al., 1985). LiP is unique in its ability to oxidize methoxylated, non-phenolic aromatic compounds of high redox potential directly (up to +1.5 V; KERSTEN et al., 1990).

WONDRACK et al. (1989) first investigated the effects of LiP on high-molecular weight fractions of brown coal. They found that coal macromolecules prepared from nitric acid-treated North Dakota lignite and a sub-bituminous German coal were partly depolymerized by *Phanerochaete chrysosporium* LiP. The coal polymer (soluble in water and N,N-dimethylformamide) was converted to smaller water-soluble fragments. Addition of veratryl alcohol, which is known to support LiP activity, forced this process.

The bleaching and depolymerization of alkali-solubilized lignite by LiP-producing cultures of *P. chrysosporium* have been reported by RALPH and CATCHESIDE (1994a, 1997a, b). The fungus converted about 85% of the coal material after 16 d incubation to a form which was not recoverable by alkali washing and acid precipitation. Extensive bleaching of coal substances coincided with the presence of extracellular LiP. The same authors demonstrated later the decolorization of native and methylated alkali-solubilized lignite preparations by LiP purified from *P. chrysosporium*. Interestingly, the LiP-catalyzed bleaching process was only accompanied by the real depolymerization of coal macromolecules (proved by GPC), when the latter were methylated with dimethyl sulfate and diazomethane. Native alkali-solubilized coal was not an appropriate substrate for LiP (RALPH and CATCHESIDE, 1997a). Methylation seems to facilitate LiP-catalyzed cleavage of bonds in coal, because otherwise phenolic groups in coal are oxidized by LiP to phenoxy radicals which tend to polymerize and counteract the degradation process. In contrast, non-phenolic aromatic groups are oxidized by LiP to aryl cation radicals which commonly decay via pathways involving C–C and C–O bond-cleavage reactions. The methylated, water-soluble coal fraction was partly converted to lower-molecular weight products, among which various methoxylated monoaromatic compounds (e.g., 2,5-dimethoxy-ethylbenzene, 3,4,5-trimethoxy-2-methylbenzoic acid), and fluoroanthene could be identified (RALPH and CATCHESIDE, 1999). However, it is questionable, whether an unpolar compound like the polyaromatic hydrocarbon fluoroanthene can be released through the action of an oxidative enzyme like LiP.

The second ligninolytic peroxidase – manganese peroxidase (EC 1.11.1.13) – was also discovered in liquid cultures of the fungus *P. chrysosporium* (KUWAHARA et al., 1984). The enzyme resembles LiP: it is extracellular, glycosylated, and contains heme as the prosthetic group (GLENN and GOLD, 1985). In contrast to other peroxidases, MnP uses Mn^{2+} and Mn^{3+} as preferred substrate and redox mediator, respectively. MnP, that is usually expressed in multiple forms, has been found in many white-rot and litter-decomposing basidiomycetes and seems, in general, to be more abundant and more widely distributed than LiP (HATAKKA, 1994, in press).

The catalytic cycle of MnP is similar to that of LiP and includes the native ferric enzyme and the reactive compounds I and II (WARIISHI et al., 1988). Significant differences, however, appear in the reductive reactions, where Mn^{2+} is a preferred electron donor. Both compound I and compound II are reduced by Mn^{2+} while the latter is oxidized to Mn^{3+}. Mn^{3+} ions are stabilized to high redox potentials via chelation with organic acids such as oxalate, malonate, malate, tartrate, or lactate (WARIISHI et al., 1992). Chelated Mn^{3+} in turn acts as diffusible redox mediator that oxidizes various aromatic compounds (WARIISHI et al., 1992; HOFRICHTER et al., 1998a) and organic acids (PEREZ and JEFFRIES, 1992; HOFRICHTER et al., 1998b). The oxidative strength of the MnP system can be considerably enhanced in the presence of appropriate co-oxidants such as thiols (FORRESTER et al., 1988; HOFRICHTER et al., 1998a) or unsaturated fatty acids, lipids and their derivatives (KAPICH et al., 1999). Moreover, MnP has been shown to produce H_2O_2 by the oxidation of glutathione and NADPH (PASZCYNSKI et al., 1985), and evidence is accumulating that MnP is able to act efficiently in the absence of external H_2O_2 by oxidizing organic acids in "oxidase-like" reactions (URZÚA et al., 1998; HOFRICHTER et al., 1998b, c).

WILLMANN and FAKOUSSA (WILLMANN, 1994; WILLMANN and FAKOUSSA, 1997a, b) have first demonstrated that MnP secretion is stimulated during the bleaching of water-soluble coal macromolecules (humic acids) by the agaric basidiomycete strain RBS 1k. Similar findings were made for Mn^{2+}-supplemented cultures of the agaric basidiomycetes *Nematoloma frowardii* and *Clitocybula dusenii* both depolymerizing coal humic substances rapidly by forming low-molecular weight fulvic acids (HOFRICHTER and FRITSCHE, 1996, 1997a). The activity of MnP increased noticeably after coal was added to the cultures indicating an inductive effect of the coal humic substances and the involvement of MnP in the depolymerization process. Further molecular biological and enzymatic studies confirmed these assumptions. Thus, the presence of coal humic substances increased the level of MnP-mRNA in both fungi, but it is still not clear, how these macromolecules take effect intracellularly (SCHEEL et al., 1999a, 2000; LUDWIG et al., 1999).

In vitro depolymerization studies using MnPs from *N. frowardii* and *C. dusenii* showed that the enzyme is indeed responsible for the depolymerization of coal humic substances (HOFRICHTER and FRITSCHE, 1997b; ZIEGENHAGEN and HOFRICHTER, 1998). Cell-free experiments showed that the depolymerization was accompanied by a drastic loss in absorbance at 450 nm (bleaching), while the formation of low-molecular weight fulvic acid-like substances – absorbing stronger at 360 nm – was observed. GPC analysis revealed that the modal molecular mass (most abundant molecular mass) decreased from 3 kDa to 0.7 kDa during the conversion of humic to fulvic acids (Fig. 4). GC-MS analysis of fulvic acids obtained from the enzymatic oxidation of coal humic acids indicated the formation of polyhydroxylated and polycarboxylated aromatic and aliphatic substances.

Pulverized native lignite, including the coal matrix (alkali insoluble, extremely high-molecular weight part of coal, see Tab. 1 in FAKOUSSA and HOFRICHTER, 1999), has recently also been found to be accessible to attack by MnP. Incubation of coal particles (approximately 2 μm in diameter) with *N. frowardii* MnP resulted in the formation of fulvic acids and small humic acid molecules (HOFRICHTER et al., 1999b).

In general, what is so special about MnP is that it acts via the real redox mediator couple Mn^{2+}/Mn^{3+}. The latter is regenerated during the depolymerization reactions and is small

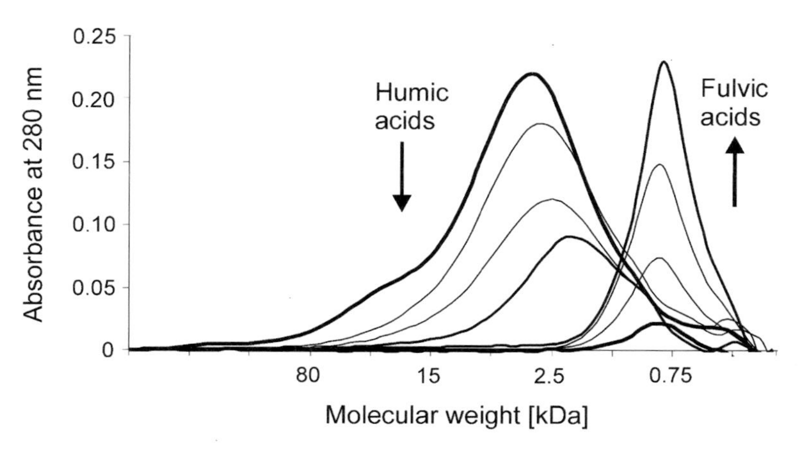

Fig. 4. Depolymerization of coal humic acids and formation of fulvic acids by MnP from *Clitocybula dusenii* in a cell-free system (modified after HOFRICHTER et al., 1999b). The reaction mixture contained: Na-malonate, $MnCl_2$, MnP, glucose, and glucose oxidase. The bold lines mark the starting and ending curves.

enough to diffuse into parts of the coal macromolecule which are not accessible to direct attack by the relatively large enzyme molecules.

In addition to fungal peroxidases, two plant peroxidases, soybean hull peroxidase (SBP) and horseradish peroxidase (HRP), have been studied for their ability to modify lignite (BLINKOVSKY et al., 1994). SBP catalyzed the partial depolymerization of a high-molecular weight coal fraction obtained from a Spanish lignite in 50% (v/v) DMF with an aqueous component consisting of acetate buffer adjusted to pH 2.2. The authors did not mention any precipitation of humic acids at this pH, probably due to the high DMF concentration. In the end, about 15% of the initial coal fraction was depolymerized to low-molecular weight products (<4 kDa). Treatment of acid-oxidized lignite with HRP resulted in an increase in coal solubility when used in hydrous organic media (dioxane, pyridine); and up to 44% of coal was converted to soluble products (SCOTT et al., 1990). However, subsequent experiments indicated that those reactions of HRP conferring increased coal solubility may not have been enzyme mediated but were attributed to the relatively high H_2O_2 concentrations applied (QUIGLEY et al., 1991).

2.2.3.2 Phenol Oxidases

Laccases (EC 1.10.3.2) are the most abundant extracellular polyphenol oxidases and they belong to the blue copper oxidases (REINHAMMAR and MALSTRÖM, 1981). Laccases are produced by most lignin-degrading basidiomycetes (HATAKKA, 1994, in press), but also by many molds and higher plants (THURSTON, 1994). Like most extracellular enzymes, laccases are glycosylated proteins which are often expressed in multiple forms. They can oxidize a variety of aromatic compounds while reducing O_2 to H_2O, a total reduction by four electrons. Laccases show a low specificity to electron-donating substrates, though differentially substituted mono-, di- and polyphenol compounds are preferred (BUSWELL and ODIER, 1987). The catalytic center of blue laccases contains four copper ions (Cu^{2+}) and their cycle resembles that of blue ascorbate oxidase and comprises several one-electron transfers between the copper atoms while O_2 is bound to the active site of the enzyme (FRATERRIGO et al., 1999; MESSERSCHMIDT et al., 1989). The action of laccase on phenolic aromatics results in the formation of phenoxy radicals, which readily undergo multiple non-enzymatic reactions leading finally both to polymerization and bond cleavages (KIRK and SHIMADA, 1985). Laccases of *Trametes versicolor* and *Pycnoporus cinnabarinus* have been shown to oxidize recalcitrant molecules with high redox potentials in the presence of appropriate "mediator substances" (ABTS, BOURBONNAIS and PAICE, 1990; 3-hydroxyanthranilic acid, EGGERT et al., 1996). Success in lignin and kraft pulp depolymerization with laccase and different redox mediators over the last years makes similar investigations into the laccase-catalyzed conversion of brown coal equally promising (CALL and MÜCKE, 1995; YOUN et al., 1996; FAKOUSSA et al., 1997).

The involvement of laccase in the bioconversion of lignite was first proposed by COHEN et al. (1987) for *Trametes* (*Coriolus*) *versicolor*, and the enzyme was thought to be the agent that is responsible for the solubilization of lignite. Further purification of the enzyme and cell-free tests, however, revealed that the solubilizing activity of laccase is negligible compared with ammonium oxalate and esterase also produced by this fungus (COHEN et al., 1990; FREDRICKSON et al., 1990).

One attempt was made to use a commercial laccase (from the phytopathogenic mold *Pyricularia oryzae*) to enhance coal solubilization in organic solvents; the enzyme preparation, however, enhanced coal solubilization only to a small extent (3–5%) in an aqueous dioxane solution (SCOTT et al., 1990).

In contrast to coal solubilization, there are clear indications that laccase is involved in the depolymerization of coal and coal humic substances. FROST (1996) has measured a correlation between laccase secretion and the decolorization and depolymerization of coal macromolecules by *Trametes versicolor* (FROST, 1996; FAKOUSSA and FROST, 1999). Laccase was also found during the bleaching of humic acids in agar plates by *Nematoloma frowardii*, and the highest activity was detected at the edge of the bleaching zones, where the depolymerizing reactions were just going on (HOFRICHTER et al., 1997a). Moreover, increased levels of laccase mRNA were detected in *N. frowardii* and other fungi when coal humic substances were present in the culture medium (SCHEEL et al., 1999a, b). *In vitro* experiments using laccase from *Pycnoporus cinnabarinus* demonstrated that the reaction system tends towards polymerizing, if appropriate mediators (e.g., ABTS, 1-hydroxybenzotriazole) are lacking. In their presence, however, substantial depolymerization of coal humic acids and pulverized coal particles was achieved (TEMP et al., 1999; GÖTZ et al., 1999).

Besides fungal laccase, tyrosinase (EC 1.14.18.1) – another extracellular, copper-containing phenoloxidase which is widely distributed among animals, plants, fungi, and actinomycetes (NAIDJA et al., 1998) – has also been tested for its coal-modifying ability. It has been reported that a tyrosinase-like enzyme from *Neurospora crassa* has coal-solubilizing activities, and that tyrosinase-negative mutants of this fungus were not able to transform lignite particles on agar to black liquids (ODOM et al., 1991; PATEL et al., 1996). Experiments with commercial mushroom tyrosinase (Sigma, S. Louis, MO) apparently confirmed these results. However, it has to be considered that this commercial enzyme preparation is obtained from soil material of the white button mushroom (*Agaricus bisporus*) and contains high amounts of soil humic substances which interfere with coal solubilization tests (SORGE and HOFRICHTER, unpublished data). Therefore, appropriate experiments using enzyme preparations which are free of humic substances, have to be performed to confirm the tyrosinase effects towards coal.

2.3 Other Approaches to Convert Coal

Coal conversion to a clean gas like methane (CH_4) by anaerobic bacteria would offer many biotechnological advantages (CRAWFORD et al., 1990; LEUSCHNER et al., 1990; ISBISTER and BARIK, 1993), however, it is anything but easy to convert a substantial fraction of native coal into methane gas for the following reasons:

(1) Anaerobic bacterial consortia are mostly complex mixed cultures and utilize only polymers consisting of identical, repeating subunits (e.g., cellulose, starch). These polymers are disassembled in a first step by extracellular hydrolytic enzymes to small molecules (e.g., monomeric sugars) that can be further assimilated in the cells. Such an extracellular step would be very difficult in the case of coal substances, because there are only few hydrolytic bonds in coal, the cleavage of which would lead to molecules (<500 Da) which can be taken up by the bacteria. Aromatic structures in coal linked by ether- and carbon–carbon bridges cannot be cleaved by anaerobic microorganisms to a greater extent (SCHINK, 1998, 2000).
(2) The relative amount of hydrogen in natural lignite and hard coals is too low to allow the substantial conversion of the carbon portion into methane.
(3) As already mentioned above, methane is anaerobically produced by complex bacterial consortia. From the economic point of view, the use of such consortia for the conversion of coal would be in any case too expensive due to media costs and too sensitive due to the complexity of regulation. Moreover to date, no *in vitro* system consisting of different stable enzymes from anaerobic bacteria has been established.

To sum up the investigations of the last decade, it can be concluded that the microbial gasification of hard coal seems almost impossible based on the present knowledge (PANOW et al., 1997). Using lower rank coals as substrates, only small fractions can be converted into methane (FAKOUSSA, AIVASIDIS, unpublished data).

In conclusion, the perspectives for an anaerobic bioconversion of lignite appear to be rather modest, too.

Another reductive approach for coal bioconversion was tested by ALEEM et al. (1991). They tried to reduce coal by applying whole cells of *Desulfovibrio desulfuricans* and *Sulfolobus brierlegi*; however, this did not lead to sufficient results. SCOTT and coworkers attempted to hydrogenate (with H_2) volatile bituminous coals by using specially modified bacterial hydrogenases stabilized in organic solvents (SCOTT et al., 1994; KAUFMANN and SCOTT, 1994). A dissolution of hard coals between 19% and 40% was achieved in such cell-free reaction systems. Lignites gave lower solubilization yields. In general, the advantages of such a process would be the generation of hydrophobic products in a single step and the independence from sensitive whole cells and expensive nutrient media.

MIKI and SATO (1994) tried to construct an artificial hydrogenase-like surface by attaching tetraphenyl porphyrin rings on silicate layers ("enzyme mimesis"). The authors were able to show an enhancement of solubility by the treatment of brown coal with such an artificial enzyme. Altogether, their experiments verified, however, that it is not easy to mimic an effective and stable catalytic center of an enzyme.

It is well known, that the greatest amount of oxygen in most lignites is bound in carboxylic groups (HÜTTINGER and MICHENFELDER, 1987). On the other hand the COOH-groups can easily be removed by thermal or chemical means (BANERJEE et al., 1989; FRISTAD et al., 1983), which is very expensive. A stable enzymatic decarboxylating activity would be cheap und would work at moderate temperatures. In preliminary experiments FAKOUSSA et al. (1999a, b) demonstrated that lignite derived substances became more hydrophobic and tended to precipitate after decarboxylation. Moreover, the specific burning value was increased and the oxygen content decreased. By using benzene carboxylic acids for screening, the authors isolated appropriate decarboxylases, which were unspecific enough to decarboxylate lignite derived humic acids to a great extent as shown by infrared spectroscopy. So far, these decarboxylases have only been found to be intracellular. Thus, it is cost intensive to isolate them, and they are not stable enough to be used for commercial purposes.

3 Use of Biotechnologically Liquefied Lignite as Chemical Feedstock for the Production of Bioplastics

Polyhydroxyalkanoic acids (PHAs) are a complex class of naturally occurring polyesters, which consist of various hydroxyalkanoic acids linked by oxo ester linkages formed between the hydroxyl of one constituent and the carboxyl group of another constituent (DOI, 1990; STEINBÜCHEL and VALENTIN, 1995). They are synthesized by a wide range of bacteria, which deposit PHAs as cytoplasmic inclusions for storage of energy and carbon (ANDERSON and DAWES, 1990; STEINBÜCHEL et al., 1995). PHAs are thermoplastic and/or elastomeric, exhibit a high degree of polymerization, consist of only the *R* stereoisomers of hydroxyalkanoic acids if the carbon atom with the hydroxyl group is chiral, and are generally biodegradable (MÜLLER and SEEBACH, 1993; JENDROSSEK et al., 1996). In addition, many of the PHAs can be produced from renewable resources (STEINBÜCHEL and FÜCHTENBUSCH, 1998; MADISON and HUISMAN, 1999). Therefore, the industry has an increasing interest in PHAs, to use them not only for the manufacturing of biodegradable packaging materials but also for various medical and pharmaceutical applications and as a source for the synthesis of enantiomeric pure chemicals (HOCKING and MARCHESSAULT, 1994).

PHAs are synthesized from many different carbon sources. In general, the carbon source must be converted into a coenzyme A thioester of a hydroxyalkanoic acid by the metabolism of the PHA accumulating bacterium. The hydroxyacyl moieties of these thioesters are then polymerized to the polyester by PHA synthases which are the key enzymes of PHA biosynthesis. The biochemistry and the molecular biology of PHA biosynthesis and of PHA synthases have been thoroughly investigated during the last years. These studies have revealed much knowledge on the physiology as well as on the enzymology, and well characterized PHA biosynthesis genes are now available from many different bacteria (REHM and STEINBÜCHEL, 1999).

3.1 Reasons to Investigate the Conversion of Coal into PHAs

There are several advantages of processes resulting in the conversion of lignite to an intracellular polymeric product. Normally, the complex structure and heterogeneous composition of coal and other features do not lend this fossil carbon source as a preferred feedstock for fermentation processes which mostly result in the formation of extracellular products. In contrast to glucose or fatty acids, which are widely used by the industry as carbon sources for microbial fermentations, the structures of lignite and other coals are rather heterogeneous and could result in the formation of several undesired by-products from the various constituents. In addition, the downstream processing, i.e., the isolation of extracellular products from the medium must be rather tedious because the product has to be separated from many components including residual components of the lignite.

On the other side, lignite and other coals are cheap and abundantly available feedstocks. In addition, PHAs offer the advantage that the product occurs intracellularly. Therefore, it can be at first easily separated from the medium components and the remaining coal liquefaction products by harvesting the cells. Subsequently, the PHAs can be released from the cells by processes established in the industry. Furthermore, the product has a uniform chemical structure allowing the conversion of a very complex substrate into a rather homogeneous product consisting of repeating units of enantiomerically pure hydroxyalkanoic acids. These arguments would also apply to some other intracellular products, such as triacylglycerols, which are accumulated to a large extent by many gram-positive bacteria belonging to the genus *Rhodococcus* and other actinomycetes (ALVAREZ et al., 1996; WÄLTERMANN et al., 2000), or polysaccharides. However, so far the biotechnological production of triacylglycerols or polysaccharides from coals has not been investigated.

3.2 Screening of Laboratory Strains which Grow on Lignite Depolymerization Products

No PHA accumulating bacteria were known which could use lignite or other coals directly as a carbon source. Recently, we have screened a wide range of PHA accumulating bacteria for their capability to use the chemically heterogeneous low-rank coal depolymerization products, which were obtained from the treatment of lignite with the fungi *Trichoderma atroviride* or *Clitocybula dusenii* or by chemical treatment of lignite, as sole carbon source for growth and the accumulation of PHAs (FÜCHTENBUSCH and STEINBÜCHEL, 1999).

Pseudomonas oleovorans turned out to be the most promising candidate. The growth of this bacterium depended strongly on the concentration of coal degradation products isolated from *T. atroviride* in the medium and an inhibitory effect of these degradation products on the growth of *P. oleovorans* was not observed at concentrations up to 2.5% (w/v). The maximum rate of conversion of the fed depolymerization products into bacterial dry mass was approximately 30% during submerse cultivation (FÜCHTENBUSCH and STEINBÜCHEL, 1999). In addition, *Pseudomonas putida*, *Rhodococcus opacus*, *R. ruber*, *Nocardia opaca*, and *N. corallina* used these depolymerization products also for growth, however, to a lesser extent. *Burkholderia cepacia* used only depolymerization products obtained by chemical treatment, whereas *R. erythropolis* and *R. fascians* did not use either lignite depolymerization product.

3.3 PHA Synthesis from Lignite Depolymerization Products by *Pseudomonas oleovorans* and *Rhodococcus ruber*

Accumulation of PHAs from coal depolymerization products was studied in detail in *P. oleovorans* and *R. ruber*. In mineral salts medium and under starvation for ammonium the wild type of *P. oleovorans* accumulated PHAs up to 8% of the cell dry weight. The accumulated PHAs represented a copolyester consisting of 3-hydroxydecanoate (3HD) as main constituent plus 3-hydroxyhexanoate (3HHx) and 3-hydroxydodecanoate (3HDD) as minor constituents. A recombinant strain of this *P. oleovorans* harboring the PHA biosynthesis operon of *Ralstonia eutropha* accumulated PHAs up to 6% of the cell dry weight. The accumulated PHAs consisted most probably of a blend of two different polyesters. One was poly(3-hydroxybutyrate) homopolyester contributing to approximately 9%, and the other was a copolyester consisting of 3HHx and 3-hydroxyoctanoate (3HO) as main constituent plus 3HD and 3HDD as minor constituents (FÜCHTENBUSCH and STEINBÜCHEL, 1999). *R. ruber* accumulated under the same conditions a copolyester consisting of 3-hydroxybutyrate (3HB) plus 3-hydroxyvalerate (3HV), however, the polyester contributed to only approximately 2 to 3% of the cell dry weight (FÜCHTENBUSCH and STEINBÜCHEL, 1999).

3.4 Isolation of Bacteria Capable to Grow on Lignite Depolymerization Products

The main bottlenecks in the synthesis and production of PHAs from depolymerization products are the scaling up of substrate production and the low conversion rate and low yield of PHAs obtained (KLEIN et al., 1999). This is mainly due to the current need to carry out the conversion in two separate steps, with the depolymerization of lignite by *T. atroviride* in the first step and the conversion of the depolymerization products into PHAs in the second step by a suitable bacterium. In addition, the isolation of the lignite depolymerization products from the hyphae of the fungus or from the medium is tedious. Therefore, it would be desirable to have microorganisms available that could convert lignite directly into PHA in a one-step process. One possibility would be to establish PHA biosynthesis in the lignite depolymerizing fungus by transfer of the genes for PHA biosynthesis, e.g., from *R. eutropha* or *P. oleovorans*. This has, however, not been done yet. Another possibility would be the identification of a bacterium that is also capable to depolymerize coal. If such a bacterium would not be able to synthesize PHAs, it would be no problem to establish PHA biosynthesis in this bacterium by genetic methods. Recently, we succeeded in the identification of such bacteria (FÜCHTENBUSCH et al., unpublished data). In mineral salts medium with untreated lignite as sole carbon source various gram-positive bacteria such as *Mycobacterium fortuitum*, *Micromonospora aurantiaca*, and three isolates belonging to the genus *Gordonia* (most probably to the species *polyisoprenivorans*) grew. Analysis of the remaining carbon source revealed significant changes in the structure of lignite indicating that growth had occurred at the expense of the carbon provided by lignite (H. SCHMIERS, B. FÜCHTENBUSCH, and A. STEINBÜCHEL, unpublished data). Interestingly, all these bacteria are capable to utilize natural rubber and also some synthetic rubbers as sole carbon sources for growth and were previously isolated as such. For this reason, these bacteria are currently also investigated in our laboratory (LINOS et al., 1999, 2000; BEREKAA et al., 2000). Since these bacteria are not able to synthesize PHAs, genetic transfer systems have to be developed allowing the transfer and expression of PHA biosynthesis genes from other bacteria.

4 Environmental Aspects of Coal Bioconversion

4.1 Sulfur in Coal and its Removal

The total sulfur content of coal consists of sulfur which is part of the coal's molecular structure (organic sulfur) and inorganic sulfidic minerals, mainly pyrite (FeS_2).

During coking and smelting processes and especially during combustion of coal the sulfur is oxidized and released mainly as SO_2, which leads to pollution of the atmosphere and water and soil ecosystems. These environmental impacts associated with the combustion of coal led to the development of new processes to control the resulting emissions. In many countries clean coal technologies have to be applied on the basis of regulations concerning fuel quality and emission standards for dust, CO_2, SO_2, and NO_x in the flue gas.

In Germany the use even of pyrite-free coal does not allow operation of large industrial power stations (>50 MW_{th}) without flue gas desulfurization because of the remaining organically bound sulfur contained in German hard coal. For economic reasons the production of low-sulfur coal, i.e., low-pyrite coal, might be preferred for small-scale combustion units.

4.1.1 Removal of Inorganic Sulfur

The inorganic sulfur species are represented mainly by sulfides. The principal example of the sulfides in coal is sulfur pyrite, in the form of pyrite. Marcasite in the form of the rhombic modification of sulfur pyrite has also been found, but its percentage in German coals is extremely low. Pyrite, the dominant iron sulfide, has been studied many times. The fine-crystalline and fine-concretionary pyrites are syngenetic and early diagenetic occurrences in a stringently anaerobic environment. These concretions may also contain sulfide minerals (galena, sphalerite, etc.).

Unlike the sulfides, sulfur also occurs fixed to the coal substance. It is then referred to as organic sulfur. The organic sulfur is covalently bound into the large complex structure of the coal and was identified mainly as sulfur compounds like thiophenols, thioethers, and dibenzothiophene. The ratio of organic to inorganic (pyritic) sulfur depends on the particular deposits and their conditions of formation and may amount to between 10 and more than 95% of total sulfur. The average ratio worldwide is around 50%. It is not known whether a correlation exists between the percentages. Research work seems to indicate that this is true at least of certain deposits (GRAY et al., 1963).

Details concerning origin and kind of sulfur in coal and its removal by mechanical, chemical and microbial methods is given by BOGENSCHNEIDER et al. (in press).

The microbial sulfide oxidation, which has already been used successfully on a commercial scale to win metals from ores (TORMA, 1977; BRIERLEY, 1978; OLSON and KELLY, 1986) was investigated intensively for the selective solubilization of pyrite associated with coal (KARGI, 1984; OLSON and BRINKMANN, 1986; MONTICELLO and FINNERTY, 1985). This microbial method of pyrite removal is claimed to offer the advantage of simple equipment, ambient temperatures and pressures, and almost no need of chemicals. For the removal of pyritic sulfur leaching processes using, e.g., autotrophic *Thiobacillus* species were developed and already scaled up into pilot plants (KLEIN, et al. 1988).

At the end of 1993, in the frame of a European joint project a pilot-scale coal desulfurization plant at Porto Torres (Sardinia) (1 t d^{-1} scale) was operated (ROSSI, 1993; LOI et al., 1994; Anonymous, 1994). The main conclusions that can be derived from these studies are:

- The removal of inorganic sulfidic minerals (pyrite and other metal sulfides) from a large variety of coals is a technically feasible process.
- The residence time in the pilot plant required for a 90% removal of the pyrite present appears to be about 50% less than suggested by laboratory shake-flask experiments.
- The maximum pulp density that does not interfere with the first-order kinetics of pyrite removal appears to be far higher than indicated in the literature. Even an almost 40% (w/v) pulp density does not reduce the specific removal rate.

Nevertheless, the economics of the technology are not (yet) favorable, mainly because of the high energy need of the reactor systems.

Thiobacillus ferrooxidans does not play a key role in the process and thus is not an appropriate model organism. In the mixed cultures from this experiment and other research, *Leptospirillum ferrooxidans* together with sulfur oxidizers like *Thiobacillus caldus* seem to be far more important in bacterial leaching of sulfidic minerals.

Although variants of the coal desulfurization process using a combination of attack on pyrite followed by froth flotation or oil agglomeration give rise to substantial sulfur removal, they are at present not of economic interest.

Approaches to large-scale biodepyritization of coal by heap leaching have been undertaken in a collaboration between the US Bureau of Mines (now liquidated) and the Pittsburgh Energy Technology Center. In one experiment using 23 t run-of-mine stoker-sized coal (<5 cm), about 50% of the pyritic sulfur was removed after 1 year of leaching. In a second experiment using 10 t cleaned coal (6–18 mm sized), about one-third of the pyritic sulfur was leached after 11 months. Some precipitation of elemental sulfur was found (HYMAN et al., 1990; ANDREWS et al., 1992; SHARP, 1992).

Basic microbiological studies on the mechanisms by which bacteria attack the extremely insoluble metal sulfide minerals have made considerable progress in recent years. It is now evident that direct enzymatic attack on metal sulfides does not occur. Instead, metal sulfides are oxidatively degraded by one or more of two different chemical pathways, depending upon the mineral. Sphalerite, galena, and chalcopyrite are attacked by Fe(III) ions and/or protons leading to oxidation via polysulfides mainly to elemental sulfur. Pyrite is attacked only by Fe(III) ions, leading to oxidation via polythionates mainly to sulfate. Furthermore, the involvement and importance of microbial extracellular polymeric substances for adhesion of bacterial cells to metal sulfides and the initiation of the (chemical) degradation became obvious (SCHIPPERS, 1998; SCHIPPERS et al., 1999). This knowledge may lead to new possibilities for improvement of desulfurization rates, e.g., by quantitative oxidation of the sulfur moiety to sulfate due to appropriate selection of microorganisms and growth conditions.

But up to now for economic reasons no commercial application has been reported (KLEIN, 1998).

4.1.2 Removal of Organosulfur Compounds

The organic sulfur in coal is covalently bound into its large complex structure. The organosulfur compounds can be described by the following classification:

- aliphatic or aromatic thiols (mercaptans, thiophenols),
- aliphatic, aromatic, or mixed sulfides (thioethers),
- aliphatic, aromatic, or mixed disulfides (dithioethers),
- heterocyclic compounds of the thiophene type (dibenzothiophene).

The content of thiols is substantially larger in high volatile than in low volatile coals. The proportion of thiophenic sulfur is greater in higher ranked coals than in lower ones.

It was generally accepted that dibenzothiophene (DBT) represents one of the frequently occurring substructures of organic sulfur compounds in coal. Therefore, in a number of studies on the biological desulfurization DBT was selected as a substrate for pre-adaptation of microbial cultures. Using these adapted microbial cultures, some authors (CHANDRA et al., 1979; GÖKCAY and YURTERI, 1983; ISBISTER and KOBYLINSKI, 1985; KLUBEK et al., 1988; RAI and REYNIERS, 1988; STEVENS and BURGESS, 1989; STONER et al., 1990; RUNNION and COMBI, 1990b) reported of relatively high desulfurization rates in coal. But as analytical estimation of sulfur species (especially of organic sulfur) is very difficult to accomplish, these results should be regarded upon critically (KLEIN et al., 1994).

Up to now a process for microbial removal of organically bound sulfur from coal was not brought into application.

4.1.2.1 Biological Mechanisms for Desulfurization of Model Compounds

C–S bond cleavage can be regarded as an important precondition for the desulfurization of fossil fuels. Consequently, biological reaction sequences which lead to a release of sulfur from the organic matrix are of great interest. The research on the reaction mechanism of biological desulfurization focused on the study of the degradation of model compounds like DBT or benzylmethylsulfide (BMS).

In principle, hydrolytic, reductive or oxidative mechanisms can be assumed. However, because of the high stability of the C–S bond a hydrolysis can only be found after previous activation of the molecule by introduction of electron withdrawing substituents as reported for the degradation of thiophene-1-carboxylate (CRIPPS, 1973). Reductive mechanisms for the desulfurization of DBT and dibenzylsulfide by methanogenic or sulfur reducing bacteria have been reported by several authors (KIM et al., 1990; KÖHLER et al., 1984; MILLER, 1992). Hydrogen serves as reactant and the contribution of a hydrogenase has been proposed.

A frequently reported strategy of bacterial attack has been described by KODAMA et al. (1973). This reaction sequence is analogous to the initial step in the pathway of naphthalene degradation and can be explained as cometabolic activity of PAH degrading bacteria even though a growth on DBT is possible by utilization of the C_3 residue of the thiophenic ring as growth substrate. A specific sulfur release could not be found in this case since formylbenzothiophene remains as dead end metabolite in the culture fluid.

A sulfur specific pathway was proposed by KILBANE (1990). This so-called "4s-pathway" is initiated by two sulfoxidations leading to DBT-1-dioxide. By the following hydrolytic cleavage dibenzyl-1-sulfonic acid is formed. In the fourth step sulfate is released and 1-hydroxydibenzyl is produced. In parts similar is the reaction pathway proposed by VAN AFFERDEN et al. (1990). A *Brevibacterium* sp. which was isolated due to its ability to use DBT as sole source of sulfur also starts with two sulfoxidation steps. In this special case, however, an angular dioxygenation follows as has been reported by ENGESSER et al. (1989) for the degradation of dibenzofuran. During rearomatization the C–S bond is cleaved. The degradation of the resulting dihydroxydibenzylsulfinic acid follows the biphenyl pathway.

4.1.2.2 Application Potential

The results obtained indicate that sulfoxidation of low molecular coal relevant compounds is possible but only in one case sulfate was released (VAN AFFERDEN, 1991).

The removal of organic sulfur from coal appears difficult since the permeation of high polymeric material into the bacterial cell is impossible. Although the results obtained until now on the field of coal biodesulfurization give an excellent contribution to the understanding of bacterial metabolism of recalcitrant sulfurorganic compounds, there is still a long way of development to an application of biological desulfurization of coal.

Significant removal of organic sulfur from native coal has not yet been demonstrated and future research in this area is not deemed practical with particulate coal.

4.1.3 Conclusion

The total sulfur in coal consists of organic sulfur and inorganic sulfidic minerals, mainly pyrite.

The biological removal of sulfidic minerals (pyrite and other metal sulfides) from a large variety of coals is a technically feasible process. Nevertheless, the economics of the technology are not (yet) favorable, mainly due to the high energy need of the reactor systems. Although variants of the coal desulfurization process using the combination of initial attack of pyrite and froth flotation or oil agglomeration give rise to substantial sulfur removal, they are at present not of economic interest. The latest scientific findings, that a direct enzymatic attack on metal sulfides does not exist, but a chemical pathway by which pyrite is oxidized via polythionates mainly to sulfate (SCHIPPERS et al., 1999), may lead to new possibilities for improvement of desulfurization rates, e.g., by quantitative oxidizing the sulfur moiety to sulfate.

The biological removal of organic sulfur from coal is not promising since the sulfur is bound to the complex chemical coal structure and, therefore, the accessibility of the bonds for microorganisms or enzymes is low. Thus, the removal of organic sulfur by biotechnological means has not yet been demonstrated.

The installed base of flue gas desulfurization technology threatens any alternative precombustion processes. Some niche markets based on certain coals may exist which can be addressed with a simple microbial system. For example, in developing countries like China, where economy does not allow to use flue gas desulfurization, the reduction of sulfur dioxide emissions by pretreatment of coal biotechnologically would mean on a global scale a serious reduction of emissions. Another immediate niche may be in the area of coal/water slurries and also in the treatment of coal fines from ponds, which is expected to be driven by environmental regulations in the near future.

4.2 Assessment of Organic Coal Conversion Products

Handling, transport, and storage of coal and coal conversion products led to pollution of the plant sites by coal-relevant compounds such as polycyclic aromatic hydrocarbons (PAH), benzene, toluene, ethyl benzene, and xylene (BTEX), phenols, cyanides, and heavy metals. Today, a great number of abandoned sites of the coal industry such as coke oven plants and gas works are contaminated by these coal-relevant compounds.

Various of these substances, e.g., PAH or benzene, give reason to strong environmental concern due to their toxic, mutagenic, or cancerogenic properties.

These environmental impacts associated with the industrial treatment of coal led to a demand for clean technologies. This is supposed to apply also to coal bioprocessing. However, it is necessary to characterize these new technologies in terms of their environmental effects. The information obtained is indispensable for the development of new clean technologies. Therefore, products of coal bioprocessing as well as the emissions from microbial coal conversion processes have to be assessed by suitable methods.

4.2.1 Products Expected from Bioconversion of Coal

Coal is a chemical and physical heterogeneous organic rock consisting of macromolecules of high molecular weight in which aromatic structures are linked together with aliphatic and ether bridges (FRANCK and COLLIN, 1968). The degradation of coal requires the cleavage of covalent bonds in the coal matrix, viz. of the ether, methylene, and ethylene linkages. In lignite, in agreement with the younger geochemical history, lignin-like structures and salt bonds in variable amounts are also present.

For biological attack on coal extremely unspecific enzymes are, therefore, required. The known mechanisms of microbial coal conversion let expect a large number of different products with aromatic structures and a wide molecular weight range (HOFRICHTER et al., 1997c, 1999; HENNING et al., 1997; RALPH and CATCHESIDE, 1994a, b). However, it is necessary to characterize this new technology with respect to the environmental effects and the toxic, mutagenic or cancerogenic properties of the products.

Therefore, the implementation of a chemical-by-chemical approach, as made possible by the development of more specific chemical and physical analytical methods, is only a tool of limited usefulness for a proper risk assessment. As recognized in the past, such an analytical approach is actually no way to protect the environment against hazardous discharges. For this reason, bioassays as essential tools for environmental risk assessment of chemical substances were denoted by the authorities of several countries (US EPA, OECD).

4.2.2 Bioassays, a Common Tool of Emission Control

As a matter of fact, in most cases the real hazard of chemical substances is not known in detail, and the global risk induced by mixes of substances is difficult to define. A complementary approach may be proposed as to the definition of hazardous substances by bioassays or biological indicators for impact assessment or hazard diagnosis for water, soil, and air. This approach based on the measurement of effective impacts on living organisms or biological activities can be used for better assessment of toxicological hazards, risks, or impacts. Up to now, toxicology has allowed to develop a large number of bioassays using specific target organisms or biological indicators. In Tab. 3 a number of bioassays used for the ecotoxicological and genotoxicological *in vivo* monitoring are listed.

Tab. 3. Bioassays for Ecotoxicological and Genotoxicological Monitoring

Test Organism	Test Parameter
Leuciscus idus (fish)	mortality
Daphnia magna	inhibition of movement
Scenedesmus subspicatus (algae)	growth inhibition
Vibrio fischeri (bacteria)	inhibition of luminescence
Vibrio fischeri (bacteria)	growth inhibition
Pseudomonas putida (bacteria)	inhibition of respiration
Pseudomonas putida (bacteria)	growth inhibition
Ames test (bacteria)	mutagenicity
UMU test (bacteria)	mutagenicity
Mutatox assay (bacteria)	mutagenicity
SOS-Chromotest (bacteria)	mutagenicity

The application of bioassays for the toxicological monitoring of industrial emissions implies different requirements concerning the test methods:

- high sensitivity for the relevant pollutants,
- standardization/reproducibility,
- interpretability of the results,
- ecological/ecotoxicological relevance.

As mentioned above, bioconversion products of coal are characterized by a various number of heterogeneous substances and a wide range of molecular weights. Up to now, the suitability of different bioassays was proved by the use of a great number of supposed as well as obtained microbial bioconversion products (KLEIN et al., 1997; SCHACHT et al., 1999). Because of the aromatic character of these substances, PAH, microbial PAH transformation products as well as humic substances were used as model compounds.

4.2.3 Ecotoxicological Testing of Coal Derived Reference Substances and Coal Bioconversion Products

Ecotoxicological investigations were carried out using *Vibrio fischeri* (bacteria, Microtox assay), *Daphnia magna*, and *Scenedesmus subspicatus* (algae) as test organisms for characterization of the above mentioned coal derived reference substances as well as bioconversion products. These organisms represent different trophic levels of aquatic ecosystems and are of ecological/ecotoxicological relevance. The choice of these organisms as well as the performance according to DIN EN-standards (Deutsches Institut für Normung, 1989, 1991a, b) assure a high degree of standardization.

The results obtained by the ecotoxicological assays show the different sensitivity of the test organisms against the reference substances (Tab. 4). The data indicate that *D. magna* is less sensitive and no toxic response was obtained using the different compounds. Otherwise, for *V. fischeri* and *S. subspicatus* significant toxic effects were found for low molecular single PAH as well as for PAH mixtures. PAHs with a water solubility below 0,1 mg L^{-1} were not detectable by *V. fischeri*. A further result was that the values for the toxicity index (EC50/EC20) for PAH are all in the same range indicating a similar mechanism of inhibition.

The investigations using the alga *S. subspicatus* as test organism show more different results for the single PAH substances in terms of the EC20-values (Tab. 4). Therefore, this bioassay allows a more distinguished interpretation of the results obtained for PAH.

Different results were obtained for the microbial PAH transformation products. Obviously, these substances have a wide range of toxic effects according to the inhibition of *V. fischeri* as characterized by different EC20-values.

Tab. 4. Ecotoxicological Characterization of Coal Derived Reference Substances and Coal Bioconversion Products by Bioassays Using *Vibrio fischeri*, *Daphnia magna*, and *Scenedesmus subspicatus*. Peat Products and Lignites were Extracted by Water and NaOH

Sample	*Vibrio fischeri*			*Daphnia magna*	*Scenedesmus subspicatus*
	DOC [mg L^{-1}]	GL20	EC20 [mg L^{-1}]	GD0	GA20
Phenanthrene	1.0	17	0.06	–	8
Fluoranthene	0.1	1	>0.1	–	12
Naphthalene	21.0	80	0.26	–	–
Acenaphthene	2.0	7	0.29	–	–
Anthracene oil	46	263	0.18	–	24
Hard coal tar	<1	3	<0.33	–	2
1,2-Naphthochinone	6.0	926	0.0065	–	1,536
1-Naphthol	35.0	48	0.73	–	256
Gentisic acid	63.0	3	21.0	–	12
Catechol	100.0	17	5.9	–	96
Hms (Fluka)	250	20	12.5	1	96
Hms (Aldrich)	1,360	128	10.6	2	256
Peat product (H_2O)	122	2	–	1	6
Peat product (NaOH)	4,800	256	18.8	4	192
Lignite type A (H_2O)	155	8	19.4	2	24
Lignite type A (NaOH)	1,540	118	13.1	2	768
Lignite type B (H_2O)	85	2	42.5	1	24
Lignite type B (NaOH)	2,760	384	7.2	3	1,536
Lignite type C (H_2O)	195	12	16.3	1	24
Lignite type C (NaOH)	3,340	512	6.5	2	1,024
I (Fulvic acids)	81	2	–	3	27.0
II (Fulvic acids)	56	2	–	1	–
III (Humic acids)	58/67	5	11.7	8	8.4
IV (Humic acids)	381/128	32	11.9	32	4.0

DOC, dissolved organic carbon; –, not determined; Hms, humic substances. Bioassay parameters: GL20, the lowest dilution factor for which the light emission is reduced by less than 20%; EC20,: the concentration of substance corresponding to the GL20 value; GD0, the lowest dilution factor for which 9/10 daphnia remained able to swim; GA20, the lowest dilution factor giving a fluorescence reduction of less than 20%.

As indicated by high EC20-values, toxicity of the humic substances extracted from peat and lignite is low. Similar results were obtained from coal bioconversion products (samples I–IV, Tab. 4) from lignite obtained by the treatment with different fungi (*Clitocybula dusenii* and *Trichoderma atroviridiae*). Samples were characterized as mixtures of different compounds similar to fulvic (sample I and II) or humic acids (sample III and IV). Results for *V. fischeri* and *S. subspicatus* show only a low toxicity, demonstrated by high EC-values.

4.2.4 Mutagenicity Testing of Coal Derived Reference Substances and Coal Bioconversion Products

Up to now, the available genotoxicological data of coal derived reference substances and coal bioconversion products have been obtained by mutagenicity testing (Ames test). Bioassays were carried out using the three different indicator strains of *Salmonella typhimurium* TA98, TA100 and TA102. The assay was performed using the relevant reference substances 2-aminoanthracene, 2-nitrofluorene, cumene hydroperoxide, and methylmethanesulfonate as positive controls and dissolution media dimethylsulfoxide (DMSO) and water as negative controls (Tab. 5). Toxicity results are presented as reversion per plate (mean values) in the presence of test samples. A mutagenic effect will be indicated if reversion frequency is significantly higher than in the negative control samples containing only DMSO or water.

Besides bioconversion products of PAH (1-naphthol, 2-naphthoate, 1,2-dihydroxynaphthalene, and 1,2-naphthochinone), humic substances as well as extracts from lignite were tested. The results showed that except for the reference substances no mutagenic effect was obtained for the tested substances (Tab. 5). Similar results were found for the coal bioconversion products of fungi (Tab. 6; Deutsches Institut für Normung, 1995).

4.2.5 Conclusions and Prospects

The investigations available up to now were focused on ecotoxicological and mutagenicity testing methods. Therefore, different bioassays with representative organisms, like *Vibrio fischeri*, *Daphnia magna*, and *Scenedesmus subspicatus*, were used for the determination of the aqueous exposure route. These organisms represent different trophical levels of aquatic ecosystems and are of ecological/ecotoxicological relevance (KANNE, 1989). All bioassays used are covered by a DIN-standard. Thus a high degree of standardization is assured. The bioassays were tested for their suitability for determination of ecotoxicological or mutagenic effects of coal-relevant substances and bioconversion products.

For classification of the ecotoxicological effect caused by microbial transformation and conversion products of coal it is necessary to compare the results to those obtained with well-known toxic and harmless compounds

Tab. 5. Mutagenicity Testing of Coal Derived Reference Substances by the Ames Test. Results for the Different Test Compounds and Mixtures are Presented as Reversions per Plate Relative to those of the Positive and Negative Control. For the Strains TA98 and TA100 Metabolic Activation of Test Substance by Addition of Microsomal Liver Fraction (S9) was Carried out

	Reversions per Plate (Mean Value)				
Sample / Tester Strain	TA98	TA98 + S9	TA100	TA100 + S9	TA102
DMSO	30	18	140	179	79
Water	33	22	157	151	83
2-Aminoanthracene	–	518	–	2,223	–
2-Nitrofluorene	580	–	–	–	–
Cumene hydroperoxide	–	–	–	–	388
Methylmethanesulfonate	–	–	735	–	–
1-Naphthol	21	55	168	161	70
2-Naphthoate	16	22	148	159	52
1,2-Dihydroxynaphthalene	22	20	119	141	130
1,2-Naphthoquinone	19	18	123	148	159
Hms Aldrich	33	35	160	188	146
Hms Fluka	18	32	154	197	110
Lignite type A (NaOH)	30	23	165	178	95
Lignite type A (H_2O)	18	17	158	171	97
Lignite type B (NaOH)	23	30	157	176	107
Lignite type B (H_2O)	43	19	171	174	115

– Not determined

Tab. 6. Mutagenicity Testing of Coal Bioconversion Products by the Ames Test. Results for the Different Test Compounds are Presented as Reversions per Plate Relative to those of the Positive and Negative Control. For the Strains TA98 and TA100 Metabolic Activation of Test Substance by Addition of Microsomal Liver Fraction (S9) was Carried out

	Reversions per Plate (Mean Value)				
Sample / Tester Strain	TA98	TA98+S9	TA100	TA100+S9	TA102
DMSO	13	11	–	–	110
Water	–	–	101	123	–
2-Aminoanthracene	–	470	–	1,254	263
2-Nitrofluorene	380	–	–	–	–
Cumene hydroperoxide	–	–	–	–	233
Methylmethanesulfonate	–	–	959	–	–
I (Fulvic acids)	15	18	122	101	100
II (Fulvic acids)	15	11	90	108	116
III (Humic acids)	10	16	148	130	–
IV (Humic acids)	17	9	130	148	–

– Not determined

with similar chemicophysical properties. All PAH compounds investigated show significant toxicological effects and confirm results presented otherwise (Roos et al., 1996). Natural humic substances which have physicochemical properties similar to those of coal bioconversion products represent the natural non-specific background toxicity. These natural substances are characterized by high EC20-values indicating a low non-specific background toxicity. Similar results, as to the ecotoxicological effects, were obtained by substances originating from water and alkaline extracts of lignite. Bioconversion products originating from the solubilization of coal by fungi show a similar pattern of ecotoxicological effects.

The data according to mutagenicity obtained by the Ames test show comparable results. No mutagenic effects were observed testing various microbial PAH transformation products, humic substances, lignite-extractable compounds, and bioconversion products. Obviously, there are significant differences compared to other investigations using other aromatic substances, e.g., PAH (Reinke et al., 1994). The results point out that high molecular weight and polarity of products effect less toxicity obviously due to less permeation of these substances into the cells.

The ecotoxicological and mutagenic data established up to now, in principle, show the suitability of bioassays used for the ecotoxicological characterization of coal bioconversion products. Future investigations have to be focused onto the suitability of further bioassays and testing of further coal bioconversion products.

5 Perspectives

Coal will remain the most important resource for energy supply, since coal accounts for about 75% of total energy resources worldwide.

For more than 20 years now, development of biologically based systems for the processing of fossil fuels has continued to be a viable topic for scientific research and commercial development on a national basis.

Bioconversion

Bioconversion of coal is an interdisciplinary area of relatively recent emergence. Its success is critically dependent on effective communication between experts in a wide range of scientific and engineering disciplines including microbiologists, molecular biologists, biochemists, coal chemist and engineers (KLEIN et al., 1999).

Despite all the difficulties and the complexity in this field of research described in this chapter, the progress achieved in the last decade is evident. Since about a decade there is already one product on the American market: a fertilizer and soil conditioner, which is made from biologically solubilized lignite. Another success is the production of "bioplastic" PHA, a comparably pure and highly value added product made by bacterial strains grown on a totally heterogeneous mixture of coal-derived humic and fulvic acids (STEINBÜCHEL and FÜCHTENBUSCH, 1997; FÜCHTENBUSCH and STEINBÜCHEL, 1999).

The recent advances in the understanding of lignite solubilization by molds as well as depolymerization by ligninolytic fungi offer new possibilities for utilizing the carbon from lower rank coals. Particularly promising biocatalysts for biotechnological applications (not only regarding coal bioconversion!) are the extracellular oxidoreductases from wood- and litter-degrading basidiomycetes. Since the biochemistry of enzymatic coal disassembly is only just emerging, it is not yet possible to predict the full range of products that might be derived from coal bioconversion. Nevertheless, since the coal humic and fulvic acids released from lignites by peroxidases appear to be assimilated by certain microbes, there should be paths to the synthesis of virtually any organic compound. Molecular biological approaches may be helpful in future to optimize such pathways. Their realization, however, requires long-term research, but some opportunities are more directly approachable such as adapting bacteria or yeasts able to accumulate useful compounds from biosolubilized coal as their carbon source.

Environmental Aspects

Removal of inorganic sulfur and some metals from coal by bioleaching is technically possible, but so far it seems that removal of organically bound sulfur is not. Thus, for coals with a high amount of organic sulfur, biodesulfurization is not economical.

Bioassays with organisms representing different trophic levels like bacteria, daphnia, and algae are suited for determination of ecotoxicological/toxicological and of mutagenic/cancerogenic effects of coal bioconversion products.

No mutagenic effects were observed testing various microbial PAH transformation products, humic substances, lignite-extractable compounds, and bioconversion products.

6 References

ALEEM, M. I. H., BHATTACHARYYA, D., HUFFMAN, G. P., KERMODE, R. I., MURTY, M. V. S. (1991), Microbial hydrogenation of coal and diphenylmethane, *Am. Chem. Soc. Div. Fuel Chem.* **36**, 53–57.

ALEKSANDROV, I. V., KOSSOV, I. I., KAMNEVA, A. I. (1988), Reclamation of solonchak soils by using modified brown coal, *Khim. Tverd. Topl. Moscow* **1**, 49–53.

ALVAREZ, H. M., MAYER, F., FABRITIUS, D., STEINBÜCHEL, A. (1996), Formation of intracytoplasmic lipid inclusions by *Rhodococcus opacus* strain PD630, *Arch. Microbiol.* **165**, 377–386.

ANDERSON, A. J., DAWES, E. A. (1990), Occurrence, metabolism, metabolic role, and industrial uses of bacterial polyhydroxyalkanoates, *Microbiol. Rev.* **54**, 450–472.

ANDREWS, G. F., STEVENS, C. J., LEEPER, S. A. (1992), Heaps as bioreactors for coal processing, in: *Proc. 14th Symp. Biotechnology for Fuels and Chemicals*, May 1992, Gatlinburg, TN.

Anonymous (1994), Biodesulfurisation of Coal, Pilot-scale R&D. Final *Report* of Contract No. Joufl-0039, CEC Joule Programme, Sub-programme: Energy from Fossil Sources: Solid Fuels.

Anonymous (2000), *Statistik der Kohlewirtschaft*. Essen.

BANERJEE, A. K., CHOUDHURY, D., CHOUDHURY, S. S. (1989), Chemical changes accompanying oxygenation of coal by air and deoxygenation of oxidized coal by thermal treatment, *Fuel* **68**, 1129–1133.

BEREKAA, M. M., LINOS, A., REICHELT, R., KELLER, U., STEINBÜCHEL, A. (2000), Effect of pretreatment of rubber material on its biodegradability by various rubber degrading bacteria, *FEMS Microbiol. Lett.* **184**, 199–206.

BLANCHETTE, R. A. (1991), Delignification by wood decay fungi, *Annu. Rev. Phytopathol.* **29**, 381–398.

BLINKOVSKY, A. M., MCELDOON, J. P., ARNOLD, J. M., DORDICK, J. S. (1994), Peroxidase-catalyzed depolymerization of coal in organic solvents, *Appl. Biochem. Biotechnol.* **49**, 153–164.

BOGENSCHNEIDER, B., JUNG, R. G., KLEIN, J. (in press), Desulfurization of coal, in: *Biopolymers* Vol. 1: *Lignin, Humic Substances and Coal* (STEINBÜCHEL, A., HOFRICHTER, M., Eds.). Weinheim: Wiley-VCH.

BOURBONNAIS, R., PAICE, M. G. (1990), Oxidation of non-phenolic substrates. An expanded role of laccase in lignin biodegradation, *FEBS Lett.* **267**, 99–102.

BRECHT, C., GOETHE, H. G., KLATT, H. J., MIDDELSCHULTE, A. REINTGES, H. et al. (1996), Jahrbuch 1996: Bergbau, Erdöl und Erdgas, Petrochemie, Elektrizität, Umweltschutz, pp. 1109–1152. Essen: Verlag Glückauf.

BRECKENRIDGE, C. R., POLMAN, J. K. (1994), Solubilization of coal by biosurfactant derived from *Candida bombicola*, *Geomicrobiol. J.* **12**, 285–288.

BRIERLEY, C. L. (1978), Bacterial leaching, *CRC Crit. Rev. Microbiol.* **6**, 207–262.

BUBLITZ, F., GÜNTHER, T., FRITSCHE, W. (1994), Screening of fungi for the biological modification of hard coal and coal derivatives, *Fuel Proc. Technol.* **40**, 347–354.

BUSWELL, J. A., ODIER, E. (1987), Lignin biodegradation, *CRC Crit. Rev. Biotechnol.* **6**, 1–60.

CALL, H. P., MÜCKE, I. (1995), The laccase mediator system (LMS) – a new concept, in: *Proc. 6th Int. Conf. Biotechnology in Pulp and Paper Industry*, Abstract O-F3, 11–15 June 1995, Vienna (Austria).

CAMPBELL, J. A., STEWART, D. L., MCCULLOUCH, M., LUCKE, R. B., BEAN, R. M. (1988), Biodegradation of coal-related model compounds, *Am. Chem. Soc. Div. Fuel Chem. Prep.* **33**, 514–523.

CATCHESIDE, D. E. A., RALPH, J. P. (1997), Biological processing of coal and carbonaceous material, in: *Proc. 9th Int. Conf. Coal Science* Vol. I (ZIEGLER, A., VAN HEEK, K. H., KLEIN, J., WANZL, W., Eds.), pp. 11–18. Essen: P & W Druck & Verlag.

CHANDRA, D., ROY, P., MISHRA, A. K., CHAKRABARTI, J. N., SENGUPTA, B. (1979), Microbial removal of organic sulfur from coal, *Fuel* **58**, 549–550.

COHEN, M. S., GABRIELE, P. D. (1982), Degradation of coal by the fungi *Polyporus versicolor* and *Poria monticola*, *Appl. Environ. Microbiol.* **44**, 23–27.

COHEN, M. S., BOWERS, W. C., ARONSON, H., GREY, E. T. (1987), Cell-free solubilization of coal by *Polyporus versicolor*, *Appl. Environ. Microbiol.* **53**, 2840–2844.

COHEN, M. S., FELDMANN, K. A., BROWN, C. S., GREY, E. T. (1990), Isolation and identification of the coal-solubilizing agent produced by *Trametes versicolor*, *Appl. Environ. Microbiol.* **56**, 3285–3290.

CRAWFORD, D. L., GUPTA, R. K. (1991), Influence of cultural parameters on the depolymerization of soluble lignite coal polymer by *Pseudomonas cepacia* DLC-07, *Res. Conserv. Recycl.* **5**, 245–254.

CRAWFORD, D. L., GUPTA, R. K., DEOBALD, L. A., ROBERTS, D. J. (1990), Biotransformation of coal and coal substructure model compounds by bacteria under aerobic and anaerobic conditions, in: *Proc. 1st Int. Symp. Biological Processing of Coal* (YUNKER, S., RHEE, K., Eds.), pp. 429–443. Electric Power Research Institute, Palo Alto, CA.

CRIPPS, R. E. (1973), The microbial metabolism of thiophen-2-carboxylate, *Biochem. J.* **134**, 353–366.

Deutsches Institut für Normung eV 38412 part 30 (1989), *Bestimmung der nicht akut giftigen Wirkung von Abwasser gegenüber Daphnien über Verdünnungsstufen*. Berlin: Beuth Verlag.

Deutsches Institut für Normung eV 38412 part 33 (1991a), *Bestimmung der nicht giftigen Wirkung von Abwasser gegenüber Grünalgen* (*Scenedesmus-Zellvermehrungs-Hemmtest*). Berlin: Beuth Verlag.

Deutsches Institut für Normung eV 38412 part 34 (1991b), *Bestimmung der Hemmwirkung von Abwasser auf die Lichtemission von Photobacterium phosphoreum*. Berlin: Beuth Verlag.

Deutsches Institut für Normung eV Entwurf UA 12 (1995), *Bestimmung des erbgutverändernden Potentials von Wasser und Abwasser mit dem Salmonella/Mikrosomen-Test*. Berlin: Beuth Verlag.

DOI, Y. (1990), *Microbial Polyesters*. New York: VCH.

DUNFORD, H. B. (1991), Horseradish peroxidase: structure and kinetic properties, in: *Peroxidases in Chemistry and Biology* (EVERSE, J., EVERSE, K. E., GISHAM, M. B., Eds.), pp. 1–24. Boca Raton, FL: CRC Press.

DURIE, R. A., LYNCH, B. M., STRENHELL, S. (1960), Comparative studies of brown coal and lignin, *Austral. J. Chem.* **13**, 156–168.

EGGERT, C., TEMP, U., ERIKSSON, K. E. L. (1996), A fungal metabolite mediates degradation of nonphenolic lignin structures and synthetic lignin by laccase, *FEBS Lett.* **391**, 144–148.

ENGESSER, K. H., STRUBEL, V., CHRISTOGLOU, K., FISCHER, P., RAST, H. G. (1989), Dioxygenolytic cleavage of aryl ether bonds: 1,10 dihydro-1,10 dihydrofluoren-9-one, novel arene dihydrodiol as evidence for angular dioxygenation of dibenzofuran, *FEMS Microbiol. Lett.* **65**, 205–210.

ENGESSER, K. H., DOHMS, C., SCHMID, A. (1994), Microbial degradation of model compounds of coal and production of metabolites with potential commercial value, *Fuel Proc. Technol.* **40**, 217–226.

FAISON, P. D. (1991), Biological coal conversions, *Crit. Rev. Biotechnol.* **11**, 347–366.

FAKOUSSA, R. M. (1981), Kohle als Substrat für Mikroorganismen: Untersuchungen zur mikrobiellen Umsetzung nativer Steinkohlen, *Thesis*, University of Bonn (Germany) (Translated as: Coal as a substrate for microorganisms: investigations of the microbial decomposition of untreated hard coal. Prepared for U.S. Department of Energy, Pittsburgh Energy Technology Center, 1987).

FAKOUSSA, R. M. (1988), Production of water-soluble coal substances by partial microbial liquefaction of untreated coal, *Resour. Conserv. Rec.* **1**, 251–260.

FAKOUSSA, R. M. (1990), Microbiological treatment of German hard coal, in: *Bioprocessing and Biotreatment of Coal* (WISE, D. L., Ed.), pp. 95–107. New York: Marcel Dekker.

FAKOUSSA, R. M. (1991), Function of active compounds in biological coal conversion. DGMK-Report No. 9106, ISBN 3-928164-26-0, pp. 187–198, Freiberg/Holzhau, Deutsche Wissenschaftliche Gesellschaft für Erdöl, Erdgas und Kohle.

FAKOUSSA, R. M. (1992), Mikroorganismen erschließen Kohle-Ressourcen, *Bioengineering* **4**, 21–28.

FAKOUSSA, R. M. (1994), The influence of different chelators on the solubilization/liquefaction of different pretreated and natural lignites, *Fuel Proc. Technol.* **40**, 183–192.

FAKOUSSA, R. M., FROST, P. (1999), *In vivo*-decolorization of coal-derived humic acids by laccase-excreting fungus *Trametes versicolor*, *Appl. Microbiol. Biotechnol.* **52**, 60–65.

FAKOUSSA, R. M., HOFRICHTER, M. (1999), Minireview: microbiology and biotechnology of coal degradation, *Appl. Microbiol. Biotechnol.* **52**, 25–40.

FAKOUSSA, R. M., TRÜPER, H. G. (1983), Kohle als mikrobielles Substrat unter aeroben Bedingungen, in: *Biotechnologie im Steinkohlebergbau*, pp. 41–50. Essen: Bergbauforschung GmbH.

FAKOUSSA, R. M., WILLMANN, G. (1991), Investigations into the mechanism of coal solubilisation/liquefaction: Chelators, in: *Proc. 2nd Int. Symp. Biological Processing of Coal* Vol. 4 (EPRI & US-Dept of Energy, Eds.), pp. 23–29. Palo Alto, CA: EPRI.

FAKOUSSA, R. M., FROST, P., SCHWÄMMLE, A. (1997), Enzymatic depolymerization of low-rank coal (lignite), in: *Proc. 9th Int. Conf. Coal Science* (ZIEGLER, A., VAN HEEK, K. H., KLEIN, J., WANZL, W., Eds.), pp. 1591–1594. Essen: P & W Verlag.

FAKOUSSA, R. M., LAMMERICH, H. P., GÖTZ, G. K. E. (1999a), Behandlung von Braunkohlebestandteilen zum Zwecke der Veredelung, *Deutsche Patentanmeldung/German Patent Application:* 24. 9. 99, Nr. 19945975.4.

FAKOUSSA, R. M., LAMMERICH, H. P., GÖTZ, G. K. E., TESCH, S. (1999b), The second step: Increasing the hydrophobicity of coal-derived humic acids enzymatically, in: *Proc. 5th Int. Symp. Biological Processing of Fossil Fuels* 26–29th Sept. 1999, Madrid, Spain (Abstract).

FISCHER, F. (1932), Biologie und Kohle, *Angew. Chem.* **45**, 185–194.

FISCHER, F., FUCHS, W. (1927a), Über das Wachstum von Schimmelpilzen auf Kohle (vorläufige Mitteilung), *Brennst. Chem.* **8**, 231–233.

FISCHER, F., FUCHS, W. (1927b), Über das Wachstum von Pilzen auf Kohle (2. Mitteilung), *Brennst. Chem.* **8**, 293–295.

FORRESTER, I. T., GRABSKI, A. C., BURGESS, R. R., LEATHAM, G. F. (1988), Manganese, Mn-dependent peroxidases, and biodegradation of lignin, *Biochem, Biophys. Res. Commun.* **157**, 992–999.

FORTUN, C., SALAS, M. L., ORTEGA, C. (1986), Response of lettuce (*Lactuca sativa*) to lignite treatments, *An. Edafol. Agrobiol.* **45/11**, 1627–1634.

FRANCK, H. G., COLLIN, G. (1968), Steinkohlenteer. Berlin: Springer-Verlag.

FRATERRIGO, T. L., MILLER, C., REINHAMMAR, B., MCMILLIN, D. R. (1999), Which copper is paramagnetic in the type 2/type 3 cluster of laccase, *J. Biol. Inorg. Chem.* **4**, 183–187.

FREDRICKSON, J. K., STEWART, D. L., CAMPBELL, J. A., POWELL, M. A., MCMULLOCH, M. et al. (1990), Biosolubilization of low-rank coal by *Trametes versicolor* siderophore-like product and other complexing agents, *J. Ind. Microbiol.* **5**, 401–406.

FRISTAD, W. E., FRY, M. A., KLANG, J. A. (1983), Persulfate/silver ion decarboxylation of carboxylic acids. Preparation of alkanes, alkenes and alcohols, *J. Org. Chem.* **48**, 3575–3577.

FRITSCHE, W., HOFRICHTER, M. (2000), Aerobic degradation by microorganisms, in: *Biotechnology* 2nd Edn. Vol. 11b (REHM, H. J., REED, G., PÜHLER, A., STADLER, P., Eds.), pp. 145–167. Weinheim: Wiley-VCH.

FRITSCHE, W., HOFRICHTER, M., ZIEGENHAGEN, D. (1999), Biodegradation of coals and lignite, in: *Biochemical Principles and Mechanisms of Biosynthesis and Biodegradation of Polymers* (STEINBÜCHEL, A., Ed.), pp. 265–272. Weinheim: Wiley-VCH.

FROST, P. J. (1996), Untersuchungen zum Mechanismus der mikrobiellen Verflüssigung Rheinischer Braunkohle durch Pilze, *Thesis*, University of Bonn, Germany.

FÜCHTENBUSCH, B., STEINBÜCHEL, A. (1999), Biosynthesis of polyhydroxyalkanoates from low-rank coal liquefaction products by *Pseudomonas oleovorans* and *Rhodococcus ruber*, *Appl. Microbiol. Biotechnol.* **52**, 91–95.

GALLE, E. (1910), Über die Selbstentzündung der Steinkohle, *Zbl. Bakt. Parasitenk. II* **28**, 461–472.

GLENN, J. K., GOLD, M. H. (1985), Purification and characterization of an extracellular Mn(II)-dependent peroxidase from the lignin-degrading basidiomycete *Phanerochaete chrysosporium*, *Arch. Biochem. Biophys.* **242**, 329–341.

GÖKCAY, C. F., YURTERI, R. N. (1983), Microbial desulfurization of lignites by a thermophilic bacterium, *Fuel* **62**, 1223–1224.

GÖTZ, G. K., SCHWÄMMLE, A., TEMP, U., EGGERT, C., FAKOUSSA, R. M. (1999), Mediator-assisted depolymerisation of brown coal by redox enzymes: Scope and limitations, in: *Proc. 5th Int. Symp. Biological Processing of Fossil Fuels* 26–29th Sept. 1999, Madrid, Spain (Abstract).

GRASSET, L., AMBLES, A. (1998), Structure of humin and humic acids from an acid soil as revealed by phase transfer catalyzed hydrolysis, *Org. Geochem.* **29**, 881–891.

GRAY, R. N., SCHAPIRO, N., COE, G. D. (1963), Distribution and forms of sulfur in a high-volatile Pittsburgh seam coal, *Trans. SME-AIME* **226**, 113–121.

GRETHLEIN, H. E. (1990), Pretreatment of lignite, in: *Bioprocessing and Biotreatment of Coal* (WISE, D. L., Ed.), pp. 73–81. New York: Marcel Dekker.

GROßE, S. A. (2000), Untersuchungen zum Chemismus der Biokonversion von Braunkohle durch kohledegradierende Deuteromyceten, *Thesis*, Bergakademie TH Freiberg, Germany.

GUPTA, R. K., SPIKER, J. K., CRAWFORD, D. L. (1988), Biotransformation of coal by ligninolytic *Streptomyces*, *Can. J. Microbiol.* **34**, 667–676.

GYORI, S. (1986), The practice and long-range development possibilities of the agricultural utilization of second rate coals, *Magy. Kem. Lapja* **45**, 176–177.

HATAKKA, A. (1994), Lignin-modifying enzymes from selected white-rot fungi: production and role in lignin degradation, *FEMS Microbiol. Rev.* **13**, 125–135.

HATAKKA, A. (in press), Biodegradation of lignin, in: *Biopolymers* Vol. 1: *Lignin, Humic Substances and Coal* (STEINBÜCHEL, A., HOFRICHTER, M., Eds.). Weinheim: Wiley-VCH.

HATCHER, P. G. (1990), Chemical structural models for coalified wood (vitrinite) in low rank coal, *Org. Geochem.* **16**, 959–970.

HAYATSU, R., WINANS, R. E., MCBETH, R. L., SCOTT, R. G., MOORE, L. P., STUDIER, M. H. (1979), Lignin-like polymers in coal, *Nature* **278**, 41–43.

HEINFLING, A., MARTÍNEZ, M. J., MARTÍNEZ, Á. T., BERGBAUER, M., SZEWZYK, U. (1998), Purification and characterization of peroxidases from the dye-decolourizing fungus *Bjerkandera adusta*, *FEMS Microbiol. Lett.* **165**, 43–50.

HENNING, K., STEFFES, H. J., FAKOUSSA, R. M. (1997), Effects on the molecular weight distribution of coal-derived humic acids studied by ultrafiltration, *Fuel Proc. Techn.* **52**, 225–237.

HOCKING, P. J., MARCHESSAULT, R. H. (1994), Biopolyesters, in: *Chemistry and Technology of Biodegradable Polymers* (GRIFFIN, G. J. L., Ed.), pp. 48–96. Glasgow: Blackie Academic.

HODEK, W. (1994), The chemical structure of coal in regard of microbiological degradation, *Fuel Proc. Technol.* **40**, 369–378.

HOFRICHTER, M., FAKOUSSA, R. (in press), Microbial degradation and modification of coal, in: *Biopolymers* Vol. 1: *Lignin, Humic Substances and Coal* (STEINBÜCHEL, A., HOFRICHTER, M., Eds.). Weinheim: Wiley-VCH.

HOFRICHTER, M., FRITSCHE, W. (1996), Depolymerization of low-rank coal by extracellular fungal enzyme systems. I. Screening for low-rank coal depolymerizing activities, *Appl. Microbiol. Biotechnol.* **46**, 220–225.

HOFRICHTER, M., FRITSCHE, W. (1997a), Depolymerization of low-rank coal by extracellular fungal enzyme systems. II. The ligninolytic enzymes of the coal-humic-acid-degrading fungus *Nematoloma frowardii* b19, *Appl. Microbiol. Biotechnol.* **47**, 419–424.

HOFRICHTER, M., FRITSCHE, W. (1997b), Depolymerization of low-rank coal by extracellular fungal enzyme systems. III. *In vitro* depolymerization of coal humic acids by a crude preparation of manganese peroxidase from the white-rot fungus *Nematoloma frowardii* b19, *Appl. Microbiol. Biotechnol.* **47**, 566–571.

HOFRICHTER, M., BUBLITZ, F., FRITSCHE, W. (1997a), Fungal attack on coal. I. Modification of hard coal by fungi, *Fuel Proc. Technol.* **52**, 43–53.

HOFRICHTER, M., BUBLITZ, F., FRITSCHE, W. (1997b), Fungal attack on coal. II. Solubilization of low-rank coal by filamentous fungi, *Fuel Proc. Technol.* **52**, 55–64.

HOFRICHTER, M., ZIEGENHAGEN, D., SORGE, S., BUBLITZ, F., FRITSCHE, W. (1997c), Enzymatic depolymerization of low-rank coal (lignite), in: *Proc. 9th Int. Conf. Coal Science* (ZIEGLER, A., VAN HEEK, K. H., KLEIN, J., WANZL, W., Eds.), pp. 1595–1598. Essen: P & W Verlag.

HOFRICHTER, M., SCHEIBNER, K., SCHNEEGAß, I., FRITSCHE, W. (1998a), Enzymatic combustion of aromatic and aliphatic compounds by manganese peroxidase from *Nematoloma frowardii*, *Appl. Environ. Microbiol.* **64**, 399–404.

HOFRICHTER, M., ZIEGENHAGEN, D., VARES, T., FRIEDRICH, M., JÄGER, M. G. et al. (1998b), Oxidative decomposition of malonic acid as basis for the action of manganese peroxidase in the absence of H_2O_2, *FEBS Lett.* **434**, 362–366.

HOFRICHTER, M., VARES, T., SCHEIBNER, K., GALKIN, S., SIPILÄ, J., HATAKKA, A. (1998c), Mineralization of synthetic lignin (DHP) by manganese peroxidases from *Nematoloma frowardii* and *Phlebia radiata*, *J. Biotechnol.* **67**, 217–228.

HOFRICHTER, M., ZIEGENHAGEN, D., SORGE, S., ULLRICH, R., BUBLITZ, F., FRITSCHE, W. (1999), Degradation of lignite (low-rank coal) by ligninolytic basidiomycetes and their manganese peroxidase system, *Appl. Microbiol. Biotechnol.* **52**, 78–84.

HÖLKER, U. (1998), Mechanismen der Verflüssigung

von rheinischer Braunkohle durch Pilze – ein Vergleich der Deuteromyceten *Fusarium oxysporum* und *Trichoderma atroviride*, *Thesis*, University of Bonn, Germany.

HÖLKER, U. (2000), Bioreaktor zur Fermentierung von festen Stoffen, *Patent Application* PCT/EP00/08929 and DE 19943853.6.

HÖLKER, U., FAKOUSSA, R. M., HÖFER, M. (1995), Growth substrates control the ability of *Fusarium oxysporum* to solubilize low-rank coal, *Appl. Microbiol. Biotechnol.* **44**, 351–355.

HÖLKER, U., LUDWIG, S., MÖNKEMANN, H., SCHEEL, T., HÖFER, M. (1997a), Different strategies of fungi to solubilize coal: a comparison of the deuteromycetes *Trichoderma atroviride* and *Fusarium oxysporum*, in: *Proc. 9th Int. Conf. Coal Science* Vol. III (ZIEGLER, A., VAN HEEK, K. H., KLEIN, J., WANZL, W., Eds.), pp. 1599–1602. Essen: P & W Druck & Verlag.

HÖLKER, U., MÖNKEMANN, H., HÖFER, M. (1997b), A system to analyze the complex physiological states of coal solubilizing fungi, *Fuel Proc. Technol.* **52**, 73–77.

HÖLKER, U., LUDWIG, S., SCHEEL, T., HÖFER, M. (1999a), Mechanisms of coal solubilization by the deuteromycetes *Trichoderma atroviride* and *Fusarium oxysporum*, *Appl. Microbiol. Biotechnol.* **52**, 57–59.

HÖLKER, U., POLSAKIEWICZ, M., SCHEEL, T., LUDWIG, S., SCHINKE-KISSING, S., HÖFER, M. (1999b), Coal as inductor of esterases in the deuteromycete *Trichoderma atroviride*, in: *Proc. 5th Int. Symp. Biological Processing of Fossil Fuels* 26–29th Sept 1999, Madrid, Spain (Abstract).

HÜTTINGER, K. J., MICHENFELDER, A. W. (1987), Molecular structure of a brown coal, *Fuel* **66**, 1164–1165.

HYMAN, D., HAMMACK, R., FINSETH, D., RHEE, K. (1990), Biologically mediated heap leaching for coal depyritization, *Proc. 1st Int. Symp. Biol. Processing of Coal*, EPRI GS-6970: 3-101-3-111, Palo Alto, CA.

ISBISTER, J. D., BARIK, S. (1990), Biogasification of low rank coals, in: *Microbial Transformations of Low Rank Coals* (CRAWFORD, D. L., Ed.), pp. 139–156. Ann Arbor, NY: CRC Press.

ISBISTER, J. D., KOBYLINSKI, E. A. (1985), Microbial desulfurization of coal, *Proc. 14th Int. Conf. Processing and Utilization of High Sulfur Coal*, Columbus, OH, pp. 627-641. Amsterdam: Elsevier.

Jahrbuch Bergbau, Erdöl und Erdgas, Petrochemie, Elektrizität und Umweltschutz (2000), 107. Jahrgang. Essen: Verlag Glückauf.

JENDROSSEK, D., SCHIRMER, A., SCHLEGEL, H. G. (1996), Biodegradation of polyhydroxyalkanoic acids, *Appl. Microbiol. Biotechnol.* **46**, 451–463.

KANNE, R. (1989), Biologische Toxizitätstests – Gegenwärtig zur Verfügung stehende Testverfahren, *UWSF-Z Umweltchem. Ökotox.* **3**, 23–26.

KAPICH, A., HOFRICHTER, M., VARES, T., HATAKKA, A. (1999), Coupling of manganese peroxidase mediated lipid peroxidation with destruction of non-phenolic lignin compounds and ^{14}C-labeled lignins, *Biochem. Biophys. Res. Comm.* **259**, 212–219.

KARGI, F. (1984), Microbial desulfurization of coal, *Biotechnol, Proc.* **3**, 241–272.

KAUFMANN, E. N., SCOTT, C. D. (1994), Liquefy coal with enzyme catalysts, *Am. Chem. Soc. Chemtech.* **4**, 27–34.

KERSTEN, P. J., KALYANARAMAN, B., HAMMEL, K. E., REINHAMMAR, B., KIRK, T. K. (1990), Comparison of lignin peroxidase, horseradish peroxidase and laccase in the oxidation of methoxybenzenes, *Biochem. J.* **268**, 475–480.

KILBANE, J. J. (1990), *Resour. Conserv. Recycl.* **3**, 69.

KIM, T. S., KIM, H. Y., KIM, B. H. (1990), Petroleum desulfurization by *Desulfovibrio desulfuricans* M6 using electrochemically supplied reduction equivalent, *Biotechnol. Lett.* **12**, 757–760.

KIRK, T. K., FARRELL, R. L. (1987), Enzymatic "combustion": the microbial degradation of lignin, *Annu. Rev. Microbiol.* **41**, 465–505.

KIRK, T. K., SHIMADA, M. (1985), Lignin biodegradation: the microorganisms involved and the physiology and biochemistry of degradation by white-rot fungi, in: *Biosynthesis and Biodegradation of Wood Components* (HIGUCHI, T., Ed.), pp. 579– 605. Orlando, FL: Academic Press.

KLEIN, J. (1998), Technological and economic aspects of coal biodesulfurisation, *Biodegradation* **9**, 293–300.

KLEIN, J., BEYER, M., VAN AFFERDEN, M., HODEK, W., PFEIFER, F. et al. (1988b), Coal in Biotechnology, in: *Biotechnology* 1st Edn., Vol. 6b (REHM, H. J., REED, G., Eds.), pp. 497–567. Weinheim: VCH.

KLEIN, J., VAN AFFERDEN, M., PFEIFER, F., SCHACHT, S. (1994), Microbial desulfurization of coal and oil, *Fuel Proc. Technol.* **40**, 297–310.

KLEIN, J., CATCHESIDE, D. E. A., FAKOUSSA, R., GAZSO, L., FRITSCHE, W. et al. (1999), Biological processing of fossil fuels – Résumé of the Bioconversion Session of ICCS'97, *Appl. Microbiol. Biotechnol.* **52**, 2–15.

KLEIN, J., PFEIFER, F., SCHACHT, S., SINDER, C. (1997), Environmental aspects of bioconversion processes, *Fuel Proc. Technol.* **52**, 17–25.

KLUBEK, B., OCHMAN, M., NABE, S., CLARK, D., ALAM, K., ABDULRASHID, N. (1988), Microbial removal of organic sulfur from coal, *Mineral Matters* **10**, 1–3.

KODAMA, K., UMEHARA, K., SHIMIZU, K., NAKATANI, S., MINODA, Y., YAMADA, K. (1973), Identification of microbial products from dibenzothiophene and its proposed oxidation pathway, *Agr. Biol. Chem.* **37**, 45–50.

KÖHLER, M., GENZ, I. L., SCHICHT, B., ECKART, V. (1984), Mikrobielle Entschwefelung von Erdöl und schweren Erdölfraktionen: 4. Mitteilung: Anaerober Abbau organischer Schwefelverbindungen des Erdöls. *Zbl. Mikrobiol.* **139**, 239–247.

KUCHER, R. V., TUROVSKII, A. A. (1977), Cultivation of *Candida tropicalis* on coal substrates, *Mikrobiologiya* **46**, 583–585.

KUWAHARA, M., GLENN, J. K., MORGAN, M. A., GOLD, M. H. (1984), Separation and characterization of two extracellular H_2O_2-dependent oxidases from ligninolytic cultures of *Phanerochaete chrysosporium*, *FEBS Lett.* **169**, 247–250.

LABORDA, F., FERNÁNDEZ, M., LUNA, N., MONISTROL, I. F. (1997), Study on the mechanisms by which microorganisms solubilize and/or liquefy Spanish coals, *Fuel Proc. Technol.* **52**, 95–107.

LABORDA, F., MONISTROL, I. F., LUNA, N., FERNÁNDEZ, M. (1999), Processes of liquefaction/solubilization of Spanish coals by microorganisms, *Appl. Microbiol. Biotechnol.* **52**, 49–56.

LEUSCHNER, A. P., LAQUIDARA, M. J., MARTEL, A. S. (1990), Biological methane production from Texas lignite, in: *Bioprocessing and Biotreatment of Coal* (WISE, D. L., Ed.), pp. 109–130. New York: Marcel Dekker.

LIESKE, R. (1929), Biologie und Kohleforschung, *Brennst. Chem.* **10**, 437–438.

LIESKE, R. (1931), Untersuchungen über die Verwendbarkeit von Kohlen als Düngemittel, *Brennst. Chem.* **12**, 81–85.

LIESKE, R., HOFMANN, E. (1928), Untersuchungen über die Mikrobiologie der Kohlen und ihrer natürlichen Lagerstätten. II. Die Mikrobiologie der Steinkohlegruben, *Brennst. Chem.* **9**, 282–285.

LINOS, A., STEINBÜCHEL, A., SPRÖER, C., KROPPENSTEDT, R. (1999), *Gordonia polyisoprenivorans* sp. nov., a rubber degrading actinomycete isolated from automobile tire, *Int. J. Syst. Bacteriol.* **49**, 1785–1791.

LINOS, A., BEREKAA, M. M., KELLER, U., REICHELT, R., SCHMITT, J. et al. (2000), Biodegradation of *cis*-1,4-polyisoprene rubbers by distinct actinomycetes: microbial strategies and detailed surface analysis, *Appl. Environ. Microbiol.* **66**, 1639–1645.

LISSNER, A., THAU, A. (1953), Die Chemie der Braunkohle. Band II, Halle: VEB Wilhelm Knapp Verlag.

LOI, G., MURA, A., TROIS, P., ROSSI, G. (1994), The Porto Torres biodepyritization pilot plant: Light and shade of one year operation, *Fuel Proc. Technol.* **40**, 261–268.

LUDWIG, S., SCHINKE-KISSING, S., SCHEEL, T., HÖFER, M., HÖLKER, U. (1999), Differential analysis of fungal extracellular enzymes induced by lignite on protein and mRNA level, in: *Proc. 5th Int. Symp. Biological Processing of Fossil Fuels* 26–29th Sept. 1999, Madrid, Spain (Abstract).

MADISON, L. L., HUISMAN, G. W. (1999), Metabolic engineering of poly(3-hydroxyalkanoates): from DNA to plastic, *Microbiol. Mol. Biol. Rev.* **63**, 21–53.

MAKA, A., SRIVASTAVA, V. J., KILBANE II, J. J., AKIN, C. (1989), Biological solubilization of untreated North Dakota lignite by a mixed bacterial and a mixed bacterial/fungal culture, *Appl. Biochem. Biotechnol.* **20/21**, 715–729.

MESSERSCHMIDT, A., ROSSI, A., LADENSTEIN, R., HUBER, R., BOLOGNESI, M., GATTI, G. et al. (1989), X-ray crystal structure of the blue oxidase ascorbate oxidase from zucchini, *J. Mol. Biol.* **205**, 513–529.

MIKI, K., SATO, Y. (1994), Biomimetic solubilization of brown coal by oxygenase model with hydrogen peroxide, *Am. Chem. Soc., Fuel Chem. Prep.* **39**, 618–622.

MILLER, K. W. (1992), Reductive desulfurization of dibenzyldisulfide, *Appl. Environ. Microbiol.*, 2176–2179.

MONISTROL, I. F., LABORDA, F. (1994), Liquefaction and/or solubilization of Spanish hard coal and coal derivatives by newly isolated microorganisms, *Fuel Proc. Technol.* **40**, 205–216.

MONTICELLO, D. J., FINNERTY, W. R. (1985), Microbial desulfurization of fossil fuels, *Ann. Rev. Microbiol.* **39**, 371–389.

MÜLLER, H. M., SEEBACH, D. (1993), Poly(hydroxyalkanoates): A fifth class of physiologically important organic biopolymers? *Angew. Chem.* **32**, 477–502.

NAIDJA, A., HUANG, P. M., BOLLAG, J.-M. (1998), Comparison of reaction products from the transformation of catechol catalyzed by birnessite or tyrosinase, *Soil Sci. Soc. Am. J.* **62**, 188–195.

ODIER, E., MOZUCH, M. D., KALYANARAMAN, B., KIRK, T. K. (1988), Ligninase-mediated phenoxy radical formation and polymerization unaffected by cellobiose:quinone oxidoreductase, *Biochemie* **70**, 847–852.

ODOM, B., COOLEY, M., MISHRA, N. C. (1991), Genetics of coal solubilization by *Neurospora crassa*, *Resour. Conserv. Rec.* **5**, 297–301.

OLSON, G. J., BRINCKMANN, F. E. (1986), Bioprocessing of coal, *Fuel* **65**, 1638–1646.

OLSON, G. J., KELLY, R. M. (1986), Microbial metal transformations: Biotechnological applications and potential, *Biotechnol. Prog.* **2**, 1–15.

OSIPOWICZ, B., JABLONSKI, L., SIEWINSKI, JASIENKO, S., RYMKIEWICZ, A. (1994), Biodegradation of hard coal and its organic extract by selected microorganisms, *Fuel* **73**, 1858–1862.

PANOW, A., FITZGERALD, J. M. P., MAINWARING, D. E. (1997), Mechanisms of biologically-mediated methane evolution from black coal, *Fuel Process Technol.* **52**, 115–125.

PASZCZYNSKI, A., HUYNH, V. B., CRAWFORD, R.

(1985), Enzymatic activities of an extracellular manganese-dependent peroxidase from *Phanerochaete chrysosporium*, *FEMS Microbiol. Lett.* **29**, 37–41.

PATEL, A., CHEN, Y. P., MISHRA, N. C. (1996), Genetics and biotechnology of *Neurospora* protein with coal solubilization activity, in: *Proc. 5th Int. Symp. Biological Processing of Fossil Fuels Madrid* (EPRI, Ed.), p. 9.

PEREZ, J., JEFFRIES, T. W. (1992), Roles of manganese and organic acids chelators in regulating lignin degradation and biosynthesis of peroxidases by *Phanerochaete chrysosporium*, *Appl. Environ. Microbiol.* **58**, 2402–2409.

POLMAN, J. K., QUIGLEY, D. R. (1991), Size exclusion chromatography of alkali-solubilized coal, *Energy Fuels* **5**, 252–253.

POLMAN, J. K., MILLER, K. S., STONER, D. L., BRECKENRIDGE, C. R. (1994a), Solubilization of bituminous and lignite coals by chemically and biologically synthesized surfactants, *J. Chem. Tech. Biotechnol.* **61**, 11–17.

POLMAN, J. K., STONER, D. L., DELEZENE-BRIGGS, K. M. (1994b), Bioconversion of coal, lignin, and dimethoxybenzyl alcohol by *Penicillium citrinum*, *J. Ind. Microbiol.* **13**, 292–299.

POTTER, M. C. (1908), Bakterien als Agentien bei der Oxidation amorpher Kohle, *Zbl. Bakteriol. Parasitenk. II* **21**, 647–665.

QUIGLEY, D. R., WEY, J. E., BRECKENRIDGE, C. R., HATCHER, H. J. (1987), Comparison of alkali and microbial solubilization of oxidized, low-rank coals, in: *Proc. Biological Treatment of Coals Workshop* (U.S. Department of Energy), p. 151. Germantown, MD.

QUIGLEY, D. R., BRECKENRIDGE, C. R., DUGAN, P. R., WARD, B. (1988a), Effect of multivalent cations found in coal on alkali- and biosolubilities, *Am. Chem. Soc., Div. Fuel Chem. Prep.* **33**, 580.

QUIGLEY, D. R., WEY, J. E., BRECKENRIDGE, C. R., STONER, D. (1988b), The Influence of pH on biological solubilization of oxidized low-rank coal, *Res. Conserv. Recycl.* **1**, 163–174.

QUIGLEY, D. R., WARD, B., CRAWFORD, D. L., HATCHER, H. J., DUGAN, P. R. (1989a), Evidence that microbially produced alkaline materials are involved in coal biosolubilization, *Appl. Biochem. Biotechnol.* **20/21**, 753–763.

QUIGLEY, D. R., BRECKENRIDGE, C. R., DUGAN, P. R. (1989b), Effects of multivalent cations on low-rank coal solubilities in alkaline solutions and microbial cultures, *Energy Fuels* **3**, 571–575.

QUIGLEY, D. R., BRECKENRIDGE, C. R., POLMAN, J. K., DUGAN, P. R. (1991), Hydrogen peroxide, peroxidases and low-rank coal, *Fuel* **70**, 581–590.

RAI, C., REYNIERS, J. P. (1988), Microbial desulfurization of coal by organisms of the genus *Pseudomonas*, *Biotechn. Prog.* 225–230.

RALPH, J. P., CATCHESIDE, D. E. A. (1993), Action of aerobic microorganisms on the macromolecular fraction of lignite, *Fuel* **72**, 1679–1686.

RALPH, J. P., CATCHESIDE, D. E. A. (1994a), Depolymerization of macromolecules from Morwell brown coal by mesophilic and thermotolerant aerobic microorganisms, *Fuel Proc. Technol.* **40**, 193–203.

RALPH, J. P., CATCHESIDE, D. E. A. (1994b), Decolourisation and depolymerisation of solubilised low-rank coal by the white-rot basidiomycete *Phanerochaete chrysosporium*, *Appl. Microbiol. Biotechnol.* **42**, 536–542.

RALPH, J. P., CATCHESIDE, D. E. A. (1996), Size-exclusion chromatography of solubilised low-rank coal, *J. Chromatogr.* **A724**, 97–105.

RALPH, J. P., CATCHESIDE, D. E. A. (1997a), Catabolism of brown coal macromolecules by the white-rot fungus *Phanerochaete chrysosporium*, in: *Proc. 9th Int. Conf. Coal Science* Vol. III (ZIEGLER, A., VAN HEEK, K. H., KLEIN, J., WANZL, W., Eds.), pp. 1603–1606. Essen: P & W Verlag.

RALPH, J. P., CATCHESIDE, D. E. A. (1997b), Transformations of low rank coal by *Phanerochaete chrysosporium* and other wood-rot fungi, *Fuel Proc. Technol.* **52**, 79–93.

RALPH, J. P., CATCHESIDE, D. E. A. (1999), Transformation of macromolecules from brown coal by lignin peroxidase, *Appl. Microbiol. Biotechnol.* **52**, 70–77.

REHM, B. H. A., STEINBÜCHEL, A. (1999), Biochemical analysis of PHA synthases and other proteins required for PHA synthesis, *Int. J. Biol. Macromol.* **25**, 3–19.

REINHAMMAR, B., MALSTRÖM, B. G. (1981), "Blue" copper-containing oxidases, in: *Copper Proteins – Metal Ions in Biology* 3 (SPIRO, T. G., Ed.), pp. 109–149. New York: John Wiley & Sons.

REINKE, M., KALNOWSKI, G., DOTT, W. (1994), Veränderung toxischer und mutagener Wirkungen beim Abbau PAK-haltiger Ölemulsionen, in: *Biologischer Abbau von polycyclischen Kohlenwasserstoffen. Schriftenreihe Biologische Abwasserreinigung* (WEIGERT, B., Ed.), *TU Berlin* **4**, pp. 197–215.

ROOS, P. H., VAN AFFERDEN, M., STROTKAMP, D., TAPPE, D., PFEIFER, F., HANSTEIN, W. G. (1996), Liver microsomal levels of cytochrome P450IA1 as biomarkers for exposure and bioavailability of soil-bound polycyclic aromatic hydrocarbons, *Arch. Environ. Contam. Toxicol.* **30**, 107–113.

ROSSI, G. (1993), Coal depyritization achievements and problems, *Fuel* **72**, 1581–1582.

RUNNION, K., COMBIE, J. D. (1990a), Thermophilic microorganisms for coal biosolubilization, *Appl. Biochem. Biotechnol.* **24/25**, 817–829.

RUNNION, K. N., COMBIE, J. D. (1990b), Microbial removal of organic sulfur from coal, *Proc. 1st Int.*

Symp. Biological Processing of Coal, Orlando, FL, EPRI GS-6970:2/62-2/76.

SCHACHT, S., SINDER, C., PFEIFER, F., KLEIN, J. (1999), Bioassays for risk assessment of coal conversion products, *Appl. Microb. Biotechnol.* **52**, 127–130.

SCHEEL, T., HÖLKER, U., LUDWIG, S., SCHINKE-KISSING, S., HÖFER, M. (1999a), Expression of laccase and manganese peroxidase in basidiomycetes in the course of humic acid degradation, in: *Proc. 5th Int. Symp. Biological Processing of Fossil Fuels* 26–29th Sept. 1999, Madrid, Spain (Abstract).

SCHEEL, T., HÖLKER, U., LUDWIG, S., HÖFER, M. (1999b), Evidence for and expression of a laccase gene in three basidiomycetes degrading humic acids, *Appl. Microbiol. Biotechnol.* **52**, 66–69.

SCHEEL, T., HÖLKER, U., LUDWIG, S., HÖFER, M. (2000), Differential expression of manganese peroxidase and laccase in white-rot fungi in the presence of manganese or aromatic compounds, *Appl. Microbiol. Biotechnol.* **54**, 686–691.

SCHINK, B. (1988), Principles and limits of anaerobic degradation – environmental and technological aspects, in: *Biology of Anaerobic Microorganisms* (ZEHNDER, A. J. B., Ed.), pp. 771–864. New York: John Wiley & Sons.

SCHINK, B. (2000), Principles of anaerobic degradation of organic compounds, in: *Biotechnology* 2nd Edn. Vol. 11b (REHM, H.-J., REED, G., PÜHLER, A., STADLER, P., Eds.), pp. 169–192. Weinheim, Wiley-VCH.

SCHIPPERS, A. (1998), Untersuchungen zur Schwefelchemie der biologischen Laugung von Metallsulfiden, *Thesis*, University of Hamburg, Germany.

SCHIPPERS, A., RHOWERDER, T., SAND, W. (1999), Intermediary sulfur compounds in pyrite oxidation: implications for bioleaching and biodepyritization of coal, *Appl. Microbiol. Biotechnol.* **52**, 104–110.

SCHOEMAKER, H. E., HARVEY, P. J., BOWEN, R. M., PALMER, J. M. (1985), On the mechanism of enzymatic lignin breakdown, *FEBS Lett.* **183**, 7–12.

SCHUMACHER, J. D., FAKOUSSA, R. (1999), Degradation of alicyclic molecules by *Rhodococcus ruber* CB4, *Appl. Microbiol. Biotechnol.* **52**, 85–90.

SCOTT, C. D., STRANDBERG, G. W., LEWIS, S. N. (1986), Microbial solubilization of coal, *Biotechnol. Prog.* **2**, 131–139.

SCOTT, C. D., WOODWARD, C. A., THOMPSON, J. E., BLANKINSHIP, S. L. (1990), Coal solubilization by enhanced enzyme activity in organic solvents, *Appl. Biochem. Biotechnol.* **24/25**, 799–815.

SCOTT, C. D., WOODWARD, C. A., SCOTT, T. C. (1994), Use of chemically modified enzymes in organic solvents for conversion of coal to liquids, *Catal. Today* **19**, 381–394.

SHARP, F. A. (1992), Biologically mediated depyritization of large sized coal, *Proc. Pittsburgh Coal Conference*, September 1992, Pittsburgh, PA.

STACH, E., MACKOWSKY, M. T., TEICHMÜLLER, M., TAYLOR, H. G., CHANDRA, D., TEICHMÜLLER, R. (1982), *Stach's Textbook of Coal Petrology* 3rd Edn., Berlin-Stuttgart: Gebrüder Bornträger.

STEFFEN, K., HOFRICHTER, M., HATAKKA, A. (2000), Mineralisation of ^{14}C-labelled synthetic lignin and ligninolytic enzyme activities of litter-decomposing basidiomycetous fungi, *Appl. Microbiol. Biotechnol.* **54**, 819–825.

STEINBÜCHEL, A., FÜCHTENBUSCH, A. (1997), PHA from coal? in: *Proc. 9th Int. Conf. Coal Science* Vol. III (ZIEGLER, A., VAN HEEK, K. H., KLEIN, J., WANZL, W., Eds.), pp. 1673–1676. Essen: P & W Druck und Verlag.

STEINBÜCHEL, A., FÜCHTENBUSCH, B. (1998), Bacterial and other biological systems for polyester production, *Trends Biotechnol.* **16**, 419–427.

STEINBÜCHEL, A., VALENTIN, H. E. (1995), Diversity of bacterial polyhydroxyalkanoic acids, *FEMS Microbiol. Lett.* **128**, 219–228.

STEINBÜCHEL, A., AERTS, K., BABEL, W., FÖLLNER, C., LIEBERGESELL, M. et al. (1995), Considerations on the structure and biochemistry of bacterial polyhydroxyalkanoic acid inclusions, *Can. J. Microbiol.* **41** (Suppl. 1), 94–105.

STEVENS, S. E., BURGESS, W. D. (1989), Microbial desulfurization of coal, *US Patent* No. 48511350.

STONER, D. I., WEY, BARRETT, K. B., JOLLY, J. G., WRIGHT, R. B., DUGAN, P. R. (1990), Modification of water-soluble coal-derived products by dibenzothiophene-degrading microorganisms, *Appl. Environ. Microbiol.* **56**, 2667–2676.

STRANDBERG, G. W., LEWIS, S. N. (1987a), Solubilization of coal by an extracellular product from *Streptomyces setonii* 75Vi2, *J. Ind. Microbiol.* **1**, 371–376.

STRANDBERG, G. W., LEWIS, S. N. (1987b), A method to enhance the microbial liquefaction of lignite coals, *Biotechnol. Bioeng. Symp.* **17**, 153–161.

SUPALUKNARI, S., LARKINS, F. P., REDLICH, P., JACKSON, W. R. (1988), An FTIR study of Australian coals: Characterization of oxygen functional groups, *Fuel Proc. Technol.* **19**, 123–140.

TEMP, U., MEYRAHN, H., EGGERT, C. (1999), Conversion of lignite with laccase-mediator-systems, in: *Proc. 5th Int. Symp. Biological Processing of Fossil Fuels* 26–29th Sept. 1999, Madrid, Spain (Abstract).

THURSTON, C. F. (1994), The structure and function of fungal laccases, *Microbiology* **140**, 19–26.

TIEN, M., KIRK, T. K. (1983), Lignin-degrading enzyme from the hymenomycete *Phanerochaete chrysosporium* Burds, *Science* **221**, 661–663.

TIEN, M., KIRK, T. K. (1984), Lignin-degrading enzyme from *Phanerochaete chrysosporium*: purification, characterization, and catalytic properties of a unique H_2O_2-requiring oxygenase, *Proc. Natl. Acad. Sci. USA* **81**, 2280–2284.

TORMA, A. E. (1977), The role of *Thiobacillus ferrooxidans* in hydrometallurgical processes, *Adv. Biochem. Eng.* **6**, 1–37.

TORZILLI, A. P., ISBISTER, J. D. (1994), Comparison of coal solubilization by bacteria and fungi, *Biodegradation* **5**, 55–62.

URZÚA, U., KERSTEN, P. J., VICUÑA, R. (1998), Manganese peroxidase-dependent oxidation of glyoxylic and oxalic acids synthesized by *Ceriporiopsis subvermispora* produces extracellular hydrogen peroxide, *Appl. Environ. Microbiol.* **64**, 68–73.

VAN AFFERDEN, M. (1991), Mikrobieller Abbau von Dibenzothiophen: Entschwefelung einer kohlerelevanten Modellverbindung, *Thesis*, University of Bonn, Germany.

VAN AFFERDEN, M., SCHACHT, S., KLEIN, J., TRÜPER, H. G. (1990), Degradation of dibenzothiophene by *Brevibacterium* sp. DO, *Arch. Microbiol.* **153**, 324–328.

WÄLTERMANN, M., LUFTMANN, H., BAUMEISTER, D., KALSCHEUER, R., STEINBÜCHEL, A. (2000), *Rhodococcus opacus* strain PD630 as a new source of high-value single-cell oil? Isolation and characterization of triacylglycerols and other storage lipids, *Microbiology* **146**, 1143–1149.

WARD, B. (1985), Lignite-degrading fungi isolated from a weathered outcrop, *Syst. Appl. Microbiol.* **6**, 236–238.

WARD, B. (1993), Quantitative measurements of coal solubilization, *Biotechnol. Techniques* **7**, 213–216.

WARD, B., SANDERS, A. (1989), Solubilization of lignites by fungi, *Proc. Electric Power Res. Inst. Symp. Biological Processing of Coal and Coal-Derived Substances*, 3/57.

WARD, B., QUIGLEY, D. R., DUGAN, P. R. (1988), Relationships between natural weathering and lignite biosolubility, *Proc. Inst. Gas Technol. Conf. Coal* **1**, 40–52.

WARIISHI, H., AKILESWARAN, L., GOLD, M. H. (1988), Manganese peroxidase from the basidiomycete *Phanerochaete chrysosporium*: spectral characterization of oxidized states and the catalytic cycle, *Biochemistry* **27**, 5365–5370.

WARIISHI, H., HUANG, J., DUNFORD, H. B., GOLD, M. H. (1991), Reactions of lignin peroxidase compounds I and II with veratryl alcohol, *J. Biol. Chem.* **266**, 20694-20699.

WARIISHI, H., VALLI, K., GOLD, M. H. (1992), Manganese(II) oxidation by manganese peroxidase from the basidiomycete *Phanerochaete chrysosporium*, *J. Biol. Chem.* **267**, 23688–23695.

WILLMANN, G. (1994), Abbau und Verwertung von verschiedenen Braunkohlen durch Pilze und Bakterien: Charakterisierung der Mikroorganismen und Abbauprodukte, *Thesis*, Universiy of Bonn, Germany.

WILLMANN, G., FAKOUSSA, R. M. (1991), Liquefaction/solubilization of brown coal by a fungus, in: *Biochemical Engineering* (REUSS, M., Ed.), pp. 429–432. Stuttgart: Fischer-Verlag.

WILLMANN, G., FAKOUSSA, R. M. (1992), Untersuchungen zum Mechanismus der Braunkohle-Solubilisierung durch Pilze. *DGMK-Bericht* Nr. 9206, ISBN 3-928164-52-X, pp. 107–116. Essen: Deutsche Wissenschaftliche Gesellschaft für Erdöl, Erdgas und Kohle.

WILLMANN, G., FAKOUSSA, R. M. (1997a), Biological bleaching of water-soluble coal macromolecules by a basidiomycete strain, *Appl. Microbiol. Biotechnol.* **47**, 95–101.

WILLMANN, G., FAKOUSSA, R. M. (1997b), Extracellular oxidative enzymes of coal-attacking fungi, *Fuel Proc. Technol.* **52**, 27–41.

WONDRACK, L., SZANTO, M., WOOD, W. A. (1989), Depolymerization of water soluble coal polymer from subbituminous coal and lignite by lignin peroxidase, *Appl. Biochem. Biotechnol.* **20/21**, 765–780.

YANG, X., ZHANG, M., LI, J., YANG, Z. (1985), Effect of ammonium nitrohumate from several materials on fertilizer superphosphate, *Yuanzineng Nongye Yingyong* **1**, 42–46.

YOUN, H. D., HAH, Y. C., KANG, S. O. (1996), Role of laccase in lignin degradation by white-rot fungi, *FEMS Lett.* **132**, 183–188.

ZIEGENHAGEN, D., HOFRICHTER, M. (1998), Degradation of humic acids by manganese peroxidase from the white-rot fungus *Clitocybula dusenii*, *J. Basic Microbiol.* **38**, 289–299.

ZIEGENHAGEN, D., HOFRICHTER, M. (2000), A simple and rapid technique to gain high amounts of manganese peroxidase with immobilized cultures of *Clitocybula dusenii*, *Appl. Microbiol. Biotechnol.* **53**, 553–557.

8 Microbial Leaching of Metals

HELMUT BRANDL
Zürich, Switzerland

1 Introduction

Future sustainable development requires measures to reduce the dependence on nonrenewable raw materials and the demand for primary resources. New resources for metals must be developed with the aid of novel technologies. in addition, improvement of alredy existing mining techniques can result in metal recovery from sources that have not been of economical interest until today. Metal-winning processes based on the activity of microorganisms offer a possibility to obtain metals from mineral resources not accessible by conventional mining (BOSECKER, 1997; BRIERLEY, 1978; BRYNER et al., 1954; TORMA and BANHEGYI, 1984). Microbes such as bacteria and fungi convert metal compounds into their water-soluble forms and are biocatalysts of these leaching processes. Additionally, applying microbiological solubilization processes, it is possible to recover metal values from industrial wastes which can serve as secondary raw materials.

2 Terminology

In general, bioleaching is a process described as being "the dissolution of metals from their mineral source by certain naturally occurring microorganisms" or "the use of microorganisms to transform elements so that the elements can be extracted from a material when water is filtered trough it" (ATLAS and BARTHA, 1997; PARKER, 1992). Additionally, the term "biooxidation" is also used (HANSFORD and MILLER, 1993). There are, however, some small differences by definition (BRIERLEY, 1997): Usually, "bioleaching" is referring to the conversion of solid metal values into their water soluble forms using microorganisms. In the case of copper, copper sulfide is microbially oxidized to copper sulfate and metal values are present in the aqueous phase. Remaining solids are discarded. "Biooxidation" describes the microbiological oxidation of host minerals which contain metal compounds of interest. As a result, metal values remain in the solid residues in a more concentrated form. In gold mining operations, biooxidation is used as a pretreatment process to (partly) remove pyrite or arsenopyrite. This process is also called "biobeneficiation" where solid materials are refined and unwanted impurities are removed (GROUDEV, 1999; STRASSER et al., 1993). The terms "biomining", "bioextraction", or "biorecovery" are also applied to describe the mobilization of elements from solid materials mediated by bacteria and fungi (HOLMES, 1991; MANDL et al., 1996; RAWLINGS, 1997; WOODS and RAWLINGS, 1989). "Biomining" concerns mostly applications of microbial metal mobilization processes in large-scale operations of mining industries for an economical metal recovery.

The area of "biohydrometallurgy" covers bioleaching or biomining processes (ROSSI, 1990). Biohydrometallurgy represents an interdisciplinary field where aspects of microbiology (especially geomicrobiology), geochemistry, biotechnology, hydrometallurgy, mineralogy, geology, chemical engineering, and mining engineering are combined. Hydrometallurgy is defined as the treatment of metals and metal-containing materials by wet processes and describes "the extraction and recovery of metals from their ores by processes in which aqueous solutions play a predominant role" (PARKER, 1992). Rarely, the term "biogeotechnology" is also used instead of biohydrometallurgy (FARBISZEWSKA et al., 1994).

3 Historical Background

One of the first reports where leaching might have been involved in the mobilization of metals is given by the Roman writer Gaius Plinius Secundus (23–79 A.D.). In his work on natural sciences, Plinius describes how copper minerals are obtained using a leaching process (KÖNIG, 1989a, b). The translation reads approximately as follows: "Chrysocolla is a liquid in the before mentioned gold mines running from the gold vein. In cold weather during the winter the sludge freezes to the hardness of pumice. It is known from experience that the most wanted [chrysocolla] is formed in copper mines, the following in silver mines. The liquid is also found in lead mines although it is of minor value. In all these mines chrysocolla is also artificially produced by slowly passing water through the mine during the winter until the month of June; subsequently, the water is evaporated in June and July. It is clearly demonstrated that chrysocolla is nothing but a decomposed vein."

The German physician and mineralogist Georgius Agricola (1494–1555) describes in his work *de re metallica* also techniques for the recovery of copper that are based on the leaching of copper-containing ores (SCHIFFNER, 1977). A woodcut from his book illustrates the (manual) transport of metal-containing leachates from mines and their evaporation in the sunlight (Fig. 1).

The Rio Tinto mines in south-western Spain are usually considered the cradle of biohydrometallurgy. These mines have been exploited since pre-Roman times for their copper, gold, and silver values. However, with respect to commercial bioleaching operations on an industrial scale, biohydrometallurgical techniques had been introduced to the Tharsis mine in Spain 10 years earlier (SALKIELD, 1987). As a consequence to the ban of open air ore roasting and its resulting atmospheric sulfur emissions in 1878 in Portugal, hydrometallurgical metal extraction has been taken into consideration in other countries more intensely. In addition to the ban, cost savings were another incentive for the development: Heap leaching techniques were assumed to reduce transportation costs and to allow the employment of locomotives and wagons for other services (SALKIELD, 1987). From 1900 on, no open air roasting of low-grade ore was conducted at the Rio Tinto mines.

Fig. 1. Woodcut from the book *de re metallica* written by Georgius Agricola (1494–1555) illustrating the manual recovery of copper-containing mine effluents which are collected in wooden basins and concentrated in the sun.

Efforts to establish bioleaching at the Rio Tinto mines had been undertaken in the beginning of the 1890s. Heaps (10 m in height) of low-grade ore (containing 0.75% Cu) were built and left for one to three years for "natural" decomposition (SALKIELD, 1987). 20 to 25% of the copper left in the heaps were recovered annually. It was calculated that approximately 200,000 t of rough ore could be treated in 1896. Although industrial leaching operations were conducted at the Rio Tinto mines for several decades, the contribution of bacteria to metal solubilization was confirmed only in 1961, when *Thiobacillus ferrooxidans* was identified in the leachates.

Early reports state that factors affecting bioleaching operations were the height of the heap, particle size, initial ore washing with acid, and temperature control to about 50 °C (SALKIELD, 1987). Another critical factor was the supply of water for the leaching heaps. Although usually acidic mine waters were used for ore processing, 4 billion liters of freshwater were required annually (SALKIELD, 1987).

Although metal leaching from mineral resources has a very long historical record (EHRLICH, 1999; ROSSI, 1990) and although the oxidation of reduced sulfur compounds and elemental sulfur resulting in the formation of sulfuric acid was demonstrated already in the 1880s (WINOGRADSKY, 1887), the oxidation of metal sulfides was not described until 1922 when mobilization of zinc from zinc sulfide was investigated (RUDOLFS, 1922; RUDOLFS and HELBRONNER, 1922). It was found that the transformation of zinc sulfide to zinc sulfate was microbially mediated. Based on these results, the economic recovery of zinc from zinc-containing ores by biological methods was proposed. In 1947, *Thiobacillus ferrooxidans* was identified as part of the microbial community found in acid mine drainage (COLMER and HINKLE, 1947). A first patent was granted in 1958 (ZIMMERLEY et al., 1958). The patent describes a cyclic process where a ferric sulfate/sulfuric acid lixiviant solution is used for metal extraction, regenerated by aeration (ferrous iron oxidation by iron-oxidizing organisms), and reused in a next leaching stage.

4 Principles of Microbial Metal Leaching

4.1 Leaching Mechanisms

Mineralytic effects of bacteria and fungi on minerals are based mainly on three principles, namely acidolysis, complexolysis, and redoxolysis. Microorganisms are able to mobilize metals by (1) the formation of organic or inorganic acids (protons); (2) oxidation and reduction reactions; and (3) the excretion of complexing agents. Sulfuric acid is the main inorganic acid found in leaching environments. It is formed by sulfur-oxidizing microorganisms such as thiobacilli. A series of organic acids are formed by bacterial (as well as fungal) metabolism resulting in organic acidolysis, complex and chelate formation (BERTHELIN, 1983). A kinetic model of the coordination chemistry of mineral solubilization has been developed which describes the dissolution of oxides by the protonation of the mineral surface as well as the surface concentration of suitable complex-forming ligands such as oxalate, malonate, citrate, and succinate (FURRER and STUMM, 1986). Proton-induced and ligand-induced mineral solubilization occurs simultaneously in the presence of ligands under acidic conditions.

4.2 Models of Leaching Mechanisms

Originally, a model with two types of mechanisms which are involved in the microbial mobilization of metals has been proposed (EWART and HUGHES, 1991; SILVERMAN and EHRLICH, 1964): (1) Microorganisms can oxidize metal sulfides by a "direct" mechanism obtaining electrons directly from the reduced minerals. Cells have to be attached to the mineral surface and a close contact is needed. The adsorption of cells to suspended mineral particles takes place within minutes or hours. This has been demonstrated using either radioactively labeled *Thiobacillus ferrooxidans* cells grown on $NaH^{14}CO_3$ or the oxidative capacity of bacteria attached to the mineral surface (ESCOBAR et al., 1996). Cells adhere selectively to mineral surfaces occupying preferentially irregularities of the surface structure (EDWARDS et al., 1999; EWART and HUGHES, 1991). In addition, a chemotactic behavior to copper, iron, or nickel ions has been demonstrated for *Leptospirillum ferrooxidans* (ACUNA et al., 1992). Genes involved in the chemotaxis were also detected in *Thiobacillus ferrooxidans* and *Thiobacillus thiooxidans* (ACUNA et al., 1992). (2) The oxidation of reduced metals through the "indirect" mechanism is mediated by ferric iron (Fe^{3+}) originating from the microbial oxidation of ferrous iron (Fe^{2+}) compounds present in the minerals. Ferric iron is an oxidizing agent and can oxidize, e.g., metal sulfides and is (chemically) reduced to ferrous iron which, in turn, can be microbially oxidized again (EWART and HUGHES, 1991). In this case, iron has a role as electron carrier. It was proposed that no direct physical contact is needed for the oxidation of iron.

In many cases it was concluded that the "direct" mechanism dominates over the "indirect" mostly due to the fact that "direct" was equated with "direct physical contact". This domination has been observed for the oxidation of covellite or pyrite in studies employing mesophilic *T. ferrooxidans* and thermophilic *Acidianus brierleyi* in bioreactors which consisted of chambers separated with dialysis membranes to avoid physical contact (LARSSON et al., 1993; POGLIANI et al., 1990). However, the attachment of microorganisms on surfaces is not an indication *per se* for the existence of a direct mechanism (EDWARDS et al., 1999). The term "contact leaching" has been introduced to indicate the importance of bacterial attachment to mineral surfaces (TRIBUTSCH, 1999).

The following equations describe the "direct" and "indirect" mechanism for the oxidation of pyrite (MURR, 1980; SAND et al., 1999):

direct:

$$2\,FeS_2 + 7\,O_2 + 2\,H_2O \xrightarrow{\text{thiobacilli}} 2\,FeSO_4 + 2\,H_2SO_4 \quad (1)$$

indirect:

$$4\,FeSO_4 + O_2 + 2\,H_2SO_4 \xrightarrow{\text{T. ferrooxidans, L. ferrooxidans}} 2\,Fe_2(SO_4)_3 + 2\,H_2O \quad (2)$$

$$FeS_2 + Fe_2(SO_4)_3 \xrightarrow{\text{chemical oxidation}} 3\,FeSO_4 + 2\,S \quad (3)$$

$$2\,S + 3\,O_2 + H_2O \xrightarrow{\text{T. thiooxidans}} 2\,H_2SO_4 \quad (4)$$

However, the model of "direct" and "indirect" metal leaching is still under discussion. Recently, this model has been revised and replaced by another one which is not dependent on the differentiation between a "direct" and an "indirect" leaching mechanisms (SAND et al., 1995, 1999). All facts have been combined and a mechanism has been developed which is characterized by the following features: (1) cells have to be attached to the minerals and in physical contact with the surface; (2) cells form and excrete exopolymers; (3) these exopolymeric cell envelopes contain ferric iron compounds which are complexed to glucuronic acid residues. These are part of the primary attack mechanism; (4) thiosulfate is formed as intermediate during the oxidation of sulfur compounds; (5) sulfur or polythionate granules are formed in the periplasmatic space or in the cell envelope.

Thiosulfate and traces of sulfite have been found as intermediates during the oxidation of sulfur (SHRIHARI et al., 1993). Sulfur granules (colloidal sulfur) have been identified as energy reserves in the exopolymeric capsule around cells of *T. ferrooxidans* during growth on synthetic pyrite films (ROJAS et al., 1995). "Footprints" of organic films containing colloidal sulfur granules are left on the mineral surface upon detachment of the bacteria.

From the existing data two "indirect" leaching mechanisms have been proposed whereas no evidence for a "direct" enzymatically mediated process has been found (SAND et al., 1999). The mineral structure is the determining factor for the prevailing type of leaching mechanism. In the "thiosulfate mechanism" thiosulfate is the main intermediate resulting from the oxidation of pyrite, molybdenite, or tungstenite. Polysulfide and elemental sulfur are the main intermediates in the "polysulfide mechanism" during the oxidation of galena, sphalerite, chalcopyrite, hauerite, orpiment, or realgar. The presence of iron(III) at the beginning of mineral degradation is an important prerequisite (SAND et al., 1999).

The following equations summarize the oxidation mechanisms (SAND et al., 1999):

Thiosulfate mechanism (found for FeS_2, MoS_2, WS_2):

$$FeS_2 + 6\,Fe^{3+} + 3\,H_2O \rightarrow S_2O_3^{2+} - 7\,Fe^{2+} + 6\,H^+ \quad (5)$$

$$S_2O_3^{2-} + 8\,Fe^{3+} + 5\,H_2O \rightarrow 2\,SO_4^{2-} + 8\,Fe^{2+} + 10\,H^+ \quad (6)$$

Polysulfide mechanism (found for PbS, $CuFeS_2$, ZnS, MnS_2, As_2S_3, As_3S_4):

$$2\,MS + 2\,Fe^{3+} + 2\,H^+ \rightarrow 2\,M^{2+} + H_2S_n + 2\,Fe^{2+} \quad (7)$$

$$H_2S_n + 2\,Fe^{3+} \rightarrow 0.25\,S_8 + 2\,Fe^{2+} + 2\,H^+ \quad (8)$$

$$0.25\,S_8 + 3\,O_2 + 2\,H_2O \rightarrow 2\,SO_4^{2-} + 4\,H^+ \quad (9)$$

Several biomolecules are involved in the aerobic respiration on reduced sulfur and iron compounds. It has been found that up to 5% of soluble proteins of *T. ferrooxidans* is made of an acid stable blue copper protein, called rusticyanin (BLAKE et al., 1993). Additionally, the iron(II) respiratory system contains a (putative) green copper protein, two types of cytochrome c, one or more types of cytochrome a, a porin, and an iron(II)-sulfate chelate (BLAKE et al., 1993). The acid stability of rusticyanin suggests that it is located in the periplasmic space. Figure 2 shows a scheme of the model which combines the electron transport sequence proposed earlier with concepts stemming from the debate on "direct"/"indirect" leaching mechanisms (BLAKE and SHUTE, 1994; BLAKE et al., 1993; HAZRA et al., 1992; SAND et al., 1995).

Some details of the metal mobilization mechanism, the importance of the presence and attachment of microorganisms and their active contribution have been demonstrated for the leaching of fly ash from municipal waste incineration (MWI) (BROMBACHER et al., 1998). Generally, several mechanisms of metal mobilization can be distinguished: (1) Contact leaching effect on the release of metals. Stock cultures of *Thiobacillus ferrooxidans* and *Thiobacillus thiooxidans* were added to ash suspensions and cells were in direct contact with the fly ash. Growth of thiobacilli might be stimulated by increased energy availability from oxidation of reduced solid particles. (2) Metal solubilization by metabolically active (enzymatic) compounds in the absence of bacterial cells. Stock cultures were filtered to obtain the cell free spent medium. This medium was used for leaching. (3) Metal solubilization by non-enzymatic extracellular metabolic products. Cell free spent medium (see 2) was autoclaved to obtain a sterile leaching solution without enzymatic activities and to evaluate the leaching ability of acid formed. (4) Leaching by fresh medium. Fresh non-inoculated and sterile medium was added to the fly ash suspension and used as control. (5) Chemical leaching due to the preparation of the ash suspension (acidification to pH 5.4). Certain elements such as, e.g., Cd or Zn might be chemically mobilized already during acidification.

MWI fly ash contains reduced copper species (chalcocite {Cu_2S} or cuprite {Cu_2O}) whereas zinc and others are present in their fully oxidized forms (BROMBACHER et al., 1998). Therefore, copper release from fly ash is directly affected and enhanced by *T. ferrooxidans*, whereas Zn, as well as Al, Cd, Cr, and Ni, are released primarily due to the acidic environment. Acidification of the fly ash pulp (chemical mobilization) led already to considerable extraction yields for Cd, Ni, and Zn and could slightly be increased using non-inoculated sterile medium as lixiviant (Fig. 3). By comparing leached amounts of copper by filtered cell free spent medium with autoclaved sterile spent medium, it was concluded that significant amounts of copper were mobilized – in contrast to other elements – by metabolic products of *T. ferrooxidans*. Leaching with cell free spent medium indicating a solubilizing mechanism due to extracellular components was significantly more effective than a leach-

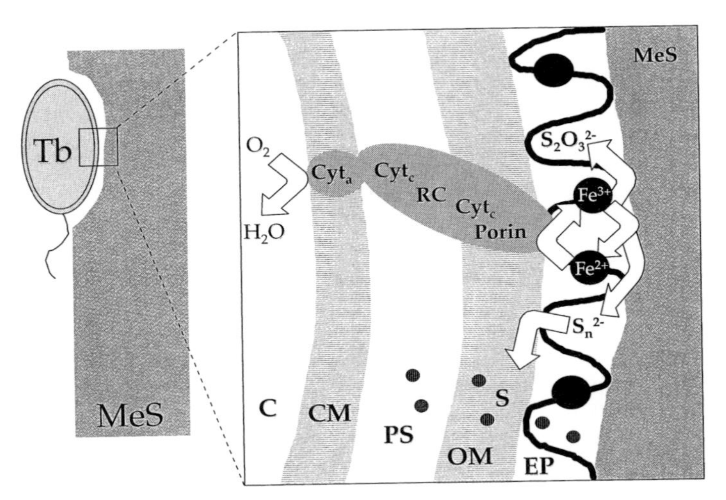

Fig. 2. Schematic mechanistic bioleaching model (after HAZRA et al., 1992; SAND et al., 1995, 1999; SCHIPPERS et al., 1996; RAWLINGS, 1999). C: cytoplasm; CM: cell membrane; PS: periplasmatic space; OM: outer membrane; EP: exopolymers; Cyt: cytochrome; RC: rusticyanin; MeS: metal sulfide

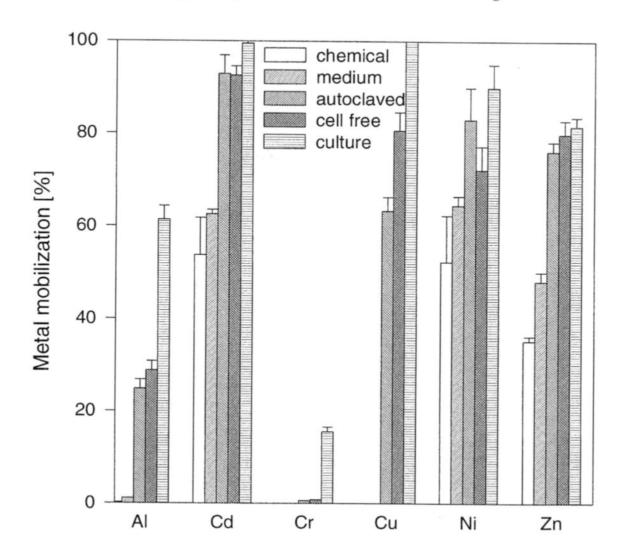

Fig. 3. Solubilized metals from fly ash originating from municipal waste incineration (in suspensions of 40 g L^{-1}) in percent of the metal amount present with different lixiviants within 8 d. All samples were incubated in triplicate. The release of metals due to acidification of the fly ash pulp is indicated as chemical mobilization (see text for explanation).

ing with autoclaved spent medium where excreted enzymes had been inactivated. It is known that several components involved in the electron transport chain of *Thiobacillus* (rusticyanin, cytochromes, iron–sulfur proteins) are located in the periplasmic space (BLAKE and SHUTE, 1994; SAND et al., 1995) and might, therefore, also be present in the cell free spent medium catalyzing oxidation of reduced metal compounds.

In many leaching environments conditions (especially iron(II) and iron(III) concentrations) vary with the duration of the leaching. This makes it difficult to assess the importance and the effect of the presence of bacteria. Using an experimental setup to maintain constant concentrations of ferrous and ferric iron, it was possible to show that in the presence of *T. ferrooxidans* rates of pyrite or zinc sulfide leaching are increased (HOLMES et al., 1999; FOWLER and CRUNDWELL, 1999; FOWLER et al., 1999).

4.3 Factors Influencing Bioleaching

Standard test methods have been developed to determine leaching rates of iron from pyrite mediated by *Thiobacillus ferrooxidans* (ASTM, 1991). An active culture of *T. ferrooxidans* is grown in a defined medium containing (in g L^{-1}): $(NH_4)_2SO_4$ (3.0); K_2HPO_4 (0.5); $MgSO_4 \cdot 7H_2O$ (0.5); KCl (1.0); $Ca(NO_3)_2 \cdot 4H_2O$ (0.01); $FeSO_4 \cdot 7H_2O$ (44.22); and 1 mL 10 N sulfuric acid (SILVERMAN and LUNDGREN, 1959). Cells are harvested, diluted, and added to pyrite suspensions with a pulp density of 20 g L^{-1}. Total soluble iron as well as sulfate formed during oxidation is periodically determined.

Metal bioleaching in acidic environments is influenced by a series of different factors (Tab. 1). Physicochemical as well as microbiological factors of the leaching environment are affecting rates and efficiencies. In addition, properties of the solids to be leached are of major importance (ACEVEDO and GENTINA, 1989; BRIERLEY, 1978; DAS et al., 1999; MURR, 1980). As examples, pulp density, pH, and particle size were identified as major factors for pyrite bioleaching by *Sulfolobus acidocaldarius* (LINDSTROM et al., 1993). Optimal conditions were 60 g L^{-1}, 1.5, and <20 μm, respectively. The influence of different parameters such as activities of the bactèria itself, source energy, mineralogical composition, pulp density, temperature, and particle size was studied for the oxidation of sphalerite by *T. ferrooxidans* (BALLESTER et al., 1989). Best zinc dissolution was obtained at low pulp densities (50 g L^{-1}), small particle sizes, and temperatures of approximately 35 °C.

Tab. 1. Factors and Parameters Influencing Bacterial Mineral Oxidation and Metal Mobilization

Factor	Parameter
Physicochemical parameters of a bioleaching environment	temperature
	pH
	redox potential
	water potential
	oxygen content and availability
	carbon dioxide content
	mass transfer
	nutrient availability
	iron(III) concentration
	light
	pressure
	surface tension
	presence of inhibitors
Microbiological parameters of a bioleaching environment	microbial diversity
	population density
	microbial activities
	spatial distribution of microorganisms
	metal tolerance
	adaptation abilities of microorganisms
Properties of the minerals to be leached	mineral type
	mineral composition
	mineral dissemination
	grain size
	surface area
	porosity
	hydrophobicity
	galvanic interactions
	formation of secondary minerals
Processing	leaching mode (*in situ*, heap, dump, or tank leaching)
	pulp density
	stirring rate (in case of tank leaching operations)
	heap geometry (in case of heap leaching)

Metal oxidation mediated by acidophilic microorganisms can be inhibited by a variety of factors such as, e.g., organic compounds, surface-active agents, solvents, or specific metals: The presence of organic compounds (yeast extract) inhibited pyrite oxidation of *T. ferrooxidans* (BACELAR-NICOLAU and JOHNSON, 1999). Certain metals present in bioleaching environments can inhibit microbial growth, therefore reducing leaching efficiencies. For instance, arsenic added to cultures inhibited *Sulfolobus acidocaldarius* grown on pyrite and *T. ferrooxidans* grown on arsenopyrite (HALLBERG et al., 1996; LAN et al., 1994). Additions of copper, nickel, uranium, or thorium adversely influenced iron(II) oxidation by *T. ferrooxidans* with uranium and thorium showing higher toxicities than copper and nickel (LEDUC et al., 1997). Silver, mercury, ruthenium, and molybdenum reduced the growth of *Sulfolobus* grown on a copper concentrate (MIER et al., 1996). Industrial biocides such as tetra-*n*-butyltin, isothiazolinones, N-dimethyl-N′-phenyl-N′-(fluorodichloro-methylthio)-sulfamide, or 2,2′-dihydroxy-5,5′-dichlorophenylmethane (dichlorophen) reduced the leaching of manganese oxides by heterotrophic microorganisms (ARIEF and MADGWICK, 1992). Biocides were externally added as selective inhibitors to suppress unwanted organisms and to improve manganese leaching efficiencies. At low concentrations of $<5\ mg\ L^{-1}$, however, manganese mobilization was increased by 20% (BOUSSIOS and MADGWICK, 1994).

Also gaseous compounds can show inhibitory effects on metal leaching: Aqueous-phase carbon dioxide at concentration $>10\ mg\ L^{-1}$ was inhibiting growth of *T. ferrooxidans* on pyrite–arsenopyrite–pyrrothite ore (NAGPAL et al., 1993). Optimal concentrations of carbon dioxide were found to be in the range of 3 to $7\ mg\ L^{-1}$. There are reports on the stimulation of bacterial leaching and the increase of leaching rates by supplementing leaching fluids with carbon dioxide (ACEVEDO et al., 1998; BRIERLEY, 1978; TORMA et al., 1972). Concentrations of 4% (v/v) carbon dioxide in the inlet gas of a fermenter showed maximum growth rates of *T. ferrooxidans*, maximum iron(II), copper, and arsenic oxidation (ACEVEDO et al., 1998).

Pulp densities of $20\ g\ L^{-1}$ delayed the onset of bioleaching of pyrite derived from coal (BALDI et al., 1992). Increasing pulp densities from 30 to $100\ g\ L^{-1}$ decreased rates of pyrite oxidation in *Sulfolobus* cultures (NGUBANE and BAECKER, 1990). For fungi such as *Aspergillus niger*, optimal pulp densities for maximum metal leaching efficiencies were found to be in the range of 30 to $40\ g\ L^{-1}$ (BOSSHARD et al., 1996). Quartz particles at pulp densities of $80\ g\ L^{-1}$ almost completely inhibited the oxidation of covellite by *T. ferrooxidans* especially in the absence of iron(II) (CURUTCHET et al., 1990).

During bioleaching processes, coprecipitation of metals with mineral phases such as jarosites can reduce leaching efficiencies (HIROYOSHI et al., 1999). In addition, the precipitation of compounds present in the leachates on the minerals to be leached can make the solid material inaccessible for bacterial leaching.

Organic solvents such as flotation or solvent extraction agents, which are added for the downstream processing of leachates from bioleaching, might also lead to inhibition problems (ACEVEDO and GENTINA, 1989). Isopropylxanthate and LIX 984 (used as flotation agent and solvent extraction agent, respectively) prevented the oxidation of pyrite and chalcopyrite by *T. ferrooxidans* (HUERTA et al., 1995). This fact is of special importance when spent leaching liquors are recycled for a reuse.

It has been demonstrated recently that the addition of small amounts of amino acids (cysteine in this case) resulted in an increased pyrite corrosion by *T. ferrooxidans* as compared to controls without additions (ROJAS-CHAPANA and TRIBUTSCH, 2000). It is suggested that the microorganisms may profit from weakening and break up of chemical bonds mediated by the formation of the cysteine–pyrite complex. This might also be the case under natural conditions by the excretion of cysteine-containing metabolites. An inexpensive alternative to increase metal recovery from ore heaps by the addition of sulfur-containing amino acids such as cysteine has been suggested (TRIBUTSCH and ROJAS-CHAPANA, 1999).

Other metabolites excreted by *Thiobacillus* might also enhance metal leaching efficiencies: Wetting agents such as mixtures of phospholipids and neutral lipids are formed by *Thiobacillus thiooxidans* (BEEBE and UMBREIT, 1971). As a consequence, growth of *T. thiooxidans* on sulfur particles is supported by the excretion of metabolites acting as biosurfactants which facilitate the oxidation of elemental sulfur. It was also hypothesized that *Thiobacillus caldus* is stimulating the growth of heterotrophic organisms in leaching environments by the excretion of organic compounds and is supporting the solubilization of solid sulfur by the formation of surface-active agents (DOPSON and LINDSTROM, 1999). Metal solubilization might also be facilitated by microbial metabolites excreted by organisms other than *Thiobacillus* which are part of microbial consortia found in bioleaching operations. Microbial surfactants, which show large differences in their chemical nature, are formed by a wide variety of microorganisms. In the presence of biosurfactants which lead to changes in the surface tension, metal desorption from solids might be enhanced resulting in an increased metal mobility in porous media. It has been suggested that this metabolic potential can be practically used in the bioremediation of metal-contaminated soils (MILLER, 1995). However, there is some evidence that surface-active compounds as well as organic solvents are inhibitory to bioleaching reactions and prevent bacterial attachment (MURR, 1980). The external addition of Tween reduced the oxidation of chalcopyrite by *T. ferrooxidans* (TORMA et al., 1976). It was concluded that the need of the microorganisms for surfactants is met by their own formation. In contrast, it was reported that the addition of Tween 80 increased the attachment of *T. ferrooxidans* on molybdenite and the oxidation of molybdenum in the absence of iron(II) (PISTACCIO et al., 1994).

4.4 Bacterial Attachment on Mineral Surfaces

It is known that the formation of extracellular polymeric substances plays an important role in the attachment of thiobacilli to mineral surfaces such as, e.g., sulfur, pyrite, or covellite. Extraction or loss of these exopolymers prevent cell attachment resulting in decreased metal leaching efficiencies (ESCOBAR et al., 1997; GEHRKE et al., 1998; POGLIANI and DONATI, 1992). It was concluded that a direct contact between bacterial cells and solid surfaces is needed and represents an important prerequisite for an effective metal mobilization (OSTROWSKI and SKLODOWSKA, 1993). Interactions between microorganisms and the mineral surface occur on two levels (BARRETT et al., 1993). The first level is a physical sorption because of electrostatic forces. Due to the low pH usually occurring in leaching environments, microbial cell envelopes are positively charged leading to electrostatic interactions with the mineral phase. The second level is characterized by chemical sorption where chemical bonds between cells and minerals might be established (e.g., disulfide bridges). In addition, extracellular metabolites are formed and excreted during this phase in the near vicinity of the attachment site (EWART and HUGHES, 1991). Low-molecular weight metabolites excreted by sulfur oxidizers include acids originating from the TCA cycle, amino acids, or ethanolamine, whereas compounds with relatively high molecular weights include lipids and phospholipids (BARRETT et al., 1993). In the presence of elemental sulfur, sulfur-oxidizing microorganisms from sewage sludge form a filamentous matrix similar to a bacterial glycocalyx suggesting the relative importance of these extracellular substances in the colonization of solid particles (BLAIS et al., 1994).

5 Microbial Diversity in Bioleaching Environments

A variety of microorganisms is found in leaching environments and has been isolated from leachates and acidic mine drainage. Although environmental conditions are usually described (from an anthropocentric view!) as being extreme and harsh due to pH values (as low as -3.6; NORDSTROM et al., 2000) and high metal concentrations (as high as 200 $g\,L^{-1}$; NORDSTROM et al., 2000), these systems can show high levels of microbial biodiversity including bacteria, fungi, and algae (LOPEZ-ARCHILLA et al., 1993). It has long been known that bacteria (*Thiobacillus* sp.), yeasts (*Rhodotorula* sp., *Trichosporon* sp.), flagellates (*Eutrepia* sp.), amoebes and protozoa are part of the microbial biocenosis found in acidic waters of a copper mine (EHRLICH, 1963). Recent detailed investigations based on molecular methods such as DNA–DNA hybridization, 16S rRNA sequencing, RCR-based methods with primers derived from rRNA sequencing, fluorescence *in situ* hybridization (FISH), or immunological techniques revealed that microbial bioleaching communities are composed of a vast variety of microorganisms resulting in complex microbial interactions and nutrient flows (such as synergism, mutualism, competition, predation) (AMARO et al., 1992; DE WULF-DURAND et al., 1997; EHRLICH, 1997; JOHNSON, 1998; EDWARDS et al., 1999). Selected organisms of these communities are given in Table 2. The composition of these communities is usually subjected to seasonal fluctuations and may vary between different mining locations (EDWARDS et al., 1999; GROUDEV and GROUDEVA, 1993). In addition, organisms are not homogeneously distributed over the whole leaching environment (CERDÁ et al., 1993).

The organism studied most is *Thiobacillus ferrooxidans*. Although this is the best known organism from acidic habitats, one may not conclude that this organism is dominant in these ecosystems. It has been found that under specific environmental conditions *Leptospirillum* sp. is even more abundant than *T. ferrooxidans* suggesting an important ecological role in the microbial community structure of bioleaching habitats (SAND, 1992; SCHRENK et al., 1998). Thiobacilli are members of the division of Proteobacteria close to the junction between the β and γ subdivision whereas leptospirilli are placed in the *Nitrospira* division (RAWLINGS, 1999). Genetic studies revealed that the role of *T. ferrooxidans* in leaching operations has probably been overestimated. Excellent reviews on the genetics of Thiobacilli and leptospirilli have been published recently (RAWLINGS, 1999; RAWLINGS and KUSANO, 1994).

Thiobacillus ferrooxidans belongs to the group of chemolithotrophic organisms. The organism is rod-shaped (usually single or in pairs), non-spore forming, gram-negative, motile, and single-pole flagellated (HORAN, 1999; KELLY and HARRISON, 1984; LEDUC and FERRONI, 1994; MURR, 1980). As carbon source, carbon dioxide is utilized. Ferrous iron is oxidized. Ammonium is used as nitrogen source. Although *T. ferrooxidans* has been characterized as being a strictly aerobic organism, it can also grow on elemental sulfur or metal sulfides under anoxic conditions using ferric iron as electron acceptor (DONATI et al., 1997; PRONK et al., 1992).

The genus *Thiobacillus* represents a versatile group of chemolithoautotrophic organisms. Optimum pH values for growth vary between 2 and 8 (Fig. 4). It has been demonstrated that sulfur-oxidizing bacteria are capable of reducing the pH of highly alkaline fly ash suspensions amended with elemental sulfur from approximately 9 to 0.5 (KREBS et al., 1999) (Fig. 5). It is likely that thiobacilli contribute to increasing acidification of leaching ecosystems in a successive mode: In the initial stages the growth of less acidophilic strains (e.g., *Thiobacillus thioparus*) is stimulated whereas during prolonged leaching the pH decreases gradually supporting growth of more acidophilic strains. This has already been observed in metal leaching from wastewater sewage sludge (BLAIS et al., 1993).

A variety of thermophilic microorganisms (especially *Sulfolobus* species) has been enriched and isolated from bioleaching environments (BRIERLEY, 1990; NEMATI et al., 2000; NORRIS and OWEN, 1993). Temperature optima for growth and metal leaching were in the range between 65 and 85 °C. Although copper extraction from mine tailings is more efficient using thermophilic instead of mesophilic organisms, extremely thermophilic microorganisms show a higher sensitivity to copper and to high pulp densities in agitated systems limiting, therefore, some practical applications (DUARTE et al., 1993; NORRIS and OWEN, 1993).

Although environmental conditions in leaching operations favor the growth and development of mesophilic, moderately thermophilic, and extremely thermophilic microbial communities, metal leaching at low temperatures has also been observed. Copper and nickel were leached from pyritic ore samples in significant amounts at 4 °C (AHONEN and TUOVINEN, 1992). However, leaching rates were lower by a factor of 30 to 50 as compared to experiments conducted at 37 °C. *T. ferrooxidans* recovered from mine waters was able to grow at 2 °C with a generation time of approximately 250 h suggesting a psychrotrophic nature of the organism (FERRONI et al., 1986). Bacterial iron mobilization has also been observed at 0 °C in ore samples obtained from Greenland (LANGDAHL and INGVORSEN, 1997). Solubilization rates at these low temperatures were still approximately 25 to 30% of the maximum values observed at 21 °C. All these findings may have a potential for practical applications in geographical areas where field operations are subjected to low temperature regimes.

A series of heterotrophic microorganisms (bacteria, fungi) is also part of microbial bioleaching communities (Tab. 2). This group of organisms uses extracellular metabolites and cell lysates from autotrophs as carbon source resulting in the removal of an inhibitory excess of carbon and stimulating, therefore, growth and iron oxidation of thiobacilli (BUTLER and KEMPTON, 1987; FOURNIER et al., 1998). In addition, several heterotrophs can also contribute to metal solubilization by the excretion of organic acids such as citrate, gluconate, oxalate, or succinate.

Tab. 2. Microbial Diversity of Acidic Bioleaching Environments and Acidic Mine Drainage: Selection of Microorganisms known to Mediate Metal Bioleaching Reactions from Ores and Minerals or known to be Part of the Microbial Consortia Found in Bioleaching Habitats

Domain	Organism	Nutrition Type	Main Leaching Agent	pH Range	pH Opt.	Temperature [°C]	Reference
Archaea	*Acidianus ambivalens*	facult. heterotrophic	sulfuric acid				Johnson (1998)
	Acidianus brierleyi	facult. heterotrophic	sulfuric acid	acidophilic	1.5–3.0	45–75	Muñoz et al. (1995)
	Acidianus infernus	facult. heterotrophic	sulfuric acid				Johnson (1998)
	Ferroplasma acidiphilum	chemolithoautotrophic	ferric iron	1.3–2.2	1.7	15–45	Golyshina et al. (2000)
	Metallosphaera prunae	chemolithoautotrophic	ferric iron, sulfuric acid				Johnson (1998)
	Metallosphaera sedula	chemolithoautotrophic	ferric iron, sulfuric acid	acidophilic		extr. thermophilic	Johnson (1998)
	Picrophilus oshimae						Johnson (1998)
	Picrophilus torridus						Johnson (1998)
	Sulfolobus acidocaldarius	chemolithoautotrophic	ferric iron, sulfuric acid	0.9–5.8	2.0–3.0	55–85	Amaro et al. (1992)
	Sulfolobus ambivalens	chemolithoautotrophic	ferric iron, sulfuric acid			extr. thermophilic	Rossi (1990)
	Sulfolobus brierleyi	chemolithoautotrophic	ferric iron, sulfuric acid			extr. thermophilic	Brierley (1977)
	Sulfolobus hakonensis	chemolithoautotrophic					Rossi (1990)
	Sulfolobus metallicus	chemolithoautotrophic					Rossi (1990)
	Sulfolobus solfataricus	chemolithoautotrophic	ferric iron, sulfuric acid			extr. thermophilic	Johnson (1998)
	Sulfolobus thermosulfidooxidans	chemolithoautotrophic	ferric iron, sulfuric acid			extr. thermophilic	Johnson (1998)
	Sulfolobus yellowstonii	chemolithoautotrophic	ferric iron, sulfuric acid			extr. thermophilic	Johnson (1998)
	Sulfurococcus mirabilis	mixotrophic	ferric iron, sulfuric acid	acidophilic		extr. thermophilic	Barrett et al. (1993), Johnson (1998)
	Sulfurococcus yellowstonii	mixotrophic	ferric iron, sulfuric acid				Johnson (1998)
	Thermoplasma acidophilum						Johnson (1998)
	Thermoplasma volcanicum						Johnson (1998)
Bacteria	*Acetobacter methanolicus*	heterotrophic	gluconate	acidophilic			Glombitza et al. (1988)
	Acidimicrobium ferrooxidans						Johnson (1998), Edwards et al. (1999)

Tab. 2. Continued

Domain	Organism	Nutrition Type	Main Leaching Agent	pH Range	pH Opt.	Temperature [°C]	Reference
	Acidiphilium angustum						EDWARDS et al. (1999)
	Acidiphilium cryptum	heterotrophic	organic acids	2.0–6.0		mesophilic	GOEBEL and STACKEBRANDT (1994)
	Acidiphilium symbioticum	heterotrophic	organic acids		3.0	mesophilic	BHATTACHARYYA et al. (1991)
	Acidobacterium capsulatum	chemoorganotrophic		3.0–6.0		mesophilic	KISHIMOTO et al. (1991)
	Acidocella sp.						JOHNSON (1998)
	Acidomonas methanolica	heterotrophic					JOHNSON (1998)
	Arthrobacter sp.	heterotrophic					BOSECKER (1993)
	Aureobacterium liquifaciens	heterotrophic					EDWARDS et al. (1999)
	Bacillus sp.	heterotrophic					CERDÁ et al. (1993), GROUDEV and GROUDEVA (1993)
	Bacillus coagulans	heterotrophic		5.4–6.0		22	BAGLIN et al. (1992)
	Bacillus licheniformis	heterotrophic				37	MOHANTY and MISHRA (1993)
	Bacillus megaterium	heterotrophic	citrate				KREBS et al. (1997)
	Bacillus polymyxa	heterotrophic					
	Chromobacterium violaceum	heterotrophic	cyanide				LAWSON et al. (1999)
	Comamonas testosteroni	heterotrophic					EDWARDS et al. (1999)
	Crenothrix sp.	facult. autotrophic	ferric iron	5.5–6.2		18–24	ROSSI (1990)
	Enterobacter agglomerans	heterotrophic		5.4–6.0		22	BAGLIN et al. (1992)
	Enterobacter cloacae	heterotrophic				22	BAGLIN et al. (1992)
	Gallionella sp.	autotrophic	ferric iron	6.4–6.8		6–25	ROSSI (1990)
	Kingella kingae						EDWARDS et al. (1999)
	Lactobacillus acidophilus	heterotrophic				37	ACHARYA et al. (1998)
	Leptospirillum ferrooxidans	chemolithoautotrophic	ferric iron		2.5–3.0	30	SAND (1992), RAWLINGS et al. (1999)
	Leptospirillum thermoferrooxidans	chemolithoautotrophic	ferric iron		1.7–1.9	45–50	BARRETT et al. (1993)
	Leptothrix discophora	facult. autotrophic	ferric iron, sulfuric acid	5.8–7.8		5–40	EDWARDS et al. (1999)
	Metallogenium sp.	heterotrophic	ferric iron	3.5–6.8	4.1		ROSSI (1990)
	Ochrobacterium anthropi	heterotrophic					EDWARDS et al. (1999)
	Propionibacterium acnes	heterotrophic				37	ACHARYA et al. (1998)
	Pseudomonas cepacia	heterotrophic		5.4–6.0		22	BAGLIN et al. (1992)

Tab. 2. Continued

Domain	Organism	Nutrition Type	Main Leaching Agent	pH Range	pH Opt.	Temperature [°C]	Reference
	Pseudomonas putida	heterotrophic	citrate, gluconate				Krebs et al. (1997)
	Psychrobacter glacincola	heterotrophic					Edwards et al. (1999)
	Serratia ficaria	heterotrophic					Edwards et al. (1999)
	Siderocapsa sp.	heterotrophic	ferric iron				Rossi (1990)
	Staphylococcus lactis	heterotrophic				37	Acharya et al. (1998)
	Stenotrophomonas maltophila	heterotrophic					Edwards et al. (1999)
	Sulfobacillus thermosulfidooxidans	chemolithoautotrophic	ferric iron, sulfuric acid	extr. acidoph.		50	Johnson (1998)
	Thermothrix thiopara	chemolithoautotrophic	sulfuric acid	neutral		60–75	Brierley (1977)
	Thiobacillus acidophilus	mixotrophic	sulfuric acid	1.5–6.0	3.0	25–30	Cerdá et al. (1993), Johnson (1998)
	Thiobacillus albertis	chemolithoautotrophic	sulfuric acid	2.0–4.5	3.5–4.0	28–30	Johnson (1998)
	Thiobacillus caldus	chemolithoautotrophic	sulfuric acid			45	Amaro et al. (1992), Dopson and Lindstrom (1999)
	Thiobacillus capsulatus	chemolithoautotrophic	sulfuric acid				Ewart and Hughes (1991)
	Thiobacillus concretivorus	chemolithoautotrophic	sulfuric acid	0.5–6.0			Rossi (1990)
	Thiobacillus delicatus	mixotrophic	sulfuric acid		5.0–7.0	25–30	Rossi (1990)
	Thiobacillus denitrificans	chemolithoautotrophic	sulfuric acid	5.0–7.0		30	Groudev and Groudeva (1993)
	Thiobacillus ferrooxidans	chemolithoautotrophic	ferric iron, sulfuric acid	1.4–6.0	2.4	28–35	Sand (1992)
	Thiobacillus intermedius	facult. heterotrophic	sulfuric acid	1.9–7.0	6.8	30	Rossi (1990)
	Thiobacillus kabobis	mixotrophic	sulfuric acid	1.8–6.0	3.0	28	Rossi (1990)
	Thiobacillus neapolitanus	chemolithoautotrophic	sulfuric acid	3.0–8.5	6.2–7.0	28	Groudev and Groudeva (1993)
	Thiobacillus novellus	chemolithoautotrophic	sulfuric acid	5.0–9.0	7.8–9.0	30	Rossi (1990)
	Thiobacillus organoparus	mixotrophic	sulfuric acid	1.5–5.0	2.5–3.0	27–30	Rossi (1990)
	Thiobacillus perometabolis	chemolithoheterotrophic	sulfuric acid	2.6–6.8	6.9	30	Rossi (1990)
	Thiobacillus prosperus	chemolithoautotrophic	sulfuric acid	1.0–4.5		23–41	Huber and Stetter (1989)
	Thiobacillus pumbophilus	chemolithoautotrophic	sulfuric acid	4.0–6.5		27	Drobner et al. (1992)
	Thiobacillus rubellus	chemolithoautotrophic	sulfuric acid		5.0–7.0	25–30	Barrett et al. (1993)

Tab. 2. Continued

Domain	Organism	Nutrition Type	Main Leaching Agent	pH Range	pH Opt.	Temperature [°C]	Reference
	Thiobacillus tepidarius	chemolithoautotrophic	sulfuric acid				Hughes and Poole (1989)
	Thiobacillus thiooxidans	chemolithoautotrophic	sulfuric acid	0.5–6.0	2.0–3.5	10–37	Sand (1992)
	Thiobacillus thioparus	chemolithoautotrophic	sulfuric acid	4.5–10.0	6.6–7.2	11–25	Blowes et al. (1998)
	Thiobacillus versutus	chemolithoautotrophic	sulfuric acid		8.0–9.0		Rossi (1990)
	Thiomonas cuprinus	facult. heterotrophic	sulfuric acid		3.0–4.0	30–36	Huber and Stetter (1990)
Eukarya	*Actinomucor* sp.	heterotrophic	succinate			27	Müller and Förster (1964)
Fungi	*Alternaria* sp.	heterotrophic	citrate, oxalate			32	Kovalenko and Malakhova (1990)
	Aspergillus awamori	heterotrophic				28	Ogurtsova et al. (1989)
	Aspergillus fumigatus	heterotrophic					Bosecker (1989)
	Aspergillus niger	heterotrophic	oxalate, citrate, gluconate, malate, tartrate, succinate			30	Dave et al. (1981), Bosecker (1987)
	Aspergillus ochraceus	heterotrophic	citrate			28	Ogurtsova et al. (1989)
	Aspergillus sp.	heterotrophic	citrate, oxalate			30	Tzeferis (1994)
	Cladosporium resinae	heterotrophic				28	Ogurtsova et al. (1989)
	Cladosporium sp.	heterotrophic					Kovalenko and Malakhova (1990)
	Coriolus versicolor	heterotrophic	oxalate				Sayer et al. (1999)
	Fusarium sp.	heterotrophic	oxalate, malate, pyruvate, oxalacetate				Bosecker (1989)

Tab. 2. Continued

Domain	Organism	Nutrition Type	Main Leaching Agent	pH Range	pH Opt.	Temperature [°C]	Reference
	Mucor racemosus	heterotrophic	citrate, succinate			27	MÜLLER and FÖRSTER (1964)
	Paecilomyces variotii	heterotrophic	citrate, oxalate				DAVE et al. (1981)
	Penicillium sp.	heterotrophic				25	GUPTA and EHRLICH (1989)
	Penicillium chrysogenum	heterotrophic				28	OGURTSOVA et al. (1989)
	Penicillium funiculosum	heterotrophic	citrate				BOSECKER (1989)
	Penicillium notatum	heterotrophic				26	KARAVAIKO et al. (1980)
	Penicillium simplicissimum	heterotrophic	citrate, oxalate, gluconate			22–30	TARASOVA et al. (1993), SILVERMAN and MUNOZ (1971)
	Rhizopus japonicus	heterotrophic					OGURTSOVA et al. (1989)
	Trichoderma lignorum	heterotrophic				24–26	AVAKYAN et al. (1981)
	Trichoderma viride	heterotrophic				32	BOROVEC (1990)
Yeasts	*Candida lipolytica*	heterotrophic				30	GROUDEV (1987)
	Rhodotorula sp.	heterotrophic					EHRLICH (1963), CERDÁ et al. (1993)
	Saccharomyces cerevisiae	heterotrophic				28	OGURTSOVA et al. (1989)
	Torulopsis sp.	heterotrophic					CERDÁ et al. (1993)
	Trichosporon	heterotrophic					EHRLICH (1963)
Algae	not identified						GROUDEV and GROUDEVA (1993)
Protozoa	not identified						GROUDEV and GROUDEVA (1993)
Amoebae	not identified						EHRLICH (1963)

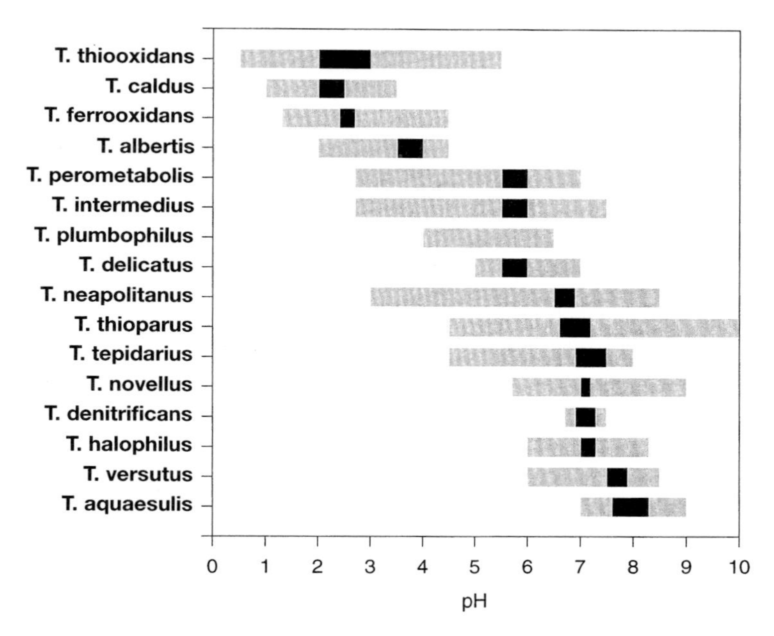

Fig. 4. Ranges of pH values for optimal growth as well as tolerance limits for *Thiobacillus* species (KELLY and HARRISON, 1984).

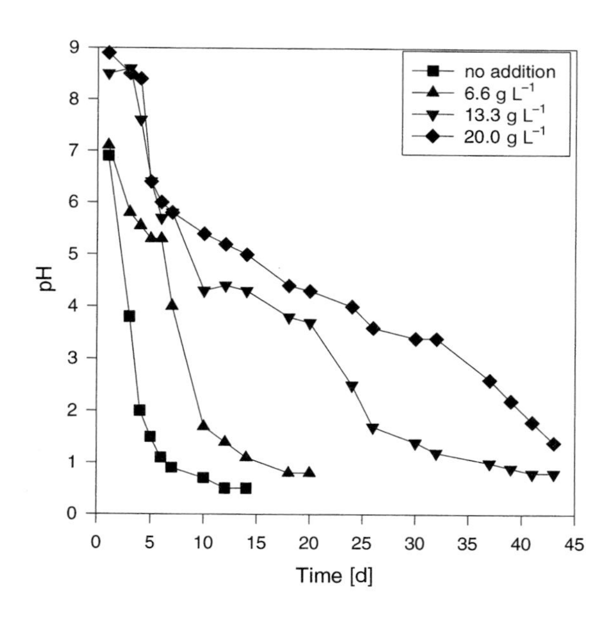

Fig. 5. Decrease of pH in fly ash suspensions (up to 20 g L^{-1}) supplemented with suspensions of sewage sludge (40 mL L^{-1}) containing elemental sulfur (10 g L^{-1}). Inoculation with a community containing equal amounts of *Thiobacillus acidophilus*, *Thiobacillus neapolitanus*, and *Thiobacillus thiooxidans* (adapted from KREBS et al., 1999).

6 Case Studies of Bioleaching Applications

6.1 Commercial-Scale Copper Ore Bioleaching

Commercial-scale leaching technologies are applied mainly in the mining industry where biohydrometallurgical technologies were first introduced to the copper industry and subsequently to the gold (OLSON, 1994; OLSON and KELLY, 1986), uranium (KHALID et al., 1993), nickel and zinc (AGATE, 1996) industry for the production of metal values from low-grade ores or from low-grade mineral resources (ANDREWS, 1990). Very recently, cobaltiferous ores are also treated on a commercial scale (D'HUGUES et al., 1999). Table 3 gives an overview of industrial bioleaching operations.

Several types of commercial-scale leaching application are in use (BROMBACHER et al., 1997): (1) *in situ* leaching where metals are solubilized and recovered from rocks without removing them mechanically from the ore body

Tab. 3. Selection of Recent and Ceased Industrial Bioleaching Operations for Gold, Nickel, Copper, and Cobalt (BRIERLEY, 1997, 1999; BRIERLEY and BRIELEY, 1999; LOAYZA et al., 1999; ROSSI, 1999)

Country	Locality, Designation	Metal	Mineral Source	Technology	Capacity [$t\ d^{-1}$]	Metal Yield [$t\ y^{-1}$]
Australia	Harbour Lights	Au	flotation concentrate	tank leaching ($160\ m^3$)	40	
	Girilambone	Cu	chalcocite	heap leaching	16,000	14,000
	Gunpowder	Cu	chalcocite, bornite	*in situ* leaching		13,000
	Maggie Hays	Ni	concentrate	tank leaching (pilot plant)		7
	Wiluna	Au	flotation concentrate	tank leaching ($480\ m^3$)	115	
	Youanmi	Au	flotation concentrate	tank leaching ($6 \cdot 480\ m^3$)	120	
Brasil	Sao Bento	Au	flotation concentrate	tank leaching ($550\ m^3$)	150	
Canada	Goldbridge	Au	pyrite, markasite, arsenopyrite	tank leaching ($225\ m^3$)	75	
Chile	Andacollo	Cu	chalcocite	heap leaching	10,000	
	Cerro Colorado	Cu	chalcocite	heap leaching	16,000	60,000
	Dos Amigos	Cu	chalcocite	heap leaching	3,000	
	Quebrada Blanca	Cu	chalcocite	heap leaching	17,300	75,000
	Zaldivar	Cu	chalcocite	heap leaching	20,000	
Ghana	Ashanti	Au	flotation concentrate	tank leaching ($6 \cdot 900\ m^3$)	960	
India	Malanjkhand	Cu	malachite, chalcocite, bornite, covellite	heap leaching	2.5	
Peru	Tamborque	Au, Ag	arsenopyrite from zinc flotation	tank leaching (pilot plant $1\ m^3$)		
South Africa	Fairview	Au	flotation concentrate	tank leaching ($90\ m^3$)	35	
Uganda	Kasese	Co	flotation concentrate	tank leaching ($4 \cdot 1{,}350\ m^3$)		1,000
USA	Carlin	Au	Au-containing sulfidic ore	heap leaching	10,000	
	Chino	Cu	chalcocite, chrysokolla	heap leaching		55,000
	San Manuel	Cu	chalcocite	*in situ* leaching		20,000

(MURR, 1980). Abandoned mines are flooded with water and left for a certain time period. Subsequently, the metal-containing solution is recovered and subjected to a metal recovery process. (2) Dump leaching which is applied for the recovery of metals from very lean ore mixed with rocks. Dumps are formed with material obtained from preparation plants in underground mines or from the stripping of open cast mines (MURR, 1980). (3) Heap leaching which consists of the build-up of run-of-mines ore heaps or of low-grade mine tailings in appropriately prepared areas. Figure 6 gives a schematic overview. The ore is pretreated by crushing to an optimal grain size, screening, or even partial roasting (BRIERLEY, 1999). In addition, the crushed ore is preconditioned with sulfuric acid. Ore heaps are formed on a lined water-impermeable ground. A system of pipes and sparging devices are placed on the ground and on top of the heaps to aerate and irrigate the ore, respectively (Fig. 7a). Often, coarse material is placed at the bottom of a heap to improve aeration. Inoculation with bacterial populations can be done using acidic mine effluents or recycled leachates from metal electrowinning (BRIERLEY, 1999). Metal-containing leachates are collected at the foot of the heap as shown for copper in Figure 7b. The pregnant solutions are concentrated and metals are extracted with solvents which are treated by an electrowinning process to obtain solid copper (Figs. 7c and d). Environmental conditions (temperature, oxygen concentration, metal recovery) in dumps or heaps for the leaching of chalcocite and pyrite have been assessed using a 2-D model (MORENO et al., 1999). (4) Vat leaching which is carried out when ore, after crushing to a suitable size, is dumped into concrete vats lined with acid-proof material. In general, there are a few vats in series, and the leaching solution percolates from a bottom opening upwards through the ore mass, and the overflow is pumped into the next vat. (5) Tank leaching which represents the most expensive and material consuming technology (BRUYNESTEYN and HACKL, 1985; ROSSI, 1999). Special installations are required for a reactor or chemostat leaching process. This process is carried out in reactors consisting of adequately stirred tanks in which the pulp to be leached resides for a defined time interval. This technique, therefore, appears to offer the best potential for leaching operations since pronouncedly higher reaction rates can be achieved as compared to other process systems.

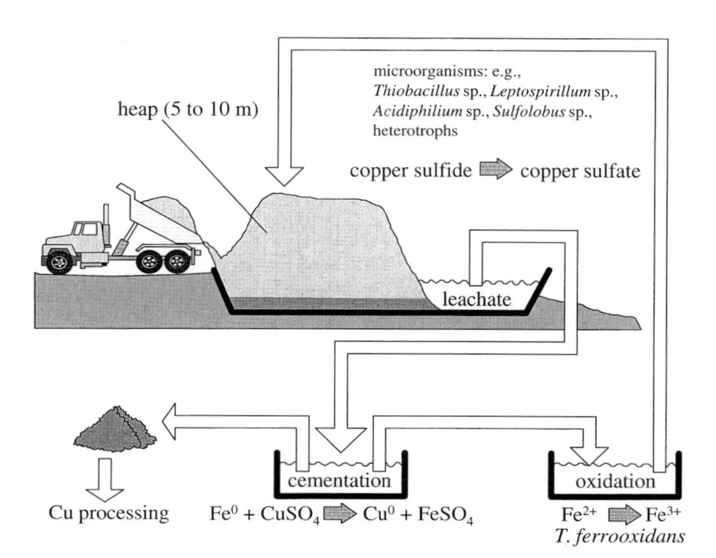

Fig. 6. Schematic overview of heap bioleaching (modified after LUGGER, 2000).

Fig. 7. Copper bioleaching at the Phelps Dodge mine (New Mexico, USA). **a** Heap formation on water-impermeable lining covered with a piping system for irrigation; **b** leachates at the foot of a heap; **c** collection of heap leachates for further treatment; **d** solid copper after electrowinning. (Courtesy of R. BACHOFEN, University of Zürich, Switzerland).

6.2 Reactor Bioleaching of Fly Ash

In addition to mining operations, biohydrometallurgy is a promising technology useful either to recover valuable metals from industrial waste materials (e.g., slag, galvanic sludge, filter dust, fly ash) or to detoxify them for a less hazardous deposition. Biohydrometallurgical processing of solid waste allows the cycling of metals similar to biogeochemical metal cycles and diminishes the demand for resources such as ores, energy, or landfill space. Fly ash from municipal waste incineration (MWI) represents a concentrate of a wide variety of toxic heavy metals (e.g., Cd, Cr, Cu, Ni). The low acute and chronic toxicity of fly ash for a variety of microorganisms and a low mutagenic effect of fly ash from MWI has been demonstrated (LAHL and STRUTH, 1993). Nevertheless, the deposition of heavy metal-containing material bears severe risks of spontaneous leaching of heavy metals due to natural weathering processes and due to uncontrolled bacterial activities (LEDIN and PEDERSEN, 1996; NOTTER, 1993; ROSSI, 1990). In the light of Agenda 21 established at the Earth Summit in Rio in 1992 there is a strong demand to support sustainable developments which include also the ecological treatment of wastes and their safe disposal. A biological metal leaching of fly ash is a very important step in this direction. Published data for biohydrometallurgical treating of fly ash or industrial wastes using bacteria or fungi showed extensive residence times for a leaching period of up to 50 d (BOSECKER, 1986; BOSSHARD et al., 1996; SCHÄFER, 1982) and most of them were only performed in shake flask experiments with small quantities of heavy metal-containing material. To optimize the process, a semi-continuous laboratory-scale leaching plant (LSLP) was constructed in order to achieve high leaching efficiencies resulting in an increased overall load of elements in the effluent in a shorter period of time as compared to batch cultures (BROMBACHER et al., 1998). A mixture of *Thiobacillus thiooxidans*, which forms a high concentration of sulfuric acid due to bacterial energy metabolism, and *Thiobacillus ferrooxidans*, which is able to oxidize reduced metal compounds resulting in an increased solubility of these metals, is used to perform the leaching experiments.

Biohydrometallurgical processing of fly ash poses severe problems especially at higher pulp densities, because of the high content of toxic metals and the saline and strongly alkaline (pH >10) environment (BROMBACHER et al., 1998). For examining the biological leaching of fly ash at high pulp densities, pulp (80 g L^{-1}) from a storage vessel was mixed with a culture of *T. ferrooxidans* and *T. thiooxidans* in such a way, that the three reaction vessels (A, B, C, each of them 1 dm^3) ran in a semi-continuous process at pulp densities of 40 g L^{-1} (BROMBACHER et al., 1998). Leaching efficiencies are in the same order as in experiments where *T. thiooxidans* was grown in batch cultures in the presence of fly ash. However, by employing a semi-continuous process, higher pulp densities can be applied resulting in increased overall loads of elements in a certain time period. Results show the potential of thiobacilli together with different microorganisms to leach substantial quantities of toxic metals from fly ash (Fig. 8). Depending on the point of view, the mobilization or bioleaching of these metals could be either a hazard for the environment (leachates from landfills or ore deposits) or a chance to reduce toxic elements and to recover valuable metals by a low-cost and low-energy level technology compared with thermal treatment (e.g., vitrification or evaporation). The experimental installation described seems to be a first promising step on the way to a pilot plant with high capacities to detoxify fly ash (for a reuse of these materials for construction purposes) and for an economical recovery of valuable metals such as zinc.

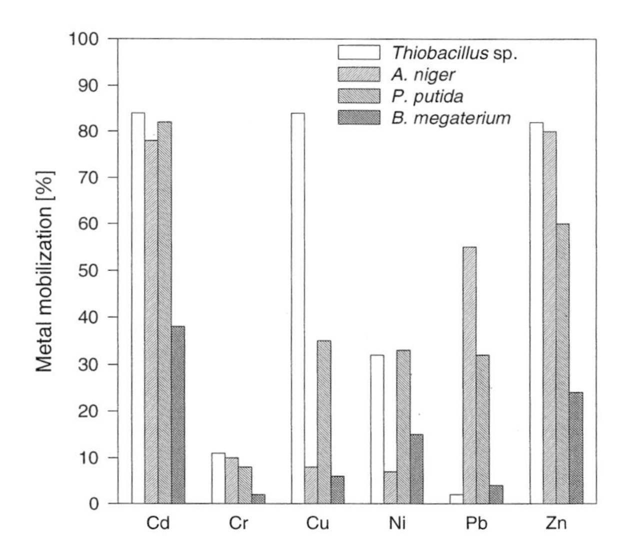

Fig. 8. Leaching of fly ash from municipal waste incineration with four different microorganisms (*Bacillus megaterium, Pseudomonas putida, Thiobacillus* sp., *Aspergillus niger*) in suspensions of 20 g L^{-1}.

Slag and ash from municipal waste incineration represent concentrates of a wide variety of elements. Some metals (e.g., Al, Zn) are present in concentrations that allow an economical metal recovery (Tab. 4), whereas certain elements (e.g., Ag, Ni, Zr) show relatively low concentrations (comparable to low-grade ores) which make a conventional technical recovery difficult. Especially in these cases, microbial processes are the technique of choice and basically the only possibility to obtain metal values from these materials. In addition, the low toxicity of these residues to bacteria has been demonstrated (BROMBACHER, 1994). Therefore, a biological metal recovery process is potentially possible. As a result, heavy metal contents are reduced to fulfill the regulations for landfilling and valuable raw materials are recovered.

In a recent study, thiobacilli, *Pseudomonas putida*, *Bacillus megaterium*, and *Aspergillus niger* were used as test organisms and incubated with fly ash obtained from municipal waste incineration (KREBS et al., 1997). Regarding thiobacilli, growth and the pH decrease were slow (1 to 3 months, depending on the amount of fly ash), but finally the pH reached a value of about 1. Elements such as cadmium, copper, or zinc were mobilized to >80% whereas others (e.g., Pb) were solubilized only by a small percentage (Fig. 8). When *Pseudomonas putida* was exposed to fly ash, citric acid – a product from the metabolism of glucose – acted with its complexing properties as leaching agent. Growth was very fast (less than one week) and the pH dropped to about 4. Highest leaching efficiencies were in the range of 60 to 80%. For fly ash amounts of >20 g L^{-1}, efficiencies were below 20% for all elements. The reason was a limitation of glucose. Therefore, the pH did not decrease further and the ratio of citrate to fly ash was lower. In the presence of fly ash, the fungus *Aspergillus niger* produced gluconic acid instead of citric acid (inhibition of enzymes of the TCA-cycle by manganese). Growth was fast (2 to 3 weeks) and the pH decreased to 3.3 resulting in leaching efficiencies of >80% for certain elements (Fig. 8). *Aspergillus niger* proved especially effective for the leaching of Pb.

6.3 Shake Flask Bioleaching of Electronic Scrap

Microbiological processes were applied to mobilize metals from electronic waste materials (BRANDL et al., 1999). Results indicate that it is possible to mobilize metals from electrical and electronic waste materials by the use of microorganisms such as bacteria (*Thiobacillus thiooxidans*, *Thiobacillus ferrooxidans*) and fungi (*Aspergillus niger*, *Penicillium simplicis-*

Tab. 4. Metal-Containing Solids Treated by Biohydrometallurgical Processes

Material	Source	Metal Content [g kg^{-1}]							
		Al	Cd	Cr	Cu	Ni	Pb	Sn	Zn
Ore	rock[a]	300	0.1	300	2	7	4	10	10
Bottom ash	municipal waste incineration	46	tr	0.4	1.5	0.1	0.7	0.3	2.5
Fly ash	municipal waste incineration	70	0.5	0.6	1	0.1	8	8	31
Dust	electronic scrap	240	0.3	0.7	80	15	20	20	25
Sludge	galvanic industries	nd	nd	26	43	105	nd	nd	166
Soil	mining activities (Zn/Pb-mine)	nd	0.1	0.2	9	0.1	25	1.6	24
Organic biomass	plants from phytoremediation	nd	0.007	nd	0.03	nd	nd	nd	0.2
Soil	earth crust[b]	72	0.0003	0.1	0.03	0.04	0.02	0.01	0.05

[a] Metal concentration which makes a recovery economically interesting.
[b] Average metal concentration.
tr: traces; nd: not determined.

simum). After a prolonged adaptation time, fungi as well as bacteria grew also at concentrations of 100 g L^{-1}. Both fungal strains were able to mobilize Cu and Sn by 65%, and Al, Ni, Pb, and Zn by more than 95%. At scrap concentrations of 5 to 10 g L^{-1} thiobacilli were able to leach more than 90% of the available Cu, Zn, Ni, and Al. Pb precipitated as $PbSO_4$ while Sn precipitated probably as SnO. For a more efficient metal mobilization a two-step leaching process is proposed where biomass growth is separated from metal leaching. Metals were precipitated and recovered as hydroxides by the stepwise addition of sodium hydroxide.

7 Economics of Metal Bioleaching

Economical factors for commercial-scale leaching applications can be divided into capital costs associated with construction and operation and maintenance costs (BROMBACHER et al., 1997). Operation costs include the running of the process equipment, the supply of reagents, services, and labor (BRUYNESTEYN et al., 1986). They are comparable to those of competing technologies (POULIN and LAWRENCE, 1996). However, capital costs are generally smaller. According to BARRETT et al. (1993), these costs depend on the methods of application and increase in the order dump < vat ≤ heap < agitated reactor.

Major items affecting costs are construction material and equipment. Equipment required for agitated bacterial oxidation processes includes conventional tanks with impellers to suspend the solids and disperse the air, compressors for air supply, thickeners for solid/liquid separation, and conventional slurry pumps for delivering and removing the slurry from the plant. Construction material needs to be acid resistant which furthermore withstands temperatures of 30 to 50 °C. In general, technical equipment for bioleaching processes is less expensive as compared to physicochemical processes. However, longer treatment times typical for bacterial leaching processes are required which compensate in part for this profit (RIEKKOLA-VANHANEN and HEIMALA, 1993; SAND et al., 1993). Other major capital costs in establishing a particular application are the services to support the process. In addition, except for the extracted metals, no saleable products are formed by the process. The acid generated is of low-grade quality and mostly contaminated with metals and salts. By-products contained in waste streams are normally not recoverable in an economical way.

An economical profile was calculated for old copper tailings in Chile containing between 0.18 to 0.26% copper (BRICENO et al., 1993). A total investment of US$ 20.5 million for the whole leaching process (repulping, desliming, dewatering, continuous heap leaching, solvent extraction, and electrowinning) for a plant capacity of 16 t Cu d^{-1} would have been necessary. The operation costs amount to US$ 0.99 per kg Cu. The annual net present value leads to US$ 2.11 million for a copper price of US$ 1.76 per kg Cu, and results in an internal rate of return of 14.2%. In comparison to ore roasting, capital costs for industrial-scale treatment plants are reduced by 12 to 20%, operating costs by 10% (BRIERLEY, 1995). Costs of 4.30 US$ per ton of ore have been calculated for a bacterial ore treatment compared to 4.10 and 4.50 US$ for pressure oxidation and roasting, respectively (BRUYNESTEYN and HACKL, 1985; BRUYNESTEYN et al., 1986; MCNULTY and THOMPSON, 1990).

Over the years, the amount of copper which has been obtained by biohydrometallurgical operations has been gradually increasing (Fig. 9). Today, between 25 and 30% of copper produced yearly is based on microbiological treatment of mineral resources. Together with other metals such as cobalt, gold, nickel, uranium, or zinc, biological metal extraction processes result in a surplus value of over 10 billion US dollars (Tab. 5).

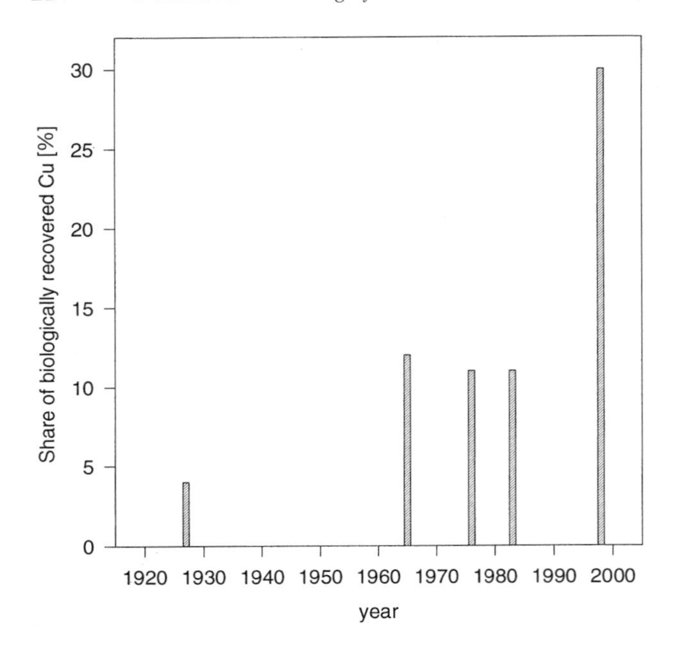

Fig. 9. Increase of microbiologically obtained copper (in percent of the annual world production).

Tab. 5. Annual World Production of Metals: Share of Microbiologically Mediated Production and Value-Added According to Market Prices from May 2000

Metal	Production [thousand t]	Year	Biological Production [%]	Actual Market Price [US$ kg^{-1}]	Value-Added [billion US$]
Cobalt	18	1996	10	60.63	0.11
Copper	11,481	1997	15–30	1.723	5.9
Gold	2.3	1997	20	10,582	4.9
Nickel	1,100	1998	2[a]	8.41	0.19
Uranium	35.6	1997	10–15	35.00	0.19
Zinc	6,963	1997	2[a]	1.09	0.15

[a] Percentage of biological production unknown. A value of 2% was assumed.

8 Perspectives of Bioleaching Technology

8.1 Heterotrophic Leaching

Future developments can be carried out in microbiological as well as technical areas, e.g., the application of thermophiles and heterotrophs for leaching operations; full characterization of microbial populations found in bioleaching systems especially regarding mechanisms and organismic interactions; combinations of bioleaching and biobeneficiation techniques; improvement of aeration systems for stirred tank applications (BRIERLEY and BRIERLEY, 1999).

As pointed out recently, the future of biohydrometallurgical applications should be oriented towards leaching by heterotrophic microorganisms, including fungi (EHRLICH, 1999). However, there are several drawbacks in pursuing this goal: Usually, conditions for thiobacilli-mediated leaching are very restrictive (extremely low pH!) and do, therefore, not require sterilization of growth media and solid materials. In contrast, leaching activities of heterotrophic organisms take place over a much broader pH range giving rise to possible contamination with an unwanted microbial flora. Thus, sterility might become a problem especially for large-scale applications. In addition, sterilization treatment is not interesting from an economic point of view (e.g., energy costs). Another economic restriction is the use of organic carbon sources required for fungal growth and acid formation. Generally, these carbon sources are a major cost factor regarding biotechnological applications (MULLIGAN et al., 1999). The availability of an inexpensive carbon source (e.g., sugar molasses, by-products from the food industry) is, therefore, a prerequisite for the development of heterotrophic leaching processes.

Advantages of using heterotrophic bacteria and fungi as mediators for metal mobilization can be deduced from metal leaching patterns that differ from those obtained by the use of sulfur-oxidizing bacteria (Fig. 8). Depending on the strains used, the growth conditions, and the metal-containing solids, certain metals (e.g., Pb, Sn) are solubilized which cannot be leached by thiobacilli. A technical (stepwise) process which combines the fungal and bacterial versatility might result in a very effective metal mobilization and increase recovery yields.

8.2 Leaching under Thermophilic Conditions

The stimulation and use of thermophilic microorganisms is a further possibility to develop novel bioleaching technologies. Parts of bioleaching heaps can reach temperatures of approximately 75 °C (BRIERLEY, 1999). Therefore, microorganisms such as *Sulfolobus*, *Acidianus*, or *Metallosphaera* species might have the potential for industrial applications (HAN and KELLY, 1998; LINDSTROM et al., 1993; KONISHI et al., 1995). In general, thermophiles have simple nutritional requirements and are often characterized by a high growth rate and fast substrate turnover. In addition, they are often more resistant to stress conditions. Problems of possible unwanted microbial contaminations are minimized due to the high cultivation temperatures.

8.3 Tapping Microbial Diversity

There is an immense potential of finding many microorganisms suitable for metal solubilization from a basically untapped resource, namely microbial species richness. Several groups of organisms which are known to mediate leaching processes are poorly investigated. As examples, cyanogenic *Chromobacter violaceum* is known to mobilize gold from gold-containing pyrite as a gold cyanide complex (LAWSON et al., 1999; SMITH and HUNT, 1985). Other microorganisms also able to form cyanide have not been investigated regarding their metal-mobilizing potential. Although it has long been known that nitrifying organisms can contribute to rock decomposition (MUNTZ, 1890) or to the biocorrosion of concrete (SAND and BOCK, 1991) by forming nitrous and nitric acid, their metal-leaching potential has basically not been investigated.

8.4 Treatment of Solid Wastes

Another broad area which has to be evaluated in the future is the microbial treatment of metal-containing waste materials from industrial processes as has been demonstrated for, e.g., fly and bottom ash from waste incineration, galvanic sludge, or sewage sludge (Tab. 6). Due to their high metal concentrations these waste materials can be considered as "artificial ores" (Tab. 4). They can serve as secondary raw materials and might reduce the demand for primary mineral resources. However, a prerequisite is the development of a process which is technically feasible and economically as well as ecologically justifiable.

8.5 Bioremediation of Metal-Contaminated Sites

Besides applications in mining operations for the recovery of base and precious metals, microbial metal mobilization methods can also find future applications for the remediation

Tab. 6. Examples of Industrial Waste Materials which are Microbially Treated for Metal Recovery (Adapted from KREBS, 1997)

Waste Material	Metals Leached	Organisms Applied	Reference
Catalysts	Cu, Pb, V, Zn	*Thiobacillus ferrooxidans*, *Bacillus* sp., *Saccharomyces cerevisiae*, *Yarrowia lipolytica*	BRIAND et al. (1999), CLARK and EHRLICH (1992), HAHN et al. (1993)
Electronic scrap	Al, Cu, Ga, Ni, Pb, Sn, Zn	*Thiobacillus thiooxidans*, *Thiobacillus ferrooxidans*, *Aspergillus niger*, *Penicillium simplicissimum*, *Bacillus* sp., *Saccharomyces cerevisiae*, *Yarrowia lipolytica*	BRANDL et al. (1999), HAHN et al. (1993)
Fly ash from coal power plant	Al	*Aspergillus niger*	SINGER et al. (1982)
Fly ash from pyrite roasting	Cu	*Thiobacillus ferrooxidans*, *Thiobacillus thiooxidans*	EBNER (1977)
Fly ash from waste incineration	Al, Cd, Cr, Cu, Fe, Mn, Ni, Pb, Zn	*Thiobacillus ferrooxidans*, *Thiobacillus thiooxidans*, *Aspergillus niger*	BOSSHARD et al. (1996), BROMBACHER et al. (1998), KREBS et al. (1999)
Galvanic sludge	Cr, Cu, Zn	*Thiobacillus* sp., *Thiobacillus thiooxidans*	BOSECKER (1986), BRUNNER and SCHINNER (1991)
Hydroxide sludge	Ni	*Thiobacillus ferrooxidans*, *Thiobacillus thiooxidans*	CHAWAKITCHAREON et al. (1999)
Red mud from bauxite extraction	Al	*Thiobacillus* sp., *Aspergillus niger*, *Penicillium simplicissimum*, *Trichoderma viride*	VACHON et al. (1994)
Semiconductor wafers	Ga	*Acidothermus* sp., *Sulfolobus* sp.	BOWERS-IRONS et al. (1993)
Sewage sludge	Cd, Cr, Cu, Mn, Ni, Pb, Zn	*Thiobacillus* sp., *Thiobacillus thioparus*	BLAIS et al. (1993), SREEKRISHNAN et al. (1993)
Slag from lead smelting	Cu, Zn	*Thiobacillus ferrooxidans*, *Thiobacillus thiooxidans*	EBNER (1977)

and decontamination of metal-contaminated sediments, soils, or sewage sludge (BOSECKER, 1999). Sulfur-oxidizing thiobacilli have been used to remove cadmium, copper, and zinc from contaminated harbor sediments (CRANE and HOLDEN, 1999). Chromium was only poorly leached. A pilot plant was operated for the removal of cadmium, cobalt, nickel, and zinc from river sediments by the stimulation of indigenous sulfur oxidizers (SEIDEL et al., 1998). Soil samples were treated with iron- and sulfur-oxidizing bacteria (especially *Thiobacillus ferrooxidans* and *Thiobacillus thiooxidans*) resulting in leaching efficiencies of >50% for, e.g., arsenic, cadmium, cobalt, chromium, copper, nickel, vanadium, or zinc (GOMEZ and BOSECKER, 1999). A combined bioleaching–bioprecipitation process has been applied for the decontamination of soils polluted with radioactive elements such as uranium or radium (GROUDEV et al., 1999). Metals in the upper soil horizon are mobilized by the stimulation of the soil microflora and transferred to deeper soil layers by drainage where metals are precipitated by the activity of anaerobic sulfate-reducing bacteria forming sulfide. *Aspergillus niger* was used to remove cadmium, chromium, copper, lead, manganese, mercury, and zinc from metal-polluted soils (WASAY et al., 1998). An effect of the soil type on the leaching efficiency was observed.

Combining both microbial mobilizing and immobilizing abilities (by biosorption) will lead to an integrated biologically based process as already demonstrated for bacterial sulfur involved in the sulfur cycle (sulfur oxidizers, sulfate reducers) (WHITE et al., 1997, 1998). Increased metal leaching efficiencies, increased metal tolerance, and (perhaps) metal selectivities might be the result of a genetic improvement of microorganisms, thus representing possible strategies to make bioleaching processes more suitable for bioremediation purposes (BOSECKER, 1999).

9 Conclusion

Microbial leaching processes can be applied for the recovery of metals from mineral resources as well as metal-containing solid wastes and metal-contaminated natural ecosystems. It is possible to recycle the leached and recovered metals and to reuse them as raw materials by metal manufacturing industries. Bioleaching allows the cycling of metals by a process close to natural biogeochemical cycles reducing the demand for non-renewable resources such as ores, energy, or landfill space. Bioleaching represents a "clean technology" process on a low-cost and low-energy level compared with some conventional mining and waste treatment techniques. Government regulations and research policies that favor "green" technologies are a key incentive for developing such processes. These processes find a wide acceptance by the public and in politics and represent innovative technologies within a proved market gap.

Note added in proof

Very recently, some species of the genus *Thiobacillus* have been reclassified and renamed. For details see KELLY and WOOD, 2000.

10 References

ACEVEDO, F., GENTINA, J. C. (1989), Process engineering aspects of the bioleaching of copper ores, *Bioproc. Eng.* **4**, 223–229.

ACEVEDO, F., GENTINA, J. C., GARCIA, N. (1998), CO_2 supply in the biooxidation of an enargite–pyrite gold concentrate, *Biotechnol. Lett.* **20**, 257–259.

ACHARYA, C., KAR, R. N., SUKLA, L. B. (1998), Leaching of chromite overburden with various native bacterial strains, *World J. Microbiol. Biotechnol.* **14**, 769–771.

ACUNA, J., ROJAS, J., AMARO, A. M., TOLEDO, H., JEREZ, C. C. (1992), Chemotaxis of *Leptospirillum ferrooxidans* and other acidophilic chemolithotrophs: comparison with the *Escherichia coli* chemosensory system, *FEMS Microbiol. Lett.* **96**, 37–42.

AGATE, A. D. (1996), Recent advances in microbial mining, *World J. Microbiol. Biotechnol.* **12**, 487–495.

AHONEN, L., TUOVINEN, O. H. (1992), Bacterial oxidation of sulfide minerals in column leaching experiments at suboptimal temperatures, *Appl. Environ. Microbiol.* **58**, 600–606.

AMARO, A. M., HALLBERG, K. B., LINDSTROM, E. B., JEREZ, C. A. (1992), An immunological assay for detection and enumeration of thermophilic biomining microorganisms, *Appl. Environ. Microbiol.* **60**, 3470–3473.

ANDREWS, G. (1990), Large-scale bioprocessing of solids, *Biotechnol. Prog.* **6**, 225–230.

ARIEF, Y. Y., MADGWICK, J. C. (1992), The effect of biocides on microaerobic bacterial biodegradation of manganese oxide, *Biorecovery* **2**, 95–106.

ASTM (1991), Standard test method for determining the rate of bioleaching of iron from pyrite by *Thiobacillus ferrooxidans* (E 1357–90). *Annual Book of ASTM Standards* 11.04., pp. 1007–1008.

ATLAS, R. M., BARTHA, R. (1997), Microbial ecology – Fundamentals and applications, 4th Edn. Menlo Park: Addison Wesley Longman.

AVAKYAN, Z. A., KARAVAIKO, G. I., MEL'NIKOVA, E. O., KRUTSKO, V. S., OSTROUSHKO, Y. I. (1981), Role of microscopic fungi in weathering of rocks and minerals from a pegmatite deposit, *Microbiology* **50**, 115–120.

BACELAR-NICOLAU, P., JOHNSON, D. B. (1999), Leaching of pyrite by acidophilic heterotrophic iron-oxidizing bacteria in pure and mixed cultures, *Appl. Environ. Microbiol.* **65**, 585–590.

BAGLIN, E. G., NOBLE, E. G., LAMPSHIRE, D. L., EISELE, J. A. (1992), Solubilization of manganese from ores by heterotrophic microorganisms, *Hydrometallurgy* **29**, 131–144.

BALDI, F., CLARK, T., POLLACK, S. S., OLSON, G. J. (1992), Leaching of pyrites of various reactivities by *Thiobacillus ferrooxidans*, *Appl. Environ. Microbiol.* **58**, 1853–1856.

Ballester, A., Blazquez, M. L., Gonzalez, F., Munoz, J. A. (1989), The influence of different variables on the bioleaching of sphalerite, *Biorecovery* **1**, 127–144.

Barrett, J., Hughes, M. N., Karavaiko, G. I., Spencer, P. A. (1993), *Metal Extraction by Bacterial Oxidation of Minerals*. New York: Ellis Horwood.

Beebe, J., Umbreit, W. W. (1971), Extracellular lipid of *Thiobacillus thiooxidans*, *J. Bacteriol.* **108**, 612–614.

Berthelin, J. (1983), Microbial weathering processes, in: *Microbial Geochemistry* (Krumbein, W. E., Ed.), pp. 223–262. Oxford: Blackwell.

Bhattacharyya, S., Chakrabarty, B. K., Das, A., Kundu, P. N., Banerjee, P. C. (1991), *Acidiphilium symbioticum* sp. nov., an acidophilic heterotrophic bacterium from *Thiobacillus ferrooxidans* cultures isolated from Indian mines, *Can. J. Microbiol.* **37**, 78–85.

Blais, J. F., Tyagi, R. D., Auclair, J. C. (1993), Bioleaching of metals from sewage sludge: microorganisms and growth kinetics, *Water Res.* **27**, 101–110.

Blais, J. F., Tyagi, R. D., Meunier, N., Auclair, J. C. (1994), The production of extracellular appendages during bacterial colonization of elemental sulfur, *Process Biochem.* **29**, 475–482.

Blake, R. C. II., Shute, E. A. (1994), Respiratory enzymes of *Thiobacillus ferrooxidans*. Kinetic properties of an acid-stable iron:rusticyanin oxidoreductase, *Biochemistry* **33**, 9220–9228.

Blake, R. C. II., Shue, E. A., Greenwood, M. M., Spencer, G. H., Ingledew, W. J. (1993), Enzymes of aerobic respiration of iron, *FEMS Microbiol. Rev.* **11**, 9–18.

Blowes, D. W., Jambor, J. L., Hanton-Fong, C. J., Lortie, L., Gould, W. D. (1998), Geochemical, mineralogical and microbiological characterization of a sulphide-bearing carbonate-rich goldmine tailings impoundment, Joutel, Québec, *Appl. Geochem.* **13**, 687–705.

Borovec, Z. (1990), Úprava bauxitu a keramických surovin bakteriemi a mikroskopickými houbami (Treatment of bauxite and ceramic raw materials by means of bacteria and microscopic fungi), *Ceram. Silikaty* **34**, 163–168 (in Czech).

Bosecker, K. (1986), Bacterial metal recovery and detoxification of industrial waste, *Biotechnol. Bioeng. Symp.* **16**, 105–120.

Bosecker, K. (1987), Microbial recycling of mineral waste products, *Acta Biotechnol.* **7**, 487–497.

Bosecker, K. (1989), Bioleaching of valuable metals from silicate ores and silicate waste products, in: *Biohydrometallurgy* (Salley, J., McGready, R. G. L., Wichlacz, P. L., Eds.), pp. 15–24. CANMET: Ottawa.

Bosecker, K. (1993), Bioleaching of silicate manganese ores, *Geomicrobiol. J.* **11**, 195–203.

Bosecker, K. (1997), Bioleaching: metal solubilization by microorganisms, *FEMS Microbiol. Rev.* **20**, 591–604.

Bosecker, K. (1999), Microbial leaching in the environmental clean-up programmes, in: *Biohydrometallurgy and the Environment Toward the Mining of the 21st Century* Vol. 9B (Amils, R., Ballester, A., Eds.), pp. 533–536. Amsterdam: Elsevier.

Bosshard, P. P., Bachofen, R., Brandl, H. (1996), Metal leaching of fly ash from municipal waste incineration by *Aspergillus niger*, *Environ. Sci. Technol.* **30**, 3066–3070.

Boussios, L., Madgwick, J. (1994), The effect of the biocide 2,2′-dihydroxy-5,5′-dichlorophenylmethane on manganese dioxide biodegradation, *Biorecovery* **2**, 227–239.

Bowers-Irons, G., Pryor, R., Bowers-Irons, T., Glass, M., Welsh, C., Blake, R. (1993), The bioliberation of gallium and associated metals from gallium arsenide ore and semiconductor wafers, in: *Biohydrometallurgical Technologies* (Torma, A. E., Wey, J. E., Lakshmanan, V. L., Eds.), pp. 335–343. Warrendale: The Minerals, Metals and Materials Society.

Brandl, H., Bosshard, R., Wegmann, M. (1999), Computer-munching microbes: Metal leaching from electronic scrap by bacteria and fungi, in: *Biohydrometallurgy and the Environment Toward the Mining of the 21st Century* Vol. 9B (Amils, R., Ballester, A., Eds.), pp. 569–576. Amsterdam: Elsevier.

Briand, L., Thomas, H., de la Vega Alonso, A., Donati, E. (1999), Vanadium recovery from solid catalysts by means of Thiobacilli action, in: *Biohydrometallurgy and the Environment Toward the Mining of the 21st Century* Vol. 9A (Amils, R., Ballester, A., Eds.), pp. 263–271. Amsterdam: Elsevier.

Briceno, H., Rossi, M. C., Montoya, R. (1993), Column bioleaching of gold and fresh copper flotation tailings from El Teniente, in: *Biohydrometallurgical Technologies* (Torma, A. E., Wey, J. E., Lakshmanan, V. I., Eds.), pp. 195–205. Warrendale: The Minerals, Metals and Materials Society.

Brierley, C. L. (1977), Thermophilic microorganisms in extraction of metals from ores, *Dev. Ind. Microbiol.* **18**, 273–284.

Brierley, C. L. (1978), Bacterial leaching, *Crit. Rev. Microbiol.* **6**, 207–261.

Brierley, C. L. (1995), Bacterial oxidation, *Eng. Min. J.* **196**, 42–44.

Brierley, C. L. (1997), Mining biotechnology: Research to commercial development and beyond, in: *Biomining: Theory, Microbes and Industrial Processes* (Rawlings, D. E., Ed.), pp. 3–17. Berlin:

Springer-Verlag.

BRIERLEY, C. L. (1999), Bacterial succession in bioheap leaching, in: *Biohydrometallurgy and the Environment Toward the Mining of the 21st Century* Vol. 9A (AMILS, R., BALLESTER, A., Eds.), pp. 91–97. Amsterdam: Elsevier.

BRIERLEY, J. A. (1990), Acidophilic thermophilic archaebacteria: potential application for metals recovery, *FEMS Microbiol. Rev.* **75**, 287–292.

BRIERLEY, J. A., BRIERLEY, C. L. (1999), Present and future commercial applications of biohydrometallurgy, in: *Biohydrometallurgy and the Environment Toward the Mining of the 21st Century* Vol. 9A (AMILS, R., BALLESTER, A., Eds.), pp. 81–89. Amsterdam: Elsevier.

BROMBACHER, C. (1994), Ökotoxikologische Untersuchung von Kehrichtverbrennungsrückständen mittels bakterieller Toxizitätstests, *Thesis*, University of Zurich, Switzerland.

BROMBACHER, C., BACHOFEN, R., BRANDL, H. (1997), Biohydrometallurgical processing of solids: A patent review, *Appl. Microbiol. Biotechnol.* **48**, 577–587.

BROMBACHER, C., BACHOFEN, R., BRANDL, H. (1998), Development of a laboratory-scale leaching plant for metal extraction from fly ash by *Thiobacillus strains*, *Appl. Environ. Microbiol.* **64**, 1237–1241.

BRUNNER, H., SCHINNER, F. (1991), Bakterielle Laugung von Galvanikschlamm aus der metallverarbeitenden Industrie, *Metall* **45**, 898–899.

BRUYNESTEYN, A., HACKL, R. P. (1985), The BIOTANKLEACH process for the treatment of refractory gold/silver concentrates, in: *Microbiological Effects on Metallurgical Processes* (CLUM, J. A., HAAS, L. A., Eds.), pp. 121–127. New York: The Metallurgical Society.

BRUYNESTEYN, A., HACKL, R. P., WRIGHT, F. (1986), The BIOTANKLEACH process, in: *Gold 100. Proc. Int. Conf. Gold.* (KING, R. P., Ed.), pp. 353–365. Johannesburg: South African Institute of Mining and Metallurgy.

BRYNER, L. C., BECK, J. V., DAVIS, D. B., WILSON, D. G. (1954), Microorganisms in leaching sulfide minerals, *Ind. Eng. Chem.* **46**, 2587–2592.

BUTLER, B. J., KEMPTON, A. G. (1987), Growth of *Thiobacillus ferrooxidans* on solid media containing heterotrophic bacteria, *J. Ind. Microbiol.* **2**, 41–45.

CERDÁ, J., GONZÁLEZ, S., RÍOS, J. M., QUINTANA, T. (1993), Uranium concentrates bioproduction in Spain: A case study, *FEMS Microbiol. Rev.* **11**, 253–260.

CHAWAKITCHAREON, P., THIRAVETYAN, P., NGEIMVIJAWAT, T. (1999), Comparison of chemical leaching and bioleaching of nickel from nickel hydroxide sludge, in: *Biohydrometallurgy and the Environment Toward the Mining of the 21st Century* Vol. 9A (AMILS, R., BALLESTER, A., Eds.), pp. 187–199. Amsterdam: Elsevier.

CLARK, T. R., EHRLICH, H. L. (1992), Copper removal from an industrial waste by bioleaching, *J. Ind. Microbiol.* **9**, 213–218.

COLMER, A. R., HINKLE, M. E. (1947), The role of microorganisms in acid mine drainage, *Science* **106**, 253–256.

CRANE, A. G., HOLDEN, P. J. (1999), Leaching of harbour sediments by estuarine iron oxidising bacteria, in: *Biohydrometallurgy and the Environment Toward the Mining of the 21st Century* Vol. 9A (AMILS, R., BALLESTER, A., Eds.), pp. 347–356. Amsterdam: Elsevier.

CURUTCHET, G., DONATI, E., TEDESCO, P. (1990), Influence of quartz in the bioleaching of covellite, *Biorecovery* **2**, 29–35.

DAS, T., AYYAPPAN, S., CHAUDKURY, G. R. (1999), Factors affecting bioleaching kinetics of sulfide ores using acidophilic micro-organisms, *Biometals* **12**, 1–10.

DAVE, S. R., NATARAJAN, K. A., BHAT, J. V. (1981), Leaching of copper and zinc from oxidised ores by fungi, *Hydrometallurgy* **7**, 235–242.

DE WULF-DURAND, P., BRYANT, L. J., SLY, L. I. (1997), PCR-mediated detection of acidophilic, bioleaching-associated bacteria, *Appl. Environ. Microbiol.* **63**, 2944–2948.

D'HUGUES, P., CEZAC, P., BATTAGLIA, F., MORIN, D. (1999), Bioleaching of a cobaltiferous pyrite at 20% solids: a continous laboratory-scale study, in: *Biohydrometallurgy and the Environment Toward the Mining of the 21st Century* Vol. 9A (AMILS, R., BALLESTER, A., Eds.), pp. 167–176. Amsterdam: Elsevier.

DONATI, E., POGLIANI, C., BOIARDI, J. L. (1997), Anaerobic leaching of covellite by *Thiobacillus ferrooxidans*, *Appl. Microbiol. Biotechnol.* **47**, 636–639.

DOPSON, M., LINDSTROM, E. B. (1999), Potential role of *Thiobacillus caldus* in arsenopyrite bioleaching, *Appl. Environ. Microbiol.* **65**, 36–40.

DROBNER, E., HUBER, H., RACHEL, R., STETTER, K. O. (1992), *Thiobacillus plumbophilus* spec. nov., a novel galena and hydrogen oxidizer, *Arch. Microbiol.* **157**, 213–217.

DUARTE, J. C., ESTRADA, P. C., PEREIRA, P. C., BEAUMONT, H. P. (1993), Thermophilic vs. mesophilic bioleaching process performance, *FEMS Microbiol. Rev.* **11**, 97–102.

EBNER, H. G. (1977), Metal extraction from industrial waste with *Thiobacilli*, in: *Conference bacterial leaching* (SCHWARTZ, W., Ed.), pp. 217–222. Weinheim: Verlag Chemie.

EDWARDS, K. J., GOEBEL, B. M., RODGERS, T. M., SCHRENK, M. O., GIHRING, T. M. et al. (1999), Geomicrobiology of pyrite (FeS_2) dissolution: Case study at Iron Mountain, California, *Geomicrobiol. J.* **16**, 155–179.

EHRLICH, H. L. (1963), Microorganisms in acid drainage from copper mine, *J. Bacteriol.* **86**, 351–352.

EHRLICH, H. L. (1997), Microbes and metals, *Appl. Microbiol. Biotechnol.* **48**, 687–692.

EHRLICH, H. L. (1999), Past, present and future of biohydrometallurgy, in: *Biohydrometallurgy and the Environment Toward the Mining of the 21st Century* (AMILS, R., BALLESTER, A., Eds.), pp. 3–12. Amsterdam: Elsevier.

ESCOBAR, B., JEDLICKI, E., WIERTZ, J., VARGAS, T. (1996), A method for evaluating the proportion of free and attached bacteria in the bioleaching of chalcopyrite with *Thiobacillus ferrooxidans*, *Hydrometallurgy* **40**, 1–10.

ESCOBAR, B., HUERTA, G., RUBIO, J. (1997), Influence of lipopolysaccharides on the attachment of *Thiobacillus ferrooxidans* to minerals, *World J. Microbiol. Biotechnol.* **13**, 593–594.

EWART, D. K., HUGHES, M. N. (1991), The extraction of metals from ores using bacteria, *Adv. Inorg. Chem.* **36**, 103–135.

FARBISZEWSKA, T., CWALINA, B., FARBISZEWSKABAJER, J., DZIERZEWICZ, Z. (1994), The use of bacterial leaching in the utilization of wastes resulting from mining and burning lignite, *Acta Biol. Cracov. Bot.* **36**, 1–9.

FERRONI, G. D., LEDUC, L. G., TODD, M. (1986), Isolation and temperature characterization of psychrophilic strains of *Thiobacillus ferrooxidans* from the environment of a uranium mine, *J. Gen. Appl. Microbiol.* **32**, 169–175.

FOURNIER, D., LEMIEUX, R., COUILLARD, D. (1998), Essential interactions between *Thiobacillus ferrooxidans* and heterotrophic microorganisms during wastewater sludge bioleaching process, *Environ. Poll.* **101**, 303–309.

FOWLER, T. A., CRUNDWELL, F. K. (1999), Leaching of zinc sulfide by *Thiobacillus ferrooxidans:* Bacterial oxidation of the sulfur product layer increases the rate of zinc sulfide dissolution at high concentrations of ferrous ions, *Appl. Environ. Microbiol.* **65**, 5285–5292.

FOWLER, T. A., HOLMES, P. R., CRUNDWELL, F. K. (1999), Mechanism of pyrite dissolution in the presence of *Thiobacillus ferrooxidans*, *Appl. Environ. Microbiol.* **65**, 2987–2993.

FURRER, G., STUMM, W. (1986), The coordination chemistry of weathering: I. Dissolution kinetics of δ-Al_2O_3 and BeO, *Geochim. Cosmochim. Acta* **50**, 1847–1860.

GEHRKE, T., TELEGDI, J., THIERRY, D., SAND, W. (1998), Importance of extracellular polymeric substances from *Thiobacillus ferrooxidans* for bioleaching, *Appl. Environ. Microbiol.* **64**, 2743–2747.

GLOMBITZA, F., ISKE, U., BULLMANN, M. (1988), Mikrobielle Laugung von Seltenen Erdelementen und Spurenelementen, *BioEngineering* **4**, 37–43.

GOEBEL, B. M., STACKEBRANDT, E. (1994), Cultural and phylogenetic analysis of mixed microbial populations found in natural and commercial bioleaching environments, *Appl. Environ. Microbiol.* **60**, 1614–1621.

GOLYSHINA, O. V., PIVOVAROVA, T. A., KARAVAIKO, G. I., KONDRAT'EVA, T. F., MOORE, E. R. B. et al. (2000), *Ferroplasma acidiphilum* gen. nov., sp. nov., an acidophilic, autotrophic, ferrous-iron-oxidizing, cell-wall-lacking, mesophilic member of the Ferroplasmaceae fam. nov., comprising a distinct lineage of the Archaea, *Int. J. Syst. Evol. Microbiol.* **50**, 997–1006.

GOMEZ, C., BOSECKER, K. (1999), Leaching of heavy metals from contaminated soil by using *Thiobacillus ferrooxidans* and *Thiobacillus thiooxidans*, *Geomicrobiol. J.* **16**, 233–244.

GROUDEV, S. N. (1987), Use of heterotrophic microorganisms in mineral biotechnology, *Acta Biotechnol.* **7**, 299–306.

GROUDEV, S. N. (1999), Biobeneficiation of mineral raw material, *Miner. Metall. Process.* **16**, 19–28.

GROUDEV, S. N., GROUDEVA, V. I. (1993), Microbial communities in four industrial copper dump leaching operations in Bulgaria, *FEMS Microbiol. Rev.* **11**, 261–268.

GROUDEV, S. N., GEORGIEV, P. S., SPASOVA, I. I., KOMNITSAS, K. (1999), Bioremediation of a soil contaminated with radioactive elements, in: *Biohydrometallurgy and the Environment Toward the Mining of the 21st Century* Vol. 9B (AMILS, R., BALLESTER, A., Eds.), pp. 627–634. Amsterdam: Elsevier.

GUPTA, A., EHRLICH, H. L. (1989), Selective and non-selective bioleaching of manganese from a manganese-containing silver ore, *J. Biotechnol.* **9**, 287–304.

HAHN, M., WILLSCHER, S., STRAUBE, G. (1993), Copper leaching from industrial wastes by heterotrophic microorganisms, in: *Biohydrometallurgical Technologies* (TORMA, A. E., WEY, J. E., LAKSMANAN, V. L., Eds.), pp. 99–108. Warrendale: The Minerals, Metals and Materials Society.

HALLBERG, K. B., SEHLIN, H. M., LINDSTRÖHM, E. B. (1996), Toxicity of arsenic during high temperature bioleaching of gold-bearing arsenical pyrite, *Appl. Microbiol. Biotechnol.* **45**, 212–216.

HAN, C. J., KELLY, R. M. (1998), Biooxidation capacity of the extremely thermoacidophilic archaeon *Metallosphaera sedula* under bioenergetic challenge, *Biotechnol. Bioeng.* **58**, 617–624.

HANSFORD, G. S., MILLER, D. M. (1993), Biooxidation of a gold-bearing pyrite-arsenopyrite concentrate, *FEMS Microbiol. Rev.* **11**, 175–182.

HAZRA, T. K., MUKHERJEA, M., MUKHERJEA, R. N. (1992), Role of rusticyanin in the electron transport process in *Thiobacillus ferrooxidans*. *Ind. J.*

Biochem. Biophys. **29**, 77–81.

HIROYOSHI, N., HIROTA, M., HIRAJIMA, T., TSUNEKAWA, M. (1999), Inhibitory effect of iron-oxidizing bacteria on ferrous-promoted chalcopyrite leaching, *Biotechnol. Bioeng.* **64**, 478–483.

HOLMES, D. S. (1991), Biorecovery of metals from mining, industrial and urban wastes, in: *Bioconversion of Waste Materials to Industrial Products* (MARTIN, A. M., Ed.), pp. 441–474. London: Elsevier.

HOLMES, P. R., FOWLER, T. A., CRUNDWELL, F. K. (1999), The mechanism of bacterial action in the leaching of pyrite by *Thiobacillus ferrooxidans*, *J. Electrochem. Soc.* **146**, 2906–2912.

HORAN, J. (1999), Acid producing potential of mine overburden: Acid mine drainage chemistry and treatment. *http://www.mines.edu/fs_home/jhoran/ch126*.

HUBER, H., STETTER, K. O. (1989), *Thiobacillus prosperus* sp. nov., represents a new group of halotolerant metal-mobilizing bacteria isolated from marine geothermal field, *Arch. Microbiol.* **151**, 479–485.

HUBER, H., STETTER, K. O. (1990), *Thiobacillus cuprinus* sp. nov., a novel facultatively organotrophic metal-mobilizing bacterium, *Appl. Environ. Microbiol.* **56**, 315–322.

HUERTA, G., ESCOBAR, B., RUBIO, J., BADILLA-OHLBAUM, R. (1995), Adverse effect of surface-active reagents on the bioleaching of pyrite and chalcopyrite by *Thiobacillus ferrooxidans*, *World J. Microbiol. Biotechnol.* **11**, 599–600.

HUGHES, M. N., POOLE, R. K. (1989), Metals and microorganisms. London: Chapman & Hall.

JOHNSON, D. B. (1998), Biodiversity and ecology of acidophilic microorganisms, *FEMS Microbiol. Rev.* **27**, 307–317.

KARAVAIKO, G. I., KRUTSKO, V. S., MEL'NIKOVA, E. O., AVAKYAN, Z. A., OSTROUSHKO, Y. I. (1980), Role of microorganisms in spodumene degradation, *Microbiology* **49**, 402–406.

KELLY, D. P., HARRISON, A. P. (1984), Genus *Thiobacillus*, in: *Bergey's Manual of Systematic Bacteriology* (BUCHANAN, R. E., GIBBONS, N. E., Eds.), pp. 1842–1859. Baltimore, MD: Williams & Wilkins.

KELLY, D. P., WOOD, A. P. (2000), Reclassification of some species of *Thiobacillus* to the newly designated genera *Acidithiobacillus* gen. nov., *Halothiobacillus* gen nov. and *Thermithiobacillus* gen. nov., *Int. J. Syst. Evol. Microbiol.* **50**, 511–516.

KHALID, A. M., ANWAR, M. A., SHEMSI, A. M., NIAZI, G., AKHTAR, K. (1993), Biohydrometallurgy of low-grade, carbonate bearing sandstone uranium ore, in: *Biohydrometallurgical Technologies* (TORMA, A. E., WEY, J. E., LAKSHMANAN, V. I., Eds.), pp. 285–292. Warrendale: The Minerals, Metals and Materials Society.

KISHIMOTO, N., KOSAKO, Y., TANO, T. (1991), *Acidobacterium capsulatum* gen. nov., sp. nov.: An acidophilic chemoorganotrophic bacterium containing menaquinone from acidic mineral environment, *Curr. Microbiol.* **22**, 1–8.

KÖNIG, R. (1989a), C. Plinii secundi – Naturalis historiae, Libri XXXVII, Volume Liber XXXIII. München: Artemis.

KÖNIG, R. (1989b), C. Plinii secundi – Naturalis historiae, Libri XXXVII, Volume Liber XXXIV. München: Artemis.

KONISHI, Y., YOSHIDA, S., ASAI, S. (1995), Bioleaching of pyrite by acidophilic thermophile *Acidianus brierleyi*, *Biotechnol. Bioeng.* **48**, 592–600.

KOVALENKO, E. V., MALAKHOVA, P. T. (1990), Microbial succession in compensated sulfide ores, *Microbiology* **59**, 227–232.

KREBS, W. (1997), Mikrobiell induzierte Mobilisierung von Metallen aus Rückständen der Kehrichtverbrennung, *Thesis*, University of Zurich, Switzerland.

KREBS, W., BROMBACHER, C., BOSSHARD, P. P., BACHOFEN, R., BRANDL, H. (1997), Microbial recovery of metals from solids, *FEMS Microbiol. Rev.* **20**, 605–617.

KREBS, W., BACHOFEN, R., BRANDL, H. (1999), Optimization of metal leaching efficiency of fly ash from municipal solid waste incineration by sulfur oxidizing bacteria, in: *Biohydrometallurgy and the Environment Toward the Mining of the 21st Century* (AMILS, R., BALLESTER, A., Eds.), pp. 377–386. Amsterdam: Elsevier.

LAHL, U., STRUTH, R. (1993), Verwertung von Müllverbrennungsschlacken aus der Sicht des Grundwasserschutzes, *Vom Wasser* **80**, 341–355.

LAN, X., JIAJUN, K., RONGQING, Q. (1994), Effect of As^{3+} and As^{5+} on *Thiobacillus ferrooxidans*: Consequences for the biological pretreatment of gold and silver-bearing arsenopyrite, *Biorecovery* **2**, 241–251.

LANGDAHL, B. R., INGVORSEN, K. (1997), Temperature characteristics of bacterial iron solubilisation and ^{14}C assimilation in naturally exposed sulfide ore material at Citronen Fjord, North Greenland (83°N), *FEMS Microbiol. Ecol.* **23**, 275–283.

LARSSON, L., OLSSON, G., HOLST, O., KARLSSON, H. T. (1993), Oxidation of pyrite by *Acidianus brierley*: Importance of close contact between the pyrite and the microorganisms, *Biotechnol. Lett.* **15**, 99–104.

LAWSON, E. N., BARKHUIZEN, M., DEW, D. W. (1999), Gold solubilisation by cyanide producing bacteria *Chromobacterium violaceum*, in: *Biohydrometallurgy and the Environment Toward the Mining of the 21st Century* (AMILS, R., BALLESTER, A., Eds.), pp. 239–246. Amsterdam: Elsevier.

LEDIN, M., PEDERSEN, K. (1996), The environmental impact of mine wastes – Roles of microorganisms and their significance in treatment of mine wastes, *Earth Sci. Rev.* **41**, 67–108.

LEDUC, L. G., FERRONI, G. D. (1994), The chemolithotrophic bacterium *Thiobacillus ferrooxidans*, *Microbiol. Rev.* **14**, 103–120.

LEDUC, L. G., FERRONI, G. D., TREVORS, J. T. (1997), Resistance to heavy metals in different strains of *Thiobacillus ferrooxidans*, *World J. Microbiol. Biotechnol.* **13**, 453–455.

LINDSTROM, E. B., WOLD, S., KETTANEH-WOLD, N., SAAF, S. (1993), Optimization of pyrite bioleaching using *Sulfolobus acidocaldarius*, *Appl. Microbiol. Biotechnol.* **38**, 702–707.

LOAYZA, C., LY, M. E., YUPANQUI, R., ROMAN, G. (1999), Laboratory biooxidation tests of arsenopyrite concentrate for the Tamborque industrial plant, in: *Biohydrometallurgy and the Environment Toward the Mining of the 21st Century* (AMILS, R., BALLESTER, A., Eds.), pp. 405–410. Amsterdam: Elsevier.

LOPEZ-ARCHILLA, A. I., MARIN, I., AMILS, R. (1993), Bioleaching and interrelated acidophilic microorganisms from Rio Tinto, Spain, *Geomicrobiol. J.* **11**, 223–233.

LUGGER, B. (2000), Biomining. *http://www.lifescience.de/bioschool/sheets/13.html.*

MANDL, M., HRBAC, D., DOCEKALOVA, H. (1996), Inhibition of iron(II) oxidation by arsenic(III,V) in *Thiobacillus ferrooxidans*: Effects on arsenopyrite bioleaching, *Biotechnol. Lett.* **18**, 333–338.

MCNULTY, T. P., THOMPSON, D. L. (1990), Economics of bioleaching, in: *Microbial Mineral Recovery* (EHRLICH, H. L., BRIERLEY, C. L., Eds.), pp. 171–182. New York: McGraw-Hill.

MIER, J. L., BALLESTER, A., GONZALEZ, F., BLAZQUEZ, M. L., GOMEZ, E. (1996), The influence of metallic ions on the activity of *Sulfolobus* BC, *J. Chem. Technol. Biotechnol.* **65**, 272–280.

MILLER, R. M. (1995), Biosurfactant-facilitated remediation of metal-contaminated soils, *Environ. Health Perspect.* **103**, 59–62.

MOHANTY, B. K., MISHRA, A. K. (1993), Scanning electron micrograph study of *Bacillus licheniformis* during bioleaching of silica from magnesite ore, *J. Appl. Bacteriol.* **74**, 184–289.

MORENO, L., MARTINEZ, J., CASAS, J. (1999), Modelling of bioleaching copper sulfide in heaps or dumps, in: *Biohydrometallurgy and the Environment Toward the Mining of the 21st Century* Vol. 9A (AMILS, R., BALLESTER, A., Eds.), pp. 443–452. Amsterdam: Elsevier.

MÜLLER, G., FÖRSTER, I. (1964), Der Einfluss mikroskopischer Bodenpilze auf die Nährstofffreisetzung aus primären Materialien, als Beitrag zur biologischen Verwitterung. II. Mitteilung, *Zbl. Bakt. II.* **118**, 594–621.

MULLIGAN, C. N., GALVEZ-CLOUTIER, R., RENAUD, N. (1999), Biological leaching of copper mine residues by *Aspergillus niger*, in: *Biohydrometallurgy and the Environment Toward the Mining of the 21st Century* Vol. 9A (AMILS, R., BALLESTER, A., Eds.), pp. 453–461. Amsterdam: Elsevier.

MUÑOZ, J. A., GONZÁLEZ, F., BLÁZQUEZ, M. L., BALLESTER, A. (1995), A study of the bioleaching of a Spanish uranium ore. Part I: A review of the bacterial leaching in the treatment of uranium ores, *Hydrometallurgy* **38**, 39–57.

MUNTZ, A. (1890), Sur la décomposition des roches et la formation de la terre arable, *C. R. Acad. Sci.* (Paris) **110**, 1370–1372.

MURR, L. E. (1980), Theory and practice of copper sulphide leaching in dumps and *in situ*, *Min. Sci. Eng.* **12**, 121–189.

NAGPAL, S., DAHLSTROM, D., OOLMAN, T. (1993), Effect of carbon dioxide concentration on the bioleaching of a pyrite-arsenopyrite ore concentrate, *Biotechnol. Bioeng.* **41**, 159–464.

NEMATI, M., LOWENADLER, J., HARRISON, S. T. L. (2000), Particle size effects in bioleaching of pyrite by acidophilic *Sulfolobus metallicus* (BC), *Appl. Microbiol. Biotechnol.* **53**, 173–179.

NGUBANE, W. T., BAECKER, A. A. W. (1990), Oxidation of gold-bearing pyrite and arsenopyrite by *Sulfolobus acidocaldarius* and *Sulfolobus* BC in airlift bioreactors, *Biorecovery* **1**, 255–269.

NORDSTROM, D. K., ALPERS, C. N., PTACEK, C. J., BLOWES, D. W. (2000), Negative pH and extremely acidic mine waters from Iron Mountain, California, *Environ. Sci. Technol.* **34**, 254–258.

NORRIS, P. R., OWEN, J. P. (1993), Mineral sulfide oxidation by enrichment cultures of novel thermoacidophilic bacteria, *FEMS Microbiol. Rev.* **11**, 51–56.

NOTTER, M. (1993), Metals and the environment (*Report* 4245). Swedish Environmental Protection Agency.

OGURTSOVA, L. V., KARAVAIKO, G. I., AVAKYAN, Z. A., KORENEVSKII, A. A. (1989), Activity of various microorganisms in extracting elements from bauxite, *Microbiology* **58**, 774–780.

OLSON, G. J. (1994), Microbial oxidation of gold ores and gold bioleaching, *FEMS Microbiol. Lett.* **119**, 1–6.

OLSON, G. J., KELLY, R. M. (1986), Microbial metal transformations: Biotechnological applications and potential, *Biotechnol. Progr.* **2**, 1–15.

OSTROWSKI, M., SKLODOWSKA, A. (1993), Bacterial and chemical leaching pattern on copper ores of sandstone and limestone type, *World J. Microbiol. Biotechnol.* **9**, 328–331.

PARKER, S. P. (1992), *Concise Encyclopedia of Science and Technology*. New York: McGraw-Hill.

PISTACCIO, L., CURUTCHET, G., DONATI, E., TEDESCO, P. (1994), Analysis of molybdenite bioleaching by *Thiobacillus ferrooxidans* in the absence of iron (II), *Biotechnol. Lett.* **16**, 189–194.

POGLIANI, C., DONATI, E. (1992), The role of exopolymers in the bioleaching of a non-ferrous metal sulfide, *J. Ind. Microbiol. Biotechnol.* **22**, 88–92.

POGLIANI, C., CURUTCHET, G., DONATI, E., TEDESCO, P. H. (1990), A need for direct contact with particle surfaces in the bacterial oxidation of covellite in the absence of a chemical lixiviant, *Biotechnol. Lett.* **12**, 515–518.

POULIN, R., LAWRENCE, R. W. (1996), Economic and environmental niches of biohydrometallurgy, *Min. Eng.* **9**, 799–810.

PRONK, J. T., DE BRUYN, J. C., KUENEN, J. G. (1992), Anaerobic growth of *Thiobacillus ferrooxidans*, *Appl. Environ. Microbiol.* **58**, 2227–2230.

RAWLINGS, D. E. (1997), *Biomining: Theory, Microbes and Industrial Processes*. Berlin: Springer-Verlag.

RAWLINGS, D. E. (1999), The molecular genetics of mesophilic, acidophilic, chemolithotrophic, iron- or sulfur-oxidizing microorganisms, in: *Biohydrometallurgy and the Environment Toward the Mining of the 21st Century* Vol. 9B (AMILS, R., BALLESTER, R., Eds.), pp. 3–20. Amsterdam: Elsevier.

RAWLINGS, D. E., KUSANO, T. (1994), Molecular genetics of *Thiobacillus ferrooxidans*. *Microbiol. Rev.* **58**, 39–55.

RAWLINGS, D. E., TRIBUTSCH, H., HANSFORD, G. S. (1999), Reasons why "*Leptospirillum*"-like species rather than *Thiobacillus ferrooxidans* are the dominant iron-oxidizing bacteria in many commercial processes for the biooxidation of pyrite and related ores, *Microbiology* **145**, 5–13.

RIEKKOLA-VANHANEN, M., HEIMALA, S. (1993), Electrochemical control in the biological leaching of sulfidic ores, in: *Biohydrometallurgical Technologies* (TORMA, A. E., WEY, J. E., LAKSHMANAN, V. I., Eds.), pp. 561–570. Warrendale: The Minerals, Metals and Materials Society.

ROJAS, J., GIERSIG, M., TRIBUTSCH, H. (1995), Sulfur colloids as temporary energy reservoirs for *Thiobacillus ferrooxidans* during pyrite oxidation, *Arch. Microbiol.* **163**, 352–356.

ROJAS-CHAPANA, J. A., TRIBUTSCH, H. (2000), Bioleaching of pyrite accelerated by cysteine, *Proc. Biochem.* **35**, 815–824.

ROSSI, G. (1990), *Biohydrometallurgy*. Hamburg: McGraw-Hill.

ROSSI, G. (1999), The design of bioreactors, in: *Biohydrometallurgy and the Environment Toward the Mining of the 21st Century* Vol. 9A (AMILS, R., BALLESTER, A., Eds.), pp. 61–80. Amsterdam: Elsevier.

RUDOLFS, W. (1922), Oxidation of iron pyrites by sulfur-oxidizing organisms and their use for making mineral phosphates available, *Soil Sci.* **14**, 135–147.

RUDOLFS, W., HELBRONNER, A. (1922), Oxidation of zinc sulfide by microorganisms, *Soil Sci.* **14**, 459–464.

SALKIELD, L. U. (1987), *A technical History of the Rio Tinto Mines: Some Notes on Exploitation from Pre-Phoenician Times to the 1950s*. London: Institution of Mining and Metallurgy.

SAND, W. (1992), Evaluation of *Leptospirillum ferrooxidans* for leaching, *Appl. Environ. Microbiol.* **58**, 85–92.

SAND, W., BOCK, E. (1991), Biodeterioration of mineral materials by microorganisms – Biogenic sulfuric and nitric acid corrosion of concrete and natural stone, *Geomicrobiol. J.* **9**, 129–138.

SAND, W., HALLMANN, R., ROHDE, K., SOBOTKE, B., WENTZIEN, S. (1993), Controlled microbiological *in situ* stope leaching of a sulfidic ore, *Appl. Microbiol. Biotechnol.* **40**, 421–426.

SAND, W., GEHRKE, T., HALLMANN, R., SCHIPPERS, A. (1995), Sulfur chemistry, biofilm, and the (in)direct attack mechanism – a critical evaluation of bacterial leaching, *Appl. Microbiol. Biotechnol.* **43**, 961–966.

SAND, W., GEHRKE, T., JOZSA, P. G., SCHIPPERS, A. (1999), Direct versus indirect bioleaching, in: *Biohydrometallurgy and the Environment Toward the Mining of the 21st Century* Vol. 9A (AMILS, R., BALLESTER, A., Eds.), pp. 27–49. Amsterdam: Elsevier.

SAYER, J. A., COTTER-HOWELLS, J. D., WATSON, C., HILLIER, S., GADD, G. M. (1999), Lead mineral transformation by fungi, *Curr. Biol.* **9**, 691–694.

SCHÄFER, W. (1982), Bakterielle Laugung von metallhaltigen Industrierückständen. In Chemietechnik. *Thesis*, University of Dortmund, Germany.

SCHIFFNER, C. (1977), *Georg Agricola: Vom Berg- und Hüttenwesen*. München: Deutscher Taschenbuch Verlag.

SCHIPPERS, A., JOZSA, P. G., SAND, W. (1996), Sulfur chemistry in bacterial leaching of pyrite, *Appl. Environ. Microbiol.* **62**, 3424–3431.

SCHRENK, M. O., EDWARDS, K. J., GOODMAN, R. M., HAMERS, R. J., BANFIELD, J. F. (1998), Distribution of *Thiobacillus ferrooxidans* and *Leptospirillum ferrooxidans*: Implications for generation of acid mine drainage, *Science* **279**, 1519–1522.

SEIDEL, H., ONDRUSCHKA, J., MORGENSTERN, P., STOTTMEISTER, U. (1998), Bioleaching of heavy metals from contaminated aquatic sediments using indigenous sulfur-oxidizing bacteria: a feasibility study, *Water Sci. Technol.* **37**, 387–394.

SHRIHARI, S. R. B., MODAK, J. M., KUMAR, R., GANDHI, K. S. (1993), Dissolution of sulfur particles by *Thiobacillus ferrooxidans*: Substrate for unattached cells, *Biotechnol. Bioeng.* **41**, 612–616.

SILVERMAN, M. P., EHRLICH, H. L. (1964), Microbial formation and degradation of minerals, *Adv.*

Appl. Microbiol. **6**, 153–206.

SILVERMAN, M. P., LUNDGREN, D. G. (1959), Studies on the chemoautotrophic iron bacterium *Ferrobacillus ferrooxidans*. I. An improved medium and harvesting procedure for securing high cell yields, *J. Bacteriol.* **77**, 642–647.

SILVERMAN, M. P., MUNOZ, E. F. (1971), Fungal leaching of titanium from rock, *Appl. Microbiol.* **22**, 923–924.

SINGER, A., NAVROT, J., SHAPIRA, R. (1982), Extraction of aluminum from fly-ash by commercial and microbiologically-produced citric acid, *Eur. J. Appl. Microbiol. Biotechnol.* **16**, 228–230.

SMITH, A. D., HUNT, R. J. (1985), Solubilisation of gold by *Chromobacterium violaceum*, *J. Chem. Tech. Biotechnol.* **35B**, 110–116.

SREEKRISHNAN, T. R., TYAGI, R. D., BLAIS, J. F., CAMPBELL, P. G. C. (1993), Kinetics of heavy metal bioleaching from sewage sludge. I. Effects of process parameters, *Water Res.* **27**, 1641–1651.

STRASSER, H., PÜMPEL, T., BRUNNER, H., SCHINNER, F. (1993), Veredelung von Quarzsand durch mikrobielle Laugung von Eisenoxid-Verunreinigungen, *Arch. Lagerstättenforsch., Geol. B.-A.* **16**, 103–107.

TARASOVA, I. I., KHAVSKI, N. N., DUDENEY, A. W. L. (1993), The effects of ultrasonics on the bioleaching of laterites, in: *Biohydrometallurgical Technologies* (TORMA, A. E., WEY, J. E., LAKSHMANAN, V. L., Eds.), pp. 357–361. Warrendale: The Minerals, Metals and Materials Society.

TORMA, A. E., BANHEGYI, I. G. (1984), Biotechnology in hydrometallurgical processes, *Trends Biotechnol.* **2**, 13–15.

TORMA, A. E., GABRA, G. G., GUAY, R., SILVER, M. (1976), Effects of surface active agents on the oxidation of chalcopyrite by *Thiobacillus ferrooxidans*, *Hydrometallurgy* **1**, 301–309.

TORMA, A. E., WALDEN, C. C., DUNCAN, D. W., BRANION, R. M. R. (1972), The effect of carbon dioxide and particle surface area on the microbiological leaching of a zinc sulphide concentrate, *Biotechnol. Bioeng.* **14**, 777–782.

TRIBUTSCH, H. (1999), Direct versus indirect bioleaching, in: *Biohydrometallurgy and the Environment Toward the Mining of the 21st Century* Vol. 9A (AMILS, R., BALLESTER, A., Eds.), pp. 51–60. Amsterdam: Elsevier.

TRIBUTSCH, H., ROJAS-CHAPANA, J. A. (1999), Verfahren zur mikrobiologischen Laugung von sulfidhaltigen Materialien und Verwendung von schwefelhaltigen Aminosäuren, *Int. Patent Application* PCT/EPGG/05272.

TZEFERIS, P. G. (1994), Fungal leaching of nickeliferous laterites, *Folia Microbiol.* **39**, 137–140.

VACHON, P., TYAGI, R. D., AUCLAIR, J. C., WILKINSON, K. J. (1994), Chemical and biological leaching of aluminum from red mud, *Environ. Sci. Technol.* **28**, 26–30.

WASAY, S. A., BARRINGTON, S. F., TOKUNGA, S. (1998), Using *Aspergillus niger* to bioremediate soils contaminated by heavy metals, *Bioremed. J.* **2**, 183–190.

WHITE, C., SAYER, J. A., GADD, G. M. (1997), Microbial solubilization and immobilization of toxic metals: key biogeochemical processes for treatment of contamination, *FEMS Microbiol. Rev.* **20**, 503–516.

WHITE, C., SHARMAN, A. K., GADD, G. M. (1998), An integrated process for the bioremediation of soil contaminated with toxic metals, *Nature Biotechnol.* **16**, 572–575.

WINOGRADSKY, S. (1887), Ueber Schwefelbacterien, *Bot. Z.* **45**, 489–610.

WOODS, D., RAWLINGS, D. E. (1989), Bacterial leaching and biomining, in: *A Revolution in Biotechnology* (MARX, J. L., Ed.), pp. 82–93. Cambridge: University of Cambridge Press.

ZIMMERLEY, S. R., WILSON, D. G., PRATER, J. D. (1958), Cyclic leaching process employing iron oxidizing bacteria, *US Patent* 2,829,964.

9 Accumulation and Transformation of Metals by Microorganisms

GEOFFREY M. GADD
Dundee, Scotland

1 Introduction

The influence of microbiological processes on contamination of the environment by toxic metals, metalloids, organometals, and radionuclides is of economic and environmental significance (GADD, 1986a, b, 1988, 1990a, b, c, 1992a, b, c, 1997, 2000a, b, c; WHITE et al., 1995; WAINWRIGHT and GADD, 1997; WHITE and GADD, 1998a, b; AVERY et al., 1998; GADD and SAYER, 2000). However, the potential of microbial processes for bioremediation may be dependent on the physical and chemical nature of the site which influences the form in which metals occur. Mineral components contain considerable quantities of metal(s) which are biologically unavailable (WAINWRIGHT and GADD, 1997; WHITE and GADD, 1998b; GADD, 1999). Certain microbial processes solubilize metals, thereby increasing their bioavailability and potential toxicity, whereas others immobilize them and reduce bioavailability (Tab. 1). The relative balance between mobilization and immobilization varies depending on the organisms involved and the physicochemical attributes of their environment. As well as being an integral component of biogeochemical cycles for metals and associated elements (EHRLICH, 1999; LEDIN, 2000), these processes may be exploited for the treatment of contaminated solid and liquid wastes (GADD, 1988, 1990a, 1992a, 1996, 1997, 2000a, b; GADD and WHITE, 1993; TOBIN et al., 1994; WHITE et al., 1997).

Table 1. Summary of Main Microbial Metal Transformations of Biotechnological Significance

$Metal_{soluble}$ ◄——►	$Metal_{insoluble}$
Autotrophic leaching	biosorption
Heterotrophic leaching	organic and inorganic precipitation
Complexation and chelation, siderophores, biomolecules	metal-binding peptides, proteins, polysaccharides, derived products, biomolecules
Methylation	transport and intracellular localization
Reductive mobilization	reductive immobilization

Metal mobilization can be achieved by autotrophic and heterotrophic leaching, chelation by microbial metabolites and siderophores, and methylation which can result in volatilization. Similarly, immobilization can result from sorption to cell components or exopolymers, transport and intracellular sequestration or precipitation as insoluble organic and inorganic compounds, e.g., oxalates (SAYER and GADD, 1997; GHARIEB et al., 1998; GADD, 1999), sulfides or phosphates (YONG and MACASKIE, 1995; WHITE and GADD, 1996a, b, 1997, 1998a). In addition, reduction of higher-valency species may result in mobilization, e.g., Mn(IV) to Mn(II), or immobilization, e.g., Cr(VI) to Cr(III) (FUJIE et al., 1994; LOVLEY, 1995; WHITE and GADD, 1998b). Solubilization provides a means for removal of the metals from solid matrices such as soils, sediments, dumps, and industrial wastes. Alternatively, immobilization may enable metals to be transformed *in situ* into insoluble forms but are particularly applicable to removing metals from aqueous solution. In relation to the latter instance, microorganisms and microbial products can be highly efficient bioaccumulators of soluble and particulate forms of metals, especially from dilute external concentrations. Such removal of potentially hazardous toxic metals and radionuclides from industrial effluents and wastewaters by microbial biomass can lead to detoxification and/or metal recovery after appropriate treatment of the loaded biomass (SHUMATE and STRANDBERG, 1985; GADD, 1988; GADD and WHITE, 1993; AVERY et al., 1998). Microorganisms can also transform certain metal, metalloid and organometallic species by methylation and dealkylation (GADD, 1993a, 2000a, b; LOVLEY, 1995; BRADY et al., 1996). Biomethylated derivatives are often volatile and may be eliminated from a system by evaporation (GADD, 1993a). Two major transformation processes are reduction of metalloid oxyanions to elemental forms, e.g., SeO_4^{2-} and SeO_3^{2-} to Se^0 and methylation of metalloids, metalloid oxyanions or organometalloids to methyl derivatives, e.g., AsO_4^{3-}, AsO_2^- and methylarsonic acid to $(CH_3)_3As$ (trimethylarsine). Such transformations modify the mobility and toxicity of metalloids, have biogeochemical significance, and are also of biotechnological potential in bioremediation (TAMAKI and FRANKENBERGER, 1992; LOVLEY, 1993, 1995; KARLSON and FRANKENBERGER, 1993; GADD, 1993a,; BRADY et al., 1996). There are indications that some microbe-based processes could be more economical than existing treatments, and some processes have been or are in commercial or field-based operation in the mining and metallurgical industries (HUTCHINS et al., 1986; GADD, 1988; GADD and WHITE, 1993). However, many areas of metal/microbe interactions remain at the laboratory scale and are unexplored in a biotechnological context. This chapter will outline some of the more important microbiological processes which are of significance in determining environmental metal mobility and which have actual or potential applications in bioremediation.

2 Metal Immobilization

2.1 Metal Accumulation

Microorganisms, including actinomycetes, cyanobacteria, and other bacteria, algae, and fungi, can accumulate metals and radionuclides from their external environment (GADD and GRIFFITHS, 1978; BORST-PAUWELS, 1981, 1989; SHUMATE and STRANDBERG, 1985; GADD, 1986a, b, 1988, 1990a,c, 1992c; TREVORS et al., 1986; BELLIVEAU et al., 1987; MANN et al., 1988; GADD and WHITE, 1989a). Amounts accumulated can be large, and a variety of mechanisms may be involved ranging from physicochemical interactions to processes dependent on cell metabolism, e.g., transport. Both living and dead cells are capable of metal accumulation and so are products produced by or derived from microbial cells such as excreted metabolites, polysaccharides, and cell wall constituents (KELLY et al., 1979; BRIERLEY et al., 1986; GADD, 1990a, 1993b, 1999; GADD and DE ROME, 1988). Uptake capacities and mechanisms may vary widely in different microorganisms, although common features do exist. Both living and dead cells are capable of metal accumulation, but there may be obvious differences in the mechanisms involved. Many metals are essential for growth and metabolism in low concentrations, e.g., Cu, Zn, Mn, whereas some have no known biological function, e.g., Au, Ag, Pb, Cd. A feature of all these and related elements is that they can be toxic towards living cells (GADD and GRIFFITHS, 1978, GADD and MOWLL, 1983; MOWLL and GADD, 1985). Thus, if the use of living cells is envisaged in any metal removal system, toxicity may lead to poisoning and inactivation (SHUMATE and STRANDBERG, 1985). However, it may be possible to separate the means of microbial propagation from the metal contacting phase or to use strains tolerant to the metal concentrations encountered. The use of dead biomass or derived products eliminates the problem of toxicity, not only from dissolved metals but also from adverse operating conditions, and the economic component of maintenance including nutrient supply. However, living cells may exhibit a wider variety of mechanisms for metal accumulation and transformations such as transport, extracellular complex formation, and precipitation. In addition, tolerance and resistance are properties found within all microbial groups (GADD and GRIFFITHS, 1978; GADD, 1986a, 1990d, 1992b; TREVORS et al., 1986; BELLIVEAU et al., 1987; REED and GADD, 1990).

In living cells, microbial metal uptake can often be simply divided into two main components. The first, which can also occur in dead cells, is metabolism-independent association with cell walls and other external surfaces. The second comprises metabolism-dependent transport across the cell membrane and transport systems of varying affinity and specificity exist. Such uptake phases may be readily observed in short-term experiments with cell suspensions of algae, heterotrophic bacteria, and yeasts if supplied with an energy source and if significant growth is precluded (GADD, 1988). In some cases intracellular uptake is a result of permeation and diffusion due to increased membrane permeability especially if toxicity is manifest. These two phases of uptake may not be seen in all microbes, and for some elements, e.g., Pb, U, Th, most accumulation in living or dead cells appears to involve surface phenomena with little or no intracellular uptake. Note that either or both phases of uptake may be enhanced or obscured by additional aspects of metabolism, such as changes in the chemical nature of the growth medium, and excretion of substances that may complex or precipitate metals, and/or various mechanisms of detoxification. Once inside cells, metal species may be bound, precipitated, localized within intracellular structures or organelles, or translocated to specific structures depending on the element concerned and the organism (MORLEY et al., 1996; GADD, 1996, 1997; WHITE et al., 1997; GADD and SAYER, 2000). While living cells may be subject to toxicity, many organisms exhibit tolerance or resistance, and it is relatively easy to isolate or engineer further resistant strains of bacteria, algae, and fungi which may be of use in some projected applications. It should be stressed that in a given microbial system several mechanisms may operate simultaneously or sequentially.

2.1.1 Biosorption

The term "adsorption" is often used to describe metabolism-independent uptake or binding of heavy metals and/or radionuclides to microbial cell walls and other extracellular surfaces. However, this may be a simplistic term when applied to complex biological systems and does not fully reflect the many interactions that may occur. Adsorption involves the accumulation or concentration of substances, the adsorbate, at a surface or interface, the adsorbent. In contrast, absorption occurs when the atoms or molecules of one phase almost uniformly penetrate into those of another phase (WEBER, 1972). There are three main types of adsorption. The first involves electrical attraction and is often called "exchange" adsorption. In relation to microbial cells, such adsorption is often simply defined as the attraction of positively charged ions to negatively charged ligands (BRIERLEY and BRIERLEY, 1983). The second kind of adsorption, where adsorbed molecules can have translational movement within the interface, is called "physical" or "ideal" adsorption and involves van der Waals forces. The third type is chemical attraction between the adsorbate and adsorbent which is called "chemical" or "activated" adsorption (WEBER, 1972). It is generally difficult to separate physical and chemical adsorption and most adsorption phenomena involve the three forms described (WEBER, 1972) and, as will be described later, may further involve crystallization and other forms of deposition (BEVERIDGE and DOYLE, 1989). The term "sorption" may include both adsorption and absorption and refers to a process where a component moves from one phase into another, preferably solid phase (WEBER, 1972). Thus, the term "biosorption" is frequently used to describe the non-directed physicochemical interactions that may occur between organic and inorganic metal/radionuclide species and cellular components (SHUMATE and STRANDBERG, 1985; GADD, 1992a, b; ZOUBOULIS et al., 1999; GUPTA et al., 2000). In living cells, metabolic activity may influence biosorption because of changes in pH, E_h, organic and inorganic nutrients and metabolites in the cellular microenvironment. Almost all biological macromolecules have some affinity for metal species with cell walls and associated materials being of the greatest significance in biosorption.

2.1.2 Biosorption by Microbial Biomass, Cell Walls, and Associated Components

Peptidoglycan carboxyl groups are the main binding site for metal cations in gram-positive bacterial cell walls with phosphate groups contributing significantly in gram-negative species (BEVERIDGE and DOYLE, 1989). Metal binding may be at least a two-stage process first involving interaction between metal ions and reactive groups followed by inorganic deposition of increased amounts of metal. This leads to the accumulation of greater than stoichiometric amounts of metals and also confirms that such biosorption is often not solely composed of ion-exchange phenomena. Metal-loaded cells of *Bacillus subtilis,* when mixed with synthetic sediment and exposed to simulated low-temperature sediment diagenesis, acted as nuclei for the formation of various crystalline metal deposits including phosphates, sulfides, and polymeric, metal-complexed organic residues (BEVERIDGE et al., 1983). The previous growth history of both gram-negative and gram-positive bacteria may affect biosorption

(BONTHRONE et al., 2000; ANDRES et al., 2000). For both *Pseudomonas fluorescens* and *B. subtilis*, copper uptake increased with the limiting nutrient in the order C<Mg<N<K with S and P occupying different positions depending on species. Such nutrient limitations had probably affected wall composition (BALDRY and DEAN, 1981). In *Pseudomonas aeruginosa*, lanthanide biosorption obeyed the Brunauer–Emmett–Teller (BET) isotherm for multilayer adsorption with uptake of La^{3+}, Eu^{3+}, and Yb^{3+} unaffected by K^+, Ca^{2+}, Cl^-, SO_4^{2-}, or NO_3^- but strongly inhibited by Al^{3+} (TEXIER et al., 2000). Ni(II) biosorption by *Pseudomonas stutzeri* followed the linearized Langmuir isotherm (RAMTEKE, 2000) while Scatchard models were established for gadolinium biosorption by *Bacillus subtilis*, *Pseudomonas aeruginosa*, *Ralstonia metallidurans*, *Mycobacterium smegmatis,* and *Saccharomyces cerevisiae* (ANDRES et al., 2000). Europium binding to *P. aeruginosa* occurred mostly through carboxyl and phosphate groups (TEXIER et al., 1999).

Chitin is an important structural component of fungal cell walls and this is an effective biosorbent for metals and radionuclides as are chitosan and other chitin derivatives (TOBIN et al., 1994; BHANOORI and VENKATESWERLU, 2000). Fungal phenolic polymers and melanins possess many potential metal-binding sites with oxygen-containing groups including carboxyl, phenolic and alcoholic hydroxyl, carbonyl and methoxyl groups being particularly important (GADD, 1993b). In *Penicillium chrysogenum*, two different Pb-binding sites were identified as carboxyl and phosphoryl groups with the former strongest but minor (5%) and the latter weaker but predominant (SARRET et al., 1999). Melanized chlamydospores of *Aureobasidium pullulans* bind more metal ions than non-pigmented yeast-like cells and mycelium, the cell wall acting as a permeability barrier (GADD and GRIFFITHS, 1980; GADD, 1984; MOWLL and GADD, 1984; GADD and MOWLL, 1985; GADD et al., 1987, 1990). Adsorption was related to the ionic radius in *Rhizopus arrhizus* for La^{3+}, Mn^{2+}, Cu^{2+}, Zn^{2+}, Cd^{2+}, Ba^{2+}, Hg^{2+}, Pb^{2+}, UO_2^{2+}, and Ag^+ but not Cr^{3+} or the alkali metal cations, Na^+, K^+, Rb^+, and Cs^+, which were not adsorbed (TOBIN et al., 1984). Copper adsorption by *Cladosporium resinae* and *Penicillium italicum* obeyed the Langmuir and Freundlich isotherms for monolayer adsorption whereas *Rhizopus arrhizus* obeyed the BET isotherm for multilayer adsorption (DE ROME and GADD, 1987). Such isotherms are widely used in studies of fungal biosorption and have some value in providing details of uptake capacities, for describing equilibrium conditions and adsorption characteristics of different species and morphologies (TSEZOS and VOLESKY, 1981; GADD and MOWLL, 1985; GADD, 1986b, 1990a; ROSS and TOWNSLEY, 1986; MORLEY and GADD, 1995; KAPOOR et al., 1999; KARAMUSHKA and GADD, 1999; FILIPOVIC-KOVACEVIC et al., 2000). However, they are simplistic and usually provide little detail on mechanistic aspects. Zinc biosorption to *R. arrhizus* was predominantly to wall chitin and chitosan (maximum sorptive capacity of 312 μmol g dry wt.$^{-1}$ compared to 213 μmol g dry wt.$^{-1}$ for the wall) (ZHOU, 1999). Sorption activity for UO_2^{2+}, Pb^{2+}, Mn^{2+}, and Co^{2+} of *Aspergillus niger* structural cell wall polysaccharides also depended on the chitin/glucan ratio (TERESHINA et al., 1999). Freeze-dried biomass of white-rot fungi was found to be the most efficient biomass type for Cd^{2+} biosorption (CIHANGIR and SAGLAM, 1999) while dry biomass of *Phanerochaete chrysosporium* sorbed mercury and alkylmercury species with an affinity of $CH_3HgCl > C_2H_5HgCl > Hg(II)$ and maximum capacities of 79, 67 and 61 mg g dry wt.$^{-1}$, respectively (SAGLAM et al., 1999).

For radionuclides like uranium and thorium, biosorption appears to be the main mechanism of uptake in fungi (GADD and WHITE, 1989b, 1990). In *R. arrhizus*, uranium biosorption involved at least three processes. Rapid, simultaneous coordination of uranium to the amine nitrogen of chitin and adsorption in the cell wall chitin structure are followed by precipitation of uranyl hydroxide at a slower rate (TSEZOS and VOLESKY, 1982). A free radical on the chitin molecule appears to be involved in the coordination of UO_2^{2+} to nitrogen (TSEZOS, 1983). Phosphate and carboxyl groups may be involved in initial binding of UO_2^{2+} to *S. cerevisiae* walls followed by deposition of uranium as needle-like fibrils on the walls with little or none being located within cells (STRANDBERG et al., 1981). Cell bound uranium comprised 10–15% of dry weight, but only about 32% of cells had measurable quantities of uranium associated with them. This means that the uranium concentration of that fraction approached 50% of dry weight, and it appeared that uranium "crystallized" on already bound molecules (STRANDBERG et al., 1981).

Over modest ranges of temperature from 4–30 °C fungal biosorption is relatively unaffected (DE ROME and GADD, 1987a). Low external pH generally decreases the rates and the extent of fungal biosorption of, e.g., Cu, Cd, Zn, Mn, and Co (GADD and MOWLL, 1985). Initial rates of uranium uptake by yeast increased above pH 2.5 (STRANDBERG et al., 1981), and maximal radium uptake by *Penicillium chrysogenum* occurred between pH 7 and 10 (TSEZOS and KELLER, 1983). Low pH can also reduce thorium uptake. At pH ≤ 2 solubility is high with Th^{4+}, the main species, but as the pH increases various hydrolysis products appear, particularly $Th(OH)_2^{2+}$, which are taken up more efficiently than Th^{4+} (TSEZOS and VOLESKY, 1981). Low pH reduces uranium uptake because of increased H_3O^+, which competes with uranium species for binding sites, and effects on the ionic species of uranium present. At pH < 2.5 the predominant cation is UO_2^{2+} but at > pH 2.5 it hydrolyzes to, e.g., $(UO_2)_2(OH)_2^{2+}$, $UO_2(OH)^+$, and $(UO_2)_3(OH)_5^+$ and there may also be complex formation with carbonates (TSEZOS and VOLESKY, 1981; STRANDBERG et al., 1981). The accompanying reduction in solubility favors adsorption (TSEZOS and VOLESKY, 1982). The presence of other anions and cations may also affect fungal biosorption by precipitation and, therefore, reducing the external free ion concentration, or competing with binding sites (TSEZOS and VOLESKY, 1982; TSEZOS, 1983; GALUN et al., 1984; MOWLL and GADD, 1984; TOWNSLEY and ROSS, 1986). The biomass concentration can also be critical. At a given equilibrium concentration of metal, fungal biomass adsorbs more metal ions at low rather than high biomass densities (DE ROME and GADD, 1987a). Contributing factors that have been suggested for such biomass concentration dependency of metal adsorption include electrostatic interactions, interference between binding sites, and reduced mixing at high cell densities (GADD and WHITE, 1989a). Certain treatments may increase fungal biosorption. Powdering exposes additional binding sites as may detergent treatment (GALUN et al., 1983; TOBIN et al., 1984; ROSS and TOWNSLEY, 1986). Removal of Pb, Cd, and Cu by alkali-treated *A. niger* biomass was greater than removal achieved by granular activated carbon (F-400) (KAPOOR et al., 1999). Benzaldehyde was an effective fixation agent for *S. cerevisiae* biomass which retained its original lead biosorption capacity (ASHKENAZY et al., 1999).

As well as dissolved metal forms, particulate material may also be adsorbed by fungal cells. Copper, lead, and zinc sulfides were adsorbed onto mycelium of *A. niger* (WAINWRIGHT and GRAYSTON, 1986) while *Mucor flavus* could adsorb lead sulfide, zinc dust, and ferric hydroxide ("ochre") from acid mine drainage (WAINWRIGHT et al., 1986). Intact, fresh mycelium and a temperature of 25 °C was best for adsorption, and the presence of a carbon source was unnecessary (WAINWRIGHT et al., 1986). The physical nature of particulate metal forms may be altered by the fungus prior to adsorption. For example, when *M. flavus* was grown with lead sulfide, particles of this compound were converted to a fine, even suspension which was then completely adsorbed (WAINWRIGHT et al., 1986).

Metabolism-independent accumulation of metals and radionuclides by algae is often rapid and reversible, and may be species-dependent (AVERY et al., 1993a; GARNHAM et al., 1994; HAMDY, 2000a, b). However, toxicity and cell death may lead to non-specific intracellular uptake (REED and GADD, 1990). The relative importance of biosorption to total uptake, and uptake capacities, may vary between different algal species. In *Ankistrodesmus braunii* and *Chlorella vulgaris*, cadmium binding to cell walls accounted for approximately 80% of total uptake. Biosorption was the major uptake component in *C. vulgaris* with respect to a wide range of other metals including uranium (NAKAJIMA et al., 1981). In common with other microbial groups, many potential binding sites occur in algal cell walls and extracellular matrices including polysaccharides, cellulose, uronic acid, and proteins (TREVORS et al., 1986; AVERY et al., 1998). Both ionic charge and covalent bonding are involved in biosorption with proteins and polysaccharides playing significant roles. Covalent bonding is likely to amino and carbonyl groups (CRIST et al., 1981). In *Scenedesmus obliquus*, the walls behave like a weakly acidic cation exchanger though cadmium, e.g., may be accumulated on *S. obliquus* walls as neutral complexes. This may occur in the marine habitat where cadmium can exist as $CdCl_3^-$ and $CdCl_2$. For algae, *in situ* metal removal from the marine environment has been postulated with one *Chlorella* strain exhibiting maximal removal of 39.4 mg Cd g dry cells^{-1} from 50 μM Cd under marine conditions (MATSUNAGA et al., 1999). Cu was preferentially adsorbed to Zn by *Cymodocea nodosa*, a brown alga (SANCHEZ et al., 1999).

The biosorption of metals and radionuclides by a variety of algae can also be represented by Freundlich and/or Langmuir isotherms confirming a linear equilibrium relationship between the metal concentration in solution and that bound to the cell surface (KHUMMONGKOL et al., 1982; GARNHAM et al., 1993). For *Chlorella vulgaris*, the order of binding capacities for living cells was $UO_2^{2+} \gg Cu^+ \gg Zn^{2+} \geq Ba^{2+} \cong Mn^{2+} > Co^{2+} \cong Cd^{2+} \geq Ni^{2+} \cong Sr^{2+}$ whereas for dead cells, the roughly similar sequence was $UO_2^{2+} \gg Cu^{2+} \gg Mn^{2+} \geq Ba^{2+} > Zn^{2+} \geq Co^{2+} \geq Cd^{2+} > Ni^{2+} > Sr^+$ (NAKAJIMA et al., 1981). For *Vaucheria* cell walls, the order of accumulation was $Cu^{2+} > Sr^+ > Zn^{2+} > Mg^{2+} > Na^+$ (CRIST et al., 1981) while for a variety of marine algae the order $Hg^{2+} > Ag^+ > Zn^{2+} > Cd^{2+}$ was obtained. Au^{3+}, Au^+ and various gold complexes can be strongly bound by algal biomass and the gold concentration may approach 10% of dry weight. Experiments suggested that Au^{3+} was reduced to Au^+ and then to elemental gold of colloidal dimensions (GREENE et al., 1986). Bacterial, algal and fungal biomass were able to remove and concentrate gold from solutions in which it occurred as a gold–cyanide complex (NIU and VOLESKY, 1999). Biosorption of Cd^{2+}, Cu^{2+}, Zn^{2+}, and Mn^{2+} is generally reduced at low pH in a variety of algae. However, this may depend on the metal concerned, and with *C. vulgaris*, Au^{3+}, Ag^+, and Hg^{2+} were strongly bound at pH 2 (DARNALL et al., 1986).

For biosorption of uranium and other radionuclides by algae, it seems that physico-chemical interactions comprise the majority of total uptake. In *Chlorella regularis* uranium uptake was optimal at pH 5 and was retarded by PO_4^{3-}, CO_3^{2-}, and HCO_3^-, but unaffected by Na^+, K^+, NH_4^+, Ca^{2+}, Mg^{2+}, Mn^{2+}, Co^{2+}, Ni^{2+}, Zn^{2+}, NO_3^- or SO_4^{2-}. UO_2^{2+} and UO_2OH^+ were the main species taken up in exchange for protons (NAKAJIMA et al., 1982; YANG and VOLESKY, 1999). Little uranium was taken up from sea water, in contrast to fresh water or decarboxylated sea water, and this was thought to be due to interference by CO_3^{2-} which results in the formation of $UO_2(CO_3)_2^{2-}$ and $UO_2(CO_3)_3^{4-}$ complexes which are not taken up by cells.

2.1.3 Biosorption by Immobilized Biomass

Both freely suspended and immobilized biomass from bacterial, cyanobacterial, algal and fungal species has received attention with immobilized systems possessing several advantages including higher mechanical strength and easier biomass/liquid separation (MACASKIE et al., 1987a; MACASKIE and DEAN, 1989; GARNHAM et al., 1992a; GUPTA et al., 2000). Living or dead biomass of all groups has been immobilized by encapsulation or cross-linking using supports which include agar, cellulose, alginates, cross-linked ethyl acrylate-ethylene glycol dimethylacrylate, polyacrylamide, polyvinyl alcohol, silica gel and cross-linking reagents such as toluene diisocyanate and glutaraldehyde (MACASKIE and DEAN, 1989; BRIERLEY, 1990; MACASKIE, 1991; TOBIN et al., 1994; TING and SUN, 2000). Immobilized living biomass has mainly taken the form of bacterial biofilms on inert supports and is used in a variety of bioreactor configurations including rotating biological contactors, fixed bed reactors, trickle filters, fluidized beds and air-lift bioreactors (GADD, 1988; MACASKIE and DEAN, 1989; GADD and WHITE, 1990, 1993; COSTLEY and WALLIS, 2000). *Streptomyces albus* was immobilized in polyacrylamide gel, crushed to 50–100 mesh size, and used in batch and column experiments. Immobilized cells had good mechanical stability and removed uranium, copper, and cobalt with some degree of selectivity, the order of efficiencies being

$UO_2^{2+} \gg Cu^{2+} > Co^{2+}$. Desorption was possible with 0.1 mol L^{-1} Na_2CO_3, and the immobilized particles were not significantly affected by five adsorption–desorption cycles (NAKAJIMA et al., 1982; NAKAJIMA and SAKAGUCHI, 1986). Polyacrylamide immobilized cells of *Citrobacter* were capable of high removal of uranium, cadmium, copper, and lead from solutions supplemented with glycerol-2-phosphate. The mechanism of uptake involved phosphatase-mediated cleavage of glycerol-2-phosphate to release HPO_4^{2-} which precipitated metal extracellularly as insoluble metal phosphate, e.g., $CdHPO_4$ (MACASKIE et al., 1987a, b; MACASKIE and DEAN, 1984, 1987). Metal could be removed from loaded cells, the biomass regenerated, and such immobilized columns could function for extended periods over a temperature range of 2–45 °C. They were unaffected by the metal used or the concentration. High levels of Cl^- and CN^- inhibited cadmium uptake probably by complex formation (MACASKIE et al., 1987b). Polyacrylamide-entrapped *Chlorella* can remove $UO_2{}^{2+}$, Au^{3+}, Cu^{2+}, Hg^{2+}, and Zn^{2+} from solution, and a selective elution scheme for recovery has been described (DARNALL et al., 1986). Particles of microbial biomass, immobilized with polymeric membranes, have been used for uranium adsorption. Such particles can be made to any size, contain only 10% inert material, and can be used in multiple cycles (TZESOS, 1986). Commercial processes have been attempted for the operation of such systems particularly for the recovery of valuable metals (BRIERLEY et al., 1986). These granules have a high mechanical strength in contrast to polyacrylamide. They have high uptake capacities, and some values obtained are (in mmol g^{-1}) 0.8, Ag; 1.9, Cd; 2.4, Cu; 2.9, Pb; 2.0, Au; and 2.1, Zn. Metal removal from dilute solutions (10–100 mg L^{-1}) exceeded 99%, and the granules were used in fixed-bed canisters or fluidized-bed reactor systems (HUTCHINS et al., 1986).

Whole cell immobilization within a polyacrylamide gel provides a useful laboratory-scale system but has low mechanical strength. Treatment with hardening agents or chemical coupling to rigid supports may result in toxicity (which may or may not be a problem), whereas electrostatic attraction is weak and affected by pH (MACASKIE and DEAN, 1987). The use of biofilms on inert matrices appears to be one way of overcoming such problems. A technique whereby *Citrobacter* is grown on solid supports as a cohesive biofilm has been described with metal removal capacities comparable to those for gel-immobilized cells (MACASKIE and DEAN, 1987). Other examples include the use of *Aspergillus oryzae* grown on reticulate foam particles (KIFF and LITTLE, 1986) and *Trichoderma reesei* packed in molochite and used for copper removal from simulated leach effluents (TOWNSLEY et al., 1986). Filamentous fungi can be grown in the form of pellets and these may have similar advantages to immobilized particles. Pellets (4 mm diameter) of *Aspergillus niger* were used for uranium removal in a fluidized-bed reactor. Such a system was more efficient than the commercial ion-exchange resin IRA-400 (YAKUBU and DUDENEY, 1986). However, there were some disadvantages. Fungal pellets broke down, causing increased resistance to liquid flow, and the similarity in specific gravity between the biomass and the liquid medium made continuous operation difficult (YAKUBU and DUDENEY, 1986).

Both living and dead biomass can be used in immobilized systems. If living cells are used, there are possibilities for the removal of other undesirable substances, as well as heavy metals and radionuclides, from wastewaters although toxicity is an important consideration. Living cells of *Pseudomonas aeruginosa* have been immobilized on particles of polyvinyl chloride (PVC) and melt blown polypropylene webs and used in batch and column systems for simultaneous denitrification and heavy metal removal from contaminated wastewater. Another system employed a mixed bacterial culture, mainly *Pseudomonas* sp. for denitrification and uranium removal. The denitrification bioreactor was operated continuously with the bacteria growing as a film on anthracite coal particles. Particles with excess cell mass, which accumulated at the upper boundary of the fluidized bed, were continuously removed, passed through a vibrating screen, and returned to the lower region of the fluidized bed to maintain optimum microbial film thickness. Excess cells from particle passage through the vibrating screen were used as a biosorbent in a separate stirred-tank reactor for uranium removal (SHUMATE et al., 1980). An alternative system relied on circulation of biosorbent particles, prepared as above, in opposite direction to uranium flow. This system resulted in a uranium reduction from 25 to 0.5 g m^{-3} with a mean liquid residence time of only 8 min (SHUMATE et al., 1980). A large-scale ($5.5 \cdot 10^6$ gallons per day) commercial process, which treats effluents from gold mining and milling, uses rotating-disc biological-contacting units to simultaneously degrade cyanide, thiocyanate, and ammonia. Heavy metals were removed by biosorption onto the microbial biofilm on the disc surfaces. Metal-loaded biomass was recovered for controlled disposal (HUTCHINS et al., 1986).

Note that many other studies have demonstrated the efficiency of metal removal by microbial biomass in a range of reactor formats and under a range of physical and chemical conditions (BUTTER et al., 1998; AL SARAJ et al., 1999; ALOYSIUS et al., 1999; MAMERI et al., 1999; PRAKASHAM et al., 1999; YIN et al., 1999; ZHOU, 1999; FIGUEIRA et al., 2000). Hollow-fiber crossflow microfiltration using *Pseudomonas aeruginosa* was used for biosorption of Pb, Cu, and Cd in single and ternary metal systems. Removal efficiencies in both single and ternary biosorption were Pb > Cu > Cd and at an influent (metal) of 200 μM and a flow rate of 350 mL h^{-1}, effluent concentrations of Cu and Pb satisfied discharge regulations. A rapid-equilibrium- and a mass-transfer model was developed which accurately described single- and multi-metal biosorption respectively (CHANG and CHEN, 1999). Dispersed air flotation was used to separate Cd-laden actinomycete biomass. At a [Cd] of 5 mg L^{-1} and 0.5 g L^{-1} freeze-dried biomass, Cd removal was virtually 100% over a wide range of pH values (KEFALA et al., 1999). *Arthrobacter* sp. was entrapped within a macro- and microporous matrix, (poly)hydroxyethyl methacrylate cross-linked with trimethylolpropane trimethacrylate. The rate of biosorption was controlled by the rate of diffusion through the shell of saturated material and a maximum Cu uptake of ~7 mg g dry $wt.^{-1}$ was observed for a resin-biomass complex containing 8% w/w biomass (VEGLIO et al., 1999a, b). *S. cerevisiae* accumulated metals when immobilized in a sol–gel matrix with the entrapment matrix contributing to overall uptake. Metal accumulation affinities of the biogel were Hg > Zn > Pb > Cd > Co, Ni > Cu (AL SARAJ et al., 1999). Immobilized *Rhizopus oligosporus* biomass was more effective than freely suspended biomass for Cd^{2+} (ALOYSIUS et al., 1999). A reliable model based on Fick's law of diffusion and the Langmuir adsorption model was able to predict experimentally determined kinetics of Ni^{2+} removal by fungal biosorbents (SUHASINI et al., 1999). Dispersed air flotation has also been applied as a separation method for harvesting metal-laden microorganisms following Cd biosorption (KEFALA et al., 1999).

2.1.4 Metal Desorption and Recovery

Application of microbial metal biosorption may depend on the ease of metal recovery for subsequent reclamation or for further containment of toxic/radioactive waste (GADD and WHITE, 1992). Non-destructive recovery may also be required for regeneration of the biomass for reuse (TSEZOS, 1984). Destructive recovery may be accomplished by pyrometallurgical treatment of the biomass or dissolution in strong acids or alkalis. The method of metal recovery depends on the ease of removal from the biomass and the value of the organism or product as a reusable entity. If cheap, waste biomass was used to take up valuable metals, then the economics of destructive recovery may be satisfactory. The choice of a recovery process also depends on the mechanism of accumulation. Metabolism-independent biosorption is often reversible and lends itself to non-destructive desorption whereas intracellular accumulation may necessitate destructive recovery. Most attention has focused on non-destructive desorption from loaded biomass (GADD and WHITE, 1992). For maximum benefit this should obviously be highly efficient, cheap, and should not cause any damage to the biomass or impairment of subsequent use.

Acidic conditions can repress biosorption and not surprisingly, acids can act as effective desorption agents. Dilute (0.1 mol L^{-1}) mineral acids, e.g., HC1, H_2SO_4, HNO_3, can remove copper and cadmium from bacteria, yeasts, and filamentous fungi and uranium from *Zoogloea ramigera, Rhizopus arrhizus,* and *Saccharomyces cerevisiae* with high efficiencies (STRANDBERG et al., 1981; TSEZOS, 1984; DE ROME and GADD, 1987a; RAMTEKE, 2000). However, at higher concentrations of acids (>1 mol L^{-1}), there may be damage to the biomass and reduction of subsequent uptake (TSEZOS, 1984). Organic complexing agents may be effective desorbers and do not have adverse effects on biomass integrity. EDTA ($\simeq 0.1$ mol L^{-1}) can remove heavy metals and uranium from bacteria, algae, yeasts, and filamentous fungi with high efficiency (HORIKOSHI et al., 1981; DUNN and BULL, 1983; RAMTEKE, 2000). For activated sludge biomass EDTA treatment was additionally useful in that differentiation was possible between surface complex formation and unexchangeable metal (LAWSON et al., 1984). In *Penicillium digitatum,* EDTA treatment improved the capacity of the biomass for subsequent re-adsorption (GALUN et al., 1983). Other organic compounds used include 8-hydroxyquinoline for copper removal and nitrilotriacetic acid (NTA) and diethylenetriamine pentacetic acid (DTPA) for cadmium removal from fungi (KUREK et al., 1982; MOWLL and GADD, 1984). For *Chlorella vulgaris* biomass, an elution scheme based on pH changes and organic binding agents has been described for selective recovery of copper, zinc, gold, and mercury. Between pH 5 and 7 a variety of metal ions were strongly bound to the cell surface and most were desorbed by lowering the pH to 2. However, Au^{3+}, Ag^{+}, and Hg^{2+} bound strongly at pH 2, and the addition of a strong ligand at differing pH values was necessary to elute these. It was found that Au^{3+} and Hg^{2+} could be selectively eluted using mercaptoethanol (DARNALL et al., 1986).

Carbonates have possibly the most commercial potential for non-destructive metal recovery. Although sodium carbonate was highly effective for uranium removal from fungal biomass, the elution pH is highly alkaline and the biomass structure may be impaired (TSEZOS, 1984). Sodium and ammonium bicarbonates were also highly efficient for uranium removal from loaded biomass and increased the adsorption capacity of the biomass after treatment. Operating pH levels are satisfactory, and there is little damage to the biomass (GALUN et al., 1983; TSEZOS, 1984). The solid:liquid ratios that can be used for bicarbonate elution systems can exceed 120:1 (mg:mL) for a 1 mol L^{-1} $NaHCO_3$ solution with $\sim 100\%$ uranium recovery and concentrations in the eluate of $\geq 1.98 \cdot 10^4$ mg L^{-1} U (TSEZOS, 1984). The presence of other ionic species may affect desorption, e.g., SO_4^{2-}-limited uranium removal from *Rhizopus arrhizus* by altering the cell wall chitin structure which led to entrapment of bound uranium (TSEZOS, 1984).

Earlier commercial promise and development of biosorption appears to have largely ceased, and there is no adoption of biosorption as a commercially viable treatment method to date. Little attention has also been given to metal-organic mixtures though equilibrium uptakes of, e.g., Ni and phenol are changed due to initial pH and other reaction components (AKSU et al., 1999). The lack of commercial development is somewhat perplexing although the lack of specificity and lower robustness of biomass-based systems compared to ion exchange resins is often cited as a reason (ECCLES, 1999).

2.1.5 Metal-Binding Proteins, Polysaccharides, Derived Products, and Other Biomolecules

A diverse range of specific and non-specific metal-binding compounds are produced by microorganisms. Non-specific metal-binding compounds range from simple organic acids and alcohols to macromolecules such as polysaccharides, humic and fulvic acids (BIRCH and BACHOFEN, 1990). Extracellular polymeric substances (EPS), a mixture of polysaccharides, mucopolysaccharides and proteins (ZINKEVICH et al., 1996) are produced by bacteria, algae and fungi and also bind significant amounts of potentially toxic metals (SCHREIBER et al., 1990; GEESEY and JANG, 1990; BEECH and CHEUNG, 1995; CHEN et al., 2000). Many such exopolymers act as polyanions under natural conditions, and negatively charged groups can interact with cationic metal/radionuclide species although uncharged polymers are also capable of binding and entrapment of insoluble forms (BEVERIDGE and DOYLE, 1989). Extracellular polysaccharides of microbial origin are able to bind metals and also adsorb or entrap particulate matter such as precipitated metal sulfides and oxides (FLEMMING, 1995; VIEIRA and MELO, 1995). One process has been developed which uses silage as a support for cyanobacterial growth. Floating mats were formed which removed metals from waters, the metal-binding process being due to large polysaccharides (>200,000 Da) produced by the cyanobacteria (BENDER et al., 1994). Bacterial extracellular polymers are strongly implicated in the removal of soluble metal ions by flocculated activated-sludge biomass, and the extraction of extracellular polymers from cultures of *Klebsiella aerogenes* and activated sludge considerably reduced the capacity of the cells and flocs to remove metals (BROWN and LESTER, 1982). Extracellular polymers are intimately involved in Cd adsorption by activated sludges (GUIBAUD et al., 1999). *Zoogloea ramigera,* common in sewage treatment, produces extracellular polysaccharides which assist flocculation and exhibit significant metal-binding properties. For instance, Cu and Cd were taken up to 0.3 and 1 g metal per g dry wt., respectively. The extracellular polymer of *K. aerogenes* bound metals in the order Cu, Cd > Co > Mn > Ni which was similar to the sequence observed for *Z. ramigera* flocs, Fe > Cu > Co > Ni (RUDD et al., 1983). Uptake was reduced at low pH and little metal was bound at pH 4.5 as compared with pH 6.8. Maximum polymer production occurred in stationary phase cultures, and these were most efficient at metal removal (RUDD et al., 1983).

Specific metal-binding compounds may be produced in response to the levels of metals present in the environment. The best known extracellular metal-binding compounds are siderophores which are low molecular weight ligands (500–1,000 Da) possessing a high affinity for Fe(III) (NEILANDS, 1981). Siderophores are produced extracellularly in response to low iron availability. They scavenge for Fe(III), and complex and solubilize it, making it available for microorganisms. Although primarily produced as a means of obtaining iron, siderophores are also able to bind other metals such as magnesium, manganese, chromium (III), gallium (III), and radionuclides such as plutonium (IV) (BIRCH and BACHOFEN, 1990). Analysis of culture media from *Pseudomonas aeruginosa,* when grown in the presence of uranium or thorium, showed that various chelating agents were produced. Some of these resembled but were not identical with iron chelating siderophores. In certain *Aspergillus* and *Rhodotorula* species, excretion of iron binding molecules is stimulated under iron limitation, and such compounds may also bind gallium (ADJIMANI and EMERY, 1987).

Other metal-binding molecules include specific, low molecular weight (6,000–10,000 Da) metal-binding proteins, metallothioneins, that are produced by animals, plants, and microorganisms in response to the presence of toxic metals (HOWE et al., 1997). Other metal binding proteins, phytochelatins and related peptides, all contain glutamic acid and cysteine at the amino-terminal position, and have been identified in plants, algae, and several microorganisms (RAUSER, 1995). Eukaryotic metallothioneins and other metal-binding peptides have been expressed in *E. coli* as fusions to membrane or membrane-associated proteins such as LamB, an outer membrane protein which functions as a coliphage surface receptor and is involved in maltose/maltodextrin transport. Such *in vivo* expression of metallothioneins provides a means of designing biomass with specific metal-binding properties (VALLS et al., 1998; CHEN et al., 1999; PAZIRANDEH et al., 1998; BANG and PAZIRANDEH, 1999). Metal-binding peptides of sequences Gly-His-His-Pro-His-Gly (HP) and Gly-Cys-Gly-Cys-Pro-Cys-Gly-Cys-Gly (CP) were engineered into LamB protein and expressed in *E. coli.* Cd^{2+} : HP and Cd^{2+} : CP were 1 : 1 and 3 : 1, respectively. Surface display of CP increased the Cd^{2+} binding ability of *E. coli* fourfold. Some competition of Cu^{2+} with Cd^{2+} for HP resulted from strong Cu^{2+} binding to HP indicating that the relative metal-binding affinities of inserted peptides and the wall to metal ion ratio were important in the design of peptide sequences and their metal specificities (KOTRBA et al., 1999). Another gene encoding for a *de novo* peptide sequence containing the metal-binding motif Cys-Gly-Cys-Cys-Gly was chemically synthesized and expressed in *E. co-*

li as a fusion with the maltose-binding protein. Such cells possessed enhanced binding of Cd^{2+} and Hg^{2+} compared to cells lacking the peptide (PAZIRANDEH et al, 1998). A hybrid protein between mouse metallothionein (MT) and the *beta* domain of the IgA protease of *Neisseria* was expressed in the outer membrane of *Pseudomonas putida* which increased metal-binding capacity three-fold (VALLS et al., 2000). The outer membrane protein, OmpC, from *E. coli* was used to display metal-binding poly-histidine peptides and this resulted in up to 6x more adsorption of Zn^{2+} and Ni^{2+} than control cells expressing the wild-type protein (CRUZ et al., 2000).

Related to the application of metal-binding molecules is the identification of genes encoding phytochelatin synthases, phytochelatins playing major roles in metal detoxification in plants and fungi. This provides molecular evidence for the role of phytochelatin molecules in metal tolerance since heterologous expression of *PCS* genes dramatically enhanced metal tolerance. Future research with microorganisms and plants may allow testing of the potential of *PCS* genes for bioremediation (CLEMENS et al., 1999). A variety of other metal-binding molecules may have future potential for metal recovery. Culture filtrates from the sulfate-reducing bacterium *Desulfococcus multivorans* exhibited copper-binding with 12 d old filtrates having a copper-binding capacity of 3.64 ± 0.33 μmol mL^{-1} with a stability constant, $\log_{10}K$ of 5.68 ± 0.64 $(n=4)$ (BRIDGE et al., 1999). A hollow fiber reactor containing an engineered *E. coli* strain expressing Hg^{2+} transport and metallothionein accumulated Hg^{2+} effectively, reducing a concentration of 2 mg L^{-1} to ~5 μg L^{-1} (CHEN et al., 1998). It can be noted that biosensors for the detection of metal bioavailability have been developed based on the direct interaction between metal-binding proteins and metal ions. Here, capacitance changes of the proteins, e.g., a synechococcal metallothionein and a mercury resistance regulatory protein, were detected in the presence of femto- to millimolar metal ion concentrations (CORBISIER et al., 1999).

Compounds derived from or produced by microorganisms may act as efficient biosorption agents. It seems that components obtained from fungal biomass have received the most attention. A variety of potential biosorption agents occur in fungal walls in considerable amounts, including mannans, glucans, phosphomannans, chitin, chitosan, and melanin. Uranium adsorption by chitin phosphate and chitosan phosphate was rapid and affected by pH and temperature in an analogous manner to whole biomass. For chitin phosphate biosorption efficiencies were in the order of $UO_2^{2+} \gg Cu^{2+} > Cd^{2+} > Mn^{2+} > Zn^{2+} > Mg^{2+} > Co^{2+} > Ni^{2+} > Ca^{2+}$, for chitosan phosphate orders were $UO_2{}^{2+} \gg Cu^{2+} > Zn^{2+} > Mn^{2+} > Co^{2+} > Ni^{2+} > Mg^{2+} > Ca^{2+}$. UO_2^{2+} and Co^{2+} were easily separated from each other using chitin phosphate and could be recovered by using sodium carbonate for desorption. Non-phosphorylated chitin and chitosan were not as effective for biosorption as phosphorylated derivatives (SAKAGUCHI and NAKAJIMA, 1982). Insoluble chitosan–glucan complexes obtained from *Aspergillus niger* were efficient metal chelating agents, while glucans possessing amino acid or sugar acid groups exhibit enhanced binding of transition metal ions (MUZZARELLI et al., 1986). Chemical modification of biomass may create derivatives with altered metal binding abilities and affinities. *A. niger* mycelium was modified by introducing additional carboxy or ethyldiamino groups with the latter substitution increasing the maximal capacity for nickel from ~70 to ~1,060 mmol kg^{-1}. An increase in maximal binding capacities for Cd, Ni, and Zn also resulted in carboxy-modified biomass (Cd 100 to 274; Ni 71 to 182; Zn 185 to 303 mmol kg^{-1}, respectively) (KRÄMER and MEISCH, 1999).

2.1.6 Transport and Intracellular Fate

Metabolism-dependent transport of metal ions may be a slower process than biosorption and can be inhibited by low temperatures, the absence of an energy source, metabolic inhibitors and uncouplers (BORST-PAUWELS, 1981; GADD, 1986a; AVERY et al., 1991, 1992a, b, 1998). Rates of uptake may also be influenced by the physiological state of cells and the nature and composition of the growth medium.

Many metals are essential for growth and metabolism, e.g., Cu, Fe, Zn, Co, and organisms possess transport systems of varying specificity for their accumulation. Non-essential metals may also be taken in via such systems (PERKINS and GADD, 1995, 1996). Integral to described transport systems are ionic gradients of, e.g., H^+ and/or K^+ across the cell membrane and the membrane potential. For fuller treatment of transport see BORST-PAUWELS (1981), TREVORS et al. (1986), GADD (1986a, 1993b), AVERY et al., 1993b, c; EIDE (1997), EIDE and GUERINOT, 1997; WINGE et al. (1998), RADISKY and KAPLAN (1999), SILVER (1998), and GADD and SAYER (2000). Most work on metal ion transport in fungi has concerned K^+ and Ca^{2+}, in view of their importance in fungal growth, metabolism, and differentiation, although the transport of other essential metal species is now receiving renewed attention (KOSMAN, 1994; GADD and LAURENCE, 1996; EIDE and GUERINOT, 1997). Work carried out at low micronutrient metal concentrations has clearly demonstrated the existence of multiple and high affinity metal transport systems in unicellular and filamentous fungi (EIDE and GUERINOT, 1997; PARKIN and ROSS, 1986; WHITE and GADD, 1987a; HOCKERTZ et al., 1987; DE ROME and GADD, 1987b; STARLING and ROSS, 1991; PILZ et al., 1981; ZHAO and EIDE, 1996a, b). *Saccharomyces cerevisiae* possesses a high affinity Mn^{2+} transport system ($K_m = 0.3\ \mu M$) functional at low Mn^{2+} concentrations (25–1,000 nM) which is of low specificity being inhibited by Mg^{2+}, Co^{2+}, Zn^{2+}, and Cd^{2+} to varying extents. At higher concentrations of $MnCl_2$ (5–200 μM), a transport system of lower affinity ($K_m = 63\ \mu M$) is available to the cell. Despite the manganese transport system of *S. cerevisiae* being relatively non-specific, transport of manganese could still occur even when there was an excess of competing divalent cations such as Mg^{2+} (GADD and LAURENCE, 1996). This is important since organisms must be able to acquire essential "trace" divalent cations even when there is an excess of other divalent cations present in their external environment. In general terms, divalent cations appear to enter cells as a result of the electrochemical gradient ($\Delta\mu_{H^+}$) generated by the activity of the plasma membrane H^+-ATPase (GADD, 1993b; BORST-PAUWELS, 1981; SANDERS, 1990; JONES and GADD, 1990) although there is some evidence that the transmembrane electrochemical K^+ gradient ($\Delta\tilde{\mu}_{K^+}$) may also be important (OKOROKOV, 1985; RAMOS et al, 1985). While the existence of a Ca^{2+}-ATPase has been described in fungi (MILLER et al. 1990) there is no evidence of a comparable mechanism for other essential divalent cations, although divalent cation efflux mechanisms have frequently been proposed (NIEUWENHUIS et al., 1981; OKOROKOV, 1985; JONES and GADD, 1990).

It should be noted that for potentially toxic inessential and essential metals, transport analysis can be complicated by toxic symptoms which can include membrane disruption, and also compartmentation in organelles like the vacuole (GADD and MOWLL, 1983; MOWLL and GADD, 1993; GADD and WHITE, 1985, 1989a, c). Even low concentrations of toxic metals can affect the structural integrity of the cell membrane and associated functions which can alter the kinetics of uptake. Progressive inhibition of H^+ extrusion could result in de-energization of the plasma membrane while increased membrane permeability results in K^+ efflux (BORST-PAUWELS, 1981; WHITE and GADD, 1987a, b). For both Zn^{2+} and Mn^{2+}, apparent low-affinity transport coincided with such effects in *S. cerevisiae* which suggested that altered kinetics were a result of toxicity (WHITE and GADD, 1987a, b; GADD and LAURENCE, 1996). In fact, if affinity constants are compared for a range of potentially toxic divalent cations, e.g., Zn^{2+}, Cu^{2+}, Cd^{2+}, Ni^{2+}, Co^{2+}, it seems that when the concentration range used is high, the calculated affinity of the transport system is low (GADD and WHITE, 1989a; KOSMAN, 1994).

Although amounts of metal species accumulated by energy-dependent transport may exceed amounts taken up by physicochemical processes, there are several exceptions where energy-dependent uptake may not be as significant a component of total uptake as general biosorption. This is particularly true for filamentous fungi and those organisms possessing extracellular polysaccharide, slime, or mucilage where high biosorptive capacities mask low rates of intracellular uptake. In addition, several examples exist where intracellular uptake is not linked with metabolism, e.g., intracellular accumulation of U in *Synechococcus elongatus* and *Pseudomonas aeruginosa* which leads to the formation of dense internal deposits (HORIKOSHI et al., 1981; STRANDBERG et al., 1981).

Once inside cells, metal ions may be compartmentalized and/or converted to more innocuous forms by binding or precipitation (PEREGO and HOWELL, 1997). In addition, intracellular metal concentrations may also be regulated by transport, including efflux mechanisms (GARNHAM et al., 1992b; GADD, 1993b; MACREADIE et al., 1994; BLAUDEZ et al., 2000; NIES, 2000). Such mechanisms are involved in normal metal homeostasis within cells but also have a role in the detoxification of potentially toxic metals. After uranium uptake, dense intracellular deposits were found in *Pseudomonas aeruginosa* although metabolism was not required (STRANDBERG et al., 1981). In cyanobacteria, aluminum and cadmium accumulated in polyphosphate bodies in *Anabaena cylindrica* and *Plectonema boryanum*, respectively (JENSEN et al., 1982; PETTERSSON et al., 1986). A *Synechococcus* sp. synthesized large quantities of an intracellular polymer that could bind nickel, the cell interior appearing highly granular (WOOD and WANG, 1983). Polyphosphate and other electron-dense deposits have also been recorded in a variety of eukaryotic algae (VYMAZAL, 1987).

The yeast cell vacuole is involved in a number of essential functions within the cell including macromolecular degradation, storage of metabolites, and cytosolic pH homeostasis (KLIONSKY et al., 1990). In addition, the vacuole has an important role in the regulation of cytosolic metal ion concentrations both for essential metabolic functions and the detoxification of potentially toxic metal ions (WHITE and GADD, 1986; GADD, 1993b, 1995; GHARIEB and GADD, 1998; LIU and CULOTTA, 1999). Metals preferentially sequestered by the vacuole include Mn^{2+} (OKOROKOV et al., 1985; GADD and LAURENCE, 1996), Fe^{2+} (BODE et al., 1995), Zn^{2+} (WHITE and GADD, 1987a), Co^{2+} (WHITE and GADD, 1986), Ca^{2+} and Sr^{2+} (OKOROKOV et al., 1985; BORST-PAUWELS, 1989; GADD, 1993b; OKOROKOV, 1994), Ni^{2+} (JOHO et al., 1995) and the monovalent cations K^+, Li^+ and Cs^+ (OKOROKOV et al., 1980; PERKINS and GADD, 1993a, b), where some may be bound to low molecular weight polyphosphates (WHITE and GADD, 1986, 1987a). In cells lacking vacuoles, there may be involvement of polyphosphate granules in metal ion sequestration. Transport of these metal cations across the vacuolar membrane has been shown to occur by cation–proton exchange driven by the vacuolar trans-membrane electrochemical pH gradient which is energized by the vacuolar H^+-ATPase (V-ATPase) (OHSUMI and ANRAKU, 1983; OKOROKOV et al., 1985; NELSON et al., 1992). The absence of a vacuole or a functional vacuolar H^+-ATPase in *Saccharomyces cerevisiae* is associated with increased sensitivity and a largely decreased capacity of the cells to accumulate Zn, Mn, Co, and Ni (RAMSAY and GADD, 1997), metals known to be mainly detoxified in the vacuole (GADD, 1993b; JOHO et al., 1995).

For Cu and Cd, intracellular detoxification in fungi appears to predominantly depend on sequestration in the cytosol by induced metal-binding molecules (MACREADIE et al., 1994; OW et al., 1994; HAYASHI and MUTOH, 1994; RAUSER, 1995). These include low molecular weight cysteine-rich proteins (metallothioneins) and peptides derived from glutathione (phytochelatins) (MACREADIE et al., 1994; OW et al., 1994; MEHRA and WINGE, 1991; WU et al., 1995; INOUHE et al., 1996). The latter peptides have the general structure of $(\gamma$Glu-Cys$)_n$-Gly where the γGlu-Cys repeating unit may extend up to 11 (OW et al., 1994). In *Schizosaccharomyces pombe* the value of n ranges from 2–5, while in *S. cerevisiae*, only an n_2 isopeptide has been observed (MACREADIE et al., 1994). As well as being termed phytochelatins, the most widely used trivial name, such peptides are also known as cadystins and metal γ-glutamyl peptides, although the chemical structure, $(\gamma EC)_nG$, is a more precise description. Although $(\gamma EC)_nG$ induction has been reported with a wide variety of metal ions, including Ag, Au, Hg, Ni, Pb, Sn, and Zn, metal binding has only been shown for a few, primarily Cd and Cu (OW et al., 1994). For Cd, two types of complexes exist in *S. pombe* and *Candida glabrata*. A low molecular weight complex consists of $(\gamma EC)_nG$ and Cd, whereas a higher molecular weight complex also contains acid-labile sulfide (MURASUGI et al.,

1983; Ow et al., 1994). The $(\gamma EC)_nG$-Cd-S^{2-} complex has a greater stability and higher Cd binding capacity than the low molecular weight complex, and has a structure consisting of a CdS crystallite core and an outer layer of $(\gamma EC)_nG$ peptides (DAMERON et al., 1989). The higher binding capacity of the sulfide-containing complex confers a greater degree of tolerance to Cd (Ow et al., 1994). In *S. pombe*, evidence has also been presented for subsequent vacuolar localization of $(\gamma EC)_nG$-Cd-S^{2-} complexes (ORTIZ et al., 1992, 1995; OW, 1993), illustrating a dynamic link between cytosolic sequestration and vacuolar compartmentation. Although the main function of *S. cerevisiae* metallothionein (yeast MT) is cellular copper homeostasis, induction and synthesis of MT as well as amplification of MT genes leads to enhanced copper resistance in both *S. cerevisiae* and *C. glabrata* (MACREADIE et al., 1994). Production of MT has been detected in both Cu- and Cd-resistant strains of *S. cerevisiae* (TOHOYAMA et al., 1995; INOUHE et al., 1996). However, it should be noted that other determinants of tolerance also occur in these and other organisms, e.g., transport phenomena and sulfide precipitation (GADD and WHITE, 1989a; Inouhe et al., 1996; YU et al., 1996), while some organisms, e.g., *Kluyveromyces lactis*, are not capable of MT or $(\gamma EC)_nG$ synthesis (MACREADIE et al., 1994). In *S. cerevisiae*, it has been shown that changes in amino acid pools can occur in response to nickel exposure with the formation of vacuolar nickel–histidine complexes being proposed as a survival mechanism (JOHO et al., 1995). Little work has been carried out on MT or $(\gamma EC)_nG$ peptides in filamentous fungi (see GADD, 1993b; GALLI et al., 1994; HOWE et al., 1997; KAMEO et al., 2000). Other mechanisms for metal immobilization within cells include precipitation by, e.g., reduction, sulfide production, or association with polyphosphate (GADD, 1990a, b, c, 1993b). Internal metal-binding proteins occur in eukaryotic algae. Metal-binding proteins may have applications in metal recovery, either inside or outside of the cells or in purified forms, especially as such proteins are able to bind valuable elements, e.g., Au and Ag, as well as those of low value, e.g., Cu, Cd, and Zn (BUTT and ECKER, 1987).

2.2 Metal Precipitation

2.2.1 Reductive Precipitation by Metal-Reducing Bacteria

Where reduction of a metal to a lower redox state occurs, mobility and toxicity may be reduced offering potential bioremediation applications. Such processes may also accompany other indirect reductive metal precipitation mechanisms, e.g., in sulfate-reducing bacterial systems where reduction of Cr(VI) can be a result of indirect reduction by Fe^{2+} and sulfide. A diverse range of microorganisms are able to use oxidized species of metallic elements, e.g., Fe(III), Cr(VI), or Mn(IV) as terminal electron acceptors (THAMDRUP, 2000). Many of these organisms can utilize more than one terminal electron acceptor including several metals or other anions such as nitrate or sulfate. Most of these organisms are anaerobic and a few are facultative anaerobes and oxygen may also be respired (LOVLEY and COATES, 1997; THAMDRUP, 2000). Fe(III) and Mn(IV) appear to be the most commonly utilized metals as terminal electron acceptors in the biosphere and metal-reducing organisms from many habitats frequently utilize both of these metals (LOVLEY, 1993; LOVLEY and COATES, 2000). However, since the solubility of both Fe and Mn is increased by bacterial reduction, and neither metal is significantly toxic, other metals are targeted in waste treatment. However, some dissimilatory Fe(III)-reducing microbes have the ability to destroy organic contaminants under anaerobic conditions, carry out reductive dechlorination, and also reduce other metals, e.g., U(VI) (LOVLEY and ANDERSON, 2000; LOVLEY and COATES, 2000). Molybdenum(VI) was reduced to molybdenum blue by a strain of *Enterobacter cloacae* which was isolated from a molybdate-polluted aquatic environment (GHANI et al., 1993). Another strain of *E. cloacae*, also isolated from a polluted habitat, was able to reduce Cr(VI) to Cr(III) under similar conditions (WANG et al., 1989) and Cr(VI) was reduced to Cr(III) and precipitated from a simulated wastewater by this organism in a bioreactor (FUJIE et al., 1994). Dissimilatory Cr(VI) reduction was also carried out by a strain of *E. coli* under both anaerobic and aerobic conditions (SHEN and WANG, 1993). Metal reduction processes may also be useful as pretreatments for other processes, e.g., the reduction of Cr(VI) compounds to Cr(III) facilitates removal by processes such as biosorption or precipitation (AKSU et al., 1999). Since aerobic or anaerobic reduction of Cr(VI) to Cr(III) is widespread in microorganisms (BADAR et al., 2000; SMITH and GADD, 2000), both *ex situ* reactor systems and *in situ* treatment options have been documented although problems may occur with Cr(VI) toxicity as well as maintenance of anaerobic conditions for anaerobic Cr(VI) reducers. However, several aerobic and anaerobic Cr(VI) reducers can use organic contaminants as electron donors for Cr(VI) reduction, which may be relevant to *in situ* treatment of mixed wastes (LOVLEY and COATES, 1997). Sulfur- and sulfate-reducing bacteria are geochemically important in reductive precipitation of toxic metals, e.g., U(VI) and Cr(VI), a process mediated by multiheme cytochrome *c* proteins. The gene encoding cytochrome c_7 from *Desulfuromonas acetooxidans* has been cloned and expressed in *Desulfovibrio desulfuricans*, the recombinant organism exhibiting enhanced expression of metal reductase activities. Such overproduction of active cytochrome c_7 could be important in fixed-enzyme reactors or in organisms with enhanced metal reductase activities for bioremediation (AUBERT, et al., 1998). *D. desulfuricans* can couple the oxidation of a range of electron donors to reduction of Tc(VII), mediated by a periplasmic dehydrogenase, which is precipitated as an insoluble oxide at cell peripheries. Resting cells, immobilized in a flowthrough membrane bioreactor accumulated substantial quantities of Tc when supplied with formate as electron donor, suggesting some potential of this organism for treatment of Tc-contaminated wastewater (LLOYD et al., 1999a). In fact, compared to an *E. coli* mutant which overexpressed the reductase, the *D. desulfuricans* enzyme exhibited a higher specific activity, reduced anion sensitivity, a better temperature optimum, and a lower K_s for formate, the electron donor (LLOYD et al., 1999b). *D. desulfuricans* can also reduce Pd(II) to cell-bound Pd(0) with hydrogen-dependent reduction being O_2-insensitive thus providing a means of Pd recovery under oxic conditions (LLOYD and MACASKIE, 1998). Another *Desulfovibrio* strain was capable of arsenate reduction by an arsenate reductase (MACY et al., 2000). Mention should be made of those sulfate-reducing bacteria like *Desulfotomaculum reducens* that share physiological properties with both sulfate- and metal-reducing groups of bacteria, and can grow with Cr(VI), Mn(IV), Fe(III), and U(IV) as sole electron acceptors (TEBO and OBRAZTSOVA, 1998). Perhaps the most promising potential application of dissimilatory biological metal reduction is uranium precipitation which is performed by a number of organisms and may have potential both in waste treatment and in concentrating uranium from low-grade sources. While U(VI) compounds are readily soluble, U(IV) compounds such as the hydroxide or carbonate have low solubility and readily form precipitates at neutral pH (SPEAR et al., 1999, 2000). A strain of *Shewanella (Alteromonas) putrefaciens* which reduced Fe(III) and Mn(IV) also reduced U(VI) to U(IV) forming a black precipitate of U(IV) carbonate (GORBY and LOVLEY, 1992; LOVLEY et al., 1993; FREDRICKSON et al., 2000). U(VI) was also reduced by the sulfate-reducing bacterium *Desulfovibrio desulfuricans* in the presence of sulfate, utilizing the electron transport chain and producing a very pure precipitate of U(IV) carbonate (LOVLEY and PHILLIPS, 1992a, b). It was also reported that *Desulfovibrio vulgaris* carried out a similar enzymic reduction of uranium(VI) (LOVLEY et al., 1993). Bacterial uranium reduction has also been combined with chemical extraction to produce a potential process for soil bioremediation (PHILLIPS et al., 1995). Note that U(VI)-reducing bacteria are ubiquitous in nature (ABDELOUAS et al., 2000). A bioremediation system has been suggested based on microbial mats with constituent microbial groups of the mat consortium immobilized in silica gel and used for removal and subsequent reduction of U(VI) to U(IV) (BENDER et al., 2000). Application of a geochemical barrier containing sulfate-reducing bacteria has also been advocated for stabilization of uranium in a low level radioactive waste disposal cell (BLOUNT, 1998).

2.2.2 Reduction of Metalloid Oxyanions

Se(VI) reduction to elemental insoluble Se(0) has been employed to remediate contaminated waters and soils. Some bacteria can use such reduction to support growth making this a natural process for *in situ* applications. Though reduction of oxyanions of As and Se can occur by different mechanisms, the most environmentally significant process is dissimilatory reduction. Oxyanions of arsenic and selenium can be used in microbial anaerobic respiration as terminal electron acceptors providing energy for growth and metabolism (MACY et al., 1989, 1996; DEMOLLDECKER and MACY, 1993; OREMLAND et al., 1989, 1999). Their reduction can be coupled to organic substrates, e.g., lactate, acetate and aromatics, with the bacteria found in a range of habitats and not confined to any specific genus. These organisms, and perhaps even the enzymes themselves, may have applications for bioremediation of selenium- and arsenic-contaminated environments (STOLZ and OREMLAND, 1999). *Sulfurospirillum barnesii* can simultaneously reduce both NO_3^- and Se(VI). Kinetic experiments with cell membranes of *S. barnesii* suggest the presence of constitutive selenate and nitrate reduction as well as an inducible, high affinity nitrate reductase in nitrate-grown cells which also has a low affinity for selenate. Simultaneous reduction of micromolar Se(VI) in the presence of millimolar nitrate suggests a role for these organisms in bioremediating nitrate-rich seleniferous wastewaters (OREMLAND et al., 1999). It has been shown that selenate does not sorb to soil components during transport in saturated soil columns, but rapidly reduces to forms that are strongly retarded. Such findings suggest that Se will be retained near the soil surface even under extreme leaching conditions (GUO et al., 1999a, b). It should be noted that the reoxidation of Se(0) is now known to be under microbial control and where alteration in physicochemical conditions leads to, e.g., drying out of Se(0)-contaminated sediments or aerobiosis of subsurface soils, Se(0) may be reoxidized to form more toxic selenite and selenate (DOWDLE and OREMLAND, 1998).

The incidental ability of a variety of microorganisms from all major groups to reduce Se(VI) and Te(VI) by additional often uncharacterized mechanisms offers additional scope for bioreactor-based approaches (GHARIEB et al, 1999; KASHIWA et al., 2000; KAPLAN et al., 2000). Such reduction of selenate [Se(VI)] and selenite [Se(IV)] to elemental selenium can be catalyzed by numerous microbial species and often results in a red precipitate deposited around cells and colonies (LOVLEY, 1993; GHARIEB et al., 1995). Reduction of TeO_3^{2-} to Te^0 is also an apparent means of detoxification found in bacteria (WALTER and TAYLOR, 1992), the Te^0 being deposited in or around cells, particularly near the cytoplasmic membrane (TAYLOR et al., 1988; MOORE and KAPLAN, 1992; BLAKE et al., 1993; LLOYD-JONES et al., 1994). In contrast to bacteria, fungal reduction of metalloids has received less attention although numerous filamentous and unicellular fungal species are capable of SeO_3^{2-} reduction to Se^0, deposited intra- and extracellularly, resulting in a red coloration of colonies (ZIEVE et al., 1985; GHARIEB et al., 1995; MORLEY et al., 1996). Fungal reduction of TeO_3^{2-} to Te^0 results in black or gray colonies (SMITH, 1974). Transmission electron microscopy reveals the presence of large black granules, apparently in vacuoles, which corresponds to the reduction of tellurite to amorphous elemental tellurium (GHARIEB et al., 1999).

2.2.3 Metal Precipitation by Sulfate-Reducing Bacteria

The sulfate-reducing bacteria (SRB) are strictly anaerobic heterotrophic bacteria commonly found in environments where oxygen is excluded and where carbon substrates and sulfate are available (CASTRO et al., 2000a). SRB are largely mesophilic with regard to their temperature requirements (POSTGATE, 1984; BARNES et al., 1991; WHITE and GADD, 1996a, b) although thermophilic strains have been recovered from, e.g., hydrothermal vents (PRIEUR et al., 1995). Sulfate-reducing bacteria are almost entirely neutrophilic with maximum growth obtained in the range pH 6–8 (POSTGATE, 1984). However, some isolates can grow in moderately acid conditions such as mine and surface waters where the bulk phase pH is in the range 3–4. In these environments the sulfate-reducing bacteria are found in sediments and their apparent acid tolerance is derived from the existence of more neutral microenvironments which are maintained by the buffering effect resulting from the low dissociation of H_2S (HEDIN and NAIRN, 1991; WHITE and GADD, 1996a). Sulfate-reducing bacteria utilize an energy metabolism in which the oxidation of organic compounds or hydrogen is coupled to the reduction of sulfate as the terminal electron acceptor, producing sulfide. Bacterial sulfate reduction results in the formation of sulfide which may reach significant concentrations in sediments or chemostat cultures (WHITE and GADD, 1996a, b). Although low concentrations (e.g., 2–5 mM) of sulfide benefit SRB growth by ensuring a low E_h, high concentrations of sulfide are inhibitory (POSTGATE, 1984; MCCARTNEY and OLESZKIEVICZ, 1991). A sulfide concentration of 16.1 mM was toxic to an SRB culture derived from an anaerobic treatment plant (REIS et al., 1992). However, such sulfide concentrations are not generally encountered due to precipitation of sulfide with metals. With the exception of the alkali and alkaline-earth metals, metal sulfides are essentially insoluble and the resultant precipitation of sulfides has been observed to protect SRB against metal toxicity (LAWRENCE and MCCARTY, 1965; POSTGATE, 1984): and metals similarly protect the organisms against sulfide toxicity.

The solubility products of most heavy metal sulfides are very low, in the range of $4.65 \cdot 10^{-14}$ (Mn) to $6.44 \cdot 10^{-53}$ (Hg) (CHANG, 1993) so that even a moderate output of sulfide can remove metals to levels permitted in the environment (CRATHORNE and DOBBS, 1990; TAYLOR and MCLEAN, 1992) with metal removal being directly related to sulfide production (WHITE and GADD, 1996a). Sulfate-reducing bacteria can also create extremely reducing conditions which can chemically reduce metals such as uranium(VI) (SPEAR et al., 2000). In addition, sulfate reduction partially eliminates acidity from the system as a result of the shift in equilibrium when sulfate (dissociated) is converted to sulfide (largely protonated) (WHITE and GADD, 1996a). This can result in the further precipitation of metals such as copper or aluminium as hydroxides as well as increasing the efficiency of sulfide precipitation. The sulfide produced from sulfate reduction plays a major role in metal sulfide immobilization in sediments but has also been applied to bioremediation of metals in waters and leachates.

Regarding other organisms, the thiosulfate reductase gene from *Salmonella typhimurium* has been expressed in *E. coli*. This resulted in sulfide production from inorganic thiosulfate which precipitated metals as metal–sulfide complexes (BANG et al., 2000). A Cd-resistant *Klebsiella planticola* also precipitated significant amounts of cadmium sulfide when grown in thiosulfate-containing medium (SHARMA et al., 2000). As an alternative to anaerobic sulfate reduction, a novel aerobic sulfate reduction pathway has been engineered for sulfide production. The assimilatory SO_4^{2-}-reduction pathway was redirected to cysteine production which was converted to S^{2-} by cysteine desulfydrase, leading to CdS precipitation on bacterial cell surfaces (WANG et al., 2000).

2.2.4 Processes Utilizing Metal Sulfide Precipitation

Acid mine drainage occurs through the activities of sulfur- and iron-oxidizing bacteria and, due to the quantities of sulfate available, sulfate reduction is an important process controlling the efflux of metals and acidity in mine effluents (FORTIN et al., 1995; LEDIN and PEDERSEN, 1996; SCHIPPERS et al., 1996). Sulfate-reduction could, therefore, provide both *in situ* (UHRIE et al., 1996) and *ex situ* metal removal from such waters (HAMMACK and EDENBORN, 1992; LYEW et al., 1994; CHRISTENSEN et al., 1996) and contribute to the removal of metals and acidity in artificial and natural wetlands (HEDIN and NAIRN, 1991; PERRY, 1995) although other mechanisms such as biosorption may be involved (VILE andWIEDER, 1993; KARATHANASIS and THOMPSON, 1995).

Large-scale bioreactors have also been developed using bacterial sulfate reduction for treating metal-contaminated waters. A pilot-scale study used either $3 \cdot 200$ L or single

4,500 L fixed-bed vessels filled with spent mushroom compost at residence times between 9 and 17 d. Both systems raised the pH from 3.0–3.5 to 6–7 and removed almost all of the metals (Al, Zn, Cu, Ni) in the inflow and approximately 20% of the sulfate present was also reduced (DVORAK et al., 1992). The most extensive use to date of sulfate-reducing bacteria is in the treatment of contaminated groundwater at the Budelco zinc-smelting works at Budel-Dorplein in The Netherlands. The pilot plant comprised a purpose-designed 9 m^3 stainless steel sludge-blanket reactor using sulfate-reducing bacteria and was developed by Shell Research Ltd. and Budelco B.V. This plant successfully removed toxic metals (primarily Zn) and sulfate from contaminated groundwater at the long-standing smelter site by precipitation as metal sulfides. The reactor used an undefined consortium of sulfate-reducing bacteria with ethanol as the growth substrate. It was capable of tolerating a wide range of inflow pH and operating temperatures, and yielded outflow metal concentrations below the ppb range. Methanogenic bacteria in the consortium also removed the acetate produced by sulfate-reducing bacteria, leaving an effluent with an acceptably low BOD (BARNES et al., 1991, 1994). The process has since been expanded to commercial scale using a 1,800 m^3 bioreactor built by Paques B.V. (Netherlands) and has been in operation since 1992. In one integrated microbial process for the bioremediation of soil contaminated with toxic metals, sulfur-oxidizing bacteria were used to leach metals from contaminated soils by breakdown of minerals and liberation of acid-labile sorbed or chemical forms (Fig. 1) (WHITE et al., 1998). A large proportion of the acidity and almost the entirety of released metals were removed using bacterial sulfate reduction in an internal sedimentation reactor. This contained a mixed, undefined culture of sulfate-reducing bacteria produced by combining a number of metal-tolerant enrichment cultures from different environmental origins (WHITE and GADD, 1996a). The combination of soil bioleaching followed by separate bioprecipitation of leached metals by SRB, proved to be effective in removing and concentrating a range of metals including Zn, Cu, and Cd from metal-contaminated soil. The resultant solids, once separated, represented an approximate 200-fold reduction in volume compared to the soil while the low toxic metal content of the clarified liquor produced would permit discharge or allow recycling of the liquor to the bioleaching stage and permit water conservation during process operation. Where soil and water conditions are appropriate, the soil-leaching component of such an integrated process could possibly be carried out *in situ* although conditions allowing this would be prone to uncertainty. In other cases, a range of *ex situ* options, including heap leaching or slurry bioreactors, could be used to solubilize metals and feed the bioprecipitation reactor (WHITE et al., 1998; WHITE and GADD, 1997).

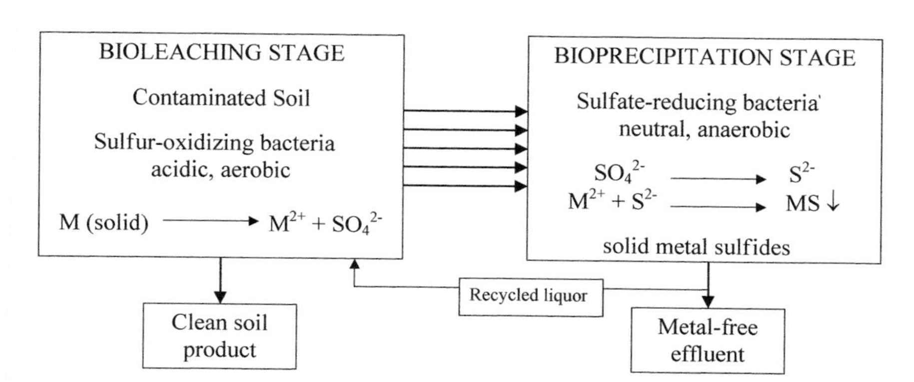

Fig. 1. Simple diagram of integrated microbial process for the bioremediation of soil contaminated with toxic metals. Bioleaching by sulfur-oxidizing bacteria is followed by precipitation of leachate metals as insoluble sulfides by sulfate-reducing bacteria (see WHITE et al., 1998).

Sulfate-reducing bacterial biofilm reactors may offer a means of process intensification and entrap or precipitate metals, e.g., Cu and Cd, at the biofilm surface (WHITE and GADD, 1998a, b, 2000). Mixed SRB cultures were more effective than pure cultures for metal removal, which was enhanced by the production of exopolymers (WHITE and GADD, 1998a, 2000; CHEN et al., 2000; JALALI and BALDWIN, 2000). Anaerobic solid substrate bioreactors have also been used in laboratory treatment of acid mine drainage with sulfate-reducing activity being an important component of the overall removal process which exceeded 99% for Cu and Cd (DRURY, 1999). Note also that a complex array of microbial, plant and soil interactions, including sulfate reduction, contribute to metal immobilization and/or removal from contaminated wetlands (WEBB et al., 1998; O'SULLIVAN et al, 1999), while SRB are also capable of mercury methylation (KING et al., 2000) and general biosorptive interactions (HARD et al., 1999).

2.2.5 Metal Precipitation as Phosphates

In this process, metal accumulation by bacterial biomass is mediated by a phosphatase, induced during metal-free growth, which liberates inorganic phosphate from an organic phosphate donor molecule, e.g., glycerol 2-phosphate. Metals/radionuclides are then precipitated as phosphates on the biomass (MACASKIE and DEAN, 1989; MACASKIE, 1991; MACASKIE et al., 1987a, b, 2000). Most work has been carried out with a *Citrobacter* sp. and a range of bioreactors, including those using immobilized biofilms, have been described (MACASKIE et al., 1994; TOLLEY et al., 1995). Microbially enhanced chemisorption where biologically produced phosphate (Pi) precipitates metals as phosphates is effective for a range of metals as well as radionuclides with both accumulative and chemisorptive mechanisms contributing to the overall process. Zirconium was mineralized by a *Citrobacter* sp. to a mixture of $Zr(HPO_4)_2$ and hydrated zirconia (ZrO_2): biomineralization of uranium as HUO_2PO_4 was repressed by zirconium in the presence of excess Pi. Cell-bound HUO_2PO_4 facilitated Ni^{2+} removal by intercalative ion exchange into the polycrystalline lattice and also promoted Zr removal (BASNAKOVA and MACASKIE, 1999). Similar phenomena were enhanced in biofilm-immobilized cells in an air-lift bioreactor: the uranyl phosphate accumulated by the bacterial biomass has potential as a bio-inorganic ion exchanger (FINLAY et al., 1999). Exocellular lipopolysaccharide (LPS) and the associated phosphate play an important role in nucleation and development of high loads of LPS-bound uranyl phosphate (BONTHRONE et al., 2000; MACASKIE et al., 2000). The *Citrobacter* acid phosphatase exists in two isoforms (CPI and CPII) with differing stabilities which may relate to their suitability for metal phosphate precipitation (JEONG and MACASKIE, 1999). An *Acinetobacter* sp. was isolated from a wastewater treatment plant operating a biological phosphate removal process. Polyphosphate was degraded under anaerobic conditions in the presence of Cd or UO_2^{2+}, the latter stimulating the reaction which produced free Pi. Cd was associated with phosphate-containing intracellular granules (BOSWELL et al., 1999). In *E. coli*, manipulation of polyphosphate metabolism by overexpression of the genes for polyphosphate kinase (*ppk*) and polyphosphatase (*ppx*) can result in lower intracellular polyphosphate, phosphate secretion, and increased metal tolerance (KEASLING et al., 1998).

2.2.6 High Gradient Magnetic Separation (HGMS)

This is a technique for metal removal from solution using bacteria rendered susceptible to magnetic fields. "Non-magnetic" bacteria can be made magnetic by precipitation of metal phosphates (aerobic) or sulfides (anaerobic) on their surfaces. For those organisms producing iron sulfide, this compound is not only magnetic but also an effective adsorbent for metallic elements (WATSON and ELLWOOD, 1994; WATSON et al., 1995, 2000). Solutions treated by HGMS can have very low residual metal levels remaining in solution (WATSON and ELLWOOD, 1994; WATSON et al., 1996).

3 Metal Mobilization

3.1 Autotrophic Leaching

Most autotrophic leaching is carried out by chemolithotrophic, acidophilic bacteria which fix carbon dioxide and obtain energy from the oxidation of ferrous iron or reduced sulfur compounds which results in the solubilization of metals because of the production of Fe(III) and H_2SO_4. The microorganisms involved include sulfur-oxidizing bacteria, e.g *Thiobacillus thiooxidans*, iron- and sulfur-oxidizing bacteria, e.g *Thiobacillus ferrooxidans* and iron-oxidizing bacteria, e.g *Leptospirillum ferrooxidans* (EWART and HUGHES, 1991; BOSECKER, 1997). Both the *Thiobacillus* species are able to oxidize inorganic sulfur and metal sulfides such as pyrite (BALDI et al., 1992). *T. ferrooxidans* and *L. ferrooxidans* are able to oxidize soluble ferrous iron producing ferric iron which can then indirectly solubilize metal sulfides (LUTHER, 1987; BOSECKER, 1997).

While most interest in leaching of mineral ores by acidophilic sulfur-oxidizing bacteria arises from a hydrometallurgical perspective, leaching of contaminating metals from soils and other matrices is also possible (WHITE et al., 1998). In fact, bioleaching using elemental sulfur as substrate can be better than sulfuric acid treatment for metal solubilization from contaminated aquatic sediments (SEIDEL et al., 1998). Iron- and sulfur-oxidizing bacteria, *Thiobacillus ferrooxidans* and *Thiobacillus thiooxidans*, were enriched from contaminated soil and were able to leach >50% of the metals present (As, Cd, Co, Cu, Ni, V, Zn, B, Be). Strains of *T. ferrooxidans* were able to remove all of the Cd, Co, Cu, and Ni (GOMEZ and BOSECKER, 1999). *T. ferrooxidans* has also been used to treat air pollution control residues (APCR: fly ash and used lime). Although viability of the thiobacilli was poor, removal of as much as 95% of Cd (from APCR of ~270 mg Cd kg^{-1}) was achieved with ~69% Pb removal (from APCR of ~5 g Pb kg^{-1}) and similar removals for Zn and Cu (MERCIER et al., 1999). Autotrophic production of sulfuric acid by *Thiobacillus* species has been used to solubilize metals from sewage sludge, thus enabling separation from the sludge which can then be used as a fertilizer (SREEKRISHNAN and TYAGI, 1994). Autotrophic leaching with *T. ferrooxidans* required acidification of the sludge to approximately pH 4.0 before adequate growth of the bacteria could occur although sulfur-oxidizing *Thiobacillus* sp. have been isolated which were able to grow at pH 7.0 making preliminary acidification unnecessary (SREEKRISHNAN and TYAGI, 1994). The elemental sulfur which acts as an energy source for the process can be of either chemical or biological origin (TICHY et al., 1994) and it has been shown that solid lumps of sulfur are as effective as powder. The use of sulfur as rods or similar may allow unused sulfur to be removed and, therefore, prevent wastage of sulfur and acidification of the sludge after disposal (RAVISHANKER et al., 1994). Simultaneous sewage sludge digestion and metal leaching under acidic conditions (pH 2.0–2.5) has an advantage over conventional aerobic and anaerobic digestion in that the acidity leads to a decrease in pathogenic microorganisms as well as the digestion of the sludge and removal of toxic metals (BLAIS et al., 1997).

Autotrophic leaching has been used to remediate other metal-contaminated solid materials including soil (ZAGURY et al., 1994). In a two-stage process for the bioremediation of metal contaminated soil, a mixture of sulfur-oxidizing bacteria, growing on elemental sulfur, was used to acidify the soil and solubilize toxic metals before treatment of the metal-contaminated liquid effluent using sulfate-reducing bacteria (WHITE et al., 1997, 1998). Laboratory studies have also been conducted into the bioremediation of red mud, the main waste product of Al extraction from bauxite, by autotrophic leaching. Various leaching methods, both chemical and biological, were used to extract aluminum from the red mud. Of the biological processes, autotrophic leaching by adapted indigenous *Thiobacillus* spp. was found to be more efficient than heterotrophic leaching by a range of fungal strains. Although this process was developed to increase the amount of aluminium obtained from bauxite, it also decreased the toxicity of the waste product (VACHON et al., 1994). Note also that the metals in leachates can be subsequently removed using bacterial sulfate reduction where the soluble metals are precipitated as insoluble sulfides (WHITE et al., 1998).

3.2 Heterotrophic Leaching

Chemoautotrophic acidophilic bacteria such as *Thiobacillus* species do not need a supplied carbon source and have a high acidification capacity. Hence, in an industrial context, most biological leaching (*autotrophic leaching*) has been carried out using such microorganisms (RAWLINGS and SILVER, 1995). However, their acidophilic nature may make them unable to tolerate the higher pH values of many industrial wastes (BURGSTALLER and SCHINNER, 1993). There may be additional advantages to using fungi for leaching purposes in that they are more easily manipulated in bioreactors than thiobacilli and by altering growth conditions, can be induced to produce high concentrations of organic acids (MATTEY, 1992; BURGSTALLER and SCHINNER, 1993; STRASSER et al., 1994). Furthermore, many species can tolerate high concentrations of metals (GADD, 1993a) and can grow in both low and high pH environments (BURGSTALLER and SCHINNER, 1993). The incidence of solubilizing ability in natural soil isolates of fungi is high, and approximately one third of soil isolates screened (including isolates from lead- and nickel-contaminated soils) were able to solubilize at least one of three test metal compounds [$Co_3(PO_4)_2$, ZnO, and $Zn_3(PO_4)_2$]; 11% were able to solubilize all three (SAYER et al., 1995).

Heterotrophic leaching by fungi occurs as a result of several processes, including the efflux of protons from hyphae and the production of siderophores (in the case of iron), but in most fungal strains, leaching occurs mainly by the production of organic acids (BURGSTALLER and SCHINNER, 1993; SAYER and GADD, 1997; DIXON-HARDY et al., 1998). Solubilization of insoluble metal compounds results from protonation of the anion of the compound (HUGHES and POOLE, 1991). The production of organic acids provides both a source of protons for solubilization and a metal-chelating anion to complex the metal cation, with complexation being dependent on the relative concentrations of the anions and metals in solution, pH, and the stability constants of the various complexes (DENÊVRE et al., 1996). Citrate and oxalate anions can form stable complexes with a large number of metals (GADD, 1999). Most metal citrates are highly mobile and are not readily degraded, degradation depending on the type of complex formed rather than the toxicity of the metal cation involved. Hence, the presence of citric acid in the terrestrial environment will leach potentially toxic metals from soil (FRANCIS et al., 1992). Oxalic acid can act as a leaching agent for those metals that form soluble oxalate complexes, including Al, Fe, and Li (STRASSER et al., 1994).

Many species of fungi are able to leach metals from industrial wastes and by-products, low grade ores (BURGSTALLER and SCHINNER, 1993), and metal-bearing minerals (TZEFERIS et al., 1994; DREVER and STILLINGS, 1997; GHARIEB and GADD, 1999; CASTRO et al., 2000b; STERFLINGER, 2000). For example, laboratory scale heterotrophic leaching of Ni and Co from low grade laterite ore has been carried out using fungi (TZEFERIS et al., 1994). 55–60% Ni was leached from the ore in the presence of strains of *Aspergillus* and *Penicillium*, and 70% was leached at high temperature (95 °C) by application of metabolic products obtained from cultivation of the fungal strains at 30 °C in a glucose and sucrose medium (TZEFERIS et al., 1994). *Trichoderma harzianum* can solubilize MnO_2, Fe_2O_3, metallic zinc, and rock (mostly Ca) phosphate with both chelation and reduction involved in oxide solubilization (ALTOMARE et al., 1999). A strain of *Penicillium simplicissimum,* isolated from a metal-contaminated site, has been used successfully to leach Zn from insoluble ZnO contained in industrial filter dust. This fungus only developed the ability to produce citric acid (>100 mM) in the presence of the filter dust (SCHINNER and BURGSTALLER, 1989; FRANZ et al., 1991, 1993). Culture filtrates from *Aspergillus niger* have also been used to leach Cu, Ni, and Co from copper converter slag (SUKLA et al., 1992). Al has been leached from red mud (the waste product of the extraction of Al from bauxite) with various fungal strains and adapted thiobacilli. The thiobacilli were best, with the most efficient of the fungal strains being *P. simplicissimum*, the fungal-derived acids (mainly citric) having a much greater ability to leach Al than pure citric acid (VACHON et al., 1994). *A. niger* is able to solubilize a wide range of insoluble metal compounds, including phosphates, sulfides, and mineral ores such as cuprite (CuS) (SAYER et al., 1995, 1997; SAYER and GADD, 1997). This fungus, a prolific organic acid producer, will produce organic acids and acidify in its surrounding medium whether a metal compound is present or not (BURGSTALLER and SCHINNER, 1993; SAYER and GADD, 1997). Fungal growth media can be manipulated, e.g., by changing the N and P balance or pH, to maximize organic acid production. It is known, e.g., that a deficiency of manganese (less than 10^{-8} M) in the growth medium leads to the production of large amounts of citric acid by *A. niger* (MEIXNER et al., 1985), and typical concentrations of citric acid produced industrially by this fungus can reach 600 mM (MATTEY, 1992). Oxalic acid production can be manipulated to yield concentrations of up to 200 mM on low cost carbon sources with the optimum pH for oxalic acid production being around neutrality (STRASSER et al., 1994). A heterotrophic mixed culture has been employed for leaching manganiferous minerals. This culture was capable of reducing MnO_2, with the process being of potential for the treatment of materials not treatable by conventional processes (VEGLIO, 1996). Cd, Zn, Cu, Pb, and Al have been leached from municipal waste fly ash using *A. niger*. A two-stage process was used, with *A. niger* being cultured and the citric acid from the culture used as a leaching agent. The environmental quality of the fly ash residue was deemed suitable for use in the construction industry (BOSSHARD et al., 1996). Effective leaching of a variety of other wastes has been demonstrated, e.g., electronic waste materials (BRANDL et al., 1999; GADD, 1999). For this, a two-stage process was proposed where initial growth of solubilizing organisms was followed by separate contact of electronic scrap with metabolite-containing medium (BRANDL et al., 1999). Another method for treatment of metal-contaminated sandy soil relies on siderophore-mediated metal solubilization by *Alcaligenes eutrophus*. Solubilized metals are adsorbed to the biomass and/or precipitated, with biomass separated from a soil slurry by a flocculation process. This process resulted in a complete decrease in the bioavailability of Cd, Zn, and Pb (DIELS et al., 1999).

Heterotrophic solubilization can have consequences for other remedial treatments for contaminated soils. Pyromorphite $[Pb_5(PO_4)_3Cl]$ is a stable lead mineral and can form in urban and industrially contaminated soils. Such insolubility reduces lead bioavailability and the formation of pyromorphite has been suggested as a remediation technique for lead-contaminated land, if necessary by means of phosphate addition. However, pyromorphite can be solubilized by phosphate-solubilizing fungi, e.g., *A. niger*, and plants grown with pyromorphite as a sole phosphorus source accumulated both P and Pb (SAYER et al., 1999). Further, during the fungal transformation of pyromorphite, the biogenic production of lead oxalate dihydrate was observed for the first time (SAYER et al., 1999). This study emphasizes the importance of considering microbial processes in developing remediation techniques for metal-contaminated soils.

Related to heterotrophic solubilization is fungal translocation of, e.g., Cs, Zn, and Cd, which can lead to concentration of metals and radionuclides in specific regions of the mycelium and/or in fruiting bodies. Whether the concentration factors observed *in vitro* can be reproduced in the field and whether such amounts can contribute to soil bioremediation remains a topic for further investigation (GRAY, 1998). Some parallels may be drawn with the phytoremediatory approach to soil contamination. Other possibilities relate to the use of copper-tolerant wood decay fungi to degrade copper-treated wood products although results obtained so far have shown only small effects on the copper content of wood before or after decay (DE GROOT and WOODWARD, 1998). Metal-tolerant bacteria have been isolated which are capable of removing copper, chromium, and arsenic from chromated copper arsenate-treated waste wood (CLAUSEN, 2000).

3.3 Reductive Mobilization

Fe(III) and Mn(IV) oxides absorb metals strongly and this may hinder metal extraction from contaminated soils. Microbial reduction of Fe(III) and Mn(IV) may be one way for releasing such metals and this process may be enhanced with the addition of humic materials, or related compounds. Such compounds may also act as electron shuttles for, e.g., U(VI) and Cr(VI), converting them to less soluble forms, especially if located in tight pore spaces where microorganisms cannot enter (LOVLEY and COATES, 1997). Bacterial Fe(III) reduction resulted in release of, e.g., Mn and Co, from goethite where 5% of the iron was substituted by these metals (BOUSSERRHINE et al., 1999). Iron-reducing bacterial strains solubilized 40% of the Pu present in contaminated soils within 6–7 d (RUSIN et al., 1993) and both iron- and sulfate-reducing bacteria were able to solubilize Ra from uranium mine tailings, with solubilization occurring largely by disruption of reducible host minerals (LANDA and GRAY, 1995).

Key microbial transformations of inorganic Hg^{2+} include reduction and methylation (see below). The mechanism of bacterial Hg^{2+} resistance is enzymic reduction of Hg^{2+} to non-toxic volatile Hg^0 by mercuric reductase. Hg^{2+} may also arise from the action of organomercurial lyase on organomercurials (SILVER, 1996). Since Hg^0 is volatile, this could provide one means of mercury removal (BARKAY et al., 1991; BARKAY and TURNER, 1992; BRUNKE et al., 1993). Mercuric reductase, from a recombinant *E. coli* strain, has been immobilized on a chemically modified diatomaceous earth support with immobilization enhancing stability and reusability: maximal activity was 1.2 nmol Hg mg^{-1} protein s^{-1} at an initial $[Hg^{2+}]$ of 50 μmol dm^{-3} (CHANG et al., 1999). A strain of *Pseudomonas putida* was able to remove over 90% of Hg (40 mg L^{-1}) in 24 h (OKINO et al., 2000). Recent work has used genetically engineered strains of *E. coli* that express *mer* and glutathione S-transferase genes that enable growth in high concentrations of $HgCl_2$ (30 μg mL^{-1}) and volatilization of part of the supplied mercury (~250 μg g dry $wt.^{-1}$) (CURSINO et al., 2000).

3.4 Methylation and Dealkylation of Metalloids and Organometals

Microbial methylation of metalloids to yield volatile derivatives, e.g., dimethylselenide or trimethylarsine, can be effected by a variety of bacteria, algae, and fungi (KARLSON and FRANKENBERGER, 1993; GADD, 1993a). Selenium methylation appears to involve transfer of methyl groups as carbonium (CH_3^+) ions via the S-adenosyl methionine system. Less work has been carried out on tellurium methylation by fungi although there is evidence of dimethyltelluride and dimethylditelluride production (KARLSON and FRANKENBERGER, 1993). Different mechanisms of tellurium transformations are species-dependent and can be influenced by physicochemical conditions which can affect tellurium speciation into soluble and insoluble forms, and bioaccumulation. Extremely small amounts of Te were volatilized by a *Fusarium* sp., perhaps precluding a role in detoxification (GHARIEB et al., 1999). Several bacterial and fungal species have been shown to methylate arsenic compounds such as arsenate [As(V), AsO_4^{3-}], arsenite [As(III), AsO_2^-], and methylarsonic acid [$CH_3H_2AsO_3$] to volatile dimethyl- [$(CH_3)_2HAs$] or trimethylarsine [$(CH_3)_3As$] (see TAMAKI and FRANKENBERGER, 1992). Volatile dimethyl selenide (DMSe) can form naturally from inorganic selenium species in soils and this can regulate geochemical cycling of Se and therefore influence selenium bioremediation. Many environmental and soil factors, e.g., organic amendments and frequent tillage, can be optimized to increase diffusive transport through soil and enhance volatilization (GUO et al., 1999a, b; STORK et al., 1999; ZHANG and FRANKENBERGER, 1999). However, in moist or flooded soil, less DMSe is lost to the atmosphere with a significant proportion of DMSe being converted to non-volatile forms of Se. Oxidized dimethylated Se (ODMSe), including dimethyl selenoxide (DMSeO) and/or dimethyl selenone ($DMSeO_2$), was the dominant form of non-volatile Se, and accounted for up to 90% of Se in flooded soils. Such transformation of DMSe can be responsible for the low Se volatilization rate in flooded soils and sediments (ZHANG et al., 1999). By determining optimum conditions which stimulate Se volatilization, it may be possible to design strategies for bioremediation of seleniferous waters (DUNGAN and FRANKENBERGER, 2000). Microorganisms enhance arsenic mobilization from soil by up to 24% compared to formaldehyde-treated controls with only low formation of dissolved methylated arsenic species (TURPEINEN et al., 1999). Other bacteria are capable of the oxidation of As(III) to As(V) (WEEGER et al., 1999). Methylation of inorganic Hg^{2+} leads to the formation of more toxic volatile derivatives: the bioremediation potential of this process has not been explored in detail as is the case for other metals and metalloids (besides selenium) capable of being methylated, e.g., As, Sn, and Pb (GADD, 1993a; MICHALKE et al., 2000).

As well as organomercurials (see above), other organometals may be microbially degraded. Organoarsenicals can be demethylated by bacteria, while organotin degradation involves sequential removal of organic groups from the tin atom which results in a reduction in toxicity (GADD, 1993b). Such mechanisms and interaction with the bioremediation possibilities described previously may provide a means of detoxification (GADD, 1993b).

3.5 Metalloid Transformations and Bioremediation

OREMLAND et al. (1990, 1991) described *in situ* removal of SeO_4^{2-}, by reduction to Se^0, by sediment bacteria in agricultural drainage regions of Nevada. Flooding of exposed sediments at Kesterson Reservoir with water (to create anoxic conditions) resulted in reduction and immobilization of large quantities of selenium that was present in the sediments (LONG et al., 1990). Microbial methylation of selenium, resulting in volatilization, has also been used for *in situ* bioremediation of selenium-containing land and water at Kesterson Reservoir, California (THOMPSON-EAGLE and FRANKENBERGER, 1992). Selenium volatilization from soil was enhanced by optimizing soil moisture, particle size and mixing while in waters it was stimulated by the growth phase, salinity, pH, and selenium concentration (FRANKENBERGER and KARLSON, 1994, 1995; RAEL and FRANKENBERGER, 1996). The selenium-contaminated agricultural drainage water was evaporated to dryness until the sediment selenium concentration approached 100 mg Se kg^{-1} dry weight. Conditions such as carbon source, moisture, temperature, and aeration were then optimized for selenium volatilization and the process continued until selenium levels in sediments declined to acceptable levels (THOMPSON-EAGLE and FRANKENBERGER, 1992). Some potential for *ex situ* treatment of selenium-contaminated waters has also been demonstrated (MACY et al., 1989; GERHARDT et al., 1991).

4 Conclusions

Microorganisms play important roles in the environmental fate of toxic metals and metalloids with physicochemical and biological mechanisms effecting transformations between soluble and insoluble phases. Such mechanisms are important components of natural biogeochemical cycles with some processes being of potential application to the treatment of contaminated materials. Although the biotechnological potential of most of these processes has only been explored at a laboratory scale, some mechanisms, notably bioleaching, biosorption, and precipitation, have been employed at a commercial scale. Of these, autotrophic leaching is an established major process in mineral extraction but has also been applied to the treatment of contaminated land. There have been several attempts to commercialize biosorption using microbial biomass but success has been limited, primarily due to competition with commercially produced ion exchange media. As a process for immobilizing metals, precipitation of metals as sulfides has achieved large-scale application, and this holds out promise of further commercial development. Exploitation of other biological processes will undoubtedly depend on a number of scientific, economic and political factors, including the availability of a market.

Acknowledgements
The author gratefully acknowledges financial support for his own work from the Biotechnology and Biological Sciences Research Council, the Natural Environment Research Council, and the Royal Societies of London and Edinburgh.

5 References

ABDELOUAS, A., LUTZE, W., GONG, W. L., NUTTALL, E. H., STRIETELMEIER, B. A. et al. (2000), Biological reduction of uranium in groundwater and subsurface soil, *Sci. Total Environ.* **250**, 21–35.

ADJIMANI, J. P., EMERY, T. (1987), Iron uptake in *Mycelia sterilia* EP-76, *J. Bacteriol.* **169**, 3664–3668.

AKSU, Z., AKPINAR, D., KABASAKAL, E., KOSE, B. (1999), Simultaneous biosorption of phenol and nickel (II) from binary mixtures onto dried aerobic activated sludge, *Process Biochem.* **35**, 301–308.

ALOYSIUS, R., KARIM, M. I. A., ARIFF, A. B. (1999), The mechanisms of cadmium removal from aqueous solution by nonmetabolizing free and immobilized live biomass of *Rhizopus oligosporus*, *World J. Microbiol. Biotechnol.* **15**, 571–578.

AL SARAJ, M., ABDELLATIF, M. S., EL NAHAL, I., BARAKA, R. (1999), Bioaccumulation of some hazardous metals by sol–gel entrapped microorganisms, *J. Non-Crystalline Solids* **248**, 137–140.

ALTOMARE, C., NORVELL, W. A., BJÖRKMAN, T., HARMAN, G. E. (1999), Solubilization of phosphates and micronutrients by the plant-growth-promoting and biocontrol fungus *Trichoderma harzianum* Rifai 1295-22, *Appl. Environ. Microbiol.* **65**, 2926–2933.

ANDRES, Y., THOUAND, G., BOUALAM, M., MERGEAY, M. (2000), Factors influencing the biosorption of gadolinium by microorganisms and its mobilisation from sand, *Appl. Microbiol. Biotechnol.* **54**, 262–267.

ASHKENAZY, R., YANNAI, S., RAHMAN, R., RABINOVITZ, E., GOTTLIEB, L. (1999), Fixation of spent *Saccharomyces cerevisiae* biomass for lead sorption, *Appl. Microbiol. Biotechnol.* **52**, 608–611.

AUBERT, C., LOJOU, E., BIANCO, P., ROUSSET, M., DURAND, M.-C. et al. (1998), The *Desulfuromonas acetoxidans* triheme cytochrome c_7 produced in *Desulfovibrio desulfuricans* retains its metal reductase activity, *Appl. Environ. Microbiol.* **64**, 1308–1312.

AVERY, S. V., CODD, G. A., GADD, G. M. (1991), Caesium accumulation and interactions with other monovalent cations in the cyanobacterium *Synechocystis* PCC 6803, *J. Gen. Microbiol.* **137**, 405–413.

AVERY, S. V., CODD, G. A., GADD, G. M. (1992a), Caesium transport in the cyanobacterium *Anabaena variabilis*: kinetics and evidence for uptake via ammonium transport systems, *FEMS Microbiol. Lett.* **95**, 253–258.

AVERY, S. V., CODD, G. A., GADD, G. M. (1992b), Interactions of caesium with cyanobacteria and microalgae, in: *Impact of Heavy Metals on the Environment*, (VERNET, J-P., Ed.), pp. 133–182. Amsterdam: Elsevier Science Publishers B. V.

AVERY, S. V., CODD, G. A., GADD, G. M. (1993a), Biosorption of tributyltin and other organotin compounds by cyanobacteria and microalgae, *Appl. Microbiol. Biotechnol.* **39**, 812–817.

AVERY, S. V., CODD, G. A., GADD, G. M. (1993b), Salt stimulation of caesium accumulation by *Chlorella salina*: potential relevance to the development of a biological Cs-removal process, *J. Gen. Microbiol.* **139**, 2239–2244.

AVERY, S. V., CODD, G. A., GADD, G. M. (1993c), Transport kinetics, cation inhibition and intracellular location of accumulated caesium in the green microalga *Chlorella salina*, *J. Gen. Microbiol.* **139**, 827–834.

AVERY, S. V., CODD, G. A., GADD, G. M. (1998), Microalgal removal of organic and inorganic metal species from aqueous solution, in: *Wastewater Treatment with Algae* (WONG, Y.-S., TAM, N. F. Y., Eds), pp. 55–72. Berlin, Heidelberg, and Landes Bioscience Georgetown, TX: Springer-Verlag.

BADAR, U., AHMED, N., BESWICK, A. J., PATTANAPIPITPAISAL, P., MACASKIE, L. E. (2000), Reduction of chromate by microorganisms isolated from metal contaminated sites of Karachi, Pakistan, *Biotechnol. Lett.* **22**, 829–836.

BALDI, F., CLARK, T., POLLACK, S. S., OLSON, G. J., (1992), Leaching of pyrites of various reactivities by *Thiobacillus ferrooxidans*, *Appl. Environ. Microbiol.* **58**, 1853–1856.

BALDRY, M. G. C., DEAN, A. C. R. (1981), Environmental change and copper uptake by *Bacillus subtilis* subsp *niger* and by *Pseudomonas fluorescens*, *Biotechnol. Lett.* **3**, 142.

BANG, S. S., PAZIRANDEH, M. (1999), Physical properties and heavy metal uptake of encapsulated *Escherichia coli* expressing a metal binding gene (NCP), *J. Microencapsul.* **16**, 489–499.

BANG, S. W., CLARKE, D. S., KEASLING, J. D. (2000), Cadmium, lead and zinc removal by expression of the thiosulfate reductase gene from *Salmonella typhimurium* in *Escherichia coli*, *Biotechnol. Lett.* **22**, 1331–1335.

BARKAY, T., TURNER, R. (1992), Biological removal of Hg(II) from a contaminated fresh-water pond, *Abstr. Pap. Am. Chem. Soc.* **203**, 166.

BARKAY, T., TURNER, R. R., VANDENBROOK, A., LIEBERT, C. (1991), The relationship of Hg(II) volatization from a fresh-water pond to the abundance of MER-genes in the gene pool of indigenous microbial communities, *Microbial. Ecol.* **21**, 151–161.

BARNES, L. J., JANSSEN, F. J., SHERREN, J., VERSTEEGH, J. H., KOCH, R. O. et al. (1991), A new process for the microbial removal of sulphate and heavy metals from contaminated waters extracted by a geohydrological control system, *Trans. Inst. Chem. Eng.* **69**, 184–186.

BARNES, L. J., SCHEEREN, P. J. M., BUISMAN, C. J. N. (1994), Microbial removal of heavy metals and sulphate from contaminated groundwaters, in: *Emerging Technology for Bioremediation of Metals* (MEANS J. L., HINCHEE R. E., Eds.), pp. 38–49, Boca Raton, FL: Lewis Publishers.

BASNAKOVA, G., MACASKIE, L. E. (1999), Accumulation of zirconium and nickel by *Citrobacter* sp., *J. Chem. Technol. Biotechnol.* **74**, 509–514.

BEECH, I. B., CHEUNG, C. W. S. (1995), Interactions of exopolymers produced by sulphate-reducing bacteria with metal ions, *Int. Biodet. Biodeg.* **35**, 59–72.

BELLIVEAU, B. H., STARODUB, M. E., COTTER, C., TREVORS, J. T. (1987), Metal resistance and accumulation in bacteria, *Biotechnol. Adv.* **5**, 101.

BENDER, J., DUFF, M. C., PHILLIPS, P., HILL, M. (2000), Bioremediation and bioreduction of dissolved U(VI) by microbial mat consortium supported on silica gel particles, *Environ. Sci. Technol.* **34**, 3235–3241.

BEVERIDGE, T. J., DOYLE, R. J. (1989), *Metal Ions and Bacteria*. New York: John Wiley & Sons.

BEVERIDGE, T. J., MELOCHE, J. D., FYFE, W. S., MURRAY, R. G. E. (1983), Diagenesis of metals chemically complexed to bacteria – laboratory formation of metal phosphates, sulfides, and organic condensates in artificial sediments, *Appl. Environ. Microbiol.* **45**, 1094–1108.

BHANOORI, M., VENKATESWERLU, G. (2000), *In vivo* chitin–cadmium complexation in cell wall of *Neurospora crassa*, *Biochim. Biophys. Acta* **1523**, 21–28.

BIRCH, L., BACHOFEN, R. (1990), Complexing agents from microorganisms, *Experientia* **46**, 827–834.

BLAIS, J. F., MEUNIER, N., TYAGI, R. D. (1997), Simultaneous sewage sludge digestion and metal leaching at controlled pH, *Environ. Technol.* **18**, 499–508.

BLAKE, R. C., CHOATE, D. M., BARDHAN, S., REVIS, N., BARTON, L. L. et al. (1993), Chemical transformation of toxic metals by a *Pseudomonas* strain from a toxic-waste site, *Environ. Toxicol. Chem.* **12**, 1365–1376.

BLAUDEZ, D., BUTTON, B., CHALOT, M. (2000), Cadmium uptake and subcellular compartmentation in the ectomycorrhizal fungus *Paxillus involutus*, *Microbiology* **146**, 1109–1117.

BLOUNT, J. G. (1998), Physicochemical and biogeochemical stabilization of uranium in a low level radioactive waste disposal cell, *Environ. Eng. Geosci.* **4**, 491–502.

BODE, H-P., DUMSCHAT, M., GAROTTI, S., FUHRMANN. G. F. (1995), Iron sequestration by the yeast vacuole. A study with vacuolar mutants of *Saccharomyces cerevisiae*, *Eur. J. Biochem.* **228**, 337–342.

BONTHRONE, K. M., QUARMBY, J., HEWITT, C. J., ALLAN, V. J. M., PATERSON-BEEDLE, M. et al. (2000), The effect of the growth medium on the composition of metal binding behaviour of the extracellular polymeric material of a metal-accumulating *Citrobacter* sp., *Environ. Technol.* **21**, 123–134.

BORST-PAUWELS, G. W. F. H. (1981), Ion transport in yeast, *Biochim. Biophys. Acta* **650**, 149–156.

BORST-PAUWELS, G. W. F. H. (1989), Ion transport in yeast including lipophilic ions, *Methods Enzymol.* **174**, 603–616.

BOSECKER, K. (1997), Bioleaching: metal solubilization by microorganisms, *FEMS Microbiol. Rev.* **20**, 591–604.

BOSSHARD, P. P., BACHOFEN, R., BRANDL, H. (1996), Metal leaching of fly-ash from municipal waste incineration by *Aspergillus niger, Environ. Sci. Technol.* **30**, 3066–3070.

BOSWELL, C. D., DICK, R. E., MACASKIE, L. E. (1999), The effect of heavy metals and other environmental conditions on the anaerobic phosphate metabolism of *Acinetobacter johnsonii*, *Microbiology* **145**,1711–1720.

BOUSSERRHINE, N., GASSER, U. G., JEANROY, E., BERTHELIN, J. (1999), Bacterial and chemical reductive dissolution of Mn-, Co-, Cr-, and Al-substituted geothites, *Geomicrobiol. J.* **16**, 245–258.

BRADY, J. M., TOBIN, J. T., GADD, G. M. (1996), Volatilization of selenite in an aqueous medium by a *Penicillium* species, *Mycol. Res.* **100**, 955–961.

BRANDL, H., BOSSHARD, R., WEGMANN, M. (1999), Computer-munching microbes: metal leaching from electronic scrap by bacteria and fungi, *Proc. Metallurgy* **9B**, 569–576.

BRIDGE, T. A. M., WHITE, C., GADD, G. M. (1999), Extracellular metal-binding activity of the sulphate-reducing bacterium *Desulfococcus multivorans*, *Microbiology* **145**, 2987–2995.

BRIERLEY, C. L. (1990), Bioremediation of metal-contaminated surface and groundwaters, *Geomicrobiol. J.* **8**, 201–223.

BRIERLEY, J. A., BRIERLEY, C. L. (1983), Biological accumulation of some heavy metals – biotechnological applications, in: *Biomineralization and Biological Metal Accumulation*, (WESTBROEK, P., DE JONG, E. W., Eds.), pp. 499–509. Dordrecht: Reidel Publishers.

BRIERLEY, J. A., GOYAK, G. M., BRIERLEY, C. L. (1986), Considerations for commercial use of natural products for metals recovery, in: *Immobilisation of Ions by Bio-sorption* (ECCLES, H., HUNT, S., Eds.), pp. 105–117. Chichester: Ellis Horwood.

BROWN, M. J., LESTER, J. N. (1982), Role of bacterial extracellular polymers in metal uptake in pure bacterial culture and activated-sludge. 1. Effects of metal concentration, *Water Res.* **16**, 1539–1548.

BRUNKE, M., DECKWER, W. D., FRISCHMUTH, A., HORN, J. M., LUNSDORF, H. et al., (1993), Microbial retention of mercury from waste streams in a laboratory column containing merA gene bacteria, *FEMS Microbiol. Rev.* **11**, 145–152.

BURGSTALLER, W., SCHINNER, F. (1993), Leaching of metals with fungi, *J. Biotechnol.* **27**, 91–116.

BUTT, T. R., ECKER, D. J. (1987), Gene synthesis, expression, structures, and functional activities of site-specific mutants of ubiquitin, *Microbiol. Rev.* **51**, 351.

BUTTER, T. J., EVISON, L. M., HANCOCK, I. C., HOLLAND, F. S. (1998), The kinetics of metal uptake by microbial biomass: implications for the design of a biosorption reactor, *Water Sci. Technol.* **38**, 279–286.

CASTRO, H. F., WILLIAMS, N. H., OGRAM, A. (2000a), Phylogeny of sulfate-reducing bacteria, *FEMS Microbiol. Ecol.* **31**, 1–9.

CASTRO, I. M., FIETTO, J. L. R., VIEIRA, R. X., TROPIA, M. J. M., CAMPOS, L. M. M. et al. (2000b), Bioleaching of zinc and nickel from silicates using *Aspergillus niger* cultures, *Hydrometallurgy* **57**, 39–49.

CHANG, J. C. (1993), Solubility product constants, in: *CRC Handbook of Chemistry and Physics* (LIDE, D. R., ed.), pp. 8–39. Boca Raton, FL: CRC Press.

CHANG, J.-S., CHEN, C.-C. (1999), Biosorption of lead, copper and cadmium with continuous hollow-fiber microfiltration processes, *Sep. Sci. Technol.* **34**, 1607–1627.

CHANG, J. S., HWANG, Y. P., FONG, Y. M., LIN, P. J. (1999), Detoxification of mercury by immobilized mercuric reductase, *J. Chem. Technol. Biotechnol.* **74**, 965–973.

CHEN, S. L., KIM, E. K., SHULER, M. L., WILSON, D. B. (1998), Hg^{2+} removal by genetically engineered *Escherichia coli* in a hollow fiber bioreactor, *Biotechnol. Prog.* **14**, 667–671.

CHEN, W., BRÜHLMANN, F., RICHINS, R. D., MULCHANDANI, A. (1999), Engineering of improved microbes and enzymes for bioremediation, *Curr. Opin. Biotechnol.* **10**, 137–141.

CHEN, B. Y., UTGIKAR, V. P., HARMON, S. M., TABAK, H. H., BISHOP, D. F. et al. (2000), Studies on biosorption of zinc(II) and copper(II) on *Desulfovibrio desulfuricans*, *Int. Biodeter. Biodegrad.* **46**, 11–18.

CHRISTENSEN, B., LAAKE, M., LIEN, T. (1996), Treatment of acid-mine water by sulfate-reducing bacteria – results from a bench-scale experiment, *Water Res.* **30**, 1617–1624.

CIHANGIR, N., SAGLAM, N. (1999), Removal of cadmium by *Pleurotus sajor-caju* basidiomycetes, *Acta Biotechnol.* **19**, 171–177.

CLAUSEN, C. A. (2000), Isolating metal-tolerant bacteria capable of removing copper, chromium, and arsenic from treated wood, *Waste Manage. Res.* **18**, 264–268.

CLEMENS, S., KIM, E. J., NEUMANN, D., SCHROEDER, J. I. (1999), Tolerance to toxic metals by a gene family of phytochelatin synthases from plants and yeast, *EMBO J.* **18**, 3325–3333.

CORBISIER, P., VAN DER LEILIE, D., BORREMANS, B., PROVOOST, A., DE LORENZO, V. (1999), Whole cell- and protein-based biosensors for the detection of bioavailable heavy metals in environmental samples, *Analyt. Chim. Acta* **387**, 235–244.

COSTLEY, S. C., WALLIS, F. M. (2000), Effect of flow rate on heavy metal accumulation by rotating biological contactor (RBC) biofilms, *J. Ind. Microbiol. Biotechnol.* **24**, 244–250.

CRATHORNE, B., DOBBS, A. J. (1990), Chemical pollution of the aquatic environment by priority pollutants and its control, in: *Pollution: Causes, Effects and Control* (HARRISON, R. M., Ed.), pp. 1–18. Cambridge: The Royal Society of Chemistry.

CRIST, R. H., OBERHOLSER, K., SHANK, N., NGUYEN, M. (1981), Nature of bonding between metallic-ions and algal cell-walls, *Environ. Sci. Technol.* **15**, 1212–1217.

CRUZ, N., LE BORGNE, S., HERNANDEZ-CHAVEZ, G., GOSSET, G., VALLE, F. (2000), Engineering the *Escherichia coli* outer membrane protein OmpC for metal bioadsorption, *Biotechnol. Lett.* **22**, 623–629.

CURSINO, L., MATTOS, S. V. M., AZEVEDO, V., GALARZA, F., BUCKER, D. H. et al. (2000), Capacity of mercury volatilization by mer (from *Escherichia coli*) and glutathione S-transferase (from *Schistosoma mansoni*) genes cloned in *Escherichia coli*, *Sci. Total Environ.* **261**, 109–113.

DAMERON, C. T., REESE, R. N., MEHRA, R. K., KORTAN, A. R., CARROL, P. J. et al. (1989), Biosynthesis of cadmium sulfide quantum semiconductor crystallites, *Nature* **338**, 596–597.

DARNALL, D. W., GREENE, B., HENZEL, M. T., HOSEA, J. M., MCPHERSON, R. A. et al. (1986), Selective recovery of gold and other metal-ions from an algal biomass, *Environ. Sci. Technol.* **20**, 206.

DEGROOT, R. C., WOODWARD, B. (1998), *Wolfiporia cocos* – a potential agent for composting or bioprocessing Douglas-fir wood treated with copper-based preservatives, *Material und Organismen*, **32**, 195–215.

DEMOLLDECKER, H., MACY, J. M., (1993), The periplasmic nitrite reductase of *Thaueria selenatis* may catalyze the reduction of selenite to elemental selenium, *Arch. Microbiol.* **160**, 241–247.

DENÊVRE, O., GARBAYE, J., BOTTON, B. (1996), Release of complexing organic acids by rhizosphere fungi as a factor in Norway Spruce yellowing in acidic soils, *Mycol. Res.* **100**, 1367–1374.

DE ROME, L., GADD, G. M. (1987a), Copper adsorption by *Rhizopus arrhizus, Cladosporium resinae* and *Penicillium italicum*, *Appl. Microbiol. Biotechnol.* **26**, 84–90.

DE ROME, L., GADD, G. M. (1987b), Measurement of copper uptake in *Saccharomyces cerevisiae* using a Cu^{2+}-selective electrode, *FEMS Microbiol. Lett.* **43**, 283–287.

DE ROME, L., GADD, G. M. (1991), Use of pelleted and immobilized yeast and fungal biomass for heavy metal and radionuclide recovery, *J. Ind. Microbiol.* **7**, 97–104.

DIELS, L., DESMET, M., HOOYBERGHS, L., CORBISIER, P. (1999), Heavy metals bioremediation of soil, *Mol. Biotechnol.* **12**, 149–158.

DIXON-HARDY, J. E., KARAMUSHKA, V. I., GRUZINA, T. G., NIKOVSKA, G. N., SAYER, J. A. et al. (1998), Influence of the carbon, nitrogen and phosphorus source on the solubilization of insoluble metal compounds by *Aspergillus niger*, *Mycol. Res.* **102**, 1050–1054.

DOWDLE, P. R., OREMLAND, R. S. (1998), Microbial oxidation of elemental selenium in soil slurries and bacterial cultures, *Environ. Sci. Technol.* **32**, 3749–3755.

DREVER, J. I., STILLINGS, L. L. (1997), The role of organic acids in mineral weathering, *Colloids Surf.* **120**, 167–181.

DRURY, W. J. (1999), Treatment of acid mine drainage with anaerobic solid-substrate reactors, *Water Environ. Res.* **71**, 1244–1250.

DUNGAN, R. S., FRANKENBERGER, W. T. (2000), Factors affecting the volatilization of dimethylselenide by *Enterobacter cloacae* SLD1a-1, *Soil Biol. Biochem.* **32**, 1353–1358.

DUNN, G. M., BULL, A. T. (1983), Bioaccumulation of copper by a defined community of activated-sludge bacteria, *Eur. J. Appl. Microbiol. Biotechnol.* **17**, 30.

DVORAK, D. H., HEDIN, R. S., EDENBORN, H. M., MCINTIRE, P. E. (1992), Treatment of metal-contaminated water using bacterial sulfate reduction – results from pilot-scale reactors, *Biotechnol. Bioeng.* **40**, 609–616.

ECCLES, H. (1999), Treatment of metal-contaminated wastes: why select a biological process? *Trends Biotechnol.* **17**, 462–465.

EHRLICH, H. L. (1999), Microbes as geologic agents: their role in mineral formation, *Geomicrobiol. J.* **16**, 135–153.

EIDE, D. (1997), Molecular biology of iron and zinc uptake in eukaryotes, *Curr. Opin. Cell. Biol.* **9**, 573–577.

EIDE, D., GUERINOT, M. L. (1997), Metal ion uptake in eukaryotes, *Am. Soc. Microbiol. News* **63**, 199–205.

EWART, D. K., HUGHES, M. N. (1991), The extraction of metals from ores using bacteria, *Adv. Inorg. Chem.* **36**, 103–135.

FIGUEIRA, M. M., VOLESKY, B., AZARIAN, K., CIMINELLI, S. T. (2000), Biosorption column performance with a metal mixture, *Environ. Sci. Technol.* **34**, 4320–4326.

FILIPOVIC-KOVACEVIC, Z., SIPOS, L., BRISKI, F. (2000), Biosorption of chromium, copper, nickel and zinc ions onto fungal pellets of *Aspergillus niger* 405 from aqueous solutions, *Food Technol. Biotechnol.* **38**, 211–216.

FINLAY, J. A., ALLAN, V. J. M., CONNER, A., CALLOW, M. E., BASNAKOVA, G. et al. (1999), Phosphate release and heavy metal accumulation by biofilm-immobilized and chemically-coupled cells of a *Citrobacter* sp. pre-grown in continuous culture, *Biotechnol. Bioeng.* **63**, 87–97.

FLEMMING, H.-K. (1995), Sorption sites in biofilms, *Wat. Sci. Technol.* **33**, 27–33.

FORTIN, D., DAVIS, B., SOUTHAM, G., BEVERIDGE, T. J. (1995), Biogeochemical phenomena induced by bacteria within sulphidic mine tailings, *J. Ind. Microbiol.* **14**, 178–185.

FRANCIS, A. J., DODGE, C. J., GILLOW, J. B. (1992), Biodegradation of metal citrate complexes and implications for toxic metal mobility, *Nature* **356**, 140–142.

FRANKENBERGER, W. T., KARLSON, U. (1994), Soil-management factors affecting volatization of selenium from dewatered sediments, *Geomicrobiol. J.* **12**, 265–278.

FRANKENBERGER, W. T., KARLSON, U. (1995), Volatization of selenium from a dewatered seleniferous sediment – a field-study, *J. Ind. Microbiol.* **14**, 226–232.

FRANZ, A., BURGSTALLER, W., SCHINNER, F. (1991), Leaching with *Penicillium simplicissimum*: influence of metals and buffers on proton extrusion and citric acid production, *Appl. Environ. Microbiol.* **57**, 769–774.

FRANZ, A., BURGSTALLER, W., MULLER, B., SCHINNER, F. (1993), Influence of medium components and metabolic inhibitors on citric acid production by *Penicillium simplicissimum, J. Gen. Microbiol.* **139**, 2101–2107.

FREDRICKSON, J. K., ZACHARA, J. M., KENNEDY, D. W., DUFF, M. C., GORBY, Y. A. et al. (2000), Reduction of U(VI) in geothite (*alpha*-FeOOH) suspensions by a dissimilatory metal-reducing bacterium, *Geochim. Cosmochim. Acta* **64**, 3085–3098.

FUJIE, K., TSUCHIDA, T., URANO, K., OHTAKE, H. (1994), Development of a bioreactor system for the treatment of chromate wastewater using *Enterobacter cloacae* HO1, *Water Sci. Technol.* **30**, 235–243.

GADD, G. M. (1984), Effect of copper on *Aureobasidium pullulans* in solid medium: adaptation not necessary for tolerant behaviour, *Trans. Brit. Mycol. Soc.* **82**, 546–549.

GADD, G. M. (1986a), Fungal responses towards heavy metals, in: *Microbes in Extreme Environments* (HERBERT, R. A., CODD, G. A., Eds.), pp. 83–110. London: Academic Press.

GADD, G. M. (1986b), The uptake of heavy metals by fungi and yeasts: the chemistry and physiology of the process and applications for biotechnology, in: *Immobilisation of Ions by Bio-sorption*, (ECCLES, H., HUNT, S., Eds), pp. 135–147. Chichester: Ellis Horwood.

GADD, G. M. (1988), Accumulation of metals by microorganisms and algae, in: *Biotechnology 1st Edn.* Vol. 6b (REHM, H.-J., REED, G., Eds.), pp. 401–433. Weinheim: VCH.

GADD, G. M. (1990a), Fungi and yeasts for metal binding, in: *Microbial Mineral Recovery* (EHRLICH, H., BRIERLEY, C. L., Eds), pp. 249–275. New York: McGraw-Hill.

GADD, G. M. (1990b), Heavy metal accumulation by bacteria and other microorganisms, *Experientia* **46**, 834–840

GADD, G. M. (1990c), Biosorption, *Chem. Ind.* **13**, 421–426.

GADD, G. M. (1990d), Metal tolerance, in: *Microbiology of Extreme Environments*, (EDWARDS, C., Ed.), pp. 178–210. Milton Keynes: Open University Press.

GADD, G. M. (1992a), Microbial control of heavy metal pollution, in: *Microbial Control of Pollution*, (FRY, J. C., GADD, G. M., HERBERT, R. A., JONES, C. W., WATSON-CRAIK, I., Eds.), pp. 59–88. Cambridge: Cambridge University Press.

GADD, G. M. (1992b), Molecular biology and biotechnology of microbial interactions with organic and inorganic heavy metal compounds, in: *Molecular Biology and Biotechnology of Extremophiles* (HERBERT, R. A., SHARP, R. J., Eds.), pp. 225–257. Glasgow: Blackie and Sons.

GADD, G. M. (1992c), Metals and microorganisms: a problem of definition (invited article), *FEMS Microbiol. Lett.* **100**, 197–204.

GADD, G. M. (1993a), Microbial formation and transformation of organometallic and organometalloid compounds, *FEMS Microbiol. Rev.* **11**, 297–316.

GADD, G. M. (1993b), Interactions of fungi with toxic metals, *New Phytol.* **124**, 25–60.

GADD, G. M. (1995), Signal transduction in fungi, in: *The Growing Fungus* (GOW, N. A. R, GADD, G. M., Eds.), pp. 183–210. London: Chapman & Hall.

GADD, G. M. (1996), Roles of microorganisms in the environmental fate of radionuclides, *Endeavour* **20**, 150–156.

GADD, G. M. (1997), Roles of microorganisms in the environmental fate of radionuclides, in: *CIBA Foundation Symposium 203: Health Impacts of Large Releases of Radionuclides*, (LAKE, J. V., BOCK, G. R., CARDEW, G., Eds.), pp. 94–108. Chichester: John Wiley & Sons.

GADD, G. M. (1999), Fungal production of citric and oxalic acid: importance in metal speciation, physiology and biogeochemical processes, *Adv. Microb. Physiol.* **41**, 47–92.

GADD, G. M. (2000a), Bioremedial potential of microbial mechanisms of metal mobilization and immobilization, *Curr. Opin. Biotechnol.* **11**, 271–279.

GADD, G. M. (2000b), Heavy metal pollutants: environmental and biotechnological aspects, in: *The Encyclopedia of Microbiology*, 2nd edn. (LEDERBERG, J., Ed.), pp. 607–617. San Diego, CA: Academic Press.

GADD, G. M. (2000c), Heterotrophic solubilization of metal-bearing minerals by fungi, in: *Environmental Mineralogy: Microbial Interactions, Anthropogenic Influences, Contaminated Land and Waste Management*, (COTTER-HOWELLS, J. D., BATCHELDER, M., CAMPBELL, L., VALSAMI-JONES, E., Eds.), pp. 21–39. London: Mineralogical Society.

GADD, G. M., DE ROME, L. (1988), Biosorption of copper by fungal melanin, *Appl. Microbiol. Biotechnol.* **29**, 610–617.

GADD, G. M., GRIFFITHS, A. J. (1978), Microorganisms and heavy metal toxicity, *Microb. Ecol.* **4**, 303–317.

GADD, G. M., GRIFFITHS, A. J. (1980), Effect of copper on morphology of *Aureobasidium pullulans*, *Trans. Brit. Mycol. Soc.* **74**, 387–392.

GADD, G. M., LAURENCE, O. S. (1996), Demonstration of high-affinity Mn^{2+} uptake in *Saccharomyces cerevisiae* – specificity and kinetics, *Microbiology*, **142**, 1159–1167.

GADD, G. M., MOWLL, J. L. (1983), The relationship between cadmium uptake, potassium release and viability in *Saccharomyces cerevisiae*, *FEMS Microbiol. Lett.* **16**, 45–48.

GADD, G. M., MOWLL, J. L. (1985), Copper uptake by yeast-like cells, hyphae, and chlamydospores of *Aureobasidium pullulans*, *Exp. Mycol.* **9**, 230.

GADD, G. M., SAYER, G. M. (2000), Fungal transformations of metals and metalloids, in: *Environmental Microbe–Metal Interactions*, (LOVLEY, D. R., Ed.), pp. 237–256. Washington, DC: American Society for Microbiology.

GADD, G. M., WHITE, C. (1985), Copper uptake by *Penicillium ochro-chloron*: influence of pH on toxicity and demonstration of energy-dependent copper influx using protoplasts, *J. Gen. Microbiol.* **131**, 1875–1879.

GADD, G. M., WHITE, C. (1989a), Heavy metal and radionuclide accumulation and toxicity in fungi and yeasts, in: *Metal–Microbe Interactions* (POOLE, R. K., GADD, G. M., Eds.), pp. 19–38. Oxford: IRL Press.

GADD, G. M., WHITE, C. (1989b), The removal of thorium from simulated acid process streams by fungal biomass, *Biotechnol. Bioeng.* **33**, 592–597.

GADD, G. M., WHITE, C. (1989c), Uptake and intracellular distribution of thorium in *Saccharomyces cerevisiae*, *Environ. Pollut.* **61**, 187–197.

GADD, G. M., WHITE, C. (1990), Biosorption of radionuclides by yeast and fungal biomass, *J. Chem. Technol. Biotechnol.* **49**, 331–343.

GADD, G. M., WHITE, C. (1992), Removal of thorium from simulated acid process streams by fungal biomass: potential for thorium desorption and reuse of biomass and desorbent, *J. Chem. Technol. Biotechnol.* **55**, 39–44.

GADD, G. M., WHITE, C. (1993), Microbial treatment of metal pollution – a working biotechnology? *Trends Biotechnol.* **11**, 353–359.

GADD, G. M., WHITE, C., MOWLL, J. L. (1987), Heavy metal uptake by intact cells and protoplasts of *Aureobasidium pullulans*, *FEMS Microbiol. Ecol.* **45**, 261–267.

GADD, G. M., GRAY, D. J., NEWBY, P. J. (1990), Role of melanin in fungal biosorption of tributyltin chloride, *Appl. Microbiol. Biotechnol.* **34**, 116–121.

GALLI, U., SCHUEPP, H., BRUNOLD, C. (1994), Heavy metal binding by mycorrhizal fungi, *Physiol. Plant.* **92**, 364–368.

GALUN, M., KELLER, P., FELDSTEIN, H., GALUN, E., SIEGEL, S. et al. (1983), Recovery of uranium (VI) from solution using fungi. 2. Release from uranium-loaded *Penicillium* biomass, *Water Air Soil Pollut.* **20**, 277.

GALUN, M., KELLER, P., MALKI, D., FELDSTEIN, H., GALUN, E., SIEGEL, S. et al. (1984), Removal of uranium (VI) from solution by fungal biomass – inhibition by iron, *Water Air Soil Pollut.* **21**, 411.

GARNHAM, G. W., CODD, G. A., GADD, G. M. (1992a), Accumulation of cobalt, zinc and manganese by the estuarine green microalga *Chlorella salina* immobilized in alginate microbeads, *Environ. Sci. Technol.* **26**, 1764–1770.

GARNHAM, G. W., CODD, G. A., GADD, G. M. (1992b), Kinetics of uptake and intracellular location of cobalt, manganese and zinc in the estuarine green alga, *Chlorella salina*, *Appl. Microbiol. Biotechnol.* **37**, 270–276.

GARNHAM, G. W., CODD, G. A., GADD, G. M. (1993), Uptake of cobalt and caesium by microalgal- and cyanobacterial-clay mixtures, *Microb. Ecol.* **25**, 71–82.

GARNHAM, G. W., AVERY, S. V., CODD, G. A., GADD, G. M. (1994), Interactions of microalgae and cyanobacteria with toxic metals and radionuclides: physiology and environmental implications, in: *Changes in Fluxes in Estuaries – Implications from Science to Management*, (DYER, K. R., ORTH, R. J., Eds.), pp. 289–293. Fredensborg, Denmark: Olsen and Olsen.

GEESEY, G., JANG, L. (1990), Extracellular polymers for metal binding, in: *Microbial Mineral Recovery* (EHRLICH, H. L., BRIERLEY, C. L., Eds.), pp. 223–275. New York: McGraw-Hill.

GERHARDT, M. B., GREEN, F. B., NEWMAN, R. D., LUNDQUIST, T. J., TRESAN, R. B. et al. (1991), Removal of selenium using a novel algal bacterial process, *Res. J. Water Pollut. Control Fed.* **63**, 799–805.

GHANI, B., TAKAI, M., HISHAM, N. Z., KISHIMOTO, N., ISMAIL, A. K. et al. (1993), Isolation and characterisation of a Mo^{6+} reducing bacterium, *Appl. Environ. Microbiol.* **59**, 1176–1180.

GHARIEB, M. M., GADD, G. M. (1998), Evidence for the involvement of vacuolar activity in metal(loid) tolerance: vacuolar-lacking and -defective mutants of *Saccharomyces cerevisiae* display higher sensitivity to chromate, tellurite and selenite, *BioMetals* **11**, 101–106.

GHARIEB, M. M., GADD, G. M. (1999), Influence of nitrogen source on the solubilization of natural gypsum ($CaSO_4 \cdot 2H_2O$) and the formation of calcium oxalate by different oxalic and citric acid-producing fungi, *Mycol. Res.* **103**, 473–481.

GHARIEB, M. M., WILKINSON, S. C., GADD, G. M. (1995), Reduction of selenium oxyanions by unicellular, polymorphic and filamentous fungi: cellular location of reduced selenium and implications for tolerance, *J. Ind. Microbiol.* **14**, 300–311.

GHARIEB, M. M., SAYER, J. A., GADD, G. M. (1998), Solubilization of natural gypsum ($CaSO_4 \cdot 2H_2O$) and the formation of calcium oxalate by *Aspergillus niger* and *Serpula himantioides*, *Mycol. Res.* **102**, 825–830.

GHARIEB, M. M., KIERANS, M., GADD, G. M. (1999), Transformation and tolerance of tellurite by filamentous fungi: accumulation, reduction and volatilization, *Mycol. Res.* **103**, 299–305.

GOMEZ, C., BOSECKER, K. (1999), Leaching heavy metals from contaminated soil by using *Thiobacillus ferrooxidans* or *Thiobacillus thiooxidans*, *Geomicrobiol. J.* **16**, 233–244.

GORBY, Y., LOVLEY, D. R. (1992), Enzymatic uranium precipitation, *Environ. Sci. Technol.* **26**, 205–207.

GRAY, S. N. (1998), Fungi as potential bioremediation agents in soil contaminated with heavy or radioactive metals, *Biochem. Soc. Trans.* **26**, 666–670.

GREENE, B., HOSEA, M., MCPHERSON, R., HENZL, M., ALEXANDER, M. D. et al. (1986), Interaction of gold(I) and gold(III) complexes with algal biomass, *Environ. Sci. Technol.* **20**, 627.

GUIBAUD, G., BAUDU, M., DOLLET, P., CONDAT, M. L., DAGOT, C. (1999), Role of extracellular polymers in cadmium adsorption by activated sludges, *Environ. Technol.* **20**, 1045–1054.

GUO, L., FRANKENBERGER, W. T., JURY, W. A. (1999a), Adsorption and degradation of dimethyl selenide in soil, *Environ. Sci. Technol.* **33**, 2934–2938.

GUO, L., FRANKENBERGER, W. T., JURY, W. A. (1999b), Evaluation of simultaneous reduction and transport of selenium, *Water Resources Res.* **35**, 663–669.

GUPTA, R., AHUJA, P., KHAN, S., SAXENA, R. K., MOHAPATRA, H. (2000), Microbial biosorbents: meeting challenges of heavy metal pollution in aqueous solutions, *Curr. Sci.* **78**, 967–973.

HAMDY, A. A. (2000a), Biosorption of heavy metals by marine algae, *Curr. Microbiol.* **41**, 232–238.

HAMDY, A. A. (2000b), Removal of Pb^{2+} by biomass of marine algae, *Curr. Microbiol.* **41**, 239–245.

HAMMACK, R. W., EDENBORN, H. M. (1992), The removal of nickel from mine waters using bacterial sulphate-reduction, *Appl. Microbiol. Biotechnol.* **37**, 674–678.

HARD, B. C., WALTHER, C., BABEL, W. (1999), Sorption of aluminum by sulfate-reducing bacteria isolated from uranium mine tailings, *Geomicrobiol. J.* **16**, 267–275.

HAYASHI, Y., MUTOH, N. (1994), Cadystin (phytochelatin) in fungi, in: *Metal Ions in Fungi* (WINKELMANN, G., WINGE, D. R., Eds.), pp. 311–337. New York: Marcel Dekker.

HEDIN, R. S., NAIRN, R. W. (1991), Contaminant removal capabilities of wetlands constructed to treat coal mine drainage, in: *Proc. Int. Symp. Constructed Wetlands for Water-Quality Improvement* (MOSHIRI, G. A., Ed.), pp. 187–195. Chelsea, MI: Lewis Publishers.

HOCKERTZ, S., SCHMID, J., AULING, G. (1987), A specific transport system for manganese in the filamentous fungus *Aspergillus niger*, *J. Gen. Microbiol.* **133**, 3513–3519.

HORIKOSHI, T., NAKAJIMA, A., SAKAGUCHI, T. (1981), Studies on the accumulation of heavy metal elements in biological systems. XIX. Accumulation of uranium by microorganisms, *Eur. J. Appl. Microbiol. Biotechnol.* **12**, 90–96.

HOWE, R., EVANS, R. L., KETTERIDGE, S. W. (1997), Copper binding proteins in ectomycorrhizal fungi, *New Phytol.* **135**, 123–131.

HUGHES, M. N., POOLE, R. K. (1991), Metal speciation and microbial growth – the hard (and soft) facts, *J. Gen. Microbiol.* **137**, 725–734.

HUTCHINS, S. R., DAVIDSON, M. S., BRIERLEY, J. A., BRIERLEY, C. L. (1986), Microorganisms in reclamation of metals, *Annu. Rev. Microbiol.* **40**, 311.

INOUHE, M., SUMIYOSHI, M., TOHOYAMA, H., JOHO, M. (1996), Resistance to cadmium ions and formation of a cadmium-binding complex in various wild-type yeasts, *Plant Cell Physiol.* **37**, 341–346.

JALALI, K., BALDWIN, S. G. (2000), The role of sulphate reducing bacteria in copper removal from aqueous sulphate solutions, *Water Res.* **34**, 797–806.

JENSEN, T. E., BAXTER, M., RACHLIN, J. W., JANI, V. (1982), Uptake of heavy metals by *Plectonema boryanum* (Cyanophyceae) into cellular components, especially polyphosphate bodies – an X-ray-energy dispersive study, *Environ. Pollut.* **27**, 119.

JEONG, B. C., MACASKIE, L. E. (1999), Production of two phosphatases by a *Citrobacter* sp. grown in batch and continuous culture, *Enzyme Microb. Technol.* **24**, 218–224.

JOHO, M., INOUHE, M., TOHOYAMA, H., MURAYAMA. T. (1995). Nickel resistance mechanisms in yeasts and other fungi. *J. Ind. Microbiol.* **14**, 164–168.

JONES, R. P., GADD, G. M. (1990), Ionic nutrition of yeast – physiological mechanisms involved and implications for biotechnology, *Enzyme Microb. Technol.* **12**, 402–418.

KAMEO, S., IWAHASHI, H., KOJIMA, Y., SATOH, H. (2000), Induction of metallothioneins in the heavy metal resistant fungus *Beauveria bassiana* exposed to copper or cadmium, *Analysis* **28**, 382–385.

KAPLAN, D., WILHELM, R., ABELIOVICH, A. (2000), Interdependent environmental factors controlling nitrification in waters, *Water Sci. Technol.* **42**, 167–172.

KAPOOR, A., VIRARAGHAVAN, T., CULLIMORE, D. R. (1999), Removal of heavy metals using the fungus *Aspergillus niger*, *Bioresource Technol.* **70**, 95–104.

KARAMUSHKA, V. I., GADD, G. M. (1999), Interactions of *Saccharomyces cerevisiae* with gold: toxicity and accumulation, *BioMetals* **12**, 289–294.

KARATHANASIS, A. D., THOMPSON, Y. L. (1995), Mineralogy of iron precipitates in a constructed acid-mine drainage wetland, *Soil Sci. Soc. Am. J.* **59**, 1773–1781.

KARLSON, U., FRANKENBERGER, W. T. (1993), Biological alkylation of selenium and tellurium, in: *Metal Ions in Biological Systems* (SIGEL, H., SIGEL, A., Eds.), pp. 185–227. New York: Marcel Dekker.

KASHIWA, M., NISHIMOTO, S., TAKAHASHI, K., IKE, M., FUJITA, M. (2000), Factors affecting soluble selenium removal by a selenate-reducing bacterium *Bacillus* sp. SF-1, *J. Biosci. Bioeng.* **89**, 528–533.

KEASLING, J. D., VAN DIEN, S. J., PRAMANIK, J. (1998), Engineering polyphosphate metabolism in *Escherichia coli*: implications for bioremediation of inorganic contaminants, *Biotechnol. Bioeng.* **58**, 231–239.

KEFALA, M. I., ZOUBOULIS, A. I., MATIS, K. A. (1999), Biosorption of cadmium ions by Actinomycetes and separation by flotation, *Environ. Pollut.* **104**, 283–293.

KELLY, D. P., NORRIS, P. R., BRIERLEY, C. L. (1979), Microbiological methods for the extraction and recovery of metals, in: *Microbial Technology: Current State, Future Prospects* (BULL, A. T., ELLWOOD, D. C., RATLEDGE, C., Eds.), pp. 263–308. Cambridge: Cambridge University Press.

KHUMMONGKOL, D., CANTERFORD, G. S., FRYER, C. (1982), Accumulation of heavy-metals in unicellular algae, *Biotechnol. Bioeng.* **24**, 2643.

KIFF, R. J., LITTLE, D. R. (1986), Biosorption of heavy metals by immobilized fungal biomass, in: *Immobilisation of Ions by Biosorption*, (ECCLES, H., HUNT, S., Eds.), pp. 71– 80. Chichester: Ellis Horwood.

KING, J. K., KOSTKA, J. E., FRISCHER, M. E., SAUNDERS, F. M. (2000), Sulfate-reducing bacteria methylate mercury at variable rates in pure culture and in marine sediments, *Appl. Environ. Microbiol.* **66,** 2430–2437.

KLIONSKY, D. J., HERMAN, P. K., EMR, S. D. (1990), The fungal vacuole: composition, function and biogenesis, *Microbiol. Rev.* **54**, 266–292.

KOSMAN, D. J. (1994), Transition metal ion uptake in yeasts and filamentous fungi, in: *Metal Ions in Fungi* (WINKELMANN, G., WINGE, D. R., Eds.), pp. 1–38. New York: Marcel Dekker.

KOTRBA, P., DOLEČKOVÁ, L., DE LORENZO, V., RUML, T. (1999), Enhanced bioaccumulation of heavy metal ions by bacterial cells due to surface display of short metal binding peptides, *Appl. Environ. Microbiol.* **65**, 1092–1098.

KRÄMER, M., MEISCH, H.-U. (1999), New metal-binding ethyldiamino- and dicarboxy-products from *Aspergillus niger* industrial wastes, *BioMetals* **12**, 241–246.

KUREK, E., CZABAN, J., BOLLAG, J. M. (1982), Sorption of cadmium by microorganisms in competition with other soil constituents, *Appl. Environ. Microbiol.* **43**, 1011.

LANDA, E. R., GRAY, J. R. (1995), US Geological Survey – results on the environmental fate of uranium mining and milling wastes, *J. Ind. Microbiol.* **26**, 19–31.

LAWRENCE, A. W., MCCARTY, P. L. (1965), The role of sulphide in preventing heavy metal toxicity in anaerobic treatment, *J. Water Pollut. Control Fed.* **37**, 392–406.

LAWSON, P. S., STERRITT, R. M., LESTER, J. N. (1984), Adsorption and complexation mechanisms of heavy-metal uptake in activated-sludge, *J. Chem. Technol. Biotechnol.* **34B**, 253.

LEDIN, M. (2000), Accumulation of metals by microorganisms – processes and importance for soil systems, *Earth Sci. Rev.* **51**, 1–31.

LEDIN, M., PEDERSEN, K. (1996), The environmental impact of mine wastes – roles of microorganisms and their significance in the treatment of mine wastes, *Earth Sci. Rev.* **41**, 67–108.

LIU, X. F., CULOTTA, V. C. (1999), Mutational analysis of *Saccharomyces cerevisiae* Smf1p, a member of the Nramp family of metal transporters, *J. Mol. Biol.* **289**, 885–891.

LLOYD, J. R., MACASKIE, L. E. (1998), Enzymatic recovery of elemental palladium using sulfate-reducing bacteria, *Appl. Environ. Microbiol.* **64**, 4607–4609.

LLOYD, J. R., RIDLEY, J., KHIZNIAK, T., LYALIKOVA, N. N., MACASKIE, L. E. (1999a), Reduction of technetium by *Desulfovibrio desulfuricans*: biocatalyst characterization and use in a flow-through bioreactor, *Appl. Environ. Microbiol.* **65**, 2691–2696.

LLOYD, J. R., THOMAS, G. H., FINLAY, J. A., COLE, J. A., MACASKIE, L. E. (1999b), Microbial reduction of technetium by *Escherichia coli* and *Desulfovibrio desulfuricans*: enhancement via the use of high activity strains and effect of process parameters, *Biotechnol. Bioeng.* **66**, 122–130.

LLOYD-JONES, G., OSBORN, A. M., RITCHIE, D. A., STRIKE, P., HOBMAN, J. L. et al. (1994), Accumulation and intracellular fate of tellurite in tellurite-resistant *Escherichia coli* – a model for the mechanism of resistance, *FEMS Microbiol. Lett.* **118**, 113–119.

LONG, R. H. B., BENSON, S. M., TOKUNAGA, T. K., YEE, A. (1990), Selenium immobilization in a pond sediment at Kesterton Reservoir, *J. Environ. Qual.* **19**, 302–311.

LOVLEY D. R. (1993), Dissimilatory metal reduction, *Ann. Rev. Microbiol.* **47**, 263–290.

LOVLEY, D. R. (1995), Bioremediation of organic and metal contaminants with dissimilatory metal reduction, *J. Ind. Microbiol.* **14**, 85–93.

LOVLEY, D. R., ANDERSON, R. T. (2000), Influence of dissimilatory metal reduction on fate of organic and metal contaminants in the subsurface, *Hydrogeol. J.* **8**, 77–88.

LOVLEY, D. R., COATES, J. D. (1997), Bioremediation of metal contamination, *Curr. Opin. Biotechnol.* **8**, 285–289.

LOVLEY, D. R., COATES, J. D. (2000), Novel forms of anaerobic respiration of environmental relevance, *Curr. Opin. Microbiol.* **3**, 252–256.

LOVLEY, D. R., PHILLIPS, E. J. P. (1992a), Bioremediation of uranium contamination with enzymatic uranium reduction, *Environ. Sci. Technol.* **26**, 2228–2234.

LOVLEY, D. R., PHILLIPS, E. J. P. (1992b), Reduction of uranium by *Desulfovibrio desulfuricans*, *Appl. Environ. Microbiol.* **58**, 850–856.

LOVLEY, D. R., GIOVANNONI, S. J., WHITE, D. C., CHAMPINE, J. E., PHILLIPS, E. J. P. et al. (1993), *Geobacter metallireducens* gen. nov. sp. nov., a microorganism capable of coupling the complete oxidation of organic compounds to the reduction of iron and other metals, *Arch. Microbiol.* **159**, 336–344.

LUTHER, G. W. (1987), Pyrite oxidation and reduction: molecular orbital theory considerations, *Geochim. Cosmochim. Acta* **51**, 3193–3199.

LYEW, D., KNOWLES, R., SHEPPARD, J. (1994), The biological treatment of acid-mine drainage under continuous-flow conditions in a reactor, *Proc. Safety Environ. Prot.* **72**, 42–47.

MACASKIE, L. E. (1991), The application of biotechnology to the treatment of wastes produced by the nuclear fuel cycle – biodegradation and bioaccumulation as a means of treating radionuclide-containing streams, *Crit. Rev. Biotechnol.* **11**, 41–112.

MACASKIE, L. E., DEAN, A. C. R. (1984), Cadmium accumulation by a *Citrobacter* sp., *J. Gen. Microbiol.* **130**, 53.

MACASKIE, L. E., DEAN, A. C. R. (1987), Use of immobilized biofilm of *Citrobacter* sp. for the removal of uranium and lead from aqueous flows, *Enzyme Microb. Technol.* **9**, 2.

MACASKIE, L. E., DEAN, A. C. R., (1989), Microbial metabolism, desolubilization and deposition of heavy metals: uptake by immobilized cells and application to the treatment of liquid wastes, in: *Biological Waste Treatment* (MIZRAHI, A., Ed.), pp. 150–201. New York: Alan R. Liss.

MACASKIE, L. E., WATES, J. M., DEAN, A. C. R. (1987a), Cadmium accumulation by a *Citrobacter* sp. immobilized on gel and solid supports – applicability to the treatment of liquid wastes containing heavy-metal cations, *Biotechnol. Bioeng.* **30**, 66.

MACASKIE, L. E., DEAN, A. C. R., CHEETHAM, A. K., JAKEMAN, R. J. B., SKARNULIS, A. J. (1987b), Cadmium accumulation by a *Citrobacter* sp. – the chemical nature of the accumulated metal precipitate and its location on the bacterial cells, *J. Gen. Microbiol.* **133**, 539.

MACASKIE, L. E., JEONG, B. C., TOLLEY, M. R. (1994), Enzymically-accelerated biomineralization of heavy metals: application to the removal of americium and plutonium from aqueous flows, *FEMS Microbiol. Rev.* **14**, 351–368.

MACASKIE, L. E., BONTHRONE, K. M., YONG, P., GODDARD, D. T. (2000), Enzymatically mediated bioprecipitation of uranium by a *Citrobacter* sp.: a concerted role for exocellular lipopolysaccharide and associated phosphatase in biomineral formation, *Microbiology* **146**, 1855–1867.

MACREADIE, I. G., SEWELL, A. K., WINGE, D. R. (1994), Metal ion resistance and the role of metallothionein in yeast, in: *Metal Ions in Fungi* (WINKELMANN, G., WINGE, D. R., Eds.), pp. 279–310. New York: Marcel Dekker.

MACY, J. M., MICHEL, T. A., KIRSCH, D. G. (1989), Selenate reduction by a new *Pseudomonas* species: a new mode of anaerobic respiration, *FEMS Microbiol. Lett.* **61**, 195–198.

MACY, J. M., LAWSON, S., DEMOLLDECKER, H. (1993), Bioremediation of selenium oxyanions in San Joaquin drainage water using *Thaueria selenatis* in a biological reactor system, *Appl. Microbiol. Biotechnol.* **40**, 588–594.

MACY, J. M., NUNAN, K., HAGEN, K. D., DIXON, D. R., HARBOUR, P. J. et al. (1996), *Chrysiogenes arsenatis* gen. nov., sp. nov.; a new arsenate-respiring bacterium isolated from gold mine wastewater, *Int. J. Syst. Bacteriol.* **46**, 1153–1157.

MACY, J. M., SANTINI, J. M., PAULING, B. V., O'NEILL, A. H., SLY, L. I. (2000), Two new arsenate/sulfate-reducing bacteria: mechanisms of arsenate reduction, *Arch. Microbiol.* **173**, 49–57.

MAMERI, M., BOUDRIES, N., ADDOUR, L., BELHOCINE, D., LOUNICI, H. et al. (1999), Batch zinc biosorption by a bacterial non-living *Streptomyces rimosus* biomass, *Water Res.* **33**, 1347–1354.

MANN, H., FYFE, W. S., KERRICH, R. (1988), The chemical content of algae and waters – bioconcentration, *Tox. Assess.* **3**, 1–16.

MATSUNAGA, T., TAKEYAMA, H., NAKAO, T., YAMAZAWA, A. (1999), Screening of marine microalgae for bioremediation of cadmium-polluted seawater, *J. Biotechnol.* **70**, 33–38.

MATTEY, M. (1992), The production of organic acids, *Crit. Rev. Biotechnol.* **12**, 87–132.

MCCARTNEY, D. M., OLESZKIEVICZ, J. A. (1991), Sulfide inhibition of anaerobic degradation of lactate and acetate, *Water Res.* **25**, 203–209.

MEHRA, R. K., WINGE, D. R. (1991), Metal ion resistance in fungi: molecular mechanisms and their related expression, *J. Cell. Biochem.* **45**, 30–40.

MEIXNER, O., MISCHACK, H., KUBICEK, C. P., ROHR, M. (1985), Effect of manganese deficiency on plasma-membrane lipid composition and glucose uptake in *Aspergillus niger*, *FEMS Microbiol. Lett.* **26**, 271–274.

MERCIER, G., CHARTIER, M., COUILLARD, D., BLAIS, J. F. (1999), Decontamination of fly ash and used lime from municipal waste, *Environ. Manage.* **24**, 517–528.

MICHALKE, K., WICKENHEISER, E. B., MEHRING, M., HIRNER, A. V., HENSEL, R. (2000), Production of volatile derivatives of metal(loids)s by microflora involved in anaerobic digestion of sewage sludge, *Appl. Environ. Microbiol.* **66**, 2791–2796.

MILLER, A. J., VOGG, G., SANDERS, D. (1990), Cytosolic calcium homeostasis in fungi: roles of plasma membrane transport and intracellular sequestration of calcium, *Proc. Natl. Acad. Sci. USA* **87**, 9348–9352.

MOORE, M. D., KAPLAN, S. (1992), Identification of intrinsic high-level resistance to rare-earth-oxides and oxyanions in members of the class Proteobacteria – characterization of tellurite, selenite, and *Rhodium sesquioxide* reduction in rhodobacter sphaeroides, *J. Bacteriol.* **174**, 1505–1514.

MORLEY, G. F., GADD, G. M. (1995), Sorption of toxic metals by fungi and clay minerals, *Mycol. Res.* **99**, 1429–1438.

MORLEY, G. F., SAYER, J., WILKINSON, S., GHARIEB, M., GADD, G. M. (1996), Fungal sequestration, solubilization and transformation of toxic metals, in: *Fungi and Environmental Change* (FRANKLAND, J. C., MAGAN, N., GADD, G. M., Eds.), pp. 235–256. Cambridge: Cambridge University Press.

MOWLL, J. L., GADD, G. M. (1983), Zinc uptake and toxicity in the yeasts *Sporobolomyces roseus* and *Saccharomyces cerevisiae*, *J. Gen. Microbiol.* **129**, 3421–3425.

MOWLL, J. L., GADD, G. M. (1984), Cadmium uptake by *Aureobasidium pullulans*, *J. Gen. Microbiol.* **130**, 279–284.

MOWLL, J. L., GADD, G. M. (1985), The effect of vehicular lead pollution on phylloplane mycoflora, *Trans. Brit. Mycol. Soc.* **84**, 685–689.

MURASUGI, A., WADA, C., HAYASHI, Y. (1983), Occurrence of acid labile sulfide in cadmium binding peptide I from fission yeast, *J. Biochem.* **93**, 661–664.

MUZZARELLI, R. A. A., BREGANI, F., SIGON, F. (1986), Chelating capacities of amino acid glucans and sugar acid glucans derived from chitosan, in: *Immobilisation of Ions by Biosorption* (ECCLES, H., HUNT, S., Eds.), pp. 173–182. Chichester: Ellis Horwood.

NAKAJIMA, A., SAKAGUCHI, T. (1986), Selective accumulation of heavy metals by microorganisms, *Appl. Microbiol. Biotechnol.* **24**, 59.

NAKAJIMA, A., HORIKOSHI, T., SAKAGUCHI, T. (1981), Studies on the accumulation of heavy metal elements in biological systems. 17. Selective accumulation of heavy metal ions by *Chlorella regularis*, *Eur. J. Appl. Microbiol. Biotechnol.* **12**, 76–83.

NAKAJIMA, A., HORIKOSHI, T., SAKAGUCHI, T. (1982), Studies on the accumulation of heavy metal elements in biological systems. 21. Recovery of uranium by immobilized microorganisms, *Eur. J. Appl. Microbiol. Biotechnol.* **16**, 88–91.

NEILANDS, J. B. (1981), Microbial iron compounds, *Ann. Rev. Biochem.* **50**, 715–731.

NELSON, N., BELTRAN, C., SUPEK, F., NELSON, H. (1992), Cell biology and evolution of proton pumps, *Cell. Physiol. Biochem.* **2**, 150–158.

NIES, D. J. (2000), Heavy metal-resistant bacteria as extremophiles: molecular physiology and biotechnological use of *Ralstonia* sp. CH34, *Extremophiles* **4**, 77–82.

NIEUWENHUIS, B. J. W. M., WEIJERS, A. G. M., BORST-PAUWELS, G. W. F. H. (1981), Uptake and accumulation of Mn^{2+} and Sr^{2+} in *Saccharomyces cerevisiae*, *Biochim. Biophys. Acta* **649**, 83–88.

NIU, H., VOLESKY, B. (1999), Characteristics of gold biosorption from cyanide solution, *J. Chem. Technol. Biotechnol.* **74**, 778–784.

OHSUMI, Y., ANRAKU, Y. (1983), Calcium transport driven by a proton motive force in vacuolar membrane vesicles of *Saccharomyces cerevisiae*, *J. Biol. Chem.* **41**, 17–22.

OKINO, S., IWASAKI, K., YAGI, O., TANAKA, H. (2000), Development of a biological mercury removal–recovery system, *Biotechnol. Lett.* **22**, 783–788.

OKOROKOV, L. A. (1985), Main mechanisms of ion transport and regulation of ion concentrations in the yeast cytoplasm, in: *Environmental Regulation of Microbial Metabolism* (KULAEV, I. S., DAWES, E. A., TEMPEST, D. W., Eds.), pp. 463–472. London: Academic Press.

OKOROKOV, L. A. (1994), Several compartments of *Saccharomyces cerevisiae* are equipped with Ca^{2+} ATPase(s), *FEMS Microbiol. Lett.* **117**, 311–318.

OKOROKOV, L. A., LICHKO, L. P., KULAEV, I. S. (1980), Vacuoles: main compartments of potassium, magnesium and phosphate in *Saccharomyces carlsbergensis* cells, *J. Bacteriol.* **144**, 661–665.

OKOROKOV, L. A., KULAKOVSKAYA, T. V., LICHKO, L. P., POLOROTOVA, E. V. (1985), H^+/ion antiport as the principal mechanism of transport systems in the vacuolar membrane of the yeast *Saccharomyces carlsbergensis*, *FEBS Lett.* **192**, 303–306.

OREMLAND, R. S., HOLLIBAUGH, J. T., MAEST, A. S., PRESSER, T. S., MILLER, L. G. et al. (1989), Selenate reduction to elemental selenium by anaerobic bacteria in sediments and culture: biogeochemical significance of a novel sulfate-independent respiration, *Appl. Environ. Microbiol.* **55**, 2333–2343.

OREMLAND, R. S., STEINBERG, N. A., MAEST, A. S., MILLER, L. G., HOLLIBAUGH, J. T. (1990), Measurement of *in situ* rates of selenate removal by dissimilatory bacterial reduction in sediments, *Environ. Sci. Technol.* **24**, 1157–1164.

OREMLAND, R. S., STEINBERG, N. A., PRESSER, T. S., MILLER, L. G. (1991), *In situ* bacterial selenate reduction in the agricultural drainage systems of Western Nevada, *Appl. Environ. Microbiol.* **57**, 615–617.

OREMLAND, R. S., SWITZER BLUM, J., BURNS BINDI, A., DOWDLE, P. R., HERBEL, M. et al. (1999), Simultaneous reduction of nitrate and selenate by cell suspensions of selenium-respiring bacteria, *Appl. Environ. Microbiol.* **65**, 4385–4392.

ORTIZ, D. F., KREPPEL, D. F., SPEISER, D. M., SCHEEL, G., MCDONALD, G. et al. (1992), Heavy metal tolerance in the fission yeast requires an ATP-binding cassette-type vacuolar membrane transporter, *EMBO J.* **11**, 3491–3499.

ORTIZ, D. F., RUSCITTI, T., MCCUE, K. F., OW, D. W. (1995), Transport of metal-binding peptides by HMT1, a fission yeast ABC-type vacuolar membrane protein, *J. Biol. Chem.* **270**, 4721–4728.

O'SULLIVAN, A. D., MCCABE, O. M., MURRAY, D. A., OTTE, M. L. (1999), Wetlands for rehabilitation of metal mine wastes, *Biol. Environ. Proc. R. Irish Acad.* **99b**, 11–17.

OW, D. W. (1993), Phytochelatin-mediated cadmium tolerance in *Schizosaccharomyces pombe*, *In Vitro Cell. Dev. Biol. Plant* **29P**, 13–219.

OW, D. W., ORTIZ, D. F., SPEISER, D. M., MCCUE, K. F. (1994), Molecular genetic analysis of cadmium tolerance in *Schizosaccharomyces pombe*, in: *Metal Ions in Fungi* (WINKELMANN, G., WINGE, D. R., Eds.), pp. 339–359. New York: Marcel Dekker.

PARKIN, M. J., ROSS, I. S. (1986), The specific uptake of manganese in the yeast *Candida utilis*, *J. Gen. Microbiol*, **132**, 2155–2160.

PAZIRANDEH, M., WELLS, B. M., RYAN, R. L. (1998). Development of bacterium-based heavy metal biosorbents: enhanced uptake of cadmium and mercury by *Escherichia coli* expressing a metal binding motif, *Appl. Environ. Microbiol.* **64**, 4068–4072.

PEREGO, P., HOWELL, S. B. (1997), Molecular mechanisms controlling sensitivity to toxic metal ions in yeast, *Toxicol. Appl. Pharmacol.* **147**, 312–318.

PERKINS, J., GADD, G. M. (1993a), Accumulation and intracellular compartmentation of lithium ions in *Saccharomyces cerevisiae*, *FEMS Microbiol. Lett.* **107**, 255–260.

PERKINS, J., GADD, G. M. (1993b), Caesium toxicity, accumulation and intracellular localization in yeasts, *Mycol. Res.* **97**, 717–724.

PERKINS, J., GADD, G. M. (1995), The influence of pH and external K^+ concentration on caesium toxicity and accumulation in *Escherichia coli* and *Bacillus subtilis*, *J. Ind. Microbiol.* **14**, 218–225.

PERKINS, J., GADD, G. M. (1996), Interactions of Cs^+ and other monovalent cations (Li^+, Na^+, Rb^+, NH_4^+) with K^+-dependent pyruvate kinase and malate dehydrogenase from the yeasts *Rhodotorula rubra* and *Saccharomyces cerevisiae*, *Mycol. Res.* **100**, 449–454.

PERRY, K. A. (1995), Sulfate-reducing bacteria and immobilization of metals, *Mar. Geores. Geotechnol.* **13**, 33–39.

PETTERSSON, A., HALLBOM, L., BERGMAN, B. (1986), Aluminum uptake by *Anabaena cylindrica*, *J. Gen. Microbiol.* **132**, 1771–1774.

PHILLIPS, E. J. P., LANDA, E. R., LOVLEY, D. R. (1995), Remediation of uranium contaminated soils with bicarbonate extraction and microbial U(VI) reduction, *J. Ind. Microbiol.* **14**, 203–207.

PILZ, F., AULING, G., STEPHAN, D., RAU, U., WAGNER, F. (1981), A high affinity Zn^{2+} uptake system controls growth and biosynthesis of an extracellular, branched β-1,3-β-1,6–glucan in *Sclerotium rolfsii* ATCC 15205, *Exp. Mycol.* **15**, 181–192.

POSTGATE, J. R. (1984), *The Sulphate-Reducing Bacteria*. Cambridge: Cambridge University Press.

PRAKASHAM, R. S., SHENO MERRIE, J., SHEELA, R., SASWATHI, N., RAMAKRISHNA, S. V. (1999), Biosorption of chromium VI by free and immobilized *Rhizopus arrhizus*, *Environ. Poll.* **104**, 421–427.

PRIEUR, D., ERAUSO, G., JEANTHON, C. (1995), Hyperthermophilic life at deep-sea hydrothermal vents, *Planet. Space Sci.* **43**, 115–122.

RADISKY, D., KAPLAN, J. (1999), Regulation of transition metal transport across the yeast plasma membrane, *J. Biol Chem.* **274**, 4481–4484.

RAEL, R. M., FRANKENBERGER, W. T. (1996), Influence of pH, salinity, and selenium on the growth of *Aeromonas veronii* in evaporation agricultural drainage water, *Water Res.* **30**, 422–430.

RAMOS, S., PENA, P., VALLE, E., BERGILLOS, L., PARRA, F. et al. (1985), Coupling of protons and potassium gradients in yeast, in: *Environmental Regulation of Microbial Metabolism* (KULAEV, I. S., DAWES, E. A., TEMPEST, D. W., Eds.), pp. 351–357. London: Academic Press.

RAMSAY, L. M., GADD, G. M. (1997), Mutants of *Saccharomyces cerevisiae* defective in vacuolar function confirm a role for the vacuole in toxic metal ion detoxification, *FEMS Microbiol. Lett.* **152**, 293–298.

RAMTEKE, P. W. (2000), Biosorption of nickel (II) by *Pseudomonas stutzeri*, *J. Environ. Biol.* **21**, 219–221.

RAUSER, W. E. (1995), Phytochelatins and related peptides, *Plant Physiol.* **109**, 1141–1149.

RAVISHANKAR, B. R., BLAIS, J. F., BENMOUSSA, H., TYAGI, R. D. (1994), Bioleaching of metals from sewage sludge: elemental sulfur recovery, *J. Environ. Eng.* **120**, 462–470.

RAWLINGS, D. E., SILVER, S. (1995), Mining with microbes, *Biotechnology* **13**, 773–778.

REED, R. H., GADD, G. M. (1990), Metal tolerance in eukaryotic and prokaryotic algae, in: *Heavy Metal Tolerance in Plants: Evolutionary Aspects* (SHAW, A. J., Ed.), pp. 105–118. Boca Raton, FL: CRC Press.

REIS, M. A. M., ALMEIDA, J. S., LEMOS, P. C., CARRONDO, M. J. T. (1992), Effect of hydrogen sulfide on the growth of sulfate-reducing bacteria, *Biotechnol. Bioeng.* **40**, 593–600.

ROSS, I. S., TOWNSLEY, C. C. (1986), The uptake of heavy metals by filamentous fungi, in: *Immobilisation of Ions by Biosorption* (ECCLES, H., HUNT, S., Eds.), pp. 49–58. Chichester: Ellis Horwood.

RUDD, T., STERRITT, R. M., LESTER, J. N. (1983), Stability-constants and complexation capacities of complexes formed between heavy metals and extracellular polymers from activated sludge, *J. Chem. Technol. Biotechnol.* **33A**, 374–380.

RUSIN, P. A., SHARP, J. E., ODEN, K. L., ARNOLD, R. G., SINCLAIR, N. A. (1993), Isolation and physiology of a manganese-reducing *Bacillus polymyxa* from an Oligocene silver-bearing ore and sediment with reference to Precambrian biogeochemistry, *Precambrian Res.* **61**, 231–240.

SAGLAM, N., SAY, R., DENIZLI, A., PATIR, S., ARICA, M. Y. (1999), Biosorption of inorganic mercury and alkylmercury species onto *Phanerochaete chrysosporium* mycelium, *Process Biochem.* **34**, 725–730.

SAKAGUCHI, T., NAKAJIMA, A. (1982), Recovery of uransium by chitin phosphate and chitosan phosphate, in: *Chitin and Chitosan* (MLRANO, S., TOKURA, S., Eds.), pp. 177–182. Tottori: Japanese Society Chitin and Chitosan.

SANCHEZ, A., BALLESTER, A., BLAZQUEZ, M. L., GONZALEZ, F., MUNOZ, J. et al. (1999), Biosorption of copper and zinc by *Cymodocea nodosa*, *FEMS Microbiol. Rev.* **23**, 527–536.

SANDERS, D. (1990), Kinetic modelling of plant and fungal membrane transport systems, *Ann. Rev. Plant Physiol. Plant Mol. Biol.* **41**, 77–107.

SARRET, G., MANCEAU, A., SPADINI, L., ROUX, J. C., HAZEMANN, J. L. et al. (1999), Structural determination of Pb binding sites of *Penicillium chrysogenum* cell walls by EXAFS spectroscopy and solution chemistry, *J. Synchrotron. Radiat.* **6**, 414–416.

SAYER, J. A., GADD, G. M. (1997), Solubilization and transformation of insoluble metal compounds to insoluble metal oxalates by *Aspergillus niger*, *Mycol. Res.* **101**, 653–661.

SAYER, J. A., RAGGETT, S. L., GADD, G. M. (1995), Solubilization of insoluble metal compounds by soil fungi: development of a screening method for solubilizing ability and metal tolerance, *Mycol. Res.* **99**, 987–993.

SAYER, J. A., KIERANS, M., GADD, G. M. (1997), Solubilization of some naturally occurring metal-bearing minerals, limescale and lead phosphate by *Aspergillus niger*, *FEMS Microbiol. Lett.* **154**, 29–35.

SAYER, J. A., COTTER-HOWELLS, J. D., WATSON, C., HILLIER, S., GADD, G. M. (1999), Lead mineral transformation by fungi, *Curr. Biol.* **9**, 691–694.

SCHINNER, F., BURGSTALLER, W. (1989), Extraction of zinc from an industrial waste by a *Penicillium* sp., *Appl. Environ. Microbiol.* **55**, 1153–1156.

SCHIPPERS, A., VON REGE, H., SAND, W. (1996), Impact of microbial diversity and sulphur chemistry on safeguarding sulphidic mine waste, *Minerals Eng.* **9**, 1069–1079.

SCHREIBER, D. R., MILLERO, F. J., GORDON, A. S. (1990), Production of an extracellular copper-binding compound by the heterotrophic marine bacterium *Vibrio alginolyticus*, *Mar. Chem.* **28**, 275–284.

SEIDEL, H., ONDRUSCHKA, J., MORGENSTERN, P., STOTTMEISTER, U. (1998), Bioleaching of heavy metals from contaminated aquatic sediments using indigenous sulfur-oxidizing bacteria: a feasibility study, *Water Sci. Technol.* **37**, 387–394.

SHARMA, P. K., BALKWILL, D. L., FRENKEL, A., VAIRAVAMURTHY, M. A. (2000), A new *Klebsiella planticola* strain (Cd^{-1}) grows anaerobically at high cadmium concentrations and precipitates cadmium sulfide, *Appl. Environ. Microbiol.* **66**, 3083–3087.

SHEN, H., WANG, Y.-T. (1993), Characterization of enzymatic reduction of hexavalent chromium by *Escherichia coli*, ATCC 33456, *Appl. Environ. Microbiol.* **59**, 3771–3777.

SHUMATE, S. E., STRANDBERG, G. W. (1985), Accumulation of metals by microbial cells, in: *Comprehensive Biotechnology*, (MOO-YOUNG, M., ROBINSON, C. N., HOWELL, J. A., Eds.), pp. 235–247. New York: Pergamon Press.

SHUMATE, S. E., STRANDBERG, G. W., MCWHIRTER, D. A., PARRON, J. R., BO-GACKI, G. M. et al. (1980), Separation of heavy metals from aqueous solutions using "biosorbents" – development of contracting devices for uranium removal, *Biotechnol. Bioeng. Symp.* **10**, 27–34.

SILVER, S. (1996), Bacterial resistances to toxic metal ions – a review, *Gene* **179**, 9–19.

SILVER, S. (1998), Genes for all metals – a bacterial view of the periodic table, *J. Ind. Microbiol. Biotechnol.* **20**, 1–12.

SMITH, D. G. (1974), Tellurite reduction in *Schizosaccharomyces pombe*, *J. Gen. Microbiol.* **83**, 389–392.

SMITH, W. L., GADD, G. M. (2000), Reduction and precipitation of chromate by mixed culture sulphate-reducing bacterial biofilms, *J. Appl. Microbiol.* **88**, 983–991.

SPEAR, J. R., FIGUEROA, L. A., HONEYMAN, B. D. (1999), Modeling the removal of uranium U(VI) from aqueous solutions in the presence of sulfate reducing bacteria, *Environ. Sci. Technol.* **33**, 2667–2375.

SPEAR, J. R., FIGUEROA, L. A., HONEYMAN, B. D. (2000), Modeling reduction of uranium U(VI) under variable sulfate concentrations by sulfate-reducing bacteria, *Appl. Environ. Microbiol.* **66**, 3711–3721.

SREEKRISHNAN, T. R., TYAGI, R. D. (1994), Heavy metal leaching from sewage sludges: a techno-economic evaluation of the process options, *Environ. Technol.* **15**, 531–543.

STARLING, A. P., ROSS, I. S. (1991), Uptake of zinc by *Penicillium notatum*, *Mycol. Res.* **95**, 712–714.

STERFLINGER, K. (2000), Fungi as geologic agents, *Geomicrobiol. J.* **17**, 97–124.

STOLZ, J. F., OREMLAND, R. S. (1999), Bacterial respiration of arsenic and selenium, *FEMS Microbiol. Rev.* **23**, 615–627.

STORK, A., JURY, W. A., FRANKENBERGER, W. T. (1999), Accelerated volatilization rates of selenium from different soils, *Biol. Trace Element Res.* **69**, 217–234.

Strandberg, G. W., Shumate, S. E., Parrott, J. R. (1981), Microbial cells as biosorbents for heavy metals – accumulation of uranium by *Saccharomyces cerevisiae* and *Pseudomonas aeruginosa*, *Appl. Environ. Microbiol.* **41**, 237–245.

Strasser, H., Burgstaller, W., Schinner, F. (1994), High yield production of oxalic acid for metal leaching purposes by *Aspergillus niger*, *FEMS Microbiol. Lett.* **119**, 365–370.

Suhasini, I. P., Sriram, G., Asolekar, S. R., Sureshkumar, G. K. (1999), Nickel biosorption from aqueous systems: studies on single and multimetal equilibria, kinetics, and recovery, *Sep. Sci. Technol.* **34**, 2761–2779.

Sukla, L. B., Kar, R. N., Panchanadikar, V. (1992), Leaching of copper converter slag with *Aspergillus niger* culture filtrate, *Biometals* **5**, 169–172.

Tamaki, S., Frankenberger, W. T. (1992), Environmental biochemistry of arsenic, *Rev. Environ. Contam. Toxicol.* **124**, 79–110.

Taylor, M. R. G., McLean, R. A. N. (1992), Overview of clean-up methods for contaminated sites, *J. Inst. Water Environ. Management* **6**, 408–417.

Taylor, D. E., Walter, E. G., Sherburne, R., Bazettjones, D. P. (1988), Structure and location of tellurium deposited in *Escherichia coli* cells harboring tellurite resistance plasmids, *J. Ultrastruct. Mol. Struct. Res.* **99**, 18–26.

Tebo, B. M., Obraztsova, A. Y. (1998), Sulfate-reducing bacterium grows with Cr(VI), U(VI), Mn(IV), and Fe(III) as electron acceptors, *FEMS Microbiol. Lett.* **162**, 193–198.

Tereshina, V. M., Marin, A. P., Kosyakov, V. N., Kozlov, V. P., Feofilova, E. P. (1999), Different metal sorption capacities of cell wall polysaccharides of *Aspergillus niger*, *Appl. Biochem. Microbiol.* **35**, 389–392.

Texier, A.-C., Andres, Y., Le Cloirec, P. (1999), Selective biosorption of lanthanide (La, Eu, Yb) ions by *Pseudomonas aeruginosa*, *Environ. Sci. Technol.* **33**, 489–495.

Texier, A.-C., Andres, Y., Illemassene, M., Le Cloirec, P. (2000), Characterization of lanthanide ions binding sites in the cell wall of *Pseudomonas aeruginosa*, *Environ. Sci. Technol.* **34**, 610–615.

Thamdrup, B. (2000), Bacterial manganese and iron reduction in aquatic sediments, *Adv. Microb. Ecol.* **16**, 41–84.

Thompson-Eagle, E. T., Frankenberger, W. T. (1992), Bioremediation of soils contaminated with selenium, in: *Advances in Soil Science* (Lal, R., Stewart, B. A., Eds.), pp. 261–309. New York: Springer-Verlag.

Tichy, R., Janssen, A., Grotenhuis, J. T. C., Lettinga, G., Rulkens, W. H. (1994), Possibilities for using biologically-produced sulphur for cultivation of thiobacilli with respect to bioleaching processes, *Bioresource Technol.* **48**, 221–227.

Ting, Y. P., Sun, G. (2000), Use of polyvinyl alcohol as a cell immobilization matrix for copper biosorption by yeast cells, *J. Chem. Technol. Biotechnol.* **75**, 541–546.

Tobin, J. M., Cooper, D. G., Neufeld, R. J. (1984), Uptake of metal ions by *Rhizopus arrhizus* biomass, *Appl. Environ. Microbiol.* **47**, 821.

Tobin, J. M., White, C., Gadd, G. M. (1994), Metal accumulation by fungi – applications in environmental biotechnology, *J. Ind. Microbiol.* **13**, 126–130.

Tohoyama, H., Inouhe, M., Joho, M., Murayama, T. (1995), Production of metallothionein in copper-resistant and cadmium-resistant strains of *Saccharomyces cerevisiae*, *J. Ind. Microbiol.* **14**, 126–131.

Tolley, M. R., Strachan, L. F., Macaskie, L. E. (1995), Lanthanum accumulation from acidic solutions using a *Citrobacter* sp. immobilized in a flow-through bioreactor, *J. Ind. Microbiol.* **14**, 271–280.

Townsley, C. C., Ross, I. S. (1986), Copper uptake in *Aspergillus niger* during batch growth and in non-growing mycelial suspensions, *Exp. Mycol.* **10**, 281–288.

Townsley, C. C., Ross, I. S., Atkins, A. S. (1986), Copper removal from a simulated leach effluent using the filamentous fungus *Trichoderma viride*, in: *Immobilisation of Ions by Biosorption* (Eccles, H., Hunt, S., Eds.), pp. 159–170. Chichester: Ellis Horwood.

Trevors, J. T., Stratton, G. W., Gadd, G. M. (1986), Cadmium transport, resistance and toxicity in algae, bacteria and fungi, *Can. J. Microbiol.* **32**, 447–464.

Tsezos, M. (1983), The role of chitin in uranium adsorption by *Rhizopus arrhizus*, *Biotechnol. Bioeng.* **25**, 2025.

Tsezos, M. (1984), Recovery of uranium from biological adsorbents – desorption equilibrium, *Biotechnol. Bioeng.* **26**, 973.

Tsezos, M. (1986), Adsorption by microbial biomass as a process for removal of ions from process or waste solutions, in: *Immobilisation of Ions by Biosorption* (Eccles, H., Hunt, S., Eds.), pp. 201–218. Chichester: Ellis Horwood.

Tsezos, M., Keller, D. M. (1983), Adsorption of RA-226 by biological origin absorbents, *Biotechnol. Bioeng.* **25**, 201–215.

Tsezos, M., Volesky, B. (1981), Biosorption of uranium and thorium, *Biotechnol. Bioeng.* **23**, 583.

Tsezos, M., Volesky, B. (1982), The mechanism of uranium biosorption by *Rhizopus arrhizus*, *Biotechnol. Bioeng.* **24**, 385.

Turpeinen, R., Pantsarkallio, M., Haggblom, M., Kairesalo, T. (1999), Influence of microbes on the mobilization, toxicity and biomethylation of arsenic in soil, *Sci. Total Environ.* **236**, 173–180.

TZEFERIS, P. G., AGATZINI, S., NERANTZIS, E. T. (1994), Mineral leaching of non-sulphide nickel ores using heterotrophic microorganisms, *Lett. Appl. Microbiol.* **18**, 209–213.

UHRIE, J. L., DREVER, J. I., COLBERG, P. J. S., NESBITT, C. C. (1996), *In situ* immobilization of heavy metals associated with uranium leach mines by bacterial sulphate reduction, *Hydrometallurgy* **43**, 231–39.

VACHON, P., TYAGI, R. D., AUCLAIR, J. C., WILKINSON, K. J. (1994), Chemical and biological leaching of aluminium from red mud, *Environ. Sci. Technol.* **28**, 26–30.

VALLS, M., GONZÁLEZ-DUARTE, R., ATRIAN, S., DE LORENZO, V. (1998), Bioaccumulation of heavy metals with protein fusions of metallothionein to bacterial OMPs, *Biochimie* **80**, 855–861.

VALLS, M., DE LORENZO, V., GONZÁLEZ-DUARTE, R., ATRIAN S. (2000), Engineering outer-membrane proteins in *Pseudomonas putida* for enhanced heavy-metal bioadsorption, *J. Inorg. Biochem.* **79**, 219–223.

VEGLIO, F. (1996), The optimisation of manganese-dioxide bioleaching media by fractional factorial experiments, *Proc. Biochem.* **31**, 773–785.

VEGLIO, F., BEOLCHINI, F., GASBARRO, A., TORO, L. (1999a), *Arthrobacter* sp. as a copper biosorbing material: ionic characterisation of the biomass and its use entrapped in a poly-hema matrix, *Chem. Biochem. Eng. Quart.* **13**, 9–14.

VEGLIO, F., BEOLCHINI, F., BOARO, M., LORA, S., CORAIN, B., TORO, L. (1999b), Poly(hydroxyethyl methacrylate) resins as supports for copper (II) biosorption with *Arthrobacter* sp.: matrix nanomorphology and sorption performances, *Proc. Biochem.* **34**, 367–373.

VIEIRA, M. J., MELO, L. F. (1995), Effect of clay particles on the behaviour of biofilms formed by *Pseudomonas fluorescens*, *Wat. Sci. Technol.* **32**, 45–52.

VILE, M. A., WIEDER, R. K. (1993), Alkalinity generation by Fe(III) reduction versus sulfate reduction in wetlands constructed for acid-mine drainage treatment, *Water Air Soil Pollut.* **69**, 425–441

VYMAZAL, J. (1987), Toxicity and accumulation of cadmium with respect to algae and cyanobacteria – a review, *Tox. Assess.* **2**, 387.

WAINWRIGHT, M., GADD, G. M. (1997), Industrial pollutants, in: *The Mycota*, Vol. IV: *Environmental and Microbial Relationships* (WICKLOW, D. T., SODERSTROM, B., Eds.), pp. 86–97. Berlin: Springer-Verlag.

WAINWRIGHT, M., GRAYSTON, S. J. (1986), Oxidation of heavy-metal sulphides by *Aspergillus niger* and *Trichoderma harzianum*, *Trans. Br. Mycol. Soc.* **86**, 269.

WAINWRIGHT, M., GRAYSTON, S. J., DE JONG, P. (1986), Adsorption of insoluble compounds by mycelium of the fungus *Mucor flavus*, *Enzyme Microb. Technol.* **8**, 597.

WALTER, E. G., TAYLOR, D. E. (1992), Plasmid-mediated resistance to tellurite: expressed and cryptic, *Plasmid* **27**, 52–64.

WANG, P,-C., MORI, T., KOMORI, K., SASATU, M., TODA, K. et al. (1989), Isolation and characterisation of an *Enterobacter cloacae* strain that reduces hexavalent chromium under anaerobic conditions, *Appl. Environ. Microbiol.* **55**, 1665–1669.

WANG, C. L., MARATUKULAM, P. D., LUM, A. M., CLARK, D. S., KEASLING, J. D. (2000), Metabolic engineering of an aerobic sulfate reduction pathway and its application to precipitation of cadmium on the cell surface, *Appl. Environ. Microbiol.* **66**, 4497–4502.

WATSON, J. H. P., ELLWOOD, D. C. (1994), Biomagnetic separation and extraction process for heavy metals from solution, *Minerals Eng.* **7**, 1017–1028.

WATSON, J. H. P., ELLWOOD, D. C., DENG, Q. X., MIKHALOVSKY, S., HAYTER, C. E. et al. (1995), Heavy metal adsorption on bacterially-produced FeS, *Minerals Eng.* **8**, 1097–1108.

WATSON, J. H. P., ELLWOOD, D. C., DUGGLEBY, C. J. (1996), A chemostat with magnetic feedback for the growth of sulphate-reducing bacteria and its application to the removal and recovery of heavy metals from solution, *Minerals Eng.* **9**, 937–983.

WATSON, J. H. P., CRESSEY, B. A., ROBERTS, A. P., ELLWOOD, D. C., CHARNOCK, J. M. et al. (2000), Structural and magnetic studies on heavy-metal-adsorbing iron sulphide nanoparticles produced by sulphate-reducing bacteria, *J. Magn. Magn. Mater.* **214**, 13–30.

WEBB, J. S., MCGINNESS, S., LAPPIN-SCOTT, H. M. (1998), Metal removal by sulphate-reducing bacteria from natural and constructed wetlands, *J. Appl. Microbiol.* **84**, 240–248.

WEBER, W. J. (1972), Adsorption, in: *Physico-Chemical Processes for Water Quality Control* (WEBER, W. J., Ed.), pp. 199–259. New York: John Wiley & Sons.

WEEGER, W., LIEVREMONT, D., PERRET, M., LAGARDE, F., HUBERT, J. C. et al. (1999), Oxidation of arsenite to arsenate by a bacterium isolated from an aquatic environment, *BioMetals* **12**,141–149.

WHITE, C., GADD, G. M. (1986), Uptake and cellular distribution of copper, cobalt and cadmium in strains of *Saccharomyces cerevisiae* cultured on elevated concentrations of these metals, *FEMS Microbiol. Ecol.* **38**, 277–283.

WHITE, C., GADD, G. M. (1987a), The uptake and cellular distribution of zinc in *Saccharomyces cerevisiae, J. Gen. Microbiol.* **133**, 727–737.

WHITE, C., GADD, G. M. (1987b), Inhibition of H^+ efflux and induction of K^+ efflux in yeast by heavy metals, *Tox. Assess.* **2**, 437–444.

WHITE, C., GADD, G. M. (1996a), Mixed sulphate-reducing bacterial cultures for bioprecipitation of toxic metals: factorial and response-surface analysis of the effects of dilution rate, sulphate and substrate concentration, *Microbiology* **142**, 2197–2205.

WHITE, C., GADD, G. M. (1996b), A comparison of carbon/energy and complex nitrogen sources for bacterial sulphate reduction: potential applications to bioprecipitation of toxic metals as sulphides, *J. Ind. Microbiol.* **17**, 116–123.

WHITE, C., GADD, G. M. (1997), An internal sedimentation bioreactor for laboratory-scale removal of toxic metals from soil leachates using biogenic sulphide precipitation, *J. Ind. Microbiol. Biotechnol.* **18**, 414–421.

WHITE, C., GADD, G. M. (1998a), Accumulation and effects of cadmium on sulphate-reducing bacterial biofilms, *Microbiology* **144**, 1407–1415.

WHITE, C., GADD, G. M. (1998b), Reduction of metal cations and oxyanions by anaerobic and metal-resistant microorganisms: chemistry, physiology and potential for the control and bioremediation of toxic metal pollution, in: *Extremophiles: Microbial Life in Extreme Environments*, (HORIKOSHI, K., GRANT, W. D., Eds.), pp. 233–254. New York: John Wiley & Sons.

WHITE, C., GADD, G. M. (2000), Copper accumulation by sulphate-reducing bacterial biofilms and effects on growth, *FEMS Microbiol. Lett.* **183**, 313–318.

WHITE, C., WILKINSON, S. C., GADD, G. M. (1995), The role of microorganisms in biosorption of toxic metals and radionuclides, *Int. Biodeter. Biodegrad.* **35**, 17–40.

WHITE, C., SAYER, J. A., GADD, G. M. (1997), Microbial solubilization and immobilization of toxic metals: key biogeochemical processes for treatment of contamination, *FEMS Microbiol. Rev.* **20**, 503–516.

WHITE, C., SHARMAN, A. K., GADD, G. M. (1998), An integrated microbial process for the bioremediation of soil contaminated with toxic metals, *Nature Biotechnol.* **16**, 572–575.

WINGE, D. R., JENSEN, L. T., SRINIVASAN, C. (1998), Metal ion regulation of gene expression in yeast, *Curr. Opin. Chem. Biol.* **2**, 216–221.

WOOD, J. M., WANG, H.-K. (1983), Microbial resistance to heavy metals, *Environ. Sci. Technol.* **17**, 582.

WU, J. S., SUNG, H. Y., JUANG, R. J. (1995), Transformation of cadmium-binding complexes during cadmium sequestration in fission yeast, *Biochem. Mol. Biol. Int.* **36**, 1169–1175.

YAKUBU, N. A., DUDENEY, A. W. L. (1986), Biosorption of uranium with *Aspergillus niger*, in: *Immobilisation of Ions by Biosorption* (ECCLES, H., HUNT, S., Eds.), pp. 183–200. Chichester: Ellis Horwood.

YANG, J. B., VOLESKY, B. (1999), Modeling uranium-proton ion exchange in biosorption, *Environ. Sci. Technol.* **33**, 4079–4085.

YIN, P. H., YU, Q. M., JIN, B., LING, Z. (1999), Biosorption removal of cadmium from aqueous solution by using pretreated fungal biomass cultured from starch wastewater, *Water Res.* **33**, 1960–1963.

YONG, P., MACASKIE, L. E. (1995), Enhancement of uranium bioaccumulation by a *Citrobacter* sp. via enzymically-mediated growth of polycrystalline $NH_4UO_2PO_4$, *J. Chem. Technol. Biotechnol.* **63**, 101–108.

YU, W., FARRELL, R. A., STILLMAN, D. J., WINGE, D. R. (1996), Identification of SLF1 as a new copper homeostasis gene involved in copper sulfide mineralization in *Saccharomyces cerevisiae*, *Mol. Cell. Biol.* **16**, 2464–2472.

ZAGURY, G. J., NARASASIAH, K. S., TYAGI, R. D. (1994), Adaptation of indigenous iron-oxidizing bacteria for bioleaching of heavy metals in contaminated soils, *Environ. Technol.* **15**, 517–530.

ZHANG, Y. Q., FRANKENBERGER, W. T. (1999), Effects of soil moisture, depth, and organic amendments on selenium volatilization, *J. Environ. Qual.* **28**, 1321–1326.

ZHANG, Y. Q., FRANKENBERGER, W. T., MOORE, J. N. (1999), Effect of soil moisture on dimethylselenide transport and transformation to non-volatile selenium, *Environ. Sci. Technol.* **33**, 3415–3420.

ZHAO, H., EIDE, D. (1996a), The yeast *ZRT1* gene encodes the zinc transporter of a high affinity uptake system induced by zinc limitation, *Proc. Natl. Acad. Sci. USA* **93**, 2454–2458.

ZHAO, H., EIDE, D. (1996b), The *ZRT2* gene encodes the low affinity zinc transporter in *Saccharomyces cerevisiae*, *J. Biol. Chem.* **271**, 23203–23210.

ZHOU, J. L. (1999), Zn biosorption by *Rhizopus arrhizus* and other fungi, *Appl. Microbiol. Biotechnol.* **51**, 686–693.

ZIEVE, R., ANSELL, P. J., YOUNG, T. W. K., PETERSON, P. J. (1985), Selenium volatilization by *Mortierella* species, *Trans. Brit. Mycol. Soc.* **84**, 177–179.

ZINKEVICH, V., BOGDARINA, I., KANG, H., HILL, M. A. W., TAPPER, R. et al. (1996), Characterization of exopolymers produced by different isolates of marine sulphate-reducing bacteria, *Int. Biodet. Biodeg.* **37**, 163–172.

ZOUBOULIS, A. I., ROUSOU, E. G., MATIS, K. A., HANCOCK, I. C. (1999), Removal of toxic metals from aqueous mixtures. Part 1: Biosorption, *J. Chem. Technol. Biotechnol.* **74**, 429–436.

10 Microbial Corrosion and its Inhibition

WOLFGANG SAND
Hamburg, Germany

1 Introduction

Microbial corrosion or biocorrosion is known in the literature under several terms, of which microbial corrosion (or MIC = microbially influenced corrosion) and biodeterioration are the best known. Besides these, also the terms biodegradation, biomineralization, bioerosion, and bioleaching exist.

Although these terms are often used as synonyms, all of them possess a slightly different meaning, which leads consequently to confusion. The problem of these definitions results from their generation: materials scientists dealing with corrosion of metals and/or other materials use the term microbial or biocorrosion. The microbiologist uses biodeterioration to indicate the biological destruction of a material and, besides, the term biodegradation, if the material is microbiologically degraded to serve as a substrate (nutrient). The geologist/soil scientist uses bioerosion referring to the importance of microorganisms for the generation of soil. Also biomineralization is used to indicate that organic materials are fully degraded to inorganic compounds like CO_2 and H_2O. Finally, the hydrometallurgist uses bioleaching to indicate that microorganisms dissolve metal sulfides and make it possible to win precious metals out of resources, which would otherwise not be economically usable.

It becomes obvious from these explanations that microbial corrosion or biodeterioration is related to materials in the technical sense. The current definition of corrosion according to DIN 50900 is related to metals only. It defines that corrosion in case of metals is a reaction of the material with its environment, causing a measurable change of the material itself. This change may result in a damage. In most cases the reaction is of an electrochemical nature. However, chemical and/or processes of chemical or metal physical nature may occur. Corrosion damage is defined as an impairment of the function of a metallic construction element and/or a whole system (metal/medium) by corrosion. This definition does not include any type of microbiological involvement (MIC). Thus, all processes concerning equipment, constructional materials, hydrocarbon reservoirs, etc., will be dealt with, whereas processes concerning food/foodstuff, cosmetics, pharmaceuticals, and xenobiotics are not the subject of this review.

In general, biodeterioration seems to be the term which incorporates all the processes to be explained. Therefore, it shall be used throughout, although biocorrosion (MIC), especially in the technical field, is a well established expression. However, in case of organic materials like wood, plastics, or hydrocarbons one cannot speak of corrosion, since it is basically a destruction/degradation of the polymeric compounds to the small basic units, which consequently become metabolizable.

The interest in biodeterioration is a result of its economic impact. In pipeline and building maintenance, water systems, air conditioning, etc., the costs caused by biodeterioration are estimated to range annually in double-digit numbers of billions of Euros. If only some of these processes could successfully be described and consequently, be prevented, considerable economical gains would result. Already the problems of definition indicate that biodeterioration is a highly complex field requiring intensive interdisciplinary (research) efforts. Besides microbiology, disciplines like materials science, (electro)chemistry, physics, engineering, etc., need to become involved to find successful solutions. From the microbiological side the best way to describe the requirements is the term applied microbial ecology, which indicates that a profound knowledge of microorganisms, pathways of metabolism, interactions with the habitat/environment, etc., are an absolute requirement. Besides, knowledge on modern detection methods based on physical and molecular techniques is also a must. Due to the size of microorganisms it is always a problem to verify the involvement of microorganisms in deterioration processes. Especially colleagues, who deal with objects in the range of tons or kilotons, often have problems in accepting that microorganisms may contribute substantially to the deterioration of pipelines or buildings/monuments. Because of the size of a microorganism (either a ball of 1 μm diameter or a cylinder of 1 μm length and up to 1 μm diameter) a single cell is without any importance. To become deleterious, many cells have to act jointly. This means that a coordinated action of at least 10^7 cells is required to produce measurable effects and to become detectable, e.g., by light microscopy. The effect of their metabolism is of chemical nature, and later on, may also influence physical processes (see freeze–thaw effect).

According to an US estimate the contribution of microorganisms to the deterioration of materials as a whole may be in the range of 30%. This figure indicates how important biodeterioration is from an economic point of view.

2 History of Biodeterioration

Biodeterioration as such is a rather new subject of research. It slowly becomes recognized by the "classical" technical disciplines (and accepted as a reality). This is largely due to the missing education of, e.g., engineers at the university. Although on a worldwide scale quite a few people are actively dealing with biodeterioration, up to now none of the universities have an institute or a chair exclusively for this subject. Publications are also widespread in very different types of journals, besides that the publications are often lacking from a scientific point of view necessary data for sound explanations. In addition, the information is somewhat separated into microbiological and technical papers, which are either read by the engineer or the microbiologist. Textbooks on biodeterioration are very rare, although in the last years at least in corrosion books chapters about biodeterioration are included. This will hopefully result in an increased awareness by the responsible people for such kinds of problems. Nowadays biodeterioration is considered to play a role, only if all classical types of (corrosion) failure explanations do not work.

This is somewhat astonishing, since mankind has used already thousands of years measures to protect their buildings, properties, or food against microbial deterioration. However, these were empirical measures, to be applied without causal explanation. To be mentioned are

- Drying: reduction of water activity
- Salt addition: reduction of water activity
- Sugar addition: reduction of water activity
- Boiling: killing of vegetative cells of microorganisms
- Smoking: addition of antimicrobials
- Fermentation: acidification to kill microorganisms and make the product hostile for further colonization
- Addition of hops (beer): introduction of antimicrobials like lumulin and lupulin
- To char lumber: to kill vegetative cells and retain an almost non-degradable surface
- Filtration: removal of microorganisms

These ancient techniques already comprise in general all strategies to fight biodeterioration. At present, the use of antimicrobials seems to be the major way, however, this shall certainly change due to environmental legislation and its requirements. Often the effect of such compounds, like tributyltin, is highly deleterious to the environment outcompeting the positive effects.

The first known documentation of a biodeterioration is in the third book of Moses of the Bible. There it is explained that walls of a house, which are overgrown by lichens, cannot be saved. They have to be taken down and the stones affected may not be used for further constructions.

In the **Roman Empire** (BC) the famous galleys were covered on their outside by lead coatings. This lead was kept in place by copper nails. To avoid corrosion elements between the two metals, the copper nails were also covered by lead. No further documentation about cases exists from the following centuries.

In 1870 the first report indicated that lumber was destroyed by the action of (wood-degrading) fungi. Previously it was believed that the deterioration was a result of a non-understood interaction between water and dead wood substance.

In 1884 the famous microbiologist ROBERT KOCH published his postulates (MADIGAN et al., 2000). Although he developed them for diseases, they are also fully applicable to biodeterioration cases. In the first step the disease-causing microorganism needs to be isolated. In the second step this organism must be identified and characterized. In the third step a reinfection with this microorganism shall produce the same disease as previously experienced and in the fourth step the same organism must be reisolated.

In 1885 started the use of $CuCO_3$ in vineyards. Also, copper was used for ship hulls to reduce fouling, and thus, avoid speed reduction because of increased friction.

In 1900 Creosote from coal tar was used for lumber protection.

In 1934 VAN DER VLUGT and von WOLZOGEN-KÜHR described the involvement of sulfate-reducing bacteria in anaerobic iron corrosion and developed the theory of cathodic depolarization.

In 1947 *Thiobacillus* (now: *Acidithiobacillus*) *ferrooxidans* was detected in Acid Mine/Rock Drainage as the causal agent for metal sulfide dissolution/sulfuric acid generation plus heavy metal pollution.

In 1962 the organization for economic development, OECD, founded an international research group for biodeterioration cases.

In 1965 in the USA a center for prevention of deterioration was founded, in order to do research on the influence of a humid climate on the function and reliability of electronic compounds. This was a result of the problems the US forces encountered in the World War II in the Pacific. Also in this year the first international symposium was held in London on the subject: microbial deterioration in the tropics.

In 1968 the first international biodeterioration symposium in Southampton took place.

In 1970 in Britain the Biodeterioration Society, BS, was founded. Almost at the same time the International Biodeterioration Research Group, IBRG, started its activity. Whereas the BS is concerned with all types and materials of biodeterioration, the IBRG is focused on man-made organic materials like plastics, resins, hydrocarbons, etc.

In 1984 the Panamerican Biodeterioration Society started activity. All these research groups finally founded **in 1987** the International Biodeterioration Association as head organization.

Since this date in many countries research groups have been organized, which deal with biodeterioration problems. For example, in Spain, in France, or in Germany active groups dealing with biodeterioration could be established. Often this was a consequence of problems and of financing. However, it still remains a matter of interest of individuals, to keep the knowledge and expertise available. In Europe, the European Union, EU, started to recognize the importance of biodeterioration in the late 1980s. Since 1992 in the framework of COST (Cooperation on Science and Technology in Europe) two actions were already supported by the EU dealing with the impact microorganisms have on materials (COST 511 and COST 520).

3 Microorganisms

Generally, all types of microorganisms can be involved in cases of biodeterioration. Thus, bacteria and cyanobacteria as prokaryota, algae (green, red, brown), lichens, yeasts, and fungi as eukaryotes may occur. All of them contribute usually and jointly in complex interactions by various mechanisms to biodeterioration.

It is important to take into account that prokaryotes and eukaryotes may occur jointly, because some countermeasures may work only for one or the other group (like antibiotics).

3.1 Bacteria and Cyanobacteria

These two groups essentially comprise the prokaryotes. Morphologically they are characterized by a few forms: rods, cocci, spirillae, and variable (pleomorphic). In some cases (fungus-like) mycelium-like structures occur.

In contrast to the low variability in morphology, an enormous versatility exists concerning metabolism. Except a few man-made materials (polymers and xenobiotics) all others are biologically degradable by microorganisms. This statement refers to organic materials, however, mineral and metallic materials may also be used sometimes as an energy source, sometimes as reactant for metabolic intermediates and/or end products, and sometimes as support for attachment. Besides this versatility, there is an enormous potential in reactivity. This is due to the surface area available for chemical reactions and to the rapid growth. Often doubling times under optimal conditions of less than one hour exist, with the lowest value known for *Escherichia coli* of about 20 min. Thus, if favorable conditions exist, microorganisms multiply exponentially and become possibly endangering.

The surface area of microorganisms is also an important factor. Under the assumption that a cubic cell has a volume of 1 μm^3, the surface area amounts to $6 \cdot 10^{-12}$ m^2 and the volume to 10^{-18} m^3. Thus, 10^{12} cells have a surface of 6 m^2 and a volume of 10^{-6} m^3 (=1 mL) only. This explains, why microorganisms may become highly dangerous: it is a result of the enormous surface, which can take up and metabolize nutrients. As an example the uptake rates for oxygen shall be compared for prokaryotes and eukaryotes. *Azotobacter* is able to use oxygen at 2,000 μL mg^{-1} dry weight per hour. In contrast, active mammalian cells in the liver or kidney can use oxygen only at 10 μL to 20 μL O_2 mg^{-1} dry weight per hour, and plant roots or leaves show rates of 0.5 μL to 4 μL O_2 mg^{-1} dry weight per hour. On the other hand, the size of bacteria makes the detection quite difficult, since simple visual inspection except in cases of mass growth with intensive turbidity and macrocolony formation does not detect these little living beings. For this purpose physical and/or chemical/physiological techniques need to be applied. Physical techniques comprise all types of microscopy (light, UV, IR, atomic force, electron, etc.). These are direct detection techniques. However, to be able to see microorganisms at least in case of light microscopy the presence of 10^6 cells mL^{-1} is necessary. Chemical/physiological techniques are always indirect detection techniques. They require the growth of an organism to become detectable, since at least 10^6 cells are required to exert a measurable, chemical effect like the oxidation (consumption) of a substrate like glucose or the formation of end products like sulfate. This may be achieved via plate count techniques (the most widely used technique) using solidified nutrient solutions, most-probable-number techniques with dilution steps, turbidity measurements, ATP content, fatty acid content and composition, chinones, immunology, PCR-technique, metabolic heat dissipation (microcalorimetry), pH and/or redox potential, etc. In all cases the nutrient solution used exerts a certain selective pressure on the microorganisms present, meaning that only the ones, which are optimally able to grow under the conditions applied, will proliferate and, thus, become detectable. For this reason, all techniques using selective pressure will only yield a section of the microflora present and, thus, never represent the full spectrum. In addition, so-called viable, but non-culturable cells, VBNC, exist. These are to the best of the existing knowledge dormant, size-reduced cells with, consequently, less nucleic acid, enzymes, etc., than normal cells. The VBNC cells are metabolically inactive and often need up to now unknown external metabolites and/or physical or chemical treatments to become resuscitated and active. Recently it has been shown by Szewzyk (2000) that hydrolactones, compounds used in quorum sensing (cell-to-cell signaling) function in revitalization. Whether this may be generalized, needs to be demonstrated. However, due to the reduced size these cells, often less than 0.1 μm in diameter, escape detection by filtration, etc.

An important feature of a prokaryotic cell is its structure and the organization within the cytoplasm. Based on this knowledge some successful measures may be developed for fighting the microbial attack on materials (as was done in medicine). Prokaryotes, according to their name, do not possess a separate compartment for their nucleic acid DNA (hence the name prokaryon). They also do not have other distinct cell organelles like plastids, chloroplasts, and mitochondria or any compartments within their cells, although intracytoplasmic membranes, stacks of membranes, and gas vacuoles exist. The flagella as means for motility consist of one or a bundle of fibrillae, which are not organized according to the 9+2 rule of eukaryotes but have a diameter of 14 nm. The material is flagellin, a protein, whose basic units have a molecular weight of 40 kDa, and are helically organized. The velocity of bacteria may in case of *Vibrio cholerae* reach up to 200 μm s^{-1}, which is equivalent to 60 times its own size. The ribosomes, place of protein production, consist of two subunits of 30 S and 50 S (Svedberg) units, in contrast to eukaryotic ones with 40 S and 60 S units. The cell wall contains chemical compounds, which are only rarely found in eukaryotes, namely N-acetyl-muramic acid, D-glucose, *meso*-diamino-pimelic acid, and D-alanine. The boundary layer of the cytoplasm, the cytoplasmic membrane, is supported on the outside by the murein sacculus. It contains the above mentioned compounds including N-acetylglucosamine. The latter is linked through β-1,4-glycosidic bonding to N-acetyl-muraminic acid and both constitute the backbone of the sacculus. On the outside of the sacculus follows the periplasmic space. It is limited by the outer membrane, which is like the cytoplasmic one a bilayer containing lipid molecules and proteins in various forms (buried or attached to the periplasmic space or, like porins, transmembranous) plus the extracellular polymeric substances, EPS (or formerly lipopolysaccharides). These EPS play a pivotal role in infection and also in biodeterioration. They are responsible not only for the attachment of cells to several surfaces and the localized growth, but also for the degradation of extracellular insoluble substrates. The EPS consist of sugars, uronic acids, proteins, lipids, nucleic acids, and complexed metals (cations). The sugars may be substituted by short chain acids like acetic acid. Together with the uronic acids they are responsible for (heavy) metal binding. The proteins may result from cell decay, but it becomes more and more clear that the EPS contain deliberately excreted proteins = exoenzymes, which catalyze processes like oxygen reduction, H_2O_2 degradation, and depolymerizations like protein to amino acids (proteinase) or cellulose to sugar (cellulase). Thus, the EPS are not merely responsible for attachment, surface colonization, and extracellular substrate degradation, but comprise an extension of the action radius of the cells far beyond their cell wall. Accordingly, in the EPS space conditions may exist, of which we have no information, but which are responsible for the often rapid degradation of a substrate, etc. For example, the bacterium *Acidithiobacillus ferrooxidans* contains in its EPS complexed iron(III) ions in a concentration of up to 50 g L^{-1}. Even though the biotope has a pH of 2, this concentration is far beyond one which is chemically stable in solution. Only due to the complexation/chelation by glucuronic acid residues with a stoichiometry of two acid residues to one iron(III) ion this concentration is achieved. Interestingly, also *Leptospirillum ferrooxidans*, a totally unrelated bacterium, but also able to live in the same biotope by catalyzing the same reactions, contains these complexed iron(III) ions in a similar concentration. The latter is a good example for the fact that environmental conditions enforce convergent evolutions.

3.2 Algae

Algae are eukaryotes living as primary producers on the conversion of solar energy (radiation) to cellular carbon. According to their pigmentation green, red, and brown species are differentiated. The formerly blue-green algae are prokaryotes and, due to their cellular organization, have to be considered as bacteria. Algae occur ubiquitously and some species are among the largest living organisms on earth like some of the brown algae (up to 100 m). As indicated above, they are usually phototrophs, but also mixotrophs and saprophytes occur. Their cells are with a minimum of 10 μm considerably larger than the ones of bacteria. They may contain chlorophylls of the a, b, c, and d types, besides phycobilines, carotenes, and xanthophylls. The cell is a truly eukaryotic cell with a distinct nucleus and all organelles (plastids, mitochondria, endoplasmic reticulum, etc.). The ribosomes contain 40 S and 60 S subunits. The cell wall usually contains polysaccharides. In case of cell division, mitosis and meiosis occur. The flagella are constructed according to the (9+2) rule: 9 peripheral double filaments plus 2 inner single filaments. In the growth and multiplication cycles phases with haploid and diploid type are known. Some species are able to fix nitrogen. The relevance for biodeterioration results from the fact that they are primary producers meaning a source of organic nutrient, that they excrete EPS, and that they produce extracellular organic acids and CO_2 (carbonic acid).

3.3 Lichens

Lichens are eukaryotic plants resulting from a symbiosis of a fungus with an alga or a cyanobacterium. In this symbiosis usually Ascomycetes occur, only rarely Basidiomycetes are found. Due to their symbiotic form of life, unique metabolic products are formed and to some extent excreted like the lichenic acids, aliphatic acids, deprides, quinones, dibenzofurane derivatives, etc. Like algae, some species grow endolithically. Cell multiplication occurs only vegetatively. Lichens are known to be primary colonizers of many substrata including extreme habitats. They are able to withstand temperatures of up to 70°C and several months of desiccation.

3.4 Fungi

Another extremely important group of microorganisms for biodeterioration are the fungi, eukaryotic organisms, which are in general saprophytes. Thus, many parasitic (pathogenic) forms exist and are a constant danger for living organisms. The fungal cell wall in general consists of chitin, a polymer of β-1,3-glycosidically linked N-acetyl-glucosamine. The vegetative forms of fungi are cells, hyphae, and mycelia. The latter may be branched or unbranched, with or without septa, and with one or many nuclei. The growth in case of hyphae often occurs apically. The mycelium may grow in, on, and above the surface of materials. Hyphae are usually at least 5 µm in diameter. Multiplication may occur sexually and asexually indicating the presence of mitosis and meiosis like in other eukaryotes.

Fungi have a strong secondary metabolism, although the meaning of this metabolism remains under discussion. Even processes like the production and excretion of antibiotics seem to be ambiguous. It seems that the secondary metabolism becomes important at the end of growth, when substrates are depleted, and when the cells need to get rid of metabolic intermediates. Fungi are the first organisms mankind has used in biotechnological processes for food production (beer, honey wine, bread, silage, etc.). Also the first countermeasures known against biodeterioration are directed against fungi: the decay of lumber was inhibited by charring it in the fire. In this way lignin- and cellulose-degrading fungi were killed and a recolonization became impossible.

Fungi consist of four main groups. All of them contain many organisms important for biodeterioration processes: Myxomycetes, Phycomycetes, Ascomycetes, and Basidiomycetes. The latter two comprise together 30% of all known fungal species. Another 30% of all species are categorized as fungi imperfecti, since no sexual multiplication has been observed for these species. They probably belong in their majority to the Ascomycetes. Species of utmost importance like *Candida* and *Rhodotortula* are fungi imperfecti. The importance of fungi results from the production of organic acids, alcohols, and solvents, EPS, CO_2, and the enormous degradative potential for organic compounds.

4 Microbial Growth

There are several factors influencing the growth of microorganisms, among these are availability of water, minerals, nutrients, oxygen or other gases, temperature, redox potential, and pH. All these factors influence the composition of the microbial biocenosis in a habitat. Each factor has a broad range which microorganisms can tolerate, mostly separated into more narrow areas for different kinds of microorganisms.

4.1 Water

The most important factor for microbial growth is the availability of water. The measure for availability is the water activity, a_w. It is given as the quotient of the vapor pressure of a solution containing dissolved compounds, P, divided by the vapor pressure of distilled water at the same temperature, P_0:

$$a_w = \frac{P}{P_0} \qquad (1)$$

A few examples are given in Tab. 1. It becomes obvious that bacteria and especially fungi can

Tab. 1. Minimum Water Potential for Growth of Bacteria and Fungi (STOLP, 1988)

a_w	Bar	MPa (25 °C)	Organism
0.96 to 0.94	−56 to −85	−5.6 to −8.5	gram-negative rods
0.93 to 0.90	−100 to −145	−10.0 to −14.5	bacilli
0.95 to 0.83	−71 to −257	−7.1 to −25.7	gram-positive cocci
0.83 to 0.75	−257 to −396	−25.7 to −39.6	halophiles
0.80 to 0.60	−307 to −703	−30.7 to −70.3	xerophytic fungi
0.71	−472	−47.2	*Aspergillus glaucus*
0.61	−681	−68.1	*Xeromyces bisporus*

resist "dryness" to a considerably larger extent than plants, animals, and especially human beings. Whereas humans already become restricted at a_w-values of 0.999, bacteria can resist easily. The a_w-values for bacteria are in the range of 0.9–0.8, with the lowest value found for halophiles (in salines) of 0.75. Fungi are even more resistant. Their lowest value is about 0.6. An example is the extremely dangerous *Xeromyces bisporus*, which grows in the lumber construction of roofs. For the degradation of the lumber the fungus needs water. Thus, it sends its hyphae often over distances of more than 30 m into the cellar/soil for water acquisition. On its way the hyphae are able to penetrate stone, mortar, or even concrete. Consequently, if a roof is infected with this fungus, all lumber has to be exchanged plus parts of the (mineral) walls, etc., to avoid reinfection. This makes such an infection very expensive for the owner. Besides, due to the danger of infection of neighboring buildings such cases have often to be notified to the local authorities.

4.2 Minerals

(Micro)organisms need for their metabolic processes a variety of minerals. To be mentioned are N, S, P, K, Ca, Mg, and Fe, which are needed in quite large amounts, and the trace minerals like Mn, Mo, Zn, Cu, Co, Ni, V, B, Cl, Na, Si, and Wo. The latter are used for the complementation of specific enzymes, electron transfer compounds, etc., as central atoms of prosthetic groups. Besides, some microorganisms need a supply of vitamins for growth. The need for these minerals and vitamins can be very specific and varies from one microorganism to another. In case of countermeasures against specific deterioration causing bacteria this need may allow for very efficient strategies.

4.3 Temperature

Microbial growth is limited by the availability of water and the stability of the cellular compounds like membranes, enzymes, etc. Thus, it is difficult to define a lower limit, since highly concentrated salt solutions have, even after being frozen at −30 °C, niches with liquid water. In these niches a few adapted microorganisms may survive, as research in Siberia and Arctic/Antarctica has shown. Consequently, it is difficult to determine an absolute lowest value. The upper known value for life is at present at 116 °C, temperatures found in the deep sea at hydrothermal vents. Around 130 °C the spontaneous decomposition of nucleic acids and proteins starts, consequently life should not be expected at temperatures considerably above the current limit of 116 °C. The term to describe the various temperature ranges in which microorganisms grow, are the following:

- Psychro- or cryophiles: temperature range <0 °C up to 20 °C
- Psychrotrophs: temperature range 0 °C to 30 °C
- Mesophiles: temperature range 20 °C to 45 °C
- Moderate thermophiles: temperature range 35 °C to 55 °C

- Thermophiles: temperature range 50 °C to 85 °C
- Extreme thermophiles: temperature range 75 °C to 95 °C
- Hyperthermophiles: temperature range >90 °C.

4.4 Oxygen Usage

Microorganisms may also be distinguished by their oxygen demand/usage. Strict aerobes have a fully oxygen-dependent metabolism. Facultative aerobes may use oxygen, but are also able to use other electron acceptors like inorganic Fe(III), Mn(IV), or organic compounds (succinate, pyruvate, etc.). The same is valid for facultative anaerobes, only that these organisms grow in contrast to the previously mentioned ones preferably under anaerobic conditions. Finally, the strict anaerobes need to be mentioned. These are organisms, for which oxygen is toxic and cannot be used for any type of metabolism.

4.5 Redox Potential

The redox potential is an indication of the oxidative/reductive potential of anions and cations solution. The value does not determine the growth of microorganisms, but allows to deduct the metabolic type to be present. Microbial life occurs in the range of (roughly) −500 mV and +800 mV. At pH 7 and standard conditions these values resemble the potentials of H_2 and O_2, respectively. All metabolic processes take place in the range between these upper and lower limiting values (biological oxygen/hydrogen reaction). As indicated in Tab. 2, the redox potential can be assigned to various types of metabolism. An addition of these compounds to the medium consequently will result in the modification of the dominant type of metabolism. This is simply a consequence of thermodynamics, since the larger the difference between electron donor and acceptor, the more energy is available for the growth of a microorganism. Consequently, the organisms able to use oxidants (electron acceptors) with a more positive redox potential grow better and, thus, outcompete those growing only with more reduced electron acceptors.

4.6 pH

The pH is an important parameter for description and limitation of microbial growth. Basically, three main categories need to be distinguished. At high proton strength/concentration acidophiles grow. They are subdivided into strong and moderate acidophiles with pH

Tab. 2. Thermodynamic Sequence for Reduction of Inorganic Substances (STEVENSON, 1986)

Reaction	E_H (pH 7.0)
Reduction of O_2 to H_2O	
$O_2 + 4H^+ + 4e^- \leftrightarrow 2H_2O$	816 mV
Reduction of NO_3^- to NO_2^-	
$NO_3^- + 2H^+ + 2e^- \leftrightarrow NO_2^- + H_2O$	421 mV
Reduction of Mn^{4+} to Mn^{2+}	
$MnO_2 + 4H^+ + 2e^- \leftrightarrow Mn^{2+} + 2H_2O$	396 mV
Reduction of Fe^{3+} to Fe^{2+}	
$Fe(OH)_3 + 3H^+ + e^- \leftrightarrow Fe^{2+} + 3H_2O$	−182 mV
Reduction of SO_4^{2-} to H_2S	
$SO_4^{2-} + 10H^+ + 8e^- \leftrightarrow H_2S + 4H_2O$	−215 mV
Reduction of CO_2 to CH_4	
$CO_2 + 8H^+ + 8e^- \leftrightarrow CH_4 + 2H_2O$	−244 mV

ranges of <0 to 3 and 2 to 5, respectively. The second category are the neutrophiles growing between pH 5 and 9. Finally, at low proton concentrations the alkaliphiles take over. These organisms are able to grow even at a pH of 12, as has been found in alkaline lakes (Lake Natron in Africa). Interestingly, fungi are more resistant to pH extremes than bacteria. Organisms like *Aspergillus niger* have been shown to occur at pH 2 and pH 10. However, generally at these extreme values specialized bacteria are the dominant microorganisms. Furthermore, in most cases fungi prefer slightly acidic conditions, whereas most bacteria thrive under neutral to slightly alkaline ones.

4.7 Carbon- and Energy Source

An important differentiation of microorganisms results from their source of carbon and/or energy (electrons). As shown in Tab. 3, several types need to be taken into account. If the energy results from solar radiation, the organism is called phototroph, whereas any chemical reaction results in chemotrophy. In case of chemotrophy, again two sources need to be differentiated. In case of an inorganic electron donor the term is lithotrophy, whereas in case of an organic one organotrophy results. Finally, the carbon source may be inorganic CO_2 from the air with the resulting term autotrophy or the carbon source may be another organic compound causing heterotrophy. By combination of these terms, the metabolism of a strain may quite precisely be described.

4.8 Biofilm and Biofouling

A biofilm is an assembly of microbial cells, living or dead, embedded in extracellular polymeric substances (EPS). In addition, organic and inorganic particles may also be present. Biofouling is a technical expression indicating that fouling took place on a technical surface, which is caused by living organisms. In the consequence of this fouling process, technical characterization of materials and/or apparatus are changed. Osmosis membranes do not function anymore due to biofilm/biofouling meaning that a film has formed, which clogs the pores causing an increased pressure to be applied. Heat exchangers loose their exchange capacity by biofouling, because of an insulating biofilm. Even corrosion may result from the formation of biofilm/biofouling, due to the formation of aeration/corrosion cells under biofilm covered and non-covered areas (cathode/anode). The primary process is caused by the adhesion of microorganisms to materials surfaces. As previously discussed, the EPS confer to microorganisms a means for attachment to surfaces of materials and, besides, seem to be involved in the degradation of insoluble substrates. Recent ecological research has indicated that about 90% of all microorganisms in the natural environment occur attached to surfaces. Only a small percentage seems to be free-living as planktonic organisms. The reasons for this behavior are the prevailing nutrient conditions. Under natural conditions a surplus of nutrients only extremely rarely exists. Thus, the natural nutritional state of a population is hunger. Only under artificial, man-made con-

Tab. 3. Catchwords for Description of Bacterial Metabolism

Term[a]	Energy Source	Hydrogen Source	Carbon Source
Phototrophy	light		
Chemotrophy	chemical reaction		
Lithotrophy		inorganic substance	
Organotrophy		organic substance	
Autotrophy			carbon dioxide
Heterotrophy			organic substance

[a] Combinations of these terms are possible like chemolithoautotrophy.

ditions this may be changed (pollution, biotechnology, etc.) and will result in totally different populations and behavior of these. In case of a hunger, meaning a shortage of nutrient, it is much more advantageous for a microorganism to remain at one site (attached) and not to use energy for movement than to actively search for nutrients. Besides, due to turbulence the available nutrients may be transported to the site of the attached organism. Consequently, the attached mode of growth is considerably more advantageous than the planktonic one. Further benefits of attached growth are the protection a biofilm may confer against desiccation, action of biocides, and/or other toxic compounds, grazing of protozoa (amoebae), and the possibilities for exchange of genetic material for adoption of new metabolic capabilities (lateral gene transfer).

It has also become obvious in the last years that the biofilm mode of growth confers to the cells an enlargement of their radius of action. For example, it is known for a long time that exoenzymes are responsible for the cleavage of polymers like cellulose, lignin, or protein. This fact was never connected with biofilm/biofouling. Recent evidence indicates that the exoenzymes are deliberately excreted and, somehow, in a yet unknown way, the cells collect the cleavage products. Thus, their radius of action becomes considerably enlarged, as has been shown for the process of bioleaching of pyrite (which can also be called a biocorrosion of a metal sulfide). Consequently, leaching bacteria are mainly found to be attached to surfaces of metal sulfides. This principle may in general be valid for organisms growing at the expense of solid substrates. Therefore also sulfate reducing bacteria, SRB, need to be included in this scheme, since in a biofilm they are protected against oxygen and, furthermore, the Fe(II) ions are delivered directly to the cells, to react with the metabolic end product H_2S for detoxification as insoluble iron sulfides, FeS and FeS_2. Explanations will be given in the section on metal biocorrosion (Sect. 9). Further reading on this important subject can be found in Sect. 15, since the scope of this review is considerably more general and biofilm is only one aspect therein.

Summarizing this section it has to be pointed out that the precise knowledge of all these growth parameters is of utmost importance in case of countermeasures to be applied against biodeterioration. Although never pure cultures are responsible for such problems, a change in the parameters discussed above often results in a substantial reduction of the problem. Therefore, a careful examination is needed to allow for a successful measure. Even more, it might help avoid the often recommended chemical weapon biocide usage, which is always detrimental to the environment.

5 Selected Metabolic Circles

5.1 Sulfur Cycle

Sulfur is an element occurring in many variations on earth. Its most frequent oxidation states are −2, 0, +2, +4, and +6, although due to the complex chemistry many intermediate values may occur, e.g., in polysulfides or polythionates. Consequently, the sulfur chemistry, which is of utmost importance in many cases of biocorrosion of metal and of many minerals, is often extremely difficult and not yet totally analyzed. In the crust of the earth sulfur occurs with a mean value of 520 mg kg^{-1}, in (sweet) water with 4 mg kg^{-1}, and in seawater with 905 mg kg^{-1}. The most abundant forms of sulfur in soil are gypsum, $CaSO_4 \cdot 2H_2O$, epsomite, $MgSO_4 \cdot 7H_2O$, sphalerite, ZnS, chalcopyrite, $CuFeS_2$, pyrite/marcasite, FeS_2, sulfide, H_2S or HS^-, and methylmercaptan, CH_3SH. Sulfide is the precursor of the other sulfides. It is generated either by protein degradation from the amino acids cysteine/cystine, which are degraded aerobically by serine sulfhydrase to serine and H_2S or by cysteine desulfhydrase, which produces H_2S, NH_3, and pyruvate anaerobically, or by sulfate reduction (not only of sulfate but also of sulfite, thiosulfate, polythionates, and elemental sulfur). Methylmercaptan results from the degradation of the amino acid methionine. The products are α-ketoglutarate, ammonia, and methylmercaptan. Microorganisms able to perform this degradation are *Clostridium* and *Pseudomonas* species.

The biological sulfur assimilation, especially for the formation of amino acids, usually proceeds via the reduction of sulfate to sulfide and the incorporation of the latter via serine to form cysteine. Organisms, which are able to perform these reactions, include *Salmonella typhimurium*, *Bacillus subtilis*, *Micrococcus luteus*, *Enterobacter aerogenes*, and *Aspergillus niger*. The biological sulfur dissimilation (or sulfur respiration) is in contrast from the point of mass turnover a much more important process than the assimilation. It is performed by sulfate reducing bacteria, which take the sulfate ion as electron acceptor reducing the sulfur (+6) to sulfide (−2) and the oxygen to water. In this way they are able to oxidize anaerobically organic compounds and hydrogen and to grow. The reactions of this process are well known and can be found in any microbiological textbook. Besides SRB other bacteria are also able to reduce oxidized sulfur compounds. *Clostridium pasteurianum*, e.g., has a sulfite reductase, which takes sulfite, but also NO_2^-, SeO_3^-, and NH_2OH as electron acceptors. For sulfite the end product may be sulfide. Other ways of dissimilation are the disproportionation of thiosulfate or sulfite as effected by *Desulfovibrio sulfodismutans*. Thiosulfate is oxidized in part to sulfate and a proton and is reduced in part to sulfide. 3 Mol of sulfite are oxidized to 3 Mol of sulfate and the electrons of this oxidation are used to reduce another mol of sulfite to sulfide. Both reactions are at the end exergonic: $\Delta G' = -21.9$ kJ mol^{-1} and -58 kJ mol^{-1}, respectively. Sulfide may again be oxidized, aerobically or anaerobically. In the latter case, which is mainly effected by photosynthetic bacteria and cyanobacteria, elemental sulfur results. A consequence of its formation and accumulation under these conditions is the formation of a sulfur deposit, which in the history of the earth has occurred at several places. These biogenic deposits can be distinguished from vulcanian ones by the isotope ratio of S^{32} to S^{34}. The ratio is determined according to the following equation:

$$\delta S^{34} = \frac{(S^{34} \div S^{32})_{\text{sample}} - (S^{34} \div S^{32})_{\text{meteorite}}}{(S^{34} \div S^{32})_{\text{meteorite}}} \qquad (2)$$

It is known that several enzymes prefer the lighter isotope to the heavier one. Consequently, by determination of the S^{32}/S^{34} ratio it becomes obvious, whether a deposit is of biogenic or magmatic origin. For this purpose the isotope ratios of carbon C^{12}/C^{13}, nitrogen N^{14}/N^{15}, or oxygen O^{16}/O^{18} may also be used. The reference standard is derived from meteoritic material, which is supposed to be free of any biological alterations.

For sulfur the ratio of S^{32}/S^{34} in meteorites is 22.22. The isotope S^{32} makes up about 95.1% of the sulfur, whereas S^{34} is present to about 4.2%. The other isotopes like S^{35}, etc., account for the remaining 0.7% of the total. Sulfur deposits exhibit ratios between 21.3 and 23.2 indicating that the ones with the value higher than the standard are most probably of biological origin.

Sulfide is especially a deleterious compound for metals and here for iron. It reacts at anodic dissolution sites with the ferrous ions to iron sulfide, which due to its low solubility product precipitates out of solution. Since iron sulfide is a semiconductor, its precipitate enlarges the area of the cathode. That causes in turn an improvement of the anodic dissolution process. Consequently, an increase of corrosion results.

The mostly aerobic oxidation of sulfide(s) results in the formation of sulfuric acid. The reaction sequence, as shown below, is by no means complete nor is it applicable to all microorganisms. There exists such a diversity in the pathways of sulfur compound oxidation that it is impossible within the scope of this review to give a complete overview (see Sect. 15). Basically sulfide is oxidized via sulfite and polythionates to sulfate. It is evident from Fig. 1 that the polythionates play an important role in this process. The decisive role is played by sulfanemonosulfonic acid, an unstable intermediate. Its degradation products include also elemental sulfur (not included above). Thus, in the course of this oxidation intermediate sulfur compounds may occur, which cannot all be mentioned here. The most important are the polysulfides, elemental sulfur, polythionates, and sulfite. The oxidation is effected by various microorganisms and occurs irrespective of pH and temperature. Gram-positive and gram-negative eubacteria, archaebacteria, and fungi are able to oxidize reduced

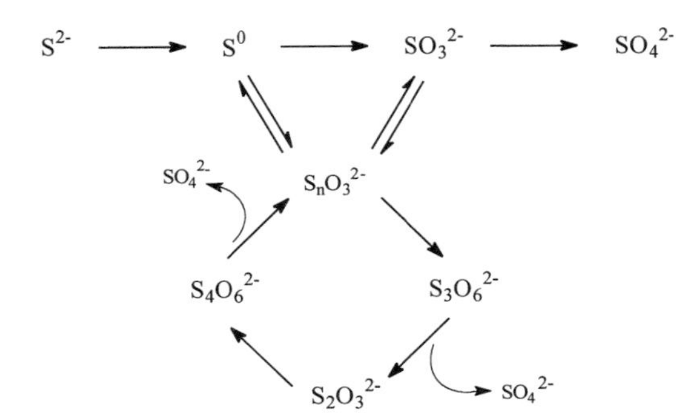

Fig. 1. Simplified scheme of the pathway of sulfide oxidation (adapted from SCHIPPERS et al., 1996).

sulfur compounds. The chemistry of this oxidation is in many cases very complex and not known in all details. There are also quite a few organisms known, which are able to oxidize sulfur compounds anaerobically. Furthermore, some organisms store elemental sulfur in globules intermittently inside or outside of their cells. Bacteria performing sulfur compound oxidation aerobically are thiobacilli, sulfobacilli, *Sulfolobus*, *Beggiatoa*, *Acidianus*, *Thiothrix*, *Thiovolum*, *Thiobacterium*, and many more.

Beggiatoa oxidizes sulfur compounds incompletely and stores elemental sulfur in globules intracellularly. The genus *Ectothiorhodospira* forms extracellular sulfur globules. It has recently been shown that these globules do not consist exclusively of S^8-sulfur rings, but also of S^6 and S^7 rings, besides polysulfides and polythionates of varying chain lengths. Anaerobic oxidation of sulfur compounds may be effected by photosynthetic bacteria such as *Ectothiorhodospira*, *Chromatium*, *Rhodopseudomonas*, etc., or chemolithotrophs like *Thiobacillus denitrificans* or the recently discovered, gigantic bacterium *Thiomargarita namibiensis* (SCHULZ et al., 1999) and others. With the phototrophs the end products are often only elemental sulfur, with *Thiobacillus* it is sulfuric acid. *Thiobacillus denitrificans* uses as electron acceptor nitrate, whereas *Acidithiobacillus thiooxidans* and *Acidithiobacillus ferrooxidans* have been shown to be able to use also iron(III) ions.

All of these sulfur compounds including elemental sulfur have been shown to be highly corrosive for metals like iron or copper. The mode of action is still not clear. The compounds, which usually are present as anions (except for elemental sulfur), may either migrate like chloride anions to anodic sites, and, thus, cause by comigration of protons a local acidification, or they can be reduced at a cathodic site to sulfide, which then can form with iron(II) ions the already described FeS precipitate with all the inherent consequences. In recent work it could be demonstrated that the occurrence of the compounds [sulfate and chromium reducible sulfur compounds (pyrite); VON RÈGE, 1999; VON RÈGE and SAND, 1999] exclusively result from bacterial activity and, thus, can be used as markers for microbially influenced corrosion, MIC.

5.2 Nitrogen Cycle

Nitrogen compounds also exert an important effect on materials. Especially ammonia, nitrous and nitric acid, besides the nitric oxides are well-known corrosive compounds for many materials. Nitrogen occurs in the biosphere in the inorganic forms of N_2 gas, NO_x gases, nitrous and nitric acid, and ammonia. Organic forms of nitrogen are urea, amino acids, peptides, proteins, amines, amides, nucleic acids, and humic substances. The degradation of the organic nitrogen compounds usually gives rise to compounds like urea and finally ammonia in a process called ammonification. Many bacteria from many groups may be involved. Ammonia is then oxidized via nitrous acid to nitric acid in a process called nitrification. In this process mainly two groups of bacteria participate. The ammonia oxidizers like *Nitrosomonas* convert ammonia via several steps to nitrous acid. Intermediate products are hydroxylamine and NO. The second group, the nitrite oxidizers like *Nitrobacter*, converts the nitrous acid into nitric acid. Although these groups are phylogenetically not related, they usually occur jointly in their habitats. Recent research indicated that neither *Nitrosomonas* nor *Nitrobacter* are the dominant nitrifying bacteria. Rather *Nitrosovibrio* and *Nitrospira* seem to be the dominant ones.

Starting with their end product nitrate a process called denitrification ensues, which under anaerobic conditions causes the formation of N_2 gas via the intermediates nitrite, NO, and N_2O. Again, a many organisms are able to perform this process. Besides, for the use of nitrate as a nitrogen source for cell compound synthesis like amino acids an enzymatic process called nitrate ammonification exists, which causes the reduction of nitrate to ammonia. However, from the point of mass turnover, this latter process is negligible. All organisms, which need to grow at the expense of nitrate as a nitrogen source, possess the necessary enzymes for this process. Finally, to close the cycle, nitrogen gas may be reduced by an enzyme system, nitrogenase, to ammonia. The system is unique for bacteria. It activates the inert triple bond of N_2 molecules by a high energy consumption. On a world wide scale, about 50% of all nitrogen is introduced into the biosphere by this process. The remaining amount comes from industrial ammonia production and processes like lightning or vulcanism.

5.3 Carbon Cycle

In this context only the turnover of the inorganic carbon compounds shall be discussed. Organic carbon compounds of importance are all organic acids and organic solvents. The description of their formation and degradation, basically the whole science of biochemistry, is clearly beyond the scope of this review. The majority of carbon occurs in form of either inorganic deposits like limestone and other carbonaceous rocks or as organic deposits like oil, gas, coal, etc. Less than 0.1% of the amount of these deposits is contained in the oceans in the form of dissolved CO_2. About 0.01% of carbon occurs in the form of dissolved organic compounds and in sediments and soils. The living biomass and the amount of atmospheric CO_2 are almost negligible, if compared with the aforementioned deposits. CO_2 is an important buffer system for pH. According to the following equation, it keeps the pH on this planet mostly in the neutral range, since HCO_3^- is the preferred form.

$$H_2O + CO_2 \leftrightarrow HCO_3^- + H^+ \leftrightarrow 2\,H^+ + CO_3^{2-} \quad (3)$$

The balance may be temporarily changed by proton or hydroxide ion introduction, but due to the atmospheric buffer this will be revised in the long term. CO_2, besides the bicarbonate anion, has due to this interchange a serious impact on materials consisting of carbonates and alkaline substances. Concrete is a good example. In the presence of CO_2 and water $Ca(OH)_2$ reacts with the carbonate anion to calcium carbonate and further to calcium bicarbonate, which finally causes the dissolution of concrete. Similar problems exist with sedimentary stones like carbonaceous sandstone, etc., and also with marble. High concentrations of CO_2, occurring possibly under microbial colonies in the course of metabolization of organic compounds, cause the dissolution of this material. This is one of the reasons, why marble on graves often shows signs of deterioration.

5.4 Iron Cycle

Iron as the most abundant element on earth is in many ways involved in the biosphere. It mostly occurs in the oxidation states +2 and +3, however, especially in human environments metallic iron is also available. Especially the latter is of interest for biodeterioration processes, since it comprises the valuable materials cast iron and steel (alloyed and non-alloyed). Iron reacts with acids by formation of iron(II) ions and hydrogen:

$$Fe + 2\,H^+ \rightarrow Fe^{2+} + H_2 \quad (4)$$

The acids can be biogenic like sulfuric or nitric acid. Another mechanism results from the attack of iron(III) ions:

$$Fe + 2\,Fe^{3+} \rightarrow 3\,Fe^{2+} \quad (5)$$

In this way metallic iron is corroded. Biogenic hydrogen sulfide has an especially negative effect on iron materials, since it forms highly insoluble complexes like FeS and/or FeS_2. These complexes, due to their semiconductor properties, allow the facilitated transfer of electrons to react with oxygen or other electron acceptors. Considerable corrosion problems are the consequence.

The main function of iron in the biosphere is the transfer of electrons. Switching from iron(II) ions to iron(III) ions and back allows the delivery and/or uptake of electrons for many biochemical processes. Main compounds involved are the heme groups of, e.g., cytochromes and iron–sulfur proteins. The nitrogen fixing enzyme nitrogenase contains an iron–molybdenum cofactor.

Iron(II) ions are an important source of energy for many microorganisms. Especially under acidic conditions (at pH above 3.5 iron(II) ions autoxidize rapidly) bacteria like *Acidithiobacillus ferrooxidans* and *Leptospirillum ferrooxidans* oxidize iron(II) to iron(III) ions and use the energy for metabolic purposes. In this context it is interesting to note that *Leptospirillum ferrooxidans* is probably to date the only organism known to be restricted to one substrate, the iron(II) ions. Other acidophilic microorganisms, however, growing at higher temperatures, are the gram-positive *Sulfobacillus* and the archaea *Acidianus*, etc. At slightly acidic conditions *Metallogenium* is effecting the oxidation of iron(II) ions. However, little is known about the growth conditions, since in its optimal pH range 3.5 to 5.5 due to autoxidation only little iron(II) ions can remain available (in solution). At neutral pH *Gallionella ferruginea* effects the oxidation of iron(II) ions. Typically this bacterium occurs in seeps containing iron(III) compounds. *Gallionella* manages to oxidize iron(II) ions at neutral pH in a microenvironment, which, due to its reduced O_2 concentration of only 1 mg L^{-1}, allows for about 20 mg L^{-1} of iron(II) ions to remain in solution. Another organism occurring in such an environment is *Leptothrix*, a species of the genus *Sphaerotilus*. However, about its mode of growth very little is known.

Iron(III) ions, in contrast to iron(II) ions, serve as electron acceptors and thus allow for the anaerobic oxidation of other, mostly organic compounds. The process is called anaerobic (iron)respiration. Resent research has shown that it is the major process for the mineralization of organic matter in sediments (LOVLEY, 1991), etc. Many bacteria and fungi are able to perform this reaction. This process is of significance, since in the course of degradation of organic compounds organic acids are often formed. These may contribute in many ways to biodeterioration. Besides, CO_2 is another product. Its significance has been discussed above.

6 Overview of Microbial Deterioration Mechanisms

Usually, biodeterioration of materials is the result of combined chemical, physical, and microbiological forces. Due to these interactions the cause of the process is often difficult to elucidate. According to KOCH's postulates the reinfection and generation of the same deterioration phenomena is a requirement. Therefore, a process simulation under conditions as natural as possible is often a requirement. It might allow the interacting forces to be differentiated. Biodeterioration is in nearly all cases a process occurring at interfaces between solid (including liquid) materials and water or gas (atmosphere). In the latter case water in the form of humidity has to be present to allow microbial life. Consequently, growth forms like microcolonies and biofilm play a pivotal role in such processes.

Organic as well as inorganic materials may be attacked and deteriorated by microorganisms. The types of deterioration may range from simple color changes to the formation of noxious gases or caustic compounds up to the total destruction of a single, a mixed, or a composite material. Depending on the function of a given piece of material the microbially influenced deterioration may be of considerable impact. For instance, the microbial attack on a historical church window may change the aesthetic impression so much that it must be considered as totally lost. Similar negative effects of microbial growth may occur on surfaces of materials like concrete, stone, or plastic. Also in case of metals, a microbial attack may cause an effect ranging from a deteriorated product to an endangerment of the environment.

Members of all groups of microorganisms may be involved in microbiologically influenced corrosion processes. The most important are bacteria, algae, lichens, and fungi. Protozoa contribute by grazing, i.e., by devouring of microorganisms. They regulate and maintain a stable microbial population. Microbiologically influenced deterioration of materials is a result of a complex interaction between different microorganisms. A single species (a pure culture) is never the only one responsible for a deterioration. Frequently, the accompanying flora permits the organisms causing the real deterioration to grow by creating optimum conditions. For instance, if organic compounds are available the accompanying flora consumes the oxygen and forms short chain organic acids. These acids and the lack of oxygen allow sulfate reducing bacteria to grow. As a consequence, hydrogen sulfide (H_2S) is produced which reacts with ferrous iron-releasing materials to form insoluble iron sulfide (FeS). Thus, these materials are deteriorated.

Despite the variety of microorganisms involved, the most important microbiological deterioration mechanisms can be classified by seven main categories.

6.1 Physical Presence

The physical presence of (microbial) cells may already lead to harmful effects. Especially the development in the production of electronic compounds like microchips caused the physical presence of cells to be a problem. Due to the reduced size even bacterial cells of 0.5 μm diameter are able to connect two electronic conductors. Structural elements of 4MB-chips are as small as 0.5 μm and have a critical size for lateral defects of 0.15 μm. Even microcells are able to overlap such distances.

6.2 Deterioration by Inorganic Acids

Some specialized microorganisms excrete a strong inorganic acid as an intermediate or end product of metabolism like sulfuric acid, H_2SO_4, or nitric acid, HNO_3 (Bock and Sand, 1993; Milde et al., 1983). Other biogenic acids are sulfurous acid, H_2SO_3, nitrous acid, HNO_2, and carbonic acid, H_2CO_3. Sulfurous and sulfuric acid are especially produced by bacteria of the genus *Thiobacillus*, but other bacteria like *Thiothrix* sp. or *Beggiatoa* are capable as well. In addition, some fungi are able to oxidize sulfur to sulfuric acid. However, a fungus like *Aureobasidium pullulans* can not use the energy from sulfur oxidation for growth. This is called co-oxidation. If materials are susceptible to an acid like metals or cement-bound constructional materials, growth of these specialized microorganisms and the concomitant excretion of acidic metabolites will cause a deterioration of the material. The following equations describe the reaction of calcium hydroxide, the binding component of cement, with (biogenic) sulfuric acid. The reaction product is calcium sulfate, i.e., gypsum:

$$Ca(OH)_2 + H_2SO_4 \rightarrow CaSO_4 + 2\,H_2O \qquad (6)$$

$$CaSiO_3 + H_2SO_4 \rightarrow CaSO_4 + H_2SiO_3 \qquad (7)$$

Simultaneously, the solubility of the calcium compound changes from low to moderately or fairly soluble. The mechanical strength of a cementitious element is based on the felted needles of the hydrated calcium hydroxide crystals, whereas the gypsum is not felted and is solubilized by water. A loss of the binder results in the loss of building material, which finally may end with a total loss of the building. Reports exist on concrete sewers where the pipes were totally degraded by biogenic sulfuric acid. The sewage flowed through the surrounding, solidified soil. Another report claimed a traffic accident of a tank lorry. It occurred at the site of a collapsing road because of a totally degraded sewer pipe. At sites with a strong attack by biogenic sulfuric acid corrosion pH values below 1 may occur. Thiobacilli are extremely acid tolerant and can resist these pH values. Especially the abundance of the species *Thiobacillus thiooxidans*, the most acid resistant *Thiobacillus*, indicates a strong attack on materials (Milde et al., 1983).

A similar damage is caused by biogenic nitric acid attack. Nitric acid as well as nitrous acid jointly occur. The formation of these acids is caused by bacteria of the groups of ammonia and nitrite oxidizers. However, heterotrophic nitrifiers need to be mentioned as well. The latter organisms oxidize nitrogen compounds to nitrate as well, but, like fungi in case of sulfur oxidation, are not able to use the energy evolved (co-oxidation).

Nitric acid reacts like sulfuric acid with calcareous materials dissolving the calcareous components. An important difference to biogenic sulfuric acid corrosion is the good solubility of nitrates. Thus, these salts do not necessarily accumulate at the site of their formation. Furthermore, at sites with biogenic nitric acid attack neutral pH-values are measured. Nitrifiers are inhibited by low concentrations of acids.

Both groups of microorganisms, thiobacilli and nitrifiers, use a similar strategy while deteriorating materials. They actively move into regions with optimum pH. For thiobacilli regions of neutral to moderately acidic, for nitrifiers regions of neutral to slightly alkaline pH are optimal. As a result, the microbial attack proceeds to the depth of the materials where enough buffering capacity remains (pH gradient).

The third inorganic acid is carbonic acid, H_2CO_3. It is excreted by all living beings. Whenever organic compounds are degraded, carbon dioxide, CO_2, is the end product. CO_2 is in equilibrium with hydrogen carbonate and carbonate when dissolved in water. Carbonic acid, especially in high concentrations, poses a threat to calcareous materials. It reacts like the other acids with calcium carbonate, $CaCO_3$, and calcium hydroxide transforming them into soluble calcium hydrogen carbonate, $Ca(HCO_3)_2$ (aggressive carbonic acid). Another product of the reaction between CO_2 and $Ca(OH)_2$ is $CaCO_3$. This process is known as carbonatization. It causes a pH drop in the surfacial water layer. In the case of cement the pH decreases from 12.5 to 8.5. A material with a pH of 12.5 is protected against microbial growth. At this pH almost all organisms are killed by saponification (alkaline hydrolysis), whereas a pH of 8.5 is adequate for microbial growth.

6.3 Deterioration by Organic Acids

Almost all microorganisms either temporarily or permanently excrete organic acids. As a rule, this is caused by unbalanced growth. Some steps of metabolism are rate determining (bottleneck). Thus, an intermediate compound may intracellularly accumulate. In this case it might be advantageous for the organism to avoid the accumulation of, e.g., acidifying compounds by excreting them. Sometimes fungi deliberately excrete organic acids to solubilize essential cations.

Basically, the action of organic acids is similar to that of inorganic acids. Strong and weak acids exist, for instance, acetic, gluconic, oxalic, citric, malic, succinic acid, and many others like sugar- or amino acids. All acids occurring in the general metabolism may be encountered. Besides the solubilizing effect of the acid, complexation is another type of action of organic acids. It will be discussed below. An attack of biogenic acids on materials may be difficult to detect, since these acids are metabolic end products of some bacteria but may be used by others. A detection of the action of organic acids often is only possible by effects like recrystallization. Only oxalic acid is mentioned as an exception. It comparably is of low solubility, especially after reaction with calcium to calcium oxalate. This compound may be used as a marker of an attack by biogenic organic acids.

6.4 Deterioration by Complexation

Besides the acid attack organic acids act by the complexation of cations. These complexes are chemically relatively stable. They allow microorganisms to overcome a temporary shortage. Additionally, insoluble compounds are attacked and dissolved by complexation. Into this group belong compounds like siderophores, which are excreted by microorganisms for the uptake of essential trace elements. Best known are those, which are excreted for iron uptake, e.g., by pathogenic bacteria. In the host, iron is an essential element, e.g., for hemoglobin formation. Microorganisms need iron for cytochrome formation, etc. Thus, high affinity compounds are produced by pathogenic microorganisms to ensure their iron supply. Some anti-pathogenic strategies use this fact by reducing the iron concentration in the serum. Consequently, these pathogenic bacteria are deprived of essential iron.

6.5 Deterioration by Organic Solvents

Many microorganisms can metabolize organic compounds under anaerobic conditions. If a suitable electron acceptor like nitrate or trivalent iron is not available. A mineralization is not possible and a fermentation results. Hydrogen (as redox equivalent) is transferred from one to another organic compound. By this transfer microbial growth is possible because of a substrate phosphorylation. As a result another organic compound is produced. Often these compounds are organic acids, as mentioned above, or organic solvents like ethanol, propanol, butanol, etc. They can react with organic materials of natural and/or synthetic origin and cause a swelling or a partial up to a total dissolution, etc. This finally results in biodeterioration.

6.6 Deterioration by Salt Stress

Anions of organic and inorganic acids react with cations like calcium from alkaline binding materials or with other metal cations to salts and water. If these salts remain within a porous system like natural stone, the water content of the system may rise because salts generally are hygroscopic. As a consequence the physical freeze–thaw attack will be enhanced. The formation of ice causes swelling attacks in the pore system of the material. If a desiccation occurs a crystallization of the salts results due to the solubility product. An especially deleterious salt in the case of concrete is ettringite, a gypsum-related salt. Due to the large amount of crystal water (as seen in its formula $Ca_6Al_2(SO_4)_3(OH)_{12} \cdot 24\ H_2O$) its volume is so large that it causes an extremely harmful swelling attack on concrete in sulfate-containing environments. Due to the growth of crystals another type of swelling attack is caused. A serious splitting off of facades may be the consequence.

6.7 Deterioration by H_2S, NO, and NO_2, or CO

Hydrogen sulfide (H_2S) is produced mainly by anaerobic biofilm forming microorganisms, the SRB (HAMILTON, 1985). They reduce oxidized sulfur compounds as sulfate, sulfite, thiosulfate, or sulfur to hydrogen sulfide. The SRB have an oxidative metabolism. They produce CO_2 by oxidizing short chain organic acids like lactate, acetate, or propionate that originate from fermentation processes (see biofilm). Electrons resulting from the oxidation of these organic compounds are used to reduce oxidized sulfur compounds. Hydrogen sulfide is formed and excreted into the medium. The oxygen chemically bound in the sulfur compounds is liberated as water. In contrast to denitrification, which is comparable to oxygen reduction, the reduction of sulfate yields only little energy for microbial growth. Sulfate as the most important sulfur compound ubiquitously occurs in water. Especially the high sulfate content of seawater favors the growth of SRE. It was believed that hydrogen sulfide was produced only under anaerobic conditions because of the toxicity of oxygen to SRB. This statement has to be re-evaluated. Recently it was shown that some SRB species tolerate oxygen and even are able to perform an oxidative metabolism with a low oxygen partial pressure (DILLING and CYPIONKA, 1991). Furthermore, it has recently been demonstrated that several species of SRB are able to use iron(III) ions as (preferred) electron acceptor. Obviously, these bacteria are quite versatile in the choice of their electron acceptor. H_2S is produced as well under aerobic conditions. The first step in the degradation of the sulfur containing amino acid cysteine (and its dimer cystine) is a splitting off of the thiol group as H_2S. This reaction considerably contributes to the H_2S pool in wastewater biotopes containing high amounts of amino acids. Methylotrophic bacteria metabolize other sulfur compounds like methylmercaptan, $CH_3–SH$, or dimethylsulfide, $CH_3–S–CH_3$, producing H_2S as well.

Hydrogen sulfide is a weak acid and reacts with cations forming sulfides. In the case of metal ions the corresponding sulfides are generally very little soluble and, thus, precipitate. Because in metal corrosion the pool of metal ions is always replenished by physicochemical processes, a continuous H_2S formation by SRB may result in a strong corrosion. The corrosive effect is enhanced by precipitated iron sulfide. This amplifies the depolarization of the metal surface and favors the formation of local corrosion cells. Furthermore, most SRB possess the enzyme hydrogenase. The enzyme permits SRB to use cathodically produced hydrogen, H_2. Hydrogen causes a cathodic passivation (protection) of a metal surface by cathodic polarization. It is believed that the hydrogenase uses the hydrogen, which would result in a depolarization. However, this ability is the subject of scientific discussion.

Another important role for hydrogen sulfide is its function as nutrient for sulfur oxidizing bacteria. In sewage pipelines it is the most important source of biogenic sulfuric acid (causing biogenic sulfuric acid corrosion).

The nitrogen oxides NO and NO_2 occur in soil, in water, in the atmosphere, and, NO in living organisms. They predominantly result by combustion processes. In living beings NO functions as a neurotransmitter explaining the toxicity of atmospheric NO to living beings. Biogenic NO gas may be an intermediate or final product of metabolic processes like nitrification and denitrification. Chemodenitrification may as well be a source of NO, e.g., by a chemical reaction of nitrite with sulfur dioxide. This reaction occurs wherever SO_2 (from combustion of gas, oil, coal, or wood) coincides with (biogenic) nitrite, NO_2^-. NO itself seems to be without detrimental effect on materials. It can be oxidized by microorganisms to nitrous and nitric acid. Therefore, NO can be considered as a precursor of a deteriorating agent. Like oxygen NO is only slightly water soluble (about 8 mg L^{-1}). It is quickly transformed into NO_2 by UV radiation. However, NO_2 is highly water soluble. By dissolution NO_2 reacts to nitrous and nitric acid. It is the anhydride of these two acids. Consequently, NO_2 is a deteriorating agent if humidity is available. It is also a strongly oxidizing agent. The influence of the two gases, NO and NO_2, on the atmosphere and on the ozone layer cannot be discussed here. Carbon monoxide, CO, may indirectly cause deleterious effects. Since it serves methanogenic bacteria as carbon source, organic acids like acetic or pyruvic acid may be formed. Besides, it can be reduced as electron acceptor to methane. CO oxidizing bacteria produce CO_2, which then may act as described above for carbonic acid.

6.8 Deterioration by Biological Growth on Surfaces – Biofouling and Biofilm

When growing on or in materials, microorganisms often excrete extracellular polymeric substances (EPS). The EPS facilitate the attachment of the cells to the surface of materials and embed the cells in a slime layer. The layer protects the cells against desiccation, against the action of toxic agents like biocides or heavy metals, against shearing forces, as well as against natural enemies like "grazing" protozoa. Consequently, if biofilm forming microorganisms need to be removed the concentration of the poison needs to be increased. Often the concentrations have to be increased by one or more orders of magnitude.

A biofilm can cause a deterioration, e.g., in the case of porous materials like natural stone or concrete. The EPS clog the pores and seal them towards the exterior. Thus, the diffusion of water vapor is inhibited and desiccation is retarded. Another consequence is a reduced accessibility for stone consolidating substances. The EPS contain, besides carbohydrates, ionic compounds like sugar acids (uronic acids), glycoproteins, amino acids, etc. These can additionally act as ion exchangers able to fix salts. The salts are hydrated and cause an increased water content. As mentioned above, an increased water content results in an increased susceptibility for a physical attack like the freeze–thaw attack.

Another possibly deleterious effect of a biofilm is the creation of anaerobic niches on the surface of materials. An aerobic metabolism in the well-aerated outer biofilm layer causes oxygen depletion. In the inner layer anaerobic metabolic processes like nitrate reduction, fermentation, and, finally, sulfate reduction take place. The consequences have been discussed above.

6.9 Deterioration by Exoenzymes and Emulsifying Agents

Microorganisms generally attack organic, biologically degradable materials by a similar strategy. Insoluble high molecular weight compounds are degraded into soluble, low molecular weight compounds by excreted exoenzymes. These enzymes act as depolymerases. In many cases an adhesion of the microorganisms to the surface of the material and the generation of a biofilm is required. The materials, which may be attacked, consist of metabolizable compounds. To mention are wood and wood containing materials, paper and pasteboard, leather, wool and cotton, some sorts of plastics, and hydrocarbons, waxes, adhesives, lacquers, paints, and lubricants.

As an example, the degradation of cellulose will be explained. Cellulose may be aerobically and anaerobically degraded. Cellulases decompose cellulose into the basic components cellobiose (two molecules of D-glucose linked α-1,4-glycosidically) and, finally, glucose (Eq. 8). The original molecule contains about 14,000 molecules of glucose and, consequently, is insoluble. Cellobiose and glucose are soluble, metabolizable compounds.

$$\text{Cellulose} \xrightarrow{\text{Cellulases}} \text{Cellobiose and/or Glucose} \quad (8)$$

The degradation of other polymers proceeds similarly. Depolymerases transform polymers into small, metabolizable mono-, di-, and trimers. In the case of natural materials and their derivatives microorganisms are presumably able to degrade and mineralize them.

This potential is called "microbial omnipotence". A different situation arises for polymeric compounds that do not occur in nature like polyether, polystyrene, polymethacrylate, etc. Microorganisms do not possess the necessary enzymes for degradation because of an unfamiliar chemical bonding. Therefore, these compounds are called xenobiotics or foreign compounds (in the biosphere). They do not occur sufficiently long in the living world to induce an evolution of appropriate degradative enzymes. For example, the experience of more than 70 years demonstrates that polymethacrylate (Plexiglass) is not microbially degradable. Recently it has been shown that besides these polymer-degrading exoenzymes others exist, which are able to react with dissolved oxygen and cause its reduction to water. These exoenzymes are probably excreted for metabolic purposes, although it is totally unknown up to now, how they function and how they are connected to cell metabolism. The evidence for their function is nevertheless quite strong, and several research apers demonstrated their effect and/or their inhibitation (by azide addition – VON RÈGE and SAND, 1999; SCOTTO and LAI, 1998).

Besides exoenzymes, microorganisms excrete compounds able to hydrophilize hydrophobic surfaces of materials. These substances are detergents or emulsifiers. For example, sulfur oxidizing bacteria excrete phospholipids (SHIVELY and BENSON, 1967) to hydrophilize hydrophobic sulfur or pyrite. Because of these phospholipids the solubility of sulfur in water increases from 5 to 20,000 μg L^{-1} (STEUDEL and HOLDT, 1988). The consequence is an increased biological degradability.

Summarizing these mechanisms it is obvious that countermeasures need a profound knowledge of the contribution of microorganisms to (bio)deterioration processes.

7 Countermeasures for Microbial Growth

Measures against microbial growth are numerous and chemical, physical, and/or biological weapons are used, sometimes in combination. The most important terms in this respect are bacteriostatic or bactericidal effect, sterilization, and conservation.

Bacteriostatic effect means that bacteria become due to the action of a compound or a physical influence growth-inhibited. However, the cells may remain alive. Bactericidal effect indicates that bacterial cells are killed by the action of a chemical compound and/or a physical influence. Sterilization is used for the removal of all organisms either by a physical process like filtration or by an action of a chemical compound or a physical effect.

Conservation finally is a term applied to food, etc., which becomes preserved by some kind of chemical or physical treatment, but may still contain some living cells (which do not multiply) and/or spores, etc.

The knowledge about countermeasures to reduce microbial growth seems to be as old as mankind. Any storage of food or the development of housing required the use of such principles, which were inhibitory to microorganisms. Processes like smoking, carbonizing, fermenting, etc., can be traced back into the early times of human development. The first written evidence in the European civilization appears in Greece. At about 1,000 years BC it was described that the burning of sulfur was used for deodorizing and disinfection. Obviously the oxidizing effect of SO_2 gas was used. In the middle ages at about 1,200 years AC letters were smoked to fight the transmission of pest and other diseases. An important development resulted in the 19th century by Lister, who recognized the disinfecting capacity of phenol. Lister introduced phenol into hospitals for room disinfection by aerosol formation and, at the same time, introduced the sterilization of bed cloths, etc., by applying formaldehyde treatment in pressure chambers. In 1949, Kaye and Phillips discovered the high reactivity of ethylene oxide and used it for the sterilization of medical equipment, bandaging materials, etc. The latter materials are often heat susceptible and were, thus, not sterilizable by autoclave treatment, etc. However, ethylene oxide is a cancerogenic compound and its application requires intensive measures to ensure its quantitative removal prior to the use of treated materials. For this reason, sterilization of sensitive equipment is also done by cobalt-60 (radioactive) bombardment. However, not all equipment withstands this type of treatment. At present, the ratio of ethylene oxide to cobalt-60 treatment is about 1 : 1.

The general requirements for sterilization agents are the following:

- highest effectiveness,
- rapid effect,
- good penetration,
- good tolerability,
- non-toxic,
- stable against organic compounds,
- versatility in use,
- controllability of effect,
- low price.

From this list it is clear that it is almost impossible to fulfill simultaneously all of these requirements. Currently, more than 800 chemical compounds are available on the market to deal with microorganisms. In addition these compounds are mixed with each other in various combinations and concentrations. Consequently, an immense number of products is available, of which effect and safety are not in all cases guaranteeable. For this reason the Commission of the European Communities has implemented in 1998 a directive "*European Biocidal Products Directive* (98/8/EC)" (123, Amtsblatt der Europäischen Gemeinschaften, Vol. 41, 1998, pp. 1–63) aiming at a systematic approach to register, categorize, and scrutinize the application of biocides for the various fields of application. The main idea is the protection of the consumers by increasing the knowledge about chemistry, application, toxicity, etc., of such compounds. It should never be forgotten that in order to be effective against living organisms like bacteria or fungi biocides must have a toxic effect. Consequently, other living beings including humans can be harmed in case of a faulty application.

Countermeasures for microorganisms have to be effective in three different types of growth media/conditions, air/atmosphere, water/solution, and soil/solid material.

7.1 Microorganisms in the Atmosphere

Microbial cells also occur in the atmosphere, inside and outside of buildings. The majority of microorganisms occurring in the air are gram-positives. Gram-negatives are found only to 0.1%. Pathogenic bacteria comprise 0.4%. Molds also occur to a significant amount. The primary source for the occurrence of these microorganisms are dust and aerosol particles in the air. Anthropogenic sources, which also contribute, are breathing, coughing, sneezing, etc. In addition, many microorganisms proliferate by expelling spores into the atmosphere, once maturation has finished. Another source of contamination is the dust in clothing, furniture, etc. Primary means to reduce the danger of contamination are consequently equipment like laminar-flow-benches, inoculation rooms (with UV light), and special clothing.

Air-bound microorganisms may pose a serious threat to humans. Especially air-conditioning installations produce aerosols containing microorganisms. This is one of the reasons why so many people have health problems in fully climatized buildings. For comparison, the number of colony-forming units (CFU) in the air outside of buildings ranges between 100 and 500 CFU m^{-3}. In an operating theatre only 10–70 CFU m^{-3} are allowed, which consequently requires the installation of filter systems. Especially critical conditions develop, if the temperature decreases below the dew point. In the ensuing process of condensate formation aerosol droplets are also included. This may lead to a massive accumulation of microorganisms. Consequently, intensive microbial growth may start fom the affected area. Examples for such occurrences are found in everyday life. Behind cupboards, etc., where heat bridges cause the wall to adopt the temperature of the outside, especially in winter cold spots develop. At these sites water vapor condenses and allows microorganisms to grow. Usually, black spots on paint or wallpaper develop, which is not only unwanted because of esthetical reasons, but also may be a hygienic problem. Besides malodors also spores may be emitted into the atmosphere possibly causing diseases like allergy. Often forgotten are sources like house animals, pets, and insects.

7.2 Microorganisms in Water

Aqueous solutions of whatever type are one of the main sources for microorganisms. For this reason big efforts are done worldwide to produce drinking or process water containing acceptable levels of microorganisms. Since water is crucial to all living beings, it is considered in many countries a nutrient with strict regulations. According to the German regulations drinking water must not contain more than 100 CFU per mL after incubation of a sample for 48 h at 20 °C. Microorganisms like *E. coli* must not be detectable at all, since this would indicate contamination with fecal waters. The conductivity should be below 2,000 μS cm^{-1}. To achieve this purity, many physical and chemical treatments are in use like UV, ozone, heating, oxidation, filtration, etc.

7.3 Microorganisms in Solid Matrices

Soils and sediments as well as products may contain microorganisms. The first two mentioned are one of the main sources of microorganisms. These microorganisms are the ones, which are most difficult to remove or to kill. This results from the fact that the surrounding matter needs to be treated together with the microorganisms to remove the latter. Especially in the case of equipment, machines, installations, buildings, etc., this is usually impossible. Many materials do not allow a thorough treatment due to their properties. In such cases only a reduction in the number of microorganisms is possible.

7.4 Physical Treatment

Several ways of physical measures exist to reduce the number of microorganisms. Widely used is the application of humid heat. The most gentle treatment is pasteurization. Either at 61.5 °C or 71 °C for 30 min or 15 s, respectively, mainly food and other susceptible materials are treated. These measures aim at killing living cells, mainly pathogens. It is also applied in continuous processes. In such cases the temperature is kept for some seconds at 80–85 °C. Tyndallization requires that the material/liquid, etc., is heated up to 70–100 °C for several times, typically at least two times, with intervals. In these intervals with ambient temperatures of 20–25 °C for 16–24 h it is hoped that spores will germinate and will become susceptible to the next heat period. Boiling/steaming at 100 °C is another method to reduce the microbial load of materials. This technique is often used to preserve food and materials, which do not tolerate higher temperatures. It requires typically one up to several hours. The use of the autoclave with pressurized steam is the first technique allowing a thorough sterilization of materials. Usually at temperatures up to 121 °C, which means a pressure of up to 2 bar, materials are incubated for periods of several minutes up to some hours. This treatment ensures that a germ-free product results. However, many materials are not able to withstand these conditions. To ensure the success of autoclave treatment, thermo- and/or bioindicators are often used. They indicate the temperature, which was achieved, or they contain heat-resistant spores and, thus, show, whether the latter have been killed. A variation of this process is the so-called UHT process or uperization, which is often used for milk, dairy products, and liquids like juices. UHT stands for ultra-high-temperature treatment and means that in a continuous flow-through process the liquid is heated to 130–150 °C for a few seconds. Microwaves are increasingly used for the sterilization of materials. In an electric field the water molecules/dipoles are energized by the apparatus. The energy finally results in heat, which kills microorganisms. Almost without any apparatus the well-known technique of desiccation or drying functions. Either solar radiation or the heat from sources like stoves is used to reduce the available water from materials and food. The water availability is reduced by this extraction and the water activity becomes too low to support life. More fierce is the direct application of dry heat to materials. Only few materials, mostly metals and ceramics, are able to withstand these measures. Metals may be annealed, which means a treatment at temperatures above 500 °C. Some materials may be burnt to kill microorganisms. The least efficient is the flaming of materials by their immersion into ethanol followed by burning the ethanol. Usually the time required for the killing of spores under these conditions (100 s at 300 °C) is not achieved. Dry heat is another measure to kill microorganisms. However, in contrast to humid hot air (pressurized) increased incubation times and periods are needed. For example, the spores of *Clostridium sporogenes* need at 180 °C an incubation time of 15 s, whereas in an autoclave at 121 °C a time of 15 min is sufficient. This means that dry heat is considerably more problematic for many materials. Sterilization by application of ionizing radiation is a widely used technique for susceptible materials. A disadvantage of radiation is the fact that solid materials are often impenetrable or reduce the X-rays so strongly that only the surface of a material is sterilized, not the interior part. Thus, only a few materials can be treated by this technique. Furthermore, especially polymers may pose a problem, since the chemical bonds may be unstable. The radiation source is mainly radioactive cobalt: Co^{60}. But also Cs^{137} is used sometimes. In special plants with high safety requirements plastic articles, solutions, etc., are sterilized by doses of 1 up to 20 Megarad. Doses below 1 Megarad kill only vegetative cells, spores remain alive. This would be a radiation-pasteurization. It is applied in some countries also to food. However, there is still some discussion continuing about the possible negative effects of this treatment due to the formation of radicals in the food. A second form are the electron rays. They are in use for about 30 years. The product is transported through an "electron curtain". The electrons are less damaging than ionizing radiation. However, their penetration depth is also reduced. Another form of radiation is UV radiation. UV light of wavelengths between 200 nm and 315 nm is used for this purpose. Especially the wavelengths 260 nm and 280 nm are important, because at these wavelengths nucleic acids show resonance to UV light. Consequently, by UV radiation they become energized and may break or form chemical alterations. UV ionizing radiation may only be used in air or liquid, not with solid materials. It is, thus, a measure restricted to surface application. For liquid materials filtration is often the method of choice. Depending on the size, filtration allows to separate microorganisms, viruses, and sometimes even molecules from a liquid. To sterilize water by filtration, filters with a pore diameter of 0.45 μm or even 0.2 μm are generally used. The size is determined by Hg porosimetry. However, the size of the pores is always only a medium value. Pores of larger and smaller diameter also occur. Consequently, some cells might pass such filters. For this reason the effectivity of filters is determined by test filtrations with solutions of about 100 L containing defined concentrations of microbial cells, e.g., of *Pseu-*

domonas diminuta. For drinking water it is required that 100 L water containing 10^6 cells mL^{-1} have to pass the filter within 18 min with the filtrate having only 13 cells L^{-1}. In case of 10^3 cells mL^{-1} the filtrate has to be obtained within 4 min and may not contain more than 4 cells L^{-1}. Another problem are the so-called ultra-micro-cells. These cells are formed in periods of starvation and may have cell sizes of 0.1 μm and below. Since they are still viable, they pose a serious threat, e.g., in the case of sterile pharmaceuticals. Furthermore, if filters are used in a continuous mode, microorganisms may settle on the surface and/or clog or even penetrate the filter itself. A filtration of high importance for providing drinking water is reverse osmosis, which is used in arid areas, where only saline water is available or under extreme conditions, where a total water recycling is necessary like a space mission. These are semipermeable membranes with an exclusion limit for molecules above 300 Da. By pumps a pressure is build up on one side of the membrane of 60 bar and more, which allows the water to permeate through the membrane, however, the salt is kept back. In many areas worldwide the drinking water supply is produced in this way. Ultrafiltration got its name from the filters, which have only pores of 0.05 to 0.01 μm diameter.

If a material cannot be treated by the previously discussed procedures, or because of economic reasons, the reduction of the amount of available water (water withdrawal), the water activity, a_w, may be an alternative. Procedures for withdrawal are drying, salt addition, sugar addition, or freeze-drying. Cooling below $-30\,°C$ or even in liquid nitrogen ($-197\,°C$) is another technique for preservation.

7.5 Chemical Treatment

For a chemical treatment of materials several groups of compounds may be considered: biocidal gases, heavy metals, antibiotics, biocides, water repellents, and coatings. Biocidal gases need the direct contact with the material to be treated and, hence, the microorganism to be removed. The gases usually fail to function sufficiently in case of porous materials (in the pores), in case of humidity, if materials are covered by a crust of, e.g., detritus, or if the materials are reactive by themselves. Formaldehyde, although cancerogenic, is in use since 1937. It is still the only proved gas for application at low temperatures in hospitals. Glutardialdehyde is used since 1975. It may cause corrosion with some materials. Ethylene oxide is cancerogenic and mutagenic. However, due to its high reactivity, it is still in use for susceptible materials. β-Propiolactone is cancerogenic and almost out of use. Ozone, highly toxic and reactive, is non-carcinogenic and non-mutagenic. In spite of its high reactivity it is still in use, although materials like rubber and a few metals are destroyed by its action. Carbon dioxide is used at high pressures to conserve food (8 bar for fruit juices). If increased pressures are used, CO_2 becomes toxic for microorganisms: 2,000 bar for 7 h at 60 °C kills even spores. Halogens like Cl_2, J_2, Br_2, ClO_2, and NaOCl are also in use. A modern process is using H_2O_2 vapor (plasma sterilization). The American FDA adopted it for general use in 1983. Peracetic acid as oxidant is another chemical compound. Antibiotics, although intensively used in medicine, are usually not usable in technical environments. These compounds are of biological origin, meaning that they are biodegradable, and often unstable under environmental conditions. Smoking as one of the oldest ways of humans to preserve food and/or materials is based on the effect of an introduction of antimicrobial compounds into materials. Acidification, either naturally by fermentation or by acid addition, is another traditional way of preservation. It is based on the fact that most spores do not germinate at a pH below 4. Preservatives are available in a large amount of different types and combinations. They are mainly used for food, pharmaceuticals, cosmetics, and special products like paints, hydraulic fluids, cutting fluids, etc. The main groups are alcoholic compounds, acids, PHB esters, phenol derivates, quaternary ammonium compounds (quats), chlorhexidines, organomercury compounds, silver salts, etc. Due to the vast amount of available products, it is impossible to list them at this place. The interested reader is referred to publications dealing with food preservation, etc.

It needs to be noted, however, that in case of biofilm/biofouling the recommended dosages are often far to low, due to the protection supplied by the EPS (VON RÈGE and SAND, 1998). Heavy metals are only exceptionally used, due to their non-degradability. However, a few applications remain, e.g., for mercury salts in the preservation of seeds. Antibiotics must only be used in medical cases. Besides, these compounds are usually labile and provide no long-term effect. Water repellents and coatings are aimed at an avoidance of water access to a material by either providing a hydrophobic surface or a physical separation layer, respectively.

7.6 Constructive Measures

In case of equipment or buildings constructive measures may help to avoid problems of biodeterioration. The principle to be applied while constructing is "to build a hostile environment" for microbes. This can be achieved in several ways. Very important is the removal of water. This can be done by drying, removal or avoiding heat/cold bridges, hydrophobization, coating, and ventilation. In all these measures the local microclimate should be considered. The use of appropriate materials for construction is another point on the agenda. Materials should be tested for their resistance against a biological attack under realistic conditions and prior to use (simulation!). Furthermore, in the process of construction care should be taken to avoid inaccessible areas, because the possibility of cleaning a machine is important. Besides, the surface roughness of materials needs to be considered. Finally, as a secondary treatment coating, hydrophobization, or the application of foils may be the measure of choice.

7.7 Biological Measures

The use of biological compounds like antibiotics has already been included in the chemical measures. Other measures play no role up to now. It may be possible in the future that grazing microorganisms, especially if genetically modified, may be used to remove populations of detrimental microorganisms. Also the use of phages and bdellovibrios may have some potential. However, these biomeasures suffer from serious problems in cases with biofilm, since the EPS provide some protection. It also remains unclear, whether the often discussed effect of covering a surface with harmless microbes "protective EPS" can really prevent biodeterioration. Besides, the principle of this biological protection is totally unknown.

8 Biofilm – Biofouling

The majority of microorganisms grows attached to surfaces in a sessile way (COSTERTON et al., 1987; WINGENDER and FLEMMING, 1999). This has only recently been accepted, since microbiologists are used to deal with planktonic organisms. This simply results from the fact that the study of physiology, systematics, ultrastructure, genetics, etc., is done using rich nutrient solutions to obtain rapidly a sufficient cell mass for study. Also, for biotechnological purposes usually high nutrient concentrations are applied. Consequently, cells do not need to attach due to the surplus of nutrient. Only when the nutrient becomes limiting, some attachment may occur. However, the decline in cell numbers and/or protein concentration is often misunderstood and considered as population death in a stationary phase. As pointed out previously, about 90% to 95% of all microorganisms on a worldwide scale occur attached to surfaces. The latter is also called substratum. This term needs to be distinguished from the term substrate, which means nutrient. However, for confusion, many cases exist, where substrate and substratum are identical. Such cases are organic materials like wood, hydrocarbons, sulfur, metal sulfides, etc.

The sessile bacteria, once they start to multiply, form microcolonies and, by further proliferation, finally a biofilm. Biofilm means that living cells from various groups of microorganisms live together in a matrix of EPS, in which detritus besides organic and inorganic particles may also be embedded as is shown in Fig. 2. The main component of a biofilm is water. It makes up to 95–99% of its whole mass.

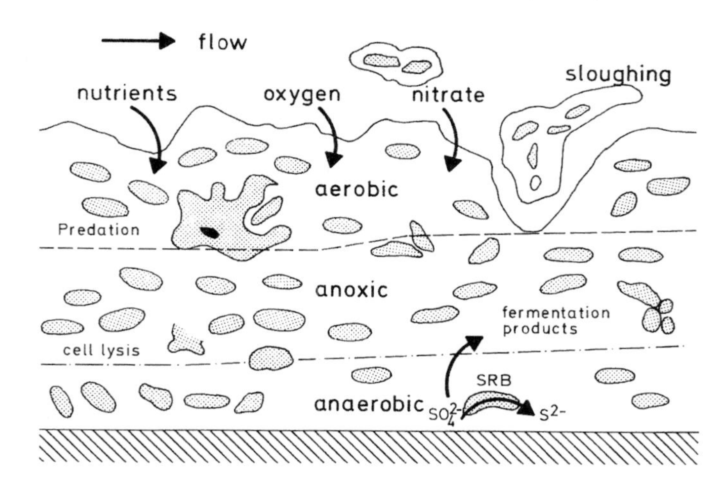

Fig. 2. A schematic representation of a mature biofilm showing various gradients and community interactions (MARSHALL and BLAINEY, 1991).

EPS, cells, and detritus/particles are, thus, negligible in quantity. In technical environments the term biofouling is used. This term clearly differentiates biofouling, which means accumulation and metabolism of microorganisms on technical surfaces impairing the proper function of a piece of equipment, from other types of fouling. Chemical fouling is the surface coverage by chemical polymerization and/or corrosion products, freeze-fouling indicates surface coverage by frozen liquids/solids, particle fouling refers to a coverage by particles, and precipitate fouling indicates the coverage by (chemical) precipitates.

The most difficult process to understand is the way the primary attachment of microbial cells to material surfaces is achieved (BUSSCHER and WEERKAMP, 1987). Bacteria usually have a negative surface charge, which by complexation with heavy metals may be neutralized or even converted to positive values. The surface is due to the composition of the EPS more or less hydrophobic (VAN LOOSDRECHT and ZEHNDER, 1990), allowing hydrophobic interactions. However, if the EPS contain substitutions with acetyl groups or if amino acids or uronic acids are present, they may become even hydrophilic. In physical chemistry, bacteria are considered due to their size as stable colloidal suspensions and respective laws describing their behavior are applied, although bacteria due to mobility might behave quite differently. Microorganisms may come into contact with surfaces by several modes. The most obvious one is a current or flow, which may be turbulent or laminar. In this case clearly many possibilities exist to randomly come into contact with a surface. The second possibility is via diffusion. This mode considers the cells to be passive particles. The third mode also uses passive particles in the process of sedimentation. The fourth way, using the colloidal model, is Brownian movement, which randomly would bring cells into surface contact. The fifth mode is, in contrast to all previous ones, assuming that microbial cells actively swim to the sites of a surface, where they intend to settle. The mechanism called chemotaxis is one of the most studied regulatory systems in microorganisms. By sensing with special receptor proteins a source of food or a sink for inhibitory, metabolic products like sulfide [in sulfate reducing bacteria, because iron(II) ions form highly insoluble iron sulfide precipitates] or sulfuric acid in sulfur oxidizing thiobacilli [reacts with alkaline compounds like $Ca(OH)_2$ in concrete to gypsum and, thus, becomes neutralized] microorganisms are able to deliberately find the most appropriate sites for growth. Chemotaxis seems to play in many cases an up to now grossly underestimated role. It is known (D. C. WHITE, personal communication) that bacteria (SRB) settle on steel plates at local anodic sites, not at cathodes.

This makes sense, because at the anodes the dissolution of iron [Fe(II) ions] takes place. The current opinion is that SRB are able to sense the anodes as sites of dissolution and, by their settlement, stabilize the anodes, whereas without bacteria the anodes would repassivate and new ones might occur at other sites.

An important factor in attachment is also the surface characteristics of materials (BUSSCHER and WEERKAMP, 1987). This means the surface charge (positive or negative), free surface energy (hydrophilic or hydrophobic), and surface roughness (if valleys of 0.5 μm or more occur, bacteria are protected against shear forces and cannot be removed by a strong flow).

The attachment process, as it is nowadays understood, consists of several phases, which are mainly defined by microscopical observation. The whole process starts with the conditioning film. This term means the spontaneous adsorption of macromolecules by several mechanisms to the surface of a material. The consequences are a change in surface charge and surface energy. Besides, approaching microorganisms encounter a chemically modified surface. If a microbial cell comes by whatever mechanism near to such a surface and adheres, the attachment is considered to be reversible (reversible adhesion). The cells may still have Brownian movement and can be removed by flowing water. They are also able to spontaneously detach. However, some microorganisms become immediately irreversibly attached. For the explanation of the primary adhesion and the factors controlling this phase several theories have been developed. According to the theory of Derjaguin, Landau, Verway, and Overbeek (DLVO) the cells remain reversibly attached, because the negative charges of the cell envelope and of the material allow only a limited approach with a remaining distance of 50–100 nm. Positive, attracting forces are hydrogen bridges (VAN DER WAALS) and hydrophobic interactions. Furthermore, the formation of polymer bridges between cell and substratum starts. These bridges may be pili, fimbriae, other proteins, and EPS components. This stadium merges into the phase of irreversible adhesion. Now cells are immobile without any type of movement and cannot be washed away by water flow (rinsing). It is believed that the polymer bridges are mainly responsible for this phase. However, not all organisms are able to proceed into this phase. Some microorganisms remain reversibly attached, which also depends on the type of material. The forces acting between cell and substratum are now chemical forces: covalent bonds, electrostatic forces, hydrogen bridges, dipole interactions of molecules, and hydrophobic interactions.

According to the chemical nature of a substratum, microorganisms adapt their EPS (as the main compounds to interact) to the type of substratum they are faced with. This is achieved by a variation of the chemical composition of the EPS (more hydrophilic or hydrophobic constituents, substituents, etc.). In case of *Acidithiobacillus ferrooxidans* it could be shown that cells grown on the dissolved substrate ferrous sulfate contain more sugars and uronic acids in their EPS than sulfur-grown cells. The latter contain in contrast more lipids in their EPS than the previously mentioned cells. By this adaptation attachment and degradation of substrate and substratum are regulated. The attached state contains the lowest free energy. The change occurring in the process of adhesion is given by the following equation:

$$\Delta F_{\mathrm{adh}} = \gamma BS - \gamma BL - \gamma SL \tag{9}$$

where ΔF_{adh} is the change of free energy due to adhesion of cells, γBS is the interfacial tension between bacterium and substratum, γBL is the interfacial tension between bacterium and liquid, and γSL is the interfacial tension between substratum and liquid.

Although bacteria are irreversibly adhering to surfaces, situations exist, where a detachment might occur. The most obvious mechanism results from growth, when daughter cells are formed. Other mechanisms of detachment may result from a change in the exopolymer composition during growth or an addition of surface active compounds changing the surface characteristics of the materials under the biofilm. Also the metabolic state, e.g., inactivity, might cause detachment. Exoenzymes, which cleave enzymatically the bridging EPS molecules, can also cause a detachment of cells.

From the point of biodeterioration it needs to be pointed out that all problems connected with adhesion and detachment of cells are of exceptional importance. For this reason considerable activity is orientated at finding ways to prevent adhesion of microorganisms. Up to now the chemical compounds, which are in use, produce insufficient results. Even highly toxic paints containing tributyltin compounds have only a temporary effect. Obviously resistant organisms are always present, and once the surface has been covered by these, others start to proliferate, too. Only the clarification of adhesion phenomena at the molecular level seems to be promising. It is known from nature that some possibilities exist. Crustaceans manage to keep their shell free of surficial growth. Electron micrographs indicate that the surface consists of ultrathin needles, which prevent adhesion. Such strategies, besides more detailed work on chemotactic mechanisms may help solve this problem in the future. The field of biofilm/biofouling (FLEMMING, 2000) is too large to be extensively treated in this overview. Interested readers should refer to several excellent books some of which have been published recently.

9 Biocorrosion of Metals

The biocorrosion of metals is a serious economic problem. The economic value of biocorrosion must amount on a worldwide scale to a few billions of Euros annually. Estimations for metal corrosion assume a loss of the gross national product, GNP, of 4% annually because of this process. If by improved diagnosis, materials selection, modification of environmental conditions, etc., a part of this sum could be saved, a considerable amount of money would become available for other, more useful purposes.

The first notification of the metal corrosion process results from PLATO, who lived in the year 427 until 347 BC. He described rust as the soil-like substance forming from iron metal. Around the year 1600 AC the famous scientist GEORG AGRICOLA wrote his book "*De fossilia naturum*", where he adopted PLATO's definition. In the year 1667 the term corrosion was born in England. German scientists adopted the term and in 1785 described it as a chemical process. Once this state had been achieved, physical and chemical research started to understand the corrosion process. Still nowadays gaps in knowledge exist, especially in biocorrosion processes.

Metals are usually separated, due to the amount of production, into iron and non-iron metals. The non-iron metals are subdivided according to their specific density into light and heavy metals with values of below 4.5 g cm^{-3} and above, respectively. Besides pure metals also a large amount of alloys exists. The resistance against (bio)corrosion is in general determined by the electrochemical potential of the metals/alloys. This means that metals in an aqueous environment are able to take up or to emit electrons forming either reduced or oxidized states. Reference point for this tendency is the standard hydrogen electrode (SHE) at 25 °C, 1,013 mbar, and an ion concentration of 1.00 mol L^{-1}. However, some metals tend to form passive layers on their surface (oxides), a process which results in a more noble surface potential than the original metal exhibited. Consequently, their tendency to take up or emit electrons is changed in a way that they are able to better resist environmental influences. However, these metals are especially susceptible to localized corrosion as occurring in pits or crevices.

(Bio)corrosion may be measured by applying Faraday's law:

$$m = \frac{M}{z \cdot F} \cdot I \cdot t \tag{10}$$

where m is the amount of electrochemically converted (lost) mass (in g), M is the molar mass (in g mol^{-1}), z is the charge number of the electrode reaction, F is the Faraday constant ($F = 96.487$ A·s mol^{-1} for $z = 1$; A = Ampere), I is the electric current (in A), and t is the time (in s). The amount of current, which flows in a corrosion element, is proportional to the amount of dissolved metal. Consequently, the current density allows to determine the amount of metal lost. For iron the following rule exists: 1 mA cm^{-2} (current density) $\cong$ 250 g Fe lost m^{-2} d^{-1} (mass loss rate) $\cong$ 11.6 mm a^{-1} (loss of plate thickness).

Microbial corrosion of metals/alloys deals with the effect, which microbial metabolites exert on the electrochemical reactions. It means that the anodic and cathodic processes become enhanced, however, not that microorganisms are using *de novo* reactions with metallic materials (WIDDEL, 1990).

Two types of corrosion need to be distinguished: localized and flat-spread corrosion. Localized corrosion develops preferentially under aeration or concentration elements. This may result from a microcolony, under which due to its oxygen consumption by metabolism anaerobic zones develop, whereas the adjacent surface without microcolony remains aerobic. Consequently, under the microcolony an anode is formed and the adjacent area becomes cathodic. Dissolution of the metal may result. Also iron(III) hydroxide precipitates (tubercles) often cause localized corrosion, although the tubercles may be only an indication of the corrosion process, not the reason. Flat-spread corrosion phenomena have often been found to be associated with SRB under biofilms and/or the presence of acid-forming microorganisms. SRB promote corrosion by the formation of their end product sulfide, which reacts with the anodically produced iron(II) ions to nearly insoluble iron sulfide. The latter acts, if in contact with the metal as an enlargement of the cathode, thus enhancing the corrosion.

Several groups of microorganisms are of special importance for metal biocorrosion. The aerobic sulfur oxidizing bacteria like thiobacilli excrete sulfuric acid, which is corrosive to many metals. Nitrifying bacteria form nitric acid, also a corrosive metabolite. SRB form sulfide, the corrosivity has been described above. Clostridia can contribute to sulfide formation by reducing sulfite, etc. Chemoorganotrophic bacteria, besides forming microcolonies and, hence, aeration cells, may contribute by degrading organic compounds to produce sulfides, ammonia, and/or organic acids (fermentation). The latter are nutrients for SRB, whereas the former compounds may be converted to inorganic acids by the previously mentioned organisms. Besides, NH_3 can cause corrosion cracking in case of copper stress. The so-called iron bacteria like *Gallionella* and *Leptothrix* may contribute, because of the formation of rust precipitates. Besides creating aeration cells, the iron(III) compounds may directly react with the metal and oxidize it according to the following equation:

$$Fe^0 + 2\,Fe^{3+} \rightarrow 3\,Fe^{2+} \tag{11}$$

The most important and most dangerous biocorrosion is caused by SRB, especially under conditions with a temporary oxygen availability. The classical equations for metal corrosion in aquieous environments are the following:

$$\text{Anode:} \quad Fe \rightarrow Fe^{2+} + 2\,e^- \tag{12}$$

$$\text{Cathode:} \quad 2\,H_2O + 2\,e^- \rightarrow 2\,H + 2\,OH^- \tag{13}$$

As a result, iron(II) hydroxide and hydrogen are formed. The hydrogen is adsorbed by the metal as atomic hydrogen. By this adsorption the metal dissolution comes to an end after a certain amount, which is metal specific, due to a generation of an excess voltage (cathodic polarization). This voltage causes the inhibition of further water reduction by the amount of hydroxide ions and the adsorbed hydrogen.

If oxygen is present in the system, the following reaction takes place:

$$^1/_2\,O_2 + 2\,e^- + H_2O \rightarrow 2\,OH^- \tag{14}$$

The metal dissolution can continue since in this reaction the electrons are consumed (cathodic depolarization) and hydrogen is not produced (IVERSON, 1966). This case is called oxygen corrosion.

In the presence of HS^- the following equation is of importance:

$$2\,HS^- + 2\,e^- \rightarrow H_2 + 2\,S^{2-} \quad \text{(analogous to water reduction)} \tag{15}$$

This equation is from an energetic point of view more favorable than the reduction of water. The final reaction would be in this case:

$$Fe^{2+} + S^{2-} \rightarrow FeS\downarrow \tag{16}$$

Iron sulfide, FeS, is a semiconductor and behaves cathodic to iron. Thus, the cathodic area becomes enlarged. This produces more surface area, where the reduction of O_2, H_2O, or H_2S can take place. In this case, the corrosion continues in spite of hydrogen production, because the dissolution product, the iron(II) ion, is constantly removed from equilibrium by the ensuing precipitation as iron sulfide.

The fate of the adsorbed, atomic hydrogen is unclear. Interestingly, many SRB possess a hydrogenase (PANKHANIA et al., 1986). It is unknown, however, and subject to scientific debate, whether this enzyme can extract atomic hydrogen from the metal and use it. Hydrogenase obviously plays no role in the corrosion of metals. Experiments with SRB possessing and non-possessing a hydrogenase demonstrated that a correlation between hydrogenase and its activity with the corrosion rate does is not exist. Also the H_2S production of SRB seems not to be the corrosion causing mechanism, since H_2S production by SRB in a medium devoid of any dissolved iron compounds demonstrated a corrosion rate of almost zero.

Probably the mechanism of metal corrosion by SRB is analogous to the one which causes metal sulfide dissolution (leaching) by *Acidithiobacillus ferrooxidans*. Recent work demonstrated that the EPS of *Desulfovibrio* contain iron(III) ions, as do the EPS of *Acidithiobacillus ferrooxidans* (GEHRKE et al., 1998). The latter uses the iron(III) ions, which are complexed in its EPS via uronic acids, for a dissolutive attack on the metal sulfide. *Desulfovibrio* with its iron(III) ions could take up electrons reducing the iron(III) to iron(II) ions. Thus, the excess voltage would be reduced and a further dissolution of the metal at the anode could take place. Besides, the iron(II) ions could react with the biogenic H_2S to FeS precipitates. By this hypothetical mechanism the corrosion could continue as long as *Desulfovibrio* has organic nutrients and sulfate in the medium.

The classical theory for the explanation of the action of SRB in metal biocorrosion originates from 1934 and was given by VON WOLZOGEN-KÜHR and VAN DER VLUGT (1934). These authors summarized the role of SRB as follows:

Anodic reaction:
$$4\,Fe \leftrightharpoons 4\,Fe^{2+} + 8\,e^- \tag{17}$$

Water reaction:
$$8\,H_2O \leftrightharpoons 8\,H^+ + 8\,OH^- \tag{18}$$

Cathodic reaction:
$$8\,H^+ + 8\,e^- \leftrightharpoons 8\,H \text{ (atomic hydrogen)} \tag{19}$$

Hydrogenase of SRB:
$$8\,H + SO_4^{2-} + 2\,H^+ \leftrightharpoons 4\,H_2O + H_2S \tag{20}$$

Sulfide precipitation:
$$Fe^{2+} + H_2S \leftrightharpoons FeS + 2\,H^+ \tag{21}$$

Hydroxide precipitation:
$$3\,Fe^{2+} + 6\,OH^- \leftrightharpoons 3\,Fe(OH)_2 \tag{22}$$

The equations are in many cases similar or even identical with the previously discussed ones. The main difference is the assumption that the atomic hydrogen is used for the reduction of sulfate using the hydrogenase system. Although theoretically possible, it remains questionable since direct evidence is missing (see above, paragraph on hydrogenase). It is more probable that the reduction of sulfate is a consequence of the oxidation of organic compounds like lactate, etc., which are available in the biofilm due to the association with facultative anaerobes with a fermentative metabolism. What VON WOLZOGEN-KÜHR and VAN DER VLUGT did not include in their model, is the role of the EPS containing complexed iron(III) ions. However, their model is nowadays still valid except the point at issue concerning hydrogenase and the use of atomic hydrogen for sulfate reduction.

As stated in the beginning, especially conditions with changing oxygen availability have been shown to be the most detrimental ones. This was demonstrated in several laboratory experiments, but also on site. Steel pilings in harbors exhibited the highest corrosion rate at a depth, which normally remains below the water level, but which was not the deepest part near the ground (Fig. 3). The reason for this phenomenon results from the chemistry of sulfur compounds. Iron sulfide under partially oxidizing conditions, i.e., if oxygen or iron(III) ions are present, transforms to either pyrite or to elemental sulfur plus iron ions. Elemental sulfur may be oxidized by sulfur oxidizing bacteria and fungi to sulfuric acid with a lot of intermediary products like sulfite, thiosulfate,

Fig. 3. Overview of a steel piling structure at Norddeich harbor, Germany, with typical pitting attack at the low-water level.

etc. Pyrite is decomposed by iron(III) ions chemically and also biologically via thiosulfate and polythionates to sulfuric acid. All these intermediary sulfur compounds including elemental sulfur may react with metals in the sense that they are cathodically reduced (most likely to sulfides) and, thus, keep the corrosion process ongoing, as shown in Fig. 4 (VON RÈGE and SAND, 1999).

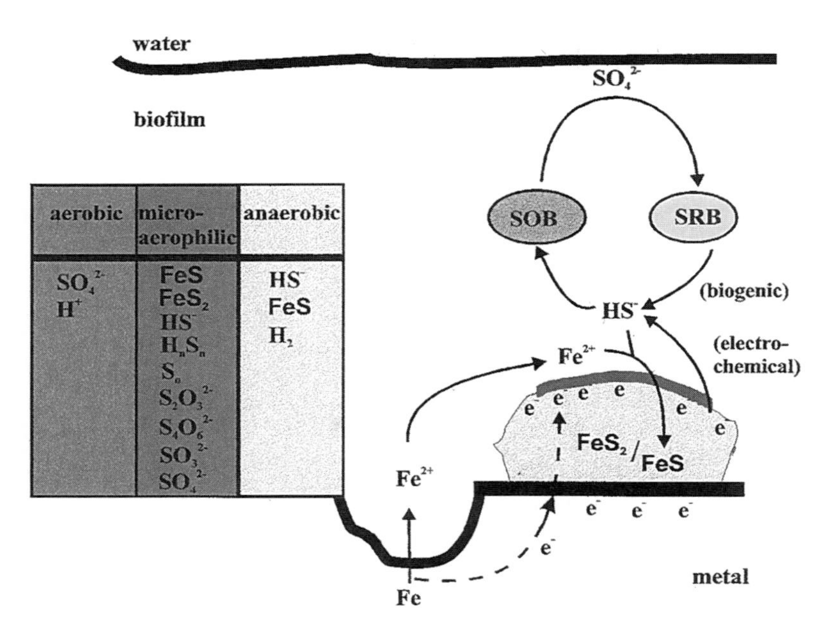

Fig. 4. Model for the metal corrosion in the presence of microorganisms of the sulfur cycle (VON RÈGE and SAND, 1999). Under aerobic conditions (left box) sulfur/-compounds are oxidized to sulfuric acid by SOB (sulfur/-compound oxidizing bacteria). Under anaerobic conditions SRB reduce sulfate to HS^-, which reacts with Fe^{2+} ions to give FeS (right box). Under microaerophilic conditions FeS and FeS_2 are chemically degraded and a variety of sulfur/-compounds is formed (middle box). These compounds are cathodically reduced on the surface of the conductive FeS/FeS_2-layer, strongly accelerating the corrosion process. Sulfur/-compounds are delivered by the biogenic sulfur cycle.

Consequently, all these compounds are known for a long time to be highly corrosive for steels, etc. This finding explains, why providing conditions with intermittent oxygenation are so detrimental for metals. Recently, it was demonstrated that the compounds CRS (chromium reducible sulfides – pyrite) and sulfate in the surface layer are bioindicators for the strength of corrosion, since their amount was directly correlated with the corrosion rate. As an example for the importance of oxygen, the sulfur cycle on the cited steel sheet pilings in harbors is shown in Fig. 5.

At the low water level, oxygen may penetrate the biofilm on the pilings and cause oxidizing conditions. Consequently, FeS and FeS_2 (pyrite) are degraded to sulfur or polythionates, respectively. These compounds may either be directly electrochemically reduced (acting as cathode) or further oxidized by sulfur oxidizing bacteria to sulfuric acid (→attack). If the water reaches its highest level, the oxygen supply from the water is too limited to allow the development of oxidizing conditions in the biofilm. SRB become active and reduce the sulfate ions to sulfide, which may react with iron ions. This cycle is driven by the corrosion of the iron sheets and the availability of organic compounds in the biofilm. The latter can result from photosynthetic microorganisms as primary producers and/or from pollution of the water, which in harbors is a common phenomenon. In some samples SRB were consequently detectable in the inner layer near the steel, and sulfur oxidizers at the surface of the biofilm, a finding in accordance with the hypothesis of this corrosion process.

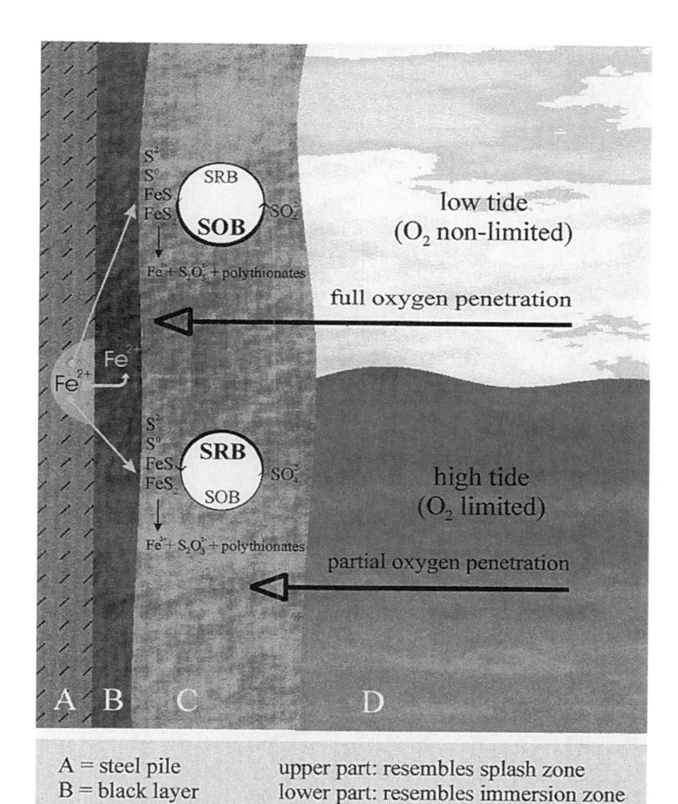

Fig. 5. Schematic summary of MIC on marine steel piling structures at the low-water level.

Another type of metal corrosion is caused by manganese oxidizing bacteria. Although these bacteria are known for quite a while, their importance has been recognized only in the last years. Similar to sulfate reducers manganese oxidizing bacteria settle on the surface of metals in aqueous environments. If manganese(II) ions are present in the water, these bacteria oxidize them to manganese(IV) ions, which form rapidly with dissolved oxygen the highly insoluble manganese dioxide, MnO_2. MnO_2 precipitates, if in contact with a metal, especially an iron-containing one like steel, take up the electrons from the anodic dissolution and, thus get reduced again to manganese(II) ions. The latter may be reoxidized by the manganese oxidizers. As a consequence, a manganese cycle exists, which is driven by the dissolution of the iron. Furthermore, a serious ennoblement is caused (redox potential of $+600\ mV_{SHE}$ and more), which renders even highly alloyed steels susceptible to pitting corrosion (SCOTTO and LAI, 1998). The following equations apply:

Anodic reaction:

$$Fe \rightarrow Fe^{2+} + 2e^- \quad (23)$$

Cathodic reaction:

$$MnO_2 + 4H^+ + 2e^- \rightarrow Mn^{2+} + 2H_2O \quad (24)$$

Manganese(II) ions remain either soluble or form under reduced oxygen partial pressure MnO or in presence of CO_2 (H_2CO_3) the solid $MnCO_3$. Both conditions are present in case of biofilm/microcolonies, as Fig. 6 shows. The consequence is the dissolution of the metal in the presented case of iron. This cycle may cause serious corrosion problems, even with highly alloyed steels. It is unclear up to now, which species are causally involved in the process. It is also unknown, whether these bacteria are able to live lithoautotrophically by manganese(II) ion oxidation. The oxidation yields only little energy for metabolic purposes.

For protection against biocorrosion many techniques have been developed. First of all measures concerning the construction and the choice of materials need to be mentioned. If non- or low-alloyed steels fail to function, often high-alloyed types may help. This means that their passivity is improved by high chromium additions. Coatings also are a frequently used technique. In this way the material becomes separated from the aggressive environment. In some special cases the aggressive medium itself may also be modified to produce a less aggressive one. An electrochemical method of protection is the use of sacrificial anodes. These anodes are made of a non-noble metal like magnesium and dissolve easily, thereby liberating electrons via a conductive bridge to the metal to be protected. As a consequence, the valuable metal becomes cathodic and protected. A similar technique is cathodic protection of equipment by applying electric current from an external source like a generator, not from a sacrificial anode. Due to the electric power the object to be protected becomes cathodic and, thus, is protected. The latter two techniques are applied in water circulation systems, heat generators, pipelines, steel pilings in harbors, etc.

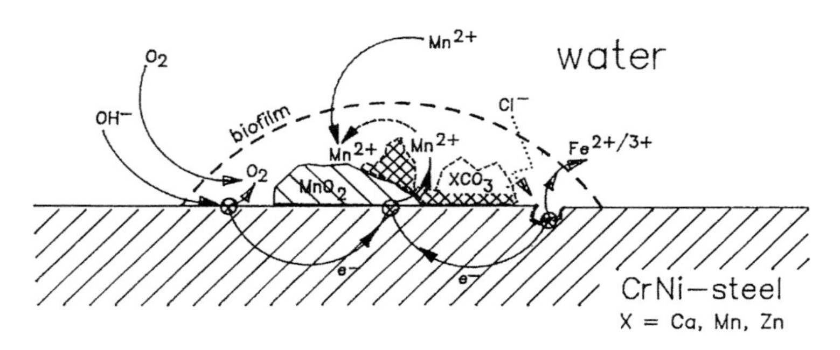

Fig. 6. Mechanism for the initiation of microbial influenced pitting corrosion (LINHARDT, 1996).

A chemical way of protection is the use of corrosion inhibitors. These are chemical compounds, interacting with the anodes or cathodes on the metal surface or in case of biocorrosion specifically inhibit microorganisms. A vast diversity of such compounds is on the market. This is not the place to discuss their chemistry and efficiency. For this purpose an unsurpassable amount of technical literature is available.

10 Biodeterioration of Mineral Materials

Mineral materials consist of four main groups. These are natural stone, concrete, ceramics, and glass.

Natural stone may be subdivided into magmatic stone (granite, basalt, andesite, etc.), sedimentary stone (gypsum, limestone, sandstone, shale, etc.), and metamorphic stone (quartzite, marble, etc.). Concrete consists of water, filler (sand, gravel, etc.), and hydraulic binder (cement, gypsum, lime, sulfur, etc.). Ceramics are mainly burnt-clay products. Glass is an inorganic, non-crystalline material (undercooled liquid).

All of the materials are susceptible to some attack by microorganisms (BOCK and SAND, 1993; SAND et al., 1989). Especially the acid excretion besides the chelation effect and biofilm plus salt stress need to be mentioned. Consequently, almost all microorganisms may contribute to a biodeterioration of mineral materials (BERTHELIN, 1983). It is often difficult to determine the effect microorganisms have in the (bio)deterioration process, because physical and chemical stress factors are also involved. A few examples are finally available, where the causality has been cleared. The best understood one is the so-called biogenic sulfuric acid corrosion of concrete in sewage pipelines and works.

The term biogenic sulfuric acid corrosion is used for an attack of biologically produced sulfuric acid on cement-bound materials. The acid is excreted by lithotrophic sulfur and sulfur compounds oxidizing bacteria as end product of their metabolism.

The corrosion occurs mainly in sewage pipelines installations made of cement-bound concrete. In addition, cement-bound mortar (between bricks) is also destroyed. In general, all materials are destroyed that may react (chemically) with sulfuric acid.

Beyond the scope of this report are working materials that themselves may be used as substrate by microorganisms like some resins, sulfur-bound concrete, natural materials, etc. Even though these materials are of great public interest, the underlying corrosion mechanism is different from the one for biogenic sulfuric acid corrosion and, thus, clearly differentiates them from the materials concerned here. The fact that concrete sewers carrying sulfide-bearing sewage were subject to a rapid and extensive destruction was observed as early as 1900 (OLMSTEAD and HAMLIN, 1900). It was reported that the destroyed concrete was highly acidic due to the presence of sulfuric acid. Although in the following decades the idea was discussed several times that this sulfuric acid could have been produced by bacteria, it was generally thought that the corrosion process was due essentially to chemical transformations by which H_2S in the sewer air was converted to sulfate. The following extract from the authoritative book on cement chemistry written by LEA and DESCH in the year 1936 is indicative: "It seems probable that the calcium sulfide or hydrosulfide first produced by the action of the hydrogen sulfide on the concrete, becomes converted in the presence of oxygen to calcium sulfide and sulfate" (LEA and DESCH, 1936).

In 1945, PARKER published the discovery of strongly acid-forming bacteria from samples of corroded concrete sewage pipelines (PARKER, 1945). He called them *Thiobacillus concretivorus*, a synonym to *Acidithiobacillus thiooxidans* (correct name). Since this discovery, the understanding began to develop that this type of corrosion is caused by the action of sulfur-oxidizing bacteria belonging mainly to the genus *Thiobacillus*.

Since then, several reports have been published elucidating the mechanism of the biogenic sulfuric acid corrosion in sewage pipelines (MILDE et al., 1983; SAND et al., 1983, 1989; SAND, 1987). Summarizing the literature, it becomes clear that the corrosion is a result of bacterial activity. It is to be assigned to the sulfur cycle, which in contrast to the carbon cycle (global warming), attracts considerably less public attention. The main source for biogenic sulfuric acid is the sulfur compound hydrogen sulfide, H_2S. It is produced by microorganisms that live in the sewage, in the mud at the bottom of the pipelines, and in the slime layer coating the surfaces of sewage pipelines above and below the water level. The layer may be called a biofilm. Under anaerobic conditions, SRB are active in these habitats, reducing oxidized sulfur compounds to H_2S. As H_2S is a weak acid, it remains dissolved in the sewage under neutral or alkaline conditions. If the pH value decreases, e.g., due to acidification, H_2S is emitted into the sewers atmosphere. Turbulence in the sewage flow is another reason for H_2S emission. Even at comparably low concentrations H_2S escapes into the gas phase (stripping). Once H_2S has reached the atmosphere it may react with oxygen to elemental sulfur that is deposited on the walls. In fact, this reaction may be accelerated catalytically by an alkaline surface. Sulfur is a good substrate for thiobacilli. By their metabolism, sulfur is oxidized to sulfuric acid. The energy obtained is used for CO_2 fixation for cell mass production. Fig. 7 gives an overview of the sulfur cycle.

The end product sulfate may be used again by SRBs as a sink for electrons producing sulfide. Thus, the sulfur cycle is closed. As a side reaction, thiosulfate may also occur in sewage pipelines. Sulfur reacts chemically with sulfur dioxide, SO_2, to form thiosulfate. The thiobacilli themselves are one source for sulfur dioxide. During sulfur oxidation, a strain of *Acidithiobacillus thiooxidans* produced measurable quantities of SO_2 probably indicating an overflow of the primary sulfur oxidation as compared with the sulfite oxidation. All these compounds have been detected in sewage pipelines (SAND, 1987).

The H_2S production often seems to be the limiting factor, as H_2S is the substrate for conversion into sulfuric acid. However, man influences H_2S production in several ways. Most important is the temperature of the sewage itself. Because of an increasing use of hot water (for dishwashers, washing machines, showers, baths), the temperature of the sewage increased by several degrees celsius during the last decades. Estimations range up to 10 °C. It is well known that microorganisms convert their substrate at high temperatures more rapidly than at low temperatures, although the rate does not double, as in chemistry. Thus, increased acidification is caused. The second reason for increased H_2S production is the increased use of detergents containing sulfates and the increase in protein consumption. Whereas sulfates may be used by SRBs in the way described above, protein is reaching the sewage after digestion of food. It is degraded into amino acids. Two amino acids, methionine and cysteine/cystine, contain sulfur atoms. The degradation of methionine usually yields methylmercaptan, CH_3SH (methanethiol), whereas cysteine gives rise to H_2S under aerobic and anaerobic conditions. Thus, H_2S may be produced aerobically as well. The third reason for an increase in H_2S production are the spreading cities. Because people tend to live in green suburbs, far outside of town, long sewage networks have to be installed for treating household sewage. Due to the long distances, especially in flat areas, the sewage has long residence times until it reaches the treatment plant. As a consequence, a lot of degradation takes place during the flow time, producing volatile sulfur compounds like H_2S. Often pressure pipes are used to reduce residence time. However, if these pipes are not ventilated and aerated, e.g., by pressurized air, they tend to develop anaerobic conditions, giving rise to H_2S formation again. Summarizing, it becomes obvious that the human way of life influences the sulfur cycle and its reactions finally causing an increased possibility of attack on working materials by biogenic sulfuric acid.

10.1 Concrete Systems

Ordinary concrete systems are very sensitive to the production of sulfuric acid. Concrete consists of a hydraulic mixture of cement fines with coarse aggregates, sand, and water. The cement first dissolves into water, then precipitates, leading to formation of hydrates. The interpenetration of these hydrates leads to setting and final hardening, giving the structure

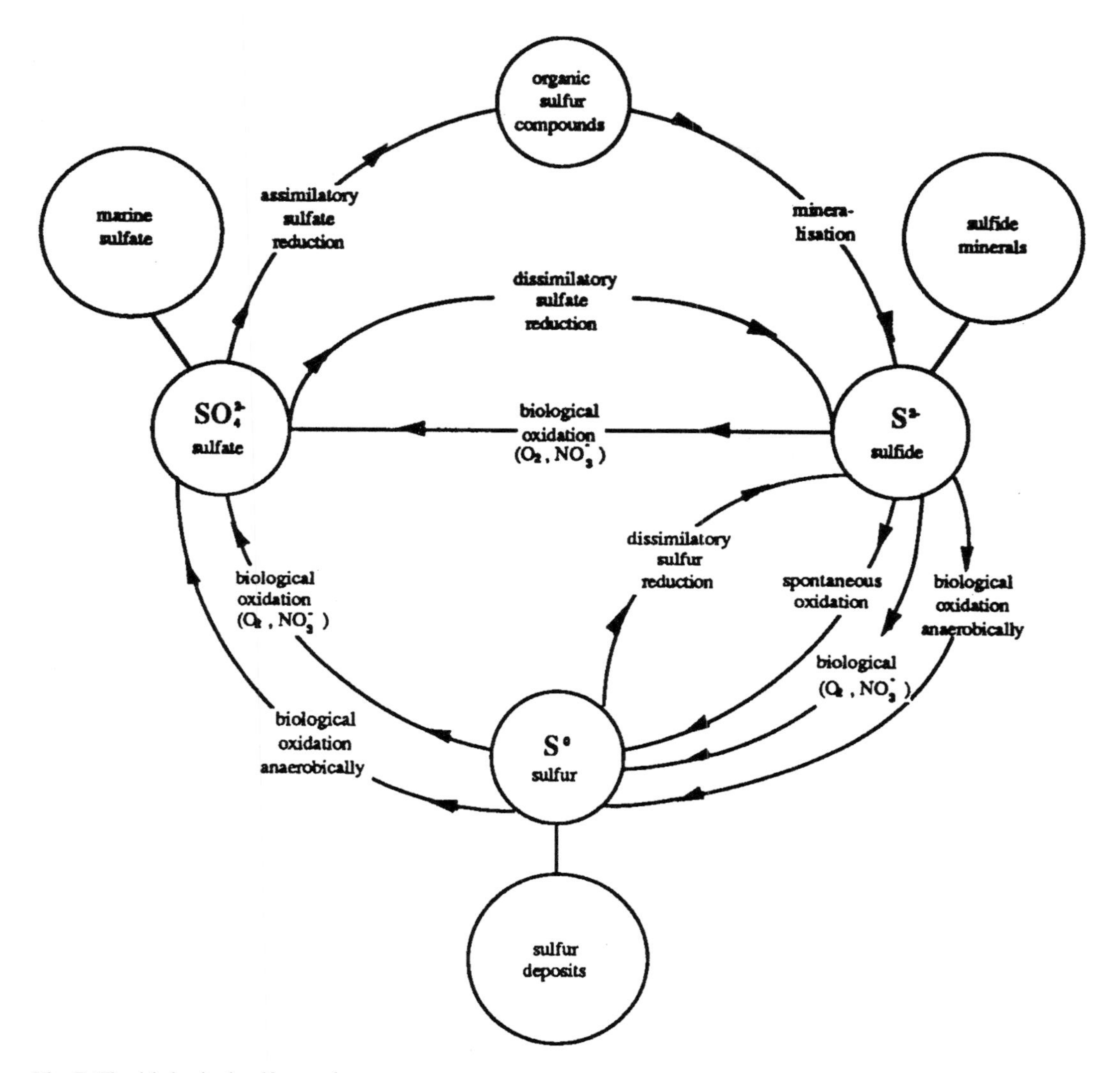

Fig. 7. The biological sulfur cycle.

its mechanical strength. The most common cement material is portland cement. It is mainly composed of the following mineral anhydrates: tricalcium silicate ($3\,CaO \cdot SiO_2$), dicalcium silicate ($2\,CaO \cdot SiO_2$), tricalcium aluminate ($3\,CaO \cdot Al_2O_3$), and tetracalcium ferro aluminate ($4\,CaO \cdot Fe_2O_3 \cdot Al_2O_3$). These anhydrates react with water to form a number of hydrated oxides, including high amounts of calcium hydroxide $Ca(OH)_2$, which is water soluble. In addition to silica, alumina, and iron compounds, cement also contains small amounts of adventitious compounds such as sodium, potassium, and magnesium oxides, phosphates, and sulfates, which are also water soluble. In a normal clean atmosphere, any moisture on a concrete surface would have dissolved in it small quantities of these compounds, as well as some CO_2 and oxygen. In polluted city atmospheres, the air contains SO_2, ammonia, and oxides of nitrogen. These will also be dissolved in the moisture.

The alkaline hydrates react on sulfuric acid attack by forming calcium sulfates such as gypsum ($CaSO_4$) and ettringite ($6\,CaO \cdot Al_2O_3 \cdot 3\,SO_4 \cdot 32\,H_2O$). This is all the more true, since $Ca(OH)_2$ is water soluble and especially reactive even deep inside the concrete structure.

10.2 Corrosion Tests

In order to help in predicting the behavior of pipe materials to this phenomenon, tests are often conducted by materials scientists immersing cement-bound test materials in sulfuric acid. With this type of test it is conceived that the cement-bound material is destroyed by a purely chemical reaction with the acid. The nature of the acid and its concentration may vary from assay to assay. The acidic solution is renewed from time to time, and the loss of weight due to chemical reaction is measured. The corrosion rate is then calculated from these values.

However, the experience of many years indicates that corrosion does not occur uniformly in sewage pipelines. From one joint to another, a pipe may be severely corroded, whereas the adjacent pipe (of the same material) exhibits only negligible corrosion. The reason for these differences is not known and, thus, predictions about the possible rate of corrosion of materials in sewage environments remain difficult.

Experiments in concentrated sulfuric acid with various types of concrete (pH 1–2) resulted in corrosion rates of similar order, though using different materials. All samples but one lost 50% of their initial weight between the 8th and the 16th week. Taking into account the expected service life of pipes of 50–80 years, the differences observed have little significance. These results are in contradiction to the field experience, where at severely endangered pipes of the sewage system pH values between 1 and 2, sometimes even below 1 occur, although the grade of corrosion shows considerable variation. As a consequence, besides the action of sulfuric acid, there must be another factor determining the resistance of a cement-bound material against biogenic sulfuric acid attack. Obviously corrosion testing based upon immersion in "simple" sulfuric acid is misleading and not appropriate when it comes to simulating the specific case of biogenic sulfuric acid corrosion. The reason is obvious: bacteria produce the sulfuric acid. Because they are living organisms, they may thrive in suitable habitats, whereas in others their growth is restricted. This means that the interactions between microorganisms and substratum must play an important role in the corrosion of cement-bound materials. This is to be taken into account in the testing procedure.

10.3 Tests in a Simulation Chamber

Because of the apparent lack of an appropriate testing method, which would include microorganisms as the main corrosion-causing agents, a simulation chamber was constructed that allowed to remodel the conditions of a sewage pipeline. In this chamber the conditions for the main corrosion-causing agent, the sulfuric acid-producing bacteria were optimized (SAND et al., 1983). Thus, a possibility for the testing of materials was created that included the interactions between bacteria and substratum. The first testings conducted with the experimental chamber confirmed the importance of the bacteria, and gave results that were consistent with on-site observations: two different materials may exhibit similar resistance to acid attack according to immersion tests, though having entirely different resistance to bacterially influenced corrosion. This is illustrated by results shown in Fig. 8. Two test blocks, one made of a blast-furnace cement, the other made of a portland cement, are shown. After having passed two simulation experiments, some difference is obvious. In this particular case, the portland cement-based formulation gave evidence of better resistance, though the purely chemical testing with sulfuric acid had caused insignificant differences.

Field experiments were conducted with portland cement-based concrete blocks in order to determine how much the simulation experiments were accelerated compared to natural conditions in Hamburg's most severely endangered sewers. Because the testing conditions are optimal for the bacteria in the experimental chamber, the test run produced corrosion on cement-bound concrete test blocks.

Fig. 8. Comparative testing of two test blocks of a portland cement (left side) and a blast-furnace cement (right side) (SAND et al., 1994).

That needed more than eight times as long as under natural conditions in the aggressive atmosphere of a sewage pipeline (SAND, 1987). Fig. 9 gives an impression of the simulation chamber for reproduction of partially filled sewage pipelines. Fig. 10 shows the diagram of the apparatus. The chamber has a volume of about 1 m^3. The walls and the bottom are thermostated separately, and the bottom is usually 2–3 °C warmer than the walls. The chamber contains about 10 cm of water. Due to the elevated temperature, water evaporates and condenses at the top lid and the walls, producing a high humidity in the gas space (usually above 98% relative humidity). The temperature is kept at 30 °C. As a nutrient for the sulfur-oxidizing bacteria, H_2S gas is used. The concentration in the gas phase amounted to 10 ± 5 ppm H_2S. At both sides of the lid a nozzle is installed. The nozzles are used for spraying a salt solution into the chamber that contains a supply of nitrogen and phosphorus for the bacteria (to avoid any shortage). Secondly, the nozzles are used for inoculation of the test blocks with bacteria. Several strains of the species *Acidithiobacillus thiooxidans*, *Halothiobacillus neapolitanus*, *Thiomonas intermedia*, and *Thiobacillus novellus* (originating from samples of corroded sewers in Hamburg) are grown as mass cultures in the laboratory, harvested, and sprayed as an aerosol (fog) into the chamber. A part of the aerosol settles on the surface of the test blocks. The results of 10 experiments show that by this inoculation method, cell densities of more than 10^6 cells cm^{-2} are usually achieved. Compared with natural conditions this equals a cell count that occurs only at highly endangered parts of the sewage system (MILDE et al., 1983). Thus, a start of the experiment without lag phase is

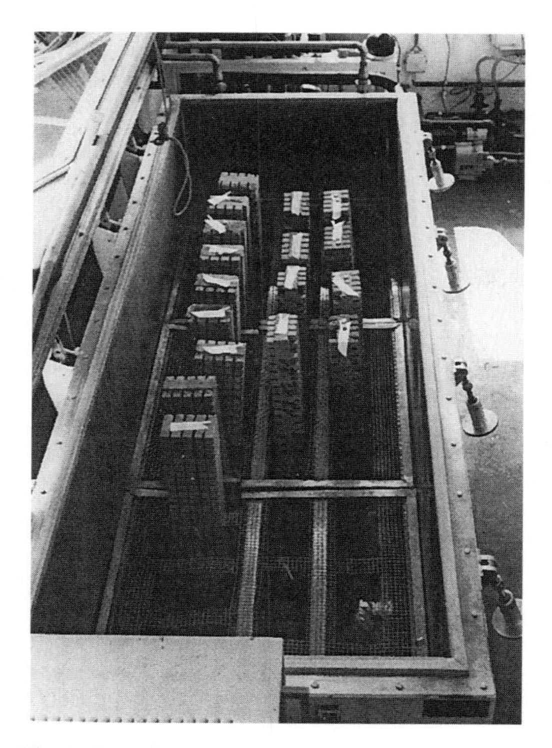

Fig. 9. Overview of the simulation chamber (SAND et al., 1994).

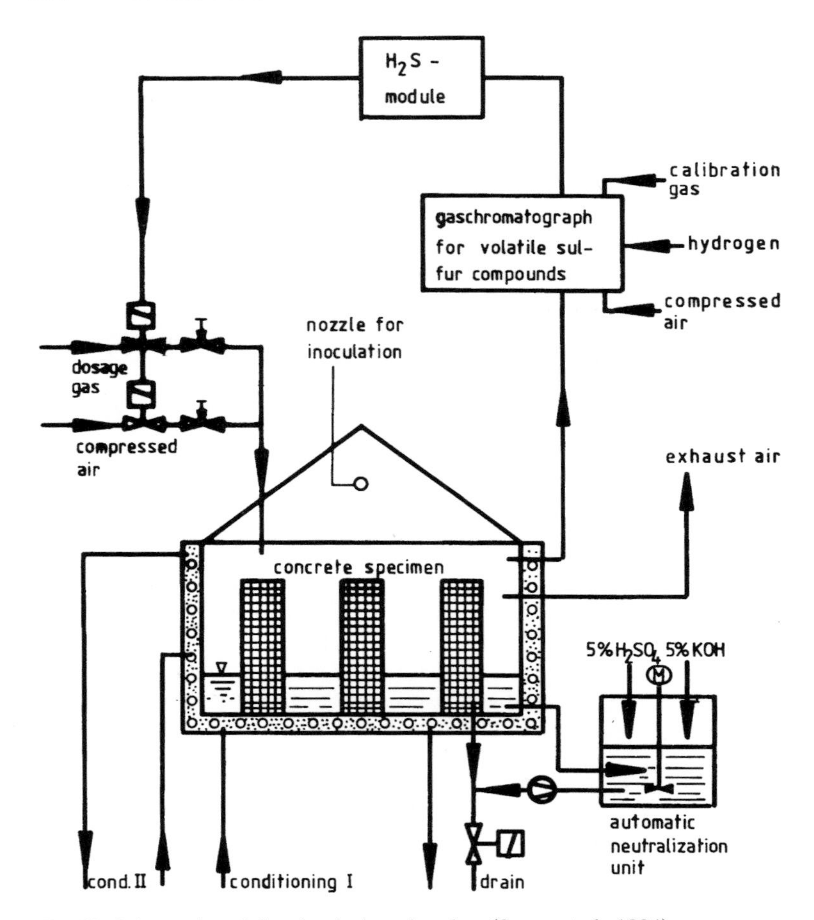

Fig. 10. Schematics of the simulation chamber (SAND et al., 1994).

achieved. The inoculation is done during about 90 d. A "resting" period of about 30 d follows, after which sampling for evaluation of resistance against biogenic sulfuric acid attack starts.

Because of the biogenic sulfuric acid the pH value of the surface water dropped from above 10 (fresh concrete) to less than pH 5 during the inoculation period (100 d). During the following incubation, the pH decreased further to a range of about 1 and 1.5 and remained with minor deviations in this range. The cell counts of thiobacilli started between 10^6 and 10^7 cells cm^{-2} test block surface. Although some deviations occurred, they remained at this level throughout the experiment. The weight loss measurements indicate a low loss after inoculation and 40% weight loss after 240 d until the end. It is evident from these data that such a test material is not suitable for the construction of sewage pipelines. Owing to the accelerated-simulation test, this result was achieved within 1 year, whereas at least 8 years would have been necessary to obtain the same information on site.

10.4 Conclusions

Corrosion tests have been generally conducted in the past with little attention paid to the role of the bacteria in the process. This indeed helped classify the materials according to their chemical resistance to different acids with varying conditions. However, it was misleading, when it came to predict the behavior of materials in natural conditions.

Due to optimal breeding conditions, the chamber proved to work as an accelerated test, and to simulate within a year the corrosion results observed after 8 years in the most aggressive part of the Hamburg sewer system. Thus, this simulation apparatus is an excellent tool to sort out rapidly appropriate cementitious materials for use in sewage systems. The city of Hamburg's sanitation authority requires this test for any new material to be used in the city's sewage network.

It is now an understood fact that natural phenomena require expertise from different people. Biogenic sulfuric acid corrosion is an excellent example, where progress could be done through combined work of complementary expertises, such as mineral chemistry, materials science, and microbiology.

A similar example exists for biogenic nitric acid corrosion, as it occurs, e.g., in cooling towers and also to some extent in sewage pipelines and works. Whenever ammonium is available, e.g., from decaying organic matter or as an addition to cooling water nitrifiers start to colonize concrete and to produce nitric acid. This strong inorganic acid dissolves the cementitious binding material like $Ca(OH)_2$ to form calcium nitrates. Since these are highly soluble salts, they are washed off the surface. Finally, only the gravel/filler remains. Thils biodeterioration could also be reproduced by simulation experiments, which allowed the development of a material test.

For the involvement of biodeterioration in the decay of natural stone a simulation system has also been developed. Under the complex conditions of an urban, anthropogenically polluted atmosphere (SO_2, NH_3, NO_x) natural stone was inoculated with indigenous bacteria and incubated in a model atmosphere. The highly complex test system allowed to recognize the impact of the microorganisms on the process. In the presence of nitrifying bacteria, which feed on ammonia salts and NO_x gases of the atmosphere, SO_2 becomes considerably more rapidly oxidized to SO_3 than in sandstone without these bacteria. Consequently, the nitrifying bacteria enhanced the SO_2 attack on sandstone (by chemical reaction between nitrite and SO_2 to NO and SO_3). This is one of the few examples, where the interactions between microorganisms and chemical reactions could be clarified.

In any case, the summary of these experiments is that microorganisms interact with the substrate/substratum. This interaction, since it occurs usually in biofilms and/or microcolonies, cannot be reproduced by purely chemical or physical testing.

11 Biodeterioration of Organic Materials

Organic materials by far outnumber the previously discussed ones of metallic and mineral origin. Organic materials, with very few exceptions, may serve microorganisms as a substrate and also as a substratum. In this respect they are clearly different from the previous two groups.

Three main groups need to be distinguished: natural organic materials, synthetic organic materials, and miscellaneous organic compounds.

11.1 Natural Organic Materials

The origin of natural organic materials is in its majority wood. Materials like lumber, cellulose, paper, cardboard, furniture wood, limewood, etc., are produced. A second important source is leather. It is produced from hides. Parchment and furs are of the same origin. Finally, textiles need to be mentioned. They may be made out of cellulose (cotton, hemp, jute, sisal) or protein (wool, silk) as basic compounds.

All of these materials have been generated by organisms, usually higher plants, in the course of their life cycle from the CO_2 in the atmosphere. Consequently, all these materials are biologically degradable. In this context the term microbiological omnipotence is used indicating the potential of microorganisms to achieve this goal.

Wood is used as energy source, as raw material, and for constructional purposes. It consists roughly of 40–50% cellulose, 15–35% hemicelluloses (polyoses), and 20–35% lignin. Besides, some minor compounds present are terpenes, waxes, and tanning agents (1–3% of total), inorganic compounds (0.1–0.5% of total). In case of tropic wood the amount of extractable compounds (terpenes, etc.) can increase up to 15% of the total. These extractable compounds are responsible mainly for the resistance against pathogens like fungi, insects, etc.

The basic unit of a cellulose fiber is the glucose molecule, which is β-1,4 glycosidically connected. In wood, chain lengths of 10,000 up to 14,000 occur. Cellulose is the backbone of the cell walls.

Polyoses are branched polysaccharides with substitutions and a chain length of 50–200 units. Coniferous wood has as its main polyose galactoglucomannan, whereas in hardwood 4-O-methylglucuronoxylan is the main component. The polyoses are the connecting compounds between the cellulose and the lignin in the cell wall. They are also responsible for the swelling and shrinking observed on wood in the course of storage for preparation.

Lignin is a complex macromolecule and functions as enforcement of the cell wall. Coniferous wood contains mainly guajacylpropane units, whereas hardwood is composed of guajacylpropane and syringylpropane units.

Besides the three mentioned main compounds wood may also contain starch, sugars, pectins, proteins, and lipids. Especially these compounds serve as nutrient source for deteriogens.

The biodeterioration of wood is mainly caused by fungi. Bacteria are of minor importance. According to the optical appearance several types of decay are distinguished. Molds cover the wood and penetrate only up to 1 mm depth. These fungi belong to the groups Zygomycetes, Ascomycetes, and fungi imperfecti and comprise about 100,000 species. They live mainly of compounds like starch and do not degrade the basic structure, the cell walls, consisting of cellulose and lignin. The second group are the fungi, which cause a bluish stain of the wood. About 100 known species of Ascomycetes and Deuteromycetes cause this type of biodeterioration. They also do not degrade the cell walls, but cause the stain because of the growth of their hyphae within the radially organized, nutrient rich wood cells. This type of biodeterioration is usually a consequence of an exceedingly humid storage after harvest. It may also occur, if matured wood becomes wet again for long periods of time. These fungi are even able to penetrate a coating and grow within the underlying wood. The third group of wood-destroying fungi, the mold-rot causing groups, need high humidity, e.g., contact of the wood with water (soil, river), and cause in such cases considerable losses of mass and strength. The members come from the groups of Ascomycetes and Deuteromycetes and are able to degrade the cellulose backbone of the cells. Consequently, the wood changes its color to become dark-brownish and shiny. Hardwood is usually more susceptible to an attack by these fungi than coniferous wood. The fourth group causes the so-called white-rot. This type of biodeterioration causes a serious damage by degradation of cellulose and lignin. The main group belongs to the wood-destroying Basidiomycetes. As a result of their action wood becomes lightweight, weak, and fiber- or sponge-like. The fifth group also consists of Basidiomycetes. These fungi, to which the most dangerous species especially for roof constructions belongs, the dry-rot fungus, degrade cellulose and the polyoses. Lignin remains, wherefrom their name was derived: brown-rot fungi. In the final state of the attack the wood may be rubbed to powder by fingers. These fungi may also occur upon prolonged storage in sawmills, etc., and with living trees. Consequently, this group is of considerable economic importance.

11.1.1 Laminated Wood

Wooden chips of varying size and quality are glued together with compounds like urea-formaldehyde resins under pressures of 15–40 bar at 140–200 °C to produce plates of laminated wood. The advantage of this type of wood is the price, the reduced problem of shrinkage and distortion, especially when wetted with water, and the reduced degradability. The latter often results besides the gluing compounds from additions of chemicals like phenol, cresol, or isocyanate to the resin. However, if humidity is constantly present, problems may occur caused by molds, mold-rot fungi, and white-rot fungi.

11.1.2 Paper and Cardboard

These two materials are differentiated according to their thickness and specific weight. A product of more than 1 mm thickness and above 225 g m^{-2} is called cardboard. In the course of production wood fibers, either from raw wood or from recycled paper/cardboard, are dewatered and pressed. The fibers get matted and cause the coherence of the product. Besides, glues, resins, colorants, antifoaming compounds, fillers, and biocides are added. The process of production suffers in most plants from severe microbial problems. It starts with the raw material, which in the case of recycled paper and cardboard may already be seriously contaminated with microorganisms. Starting with the process of pulp production microorganisms find optimal growth conditions. In an aqueous environment with a temperature of 30–50 °C and an almost unlimited substrate supply (due to the wood fibers and the additions) considerable growth of aerobic and anaerobic species occurs. The only constraint is the relatively short time of residence, which in most cases is below 8 h. Consequently, the wood fiber is rarely degraded, but the dissolvable contents of the wood like starch besides the degradable additions serve as nutrient source. The large amount of available nutrients results in problems like malodor due to fermentation of starch to organic acids, etc., and due to sulfate reduction, like metal corrosion in the paper machine, because of sulfide formation, and like the standstill of machines, since the paper film broke due to inclusions of high amounts of bacterial slime mass into the fiber matrix. The slimes may be part of the biofilm (sloughing off) and/or may result from the bacterial production of dextrans or levans in the course of starch degradation. Consequently, until the machine is running again, a costly time lapse occurs, which is usually in the rang of 10,000–20,000 Euros per event. Considerable efforts are done to achieve an uninterrupted production.

For this purpose large amounts of biocides and dispersants besides enzymes for slime degradation are often added. The problem of microbial growth in paper production has been aggravated due to the generation of closed water cycles in paper mills. Consequently, the concentration of organic compounds in the production waters has increased considerably. This necessarily resulted in an increased use of antibacterial compounds to ensure an economic (undisturbed) production of paper products.

11.1.3 Leather

Rawhide contains 33% protein, 1.5% lipids, and 65% water. By removal of hair and surface layers the leather hide is produced, which consists finally only of matted collagen fibers. By tanning either with chromium (80% of all leather products, CrO_2 concentration 4–7%) or with plant extracts (20% of all leather products, tanning concentration 20–35%) and fat addition the final product is made. It contains 5–15% fat, 10–14% water, and has a pH of 4–4.5. Since leather is basically a protein, in spite of its protection by the heavy metal chromium or the tanning, bacteria, molds, and yeasts are able to degrade it. The effects range from discoloration via a mat finish to total destruction. Countermeasures are protection from humidity and surface protection by waxing, etc.

11.1.4 Parchment Paper

Parchment paper is basically similar to leather. It consists of matted collagen fibers. However, the process of tanning has been omitted. Consequently, it may be easily degraded, especially by collagen degrading bacteria. The only possibility for protection is the storage under a strictly controlled atmosphere. Especially for museums this poses a serious problem, since many historic documents have been written on parchment paper.

11.1.5 Textiles

The raw materials used for textile production consist of plant or animal fibers. The former are mostly produced from cotton fibers, besides some jute and linen. The fibers consist mainly of cellulose fibrils. In cotton the cellulose amount is about 80% and in jute about 60%. Cotton fibers are used for about 50% of all textiles. Wool comprises about 10%. The wool fiber (hair) is basically the protein keratin (a sulfur containing protein). Besides wool, silk is also a protein fiber used for textile production. Consequently, cotton, wool, and silk are biodegradable. Bacteria and fungi can cause serious problems, if enough humidity is present.

11.2 Synthetic Organic Materials

The main constituents of synthetic organic (plastic) materials are organic macromolecules, which are formed synthetically or by a modification of naturally occurring ones. In order to achieve the desired technical properties, a variety of chemical compounds is added to the macromolecular backbone. Unvulcanized natural and synthetic caoutchouc (for rubber production), besides sealants like silicones, polysulfides, epoxides, polyurethanes, etc., belong also to this group.

Rubber is a product resulting from the vulcanization of either natural or synthetic caoutchouc, which preserves the elasticity for a long time. Natural caoutchouc is a high molecular hydrocarbon, whose basic units 2-methyl-1,3-butadiene (isoprene molecules) are *cis*-1,4-linked to form a macromolecule of about 1,500 units. Synthetic caoutchouc consists mainly of polybutadiene. In case of natural rubber, bacteria and fungi are able to attack the resulting products thereby reducing elasticity and tensile strength. Synthetic rubber is, due to its origin very resistant against biodeterioration. However, in case of additives used in the course of production products of synthetic rubber may become partly susceptible to microbial attack. The following groups of compounds may be added: vulcanization accelerators, activators, plasticizers, fillers, pigments, protectants, etc. Many of these groups contain biodegradable substances, which serve microorganisms as nutrient source (xanthates, mineral oil, paraffins, etc.).

Synthetic materials of natural origin may be derived from casein or cellulose (cellulose-nitrate, -acetate, -ether, etc.). The casein-based products are duroplasts, meaning that they cannot be formed, once they are fully polymerized. In contrast, the thermoplasts, to which the modified cellulose products belong, can be formed also later on by heating up and pressing into a desired form. Modified cellulose is in general biodegradable.

Based on PHB (poly-β-hydroxybutyric acid), which is formed by several bacteria with the function of an energy reserve polymer, a natural polymeric material has been developed. This PHB and derivatives of it are used for the production of biologically degradable "plastic" bottles, etc. However, serious problems exist concerning the shelf life of such bottles. Often periods of several months are encountered under adverse conditions, i.e., temperature and/or humidity. Consequently, because of the biodegradability either from the inside or from the outside or both microorganisms start to grow and destroy the PHB material. In addition, due to the necessity of including a microbial fermentation process for production of the polymer, the price is still two times as high as that of those polymers, which are directly produced from raw materials like oil or gas. Consequently, the application is still quite limited.

Synthetic materials resulting from chemical synthesis comprise three main groups: the polycondensates, the polymerizates, and the polyadducts. The polycondensates and the polyadducts consist each of two subgroups, duroplasts and thermoplasts, whereas the polymerizates are all thermoplasts. Tab. 4 shows the main materials, which fall into these (sub)-groups. This review is not the place to consider the chemistry of all of these compounds. For further reading literature on organic polymer chemistry is recommended. In the context of biodeterioration of these materials the basic units of the most important synthetic polymers and possibilities for microbial attack shall be discussed.

Tab. 4. List of some Important Synthetic Polymers (HEITZ et al., 1992)

Polycondensates		Polymerizates	Polyadducts	
Duroplasts	Thermoplasts	Thermoplasts	Duroplasts	Thermoplasts
Phenolic resins	polyamides	polyethylene	epoxidic resins	linear polyurethanes
Urea resins	polycarbonates	polypropylene	cross-linked polyurethanes	chlorinated polyethers
Thiourea resins	polyesters	poly-1-butene		
Melamine resins	polyphenylene oxide	poly-4-methyl-1-pentene ionomers		
Unsaturated polyester resins	polysulfones			
Alkylic resins	polyvinylacetal	polyvinylchloride		
Allyl resins		polyvinylidenechloride		
Silicones		polymethylmethacrylate		
Polyimide		polyacrylonitrile		
Poly-benzimid-azole		polystyrene		
		polyacetal		
		fluor synthetics		
		polyvinyl alcohol		
		polyvinyl acetate		
		poly-*p*-xylylene		

- Polyurethanes
  $-OCN-\text{diisocyanate}-\underset{H}{N}-\underset{O}{\overset{\|}{C}}-O-\text{polyester or polyether}-$
  with a polymer weight of 10,000 and more are in general biodegradable, although some differences exist between the polyurethane ethers, which are more resistant, and the polyesters, which are less resistant, due to the wide occurrence of esterase enzymes in microorganisms.
- Polyethylenes with the basic unit $-CH_2-CH_2-$ and a polymer weight starting with 1,500 up to 10,000 and more are generally quite resistant to a microbial attack. A problem may result from microbial pigments, which may dissolve in the material and cause unwanted discolorations.
- Polypropylene $-CH_2-\underset{CH_3}{CH}-$ with a polymer weight of 25,000 up to 500,000 is like polyethylene resistant to microbial attack. Even a fungal surface growth remains without changed properties, besides that some discolorations may occur.
- Polymethacrylate (Plexiglass) $-CH_2-C(H_3C)(COOCH_3)-$ with a polymer weight of 500,000 up to 1,000,000 is considered as biologically non-degradable, if no plasticizers have been added. In the latter case, an attack by fungi may cause an impairment of mechanical properties.
- Polyamide $-\underset{H}{N}-(CH_2)_n-\underset{O}{\overset{\|}{C}}-$ with a polymer weight of 10,000 up to 100,000 comprises materials, which are well degradable, and others, which are almost non-degradable. The degradability is dependent on the chemical substituents, which may be added to the basic unit. Basically, the amide group is susceptible to enzymatic attack by hydrolysis.

- Polystyrene $—CH_2—CH—$ with a phenyl group (benzene ring attached to CH) polymer weight of 200,000 up to 300,000 is considered to be resistant against microbial attack. Some fungal growth may occur in cases of cracks in the surface.
- Polyvinylchloride, $—CH_2—CH(Cl)—$ the basis of PVC, with a polymer weight of 30,000 up to 520,000 is considered to be resistant against microbial attack. However, due to many fields of application, PVC contains many additives, which are to a large extent biodegradable.
- Polytetrafluoroethylene (PTFE) with the basic unit $—CF_2—CF_2—$ and a polymer weight of 500,000 up to 5,000,000 is very resistant against microbial attack. No evidence exists concerning any biodeterioration problem.

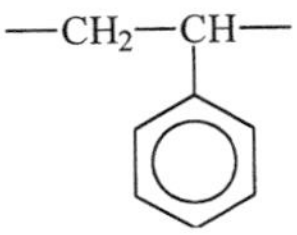

In contrast, polyesters are usually well biodegradable, except phthalic acid esters, which are resistant. Many microorganisms possess hydrolytic enzymes, esterases, which can cleave the ester bond of the polymer into the basic unit $—CH_2—C(=O)—O—$.

Additives are used to achieve the technical properties needed. The main groups are plasticizers (up to 60%), stabilizers, antioxidants (up to 2%), antistatic compounds, photoprotectants (up to 1%), flame proof compounds (up to 30%), microbicides (up to 5%), polymerization accelerators, pigments (up to 5%), fillers (up to 30 %), etc. These groups, especially the plasticizers, contain a variety of biodegradable compounds. Consequently, their addition in the course of production might render a synthetic material susceptible to biodeterioration, even though the backbone remains intact.

Deterioration problems of plastic materials result from the fact that they serve as nutrient source for microorganisms. Degradation is effected by exoenzymes, which depolymerize the polymer into biodegradable units. But also microbial metabolites can cause harm, if they react chemically with the polymer. In some instances, organic acids and solvents are excreted by microorganisms. Some of these compounds are known to react with some polymers. The effects are mainly embrittlement and/or reduced bending or tensile strength. The biodeterioration under atmospheric conditions is mainly caused by fungi, whereas in soil or water bacteria are the responsible agents.

The protection of synthetic materials may be achieved by technological measures (polymer chemistry), microbicide addition, and measures dealing with the environmental conditions, in which these materials are used (humidity, pH, O_2, temperature, etc.).

11.3 Miscellaneous Organic Materials

This group contains a diversity of materials for various applications. These are paints, varnishes/lacquers, plasters, adhesives/glues, and hydrocarbons.

Paints are used widely for protective and/or decorative purposes. Fronts and the interior of buildings as well as wet areas (in soil, on water, etc.) are covered by paints. Depending on the type the paint is either water-based (acrylate) or solvent-based (resin paint). Besides, additives like chalk, cellulose, casein, pigments, thickeners like carboxymethyl cellulose etc., are added. In the course of production these compounds are mixed and filled as paint in cans for distribution. Especially the water-based paints are endangered by biodeterioration. Problems may arise from the quality of the products (contamination of raw materials), the water (water-based paints), the production process (antiseptic mixing), and the packaging in (small) cans (contamination of air, packaging material, etc.). Besides, the ready-to-use product is often not used immediately. It may be stored in shelves for periods of more than a year facing heat or cold climatic conditions. Problems with biodeterioration are quite likely to arise. Typical problems encountered are discolorations of the product, coagulation, breaking of the dispersion (separation into water phase and organic phase), viscosity change, pH change, gas production (H_2S, NH_3, H_2, CO_2), and malodors (H_2S, NH_3, amines, butyric acid, etc.). Countermeasures are production hygiene, quality of raw materials, and biocide addition.

Plasters are used inside and outside of buildings for protective and/or decorative purposes. Often plasters comprise rough dispersions of resins with mineral particles like sand. Due to the environment surface growth may occur, besides a degradation of the constituents. Consequently discolorations up to a loss of function may result.

Varnishes/lacquers are materials similar to paints. They are usually used in thin layers for materials protection. The problems encountered are similar to those of paints.

Adhesives/glues are according to the DIN definition 16920 "non-metallic materials, which can join pieces of materials by adhesion and cohesion". Several types of adhesives need to be distinguished. The gluing compounds can be of natural (animals, plants) or synthetic origin and may contain water as solvent. Paste is a low concentrated, highly viscous, water containing product. Dispersion adhesives are organic compounds dispersed in a liquid, in which the basic compounds are insoluble. Solution adhesives are organic compounds dissolved in an organic solvent. Melting adhesives are at room temperature solid compounds without any solvent. To become usable, they have to be heated (melt). Reaction adhesives cause the joining effect by a chemical reaction (polymerization), once they are mixed or get into contact with air (oxygen). Adhesive mortars contain cement, an organic compound, and water. Contact adhesives are adhesive films, which appear visually to be dry. By joining the materials under application of pressure the joining takes place. These materials consist of a basic compound (responsible for joining), solvents, dispersants, dilution, thickeners, resins, filler, plasticizers, hardeners, accelerators, etc. The basic joining compound may be a natural compound or of synthetic origin. Most of these compounds are susceptible to microbial attack. Consequences are changes in viscosity, surface growth, discoloration, and loss of cohesion (disintegration of the material).

Hydrocarbons can be either gases, solids, or liquids. The origin is mainly raw oil, but coal, oil shale, and tar sand may be also used for the production of these compounds. In the process of fractionation in the refinery, various products are separated (alkanes, isoalkanes, cycloalkanes, and aromatic compounds including N-, S-, or O-heterocycles). Since hydrocarbons are of natural origin (incomplete plant decay) microorganisms are able to degrade these compounds (ATLAS, 1981), if oxygen (or another electron acceptor like NO_3^-) and water are available. Due to their chemical nature, hydrocarbons are in general very little soluble in water. For degradation microorganisms need to improve the solubility. Therefore they excrete emulsifying agents into the medium. Serious problems may result from hydrocarbon biodeterioration. These compounds and/or their storage tanks very often contain some water (0.1–2%). Microbial growth can take place and may result in corrosion problems (organic acids), the clogging of tubings, etc., and the formation of malodors. Especially in the case of airplanes a microbial growth in tanks must be avoided, because of the existing danger that mycelia can interfere with the kerosene supply by blocking pumps and/or tubings. Storage tanks embedded in soil may become leaky due to microbial corrosion, which would cause serious soil pollution.

Hydrocarbons are also used in hydraulic fluids, cooling liquids, or lubricants. These are often not pure hydrocarbons, but oil in water emulsions. In case of metal-working fluids about 100,000 t a^{-1} are consumed. Due to the presence of hydrocarbons these compounds/materials are susceptible to biodeterioration and may loose their function. Besides, especially metal-working fluids are used in such a way that aerosols are formed. The problem exists that pathogenic bacteria might be contained and, thus, become distributed among workers. Consequently, microbicides have to be added, besides physical measures to be applied.

12 Useful Biodeterioration

Depending upon the point of view biodeterioration may also be a useful process. Three selected examples shall briefly be mentioned: bacterial leaching of low-grade ores, kaolin purification, and microbial enhanced oil recovery.

Bacterial leaching is applied to extract valuable metals out of low-grade ores, which cannot be economically processed by conventional means like flotation and roasting. Specialized lithoautotrophic bacteria like *Acidithiobacillus ferrooxidans*, *Acidithiobacillus thiooxidans*, *Leptospirillum ferrooxidans*, and many others are able to degrade metal sulfides to sulfuric acid, which contains the metal cations in a dissolved state. The process of metal sulfide dissolution is effected by an attack of iron(III) ions and/or protons (according to the mineralogical type of sulfide). The iron(III) ions and protons are brought into contact with the metal sulfide surface via EPS, which contain the iron(III) ions in a complexed state (two glucuronic acid residues per iron(III) ion). From electrochemistry it is known that tunneling of electrons is possible up to 20 nm, which is exactly the distance between the bacterial outer membrane and the metal sulfide surface. This distance/space is filled by the EPS of the attached cells containing the complexed iron(III) ions. The latter may become reduced by the tunneling effect and, thus, become mobile [from complex chemistry it is known that iron(II) ions are considerably less firmly bound than iron(III) ions]. These iron ions will be oxidized by the iron oxidizing system of the cells for bacterial energy generation. The oxidized iron ions may enter the cycle again. Consequently, bacterial leaching can be seen as a wanted biocorrosion of a metal sulfide. The economic value of bioleaching is huge: copper is produced on a worldwide scale of about 20% by bioleaching, gold production may depend to the same amount on bioleaching. Currently, the largest bioreactors are built for gold and cobalt winning. Up to six tanks of 1,000 m^3 volume each are connected to process gold containing ores (Fig. 11).

Besides, the metals uranium, nickel, cobalt, zinc, manganese, silver, and molybdenum are increasingly produced by bioleaching. Above the positive aspects serious environmental problems connected with bioleaching may not be forgotten. A serious problem results from mining waste (resulting from metal mining, coal and lignite mining, smelter and power plant storage), which is called acid rock/mine drainage (COLMER and HINKLE, 1947). Since the same processes function if metal sulfides are exposed to the environment without the intention to recover the dissolved metal ions, severe pollution can result (Fig. 12). Large

Fig. 11. Stirred biooxidation tanks of the Ashanti gold recovery plant in Obuasi, Ghana, for the bioleaching of refractory ores prior to cyanidation (courtesy of D. DEW).

Fig. 12. Acid rock drainage at Rosia Poieni, Romania.

areas may be polluted by sulfuric acid containing metal cations and even derive their name from this pollution (e.g., Rio Tinto, Spain).

Kaolin purification has already been exploited by Chinese potters in the millennium B.C. to produce a white shining porcelain. These potters used biological processes like bioleaching of metal sulfides and the complexation of metal cations by organic acids (produced by fermentation) to remove metal compounds from kaolin and clay. Besides, the physical properties of the raw materials were improved. Basically, bioleaching was used to dissolve the metal sulfides FeS and FeS_2, which are often encountered in kaolin and clay. The iron sulfides upon burning cause the product to be of brownish appearance. Fermentation for production of organic acids was also used, since these acids complex the unwanted metals and, thus, remove them. For this purpose clay, kaolin, etc., were mixed with manure and incubated for several years.

Microbial enhanced oil recovery, MEOR, is used to increase the amount of oil produced from wells, which are almost exhausted (Jack, 1988). The principles used are *in situ* microbial gas production (CO_2, H_2, H_2S, N_2) to increase the pressure in the formation by introduction of the respective microbes, introduction of microorganisms, which produce *in situ* emulsifying agents for a viscosity reduction of the remaining oil, the introduction of EPS producing microorganisms into the formation to clog the pores, through which the preferential flow takes place, in order to enforce other, up to now not used pathways to open and allow the extraction of the oil (Geesey et al., 1987). The latter may be also achieved by using ultramicro-cells, which are optimally suited because of their reduced diameter (below 0.2 μm) to enter small pores and, by starting to grow, clog these efficiently and consequently divert the flow.

13 Synopsis

The aim of this review is to demonstrate that microorganisms, although of microscopic size, can interfere in many ways with man-made materials, either of natural or of synthetic origin. All microorganisms, bacteria, cyanobacteria, algae, protozoa, lichen, and fungi, are able to participate. Nevertheless, the amount of mechanisms causing biodeterioration is limited to less than ten. These become effective by chemical and/or physical reactions and do not comprise any *de novo* mode of action. This always causes the problem of detection (that it is a microbial problem) and differentiation (different from chemical or physical effects only). The reality is at present that except in cases, where the impact of microorganisms has been acknowledged, all types of physical and/or chemical explanations are tried. Only when these (repeatedly) fail to explain the process, the responsible people start to think about the possibility that microorganisms could be involved in their problem. The second problem is directly evolving out of this situation: countermeasures are tested in laboratory experiments by using the chemical and/or physical stress on materials only. The idea that microorganisms are involved may be accepted, but their effect is considered to be of physical and/or chemical nature only. Although this is basically true, something is forgotten: microorganisms are living beings, which like others try to find (actively) the most suitable environment for growth. Thus, via some very complex systems they interact with the environment. Furthermore, it is unknown that the extracellular polymeric compounds, which surround the cells, play a pivotal role in biodeterioration processes. All evidence available up to now indicates that in the tiny space between the cell wall and the substratum (material), i.e., in dimensions of 10 nm to 30 nm, the processes take place, which are of interest for us. However, no techniques are available up to now to detect and analyze reactions in this nanometer range. A serious mistake, which often occurs, is the use of homogenized samples for analysis of (corrosive) parameters. One gram of sample can never reflect the values in the few nanometers between cell and substratum. Hopefully microelectrodes, AFM (atomic force microscopy), ESCA (electron spectroscopy for chemical analysis), and other techniques will allow in the future to elucidate the processes, which cause so much damage to materials and waste a considerable amount of the gross national product. Up to now the only way to improve materials is the use of simulation equipment. If the processes are known, materials may be selected by this technique, which are more resistant against the biogenic attack than the currently used ones. Consequently, to be successful, the interaction between scientists from various specifications is needed: (electro-) chemists, physicists, engineers, materials scientists, and microbiologists.

Acknowledgement

This chapter is the result of practical experience, the continuous exchange of information in the DECHEMA working group "Microbial Deterioration of Materials and Protection" and in the GfKORR group "Microbial Corrosion", and teaching of students. I especially appreciate the cooperation of T. Gehrke, T. Rohwerder, K. Kinzler, and G. Meyer. Without their support this manuscript could not have been finished.

14 References

Atlas, R. M. (1981), Microbial degradation of petroleum hydrocarbons: an environmental perspective, *Microbiol. Rev.* **45**, 180–209.

Berthelin, J. (1983), Microbial weathering processes, in: *Microbial Geochemistry* (Krumbein, W. E., Ed.), pp. 223–262. Oxford: Blackwell Scientific Publications.

Bock, E., Sand, W. (1993), A review: the microbiology of masonry biodeterioration, *J. Appl. Bacteriol.* **74**, 503–514.

Busscher, H. J., Weerkamp, A. H. (1987), Specific and nonspecific interactions in bacterial adhesion to solid substrata, *FEMS Microbiol. Rev.* **46**, 165–173.

Colmer, A. R., Hinkle, M. E. (1947), The role of microorganisms in acid mine drainage: a preliminary report, *Science* **106**, 253.

Costerton, J. W., Cheng, K. J., Geesey, G. G., Ladd, T. I., Nickel, J. C. et al. (1987), Bacterial biofilms in nature and disease, *Am. Rev. Microbiol.* **41**, 435–464.

Dilling, W., Cypionka, H. (1991), Aerobic respiration in sulfate-reducing bacteria, *FEMS Microbiol. Lett.* **71**, 123–128.

Flemming, H.-C. (2000), Biofilms in biofiltration, in: *Biotechnology* 2nd Edn., Vol. IIc (Rehm, H.-J., Reed, G., Pühler, A., Stadler, P., Eds.), pp. 445–455. Weinheim: Wiley-VCH.

Geesey, G. G., Mittelman, M. W., Lieu, V. T. (1987), Evaluation of slime-producing bacteria in oil field core flood experiments, *Appl. Environ. Microbiol.* **53**, 278–283.

Gehrke, T., Telegdi, J., Thierry, D., Sand, W. (1998), Importance of extracellular polymeric substances from *Thiobacillus ferrooxidans* for bioleaching, *Appl. Environ. Microbiol.* **64**, 2743–2747.

Hamilton, W. A. (1985), Sulphate-reducing bacteria and anaerobic corrosion, *Annu. Rev. Microbiol.* **39**, 195–217.

Heitz, E., Mercer, A. D., Sand, W., Tiller, A. K. (Eds.) (1992), *Microbial Degradation of Materials – and Methods of Protection*, European Federation of Corrosion Publications Number 9. London: The Institute of Materials.

IVERSON, W. P. (1966), Direct evidence for the cathodic depolarization theory of bacterial corrosion, *Science* **151**, 986–988.

JACK, T. R. (1988), Microbially enhanced oil recovery, *Biorecovery* **1**, 59–73.

KAYE, S., PHILLIPS, C. R. (1949), *Am. J. Hyg.* **50**, 296–298.

LEA, F. M., DESCH, C. H. (1936), *The Chemistry of Cement and Concrete*. London: Edward Arnold.

LINHARDT, P. (1996), Failure of chromium-nickel steel in a hydroelectric power plant by manganese-oxidizing bacteria, in: *Microbially Influenced Corrosion of Materials* (HEITZ, E., FLEMMING, H.-C., SAND, W., Eds.), pp. 221–230. Berlin: Springer-Verlag.

LOVLEY, D. R. (1991), Dissimilatory Fe(III) and Mn(IV) reduction, *Microbiol. Rev.* **55**, 259–287.

MARSHALL, K. C., BLAINEY, L. (1991), Role of bacterial adhesion in biofilm formation and biocorrosion, in: *Biofouling and Biocorrosion in Industrial Water Systems* (FLEMMING, H.-C., GEESEY, G. G., Eds.), pp. 29–46. Berlin: Springer-Verlag.

MILDE, K., SAND, W., WOLFF, W., BOCK, E. (1983), Thiobacilli of the corroded concrete walls of the Hamburg sewer system, *J. Gen. Microbiol.* **129**, 1327–1333.

OLMSTEAD, W. M., HAMLIN, H. (1900), Converting portions of the Los Angeles outfall sewer into a septic tank, *Eng. News* **44**, 317–318.

PANKHANIA, I. P., MOOSAVI, A. N., HAMILTON, W. A. (1986), Utilization of cathodic hydrogen by *Desulfovibrio vulgaris* (Hildenborough), *J. Gen. Microbiol.* **132**, 3357–3365.

PARKER, C. D. (1945), The corrosion of concrete. I. The isolation of a species of bacterium associated with the corrosion of concrete exposed to atmospheres containing hydrogen sulphide, *Austr. J. Exper. Biol. Med. Sci.* **23**, 81–90.

SAND, W. (1987), Importance of hydrogen sulfide, thiosulfate, and methylmercaptan for growth of thiobacilli during simulation of concrete corrosion, *Appl. Environ. Microbiol.* **53**, 1645–1648.

SAND, W., MILDE, K., BOCK, E. (1983), Simulation of concrete corrosion in a strictly controlled H_2S-breeding chamber, in: *Recent Progress in Biohydrometallurgy* (ROSSI, G., TORMA, A. E., Eds.), pp. 667–677. Iglesias: Associazione Mineraria Sarda.

SAND, W., AHLERS, B., KRAUSE-KUPSCH, T., MEINCKE, M., KRIEG, E. et al. (1989), Mikroorganismen und ihre Bedeutung für die Zerstörung von mineralischen Baustoffen, *UWSF – Z. Umweltchem. Ökotox.* **3**, 36–40.

SAND, W., DUMAS, T., MARCDARGENT, S. (1994), Accelerated biogenic sulfuric-acid corrosion test for evaluating the performance of calcium-aluminate based concrete in sewage applications, in: *Microbiologically Influenced Corrosion Testing*, ASTM STP 1232 (KEARNS, J. R., LITTLE, B. J., Eds.), pp. 234–249. Philadelphia, PA: ASTM.

SCHIPPERS, A., JOZSA, P.-G., SAND, W. (1996), Sulfur chemistry in bacterial leaching of pyrite, *Appl. Environ. Microbiol.* **62**, 3424–3431.

SCHULZ, H. N., BRINKHOFF, T., FERDELMAN, T. G., HERNÁNDEZ MARINÉ, M., TESKE, A., JØRGENSEN, B. B. (1999), Dense populations of a giant sulfur bacterium in Namibian shelf sediments, *Science* **284**, 493–495.

SCOTTO, V., LAI, M. E. (1998), The ennoblement of stainless steels in seawater: a likely explanation coming from the field, *Corr. Sci.* **40**, 1007–1018.

SHIVELY, J. M., BENSON, A. A. (1967), Phospholipids of *Thiobacillus thiooxidans*, *J. Bacteriol.* **94**, 1679–1683.

STEUDEL, R., HOLDT, G. (1988), Solubilization of elemental sulfur in water by cationic and anionic surfactants, *Angew. Chem.* (*Int. Edn. Engl.*) **27**, 1358–1359.

STEVENSON, F. J. (1986), *Cycles of Soil*. New York: John Wiley.

STOLP, H. (1988), *Microbial Ecology: Organisms, Habitats, Activities*. Cambridge: Cambridge University Press.

SZEWZYK (2000), in: ASM Conf. of Biofilms 2000, Poster 163. Washington, DC: ASM.

VAN LOOSDRECHT, M. C. M., ZEHNDER, A. J. B. (1990), Energetics of bacterial adhesion, *Experientia* **46**, 817–822.

VON RÈGE, H. (1999), Bedeutung von Mikroorganismen des Schwefelkreislaufes für die Korrosion von Metallen, *Thesis*, University of Hamburg, Germany.

VON RÈGE, H., SAND, W. (1998), Evaluation of biocide efficacy by microcalorimetric determination of microbial activity in biofilms, *J. Microbiol. Methods* **33**, 227–235.

VON RÈGE, H., SAND, W. (1999), Importance of biogenic iron–sulfur compounds for metal-MIC, in: *Solution of Corrosion Problems in Advanced Technologies* (SCHMITT, G., SCHÜTZE, M., Eds.), pp. 85–88. Frankfurt: DECHEMA e.V.

VON WOLZOGEN-KÜHR, G. A. H., VAN DER VLUGT, L. R. (1934), De graphiteering van gietijzer ais electrobiochemisch proces in anaerobe gronden, *Water* **18**, 147–165.

WIDDEL, F. (1990), Mikrobielle Korrosion, in: *Jahrbuch Biotechnologie*, Vol. 3 (PRÄVE, P., SCHLINGMANN, M., CRUEGER, W., ESSER, K., THAUER, R., WAGNER, F., Eds.), pp. 277–318. München: Carl Hanser.

WINGENDER, J., FLEMMING, H.-C. (1999), Autoaggregation of microorganisms: Flocs and biofilms, in: *Biotechnology* 2nd Edn., Vol 11a (REHM, H.-J., REED, G., PÜHLER, A., STADLER, P., Eds.), pp. 65–83. Weinheim: Wiley-VCH.

15 Further Reading

The interested reader may find articles about biodeterioration in many journals and books. Problematic is the fact that microbiologists prefer to publish in their journals (for scientific merit), but these are seldom read by other scientists. The same is valid for (electro)chemists and other scientists. In addition, to complicate the situation, many articles appear in specialized, practical journals aiming specifically at one industrial branch. For this reason only citations for a few selected books and journals can be given.

Books

Angell, P. (Ed.) (1995), *International Conference on Microbially Influenced Corrosion*. Houston, TX: NACE International.

Borenstein, S. W. (1994), *Microbiologically Influenced Corrosion Handbook*. New York: Woodhead Publishing.

Brill, H. (Ed.) (1995), *Mikrobielle Materialzerstörung und Materialschutz: Schädigungsmechanismen und Schutzmaßnahmen*. Jena: Gustav Fischer.

Characklis, W. G., Wilderer, P. A. (Eds.) (1989), *Structure and Function of Biofilms*. Chichester: John Wiley & Sons.

Dexter, S. C. (Ed.) (1986), *Biologically Induced Corrosion*, NACE 8. Houston, TX: NACE International.

Doyle, R. J. (Ed.) (1999), *Methods in Enzymology*, Vol. 310, *Biofilms*. San Diego, CA: Academic Press.

Doyle, R. J., Rosenberg, M. (Eds.) (1990), *Microbial Cell Surface Hydrophobicity*. Washington, DC: ASM.

Ehrlich, H. L. (1996), *Geomicrobiology*, 3rd Edn. New York: Marcel Dekker.

Flemming, H.-C., Geesey, G. G. (Eds.) (1991), *Biofouling and Biocorrosion in Industrial Water Systems*. Berlin: Springer-Verlag.

Flemming, H.-C., Szewzyk, U., Griebe, T. (Eds.) (2000), *Investigation of Biofilms*. Lancaster: Technomic.

Fletcher, M. (1996), *Bacterial Adhesion*. New York: Wiley-Liss.

Heitz, E., Mercer, A. D., Sand, W., Tiller, A. K. (Eds.) (1992), *Microbial Degradation of Materials – and Methods of Protection*, European Federation of Corrosion Publications Number 9. London: The Institute of Materials.

Heitz, E., Flemming, H.-C., Sand, W. (1996), *Microbially Influenced Corrosion of Materials*. Berlin: Springer-Verlag.

Hill, E. C., Shennan, J. L., Watkinson, R. J. (Eds.) (1987), *Microbial Problems in the Offshore Oil Industry*. Chichester: John Wiley & Sons.

Kearns, J. R., Little, B. J. (Eds.) (1994), *Microbiologically Influenced Corrosion Testing*, ASTM STP 1232. Philadelphia, PA: ASTM.

Kobrin, G. (Ed.) (1993), *Microbiologically Influenced Corrosion*. Houston, TX: NACE International.

Madigan, M. T., Martinko, J. M., Parker, J. (Eds.) (2000), *Brock Biology of Microorganisms* 9th Edn. New Jersey: Prentice Hall.

Rose, A. H. (Ed.) (1981), *Microbial Biodeterioration*. London: Academic Press.

Rossi, G. (1990), *Biohydrometallurgy*. Hamburg: McGraw-Hill.

Sand, W., Kreysa, G. (Eds.) (1996), *Biodeterioration and Biodegradation*, DECHEMA-Monographie 133. Weinheim: VCH.

Schmidt, R. (1999), *Werkstoffverhalten in biologischen Systemen*. Berlin: Springer-Verlag.

Stolp, H. (1988), *Microbial Ecology: Organisms, Habitats, Activities*. Cambridge: Cambridge University Press.

Tiller, A. K., Sequeira, C. A. C. (Eds.) (1995), *Microbial Corrosion*. London: The Institute of Materials.

Wallhäußer, K. H. (1988), *Praxis der Sterilisation – Desinfektion – Konservierung*. Stuttgart: Thieme.

Wingender, J., Neu, T., Flemming, H.-C. (Eds.) (1999), *Microbial Extracellular Polymeric Substances*. Berlin: Springer-Verlag.

Journals

International Biodeterioration and Biodegradation, Elsevier

Materials and Corrosion, Wiley-VCH

Corrosion, NACE International

Materials Performance, NACE International

Biofouling, Harwood Academic Publishers

Biodegradation, Kluwer Academic Publishers

Water Research, Pergamon Press

Special Substances

11 Vitamins and Related Compounds: Microbial Production

SAKAYU SHIMIZU
Kyoto, Japan

1 Introduction

Vitamins are defined as essential micronutrients that are not synthesized by mammals. Most vitamins are essential for the metabolism of all living organisms, and they are synthesized by microorganisms and plants. Coenzymes (and/or prosthetic groups) are defined as organic compounds with low molecular weight that are required to show enzyme activity by binding with their apoenzymes. Many coenzymes are biosynthesized from vitamins and contain a nucleotide (or nucleoside) moiety in their molecules. Besides their functions as vitamins and coenzymes, most of vitamins and coenzymes have been shown to have various other biofunctions. Accordingly, it is more appropriate to understand both as effective biofactors (see FRIEDRICH, 1988 for basic information).

Most vitamins and related compounds are now industrially produced and widely used as food or feed additives, medical or therapeutic agents, health aids, cosmetic and technical aids, and so on. Thus, vitamins and related compounds are important products for which many biotechnological production processes (i.e., fermentation and microbial/enzymatic transformation) as well as organic chemical synthetic ones have been reported; some of them are now applied for large-scale production. Industrial production methodology, annual production amounts, and fields of application for these vitamins and related compounds are summarized in Tabs. 1–3.

In this chapter, some of the vitamins and related compounds are described from the viewpoint of their microbial production. Previous reviews, from a similar viewpoint, may be useful for further information (DE BAETS et al., 2000; EGGERSDORFER et al., 1996; FLORENT, 1986; SHIMIZU and YAMADA, 1986; VANDAMME, 1989).

2 Water-Soluble Vitamins

2.1 Riboflavin (Vitamin B_2) and Related Coenzymes

Riboflavin is used for human nutrition and therapy and as an animal feed additive. The crude concentrated form is also used for feed. It is produced by both synthetic and fermentation processes [major producers, Hoffmann-La Roche (Switzerland), BASF (Germany), ADM (USA), Takeda (Japan)]. The current world production of riboflavin is about 2,400 t a^{-1}, of which 75% is for feed additive and the remaining for food and pharmaceuticals. Two closely related ascomycete fungi, *Eremothecium ashbyii* and *Ashbya gossypii*, are mainly used for the industrial production (OZBAS and KUTSAL, 1986; STAHMANN et al., 2000). Yields much higher than 10 g of riboflavin per liter of culture broth are obtained in a sterile aerobic submerged fermentation with a nutrient medium containing molasses or plant oil as a major carbon source. Yeasts (*Candida flaeri*, *C. famata*, etc.) and bacteria can also be used for the practical production. Riboflavin production by genetically engineered *Bacillus subtilis* and *Corynebacterium ammoniagenes* which overexpress genes of the enzymes involved in riboflavin biosynthesis reach 4.5 g L^{-1} and 17.4 g L^{-1}, respectively (KOIZUMI et al., 1996; PERKINS et al., 1999). D-Ribose is used as the starting material in the chemical production processes, in which it is transformed to riboflavin in three steps. D-Ribose is obtained directly from glucose by fermentation with a genetically engineered *Bacillus* strain which is transketolase-defective and overexpresses the gluconate operon (DE WULF and VANDAMME, 1997).

Flavin mononucleotide (FMN), a coenzyme form of riboflavin, is synthesized from riboflavin by chemical phosphorylation, after which FMN is crystallized as the diethanolamine salt to separate isomeric riboflavin phosphates and unreacted riboflavin.

The other coenzyme form of riboflavin, flavin adenine dinucleotide (FAD), is used in pharmaceutical and neutraceutical applications. Several tons of FAD are annually pro-

Tab. 1. Industrial Production of Vitamins and Coenzymes

Compound	Production Method			World Production [$t\ a^{-1}$]		Use
	Biotechnological	Chemical	Extraction	1980s[a]	1990s[b]	
Thiamin (B_1)		+		1,700	4,200	food, pharmaceutical
Riboflavin (B_2)	+			2,000	2,400	feed, pharmaceutical
FAD	+	+			10	pharmaceutical
Nicotinic acid, nicotinamide	+	+		8,500	22,000	feed, food, pharmaceutical
NAD, NADP	+					technical
Pantothenic acid	+[c]	+		5,000	7,000	feed, food, pharmaceutical
Coenzyme A	+					technical, neutraceutical
Pyridoxine (B_6)		+		1,600	2,550	feed, food, pharmaceutical
Biotin	(+)[d]	+		2.7	25	feed, pharmaceutical
Folic acid		+		100	400	feed, food, pharmaceutical
Vitamin B_{12}	+			12	10	feed, food, pharmaceutical
Vitamin C	+[c]			40,000	60,000	feed, food, pharmaceutical
ATP	+					pharmaceutical, technical
S-Adenosyl-methionine	+					pharmaceutical, nutraceutical
Lipoic acid		+				pharmaceutical
Pyrroloquinoline quinone	+	+				technical
Vitamin A		+		2,500	2,700	feed, food, pharmaceutical
β-Carotene	+		+	100	400	feed, food
Ergosterol	+			25	38	feed, food
Vitamin D_3		+	+		5,000	feed, food
α-Tocopherol (E)	(+)[d]	+	+	6,800	22,000	feed, food, pharmaceutical, nutraceutical
PUFAs[e]	+		+			feed, food, pharmaceutical, nutraceutical
Phylloquinone (K_1)		+			3.5	pharmaceutical
Menaquinone (K_2)		+			500	pharmaceutical
Ubiquinone-10	+					feed, food, pharmaceutical

[a] Values were taken from FLORENT (1986).
[b] Values were taken from EGGERSDORFER et al. (1996).
[c] Hybrid of microbial and chemical reactions.
[d] Parentheses indicate pilot scale process.
[e] PUFAs, polyunsaturated fatty acids.

duced by chemical synthesis or by microbial transformation. The latter uses FMN and adenosine 5′-triphosphate (ATP) as the substrates and *C. ammoniagenes* cells as a source of FMN adenylyltransferase. In this transformation, ATP is generated from adenine and phosphoribosyl pyrophosphate is *de novo* synthesized from glucose by the same organism (see Sect. 2.8). In a similar fashion using the *C. ammoniagenes* ATP generating system, genetically engineered strains of *Escherichia coli* which overexpress flavokinase, and FMN adenylyltransferase can be used as the catalyst in the transformation from riboflavin (KITATSUJI et al., 1992) (Fig. 1).

Tab. 2. Microbial and Enzymatic Processes for the Production of Water-Soluble Vitamins and Coenzymes

Vitamin, Coenzyme	Enzyme (Microorganism)	Method
Vitamin C (2-Keto-L-gulonic acid)	2,5-diketo-D-gulonic acid reductase (*Corynebacterium* sp.)	enzymatic conversion of 2,5-diketo-D-gluconate obtained through fermentative process to 2-keto-L-gulonic, followed by chemical conversion to L-ascorbic acid
Biotin	fermentation (*Serratia marcescens*)	fermentative production from glucose by a genetically engineered bacterium
	multiple enzyme system (*Bacillus sphaericus*)	conversion from diaminopimelic acid using the biotin biosynthesis enzyme system of a mutant of *B. sphaericus*
Pantothenic acid (D-Pantoic acid)	lactonohydrolase (*Fusarium oxysporum*)	resolution of D,L-pantolactone to D-pantoic acid and L-pantolactone by stereoselective hydrolysis
Coenzyme A	multiple enzyme system (*Brevibacterium ammoniagenes*)	conversion by enzymatic coupling of ATP-generating system and coenzyme A biosynthesis system of *B. ammoniagenes* (parent strain or mutant) with D-pantothenic acid, L-cysteine, and AMP (or adenosine, adenine, etc.) as substrates
Nicotinamide	nitrile hydratase (*Rhodococcus rhodochrous*)	hydration of 3-cyanopyridine
Nicotinic acid	nitrilase (*Rhodococcus rhodochrous*)	hydrolysis of 3-cyanopyridine to form corresponding acid (nicotinic acid) and ammonia
NAD	multiple enzyme system (*Corynebacterium ammoniagenes*)	conversion by enzymatic coupling of ATP-generating system and NAD biosynthesis enzymes of *B. ammoniagenes* with adenine and nicotinamide as substrates
NADP	NAD kinase (*Brevibacterium* sp., *Corynebacterium* sp., etc.)	phosphorylation of NAD with ATP as the phosphate group donor
NADH	formic acid dehydrogenase (*Arthrobacter* sp., *Candida boidinii*, etc.)	reduction of NAD with formic acid as the hydrogen donor
NADPH	glucose dehydrogenase (*Bacillus* sp., *Gluconobacter* sp., etc.)	reduction of NADP with glucvose as the hydrogen donor
Riboflavin	fermentation (*Eremothecium ashbyii*, *Ashbya gossypii*, *Bacillus* sp., etc.)	fermentative production from glucose

Tab. 2. Continued

Vitamin, Coenzyme	Enzyme (Microorganism)	Method
FAD	FAD synthetase (*Corynebacterium* sp., *Arthrobacter* sp., etc.)	enzymatic pyrophosphorylation of ATP and flavin mononucleotide synthesized chemically
ATP	multiple enzyme system (baker's yeast, methylotrophic yeasts, *Corynebacterium ammoniagenes*, etc.)	ribotidization of adenine (or adenosine) under coupling of the glycolysis system or methanol oxidation system
S-Adenosylmethionine	S-adenosylmethionine synthetase (*Saccharomyces saké*)	conversion of L-methionine by *S. saké* mutant
S-Adenosylhomocysteine	S-adenosylhomocysteine hydrolase (*Alcaligenes faecalis*)	condensation of adenosine and homocysteine
L-Carnitine	β-oxidation-like enzymes (*Agrobacterium* sp.) aldehyde reductase (*Sporobolomyces salmonicolor*)	conversion of butyrobetaine to L-carnitine enzymatic asymmetric reduction of 4-chloroacetoacetic acid ester to *R*-(−)-3-hydroxy-4-chlorobutanoic acid ester, followed by its chemical conversion to L-carnitine
Pyridoxal-5′-phosphate	pyridoxamine oxidase (*Pseudomonas* sp.)	oxidation of chemically synthesized pyridoxine-5′-phosphate
CDP-choline, GDP-glucose, etc.	CDP-choline pyrophosphorylase, NDP-glucose pyrophosphorylase, etc. (yeasts, etc.)	pyrophosphoric acid condensation of choline (or glucose, etc.) and nucleotide triphosphate (or the corresponding nucleoside)
Vitamin B_{12}	fermentation (*Propionibacterium shermanii*, *Pseudomonas denitrificans*, etc.)	fermentative production from glucose
Pyrroloquinoline quinone (PQQ)	fermentation (methanol-utilizing bacterium)	fermentative production from methanol

2.2 Nicotinic Acid, Nicotinamide, and Related Coenzymes

The world production of nicotinic acid and nicotinamide is estimated to be 22,000 t a^{-1} [major producers, BASF, Lonza (Switzerland) and Degussa (Germany)]. The major use (ca. 75%) is for animal nutrition and the remaining for food enrichment and pharmaceutical application. Chemical processes involving oxidation of 5-ethyl-2-methylpyridine or total hydrolysis of 3-cyanopyridine are used for nicotinic acid production. Bacterial nitrilase has been shown to be useful for the same purpose (Fig. 2a). For example, 3-cyanopyridine is almost stoichiometrically converted to nicotinic acid (172 g L^{-1}) on incubation with the nitrilase-overexpressed *Rhodococcus rhodochrous* J1 cells (NAGASAWA and YAMADA, 1989). The same *R. rhodochrous* enzyme can be used for the production of *p*-aminobenzoic acid from *p*-aminobenzonitrile.

Tab. 3. Microbial and Enzymatic Processes for the Production of Fat-Soluble Vitamins

Vitamin	Enzyme (Microorganism)	Method
Vitamin E and K_1 side chains	multiple enzyme system (*Geotrichum candidum*)	enzymatic conversion from (*E*)-3-(1′,3′-dioxolane-2′-yl)-2-butene-1-ol
[(*S*)-2-methyl-γ-butyrolactone] [(*S*)-3-methyl-γ-butyrolactone]	reductase bakers' yeast, (*Geotrichum* sp., etc.)	asymmetric reduction of ethyl-4,4-dimethoxy-3-methylcrotonate
[(*S*)- or (*R*)-β-hydroxy-isobutyric acid]	multiple enzyme system (*Candida* sp., etc.)	stereoselective oxidation of isobutyric acid
Vitamin K_2	multiple enzyme system (*Flavobacterium* sp.)	conversion of quinone- and side chain-precursors to the vitamin
Arachidonic acid	fermentation (*Mortierella alpina*)	fermentative production from glucose
Dihomo-γ-linolenic acid	fermentation (*Mortierella alpina*)	fermentative production from glucose by a Δ5-desaturase-defective mutant
Mead acid	fermentation (*Mortierella alpina*)	fermentative production from glucose by a Δ12-desaturase-defective mutant
Eicosapentaenoic acid	multiple enzyme system (*Mortierella alpina*)	Δ17-desaturation of arachidonic acid or conversion from α-linolenic acid

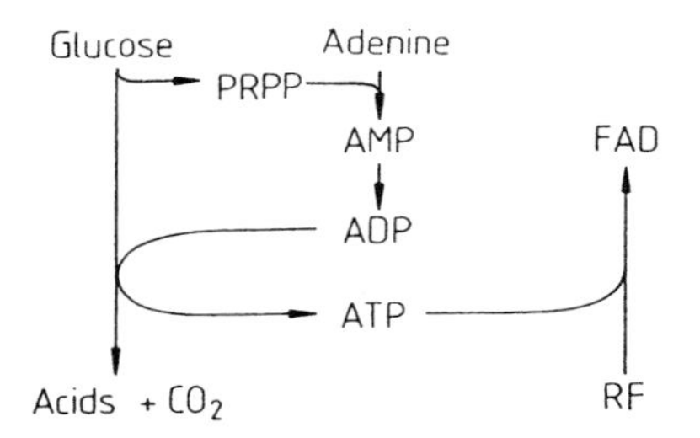

Fig. 1. Schematic representation for the FAD production from riboflavin (RF) coupled with bacterial ATP-generating system (see also Fig. 8).

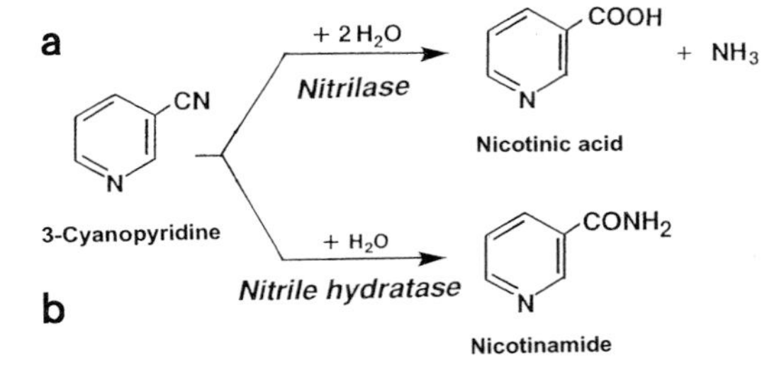

Fig. 2. Transformation of 3-cyanopyridine to nicotinic acid by nitrilase **a** and nicotinamide by nitrile hydratase **b**.

Nicotinamide is available from partial hydrolysis of 3-cyanopyridine, which is performed by both chemical and enzymatic processes. The enzymatic process uses nitrile hydratase as the catalyst (Fig. 2b). This novel enzyme catalyzing a simple hydration reaction was discovered as one of the responsible enzymes for the two-step transformation of nitriles to acids via amides (ASANO et al., 1980). Extensive studies of this enzyme as well as screening of the enzyme from a variety of microbial strains revealed the presence of several different types of nitrile hydratases, especially Co- and Fe-containing enzymes, in various bacteria (KOBAYASHI et al., in press). The Co-containing enzyme from *R. rhodochrous* J1 hydrates various kinds of aliphatic and aromatic nitriles to the corresponding amides and has been shown to be useful for the production of useful amides (YAMADA and KOBAYASHI, 1996). For example, using the bacterial cells containing highly elevated amounts of this enzyme exceeding 50% of the total cellular proteins, 1.23 kg of 3-cyanopyridine suspended in 1 liter of water are stoichiometrically converted to 1.46 kg of nicotinamide crystals (NAGASAWA and YAMADA, 1989). Based on these studies, Lonza (Switzerland) has constructed a plant for the commercial production of nicotinamide in China in 1997. This enzymatic process surpasses the chemical process in regard to several points such as stoichiometric conversion of high concentration of the substrate and the quality of the product actually with zero contents of byproducts. The same enzyme has been used for the industrial production of acrylamide from acrylonitrile by Nitto (Japan) since 1991.

Nicotinamide adenine dinucleotide (NAD) is used in pharmaceutical application and as a reagent for clinical analysis. Nicotinamide adenine dinucleotide phosphate (NADP) is also used for analysis. Practical production of NAD is carried out by extraction. Yeasts such as *Saccharomyces cerevisiae* are favorable sources of NAD. It is also produced by microbial transformation utilizing the salvage pathway for the biosynthesis of NAD from nicotinamide (or nicotinic acid) and ATP. On cultivation of *Corynebacterium ammoniagenes* with the precursors, nicotinamide and adenine, the amount of NAD in the medium reaches 2.3 mg mL^{-1} (NAKAYAMA et al., 1968). For the mechanism involved in the transformation it has been suggested that both precursors are first ribotidated to nicotinamide monophosphate and ATP, respectively, which are then converted to NAD by pyrophosphorylation (for ATP generation, see Sect. 2.8). NADP can be prepared by enzymatic phosphorylation. Reduced forms of these coenzymes, NADH and NADPH, can be obtained by both chemical and enzymatic methods. The latter uses formate dehydrogenase from methanol-utilizing yeasts for NADH. Glucose dehydrogenase from *Bacillus* sp. is also used for both NADH and NADPH. *In situ* regeneration of these coenzymes is currently attracting more attention for the production of chiral alcohols from prochiral carbonyl compounds with carbonyl reductases (Fig. 3). For example, genetically engineered *E. coli* cells overexpressing glucose dehydrogenase from *Bacillus* sp. and aldehyde reductase from *Sporobolomyces salmonicolor* effectively catalyze stereospecific reduction of ethyl 4-chloro-3-oxobutanoate to ethyl $R(-)$-4-chloro-3-hydroxybutanoate in the presence of glucose and a catalytic amount of NADP (SHIMIZU and KATAOKA, 1999a; SHIMIZU et al., 1997).

2.3 Pantothenic Acid and Coenzyme A

About 6,000 t of calcium D-pantothenate are produced annually. It is mainly used as an animal feed additive (80%). It is also used for pharmaceutical, health care and food products. D-Pantothenyl alcohol (1,000 t a^{-1}) is also used for the same purposes. The commercial production process involves reactions yielding racemic pantolactone from isobutyraldehyde, formaldehyde, and cyanide, optical resolution of the racemic pantolactone to D-pantolactone, and condensation of D-pantolactone with β-alanine to form D-pantothenic acid. 3-Aminopropanol is used for D-pantothenyl alcohol [major producers, Hoffmann-La Roche, Fuji (Japan), and BASF]. The conventional optical resolution which requires expensive alkaloids as resolving agents is troublesome. Recently, an efficient enzymatic method has been introduced into this optical resolution step (SHIMIZU et al., 1997). This enzymatic resolution uses a novel fungal enzyme, lactonohydrolase, as the catalyst. The enzyme catalyzes stereospecific hydrolysis of various kinds of lactones. D-Pantolactone is a favorable substrate of this enzyme, but the L-enantiomer is not hydrolyzed

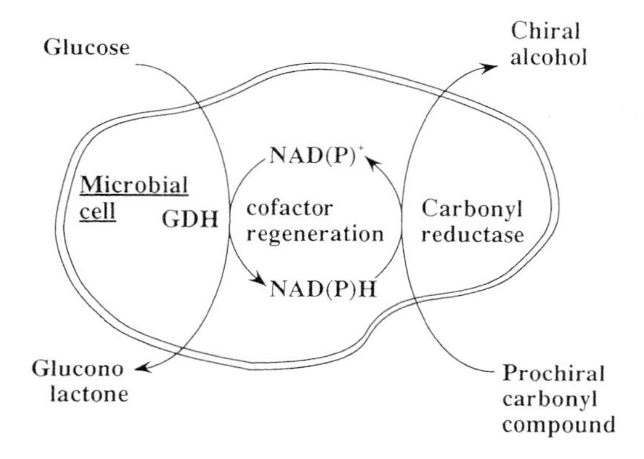

Fig. 3. *In situ* NAD(P)H regeneration with glucose dehydrogenase (GDH) for the stereospecific reduction of prochiral carbonyl compounds to chiral alcohols.

Lactonase

DL-PL D-PA L-PL

Fig. 4. Principle of the optical resolution of D,L-pantolactone by fungal lactonase. PL, pantolactone; PA, pantoic acid.

at all (SHIMIZU et al., 1992). Thus, the racemic mixture can be separated into D-pantoic acid and L-pantolactone (Fig. 4). As this lactonase reaction is an intermolecular ester bond hydrolysis, the pantolactone as the substrate needs not to be modified for resolution, which is one of the practical advantages of the use of this enzyme. Several filamentous fungi of the genera *Fusarium*, *Gibberella*, and *Cylindrocarpon* show high activity of this enzyme. On incubation with *Fusarium oxysporum* cells for 24 h at pH 7.0, D-pantolactone in a racemic mixture (700 g L^{-1}) is almost completely hydrolyzed to D-pantoic acid (96% ee) (KATAOKA et al., 1995a, b). Practically, this stereospecific hydrolysis is carried out by *F. oxysporum* cells immobilized with calcium alginate gels. When the immobilized cells were incubated in a racemic mixture (350 g L^{-1}) for 21 h at 30 °C, 90–95% of the D-pantolactone was hydrolyzed to D-pantoic acid (90–97% ee). After repeated reaction for 180 times (i.e., 180 d), the immobilized cells retained about 90% of their initial activity (SHIMIZU and KATAOKA, 1996, 1999b; SHIMIZU et al., 1997). The overall process for this enzymatic resolution is compared with the conventional chemical process in Fig. 5. The enzymatic process can skip several tedious steps which are necessary in the chemical resolution. Based on these studies, Fuji (Japan) changed over their chemical resolution with this enzymatic resolution in 1999.

Several enzymatic methods to skip this resolution step have also been reported. The two-step chemicoenzymatic method, which involves a one-pot synthesis of ketopantolactone and its stereospecific reduction to D-pantolactone (SHIMIZU and YAMADA, 1989b; SHIMIZU et al., 1997) is practically promising. The chemical synthesis is performed in one step from isobutyraldehyde, sodium methoxide, diethyl oxalate, and formalin at room temperature

Chemical resolution

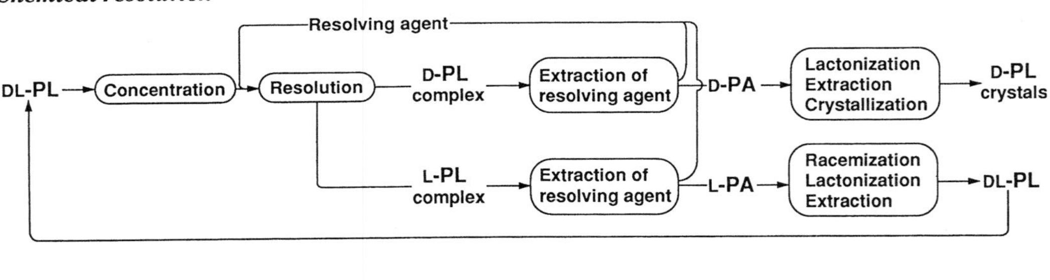

Enzymatic resolution

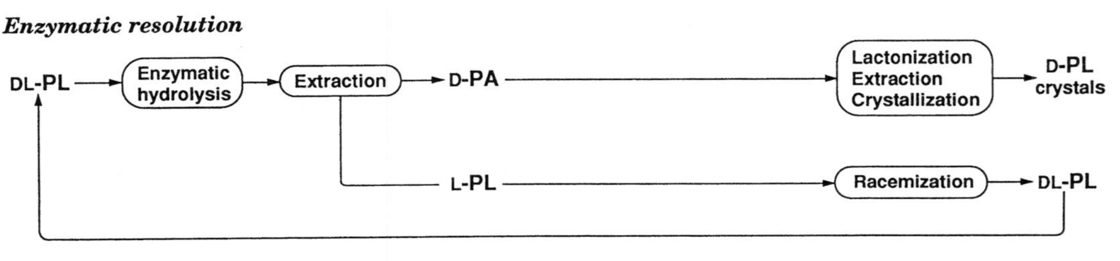

Fig. 5. Comparison of chemical and enzymatic methods for the optical resolution of D,L-pantolactone. For abbreviations, see Fig. 4.

with a yield of 81%. The bioreduction is performed in the presence of glucose as an energy source for the reduction and *Candida parapsilosis* cells with high carbonyl reductase activity as the catalyst. In this bioreduction, ketopantolactone is stoichiometrically converted to D-pantolactone (90 g L^{-1}, 94% ee) with a molar yield of 100% (Hata et al., 1987). Alternatively, ketopantoic acid, which is easily obtained from ketopantolactone by spontaneous hydrolysis under mild alkaline conditions, can be used as the substrate for the stereospecific bioreduction. In this case, *Agrobacterium* sp. cells with high activity of ketopantoic acid reductase are used as the catalyst. The yield of D-pantoic acid was 119 g L^{-1} (molar yield, 90%; optical purity, 98% ee) (Kataoka et al., 1990). The chemical step can be replaced by an enzymatic one using L-pantolactone dehydrogenase of *Nocardia asteroides*. The enzyme specifically oxidizes the L-isomer in a racemic pantolactone mixture to ketopantolactone, which is then converted to D-pantolactone or D-pantoic acid by the above mentioned reduction with *C. parapsilosis* or *Agrobacterium* sp., respectively (Kataoka et al., 1991a, b; Shimizu et al., 1987) (Fig. 6).

A direct fermentation process for D-pantoic acid and/or D-pantothenic acid is also promising. A genetically engineered strain of *E. coli* overexpressing pantothenic acid biosynthesis enzymes produced 65 g L^{-1} D-pantothenic acid from glucose upon addition of β-alanine as a precursor (Hikichi et al., 1993).

Coenzyme A (CoA) is used as an analytical reagent and for pharmaceutical, neutraceutical, and cosmetic applications. A successful microbial transformation method uses *Brevibacterium ammoniagenes* cells, in which all five enzymes necessary for the biosynthesis of CoA from D-pantothenic acid, L-cysteine, and ATP abundantly occur, as the catalyst, and these three precursors as the substrates (Shimizu and Yamada, 1986, 1989b). On cultivation of the bacterium in a medium containing glucose (10%), D-pantothenic acid, L-cysteine and AMP (or adenine), from which ATP is effectively generated by the same bacterium (see Sect. 2.8), the yield of CoA was 3–6 g L^{-1}. Higher yields (ca. 20 g L^{-1}) were obtained when 4′-phosphopantothenic acid was used in place of D-pantothenic acid or oxypantetheine-resistant mutants which were free from the feedback inhibition of pantothenate kinase by CoA were used as the catalyst (Shimizu et al., 1984). In a similar manner, all the intermediates involved in the biosynthesis of CoA from D-pantothenic acid, i.e., 4′-phosphopantothenic acid, 4′-phosphopantothenoylcysteine, 4′-phosphopantetheine, and 3′-dephospho-CoA, can be prepared (Shimizu and Yamada, 1986, 1989b).

2.4 Pyridoxine (Vitamin B_6)

Vitamin B_6 compounds, mainly pyridoxine and pyridoxal 5′-phosphate, are exclusively produced by chemical synthesis [ca. 2,500 t a^{-1}; major producers, Takeda, Hoffmann-La Roche, Fuji/Daiichi (Japan)]. They have many pharmaceutical and feed/food applications. Recent chemical and molecular biology studies revealed that 1-deoxy-D-xylulose and 4-hydroxy-L-threonine are the precursors for the biosynthesis of pyridoxine (Tazoe et al., 2000), but its complete biosynthetic pathway is not known in detail. Screening for vitamin B_6 producers among microorganisms found several potential strains, such as *Klebsiella* sp., *Flavobacterium* sp., *Pichia guilliermondii*, *Bacillus subtilis*, *Rhizobium meliloti* and so on.

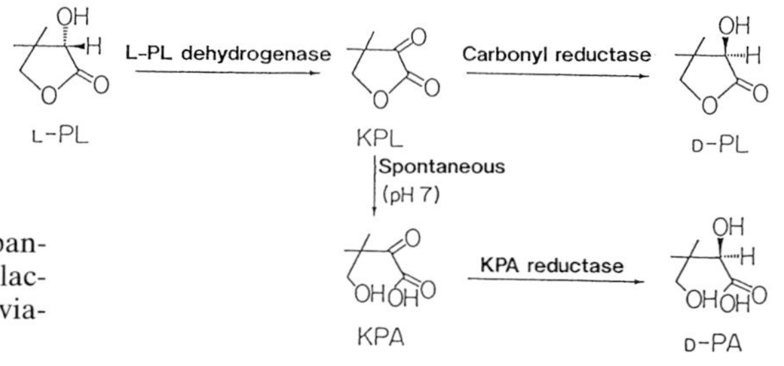

Fig. 6. Enzymatic routes for the synthesis of D-pantolactone via ketopantolactone. KPL, ketopantolactone; KPA, ketopantoic acid. For other abbreviations, see Fig. 4.

The best producers are *Flavobacterium* sp. (20 mg L^{-1}) (TANI, 1989c) and *Rhizobium meliloti* (84 mg L^{-1}) (TAZOE et al., 1999), but they are still not useful for industrial fermentation (for a review see TANI, 1989c).

2.5 Biotin

Biotin is widely used as a pharmaceutical and a feed additive. Commercial production of biotin exclusively depends on the chemical process [25 t a^{-1}; major producers, Hoffmann-La Roche, Tanabe (Japan) and Sumitomo (Japan)]. The extensive studies of biotin biosynthesis in microorganisms have revealed the details of the first three steps from pimeloyl-CoA to dethiobiotin. The last step, the introduction of sulfur into dethiobiotin to form biotin, remains enzymatically unsolved at present (IZUMI and YAMADA, 1989). The strong feedback inhibition/repression in the biotin biosynthesis makes biotin overproduction difficult. Several regulatory mutants of *Bacillus sphaericus* and *Serratia marcescens* have been selected among the mutants resistant to biotin analogs such as acidomycin to improve biotin production capacity. *S. marcescens* mutants were further improved by using recombinant DNA technology. The current production level of 230 mg L^{-1} by *S. marcescens* fermentation (MASUDA et al., 1995; SAKURAI et al., 1993) is still not satisfactory for the industrial application.

2.6 Vitamin B_{12}

Vitamin B_{12} is used for pharmaceutical products and as a supplement for food, beverages, and animal feed. It is produced exclusively by fermentation with bacterial strains of *Propionibacterium* or *Pseudomonas* genera [10 t a^{-1}; major producer, Rhône-Poulenc (France)]. In the past, it was also obtained as a by-product of antibiotic production with *Streptomyces* strains, however this process is no longer economical. The fermentation production of vitamin B_{12} with *Propionibacterium shermanii* is carried out in two stages; the first aerobic stage promoting bacterial growth and production of cobinamides intermediates, and the second aerobic stage promoting 5,6-dimethylbenzimidazol production, its coupling with the cobinamide, 5′-deoxyadenosylcobinamide guanosine diphosphate, to form 5′-deoxyadenosylcobalamin phosphate, and dephosphorylation to 5′-deoxyadenosylcobalamin, the coenzyme form of vitamin B_{12}. The amount of vitamin B_{12} production by the bacterium on cultivation in a medium containing corn steep liquor and glucose with $CoCl_2$ (20 mg L^{-1}) is 25–40 mg L^{-1} (FLORENT, 1986). In contrast to the fermentation by propionibacteria, the production by *Pseudomonas* bacteria parallels with growth under the aerobic conditions. The production by *Pseudomonas denitrificans* on cultivation in a medium containing sugar beet molasses as a major carbon source with addition of 5,6-dimethylbenzimidazol (25 mg L^{-1}) is up to 150 mg L^{-1} (SPALLA et al., 1989). Additions of the intermediates such as glycine, threonine, δ-aminolevulinic acid, and aminopropanol stimulate this bacterial vitamin B_{12} production. Betaine and choline also have stimulatory effects. Although these bacteria produce the coenzyme form, 5-deoxyadenosylcobalamin (coenzyme B_{12}), the vitamin is usually isolated as the cyanide form because of instability of the coenzyme. Several vitamin B_{12} derivatives, such as 5′-adenosylcobalamin and hydroxycobalamin and so on can also be produced by fermentation (for a review, see SPALLA et al., 1989).

2.7 L-Ascorbic Acid (Vitamin C)

Total world production of L-ascorbic acid is estimated to be 60,000–70,000 t a^{-1} [major producers, Hoffmann-La Roche, Dalry (UK), Belvidere (USA), Takeda, etc.]. It is used in pharmaceutical products and feed, and as an antioxidant for food. The commercial production method is based on Reichstein–Grüssner synthesis in which L-ascorbic acid is produced from glucose by a five-step reaction involving reduction of D-glucose to D-sorbitol, oxidation of D-sorbitol to L-sorbose, condensation of L-sorbose with acetone to form diacetone L-sorbose, oxidation of diacetone L-sorbose to L-ascorbic acid via 2-keto-L-gulonic acid (Fig. 7). The second step, oxidation of D-sorbitol to L-sorbose, is performed by microbial oxidation. The most frequently used strains are

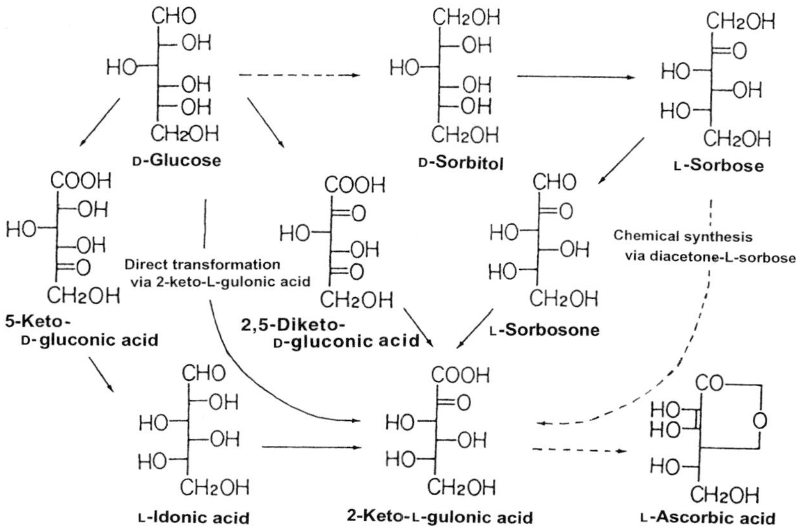

Fig. 7. Various routes for the production of L-ascorbic acid via 2-keto-L-gulonic acid. Solid and broken lines indicate microbial and chemical reactions, respectively.

those of *Gluconobacter oxydans*, which convert D-sorbitol to L-sorbose with more than 90% yield (BOUDRANT, 1990; DELIC et al., 1989).

Various alternative routes have been investigated. Most of them, as well as the Reichstein–Grüssner route, involve 2-keto-L-gulonic acid as a key intermediate (see Fig. 7). The chemical oxidation of L-sorbose to 2-keto-L-gulonic acid via diacetone sorbose can be replaced by enzymatic transformation by genetically engineered strains of *G. oxydans* (L-sorbosone route), in which L-sorbose is oxidized to 2-keto-L-gulonate via L-sorbosone by sorbose dehydrogenase and sorbosone dehydrogenase. The production of 2-keto-L-gulonic acid by these recombinant strains is 100–130 mg mL^{-1} from 150 mg mL^{-1} L-sorbose (SAITO et al., 1997). The 5-keto-D-gluconic acid route involves oxidation of D-glucose by *Acetobacter suboxydans* to 5-keto-D-gluconic acid, followed by chemical hydrogenation to a mixture of D-gluconic acid and L-idonic acid. Oxidation of L-idonic acid with *Pseudomonas* sp. gives 2-keto-L-gulonic acid. Although both microbial oxidation steps give high conversion yields (90–100%), nonselectivity of chemical reduction to L-idonic acid seems to be the reason that little attention has been paid to further development of this route. The 2,5-diketo-L-gulonic acid route uses two different microbial reactions to form 2-keto-L-gulonic acid. D-Glucose is oxidized to 2,5-diketo-L-gulonic acid with *Erwinia* sp. in the first step, and then it is reduced to 2-keto-L-gulonic acid with *Corynebacterium* sp. in the second step (SONEYAMA et al., 1982). This is one of the efficient processes for practical purposes, because each bioconversion can be performed efficiently and all the chemical steps before the 2-keto-L-gulonic acid production can be eliminated. This route was further improved by combining the relevant traits of each bacterium into a single organism by recombinant DNA technology. Thus, D-glucose is directly converted to a key intermediate, 2-keto-L-gulonic acid, by using such kinds of genetically engineered strains (ANDERSEN et al., 1985; GRINDLEY et al., 1988).

L-Ascorbic acid 2-phosphate is a more stable derivative of L-ascorbic acid and is used for health care products and aquaculture feed. *Sphingomonas trueperi* cultivated in an appropriate medium produces L-ascorbic acid 2-

phosphate on addition of L-ascorbic acid and pyrophosphate. The phosphorylation is regiospecific at the 2-position hydroxyl group of L-ascorbic acid and proceeds efficiently on addition of a surfactant, polyoxyethylene stearylamine, and xylene in the medium (KOIZUMI et al., 1990).

2.8 Adenosine Triphosphate and Related Nucleotides

Adenosine 5′-triphosphate (ATP) is used for human therapy. It is also used as a technical reagent for analyses. Several effective production methods have been developed (for reviews see SHIMIZU and YAMADA, 1986 and TANI, 1989d); phosphorylation of adenosine or adenosine 5′-monophosphate (AMP) through glycolysis by yeasts using glucose as the energy donor (Fig. 8a), ribotidation of adenine through the salvage pathway to adenosine followed by phosphorylation with *Corynebacterium ammoniagenes* (Fig. 8b), and phosphorylation of adenosine through the oxidative phosphorylation system of methylotrophic yeasts using methanol as the energy donor (Fig. 8c).

ATP production by the yeast phosphorylation through glycolysis was developed based on the observation that dried cells of bakers' yeast convert added AMP or adenosine to ATP. For high yield production, high concentrations of glucose (300 mg mL^{-1}) and inorganic phosphate (250–300 mM) are necessary. A variety of yeasts other than bakers' yeast have been reported to produce ATP under the similar reaction conditions. For example, dried cells of brewers' yeast converted 300 mmol adenosine to 190 mmol ATP and 20 mmol ADP in 1 liter of reaction mixture in 8 h at 28 °C (TOCHIKURA et al., 1976). A strain of *Saccharomyces cerevisiae* overexpressing the hexokinase gene converted adenosine (130 mM) to ATP stoichiometrically (FUKUDA et al., 1984). This ATP-generating system can be used for the preparation of a variety of sugar nucleosides, CDP-choline, and so on (see SHIMIZU and YAMADA 1986, 1989d).

ATP production through bacterial ribotidation of adenine has been commercially operated since 1983. On incubation of permeabilized cells of *C. ammoniagenes* with adenine for 13 h, the amount of ATP produced was more than 37 g L^{-1}. *De novo* synthesis of phosphoribosyl pyrophosphate from glucose, ribotidation of adenine with phosphoribosyl pyrophosphate to AMP, and generation of ATP by adenylate kinase coupled with glycolysis energy are involved in the reaction mechanism (FUJIO and FURUYA, 1983; MARUYAMA and FUJIO, 1999). This ATP generation system has been used for the production of many compounds (i.e., FAD, NAD, CDP-choline, etc.) which require ATP for their biosynthesis (FUJIO et al., 1998; see also Fig. 1 and Fig. 8b).

Various methylotrophic yeasts can be used for the production of ATP through the oxidative phosphorylation system (TANI, 1989d). The system is based on the conversion of free energy of methanol to ATP through oxidative phos-

a

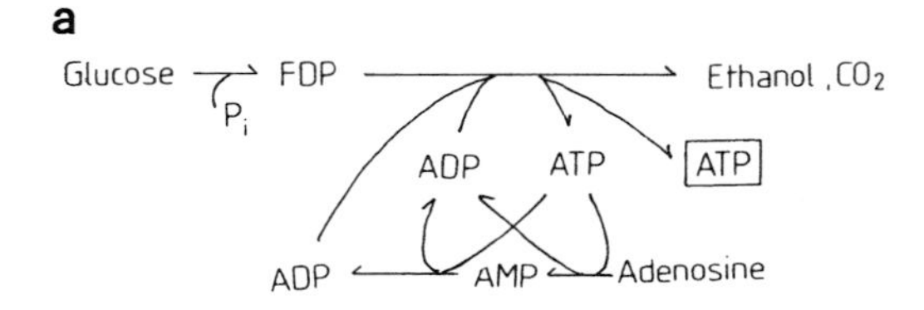

b

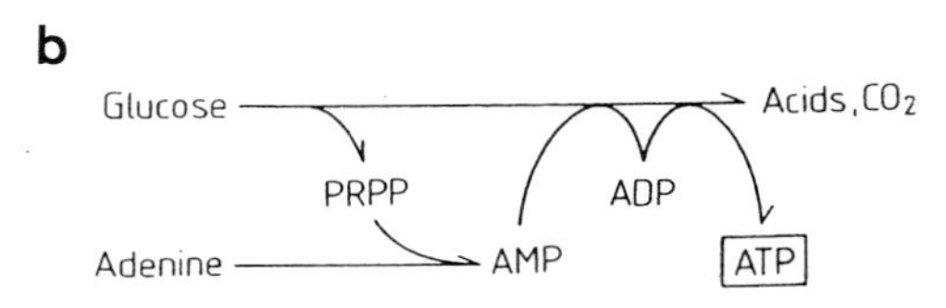

c

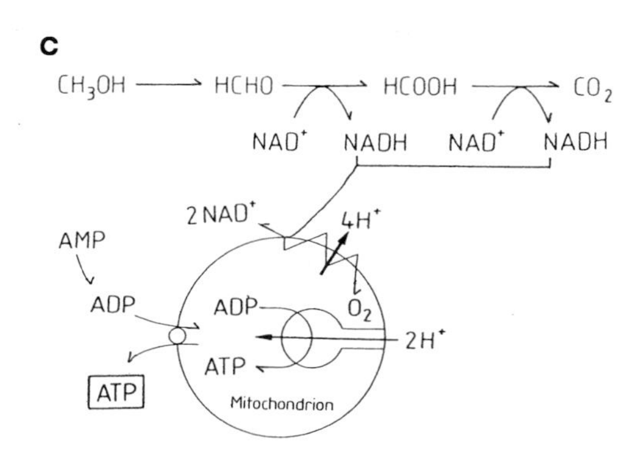

Fig. 8. Schematic representation of several ATP generating systems. Yeast glycolysis **a**, bacterial ribotidation of adenine **b**, and oxidative phosphorylation of methylotrophic yeasts **c**.

phorylation in respiration as shown in Fig. 8c. *Candida boidinii* genetically enriched with adenylate kinase, when incubated in a reaction mixture with successive addition of adenosine, has been reported to accumulate ATP efficiently (117 g L^{-1}; molar yield based on added adenosine, 78%) (SAKAI et al., 1994).

2.9 S-Adenosylmethionine and Related Nucleosides

S-Adenosylmethionine (AdoMet) has pharmaceutical and nutraceutical applications. It is produced exclusively by fermentation with yeasts (major producer, BASF). A variety of microorganisms have been reported to accumulate AdoMet intracellularly on cultivation in the presence of L-methionine. The best producers are strains of *Saccharomyces* yeasts. Especially several strains of *Saccharomyces saké* show high activities of methionine adenosyltransferase, the responsible enzyme for the conversion of L-methionine to AdoMet, suggesting that they are potent producers for the practical purpose. The production of AdoMet by *S. saké* Kyokai No. 6 is over 11 g L^{-1} on cultivation with sucrose (100 g L^{-1}) and L-methionine (10 g L^{-1}) and other nutrients. The value corresponds to 250–300 mg g^{-1} dried cells (SHIOZAKI et al., 1986). The storage place for AdoMet in the cells is the vacuoles in which it is stably present. As leakage of AdoMet from the vacuoles leads to its rapid decomposition, thus vacuolar capacity and integrity, in addition to AdoMet productivity, are also important factors for the satisfactory large-amount production of AdoMet. The usual large-scale separation method involves extraction of AdoMet from yeast cells by perchloric acid, decolorization of the extract by synthetic adsorbent, separation of AdoMet from the compounds having positive electric charges, such as adenosine, adenine, methylthioadenosine, and so on, by cationic ion-exchange resin, and salt formation with *p*-toluenesulfonate or *p*-toluenesulfonate/HSO_4^-. Some commercially available preparations of AdoMet of reagent grade contain 10–20% $S(+)$-enantiomer, the source of which is unknown (for reviews see SHIMIZU and YAMADA, 1984, 1986, 1989c).

S-Adenosylhomocysteine (AdoHcy) is the product of biological transmethylation reactions in which AdoMet is the methyl donor. AdoHcy, as well as AdoMet, has been suggested to have high potential in pharmaceutical and chemotherapeutic sectors. A practical method for AdoHcy production using the bacterial AdoHcy hydrolase reaction has been developed. Washed cells of *Alcaligenes faecalis* with a high content of AdoHcy hydrolase, on incubation in a reaction mixture containing L-homocysteine and adenosine at 37 °C for 40 h, gave AdoHcy stoichiometrically (70–80 g L^{-1}). Essentially the same yield could be obtained from D,L-homocysteine and adenosine on incubation with *A. faecalis* cells and a small amount of *Pseudomonas putida* cells as the source of racemase (SHIMIZU and YAMADA, 1984, 1989c; SHIMIZU et al., 1986a, b).

2.10 Miscellaneous

Thiamin (vitamin B_1) is used in pharmaceuticals and foods. Total annual production is estimated to be 4,200 t (major producers, Hoffmann-La Roche, Takeda). It is exclusively produced chemically. A mutant of *Saccharomyces cerevisiae* excreting thiamin (200 mg L^{-1}) has been isolated (HAJ-AHMAD et al., 1992).

Folic acid is mainly used as an animal feed additive (80%). It also has pharmaceutical and food applications (20%). Total annual production by chemical process is estimated to be 400 t (major producers, Hoffmann-La Roche, Takeda, Sumitomo). Although each step involved in the bacterial biosynthesis from the precursors, guanosine 5′-triphosphate, *p*-aminobenzoic acid, is well elucidated, potent microorganisms useful for practical purposes have not been obtained (SLOCK et al., 1990; LACKS et al., 1995).

Lipoic acid is used in pharmaceutical application. It is exclusively produced by a chemical process. *Streptococcus faecalis*, *Escherichia coli*, and *Saccharomyces cerevisiae* have been shown to produce small amounts of this vitamin (BRODY et al., 1997; JOHNSON and COLLINS, 1973). Microorganisms with high amounts generally exhibit high activities of pyruvate and α-ketoglutarate dehydrogenase complexes (HERBERT and GUEST, 1975).

Pyrroloquinoline quinone (PQQ), which was originally found as a coenzyme of quinoprotein oxidoreductase in methylotrophic bacteria, is widely distributed among living organisms and plays a role in oxidation–reduction reactions. It has a growth-promoting effect on several bacteria. A small amount is commercially produced both by fermentation and chemical synthesis [producers, Mitsubishi Gas Chemical (Japan) and Ube (Japan)] and is used as a technical reagent. *Methylobacillus glycogenes* produces about 10 mg L^{-1} of PQQ extracellularly.

3 Fat-Soluble Vitamins

3.1 Vitamin A (Retinoids) and β-Carotene (Provitamin A)

Vitamin A (for chemical structure, see Fig. 9), a group of monocyclic diterpenes with similar biological activity, occurs only in animals. It is produced chemically on a large scale (2,700 t a^{-1}; 80% for animal feed, 20% for food and pharmaceuticals; major producer, Hoffmann-La Roche, BASF, Rhône-Poulenc). β-Carotene (for chemical structure, see Fig. 9) is one of the members of carotenes, a large group of isomeric unsaturated hydrocarbons with conjugated *trans* double bonds, four methyl branches, and a β-ionone ring at one end. It is synthesized through the so-called carotenoid biosynthesis pathway in microorganisms and plants (Fig. 10). β-Carotene itself is inactive as a vitamin, but is oxidatively degraded into an active form, vitamin A, in animals after adsorption from the diet. It is used as an antioxidant, pigment in food, and an animal feed additive, and also used in pharmaceutical and cosmetic sectors. The major industrial process is based on chemical synthesis (400 t a^{-1}; major producer, Hoffmann-La Roche, BASF). Efficient bioprocesses by cultivating *Dunaliella* green microalgae in salt ponds or lakes are also used in Australia, USA, Israel, and so on. The β-carotene content of more than 100 mg g^{-1} dry cells can be obtained on cultivation of the algae under the nitrogen-limited conditions of 20–30% salts and 4–5% carbon dioxide at 10,000 lux and 25–27 °C for 3 months in a bench-scale fermenter. The resultant cells contain small amounts of α-carotene, cryptoxanthin, zeaxanthin, and luteine other than β-carotene (BOROWITZKA and BOROWITZKA, 1988). The dried cells or crude extracts with plant oil are used mainly as a feed additive. Another industrial bioprocess, which is used in Russia, is the fermentation with a fungus *Blakeslea trispora*. β-Carotene production by this fungus is strongly affected to sex-

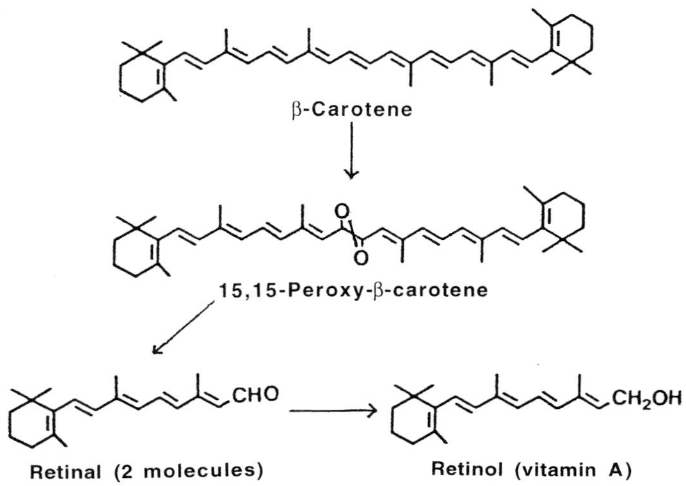

β-Carotene

15,15-Peroxy-β-carotene

CHO

Retinal (2 molecules)

CH_2OH

Retinol (vitamin A)

Fig. 9. Transformation of β-carotene to retinol.

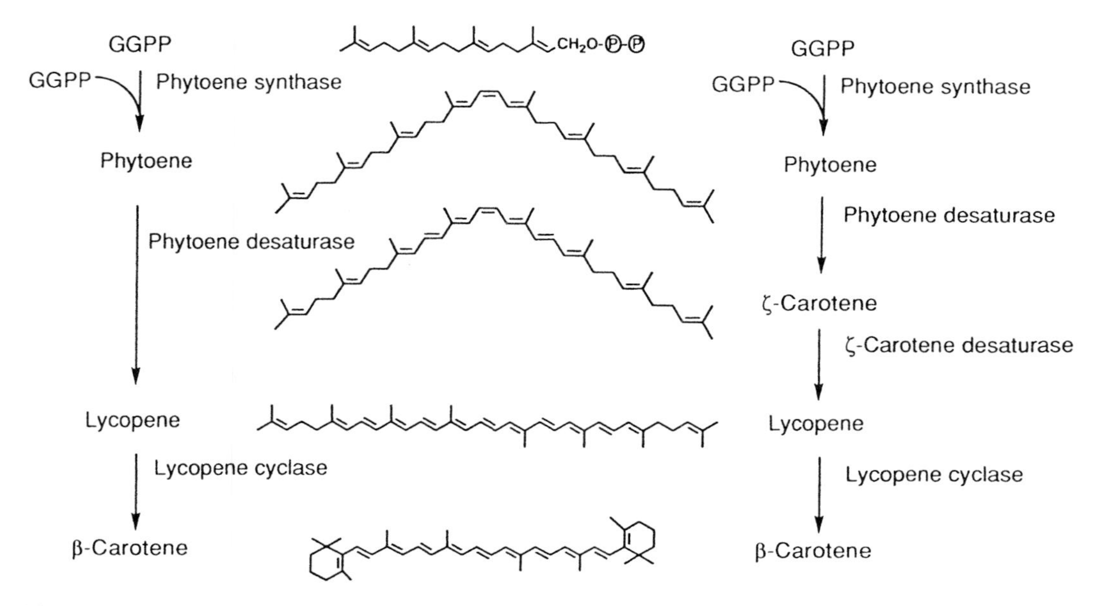

Fig. 10. Biosynthetic pathways for *β*-carotene from geranylgeranyl pyrophosphate (GGPP) in microorganisms (left) and plants (right).

ual interaction between the two mating types. A 7 d fermentation with a mutant strain of the fungus gave more than 7 g L^{-1} of *β*-carotene, the value corresponding to 200 mg g^{-1} dried cells. The resultant cells are used as a feed additive or a source of purified *β*-carotene. Several Mucorales fungi are also known as *β*-carotene producers. However, these are not satisfactory for practical purposes compared to *Blakeslea* strains (for a review see BOROWITZKA and BOROWITZKA, 1989).

Genetically engineered microorganisms in which the carotenoid biosynthetic pathway was conferred and enriched can also be used for the practical production. The transfer of the bacterial carotenoid biosynthesis genes from *Erwinia uredovora* and *Agrobacterium aurantiacum* into yeasts such as *Candida utilis* and *Saccharomyces cerevisiae* has succeeded. The resultant recombinant yeasts are a good source of *β*-carotene (0.1–0.4 mg g^{-1} dried cells) (MIURA et al., 1998). These recombinant strains may be advantageous over algal strains, because algal production requires larger space and longer cultivation time and is sometimes affected by climatic conditions.

3.2 Vitamin D

Vitamins D_3 and D_2 are mainly used for food and feed. They are also used for antirachitic treatment. These vitamin D compounds (see Fig. 11 for chemical structures) are manufactured from sterols (wool grease cholesterol for vitamin D_3 and ergosterol for vitamin D_2) by the essentially same chemical synthesis including ultraviolet irradiation on an industrial scale [38 t a^{-1}; major producer, Solvay-Duphar (The Netherlands), Hoffmann-La Roche, BASF]. The fish oils, the natural source of vitamin D_3 compounds, are only used for preparation of vitamin concentrates. Ergosterol can be extracted from food yeasts, such as *Saccharomyces cerevisiae*, *Candida utilis*, and so on, in which the content of ergosterol is 10–30 mg g^{-1} dried cells. A strain of *S. cerevisiae* has been reported to accumulate 3.76 g L^{-1} ergosterol (100 mg g^{-1} dried cells) on cultivation in a medium containing inverted blackstrap molasses and corn steep liquor at 28 °C for 96 h (DULANEY et al., 1954). Several filamentous fungi of *Trichodelma*, *Cephalosporium*, and *Fusarium* are also potential

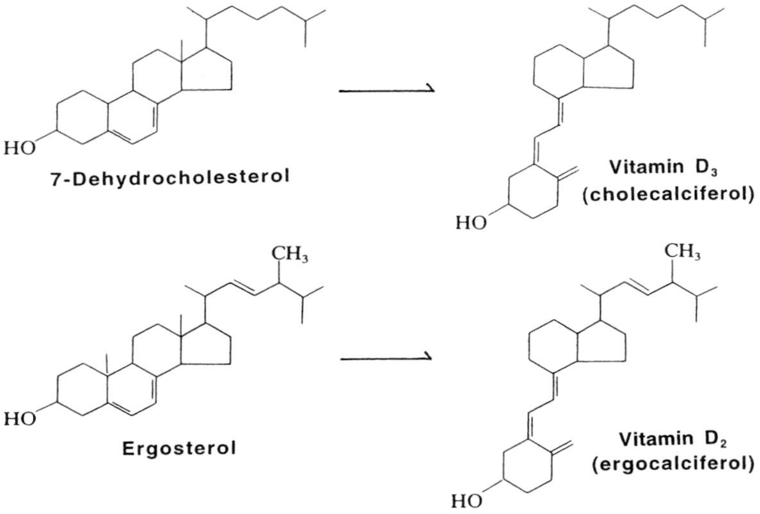

Fig. 11. Transformation of vitamin D compounds.

producers of ergosterol, however, these are lower producers compared to the *S. cerevisiae* strain (for a review, see MARGALITH, 1989).

3.3 Tocopherols (Vitamin E)

Tocopherols are used mainly as an antioxidant in clinical, nutraceutical and nutritional applications. It is also used as an additive in animal feed and cosmetics. Commercial production is based on chemical synthesis which gives a racemic mixture of α-tocopherol [20,000 t a^{-1}; major producers, Eastman-Kodak (USA), Eizai (Japan), Hoffmann-La Roche]. A mixture of tocopherols extracted from vegetable oils is also commercially produced (2,000 t a^{-1}). Photoheterotrophically cultivated *Euglena gracilis* has been suggested as a potent producer of tocopherols (TANI, 1989a). The chiral C_{14} side chain of tocopherols can be obtained by coupling of *S*- or *R*-3-hydroxyisobutyric acid; practical production of these chiral C_4 units by biocatalytic oxidation of isobutyric acid has been developed (Fig. 12) (LEUENBERGER, 1985).

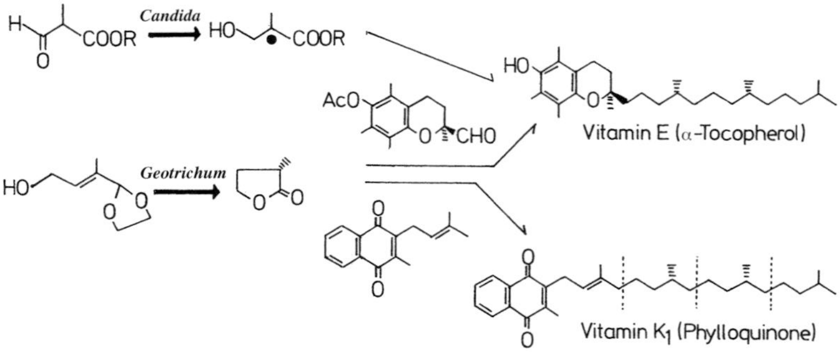

Fig. 12. Chemicoenzymatic routes for the synthesis of optically active vitamin E (top) and vitamin K_1 (bottom).

3.4 Polyunsaturated Fatty Acids (Vitamin F Group)

Several C_{18} and C_{20} polyunsaturated fatty acids (PUFAs) exhibit unique biological activities, such as lowering the plasma cholesterol level and preventing thrombosis, and are essential in human nutrition. According to the recent definition given for vitamins, they are not included in a group of vitamin compounds, but sometimes categorized as vitamin-like compounds. Among these series of unique fatty acids, γ-linoleic acid (GLA), arachidonic acid (AA), eicosapentaenoic acid (EPA), and docosahexaenoic acid (DHA) are commercially produced. Because food sources rich in these PUFAs are limited to a few seed oils (GLA) and fish oils (EPA and DHA), screening has been undertaken for alternative sources of these PUFAs in microorganisms. Several Mucorales fungi, such as *Mortierella isabellina* and *Mucor circinelloides*, have been selected as potent producers of triacylglycerol containing GLA. They accumulate more than 5 g L^{-1} GLA on cultivation in a medium containing glucose or molasses. Based on these basic observations, efficient fermentation processes have been performed on industrial scale (KENDRICK and RATLEDGE, 1992). For C_{20} PUFAs, *Mortierella alpina* 1S-4 and related filamentous fungi have been isolated. *M. alpina* 1S-4 produces 30–60 g cells as dry weight per liter in large-scale fermentation involving intermittent feeding of glucose for 7–10 d. One gram of dried cells contains 600 mg lipid (triacylglycerol) that consists of AA (40–70%; ca. 13 g L^{-1}) (HIGASHIYAMA et al., 1998a, b). Based on these observations, AA-rich oil is now commercially produced and is used as an additive for formulated milk for infants. A variety of mutants defective in desaturases in the biosynthesis of AA have been isolated; they are useful for the practical production of all C_{20} PUFAs (i.e., dihomo-γ-linoleic acid, EPA, Mead acid, etc.) of n-6, n-3 and n-9 pathways (Fig. 13) (for reviews, see CERTICK et al., 1998; SHIMIZU and OGAWA, 1999; SHIMIZU and YAMADA, 1989a; SHIMIZU et al., 1997).

3.5 Vitamin K Compounds

All vitamin K compounds, phylloquinone (vitamin K_1; Fig. 14), menaquinones (vitamin K_2; Fig. 14), and menadione (vitamin K_3), are

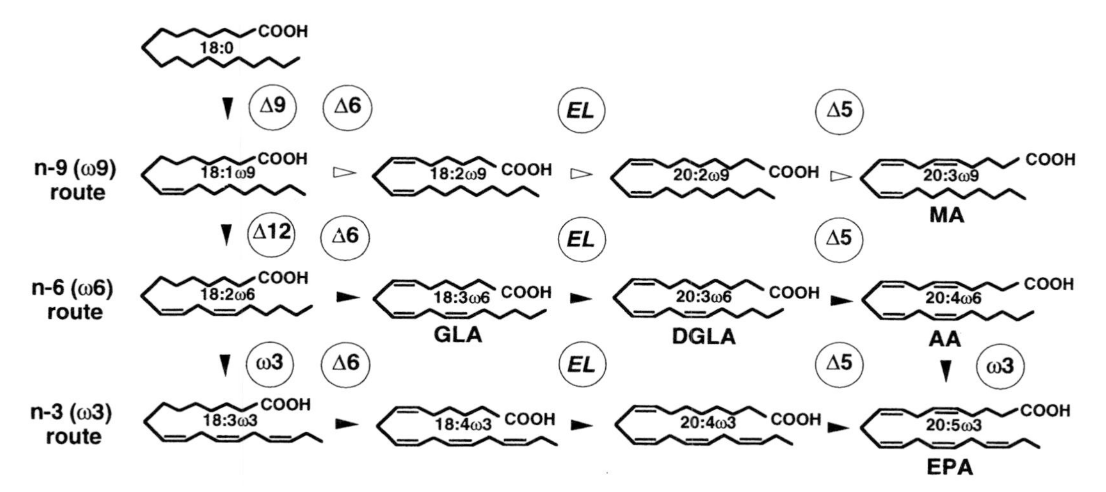

Fig. 13. Pathways for the biosynthesis of C_{20} polyunsaturated fatty acids from C_{18} fatty acids by a fungus *Mortierella alpina* 1S-4 and its desaturase-defective mutants. Open arrows show paths through which n-9 (ω9) fatty acids are formed by Δ12 desaturase-defective mutants.
MA, Mead acid; GLA, γ-linolenic acid; DGLA, dihomo-γ-linolenic acid; AA, arachidonic acid; EPA eicosa pentaenoic acid; EL, elongation enzyme; open circle with Δn or ωn, Δn or ωn desaturase.

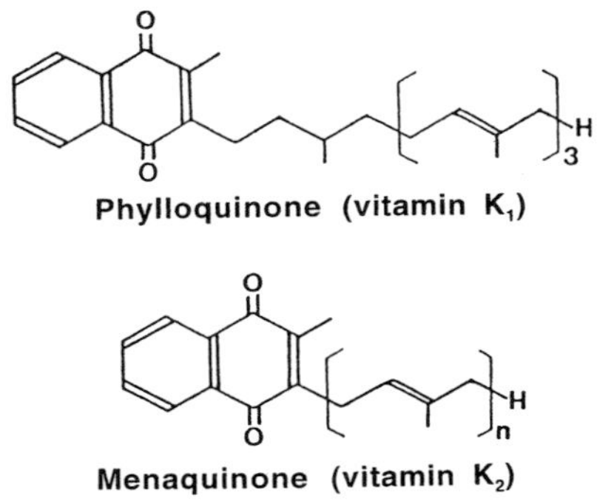

Phylloquinone (vitamin K_1)

Menaquinone (vitamin K_2)

Fig. 14. Chemical structures of vitamin K compounds.

produced by chemical processes. Total annual production of phylloquinone is estimated to be about 3.5 t a^{-1} (major producers, Hoffmann-La Roche, Eizai). It is mainly used for clinical applications, such as those for bleeding symptoms. The chiral side chain of phylloquinone can be obtained by biotransformation (see Fig. 12). About 500 t of menadione (or derivatives) are produced annually and are used as an animal feed additive [major producer, Vanetta (Italy)]. It does not occur in nature and is not active as a vitamin, but it can be converted to menaquinone-4 in animals after adsorption from the diet. Menaquinone-4 is used for the prevention of hemorrhagic disease of the newborn in Japan (for a review, see TANI, 1989b).

Plants and bacteria have been found to produce phylloquinone and menaquinones, respectively, through the biosynthetic pathways for these vitamins involving formation of the naphthoquinone ring from shikimate via chorismate and formation of a phytyl or polyprenyl side chain a *flavobacterium* sp. was obtained as a potent producer of menaquinone-6 through extensive screening. A mutant of this bacterium, which was resistant to both 1-hydroxy-2-naphthoate (a specific inhibitor of menaquinone biosynthesis) and sulfapyridine, has been reported to produce menaquinone-4 extracellularly (240 mg L^{-1}) and menaquinone-6 intracellularly (40 mg L^{-1} or 2.7 mg g^{-1} dried cells) on cultivation with glycerol in the presence of a detergent, polyoxyethylene oleyl ether, for 6 d (TAGUCHI et al., 1989).

3.6 Ubiquinone Q (Coenzyme Q)

Ubiquinone Q is a group of 2,3-dimethyl-5-methyl benzoquinone compounds with an isoprenoid side chain. They are widely distributed in organisms. There are various types which are designated according to the number of isoprene units in the side chain. Among these homologs, ubiquinone-10 is industrially manufactured by fermentation [major producer, Kanegafuchi (Japan)] and used for pharmaceutical and nutraceutical applications. Efficient fermentation processes for ubiquinone-10 using a yeast, *Rhodotorula mucilaginosa*, a methanol-utilizing bacterium, *Protaminobacter rubber* (NATORI and NAGASAKI, 1978), and so on, have been established, but most information is restricted to the patent literature. *Agrobacterium tumefaciens*, *Paracoccus denitrificans*, and *Rhodobacter sphaeroides* are also known as potent producers (YOSHIDA et al., 1998).

4 Concluding Remarks

Many vitamins and related compounds have become large-scale products in concert with the constantly increasing demand (see Tab. 1). Fields of vitamin applications and economic importance of vitamins have increased as well. At present, the annual worldwide market value for vitamins as a bulk product is estimated to be ca. $30 \cdot 10^9$ US$, the amount being about 3 times increased compared to those of the 1980s. A half is in feed, the remaining in pharmaceuticals (30%) and in food (20%).

A broad range of application now exists for vitamins and related compounds for feed, food, pharmaceutical, nutraceutical, cosmetics and technical purposes. Nutraceutical products containing S-adenosylmethionine, e.g., are currently on the market in USA. The fungal oil containing arachidonic acid is also used as an additive for infant milk.

Many vitamins and related compounds are now produced on an industrial scale by biotechnological processes, although chemical synthesis is still the dominant method for several vitamins. Compared to chemical reactions, those involved in fermentation and microbial/enzymatic transformation have many superior properties for the practical purposes. One of those is their excellent specificity. Most reactions involve only one compound or a few that are structurally related, and proceed by distinguishing almost completely between stereoisomers or regioisomers. This property is particularly useful to obtain biologically and optically active complex compounds like many vitamins (for reviews, see YAMADA and SHIMIZU, 1987; SHIMIZU et al., 1997; OGAWA and SHIMIZU, 1999). A current successful example is the introduction of the stereospecific hydrolysis reaction by a novel fungal lactonase to the optical resolution of racemic pantolactone involved in the synthesis of D-pantothenic acid. It has been shown that this enzymatic resolution is highly satisfactory not only economically but also environmentally (water, −49%; CO_2, −30%; and BOD, −62%; comparing the former chemical resolution). The nitrile hydratase catalyzed hydration reaction of 3-cyanopyridine to nicotinamide is also a good example. The reaction gives only nicotinamide, no nicotinic acid being formed, which is fairly difficult by chemical way. Also, it should be emphasized that microorganisms performing these unique reactions have been discovered through extensive screening among a wide variety of microorganisms, which show unusual diversity and versatility. In this respect, further exploration of potential microbial abilities should become a key not only in the vitamin production but also in the production of other biologically and chemically useful compounds (for a review, see OGAWA and SHIMIZU, 1999).

Acknowledgement:
The author thanks Dr. T. HOSHINO, Nippon Roche Research Center, for suggestions.

5 References

ANDERSON, S., MARKS, R., LAZARUS, R., MILLER, J., STAFFORD, K. et al. (1985), Production of 2-keto-L-gulonate, an intermediate in L-ascorbate synthesis, by a genetically modified *Erwinia herbicola*, *Science* **230**, 144–149.

ASANO, Y., TANI, Y., YAMADA, H. (1980), A new enzyme, "nitrile hydratase" which degrades acetonitrile in combination with amidase, *Agric. Biol. Chem.* **44**, 2251–2252.

BOROWITZKA, L. J., BOROWITZKA, M. A. (Eds.) (1988), *Microalgal Biotechnology*. Cambridge: Cambridge University Press.

BOROWITZKA, L. J., BOROWITZKA, M. A. (1989), β-Carotene (provitamin A) production with algae, in: *Biotechnology of Vitamins, Pigments and Growth Factors* (VANDAMME, E. J., Ed.), pp. 15–26. London: Elsevier Applied Science.

BOUDRANT, J. (1990) Microbial processes for ascorbic acid biosynthesis: a review, *Enzyme Microb. Technol.* **12**, 322–329.

BRODY, S., OH, C., HOJA, U., SCHWEIZER, E. (1997), Mitochondrial acyl carrier protein is involved in lipoic acid synthesis in *Saccharomyces cerevisiae*, *FEBS Lett.* **408**, 217–220.

CERTICK, M., SAKURADANI, E., SHIMIZU, S. (1998), Desaturase-defective fungal mutants: useful tools for the regulation and overproduction of polyunsaturated fatty acids, *Trends Biotechnol.* **16**, 500–505.

DE BAETS, S., VANDEDRINCK, S., VANDAMME, E. J. (2000), Vitamins and related biofactors: microbial production, in: *Encyclopedia of Microbiology*, 2nd Edn. (LEDERBERG, J., Ed.), pp. 837–853. London: Academic Press.

DELIĆ, V., ŠUNIĆ, D., VLAŠIĆ, D. (1989), Microbial reactions for the synthesis of vitamin C (L-ascorbic acid), in: *Biotechnology of Vitamins, Pigments and Growth Factors* (VANDAMME, E. J., Ed.), pp. 299–334. London: Elsevier Applied Science.

DE WULF, P., VANDAMME, E. J. (1997), Microbial synthesis of D-ribose: metabolic deregulation and fermentation process, *Adv. Appl. Microbiol.* **44**, 167–214.

DULANEY, E. L., STAPLEY, E. O., SIMPF, K. (1954), Studies on ergosterol production by yeasts, *Appl. Microbiol.* **2**, 371–379.

EGGERSDORFER, M., ADAM, G., JOHN, M., HAHNLEIN, W., LABLER, L. et al. (1996), Vitamins, in: *Ullmann's Encyclopedia of Industrial Chemistry*, 5th Edn., pp. 443–613. Weinheim: VCH.

FLORENT, J. (1986), Vitamins, in: *Biotechnology*, 1st Edn., Vol. 4 (PAPE, H., REHM, H.-J., Eds.), pp. 114–158. Weinheim: VCH.

FRIEDRICH, W. (1988), *Vitamins*. Berlin: Walter de Gruyter.

FUJIO, T., FURUYA, A. (1983), Production of ATP from adenine by *Brevibacterium ammoniagenes*, *J. Ferment. Technol.* **61**, 261–267.

FUJIO, T., MARUYAMA, A. (1997), Enzymatic production of pyrimidine nucleotides using *Corynebacterium ammoniagenes* cells and recombinant *E. coli* cells: enzymatic production of CDP-choline from orotic acid and choline chloride, *Biosci. Biotech. Biochem.* **61**, 956–959.

FUJIO, T., MARUYAMA, A., MORI, H. (1998), New production methods for useful substances using ATP regenerating system, *Biosci. Ind.* **56**, 737–742.

FUKUDA, Y., YAMAGUCHI, S., HASHIMOTO, H., SHIMOSAKA, M., KIMURA, A. (1984), Cloning of glucose phosphorylating genes in *S. cerevisiae* by the KU-method and application to ATP production, *Agric. Biol. Chem.* **48**, 2877–2881.

GRINDLEY, J. F., PAYTON, M. A., VAN DE POL, H., HARDY, K. G. (1988), Conversion of glucose to 2-keto-L-gluconate, an intermediate in L-ascorbate synthesis, by a recombinant strain of *Erwinia citreus*, *Appl. Environ. Microbiol.* **54**, 1770–1775.

HAJ-AHMAD, Y., BILINSKI, C. A., RUSSEL, I., STEWART, G. G. (1992), Thiamin secretion in yeast, *Can. J. Microbiol.* **38**, 1156–1161.

HATA, H., SHIMIZU, S., YAMADA, H. (1987), Enzymatic production of D-(−)-pantoyl lactone from ketopantoyl lactone, *Agric. Biol. Chem.* **51**, 3011–3016.

HERBERT, A. A., GUEST, J. R. (1975), Lipoic acid content of *Escherichia coli* and other microorganisms, *Arch. Microbiol.* **106**, 259–266.

HIGASHIYAMA, K., YAGUCHI, T., AKIMOTO, K., FUJIKAWA, S., SHIMIZU, S. (1998a), Enhancement of arachidonic acid production by *Mortierella alpina* 1S-4, *J. Am. Oil Chem. Soc.* **75**, 1501–1505.

HIGASHIYAMA, K., YAGUCHI, T., AKIMOTO, K., FUJIKAWA, S., SHIMIZU, S. (1998b), Effects of mineral addition on the growth morphology and arachidonic acid production by *Mortierella alpina* 1S-4, *J. Am. Oil Chem. Soc.* **75**, 1815–1819.

HIKICHI, Y., MIKI, K., MORIYA, T., KASHIWA, K., YAMAGUCHI, T., NOGAMI, A. (1993), Microbial production of D-pantothenic acid, *Nippon Nogeikaagaku-kaishi* **67**, 294 (abstract in Japanese).

IZUMI, Y., YAMADA, H. (1989), Microbial production of biotin, in: *Biotechnology of Vitamins, Pigments and Growth Factors* (VANDAMME, E. J., Ed.), pp. 231–256. London: Elsevier Applied Science.

JOHNSON, M. G., COLLINS, E. B. (1973), Synthesis of lipoic acid by *Streptococcus faecalis* 10C1 and end products produced anaerobically from low concentrations of glucose, *J. Gen. Microb.* **78**, 47–55.

KATAOKA, M., SHIMIZU, S., YAMADA, H. (1990), Novel enzymatic production of D-(−)-pantoyl lactone through the stereospecific reduction of ketopantoic acid, *Agric. Biol. Chem.* **54**, 177–182.

KATAOKA, M., SHIMIZU, S., YAMADA, H. (1991a), Purification and characterization of a novel FMN-dependent enzyme, membrane-bound L-(+)-pantoyl lactone dehydrogenase from *Nocardia asteroides*, *Eur. J. Biochem.* **204**, 799–806.

KATAOKA, M., SHIMIZU, S., YAMADA, H. (1991b), Stereospecific conversion of a racemic pantoyl lactone to D-(−)-pantoyl lactone through microbial oxidation and reduction reactions, *Recl. Trav. Chim. Pays-Bas* **110**, 155–157.

KATAOKA, M., SHIMIZU, K., SAKAMOTO, K., YAMADA, H., SHIMIZU, S. (1995a), Optical resolution of racemic pantolactone with a novel fungal enzyme, lactonohydrolase, *Appl. Microbiol. Biotechnol.* **59**, 2292–2294.

KATAOKA, M., SHIMIZU, K., SAKAMOTO, K., YAMADA, H., SHIMIZU, S. (1995b), Lactonohydrolase-catalyzed optical resolution of pantoyl lactone: selection of a potent enzyme producer and optimization of culture and reaction conditions for practical resolution, *Appl. Microbiol. Biotechnol.* **44**, 333–338.

KENDRICK, A. J., RATLEDGE, C. (1992), Microbial polyunsaturated fatty acids of potential commercial interest, *SIM Ind. Microbiol. News* **42**, 59–65.

KITATSUJI, K., ISHINO, S., TESHIBA, S., ARIMOTO, M. (1992), *Japanese Open Patent* H5-304975.

KOBAYASHI, M., SHIMIZU, S., YAMADA, H. (in press), Hydratases involved in nitrile conversion: screening, characterization and application, *Chem. Records* **1**.

KOIZUMI, S., MARUYAMA, A., FUJIO, T. (1990) Purification and characterization of ascorbic acid phosphorylating enzyme from *Pseudomonas azotocolligans*, *Agric. Biol. Chem.* **54**, 3235–3239.

KOIZUMI, S., YONETANI, Y., MARUYAMA, A., TESHIBA, S. (1996), Riboflavin fermentation: construction of the plasmids with high expression of riboflavin biosynthesis genes, *Nipponnogeikagaku-kaishi* **70**, 286 (abstract in Japanese).

LACKS, A. S., GREENBERG, B., LOPEZ, P. (1995), A cluster of four genes encoding enzymes for five steps in the folate biosynthetic pathway of *Streptococcus pneumoniae*, *J. Bacteriol.* **177**, 66–74.

LEUENBERGER, H. G. W. (1985), Microbiologically catalyzed reaction steps in the field of vitamin and carotenoid synthesis, in: *Biocatalysis in Organic Synthesis* (TRAMPER, J., VAN DER PLAS, M. C., LINKO, P., Eds.), pp. 99–118. Amsterdam: Elsevier.

MARGALITH, P. (1989), Vitamin D: the biotechnology of ergosterol, in: *Biotechnology of Vitamins, Pigments and Growth Factors* (VANDAMME, E. J., Ed.), pp. 81–93. London: Elsevier Applied Science.

MASUDA, M., TAKAHASHI, N., SAKURAI, N., YANAGIYA, S., KOMATSUBARA, S., TOSA, T. (1995), Further improvement of D-biotin production by a recombinant strain of *Serratia marcescens*, *Process Biochem.* **30**, 553–562.

MIURA, Y., KONDO, K., SAITO, T., SHIMADA, H., FARASER, P. D., MISAWA, N. (1998), Production of the carotenoids lycopene, β-carotene and astaxanthin in the food yeast *Candida utilis*, *Appl. Environ. Microbiol.* **64**, 1226–1229.

NAGASAWA, T., YAMADA, H. (1989), Microbial transformations of nitriles, *Trends Biotechnol.* **7**, 153–158.

NAKAYAMA, K., SATO, Z., TANAKA, H., KINOSHITA, S. (1968), Production of nucleic acid-related substances by fermentation processes, part XVII: production of NAD and nicotinic acid mononucleotide with *Brevibacterium ammoniagenes*, *Agric. Biol. Chem.* **32**, 1331–1336.

NATORI, Y., NAGASAKI, T. (1978), Occurrence of coenzyme Q_{12} and 13 in facultative methanol-utilizing bacteria, *Agric. Biol. Chem.* **44**, 2105–2110.

OGAWA, J., SHIMIZU, S. (1999), Microbial enzymes: new industrial applications from traditional screening methods, *Trends Biotechnol.* **17**, 13–21.

OZBAS, T., KUTSAL, T. (1986), Comparative study of riboflavin production from two microorganisms: *Eremothecium ashbyii* and *Ashbya gossypii*, *Enzyme Microb. Technol.* **8**, 593–596.

PERKINS, J. B., SLOMA, A., HERMANN, T., THERIAULT, K., ZACHGO, E. et al. (1999), Genetic engineering of *Bacillus subtilis* for the commercial production of riboflavin, *J. Ind. Microbiol. Biotechnol.* **22**, 8–18.

SAITO, Y., ISHII, Y., HAYASHI, M., IMANO, Y., AKASHI, T., YOSHIKAWA, K. et al. (1997), Cloning of genes coding for L-sorbose and L-sorbosone dehydrogenases from *Gluconobacter oxydans* and microbial production of 2-keto-L-gulonate, a precursor of L-ascorbic acid, in a recombinant *G. oxydans* strain, *Appl. Environ. Microbiol.* **63**, 454–460.

SAKAI, Y., ROGI, T., YONEHARA, T., KATO, N., TANI, Y. (1994), High-level ATP production by a genetically engineered *Candida* yeast, *Bio/Technology* **12**, 291–293.

SAKURAI, N., MIYAZAKI, A., MASUDA, M., KOMATSUBARA, S., TOSA, T. (1993), Molecular breeding of a biotin-hyperproducing *Serratia marcescens* strain, *Appl. Environ. Microbiol.* **59**, 3225–3232.

SHIMIZU, S., KATAOKA, M. (1996), Optical resolution of pantolactone by a novel fungal enzyme, lactonohydrolase, *Enzyme Eng.* **8**, 650–658.

SHIMIZU, S., KATAOKA, M. (1999a), Production of chiral C3- and C4-units by microbial enzymes, *Adv. Biochem. Eng. Biotechnol.* **63**, 109–123.

SHIMIZU, S., KATAOKA, M. (1999b), Lactonohydrolase, in: *Encyclopedia of Bioprocess Technology: Fermentation, Biocatalysis, and Bioseparation* (FLICKINGER, M. C., DREW, S. W., Eds.), pp. 1571–1577. New York: John Wiley & Sons.

SHIMIZU, S., OGAWA, J. (1999), Oils, microbial production, in: *Encyclopedia of Bioprocess Technology: Fermentation, Biocatalysis, and Bioseparation* (FLICKINGER, M. C., DREW, S. W., Eds.), pp. 1839–1851. New York: John Wiley & Sons.

SHIMIZU, S., YAMADA, Y. (1984), Microbial and enzymatic processes for the production of pharmacologically important nucleosides, *Trends Biotechnol.* **2**, 137–141.

SHIMIZU, S., YAMADA, Y. (1986), Coenzymes, in: *Biotechnology*, Vol. 4 (PAPE, H., REHM, H.-J., Eds.), pp. 159–184. Weinheim: VCH.

SHIMIZU, S., YAMADA, H. (1989a), Microbial production of polyunsaturated fatty acids (vitamin F group), in: *Biotechnology of Vitamins, Pigments and Growth Factors* (VANDAMME, E. J., Ed.), pp. 105–136. London: Elsevier Applied Science.

SHIMIZU, S., YAMADA, H. (1989b), Pantothenic acid (vitamin B_5), coenzyme A and related compounds, in: *Biotechnology of Vitamins, Pigments and Growth Factors* (VANDAMME, E. J., Ed.), pp. 199–220. London: Elsevier Applied Science.

SHIMIZU, S., YAMADA, H. (1989c), Adenosylmethionine, adenosylhomocysteine and related nucleosides, in: *Biotechnology of Vitamins, Pigments and Growth Factors* (VANDAMME, E. J., Ed.), pp. 351–372. London: Elsevier Applied Science.

SHIMIZU, S., YAMADA, H. (1989d), Other vitamin-related coenzymes, in: *Biotechnology of Vitamins, Pigments and Growth Factors* (VANDAMME, E. J., Ed.), pp. 373–382. London: Elsevier Applied Science.

SHIMIZU, S., ESUMI, A., KOMAKI, R., YAMADA, H. (1984), Production of coenzyme A by a mutant of *Brevibacterium ammoniagenes* resistant to oxypantetheine, *Appl. Environ. Microbiol.* **48**, 1118–1122.

SHIMIZU, S., SHIOZAKI, S., YAMADA, H. (1986a), High yield production of *S*-adenosyl-L-homocysteine with microbial cells as the catalyst, *J. Biotechnol.* **4**, 81–90.

SHIMIZU, S., SHIOZAKI, S., YAMADA, H. (1986b), Production of S-adenosyl-L-homocysteine by bacterial cells with a high content of S-adenosylhomocysteine hydrolase: utilization of a racemic mixture of homocysteine as the substrate, *J. Biotechnol.* **4**, 91–100.

SHIMIZU, S., HATTORI, S., HATA, H., YAMADA, H. (1987), Stereoselective enzymatic oxidation and reduction system for the production of D-(−)-pantoyl lactone from a racemic mixture of pantoyl lactone, *Enzyme Microb. Technol.* **9**, 411–416.

SHIMIZU, S., KATAOKA, M., SHIMIZU, K., HIRAKATA, M., SAKAMOTO, K., YAMADA, H. (1992), Purification and characterization of a novel lactonohydrolase, catalyzing the hydrolysis of aldonate lactones and aromatic lactones, from *Fusarium oxysporum*, *Eur. J. Biochem.* **209**, 383–390.

SHIMIZU, S., OGAWA, J., KATAOKA, M., KOBAYASHI, M. (1997), Screening of novel microbial enzymes for the production of biologically and chemically useful compounds, *Adv. Biochem. Eng. Biotechnol.* **58**, 46–87.

SHIOZAKI, S., SHIMIZU, S., YAMADA, H. (1986), Production of S-adenosyl-L-methionine by *Saccharomyces saké*, *J. Biotechnol.* **4**, 345–354.

SLOCK, J., STAHLY, D. P., HAN, C.-Y., SIX, R. W., CRAWFORD, I. P. (1990), An apparent *Bacillus subtilis* folic acid biosynthetic operon containing *pab*, an amphibolic *trpG* gene, a third gene required for synthesis of *p*-aminobenzoic acid, and the dihydropteroate synthase gene, *J. Bacteriol.* **172**, 7211–7226.

SONEYAMA, T., TANI, H., MATSUDA, K., KAGEYAMA, B., TANIMOTO, M. et al. (1982), Production of 2-keto-L-gulonic acid from D-glucose by two-stage fermentation, *Appl. Environ. Microbiol.* **43**, 1064–1069.

SPALLA, C., GREIN, A., GAROFANO, L., FERNI, G. (1989), Microbial production of vitamin B_{12}, in: *Biotechnology of Vitamins, Pigments and Growth Factors* (VANDAMME, E. J., Ed.), pp. 257–284. London: Elsevier Applied Science.

STAHMANN, K.-P., REVELTA, J. L., SEULBERGER, H. (2000), Three bacterial processes using *Ashbya gossypii*, *Candida famata*, or *Bacillus subtilis* compete with chemical riboflavin production, *Appl. Microbiol. Biotechnol.* **53**, 509–516.

TAGUCHI, H., SHIBATA, T., DUANGMANEE, C., TANI, Y. (1989), Menaquinone-4 production by a sulfonamide-resistant mutant of *Flavobacterium* sp. 238-7, *Agric. Biol. Chem.* **53**, 3017–3023.

TANI, Y. (1989a), Algal and microbial production of vitamin E, in: *Biotechnology of Vitamins, Pigments and Growth Factors* (VANDAMME, E. J., Ed.), pp. 95–104. London: Elsevier Applied Science.

TANI, Y. (1989b), Microbial production of vitamin K_2 (menaquinone) and vitamin K_1 (phylloquinone), in: *Biotechnology of Vitamins, Pigments and Growth Factors* (VANDAMME, E. J., Ed.), pp. 123–136. London: Elsevier Applied Science.

TANI, Y. (1989c), Microbial production of vitamin B_6 derivatives, in: *Biotechnology of Vitamins, Pigments and Growth Factors* (VANDAMME, E. J., Ed.), pp. 221–230. London: Elsevier Applied Science.

TANI, Y. (1989d), Microbial production of ATP, in: *Biotechnology of Vitamins, Pigments and Growth Factors* (VANDAMME, E. J., Ed.), pp. 337–350. London: Elsevier Applied Science.

TAZOE, M., ICHIKAWA, K., HOSHINO, T. (1999), Production of vitamin B_6 in *Rhizobium*, *Agric. Biol. Chem.* **63**, 1378–1382.

TAZOE, M., ICHIKAWA, K., HOSHINO, T. (2000), Biosynthesis of vitamin B_6 in *Rhizobium*, *J. Biol. Chem.* **275**, 11300–11305.

TOCHIKURA, T., KARIYA, Y., YANO, T., TACHIKI, T., KIMURA, H. (1976), Fermentative production by means of coupling by energy transfer, in: *Abstracts 5th Int. Fermentation Symp.*, Berlin, p. 441.

VANDAMME, E. J. (1989) (Ed.), *Biotechnology of Vitamins, Pigments and Growth Factors*. London: Elsevier Applied Science.

YAMADA, H., KOBAYASHI, M. (1996), Nitrile hydratase and its application to industrial production of acrylamide, *Biosci. Biotech. Biochem.* **60**, 1391–1400.

YAMADA, H., SHIMIZU, S. (1987), Microbial processes for the production of biologically and chemically useful compounds, *Angew. Chem.* (*Int. Eng. Edn.*) **27**, 622–642.

YOSHIDA, H., KOTANI, Y., OCHIAI, K., ARAKI, K. (1998), Production of ubiquinone-10 using bacteria, *J. Gen. Appl. Microbiol.* **44**, 19–26.

12 Biochemistry of Polyketide Synthases

RAJESH S. GOKHALE
DIPIKA TUTEJA
New Delhi, India

1 Introduction

Polyketides are a group of secondary metabolites exhibiting remarkable diversity both in terms of their structure and function. These metabolites are ubiquitous in distribution and have been reported from organisms as diverse as bacteria, fungi, plants, insects, dinoflagellates, mollusks, and sponges (O'HAGAN, 1991). The wide spectrum of activity of polyketides makes them economically, clinically, and industrially most sought after biomolecules (Fig. 1). Polyketides, unlike the names of most other classes of chemical compounds, are united by the way they are synthesized during the biosynthetic pathway. Such biosynthetic mechanisms possess enormous potential and can generate strikingly different chemical end products. The carbon skeleton of polyketides is biosynthesized by step-wise decarboxylative condensation of small carboxylic acid thioesters, a process that closely parallels fatty acid synthesis (HOPWOOD and SHERMAN, 1990). However, in contrast to fatty acid synthases (FASs), polyketide synthases (PKSs) generate an enormous variety of different products. This is achieved by using a broad palette of primers and extender units, by varying the degree of processing after each condensation step, and by controlling the stereochemistry of the reduction of the intermediate β-ketoacyl thioesters. Virtually, all polyketides are modified during or after their synthesis, e.g., through hydroxylation, reduction, or epoxidation. Many polyketides cyclize to form either lactone rings (e.g., macrolides, polyenes) or aromatic rings (e.g., the anthracyclines), or to form multiple rings as in polyethers (see the thematic issue of *Chemical Reviews*, 1997, Vol. 97).

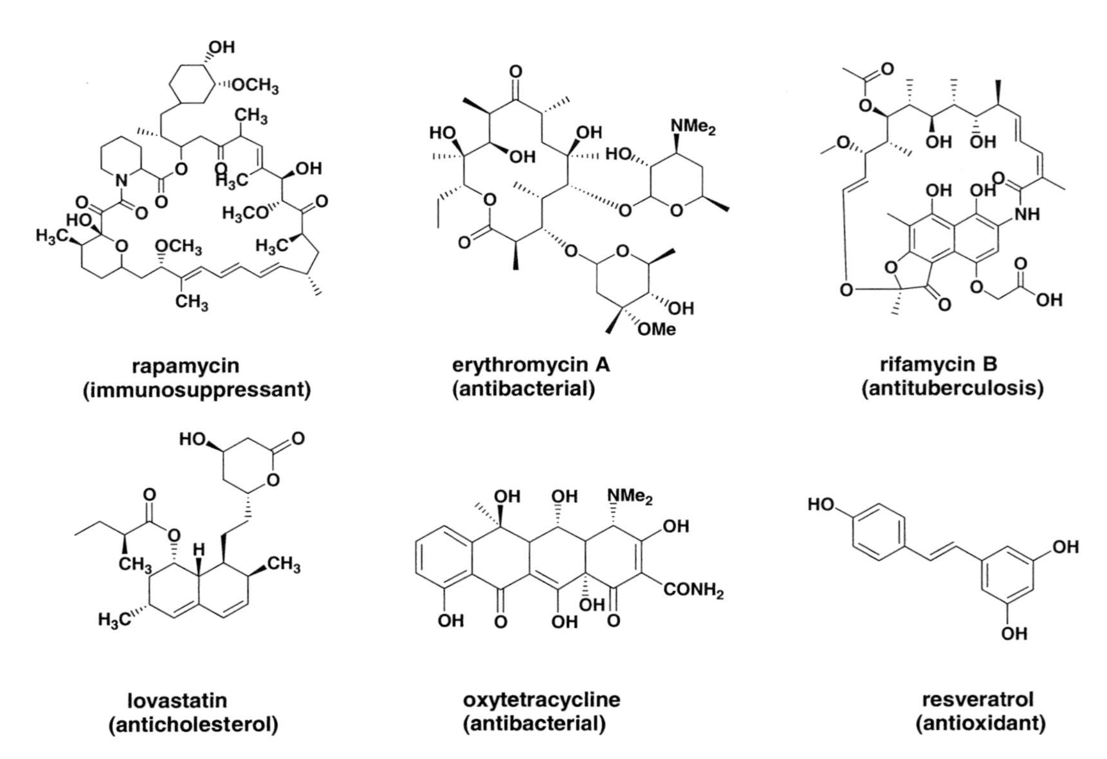

Fig. 1. Chemical structures of some of the polyketides. The functions of these metabolites are shown in parentheses.

The last two decades have witnessed an enormous expansion in our understanding of the polyketide biosynthetic machinery. Although organic chemists have primarily conducted the initial research, the advent of geneticist and molecular biologist has completely transformed this field. The genetic manipulation techniques established thereof have resulted in the identification, cloning, sequencing, and functional analysis of several polyketide biosynthetic pathways (HOPWOOD, 1997). Interestingly, the classical enzymological approaches, which have usually provided leads in other biological fields, failed to characterize PKSs, probably due to their multisubunit, multienzyme protein assembly that hindered establishment of a suitable enzymatic assay (BENTLEY and BENNETT, 1999). Moreover, since the isolation of these enzymes systems from various fungi and *Streptomyces* was also technically demanding, it dissuaded a number of biochemists from accepting the challenge. It is, therefore, not surprising that the present insights into the understanding of the polyketide biosynthetic machinery have primarily emerged from the genetic "mix-and-match" approach that involves combining different PKS genes. These empirical gene fusion approaches have led to the biosynthesis of diverse "unnatural" natural products revealing the versatility and combinatorial potential of PKSs. With our advancement in the understanding of the molecular mechanistic basis of polyketide machinery, it should be possible to rationally alter the PKS genes. Recently, there have been a number of excellent review articles highlighting enormous potential for producing novel polyketides, both singly and in libraries (LEADLAY, 1997; HUTCHINSON, 1998; KATZ and MCDANIEL, 1999; KHOSLA et al., 1999; LAL et al., 2000). This chapter assimilates our current biochemical understanding of polyketide biosynthesis and discusses how mechanistic and structural studies of PKSs would play a decisive role in the developing a technology of polyketide libraries.

2 Polyketide Synthases and Assembly of Polyketides

Polyketides are produced by sequential reactions catalyzed by a collection of enzyme activities called polyketide synthases. These are large (100 to 1,000 kDa) multienzyme systems that contain coordinated groups of active sites. The biosynthetic origin of polyketides first became possible with the availability of radioactive isotopes and then by stable NMR isotopes of carbon and hydrogen (O'HAGAN, 1993; BENTLEY, 1999). From these studies it was evident that the assembly of polyketides resembled fatty acid biosynthesis. The cloning and sequencing of the structural genes for a variety of PKSs further established their analogy with FASs. Since then four different types of PKSs have been discovered in the microbial world. Type I synthases, which are analogous to vertebrate fatty acid synthases (WITKOWSKI et al., 1991; HOPWOOD and SHERMAN, 1990), are typically involved in the biosynthesis of fungal polyketides such as 6-methylsalicylic acid (BECK et al., 1990) and aflatoxin (BROWN et al., 1996). These PKSs are large multidomain proteins carrying all the active sites required for polyketide biosynthesis. In contrast, in the type II PKS system, similar active sites are distributed over several smaller polypeptides. Type II synthases catalyze the formation of compounds that require aromatization and cyclization, but not extensive reduction or reduction/dehydration cycles. These PKSs are analogous to bacterial fatty acid synthases and are involved in the biosynthesis of bacterial aromatic natural products such as actinorhodin, tetracenomycin, and doxorubicin (FERNANDEZ-MORENO et al., 1992; BIBB et al., 1989; GRIMM et al., 1994). A type II synthase protects the poly-β-ketone intermediates during the chain elongation and then presents the completed chain for further processing and cyclization. Recently, chalcone synthases that were believed to be specific to the plant kingdom, were identified in the gram-positive, soil-living filamentous bacterium *Streptomyces griseus* (FUNA et al., 1999). Chalcone synthase-like proteins are comparatively small proteins with a single polypeptide chain and are involved in the biosynthesis of precursors for flavonoids. Unlike all other PKSs, these proteins do not have a phosphopantetheinyl (P-pant) arm on which the growing polyketide chains are tethered (FERRER et al., 1999; SCHRODER, 1999). Modular PKSs constitute another class of polyketide synthases. These proteins have a unique architecture involving multiple copies of active sites that are present on exceptionally large multifunctional proteins. The multiple copies of active sites are organized into coordinated groups termed modules. Each module is responsible for the catalysis of one cycle of polyketide chain elongation and associated functional group modifications (CAFFREY et al., 1992). Modular PKSs are responsible for the biosynthesis of polyketide precursors of a wide variety of metabolites including antibiotics such as erythromycin (CORTES et al., 1990; DONADIO et al., 1991), epothilone (JULIEN et al., 2000), rapamycin (SCHWECKE et al., 1995), and rifamycin (AUGUST et al., 1998; SCHUPP et al., 1998). The order of biosynthetic modules from NH_2- to the COOH-terminus on each PKS and the number and type of catalytic domains within each module determine the order of structural and functional elements in the resulting natural product.

The catalytic sites involved in various steps of polyketide biosynthesis are: acyl transferase (AT), keto synthase (KS), acyl carrier protein (ACP), keto reductase (KR), enoyl reductase (ER), and dehydratase (DH). Although one or more of these active sites may be added/excluded depending upon the PKS type, these systems share a common general mechanism of chain assembly. The 2-carbon unit building blocks are systematically combined by decarboxylative Claisen-like condensations. The specificity for the 2-carbon building block resides in the AT domain, whereas KS and ACP together constitute one catalytic center. Other catalytic domains (KR, DH, and ER) dictate the oxidation states of the odd numbered carbon of each 2-carbon unit of the polyketide. The basic chemical reactions performed by various catalytic centers in polyketide synthases are illustrated in Fig. 2. In the following sections we have discussed our present molecular mechanistic understanding of all four types of polyketide synthases.

3 Type I Iterative Polyketide Synthases

This class of PKSs is involved in the biosynthesis of both polycyclic aromatic compounds such as 1,3,6,8-tetrahydroxynaphthalene and 6-methylsalicylic acid (6-MSA) as well as non-aromatic reduced compounds such as lovastatin, brefeldin A, and T-toxin (CARRERAS et al., 1997). 6-Methylsalicylic acid synthase (MSAS), a fungal polyketide synthase from *Penicillium patulum*, is perhaps the simplest polyketide synthase that exhibits typical characteristics of this family of multifunctional enzymes – a large multi domain protein, specificity towards acetyl-CoA and malonyl-CoA substrates, a high degree of control on chain length, and regiospecific ketoreduction. The 6-MSAS gene was isolated and cloned by antibody screening of a *P. patulum* DNA expression library (BECK et al., 1990). The DNA sequence revealed a single open reading frame that was interrupted by a short intron. The four catalytic sites identified in the sequence of encoded protein, KS, AT, KR, and ACP domains, resembled colinearly with rat FAS, which had provided clue to the relationship between these two synthases (SMITH, 1994).

6-MSAS codes for the biosynthesis of 6-MSA from one molecule of acetyl-CoA and three molecules of malonyl-CoA, which is then glycosylated to produce the antibiotic patulin (MARTIN and DEMAIN, 1978). The mechanism of biosynthesis involves initiation with an acetyl group, which is then extended to a tetraketide via three decarboxylative condensations with malonyl-CoA extender units. The NADPH is specifically used for β-ketoreduction after the formation of a C-6 triketide intermediate. In the absence of NADPH, the enzyme-bound triketide intermediate fails to react with the third malonyl-CoA extender unit, and instead it cyclizes to form the triacetic acid lactone (SPENCER and JORDAN, 1992). The thioesterase domain, which is usually involved in the release of polyketide chains, is not present in 6-MSAS. It is, therefore, unclear as to how 6-MSAS catalyzes the cleavage of the thioester bond to its final intermediate. 6-MSAS has now been expressed in a number of

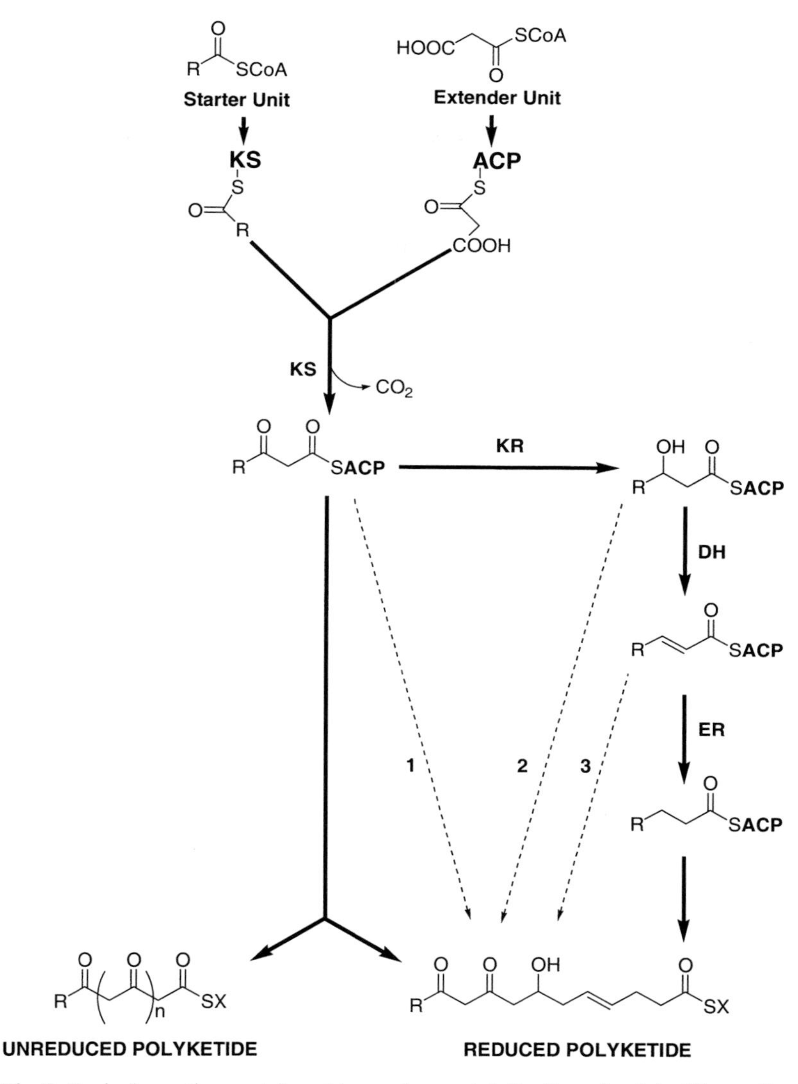

Fig. 2. Typical reactions catalyzed by various catalytic sites of polyketide synthases.

heterologous hosts such as *Streptomyces coelicolor* (BEDFORD et al., 1995), *Saccharomyces cerevisiae*, and *Escherichia coli* (KEALEY et al., 1998). Manipulation of the 6-MSAS gene should lead to the analysis of functions attributed to the regions of the multifunctional enzyme. Recent studies with another type I iterative PKS have revealed a novel mechanism of controlling regiospecific reduction.

3.1 Lovastatin Biosynthesis and Substrate Discrimination during Chain Assembly

Lovastatin is a medically important anti-hypercholesterolemic drug by its ability to inhibit hydroxymethylglutaryl-coenzyme A reductase (ALBERTS et al., 1980). It is a secondary metabolite from the filamentous fungus *Aspergillus terreus*. Lovastatin was suggested to be biosynthesized by a type I iterative PKS assembly. From the lovastatin structure it was not clear how a single set of activities could give rise to a variable state of reduction and dehydration during each of the chain-extending condensation steps. Dihydromonacolin L was the first recognized intermediate in the lovastatin biosynthetic pathway and it could be deduced that this molecule could be derived from 1 acetate- and 8 malonyl-CoA units, although the variable state of oxidation states was unclear. Recent studies with the lovastatin biosynthetic machinery have revealed new features in polyketide biosynthesis (HENDRICKSON et al., 1999; KENNEDY et al., 1999). The DNA sequencing and analysis of the lovastatin gene cluster revealed 18 potential genes which spanned 64 kb in *A. terreus*. Four of the genes are essential for the biosynthesis of lovastatin (Fig. 3): (1) *lov B* that codes for lovastatin nonaketide synthase (LNKS), (2) *lov F* that codes for lovastatin diketide synthase (LDKS), (3) *lov C* that is probably an enoyl reductase, and (4) *lov D* was identified to be a transesterase. Other genes from this gene cluster are probably involved in the regulation, resistance, and transportation of this molecule.

The LNKS is a type I iterative PKS and is involved in the biosynthesis of the nonaketide skeleton. The relative positions of its functional domains are primarily similar to rat fatty acid synthase (AMY et al., 1989) and to *P. patulum* 6-MSAS (BECK et al., 1990). Whereas domains such as KS, AT, DH, KR, and ACP are very similar to their counterparts from other enzymes, the ER domain sequence lacks the signature motif. This is unexpected because several of the carbons in the intermediate dihydromonacolin L are fully reduced. Two additional domains have also been identified in the LNKS sequence. A methyl transferase (MeT) domain is positioned between DH and inactive ER domains. MeT domains are usually reported in non-ribosomal polypeptide synthetases (NRPSs), where these enzymes carry out N-methylations (MARAHIEL et al., 1997). The other additional domain of LNKS is present at the C-terminus and is homologous to the elongation domains of NRPS. The role of this domain in lovastatin biosynthesis remains unclear. Interestingly, similar condensation domains have been identified in PKS proteins of *Mycobacterium tuberculosis* (KOLATTUKUDY et al., 1997; FITZMAURICE and KOLATTUKUDY, 1997; COLE et al., 1998). For identifying the product biosynthesized by LNKS, the *lov B* gene was expressed in a heterologous system under the control of the *alcA* promoter (KENNEDY et al., 1999). The recombinant strain of *Aspergillus nidulans* produced two truncated products. These shunt products had a shorter carbon chain and a lower degree of reduction than dihydromonacolin L. In order to investigate if this was due to the lack of some protein components, various genes from the lovastatin gene cluster were expressed together with *lov B* in *A. nidulans*. Expression of *lov C* gene was able to rescue the aberrant behavior of *lov B* and resulted in the biosynthesis of monacolin J. The *lov C* gene encoded a protein of 363 amino acids and had some homology to the ER domains of PKSs. Although no direct proof is available, it has been suggested that *lov C* brings about the three enoyl reductions that are necessary for dihydromonacolin L production. Further studies have provided strong evidence in support of a specific functional interaction between LNKS and *lov C* protein. Probably, it is this interaction that controls the substrate discrimination by performing ER activity at tetra-, penta-, and heptaketide stages. Remarkably, whereas LNKS could recognize when to methylate the polyketide intermediate, it apparently was not able to discriminate upon failure of one ER reaction. Similar regiospecificity in reduction was also observed for the last two iterative steps of dihydromonacolin L biosynthesis by LNKS, where only the KR domain is used. The second PKS gene, *lov F*, was identified to be non-iterative lovastatin diketide synthase (LDKS) and was shown to be involved in the biosynthesis of the 2-methyl butyryl side chain. The organization of domains in LDKS is similar to LNKS – KS, AT, DH, MeT, ER, KR, ACP domains. Also like LNKS, LDKS contains the MeT domain, although the condensation domain of NRPS is absent. Moreover the ER domain here resembles closely the other active ER domains from other PKSs. Since 2-methyl butyrate is synthesized independently of the nonaketide backbone, another gene *lov D* was

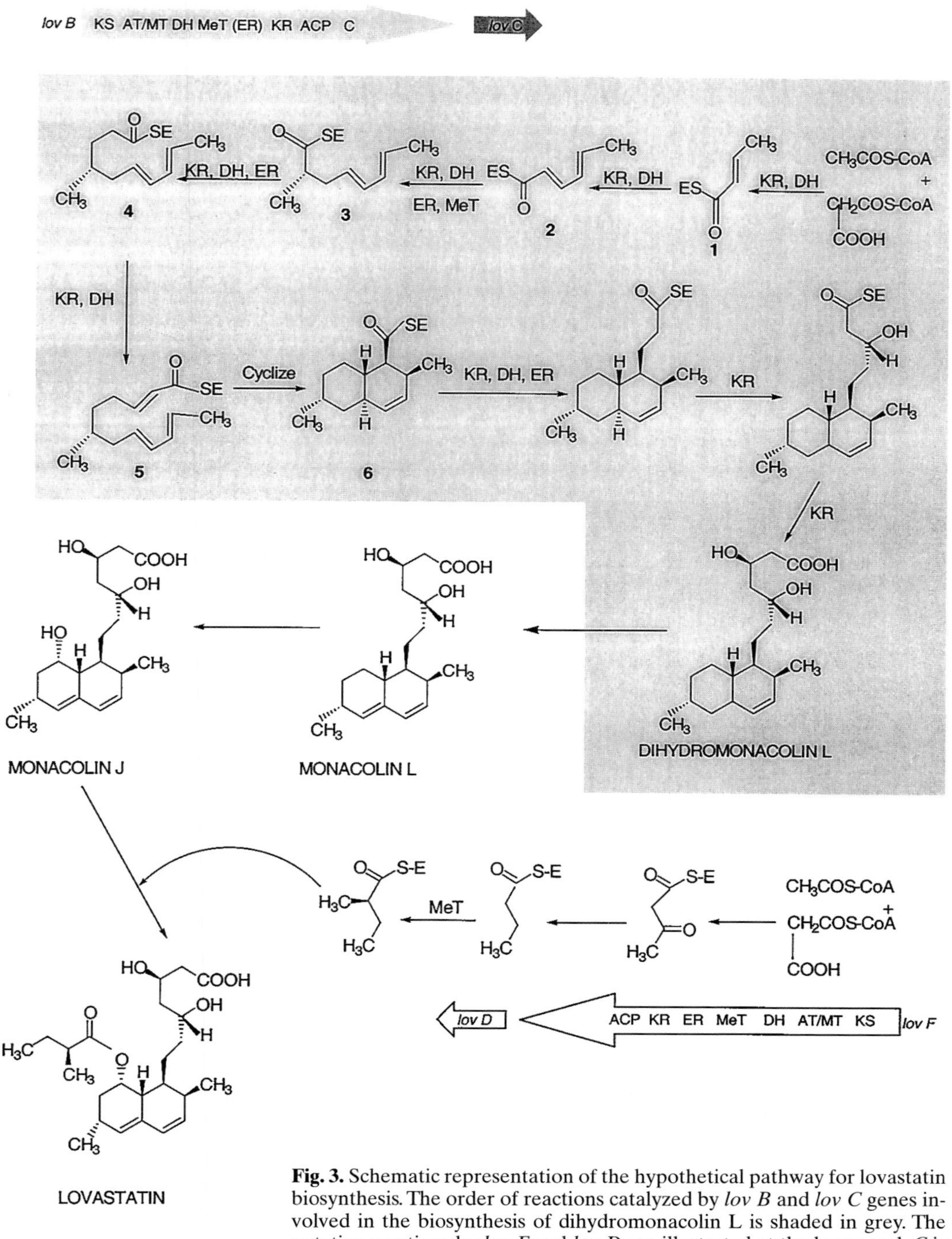

Fig. 3. Schematic representation of the hypothetical pathway for lovastatin biosynthesis. The order of reactions catalyzed by *lov B* and *lov C* genes involved in the biosynthesis of dihydromonacolin L is shaded in grey. The putative reactions by *lov F* and *lov D* are illustrated at the lower end. *C* is the condensation domain of non-ribosomal polypeptide synthases.

identified from this study, and was suggested to be involved in connecting two polyketide components of lovastatin PKS (KENNEDY et al., 1999).

Lovastatin biosynthesis demonstrates a wonderful example where protein interaction has been utilized to facilitate a remarkable discriminatory power in catalytic activity. Although much less is known about their molecular recognition powers, these enzymes can be coupled to other PKS systems to generate novel polyketides. If the discrimination for the extender units can be designed in these systems, it may even be possible to replace very large modular PKS by these relatively smaller PKSs.

4 Type II Iterative Polyketide Synthase

Type II polyketide synthases are a family of bacterial PKSs related to type II fatty acid synthases found in bacteria and plants (CARRERAS et al., 1997; HOPWOOD, 1997). They catalyze the biosynthesis of a broad range of polyfunctional aromatic natural products. These PKSs contain a single set of iteratively used active sites carried on separate proteins (BIBB et al., 1989; FERNANDEZ-MORENO et al., 1992; GRIMM et al., 1994). Two examples of bacterial aromatic polyketide pathways, the actinorhodin and the tetracenomycin pathways, are shown in Fig. 4. In each case the enzymatic assembly consists of "minimal" PKS and other auxiliary subunits. The minimal PKS is composed of four subunits: ketosynthase (KS), chain length factor (CLF), acyl carrier protein (ACP), and possibly a malonyl-CoA:ACP transacylase (MAT) (CARRERAS and KHOSLA, 1998; BAO et al., 1998). A model for the sequence of reactions catalyzed by the minimal PKS is shown in Fig. 5. ACP requires a P-pant arm for catalysis, which is post-translationally carried out by P-pant transferases (LAMBALOT et al., 1996; GEHRING et al., 1997). Malonyl units are transferred from malonyl-CoA onto the P-pant arm of ACP by MAT. Repeated decarboxylative condensations occur between the ACP-bound nucleophilic extender units and the KS-bound electrophilic growing chains. This gives rise to the poly-β-ketone backbone of specified length. Poly-β-ketone products formed as intermediates by the synthetic activity of minimal PKSs are extremely labile. Downstream auxiliary enzymes, such as ketoreductase (KR), aromatase, and cyclase (ARO/CYC) process these intermediates, all of which possess a high degree of regiospecificity.

The CLF has high sequence similarity to the KS, and it associates tightly to form a heterodimer. Although the exact role of CLF is not clear, it has been shown to play an important role in chain length determination (MCDANIEL et al., 1993, 1995). A recent report suggests that CLF is also specifically involved in polyketide chain initiation and that this factor has specific decarboxylase activity towards malonyl-ACP (BISANG et al., 1999). There is also some debate about the requirement of the MAT for minimal PKS activity. There are reports in the literature, where it has been demonstrated that type II PKS ACPs can catalyze self-malonylation and that they do not require MAT (HITCHMAN et al., 1998; MATHARU et al., 1998).

In the following subsections we will discuss our current understanding of the bacterial aromatic PKS subunits.

4.1 Minimal PKS

A central component of aromatic PKS is the β-ketoacyl synthase-chain length factor (KS-CLF) heterodimer. In the presence of an acyl carrier protein (ACP) and a malonyl-CoA:ACP malonyl transferase (MAT), this enzyme synthesizes a polyketide chain of defined length from malonyl-CoA. The dependence of the rate of polyketide synthesis on the concentration of individual protein components was investigated for *act* minimal PKS by studying its steady-state kinetic parameters (DREIER et al., 1999). It was observed that ACP was needed in substantial amounts (much more than the stoichiometry of KS/CLF) for maximal activity. This result suggested that the ACP domain might be getting recycled during the overall catalytic cycle. Also the apo ACP was found to competitively inhibit activity of

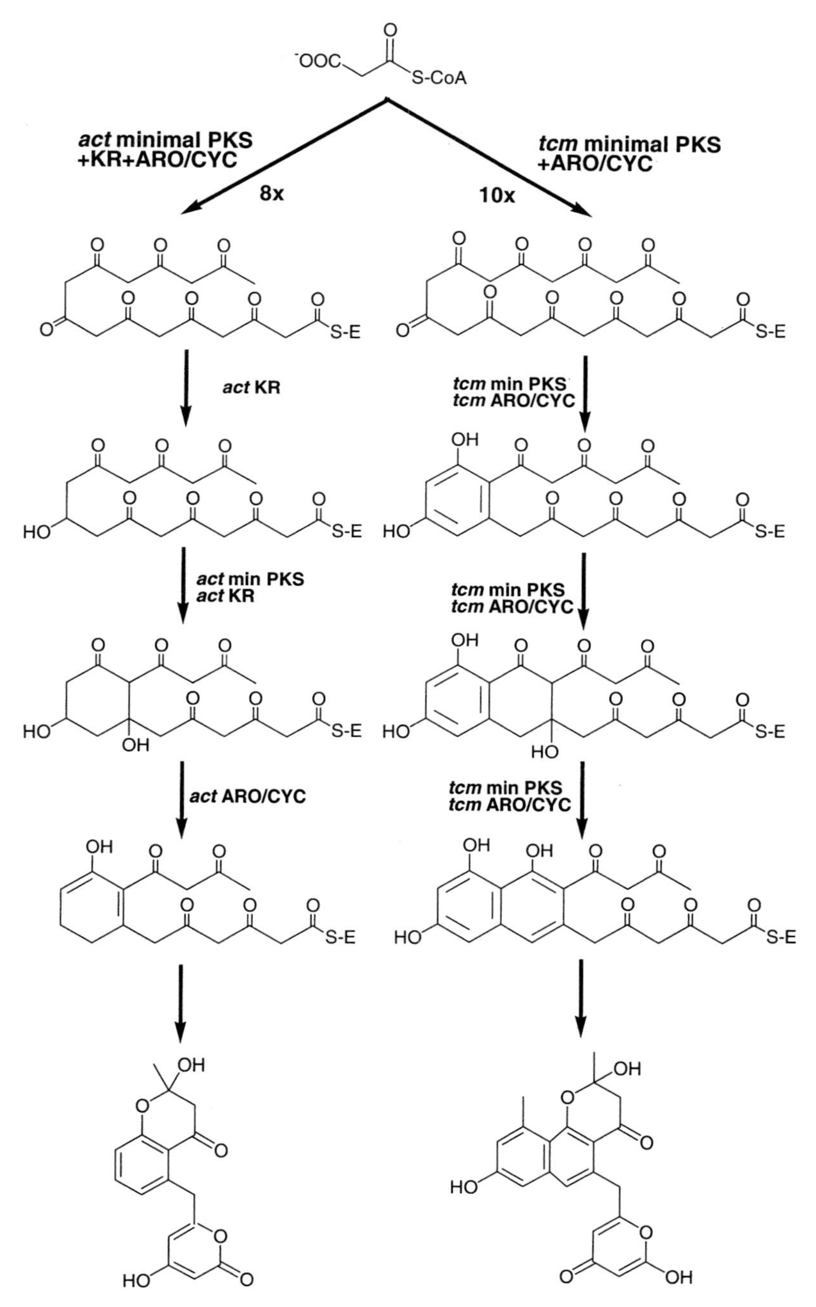

Fig. 4. The actinorhodin (*act*) and tetracenomycin (*tcm*) pathways for biosynthesis of aromatic polyketides. Whereas, *act* minimal PKS uses 8 malonyl-CoA units to form 16-carbon chain, the *tcm* minimal PKS catalyzes a 20-carbon chain from 10 malonyl-CoA units. The catalytic functions of auxiliary subunits in these two pathways are also illustrated.

minimal PKS, demonstrating the importance of protein–protein interactions between the polypeptide moiety of the ACP and the remainder of the minimal PKS. Two models of biosynthesis have been suggested:

(1) a "static" model in which ACP KS/CLF forms a ternary complex during the synthesis of the complete backbone, and
(2) a "dynamic" model in which the ternary complex dissociates following every round of condensation (depicted in Fig. 5).

A detailed investigation of the role of the actinorhodin KS-CLF in priming, elongation, and termination was carried out by using the wild-type enzyme and selected mutants that probe key steps in the overall catalytic cycle (DREIER, 2000). Under conditions reflecting steady-state turnover of the PKS, a unique acyl-ACP intermediate was detected that carried a long, possibly full-length, acyl chain. This species was not detected when C169S, H309A, K341A, and H346A mutants of the KS were used in the minimal PKS assay. These four residues are conserved in all known ketosynthases and all the four mutants were blocked in the early steps of the catalytic cycle. Among these mutants, C169S was efficient in decarboxylation of malonyl-ACP, and other mutants H309A, K341A, and H346A were unable to catalyze decarboxylation. The transfer of label from [^{14}C]-malonyl-ACP to the active site nucleophile of Cys169 in the KS domain was detected for the wild-type enzyme and for the C169S and K341A mutants, but not for the H309A mutant and only very weakly for the H346A mutant. Based on these results and its homology with the KASII ketosynthase homodimer, whose crystal structure has been determined (HUANG et al., 1998), the role of these conserved residues was proposed. Cys169 and His346 were proposed to form a catalytic dyad for acyl chain attachment. His309 was proposed to be required for positioning the malonyl-ACP in the active site and for supporting the carbanion formation by interacting with the thioester carbonyl. Lys341 was suggested to enhance the rate of malonyl-ACP decarboxylation via electrostatic interaction. CLF

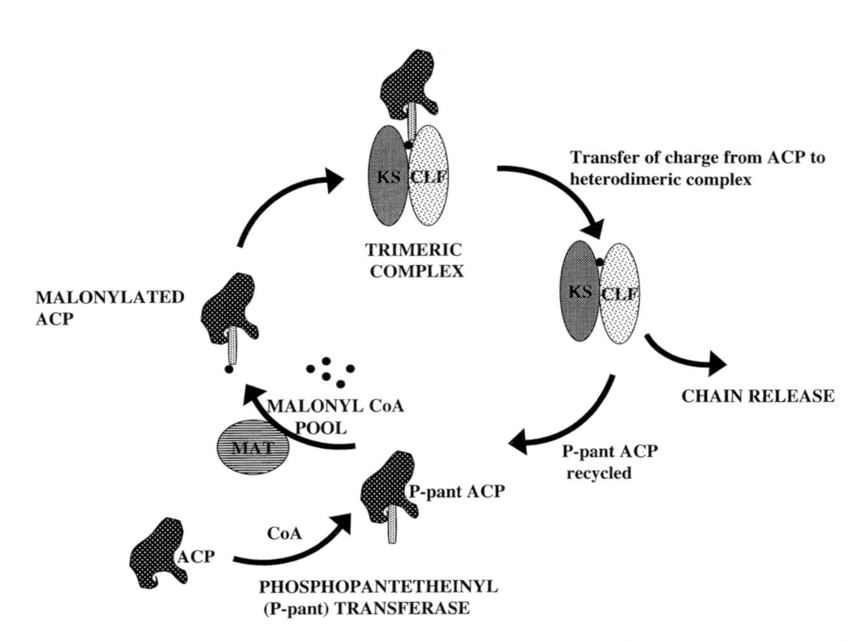

Fig. 5. A cartoon depicting the hypothetical "dynamic" model of aromatic polyketide synthase.

mutagenic studies with a universally conserved Ser145 residue and Gln171 were not found essential for PKS activity (DREIER, 2000). There is some debate on the role of MAT in aromatic polyketide synthesis. HITCHMAN et al. (1998) has suggested that ACP is capable of transferring malonyl groups from malonyl-CoA in the absence of MAT. However, several other reports have suggested that at the physiologically relevant concentrations of individual proteins, the MAT dependent pathway for malonyl-S-ACP is dominant (BAO et al., 1998; CARRERAS and KHOSLA, 1998; DREIER et al., 1999). All these mechanistic studies are supportive of a dynamic model, where the ACP and the KS domains would dissociate after each C–C bond forming event, and that the newly extended acyl chain is transferred back from the ACP pantetheine to the KS cysteine before dissociation can occur.

4.2 Auxiliary Units (Ketoreductase and Aromatase/Cyclase)

KRs are found in many aromatic PKS gene clusters and catalyze regiospecific reduction of the C-9 carbonyl of the poly-β-ketone backbone (BARTEL et al., 1990). KR domains perform their function regardless of the chain length. The *act* KR has been shown to carry out reduction of C-9 carbonyl of polyketides ranging from octaketide to dodecaketide (BARTEL et al., 1990; MCDANIEL et al., 1993). ARO/CYC occur in two architectural forms and they have been shown to play an important role in the formation of the first aromatic ring (BIBB et al., 1994; ZAWADA and KHOSLA, 1997). Didomain ARO/CYC (such as from *act*) catalyzes the aromatization of the first six-membered carbocyclic ring derived from polyketide backbones that have undergone C-9 reduction. The specificity of monodomain ARO/CYCs is not clear. Recently, several auxiliary PKS components including the actinorhodin ketoreductase (*act* KR), the griseus aromatase/cyclase (*gris* ARO/CYC), and the tetracenomycin aromatase/cyclase (*tcm* ARO/CYC) have been expressed independently in *E. coli* and reconstituted along with minimal PKS for detailed mechanistic analysis of the multienzyme system (ZAWADA and KHOSLA, 1999). The polyketide products of reconstituted *act* and *tcm* PKSs were identical to those identified in previous *in vivo* studies (MCDANIEL et al., 1993). Steady-state kinetic analysis revealed that the extended PKS comprised of the *act* minimal PKS, the *act* KR, and the *gris* ARO/CYC had a higher turnover number than the *act* minimal PKS plus the *act* KR or the *act* minimal PKS by itself. This was surprising and the reasons for such changes in kinetic behavior can be attributed to the following two reasons. Firstly, it may be possible that addition of ARO/CYC to minimal PKS leads to favorable protein–protein interactions, which performs the chemistry optimally. However, such a stable protein complex was not detected between minimal and auxiliary PKS components. Secondly, the product release may be more efficient after aromatization reaction, which in turn would indicate that chain release is a rate-limiting step in aromatic polyketide biosynthesis. A similar notion has also been postulated by DREIER (2000) in the case of minimal PKS. Another independent reconstitution experiment with *tcm* ARO/CYC to the *tcm* minimal PKS had also resulted in an increased overall rate (ZAWADA and KHOSLA, 1999). The absence of the identification of a stable complex suggests that intermediates are channeled between the various subunits in these multienzyme systems.

Previous genetic-based approaches had provided useful insights into molecular recognition features of bacterial aromatic PKSs (MCDANIEL et al., 1995). These studies had ascertained a preliminary set of rules for the rational manipulation of chain synthesis, reduction of keto groups, and early cyclization steps. The advent of a reconstituted cell-free systems could provide fundamentally a new perception into the importance of protein–protein interaction in regulating type II PKS function and specificity. Major outstanding challenges like the precise sequence of reactions can be tackled by developing alternative fluorescent assays, where kinetics of different substrates could be monitored simultaneously. The identification of three-dimensional structures of these or their homologous proteins from FASs should offer an attractive opportunity to unravel the mysteries of these PKSs by structure-based approaches (HE et al., 2000).

5 Chalcone-Like Synthases

Chalcone-like synthases are a superfamily of PKS that have been believed to be specific to the plant kingdom (SCHRODER and SCHRODER, 1990; DIXON, 1999). Chalcone synthases (CHSs) are involved in the biosynthesis of chalcones that are precursors of thousands of secondary plant products. These products have important functions as flower pigments (insect attraction), phytoalexins (defense against pathogens), signals for interaction with rhizobia (formation of nitrogen-fixing nodules), mediators of fertility, and protectors against damaging effects of ultraviolet radiation (SCHRODER, 1999). Even though there are significant differences between the biochemistry of chalcone synthases and other types of PKSs (LANZ et al., 1991; TROPF et al., 1995), CHSs are grouped together with other PKSs because of their similar mechanism of biosynthesis, which involves decarboxylative 2-carbon condensation. Reactions catalyzed by two different chalcone synthase-like proteins are illustrated in Fig. 6. CHS is a homodimeric iterative poly-

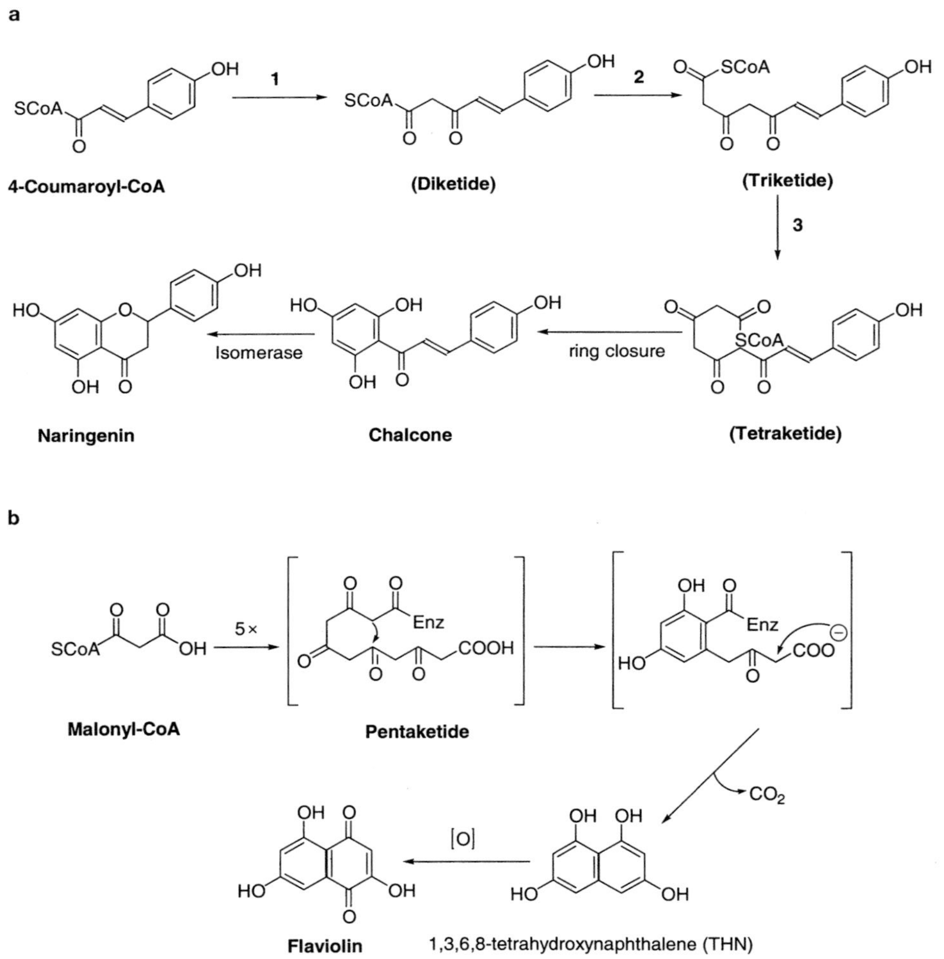

Fig. 6. Reactions catalyzed by two different chalcone synthase-like proteins: **a** CHS, **b** *rppA*.

ketide synthase, with a modest-size polypeptide chain (monomer $Mr \sim 42$ kDa) that performs consecutive elongation reactions at two independent active sites. These enzymes are not phosphopantetheinylated and use free CoA esters as substrates. CHS thus functions as a unimodular PKS that performs a series of decarboxylation, condensation, cyclization and aromatization reactions at a single active site. Plant-specific CHSs use coenzyme-A (CoA) esters from the phenylpropanoid pathway and malonyl-CoA to synthesize chalcones. Most CHSs use *p*-coumaroyl-CoA as a starter unit for synthesis, however, alternate CoA thioesters of acetate, butyrate, hexanoate, benzoate, cinnamoate, and phenylpropionate are also acceptable (FLIEGMANN et al., 1992). Recent studies have identified similar genes in *Streptomyces* as well as in *Mycobacterium*, although their role in these microorganisms is not clear (COLE et al., 1998; FUNA et al., 1999).

5.1 Molecular Basis of Chalcone Biosynthesis

A large number of CHS-related genes have been cloned and sequenced, and their initial sequence comparison has suggested that CHSs are phylogenetically distinct from all other groups of PKSs and FASs. These include stilbene synthase (STS) (YAMAGUCHI et al., 1999), acridone synthase (ACS) (LUKACIN et al., 1999), and phlorisovalerophenone synthase (VPS) (PANIEGO et al., 1999). All these enzymes share a common chemical mechanism, but differ from CHS in their substrate specificity and their cyclization reactions. STS, e.g., condenses one coumaryl-CoA and three malonyl-CoA molecules, like CHS, but synthesizes resveratrol through a structurally distinct cyclization intermediate. Similarly, VPS uses isovaleryl-CoA and three molecules of malonyl-CoA to form phlorisovalerophenone, an intermediate in the biosynthesis of hop bitter acids.

The three-dimensional crystal structure of recombinant alfalfa CHS along with a number of other co-complex structures with substrate and product analogs have provided a framework for understanding the biosynthesis of plant polyketides (FERRER et al., 1999). The CHS enzyme crystallized as a symmetrical dimer, such that the two-fold axis of the dimer overlapped with the crystallographic 2-fold axis. The CHS monomer consisted of two structural domains. An upper domain exhibited an $\alpha\beta\alpha\beta\alpha$ pseudo-symmetric motif observed in thiolase (MATHIEU et al., 1994) from *Saccharomyces cerevisae* and in FAS β-ketoacyl synthase II from *E. coli* (HUANG et al., 1998). Such a domain in these enzymes carries out a similar function that involves shuttling CoA intermediates. The lower domain is significantly different in all these three related proteins and probably accommodates polyketide reaction intermediates of different size and specificity. The CHS-complex crystal structures of bound CoA thioesters and product analogs occupied both the active sites suggesting that the homodimer contains two functional independent active sites. Each of these active sites consisted entirely of residues from its own monomer, except Met137, which was contributed from the other subunit. Remarkably, very few chemically reactive residues were present in its active site. Crystal structure studies have identified four residues, Cys164, Phe215, His303, and Asn336 that lie at the active site pocket of CHS. These residues are found to be conserved in all CHS-related enzymes (FERRER et al., 1999).

5.2 Mechanism of Biosynthesis

The catalytic cycle in chalcone synthases begins with the binding of the *p*-coumaroyl-CoA at the active site of nucleophile Cys164. During this process coumaroyl thioester is formed at Cys164 and the coenzyme dissociates from the enzyme. The first extender unit malonyl-CoA then positions itself in the binding pocket. It is to be noted that CHS-like enzymes lack the traditional phophopantetheine arm, and thus malonyl-CoA is not covalently attached to the protein. Decarboxylation of malonyl-CoA leads to the formation of a carbanion that is stabilized by the keto–enol tautomerism. This carbanion attacks the electrophilic coumaroyl thioester, which transfers the coumaroyl moiety onto the acetyl group of CoA thioester. This diketide intermediate then gets loaded onto Cys164 releasing the free CoA. Similarly two more condensation cycles take place to form the tetraketide. It is not unambiguously assigned whether cyclization takes place when the tetraketide is bound on the Cys164 or when it is present in a CoA form. The cyclization involves an intramolecular Claisen condensation. The nucleophilic methylene group nearest to the coumaroyl moiety attacks the carbonyl carbon of the thioester. Ring closure proceeds through an internal proton transfer from the nucleophilic carbon to carbonyl oxygen. Cyclization in the case of STS may involve nucleophilic attack of the methylene group nearest to the thioester linkage to Cys164 on the carbonyl carbon of the coumaroyl moiety. In STS cyclization it is likely that this intermediate is covalently attached to STS. Completion of the reaction sequence would require hydrolysis from Cys164 and an additional decarboxylation step before formation of resveratol (FERRER et al., 1999).

Recent mechanistic studies with CHS have explored the nucleophilicity of the catalytic residue Cys164 (JEZ and NOEL, 2000; SUH et al., 2000). This study has demonstrated that the pK_a of Cys164 is in the acidic range ($pK_a=5.50$), such that the thiolate anion is present at the physiological pH in the CHS active site. The proximity of His303 to Cys164 in the CHS structure suggests that histidine might stabilize the thiolate anion. Mutational analyses of His303 with Gln and Ala provided consistent results and changed the pK_a of Cys164's thiol to 6.61 and 7.62, respectively (JEZ and NOEL, 2000). This suggests that a stable thiolate–imidazolium ion pair lowers the pK_a and maintains the nucleophile thiolate required for loading reactions. Interestingly, the catalytic residues Cys164 and His303 are conserved not only among various CHS-like proteins but also are present in other PKSs, like DEBS, *act* PKS, and FASs. It is likely that this thiolate–imidazolium ion pair may play an important role in the catalysis of all these enzymes. Indeed, mechanistic analysis of *act* PKS has suggested that Cys–His is an essential component of the catalytic machinery (discussed in Sect. 4.1).

6 Modular Polyketide Synthase

The identification of the modular polyketide machinery has provided tremendous excitement in this field. Modular polyketides catalyze biosynthesis of macrolides and other macrocyclic polyketides that have medicinal properties. For example, erythromycin (CORTES et al., 1990; DONADIO et al., 1991), rapamycin (SCHWECKE et al., 1995), rifamycin (AUGUST et al., 1998; SCHUPP et al., 1998), spiramycin (TANG et al., 1991), FK506 (WU et al., 2000a), and avermectin (IKEDA and OMURA, 1998). Modular PKSs are found both in gram-positive as well as gram-negative bacteria, but they are abundant in the group Actinomycetes. Although minor variations have been encountered in content and organization of modular polyketides their fundamental features and chemistry of assembly are the same.

6.1 Mechanism of Polyketide Chain Assembly

The erythromycin-producing polyketide synthase is the paradigm for all modular type I PKSs. Erythromycin is composed of the 14-membered polyketide-derived macrocyclic lactone ring, 6-deoxyerythronolide B (6-dEB), to which two deoxy sugars are attached. The synthesis of 6dEB from one propionyl and six methylmalonyl units requires six consecutive cycles of chain extension. The catalytic activities of the multienzyme 6-deoxyerythronolide B synthase (DEBS) are organized in six modules such that an individual module is responsible for a single round of condensation and associated reduction reactions (Fig. 7). Each of these modules are dimers and DEBS thus forms a hexameric protein complex ($\alpha_2\beta_2\gamma_2$) of three polypeptide chains DEBS1, DEBS2, and DEBS3. During the entire course of biosynthesis, the various intermediates are covalently bound as acyl thioesters to the acyl carrier protein (ACP) and ketosynthase (KS) domains of the relevant module. ACP domains are post-translationally modified at a conserved serine residue by phosphopantetheinyla-

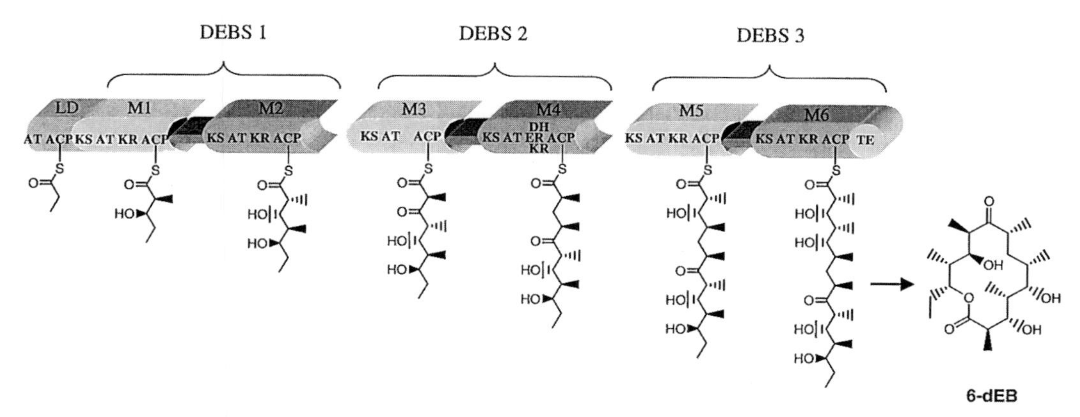

Fig. 7. Modular organization of 6-deoxyerythronolide B synthase (DEBS), which catalyzes the biosynthesis of 6-dEB.

tion. The P-pant group is thought to serve as a flexible tether for both monomeric units and the growing acyl chains as it passes from one module to another. The loading domain (LD), which is posted at the start of the PKS enzyme assembly, is followed by an appropriate number of chain-extending modules in correct sequence, each carrying the relevant complement of reductive sites.

Polyketide synthesis is initiated by transferring the starter unit from the loading domain onto the active site cysteine residue of the KS domain of module 1. This bound starter unit carries out the first condensation reaction with the extender unit, which is attached to the P-pant arm of the ACP domain of the same module. The energy for condensation is supplied by the exergonic decarboxylation of the extender unit. The growing chain that is linked to the ACP domain thus bears the β-carbonyl groups that are either left unreduced or are reduced by the stereospecific conversion into a hydroxyl, olefin, or methylene functionality. The full reduction of β-carbonyl is a three-step process requiring three distinct enzymatic functions *viz*., (1) ketoreduction (KR) to produce the secondary alcohol, where an electron is supplied by NADPH, followed by (2) dehydration (DH), leaving the α,β-unsaturation in the polyketide chain, and finally (3) enoyl reduction (ER) which utilizes NADPH as electron donor, to result in the production of a methylene function at the β-carbon. Following condensation and reduction, the growing chain is transferred from the ACP to the reactive cysteine of the next KS domain, where the next chain elongation step is carried out. This cycle is repeated and the number of elongation cycles is equal to the number of modules present in the particular PKS. The polyketide chain is released from the PKS machinery by the "off-loading" domain such as "thioesterase" (TE), which is present at the end of the PKS assembly line. The TE domain probably also plays a role in the release and cyclization of the polyketide chain. Thus, the final chain length, the level of ketoreduction, the stereochemistry, and the initial cyclization pattern of the full-length polyketide chain are determined by the PKS. Following the formation of the initial cyclized polyketide, additional enzymes can transform the carbon skeleton with modifications such as reduction, glycosylations, and methylations. Such enzymes have been referred to as "tailoring enzymes" in the polyketide biosynthesis. The following sections discuss our present understanding of a modular machinery in terms of hierarchy in organization and structural and functional specificity and tolerance.

6.2 Hierarchical Organization of Modular Polyketide Synthases

The sequence of erythromycin PKS had provided the first evidence for its assembly-line programing of complex PKSs (CORTES et al.,

1990; DONADIO et al., 1991). Apart from the identification of six modules with six putative catalytic sites of condensations for building the heptaketide backbone, all the other features like reductive domains, loading domain, and off-loading domains were positioned appropriately. Since then a large number of PKSs has been identified with modular architecture. This structural modularity has been substantiated by a number of functional studies involving domain inactivation (DONADIO et al., 1993), substitution (OLIYNYK et al., 1996; LAU et al., 1999; MARSDEN et al., 1998; MCDANIEL et al., 1999) and addition (MCDANIEL et al., 1997, KAO et al., 1997, 1998), and also recently with module swapping (GOKHALE et al., 1999b, RANGANATHAN et al., 1999; TANG et al., 2000) experiments. All these studies have suggested that the biosynthesis of complex polyketides requires the interplay of a panoply of hierarchical structures. These hierarchically disposed structures are intrinsic components of modular PKSs and include modules, domains, and linkers. Modules occupy the first level in the hierarchy and are a collection of active sites. The number, sequence, and composition of modules largely specify a polyketide structure. Each module is responsible for one round of chain elongation and associated reduction reactions. Catalytic centers within the modules are called domains and these form the second level of hierarchy. In contrast to structural folding domains, the domains within modules have been defined based on their function. ACP, KS, and AT are core domains, whereas KR, DH, ER are accessory reductive domains. The specificity of these domains dictates the final outcome of the complex polyketide. Regions connecting these domains, referred to as "linkers", are probably important for the structural integrity and establish cooperation between various active sites during the catalytic process. These linkers form the last tier in the hierarchical setup of polyketide machinery. Various polypeptide chains of the multienzyme complex are specifically organized such that the growing polyketide chain transverses through a unique path. Recent studies suggest that these linkers might play an important role in the overall organization of PKSs (GOKHALE et al., 1999b; GOKHALE and KHOSLA, 2000). It is interesting to note that in polyketide biosynthesis the catalyst itself is modularized, while in the template-based biosynthesis of proteins and nucleic acids, the template is modular in nature. In order to exploit the catalytic versatility of these modular enzymatic assemblies, it is important that we understand all these levels of topological organization. Surprisingly, very few studies have directly addressed molecular recognition and mechanistic aspects of modular enzymes. Instead most of the knowledge about modular PKSs emanates from domain and module shuffling experiments. In the next three sections we will discuss our present understanding of these three tiers of the modular PKS machinery.

6.2.1 Properties of Modules

Since the discovery of the modular architecture of certain PKSs, several structural and functional studies have validated the notion of module being a "unit" of polyketide biosynthesis. Each module is folded into a homodimer such that there are two sets of active sites for polyketide biosynthesis (CAFFREY et al., 1992; PIEPER et al., 1995; STAUNTON et al., 1996). Although the functional independence of one set of catalytic site from the other is established, it is not known whether both the catalytic sites are functional simultaneously. It is also not clear whether all modules in a complex operate independently of each other or only one chain is loaded at any given time in the enzymatic assembly. Experimental support for the processive nature of biosynthesis has come from two different experiments. Firstly, the genetic manipulation of DEBS has demonstrated that upstream modules can function independently of downstream modules (CORTES et al., 1995; KAO et al., 1994, 1995). Secondly, inactivation of the releasing enzyme in rifamycin PKS leads to the formation of products of all intermediate chain lengths, suggesting that each ACP is occupied by a product of the corresponding module (YU et al., 1999). We here enumerate several experiments that have suggested that modules are structurally and functionally independent folding units.

6.2.1.1 Structural Studies of Erythromycin PKS

The homodimeric structure of modules was first investigated by limited proteolysis studies of DEBS proteins (STAUNTON et al., 1996). Each DEBS protein consists of two modules. The proteolytic cleavage pattern of all DEBS proteins was specific and partial digestions resulted in the release of dimeric modules. All the module fragments gave a single N-terminal sequences demonstrating that the dimeric multienzyme must have an element of symmetry that leads to identical cuts in both subunits. Further extended cleavage occurred between domain boundaries predicted on the basis of sequence alignments (APARICIO et al., 1994). Based on these proteolytic data and the sequential arrangement of domains in the module a "double helical" model was proposed, such that each module could exist autonomously (STAUNTON et al., 1996). The functional topology and organization of various domains in a module has also been studied by *in vitro* complementation analyses (KAO et al., 1996). In this approach two inactive PKSs carrying mutations in different functional domains were reconstituted to form a catalytically active heterodimer. Complementation studies between various combinations of inactivated KS1, KS2, ACP1, and ACP2 domains of DEBS1-TE confirmed that there are two sets of independent catalytic sites. These studies also suggested that KS and ACP from different subunits constitute a catalytic center and that the chain transfer from one module to the next module takes place on the same polypeptide chain. Similar complementation of AT2 null DEBS1-TE protein with KS1 and KS2 inactivated DEBS1-TE protein, respectively, revealed that the AT domain could be shared between two clusters of active sites within the same dimeric module (GOKHALE et al., 1998). In the absence of a three-dimensional structure of these proteins, such biochemical experiments have provided a crude-working model of modular PKSs.

6.2.1.2 Expression of Individual Modules and their Substrate Specificity

The genetic evidence for the independent functional existence of an isolated single modules was provided by the construction of a trimodular derivative of DEBS (KAO et al., 1996). The DEBS1-Module3-TE was engineered such that the DEBS2 protein was spliced and the TE domain was pasted at the end of ACP3. Recently, several single modules of DEBS have been heterologously expressed in *E. coli* (GOKHALE et al., 1999b). Since the host's P-pant transferase does not post-translationally modify ACP domains of PKS modules, surfactin P-pant transferase (*sfp*) was co-expressed with all modules. The N-terminal modules did not require any modification for functional expression. The TE domain was pasted at the end of these modules for catalytic turnover. The C-terminal modules in contrast, were functionally expressed after changing the N-terminal intra-polypeptide linker residues with the inter-polypeptide linker residues (for details see Sect. 6.2.3).

The specificity and flexibility of individual modules were tested by using a an N-acetylcysteamine (NAC) thioester-diketide (I) along with methyl malonyl-CoA and NADPH. The Michaelis constant (K_M) for this diketide substrate for modules 2, 3, 5, and 6 of DEBS was within the same order of magnitude for each module (GOKHALE et al., 1999b). This result was surprising as the same diketide during the *in vivo* feeding experiments would specifically be incorporated at module 2, and all other modules were apparently incapable of accepting and elongating this substrate. The reason for this apparent paradox is not clear. It has been hypothesized that kinetic channeling in a modular PKS ensures a high level of overall biosynthetic selectivity, even though each module is relatively flexible in its specificity (GOKHALE and KHOSLA, 2000). A recent study has probed the specificity of individual DEBS modules with varying stereochemistry at α- and β-positions of diketide and also with another substrate having a longer chain length of the carbon backbone (WU et al., 2000b). These studies have confirmed that DEBS modules are tolerant towards diverse incoming acyl chains. All modules distinctly favored *syn* diketides over *anti* diketides confirming that there is a high degree of similarity in the specificity of all DEBS modules, despite the diverse structural features of their natural substrates. This lack of correlation between the optimal substrates and their natural substrates for the modules certainly needs a detailed investigation. It will be interesting in future to probe the intrinsic preference for modules from other PKS systems.

6.2.1.3 Module Swaps

The ultimate proof of the existence of module as structural and functional unit has been provided by the success of module swap experiments. By maintaining appropriate linker residues at the ends of modules that are probably important in establishing protein–protein interactions between heterologous modules (for details see Sect. 6.2.3), the transfer of biosynthetic intermediates between unnaturally linked modules was achieved. The expression of bimodular *ery* M1-*rif* M5-TE synthesized the expected triketide lactone. In the context of rifamycin PKS, module 5 (Fig. 8a) accepts completely different acyl chains. As a more demanding test this bimodular construct was co-expressed with DEBS2 and DEBS3, after appropriate C-terminal engineering of *rif* M5 (Fig. 8b). Expression of this construct in the recombinant *Streptomyces coelicolor* CH999 strain produced good levels of expected 6-dEB (GOKHALE et al., 1999b). These experiments, therefore, have demonstrated that it is feasible to rewire PKSs by swapping modules.

6.2.2 Properties of Domains

From the sequence homology studies it was clear that the arrangement of domains in a module was analogous to FASs. Also the cor-

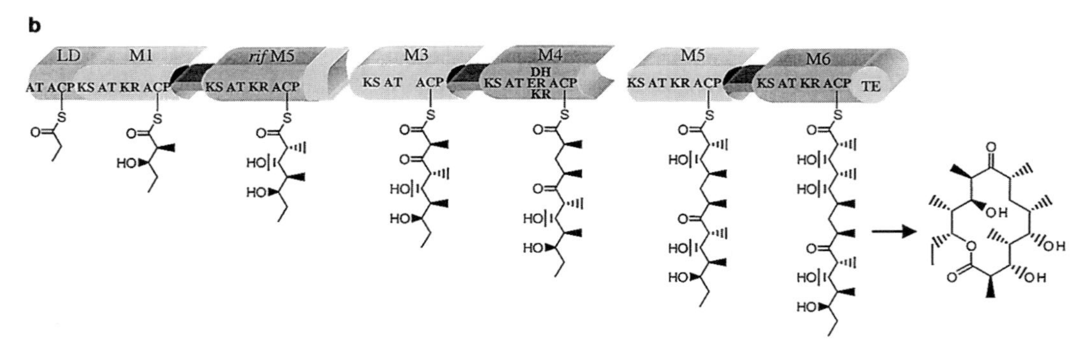

Fig. 8. Heterologous modular fusions between *ery* and *rif* PKS gene clusters. **a** The *ery* M1–M2 intra-polypeptide linker was maintained in this construct. **b** This construct demonstrated the successful inter-polypeptide chain transfer. Here, in addition to retaining *ery* M1–M2 linker, the inter-polypeptide linker that are naturally present between *ery* M2 and M3 (shown in lighter shade after *rif* M5) were also maintained.

relation between the functionality of the growing chain with the nature of domain present in the module had suggested that these domains would be used in a linear order. Despite this there was a possibility that various domains of a module might fold in such a manner that a linear set of domains would not form a functional module. The first experimental proof for the linear use of the individual domains came from the domain inactivation experiments carried out by KATZ and coworkers (DONADIO et al., 1993). They disrupted KR5 and ER4 domains in DEBS to obtain the corresponding 5-keto- and 6,7-dehydro analogs of 6-dEB, respectively. These studies also provided the first evidence for the fact that the downstream domains of module 5, module 6, and TE domain were able to accept these altered substrates and performed their normal catalytic function. Since then a large number of studies involving domain inactivation, shuffling, addition and replacement studies have been carried out (KHOSLA et al., 1999). These studies have shown that domains are "mobile" and can be utilized to engineer novel polyketides (McDANIEL et al., 1999; XUE et al., 1999). McDANIEL et al. (1999) have combined this potential to generate more than 50 different 6-dEB analogs by substituting ATs and β-carbon processing domains of DEBS with counterparts from the rapamycin PKS that encode alternative substrate specificities and β-carbon reduction/dehydration activities. All these studies have reinforced the view that domains are structurally and functionally a specific entity. Surprisingly, apart from the TE domain none of the other individual domains have been independently expressed, and, therefore, all the specificity and tolerance studies for each type of domain have been carried out in the context of the complete module. Presently, all the reactions involved in polyketide biosynthesis, except epimerase activity, have been specifically assigned to individual domains. A detailed account of tolerance and specificity of individual domains has been recently reviewed by KHOSLA et al. (1999). We will here briefly summarize the properties of individual active sites.

6.2.2.1 Specificity of Ketosynthase Domain

The assembly-line biosynthesis of complex polyketides suggests that the KS domain may be playing at least two crucial roles: (1) Substrate recognition during inter-modular chain transfer and (2) specificity in the catalytic steps of the chain condensation reaction. A number of studies with DEBS as a model system has suggested that there is a broad substrate permissibility for various KS domains (TSUKAMOTO et al., 1996; JACOBSEN et al., 1997, 1998). It is, therefore, surprising that naturally occurring PKSs typically produce a single product or a small group of closely related products. Initial studies explored the molecular recognition properties of KS2 by incubating series of diketide substrates to the KS1-inactivated-DEBS1-TE mutant (CHUCK et al., 1997). Various features in diketide substrate, such as chain length, stereochemistry, and functional groups were varied. The specificity was studied by measuring the steady-state kinetic parameter for the turnover of diketide-SNAC analogs into their corresponding triketide lactones. Diketides with varying chain length and with natural C-2 stereochemistry were acceptable to KS2. The removal of C-3 oxygen from the diketide substrate was also tolerated. The influence of other domains (KR, TE) were ruled out in this study. CHUCK et al. (1977) showed that the inactive analogs did not inhibit biosynthesis, which implicated that inactive diketides do not compete for the KS acylation step and, therefore, the overall kinetic parameters exhibited properties of the KS domain. In an *in vivo* "chemo-biosynthesis" approach chemically synthesized analogs were incorporated in these complex polyketide structures (JACOBSEN et al., 1997). A genetic block was introduced in the first condensation step of the DEBS pathway. Microorganisms harboring this modified PKS were then fed with analogs of the second module intermediates. These studies have demonstrated the remarkable tolerance and specificity of DEBS. In a rather remarkable example a long-chain diketide with native (2*S*,3*R*) stereochemistry was specifically incorporated *in vivo* to synthesize an interesting 6-dEB analog, as shown in

Fig. 9 (HUNZIKER et al., 1999). Similar flexibility in incorporation of unusual precursors was not completely successful with rifamycin PKS. Incorporation of certain 3-amino-5-hydroxy benzoic acid (AHBA) analogs in a genetically modified rifamycin strain *Amycolatopsis mediterranei* with a block in the AHBA pathway, resulted in the synthesis of truncated tetraketide analogs (HUNZIKER et al., 1998). These tetraketide products were intermediates released after module 3, which indicated that the fourth module of the rifamycin PKS shows absolute discrimination. It is also possible that a downstream block might have led to the accumulation of stable tetraketides. As discussed in Sect. 6.2.1.2, a number of diketide variants has been tested on isolated modules. All these studies have also re-imposed the flexibility of DEBS KS domains in accepting substrates.

6.2.2.2 Specificity of Acyl Transferase Domains

Typically, there are two distinct types of AT domains in modular polyketide synthases. The loading AT (AT_L) domain is involved in charging the PKS assembly with the primer unit and usually has a broad substrate specificity. In contrast, the individual extender AT domains present within each of the modules exhibit strict structural as well as stereochemical specificity. In the case of DEBS, all chain extending AT domains are specific for (2*S*)-methylmalonyl-CoA (WEISSMAN et al., 1998; PIEPER et al., 1996). The 14 modules of rapamycin PKS have half of its modules specific for malonyl-CoA and the other half for methylmalonyl-CoA (HAYDOCK et al., 1995). Several research

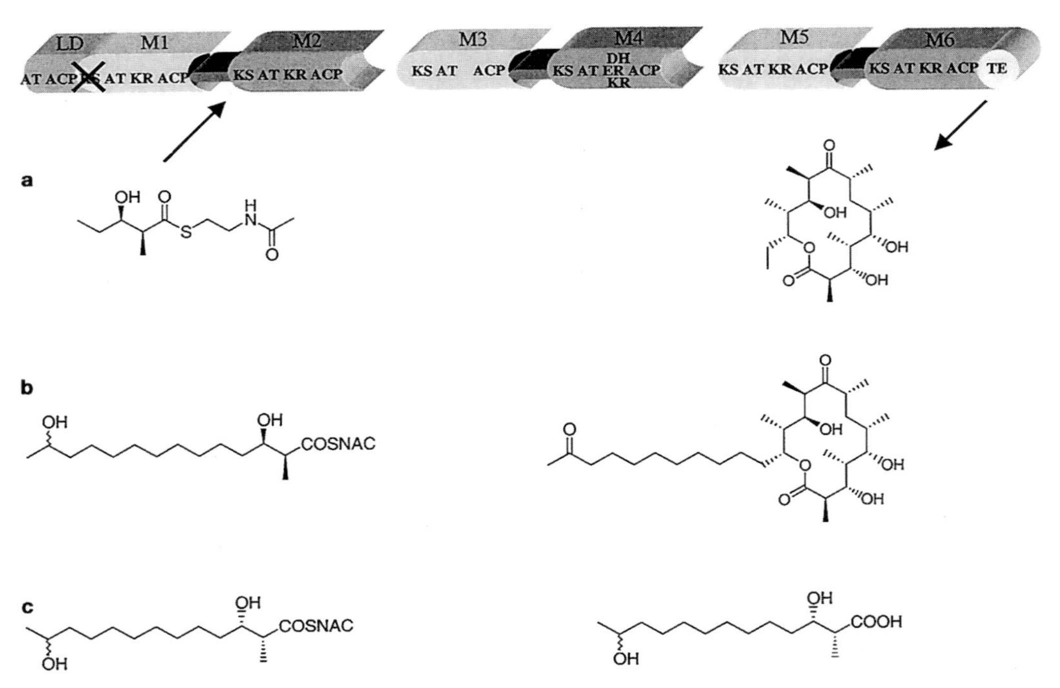

Fig. 9. Schematic representation of the "chemo-biosynthetic" approach. The KS1 domain of DEBS is inactivated and intermediates are fed in the growing cultures. **a** Natural diketide intermediate is loaded at module 2 to synthesize 6-dEB. **b** Long-chain analog of diketide with (2*S*,3*R*) stereochemistry was incorporated to produce aglycone. **c** Substrate with (2*R*,3*S*) stereochemistry did not produce any cyclic product and only resulted in hydrolysis of substrate. In the **b** and **c** panel, the oxidation of the distal hydroxyl occurred during the fermentation process and was probably not catalyzed by PKS.

groups have taken the advantage of the intrinsic AT specificity from different PKSs to alter the specificity for chain elongation (OLIYNYK et al., 1996; RUAN et al., 1997; LAU et al., 1999). Sequence comparison of different AT domains from various PKSs clearly aligns into two families with malonyl and methylmalonyl substrate specificity (HAYDOCK et al., 1995). Even though this can be used to predict the specificity of AT domains with reasonable confidence, it has not been possible experimentally to change AT specificity by swapping these conserved signature sequences. LAU et al. (1999) studied a series of hybrid mutants with chimeric AT domains, which were constructed by fragments of RAPS AT2 domain (specific for malonyl-CoA) with systematic replacement by DEBS AT2 (which uses methylmalonyl-CoA). These chimeric ATs were placed in the context of DEBS system at AT2 position and then the production of 6-dEB was monitored. This study has identified a small region in the AT protein sequence that seems to play an important role in discriminating malonyl versus methylmalonyl-CoA specificity. Intriguingly, this region shows low sequence conservation across all known AT domains.

Several studies have measured the substrate specificity of loading AT domain in the context of modular PKS (MARSDEN et al., 1994; WEISMANN et al., 1995; PIEPER et al., 1996, 1997). The DEBS loading domain has been swapped with the loading domain from avermectin PKS to synthesize branched chain erythromycin analogs (MARSDEN et al., 1998). The primer unit specificity in rifamycin biosynthesis has also been examined by feeding 3-amino-5-hydroxy benzoic acid (AHBA) analogs to a non-producing mutant of *Amycolatopsis mediterranei*, as discussed earlier. Recently, the loading didomain of DEBS has been expressed in *E. coli* (LAU et al., 2000). The loading ACP (ACP_L) domain was phosphopantetheinylated by *sfp* P-pant transferase. Competition experiments were performed with the holo-protein to determine relative rates of incorporation of a variety of unnatural substrates, in the presence of comparable concentrations of labeled acetyl-CoA. This study has suggested that AT_L of DEBS has greatest preference for unbranched alkyl chains, although it is tolerant towards branched alkyl chains and olefinic functional groups, while the bulkier polar and charged groups are not preferred.

6.2.2.3 Specificity of Ketoreductase Domain

Reduced polyketides are known to have hydroxyl groups of both D- and L-configuration. The cofactor stereochemistry of NADPH-mediated reduction of acyl chains has been tested for both KR1 and KR2 domains of DEBS. Even though these two domains generate hydroxyl groups with D- and L-configuration, respectively, the hydride is donated from the *pro-S* face of NADPH (MCPHERSON et al., 1998). It has now been established that the stereochemistry of β-ketoacyl-ACP reduction is an intrinsic property of the KR domain and does not depend on other factors. Replacing DEBS KR2 with either RAPS KR5 or KR2, both of which have different stereochemical preferences, reversed the stereochemistry (KAO et al., 1998). In another study a "minimal PKS" was engineered by fusing KS from module 1 to AT2 in DEBS (BOHM et al., 1998). This study suggested that the KR domain selectively reduces one diastereomeric intermediate (here 2*R*-diastereomer of 2-methyl-3-ketopentanoyl intermediate), and thus KR domains could influence the stereochemistry of methyl-branched centers in polyketide products.

6.2.2.4 Specificity of Acyl-Carrier Protein

The ACP domain in PKS is post-translationally modified by phosphopantetheinylation, which is approximately 20 Å in length (SMITH, 1994; LAMBALOT et al., 1996). The swinging arm motion of the phosphopantetheine arm has long been accepted as a key factor in facilitating coupling of various partial reactions of FASs and PKSs. Fluorescence energy transfer experiments in FASs had indicated that the distance between the arm and the catalytic active site is much longer than 20 Å, and, therefore, a concept of domain mobility was proposed (CHANG and HAMMES, 1990). Such motions involving domains of both FAS and PKS have not been defined and the motion of the ACP arm remains unclear. Also it is not clear whether these movements are coordinated with catalytic reactions or are random in nature. Various domain-swapping experiments in DEBS have suggested that ACP does not possess intrinsic specificity. However, it must be noted that the ACP domain must interact productively with two distinct KS domains in order to catalyze respective specific reactions. During its intra-modular stroke the extender unit on the P-pant arm of ACP provides the nucleophilic component of the decarboxylative condensation reaction, whereas the intermodular stroke results in polyketide chain transfer from ACP (n) to the downstream KS (n+1) domain. The importance of specific protein interactions between KS and ACP domains is not clear.

6.2.2.5 Specificity of Thioesterase Domain

The "mobility" of the TE domain has been substantiated from a number of studies. Genetic studies have suggested that the TE domain from DEBS can interact with all ACP domains of DEBS to carry out chain release and cyclization (KAO et al., 1995; CORTES et al., 1995). Another study has shown that the triketide chain from DEBS is released much more efficiently when the TE domain is covalently bound to DEBS rather than DEBS1 and TE existing as two independent polypeptide chains (GOKHALE et al., 1999a). Together, these two studies have suggested that the recognition between ACP and TE domain is not very specific and the *cis* TE effect may be primarily due to its "effective concentration". The recombinant TE domain from DEBS has been expressed in *E. coli* and its substrate specificity has been evaluated in two independent studies (WEISMANN et al., 1998; GOKHALE et al., 1999a). In both of these studies the isolated TE domain failed to synthesize lactone bond formation in the activated (ω-1) and ω-hydroxycarboxylic acids. The TE domain instead hydrolyzed these nitrophenyl ester- and N-acetylcysteamine (NAC) thioester-substrates. The hydrolytic specificity of the TE domain suggested a marked kinetic preference for longer chain substrates and for (2*S*,3*R*)-NAC substrates (GOKHALE et al., 1999a). Recently, the excised TE domain from tyrocidine synthetase has been shown to efficiently catalyze cyclization of a decapeptide thioester to form the antibiotic tyrocidine A (TRAUGER et al., 2000). It can also catalyze pentapeptide thioester dimerization followed by cyclization to form the antibiotic gramicidin S. By systematically varying the decapeptide thioester substrate and comparing cyclization rates, this study has shown that only two residues (one near each end of the decapeptide) are critical for cyclization. TE domains in general, appear to have a relaxed specificity and they should be able to cyclize and release a broad range of new substrates and products produced by engineered enzymatic assembly lines.

6.2.3 Properties of Linkers

Modular architecture of proteins is widespread in biological macromolecules. Recent advances have established the crucial role of linkers in establishing structural and functional assembly of these multimodular proteins. It is believed that these dynamic linkers establish communication by directing the correlated movements of various domains (GOKHALE and KHOSLA, 2000). Even though the modularity in PKS has been highlighted by several reports of domain and module swapping experiments, the mechanisms that guide the biosynthetic intermediates from one active site to another remain poorly understood. It has recently been shown that appropriate engineering of N- and C-terminal ends plays an important role in the assembly of functional modules or domains. Two categories of linkers have been postulated: linkers that connect covalently connected modules (intra-polypeptide linkers, e.g., between modules 1 and 2 of DEBS) and linkers between modules that are present on two different polypeptides (inter-polypeptide linkers, e.g., between modules 2 and 3 of DEBS) (GOKHALE et al., 1999b). There is probably another category of linkers that connect different domains in a module and can be referred to as intra-modular linkers. Although the role of linkers is not completely understood, here we would argue for their structural and functional role in polyketide assembly.

6.2.3.1 Intra-Polypeptide Linkers

It has been hypothesized that communication between two modules is established through these linker residues. Polyketide chain transfer involves a thioester exchange reaction. Thermodynamically, it is feasible for these acyl chains to equilibrate between ACP and KS domains, particularly in the absence of an extender unit. Previous studies have shown that the PKS chain cannot move back from KS (n+1) to ACP (n) in DEBS. Since the transfer of polypeptide chains between adjacent modules is unidirectional and takes place from the ACP domain of the upstream (n) module to the KS of the downstream (n+1) module, a model for linker-mediated chain transfer has been proposed (Fig. 10). According to this model linkers control the conformational changes required to allow a given ACP to switch interactions between KS of its own module and that of the next (GOKHALE and KHOSLA, 2000). It is possible that the catalytic site KS2 is coupled to ACP1 by inter-polypeptide linker. The presence of substrate in the active site of the KS domain may modify its interaction with the linker, which in turn would uncouple ACP1–KS2 interaction. Interestingly a number of proteins are regulated through an active site. This is very similar to a well-characterized SH-2 kinase linker that couples SH3 domain to the kinase domains in *Src*-homology proteins (XUE et al., 1999; RIGGS and SMITHGALL, 1999). Recent results from RANGANATHAN et al. (1999) have provided support for this hypothesis. They have shown successfully that intermodular fusions can be constructed by preserving ACP-KS didomain that spans the junctions between naturally occurring modules. Amino acid sequence analysis of the interpolypeptide linkers in DEBS and related PKSs has shown that these linkers contain one or more proline residues. It is possible that by forming a conformationally sensitive turn these segments may serve to align two modules, or in analogy with SH2-kinase linker, these linkers residues may form a polyproline helix (XU et al., 1999; SCHINDLER et al., 1999), which might be responsible for providing cooperative interactions that couple and uncouple two modules by hinge movement.

6.2.3.2 Inter-Polypeptide Linkers

Since modules are both structurally and functionally independent, it is likely that these inter-polypeptide linkers are primarily involved in protein–protein recognition. An indication of weaker interaction between DEBS polypeptide chains had come from gel filtration studies, where DEBS did not elute as one complex (PIEPER et al., 1996). This was confirmed from the *in vitro* study, where the dissociation constant (K_D) for the interaction between DEBS1 and module3-TE was estimated

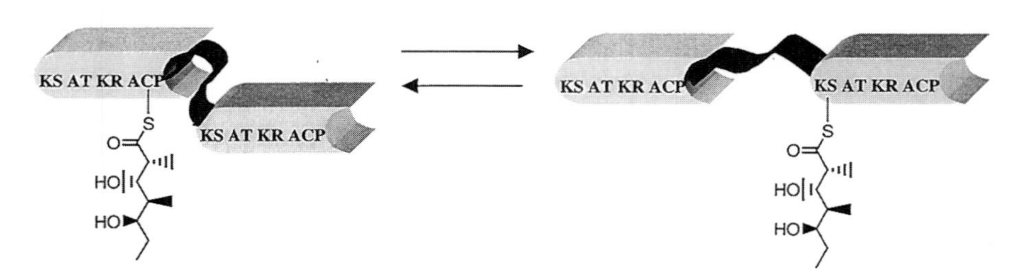

Fig. 10. A model for linker-mediated chain transfer (linker is shown in black ribbon). The catalytic site of KS is coupled to ACP by intra-polypeptide linker. The binding of substrates modulates protein interactions through the linker and thereby regulates vectorial transfer of polyketide chains.

to be ~2.6 μM (GOKHALE et al., 1999a). It is, therefore, likely that interfacial interactions in these proteins do not extend over large protein surfaces. This was further validated by constructing a hybrid synthase, wherein rifamycin PKS module 5 could communicate with DEBS2 by transferring a relatively short linker from the end of DEBS1 (GOKHALE et al., 1999b). Amino acid sequence analysis studies with inter-polypeptide linkers (sequences that are present upstream of the N-terminal module and downstream of the C-terminal module) have suggested propensity to form α-helical coiled-coils (GOKHALE and KHOSLA, 2000). Because modules are both structurally and functionally independent, it has been proposed that all amino-terminal and carboxy-terminal modules possess homodimeric coiled-coils (Fig. 11). Inter-modular interactions might then be facilitated through heterodimeric or four-stranded coiled-coils. It has been established that coiled-coil is one of nature's favorite way to establish interactions. Several studies have proved that a few critical residues are sufficient to dictate the specificity in coiled-coiled proteins, as illustrated by GCN4 (HARBURY et al., 1993) and Fos-Jun (SCHUERMANN et al., 1991).

6.2.3.3 Intra-Modular Linkers

The various regions between domains called intra-modular linkers have been thought to play a passive but vital structural role in maintaining a correct topology for optimal chemistry within PKS modules. Many of these structural regions can be identified by short sections of amino acids rich in alanine, proline, and charged residues. These are around 30 to 35 residues long and are thought to serve as linkers between folded domains (STAUNTON and WILKINS, 1997). There are also some longer sequences in all modular PKSs, and these sequences have not been associated with any specific catalytic role. Several successdul domain swapping experiments and module fusion experiments using intra-modular junctions have preserved these regions (TANG et al., 1999).

Detailed investigations of the various levels of the biosynthetic modular machinery would render exploitation of this remarkable power of catalysis.

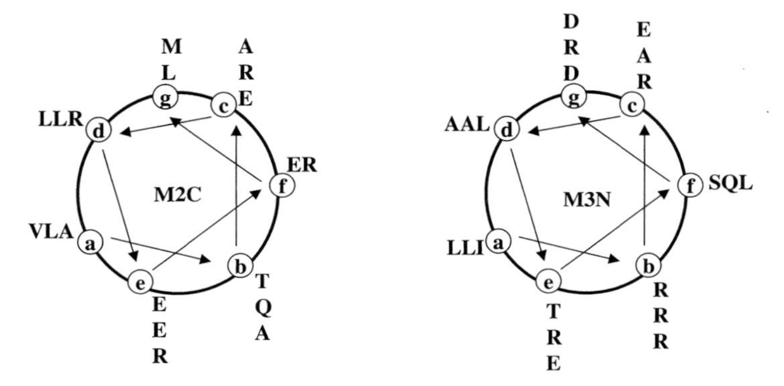

Fig. 11. Helical wheel diagrams to illustrate the postulated coiled-coil nature of inter-polypeptide linkers. All the hydrophobic amino acid residues are clustered on the same face of the helix (depicted by "a" and "d" here). The other face of the helix contains hydrophilic residues. M2C is the C-terminal linker of module 2 of *ery* PKS and M3N is the linker present at the N-terminal end of the module 3.

7 Crosstalk between Polyketide Synthases and Fatty Acid Synthases

In nature, the formation of fatty acids and polyketides is catalyzed by structurally and mechanistically and probably also evolutionary related enzymes (HOPWOOD, 1997). It is remarkable that in contrast to fatty acids, polyketides exhibit such a wide diversity of structure and function from a similar set of catalytic activities. Fatty acid synthases (FASs) catalyze the biosynthesis of fatty acids, which are primary metabolites and components of cellular membranes in all living systems (WAKIL, 1989; SMITH, 1994). Polyketide synthases (PKSs) catalyze the formation of polyketide natural products, which are secondary metabolites that have been a rich source of commercially important antibiotics and other therapeutics for the pharmaceutical industry (O'HAGAN, 1991). Traditionally, the biosynthetic machinery for fatty acids and polyketides has been studied independently, but recent evidence suggests that these enzyme systems may regularly interact in many organisms to generate hybrid molecules of fatty acids and polyketides.

Several examples of interactions between distinct biosynthetic pathways have been reported in the literature. For instance, the *Rhizobium* bacteria and various legume plants form symbiotic relationships in which bacterial infection leads to the formation of nitrogen-fixing root nodules. Bacterial lipooligosaccharide signaling molecules are involved in identifying the correct legume host for infection, and appear to be synthesized through an interaction between the bacterial FASs and *nod* gene products (SPAINK et al., 1991). Moreover, specificity determinants for different signaling molecules appear to include both the length and pattern of unsaturation of the fatty acid chains, indicating that a variety of interactions between these enzyme systems must exist. Another example involves the biosynthesis of complex lipids by the mycobacteria (KOLATTUKUDY et al., 1997). The genome sequences of *Mycobacterium tuberculosis and Mycobacterium leprae* include both fatty acid and polyketide synthase gene clusters, and recent evidence suggests that the formation of complex lipids in mycobacteria requires the action of PKSs (KOLATTUKUDY et al., 1997; REINALDO et al., 1999; COX et al., 1999). Finally, there is evidence of interactions between FASs and PKSs in *Streptomyces* bacteria, in particular via the sharing of identical enzymatic subunits (KIM et al., 1994; SUMMERS et al., 1995; REVILL et al., 1996). These examples are likely to be only a few of many interactions that exist between distinct biosynthetic pathways within a single organism.

7.1 FAS/PKS Crosstalk in *Streptomyces*

The *Streptomyces* bacteria are developmentally complex organisms, which form colonies that sporulate and produce antibiotics in response to nutrient limitation. The biosynthetic pathways that lead to antibiotic production have been extensively studied in *Streptomyces*, and have contributed to significant insights into the programing of PKSs (reviewed by HOPWOOD, 1997). Evidence for crosstalk among biosynthetic pathways within *Streptomyces* includes two distinct PKSs that catalyze the synthesis of a spore pigment molecule and the antibiotic actinorhodin. Artificial complementation of any of the spore pigment PKS subunits by the corresponding actinorhodin PKS subunits has been shown (KIM et al., 1994; REVILL et al., 1996). These studies suggested that although the two sets of PKS subunits were capable of biochemical interactions, such crosstalk is prevented during normal growth by the regulated expression of these gene sets during different stages of the *Streptomyces* life cycle. Furthermore, the malonyl acyl transferase (MAT) FAS subunit has been shown to be shared between the two systems (REVILL et al., 1995, SUMMERS et al., 1995). *In vitro* reconstitution of purified actinorhodin PKS genes unambiguously showed that this MAT was absolutely essential for the catalytic turnover (CARRERAS and KHOSLA, 1998). Similarly, purified phosphopantetheinyl transferase from fatty acid synthase was capable of post-translationally modifying Type II PKS acyl carrier proteins (CARRERAS et al., 1997).

7.2 FAS/PKS Crosstalk in *Mycobacterium*

Mycobacteria are classified in the phylogeny of the Actinomycetes, along with the *Streptomyces* bacteria. Interestingly, these two actinomycete genera have received immense attention due to their contrasting effects on human society. Whereas *Streptomyces* have provided a rich source of antibiotics and other therapeutic products for human diseases, *Mycobacterium tuberculosis* and *Mycobacterium leprae* have been two of humankind's greatest scourges (YOUNG and COLE, 1993). Like the streptomycetes, mycobacteria are developmentally complex organisms. They are thought to alter their cell type during infection and pathogenesis, and are capable of achieving a dormant state following a shift to anaerobic conditions *in vitro*, which may be an important feature for infection and pathogenesis (PARRISH et al., 1998). Polyketide macrolides called mycolactones have been recently isolated from an extracellular mycobacterial species, *Mycobacterium ulcerans* (GEORGE et al., 1999). Mycolactones in guinea pigs have been shown to produce similar histopathological changes as *M. ulcerans* infection (GEORGE et al., 1999). Genome sequencing of *M. tuberculosis* and *M. leprae* has identified a number of genes homologous to PKSs (COLE et al., 1998).

Although a polyketide product has not yet been isolated from *M. tuberculosis*, two recent studies have implicated some of its PKS genes in the biosynthesis of complex lipids (REINALDO et al., 1999; COX et al., 1999). JACOBS and coworkers have suggested a pathway for the biosynthesis of phthiocerol dimycocerosate (PDIM), a lipid exclusively found in pathogenic mycobacterial species. PDIM is probably synthesized by the combined action of (1) the PKS-like "*Pps*" enzymes, (2) an iterative mycocerosic acid synthase, *Mas*, (3) at least one acyl-CoA synthase, and (4) membrane transport proteins such as *MmpL7* (COX et al., 1999). Previously, KOLATTUKUDY and coworkers had postulated a similar mechanism for the biosynthesis of the lipids phthiocerol, phenolphthiocerol, and mycoside B (KOLATTUKUDY et al., 1997; AZAD et al., 1997). Together, these studies suggest that the biosynthesis of complex lipids in *M. tuberculosis* is a complex multistep process involving a number of enzyme systems. Furthermore, it is believed that facile crosstalk exists between and within the FASs and PKS-like enzymes, including the shuttling of elongating lipid chains between these enzymatic assemblies (AZAD et al., 1996; FITZMAURICE and KOLATTUKUDY, 1998; COX et al., 1999). However, the mechanisms by which acyl chains are transferred from one enzymatic complex to another are not clear, since all the biosynthetic intermediates are covalently sequestered as acyl thioesters and cannot diffuse between the enzymatic assemblies. There are reports that an acyl-CoA synthase, an enzyme normally involved in fatty acid degradation, may be involved in this process (FITZMAURICE and KOLATTUKUDY, 1998; COX et al., 1999).

The identification of polyketides and PKSs from pathogenic organisms like mycobacteria and the examples of interaction between FASs and PKSs have provided new dimension to this field.

8 Conclusions

Polyketides are produced by sequential reactions catalyzed by a collection of catalytic activities termed polyketide synthase. This enzymatic machinery represents a remarkable example of nature's ways to build complex biomolecules. Although rapid progress has been achieved in our understanding of the programing of polyketide biosynthesis, a better understanding of PKS structure and mechanisms should provide important leads for developing the technology to generate a vast repertoire of bioactive natural products.

Acknowledgements

We thank Ms. PREETI SAXENA for her assistance in drawing figures and proof reading the text.

9 References

ALBERTS, A. W., CHEN, J., KURON, G., HUNT, V., HUFF, J. et al. (1980), Mevinolin: a highly potent competitive inhibitor of hydroxymethylglutaryl-coenzyme A reductase and a cholesterol-lowering agent, *Proc. Natl. Acad. Sci. USA* **77**, 3957–3961.

AMY, C., WITKOWSKI, A., NAGGERT, J., WILLIAMS, B., RANDHAWA, Z., SMITH, S. (1989), Molecular cloning and sequencing of cDNA's encoding the entire rat fatty acid synthase, *Proc. Natl. Acad. Sci. USA* **86**, 3114–3118.

APARICIO, J. F., CAFFREY, P., MARSDEN, A. F., STAUNTON, J., LEADLAY, P. F. (1994), Limited proteolysis and active-site studies of the first multienzyme component of the erythromycin-producing polyketide synthase, *J. Biol. Chem.* **269**, 8524–8528.

AUGUST, P. R., TANG, L., YOON, Y. J., NING, S., MULLER, R. et al. (1998), Biosynthesis of the ansamycin antibiotic rifamycin: deductions from the molecular analysis of the rif biosynthetic gene cluster of *Amycolatopsis mediterranei* S699, *Chem. Biol.* **5**, 69–79.

AZAD, K. A., SIRAKOVA, T. D., FERNANDES, N. D., KOLATTUKUDY, P. E. (1996), Gene knockout reveals a novel gene cluster for the synthesis of a class of cell wall lipids unique to pathogenic *Mycobacteria*, *J. Biol. Chem.* **272**, 16741–16745.

AZAD, A. K., SIRAKOVA, T. D., FERNANDES, N. D., KOLATTUKUDY, P. E. (1997), Gene knockout reveals a novel gene cluster for the synthesis of a class of cell wall lipids unique to pathogenic *Mycobacteria*, *J. Biol. Chem.* **272**, 16741–16745.

BAO, W., WENDT-PIENKOWSKI, E., HUTCHINSON, C. R. (1998), Reconstitution of the iterative type II polyketide synthase for tetracenomycin F2 biosynthesis, *Biochemistry* **37**, 8132–8138.

BARTEL, P. L., ZHU, C. B., LAMPEL, J. S., DOSCH, D. C., CONNORS, N. C. et al. (1990), Biosynthesis of anthraquinones by interspecies cloning of actinorhodin biosynthesis genes in streptomycetes: clarification of actinorhodin gene functions, *J. Bacteriol.* **172**, 4816–4826.

BECK, J., RIPKA, S., SIEGNER, A., SCHILTZ, E., SCHWEIZER, E. (1990), The multifunctional 6-methylsalicylic acid synthase gene of *Penicillium patulum*. Its gene structure relative to that of other polyketide synthases, *Eur. J. Biochem.* **192**, 487–498.

BEDFORD, D. J., SCHWEIZER, E., HOPWOOD, D. A., KHOSLA, C. (1995), Expression of functional fungal polyketide synthase in the bacterium *Streptomyces coelicolor* A3(2), *J. Bacteriol.* **177**, 4544–4548.

BENTLEY, R. (1999), Secondary metabolite biosynthesis: the first century, *Crit. Rev. Biotechnol.* **19**, 1–40.

BENTLEY, R., BENNETT, J. W. (1999), Constructing polyketides: from Collie to combinatorial biosynthesis, *Annu. Rev. Microbiol.* **53**, 411–446.

BIBB, M. J., BIRO, S., MOTAMEDI, H., COLLINS, J. F., HUTCHINSON, C. R. (1989), Analysis of the nucleotide sequence of the *Streptomyces glaucescens tcmI* genes provides key information about the enzymology of polyketide antibiotic biosynthesis, *EMBO J.* **8**, 2727–2736.

BIBB, M. J., SHERMAN, D. H., OMURA, S., HOPWOOD, D. A. (1994), Cloning, sequencing and deduced functions of a cluster of *Streptomyces* genes probably encoding biosynthesis of the polyketide antibiotic frenolicin, *Gene* **142**, 31–39.

BISANG, C., LONG, P. F., CORTES, J., WESTCOTT, J., CROSBY, J. et al. (1999), A chain initiation factor common to both modular and aromatic polyketide synthases, *Nature* **401**, 502–505.

BOHM, I., HOLZBAUR, I. E., HANEFELD, U., CORTES, J., STAUNTON, J., LEADLAY, P. F. (1998), Engineering of a minimal modular polyketide synthase, and targeted alteration of the stereospecificity of polyketide chain extension, *Chem. Biol.* **5**, 407–412.

BROWN, D. W., YU, J. H., KELKAR, H. S., FERNANDES, M., NESBITT, T. C. et al. (1996), Twenty-five coregulated transcripts define a sterigmatocystin gene cluster in *Aspergillus nidulans*, *Proc. Natl. Acad. Sci. USA* **93**, 1418–1422.

CAFFREY, P., BEVITT, D. J., STAUNTON, J., LEADLAY, P. F. (1992), Identification of DEBS 1, DEBS 2 and DEBS 3, the multienzyme polypeptides of the erythromycin-producing polyketide synthase from *Saccharopolyspora erythraea*, *FEBS Lett.* **304**, 225–228.

CARRERAS, C. W., KHOSLA, C. (1998), Purification and *in vitro* reconstitution of the essential protein components of an aromatic polyketide synthase, *Biochemistry* **37**, 2084–2088.

CARRERAS, C., PIEPER, R., KHOSLA, C. (1997), The chemistry and biology of fatty acid, polyketide and non-ribosomal peptide biosynthesis, *Topics Curr. Chem.* **188**, 85–126.

CHANG, S. I., HAMMES, G. G. (1990), Structure and mechanism of action of a multifunctional enzyme: Fatty acid synthase, *Am. Chem. Soc.* **23**, 363–369.

CHUCK, J. A., MCPHERSON, M., HUANG, H., JACOBSEN, J. R., KHOSLA, C., CANE, D. E. (1997), Molecular recognition of diketide substrates by a beta-ketoacyl-acyl carrier protein synthase domain within a bimodular polyketide synthase, *Chem. Biol.* **4**, 757–766.

COLE, S. T., BROSCH, R., PARKHILL, J., GARNIER, T., CHURCHER, C. et al. (1998), Deciphering the biology of *Mycobacterium tuberculosis* from the complete genome sequence, *Nature* **393**, 537–544.

CORTES, J., HAYDOCK, S. F., ROBERTS, G. A., BEVITT, D. J., LEADLAY, P. F. (1990), An unusually large multifunctional polypeptide in the erythromycin-producing polyketide synthase of *Saccharopolyspora erythraea*, *Nature* **348**, 176–178.

CORTES, J., WIESMANN, K. E., ROBERTS, G. A., BROWN, M. J., STAUNTON, J., LEADLAY, P. F. (1995), Repositioning of a domain in a modular polyketide synthase to promote specific chain cleavage, *Science* **268**, 1487–1489.

COX, J. S., CHEN, B., MCNEIL, M., JACOBS, W. R. Jr. (1999), Complex lipid determines tissue specific replication of *Mycobacterium tuberculosis* in mice, *Nature* **402**, 79–83.

DIXON, R. A. (1999), Plant natural products: the molecular genetic basis of biosynthetic diversity, *Curr. Opin. Biotechnol.* **10**, 192–197.

DONADIO, S., STAVER, M. J., MCALPINE, J. B., SWANSON, S. J., KATZ, L. (1991), Modular organization of genes required for complex polyketide biosynthesis, *Science* **252**, 675–679.

DONADIO, S., MCALPINE, J. B., SHELDON, P. J., JACKSON, M., KATZ, L. (1993), An erythromycin analog produced by reprogramming of polyketide synthesis, *Proc. Natl. Acad. Sci. USA* **90**, 7119–7123.

DREIER, J. (2000), Mechanistic analysis of a type II polyketide synthase. Role of conserved residues in the beta-ketoacyl synthase-chain length factor heterodimer, *Biochemistry* **39**, 2088–2095.

DREIER, J., SHAH, A. N., KHOSLA, C. (1999), Kinetic analysis of the actinorhodin aromatic polyketide synthase, *J. Biol. Chem.* **274**, 25108–25112.

FERNANDEZ-MORENO, M. A., MARTINEZ, E., BOTO, L., HOPWOOD, D. A., MALPARTIDA, F. (1992), Nucleotide sequence and deduced functions of a set of cotranscribed genes of *Streptomyces coelicolor* A3(2), including the polyketide synthase for the antibiotic actinorhodin, *J. Biol. Chem.* **267**, 19278–19290.

FERRER, J. L., JEZ, J. M., BOWMAN, M. E., DIXON, R. A., NOEL, J. P. (1999), Structure of chalcone synthase and the molecular basis of plant polyketide biosynthesis, *Nature Struct. Biol.* **6**, 775–784.

FITZMAURICE, A. M., KOLATTUKUDY, P. E. (1997), Cloning, sequencing, and characterization of a fatty acid synthase encoding gene from *Mycobacterium tuberculosis* var. *bovis* BCG, *Gene* **170**, 95–99.

FITZMAURICE, A. N., KOLATTUKUDY, P. E. (1998), An acyl-CoA synthase (acoas) gene adjacent to the mycocerosic acid synthase (*mas*) locus is necessary for mycocerosyl lipid synthesis in *Mycobacterium tuberculosis* var. *bovis* BCG, *J. Biol. Chem.* **273**, 8033–8039.

FLIEGMANN, J., SCHRODER, G., SCHANZ, S., BRITSCH, L., SCHRODER, J. (1992), Molecular analysis of chalcone and dihydropinosylvin synthase from Scots pine (*Pinus sylvestris*), and differential regulation of these and related enzyme activities in stressed plants, *Plant Mol. Biol.* **18**, 489–503.

FUNA, N., OHNISHI, Y., FUJII, I., SHIBUYA, M., EBIZUKA, Y., HORINOUCHI, S. (1999), A new pathway for polyketide synthesis in microorganisms, *Nature* **400**, 897–899.

GEHRING, A. M., LAMBALOT, R. H., VOGEL, K. W., DRUECKHAMMER, D. G., WALSH, C. T. (1997), Ability of *Streptomyces* spp. acyl carrier proteins and coenzyme A analogs to serve as substrates *in vitro* for *E. coli* holo-ACP synthase, *Chem. Biol.* **4**, 17–24.

GEORGE, K. M., CHATTERJEE, D., GUNAWARDANA, G., WETY, D., HAYMAN, J. et al. (1999), Mycolactone: A polyketide toxin from *Mycobacterium ulcerans* required for virulence, *Science* **283**, 854–857.

GOKHALE, R. S., KHOSLA, C. (2000), Role of linkers in communication between protein modules, *Curr. Opin. Chem. Biol.* **4**, 22–27.

GOKHALE, R. S., LAU, J., CANE, D. E., KHOSLA, C. (1998), Functional orientation of the acyltransferase domain in a module of the erythromycin polyketide synthase, *Biochemistry* **37**, 2524–2528.

GOKHALE, R. S., HUNZIKER, D., CANE, D. E., KHOSLA, C. (1999a), Mechanism and specificity of the terminal thioesterase domain from the erythromycin polyketide synthase, *Chem. Biol.* **6**, 117–125.

GOKHALE, R. S., TSUJI, S. Y., CANE, D. E., KHOSLA, C. (1999b), Dissecting and exploiting intermodular communication in polyketide synthases, *Science* **284**, 482–485.

GRIMM, A., MADDURI, K., ALI, A., HUTCHINSON, C. R. (1994), Characterization of the *Streptomyces peucetius* ATCC 29050 genes encoding doxorubicin polyketide synthase, *Gene* **151**, 1–10.

HARBURY, P. B., ZHANG, T., KIM, P. S., ALBER, T. (1993), A switch between two-, three- and four-stranded coiled coils in GCN 4 leucine zipper mutants, *Science* **262**, 1401–1407.

HAYDOCK, S. F., APARICIO, J. F., MOLNAR, I., SCHWECKE, T., KHAW, L. E. et al. (1995), Divergent sequence motifs correlated with the substrate specificity of (methyl), malonyl-CoA : acyl carrier protein transacylase domains in modular polyketide synthases, *FEBS Lett.* **374**, 246–248.

HE, M., VAROGLU, M., SHERMAN, D. H. (2000), Structural modeling and site-directed mutagenesis of the actinorhodin beta-ketoacyl-acyl carrier protein synthase, *J. Bacteriol.* **182**, 2619–2623.

HENDRICKSON, L., DAVIS, C. R., ROACH, C., NGUYEN, D. K., ALDRICH, T. et al. (1999), Lovastatin biosynthesis in *Aspergillus terreus*: characterization of blocked mutants, enzyme activities and a multifunctional polyketide synthase gene, *Chem. Biol.* **6**, 429–439.

HITCHMAN, T. S., CROSBY, J., BYROM, K. J., COX, R. J., SIMPSON, T. J. (1998), Catalytic self-acylation of type II polyketide synthase acyl carrier proteins, *Chem. Biol.* **5**, 35–47.

HOPWOOD, D. (1997), Genetic contributions to understanding polyketide synthases, *Chem. Rev.* **97**, 2465–2497.

HOPWOOD, D. A., SHERMAN, D. H. (1990), Molecular genetics of polyketides and its comparison to fatty acid biosynthesis, *Annu. Rev. Genet.* **24**, 37–66.

HUANG, W., JIA, J., EDWARDS, P., DEHESH, K., SCHNEIDER, G., LINDQVIST, Y. (1998), Crystal structure of beta-ketoacyl-acyl carrier protein synthase II from *E. coli* reveals the molecular architecture of condensing enzymes, *EMBO J.* **17**, 1183–1191.

HUNZIKER, D., YU, T.-W., HUTCHINSON, R., FLOSS, H. G., KHOSLA, C. (1998), Primer unit specificity in rifamycin biosynthesis principally resides in the later stages of the biosynthetic pathway, *J. Am. Chem. Soc.* **120**, 1092–1093.

HUNZIKER, D., WU, N., KENOSHITA, K., CANE, D. E., KHOSLA, C. (1999), Precursor directed biosynthesis of novel 6-deoxyerythronolide B analogs containing non-natural oxygen substituents and reactive functionalities, *Tetrahedron Lett.* **40**, 635–638.

HUTCHINSON, C. R. (1998), Combinatorial biosynthesis for new drug discovery, *Curr. Opin. Microbiol.* **1**, 319–329.

IKEDA, H., OMURA, S. (1998), Combinatorial biosynthesis: engineered biosynthesis of polyketide compounds, *Tanpakushitsu Kakusan Koso* **43**, 1265–1277 (in Japanese).

JACOBSEN, J. R., HUTCHINSON, C. R., CANE, D. E., KHOSLA, C. (1997), Precursor-directed biosynthesis of erythromycin analogs by an engineered polyketide synthase, *Science* **277**, 367–369.

JACOBSEN, J. R., KEATINGE-CLAY, A. T., CANE, D. E., KHOSLA, C. (1998), Precursor-directed biosynthesis of 12-ethyl erythromycin, *Bioorg. Med. Chem.* **6**, 1171–1177.

JEZ, J. M., NOEL, J. P. (2000), Mechanism of chalcone synthase: *p*Ka of the catalytic cysteine and the role of the conserved histidine in a plant polyketide synthase, *J. Biol. Chem.* **275**, 39640– 39646.

JULIEN, B., SHAH, S., ZIERMANN, R., GOLDMAN, R., KATZ, L., KHOSLA, C. (2000), Isolation and characterization of the epothilone biosynthetic gene cluster from *Sorangium cellulosum*, *Gene* **249**, 153–160.

KAO, C. M., LUO, G., KATZ, L., CANE, D., E., KHOSLA, C. (1994), Engineered biosynthesis of a triketide lactone from an incomplete modular polyketide synthase, *J. Am. Chem. Soc.* **116**, 11612–11613.

KAO, C. M., LUO, G., KATZ, L., CANE, D. E., KHOSLA, C. (1995), Manipulation of macrolide ring size by directed mutagenesis of a modular polyketide synthase, *J. Am. Chem. Soc.* **117**, 9105–9106.

KAO, C. M., PIEPER, R., CANE, D. E., KHOSLA, C. (1996), Evidence for two catalytically independent clusters of active sites in a functional modular polyketide synthase, *Biochemistry* **35**, 12363–12368.

KAO, C. M., MCPHERSON, M., MCDANIEL, R., FU, H., CANE, D. E., KHOSLA, C. (1997), Gain of function mutagenesis of the polyketide synthase. 2. Engineered biosynthesis of an eight membered ring tetraketide lactone, *J. Am. Chem. Soc.* **119**, 11339–11340.

KAO, C. M., MCPHERSON, M., MCDANIEL, R., FU, H., CANE, D. E., KHOSLA, C. (1998), Alcohol stereochemistry in polyketide backbone is controlled by the β-ketoreductase domains of modular polyketide synthases, *J. Am. Chem. Soc.* **120**, 2478–2479.

KATZ, L., MCDANIEL, R. (1999), Novel macrolides through genetic engineering, *Med. Res. Rev.* **19**, 543–558.

KEALEY, J. T., LIU, L., SANTI, D. V., BETLACH, M. C., BARR, P. J. (1998), Production of a polyketide natural product in nonpolyketide-producing prokaryotic and eukaryotic hosts, *Proc. Natl. Acad. Sci. USA* **95**, 505–509.

KENNEDY, J., AUCLAIR, K., KENDREW, S. G., PARK, C., VEDERAS, J. C., HUTCHINSON, C. R. (1999), Modulation of polyketide synthase activity by accessory proteins during lovastatin biosynthesis, *Science* **284**, 1368–1372.

KHOSLA, C., GOKHALE, R. S., JACOBSEN, J. R., CANE, D. E. (1999), Tolerance and specificity of polyketide synthases, *Annu. Rev. Biochem.* **68**, 219–253.

KIM, E.-S., HOPWOOD, D. A., SHERMAN, D. H. (1994), Analysis of type II polyketide β-ketoacyl synthase specificity in *Streptomyces coelicolor* A3(2) by *trans* complementation of actinorhodin synthase mutants, *J. Bacteriol.* **176**, 1801–1804.

KOLATTUKUDY, P. E., FERNANDES, N. D., AZAD, A. K., FITZMAURICE, A. M., SIRAKOVA, T. D. (1997), Biochemistry and molecular genetics of cell-wall lipid biosynthesis in Mycobacteria, *Mol. Microbiol.* **24**, 263–270.

LAL, R., KUMARI, R., KAUR, H., KHANNA, R., DHINGRA, N., TUTEJA, D. (2000), Regulation and manipulation of the gene clusters encoding type-I PKSs, *Trends Biotechnol.* **18**, 264–274.

LAMBALOT, R. H., GEHRING, A. M., FLUGEL, R. S., ZUBER, P., LACELLE, M. et al. (1996), A new enzyme superfamily – the phosphopantetheinyl transferases, *Chem. Biol.* **3**, 923–936.

LANZ, T., TROPF, S., MARNER, F. J., SCHRODER, J., SCHRODER, G. (1991), The role of cysteines in polyketide synthases. Site-directed mutagenesis of resveratrol and chalcone synthases, two key enzymes in different plant-specific pathways, *J. Biol. Chem.* **266**, 9971–9976.

LAU, J., FU, H., CANE, D. E., KHOSLA, C. (1999), Dissecting the role of acyltransferase domains of modular polyketide synthases in the choice and stereochemical fate of extender units, *Biochemistry* **38**, 1643–1651.

LAU, J., CANE, D. E., KHOSLA, C. (2000), Substrate specificity of the loading didomain of the erythromycin polyketide synthase, *Biochemistry* **39**, 10514–10520.

LEADLAY, P. F. (1997), Combinatorial approaches to polyketide biosynthesis, *Curr. Opin. Chem. Biol.* **2**, 162–168.

LUKACIN, R., SPRINGOB, K., URBANKE, C., ERNWEIN, C., SCHRODER, G. et al. (1999), Native acridone synthases I and II from *Ruta graveolens* L. form homodimers, *FEBS Lett.* **448**, 135–140.

MARAHIEL, M., STACHELHAUS, T., MOOTZ, H. D. (1997), Modular peptide synthases involved in non-ribosomal peptide synthesis, *Chem. Rev.* **97**, 2651–2673.

MARSDEN, A. F., CAFFREY, P., APARICIO, J. F., LOUGHRAN, M. S., STAUNTON, J., LEADLAY, P. F. (1994), Stereospecific acyl transfers on the erythromycin-producing polyketide synthase, *Science* **263**, 378–380.

MARSDEN, A. F., WILKINSON, B., CORTES, J., DUNSTER, N. J., STAUNTON, J., LEADLAY, P. F. (1998), Engineering broader specificity into an antibiotic-producing polyketide synthase, *Science* **279**, 199–202.

MARTIN, J. F., DEMAIN, D. L. (1978), *The Filamentous Fungi*. London: Edward Arnold.

MATHARU, A. L., COX, R. J., CROSBY, J., BYROM, K. J., SIMPSON, T. J. (1998), MCAT is not required for *in vitro* polyketide synthesis in a minimal actinorhodin polyketide synthase from *Streptomyces coelicolor*, *Chem. Biol.* **5**, 699–711.

MATHIEU, M., ZEELEN, J. P., PAUPTIT, R. A., ERDMANN, R., KUNAU, W. H., WIERENGA, R. K. (1994), The 2.8 Å crystal structure of peroxisomal 3-ketoacyl CoA thiolase of *Saccharomyces cerevisiae*, *Structure* **2**, 797–808.

MCDANIEL, R., EBERT-KHOSLA, S., HOPWOOD, D. A., KHOSLA, C. (1993), Engineered biosynthesis of novel polyketides, *Science* **262**, 1546–1550.

MCDANIEL, R., EBERT-KHOSLA, S., HOPWOOD, D. A., KHOSLA, C. (1995), Rational design of aromatic polyketide natural products by recombinant assembly of enzymatic subunits, *Nature* **375**, 549–554.

MCDANIEL, R., KAO, C. M., HWANG, S. J., KHOSLA, C. (1997), Engineered intermodular and intramodular polyketide synthase fusions, *Chem. Biol.* **4**, 667–674.

MCDANIEL, R., THAMCHAIPENET, A., GUSTAFSSON, C., FU, H., BETLACH, M., ASHLEY, G. (1999), Multiple genetic modifications of the erythromycin polyketide synthase to produce a library of novel "unnatural" natural products, *Proc. Natl. Acad. Sci. USA* **96**, 1846–1851.

MCPHERSON, M., KHOSLA, C., CANE, D. E. (1998), Erythromycin biosynthesis: The β-ketoreductase domains catalyze the stereospecific transfer of the 4-pro-S hydride of NADPH, *J. Am. Chem. Soc.* **13**, 3267–3268.

O'HAGAN, D. (1991), *The Polyketide Metabolites*. Chichester: Ellis Horwood.

O'HAGAN, D. (1993), Biosynthesis of fatty acid and polyketide metabolites, *Nat. Prod. Rep.* **10**, 593–624.

OLIYNYK, M., BROWN, M. J., CORTES, J., STAUNTON, J., LEADLAY, P. F. (1996), A hybrid modular polyketide synthase obtained by domain swapping, *Chem. Biol.* **3**, 833–839.

PANIEGO, N. B., ZUURBIER, K. W., FUNG, S. Y., VAN DER HEIJDEN, R., SCHEFFER, J. J., VERPOORTE, R. (1999), Phlorisovalerophenone synthase, a novel polyketide synthase from hop (*Humulus lupulus* L.), cones, *Eur. J. Biochem.* **262**, 612–616.

PARRISH, N. M., DICK, J. D., BISHAI, W. R. (1998) Mechanisms of latency in *Mycobacterium tuberculosis*, *Trends Microbiol.* **6**, 107–112.

PIEPER, R., LUO, G., CANE, D. E., KHOSLA, C. (1995), Cell-free synthesis of polyketides by recombinant erythromycin polyketide synthases, *Nature* **378**, 263–266.

PIEPER, R., EBERT-KHOSLA, S., CANE, D., KHOSLA, C. (1996), Erythromycin biosynthesis: kinetic studies on a fully active modular polyketide synthase using natural and unnatural substrates, *Biochemistry* **35**, 2054–2060.

PIEPER, R., GOKHALE, R. S., LUO, G., CANE, D. E., KHOSLA, C. (1997), Purification and characterization of bimodular and trimodular derivatives of the erythromycin polyketide synthase, *Biochemistry* **36**, 1846–1851.

RANGANATHAN, A., TIMONEY, M., BYCROFT, M., CORTES, J., THOMAS, I. P. et al. (1999), Knowledge-based design of bimodular and trimodular polyketide synthases based on domain and module swaps: a route to simple statin analogues, *Chem. Biol.* **6**, 731–741.

REINALDO, L., ENSERGUEIX, D., PEREZ, E., GICQUEL, B., GUILHOT, C. (1999), Identification of a virulence gene cluster of *Mycobacterium tuberculosis* by signature-tagged transposon mutagenesis, *Mol. Microbiol.* **34**, 257–267.

REVILL, W. P., BIBB, M. J., HOPWOOD, D. (1995), Purification of malonyltransferase from *Streptomyces coelicolor* A3(2) and analysis of its genetic determinant, *J. Bacteriol.* **177**, 3946–3952.

REVILL, W. P., BIBB, M. J., HOPWOOD, D. (1996), Relationship between fatty and polyketide synthases from *Streptomyces coelicolor* A3(2): Characterization of the fatty acid synthase acyl carrier protein, *J. Bacteriol.* **178**, 5660–5667.

RIGGS, S. D., SMITHGALL, T. E. (1999), SH2-kinase linker mutations release Hck tyrosine kinase and transforming activities in rat-2 fibroblasts, *J. Biol. Chem.* **274**, 26579–26583.

RUAN, X., STASSI, D., LAX, S. A., KATZ, L. (1997), A second type-I PKS gene cluster isolated from *Streptomyces hygroscopicus* ATCC 29253, a rapamycin-producing strain, *Gene* **203**, 1–9.

SCHINDLER, T., SICHERI, F., PICO, A., GAZIT, A., LEVITZKI, J. (1999), Crystal structure of Hck in complex with Src family-selective tyrosine kinase inhibitor, *J. Mol. Cell* **5**, 639–648.

SCHRODER, J. (1999), Probing plant polyketide synthesis, *Nature Struct. Biol.* **6**, 714–716.

SCHRODER, J., SCHRODER, G. (1990), Stilbene and chalcone synthases: related enzymes with key functions in plant-specific pathways, *Z. Naturforsch.* [C] **45**, 1–8.

SCHUERMANN, M., HUNTER, J. B., HENNIG, G., MULLER, R. (1991), Non-leucine residues in the leucine repeats of Fos and Jun contribute to the stability and determine the specificity of dimerization, *Nucleic Acids Res.* **19**, 739–746.

SCHUPP, T., TOUPET, C., ENGEL, N., GOFF, S. (1998), Cloning and sequence analysis of the putative rifamycin polyketide synthase gene cluster from *Amycolatopsis mediterranei*, *FEMS Microbiol. Lett.* **159**, 201–207.

SCHWECKE, T., APARICIO, J. F., MOLNAR, I., KONIG, A., KHAW, L. E. et al. (1995), The biosynthetic gene cluster for the polyketide immunosuppressant rapamycin, *Proc. Natl. Acad. Sci. USA* **92**, 7839–7843.

SMITH, S. (1994), The animal fatty acid synthase: one gene, one polypeptide, seven enzymes, *FASEB J.* **8**, 1248–1258.

SPAINK, H. P., SHEELEY, D. M., VAN BRUSSEL, A. A., GLUSHKA, J., YORK, W. S. et al. (1991), A novel highly unsaturated fatty acid moiety of lipo-oligosaccharide signals determines host specificity of *Rhizobium*, *Nature* **354**, 125–130.

SPENCER, J. B., JORDAN, P. M. (1992), Purification and properties of 6-methylsalicylic acid synthase from *Penicillium patulum*, *Biochem. J.* **288**, 839–846.

STAUNTON, J., WILKINS, B. (1997), Biosynthesis of erythromycin and rapamycin, *Chem Rev.* **97**, 2611–2629.

STAUNTON, J., CAFFREY, P., APARICIO, J. F., ROBERTS, G. A., BETHELL, S. S., LEADLAY, P. F. (1996), Evidence for a double-helical structure for modular polyketide synthases, *Nature Struct. Biol.* **3**, 188–192.

SUH, D. Y., KAGAMI, J., FUKUMA, K., SANKAWA, U. (2000), Evidence for catalytic cysteine-histidine dyad in chalcone synthase, *Biochem. Biophys. Res. Commun.* **275**, 725–730.

SUMMERS, R. G., ALI, A., SHEN, B., WESSEL, W. A., HUTCHINSON, C. R. (1995), Malonyl-coenzyme A: Acyl carrier protein acyltransferase of *Streptomyces glaucescens*: A possible link between fatty acid and polyketide synthesis, *Biochemistry* **34**, 9389–9402

TANG, L., FU, H., MCDANIEL, R. (2000), Formation of functional heterologous complexes using subunits from the picromycin, erythromycin and oleandomycin polyketide synthases, *Chem. Biol.* **7**, 77–84.

TANG, L., WANG, Y. G., ZHU, X. W. (1991), Cloning and expression of spiramycin polyketide synthase genes and resistance genes from *S. spiramyceticus* U-1941, *Chin. J. Biotechnol.* **7**, 33–42.

TANG, L., FU, H., BETLACH, M. C., MCDANIEL, R. (1999), Elucidating the mechanism of chain termination switching in the picromycin/methymycin polyketide synthase, *Chem. Biol.* **6**, 553–558.

TRAUGER, J. W., KOHLI, R. M., MOOTZ, H. D., MARAHIEL, M. A., WALSH, C. T. (2000), Peptide cyclization catalysed by the thioesterase domain of tyrocidine synthetase, *Nature* **407**, 215–218.

TROPF, S., KARCHER, B., SCHRODER, G., SCHRODER, J. (1995), Reaction mechanisms of homodimeric plant polyketide synthase (stilbenes and chalcone synthase), A single active site for the condensing reaction is sufficient for synthesis of stilbenes, chalcones, and 6′-deoxychalcones, *J. Biol. Chem.* **270**, 7922–7928.

TSUKAMOTO, N., CHUCK, J. A., LUO, G., KAO, C. M., KHOSLA, C., CANE, D. E. (1996), 6-deoxy-erythronolide B synthase 1 is specifically acylated by a diketide intermediate at the beta-ketoacyl-acyl carrier protein synthase domain of module 2, *Biochemistry* **35**, 15244–15248.

WAKIL, S. J. (1989), Fatty acid synthase, a proficient multifunctional enzyme, *Biochemistry* **28**, 4523–4530.

WEISMANN, K. J., CORTES, J., BROWN, M. J., CUTTER, A. L., STAUNTON, J., LEADLAY, P. F. (1995), Polyketide synthesis *in vitro* on a modular polyketide synthase, *Chem. Biol.* **2**, 583–589.

WEISMAN, K. J., BYCROFT, M., STAUNTON, J., LEADLAY, P. F. (1998), Origin of starter units for erythromycin biosynthesis, *Biochemistry* **37**, 11012–11017.

WITKOWSKI, A., RANGAN, V. S., RANDHAWA, Z. I., AMY, C. M., SMITH, S. (1991), Structural organization of the multifunctional animal fatty acid synthase, *Eur. J. Biochem.* **198**, 571–579.

WU, K., CHUNG, L., REVILL, W. P., KATZ, L., REEVES, C. D. (2000a), The FK520 gene cluster of *Streptomyces hygroscopicus* var. *ascomyceticus* (ATCC 14891), contains genes for biosynthesis of unusual polyketide extender units, *Gene* **251**, 81–90.

WU, N., KUDO, F., CANE, D. E., KHOSLA, C. (2000b), Analysis of the molecular recognition features of individual modules derived from the erythromycin polyketide synthase, *J. Am. Chem. Soc.* **122**, 4847–4852.

Xu, W., Doshi, A., Lei, M., Eck, M. J., Harrison, S. C. (1999), Crystal structures of c-Src reveal features of its autoinhibitory mechanism, *Mol. Cell* **5**, 629–638.

Xue, Q., Ashley, G., Hutchinson, C. R., Santi, D. V. (1999), A multiplasmid approach to preparing large libraries of polyketides, *Proc. Natl. Acad. Sci. USA* **96**, 11740–11745.

Yamaguchi, T., Kurosaki, F., Suh, D. Y., Sankawa, U., Nishioka, M. et al. (1999), Cross-reaction of chalcone synthase and stilbene synthase overexpressed in *Escherichia coli*, *FEBS Lett.* **460**, 457–461.

Young, D. B., Cole, S. T. (1993), Leprosy, tuberculosis and the new genetics, *J. Bacteriol.* **175**, 1–6.

Yu, T. W., Shen, Y., Doi-Katayama, Y., Tang, L., Park, C. et al. (1999), Direct evidence that the rifamycin polyketide synthase assembles polyketide chains processively, *Proc. Natl. Acad. Sci. USA* **96**, 9051–9056.

Zawada, R. J., Khosla, C. (1997), Domain analysis of the molecular recognition features of aromatic polyketide synthase subunits, *J. Biol. Chem.* **272**, 16184–16188.

Zawada, R. J., Khosla, C. (1999), Heterologous expression, purification, reconstitution and kinetic analysis of an extended type II polyketide synthase, *Chem. Biol.* **6**, 607–615.

13 Biotechnological Production of Terpenoid Flavor and Fragrance Compounds

JENS SCHRADER
Frankfurt am Main, Germany

RALF G. BERGER
Hannover, Germany

1 Introduction

Terpene hydrocarbons and their oxyfunctionalized derivatives, the terpenoids, are the most diverse class of substances in nature (HILL, 1993). Whereas terpenoids are extensively applied in industry as flavor and fragrance compounds, the hydrocarbons are separated from their natural sources (essential oils): not only do these substances generally contribute little to flavor and fragrance, but by autoxidation and polymerization they also cause undesirable off-flavors and precipitations.

In the flavor and fragrance industry terpenoids are now used in a large number of perfumery and cosmetics products (particularly flowery fragrances), foodstuffs (citrus and peppermint flavors), and also in many household and pharmaceutical products. In view of the increasing amount of knowledge that is now accumulating on the supplementary bioactivities of terpenoids, it may be assumed that research and development of this class of substances will receive further impetus (see Sect. 4). The essential oil content of plants is, however, low, with concentrations of <0.1 to 5%, and the oils are complex. Commercial extraction of minor compounds is only in rare cases economically viable. To cover demand, >80% of flavor and fragrance substances worldwide, which includes many terpenoids, are chemically synthesized (LUCKNER, 1990). Terpene hydrocarbons, e.g., (+)-limonene and the pinenes, are the main components of some essential oils which occur in large quantities (e.g., >90% in orange oil), and are, thus, ideal starter substances for the chemical synthesis of terpenoids. Both the flavor and the fragrance sectors, however, are interested in replacing chemically synthesized ("nature identical") substances by biotechnologically produced (natural) ones. On the one hand chemical syntheses, requiring the use of environmentally harmful reagents and process conditions could, therefore, either be partly or totally substituted. On the other hand the advantages of biocatalysis permit the generation of target products which cannot be achieved by chemical means or only at great expense (see Tab. 1).

Microorganisms (co-)catabolize terpene carbohydrates to a large number of commercially interesting flavor and fragrance substances, as documented by numerous scientific publications on this topic during the past decades. Besides individual types of bacteria that degrade terpenes, e.g., *Pseudomonas*, *Rhodococcus*, *Bacillus*, and many deuteromycetes, e.g., aspergilli and penicillia, it is above all the higher fungi of the ascomycetes and basidiomycetes which have a marked capacity for *de novo* biosynthesis and for the biotransformation and bioconversion of terpene precursors.

The incidence of a transformation reaction of a multifunctional precursor depends on the species of microorganism and its cultivation conditions. But even different strains of the same species cultivated under the same conditions can produce different metabolites and

Tab. 1. Advantages of Natural Flavor and Fragrance Compounds and their Biocatalytic Synthesis

Natural Flavor and Fragrance Compounds	Biocatalysis
Higher product quality (enantiomeric purity, less by-products)	high specificities (chemo, regio, stereo)
Environmentally friendly, sustainable production (renewable raw materials, low energy consumption)	ecologically compatible agents and reaction conditions
Better accepted by the consumer (increased demand for "natural" products)	multistep syntheses "in one" possible (whole cell bioconversion)
Special or "new" products not, or only at great expense, accessible by chemical means	functionalization of chemically inert carbons, selective modification of one functional group in multifunctional molecules

show different time profiles of product concentration (ABRAHAM et al., 1995). Therefore, the transformation site in the terpenoid molecule cannot seriously be predicted and has to be investigated empirically in most cases, as illustrated by KRASNOBAJEW (1984) in the First Edition of *Biotechnology*. Nevertheless, in terpenoid transformation a manageable number of enzyme reactions are frequently found due to the uniform basic terpenoid structure which derives from the general terpene biosynthesis principle based on five-carbon (isoprene) units (LITTLE and CROTEAU, 1999). Reaction sites which are highly susceptible to chemical reagents are also easily attacked by microorganisms. Enzyme systems are able to regioselectively transform multifunctional terpenoid substrates at specific sites under mild conditions, whereas chemical processes more often suffer from too stringent reaction parameters. Furthermore, even non-activated, chemically inert carbon atoms can be functionalized by enzymatic reactions. The enzymes and corresponding reactions involved in the microbial transformation of monoterpenoids can serve as model systems for the illustration of terpenoid bioconversion in general and have been extensively compiled recently (VAN DER WERF et al., 1997). Tab. 2 gives a brief summary of this compilation and presents some examples of reactions illustrated in this chapter.

In the field of microbial terpenoid production, basic microbiological and biotechnological methods, such as screening of producing strains, media optimization, and cultivation techniques, do not significantly differ from standard procedures. It suffices, therefore, to refer to the general, bioprocessing-oriented volumes of the 2nd Edition of *Biotechnology* (Volumes 1, 3, and 9). However, among the approximately 60 to 100 flavor and fragrance compounds produced biotechnologically on an industrial scale, there are few terpenoids despite their often unique organoleptic properties, the growing demand, and the unstable supply situation from the classical (frequently

Tab. 2. Microbial Transformation Reactions Involved in the Bioconversion of Terpenes and Terpenoids[a] (adapted from VAN DER WERF et al., 1997)

Reaction	Enzyme	Exemplary Reactions Involving this Type of Enzyme
Hydroxylation	monooxygenase	**(12)** → **(17)** **(40)** → **(113)**
Epoxidation	monooxygenase	**(2)** → **(87)** **(21)** → **(25)** **(12)** → **(98)**
Interconversion of alcohols to aldehydes/ketones	alcohol dehydrogenase	**(1)** → **(2)** **(112)** → **(40)**
Hydrogenation of unsaturated double bonds	reductase	**(1)** → **(4)**
Lactonization	monooxygenase catalyzing Baeyer-Villiger reaction, lactonase	**(12)** → **(15)** **(57)** → **(58)**
Hydrolysis and synthesis of monoterpenyl esters	lipase, esterase	**(39)** → **(40)** **(90)** → **(115)**
Ring opening and cyclization	lyase/synthase and not characterized enzyme	**(45)** → **(48, 52, 117, 118)** **(12)** → **(99–102)**
Hydration of unsaturated double bonds	hydratase	**(21)** → **(38)** **(105)** → **(107)**
Allylic rearrangement	not characterized	**(22)** → **(36)** **(94)** → **(12)**
Racemization	racemase, epimerase	–

[a] The numbers refer to the corresponding molecules depicted in the figures of this chapter.

overseas) sources. The main reasons lie in the chemical nature of the substrates and the target molecules: the low water solubility, high volatility and cytotoxicity of terpenes and terpenoids impede conventional bioprocesses. Moreover, due to microbial metabolic versatility one precursor molecule is very often converted into a wide variety of derivatives, partly representing useless or not readily separable by-products. For this reason, wherever possible this contribution stresses process features that reduce or overcome the limitations mentioned and could thus serve as prototypes for the bioconversion of terpenes.

Furthermore, this chapter does not report on *de novo* syntheses, but only on bioconversions and biotransformations of terpenes and terpenoids as precursors as they alone permit industrially interesting yields of individual target products which dominate the product spectrum and, therefore, facilitate the downstream processing. As a consequence, priority is given to work on organisms, such as bacteria and fungi, which are fast-growing and easy to process by biotechnological means or, in the case of esterifications, to systems based on enzymes. The enzymatic release of terpenoids from terpenyl glucoside precursors by hydrolysis, which is a further commercially interesting field of application, is not reported in this chapter, as the aim of hydrolysis is primarily the flavor enrichment of complex fermentation products from plant raw materials and not the production of a distinct natural aroma compound (VASSEROT et al., 1995; GUEGEN et al., 1997, 1998; GANGA et al., 1999).

Plant cell cultures, on account of their metabolic and physiological proximity to the natural process of aroma synthesis, have great potential, particularly in this application area. *In vitro* cultured plant cells may provide valuable knowledge on genetics and metabolism, but hitherto no industrially relevant yields of individual flavors have been reported. Therefore, in this contribution basic research results are mentioned only in cases where interesting aroma compounds could be generated that are potentially of commercial interest and have not previously – or only in rare instances – been cited.

Section 2 of this chapter gives an overview of previous research activities by summarizing the chapter "Terpenoids" of the First Edition of *Biotechnology* (see Tab. 3) and compiling a table of selected works from the mid-1980s to today (see Tab. 4). Section 3 presents more recent research activities classified by terpenoid structure, while an outlook on future developments concludes this contribution (Sect. 4).

2 Previous Research Activities

Tab. 3 summarizes the chapter "Terpenoids" published in the First Edition of *Biotechnology* (KRASNOBAJEW, 1984), with a focus on the microbial generation of mono- and sesquiterpenoids with flavor and fragrance impact.

The wish to modify and diversify terpenoid structures by biocatalysis may arise from biochemical, microbiological, hygienic, genetic or bioprocessing aspects, or others. As a result, information in this area is scattered over many books and journals from quite different fields. Since the mid-1980s a considerable number of monographs, multi-authored works, and conference proceedings have focused on the biotechnological production of flavors and aromas (Tab. 4). All of these publications include chapters on terpene biochemistry and biogeneration which will guide the interested reader. From the books selected a wealth of references to original work and reviews on biotransformation can be derived, and also more general information on analysis, occurrence, precursors, and commercial aspects of flavors. Many of these data can be transferred and applied to bioprocesses for terpenoid flavors as well.

Tab. 3. Selection of Terpenoid Bioconversions and Biotransformations with Flavor and Fragrance Impact Cited in a Review of V. KRASNOBAJEW, published in the 1st Edition of this Series (KRASNOBAJEW, 1984)

Precursor(s) / Reference	Product(s) / Remarks				
1 Citronellal	2 Citronellol	3 Citronellic acid	4 Dihydro-citronellol	5 3,7-Dimethyl-1,7-octane-diol	6 Menthol
JOGLEKAR, S. S., DHAVALIKAR, R. S. (1969), *Appl. Microbiol.* **18**, 1084.	*Pseudomonas aeruginosa*; 0.35% (w/w) (**1**); yields: 65% (**3**), 1.7% (**5**), 0.75% (**6**), 0.6% (**2**), 0.6% (**4**); after 96 h 20% of (**1**) remained unreacted				
Takasago Perfumery Co. Ltd. (1970). *Jpn. Patent* 73 161 91.	*Candida reukaufii* AHU 3032; enantioselective reduction of (−)-(**1**) in racemic (**1**) to (−)-(**2**) with an optical purity of more than 80% after several hours				
YAMAGUCHI, Y., KOMATSU, A., MOROE, T. (1976), *J. Agric. Chem. Soc. (Japan)* **50**, 443.	diverse yeasts; stereoselective reduction of (±)-(**1**) to (−)-(**2**); efficient regeneration of stoichiometrically necessary cofactors for microbial redox systems will be key to industrial success				
BABIČKA, J., VOLF, J. (1955), *Czech. Patent* 84 320.	*Penicillium digitatum* Saccardo var. *italica*; selective cyclization of (**1**) to (**6**); yield: 93%; isopulegol and pulegol as potential intermediates; surface culture; incubation time 28 d; no determination of optical purity				
7 Citral	8 Geranic acid	9 2,6-Dimethyl-8-hydroxy-7-oxo-2-octene	10 6-Methyl-5-heptanoic acid	11 3-Methyl-2-butenoic acid	
JOGLEKAR, S. S., DHAVALIKAR, R. S. (1969). *Appl. Microbiol.* **18**, 1084.	*Pseudomonas aeruginosa*; 0.35% (w/w) (**7**); yields: 62% (**8**), 1% (**11**), 0.8% (**9**), 0.5% (**10**)				

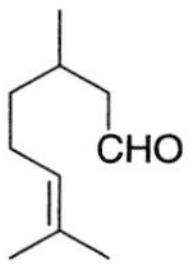

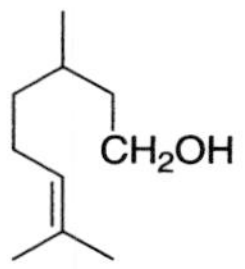

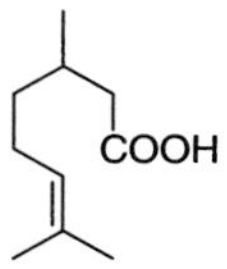

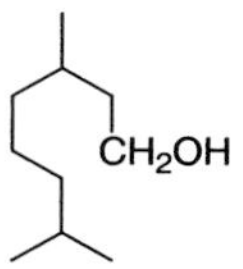

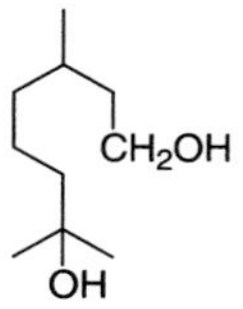

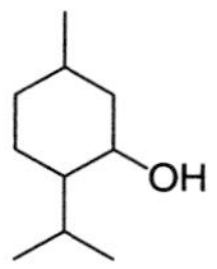

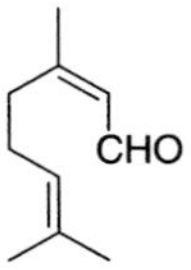

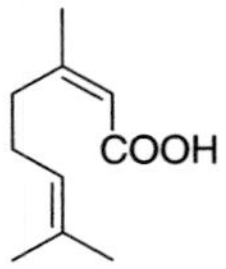

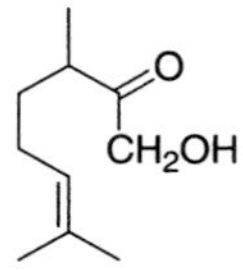

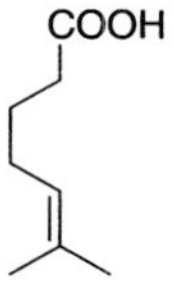

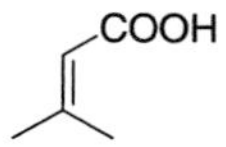

Tab. 3. Continued

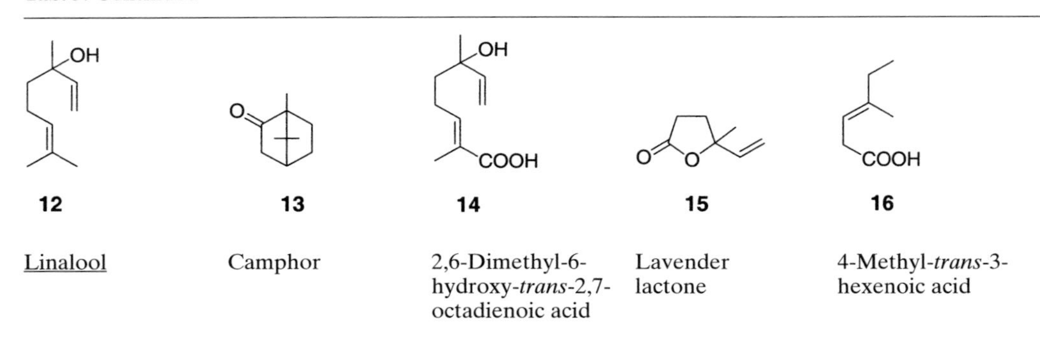

12	13	14	15	16
Linalool	Camphor	2,6-Dimethyl-6-hydroxy-*trans*-2,7-octadienoic acid	Lavender lactone	4-Methyl-*trans*-3-hexenoic acid

MIZUTANI, S., HAYASHI, T., UEDA, H., TATSUMI, C. (1971), *J. Agric. Chem. Soc. (Japan)* **45**, 368. — *Pseudomonas mallei*; conversion of (**12**) yielded a mixture of the metabolites (**13**), (**14**), (**15**), and (**16**)

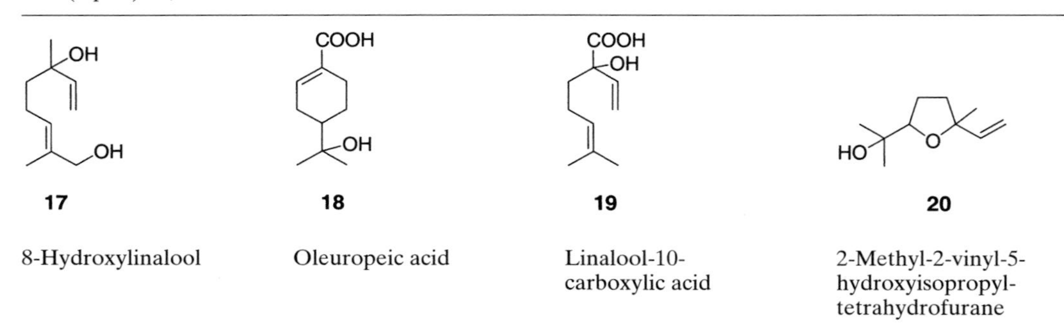

17	18	19	20
8-Hydroxylinalool	Oleuropeic acid	Linalool-10-carboxylic acid	2-Methyl-2-vinyl-5-hydroxyisopropyl-tetrahydrofurane

DEVI, J. R., BHATTACHARYYA, P. K. (1977), *Indian J. Biochem. Biophys.* **14**, 359. — *Pseudomonas incognita*; isolated by enrichment culture with (**12**) as sole carbon source; conversion of (**12**) yielded a mixture of the metabolites (**14**), (**15**), (**17**), (**18**), (**19**), and (**20**)

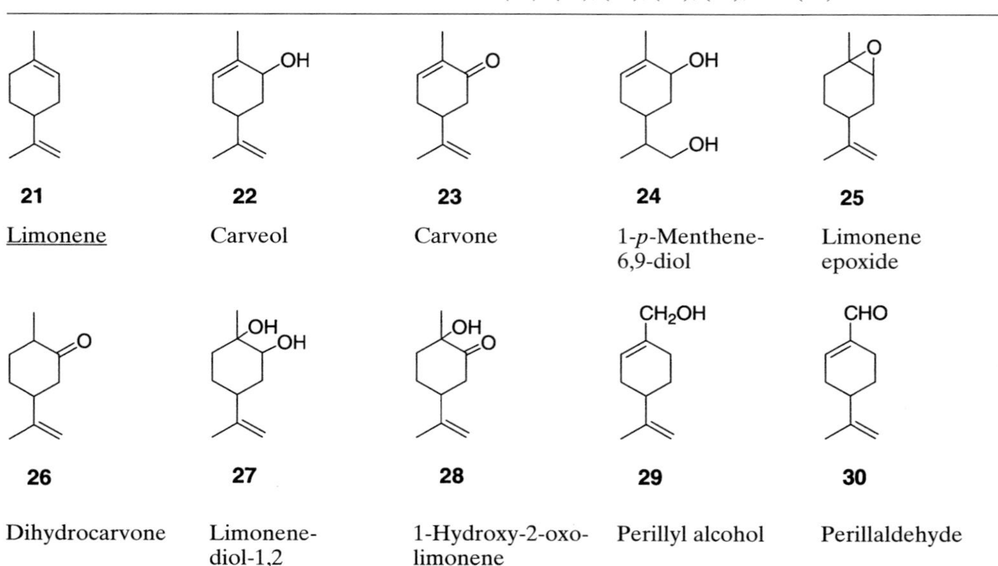

21	22	23	24	25
Limonene	Carveol	Carvone	1-*p*-Menthene-6,9-diol	Limonene epoxide

26	27	28	29	30
Dihydrocarvone	Limonene-diol-1,2	1-Hydroxy-2-oxo-limonene	Perillyl alcohol	Perillaldehyde

Tab. 3. Continued

31	32	33	34	35
Perillic acid	4,9-Dihydroxy-1-*p*-menthene-7-oic acid	2-Hydroxy-8-*p*-menthene-7-oic acid	2-Oxo-8-*p*-menthene-7-oic acid	Isopropenyl-pimelic acid

Reference	Remarks
DHAVALIKAR, R. S., RANGACHARI, P. N., BHATTACHARYYA, P. K. (1966), *Indian J. Biochem.* **3**, 158.	Proposal of three limonene degradation pathways with a soil pseudomonad, isolated by enrichment culture on (**21**); from (**21**) to both (+)-(**23**) and (**24**) via (+)-*cis*-(**22**); from (**21**) to (+)-(**26**) via (**25**) and to (**28**) via (**27**); main pathway from (**21**) to (**29**) to (**30**) to (**31**), further degradation from (**31**) to (**32**) and from (**31**) to (**35**) via (**33**) and (**34**)

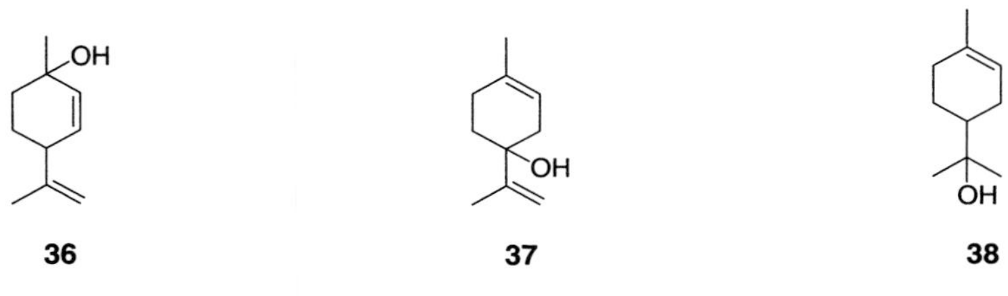

36	37	38
p-Mentha-2,8-diene-1-ol	*p*-Mentha-1,8-diene-4-ol	α-Terpineol

Reference	Remarks
BOWEN, E. R. (1975), *Proc. Fla. State Hortic. Soc.* **88**, 304.	*Penicillium digitatum* and *P. italicum*; isolated from overripe oranges by enrichment culture with 1% (**21**) as sole carbon source; inhibiting effect above 0.5% (v/v) (**21**), both fungi generated the same composition of metabolites; *P. italicum* yielded 26% *cis/trans*-(**22**), 18% *cis/trans*-(**36**), 6% (**23**), 4% (**37**), 3% (**29**), and 3% (**27**)
MUKHERJEE, B. B., KRAIDMAN, G., HILL, I. D. (1973), *Appl. Microbiol.* **25**, 447.	*Cladosporium* sp.; one main transformation product from (**21**): 1.5 g *trans*-(**27**) and 0.2 g *cis*-(**27**) from 1 L growth medium
KRAIDMAN, G., MUKHERJEE, B. B., HILL, I. D. (1969), *Bacteriol. Proc.* **69**, 63.	*Cladosporium* sp.; biotransformation of (+)-(**21**) to 1 $g\ L^{-1}$ (**38**) in a mineral medium with (+)-(**21**) as sole carbon source

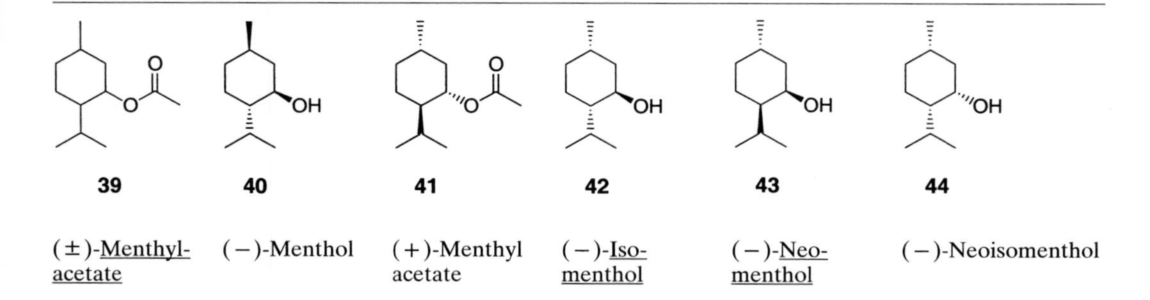

39	40	41	42	43	44
(±)-Menthyl-acetate	(−)-Menthol	(+)-Menthyl acetate	(−)-Iso-menthol	(−)-Neo-menthol	(−)-Neoisomenthol

Tab. 3. Continued

MOROE, T., HATTORI, S., KOMATSU, A., YAMAGUCHI, Y. (1971). (Takasago Perfumery Co.) *W. Ger. Patent* DT 2 036 875. Jap. prior. 1969, also *U.S. Patent* 3 607 651 (1971).	*Trichoderma reesei* AHU 9 485; approximately 20 L fermentation; medium: 2% glucose, 0.5% KH_2PO_4, 0.7% yeast extract; airflow: 5 L min^{-1}; 24 h incubation at 27 °C and 300 r.p.m; dosage of 200 g of the acetates of a racemic (±)-mixture consisting of the isomers (**40**), (**42**), and (**43**) (of which the (−)-isomers are depicted above together with the fourth menthol-isomer (**44**)); incubation for additional 24 h; product isolation by steam distillation, evaporation, chromatography, esterification, recrystallization, and hydrolysis yielded approximately 23 g of pure (**40**)
YAMAGUCHI, Y., KOMATSU, A., MOROE, T. (1977), *J. Agric. Chem. Soc.* (*Japan*) **51**, 411.	*Rhodotorula mucilaginosa* AHU 3 243; asymmetric hydrolysis of (**39**) on industrial scale, high optical selectivity and stereospecifity; 70% increase of esterase activity by UV-induced mutagenesis; 30% (**39**) per liter culture medium yielded 44.4 g of (**40**) after 24 h; detailed study of parameters affecting hydrolysis
WATANABE, Y., INAGAKI, T. (1977), (Nippon Terpene Chemical Co., Ltd.), *Japan. Patent* 77 122 690.	*Arginomonas non-fermentans* FERM-P-1 924; patent describing the large-scale biochemical production of (**40**) by asymmetric hydrolysis of the monochloroacetates of (±)-menthol; increase of reaction velocity and decrease of wastewater by using the α-halo carboxylic acid

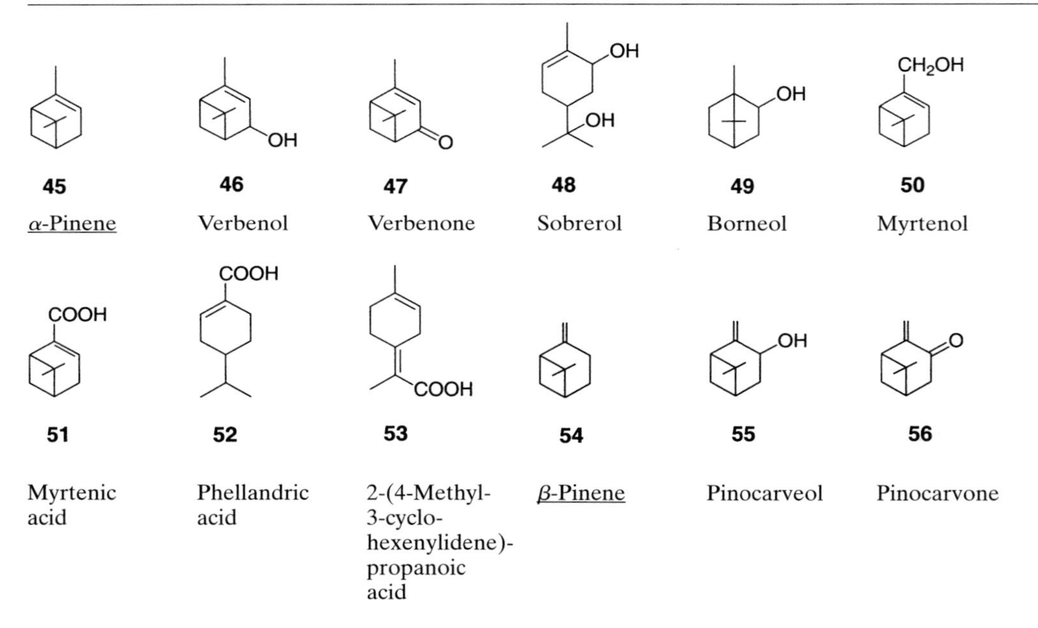

PREMA, B. R., BHATTACHARYYA, P. K. (1962), *Appl. Microbiol* **10**, 524.	*Aspergillus niger* NCIM 612; biotransformation of (**45**) to (+)-*cis*-(**46**), (+)-(**47**), and (+)-*trans*-(**48**); inhibiting effect above 0.6% (**45**); dosage of 0.5 mL (**45**) to 100 mL shake flask culture, 4–8 h incubation; 10 shake flasks yielded 0.88 g of oxygenated products while 3.15 g (**45**) remained unreacted
SHUKLA, O. P., MOHOLAY, M. N., BHATTACHARYYA, P. K. (1968), *Indian J. Biochem.* **5**, 79.	*Pseudomonas* sp. (PL-strain); the soil-isolated strain grew on (**45**) as sole carbon source; biotransformation of (**46**) and (**55**) yielded similar complex metabolite mixtures, e.g., (+)-(**49**), (**50**), (**51**), (**52**), but not (**46**) or (**47**)

Tab. 3. Continued

Nissan Chemical Ind. KK (1981), *Japan. Patent* J5-6 124-388. (Prior. date March 3, 1980).	*Pseudomonas maltophilia* FERM-P-5 420; patent claiming the biotransformation by oxidative ring cleavage of (**45**) to (**53**); starting from 75 mL (**45**) a yield of 2 g (**53**) was achieved
BHATTACHARYYA, P. K., GANAPATHY, K. (1965), *Ind. J. Biochem.* **2**, 137.	*Aspergillus niger* NCIM 612; biotransformation of (**54**) to (**55**), (**56**), and (**50**) by preferably oxidizing in the allylic position

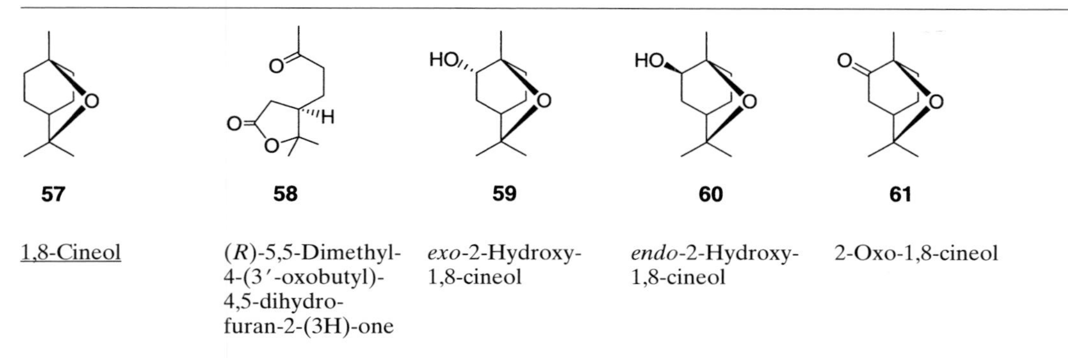

MACRAE, J. C., ALBERTS, V., CARMAN, R. M., SHAW, J. M. (1979), *Aust. J. Chem.* **32**, 917.	Bacterium closely related to *Pseudomonas flava*; strain UQM-1 742 isolated from eucalyptus leaves by enrichment culture on mineral medium with (**57**) as sole carbon source; toxic effect above 0.5 g L^{-1} (**57**); shake cultures with 0.5 g L^{-1} (**57**) yielded 14% (**61**), 10% (**58**), 5% (**59**), and 2% (**60**), while 68% (**57**) remained unreacted

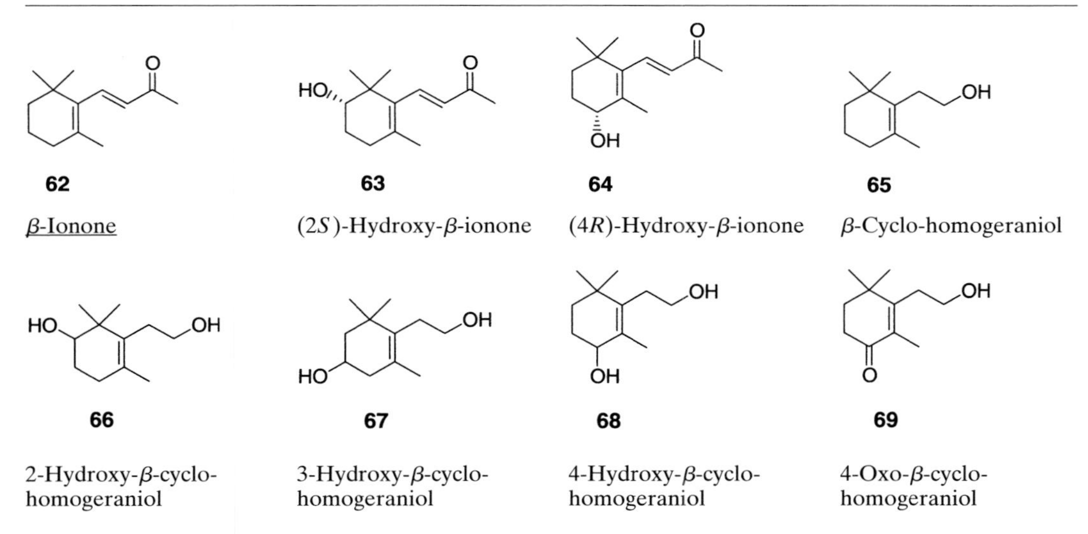

MIKAMI, Y., WATANABE, E., FUKUNAGA, Y., KISAKI, T. (1978), *Agric. Biol. Chem.* **42**, 1075.	*Aspergillus niger* JTS 191; biotransformation of (**62**) yielded two main hydroxylation products (**63**) and (**64**) being very effective for tobacco flavoring at ppm level

Tab. 3. Continued

KRASNOBAJEW, V., HELMLINGER, D. (1982). *Helv. Chim. Acta* **65**, 1590.	*Lasiodiplodia theobromae* ATCC 28 570; due to inhibiting effect of (**62**) pregrown mycelia at the end of log phase were used; 75 L scale, dry biomass 20 g L^{-1}; 28 °C, 800 r.p.m.; biotransformation proceeded by continuously feeding (**62**), of which the concentration in the metabolite mixture never exceeded 10%, and yielded (**65**) as main product; (**66**), (**67**), (**68**), and (**69**) besides other products were identified as minor components; after 4 d 400 g crude product mixture with a tobacco-like flavor was obtained

70 β-Damascone

71 4-Hydroxy-β-damascone

72 2-Hydroxy-β-damascone

73 10-Hydroxy-β-damascone

74 4,10-Dihydroxy-β-damascone

HELMLINGER, D., KRASNOBAJEW, V. (1980), (L. Givaudan & Cie S.A.) *Swiss Patent* 3966.80.	*Aspergillus niger* IFO 8541; biotransformation of (**70**) to (**71**) as main product and (**72**) as by-product; metabolite mixture showed excellent tobacco flavoring properties
HELMLINGER, D., KRASNOBAJEW, V., RYTKOENEN, S., STAUCH, W. (1981), *Abstr. 2nd Eur. Symposium on Organic Chemistry*, Stresa; p. 333.	*Lasiodiplodia theobromae* ATCC 28 570; biotransformation of (**70**) to (**71**) and (**72**); (**73**), (**74**) as minor products

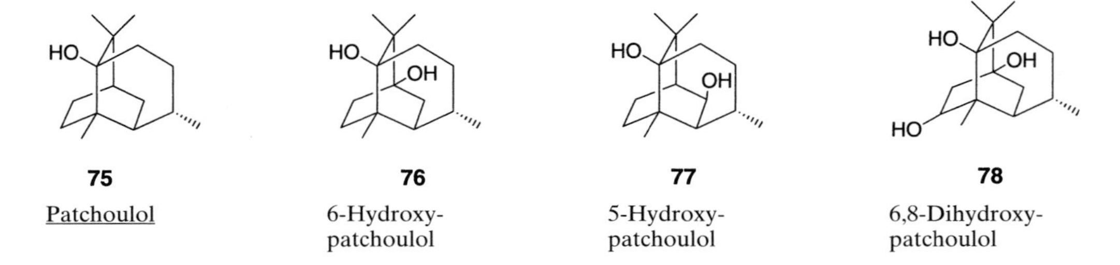

75 Patchoulol

76 6-Hydroxy-patchoulol

77 5-Hydroxy-patchoulol

78 6,8-Dihydroxy-patchoulol

Tab. 3. Continued

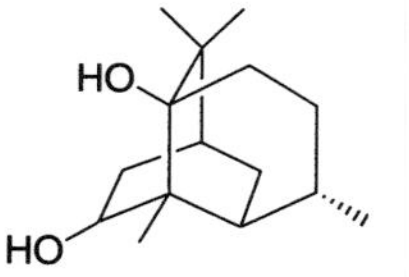

79 8-Hydroxy-patchoulol	**80** 10-Hydroxy-patchoulol	**81** Norpatchoulenol

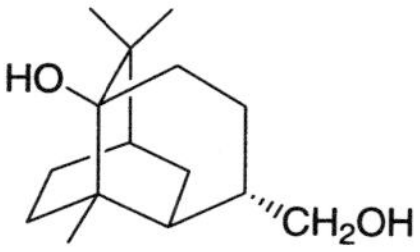

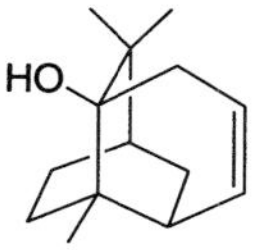

BECHER, E., SCHÜEP, W., MATZINGER, P. K., TEISSEIRE, P. J., EHRET, C. et al. (1981), (Fa. Hoffmann-La Roche & Co.). *Br. Patent* 1 586 759. (Prior. date: Sept. 2, 1976)	Different metabolites were the main biotransformation products from (**75**) depending on the employed fungi: *Gliocladium roseum* NRR 8194: (**76**); *Fusarium lycopersici*: (**77**); *Aspergillus niger* ATCC 11 393; (**78**); *Curvularia lunata* NRRL 2380: (**79**); different fungi yielded (**80**) as by-product, which is of interest as it can be chemically converted to the character impact compound of patchouli essential oil (**81**); 10 shake flasks (500 mL) with 100 mL culture broth each [5% Dextrin, 1% Pharma media (Traders Oil Mill Co., Texas), 0.05% KCl, 0.1% KH_2PO_4, and 0.1% $NaNO_3$] were inoculated with a spore suspension of *Paecilomyces carneus* FERM P-3 797; incubation for 3 d at 26.5 °C and 180 r.p.m.; thereafter, starting from 1 g (**75**), 260 mg crystalline (**80**) was obtained after 7 d of cultivation and isolation by organic solvent extractions, vacuum evaporation, chromatography, and crystallizations

Tab. 4. Books Covering Multiple Aspects of the Biogeneration of Flavors and Fragrances

Year	Author, Editor(s)	Source
1986	PARLIMENT, T. H., CROTEAU, R. (Eds.)	*Biogeneration of Aromas*, ACS Symp. Ser. no. 317. Washington, DC: ACS
1987	SCHREIER, P. (Ed.)	*Bioflavour '87*. Berlin: deGruyter
1992	PATTERSON, R. L. S., CHARLWOOD, B. V., MACLEOD, G., WILLIAMS, A. A. (Eds.)	*Bioformation of Flavours*. Cambridge: Royal Soc. Chem.
1992	TERANISHI, R., TAKEOKA, G. R., GÜNTHERT, M. (Eds.)	*Flavor Precursors*, ACS Symp. Ser. no. 490. Washington, DC: ACS
1993	SCHREIER, P., WINTERHALTER, P. (Eds.)	*Progress in Flavour Precursor Studies*. Carol Stream: Allured
1994	GABELMAN, A. (Ed.)	*Bioprocess Production of Flavor, Fragrance, and Color Ingredients*. New York: John Wiley & Sons, Inc.
1995	ÉTIÉVANT, P., SCHREIER, P. (Eds.)	*Bioflavour 95*. Paris: Institut National de la Recherche Agronomique (INRA)
1995	BERGER, R. G.	*Aroma Biotechnology*. Berlin: Springer-Verlag
1996	TAKEOKA, G. R., TERANISHI, R., WILLIAMS, P. J., KOBAYASHI, A. (Eds.)	*Biotechnology for Improved Foods and Flavors*, ACS Symp. Ser. no. 637. Washington, DC: ACS
1997	BERGER, R. G. (Ed.)	Biotechnology of Aroma Compounds, in: *Advances in Biochemical Engineering & Biotechnology* (Vol. 55). Berlin: Springer-Verlag
1999	SWIFT, K. A. D. (Ed.)	*Current Topics in Flavours and Fragrances*. Dordrecht: Kluwer
2000	SCHIEBERLE, P., ENGEL, K.-H. (Eds.)	*Frontiers of Flavour Science*. Garching: Deutsche Forschungsanstalt für Lebensmittelchemie

3 Biotechnological Production of Terpenoid Flavor and Fragrance Compounds

3.1 Acyclic Monoterpenoids

3.1.1 Myrcene

The unsaturated monoterpene-triene β-myrcene (**82**), a constituent of the terpene carbohydrate fraction of many essential oils, is prone to polymerization by its 1,3-diene structure and has, therefore, not been used very often as starting material for biotransformation studies (e.g., NARUSHIMA et al., 1981; YAMAZAKI et al., 1988). Screening experiments and bioconversion pathway studies of BUSMANN and BERGER (1994c) illustrated the vast metabolic potential of basidiomycetes to convert myrcene into oxygenated monoterpenoids of flavor and fragrance value. Strains of *Ganoderma applanatum*, *Pleurotus flabellatus*, and *Pleurotus sajor-caju* were found to be most suitable for oxyfunctionalizations. The fresh-flowery myrcenol (**83**) was the major component of a broad spectrum of acyclic and monocyclic products (Fig. 1). On the basis of the metabolites identified, different metabolic pathways leading to acyclic and monocyclic products were proposed, involving allylic hydroxylation, oxygenation, and hydration of double bonds, oxidation of alcohol to carbonyl groups and, in the case of cyclization, intramolecular formation of carbon–carbon bonds. Due to product diversity and low concentrations, further improvements, e.g., by screening for single product accumulating strains or by immobilization, are necessary to establish more refined bioconversion systems. IURESCIA et al. (1999) identified and sequenced myrcene catabolism genes from *Pseudomonas* sp., which was isolated from the sediment of the River Rhine by enrichment culture techniques using myrcene as the sole carbon and energy source. By transposon mutagenesis, a myrcene-negative mutant was obtained, which accumulated myrcen-8-ol (**84**) as the sole biotransformation product. Based on DNA sequencing and biotransformation data, a pathway for myrcene catabolism in *Pseudomonas* sp. was proposed (Fig. 2).

As many catabolic enzymes have a low substrate specificity, the authors suggested that the *Pseudomonas* mutant or the genes isolated might be useful for the production of fine chemicals, in general.

3.1.2 Citronellol, Citronellal

The monoterpene alcohol citronellol (**2**) is widely used as a flavor and fragrance compound for citrus and floral compositions, so that different methods for its chemical synthesis have been developed. The first biotechnological approach for its production was described as early as 1915 (MAYER and NEUBERG, 1915) by yeast-catalyzed reduction of citronellal. As $(-)$-citronellol is the enantiomer preferred by the flavor industry, biotechnological attempts at its enantioselective generation from $(\pm)$-citronellal have been made since the 1970s (see Tab. 3). Some of the more recent attempts in the field of citronellol transformation, based on bacteria, yeasts, and fungi and following totally different bioprocess strategies, are mentioned in the following. In the 1980s, the group of SCHREIER launched a series of publications on the biotransformation of acyclic monoterpenols by *Botrytis cinerea*, the noble rot fungus of grapes which is important on account of both its plant pathogenicity and its terpenoid transformation capability (BOCK et al., 1986, 1988; BRUNERIE et al., 1987). Bioconversion of citronellol (**2**) indicated a strong dependence of the product spectrum on the particular strain and medium used. 2,6-Dimethyl-1,8-octanediol (**85**) and (*E*)-2,6-dimethyl-2-octene-1,8-diol (**86**) were identified as the main metabolites, thus indicating ω-hydroxylation and double bond hydrogenation as preferred reactions (BRUNERIE et al., 1987) (Fig. 3).

Bioconversion of citronellol (**2**) by the basidiomycete *Cystoderma carcharias* yielded 3,7-dimethyl-1,6,7-octane-triol (**88**) as the main product (ONKEN and BERGER, 1999a), whereas among the minor products, consisting of different diols, *cis*/*trans*-rose oxide (**89**), a flavor-impact compound, was also observed (Fig. 3).

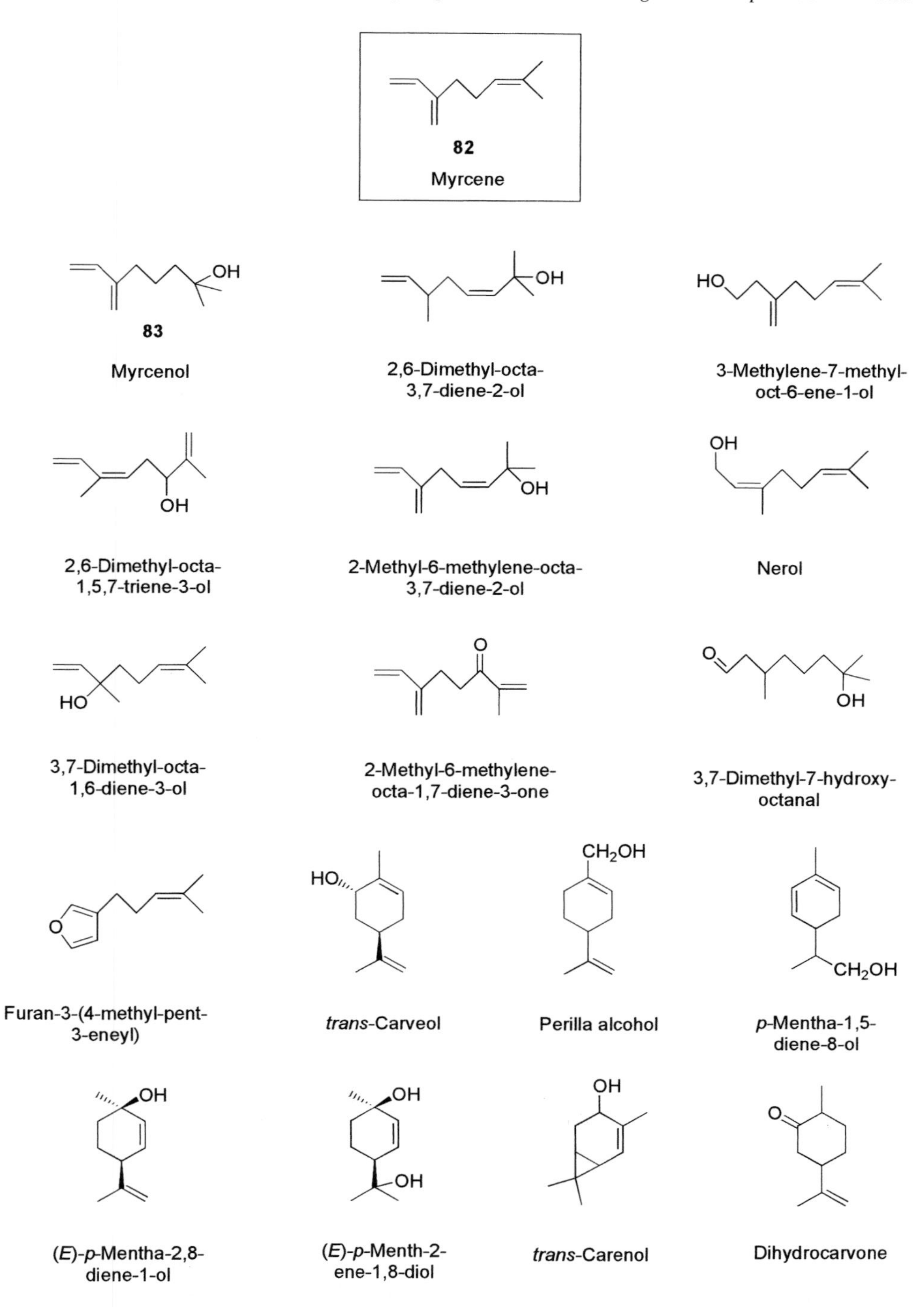

Fig. 1. Conversion products of myrcene by different strains of basidiomycetes (BUSMANN and BERGER, 1994c).

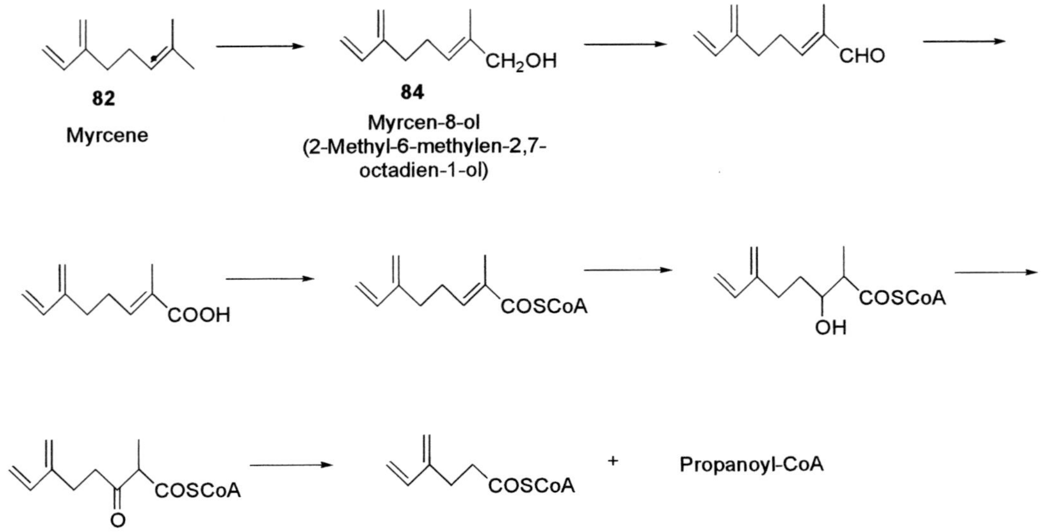

Fig. 2. Postulated catabolic pathway of myrcene by *Pseudomonas* sp. (IURESCIA et al., 1999).

In a 2 L bioreactor process, production rates of up to 0.15 g L^{-1} d^{-1} were reached by sequentially feeding amounts of up to 400 µL citronellol to the culture broth, affording a final terpene triol concentration of 0.87 g L^{-1} (52% conversion rate) after 11 d. In contrast to direct aeration, aeration via a membrane unit consisting of a coiled hydrophobic microporous polypropylene capillary membrane (0.2 µm pore size, 1,272 cm^2 total surface) led to a high gas-exchange surface. By this means, oxygen saturation, growth and transformation rates were increased. As relatively low volumetric air flow rates were necessary, the loss of the volatile substrates by way of the exhaust air was reduced.

After 5 d the bioprocess performance was increasingly hindered by partial blocking of the membrane pores, thus decreasing the oxygen supply.

As a result of inhibition studies, it was postulated that a cytochrome P-450 monooxygenase was involved in the first oxyfunctionalizing step. Epoxide hydrolase and glutathione-S-transferase or UDP-glucuronosyltransferase may be active enzymes in the subsequent transformation steps.

DEMYTTENAERE (1998) reported on sporulated surface cultures of *Aspergillus niger*, by which citronellol was partially metabolized to *cis/trans*-rose oxide, whereas citronellal was almost completely reduced to citronellol. Isopulegol, isoisopulegol, menthone, isomenthone, limonene, and α-terpineol were also found as minor products.

ODA et al. (1995) succeeded in generating (+)-citronellol by enantioselective oxidation of racemic citronellol to (−)-citronellic acid using an interface bioreactor which enables microorganisms to grow on the interface between a hydrophilic carrier (nutrient agar) and a hydrophobic organic solvent layer (*n*-decane). The main benefit of this method was the fact that, obviously, the cytotoxic effects of the substrates and/or products were reduced compared with a conventional biphasic, aqueous–organic system. After precultivating *Hansenula saturnus* IFO 0809 in the presence of filter pads swollen and coated with nutrient agar, a biofilm formed. The thus coated filter pads were arranged as parallel plates in a stirred and aerated vessel and were covered by 1.1 L of a 4% solution of (±)-citronellol in decane. After 12 d, (−)-citronellic acid (82% ee) and (+)-

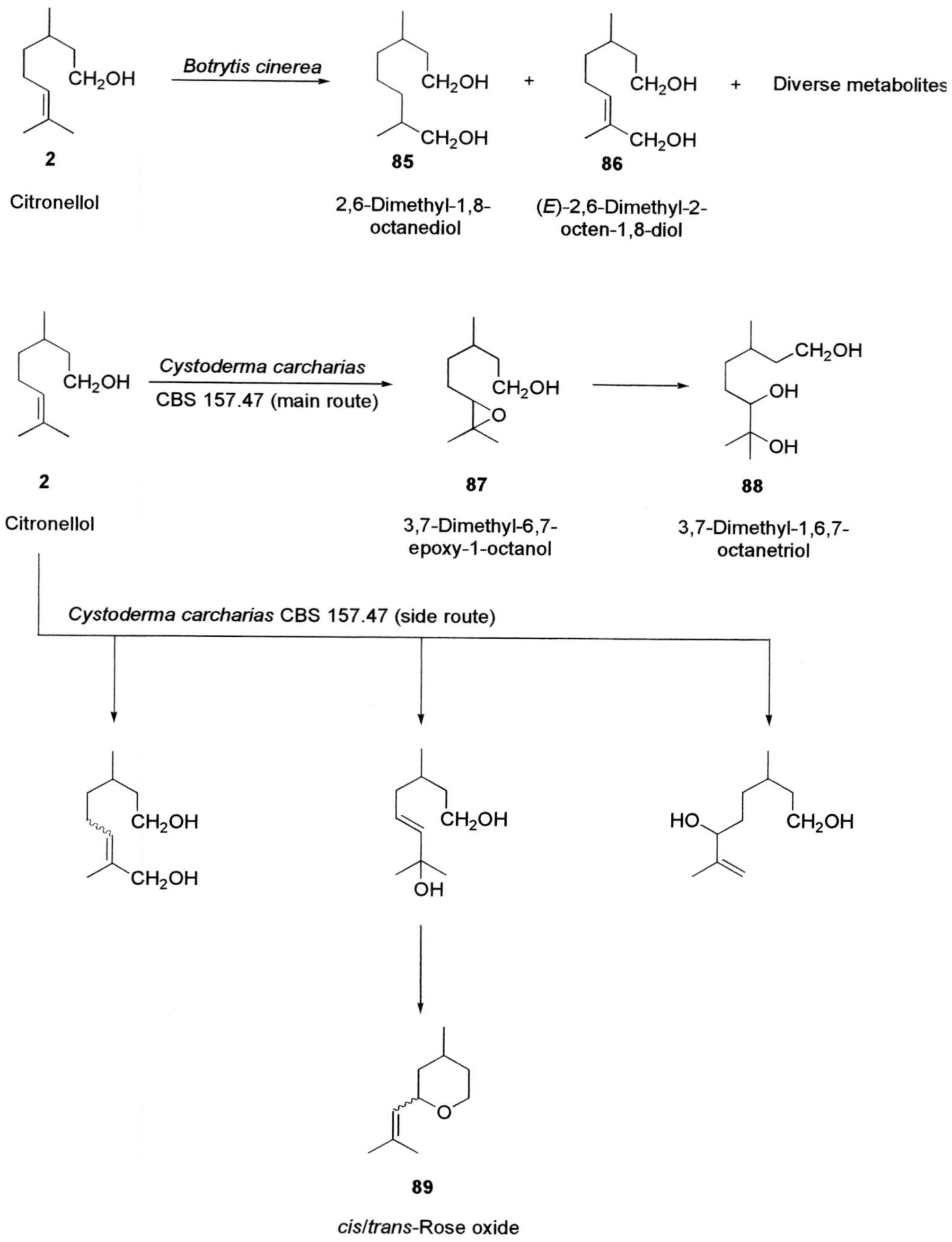

Fig. 3. Microbial conversions of citronellol (BRUNERIE et al., 1987; ONKEN and BERGER, 1999a).

citronellol (77% ee) had accumulated at 18.0 and 17.8 g L^{-1}, respectively, in the organic solvent. In a further approach using an interface bioreactor, *Rhodococcus equi* IFO 3730 preferentially oxidized racemic citronellol to (+)-citronellal (80% ee), while transforming racemic citronellal to the (−)-enantiomer of citronellic acid (ODA et al., 1996a). High product concentrations were achieved by precultivating *Geotrichum candidum* JCM 01747 on an agar plate for one day, which was subsequently covered by a 10% solution of (±)-citronellol in decane and shaken moderately. Despite its strong toxicity, 65 g L^{-1} of citronellic acid accumulated in 13 d, albeit with low optical purity.

3.1.3 Geraniol, Nerol, Linalool, Citral

To perform the biotransformation of monoterpenoids to volatile products DEMYTTENAERE and DE POOTER (1996) recommended the use of sporulated surface cultures of filamentous fungi. By means of a saturated gas phase, the volatile substrates can be directly dosed to the spores, which generally seem to be longer active in the presence of and more tolerant to higher concentrations of the cytotoxic substances than viable cells. Thus, *Penicillium italicum* transformed both geraniol (**90**) and nerol into 6-methyl-5-hepten-2-one (methylheptenone) (**93**) as the main product (97% per GC), which was first described by VANDENBERGH (1989) as a novel, fruity geraniol transformation product obtained from *Pseudomonas putida*. Using a closed system, the *Penicillium* transformation product, released into the gas phase, was concentrated by periodical headspace sampling. By this means, 670 μL methylheptenone were collected from a 300 mL cylindrical flask over a period of two months. Obviously, product adsorption to the agar layer lining the inner wall of the flask was the reason why the molar yield did not exceed 76%. Based on previous data on the bacterial degradation of geraniol, an analogous pathway for the fungus via a triol (**91**) and a dihydroxyketone (**92**) was suggested (Fig. 4).

Methylheptenone was also the main product in biotransformation experiments with *Penicillium digitatum*, starting from both the alcohol nerol and the aldehyde citral, of which the latter is a racemate consisting of the *cis* and *trans* isomers geranial and neral (DEMYTTENAERE and DE POOTER, 1998). With *Aspergillus niger* strains isolated from different plant materials, such as a cypress branch or an African *Welwitschia* plant, the main products obtained from nerol (**94**) were linalool (**12**) and α-terpineol (**38**) (up to a yield of 16%), whereas lower yields consisted of limonene (**21**), methylheptenone (**93**), and 2,6,6-trimethyl-2-vinyltetrahydropyran (**95**) (DEMYTTENAERE and DE KIMPE, 1999). To facilitate monitoring of biotransformation product profiles, the authors recommended solid phase micro-extraction (SPME) coupled with small-scale surface cultures in appropriate vials as a sensitive method of sample preparation, which, in contrast to that of dynamic headspace purge and trap, had the advantage of being quick and solvent-free.

WHITEHEAD et al. (1995) reported on the wide variety of possible biotransformations using commercial bakers' yeast (*Saccharomyces cerevisiae*), which has led to patented, industrial-scale bioprocesses for the production of short-chain acids from natural raw materials (WHITEHEAD and OHLEYER, 1997). When used as building blocks for aliphatic esters, the acids themselves are of great importance to the flavor industry, particularly as they can be labeled "natural". Geraniol (**90**) was enantioselectively oxidized to *E*-geranic acid (**96**) (85%) and (+)-citronellic acid (**97**) (15%). Under aerobic conditions and at alkaline pH, geranic acid reached a maximum concentration of 3.6 g L^{-1} after an incubation time of 48 h. Due to the identification of metabolites, an NADH-depending pathway from geraniol via geranial, neral, and citronellal to citronellol was proposed which branched off from citronellal to citronellic acid (Fig. 4).

KING and DICKINSON (2000) studied the biotransformation characteristics of different yeasts, as these organisms may be responsible for monoterpenoid transformation during beer and wine fermentation. Besides the expected reductions, also translocations, *cis*/*trans* isomerizations, cyclizations, and hydroxylations were found as reactions catalyzed by the three yeast species tested: *Saccharomyces cerevisiae*, *Kluyveromyces lactis*, and *Torulaspora*

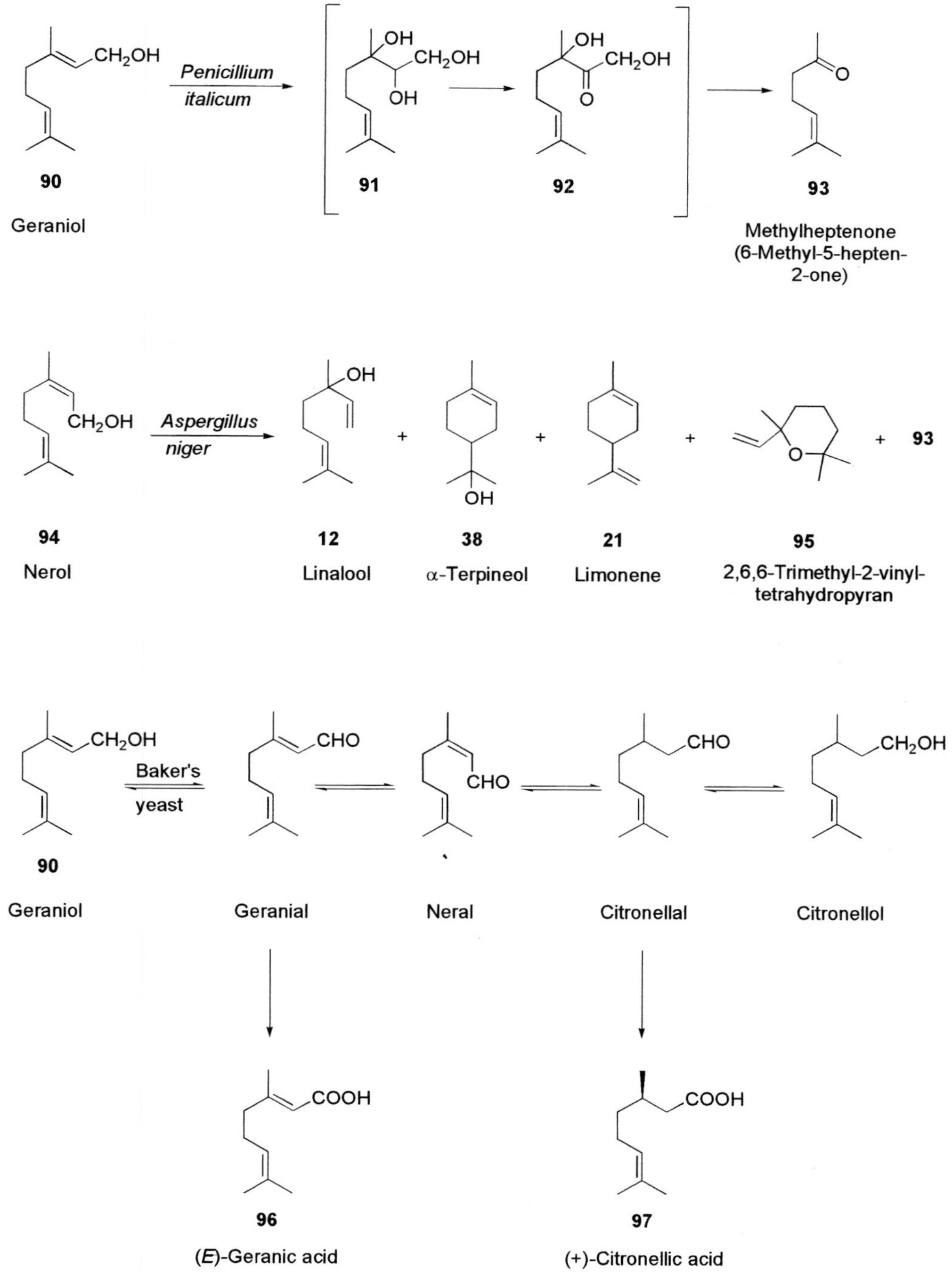

Fig. 4. Conversion of geraniol and nerol by various microorganisms (DEMYTTENAERE and DE POOTER, 1996; DEMYTTENAERE and DE KIMPE, 1999; WHITEHEAD et al., 1995).

delbrueckii. For instance, geraniol and nerol were transformed to linalool; nerol and linalool were transformed into monocyclic α-terpineol and the latter was further transformed into the diol terpin hydrate. Only small amounts of products in the milligram per liter range accumulated under the conditions used.

As outlined in Tab. 3, bacterial degradation of linalool (**12**), which occurs in many essential oils, preferably leads to a mixture of oxidation and hydroxylation products. In the following years different papers dealt with the ability of fungi to convert linalool (e.g., BOCK et al., 1986; MADYASTHA and MURTHY, 1988a, b). Starting from (±)-linalool, furanoid and pyranoid metabolites were found to be the main products from submerged shake cultures of *Aspergillus niger* ATCC 9142 (DEMYTTENAERE and WILLEMEN, 1998). This strain produced a mixture of *cis*- and *trans*-furanoid linalool oxide (**99** and **100**) (yield 15–24%) and *cis*- and *trans*-pyranoid linalool oxide (**101** and **102**) (yield 5–9%) from the racemic substrate. Biotransformation of (*R*)-(−)-linalool yielded almost pure *trans*-furanoid and *trans*-pyranoid linalool oxide (ee >95%). A reaction route via the epoxide (**98**) was proposed (Fig. 5).

Pure linalool oxides, which, for instance, are of interest for lavender notes in perfumery, have rarely been found as biotransformation products so far (e.g., DAVID and VESCHAMBRE, 1984; ABRAHAM et al., 1990).

Recently, HARDER and coworkers illustrated the existence and broad substrate range of bacterial terpene metabolism under anaerobic and anoxic conditions in a series of publications (HARDER and PROBIAN, 1995; FOß and HARDER, 1997; HARDER and FOß, 1999; HEYEN and HARDER, 2000). Novel monoterpene degradation pathways were postulated. For instance, with the transformation of linalool to geraniol and *vice versa* by a self-isolated denitrifying bacterium, a reversible and regioselective enzymatic rearrangement of an allylic alcohol was reported for the first time (FOß and HARDER, 1997). Most of the products documented represented transient metabolites of a complete substrate mineralization. Nerol, which was almost stoichiometrically oxidized to neral, was the exception.

3.2 Monocyclic Monoterpenoids

3.2.1 Limonene

Limonene (**21**) is the major constituent of citrus essential oils and a by-product of the citrus processing industry, amounting to about

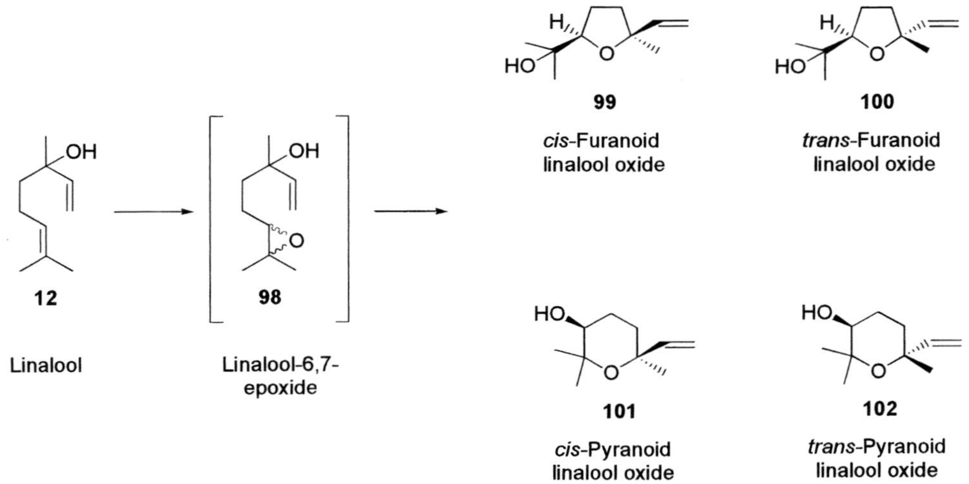

12 Linalool

98 Linalool-6,7-epoxide

99 *cis*-Furanoid linalool oxide

100 *trans*-Furanoid linalool oxide

101 *cis*-Pyranoid linalool oxide

102 *trans*-Pyranoid linalool oxide

Fig. 5. Conversion of linalool by *Aspergillus niger* ATCC 9142 (DEMYTTENAERE and WILLEMEN, 1998).

50 million kg a^{-1} (BRADDOCK and CADWALLADER, 1995). Together with the pinenes, limonene is the cheapest and most readily available monoterpene in nature. Consequently, limonene is a favorable starting material for the manufacture of a variety of terpenoid flavor and fragrance compounds in the chemical industry and is furthermore probably the best studied monoterpene in biotransformation research (for reviews, see GABRIELYAN et al., 1992; BRADDOCK and CADWALLADER, 1995). Fig. 6 illustrates five different microbial transformation pathways, starting from limonene and leading to more valuable flavor compounds, which have been proposed during the past decades by many authors (VAN DER WERF et al., 1999, and references cited therein). Recently, VAN DER WERF et al. (1999) confirmed one of these routes by biochemical studies with a strain of *Rhodococcus erythropolis* which they isolated from a freshwater sediment sample and grew on limonene as the sole source of carbon and energy. In contrast to *Pseudomonas putida*, which degrades limonene by way of the perillyl alcohol route, *R. erythropolis* starts with epoxidation at the 1,2 double bond forming limonene-1,2-epoxide (**25**) (see Fig. 6), while further mineralization proceeds by way of limonene-1,2-diol (**27**) and 1-hydroxy-2-oxolimonene (**28**) and a lactone (**103**), which spontaneously rearranges to form 3-isopropenyl-6-oxoheptanoate (**104**) and thus probably directs the metabolites to the β-oxidation pathway. As the opposite enantiomers were found in the degradation pathway of (+)- and (−)-limonene and the accompanying enzyme activities occurred in the same cell extract fractions after chromatography, the authors suggested that the enzymes involved converted both enantiomers stereospecifically.

Although the microbial generation of a multitude of metabolites seemed to be the rule for biotransformation studies (see Tab. 3), ABRAHAM et al. (1985) succeeded in producing pure (1*S*,2*S*,4*R*)-*p*-menth-8-ene-1,2-diol (**106**) from (+)-limonene (**105**) as the main product using the fungus *Corynespora cassiicola* DSM 62475 (Fig. 7). The diol itself has no flavor impact, but may be useful as a starting material for further syntheses. From the viewpoint of productivity, this practical approach represents one of the most outstanding terpene biotransformations reported so far. In a 70 L fed-batch process 1,300 g substrate was transformed into 900 g pure product during 96 h of fermentation. The process efficiency may be a result of special feeding strategies: glucose and limonene levels were controlled on line by analyzing the CO_2 and the volatile precursor concentration in the exhaust air, respectively. The same research group also reported on the stereospecific transformation of (+)-limonene (**105**) into optically pure (+)-α-terpineol (**107**) as the main product (STUMPF et al., 1984), which has a lilac scent and is one of the most frequently used fragrance compounds (Fig. 7). After precultivating *Penicillium digitatum* DSM 62840 in shake flasks for approximately 3 d, 0.2 mL (+)-limonene was added to 400 mL culture and was completely consumed over a 24 h period while α-terpineol formed with a yield of 46%.

By using mycelia of *Penicillium digitatum* NRLL 1202, TAN et al. (1998) demonstrated enantioselective and enantiospecific bioconversion of racemic limonene to (+)-α-terpineol (**107**) (Fig. 7). Bioconversion activity increased up to 12-fold after sequential substrate induction. By this means, after 96 h a yield of about 3.2 g L^{-1} terpineol could be achieved, albeit on an analytical scale of 5 mL reaction mixes. On account of enzyme induction and other characteristics, such as inhibition by the iron-chelating agent phenanthroline, the authors concluded that a multi-protein component cytochrome P-450-dependent monooxygenase must be responsible for the bioconversion analogous to bacteria, such as *Pseudomonas* sp., of which different P-450-dependent terpene monooxygenases had already been successfully purified and characterized (VAN DER WERF et al., 1997). TAN and DAY (1998a) used Ca alginate immobilized *Penicillium digitatum* mycelia in an air-lift bioreactor process with 500 mL culture in batch, fed-batch and continuous mode. A yield of about 13 mg α-terpineol (g beads)$^{-1}$ d^{-1}, corresponding to a molar conversion of 45%, was obtained at a dissolved oxygen level of 50.0 µmol L^{-1}. An economic assessment showed that the productivity of the process has to be further improved to compete with chemical synthesis. Upon an additional screening of 22 organic

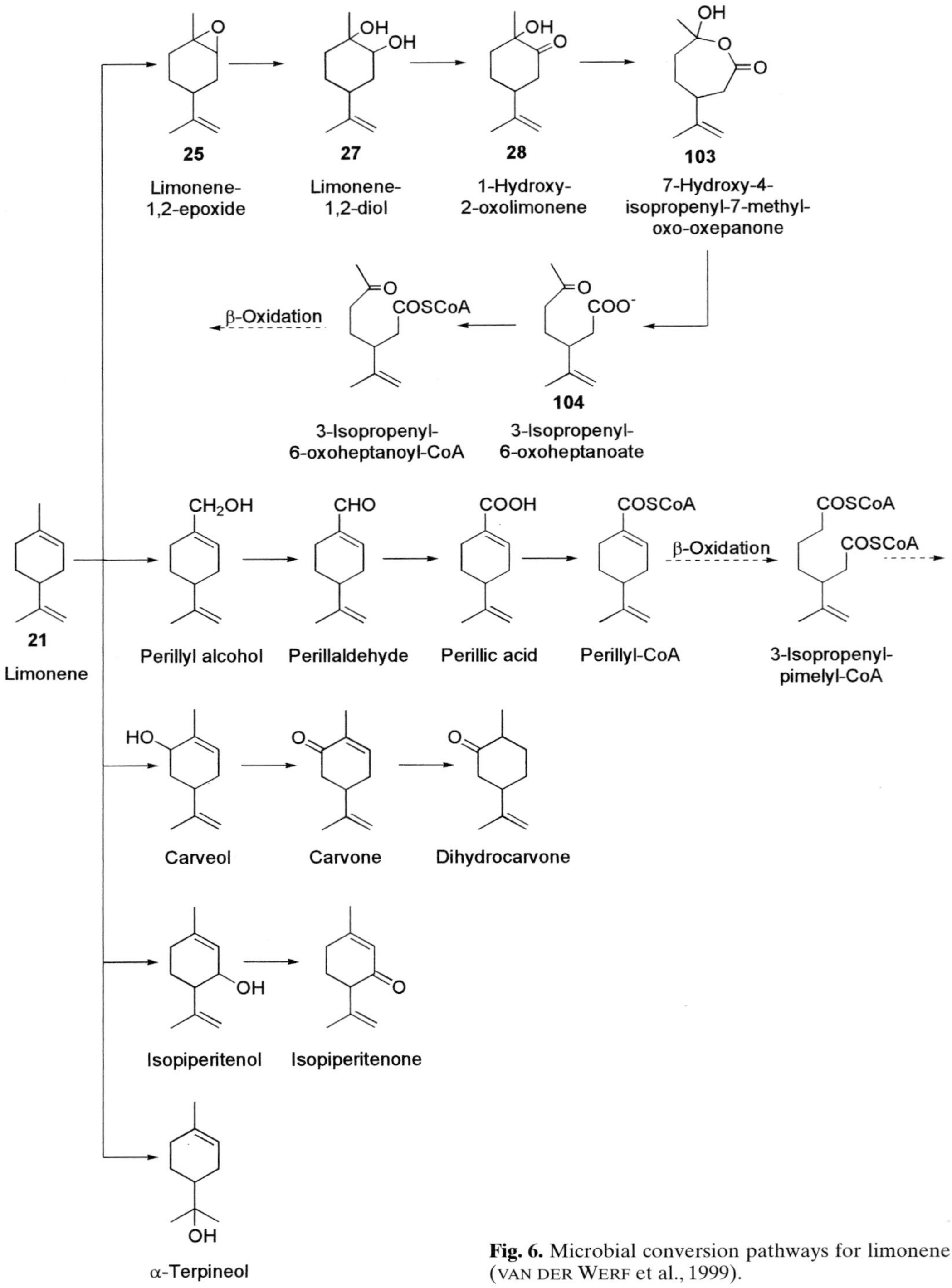

25
Limonene-1,2-epoxide

27
Limonene-1,2-diol

28
1-Hydroxy-2-oxolimonene

103
7-Hydroxy-4-isopropenyl-7-methyl-oxo-oxepanone

β-Oxidation

3-Isopropenyl-6-oxoheptanoyl-CoA

104
3-Isopropenyl-6-oxoheptanoate

21
Limonene

Perillyl alcohol

Perillaldehyde

Perillic acid

Perillyl-CoA

β-Oxidation

3-Isopropenyl-pimelyl-CoA

Carveol

Carvone

Dihydrocarvone

Isopiperitenol

Isopiperitenone

α-Terpineol

Fig. 6. Microbial conversion pathways for limonene (VAN DER WERF et al., 1999).

solvents the same authors suggested that the use of organic cosolvent aids bioconversion activity (TAN and DAY, 1998b). With immobilized cells, a two-fold increase in bioconversion yield could be achieved in detergent reversed micelle systems [0.1% (v/v) of Tween 80 or Triton 100-X], whereas water–organic cosolvent single-phase systems with dioctyl phthalate or ethyl decanoate [1.5% (v/v)] raised the conversion rate 2.2-fold with free cells. The positive effect of reverse micelles indicates that in the case of highly hydrophobic substrates, such as terpenoids, immobilization with lipophilic gel matrices is basically more effective than with hydrophilic ones, such as Ca alginate.

To decrease the cytotoxic effects of terpenes and terpenoids on microorganisms, SPEELMANS et al. (1998) followed a completely different strategy: by an enrichment technique using 5% (+)-limonene (v/v) added directly to a mineral medium, solvent-tolerant and solvent-resistant Gram-negative bacteria were selected. Subsequently, in a biphasic liquid mineral medium, oversaturated with limonene and with added glycerol as the cosubstrate, the strain with the greatest amount of accumulated products was identified as *Pseudomonas putida* and termed type GS1 (DSM 12264). 150 mM (+)-limonene (**105**), 50 mM glycerol, pH 7.0, and 30–34 °C were found to be optimal conditions for bioconversion experiments in shake flasks allowing the generation of up to 18 mM (3.0 g L^{-1}) (+)-perillic acid (**108**) after 120 h, whereas 48 h of fermentation were already sufficient to obtain 80% of the maximum concentration (Fig. 7). The fact that only one single product was formed that was not further degraded makes this biotechnological process commercially interesting. The authors announced that perillyl alcohol and perillaldehyde, the more volatile precursors of the acid, were the target molecules of further screening experiments with solvent-tolerant strains.

In a series of publications the development of a process was reported starting with the screening of terpene catabolizing *Bacillus* species via cloning and expression of a (+)-limonene degradation pathway of *Bacillus stearothermophilus* in *Escherichia coli* to proposing a simple two-phase whole cell bioreactor model using the recombinant organism (CHANG and ORIEL, 1994; CHANG et al., 1995; SAVITHIRY et al., 1997): at the beginning, a *Bacillus stearothermophilus* strain termed BR388, which grew at an optimum near 55 °C, was isolated from orange peel by an enrichment culture with (+)-limonene (**105**) as the sole carbon source. During bioconversion studies with added yeast extract to enhance growth and cometabolic activity, perillyl alcohol (**29**) was identified as the major product (Fig. 7), whereas α-terpineol (**38**) and perillaldehyde (**30**) were minor products (CHANG and ORIEL, 1994). From this thermophilic strain a 9.6 kb chromosomal fragment, responsible for the limonene degradation pathway, was cloned into *E. coli* (CHANG et al., 1995) which was the first reported cloning of a functional microbial monoterpene degradation pathway. It was suggested that the use of a recombinant mesophilic microorganism carrying foreign genes and encoding thermostable, robust enzymes, could help to eliminate undesired host enzyme activities by processing the desired bioconversion at an elevated temperature. The identification of the products α-terpineol (**38**), the perillyl derivatives, as well as carveol and carvone (**23**), showed that limonene hydratase, methyl oxidase, and allylic hydroxylase activities were coded on the cloned fragment (SAVITHIRY et al., 1997) (Fig. 7). The perilla route was favored as the growth-supporting pathway for the parent *Bacillus* strain. By contrast, for a *Pseudomonas alcaligenes* strain, TEUNISSEN and DE BONT (1995) found that α-terpineol was an important intermediate for (+)-limonene degradation. The fact that a biphasic medium with an aqueous and a neat limonene phase was successfully applied in the bioconversion of the monoterpene, thus indicating enhanced microbial tolerance to the cytotoxic substrate, corroborates the strategy of targeting eubacterial thermophilic enzyme activities for gene transfer to tailored hosts. The abovementioned process development was accompanied by a series of patents; the most recent one also claimed the aspect of integrated product recovery by the separate limonene phase, which acts both as a substrate and as an extractant (CHANG and ORIEL, 1996, 1997; ORIEL et al., 1997; SAVITHIRY and ORIEL, 1998).

(+)-Carvone (caraway flavor) and (−)-carvone (spearmint flavor) are important aroma

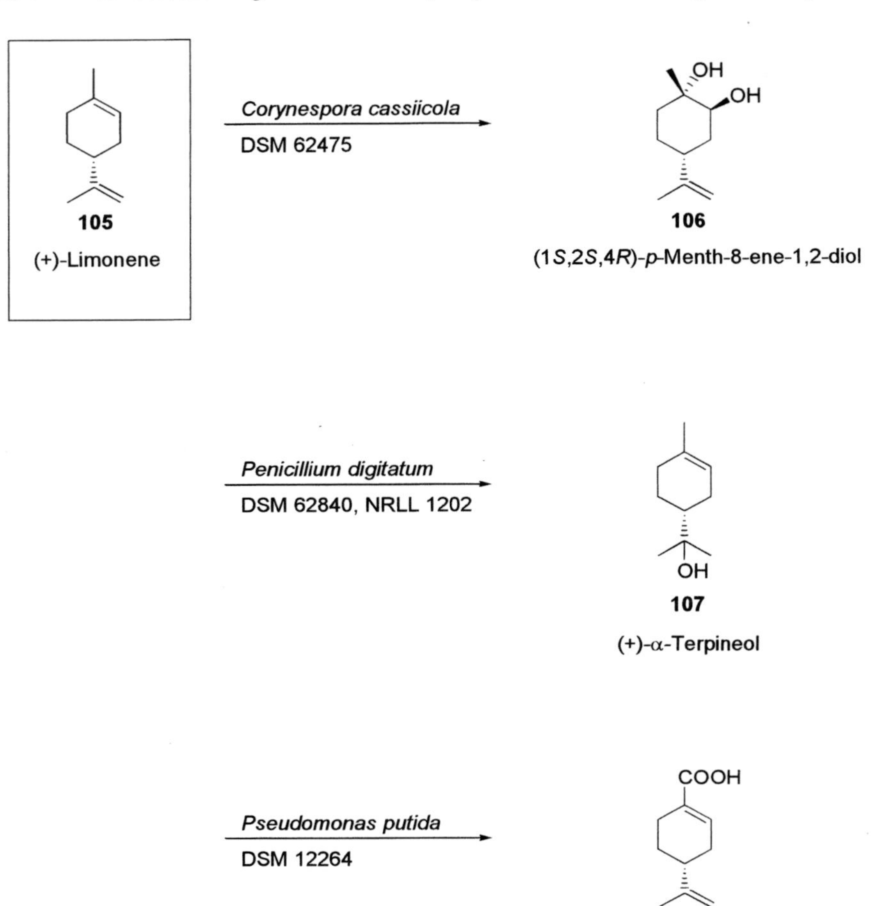

Fig. 7. Selected main products of the microbial conversion of (+)-limonene.

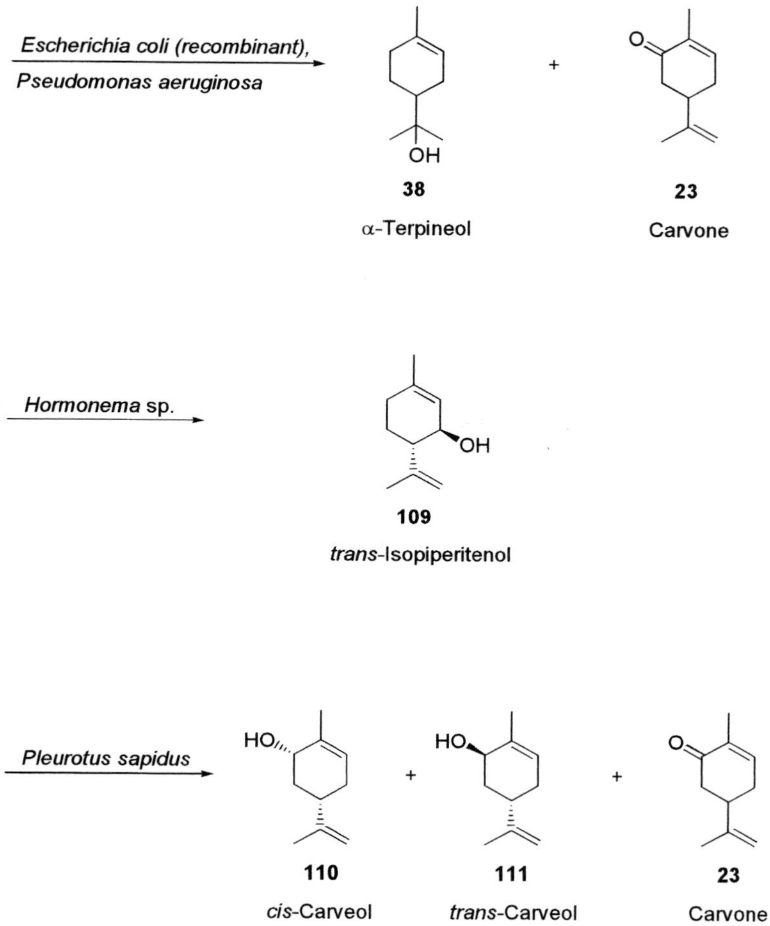

Escherichia coli (recombinant), Pseudomonas aeruginosa → 38 α-Terpineol + 23 Carvone

Hormonema sp. → 109 trans-Isopiperitenol

Pleurotus sapidus → 110 cis-Carveol + 111 trans-Carveol + 23 Carvone

Fig. 7. Continued.

compounds for foods and beverages, but their microbial generation from inexpensive terpene precursors in commercially interesting amounts remains a great challenge, as documented by the small number of capable microorganisms and low product yields (a few milligrams per liter range) which have been reported so far (see Tab. 3; DHAVALIKAR and BHATTACHARYYA, 1966; RHODES and WINSKILL, 1985; KIESLICH et al., 1986).

Recently, VAN DER WERF and DE BONT (1998) screened bacterial strains capable of growing on carvone as the sole carbon source and found 120 Gram-positive strains, however, none of them was able to transform limonene into significant amounts of carvone. By contrast, ACOSTA et al. (1996) found a *Pseudomonas aeruginosa* strain capable of converting (+)-limonene (**105**) into carvone (**23**) and α-terpineol (**38**) as the main products (Fig. 7). Final product concentrations of up to 0.63 g L^{-1} carvone and 0.24 g L^{-1} α-terpineol were achieved at 37 °C in 200 mL shake flasks, albeit requiring a long incubation time of at least 13 d. Interestingly, main product formation occurred primarily between the 9th and 13th days of cultivation, whereas further incubation did not effect the degradation of these products.

Although yeasts are well-known as useful biocatalysts for reduction reactions, the use of yeasts for hydroxylation purposes has been neglected during recent years. In their search for (+)-limonene and (−)-piperitone transformation capabilities, VAN RENSBURG et al. (1997) did not focus on prokaryotic but on eukaryotic organisms and performed a screening with 100 isolates of yeasts and yeast-like fungi, isolated from monoterpene-rich environments and 27 additional strains from a cell culture collection. (−)-Piperitone, the main constituent of *Eucalyptus dives* oils, was transformed to up to 0.2 g L^{-1} (4*R*,6*S*)-*trans*-6-hydroxy-piperitone by *Trichosporon* spp. and to 0.3 g L^{-1} (4*R*,6*R*)-*cis*-6-hydroxy-piperitone by *Hormonema* sp., besides further products with other non-conventional yeasts (VAN DYK et al., 1998a). This highly capable black yeast, identified as a *Hormonema* sp., termed UOFS Y-0067, also transformed (+)-limonene, α-pinene, and β-pinene into hydroxylation products in comparatively high yields (VAN DYK et al., 1998b) (for pinene transformation, see Sect. 3.4.1). The black yeast transformed (+)-limonene (**105**) into *trans*-isopiperitenol (**109**), a biotransformation reported for the first time (Fig. 7). A product concentration of 0.5 g L^{-1} was achieved after 12 h of incubation in shake flasks (31% molar conversion), obviously prevented from further increasing by cytotoxicity of the substrate and probably of the product, too. Unfortunately, the product concentration of *trans*-isopiperitenol, which can easily be converted into the value-added (−)-menthol by chemical hydrogenation, was not always reproducible. It was suggested that the variation in product concentration was related to the morphological mutability, as documented by the varying macroscopic appearance (production of extracellular slime and pigments). As a consequence, it is questionable whether commercialization based on an organism such as this *Hormonema* sp. will be easily feasible.

ONKEN and BERGER (1999b) focused on higher fungi which are known to transform monoterpenes into oxyfunctionalized compounds (BUSMANN and BERGER, 1994b, c), and investigated the effects of (+)-limonene on submerged cultures of the basidiomycete *Pleurotus sapidus*. By means of precultivation in the presence of small amounts of (+)-limonene (**105**), fed via the gas phase, the concentration of *cis*/*trans*-carveol (**110** and **111**) doubled and the concentration of carvone (**23**) increased by a factor of 3–4, to yield total product concentrations of >0.1 g L^{-1} (Fig. 7).

3.2.2 Menthol, Menthone

When (−)-menthol (**40**) has to be prepared chemically from optically inactive raw materials, the main problem is its separation from up to eight stereoisomers (MIKAMI, 1988) (see Tab. 3). Therefore, intensive research activities have addressed the enantioselective generation of (−)-menthol by using biotechnology. The most promising attempts so far are based on biocatalytic asymmetric hydrolysis of (±)-menthyl esters, which went into industrial-scale production in the 1970s (see Tab. 3). As these transformations are discussed in Sect. 3.3.3, the present section reports another interesting approach which focuses on the extension and fortification of plant (−)-menthol biosynthesis by microbial systems. The peppermint plant, *Mentha piperita*, the main source of (−)-menthol, contains the highest level of essential oil during the initial flowering stage; although at this stage the volatile fraction consists almost completely of (−)-menthone (**112**), in the subsequent growth phase until full bloom only 40% is converted to (−)-menthol (KEMPLER, 1983; WELSH et al., 1989). Therefore, harvesting the peppermint plants at their (−)-menthone peak and subsequent biocatalytic reduction to (−)-menthol should be a commercially interesting strategy. KISE and HAYASHIDA (1990) established a continuously operating two-phase system reactor with NADH-dependent menthone reductase from *Cellulomonas turbata* immobilized onto activated carbon as the biocatalyst. Cofactor recycling was performed converting methyl isobutyl carbinol to methyl isobutyl ketone by the same enzyme (Fig. 8). The product mixture was separated by retaining the immobilized enzyme and the cofactor by a hydrophobic microfiltration membrane. An average (−)-menthol productivity of 36.5 g L^{-1} d^{-1}

was achieved during 270 h of steady state conditions, whereas the enzyme's half-life was 607 h. To separate (−)-menthol from the mixture, two distillation steps were necessary during downstream processing. Though productivity and specific enzyme activity have to be further improved for potential industrial application, one important advantage is that in contrast to the kinetic resolution of chemically synthesized (+)-menthyl esters the (−)-menthol generated by microbial transformation of plant (−)-menthone can be declared "natural".

(+)-Menthol and (−)-menthol have also been taken as substrates for biotransformation studies (ASAKAWA et al., 1991). Using *Aspergillus niger* and *Aspergillus cellulosae*, the oxyfunctionalizing of non-activated carbon atoms of menthol by cometabolism was demonstrated. After static incubation of pregrown 40 mL cultures, biotransformation was started by adding up to 34 mg substrate. After 7 to 10 d, a series of up to 7 hydroxylation products was identified as metabolites derived from (+)- and (−)-menthol. Terpenoids with more than one hydroxy group do not in general contribute to the flavor of a composition, but the authors emphasized that 8-hydroxymenthol (**113**), a mosquito-repelling agent, was found among the (−)-menthol hydroxylation products (Fig. 8).

3.3 Synthesis of Monoterpenyl Esters

Since the early 1980s interest has focused increasingly on the biotechnological production of value-added terpenyl esters as flavor and fragrance compounds, coupled with a rising demand by consumers for natural products in general (DE CASTRO, 1995). The number of scientific publications that emerged during this period substantiates this trend which again received further impetus at the beginning of the 1990s when higher yields were obtained with lipases in organic solvents. Much research has been, and still is being, undertaken in the field of lipase-catalyzed ester production; some more recent work on terpenoid flavors and fragrances will be documented in the following. The fact should not be overlooked that, besides all the other demands, the regulations and directives prevailing in the food industry in conjunction with "natural" declarations require substrates of natural origin, which concerns both the terpene alcohol and the acyl donor.

The main targets are the acyclic terpenols citronellol, geraniol, and linalool, of which the lower fatty acid esters, especially the most widely used acetates, are important as both fragrance and flavor substances (BAUER et al., 1997). Most of the biotechnological studies so far concentrate on the primary alcohols geraniol and citronellol, whereas the sterically less accessible tertiary alcohol linalool seems to be considerably more difficult to esterify with lipases. Therefore, in the following no example is cited of a promising biotechnological attempt to esterify linalool. Some investigators demonstrated the ability of whole microbial cells to hydrolyze sterically hindered esters, such as racemic linalyl acetate, to yield the free terpene alcohol, albeit with no (MADYASTHA and MURTHY, 1988a, b) or relatively low enantioselectivities (VAN DYK and THOMAS, 1998; OSPRIAN et al., 1996). These results should stimulate the search for appropriate enzyme isolates from the *Rhodococcus* species, which showed hydrolytic activity and proved to be the most promising microorganisms in the studies.

3.3.1 Esterification in Organic Solvents

Commercially available, free or immobilized lipases are versatile catalysts. Good yields of 90 to 100% molar conversion were obtained by performing both direct esterification (CLAON and AKOH, 1993, 1994b) and transesterification experiments (CLAON and AKOH, 1994a) with the commercial, immobilized *Candida antarctica* lipase SP435 and *n*-hexane as the organic solvent. The investigators also observed substrate inhibition at concentrations above 0.3 M acetic acid, which can be attributed to the water solubility of the short-chain fatty acid and the resulting interaction with the enzyme surface. To circumvent these drawbacks in the direct acetylation of geraniol, MOLINARI et al. (1998) used intracel-

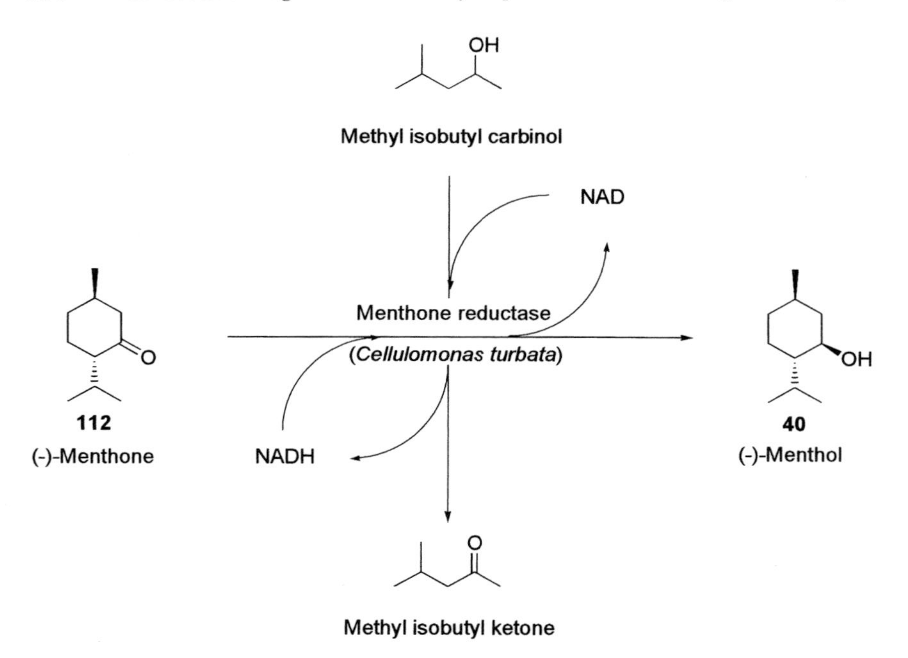

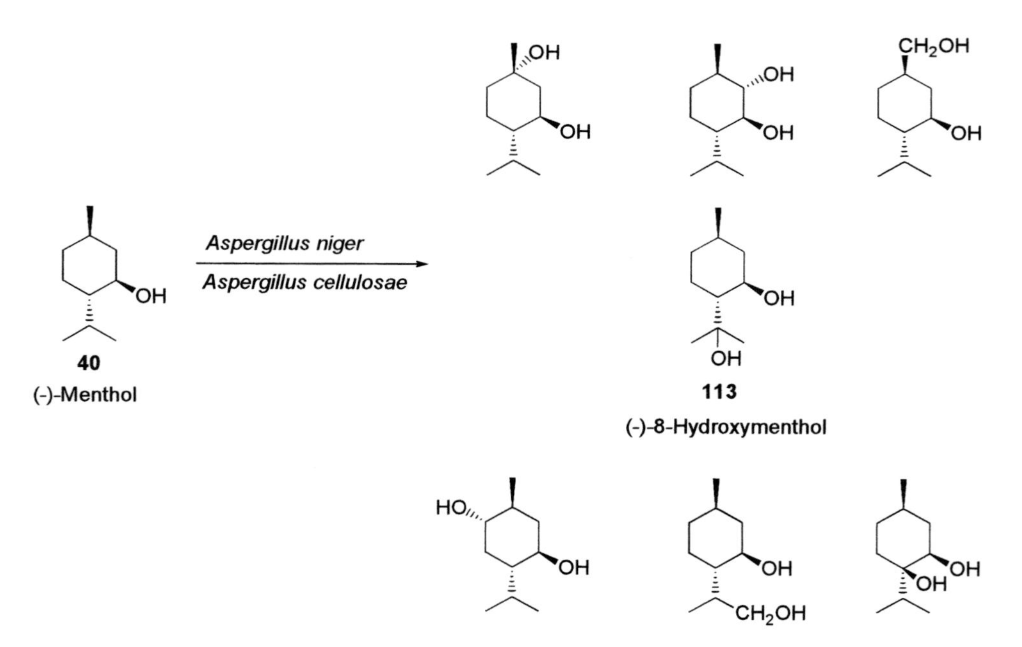

Fig. 8. Biotransformations of (–)-menthone and (–)-menthol (KISE and HAYASHIDA, 1990; ASAKAWA et al., 1991).

lular enzymes by taking dry mycelium of *Rhizopus delemar* MIM in heptane and attained geranyl acetate concentrations of up to 75 g L^{-1} after 10 d in a fed-batch approach, corresponding to a molar conversion of 55%. This requires further development prior to potential industrial application. DE CASTRO et al. (1998) chose the *Mucor miehei* lipase and acetic acid-limiting conditions in a fed-batch system to overcome the inhibiting effects of acetic acid. By this means a maximum yield of 65% conversion could be achieved after several days of incubation. HUANG et al. (1998) successfully used a surfactant-coated lipase for direct esterification of geraniol and acetic acid to prevent direct contact of the organic solvent and substrates with the lipase. From the viewpoint of the basic biochemical principle, ODA et al. (1996b) performed interesting experiments with whole cells from *Hansenula saturnus* IFO 0809 in an "interface bioreactor", consisting of nutrient agar-coated filter pads as the carrier of the microbial cell layer, covered by a layer of *n*-decane as the organic solvent (see Sect. 3.1.2). With this system, the investigators circumvented any dosage of external acyl donor by coupling *in vivo* acetyl-CoA formation and esterification with the added citronellol to yield the desired terpenyl ester. Furthermore, by feeding the less expensive citronellal, citronellyl acetate was obtained in the same way, because the microorganism reduced citronellal to citronellol prior to the esterification reaction (Fig. 9). In a subsequent screening, from a multitude of microorganisms *Pichia quercuum* IFO 0949 and *P. heedii* IFO 10019 were selected as the most promising strains for producing different esters by double and triple coupling of the aforementioned metabolic pathways (ODA and OHTA, 1997).

YEE and AKOH (1996) used acetic anhydride as a non-water producing acyl donor and a *Pseudomonas* sp. lipase (PS), which was immobilized by adsorption onto glass beads, to synthesize geranyl acetate in yields as high as 97% after 24 h at 50 °C. The results underpin the importance of enzyme immobilization both to enhance resistance to denaturation and facilitate recovery, and to enable the biocatalyst to be reused for many runs, thereby reducing the overall process costs. GONZÁLEZ-NAVARRO and BRACO (1998) described another promising tool for improving the stabil-

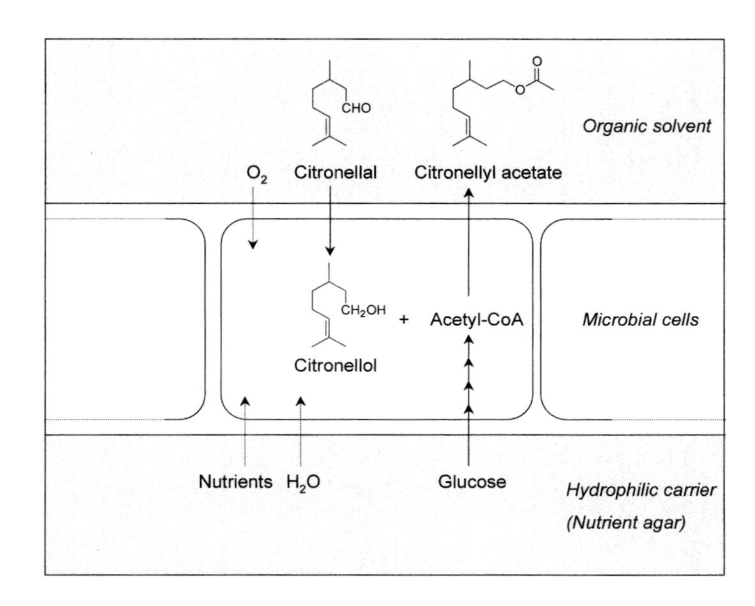

Fig. 9. *In vivo* production of terpenyl esters by whole cells of *Hansenula saturnus*, *Pichia quercuum*, and *P. heedii* (adapted from ODA et al., 1996b; ODA and OHTA, 1997).

ity and consequently non-aqueous enzymatic activity, that of interfacial activation-based molecular (bio)imprinting. For this purpose, a pre-incubation step in the presence of dioctyl sulfosuccinate and *n*-octyl β-D-glucopyranoside was performed to induce an activating conformational change in the lipase upon binding to these amphiphile interfaces. The resulting enhanced conformational rigidity was trapped by rapid freeze-drying and retained in the lyophilized powder. With this innovative technique and using the porcine pancreatic lipase and the *Candida rugosa* lipase, the esterification rate of citronellol and geraniol and also oleic acid could be increased up to 90-fold compared with the non-activated, conventionally lyophilized lipase.

Recently LABORET and PERRAUD (1999) reported on a successful process to synthesize terpenyl propanoate, butanoate, hexanoate, and acetate by direct esterification with a free *Mucor miehei* lipase in *n*-hexane. Despite confirming the inhibiting effects of acetic acid on esterification activity, they succeeded in designing a process for potential future industrial application by optimizing several reaction parameters, such as the acid: alcohol ratio, substrate and enzyme concentrations, solvent effects, kinetics, and biocatalyst reusability. For further scale-up, a triphasic 2 L fluidized bed reactor and a fairly simple way of processing were developed which permitted the production of more than 100 g of natural terpenyl ester per day even on this laboratory scale (Fig. 10).

Triacyl glycerols and fatty acid vinyl esters, which have not been used as acyl donors very extensively in terpene ester synthesis so far, seem to be very efficient reactants in transesterifications to produce geranyl and citronellyl esters, while circumventing both the release of water during the reaction and, in the case of acetate, the inhibiting effect of the free fatty acid on lipases. By this means, YEE et al. (1995, 1997) showed that a *Candida rugosa* lipase and a *Pseudomonas* sp. lipase PS yielded over 96% molar conversion after 24 h incubation time with geraniol and citronellol as the terpene alcohols and tributyrin and tricaproin as the acyl donors. By taking the transesterification of geraniol and tributyrin with lipase AY from *C. rugosa* in *n*-hexane as a model system, SHIEH et al. (1996) demonstrated the feasibility of statistical methods for experimental management to optimize the most important reaction variables (time, temperature, enzyme amount, substrate molar ratio, added water). In the biocatalysis of terpenyl acetates the use of non-natural vinyl acetate (**114**) was promising (NAKAGAWA et al., 1998; AKOH and YEE, 1998): besides eliminating the inhibiting effects by using vinyl acetate as the acyl donor, the tautomerization of vinyl alcohol (**116**), which is the direct product of transesterification, has the added advantage of rendering the process irreversible (Fig. 11).

Studying the synthesis of geranyl acetate (**115**), NAKAGAWA et al. (1998) achieved a molar conversion yield of 97.5% after a very short incubation time of 5 h with a self-isolated celite-adsorbed lipase of *Trichosporon fermentans*. No lipase inhibition by geraniol (**90**) or vinyl acetate (**114**) was evident even at 1.0 M and the optimum water content of 3% (v/v) could be raised to 10% (v/v) while only moderately decreasing the conversion yield to 80%. This might be useful for practical syntheses.

3.3.2 Solvent-Free Esterification

The advantages of using hydrocarbons as solvents in ester synthesis with lipases are well-known, e.g., the shift in thermodynamic equilibria towards synthesis, increased solubility of hydrophobic substrates, suppression of hydrolytic side reactions, low probability of microbial contaminations, etc. (FABER, 2000). Nevertheless, there are also good reasons for preferring solvent-free reaction media from the viewpoints of environmentally friendly processes, simplified downstream processing, natural products for food ingredients and, last but not least, the overall process economy (IKUSHIMA et al., 1995; CHATTERJEE and BHATTACHARYYA, 1998b). CHATTERJEE and BHATTACHARYYA (1998b) reported on laboratory-scale experiments to synthesize short and medium chain fatty acid esters of geraniol and citronellol by direct esterification in the absence of any organic solvent. Yields ranging from 96 to 99% molar conversion were achieved with a commercially available immobilized lipase of *Rhizomucor miehei* after 6 h at 55

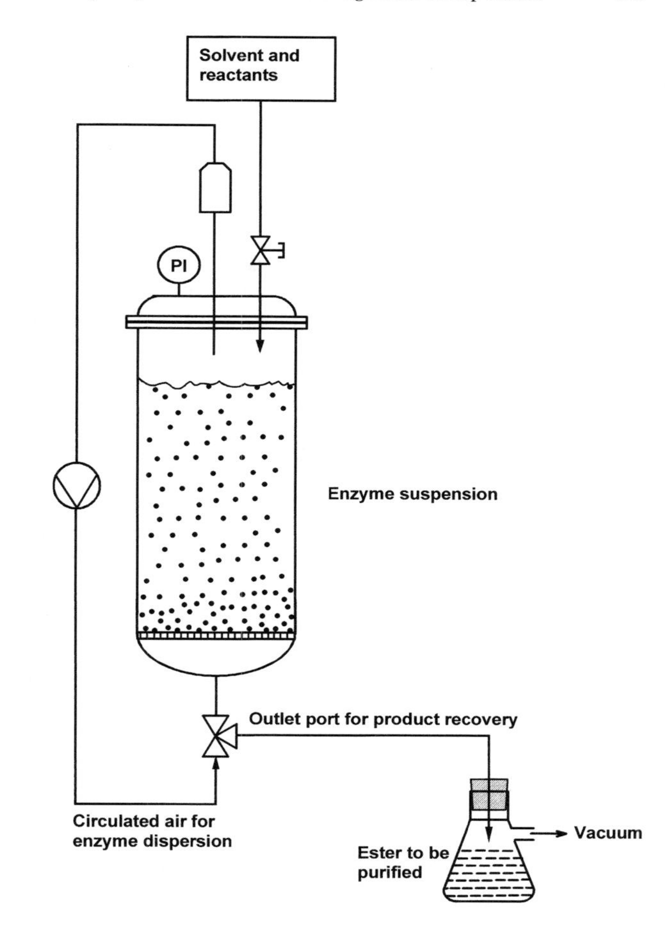

Fig. 10. Laboratory-scale production of terpenyl esters (adapted from LABORET and PERRAUD, 1999).

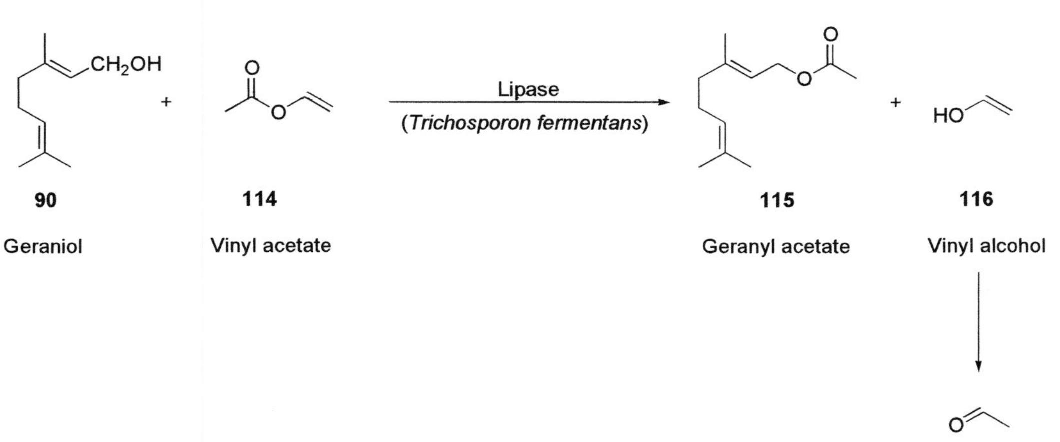

Fig. 11. Transesterification of geraniol with vinyl acetate (NAKAGAWA et al., 1998).

to 60 °C and with 0.1 M substrate solutions (terpene alcohol; hexanoate, octanoate, decanoate, and dodecanoate) while agitating the reaction medium under vacuum to remove water formed during the reaction. Using the concept of solvolysis, where the solvent itself acts as the reactant, geranyl ester yields of about 100% were obtained with immobilized *Candida antarctica* lipase, propanoate and butanoate, while acetate showed much lower conversion (OGUNTIMEIN et al., 1995). By contrast, CHATTERJEE and BHATTACHARYYA (1998a) obtained yields of 75 to 77% molar conversion to geranyl and citronellyl acetate in 8 h by transesterification with short chain alcohol acetates as the solvents and with an immobilized *Mucor miehei* lipase. They also described a procedure for recovery of pure product based on vacuum filtration, vacuum distillation, and column chromatography, which they considered to be of potential interest for an industrial process.

3.3.3 Stereoselective Esterification and Ester Hydrolysis

Whereas geraniol does not have a chiral center, citronellol occurs as a pair of two optical isomers, which slightly differ in odor. The (−)-enantiomer is said to be more pleasant than the (+)-enantiomer (BAUER et al., 1997). Due to the inherent stereochemical advantages of biological versus chemical catalysts, it may be assumed that racemic citronellol, the form mostly found in natural sources, is an ideal substrate for enantioselective esterifications. Among the citronellol esterifications described above, no details were available on the effect of chirality. The work of WANG and LINKO (1995) is an exception, as they reported on experiments for the direct esterification of citronellol with butanoic acid both with and without additional organic solvent and with different lipases. With *Candida rugosa* lipase and without the addition of organic solvent, the highest ester yields in an 18 h reaction were 98% for (+)-citronellol with 12% water content and 67% for (−)-citronellol with 18% water content. Thus, an enantiomeric preference of the lipase used was determined in a solvent-free system, whereas in an *n*-hexane system only little enantioselectivity was observed. A lipase-catalyzed total resolution of racemic (±)-citronellol was not attainable, as shown by a conversion yield of more than 65% for (±)-citronellol as the substrate. By contrast, IKUSHIMA et al. (1996) succeeded in synthesizing (−)-citronellyl ester with an optical purity of at least 98% ee from racemic citronellol and oleic acid using immobilized *Candida cylindracea* lipase (CCL) in a supercritical carbon dioxide reactor (see the patent published by the same authors, IKUSHIMA et al., 1995). Studying the relationship between the conformation of active CCL in supercritical carbon dioxide and the reactivities in the esterification reaction, it was found that, in a very limited pressure range near the critical point, drastic conformational changes in the enzyme occurred causing active sites to emerge and catalyze the stereoselective esterification, albeit at relatively low reaction rates. The patent claims not only the esterification itself, but also the hydrolysis of the (−)-citronellyl ester to recover pure (−)-citronellol. In this context it is remarkable that MICHOR et al. (1996), who conducted transesterification experiments to investigate different enzymes, racemic terpenols, and acylating reagents in supercritical carbon dioxide, found enantioselectivity only with (±)-menthol, but not with (±)-citronellol.

In contrast to citronellol, menthol has proved to be a suitable substrate in enantioselective biotransformations in general. The molecular structure of menthol shows three chiral centers, and consequently, the most important, (−)-menthol, which is used for its characteristic peppermint flavor and refreshing effect, is only one optical isomer of a group of four pairs of antipodes: (±)-menthol, (±)-neomenthol, (±)-isomenthol, and (±)-neoisomenthol (EMBERGER and HOPP, 1985). Therefore, (±)-menthol has been extensively studied for its use as a substrate in biotransformations both for enzymatic resolution of the racemic substrate to yield pure (−)-menthol and for its enantioselective esterification for the production of valuable (−)-menthyl esters (see Tab. 3).

One possibility of producing (−)-menthol on an industrial scale is the kinetic resolution of (±)-menthol, which can be synthesized by hydrogenating thymol. The use of biocatalysts for the resolution was patented as early as 1971 (see Tab. 3, MOROE et al., 1971). Subsequent publications underpin the great potential of microbial conversion in this commercially highly interesting field and document the competitiveness of biochemical methods compared with chemical ones, especially with respect to stereoselective reactions (see Tab. 3, WATANABE and INAGAKI, 1977; OMATA et al., 1981; MIKAMI, 1988, and references therein). Due to the chemically synthesized racemic terpenyl esters, which are used as substrates for the subsequent biocatalytic resolution, the (−)-menthol produced is not eligible for the premium label "natural".

Since the formation of water as a by-product is unfavorable for both the equilibrium of the esterification reaction and the enzyme stability, XU et al. (1995a) used propanoic acid anhydride as a non-water-producing acyl donor for lipase-catalyzed enantioselective esterification to obtain an efficient kinetic resolution of (±)-menthol. By continuously feeding propanoic acid anhydride into a reactor containing free *Candida cylindracea* lipase OF 360 and (±)-menthol in cyclohexane, an optical purity of more than 98% ee of the (−)-menthyl ester could be obtained. However, a batch process with a high concentration of the acid anhydride in dehydrated organic solvent did not succeed because the minimal water content necessary for the enzyme function could not be maintained due to the competing hydrolysis of the anhydride. In a subsequent attempt, the same authors established a continuously operating biotransformation with the above-mentioned system in a 120 mL reactor (XU et al., 1995b). By feeding the acid anhydride, the water content of the organic solvent was held in a range of 2–4 mM, which was optimal to maintain stable operating conditions over a reaction time of more than 50 d (Fig. 12). The free lipase suspended directly in cyclohexane proved to be an efficient biocatalyst, as the protein's insolubility in the organic solvent meant that no additional immobilization, which generally causes activity loss and generates diffusion barriers, was necessary.

An ester productivity of 0.027 μmol h^{-1} (mg lipase)$^{-1}$ caused the investigators to expect a large-scale process to be both technically and economically feasible, although they did not compile a cost-benefit analysis. In fact, the productivity was much higher than that of KAMIYA et al. (1995), who used a surfactant-coated lipase originating from *C. cylindracea* AY as the biocatalyst, though focusing on the esterification of (−)-menthol with free fatty acids in isooctane as the solvent, which, strictly

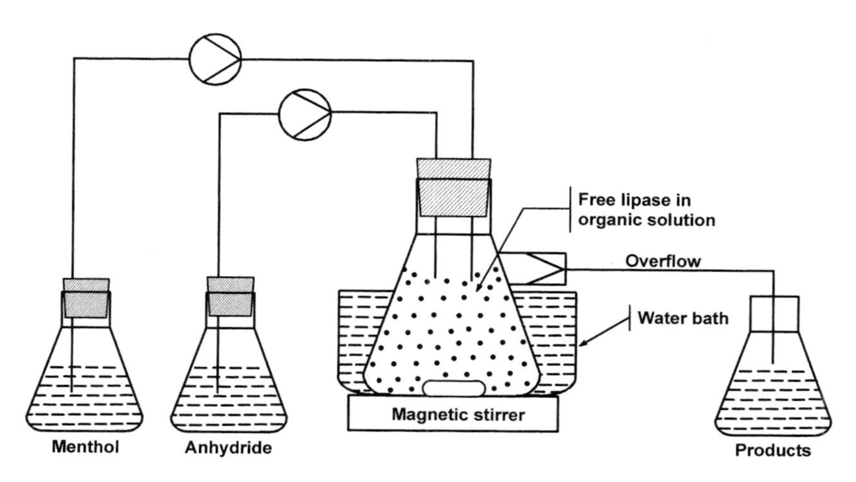

Fig. 12. Continuous laboratory-scale production of menthyl propanoate (adapted from XU et al., 1995b).

speaking, makes a comparison only partially permissible. The use of surfactant-coated enzymes in the latter case raised the reaction rate of the chosen system by more than 100 times. Thus, from a commercial point of view, such an enzyme-modifying technique should be considered to be potentially helpful in transferring promising laboratory-scale bioconversion processes into industrial applications.

3.4 Bicyclic Monoterpenoids

3.4.1 Pinenes

The pinenes are the main constituents of turpentine oil (up to 75–90%) and are also found in relevant amounts in the essential oils of non-coniferous plants, e.g., up to 12% in citrus oils (OHLOFF, 1994). 25% of the world's production of α-pinene (160,000 t) and β-pinene (26,000 t) are used as inexpensive starting material for the chemical synthesis of a wide variety of terpenoid flavor and fragrance compounds (OHLOFF, 1994; MIMOUN, 1996). Due to their broad availability and resulting low price, the pinenes also represent an ideal natural raw material for biotechnological approaches and have, therefore, been studied intensively in microbial conversion experiments since the early 1960s (BHATTACHARYYA et al., 1960; SHUKLA and BHATTACHARYYA, 1968; NARUSHIMA et al., 1982; TRUDGILL, 1994). As for most terpenes, the microbial metabolism of the pinenes often leads to diverse degradation pathways and therefore to a wide variety of products with different concentrations (see Tab. 3), thus complicating the development of bioprocesses of potential commercial interest. This effect is even more marked in the case of the structurally more complex bicyclic monoterpenes compared to acyclic and monocyclic monoterpenes (MIKAMI, 1988). For instance, RHODES and WINSKILL (1985) patented the microbial preparation of the (−)-enantiomer of carvone (**23**), the spearmint key impact compound, by bioconversion of α- or β-pinene (**45** and **54**) with the *Pseudomonas* strain NCIB 11671 which grew on pinene as the sole carbon source. The final concentrations of 13.4 mg L^{-1} after 48 h cultivation in Erlenmeyer flasks or of 7.4 mg L^{-1} in a 75 L bioreactor after 25 h were obviously too low for potential industrial use. Several metabolic pathways for conversion and degradation of α-pinene have been described, of which even the initial transformation step may be different. Some examples of initial reaction steps are allylic oxidation to form bicyclic terpenols, e.g., verbenol, the cleavage of the cyclobutane ring after initial protonation of the double bond leading to limonene and its derivatives, such as the perilla series, and the oxidation of the double bond to form α-pinene epoxide, from which biotransformations to bicyclic, monocyclic and acyclic monoterpenoids are possible (e.g., Tab. 3, PREMA and BHATTACHARYYA, 1962; SHUKLA and BHATTACHARYYA, 1968; BEST et al., 1987). For a comparison of different degradation routes, see the review by TRUDGILL (1994). The wide range of terpenoid compounds accessible by microbial conversion of α- and/or β-pinene (**45** and **54**) is illustrated in Fig. 13, which provides supplementary examples to those given in Tab. 3.

The bioconversion of α-pinene (**45**) via its epoxide to (Z)-2-methyl-5-isopropyl-2,5-hexadienal (isonovalal) (**117**), which can undergo further non-enzymatic isomerization to form the (E)-isomer (novalal) (**118**), by both *Pseudomonas fluorescens* NCIMB 11671 and the *Nocardia* sp. strain P18.3 was described by BEST et al. (1987) and GRIFFITHS et al. (1987a, b) respectively. In a patent, the production of these novel acyclic aldehydes as well as of E/Z-2-methyl-5-isopropyl-2,4-hexadienal by *P. fluorescens* NCIMB 11671 was claimed (BURFIELD et al., 1989). Due to their citrus- to woody-like organoleptic properties the novalic aldehydes were reported to be useful for perfume compositions. In an example given in the aforementioned patent, 38 g of a pale yellow oil, consisting of 92.8% isonovalal, was recovered by organic solvent extraction from a 20 L batch fermentation corresponding to a yield of 58.9%. The medium-scale bioprocess was started by adding α-pinene epoxide to a culture of *P. fluorescens*, pregrown on α-pinene and continued for 3 h. On a 2 L scale, even higher product concentrations were achieved by an adapted bioprocess design which counteracted the typical drawbacks in terpenoid transformation, such as high sub-

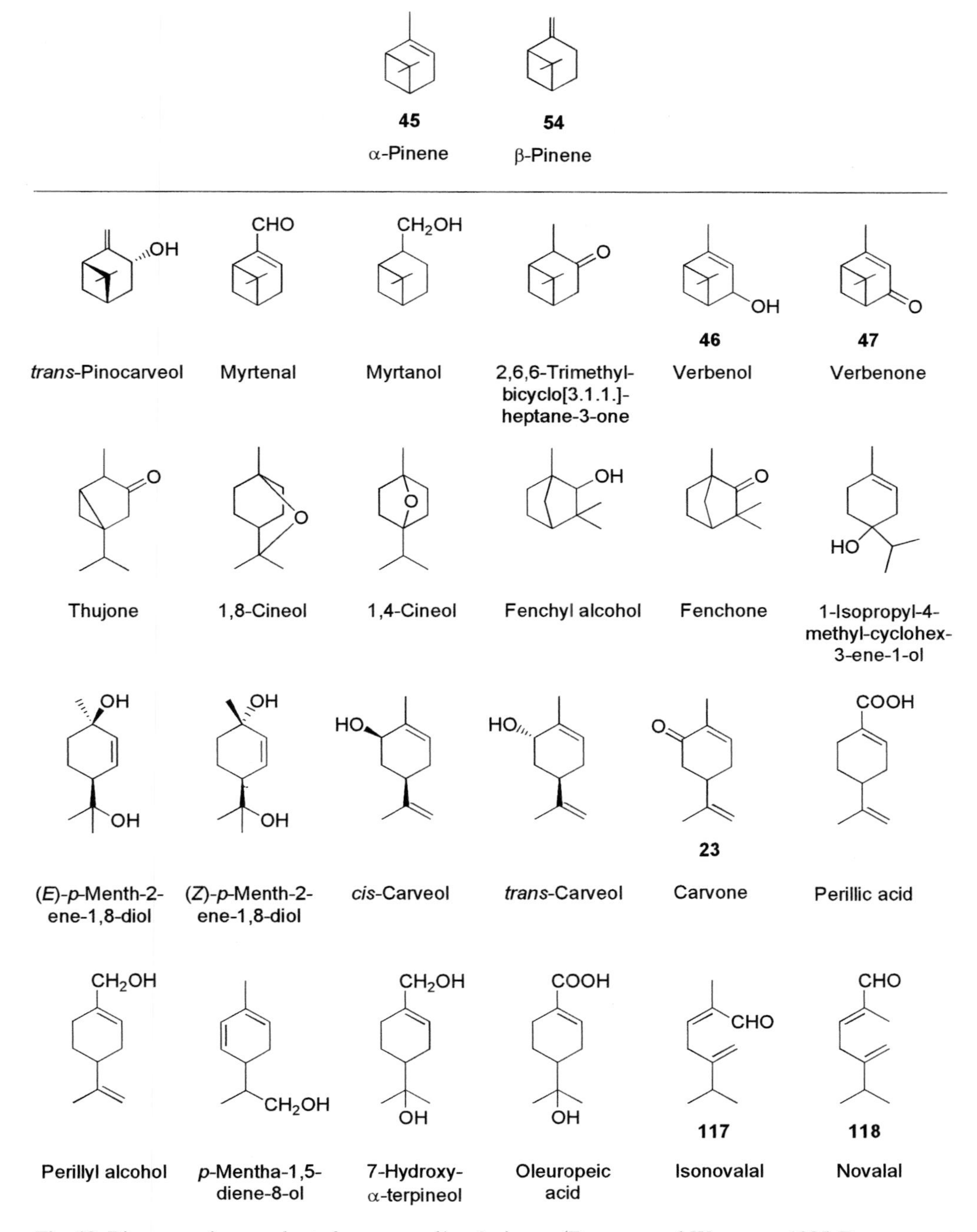

Fig. 13. Bioconversion products from α- and/or β-pinene (RHODES and WINSKILL, 1985; BURFIELD et al., 1989; MIKAMI, 1988; BUSMANN and BERGER, 1994a, b; TRUDGILL, 1994; SEITZ, 1994).

strate and product cytotoxicity, low substrate solubility, and further degradation of the desired product (BERGER et al., 1999). By using membrane aeration, fed-batch addition of α-pinene epoxide, and addition of Lewatit for *in situ* product adsorption, a final isonovalal concentration of 57 g L^{-1} (66% yield) was achieved after 5 h.

Recently, AGRAWAL et al. (1999) increased the biotransformation efficiency of *Aspergillus* sp. and *Penicillium* sp. by UV-induced mutation up to 15-fold compared with that of wild types. By this means, an *Aspergillus* sp., which had been isolated from soil, precultivated in complex medium, and transferred into phosphate buffer, transformed 200 mg L^{-1} α-pinene (**45**) into 45.6 mg L^{-1} of the camphor- and mint-like verbenol (**46**) (23% conversion yield). Furthermore, obviously no significant amounts of by-products formed; the dehydrogenase-supported oxidation of the alcohol to the corresponding aldehyde, verbenone (**47**), was depressed by the same mutation, as illustrated by parallel transformation experiments with verbenol as the sole substrate. These results suggest that the classical method of strain improvement by UV treatment is still an interesting strategy for bioprocess development, especially for application in the food industry where genetic engineering still suffers from public concern and lack of understanding. Another screening of fungal cultures led to the selection of an *Aspergillus niger* strain, which showed increased activity in the bioconversion of α-pinene to verbenone (AGRAWAL and JOSEPH, 2000). Due to its increased volatility, the ketone has a higher flavor impact than the corresponding alcohol. Besides the glucose concentration in the preculture medium, it was especially the variables in the buffer reaction mixture, such as pH, substrate concentration, and incubation period, that showed sharp maxima and therefore had to be carefully adjusted. With resting cells, pregrown until the late-exponential phase in potato-dextrose broth with 6% (w/v) glucose, 200 mg L^{-1} α-pinene (**45**) were converted into 32.8 mg L^{-1} verbenone (**47**) at pH 7.0 after 6 h of incubation. Elevated concentrations of α-pinene and prolonged incubation time effected a decrease in bioconversion yield thus indicating substrate toxicity and product degradation. No mention was made of the concentration course of verbenol, the precursor of the proposed two-step bioconversion. By contrast, VAN DYK et al. (1998b) described a mixture of verbenone and *trans*-verbenol as the main products in an α-pinene conversion using a self-isolated black yeast *Hormonema* sp. Although yielding the highest product concentrations reported so far with respect to α-pinene bioconversion experiments (0.3 g L^{-1} verbenone and 0.4 g L^{-1} *trans*-verbenol after 96 h), the unwanted morphological characteristics of the microorganism restricted further process development, as mentioned above (see Sect. 3.2.1).

3.4.2 Cineol, Carene, Camphor

Together with the pinenes, 1,8-cineol, 3-carene, and camphor represent other readily available bicyclic monoterpenes that are useful as inexpensive starting materials for bioconversions. The different biochemical pathways of microbial degradation of these terpenoids were reviewed in detail by TRUDGILL (1994). LIU and ROSAZZA (1990) described the stereospecific hydroxylation of 1,8-cineol (**57**), the chief constituent of eucalyptus oils, to *exo*-6-hydroxy-1,8-cineol (**119**) by using *Bacillus cereus* UI-1477. During 24 h, 1 g L^{-1} of the substrate was transformed to 0.74 g L^{-1} of the product, a remarkable product yield of 74% (Fig. 14). A model of the enzymatic reaction mechanisms was proposed.

(+)-3-Carene (**120**), the major component of Indian turpentine oil (SOMAN, 1995), together with (+)-2-carene, which is accessible by chemical isomerization, were the substrate for bioconversion studies using *Mycobacterium smegmatis* DSM 43061 (STUMPF et al., 1990). Products with insect-repelling properties, chaminic acids (**122**), which are difficult to obtain by chemical means, were produced as the main metabolites whereas volatile carenones (**121**), and, after cleavage of the cyclopropane ring, a monocyclic terpenol (**123**) formed as by-products (Fig. 14).

In Volume 8a of the 2nd Edition of *Biotechnology*, additional examples of recent work on microbial hydroxylations of, among others, 1,4- and 1,8-cineol are given (HOLLAND, 1998). In the same volume, the degradation pathways of

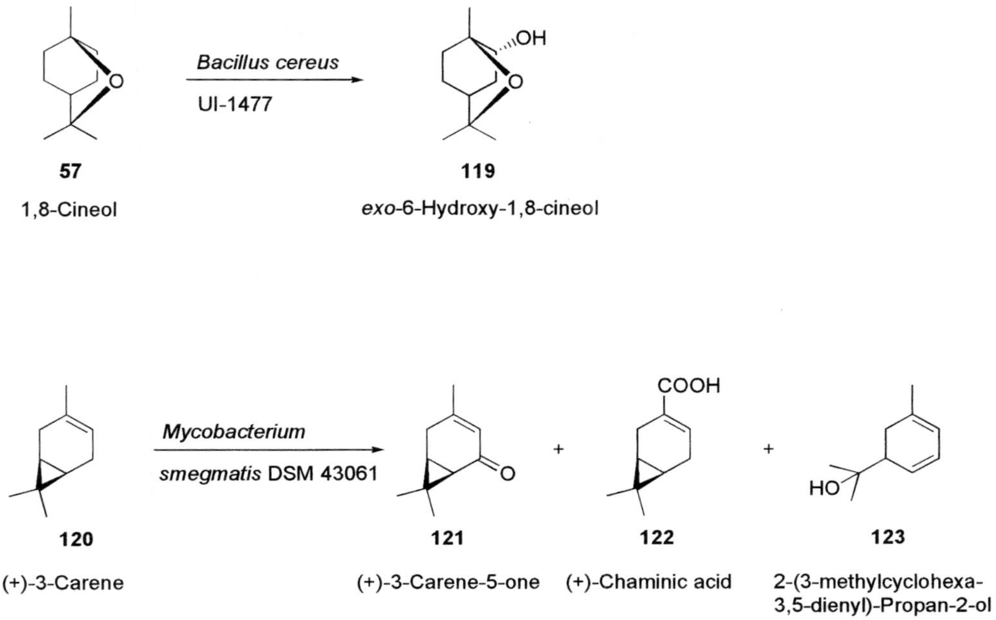

Fig. 14. Microbial conversion of 1,8-cineol and (+)-3-carene (LIU and ROSAZZA, 1990; STUMPF et al., 1990).

camphor, fenchone, and 1,8-cineol by microbial Baeyer–Villiger reactions are also described (KELLY et al., 1998).

3.5 Sesquiterpenes and C13 Norisoprenoids

Biogenetically derived from the trimeric precursor farnesyl diphosphate, sesquiterpenes constitute the structurally most diverse class of terpenoids. Sesquiterpenes play key roles in food flavors and fragrances and impart many other bioactivities. Their difficult total synthesis coupled with the abundance of non-functionalized, economically less important sesquiterpene hydrocarbons in many essential oils have stimulated a lot of research into the biotransformation of many sesquiterpene substrates from the over 70 subclasses. An impressive overview on work that had been carried out before 1990 was presented by LAMARE and FURSTOSS (1990); among the compounds biotransformed were: the acyclic representatives farnesol and nerolidol, the monocyclic humulene, germacrene and elemol, the bicyclic caryophyllene, cyperone, eudesmane, valencene, and the tricyclic patchoulol, cedrene/cedrol, santalene/santalol, (iso)/longifolene, guaioxide, liguloxide, kessanes, and other less known compounds. The following discussion will be limited to some more recent examples.

The three farnesene isomers (**124**, **125**, **126**) were fed to strains of *Arthrobacter*, *Bacillus*, *Nocardia*, and *Pseudomonas* with the aim to produce precursors of sinsensal (**127**), an orange flavor impact component (Fig. 15), but only little conversion was achieved (ABRAHAM et al., 1992a). When the diene moiety was protected by reacting with sulfur dioxide, the resulting sulfolenes were hydroxylated to ω-hydroxyfarnesene sulfones with a yield of up to 27%. Some strains discriminated the *E*- and *Z*-configurated substrates. It was suggested that the dienophile could be cleaved off after biotransformation simply by heating. Interestingly, species with known terpene transformation capabilities, such as *Diplodia gossy-*

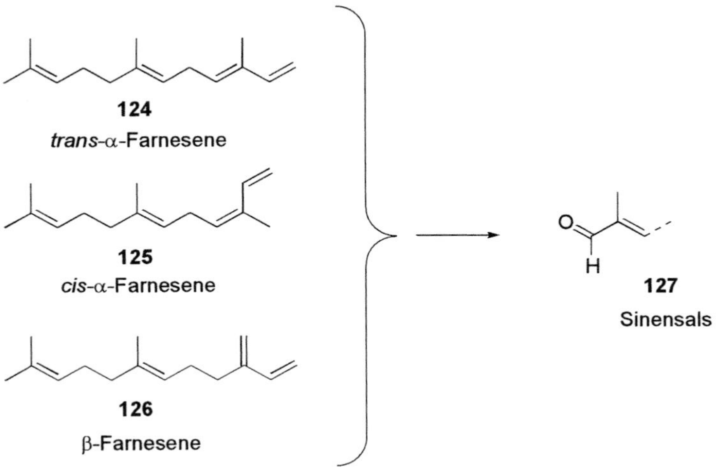

124
trans-α-Farnesene

125
cis-α-Farnesene

126
β-Farnesene

127
Sinensals

Fig. 15.

pina, did not work on the sulfone derivatives. Cell suspension cultures of *Achillea millefolium* L. ssp. *millefolium* (yarrow) converted a mixture of farnesol isomers to farnesenes and glycosylated products (FIGUEREIDO et al., 1996), reactions not observed with bacteria or fungi (LAMARE and FURSTOSS, 1990; MIYAZAWA et al., 1996; NANKAI et al., 1998).

Humulene is not a generally accepted substrate in microbial transformations. For example, strains of *Aspergillus niger*, often used in biotransformation experiments, left humulene almost unchanged. In continuation of earlier work several studies on the conversion of α-humulene were carried out using not only bacteria, but also higher fungi (ABRAHAM et al., 1989). Strains of *Diplodia gossypina*, *Corynespora cassiicola*, *Chaetomium cochliodes*, and *Mycobacterium smegmatis* produced a bewildering variety of metabolites indicating a parallel attack on different sites of the substrate molecule. Rapid epoxidation, hydroxylation, and further oxidative steps occurred. While these biotransformations yielded complex product mixtures and were, thus, less favorable for preparative use, the detection of novel bioactive principles, often with distinct stereochemistry, is a continuing driving force behind this type of work. One explanation for these observations may be that humulene in basidiomycete fungi is a versatile precursor of bicyclic and tricyclic sesquiterpenes of the illudane, hirsutane and lactarane types.

In a series of papers the group of MIYAZAWA described the conversion of cyclic sesquiterpenes using *Glomerella cingulata*, a plant pathogenic fungus. Among the substrates were (−)-globulol, (−)-α-bisabolol, (+)-cedrol, (+)-aromadendrene, (−)-alloaromadendrene, β-selinene (**128**) (MIYAZAWA et al., 1997a), and γ-gurjunene (MIYAZAWA et al., 1998 and references cited therein). Typically, *G. cingulata* oxidized the substrates at the double bond of an exomethylene group or an isopropenyl group non-stereoselectively, while the hydroxylation of a ring methylene moiety proceeded stereoselectively. This preferential attack of carbon atoms led to a smaller number of transformation products compared to the conversion of humulene (Fig. 16). However, the results of this type of biotransformation experiment are still difficult to predict: a strain of the basidiomycete *Marasmius alliaceus* hydrated the exomethylene double bond of β-bisabolene (**129**) to produce α-bisabolol (**130**) with an apple blossom-like odor, whereas other basidiomycetes, such as *Mycena pura*, *Ganoderma applanatum*, *Kuehneromyces erinaceus*, and *Trametes hirsuta*, deprenylated the substrate molecule to yield *trans*-carenol (**131**) (Fig. 17) (BUSMANN, 1994).

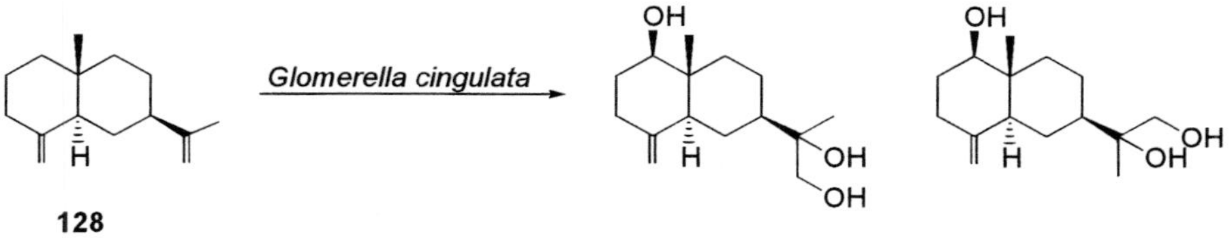

Glomerella cingulata

128

β-Selinene

Diastereomer pair of (1*S*,6*S*,9*S*,10*R*,11*RS*)-1,11,13-trihydroxy-β-Selinene

Fig. 16.

The first attempt to replace the chromium oxide catalyzed allylic oxidation of valencene (**132**) to (+)-nootkatone (**134**), a grapefruit flavor impact, by biotransformation dates back to 1973 and is described in detail by LAMARE and FURSTOSS (1990). Less well characterized strains of *Enterobacter* produced a mixture of compounds, in most of which the eremophilan skeleton was retained. Nootkatone was identified as a by-product only, although allylic/benzylic oxidations are thought to be major pathways in microbial conversion reactions (AZERAD, 1993). More recent experiments showed that a lactoperoxidase system transformed valencene to isomeric alcohols (possibly nootkatols, DRAWERT et al., 1984) and nootkatone, while a lignin peroxidase from *Phanerochaete chrysosporium* was much less efficient (WILLERSHAUSEN and GRAF, 1991). A Japanese patent used a *Rhodococcus* strain isolated from soil to perform the same reaction (OKUDA et al., 1994). Suspended photomixotrophic plant cells of *Citrus paradisi* (grapefruit) converted exogenous valencene (**132**) to 2-hydroxyvalencene (**133**) and to nootkatone (**134**) (Fig. 18), which was not further degraded within 72 h of incubation (REIL and BERGER, 1996). A comparable oxidation of (+)-limonene by endogenous enzymes of citrus peel was recently reported (LEE et al., 1999).

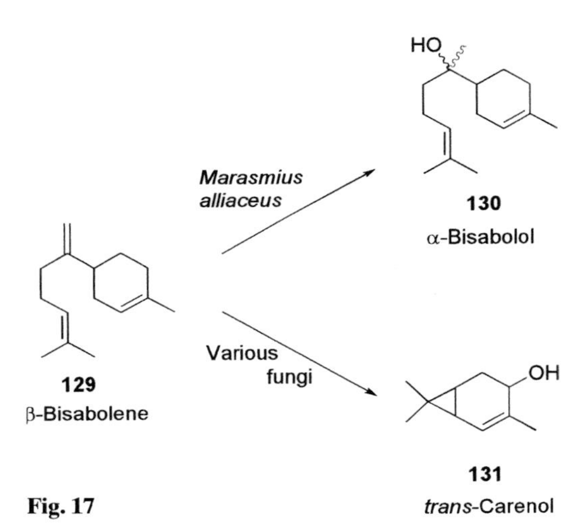

Fig. 17

The chemical basis of the antifungal, insect repelling and unique sensory properties of the essential oil of *Pogostemon patchouli* Pell. (patchouli) has been a matter of controversy.

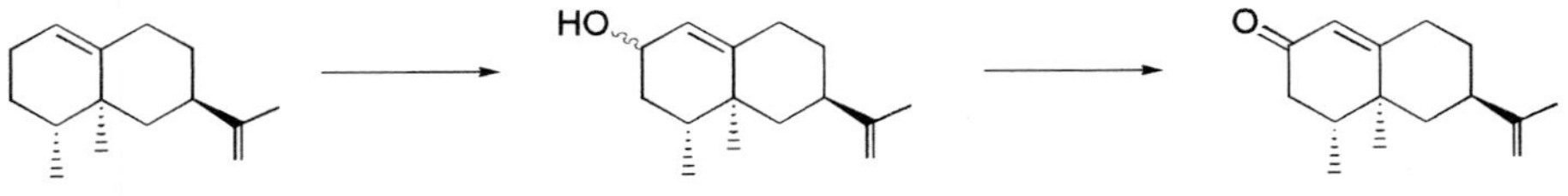

132 Valencene

133 2-Hydroxyvalencene

134 Nootkatone

Fig. 18.

It was demonstrated, however, that peracid oxidation of the hydrocarbon fraction (mainly α- and β-patchoulene) yields epoxides and other oxidation products of commercial interest. As a result of the low water solubility of the hydrocarbons, most of the experiments published deal with the conversion of the abundant tricyclic monoalcohol patchoulol (**75**). Depending on strain and conditions, varying proportions of dihydroxy and trihydroxy compounds formed. For example. when patchoulol was used to control the growth of the plant pathogen *Botrytis cinerea*, the fading activity was explained by a detoxification reaction of the fungus leading to hydroxy derivatives (Fig. 19). These in turn were regarded as substrates for subsequent degradation by *retro*-Prins, *retro*-aldol, or Baeyer-Villiger-type reactions (ALEU et al., 1999).

Juniperus virginiana L. is the botanical source of cedar wood essential oil. In close analogy to patchouli, the main hydrocarbons (−)-α-cedrene and (−)-thujopsene are of little economic importance, while (+)-cedrol, α-cedrene epoxide, and other oxyfunctionalized derivatives impart the key sensory attributes. According to earlier work (LAMARE and FURSTOSS, 1990) α-cedrene is a rather poor conversion substrate, and only TAKIGAWA et al. (1993) have reported significant yields of cedrenol and α-curcumene upon allylic hydroxylation and ring cleavage. In contrast, (+)-cedrol (**135**) (Fig. 20) was converted by a strain of *Bacillus cereus* to 2*S*-hydroxycedrol as the only detectable metabolite, while a *Streptomyces griseus* produced numerous monoalcohols, diols, ketols, and 4-oxocedrol (MAATOOQ et al., 1993). Similar mono- and dihydroxylations were obtained with calarene and globulol incubated in the presence of *Bacillus megaterium*, *Mycobacterium smegmatis*, or *Diplodia gossypina* (ABRAHAM et al., 1992b). These tricyclic sesquiterpenes were preferentially oxidized at the *exo*-orientated methyl group of the cyclopropane ring. It was noted that some of the metabolites formed were stereoisomeric to known bioactive natural compounds.

The above examples suggest that a second oxygen is usually introduced at secondary or tertiary (angular) carbons remote from the first oxo function. However, it appears easy to find biocatalysts to prove the opposite. In view of the lack of predictability of the regio- and stereoselectivity of biotransformations and the inevitably tedious and empirical screening, attempts were made to rationalize the enzymatic approach by correlating product groups or by correlating volatile products with the taxonomic position. In a study on 25 different strains of *Fusarium sambucinum* grown on wheat kernels, e.g., it was found that the strains that produced high concentrations of trichothecenes also produced volatile sesquiterpenes, such as α- and β-farnesene, β-chamigrene, β-bisabolene, longifolene, and others (JELEN et al., 1995). Fifteen strains which did not produce any trichothecenes did produce less volatile sesquiterpenes, with less chemical diversity. No oxidized sesquiterpenes from these fungi were reported. Practical experience gathered over many years was presented in a comparative study that included 40 bacterial

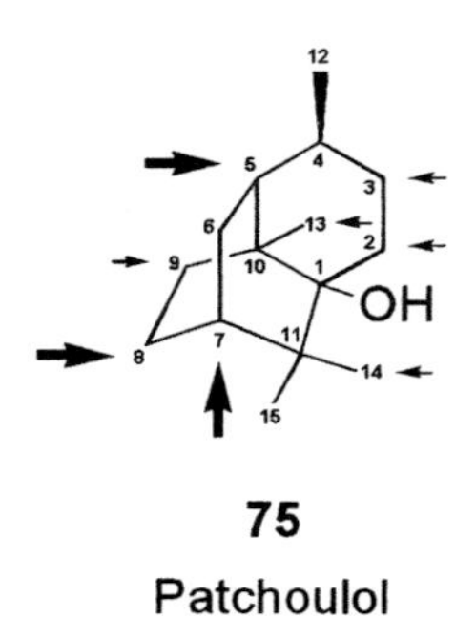

Fig. 19. Preferred sites of attack of oxygen on patchoulol, catalyzed by *Botrytis cinerea* (according to ALEU et al., 1999).

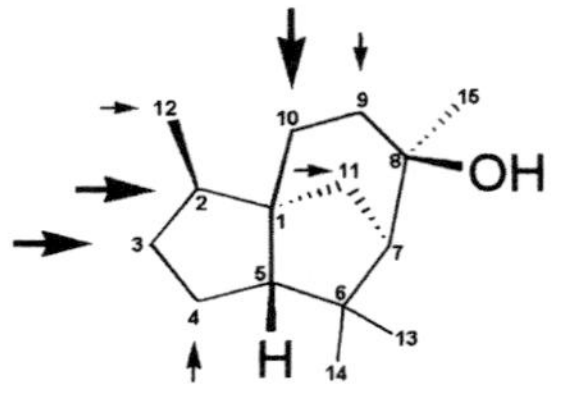

OH
H

135

(+)-Cedrol

Fig. 20. Preferred sites of attack of oxygen on cedrol, catalyzed by various microorganisms.

and 60 fungal strains (ABRAHAM, 1994). Fungal phyla and Gram-positive and Gram-negative bacteria could all be distinguished in a comprehensive statistical analysis using 12 different terpene substrates. Five microbial groups with distinct biotransformation activity were finally separated by discriminant analysis on the basis of 30 metabolites, mainly from globulol, cedrol, and 1,8-cineol. Growth rates indicated that none of the substrates were toxic to any strain under the conditions used. The author speculated that if the transport of substrate into the cell was a crucial step, different types of cell walls as associated with different taxonomy would explain the different biotransformation activity.

Some C13-norisoprenoids of the ionone/damascone family (Fig. 21) belong to the most potent fragrance compounds in nature (odor threshold in water β-ionone (**62**): 0.007 μg L^{-1}, β-damascenone (**137**): 0.009 μg L^{-1}). Fungal transformation of ionones to various hydroxy and oxo derivatives has been reported earlier. With the aim of detecting *Streptomyces* strains with powerful cytochrome P-450-dependent monooxygenases, a screening was performed on 215 strains using α- and β-ionone as a substrate (LUTZ-WAHL et al., 1998 and references cited therein). When α-ionone (**136**) was the substrate, conversion rates of six strains exceeded 90%, and the desired 3-hydroxyionone

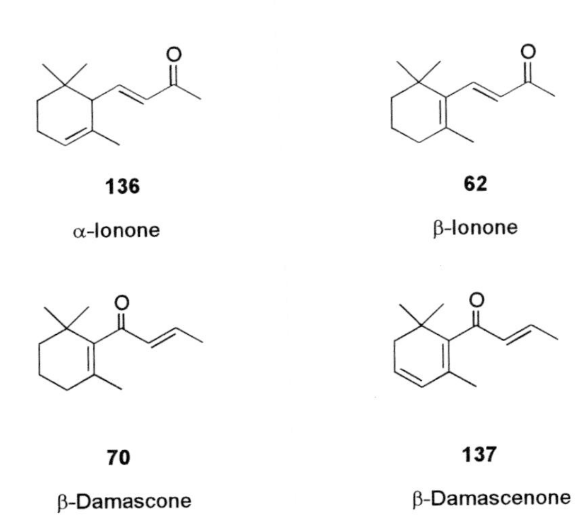

Fig. 21.

was formed without a multitude of by-products. Stereoselectivity was also remarkable with some of the strains: racemic α-ionone was transformed only into the *trans*-configurated (3*R*,6*R*)- and (3*S*,6*S*)-alcohols. An equatorial orientation of the newly introduced hydroxy function was concluded from ^{1}H, ^{1}H-NMR long-range coupling measurements. Although β-ionone was added above its solubility concentration and likewise well transformed, only little conversion to the 4-hydroxy product and no conversion to the 3-hydroxy product was found. More suitable for the transformation of β-ionone are fungal biosystems, as was demonstrated by a fed-batch process using *Aspergillus niger* strains (LARROCHE et al., 1995). From the recovery of 2- and 4-oxo and hydroxy compounds a transformation yield of close to 100% was calculated. 2.5 g L^{-1} of volatiles were collected after a cultivation time of 230 h. A dynamic model of the two-phase system was developed to show that liquid–liquid mass transfer rates were limiting (GRIVEL et al., 1999). The loss of about 70% of the added ionone substrate by air stripping was noted as a severe drawback, and the use of an organic cosolvent was suggested to reduce the solute activity. An enzymatic system for the generation of β-ionone from β-carotene consisted of xanthine oxidase and xanthine or acetaldehyde (or butanal) (BOSSER and BELIN, 1994). Degradation of β-carotene did not occur in the presence of the common substrate xanthine. Radical mechanisms and substrate solubilization were discussed as important factors. Fungal catalysis was also successfully applied to transform α- and β-damascone (**70**) to hydroxy and oxo-products (SCHOCH et al., 1991; SCHWAB et al., 1991). Depending on the composition of medium and strain of *Botrytis cinerea* used, almost full regioselectivity of oxygen insertion was achieved.

Apart from their well known sensory activities, sesquiterpenes and related compounds also show less well characterized bioactivities, such as antimicrobial, insect repelling or alternative medicine properties (CUTLER, 1995). For example, caryophyllene, δ-cadinene, and β-ionone were among the 10 most abundant volatiles of green tea, and the inhibitory activity exerted by green tea on the microflora of the oral cavity was assigned to these volatiles.

Acyclic sesquiterpenes were among the compounds named in a recent US patent that claimed the use of these phytochemicals in cancer therapy (MYERS et al., 1997). Little is still known on the enzymatic background of these bioactivities. However, first reports on a direct *in vitro* inhibition of cytochrome enzymes by sesquiterpenes (SIME et al., 2000) and by β-ionone and other terpenes (DE-OLIVEIRA et al., 1999) have appeared. A straightforward conclusion is that the low product yields usually reported for microbial conversions of sesquiterpenes are not only caused by low substrate solubility. In the classical batch processes, the substrates were administered all at once, and a simple product inhibition of the functionalizing cytochrome enzymes may have occurred rapidly. Growth inhibition and loss of cell vitality, frequently reported from bacterial and fungal strains, would then be easily explained by the inhibition of detoxification pathways; as a result, the apolar substrates would accumulate in outer and inner membrane lipids, hence rendering it impossible for the cell to further maintain the physiological gradients.

4 Future Aspects

It may be concluded that biotechnological production of terpenoid flavor and fragrance compounds, although the focus of research activities for several decades, has only found industrial application in a few cases. Recently some promising attempts have been made to overcome the inherent drawbacks of terpene and terpenoid bioconversion. The main goal always has to be an increase in product yield and ultimately in overall productivity. It can be assumed that flavor and fragrance biotechnology stands a good chance of commercial exploitation once the order of magnitude of the economic data of its competitor, chemical synthesis, has been achieved: the sustainability of environmentally friendly bioprocesses, the sensory quality of bioflavors and biofragrances, which is often superior to their chemical counterparts, and, last but not least, the declaration advantage with respect to food legislation and to the consumers' preference for natural products and processes represent values favoring biotechnology to challenge traditional flavor and fragrance manufacture. Furthermore, whole cell processes enable multistep reactions "in one", which makes this type of biocatalysis attractive for the generation of more complex compounds and a strong competitor to a costly chemical reaction sequence. Moreover, after clarifying the cell metabolism, more simple targets may allow the reduction of whole cell biocatalysts to their basic biochemical structures and active enzymes, ideally immobilized for prolonged reuse, thus providing more economically viable processes.

The scientific literature cited in this chapter illustrates the great potential of microbial systems in the field of terpenoid production. A large variety of metabolic pathways results in aroma-relevant intermediates and end products. More and more sophisticated biochemical methods leading to metabolic flux elucidation accompanied by modern flavor and fragrance analysis allow deeper insights into the physiological mechanisms. By this means "biological" parameters become achievable targets of process optimization which go beyond standard methods, such as strain screening and media optimization. As in the biopharmaceutical sector, in all biotechnology-based production processes metabolic engineering represents a research-intensive but very often straightforward strategy to improve the yield of a desired product, e.g., by selective enhancement of the corresponding pathway of a promising strain or by gene transfer into a recombinant host, tailored for bioprocessing. Thus, despite the uncertainties of the legislative situation and public perception characteristic of the food sector, the number of publications on genetic engineering of flavor and fragrance synthesis has been steadily increasing during the past few years. Some examples are given in Tab. 5.

In addition to the improvement of biological parameters, the integration of innovative techniques into bioprocess performance will play a key role in the setting up of modern, unconventional processes which are necessary to overcome the inherent drawbacks of terpene and terpenoid bioconversion and to make the transition into the commercial synthesis of "bioterpenoids". Some of these innovations

Tab. 5. Recent Genetic Engineering Approaches in Terpenoid Flavor and Fragrance Biotechnology

Target Compound(s)	Genetic Engineering	Reference
Myrcenol	identification and sequencing of myrcene catabolism genes, transposon mutagenesis (*Pseudomonas* sp.)	IURESCIA et al. (1999)
Terpenoid synthesis	overexpression of *Thermus thermophilus* and *Sulfolobus acidocaldarius* geranylgeranlyl diphosphate synthases in *E. coli*	OHTO et al. (1998, 1999)
Terpenoid synthesis	overexpression of *Bacillus stearothermophilus* farnesyl diphosphate synthase in *E. coli*	KOYAMA et al. (1994)
Geraniol, linalool, citronellol, farnesol	genetic studies and crossing experiments with *Saccharomyces cerevisiae* to increase terpenol production	CHAMBON et al. (1990), JAVELOT et al. (1991)
Sabinene, 1,8-cineol, bornyl diphosphate	isolation, sequencing and recombinant expression of DNA encoding (+)-bornyl diphosphate synthase, 1,8-cineol synthase, and (+)-sabinene synthase from common sage (*Salvia officinalis*)	CROTEAU et al. (1999)
(+)-Limonene, myrcene, (−)-pinene	isolation, sequencing and recombinant expression of DNA encoding (+)-limonene synthase, myrcene synthase, and (−)-pinene synthase from grand fir (*Abies grandis*)	BOHLMANN et al. (1999)
Perillyl alcohol, α-terpineol, perillaldehyde, carveol, carvone	cloning and expression of a (+)-limonene degradation pathway of *Bacillus stearothermophilus* in *E. coli*	ORIEL et al. (1997), SAVITHIRY et al. (1997)
Camphor derivatives	elucidation of gene regulation of camphor degradation in *Pseudomonas putida* ATCC 17453	ARAMAKI et al. (1993, 1995)

may be integrated by transfer from other biotechnological applications which are confronted with comparable constraints, such as substrate toxicity and volatility or the complexity of bioconversion and biodegradation pathways (e.g., the microbial degradation of environmentally harmful petrochemicals). Other improvements may come about as a consequence of an individual bioprocess characterization to identify weak points which could be eliminated by tailored, technical modifications. Tab. 6 summarizes whole cell terpene bioconversions from the viewpoint of technical process innovations.

Terpenoids are of increasing interest as their supplementary bioactivities, which exceed their flavor and fragrance qualities, can open new areas of application. The use of spices and herbs has played an important role in human nutrition for thousands of years and has to be put down to their antioxidative, preservative and medicinal properties besides the flavoring ones (NAKATANI, 1994; LOZA-TAVERA, 1999). The broad spectrum of biological activities of essential oils and their mainly terpenoid constituents was reviewed recently (NAKATSU et al., 2000). During the last decade, additional functions as repellents, attractants, allochemicals, and medicines have become increasingly obvious (CUTLER et al., 1996). Thus, the natural substance class of terpenoids fits in perfectly with current views on modern foodstuffs, where nutrition is being increasingly fused with convenience and health aspects. Furthermore, the purely agrochemical and pharmaceutical functions of terpenoids will widen the area of application beyond food flavorings and fragrances. For instance, several terpenoids are reported to interact with detoxifying enzymes or carcinogenic metabolic intermediates, so that even a future application in the prevention and treatment of cancer may be possible. Tab. 7 shows supplementary bioactivities of several terpenoid flavor and fragrance compounds, which are amenable by microbial processes.

Tab. 6. Some Bioprocess Modifications for Improved Terpenoid Flavor and Fragrance Production

Drawback	Technological Improvement	Reference
Substrate properties		
(Cytotoxicity, volatility, low water solubility)	on line monitoring and controlled feeding	ABRAHAM et al. (1985)
	immobilization	TAN and DAY (1998a, b)
	solvent-tolerant strains in organic–aqueous biphasic media	SPEELMANNS et al. (1998), TEUNISSEN and DE BONT (1995)
	microorganisms grown on the interface between nutrient agar and organic solvent ("interface bioreactor")	ODA et al. (1995, 1996a, b)
	sporulated surface cultures of filamentous fungi	DEMYTTENAERE and DE POOTER (1996)
	sequential feeding of non-toxic amounts of the substrate; substrate induction	TAN et al. (1998), LARROCHE et al. (1995)
	circumvention of "phase toxicity" by supplying the substrate via the gas phase	ONKEN and BERGER (1999a, b)
	strain adaptation by precultivation in the presence of the substrate	
	avoiding emulsification problems of aqueous–organic biphasic systems by using membrane bioreactors	DOIG et al. (1998), WESTGATE et al. (1998)
	decrease of loss of volatile substrate by aeration via membrane	ONKEN and BERGER (1999a)
	use of organic cosolvents	TAN and DAY (1998b)
Product properties		
(Toxicity, feedback inhibition, volatility)	continuous *in situ* adsorption to polymeric resins	KLINGENBERG and HANSSEN (1988), KRINGS et al. (1993), BERGER et al. (2000)
	in situ product recovery by organic phase/substrate phase	SAVITHIRY and ORIEL (1998), DOIG et al. (1998)
Metabolic properties		
(Low conversion yield, unwanted by-products)	recombinant microorganism with enzyme overexpression (see Tab. 5)	ORIEL et al. (1997), SAVITHIRY and ORIEL (1998)
	enrichment culture; screening	VAN RENSBURG et al. (1997)
	UV-induced mutagenesis	AGRAWAL et al. (1999)

By using plant-based raw materials and affording closed material cycles, biotechnological production processes are expected to make a significant advance towards sustainability in the synthesis of fine chemicals in the near future because they are based on renewable plant sources and catalysts and can be operated with closed mass balances. Flavor and fragrance production plays a significant role in the rapidly expanding fine chemicals industry (BARBER, 1996). Chirality is a major principle of molecular interactions in almost all areas of life. Especially in the field of terpenoid flavors and fragrances, biocatalysis, which is very often unrivalled with respect to its stereoselectivity, will undoubtedly be further intensified. Thus, both the processing and the product contribute to the uniqueness of biotechnological processes.

Tab. 7. Supplementary Bioactivities of Several Terpenoid Flavor and Fragrance Compounds

Compound(s)	Bioactivity (Potential Use)	Reference
Different terpenoids	antimicrobial	GRIFFIN et al. (1999),
Nootkatol, nootkatone	inhibiting cytochrome P450 enzymes (co-administration pharmaceutical)	SIME et al. (2000)
Limonene, perillyl alcohol, carveol, menthol, sobrerol, farnesol	antitumorigenic (chemotherapeutic agents and prodrugs)	CROWELL et al. (1996), REDDY et al. (1997), MYERS et al. (1997), LOZA-TAVERA (1999)
Citral, geraniol, linalool, cineol, menthol	antifungal, antibacterial	PATTNAIK et al. (1997)
Pulegone, carvone	inhibiting acetylcholinesterase (neuropharmacology)	MIYAZAWA et al. (1997b)
Citronellol, geraniol	inhibiting tyrosinase (anti-browning of fruits, skin-whitening)	NAKATSU et al. (2000)
Verbenol, verbenone	insect pheromones	REECE et al. (1968), RYKER and YANDELL (1983)
β-Ionone	inhibiting liver monooxygenase (anticarcinogenic agent)	DE-OLIVEIRA et al. (1999)

5 References

ABRAHAM, W.-R. (1994), Phylogenetic influences in microbial hydroxylation of terpenoids, *World J. Microbiol. Biotechnol.* **10**, 88–92.

ABRAHAM, W.-R., HOFFMANN, H. M. R., KIESLICH, K., RENG, G., STUMPF, B. (1985), Microbial transformations of some monoterpenoids and sesquiterpenoids, in: *Enzymes in Organic Synthesis. Ciba Foundation Symp. 111*, London, 15–17 May 1984 (PORTER, R., CLARK, S., Eds.), pp. 146–160. London: Pitman.

ABRAHAM, W.-R., ERNST, L., STUMPF, B., ARFMANN, H.-A. (1989), Microbial hydroxylations of bicyclic and tricyclic sesquiterpenes, *J. Ess. Oil Res.* **1**, 19–27.

ABRAHAM, W.-R., STUMPF, B., ARFMANN, H.-A. (1990), Chiral intermediates by microbial epoxidations, *J. Ess. Oil Res.* **2**, 251–257.

ABRAHAM, W.-R., ARFMANN, H.-A., GIERSCH, W. (1992a), Microbial hydroxylation of precursors of sinensal, *Z. Naturforsch.* **47c**, 851–858.

ABRAHAM, W.-R., KIESLICH, K., STUMPF, B., ERNST, L. (1992b), Microbial oxidation of tricyclic sesquiterpenoids containing a dimethylcyclopropane ring, *Phytochemistry* **31**, 3749–3755.

ABRAHAM, B., ONKEN, J., REIL, G., BERGER, R. G. (1995), Strategies toward an efficient biotechnology of aromas, in: *Flavour Perception/Aroma Evaluation, Proc. 5th Wartburg Aroma Symp.* (KRUSE, H.-P., ROTHE, M., Eds.), pp. 357–373. Bergholz-Rehbrücke: Eigenverlag Universität Potsdam.

ACOSTA, M., MAZAS, N., MEJÍAS, E., PINO, J. (1996), Obtencion de aromatizantes mediante biotransformacion del limoneno, *Alimentaria* **272**, 73–75.

AGRAWAL, R., JOSEPH, R. (2000), Bioconversion of *alpha* pinene to verbenone by resting cells of *Aspergillus niger*, *Appl. Microbiol. Biotechnol.* **53**, 335–337.

AGRAWAL, R., NAZHATH-UL-AINN, D., JOSEPH, R. (1999), Strain improvement of *Aspergillus* sp. and *Penicillium* sp. by induced mutation for biotransformation of α-pinene to verbenol, *Biotechnol. Bioeng.* **63**, 249–252.

AKOH, C. C., YEE, L. N. (1998), Lipase-catalyzed transesterification of primary terpene alcohols with vinyl esters in organic media, *J. Mol. Catal. B: Enzymatic* **4**, 149–153.

ALEU, J., HANSON, J. R., GALAN, R. H., COLLADO, I. G. (1999), Biotransformation of the fungistatic sesquiterpenoid patchoulol by *Botrytis cinerea*, *J. Nat. Prod.* **62**, 437–440.

ARAMAKI, H., SAGARA, Y., HOSOI, M., HORIUCHI, T. (1993), Evidence for autoregulation of *camR*, which encodes a repressor for the cytochrome P-450cam hydroxylase operon on the *Pseudomonas putida* CAM plasmid, *J. Bacteriol.* **175**, 7828–7833.

ARAMAKI, H., SAGARA, Y., KABATA, H., SHIMAMOTO, N., HORIUCHI, T. (1995), Purification and characterization of a *cam* repressor (CamR) for the cytochrome P-450cam hydroxylase operon on the *Pseudomonas putida* CAM plasmid, *J. Bacteriol.* **177**, 3120–3127.

ASAKAWA, Y., TAKAHASHI, H., TOYOTA, M., NOMA, Y. (1991), Biotransformation of monoterpenoids, (−)- and (+)-menthols, terpinolene and carvotanacetone by *Aspergillus niger*, *Phytochemistry* **30**, 3981–3987.

AZERAD, R. (1993), Microbiological hydroxylations: myths and realities, *Chimia* **47**, 93–96.

BARBER, M. S. (1996), Fine chemicals and biotechnology: the business and the markets, *Chimia* **50**, 438–439.

BAUER, K., GARBE, D., SURBURG, H. (1997), *Common Fragrance and Flavor Materials*, 3rd Edn. Weinheim: Wiley-VCH.

BERGER, R. G., LATZA, E., NEUSER, F., ONKEN, J. (1999), Terpenes and amino acids – progenitors of volatile flavours in microbial transformation reactions, in: *Frontiers of Flavour Science* (SCHIEBERLE, P., ENGEL, K.-H., Eds.), pp. 394–399. Garching: Deutsche Forschungsanstalt für Lebensmittelchemie.

BERGER, R. G., LATZA, E., KRINGS, U., PREUSS, A., TREFFENFELDT, W., VOLLMER, H. (2000), Process for selective enrichment and separation of aroma molecules by adsorption, *Eur. Patent* 0979806.

BEST, D. J., FLOYD, N. C., MAGALHAES, A., BURFIELD, A., RHODES, P. M. (1987), Initial enzymatic steps in the degradation of *alpha*-pinene by *Pseudomonas fluorescens* NCIMB 11671, *Biocatalysis* **1**, 147–159.

BHATTACHARYYA, P. K., PREMA, B. R., KULKARNI, B. D., PRADHAN, S. K. (1960), Microbiological transformation of terpenes: hydroxylation of α-pinene, *Nature* **187**, 689–690.

BOCK, G., BENDA, I., SCHREIER, P. (1986), Biotransformation of linalool by *Botrytis cinerea*, *J. Food Sci.* **51**, 659–662.

BOCK, G., BENDA, I., SCHREIER, P. (1988), Microbial transformation of geraniol and nerol by *Botrytis cinerea*, *Appl. Microbiol. Biotechnol.* **27**, 351–357.

BOHLMANN, J., CROTEAU, R. B., STEELE, C. L. (1999), Monoterpene synthases from grand fir (*Abies grandis*), *World Patent* 9902030.

BOSSER, A., BELIN, J. M. (1994), Synthesis of β-ionone in an aldehyde/xanthine oxidase/β-carotene system involving free radical formation, *Biotechnol. Prog.* **10**, 129–133.

BRADDOCK, R. J., CADWALLADER, K. R. (1995), Bioconversion of citrus *d*-limonene, in: *Fruit Flavors: Biogenesis, Characterization and Authentication: Proc. Symp.*, Chicago, MI: August 1993, pp. 142–148. Washington, DC: ACS.

BRUNERIE, P., BENDA, I., BOCK, G., SCHREIER, P. (1987), Bioconversion of citronellol by *Botrytis cinerea*, *Appl. Microbiol. Biotechnol.* **27**, 6–10.

BURFIELD, A. G., BEST, D. J., DAVIS, K. J. (1989), Production of 2-methyl-5-isopropylhexa-2,5-dien-1-al and of 2-methyl-5-isopropyl-hexa-2,4-dien-1-al in microorganisms, *Eur. Patent* 0304318.

BUSMANN, D. (1994), Oxofunktionalisierung und Katabolyse von Terpen-Kohlenwasserstoffen durch Basidiomyceten, *Thesis*, University of Hannover, Germany.

BUSMANN, D., BERGER, R. G. (1994a), Bioconversion of terpenoid hydrocarbons by basidiomycetes, in: *Trends in Flavour Research, Proc. 7th Weurman Flavour Res. Symp., Noordwijkerhout*, 15–18 June 1993 (MAARSE, H., VAN DER HEIJ, D. G., Eds.), pp. 503–507. Amsterdam: Elsevier Science B.V.

BUSMANN, D., BERGER, R. G. (1994b), Oxyfunctionalization of α- and β-pinene by selected basidiomycetes, *Z. Naturforsch.* **49c**, 545–552.

BUSMANN, D., BERGER, R. G. (1994c), Conversion of myrcene by submerged cultured basidiomycetes, *J. Biotechnol.* **37**, 39–43.

CHAMBON, C., LADEVEZE, V., OULMOUDEN, A., SERVOUSE, M., KARST, F. (1990), Isolation and properties of yeast mutants affected in farnesyl diphosphate synthase, *Curr. Genet.* **18**, 41–46.

CHANG, H. C., ORIEL, P. (1994), Bioproduction of perillyl alcohol and related monoterpenes by isolates of *Bacillus stearothermophilus*, *J. Food Sci.* **59**, 660–662.

CHANG, H. C., ORIEL, P. J. (1996), Preparation of perillyl compounds using *Bacillus stearothermophilus*, *U.S. Patent* 005487988A.

CHANG, H. C., ORIEL, P. J. (1997), Process and bacterial cultures for the preparation of perillyl compounds, *U.S. Patent* 5652137.

CHANG, H. C., GAGE, D. A., ORIEL, P. J. (1995), Cloning and expression of a limonene degradation pathway from *Bacillus stearothermophilus* in *Escherichia coli*, *J. Food Sci.* **60**, 551–553.

CHATTERJEE, T., BHATTACHARYYA, D. K. (1998a), Synthesis of monoterpene esters by alcoholysis reaction with *Mucor miehei* lipase in a solvent-free system, *J. Am. Oil Chem. Soc.* **75**, 651–655.

CHATTERJEE, T., BHATTACHARYYA, D. K. (1998b), Synthesis of terpene esters by an immobilized lipase in a solvent-free system, *Biotechnol. Lett.* **20**, 865–868.

CLAON, P. A., AKOH, C. C. (1993), Enzymatic synthesis of geraniol and citronellol esters by direct esterification in *n*-hexane, *Biotechnol. Lett.* **15**, 1211–1216.

CLAON, P. A., AKOH, C. C. (1994a), Lipase-catalyzed synthesis of terpene esters by transesterification in *n*-hexane, *Biotechnol. Lett.* **16**, 235–240.

CLAON, P. A., AKOH, C. C. (1994b), Effect of reaction

parameters on SP435 lipase-catalyzed synthesis of citronellyl acetate in organic solvent, *Enzyme Microb. Technol.* **16**, 835–838.

CROTEAU, R. B., WISE, M. L., KATAHIRA, E. J., SAVAGE, T. J. (1999), Monoterpene synthases from common sage (*Salvia officinalis*), *U.S. Patent* 5891697.

CROWELL, P. L., SIAR AYOUBI, A., BURKE, Y. D. (1996), Antitumorigenic effects of limonene and perillyl alcohol against pancreatic and breast cancer, *Adv. Exp. Med. Biol.* **401**, 131–136.

CUTLER, H. G. (1995), Natural product flavor compounds as potential antimicrobials, insecticides and medicinals, *Agro-Food-Industry Hi-Tech* **6**, 19–23.

CUTLER, H. G., HILL, R. A., WARD, B. G., ROHITHA, B. H., STEWART, A. (1996), Antimicrobial, insecticidal, and medicinal properties of natural product flavors and fragrances, in: *Biotechnology for Improved Foods and Flavors, ACS Symposium Series 637* (TAKEOKA, G. R., TERANISHI, R., WILLIAMS, P. J., KOBAYASHI, A., Eds.), pp. 51–66. Washington, DC: Am. Chem. Soc.

DAVID, L., VESCHAMBRE, H. (1984), Préparation d'oxydes de linalol par bioconversion, *Tetrahedron Lett.* **25**, 543–546.

DE CASTRO, H. F. (1995), Fine chemicals by biotransformation using lipases, *Quimica Nova* **18**, 544–554.

DE CASTRO, H. F., NAPOLEAO, D. A. S., OLIVEIRA, P. C. (1998), Production of citronellyl acetate in a fed-batch system using immobilized lipase, *Appl. Biochem. Biotechnol.* **70–72**, 667–675.

DE-OLIVEIRA, A. C. A. X., FIDALGO-NETO, A. A., PAUMGARTTEN, F. J. R. (1999), *In vitro* inhibition of liver monooxygenases by β-ionone, 1,8-cineole, $(-)$-menthol and terpineol, *Toxicology* **135**, 33–41.

DEMYTTENAERE, J. C. R. (1998), Biotransformation of terpenes by fungi for the production of bioflavours, *Med. Fac. Landbouww. Univ. Gent* **63/4a**, 1321–1324.

DEMYTTENAERE, J., DE KIMPE, N. (1999), Biotransformation of geraniol and nerol by sporulated surface cultures of *Aspergillus niger*, in: *Frontiers of Flavour Science* (SCHIEBERLE, P., ENGEL, K.-H., Eds.), pp. 400–404. Garching: Deutsche Forschungsanstalt für Lebensmittelchemie.

DEMYTTENAERE, J. C. R., DE POOTER , H. L. (1996), Biotransformation of geraniol and nerol by spores of *Penicillium italicum, Phytochemistry* **41**, 1079–1082.

DEMYTTENAERE, J. C. R., DE POOTER , H. L. (1998), Biotransformation of citral and nerol by spores of *Penicillium digitatum*, *Flav. Fragr. J.* **13**,173–176.

DEMYTTENAERE, J. C. R., WILLEMEN, H. M. (1998), Biotransformation of linalool to furanoid and pyranoid linalool oxides by *Aspergillus niger*, *Phytochemistry* **47**, 1029–1036.

DHAVALIKAR, R. S., BHATTACHARYYA, P. K. (1966), Microbiological transformation of terpenes: part VIII fermentation of limonene by a soil *Pseudomonas*, *Ind. J. Biochem.* **3**, 144–157.

DOIG, S. D., BOAM, A. T., LEAK, D. I., LIVINGSTON, A. G., STUCKEY, D. C. (1998), A membrane bioreactor for biotransformation of hydrophobic molecules, *Biotechnol. Bioeng.* **58**, 587–594.

DRAWERT, F., BERGER, R. G., GODELMANN, R. (1984), Regioselective biotransformation of valencene in cell suspension cultures of *Citrus*, *Plant Cell Rep.* **3**, 37–40.

EMBERGER, R., HOPP, R. (1985), Synthesis and sensory characterization of menthol enantiomers and their derivatives for the use in nature identical peppermint oils, in: *Topics in Flavour Research* (BERGER, R. G., NITZ, S., SCHREIER, P., Eds.), pp. 201–218. Hangenham: Eichhorn.

FABER, K. (2000), *Biotransformations in Organic Chemistry*, 4th Edn. Berlin: Springer-Verlag.

FIGUEREIDO, A. C., ALMENDRA, M. J., BARROSO, J. G., SCHEFFER J. J. C. (1996), Biotransformation of monoterpenes and sesquiterpenes by cell suspension cultures of *Achillea millefolium* L. ssp. *millefolium*, *Biotechnol. Lett.* **18**, 863–868.

FOß, S., HARDER, J. (1997), Microbial transformation of tertiary allylalcohol: regioselective isomerisation of linalool to geraniol without nerol formation, *FEMS Microbiol. Lett.* **149**, 71–75.

GABRIELYAN, K. A., MENYAILOVA, I. I., NAKHAPETYAN, L. A. (1992), Biocatalytic transformation of limonene (review), *Appl. Biochem. Microbiol.* **28**, 241–245.

GANGA, M. A., PINAGA, F., VALLES, S., RAMON, D., QUEROL, A. (1999), Aroma improving in microvinification processes by the use of a recombinant wine yeast strain expressing the *Aspergillus nidulans* xlnA Gene, *Int. J. Food Microbiol.* **47**, 171–178.

GONZÁLEZ-NAVARRO, H., BRACO, L. (1998), Lipase-enhanced activity in flavour ester reactions by trapping enzyme conformers in the presence of interfaces, *Biotechnol. Bioeng.* **59**, 122–127.

GRIFFIN, S. G., WYLLIE, S. G., MARKHAM, J. L., LEACH, D. N. (1999), The role of structure and molecular properties of terpenoids in determining their antimicrobial activity, *Flav. Fragr. J.* **14**, 322–332.

GRIFFITHS, E. T., BOCIEK, S. M., HARRIES, P. C., JEFFCOAT, R., SISSONS, D. J., TRUDGILL, P. W. (1987a), Bacterial metabolism of α-pinene: pathway from α-pinene oxide to acyclic metabolites in *Nocardia* sp. strain P18.3, *J. Bacteriol.* **169**, 4972–4979.

GRIFFITHS, E. T., HARRIES, P. C., JEFFCOAT, R., TRUDGILL, P. W. (1987b), Purification and properties of

α-pinene oxide lyase from *Nocardia* sp. strain P18.3, *J. Bacteriol.* **169**, 4980–4983.

GRIVEL, F., LARROCHE, C., GROS, J. B. (1999), Determination of the reaction yield during biotransformation of the volatile and chemically unstable compound β-ionone by *Aspergillus niger*, *Biotechnol. Progr.* **15**, 697–705.

GUEGUEN, Y., CHEMARDIN, P., PIEN, S., ARNAUD, A., GALZY, P. (1997), Enhancement of aromatic quality of muscat wine by the use of immobilized β-glucosidase, *J. Biotechnol.* **55**, 151–156.

GUEGEN, Y., CHEMARDIN, P., JANBON, G., ARNAUD, A., GALZY, P. (1998), Investigation of the β-glucosidases potentialities of yeast strains and application to bound aromatic terpenols liberation, in: *New Frontiers in Screening for Microbial Biocatalysis* (KIESLICH, K., VAN DER BEEK, C. P., DE BONT, J. A. M., VAN DEN TWEEL, W. J. J., Eds.), pp. 149–157. Amsterdam: Elsevier Science B.V.

HARDER, J., FOß, S. (1999), Anaerobic formation of the aromatic hydrocarbon *p*-cymene from monoterpenes by methanogenic enrichment cultures, *Geomicrobiol. J.* **16**, 295–305.

HARDER, J., PROBIAN, C. (1995), Microbial degradation of monoterpenes in the absence of molecular oxygen, *Appl. Environ. Biotechnol.* **61**, 3804–3808.

HEYEN, U., HARDER, J. (2000), Geranic acid formation, an initial reaction of anaerobic monoterpene metabolism in denitrifying *Alcaligenes defragans*, *Appl. Environ. Microbiol.* **66**, 3004–3009.

HILL, R. A. (1993), Terpenoids, in: *The Chemistry of Natural Products* (THOMSON, R. H., Ed.), pp. 106–139. London, New York: Blackie Academic & Professional; Chapman & Hall.

HOLLAND, H. L. (1998), Hydroxylation and dihydroxylation, in: *Biotechnology* 2nd Edn., Vol. 8a *Biotransformations I* (REHM, H.-J., REED, G., PÜHLER, A., STADLER, P., Eds.), pp. 475–533. Weinheim: Wiley-VCH.

HUANG, S. Y., CHANG, H. L., GOTO, M. (1998), Preparation of surfactant-coated lipase for the esterification of geraniol and acetic acid in organic solvents, *Enzyme Microb. Technol.* **22**, 552–557.

IKUSHIMA, Y., SAITO, N., HATAKEDA, K., ITO, S. (1995), Method for esterification of (*S*)-citronellol using lipase in supercritical carbon dioxide, *U.S. Patent* 5403739.

IKUSHIMA, Y., SAITO, N., HATAKEDA, K., SATO, O. (1996), Promotion of a lipase-catalyzed esterification in supercritical carbon dioxide in the near-critical region, *Chem. Eng. Sci.* **51**, 2817–2822.

IURESCIA, S., MARCONI, A. M., TOFANI, D., GAMBACORTA, A., PATERNO, A. et al. (1999), Identification and sequencing of β-myrcene catabolism genes from *Pseudomonas* sp. strain M1, *Appl. Environ. Microbiol.* **65**, 2871–2876.

JAVELOT, C., GIRARD, P., COLONNA-CECCALDI, B., VLADESCU, B. (1991), Introduction of terpene-producing ability in a wine strain of *Saccharomyces cerevisiae*, *J. Biotechnol.* **21**, 239–252.

JELEN, H. H., MIROCHA, C. J., WASOWICZ, E., KAMINSKI, E. (1995), Production of volatile sesquiterpenes by *Fusarium sambucinum* strains with different abilities to synthesize trichothecenes, *Appl. Environ. Microbiol.* **61**, 3815–3820.

KAMIYA, N., GOTO, M., NAKASHIO, F. (1995), Surfactant-coated lipase suitable for the enzymatic resolution of menthol as a biocatalyst in organic media, *Biotechnol. Prog.* **11**, 270–275.

KELLY, D. R., WAN, P. W. H., TANG, J. (1998), Flavin monooxygenases – uses as catalysts for Baeyer–Villiger ring expansion and heteroatom oxidation, in: *Biotechnology* 2nd Edn., Vol. 8a *Biotransformation I* (REHM, H.-J., REED, G. PÜHLER, A., STADLER, P., Eds.), pp. 535–587. Weinheim: Wiley-VCH.

KEMPLER, G. M. (1983), Production of flavour compounds by microorganisms, *Adv. Appl. Microbiol.* **29**, 29–51.

KIESLICH, K., ABRAHAM, W.-R., STUMPF, B., THEDE, B., WASHAUSEN, P. (1986), Transformations of terpenoids, in: *Progress in Essential Oil Research* (BRUNKE, H. J., Ed.), pp. 367–394. Berlin: Walter de Gruyter.

KING, A., DICKINSON, J. R. (2000), Biotransformation of monoterpene alcohols by *Saccharomyces cerevisiae*, *Torulaspora delbrueckii* and *Kluyveromyces lactis*, *Yeast* **16**, 499–506.

KISE, S., HAYASHIDA, M. (1990), Two-phase system membrane reactor with cofactor recycling, *J. Biotechnol.* **14**, 221–228.

KLINGENBERG, A., HANSSEN, H.-P. (1988), Enhanced production of volatile flavour compounds from yeasts by adsorber techniques. I. Model investigations, *Chem. Biochem. Eng.* **2**, 222–224.

KOYAMA, T., SAITO, K., OGURA, K., OBATA, S., TAKESHITA, A. (1994), Site-directed mutagenesis of farnesyl diphosphate synthase; effect of substitution on the three carboxyl-terminal amino acids, *Can. J. Chem.* **72**, 75–79.

KRASNOBAJEW, V. (1984), Terpenoids, in: *Biotechnology* 1st Edn., Vol. 6a *Biotransformations* (REHM, H.-J., REED, G., Eds.), pp. 97–125. Weinheim: Verlag Chemie.

KRINGS, U., KELCH, M., BERGER, R. G. (1993), Adsorbents for the recovery of aroma compounds in fermentation processes, *J. Chem. Tech. Biotechnol.* **58**, 293–299.

LABORET, F., PERRAUD, R. (1999), Lipase-catalyzed production of short-chain acids terpenyl esters of interest to the food industry, *Appl. Biochem. Biotechnol.* **82**, 185–198.

LAMARE, V., FURSTOSS, R. (1990), Bioconversion of sesquiterpenes, *Tetrahedron* **46**, 4109–4132.

LARROCHE, C., CREULY, C., GROS, J.-B. (1995), Fed-batch biotransformation of β-ionone by *Aspergillus niger*, *Appl. Microbiol. Biotechnol.* **43**, 222–227.

LEE, M.-H., LIU, H., SU, N.-W., KU, K.-L., CHOONG, Y.-M. (1999), Bioconversion of *d*-limonene to oxygenated compounds by endogenous enzymes of the citrus peel, *J. Chin. Agr. Chem. Soc.* **37**, 1–19.

LITTLE, D. B., CROTEAU, R. B. (1999), Biochemistry of essential oil terpenes. A thirty year overview, in: *Flavor Chemistry. Thirty Years of Progress* (TERANISHI, R., WICK, E. L., HORNSTEIN, I., Eds.), pp. 239–253. New York: Kluwer Academic/Plenum Publishers.

LIU, W. G., ROSAZZA, J. P. N. (1990), Stereospecific hydroxylation of 1,8-cineole using a microbial biocatalyst, *Tetrahedron Lett.* **31**, 2833–2836.

LOZA-TAVERA, H. (1999), Monoterpenes in essential oils, in: *Chemicals via Higher Plant Bioengineering* (SHAHIDI, F., KOLODZIEJCZYK, P., WHITAKER, J. R., LOPEZ MUNGUIA, A., FULLER, G., Eds.), *Advances in Experimental Medicine and Biology* (BACK, N., COHEN, I. R., KRITCHEVSKY, D., LAJTHA, A., PAOLETTI, R., Eds.), pp. 49–62. New York: Kluwer Academic/Plenum Publishers.

LUCKNER, M. (1990), *Secondary metabolism in plants and animals*. Jena: VEB Gustav Fischer Verlag.

LUTZ-WAHL, S., FISCHER, P., SCHMIDT-DANNERT, C., WOHLLEBEN, W., HAUER, B., SCHMID, R. D. (1998), Stereo- and regioselective hydroxylation of α-ionone by *Streptomyces* strains, *Appl. Environ. Microbiol.* **64**, 3878–3881.

MAATOOQ, G., EL-SHARKAWY, S., AFIFI, M. S., ROSAZZA, J. P. N. (1993), Microbial transformation of cedrol, *J. Nat. Prod.* **56**, 1039–1050.

MADYASTHA, K. M., MURTHY, N. S. R. K. (1988a), Transformation of acetates of citronellol, geraniol, and linalool by *Aspergillus niger*: regiospecific hydroxylation of citronellol by a cell-free system, *Appl. Microbiol. Biotechnol.* **28**, 324–329.

MADYASTHA, K. M., MURTHY, N. S. R. K. (1988b), Regiospecific hydroxylation of acyclic monoterpene alcohols by *Aspergillus niger*, *Tetrahedron* **29**, 579–580.

MAYER, P., NEUBERG, C. (1915), Phytochemische Reduktionen. XII. Die Umwandlung von Citronellal in Citronellol, *Biochem. Z.* **71**, 174–179.

MICHOR, H., MARR, R., GAMSE, T., SCHILLING, T., KLINGSBICHEL, E., SCHWAB, H. (1996), Enzymatic catalysis in supercritical carbon dioxide: comparison of different lipases and a novel esterase, *Biotechnol. Lett.* **18**, 79–84.

MIKAMI, Y. (1988), Microbial conversion of terpenoids, in: *Biotechnology and Genetic Engineering Reviews* Vol. 6 (RUSSELL, G. E., Ed.), pp. 271–320. Newcastle upon Tyne: Intercept.

MIMOUN, H. (1996), Catalytic opportunities in the flavor and fragrance industry, *Chimia* **50**, 620–625.

MIYAZAWA, M., NANKAI, H., KAMEOKA, H. (1996), Biotransformation of acyclic terpenoid (2*E*,6*E*)-farnesol by plant pathogenic fungus *Glomerella cingulata*, *Phytochemistry* **43**, 105–109.

MIYAZAWA, M., HONJO, Y., KAMEOKA, H. (1997a), Biotransformation of the sesquiterpenoid β-selinene using the plant pathogenic fungus *Glomerella cingulata* as a biocatalyst, *Phytochemistry* **44**, 433–436.

MIYAZAWA, M., WATANABE, H., KAMEOKA, H. (1997b), Inhibition of acetylcholinesterase activity by monoterpenoids with a *p*-menthane skeleton, *J. Agric. Food Chem.* **45**, 677–679.

MIYAZAWA, M., HONJO, Y., KAMEOKA, H. (1998), Biotransformation of the sesquiterpenoid (+)-γ-gurjunene using a plant pathogenic fungus, *Glomerella cingulata*, as a biocatalyst, *Phytochemistry* **49**, 1283–1285.

MOLINARI, F., VILLA, R., ARAGOZZINI, F. (1998), Production of geranyl acetate and other acetates by direct esterification catalyzed by mycelium of *Rhizopus delemar* in organic solvent, *Biotechnol. Lett.* **20**, 41–44.

MYERS, C. E., TREPEL, J., SAUSVILLE, E., SAMID, D., MILLER, A., CURT, G. (1997), Monoterpenes, sesquiterpenes and diterpenes as cancer therapy, *U.S. Patent* 5602184.

NAKAGAWA, H., WATANABE, S., SHIMURA, S., KIRIMURA, K., USAMI, S. (1998), Enzymatic synthesis of terpenyl esters by transesterification with fatty acid vinyl esters as acyl donors by *Trichosporon fermentans* lipase, *World J. Microbiol. Biotechnol.* **14**, 219–222.

NAKATANI, N. (1994), Antioxidative and antimicrobial constituents of herbs and spices, in: *Spices, Herbs and Edible Fungi* (CHARALAMBOUS, G., Ed.), pp. 251–271. Amsterdam: Elsevier Science B.V.

NAKATSU, T., LUPO JR., A. T., CHINN JR., J. W., KANG, R. K. L. (2000), Biological activity of essential oils and their constituents, in: *Studies in Natural Products Chemistry* Vol. 21: Bioactive Natural Products Part B (ATTA-UR-RAHMAN, H. E. J., Ed.), pp. 571–631. Amsterdam: Elsevier Science B.V.

NANKAI, H., MIYAZAWA, M., KAMEOKA, H. (1998), Biotransformation of (2*Z*,6*Z*)-farnesol by the plant pathogenic fungus *Glomerella cingulata*, *Phytochemistry* **47**, 1025–1028.

NARUSHIMA, H., OMORI, T., MINODA, Y. (1981), Microbial oxidation of β-myrcene, in: *Adv. Biotechnol.* (*Proc. Int. Ferment. Symp.*), 6th Meeting Date 1980 (VEZINA, C., SINGH, K., Eds.), pp. 525–531. Toronto: Pergamon Press.

NARUSHIMA, H., OMORI, T., MINODA, Y. (1982), Microbial transformation of α-pinene, *J. Appl. Microbiol. Biotechnol.* **16**, 174–178.

ODA, S., OHTA, H. (1997), Double coupling of acetyl coenzyme A production and microbial esterification with alcohol acetyltransferase in an interface bioreactor, *J. Ferment. Bioeng.* **83**, 423–428.

ODA, S., INADA, Y., KATO, A., MATSUDOMI, N., OHTA, H. (1995), Production of (*S*)-citronellic acid and (*R*)-citronellol with an interface bioreactor, *J. Ferment. Bioeng.* **80**, 559–564.

ODA, S., KATO, A., MATSUDOMI, N., OHTA, H. (1996a), Enantioselective oxidation of racemic citronellol with an interface bioreactor, *Biosci. Biotechnol. Biochem.* **60**, 83–87.

ODA, S., INADA, Y., KOBAYASHI, A., KATO, A., MATSUDOMI, N., OHTA, H. (1996b), Coupling of metabolism and bioconversion: microbial esterification of citronellol with acetyl coenzyme A produced via metabolism of glucose in an interface bioreactor, *Appl. Environ. Microbiol.* **62**, 2216–2220.

OGUNTIMEIN, G. B., ANDERSON, W. A., MOO-YOUNG, M. (1995), Synthesis of geraniol esters in solvent-free system catalysed by *Candida antarctica* lipase, *Biotechnol. Lett.* **17**, 77–82.

OHLOFF, G. (1994), *Scent and Fragrances*. Berlin: Springer-Verlag.

OHTO, C., NAKANE, H., HEMMI, H., OHNUMA, S., OBATA, S., NISHINO, T. (1998), Overexpression of an archaeal geranylgeranyl diphosphate synthase in *Escherichia coli* cells, *Biosci. Biotechnol. Biochem.* **62**, 1243–1246.

OHTO, C., ISHIDA, C., KOIKE-TAKESHITA, A., YOKOYAMA, K., MURAMATSU, M. et al. (1999), Gene cloning and overexpression of a geranylgeranyl diphosphate synthase of an extremely thermophilic bacterium, *Thermus thermophilus*, *Biosci. Biotechnol. Biochem.* **63**, 261–270.

OKUDA, M., SONOHARA, H., TAKIGAWA, H., TAJIMA, K., ITO, S. (1994), New microorganism and preparation of nootkatone using the microorganism, *Jpn. Patent* 06303967.

OMATA, T., IWAMOTO, N., KIMURA, T., TANAKA, A., FUKUI, S. (1981), Stereoselective hydrolysis of DL-menthol succinate by gel-entrapped *Rhodotorula minuta* var. *texensis* cells in organic solvent, *Eur. J. Appl. Microbiol. Biotechnol.* **11**, 199–204.

ONKEN, J., BERGER, R. G. (1999a), Biotransformation of citronellol by the basidiomycete *Cystoderma carcharias* in an aerated-membrane bioreactor, *Appl. Microbiol. Biotechnol.* **51**, 158–163.

ONKEN, J., BERGER, R. G. (1999b), Effects of *R*-(+)-limonene on submerged cultures of the terpene transforming basidiomycete *Pleurotus sapidus*, *J. Biotechnol.* **69**, 163–168.

ORIEL, P. J., SAVITHIRY, S., CHANG, H. C. (1997), Process for the preparation of monoterpenes using bacterium containing recombinant DNA, *U.S. Patent* 5688673.

OSPRIAN, I., STEINREIBER, A., MISCHITZ, M., FABER, K. (1996), Novel bacterial isolates for the resolution of esters of tertiary alcohols, *Biotechnol. Lett.* **18**, 1331–1334.

PATTNAIK, S., SUBRAMANYAM, V. R., BAPAJI, M., KOLE, C. R. (1997), Antibacterial and antifungal activity of aromatic constituents of essential oils, *Microbios* **89**, 39–46.

REDDY, B. S., WANG, C.-X., SAMAHA, H., LUBET, R., STELLE, V. E., KELLOFF, G. J. (1997), Chemoprevention of colon carcinogenesis by dietary perillyl alcohol, *Cancer Res.* **57**, 420–425.

REECE, C. A., RODIN, J. O., BROWNLEE, R. G., DUNCAN, W. G., SILVERSTEIN, R. M. (1968), Synthesis of the principal components of the sex attractant from male *Ips confusus* frass: 2-methyl-6-methylene-7-octen-4-ol, 2-methyl-6-methylene-2,7-octadien-4-ol, and (+)-*cis*-verbenol, *Tetrahedron* **24**, 4249–4256.

REIL, G., BERGER, R. G. (1996), Genesis of aroma compounds in phototrophic cell culture of grapefruit, *C. paradisi* cv. White Marsh, *Proc. 8th Weurman Flavour Symposium, Reading* (MOTTRAM, D. S., TAYLOR, A. J., Eds.), pp. 97–104. Cambridge: Royal Soc. Chem.

RHODES, P. M., WINSKILL, N. (1985), Microbiological process for the preparation of 1-carvone, *U.S. Patent* 4495284.

RYKER, L. C., YANDELL, K. L. (1983), Effect of verbenone on aggregation of *Dendroctonus ponderosae* Hopkins (Coleoptera, Scolytidae) to synthetic attractant, *Z. ang. Entomol.* **96**, 452–459.

SAVITHIRY, N., ORIEL, P. J. (1998), Method for production of monoterpene derivatives of limonene, *U.S. Patent* 5763237.

SAVITHIRY, N., CHEONG, T. K., ORIEL, P. (1997), Production of α-terpineol from *Escherichia coli* cells expressing thermostable limonene hydratase, *Appl. Biochem. Biotechnol.* **63–65**, 213–220.

SCHOCH, E., BENDA, I., SCHREIER, P. (1991), Bioconversion of α-damascone by *Botrytis cinerea*, *Appl. Environ. Microbiol.* **57**, 15–18.

SCHWAB, E., SCHREIER, P., BENDA, I. (1991), Biotransformation of β-damascone by *Botrytis cinerea*, *Z. Naturforsch.* Sect. C. Biosci. **46c**, 395–-397.

SEITZ, E.W. (1994), Fermentation production of pyrazines and terpenoids for flavors and fragrances, in: *Bioprocess Production of Flavor, Fragrance and Color Ingredients* (GABELMAN, A., Ed.), pp. 95–134. New York: John Wiley & Sons.

SHIEH, C.-J., AKOH, C. C., YEE, L. N. (1996), Optimized enzymatic synthesis of geranyl butyrate with lipase AY from *Candida rugosa*, *Biotechnol. Bioeng.* **51**, 371–374.

SHUKLA, O. P., BHATTACHARYYA, P. K. (1968), Microbial transformation of terpenes: part XI – pathways of degradation of α- and β-pinenes in a soil pseudomonad (PL-strain), *Ind. J. Biochem.* **5**, 92–101.

SIME, J. T., CHEETHAM, P. S. J., GRADLEY, M. L., BANISTER, N. E. (2000), Uses of sesquiterpenes for inhibiting oxidative enzymes, *U.S. Patent* 6054490.

SOMAN, R. (1995), Terpenes and derivatives industry in India, *Chem. Eng. World* **30**, 34–39.

SPEELMANS, G., BIJLSMA, A., EGGINK, G. (1998), Limonene bioconversion to high concentrations of a single and stable product, perillic acid, by a solvent-resistant *Pseudomonas putida* strain, *Appl. Microbiol. Biotechnol.* **50**, 538–544.

STUMPF, B., ABRAHAM, W.-R., KIESLICH, K. (1984), Verfahren zur Herstellung von (+)-α-Terpineol durch mikrobiologische Umwandlung von Limonen, *German Patent* 3243090 A1.

STUMPF, B., WRAY, V., KIESLICH, K. (1990), Oxidation of carenes to chaminic acids by *Mycobacterium smegmatis* DSM 43061, *Appl. Microbiol. Biotechnol.* **33**, 251–254.

TAKIGAWA, H., KUBOTA, H., SONOHARA, H., OKUDA, M., TANAKA, S. et al. (1993), Novel allylic oxidation of α-cedrene to *sec*-cedrenol by a *Rhodococcus* strain, *Appl. Environ. Microbiol.* **59**, 1336–1341.

TAN, Q., DAY, D. F. (1998a), Bioconversion of limonene to α-terpineol by immobilized *Penicillium digitatum*, *Appl. Microbiol. Biotechnol.* **49**, 96–101.

TAN, Q., DAY, D. F. (1998b), Organic co-solvent effects on the bioconversion of (*R*)-(+)-limonene to (*R*)-(+)-α-terpineol, *Proc. Biochem.* **33**, 755–761.

TAN, Q., DAY, D. F., CADWALLADER, K. R. (1998), Bioconversion of (*R*)-(+)-limonene by *P. digitatum* (NRRL 1202), *Proc. Biochem.* **33**, 29–37.

TEUNISSEN, M. J., DE BONT, J. A. M. (1995), Will terpenes be of any significance in future biotechnology?, in: *Bioflavour 95, Dijon (France)*, February 14–17, 1995 (ÉTIÉVANT, P., SCHREIER, P., Eds.), pp. 329–330. Versailles Cedex: INRA Editions.

TRUDGILL, P. W. (1994), Microbial metabolism and transformation of selected monoterpenes, in: *Biochemistry of Microbial Degradation* (RATLEDGE, C., Ed.), pp. 33–61. London: Kluwer Academic Publishers.

VANDENBERGH, P. A. (1989), Bacterial method and composition for linalool degradation, *U.S. Patent* 4800158.

VAN DYK, M. S., THOMAS, E. (1998), Hydrolysis of linalyl acetate and α-terpinyl acetate by yeasts, *Biotechnol. Lett.* **20**, 417–420.

VAN DYK, M. S., VAN RENSBURG, E., RENSBURG, I. P. B., MOLELEKI, N. (1998a), Biotransformation of monoterpenoid ketones by yeasts and yeast-like fungi, *J. Mol. Cat. B: Enzymatic* **5**, 149–154.

VAN DYK, M. S., VAN RENSBURG, E., MOLELEKI, N. (1998b), Hydroxylation of (+)-limonene, (−)-α-pinene and (−)-β-pinene by a *Hormonema* sp., *Biotechnol. Lett.* **20**, 431–436.

VAN RENSBURG, E., MOLELEKI, N., VAN DER WALT, J. P., BOTES, P. J., VAN DYK, M. S. (1997), Biotransformation of (+)-limonene and (−)-piperitone by yeasts and yeast-like fungi, *Biotechnol. Lett.* **19**, 779–782.

VAN DER WERF, M. J., DE BONT, J. A. M. (1998), Screening for microorganisms converting limonene into carvone, in: *New Frontiers in Screening for Microbial Biocatalysis* (KIESLICH, K., VAN DER BEEK, C. P., DE BONT, J. A. M., VAN DEN TWEEL, W. J. J., Eds.), pp. 231–234. Amsterdam: Elsevier Science B.V.

VAN DER WERF, M. J., DE BONT, J. A. M, LEAK, D. J. (1997), Opportunities in microbial biotransformation of monoterpenes, in: *Advances in Biochemical Engineering/Biotechnology* (SCHEPER, T., Ed.), Vol. 55: Biotechnology of Aroma Compounds (BERGER, R. G., Ed.), pp. 147–177. Berlin: Springer-Verlag.

VAN DER WERF, M. J., SWARTS, H. J., DE BONT, J. A. M (1999), *Rhodococcus erythropolis* DCL 14 contains a novel degradation pathway for limonene, *Appl. Environ. Microbiol.* **65**, 2092–2102.

VASSEROT, Y., ARNAUD, A., GALZY, P. (1995), Monoterpenol glycosides in plants and their biotechnological transformation, *Acta Biotechnol.* **15**, 77–95.

WANG, Y., LINKO, Y.-Y. (1995), Lipase-catalyzed enantiomeric synthesis of citronellyl butyrate, *J. Ferment. Bioeng.* **80**, 473–477.

WELSH, F., MURRAY, W. D., WILLIAMS, R. E. (1989), Microbiological and enzymatic production of flavor and fragrance chemicals, *Crit. Rev. Biotechnol.* **9**, 105–169.

WESTGATE, S., VAIDYA, A. M., BELL, G., HALLING, P. J. (1998), High specific activity of whole cells in an aqueous-organic two-phase membrane bioreactor, *Enzyme and Microb. Technol.* **22**, 575–577.

WHITEHEAD, I. M., OHLEYER, E. (1997), Process for the production of carboxylic acids from alcohols using *Saccharomyces*, *U.S. Patent* 5599700.

WHITEHEAD, I. M., OHLEYER, E. P., DEAN, C. (1995), Bio-oxidation reactions using baker's yeast: the industrial production of short-chain acids for use in natural flavours, in: *Bioflavour 95, Dijon (France)*, February 14–17, 1995 (ÉTIÉVANT, P., SCHREIER, P., Eds.), pp. 245–249. Versailles Cedex: INRA Editions.

WILLERSHAUSEN, H., GRAF, H. (1991), Enzymatische Transformation von Valencen zu Nootkaton, *Chemiker-Zeitung* **115**, 358–360.

XU, J.-H., KAWAMOTO, T., TANAKA, A. (1995a), Efficient kinetic resolution of *dl*-menthol by lipase-catalyzed enantioselective esterification with acid anhydride in fed-batch reactor, *Appl. Microbiol.*

Biotechnol. **43**, 402–407.
XU, J.-H., KAWAMOTO, T., TANAKA, A. (1995b), High-performance continuous operation for enantioselective esterification of menthol by use of acid anhydride and free lipase in organic solvent, *Appl. Microbiol. Biotechnol.* **43**, 639–643.
YAMAZAKI, Y., HAYASHI, Y., HORI, N., MIKAMI, Y. (1988), Microbial conversion of β-myrcene by *Aspergillus niger*, *Agric. Biol. Chem.* **52**, 2921–2922.
YEE, L. N., AKOH, C. C. (1996), Enzymatic synthesis of geranyl acetate by transesterification with acetic anhydride as acyl donor, *J. Am. Oil Chem. Soc.* **73**, 1379–1384.
YEE, L. N., AKOH, C. C., PHILLIPS, R. S. (1995), Terpene ester synthesis by lipase-catalyzed transesterification, *Biotechnol. Lett.* **17**, 67–70.
YEE, L. N., AKOH, C. C., PHILLIPS, R. S. (1997), Lipase PS-catalyzed transesterification of citronellyl butyrate and geranyl caproate: effect of reaction parameters, *J. Am. Oil Chem. Soc.* **74**, 255–260.

14 New Degradable Resins

ALEXANDER STEINBÜCHEL
Münster, Germany

1 Introduction

In Volume 6 of the Second Edition of the multivolume comprehensive treatise *Biotechnology* the state of the development of polyhydroxyalkanoates (PHA) and their applications was summarized according to the situation in the year 1995 (STEINBÜCHEL, 1996). Research was mainly driven by the little knowledge on the biosynthesis of these polyoxoesters and by the attempts to commercialize PHAs and to introduce products manufactured from PHAs on the market. Since this chapter was written, almost seven years have passed. During this time our knowledge on the biochemistry, physiology, and molecular genetics of PHA metabolism has dramatically increased, and many new biotechnological production processes were developed. In addition, many other new and interesting developments have occurred and discoveries have been made. This refers not only to PHAs, but also to many other polyesters or polymers to the production of which biotechnology can contribute. Whereas until the eighties biopolymers produced from renewable resources, exhibiting biodegradability and being used either in their natural state or in a modified state were considered as the only possible alternative to non-biodegradable plastics produced from fossil oils, the situation has now clearly changed and reality has become more diverse. New synthetic polymer resins have been developed, and polymer chemists and material scientists successfully mimicked the biodegradability of biopolymers when they designed the structures of these synthetic polymers. As a result of these efforts, synthetic polymers exhibiting excellent properties can be now produced which are biodegradable and which are readily cleaved by enzymes that in nature hydrolyze biopolymers.

2 How Does Biotechnology Contribute to the Production of Polymers?

Biotechnology can principally contribute in two different ways to the production of biopolymers. The first and more traditional way is the direct production of a biopolymer employing a suitable production organism. Various microorganisms synthesize biopolymers exhibiting interesting properties for putative technical applications either as intracellular or extracellular polymeric compounds by a fully biotechnological approach. Besides some water soluble biopolymers such as xanthan and dextran, the only biopolymers, which fulfil the criteria of a "resin" and which come close to commercialization, are polyhydroxyalkanoic acids (PHA). The new developments in the PHA field will be compiled in Sects. 3 and 4. On the other side various low molecular compounds can be produced by various microorganisms or other biotechnological approaches that are subsequently polymerized by solely chemical processes. This strategy is referred to as the semi-biotechnological approach and some of the most important biotechnological products relevant for polymer biosynthesis are listed below.

Regarding the semi-biotechnological approach, many hydroxyalkanoic acids, α,ω-dicarboxylic acids and α,ω-alkanediols are in principle suitable precursors for polymer synthesis because these are bifunctional organic compounds. They can be used as monomer alone or in combination with other compounds for polycondensations to produce polyesters, polyethers, or polyurethanes. Whereas lactic acid is currently the most interesting representative of the hydroxyalkanoic acids, succinic acid, and adipic acid are interesting α,ω-dicarboxylic acids, and 1,3-propanediol and 1,4-butanediol are interesting α,ω-alkanediols for this purpose. In addition, acrylamide has to be mentioned, which is also biotechnologically produced either by employing microorganisms or isolated nitrile hydratases. A recent review on bulk and commodity chemicals that will be provided by the biotechnology industry (WILKE, 1999) listed three chemicals that will be preferentially used as building blocks for the synthesis of biodegradable polymers. These are lactic acid, 1,3-propanediol, and succinic acid.

Lactic acid

Lactic acid is at present probably the most important biotechnological product for production of technical polymers. It is very efficiently produced by the fermentation of glucose employing homofermentative lactic acid bacteria with yields close to 100% at rates of more than 2 g lactic acid per L and h resulting in concentrations higher than 100 g L^{-1} (DATTA et al., 1995). The different processes and strategies to produce lactic acid biotechnologically were reviewed previously in this series (KASCAK et al., 1996). Cargill Dow Chemicals now anticipates to establish a 125,000 t a^{-1} plant in 2001 for the production of polylactic acid, and it anticipates a market potential of 500,000 t a^{-1} on a 10-year perspective.

1,3-Propanediol

Another compound is 1,3-propanediol, the importance of which was only recently nicely reviewed by BIEBL et al. (1999). Therefore, only a short summary is given here. 1,3-Propanediol was until the middle of the nineties produced by chemical synthesis at a price of approximately US$ 30 kg^{-1} and could not compete with diols such as 1,2-ethanediol, 1,2-propanediol, and 1,4-butanediol which were available at a price of only US$ 2 kg^{-1} from petrochemicals (MILLET, 1993). Then Shell Chemical Company and Degussa established new improved synthesis processes based on the reaction of ethylene oxide with carbon monoxide plus hydrogen (STINSON, 1995) or on hydrolysis of acrolein with subsequent catalytic hydrogenation (ARNTZ et al., 1991). Both companies announced the operation of new production plants allowing the production of 80,000 or 10,000 t 1,3-propanediol per year, respectively; later, DuPont took over the process of Degussa. 1,3-Propanediol can also be produced by biotechnological processes. For more than hundred years it is known that species of the genus *Clostridium* like *C. butyricum* and various Enterobacteriaceae like *Klebsiella pneumoniae* ferment glycerol to 1,3-propanediol. DuPont and Genencor International are currently establishing a process for the production of 1,3-propanediol from glucose, which is much cheaper than glycerol, employing genetically modified microorganisms such as *Escherichia coli* or *Saccharomyces cerevisiae* with engineered metabolism allowing the conversion of glucose into 1,3-propanediol (POTERA, 1997; GATENBY et al., 1998). On the other side, also glycerol may become a cheap carbon source as a by-product of fatty acid methylester production from oil crops, if the demand for biodiesel increases (WILKE, 1999). Therefore, 1,3-propanediol may be the first bulk chemical produced by a genetically engineered microorganism (BIEBL et al., 1999).

Succinic acid

The biotechnological production of succinic acid is less commercially developed although the market size is estimated as high as US$ $400 \cdot 10^6$ a^{-1}. Succinic acid may in addition to several other uses also provide an interesting feedstock for 1,4-butanediol and the production of adipic acid and nylon. Employing the anaerobic rumen bacterium *Actinobacillus succinogenes* (GUETTLER et al., 1999) and other strictly anaerobic bacteria, glucose may be fermented to succinic acid by anaerobic fermentations to concentrations as high as 110 g succinic acid L^{-1} (GUETTLER et al., 1996), and due to the involvement of PEP carboxykinase on a weight by weight basis, more succinic acid then glucose can be obtained. The projected price for succinic acid is of the order of US$ 0.35 kg^{-1} on a large scale (ZEIKUS et al., 1999). Succinic acid will then be an interesting commodity chemical for the production of polybutylene terephthalate resins, nylon 6,6, and linear aliphatic polyesters such as Bionolle (ZEIKUS et al., 1999). In addition, 1,4-butanediol and γ-butyrolactone and other four-carbon chemicals comprising a market size of more than $275 \cdot 10^6$ kg a^{-1} can be obtained from succinic acid, which are substrates for the fermentative production of PHAs containing 4-hydroxybutyrate as comonomer (KUNIOKA et al., 1988). Succinic acid is also used to get modified natural polymers such as for the succinylation of cellulose and starch (DIAMANTOGLON and MEYER, 1988).

Adipic acid

This organic acid is an important chemical for the production of nylon and many other products. Nylon is, e.g., produced from the condensation of hexanediamine and adipic acid. There is a worldwide demand of approximately $2 \cdot 10^6$ t of adipic acid per year. The currently used main processes for chemical production of adipic acid rely either on the oxidation of lipids or on the oxidation of cyclohexanol or cyclohexanone with HNO_3. It is estimated that only the production of adipic acid for nylon production contributes to approximately 10% of the total release of the greenhouse gas NO into the atmosphere; adipic acid production is, therefore, a highly ecocidal process. Therefore, chemists have developed more environmentally sound processes. However, there are also biotechnological processes under development. These processes are based on genetically modified microorganisms with a deregulated shikimic acid pathway and other measures allowing the overproduction of 3-dehydroshikimic acid, which is subsequently allowed to be degraded to *cis,cis*-muconic acid; the latter is then chemically reduced with hydrogen to adipic acid using platinum as catalyst (DRATHS and FROST, 1994). Many other interesting compounds could be derived from those or related strains.

Acrylamide

The chemical industry produces acrylamide on a large scale from acrylonitrile which is hydrolyzed in the presence of a copper catalyst. It is mainly used for the production of polyacrylamide and various copolymers of acrylamide. Several bacteria such as *Pseudomonas chlororaphis*, *Arthrobacter* sp., *Brevibacterium* sp., and *Rhodococcus* sp. express nitrilases, which catalyze the same reaction under much milder conditions. Consequently, acrylamide was also produced biotechnologically from acrylonitrile at a scale of more than 30,000 t a^{-1} already at the beginning of the 1990s (KOBAYASHI et al., 1992). Various strategies and improved strains have been employed since then, and meanwhile biotechnologically produced acrylamide accounts to approximately 30% of the total world production of acrylamide (BUNCH, 1998).

3 PHAs

Besides plant polyesters cutin and suberin polyhydroxyalkanoates (PHA) represent the most widespread naturally occurring polyesters in nature. They are synthesized exclusively in prokaryotes as a storage compound for energy and carbon. Approximately 150 different hydroxyalkanoic acids have been detected as constituents in PHAs (STEINBÜCHEL and VALENTIN, 1995). In addition to poly(3-hydroxybutyrate), poly(3HB), PHAs consisting of short-carbon-chain length hydroxyalkanoic acids (HA_{SCL}) comprising 3–5 carbon atoms are distinguished from PHAs consisting of medium-carbon-chain-length hydroxyalkanoic acids (HA_{MCL}) comprising 6 or more carbon atoms.

3.1 New Knowledge on PHA Synthases

At present 54 PHA synthase genes (*phaC*) from 44 different gram-negative and gram-positive eubacteria have meanwhile been cloned, and 44 of these genes have been sequenced (REHM and STEINBÜCHEL, 1999; STEINBÜCHEL and HEIN, 2001; REHM and STEINBÜCHEL, in press). Detailed information about PHA synthases from archaea or extremophilic prokaryotes is still not available. As before, still three different types of PHA synthases (I, II, and III) can be distinguished according to their subunit composition (PhaC alone or PhaC plus PhaE) and their substrate specificity (HA_{SCL} or HA_{MCL}). No fundamentally new type of PHA synthase has been discovered so far. A multiple alignment of the primary structures of all available PHA synthases showed overall identities between 8% and 96%; 8 amino acid residues are strictly conserved in any of these PHA synthases (REHM and STEINBÜCHEL, in press). Much progress has been made regarding the biochemistry of PHA synthases revealing the catalytic mechanism of this enzyme class, and the PHA synthases of *Ralstonia eutropha*, *Pseudomonas aeruginosa*, and *Allochromatium vinosum*, are currently studied in much detail. These studies revealed that PHA synthases belong to the α/β hydrolase superfamily (JIA et al., 2000). Most probably the catalytic triad proposed for members of the α/β hydrolase superfamily is made from the strictly conserved residues cysteine (Cys-319 in *R. eutropha*, Cys-296 in *P. aeruginosa*, Cys-149 in *A. vinosum*), aspartate (Asp-480 in *R. eutropha*, Asp-452 in *P. aeruginosa*, Asp-302 in *A. vinosum*), and histidine (His-508 in *R. eutropha*, His-480 in *P. aeruginosa*, His-331 in *A. vinosum*). A previously suggested catalytic mechanism related to the mechanism of fatty acid synthases and based on the provision of two thiol groups – one by the cysteine of a "lipase box"-like motif and the second by a phosphopantetheinylated strictly conserved serine residue (GERNGROSS et al., 1994) – could not be confirmed, because it was shown that a covalent modification of the serine residue by phosphopanthenylation does not occur (HOPPENSACK et al., 1999). A recently described viable colony staining method may be quite helpful to screen for PHA-positive and PHA-negative clones (SPIEKERMANN et al., 1999) to investigate PHA synthases with modified activity or substrate range. With all three enzymes mentioned above *in vitro* biosynthesis of various PHAs was obtained. Purified PHA synthases have also been combined with other enzymes, thus generating short pathways, which allowed synthesis of PHAs from substrates cheaper than hydroxyacyl-CoA thioesters and also recyclization of CoA (see Sect. 3.4).

3.2 Phasins

It is now well established that in addition to the PHA synthases (PhaC) other proteins are also bound to the surface of PHA granules. PHA granules from all bacteria have bound large amounts of small, amphiphilic proteins which cover the surface of the granules. With analogy to respective proteins, which cover the surface of triacylglycerol inclusions in plants and which are referred to as oleosins, the term "*phasin*" was proposed and is meanwhile well accepted in the published literature for these PHA granule-associated proteins in bacteria (STEINBÜCHEL et al., 1995).

Most detailed biochemical and molecular genetic studies have been done with the phasins from *Rhodococcus ruber* (PIEPER-FÜRST et al., 1994, 1995), *R. eutropha* (WIECZOREK et al., 1995; HANLEY et al., 1999), and *Paracoccus denitrificans* (MAEHARA et al., 1999), and models for the interaction of these phasins with the surface of PHA granules have been proposed. The presence of phasins during *in vivo* biosynthesis of PHA in recombinant strains of *E. coli* expressing foreign PHA biosynthesis genes exerted a negative effect on the size of the PHA granules (PIEPER-FÜRST et al., 1994) and in some cases a positive effect on the amount of PHA accumulated (MAEHARA et al., 1999). It was also shown that the addition of phasins during *in vitro* PHA biosynthesis experiments has a negative effect on the size of the granules (JOSSEK et al., 1998) and a positive effect on the activity of PHA synthases (JOSSEK and STEINBÜCHEL, 1998). These studies were paralleled with detailed electron microscopic studies principally confirming the models derived from the biochemical studies (MAYER et al., 1997). Meanwhile phasins and the corresponding genes have been identified and also partially characterized in several other bacteria.

In addition to the PHA synthases, pseudomonads belonging to the rRNA homology group I also posses various granule-associated proteins of lower molecular weight. However, the situation seems to be more complex and different from that in all other PHA accumulating bacteria, as revealed by studies of *Pseudomonas oleovorans* (STUART et al., 1995) and *P. putida* (STUART et al., 1998; VALENTIN et al., 1998), and a clear reliable model has not yet been presented.

3.3 PHA Biosynthesis Pathways

Several new pathways and enzymes providing coenzyme A thioesters as substrates to PHA synthases have been detected and analyzed recently. The enhanced knowledge of these pathways and the availability of an increasing number of genes for enzymes that could contribute to such pathways either from regular studies or the analysis of genome sequencing projects has stimulated *in vivo* metabolic engineering of new or improved pathways in various organisms. As outlined in a recent review (STEINBÜCHEL and FÜCHTENBUSCH, 1998) many of these pathways allowed the conversion of abundantly available cheap carbon sources provided by agriculture such as glucose or triacylglycerols or of CO_2 to various PHAs consisting not solely of 3-hydroxybutyrate. This will be crucial, if economically feasible processes for production of PHAs by bacterial fermentation from starch, molasses, lactose, lipids, etc., or by transgenic plants from CO_2 will be established. If PHAs other than poly(3HB) are to be produced from such carbon sources, the respective hydroxyacyl-CoA thioesters must be derived from central intermediates of the metabolism.

The following central pathways, which are present in most organisms, are suitable for this: parts of the amino acid metabolism, the citric acid cycle, fatty acid *de novo* synthesis, and fatty acid β-oxidation. The metabolism of the branched-chain amino acids valine and isoleucine and of threonine was utilized to establish the incorporation of 3-hydroxyvalerate (3HV) into PHAs. Polyesters containing 4-hydroxybutyrate (4HB) have been derived from succinyl-CoA, and different intermediates of the fatty acid metabolism were directed towards synthesis of various poly($3HA_{MCL}$). For details and for the respective literature, some recent reviews on this subject should be consulted (MADISON and HUISMAN, 1999; STEINBÜCHEL and FÜCHTENBUSCH, 1998; STEINBÜCHEL, 2001). For a functional pathway it is, however, very important to have not only expression of the required enzymes but also an appropriate flow of metabolites towards the PHA synthase as shown recently for the utilization of fatty acids for poly($3HA_{MCL}$) production in recombinant strains of *E. coli* (LANGENBACH et al., 1997; QI et al., 1997, 1998).

Two enzymes linking fatty acid metabolism and PHA biosynthesis, which are obviously specifically related to PHA metabolism, have been identified recently. One is a 3-hydroxyacyl-acyl carrier protein-coenzyme A transferase which was first identified and characterized in *Pseudomonas putida* and links fatty acid *de novo* synthesis and poly($3HA_{MCL}$) biosynthesis (REHM et al., 1998). This metabolic link seems to be present in all pseudomonads belonging to the rRNA homology group I, which are capable to synthesize poly($3HA_{MCL}$) from simple carbon sources (HOFFMANN et al., 2000a; DOI and coworkers according to a presentation at the ISBP2000 Conference in Cambridge in September 2000), and it is cryptic in *P. oleovorans* (HOFFMANN et al., 2000b). The second enzyme is a (*R*)-specific enoyl-CoA hydratase first identified in *Aeromonas caviae* (FUKUI and DOI, 1997) and later also in *P. aeruginosa* (TSUGE et al., 2000). The enzyme catalyzes the conversion of *trans*-2-enoyl-CoA to (*R*)-3-hydroxyacyl-CoA during the cultivation on fatty acids, thus linking fatty acid β-oxidation and poly($3HA_{MCL}$) biosynthesis. Both new enzymes and the corresponding genes will be probably key tools to establish poly($3HA_{MCL}$) biosynthesis heterologously in other organisms.

There still remain some pathways for the synthesis of unusual substrates to PHA synthases, which led to the incorporation of interesting constituents, to be analyzed in more detail. One example is the formation of 3-hydroxy-4-pentenoyl-CoA in a *Burkholderia cepacia* strain. This bacterium accumulates from almost any carbon source PHAs consisting of 3-hydroxy-4-pentenoate (RODRIGUES et al., 1995). However, a blend of poly(3HB) homopolyester and of a homopolyester of 3-hydroxy-4-pentenoate, instead of a copolyester of the two constituents is accumulated (VALENTIN et al., 1999a), indicating the presence of two PHA synthases with different substrate ranges. One PHA synthase was recently cloned (RODRIGUES et al., 2000a), and if this gene was disrupted by homogenotization, the cells continued to incorporate 3-hydroxy-4-pentenoate into the polyester (RODRIGUES et al., 2000b).

3.4 *In vitro* Biosynthesis of PHAs

Since the last review on PHAs was published in Volume 6 of the Second Edition of the *Biotechnology* series (STEINBÜCHEL, 1996) various systems for the *in vitro* biosynthesis of different PHAs have been established. They were extensively reviewed recently (STEINBÜCHEL, 2001); therefore, only a short summary is provided here. *In vitro* PHA biosynthesis was achieved using purified PHA synthases of *R. eutropha*, *A. vinosum*, and *P. aeruginosa* alone or together with additional auxiliary enzymes purified from various sources. The functions of the auxiliary enzymes were to be able to use other substrates than coenzyme A thioesters of hydroxyalkanoic acids and also to recycle coenzyme A which must then be used only in catalytic instead of stoichiometric amounts. The latter is important with regard to the costs of synthesizing PHAs *in vitro*. Furthermore, it is also important with respect to the kinetics, because otherwise CoA is accumulated to such high concentrations that it competitively inhibits the PHA synthase.

The currently established *in vitro* PHA synthesis systems were compiled in a very recent review (STEINBÜCHEL, 2001). Poly(3HB) was synthesized from 3-hydroxybutyryl-CoA using the PHA synthases from *R. eutropha* (GERNGROSS and MARTIN, 1995; LENZ et al., 1999) or *A. vinosum* (JOSSEK et al., 1998). Poly(3HB) was also synthesized employing the PHA synthase of *A. vinosum* in combination with the auxiliary enzymes propionyl-CoA transferase from *Clostridium propionicum* alone or in addition also with an acetyl-CoA synthetase of *Saccharomyces cerevisiae* (JOSSEK and STEINBÜCHEL, 1998). Poly(3HV) was synthesized from 3-hydroxyvaleryl-CoA using the PHA synthase of *R. eutropha* (SU et al., 2000). Homopolyesters and copolyesters of various short-carbon-chain-length poly(HA_{SCL}) such as 3HB, 4HB, and 4-hydroxyvalerate (4HV) were synthesized employing the PHA synthase of *A. vinosum* plus the butyrate kinase and the phosphotransbutyrylase of *Clostridium acetobutylicum* as auxiliary enzymes (LIU and STEINBÜCHEL, 2000a). Finally, poly(3HD) was synthesized using the PHA synthase PhaC1 of *Pseudomonas aeruginosa* plus an acyl-CoA synthetase of *Pseudomonas* sp. (QI et al., 2000).

All *in vitro* PHA biosynthesis systems mentioned above resulted in the formation of spherical PHA granules occurring in the buffer solution, and PHA material could be prepared. Although none of these *in vitro* PHA biosynthesis systems has been applied on a large or even in industrial scale, yet, some of the systems are suitable to provide amounts of the polyesters in the order of 50–100 mg, which is sufficient to investigate some properties such as crystallinity, melting temperature, glass transition temperature, etc. A further up-scale might be possible; however, it requires further R&D work. Further studies will allow a conclusion whether such cell free *in vitro* systems are suitable for a technical production of PHAs and whether some of the inherent advantages of such systems regarding the yield per volume and the ease of downstream processing can be utilized. Another peculiarity of such systems is that from these *in vitro* metabolic engineering experiments conclusions can be drawn whether *in vivo* metabolic engineering of the respective pathway is feasible and the desired pathways are functionally active, and whether it is worthwhile to initiate such work. This is because the latter are much more time consuming and require much more effort. The system for the *in vitro* biosynthesis of poly(HA_{SCL}) employing the PHA synthase of *A. vinosum* plus the butyrate kinase and the phosphotransbutyrylase of *C. acetobutylicum* (LIU and STEINBÜCHEL, 2000a), was, e.g., established in *E. coli*, allowing also the *in vivo* synthesis of the respective polyesters. The only alteration was the replacement of the *A. vinosum* PHA synthase by the more flexible *Thiocapsa pfennigii* PHA synthase (LIU and STEINBÜCHEL, 2000b; LIEBERGESELL et al., 2000).

3.5 Fermentative Production of PHAs

In the past all PHAs were only produced with wild-type strains or mutants derived from bacterial strains. Poly(3HB), poly(3HB-*co*-3HV) and poly($3HA_{MCL}$) could be produced by *Alcaligenes latus*, a process developed by the Austrian Company Chemie Linz (HRABAK, 1992), *Ralstonia eutropha*, a process developed by Imperial Chemical Industries (BYROM, 1992), or *Pseudomonas oleovorans* (JUNG et al., 2001), reaching cells densities far above 100 g or even 200 g cell dry matter per L with *A. latus* and *R. eutropha*, respectively, but significantly less with *P. oleovorans*. In a recent review (CHOI and LEE, 1999a) the cell concentrations, PHA contents of the cells, and the productivities of the different processes for biotechnological production of PHAs were compiled, and they were analyzed with regard to the economic feasibility. Recombinant DNA technology provided also other suitable genetically modified production strains. Among bacteria the main target is *E. coli*. All three polyesters can be synthesized in recombinant stains of *E. coli* if the respective PHA synthases are functionally expressed and if a carbon source is used, which can be converted into a hydroxyacyl coenzyme A thioester utilized by the PHA synthase. The review mentioned above included also analyses of data obtained for processes employing recombinant strains of *E. coli*. The authors came to the conclusion that the operating costs for a 100,000 t a^{-1} plant were cheapest using a process with *A. latus* and sucrose. This strain exhibited the highest productivity with 4.94 g poly(3HB) per L and h, and at that scale production costs of US$ 2.6 kg^{-1} poly(3HB) were calculated (CHOI and LEE, 1999a). From this *A. latus* may be superior to the recombinant strains of *E. coli*; however, *E. coli* offers much higher flexibility to engineer pathways and to further optimize the strains genetically and also to produce PHAs other than poly(3HB). In addition, downstream processing may be easier with cells of *E. coli* (CHOI and LEE, 1999b).

Many other technical or in abundant amounts available substrates are also considered for the production of PHAs. Only a few will be mentioned here. For several reasons lignite may represent another interesting carbon source for the production of PHAs (KLEIN et al., 1999). Biosynthesis of PHAs from low-rank coal liquefaction products by *P. oleovorans* and *Rhodococcus ruber* has recently been demonstrated (FÜCHTENBUSCH and STEINBÜCHEL, 1999). In addition, waste oil remaining from the production of rhamnose was shown to be suitable for the production of PHAs employing *R. eutropha* and *P. oleovorans* (FÜCHTENBUSCH et al., 2000) as the sole carbon source. Levulinic acid, a carbon source abundantly available from carbohydrates by chemical conversion, can be fermented to copolyester consisting of a high molar fraction of 4-hydroxyvalerate (SCHMACK et al., 1998; GORENFLO et al., in press).

3.6 Transgenic Plants Producing PHAs

During the last years much progress has been made to express bacterial PHA biosynthesis genes in plants and to produce poly(3HB). In addition to poly(3HB), it was also shown that poly(3HB-*co*-3HV) and PHAs consisting of various $3HA_{MCL}$ can be synthesized in transgenic plants. Although the projects on PHA production in transgenic plants pursued by Zeneca and later by Monsanto were stopped in 1994 or 1998, respectively, several smaller companies continued their R&D programs or started new programs on this aspect.

One of the most relevant achievements in this field was certainly the demonstration that poly(3HB) can be also produced in agricultural crops such as *Gossypium hirsutum* (JOHN and KELLER, 1996), *Zea mays* (HAHN et al., 1997), *Brassica napus* (VALENTIN et al., 1999b), *Nicotiana tabacum* (NAKASHITA et al., 1999), and *Solanum tuberosum* (LOTHAR WILLMITZER, unpublished results), and not only in *Arabidopsis thaliana* as shown previously (POIRIER et al., 1992). In some cases relatively high poly(3HB) yields were obtained by targeting the expression of the foreign genes to different compartments of the plant cells (NAWRATH et al., 1994). Another significant achievement and breakthrough was the establishment of biosynthetic routes for PHAs different from poly(3HB). With strategies and concepts, in most cases previously applied to bacteria for metabolic engineering, synthesic routes for poly(3HB-*co*-3HV) (SLATER et al., 1999) and also for PHAs consisting of various $3HA_{MCL}$ (MITTENDORF et al., 1998) were engineered and became reality also in transgenic plants. To obtain biosynthesis of poly(3HB-*co*-3HV) in *A. thaliana* and *B. napus*, the threonine deaminase from *E. coli* (*ilvA*) plus a β-ketothiolase (*bktB*), the acetoacetyl-CoA reductase and the PHA synthase (*phaC*) from *R. eutropha* were expressed in the chloroplasts enabling the endogenous provision of acetyl-CoA and propionyl-CoA together with enzymes of the respective host (SLATER et al., 1999). The 3HV content in the copolyester amounted up to 17 mol%, and the polyester content of the plants was up to approximately 13% of the dry matter at the maximum (VALENTIN et al., 1999b; SLATER et al., 1999). To obtain poly(3HA$_{MCL}$) biosynthesis in *A. thaliana*, the *P. aeruginosa* PHA synthase PhaC1 was expressed in the peroxisomes (MITTENDORF et al., 1998). The polyester, which was accumulated to 0.5% of the plant dry matter, contained 3-hydroxyoctanoate and 3-hydroxyoctenoate as the main constituents (together 41 mol%). Higher contents of poly(3HA$_{MCL}$) were obtained if the caproyl-acyl-carrier-protein thioesterase from *Cuphea lanceolata* was also expressed in *A. thaliana* (MITTENDORF et al., 1999). The usefulness and the favorable economics of PHA production in general and in transgenic plants in particular are intensively discussed (STEINBÜCHEL and FÜCHTENBUSCH, 1998; GERNGROSS and SLATER, 2000).

3.7 Commercialization of PHAs

The efforts to commercialize a family of poly(3HB-*co*-3HV) copolymers, which are referred to as Biopol®, have continued during the last 7 years. The research and development program, which had started at Imperial Chemical Industries (ICI) in the United Kingdom in the late 1970s, was continued after the separation of the agricultural and pharmaceutical business of ICI by the spun-off Zeneca Ltd. since 1990. Later, in 1996 the Biopol® business was acquired by Monsanto in the United States, and the research program and commercial Biopol® business was continued until the end of 1998. The major aims of Monsanto were to produce PHAs in plants and to improve the properties of PHAs for different end use applications. Several other companies were also active in this field, and despite the program was not continued by Monsanto after 1998, the other remained in the field with Metabolix Inc. and Tepha Inc. being probably the most active companies. Moreover, other companies entered the field with programs for this or related biodegradable polyesters.

At present several applications of PHAs in the medical field are being developed and have already been used in animal studies or clinical studies applied to human patients. Patches, stents, tissue engineering, tissue regeneration, drug delivery systems, prodrugs, nerve repair nutritional uses, orthopedics, and urology were the major fields in which PHAs were applied. Wound management and pericardial patches made from nonwoven poly(3HB) were successfully used (BOWALD and JOHANSSON, 1990; MALM et al., 1992a; DUVERNOY et al., 1995), as also nonwoven poly(3HB) patches in artery augmentation (MALM et al., 1994; STOCK et al., in press) or in the repair of atrial septum defects (MALM et al., 1992b). Other applications are poly(3HB) or poly(3HB-*co*-3HV) as materials for cardiovascular stents (VAN DER GIESSEN et al., 1996; SCHMITZ and BEHREND, 1997; UNVERDORBEN et al., 1998; BEHREND et al., 1998), and poly(3HB-*co*-4HB) or poly(HOH) as coatings for vascular grafts (NOISHIKI and KOMATSUZAKI, 1995; MAROIS et al., 1999, 2000) and the use of PHAs as scaffolding material in tissue engineering of, e.g., heart valves (STOCK et al., 2000; SODIAN et al., 1999, 2000; HOERSTRUP et al., 2000). Many more applications could be mentioned.

4 Polythioesters

Recently, so far unknown, sulfur-containing polythioesters were described, which were synthesized by *Ralstonia eutropha* when the cells were fed with mercaptoalkanoic acids as carbon sources (LÜTKE-EVERSLOH et al., 2001). This was the first report on a natural polymer containing sulfur in the backbone. These polymers were described as new representatives of a new class of biopolymers which are referred to as polythioesters.

4.1 Polymers Containing 3-Mercaptopropionic Acid

R. eutropha synthesized a copolymer of 3-hydroxybutyrate and 3-mercaptopropionate, poly(3HB-*co*-3MP), when 3-mercaptopropionic acid or 3,3′-thiodipropionic acid were provided as carbon sources in addition to fructose or gluconic acid under nitrogen-limited growth conditions. The peculiarity of this polymer was the occurrence of thioester linkages derived from the thiol groups of 3MP and the carboxyl groups of 3MP or 3HB, respectively, which occurred besides the common oxoester bonds of PHAs. Depending on the cultivation conditions and the feeding regime, poly(3HB-*co*-3MP) contributed up to 19% of the cellular dry weight, with a molar fraction of 3MP of up to 43%. The chemical structure of poly(3HB-*co*-3MP) was confirmed by gas chromatography/mass spectrometry, infrared spectroscopy, ^{1}H- and ^{13}C-nuclear magnetic resonance spectroscopy, and elemental sulfur analysis.

4.2 Polymers Containing 3-Mercaptobutyric Acid

Another polythioester covalently linking 3-hydroxybutyrate and 3-mercaptobutyrate, poly(3HB-*co*-3MB), was also synthesized by *R. eutropha*, when 3-mercaptobutyric acid was fed as carbon source in addition to gluconate (T. LÜTKE-EVERSLOH and A. STEINBÜCHEL, unpublished results). The polymer contributed to up to 31% of the cellular dry weight. Elemental sulfur analysis of poly(3HB-*co*-3MB) exhibited a total sulfur content of 11.7% (w/w) indicating a molar fraction of 3MB of approximately 33%. The molecular structure of this novel polymer was also confirmed by the same methods as mentioned above for poly(3HB-*co*-3MP).

5 Semisynthetic and Synthetic Polyester Resins

This section summarizes recent efforts of the chemical industry to produce various polyesters and related polymers, which currently are in part or completely based on biotechnologically produced monomers, or where in the future at least one compound that could be replaced by a biotechnologically produced monomer, provided the production of the respective compound has become economically feasible. Due to the restricted information available from the chemical companies producing these new polymers, the information provided here may not be complete and may not represent the current state of development of the respective polymer. However, since such chemically synthesized products represent an at least potentially very important sink for biotechnological products, and since most of these polymers are biodegradable, the author considered this worth mentioning in this chapter.

5.1 Biomax® (DuPont)

The interest of DuPont in the economically feasible biotechnological production of 1,3-propanediol was already outlined above. The manufacturing of the polyester poly(trimethylene terephthalate) is one use of this alkanediol (MCCOY, 1998). Biomax® is a fully biodegradable and compostable polyester consisting of terephthalate and aliphatic constituents based on the polyethylene terephthalate polyester technology at DuPont. The aliphatic comonomers make the polymer accessible to hydrolytic cleavage, and the cleavage products are then further degraded by microorganisms and converted to CO_2 and water. It is suitable for cast and blown film applications, paper coating, thermoforming, and injection molding. Examples for products manufactured from Biomax® are top and back sheets of disposable diapers, blister packs, disposable eating utensils, geotextiles, agricultural films, plant pots, coated paper products, bottles, and injection-molded objects.

5.2 Bionolle® (Showa Highpolymer Company)

In 1993 Showa Highpolymer Company announced the commercialization of a thermoplastic polyester which is based on glycols and aliphatic dicarboxylic acids such as polyethylene succinate or polybutylene succinate. All different specifications of Bionolle® are, although slowly, clearly biodegradable, and they are, therefore, suitable for the manufacturing of biodegradable and compostable packaging materials which are hitherto mainly produced from polyethylene. The company started production of this new product in 1993 by operating a 3,000 t a^{-1} plant at the beginning. Pots for plants in horticulture and compostable bags were one of the first products manufactured from this polyester.

5.3 Corterra® (Shell Chemical Company)

In 1995 Shell Chemical Company announced the commercialization of a new polyester which is based on 1,3-propanediol and terephthalic acid. The trade name of this poly(trimethylene terephthalate) is Corterra®, and it was designed in particular for fiber and textile applications. Shell started the production of Corterra® with a 4,000 t a^{-1} plant employing a new chemical process for the production of 1,3-propanediol, and recently a 80,000 t a^{-1} plant should have started to operate (Chuah et al., 1995).

5.4 Eastar Bio® (Eastman Chemical Company)

This product is a biodegradable polyester, which was introduced on the market by Eastman Chemical Company in 1997, and in 1999 it was produced at a scale of approximately 15,000 t. It can be processed on conventional polyethylene extrusion equipment and by other conventional plastic processing techniques. Compost bags, biodegradable cartons for packaging, and golf tees are only a few examples of application.

5.5 Ecoflex® (BASF AG)

Based on R&D work done at BASF in Ludwigshafen and at the Gesellschaft für Biotechnologische Forschung (GBF) in Braunschweig polyesters were developed, which were commercialized since 1998 (Witt et al., 1996). Ecoflex® is an aliphatic–aromatic copolyester made from terephthalic acid, adipic acid, and 1,4-butanediol. It is fully biodegradable and suitable for various applications such as compost bags, fast food disposables, agricultural films, hygiene films, and paper coatings. Ecoflex® can also be blended with starch in order to hydrophobize starch and make it water resistant.

5.6 Lacea® (Mitsui Chemicals Corp.)

The product from this company, which is distributed under the trade name Lacea®, is another example for a polylactide.

5.7 Natureworks™ PLA (Cargill Dow Polymers)

In a joint venture the two companies Cargill and Dow Chemicals have been focusing since 1997 on the commercialization of polylactic acid under the trade name Nature Works™ PLA. Cargill Dow Polymers has developed a low-cost continuous process which allows the production of polylactide at costs competitive with those of conventional plastics. Aqueous lactic acid is first converted into a low molecular weight PLA pre-polymer in a continuous condensation reaction, which is then converted into a mixture of lactide stereoisomers; the latter are purified by vacuum distillation. In the last step high molecular weight PLA is produced by ring-opening polymerization of the lactides in the melt. The company will extent its production capacity to 140,000 t a^{-1} by the end of 2001 in the USA and intends to construct a second production plant in Europe in the near future. This PLA is intended mainly for the use in packaging applications in high value films and rigid containers.

5.8 Resomer® (Boehringer Ingelheim AG)

Boehringer Ingelheim has developed polyesters already a long time ago on the basis of lactic acid and glycolic acid and is selling them under the generic name Resomer®. Homopolyesters of lactic acid (Resomer® L210 and Resomer® LR708) are designed for the use as resorbable implants in the field of medical devices, whereas copolyesters of lactic acid and glycolic acid (Resomer® RG503H) are raw material for the pharmaceutical industry to manufacture encapsulations of active ingredients for controlled release.

6 Polyester Amides

Biodegradable polymers mainly designed for packaging, but also for other applications are polyester amides developed by Bayer AG. BAK 1095®, e.g., is fully biodegradable and compostable. Biodegradability of the polymers is due to the occurrence of amide bonds in the polymer backbone instead of ester bonds occurring in all products mentioned in Sect. 5. It is also made mainly for applications in packaging materials and for applications in agriculture.

7 Conclusions and Outlook

PHAs, other synthetic or semisynthetic polyester resins and other resins have attracted much attention during the last years. More and more companies have successfully developed processes and products as alternatives to the conventional polymeric materials and plastics, which are not biodegradable and are manufactured from fossil resources. Moreover, several of these products are already competitive with the conventional materials with respect to performance and costs or will be probably in the near future. This reflects the market for biodegradable materials, which is significantly growing every year due to new applications as well as due to new measures and legislatives for waste disposal. The most promising advances were made with polylactides. However, encouraging developments were also made with some other polymers. PHAs have most probably the largest potential due to the incredible variability of the chemical structures that can be synthesized by bacteria and due to the flexibility of the PHA biosynthesis enzymes with regard to the broad substrate range. Moreover, due to the perspectives and impacts of molecular genetics and in particular molecular plant breeding as well as metabolic engineering, it is not unlikely that some members of PHAs can be produced at competitive costs and will, therefore, have a promising future.

8 References

ARNTZ, D., HAAS, T., MÜLLER, A., WIEGAND, N. (1991), Kinetische Untersuchungen zur Hydratisierung von Acrolein, *Chem. Ing. Tech.* **63**, 733–735.

BEHREND, D., LOOTZ, D., SCHMITZ, K. P., SCHYWALSKY, M., LABAHN, D. et al. (1998), PHB as a bioresorbable material for intravascular stents, *Am. J. Card. Transcatheter Cardiovasc. Ther.* (Abstract TCT-8, 4S).

BIEBL, H., MENZEL, K., ZENG, A.-P., DECKWER, W.-D. (1999), Microbial production of 1,3-propanediol, *Appl. Microbiol. Biotechnol.* **52**, 289–297.

BOWALD, S. F., JOHANSSON, E. G. (1990), A novel surgical material, *Eur. Patent Application* No. 0 349 505 A2.

BUNCH, A. W. (1998), Nitriles, in: *Biotechnology* Vol. 8a, 2nd Edn. (REHM, H.-J., REED, G., PÜHLER, A., STADLER, P., Eds.), pp. 277–361. Weinheim: VCH.

BYROM, D. (1992), Production of poly-β-hydroxybutyrate : poly-β-hydroxyvalerate copolymers, *FEMS Microbiol. Rev.* **103**, 247–250.

CHOI, J., LEE, S. Y. (1999a), Factors affecting the economics of polyhydroxyalkanoate production by bacterial fermentation, *Appl. Microbiol. Biotechnol.* **51**, 13–21.

CHOI, J. I., LEE, S. Y. (1999b), Efficient and economical recovery of poly(3-hydroxybutyrate) from recombinant *Escherichia coli* by simple digestion with chemicals, *Biotechnol. Bioeng.* **62**, 546–553.

CHUAH, H. H., BROWN, H. S., DALTON, P. A. (1995), Corterra poly(trimethylene terephthalate). A new performance carpet fiber, *Int. Fib. J.*, October 1995.

DATTA, R., TSAI, S.-P., BONSIGNORE, P., MOON, S.-H., FRANK, J. R. (1995), Technological and economic potential of poly(lactic acid) and lactic acid derivatives, *FEMS Microbiol. Rev.* **16**, 221–231.

DIAMANTOGLON, M., MEYER, G. (1988), Process for

the production of water-insoluble fibers of cellulose monoesters of maleic acid, succinic acid and phthalic acid, having an extremely high absorbability for water and physiological liquids, *U.S. Patent* 4,734,239.

DRATHS, K. M., FROST, J. W. (1994), Environmentally compatible synthesis of adipic acid from D-glucose, *J. Am. Chem. Soc.* **116**, 399–400.

DUVERNOY, O., MALM, T., RAMSTRÖM, J., BOWALD, S. (1995), A biodegradable patch used as a pericardial substitute after cardiac surgery: 6- and 24-month evaluation with CT, *Thorac. Cardiovasc. Surg.* **43**, 271–274.

FÜCHTENBUSCH, B., STEINBÜCHEL, A. (1999), Biosynthesis of polyhydroxyalkanoates from low-rank coal liquefaction products by *Pseudomonas oleovorans* and *Rhodococcus ruber*, *Appl. Microbiol. Biotechnol.* **52**, 91–95.

FÜCHTENBUSCH, B., WULLBRANDT, D., STEINBÜCHEL, A. (2000), Production of polyhydroxyalkanoic acids by *Ralstonia eutropha* and *Pseudomonas oleovorans* from waste oil remaining from the production of rhamnose as sole carbon source, *Appl. Microbiol. Biotechnol.* **53**, 167–172.

FUKUI, T., DOI, Y. (1997), Cloning and analysis of the poly(3-hydroxybutyrate-*co*-3-hydroxyhexanoate) biosynthesis genes of *Aeromonas caviae*, *J. Bacteriol.* **179**, 4821–4830.

GATENBY, A. A., HAYNIE, S. L., NAGARAJAN, G. (1998), Method for the production of 1,3-propanediol by recombinant organisms, *WO* 9811339 (E. I. DuPont de Nemours and Genencor International).

GERNGROSS, T. U., MARTIN, D. P. (1995), Enzyme-catalyzed synthesis of poly[(*R*)-(−)-3-hydroxybutyrate] – formation of macroscopic granules *in vitro*, *Proc. Natl. Acad. Sci. USA* **92**, 6279–6283.

GERNGROSS, T. U., SLATER, S. C. (2000), How green are green plastics, *Sci. Am.* **283**, 36–41.

GERNGROSS, T. U., SNELL, K. D., PEOPLES, O. P., SINSKEY, A. J., CSUHAI, E. et al. (1994), Overexpression and purification of the soluble polyhydroxyalkanoate synthase from *Alcaligenes eutrophus*: evidence for a required posttranslational modification for catalytic activity, *Biochemistry* **33**, 9311–9320.

GORENFLO, V., SCHMACK, G., VOGEL, R., STEINBÜCHEL, A. (in press), Development of a process for the biotechnological large-scale production of 4HV-containing polyesters and characterization of their physical and mechanical properties, *Biomacromolecules*.

GUETTLER, M. V., JAIN, M. K., RUMLER, D. (1996), Method for making succinic acid, bacterial variants for use in the process, and methods for obtaining variants, *U.S. Patent* 5,573,931.

GUETTLER, M. V., RUMLER, D., JAIN, M. K. (1999), *Actinobacillus succinogenes* sp. nov., a novel succinic acid producing strain from the bovine rumen, *Int. J. Syst. Bacteriol.* **49**, 207–216.

HAHN, J. J., ESCHENLAUER, A. C., NARROL, M. H., SOMERS, D. A., SRIENC, F. (1997), Growth kinetics, nutrient uptake, and expression of the *Alcaligenes eutrophus* poly(β-hydroxybutyrate) synthesis pathway in transgenic maize cell suspension cultures, *Biotechnol. Prog.* **13**, 347–355.

HANLEY, S. Z., PAPPIN, D. J. C., RAHMAN, D., WHITE, A. J., ELBOROUGH, K. M. et al. (1999), Re-evaluation of the primary structure of *Ralstonia eutropha* phasin and implications for polyhydroxyalkanoic acid binding, *FEBS Lett.* **447**, 99–105.

HOERSTRUP, S. P., SODIAN, R., DAEBRITZ, S., WANG, J., BACHA, E. A. et al. (2000), Functional trileaflet heart valves grown *in vitro*, *Circulation* **102**, III-44–49.

HOFFMANN, N., STEINBÜCHEL, A., REHM, B. H. A. (2000a), The *Pseudomonas aeruginosa phaG* gene product is involved in the synthesis of polyhydroxyalkanoic acids consisting of medium-chain-length constituents from non-related carbon sources, *FEMS Microbiol. Lett.* **184**, 253–260.

HOFFMANN, N., REHM, B. H. A., STEINBÜCHEL, A. (2000b), Homologous functional expression of cryptic *phaG* from *Pseudomonas oleovorans* establishes the transacylase-mediated polyhydroxyalkanoate biosynthesis pathway, *Appl. Microbiol. Biotechnol.* **54**, 665–670.

HOPPENSACK, A., REHM, B. H. A., STEINBÜCHEL, A. (1999), Analysis of 4-phosphopantetheinylation of polyhydroxybutyrate synthase from *Ralstonia eutropha*: generation of β-alanine auxotrophic Tn5 mutants and cloning of *panD* gene region, *J. Bacteriol.* **181**, 1429–1435.

HRABAK, O. (1992), Industrial production of poly-β-hydroxybutyrate, *FEMS Microbiol. Rev.* **103**, 251–256.

JIA, Y., KAPPOCK, T. J., FRICK, T., SINSKEY, A. J., STUBBE, J. (2000), Lipases provide a new mechanistic model for polyhydroxybutyrate (PHB) synthases: characterization of the functional residues in *Chromatium vinosum* PHB synthase, *Biochemistry* **39**, 3927–3936.

JOHN, M. E., KELLER, G. (1996), Metabolic pathway engineering in cotton: biosynthesis of polyhydroxybutyrate in fiber cells, *Proc. Natl. Acad. Sci. USA* **93**, 12768–12773.

JOSSEK, R., STEINBÜCHEL, A. (1998), *In vitro* synthesis of poly(3-hydroxybutyric acid) by using an enzymatic coenzyme A recycling system, *FEMS Microbiol. Lett.* **168**, 319–324.

JOSSEK, R., REICHELT, R., STEINBÜCHEL, A. (1998), *In vitro* biosynthesis of poly(3-hydroxybutyric acid) by using purified poly(hydroxyalkanoic acid) synthase of *Chromatium vinosum*, *Appl. Microbiol. Biotechnol.* **49**, 258–266.

JUNG, K., HAZENBERG, W., PRIETO, M., WITHOLT, B.

(2001), Two-stage continuous process development for the production of medium-chain-length poly(3-hydroxyalkanoates), *Biotechnol. Bioeng.* **72**, 19–24.

KASCAK, J. S., KOMINEK, J., ROEHR, M. (1996), Lactic acid, in: *Biotechnology* Vol. 6, 2nd Edn. (REHM, H.-J., REED, G., PÜHLER, A., STADLER, P., Eds.), pp. 293–306. Weinheim: VCH.

KLEIN, J., CATCHSIDE, D. E. A., FAKOUSSA, R., GAZSO, L., FRITSCHE, W. et al. (1999), Biological processing of fossil fuels. Résumé of the Bioconvension Session of ICCS '97, *Appl. Microbiol. Biotechnol.* **52**, 2–15.

KOBAYASHI, M., NAGASAWA, T., YAMADA, H. (1992), Enzymatic synthesis of acrylamide: a success story not yet over, *Trends Biotechnol.* **10**, 402–408.

KUNIOKA, M., NAKAMURA, Y., DOI, Y. (1988), New bacterial copolyesters produced in *Alcaligenes eutrophus* from organic acids, *Polym. Commun.* **29**, 174–176.

LANGENBACH, S., REHM, B. H. A., STEINBÜCHEL, A. (1997), Functional expression of the PHA synthase gene *phaC1* from *Pseudomonas aeruginosa* in *Escherichia coli* results in poly(3-hydroxyalkanoate) synthesis, *FEMS Microbiol. Lett.* **150**, 303–309.

LENZ, R. W., FARCET, C., DIJKSTRA P. J., GOODWIN, S., ZHANG, S. M. (1999), Extracellular polymerization of 3-hydroxyalkanoate monomers with the polymerase of *Alcaligenes eutrophus*, *Int. J. Biol. Macromol.* **25**, 55–60.

LIEBERGESELL, M., RAHALKAR, S., STEINBÜCHEL, A. (2000), Analysis of the *Thiocapsa pfennigii* polyhydroxyalkanoate synthase: subcloning, molecular characterization and generation of hybrid synthases with the corresponding *Chromatium vinosum* enzyme, *Appl. Microbiol. Biotechnol.* **54**, 186–194.

LIU, S.-J., STEINBÜCHEL, A. (2000a), Exploitation of butyrate kinase and phosphotransbutyrylase from *Clostridium acetobutylicum* for the *in vitro* biosynthesis of poly(3-hydroxyalkanoic acids), *Appl. Microbiol. Biotechnol.* **53**, 545–552.

LIU, S.-J., STEINBÜCHEL, A. (2000b), A novel genetically engineered pathway for synthesis of poly(hydroxyalkanoic acids) in *Escherichia coli*, *Appl. Environ. Microbiol.* **66**, 739–743.

LÜTKE-EVERSLOH, T., BERGANDER, K., LUFTMANN, H., STEINBÜCHEL, A. (2001), Biosynthesis of a new class of biopolymer: Bacterial synthesis of a sulfur containing polymer with thioester linkages, *Microbiology* **147**, 11–19.

MADISON, L. L., HUISMAN, G. W. (1999), Metabolic engineering of poly(3-hydroxyalkanoates): from DNA to plastic, *Microbiol. Mol. Biol. Rev.* **63**, 21–53.

MAEHARA, A., UEDA, S., NAKANO, H., YAMANE, T. (1999), Analyses of a polyhydroxyalkanoic acid granule-associated 16-kilodalton protein and its putative regulator in the *pha* locus of *Paracoccus denitrificans*, *J. Bacteriol.* **181**, 2914–2921.

MALM, T., BOWALD, S., BYLOCK, A., BUSCH, C. (1992a), Prevention of postoperative pericardial adhesions by closure of the pericardium with absorbable polymer patches, *J. Thorac. Cardiovasc. Surg.* **104**, 600–607.

MALM, T., BOWALD, S., KARACAGIL, S., BYLOCK, A., BUSCH, C. (1992b), A new biodegradable patch for closure of atrial septal defect, *Scand. J. Thor. Cardiovasc.* **26**, 9–14.

MALM, T., BOWALD, S., BYLOCK, A., BUSCH, C., SALDEEN, T. (1994), Enlargement of the right ventricular outflow and the pulmonary artery with a new biodegradable patch in transannular position, *Eur. Surg. Res.* **26**, 298–308.

MAROIS, Y., ZHANG, Z., VET, M., DENG, X., LENZ, R. W., GUIDOIN, R. (1999), Hydrolytic and enzymatic incubation of polyhydroxyoctanoate (PHO): A short-term *in vitro* study of a degradable bacterial polyester, *J. Biomater. Sci. Polym. Edn.* **10**, 483–499.

MAROIS, Y., ZHANG, Z., VET, M., DENG, X., LENZ, R. W., GUIDOIN, R. (2000), Bacterial polyesters for biomedical applications: *In vitro* and *in vivo* assessments of sterilization, degradation rate and biocompatibility of poly(β-hydroxyoctanoate) (PHO), in: *Synthetic Bioabsorbable Polymers for Implants* (AGRAWAL, C. M., PARR, J. E., LIN, S. T., Eds.), pp. 12–38. Scranton: ASTM.

MAYER, F., MADKOUR, M. H., PIEPER-FÜRST, U., WIECZOREK, R., LIEBERGESELL, M., STEINBÜCHEL, A. (1997), Electron microscopic observations on the macromolecular organization of the boundary layer of bacterial PHA inclusion bodies, *J. Gen. Appl. Microbiol.* **42**, 445–455.

MCCOY, M. (1998), Chemical makers try biotech paths, *Chem. Eng. News* **22**, 13–19.

MILLET, P. (1993), Retournement de la situation de la glycérine, *Informations Chimie* **345**, 102–104.

MITTENDORF, V., ROBERTSON, E. J., LEECH, R., KRÜGER, N., STEINBÜCHEL, A., POIRIER, Y. (1998), Synthesis of medium-chain-length polyhydroxyalkanoates in *Arabidopsis thaliana* using intermediates of peroxisomal fatty acid β-oxidation, *Proc. Natl. Acad. Sci. USA* **95**, 13397–13402.

MITTENDORF, V., BONGCAM, V., ALLENBACH, L., COULLEREZ, G., MARTINI, N., POIRIER, Y. (1999), Polyhydroxyalkanoate synthesis in transgenic plants as a new tool to study carbon flow through *beta*-oxidation, *Plant J.* **20**, 45–55.

NAKASHITA, H., ARAI, Y., YOSHIOKA, K., FUKUI, T., DOI, Y. et al. (1999), Production of biodegradable polyester by a transgenic tobacco, *Biosci. Biotechnol. Biochem.* **63**, 870–874.

NAWRATH, C., POIRIER, Y., SOMERVILLE, C. (1994), Targeting of the polyhydroxybutyrate biosynthetic pathway to the plastids of *Arabidopsis thaliana* results in high levels of polymer accumulation, *Proc. Natl. Acad. Sci. USA* **91**, 12760–12764.

NOISHIKI, Y., KOMATSUZAKI, S. (1995), Medical materials for soft tissue use, *Jpn. Patent Application* No. 07-275344.

PIEPER-FÜRST, U., MADKOUR, M. H., MAYER, F., STEINBÜCHEL, A. (1994), Purification and characterization of a 14-kDa protein that is bound to the surface of polyhydroxyalkanoic acid granules in *Rhodococcus ruber*, *J. Bacteriol.* **176**, 4328–4337.

PIEPER-FÜRST, U., MADKOUR, M. H., MAYER, F., STEINBÜCHEL, A. (1995), Identification of the region of a 14-kDa protein of *Rhodococcus ruber* that is responsible for the binding of this phasin to polyhydroxyalkanoic acid granules, *J. Bacteriol.* **177**, 2513–2523.

POIRIER, Y., DENNIS, D., KLOMPARENS, K., SOMERVILLE, C. R. (1992), Polyhydroxybutyrate, a biodegradable thermoplastic, produced in transgenic plants, *Science* **256**, 520–523.

POTERA, C. (1997), Genencor & DuPont create "green" polyester, *Gen. Eng. News* **17**, 17.

QI, Q., REHM, B. H. A., STEINBÜCHEL, A. (1997), Synthesis of poly(3-hydroxyalkanoates) in *Escherichia coli* expressing the PHA synthase gene *phaC2* from *Pseudomonas aeruginosa*: comparison of PhaC1 and PhaC2, *FEMS Microbiol. Lett.* **157**, 155–162.

QI, Q., STEINBÜCHEL, A., REHM, B. H. A. (1998), Metabolic routing towards polyhydroxyalkanoic acid synthesis in recombinant *Escherichia coli* (*fadR*): inhibition of fatty acid β-oxidation by acrylic acid, *FEMS Microbiol. Lett.* **167**, 89–94.

QI, Q., STEINBÜCHEL, A., REHM, B. H. A. (2000), *In vitro* synthesis of poly(3-hydroxydecanoate): purification of type II polyhydroxyalkanoate synthases PhaC1 and PhaC2 from *Pseudomonas aeruginosa* and development of an enzyme assay, *Appl. Microbiol. Biotechnol.* **54**, 37–43.

REHM, B. H. A., STEINBÜCHEL, A. (1999), Biochemical and genetic analysis of PHA synthases and other proteins required for PHA synthesis, *Int. J. Biol. Macromol.* **25**, 3–19.

REHM, B. H. A., STEINBÜCHEL, A. (in press), PHA synthases – the key enzymes of PHA synthesis, in: *Biopolymers* Vol. 3 (DOI, Y., STEINBÜCHEL, A., Eds.). Weinheim: Wiley-VCH.

REHM, B. H. A., KRÜGER, N., STEINBÜCHEL, A. (1998), A new metabolic link between fatty acid *de novo* synthesis and polyhydroxyalkanoic acid synthesis. The *phaG* gene from *Pseudomonas putida* KT2440 encodes a 3-hydroxyacyl-acyl carrier protein-coenzyme A transferase, *J. Biol. Chem.* **273**, 24044–24051.

RODRIGUES, M. F. A., DA SILVA, L. F., GOMEZ, J. G. C., VALENTIN, H. E., STEINBÜCHEL, A. (1995), Biosynthesis of poly(3-hydroxybutyric acid-*co*-3-hydroxy-4-pentenoic acid) from unrelated substrates by *Burkholderia* sp., *Appl. Microbiol. Biotechnol.* **43**, 880–886.

RODRIGUES, F. M. DE A., VALENTIN, H. E., BERGER, P. A., TRAN, M., ASRAR, J. M. et al. (2000a), Polyhydroxyalkanoate accumulation in *Burkholderia* sp.: a molecular approach to elucidate the genes involved in formation of two distinct short-chain homopolymers, *Appl. Microbiol. Biotechnol.* **53**, 453–460.

RODRIGUES, F. M. DE A., VICENTE, E. J., STEINBÜCHEL, A. (2000b), Studies on PHA accumulation in a polyhydroxyalkanoate synthase-negative mutant of *Burkholderia cepacia* generated by homogenotization, *FEMS Microbiol. Lett.* **193**, 179–185.

SCHMACK, G., GORENFLO, V., STEINBÜCHEL, A. (1998), Bioproduction and characterization of 4HV-containing polyesters, *Macromolecules* **31**, 644–649.

SCHMITZ, K.-P., BEHREND, D. (1997), Method of manufacturing intraluminal stents made of polymer material, *European Patent Application* No. 0 770 401 A2.

SLATER, S., MITSKY, T. A., HOUMIEL, K. L., HAO, M., REISER, S. E. et al. (1999), Metabolic engineering of *Arabidopsis* and *Brassica* for poly(3-hydroxybutyrate-*co*-3-hydroxyvalerate) copolymer production, *Nature Biotechnol.* **17**, 1011–1016.

SODIAN, R., STOCK, U. A., SPERLING, J. S., MARTIN, D. P., MAYER, J. E., VACANTI, J. P. (1999), Tissue engineering of a trileaflet heart valve – Early *in vitro* experiences with a combined polymer, *Tissue Eng.* **5**, 489–493.

SODIAN, R., HOERSTRUP, S. P., SPERLING, J. S., MARTIN, D. P., DEABRITZ, S. et al. (2000), Evaluation of biodegradable, three-dimensional matrices for tissue engineeering of heart valves, *ASAIO J.* **46**, 107–110.

SPIEKERMANN, P., REHM, B. H. A., KALSCHEUER, R., BAUMEISTER, D., STEINBÜCHEL, A. (1999), A sensitive, viable colony staining method using Nile Red for direct screening of bacteria that accumulate polyhydroxyalkanoic acids and other lipid storage compounds, *Arch. Microbiol.* **171**, 73–78.

STEINBÜCHEL, A. (1996), PHB and other polyhydroxyalkanoic acids, in: *Biotechnology* Vol. 6, 2nd Edn. (REHM, H.-J., REED, G., PÜHLER, A., STADLER, P., Eds.), pp. 403–464. Weinheim: VCH.

STEINBÜCHEL, A. (2001), Perspectives for biotechnological production and utilization of biopolymers: Metabolic engineering of polyhydroxyalkanoate biosynthesis pathways as a successful example, *Macromol. Biosci.* **1**, 1–24.

STEINBÜCHEL, A., FÜCHTENBUSCH, B. (1998), Bacterial and other biological systems for polyester production, *Trends Biotechnol.* **16**, 419–427.

STEINBÜCHEL, A., HEIN, S. (2001), Biochemical and

molecular basis of polyhydroxyalkanoic acids in microorganisms, in: *Biopolyesters* (Steinbüchel, A., Babel, W., Eds.), *Adv. Biochem. Eng. Biotechnol.* **71**, 81–123.

Steinbüchel, A., Valentin, H. E. (1995), Diversity of bacterial polyhydroxyalkanoic acids, *FEMS Microbiol. Lett.* **128**, 219–228.

Steinbüchel, A., Aerts, K., Babel, W., Föllner, C., Liebergesell, M. et al. (1995), Considerations on the structure and biochemistry of bacterial polyhydroxyalkanoic acid inclusions, *Can. J. Microbiol.* **41** (Suppl. 1), 94–105.

Stinson, S. C. (1995), Fine and intermediate chemical makers emphasize new projects and processes, *Chem. Eng. News* **17**, 10–14.

Stock, U. A., Sakamoto, T., Hatsuoka, S., Martin, D. P., Nagashima, M. et al. (in press), Patch augmentation of the pulmonary artery with bioabsorbable polymers and autologous cell seeding, *J. Thorac. Cardiovasc. Surg.* **120**.

Stock, U. A., Nagashima, M., Khalil, P. N., Nollert, G. D., Herden, T. et al. (2000), Tissue-engineered valved conduits in the pulmonary circulation, *J. Thorac. Cardiovasc. Surg.* **119**, 732–740.

Stuart, E. S., Lenz, R. W., Fuller, R. C. (1995), The ordered macromolecular surface of polyester inclusion bodies in *Pseudomonas oleovorans*, *Can. J. Microbiol.* **41** (Suppl. 1), 84–93.

Stuart, E. S., Tehrami, A., Valentin, H. E., Dennis, D., Lenz, R. W., Fuller, R. C. (1998), Protein organization on the PHA inclusion cytoplasmic boundary, *J. Biotechnol.* **64**, 137–144.

Su, L., Lenz, R. W., Takagi, Y., Zhang, S. M., Goodwin, S. et al. (2000), Enzymatic polymerization of (*R*)-3-hydroxyalkanoates by a bacterial polymerase, *Macromolecules* **33**, 229–231.

Tsuge, T., Fukui, T., Matsusaki, H., Taguchi, S., Kobayashi, G. et al. (2000), Molecular cloning of two (*R*)-specific enoyl-CoA hydratase genes from *Pseudomonas aeruginosa* and their use for polyhydroxyalkanoate synthesis, *FEMS Microbiol. Lett.* **184**, 193–198.

Unverdorben, M., Schywalsky, M., Labahn, D., Hartwig, S., Laenger, F. et al. (1998), Polyhydroxybutyrate (PHB) stent-experience in the rabbit, *Am. J. Card. Transcatheter Cardiovasc. Ther.* (Abstract TCT-11, 5S).

Valentin, H. E., Stuart, E. S., Fuller, R. C., Lenz, R. W., Dennis, D. (1998), Investigation of the function of proteins associated to polyhydroxyalkanoate inclusions in *Pseudomonas putida* BMO1, *J. Biotechnol.* **64**, 145–157.

Valentin, H. E., Berger, P. A., Gruys, K. J., Rodrigues, M. F. de A., Steinbüchel, A. et al. (1999a), Polyhydroxyalkanoate accumulation by *Burkholderia* sp., *Macromolecules* **32**, 7389–7395.

Valentin, H. E., Broyles, D. L., Casagrande, L. A., Colburn, S. M., Creely, W. L. et al. (1999b), PHA production, from bacteria to plants, *Int. J. Biol. Macromol.* **25**, 303–306.

van der Giessen, W. J., Lincoff, A. M., Schwartz, R. S., van Beusekom, H. M. M., Serruys, P. W. et al. (1996), Marked inflammatory sequelae to implantation of biodegradable and nonbiodegradable polymers in porcine coronary arteries, *Circulation* **94**, 1690–1697.

Wieczorek, R., Pries, A., Steinbüchel, A., Mayer, F. (1995), Analysis of a 24-kDa protein associated with the polyhydroxyalkanoic acid granules in *Alcaligenes eutrophus*, *J. Bacteriol.* **177**, 2424–2435.

Wilke, D. (1999), Chemicals from biotechnology: molecular plant genetics will challenge the chemical and the fermentation industry, *Appl. Microbiol. Biotechnol.* **52**, 135–145.

Witt, U., Müller, R.-J., Deckwer, W.-D. (1996), Evaluation of the biodegradability of copolyesters containing aromatic compounds by investigation of model oligomers, *J. Environ. Polym. Degrad.* **4**, 9–20.

Zeikus, J. G., Jain, M. K., Elankovan, P. (1999), Biotechnology of succinic acid production and markets for derived industrial products, *Appl. Microbiol. Biotechnol.* **51**, 545–552.

Biotechnology in Various Areas and Special Processes

15 Marine Biotechnology

ULRIKE LINDEQUIST
THOMAS SCHWEDER
Greifswald, Germany

1 Introduction

Marine biotechnology can be understood as the "use of marine organisms or their constituents for useful purposes in a controlled fashion" (ATTAWAY and ZABORSKY, 1993). Whereas some marine products are known for many thousands of years, marine biotechnology has increasingly attracted worldwide attention since the 1980s.

With more than 70% of the earth's surface, the oceans represent the largest habitat of the world and a prolific resource of organisms with high biological and chemical diversity. Nevertheless, research on marine biotechnology and commercialization of products are still in their infancy; the proportion of marine organisms or their products researched and used has been relatively small to date. Thus of the approximately 120,000 known natural substances less than 10% originate from marine life (JASPARS, 1998). Initially, the marine natural products field focused on metabolites from marine algae and invertebrate animals. Meanwhile, marine microorganisms attract more and more attention (FENICAL and JENSEN, 1993).

Main fields of marine biotechnology are

1. natural products from marine organisms (pharmaceuticals, enzymes, nutraceuticals, biopolymers, and other biomaterials),
2. bioremediation with marine microorganisms,
3. energy,
4. aquaculture.

Regarding these topics the ramifications of marine biotechnology reach almost every major sector of human life (ZABORSKY, 1999). Central point of this chapter are natural products from marine organisms. Within this the most productive sectors to date are biopharmaceuticals and enzymes. New reviews about marine compounds with biological activity are given by IRELAND et al. (1993), BONGHIORNI and PIETRA (1996), KÖNIG and WRIGHT (1996), JASPARS (1998), FAULKNER (2000a, b, earlier reviews since 1972 cited therein), KERR and KERR (1999), and MUNRO et al. (1999). MOORE (1999) summarized the results of biosynthetic investigations. Bioremediation and energy are treated only briefly here. Aquaculture is mentioned only in connection with other points. For marine biotechnology with microalgae, see Chapter 5, this volume.

2 Marine Diversity

The marine environment is supposed to be one of the richest biospheres on earth. This biosphere encompasses a huge thermal range (from −1,5 °C in Antarctic waters to temperatures in excess of 100 °C in shallow and 350 °C in deep hydrothermal systems), pressures ranging from 1 to over 1,000 atmospheres, nutrient variations ranging from eutrophic to oligotrophic, and extensive photic and nonphotic zones. The marine environment ranges from nutrient-rich regions to nutritionally sparse locations where only a few organisms can survive. The many habitats in the sea have created niches for the evolution of diverse forms of life, from microorganisms to mammals (BERNAN et al., 1997; COWAN, 1997; DE LONG, 1997; FENICAL, 1997).

The adaptations of organisms to growing in the ocean are fundamentally different from those in land-based organisms. Many organisms live in close associations. This leads to the necessity of producing secondary metabolites playing roles in ensuring the fitness and survival of their producers. Such functions may be intra- and interspecific signaling, deterrency of herbivores and predators, suppression of competing neighbors, inhibition of bacterial and fungal invasion, or protection against UV radiation (PROKSCH, 1999).

To date only a fraction of the types of organisms which live in the sea are known. The number of types of bacteria which live in the marine environment is estimated to be 400,000 to 3 million, the types of algae about 200,000 to 10 million. Of the 34 known phyla of animals 33 are represented in this habitat, 15 are found exclusively in the sea. Microbial diversity studies usually lag far behind macroorganism research. It is estimated that over 95% of the all existent fish species are known, but less than 5% of bacterial species (COWAN, 1997). Based on microscopic counting methods with DNA dyes and 16S rRNA analysis of marine communities it is supposed that approximately 1% of the marine bacteria have been identified. Much less is also known about symbiosis and about physiological and ecological significance of secondary metabolites.

New techniques of molecular biology, such as the DNA array technique, will give new insights into marine diversity and the interactions between marine organisms. AMANN and KÜHL (1998) suggested various approaches for *in situ* analysis of different species in mixed populations. This includes fluorescence *in situ* hybridization (FISH) and the use of microsensors. The combination of these different techniques will yield more detailed insight in the way microorganisms operate in nature. FISH with rRNA-targeted oligonucleotide probes was used to investigate the phylogenetic composition of bacterioplankton communities in several freshwater and marine samples (GLOCKNER et al., 1999). Furthermore, this technique has been applied to investigate the microbial diversity in marine sponges, such as *Aplysina cavernicola* (FRIEDRICH et al., 1999).

The revolution in the sequencing technique will certainly support this new technology, since the design of the DNA arrays requires the knowledge of the sequence of the appropriate genes. The first completely sequenced genomes of marine bacteria were those of the methanogenic bacterium *Methanococcus jannaschii* (SMITH et al., 1997) and that of the cyanobacterium *Synechocystis* spp. (NAKAMURA et al., 1998; www.kazusa.org). However, so far there are very few sequences completed (e.g., *Aeropyrum pernix*, *Thermotoga maritima*) and even fewer ongoing genome projects with marine microorganisms (*www.tigr.org*).

Proteome analysis is a powerful tool to get a more comprehensive picture of the cellular status. This approach permits an improved understanding of the means by which marine bacteria adapt to alterations in their environmental conditions. The technique is based on two-dimensional protein gel electrophoresis (2D-PAGE) used for protein separation, and MALDI-TOF mass spectrometry or N-terminal sequencing applied for protein identification. Proteome analysis could be accomplished by expression profiling with DNA arrays. Such proteome analysis has already been used for the marine bacterium *Vibrio* spp. (ÖSTLING et al., 1997). GROSS et al. (1994) used this approach to investigate the response of marine bacteria and fungi to high-pressure stress.

The first genome of an unculturable marine bacterium, the symbiont of the tube worm *Riftia pachyptila* (DISTEL et al., 1988) is currently sequenced by a Californian consortium (HORST FELBECK, personal communication). This will open a new dimension in the exploration of this deep-sea symbiontic microorganism, which seems to form a monoculture in the trophosomes of *R. pachyptila*. The knowledge of its genome sequence together with the new molecular techniques of proteome and transcriptome analysis might help to find the clue for this exclusive microbial colonization with *Riftia*. This could finally be due to the identification of new natural products possibly involved in the exclusive establishment of the bacterial symbiont in the trophosomes of *Riftia*.

3 Marine Organisms of Biotechnological Potential

3.1 Marine Microorganisms

Marine microorganisms (bacteria, fungi, protists) encompass a complex and diverse assemblage of microscopic life forms (BERNAN et al., 1997). They are main decomposers in the marine environments and provide the food for other organisms. The diversity in marine environmental conditions has facilitated a remarkable specialization of marine microorganisms. Often marine microorganisms are the original producers of many compounds isolated from sponges, ascidians, and other marine invertebrates. Marine bacteria are free-living (planktonic), commensal (epibacteria, inhabit surfaces, tissues, and internal spaces of other organisms), or symbiotic (endobacteria, live within cells or in the intracellular matter of other organisms). Generally bacteria are divided into eubacteria and archaebacteria. Most marine bacteria belong to the eubacteria. They can be subdivided into autotrophs and heterotrophs. Above all the more primitive archaebacteria inhabit extreme marine environments. They include methanobacteria (which obtain energy by converting hydrogen and carbon dioxide into methane), halobacteria (which grow in water of high salinity), sulfobacteria (which utilize sulfur compounds as a source of energy), and bacteria which use amines or metals for the generation of energy (FENICAL and JENSEN, 1993; KÖNIG and WRIGHT, 1999). The investigation of the microbial diversity of the Antarctic ocean demonstrated that up to 34% of total prokaryotic biomass is represented by archaeal species (DELONG et al., 1994).

Marine fungi are not a taxonomically-systematically, but an ecologically-physiologically defined group of fungi (Mycophyta). They facultatively grow or obligately in oceans and ocean-associated estuarine habitats. According to KOHLMEYER and KOHLMEYER (1979) obligate marine fungi are those growing and sporulating exclusively in a marine and estuarine habitat; facultative marine fungi are those from freshwater or terrestrial milieus able to grow (and possibly also to sporulate) in the marine environment. Including the non-active marine fungi, the number of mycelia-forming higher fungi is estimated to be at least 6,000 (SCHAUMANN, 1993). Marine fungi represent a considerable amount of biomass in the marine habitat. A biotechnologically important group of marine fungi are the thraustochytrids. These common marine microheterotrophs can be cultured and produce lipids containing high proportions of n-3 and n-6 PUFAs (see Sect. 5.3.2.1). On the other hand they often are contaminants in cell cultures of marine invertebrates.

Reviews about natural substances from several groups of marine microorganisms are given by FENICAL (1993), FENICAL and JENSEN (1993), KOBAYASHI and ISHIBASHI (1993), PIETRA (1997), BERNAN et al. (1997); about those from marine bacteria by JENSEN and FENICAL (1994), MIKHAILOW et al. (1995), SODE et al. (1996), and those from marine fungi by LIBERRA and LINDEQUIST (1995), and LINDEQUIST et al. (1999).

3.2 Marine Plants

The greatest part of marine vegetation are algae (seaweeds). The main divisions are the red (Rhodophyta, about 400 genera), the brown (Phaeophyta, about 200 genera), and the green (Chlorophyta) seaweeds. Another classification is those in macroalgae (macroscopic) and microalgae (visible only in the microscope). In this chapter only macroalgae are treated (for microalgae, see Chapter 5, this volume).

Algae are one of the primary producers in the sea. Using sunlight they are involved in the fixation of carbon dioxide resulting in the evolution of oxygen. In order to gather enough light for photosynthesis, seaweeds belonging to different divisions have developed different accessory pigments according to the change of color with the depth of the water. Because they lack true roots, seaweeds mostly need a solid substrate on which to set (benthic forms).

Marine phanerogams belonging to the genera *Zostera, Ruppia, Cymodeca, Posedonia*, and others are not of biotechnological interest until now.

3.3 Marine Animals

Marine animals with biotechnological potential are the sponges (Porifera), the cnidarians (Cnidaria, coelenterates), the molluscs (Mollusca), the moss animals (Bryozoa), the tunicates (Tunicata), and the crustaceans (Crustacea). All are invertebrates. Marine vertebrates like Pisces are of biotechnological importance only in aquaculture.

Sponges are primitive multicellular aquatic animals. Their body structure is simple and they do not possess organs and true tissues. Their most characteristic feature is a network consisting of chambers and channels through which water flows continuously. Sponges are attached to a solid substratum (OSINGA et al., 1998). More than 5,000 recent species are known, mostly living in a marine environment. They represent the most ancient, extant metazoan phylum (MÜLLER, 1998, 1999a). Approximately 40% of all so far known compounds of marine origin come from sponges or from associated organisms. They are the richest source of structurally different bioactive compounds among the Metazoa (SARMA et al., 1993; MÜLLER, 1999b). The discovery of spongouridin and spongothymidin in a sponge by BERGMANN and FEENEY (1951) was the starting point for this development. These nucleosides were used as a model for the development of the virostatic compounds arabinoside-A and arabinoside-C. Yet, there are several limitations to the use of sponge substances on a larger scale, as their content is usually low, the amount of starting material is very limited, or the composition of bioactive compounds in the starting material varies depending on the depth where the animals were collected (KREUTER et al., 1992; MÜLLER, 1999b). A further problem is that interesting substances may also be produced by endosymbiotic microorganisms which live in the sponge. BEWLEY et al. (1996) found four distinct cell populations consistently present in *Theonella swinhoei:* eukaryotic sponge cells, unicellular heterotrophic bacteria, unicellular cyanobacteria, filamentous heterotrophic bacteria. The macrolid swinholide A occurred only in the mixed population of unicellular heterotrophic bacteria, an anti-fungal cyclic peptide (P951) only in the filamentous heterotrophic bacteria. Bioactive substances from sponges have been reviewed by SARMA et al. (1993).

Cnidaria are the lowest metazoan group possessing an organized body structure. Classes belonging to the phylum Cnidaria with biologically highly active compounds are the Anthozoa with Gorgonaria (gorgonians, horny corals) and Alcyonaria (soft corals). At the end of the 1960s the discovery of first prostanoids in the coral *Plexaura homomalla* was one of the starting points for the investigation of marine natural substances and of prostanoids (WEINHEIMER and SPRAGGINS, 1969). Since that time marine prostanoids have attracted much attention because of their structural features and biological activity (IWASHIMA et al., 1999).

Marine molluscs (mussels, oysters, clams, snails) are ancient examples of applied marine biotechnology. Extracts from these organisms yielded the well-known Tyrian purple, an important and valuable dye during the Antique. The dying component could be identified as 6,6′-dibromindigo (BONGHIORNI and PIETRA, 1996). In view of biologically active substances *Conus* sp. are of toxicological and therapeutic interest.

Bryozoa possess a prolific chemical defense strategy. About 4,000 species are known. They are relatively primitive organisms that live in colonies. Of biotechnological interest is *Bugula neritina*. Larvae of *B. neritina* harbor bacterial symbionts which were identified as a novel species of γ-proteobacterium (HAYGOOD and DAVIDSON, 1997).

Tunicates (about 2,000 species) are the most primitive group of the phylum Chordata. Ascidiaceae (sea quirts, ascidians) are the group of tunicates best investigated chemically. They live in benthic form on the sediment and are organized in large colonies. Important classes of natural substances from ascidia are cyclopeptides, polycyclic alkaloids, indol derivatives, polysaccharides (tunicin as a skeletal substance) and sulfur-containing compounds. A review on biologically active substances from tunicates is given by BRACHER (1994).

Among the Crustacea crayfish, lobster, and shrimp are of interest for aquaculture and as producers of biopolymers and pigments.

4 Production Processes

4.1 Product Discovery

The first step of a marine biotechnological process is to find and identify new products. Main attention has been directed to the discovery of natural products with biological activity useful for medicine. Difficulties may be the overall diversity of marine organisms and habitats, the uncertainty in the identification of the source organism, the influence of habitat as well as of seasonal and geographic factors, the trace nature of some extremely bioactive compounds, and unexpected chemical functionalizations. An outline of the general approach in the discovery of biologically active products includes the following steps:

Collection and identification of the marine organisms

This is often the most important step, because the quality of the collection influences all further steps. Collection of organisms at places easy to reach (at coasts, beaches, from water surfaces) was preferred for a long time. Improvements in underwater life support systems have provided marine scientists with new possibilities for the collection from unexplored regions and depths. In addition to conventional self-contained underwater breathing apparatus (SCUBA) by using Closed Circuit Underwater Breathing Apparatus (CCUBA) it is now possible to dive to −150 m for up to 8 h (EL SAYED et al., 2000). Strategies may be to collect large numbers of highly diverse organisms in order to provide maximal taxonomic and chemical diversity, to collect specific taxonomic groups, or to focus on organisms in which due to ecological reasons production of biologically active secondary metabolites is expected. The collection must be accurately sorted, carefully documented, and voucher specimens should be prepared. All collections should be made with the informed consent of the host country or state. Sampling methods are designed to minimize environmental impact. Because of the high toxicity and irritant properties of some marine organisms they should always be handled with caution. Protective equipment such as gloves and eye protection should always be worn. The bulk sample must be frozen rapidly or stored in suitable solvent to retard bacterial degradation of the specimens (WRIGHT, 1998; FAULKNER, 2000c). Storage of collected organisms, preparation for cultivation and/or extraction and further processes depend on the aim of the study and the type of organism.

Extraction of substances

Mode and solvent of extraction determine which substances are extracted. A suitable extraction solvent is ethanol. It extracts substances of different polarity and is compatible with most bioassays. Other extraction solvents prefer lipophilic or hydrophilic substances. Another possibility would be solid-phase extractions (RIGUERA, 1997; WRIGHT, 1998).

Assay of extracts (biological screening)

Large screening programs and new screening methods exist. The screening strategy has to be adopted to the biomedical or other utility. Useful advice for screening of marine extracts is given by FAULKNER (2000c). The use of high-throughput screening methods (GRABLEY and THIERICKE, 1999) can improve the efficacy of biological tests. Methods of functional genomics allow the target identification of biologically active substances in a very specific way (SCHWEDER et al., 2000).

Isolation and identification of active compound(s); dereplication of known compounds

Chromatographic methods like column chromatography without or with pressure, countercurrent chromatography etc. are used for the isolation of active substances. Thin-layer chromatography is mainly used for control of the separation process. Structure elucidation of isolated marine compounds needs several instrumental methods including infrared (IR), ultraviolet (UV), mass spectro-metry (MS), and nuclear magnetic resonance (NMR) spectroscopy (RIGUERA, 1997; WRIGHT, 1998; EL SAYED et al., 2000).

4.2 Cultivation/Fermentation

4.2.1 Marine Microorganisms

Because of the special growth requirements only a minority of the marine microorganisms could be cultured so far. Less than 5% of the viable bacterial cells in marine samples ultimately grow under standard culture conditions (JENSEN and FENICAL, 1994; BERNAN et al., 1997), for example none of the cold-adapted archaebacteria could be cultivated so far. Marine microorganisms are adapted to saline environment, they have a quite different physiological character compared to their terrestrial counterparts, and they mostly require NaCl for optimal growth. For example, marine bacteria possess a unique type of NADH:quinone reductase, a membrane-bound enzyme which is part of the respiratory chain and requires Na^+ for its activity (YOSHIKAWA et al., 1999).

Currently most production strategies are carried out at the shake-flask level, but investigations in bioreactor engineering and fermentation protocol design are in progress in the field of marine bacteria (WEINER and MANYAK, 1999) and marine fungi (ABBANAT et al., 1998; HELMHOLZ et al., 1999). Until now only thraustochytrids and microalgae (see Chapter 5, this volume) can be cultured to produce high amounts of biomass. Cell yield and PUFA production by thraustochytrids can be varied by manipulation of physical and chemical culture parameters (LEWIS et al., 1999); media with a high C:N ratio stimulated DHA production. The use of such media in bioreactor cultivations resulted in maximal biomass, lipid, and DHA content (BOWLES et al., 1999).

New culture methods should take into account the environmental parameters associated with the habitus sampled. They should involve the nutrient composition of the media, pH, temperature, wavelength of the light, or pressure, but also the establishment of mixed cultures. Improvement of the isolation method, development of immobilization methods, mass transfer, and shear stress data of fermentations are also necessary (JENSEN and FENICAL, 1994; CHANDRASEKARAN, 1996; MARWICK et al., 1999).

4.2.2 Macroalgae

4.2.2.1 Aquaculture

Convenient methods for the production of macroalgae are harboring from natural sources and aquaculture. Present research in aquaculture is directed to species and strain selection, vegetative reproduction, methods and locations for cost-effective cultivation, timing of and methods for harvesting, and optimizing drying and storage conditions. Aims are to optimize the yields, to assure product quality, and to match product qualities for new applications. For sustaining future commercial growth it will be necessary to use modern biotechnological methods, e.g., research on genome structure and function and metabolic pathways, and to develop new culture methods (RENN, 1997).

4.2.2.2 Cell or Tissue Culture

Cell or tissue suspension cultures established from delicate and anatomically complex marine macroalgae also have the potential for the production of secondary metabolites. They are grown within bioreactor systems (stirred tank, bubble column, immobilized mesh, and tubular recycle photobioreactor systems) under controlled conditions (RORRER and MULLIKIN, 1999). Axenic culture, organic carbon source, and plant growth regulators are necessary. Progress has been made in the field of protoplast formation (allows production of hybrid and transgenic macroalgae) for *Porphyra, Gracilaria, Chondrus*, and some others, and in the field of callus or undifferentiated cell culture, e.g., for *Gelidium, Gracilaria*, and *Pterocladia* (RENN, 1997). Differentiation of excised tissues to differentiate structure types during culture could be shown, e.g., for the brown algae *Sargassum confusum* and *Undaria undarioides*. Gametophyte cell suspension cultures from the brown alga *Laminaria saccharina* produce hydroxy fatty acids (15-HETE, 13-HODTA, 13-HODE) in a yield from 100–1,000 µg product per g dry biomass. Feeding of the cultures with C18 polyunsaturated fatty acids increased the yields 2–4 times over controls (RORRER et al., 1997).

4.2.3 Marine Animals

4.2.3.1 Aquaculture

Besides harboring from natural sources there is much experience in the field of aquaculture of marine invertebrates and vertebrates for food and some technical purposes. At present, as much as 20% of the world production of fish is based on aquaculture (MEYERS, 1994), and aquaculture of molluscs and crustaceaens will be a rapidly growing global industry. Aquaculture production is constrained largely by the growth efficiency of the species produced. Biotechnological methods may open new ways for improvement of aquaculture. Examples are transgenic technologies for production of fish with improved properties, breeding of transgenic fodder plants, or cultivation of PUFA-rich microorganisms for aquaculture diets or the use of probiotics for reduction of antibiotic use on fish farms. The selection and genetic engineering of nitrifying and denitrifying bacteria could pave the way for fully enclosed, recirculating marine culture systems (LYNDON, 1999). Further possibilities are the development of new vaccines and antibiotics or the cocultivation of marine microorganisms with antimicrobial potential and the design of improved culture systems.

Aquaculture for production of biologically active substances is – with one exception (*Bugula*) – only at an experimental state. Most approaches are done with sponges. Advantages of *in situ* sponge aquacultures are low cost with respect to culture device, maintenance, and feeding of the animals, but the nutrition is not defined, and sponge growth rates and target metabolite complement and concentration per unit volume vary with season, locality, depth, and type of growing support structure. An improvement may be the use of semicontrolled conditions where additional nutrition is added to the cultures (OSINGA et al., 1998, 1999). Successful approaches of cultivation of sponge tissue samples in open systems with maintained metabolite production were done with *Geodia cydonium* (MÜLLER et al., 1999b) and *Lyssodendoryx* sp. (MUNRO et al., 1999). In the case of *G. cydonium* tissue samples of approximately 10 g were attached to the bottom of cultivation trays. The tray batteries remained in aquaria or were transferred to the vicinity of a fish and mussel farm. The trays were exposed to circulating sea water, which allows for removal of toxic compounds and supply of nutrition. Growth started after about 3 months with a doubling growth rate after 6 months. The production of cytotoxic compounds and the expression of a marker protein in cell differentiation demonstrated the intactness of the cultures. Also *Lyssodendoryx* sp. maintains its biosynthetic ability for production of halichondrin B and isohomohalichondrin B (MUNRO et al., 1999). Prospects for closed culture systems appear to be limited. Successful cultivations in closed systems were made with *Ophlitaspongia seriata* and *Pseudosuberites* (aff.) *andrewsi* (OSINGA et al., 1998, 1999). Pilot work on the production of the sponge *Suberites ficus* on cultured scallops (*Chlamys opercularis*) in Scotland has been described by ARMSTRONG et al. (1999).

In at least one case aquaculture of invertebrates is successful in producing bioactive substances and overcoming supply difficulties. Bryostatin, a complex polycyclic polyether structure, is produced by the Pacific coast bryozoan, *Bugula neritina*. At the company of Cal BioMarine Technologies (Carlsbad, CA) colonies of *B. neritina* are grown in 5,000 L tanks. Since natural colonies are sometimes destroyed by storms and not all *B. neritina* populations produce the material, aquaculture using selected high-yielding strains should solve the supply difficulties (ROUHI, 1995).

4.2.3.2 Cell or Tissue Culture

Cell culture of marine invertebrates, especially sponges, and subsequent production of compounds in bioreactors are also under investigation, but so far, mainly primary cell cultures have been established. At present (industrial) techniques to produce biomass under completely controlled conditions in bioreactors are not available (RINKEVICH, 1999). A model for a sponge bioreactor is given by OSINGA et al. (1998). A powerful novel model system to study basic mechanisms was the formation of multicellular aggregates from dissociated single cells of *Suberites domuncula*, termed primmorphs. This was achieved in seawater without addition of any further supplement. These primmorphs show an organized tissue-like structure, they are telomerase-positive and undergo DNA synthesis. They were cultured for more than 5 months and may be used as bioreactors to produce compounds from sponges in the future (CUSTODIO et al., 1998). In primary culture stimulated and nonstimulated archaecytes from the sponge *Teichaxinella morchella* produced the bioactive metabolite stevensine. Cultivation assays with other invertebrate cells were done with the tunicate *Ecteinascidia turbinata* (POMPONI et al., 1998; POMPONI, 1999).

4.3 Product Recovery

Although product recovery is the final and often cost-determining step in marine bioprocess engineering, relatively little attention has been paid to this problem. Newer aspects are discussed by VAN DER WIELEN and CABATINGAN (1999).

4.4 Prospects

4.4.1 Genetic Engineering

Current marine biotechnological applications are based on products or sequences from culturable organisms. Other ways to make use of the great biotechnological potential of the unculturable marine organisms still have to be found. Advanced techniques of DNA isolation and cloning also permit the isolation and expression of genes of unculturable organisms, especially microorganisms from different habitats. In the future, these methods could allow the production of targeted metabolites in common industrial microbes.

This concept is already being applied by different laboratories and companies. STEIN et al. (1996) used cosmid vectors to clone large (30–40 kb) fragments of genomic DNA from unculturable marine archaea. Recently, DIVERSA Corporation (San Diego, CA) has applied for a patent claiming a process allowing the construction and screening of expression libraries from nucleic acid directly isolated from the environment utilizing a fluorescence-activated cell sorter. This approach can be used to automate the screening of genes from nature either for new technical enzymes or for other specified activities.

4.4.2 Synthesis and Modification of Marine Compounds

Chemical or enzymatic synthesis offer a complementary approach to get marine natural products. Modification of the natural compound as lead substance by biotransformation or chemical methods can increase the structural variability and improve the properties of the product (ELYAKOV et al., 1994; EL SAYED et al., 2000).

5 Products of Marine Biotechnology

5.1 Pharmaceuticals

5.1.1 Cytostatic/Antitumor Compounds

Worldwide tumors range at position 3 in mortality statistics. Therefore, there is a need for new antitumor drugs with high activity and minimized side effects. In the field of potential cancer treatments most progress in research has been achieved with respect to marine active substances. Probably marine organisms need cytostatic metabolites as chemical weaps and are, therefore, a prolific resource for new drugs (SCHMITZ et al., 1993; JASPARS, 1998; FAULKNER, 2000c). Examples are summarized in Tab. 1 (see also Fig. 1), the following of which are in preclinical or clinical trials at present:

Halomon

Halomon exhibits highly differential cytotoxicity against several human tumor cell lines. Brain, renal, and colon tumor cell lines are most sensitive, while leukemia and melanoma lines are relatively less sensitive. Collections of *P. hornemannii* yielded similar monoterpenes, partly less cytotoxic and devoid of differential activity. $5 \cdot 50$ mg kg^{-1} d^{-1} halomon in an i.p./i.p. xenograft model with the highly aggressive U251 brain tumor line have shown a high percentage of apparent "cures" (40%). Structure–activity relations show that the halogen atoms at C6 and C2 are necessary, but not those at C7 (FULLER et al., 1992, 1994). The lethal dose is 200 mg kg^{-1} mouse i.v. (intravenous). After i.p. (intraperitoneal), or s.c. (subcutaneous) application into mice about half of the applied dose is bioavailable (EGORIN et al., 1996). Halomon is an interesting candidate for development as antineoplastic chemotherapeutic agent. The

Tab. 1. Compounds of Marine Origin with Cytostatic/Antitumor Activity

Source	Compound	Structure Type	Activity/Target	Reference
Bacteria				
Micromonospora marina	thiocoralline	depsipeptide	inhibition of RNA synthesis	BAZ et al., 1997
Streptomyces viridodiastaticus "litoralis"	bioxalomycins	alkaloids	inhibition of RNA and protein biosynthesis; increased survival of P388 mice; strong antibiotic activity	BERNAN et al., 1997
Agrobacterium sp.	agrochelin	thiazole alkaloid	cytotoxic, chelating to Zn^{2+}	CANEDO et al., 1999
Pelagiobacter variabilis	pelagiomicins	phenazine derivative	anticancer antibiotics	IMAMURA et al., 1997
Fungi				
Penicillium sp., from *Enteromorpha intestinalis*	penostatins, penochalasins		cytotoxic against P388 cells	NUMATA et al., 1995; TAKAHASHI et al., 1996
Aspergillus versicolor, from marine green algae		sesquiterpenoid nitrobenzoylesters	selective cytotoxic against several tumor cell lines	BELOFSKY et al., 1998
Leptosphaeria sp., from *Sargassum tortile*	leptosins	epipolysulfanyldioxopiperazinderivatives	cytotoxic against P388 cells	TAKAHASHI et al., 1994
Macroalgae				
Portieria hornemannii	halomon	monoterpene	active *in vitro* and *in vivo*	FULLER et al., 1992, 1994
Stypolopodium flabelliforme	14-keto-stypodiol-diacetate	diterpene	inhibition of microtubuli assembly	DEPIX et al., 1998
Enteromorpha prolifera	pheophytin-α and other compounds	chlorophyll-related compound	inhibit *in vitro* and *in vivo* chemically induced mouse skin tumorigenesis	HIQASHI-OKAJ et al., 1999
Cymopolia barbata	cyclocymopols	prenylated bromohydrochinons	modulators of progesterone receptors	PATHIRANA et al., 1995

Tab. 1. Continued

Source	Compound	Structure Type	Activity/Target	Reference
Ulva lactuca	ulvans	glycosaminoglycans	influence on adhesion, proliferation and differentiation of colon tumor cells	KAEFFER et al., 1999
Porifera				
Halichondria okadai, *Lyssodendoryx* sp., *Axinella* sp., *Phakellia parkeri*	halichondrins	macrocyclic lactones	against P-388 leukemia, B-16 melanoma, L-1210 leukemia cells	HIRATA and UEMURA, 1986; LITAUDON et al., 1994; MUNRO et al., 1999
Discodermia dissoluta	discodermolide	terpene	stabilizes microtubules, immunosuppressive, active against multi-drug-resistant tumors	GUNASEKERA et al., 1990; TER HAAR et al., 1996
Jaspis splendens	jaspamide	cyclodepsipeptide	inhibits actin polymerization, abandoned from clinical development	CREWS et al., 1986
Mycale sp.	peloruside A, mycalamide A, pateamine	macrolides	cytotoxic in nanomolar range	WEST et al., 2000
Aplysina aerophoba (= *Verongia* sp.)	aeroplysinin	modified bromotyrosine	protein tyrosin kinase inhibitor	FATTORUSSO et al., 1970; KREUTER et al., 1990
Cnidaria				
Eleutherobia sp.	eleutherobin	diterpene	stabilizes microtubules	LINDEL et al., 1997
Sarcodictyon roseum	sarcodictyin A	diterpene	stabilizes microtubules	D'AMBROSIO et al., 1987
Mollusca				
Dolabella auricularia	dolastatin 10 (produced by *Lyngbya majuscula*?)	peptide	effect on tubulin polymerization active *in vitro* and *in vivo*	PETTIT et al., 1993

Tab. 1. Continued

Source	Compound	Structure Type	Activity/Target	Reference
Bryozoa				
Bryozoa neritina	bryostatins	macrolides	partial agonists of protein kinase C, potent activity in human tumor xenografts	PETTIT et al., 1982
Tunicata				
Trididemnum solidum	didemnin B	cyclic peptide	active against leukemia, melanoma, immunosuppressive, antiviral, cardiotoxic, abandoned from clinical development in phase II	RINEHART et al., 1981
Aplidium albicans	dehydro-didemnin B	cyclic peptide	stronger antitumor activity than didemnin B	SAKAI et al., 1996
Ecteinascidia turbinata	ecteinascidin 743	tetrahydroisoquinoline alkaloid	active against leukemia, melanoma, ovarian and lung tumors	RINEHART et al., 1990; WRIGHT et al., 1990

main problem is the low yield of active substance (430 mg from 2–4 kg wet weight).

Halichondrin B, Isohomohalichondrin B

Halichondrin B and isohomohalichondrin B are tubulin-interacting agents. The IC 50 against B16 melanoma cells is only 0.093 ng mL^{-1} (HIRATA and UEMURA, 1986). The *in vivo* antitumor activities of these compounds were examined in human tumor models, in immunodeficient mice and rats. The compounds show distinct activities in different models. In a nude rat model for LOX bone marrow metastases, the mean lifespan was prolonged by halichondrin B to 32 d (control animals 15 d, vinblastine treated animals 17 d) (FODSTAD et al., 1996).

Discodermolide

Discodermolide has potent antiproliferative activity, also against some multidrug-resistant tumors. The relatively low yield (0.002%), the difficulty and expense of harvesting of a deep-water sponge, and the lack of practically useful synthetic routes make the further development more problematic. At present discodermolide serves as a lead compound (FAULKNER, 2000c).

Ecteinascidin

Ecteinascidin 743 shows significant *in vitro* activity against murine (L1210) leukemia (IC 50 = 0.5 ng mL^{-1}) and *in vivo* activities against P-388 lymphocytic leukemia and human mammary tumors. Ecteinascidin is a minor groove-binding, guanine-specific alkylating agent which also interacts with the microtubule network and blocks cell cycle progression at late S/G2. It is presently in clinicial trial for human cancer (RINEHART et al., 1990; VALOTI et al., 1998; TAKEBAYASHI et al., 1999). A multistep synthesis has been developed. The structurally related, totally synthetic molecule phthalascidin has been prepared and was also

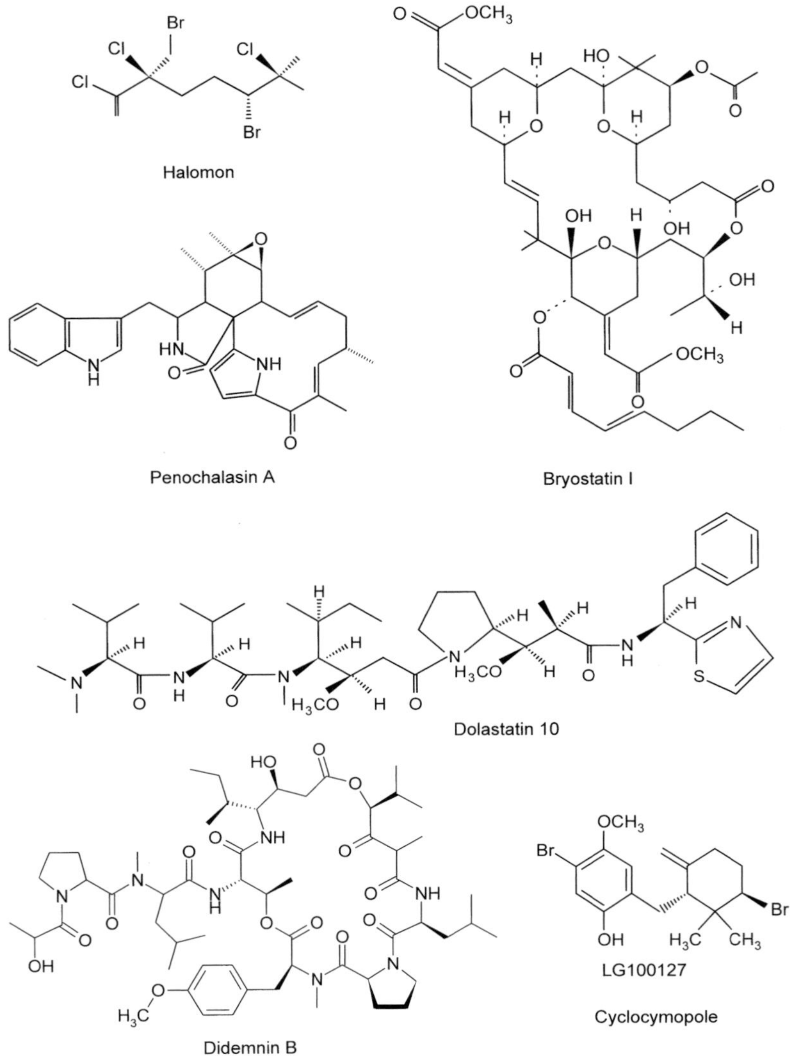

Halomon

Penochalasin A

Bryostatin I

Dolastatin 10

Didemnin B

LG100127

Cyclocymopole

Fig. 1. Cytotoxic/antitumor metabolites.

evaluated as antitumor agent (MARTINEZ et al., 1999).

Eleutherobin

Eleutherobin shows selective cytotoxicity toward breast, renal, ovarian, and lung cancer cell lines and underwent preclinical trials. Syntheses were successfully developed for both eleutherobin and sarcodictyin A (NICOLAU et al., 1997; CHEN et al., 1998; BARON et al., 1999; FAULKNER, 2000c).

Dolastatin 15

The potent antineoplastic constituent from the shell-less mollusk *Dolabella auricularia* has been selected as a lead compound for developing new antitumor drugs (FAULKNER, 2000c). Several synthetic and structurally simplified analogs (e.g., LU103793) show potent cytotoxic activity against a variety of human cancer cell lines (HU and HUANG, 1999).

Bryostatin

Bryostatin 1 acts as a partial agonist of protein kinase C. Unlike phorbol esters, it does not act as tumor promoter. It shows potent activity *in vivo* against human tumor xenografts and is currently in phase II clinical trials in the United States for the treatment of melanoma, non-Hodgkin's lymphoma, and renal cancer (FAULKNER, 2000c).

Bugula neritina also produces other bryostatins, and different populations contain different bryostatins. Two chemotypes were defined (PETTIT et al., 1996). It has been suggested that bacterial symbionts may play a role in bryostatin production (HAYGOOD and DAVIDSON, 1997; DAVIDSON and HAYGOOD, 1999). Bryostatin can be produced by aquaculture (see Sect. 4.2.3.1). The total synthesis of bryostatin 2 was also described (EVANS et al., 1999).

Dehydrodidemnin B (Aplidine)

The substance has a 5–10 times stronger cytotoxic activity against tumor cells than didemnin B and is not cardiotoxic. It can be synthesized and is in clinical studies (SAKAI et al., 1996; FAULKNER, 2000c).

It can be summarized that until now no marine cytostatic has reached the marketplace, but there are some promising candidates which will be further developed, if the problem of substance supply is solved.

5.1.2 Antiinflammatory Compounds

Marine compounds have provided several lead compounds or they are used in cosmetics and drugs as antiinflammatory compounds. Examples are given in Tab. 2 (see also Fig. 2). The following compounds are in preclinical or clinical development or on the market:

Manoalid

The sesterterpen manoalide is commercially available as an experimental tool for PLA_2 inhibition. Besides this it was introduced into clinical trials phase 1 and used as a lead substance for antiinflammatorics (FAULKNER, 2000c).

Debromohymenialdisine

Debromohymenialdisine (DBH) is under development as a therapeutic agent for osteoarthritis (FAULKNER, 2000c). It can easily be synthesized (XU et al., 1997).

Pseudopterosins

A partially purified extract of *Pseudopterogorgia elisabethae* containing pseudopterosins is used as an additive in cosmetic products by Estée Lauder. The gorgonians could be harvested from natural sources. Regrowth occurs in about 18 months (FAULKNER, 2000c).

5.1.3 Antimicrobial Compounds

There is an expanding need for new therapeutic antibiotics because of, e.g., the evolving resistance of microorganisms to existing antibiotics and the emergence of new viral diseases.

Competition for space and nutrients led to the evolution of antimicrobial defence strategies in the marine environment. Therefore, marine organisms offer a particularly rich source of new antimicrobial substances. Screening investigations showed that 35% of 400 surface-associated bacterial strains isolated from the surfaces of marine algae or invertebrates produce antimicrobial compounds. The proportion is higher than that of free living marine bacteria or soil bacteria. Many strains which normally did not produce antibiotics could be induced to do so by exposing them to small amounts of living cells, supernatants from other bacteria, or other chemicals (BURGESS et al., 1999). The initial isolation of the well-known antibiotic cephalosporin C was derived from the marine fungus *Cephalosporium acremonium* obtained from coastal sewage of the coast of Sardinia (COLWELL et al., 1985). Seaweeds and invertebrates also contain antimicrobial, cytostatic, and ichthyotoxic substances for defence against microbes and other organisms (VLACHOS et al., 1996).

Examples of antibacterial and/or antifungal and antiparasitic compounds are summarized in Tab. 3 (see also Fig. 3), examples of antiviral

Tab. 2. Compounds of Marine Origin with Antiinflammatory Activity

Source	Compound	Structure Type	Activity/Target	Reference
Bacteria				
Caulobacter sp.	thiotropocin	sulfur-containing macrolide	inhibits histamine release and hemolysis of rabbit erythrocytes	KAWANO et al., 1998
Streptomyces sp.	salinamide	bicyclic depsipeptides	anti-inflammatory in phorbol ester-induced mouse ear edema assay	TRISCHMAN et al., 1994
Fungi				
Micrascus longirostris	cathestatins	peptides	inhibition of cystein proteases	YU et al., 1996
Corollospora pulchella	pulchellalactam		inhibition of CD45 proteintyrosin-phosphatase	ALVI et al., 1998
Phoma sp., from crab shells	phomactins	sesquiterpenes	inhibition of aggregation of thrombocytes, PAF antagonists	SUGANA et al., 1991
Macroalgae				
Ulva pertussa, *Undaria pinnatifida*		fatty acids 18:4 n3	inhibition of 5-lipoxygenase and platelet aggregation	ISHIHARA et al., 1998
Ceratodyction spongiosum and symbiontic sponge *Sigmadocia symbiotica*	ceratospongamide	thiazol-containing cyclic heptapeptide	inhibition of $sPLA_2$ expression and human $sPLA_2$ promoter-based reporter	TAN et al., 2000
Porifera				
Luffariella variabilis	manoalide	sesterterpene	inhibits irreversibly PLA_2, inhibits ion channels for Ca^{2+}	DESILVA and SCHEUER, 1980; GLASER and JACOBS, 1986
Phakellia flabellata, *Hymeniacidon aldis*, *Stylotella aurantia*	debromo-hymenialdisine	alkaloid	protein kinase C inhibitor, acts against osteoarthritis	SHARMA et al., 1980; KITAGAWA et al., 1983; PATIL et al., 1997
Cnidaria				
Pseudo-pterogorgia elisabetha	pseudopterosins	glycosides	inhibit PLA_2, influence lipoxygenase, cyclooxygenase	LOOK et al., 1986; ROUSSIS et al., 1990

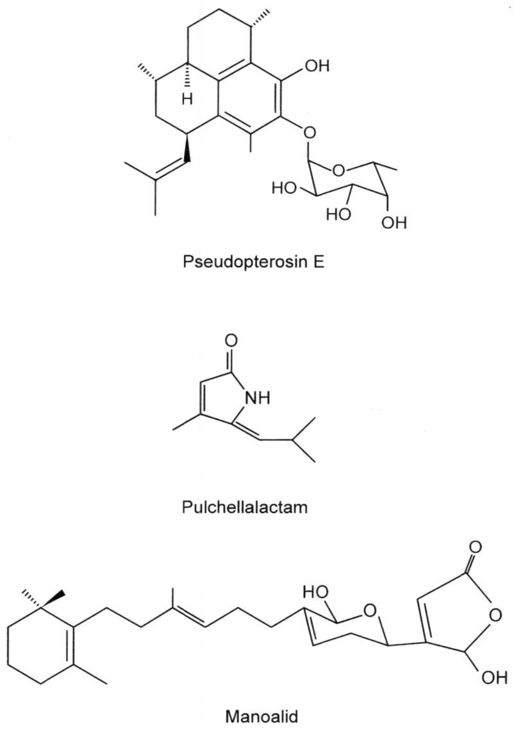

Pseudopterosin E

Pulchellalactam

Manoalid

Fig. 2. Antiinflammatory metabolites.

substances are presented in Tab. 4. It has to be noted that the antimicrobial effects were shown mostly only *in vitro*.

5.1.4 Further Biologically Active Compounds

The spectrum of biological acitivities of marine natural substances is very broad, e.g., compounds influencing the vascular system or blood coagulation/fibrinolysis or compounds with analgetic effects could be of great medical value. Examples are summarized in Tab. 5 (see also Fig. 3).

Many of the bioactive molecules isolated from marine sources are toxins with highly specific targets. Because of their strong toxicity they cannot find therapeutic application, but they may be useful as "molecular tools" for the identification of cellular and biochemical processes. For such purposes about 30 compounds of marine sources are commercially available. Examples are the neuroexcitatory amino acids kainic acid and domoic acid from algae (isolated from the Rhodophyceae *Diginea simplex* and *Chondrus armata*, in mass culture also produced by the diatom *Pseudonitzschia pungens*), the actin-targeted compounds swinholide A, bistheonellide A, and mycalolide B from sponges (*Theonella* sp., *Mycale* sp.), and the highly specific protein phosphatase inhibitor calyculin A from *Discodermia calyx* (Porifera). Kainic acid and domoic acid act as glutaminergic agonists and are used as probes in neurobiology. In Japan kainic acid is also used as an anthelminthic drug (BENTLEY, 1997). The actin-targeted substances disrupt the actin cytoskeleton of cultivated cells, stabilize actin dimers and sever actin filaments (SAITO et al., 1998). Swinholide A is commercially available as a reagent for cytoskeleton research, and calyculin A as an agent for the investigation of signal transduction processes.

5.1.5 Bone Substitutes

Skeletons of corals and some reef-forming algae consist of calcium carbonate. It can be used as a bone graft substitute in surgery. The grafts are well tolerated and become partially ossified when the calcified skeleton is resorbed (SOOST et al., 1998; ROUDIER et al., 1995). Besides this the calcium carbonate skeleton of the corals can be converted into hydroxyapatite by heating the coral to 900 °C, eliminating of the organic material, and chemically exchanging with di-ammonium phosphate under hydrothermal conditions (SIVAKUMAR et al., 1996).

5.2 Enzymes

Most of the enzymes currently on the market originate from terrestrial microorganisms. However, the remarkable diversity of marine microorganisms offers a promising source for the exploration of new or improved catalysts. The adaptation of marine microorganisms to extreme environmental conditions such as low or even very high temperature, high pressure,

Tab. 3. Compounds of Marine Origin with Antibacterial, Antifungal, or Antiparasitic Activity

Source	Compound	Structure Type/ Biogenetic Origin	Activity/Target	Reference
Bacteria				
Pseudo-alteromonas sp.	korormicin		inhibits gram-negative marine and halophilic bacteria (inhibits specific NADH:quinone reductase)	YOSHIKAWA et al., 1999
Streptomyces sp.	luisols A and B	anthraquinones	antibacterial	CHENG et al., 1999
Agrobacterium aurantiacum	hydroxyakalone		xanthin oxidase inhibitor	IZUMIDA et al., 1996
Micrococcus luteus, sponge-associated		2,4,4′-trichloro-2′-hydroxydiphenylether	antibacterial	BULTEL-PONCÉ et al., 1998
Fungi				
Corollospora maritima	corollosporine	phthalid derivative	antibacterial	LIBERRA et al., 1998
Fusarium sp., *Acremonium* sp.	halymecins	conjugates of hydroxydecanoic acids	antibacterial, antimicroalgal	CHEN et al., 1996
Exophiala pisciphila	exophilin A	polyketide	antibacterial	DOSHIDA et al., 1996
Hypoxylon oceanicum	LL-15G256	lipodepsipeptide	antifungal, inhibition of cell wall biosynthesis	ABBANAT et al., 1998; SCHLINGMANN et al., 1998
Macroalgae				
Dictyota dichotoma	dictyols	diterpenes	antibacterial, antifungal, molluscicid	SALEH et al., 1992
Dictyota dichotoma	crenulacetals	terpenes	inhibit larvae of *Polydora*	TAKIKAWA et al., 1995
Laurencia obtusa	allolaurinterol	sesquiterpene	antibacterial, antifungal, antialgal	KÖNIG and WRIGHT, 1997
Porifera				
Theonella swinhoei	theonegramide	glycopeptide	antifungal	BEWLEY and FAULKNER, 1994
Siliquaria-spongia japonica	aurantosides	polyketides	antifungal	SATA et al., 1999

Tab. 3. Continued

Source	Compound	Structure Type/ Biogenetic Origin	Activity/Target	Reference
Halichondria sp.	halichondramide	macrolide	antifungal, antimalaria	KERNAN et al., 1988
Cnidaria				
Sinularia flexibilis	flexibilid, sinulariolid	diterpenes	antibacterial	ACERET et al., 1998
Pseudo-pterogorgia elisabethae	pseudopteroxazole, secopseudo-pterazole	diterpenoidalkaloids	inhibit *Mycobacterium tuberculosis*	RODRIGUEZ et al., 1999

or nutrient limitation has evolved unique adaptation strategies. The biotechnological potential of thermophilic marine microorganisms, especially of their enzymes, has been reviewed by ANTRANIKIAN (Chapter 4, this volume).

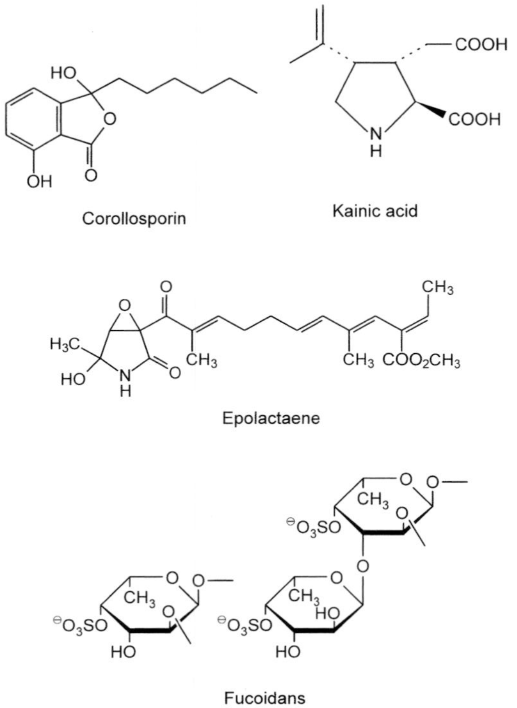

Corollosporin

Kainic acid

Epolactaene

Fucoidans

Fig. 3. Several biologically active metabolites.

90% of the ocean water is colder than 5 °C. Thus, the majority of marine organisms is cold-adapted. Reviews covering the molecular basis of cold adaptation and the biotechnological potential of psychrophilic microorganisms include those of MARSHALL (1997), RUSSELL (1998), NICHOLS et al. (1999), and GERDAY et al. (2000). The enzymes of psychrophilic microorganisms are characterized by a high specific activity at low and moderate temperature. The adaptation of psychrophilic enzymes to low temperature is due to their high thermoflexibility. The strategy for the higher flexibility of cold-adapted enzymes seems to be unique to each enzyme. A concerted action of different structural components of the enzymes determines their temperature dependent activities (GERDAY et al., 2000). The following structural properties of cold-adapted enzymes has been suggested:

1. decreased levels of prolyl and arginyl residues and increased levels of glycyl residues,
2. low degree of intramolecular interactions,
3. increased interactions with the solvent,
4. decrease in the cation or anion interaction,
5. weaker interdomain or intersubunit interaction.

The exploration and sequencing of genes of further cold-adapted enzymes and the investigation of their structural components by site-directed mutagenesis or directed evolution

Tab. 4. Compounds of Marine Origin with Antiviral Activity

Source	Compound	Structure Type	Activity/Target	Reference
Bacteria				
Pseudomonas sp.		sulfated polysaccharides	anti-HSV-1 activity	MATSUDA et al., 1999
Macroalgae				
Cryptopleura ramosa		sulfated galactan	antiviral against HSV-1, HSV-2	CARLUCCI et al., 1997a
Gigartina tenella	KMO43	sulfoquinovosyldiacyl-glycerol	inhibits retroviral reverse trans-scriptase (HIV)	OHTA et al., 1998
Gigartina skottsbergii	carrageenans and cyclized derivatives	sulfated galactans	antiviral against HSV-1	CARLUCCI et al., 1997b
Sargassum horneri		fucan sulfat	antiviral against HSV-1, HCMV, HIV	HOSHINO et al., 1998
Monostroma latissimum		sulfated rhamnan	antiviral against HSV-1, HCMV, HIV-1	LEE et al., 1999
Nothogenia fastigiata		sulfated xylomannans	antiviral against HSV, anti-coagulating	KOLENDER et al., 1997
Porifera				
Dysidea avara	avaron, avarol	sesquiterpenoid hydrochinons	inhibit retroviral reverse trans-criptase (HIV-1)	SARIN et al., 1987; LOYA and HIZI, 1990
Tunicata				
Ascidia	lamellarin-α-20-sulfate	alkaloid	inhibition of HIV-1 integrase	REDDY et al., 1999

will help to increase our understanding of the temperature adaptation of enzymes. It has been found that the amino acid sequence of enzymes from psychrophilic archaea is very similar to the enzymes of thermophilic archaea (DELONG, 1997). Thus, the discovery of the psychrophilic archaea permits a direct comparison of the sequences of these enzymes with their thermophilic counterparts.

From the biotechnological point of view many cold-adapted enzymes could replace mesophilic counterparts in the future or could help to establish even new bioprocesses under low-temperature conditions, because these enzymes

- can help to save energy (e.g., in washing processes, food processing, or bioremediation),
- save labile or volatile compounds (e.g., in biotransformations or food processing),
- prevent the growth of mesophilic contaminants at low temperatures (e.g., food processing),
- could be easily inactivated by moderate temperatures (e.g., molecular biological applications or food processing).

Examples of potential low-temperature processes are summarized in Tab. 6. However, the thermosensitivity of these enzymes is the ma-

Tab. 5. Some Further Biologically Active Compounds of Marine Origin

Source	Compound	Structure Type	Activity	Reference
Penicillium sp. (Fungi)	epolactaene	polyene	stimulates differentiation of neuroblastoma cells	KAKEYA et al., 1995
Porphyra yezoensis (Rhodophyceae)		peptides	inhibition of angiotensin I-converting enzymes	SUETSUNA, 1998
Phaeophyceae	fucoidans and other compounds	sulfated polysaccharides	inhibition of blood coagulation-like heparin	LOGEART et al., 1997
Sargassum autumnale (Phaeophyceae)	nahocols, isonahocols	prenyl ethers	endothelin antagonists potential anti-hypertonics	TSUCHIYA et al., 1998
Conus magus (Molluscs)	conotoxins	peptides	analgetics (in clinical development), inhibition of Ca^{2+} ion channels	Anonymous Ziconotide

Tab. 6. Examples for the Biotechnological Application of Psychrophilic Microorganisms and their Enzymes

Process	Enzymes	Advantage	Selected Reference
Baking	amylases	slower crust ageing	GERDAY et al. (2000)
Fruit juice production	pectinases, amylases, cellulases	keeping aroma	LEA (1995)
Milk production	β-galactosidases	removal of lactose	LAW and GOODENOUGH (1995)
Preservation	muramidases, chitinases, glucanases, glucose oxidases	protection against cold store contaminants	OHGIYA et al. (1999)
Detergents	amylases, cellulases, alkaline proteases, lipases	energy saving	OHGIYA et al. (1999)
Biotransformation	lipases, nitrile hydratases	saving of labile and volatile compounds	OHGIYA et al. (1999)
Molecular biology	phosphatase, uracil DNA glycosylase, restriction enzymes	heat inactivation	SOBEK et al. (1996)
Oil degradation, Bioremediation	whole cold-adapted bacteria and their enzyme system	*in situ* oil/pollutant degradation at ambient (low) temperature	MARGESIN and SCHINNER (1999)

jor drawback for their industrial application. Consequently, a compromise between high enzyme activity at low temperature, but moderate stability at ambient temperature is a prerequisite for their biotechnological breakthrough. By site-directed mutagenesis experiments carried out on an Antarctic subtilisin FELLER et al. (1996) demonstrated that the possibility remains to stabilize the structure of cold-adapted enzymes without affecting their high catalytic efficiency.

A further problem of the commercial application of cold-adapted thermosensitive enzymes is their difficult production with the established mesophilic expression systems. These expression systems work well at 37 °C. However, at this temperature many cold-adapted enzymes cannot correctly fold into their native functional conformation and are either degraded by host proteases or accumulate in the form of inclusion bodies. This could hinder an economical production of such enzymes. SCHWEDER and KAAN (1999) suggested a cold-inducible expression system for *Bacillus*. Such a system could be used for a biphasic production process. At the first stage *Bacillus* cells grow at an optimal temperature of 35 °C to the desired high cell density in the fermenter. At the second stage a moderate temperature down-shift to 20 °C induces the expression system and thus the overproducion of these thermosensitive enzymes in a functional conformation. Another approach has been suggested by REMAUT et al. (1999) who developed regulatable expression systems for cold-adapted enzymes in well growing psychrotrophic Antarctic bacteria based on established *E. coli*-derived expression control elements.

5.3 Nutraceuticals

5.3.1 Biomass

Because of their high nutrient value and their delicious taste, seaweed, pisces, crustaceans, and shells are used as food in many parts of the world. They are produced by harboring from natural sources or by aquaculture. Modern biotechnological methods can improve the quality of the products and increase the efficiency of the production processes (see Sect. 4.2). Edible marine products such as seaweeds are rich in proteins, PUFAs (see Sect. 5.3.2.1), dietary fiber (in seaweeds 58–76 g/100 g dry weight), vitamins (see Sect. 5.3.2.3), and some trace elements. Eating brown seaweeds can relieve the problem of insufficient iod supply (HOPPE et al., 1979; ABDUSSALAM, 1990; PAK and ARAYA, 1996). As an example the composition of one of the most important edible seaweed, *Porphyra columbina*, is mentioned: per 100 g dry weight the algae contain 7,03–11,00 g water, 16, 18–22,70 g ash, 3,18–6,41 g sodium, 1,24–1,96 g potassium, 600–836 mg magnesium, 78–276 mg phosphorus, 63–108 mg calcium, and 3,9–26,4 mg iron (FAJARDO et al., 1998).

5.3.2 Food Additives

5.3.2.1 Polyunsaturated Fatty Acids (PUFAs)

5.3.2.1.1 Importance and Producing Organisms

Most important PUFAs are eicosopentaenic acid (EPA) and docosahexaenic acid (DHA, Fig. 4). They have gained increased interest due to their beneficial properties for human health (prevention and cure of thrombosis, arteriosclerosis and related blood-circulatory diseases by reducing blood platelet aggregation and blood cholesterol) and their importance in infant development (LINKO and HAYAKAWA, 1996; GILL and VALIVETY, 1997). Long-chain PUFAs can be synthesized *de novo* from the parent essential fatty acids linoleic acid and α-linolenic acid by desaturation and chain elongation, but the overall capacity of humans to synthesize these fatty acids is relatively low. Therefore, most of the PUFAs are obtained from dietary sources, or, for a fetus or a breast fed baby, from the mother. Based on these data the FAO/WHO has recommended the inclusion of supplemental DHA (and ARA) in both pre-term and full-term infant formulas. The most important source of PUFAs, especially EPA and DHA, so far is fish

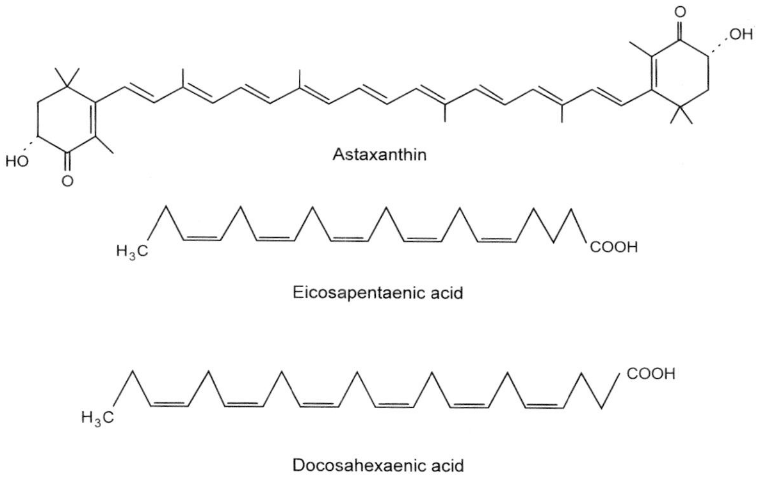

Astaxanthin

Eicosapentaenic acid

Docosahexaenic acid

Fig. 4. Food additives.

oil. However, the problem that fish stocks continue to dwindle worldwide and the disturbing taste and smell of PUFA products derived from fish oil hinder the economical mass production of these interesting marine natural products. Increasing commercial interest in these special fatty acids increases the pressure to seek new alternative PUFA producing organisms. Lower organisms such as marine fungi, bacteria, and some algae can also synthesize PUFAs. Examples are summarized in Tab. 7. Among the algae uni- and multicellular species from the Chlorophyceae, the Cryptophyceae, the Rhodophyceae are able to produce EPA, members of the Chrysophyceae and the Dinophyceae also synthesize DHA (GILL and VALIVETY, 1997; FATMA and SULTAN, 1999). However, the large-scale cultivation of photosynthetic PUFA-producing microorganisms so far seems to be economically not feasible (RATLEDGE, 1993), but heterotrophic MOs might be a more productive PUFA source, since industrial cultivation would be much easier. The heterotrophic microalga *Crypthecodinium cohnii* can accumulate up to 13% of its cell mass as lipids with 36–43% DHA (DESWAAF et al., 1999). This alga can be easily cultivated in media containing glucose, yeast extract, and sea salt at 27–30 °C with a yield of up to 1.6 g L^{-1} DHA. Recently, particular attention has been paid to a group of fungoid protists, the thraustochytrids. Members of this group, such as *Schizochytrium* or *Thraustochytrium*, are able to synthesize up to 50% of their lipids in the form of DHA (WEETE et al., 1997; YOKOCHI et al., 1998; LEWIS et al., 1999). These fungi are able to accumulate PUFAs in the cytoplasm in form of oil droplets. Although industrial cultivation of these protists is complicated (KENDRICK and RATLEDGE, 1992; LI and WARD, 1994), two patents have been filed already claiming the cultivation of thraustochytrids for the production of EPA and DHA (BARCLAY, 1992, 1994). Recently, it has been reported that a fermenter culture with a thraustochytrid identified as *Schizochytrium limacinum* obtained up to 3.3 g DHA L^{-1} d^{-1} (YAGUCHI et al., 1997).

Since 1973 it is known that marine bacteria can contain PUFAs (OLIVER and COLWELL, 1973). In contrast to fungi bacteria exclusively incorporate PUFAs in their membrane phospholipids. Although the levels of PUFAs in bacteria are not as high as in some fungi the easy cultivation of bacteria could provide an economical production of PUFAs in the future. Interestingly, most bacteria are not able to produce PUFAs (RUSSELL and NICHOLS, 1999). However, the majority of bacteria is able to take up exogeneous PUFAs when pres-

Tab. 7. Examples of EPA or DHA Synthesizing Marine Microorganisms

PUFA	Source	% of Total Lipid	Reference
EPA	**Fungi**		
	Mortierella sp.	15	KENDRICK and RATLEDGE, 1992
	Algae		
	Chlorella minutissima	20–40	HAIGH et al., 1996
	Phaeodactylum tricornutum	26–40	YONGMANITCHAI and WARD, 1991
	Nannochloropsis sp.	33	CHINI ZITELLI et al., 1999
	Bacteria		
	Shewanella gelidimarina	16	NICHOLS and RUSSELL, 1996
	Shewanella hanedai	20	BOWMAN et al., 1997
	Photobacterium profundum	5–6	ALLEN et al., 1999
DHA	**Fungi**		
	Traustochytrium sp.	25	WEETE et al., 1997
	Traustochytrium roseum	46–49	LI and WARD, 1994
	Algae		
	Phaeodactylum tricornutum	10	YONGMANITCHAI and WARD, 1991
	Isochrysis galbana	14	THOMPSON et al., 1992
	Crypthecodinium cohnii	36–43	DESWAAF et al., 1999
	Bacteria		
	Colwellia psychrerythraea	6–7	BOWMAN et al., 1998

ent in their environment (WATANABE et al., 1994). Thus, only a limited number of marine bacteria can synthesize PUFAs, such as *Photobacterium* sp. (ALLEN et al., 1999), cyanobacteria such as *Synechocystis* sp. (SAKAMOTO et al., 1994), and members of the genera *Shewanella* and *Colwellia* (NICHOLS et al., 1999).

It has been found that marine low-temperature adapted (psychrophilic or psychrotolerant) bacteria above all have the metabolic capacity to synthesize unsaturated fatty acids and to accumulate a considerable amount of EPA or DHA in their membranes. The majority of the Antarctic PUFA-producing bacteria could be associated to the genera *Colwellia* and *Shewanella* (RUSSELL and NICHOLS, 1999). It is supposed that the high ratio of unsaturated to saturated fatty acids helps to maintain membrane fluidity under the extreme marine environmental conditions of low temperature and high pressure (HAZEL and WILLIAMS, 1990). For instance, many deep-sea bacteria show a high amount of PUFAs, like EPA or DHA, in their membrane lipids. In many cases these amounts increase with increasing pressure (DELONG and YAYANOS, 1985; YAYANOS, 1995). However, a study of ALLEN et al. (1999) demonstrated that PUFAs could be exchanged by monosunsaturated fatty acids. Based on these data and due to the fact that many of the PUFA-producing bacteria discovered have been isolated from vertebrate or invertebrate sources, the authors suggested that the PUFAs synthesized by these bacteria are not essential for their survival at high pressure, but are needed as a nutritional source by their symbiotic partners.

5.3.2.1.2 Optimization of PUFA Production

Alternative production of PUFAs by means of marine organisms is under development in an increasing number of companies (GILL and VALIVETY, 1997). However, the fast growing market for DHA and EPA requires either new strong PUFA-overproducing microorganisms or a considerable optimization of established industrial microbial PUFA production processes. The following aspects have to be considered for the improvement of microbial PUFA production bioprocesses:

(1) cultivation regime:
- nutrient requirements (e.g., seawater, salt),
- temperature (PUFA production is stimulated by low temperatures),
- pressure (PUFA synthesis of barophilic MOs is stimulated by low pressure), etc.

(2) bioreactor design:
- corrosion (most of the MOs require high salt concentrations for their growth),
- light system (for the cultivation of photosynthetic MOs, like algae or cyanobacteria) (see Chapter 5, this volume),
- cooling system (for the cultivation of psychrophilic or psychrotolerant MOs),
- pressure resistance (for the cultivation of barophilic bacteria), etc.

(3) strain improvement by pathway engineering:
- improvement of the productivity by mutagenic agents or other "natural" mutation techniques,
- genetic engineering of the production strains, such as increase of the copy number of selected genes and by this overproduction of the synthesizing enzymes (i) by cloning of the appropriate genes on multi-copy plasmids or (ii) by regulation of the expression of the PUFA synthesis genes by stronger constitutive promoters,
- exclusive enrichment of selected PUFAs by mutation of genes/inhibition of enzymes.

New insights into the molecular biology of microbial PUFA synthesis enable the identification of the responsible genes and their products. The cloning of a 38 kb gene cluster from *Shewanella putrefaciens* into *Escherichia coli* and *Synechococcus* sp. resulted in successful EPA production by these microorganisms (YAZAWA, 1996; TAKEYAMA et al., 1997). PUFA genes from *Vibrio marinus*, *Photobacterium* sp., and *Shewanella* sp. have been cloned and characterized by FACCIOTTI et al. (1998). This research group found strong homologies between the PUFA genes of the three strains examined. The potential conservation of genes responsible for the synthesis of EPA and DHA offers the possibility to clone further genes from other bacteria and to investigate in more detail how these genes are regulated. Especially knowledge on the temperature- and pressure-dependent regulation of these genes could permit the uncoupling of PUFA synthesis from these environmental conditions and thus realize an overproduction of PUFAs at moderate temperature and normal (ambient) pressure in selected psychrotolerant bacteria.

5.3.2.1.3 Microbial PUFAs in Aquacultures

The fish resources worldwide are more and more exhausted, and therefore, fish and also shrimp production in aquaculture is of increasing economical importance. EPA and DHA are essential for normal growth and development of the larvae of many aquaculture species. Furthermore, because of the positive effect of PUFAs on human health, an increase of the PUFA content of aquaculture fish or shrimp would increase their market value. That is why PUFA-producing microorganisms have been considered as nutrients for aquacultures (NICHOLS et al., 1999). Especially heterotrophic PUFA-producing bacteria which are easy to cultivate in large-scale fermentation processes could be utilized in aquaculture feeding to enrich rotifers which are used as live feed for larvae with PUFAs (NICHOLS et al., 1996).

5.3.2.2 Vitamins, Pigments

Fish liver oils from *Gadus morrhua* and *Hippoglossus hippoglossus* are rich sources of the fat soluble vitamins A and D. They are used for vitamin substitution and as ointment additives for the treatment of skin diseases and wounds.

Pigments such as carotenoids, phycoerythrin, and phycocyanin are produced by algae and cyanobacteria. They are also found in marine animals such as lobsters and crabs. The compounds can be obtained biotechnologically from cultures of microalgae (see Chapter 5, this volume) or from marine bacteria (*Agrobacterium aurantiacum*) (YOKOYAMA et al., 1994). They already have a firmly established market in diagnostics, cosmetics, nutrition, and pharmacy. The biosynthetic genes for the highly oxidized carotenoid astaxanthin (3,3′-dihydroxy-β,β-carotene-4,4-dione) (Fig. 4) from the marine bacteria *Agrobacterium aurantiacum* could be expressed in *E. coli* (FRASER et al., 1997). Astaxanthin is a vitamin A precursor, a scavenger of active oxygen species, an antitumor promoter, and an agent for pigmenting cultured fish and shellfish.

5.4 Products of Marine Biotechnology for Several Purposes

5.4.1 Biopolymers

Biopolymers from marine organisms of biotechnological interest are mainly polysaccharides. In their natural environment they have a protecting function as in the case of the chitin exoskeleton of marine invertebrates and plankton or a structure forming function as in the case of alginates. Besides this they act as surface adhesives. Marine polysaccharides are produced by marine bacteria (complex polysaccharides, related extracellular substances, alginates) (WEINER, 1997), algae (agar, carrageenan, alginates), and crustaceae (chitin). Polysaccharides from marine bacteria have a very complex structure. They can be applied as underwater surface coatings, bioadhesives, for oil cleaning and viscosity reduction, as dispersing agents, in the food and textile industry, or for toxic metal bioremediation. Enhanced yields and improved properties can be achieved by using recombinant DNA modified strains of polysaccharide-exporting bacteria (WEINER et al., 1985). Other biopolymers produced by marine bacteria are polyesters and pigments.

Agar is mainly prepared from red seaweeds belonging to the genera *Gelidium*, *Gracilaria*, *Pterocladia*, and *Ahnfeltia*. Carrageenan is obtained, e.g., from *Chondrus crispus*, *Gigartina stellata*, etc. Agar and carrageenan are polygalactans consisting of D-galactopyranose, 3,6-anhydro-L-galactopyranose and their derivatives, linked by β-(1–3) and α-(1–4) linkages. Agar, agarose (the sulfate-free fraction of agar), and carrageenan have a broad spectrum of applications, e.g., as gel-forming substances in microbiology, the food industry, and cosmetics, or as mild laxatives and as auxiliary substances in galenics.

Alginates are a family of unbranched, nonrepeating copolymers consisting of β-D-mannuronic acid and α-L-guluronic acid, linked by 1–4 linkages. They are obtained commercially by harvesting brown seaweeds, mainly *Macrocystis* from the coasts of Pacific North America, New Zealand, and Australia, and *Laminaria* from the North Atlantic and Japanese coasts. Main producers are the USA, France, Great Britain, Norway, and Japan. The current annual commercial production is about 30,000 t. Fields of application are pharmacy and medicine (antacidic and hemostyptic action, immobilization of cells, sustained release of drugs, dental impression materials, reduction of body weight), cosmetics, food industry, textile printing, paper and water treatment, and other technical purposes (ONSOYEN, 1996; REHM and VALLA, 1997). Sodium alginate preparations from *Sargassum* reduce strontium absorption in animals (rats) and humans with high efficiency and virtually no toxicity. They are suitable antidotes against radiostrontium absorption on a long-term basis, when added to bread at a 6% level (VOLESKY, 1994).

Alginate biosynthesis of bacteria is particularly well-characterized in *Pseudomonas aeruginosa* and *Azotobacter vinelandii*. The alginates produced by bacteria show some structural differences from those from seaweed (acetylation of mannuronate residues). Bacterial alginates which are much more expensive than those of seaweeds may become commercial products in the future because of higher quality and of environmental aspects. It seems possible to control the structure of the final product by genetic changes. Such high-quality alginates could be useful as immunostimulants or as gel-forming agents for immobilization of cells, e.g., insulin-producing cells in alginate capsules (ONSOYEN, 1996; REHM and VALLA, 1997).

The shells of crustaceae including crabs, shrimps, and krill contain chitin (poly-β(1,4)-N-acetyl-D-glucosamine) as a structural component. This is processed industrially from crustacean shell waste. It can be prepared to dressing material for healing of animal wounds and as a textile material (HIRANO et al., 1999). Chitosan is prepared from chitin by desacetylation and can be chemically modified. These poducts are employed in several technological (immobilization of enzymes and cells, adsorbents, removal of heavy metals from water; warp sizig, printing, and dying of textiles, photography), and medical fields (wound healing, hemostasis, contact lenses, dialysis, blood expanders, dermatology, tissue substitutes, drug delivery, reduction of body weight) (COLWELL et al., 1985).

5.4.2 Compounds with Antifouling Activity

Environmental problems caused by metal-based antifouling coatings have led to a concomitant increase in interest for the development of non-toxic alternatives which interfere with the normal adhesion of marine organisms or which inhibit the settlement of potential epibionts. Antifouling compounds from marine organisms may be such an alternative (ABARZUA and JAKUBOWSKI, 1995; CLARE, 1998). About 30% of 200 investigated species were found to be active (MIKI et al., 1996). Examples are halogenated furanons from *Delisea* sp. (Rhodophyceae; DE NYS et al., 1995), halogenated monoterpenes from *Plocamium costatum* (Rhodophyceae; KÖNIG et al., 1999), and diterpenes from *Dictyota* sp. (Phaeophyceae; SCHMITT et al., 1998), and *Renilla reniformis* and *Leptogorgia virgulata* (Cnidaria; CLARE et al., 1999).

5.4.3 Antifreeze Proteins

Antifreeze proteins (AFP) have been found in many cold-adapted organisms like fishes and invertebrates. They have the ability to bind to ice and can thus inhibit its growth (FEENEY and YEH, 1998). There are two types of antifreeze proteins derived from fish: glycoproteins and non-glycoproteins. The best studied glycoproteins originate from Antarctic fish, such as the Northern fish *Boreogadus saida*. Non-glycoproteins have been isolated, e.g., from the winter flounder *Pseudopleuronectes americanus*. Antifreeze proteins are located in the blood serum of the fish and can yield up to 30 mg mL^{-1}. They can be obtained by isolation from natural sources, by synthesis, or by recombinant methods. Overexpression of genes of selected antifreeze proteins have been tested (LOEWEN et al., 1997). Cloning and overexpression in the yeast *Pichia pastoris* resulted in a yield of 30 mg L^{-1} recombinant antifreeze proteins.

Antifreeze proteins can be used in the manufacture of frozen foodstuffs, ice cream, in cloud-seeding and – via transgenic expression – in the prevention of frost damage in economic crops (FEENEY and YEH, 1998; Tab. 8).

5.4.4 Adhesive Proteins

Marine mussels attach to underwater surfaces very strongly, durable, opportunistic, and are not undermined by the presence of water. Adhesion is mediated by means of their byssus. This is an external structure consisting of several threads attached to the animal at one end and to a surface on the other. At least 4 families (*Mytilus edulis* foot proteins, Mefp-1, Mefp-2, Mefp-3, Mefp-4) of plaque proteins are involved in mussel attachment. All contain the post-translationally modified amino acid L-3,4-dihydroxyphenylalanine (L-DOPA) which seems to be important in the adsorption by its ability to form hydrogen bonds (PAPOV et al., 1995). Investigations are done for biotechnological production of mussel adhesive proteins and their use as a coatings for various purposes, e.g., for antibody immobilization on microtiter plates (JACOBS et al., 1998).

Tab. 8. Examples for the Use of Antifreeze Proteins

Application	Example	Reference
Reduction of recrystallization in frozen food	use in ice cream production	WARREN et al., 1992
	chilled and frozen meat	PAYNE and YOUNG, 1995
Storage of biological material at cold temperatures above freezing	reduced leakage from liposomes during chilling	HAYS et al., 1996
Storage of biological material with repeated freezing and thawing	better survival of pig oocytes	ARAV et al., 1993
	reduced killing of prostatic adenocarcinoma cells	KOUSHAFAR and RUBINSKY, 1997

5.4.5 Sunscreens

Many organisms inhabiting shallow water or intertidal marine environments are protected from long-term solar damage by producing UV absorbing sunscreens such as pigments (carotenoids) or mycosporine amino acids. These substances produced by algae or coral symbioses and transformed by bacteria are under development for use as a natural antioxidant in food processing and cosmetic applications (DUNLAP et al., 1998).

5.4.6 Agrochemicals

Because of the vital interest in discovering new insecticides with decreased environmental and toxicological risk and no resistance marine sources were also screened. About 40 active compounds are reported as insecticidal prototype leads from marine origin. They include polyhalogenated monoterpenes, polyhalogenated C15 metabolites, diterpenes, peptides and amino acids, phosphate esters, sulfur-containing derivatives, and macrolides (EL SAYED et al., 1997). A known example is the sulfur-containing compound nereistoxin isolated from the marine annelid *Lumbriconercis heteropoda* (OKAICHI and HASHIMOTO, 1962). It served as a model for the commercial insecticide Padan (EL SAYED, 2000).

5.4.7 Compatible Solutes

At high salt concentrations many bacteria accumulate small organic compounds which seem to function as osmoprotectants stabilizing proteins, membranes, and DNA (LIPPERT and GALINSKI, 1992). It could be found that these compatible solutes also protect whole cells against different kinds of stress such as freezing, drying, and heating. For osmoprotectants like ectoines (SEVERIN et al., 1992) biotechnological applications as stabilizers of biomolecules, as moisturizers in cosmetics, or as chiral building blocks have been proposed.

The genes for the biosynthesis of the compatible solute ectoine from the moderate halophilic *Marinococcus halophilus* could be cloned and functionally expressed in *E. coli* (LOUIS and GALINSKI, 1997). Recently, a bioprocess called "bacterial milking" has been established for the production of compatible solutes (SAUER and GALINSKI, 1998). This process is based on alternative osmotic shocks following a high cell density fermentation with the gram-negative bacterium *Halomonas elongata* allowing a continuous production of ectoines.

5.4.8 Magnetosomes

Some marine bacteria are able to orient and migrate along geomagnetic field lines. This ability is based on intracellular magnetic structures, the magnetosomes (BLAKEMORE, 1975).

The magnetosomes consist of magnetic iron mineral particles enclosed within membranes (BALKWILL et al., 1980). The mechanism of magnetite biomineralization has been intensively investigated in members of the genus *Magnetospirillum* (SCHÜLER, 1999).

Several biotechnological applications of magnetic crystals have been suggested (SCHÜLER and FRANKEL, 1999):

- Carriers for immobilization of bioactive substances (e.g., enzymes or antibodies),
- Incorporation of magnetosomes by phagocytosis or polyethylene glycol fusions into eukaryotic cells and their manipulation by magnetic fields,
- Contrast agents for magnetic resonance imaging and tumor-specific drug carriers based on intratumoral enrichment.

In comparison to synthetic magnetic particles bacterial magnetosomes are uniform and homogenous and might be better suitable for the above mentioned applications. However, as yet no application of bacterial magnetosomes has attained commercial scale. One reason for this is the problematic large-scale cultivation of these microaerobically growing bacteria. Recently, more oxygen-tolerant magnetotactic strains, such as *M. gryphiswaldense*, could be isolated. These magnetotactic strains can be cultivated more easily and could be suitable for the mass cultivation of magnetosomes (SCHÜLER and BAEUERLEIN, 1998).

5.4.9 Bioluminescence

The phenomenon of bioluminescence can be most frequently found in aquatic ecosystems. In a depth below 200–1,000 m bioluminescence is the only source of light. The best known example of bioluminescence represents the symbiotic association of the bacteria *Vibrio* spp. with the two fishes *Anomalops katoptron* and *Photoblepharon palpebratus* (HAYGOOD and DISTEL, 1993). Both fishes carry so-called light organs which accommodate the luminescent bacteria. Besides *Vibrio* there are many other luminescent bacteria, but most of these bacteria originate from marine environments (HASTINGS et al., 1985; WILSON and HASTINGS, 1998). The enzyme responsible for light production in *Vibrio* is called luciferase. This enzyme is a monooxygenase which in the presence of reduced flavin mononucleotide (FMN), molecular oxygen, and long-chain aldehydes (e.g., decanol) catalyzes the bioluminescence reaction (Fig. 5). This enzyme is commercially available and is applied together with the NAD(P)H:FMN oxidoreductase for the quantitative measurement of NAD(P)H in

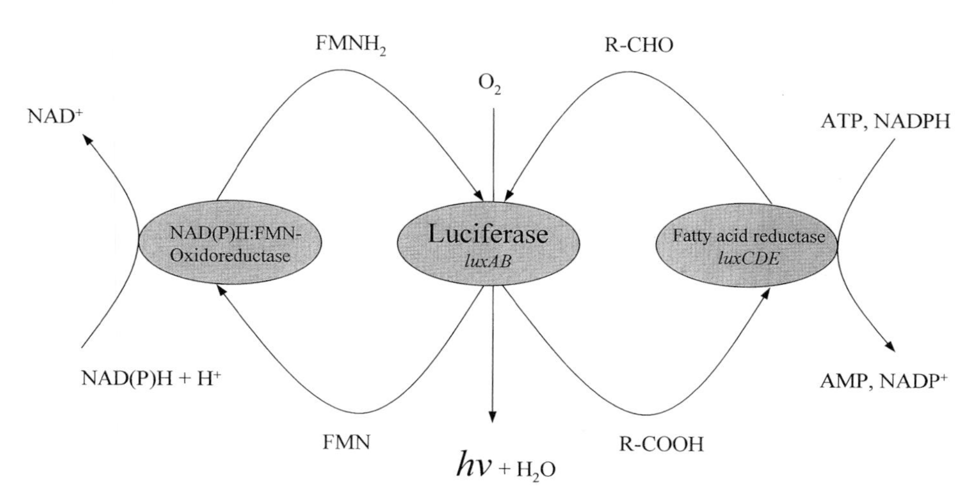

Fig. 5. Schematic representation of the bacterial bioluminescene reaction.

biosensors. Because of the unique features of the bacterial luminescence reaction, the genes *luxAB* coding for the bacterial luciferase are frequently used as reporter genes in different bacteria or eukaryotic organisms (MEIGHEN, 1993). Furthermore, intact luminescent bacteria are used as biosensors for the detection of toxins in the environment. In Germany, this approach has been introduced with *Photobacterium phosphoreum* as a classified test since 1991 (Deutsche Einheitsverfahren zur Wasser-, Abwasser- und Schlammuntersuchung, DIN 38412).

Another example of the biotechnological application of a bioluminescence-producing enzyme is green fluorescent protein (GFP). This enzyme is found in specialized organs situated in the umbrella of the jellyfish *Aequorea victoria* (JONES et al., 1999). GFP is excited by energy transfer from a photoprotein, aequorin. This reaction requires Ca^{2+} and a small chromophore, the coelenterazine as a prosthetic group for the light producing reaction. Engineered recombinant aequorin has been applied to monitor calcium signals in organelles and subcellular domains (KENDALL and BADMINTON, 1998). Furthermore, functional improvements of GFP by selected mutations resulted in various applications of this enzyme as a protein tag for the monitoring of gene expression and location. The advantage of GFP for such applications in comparison to luciferase is the independence of its enzyme reaction of molecular oxygen.

6 Bioremediation with Marine Microorganisms

6.1 Biodegradation of Marine Oil Spills

Despite rigid laws and improved technical security systems organic pollutants, especially oil spills, still are as ever a great danger to the vulnerable ecological balance of the marine environment. Because of the increasing ship traffic over the oceans it is unlikely that oil spills can be completely prevented. Despite the improvement of physical collection of the spilled oil with skimmers, biological tools allowing an effective *in situ* degradation of such pollutants are of interest. As mentioned above the majority of the marine environment is colder than 5 °C. Although the temperature of the surface water is in most cases a little higher, a temperature below 15 °C can be generally expected. Thus, microorganisms with an optimal degradation capacity at these low temperatures could achieve faster cleaning of polluted marine environments. Already in 1975 WALKER and COLWELL isolated cold-adapted petroleum-degrading microorganisms from water and sediment samples of two areas in the mid-Atlantic at environmental temperatures of 0.5–10 °C. The authors found that slower, but more extensive biodegradation with these cold-adapted bacteria occurred at 0 °C compared to 5 °C or 10 °C. They speculated that the more intensive biodegradation at lower temperature is due to the lower solubility of toxic hydrocarbons. It has been found that besides the temperature and the accessibility to degradable compounds within oil aggregates the availability of additional nutrients like nitrogen and phosphorus are the limiting factors for the effectiveness of biodegradation in marine environments (MARGESIN and SCHINNER, 1999). Although already in 1979 a patent has been filed claiming the addition of microorganisms to improve the degradation of floating oil (MOHAN et al., 1979) no such product is currently commercially available. Furthermore, there have been many proposals to use genetically engineered bacteria to enhance the oil degradation capacity, but successful systems have not yet been reported.

6.2 Biosurfactants

Biosurfactants can stimulate the rate of biodegradative removal of oil from the marine environment (PRINCE, 1997). Biosurfactants are surface-active microbial products which are produced by a variety of bacteria (SULLIVAN, 1998). YAKIMOV et al. (1999) described two biosurfactant producing, *n*-alkane-degrading marine bacteria isolated from the Ross sea in the Antarctic. These bacteria could be assigned to the genus *Rhodococcus*. During the cultivation on *n*-alkanes as sole carbon and energy source they produced extracellular and cell-bound surface-active glycolipids which reduced the surface tension of water from 72 mN m^{-1} to 32 mN m^{-1}. Recently, a new lipopeptide biosurfactant produced by a marine *Arthrobacter* strain could be identified (MORIKAWA et al., 1993). Such surfactants have a broad range of potential industrial applications, not only for enhanced oil recovery and surfactant-aided bioremediation of water-insoluble pollutants, but also for industrial processes (i.e., emulsification, phase separation, wetting foaming, emulsion stabilization, and viscosity reduction) (DESAI and BANAT, 1997; SULLIVAN, 1998).

6.3 Marine *Bacillus* Spores for Removal and Detoxification of Metals

The spores of the marine *Bacillus* sp. strain SG-1 are capable of binding a variety of heavy metals such as copper, cadmium, zinc, nickel, and cobalt or manganese (the two latter are also oxidized) (FRANCIS and TEBO, 1999). The spores can either passively adsorb these metals on the charged spore surface (biosorption), or the enzymatic activity of the outermost spore layer can directly catalyze the oxidation of divalent metals such as Mn(II) or Co(II) (direct oxidation). It has been suggested that due to the inherent physically tough nature and remarkable capacity to bind and oxidatively precipitate metals without having to sustain growth, the spores could be attractive candidates for biotechnological applications in the bioremediation of metal pollution. Furthermore, these marine spores could also be suitable for the recovery of metals from the environment.

7 Hydrogen Production

7.1 Hydrogen-Producing Marine Microorganisms

The foreseeable limitation of fossil energy resources, but also the difficulty to estimate the effect of the increasing carbon dioxide concentration in the atmosphere on our climate, enhances the pressure to find alternative sources for energy production. Some areas of the earth are rich in one of the most abundant renewable energy sources, solar energy. Photosynthetic organisms are able to collect solar energy by their biological energy conversion systems. A byproduct of the metabolism of some photosynthetic microorganisms, especially bacteria, is the powerful energy source hydrogen. Hydrogen is a renewable energy source that does not produce carbon dioxide. Hydrogen could be used as an ecologically sound energy source to partly replace the limited fossil fuels in the future (NANDI and SENGUPTA, 1998). However, so far current cost of hydrogen production cannot compete with the fossil fuels. Either an increase of the current oil price by the emergent limitation of the fossil fuel resources or a significant optimization of established microbial hydrogen producing systems would allow an economical production of this alternative energy resource.

Among the bacteria phototrophic bacteria and cyanobacteria belong to the most efficient hydrogen producers. Hydrogen production by cyanobacteria has been favored, since they are CO_2-consuming and O_2-evolving photosynthetic microorganisms which do not have sophisticated growth requirements (RAO and HALL, 1996; see Chapter 5, this volume). However, in comparison to phototrophic bacteria, the efficiency of hydrogen production by cyanobacteria or algae is lower (MIYAKE et al., 1999). Whereas cyanobacteria use water as starting component for hydrogen production, phototrophic bacteria such as the anaerobic bacterium *Rhodobacter shaeroides* RV (TSYGANKOV et al., 1993) use organic substrates, especially organic acids:

$$\text{organic acids} + H_2O + \text{sunlight} \rightarrow \times H_2 + \times CO_2$$

For this type of hydrogen production less light energy is required, but in contrast to the plant-type mechanism by cyanobacteria and algae, phototrophic bacteria produce carbon dioxide. It has been found that many of the anaerobic photosynthetic bacteria are able to decompose a variety of organic pollutants and, consequently, hydrogen production by these bacteria could be combined with organic waste treatment. The application of distillery wastewater for photoproduction of hydrogen by *R. shaeroides* has been described (SASIKALA et al., 1992).

7.2 Strategies for the Optimization of Hydrogen Production

Although the knowledge of the physiological and molecular basis of bacterial hydrogen production has remarkably increased during the last decade much more efforts are required to make microbial hydrogen production competitive compared to other energy sources. The following strategies are considered in this respect:

1. Screening of new photosynthetic microorganisms with high hydrogen production capacity,
2. Development of new or improved large-scale photobioreactor concepts (see Chapter 5, this volume),
3. Optimization of cultivation conditions (media, temperature, light),
4. Improvement of the separation and refining of the hydrogen produced,
5. Optimization of the hydrogen production capacity by an improved genetic design of selected strains (see Fig. 6),
6. Commercialization of other natural compounds produced by these microorganisms,
7. Wastewater treatment by hydrogen-producing bacteria.

The first genome of a hydrogen producing microorganism, namely from the cyanobacteria *Synechocystis* PCC 6803, has been completed (NAKAMURA et al., 1998; *www.kazusa.org*). Genetic manipulation techniques for selected hydrogen-producing bacteria have been developed (MIYAKE et al., 1999). This permit an optimization of hydrogen production by directed genetic modifications of their hydrogen metabolic machinery. Potential targets for the optimization of hydrogen production by genetically modified marine bacteria are indicated in Fig. 6.

Inactivation of genes encoding H_2 uptake enzymes (hydrogenases) and thus prevention of the recycle of hydrogen evolved by nitrogenase under N_2-fixing conditions has been suggested (HANSEL and LINDBLAD, 1998). Furthermore, an increase of the number of major hydrogen-producing enzyme complexes in

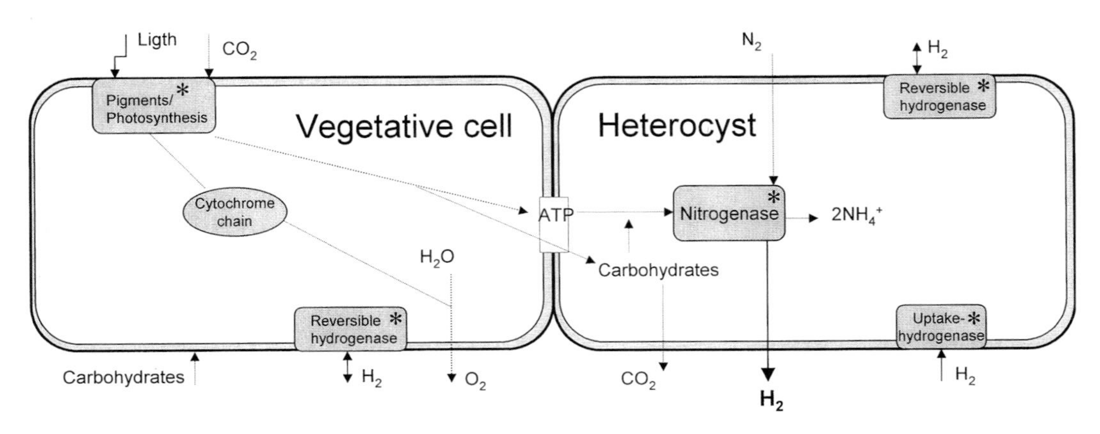

Fig. 6. Schematic representation of hydrogen metabolism in cyanobacteria. Potential targets for the optimization of hydrogen production by genetic modifications are marked with an asterisk.

nitrogen fixing cyanobacteria, nitrogenase, could also improve the level of hydrogen production (THIEL et al., 1997). Some of the hydrogen producers possess a reversible/bidirectional hydrogenase (TAMAGNINI et al., 1997). The reversible mechanism of this enzyme is not yet understood. Further investigations could help to clarify the regulation of H_2 uptake and finally suggest a strategy for the repression of this undesired enzyme function. Upregulation of genes involved in heterocyst development and thus an increased number of heterocysts could also result in improved H_2 production. Recently, a gene has been found, the product of which seems to be involved in the control of heterocyst frequency of *Anabaena* sp. (YOON and GOLDEN, 1997). The most efficient driving energy for hydrogen production comes from photosynthesis. Improvement of light energy absorption by optimization of the pigment composition could enhance the hydrogen production capacity of phototrophic bacteria such as *R. shaeroides* RV (MIYAKE et al., 1999). Finally, a decreased sensitivity of hydrogen-producing enzymes for oxygen could allow both processes oxygenic photosynthesis and H_2 production.

8 Outlook

Marine biotechnology meanwhile has a broad scientific base and attracts more and more attention also on the industrial level. It is commonly accepted that biotechnological methods often are the only possibility to overcome the problem of supply and to use the great biochemical potential of marine organisms. This potential includes the production of new biomedical compounds, of interesting enzymes for industrial and diagnostic purposes, of food- and health-promoting food additives, of numerous auxiliary materials, and in the future maybe also of energy. When evaluating the current state it seems necessary to give more emphasis to the following fields:

- Basic biology of marine organisms (diversity, taxonomy, physiology, genetics, biochemistry, interactions),
- Basic ecology of the marine habitats,
- Cultivation of marine organisms and generation of pure cultures,
- Bioprocess engineering (scale-up, genetic engineering, improvement of product purification),
- Increased involvement of the industry for commercialization.

The success of marine biotechnology will highly depend upon the intensive interdisciplinary cooperation between marine biologists and microbiologists, natural product chemists, biotechnologists, process engineers, pharmacists, etc. In this way the further development of marine biotechnology will not only offer new knowledge, products, and services, but at the same time also the possibility to structure the development and use of the marine habitat in such a way that protection and conservation of the marine and coastal environments are assured as an integral part of the global life support system.

9 References

ABARZUA, S., JAKUBOWSKI, S. (1995), Biotechnological investigation for the prevention of biofouling. I. Biological and biochemical principles for the prevention of biofouling, *Mar. Ecol.: Prog. Ser.* **123**, 301–312.

ABBANAT, D., LEIGHTON, M., MAIESE, W., JONES, E. B. G., PEARCE, C., GREENSTEIN, M. (1998), Cell wall active antifungal compounds produced by the marine fungus *Hypoxylon oceanicum* LL – 15G256. I: Taxonomy and Fermentation, *J. Antibiot.* **51**, 296–302.

ABDUSSALAM, S. (1990), Drugs from seaweeds, *Med. Hypotheses* **32**, 33–35.

ACERET, T. L., COLL, J. C., UCHIO, Y., SAMMARCO, P. W. (1998), Antimicrobial activity of the diterpens flexibilide and sinulariolide derived from *Sinularia flexibilis* Quoy and Gaimard 1833 (Coelenterata: Alcyonacea, Octocorallia), *Comp. Biochem. Physiol. C Pharmacol. Toxicol. Endocinol.* **120**, 121–126.

ALLEN, E. E., FACCIOTTI, D., BARTLETT, D. H. (1999), Monounsaturated but not polyunsaturated fatty acids are required for growth of the deep-sea bacterium *Photobacterium profundum* SS9 at high pressure and low temperature, *Appl. Environ.*

Microbiol. **65**, 1710–1720.

ALVI, K. A., CASEY, A., NAIR, B. P. (1998), Pulchellalactam: A CD45 protein tyrosin phosphatase inhibitor from the marine fungus *Corollospora pulchella*, *J. Antibiot.* **51**, 515–517.

AMANN, R., KUHL, M. (1998), *In situ* methods for assessment of microorganisms and their activities, *Curr. Opin. Microbiol.* **1**, 352–358.

Anonymous (1999), Ziconotide, *Curr. Opin. CPNS Invest. Drugs* **1**, 153–166.

ARAV, A., RUBINSKY, B., FLETCHER, G., SEREN, E. (1993), Cryogenic protection of oocytes with antifreeze proteins, *Mol. Reprod. Devel.* **36**, 488–493.

ARMSTRONG, E., MCKENZIE, J. D., GOLDSWORTHY, G. T. (1999), Aquaculture of sponges on scallops for natural products research and antifouling, *J. Biotechnol.* **70**, 163–174.

ATTAWAY, D. H., ZABORSKY, R. R. (Eds.) (1993), *Marine Biotechnology*, Vol. 1, *Pharmaceutical and Bioactive Natural Products*. New York, London: Plenum Press.

BALKWILL, D. L., MARATEA, D., BLAKEMORE, R. P. (1980), Ultrastructure of a magnetotactic *Spirillum*, *J. Bacteriol.* **141**, 1399–1408.

BARCLAY, W. R. (1992), Process for the heterotrophic production of microbial oils with high concentrations of omega-3 highly unsaturated fatty acids, *U.S. Patent* 5 130 242.

BARCLAY, W. R. (1994), Process for growing *Thraustochytrium* and *Schizochyrium* using non-chloride salts to produce a microfloral biomass having omega-3 highly unsaturated fatty acids, *U.S. Patent* 5 340 742.

BARON, A., CAPRIO, V., MANN, J. (1999), Synthesis of key intermediates for a concise and convergent approach to the marine natural product eleutherobin, *Tetrahedron Lett.* **40**, 9321–9324.

BAZ, J. P., CANEDO, L. M., FERNAN'DEZ PUENTES, J. L., SILVA ELIPE, M. V. (1997), Thiocoraline, a novel depsipeptide with antitumor activity produced by a marine *Micromonospora* II. Physico-chemical properties and structure determination, *J. Antibiot.* **50**, 738–741.

BELOFSKY, G., JENSEN, P. R., RENNER, M. K., FENICAL, W. (1998), New cytotoxic sesquiterpenoid nitrobenzoyl esters from a marine isolate of the fungus *Aspergillus versicolor*, *Tetrahedron* **54**, 1715–1724.

BENTLEY, R. (1997), Microbial secondary metabolites play important roles in medicine, prospects for discovery of new drugs, *Perspectives in Biology and Medicine* **40**, 364–394.

BERGMANN, W., FEENEY, R. J. (1951), Contribution to the study of marine sponges. 32. The nucleosides of sponges, *J. Org. Chem.* **16**, 981–987.

BERNAN, V. S., GREENSTEIN, M., MAIESE, W. M. (1997), Marine microorganisms as a source of new natural products, *Adv. Appl. Microbiol.* **43**, 57–90.

BEWLEY, C. A., FAULKNER, D. J. (1994), Theonegramide, an antifungal glycopeptide from the Philippine lithistid sponge *Theonella swinhoei*, *J. Org. Chem.* **59**, 4849–4852.

BEWLEY, C. A., HOLLAND, N. D., FAULKNER, D. J. (1996), Two classes of metabolites from *Theonella swinhoei* are localized in distinct populations of bacterial symbionts, *Experientia* **52**, 716–722.

BLAKEMORE, R. P. (1975), Magnetotactic bacteria. *Science* **190**, 377–379.

BONGHIORNI, L., PIETRA, F. (1996), Marine natural products for industrial applications, *Chem. Ind.* 54–58.

BOWLES, R. D., HUNT, A. E., BREMER, G. B., DUCHARS, M. G., EATON, R. A. (1999), Long-chain n-3 polyunsaturated fatty acid production by members of the marine protistan group the thraustochytrids: screening of isolates and optimisation of docosahexaenoic acid production, *J. Biotechnol.* **70**, 193–202.

BOWMAN, J. P., GOSINK, J. J., MCCAMMON, S. A., LEWIS, T. T., NICHOLS, D. S. et al. (1997), *Colwellia demingae* sp. nov., *Colwellia hornerae* sp. nov., *Colwellia rossensis* sp. nov. and *Colwellia psychrotropica* sp. nov.: psychrophilic Antarctic species with the ability to synthesise docosahexaenoic acid (22:6 n-3), *Int. J. Syst. Bacteriol.* **48**, 1171–1180.

BRACHER, F. (1994), Seescheiden (Ascidien) – eine neue Quelle pharmakologisch aktiver Substanzen, *Pharmazie in unserer Zeit* **23**, 147–157.

BULTEL-PONCÉ, V., DEBITUS, C., BERGE, J.-P., CERCEAU, C., GUYOT, M. (1998), Metabolites from the sponge-associated bacterium *Micrococcus luteus*, *J. Mar. Biotechnol.* **6**, 233–236.

BURGESS, J. G., JORDAN, E. M., BREGU, M., MEARNS-SPRAGG, A., BOYD, K. G. (1999), Microbial antagonism: a neglected avenue of natural products research, *J. Biotechnol.* **70**, 27–32.

CANEDO, L. M., DE LA FUENTE, J. A., GESTO, C., FERREIRO, M. J., JIMENEZ, C., RIGUERA, R. (1999), Agrochelin, a new cytotoxic alkaloid from the marine bacteria *Agrobacterium* sp., *Tetrahedron Lett.* **40**, 6841–6844.

CARLUCCI, M. J., SCOLARO, L. A., ERREA, M. I., MATULEWICZ, M. C., DAMONTE, E. B. (1997a), Antiviral activity of natural sulphated galactans on Herpes virus multiplication in cell culture, *Planta med.* **63**, 429–432.

CARLUCCI, M. J., PUJOL, C. A., CIANCIA, M., NOSEDA, M. D., MATULEWICZ, M. C. et al. (1997b), Antiherpetic and anticoagulant properties of carrageenans from the red seaweed *Gigartina skottsbergii* and their cyclized derivatives: correlation between structure and biological activity, *Int. J. Biol. Macromol.* **20**, 97–105.

CHANDRASEKARAN, M. (1996), Harnessing marine microorganisms through solid state fermentation,

J. Sci. Ind. Res. **5**, 468–471.

Chen, C., Imamura, N., Nishijima, M., Adachi, K., Sakai, M., Sano, H. (1996), Halymecins, new antimicroalgal substances produced by fungi isolated from marine algae, *J. Antibiot.* **49**, 998–1005.

Chen, X. T., Zhou, B., Bhattacharya, S. K., Gutteridge, C. E., Pettus, T. R. R., Danishefsky, S. J. (1998), The total synthesis of eleutherobin: a surprise ending, *Angew. Chem.* (Int. Edn.) **37**, 789–792.

Cheng, X. C., Jensen, P. R., Fenical, W. (1999), Luisols A and B, new aromatic tetraols produced by an estuarine marine bacterium of the genus *Streptomyces* (Actinomycetales), *J. Nat. Prod.* **62**, 608–610.

Chini Zittelli, G., Lavista, F., Bastianini, A., Rodolfi, L., Vincenzini, M., Tredici, M. R. (1999), Production of eicosapentaenoic acid by *Nannochloropsis* sp. cultures in outdoor tubular photobioreactors, *J. Biotechnol.* **79**, 299–312.

Clare, A. S. (1998), Towards nontoxic antifouling, *J. Mar. Biotechnol.* **6**, 3–6.

Clare, A. S., Rittschof, D., Gerhart, D. J., Hooper, I. R., Bonaventura, J. (1999), Antisettlement and narcotic action of analogues of diterpene marine natural product antifoulings from octocorals, *Mar. Biotechnol.* **1**, 427–436.

Colwell, R. R., Pariser, E. R., Sinskey, A. J. (Eds.) (1985), *Biotechnology of Marine Polysaccharides.* Washington, New York, London: Hemisphere Publishing Corp.

Cowan, D. A. (1997), The marine biosphere: a global resource for biotechnology. *TIBTECH* **15**, 129–130.

Crews, P., Manes, L. V., Boehler, M. (1986), Jasplakinolide, a cyclodepsipeptide from the marine sponge *Jaspis* sp., *Tetrahedron Lett.* **27**, 2797–2800.

Custodio, R. C., Prokic, I., Steffen, R., Koziol, C., Borojevic, R. et al. (1998), Primmorphs generated from dissociated cells of the sponge *Suberites domuncula*: a model system for studies of cell proliferation and cell death, *Mech. Aging Dev.* **105**, 45–59.

D'Ambrosio, M., Guerriero, A., Pietra, F. (1987), Sarcodictyin A and sarcodictyin B, novel diterpenoidic alcohols esterified by (*E*)-*N*(1)-methylurocanic acid. Isolation from the Mediterranean stolonifer *Sarcodictyon roseum*, *Helv. Chim. Acta* **70**, 2019–2027.

DaCosta, M. S., Santos, H., Galinski, E. A. (1998), An overview of the role and diversity of compatible solutes in bacteria and archaea, *Adv. Biochem. Eng. Biotechnol.* **61**, 117–153.

Davidson, S. K., Haygood, M. G. (1999), Identification of sibling species of the bryozoan *Bugula neritina* that produce different anticancer bryostatins and harbor distinct strains of the bacterial symbiont "*Candidatus Endobugula sertula*", *Biol. Bull.* **196**, 273–280.

DeNys, R., Steinberg, P. D., Willemsen, P., Dworjanyn, S. A., Gabelish, C. L., King, R. J. (1995), Broad spectrum effects of secondary metabolites from the red alga *Delisea pulchra* in antifouling assays, *Biofouling* **8**, 259–271.

DeLong, E. F. (1997), Marine microbial diversity: the tip of the iceberg, *TIBTECH* **15**, 203–207.

DeLong, E. F., Wu, K. Y., Prezelin, B. B., Jovine, R. V. (1994), High abundance of Archaea in Antarctic marine picoplankton, *Nature* **371**, 695–697.

DeLong, E. F., Yayanos, A. A. (1985), Adaptation of the membrane lipids of a deep-sea bacterium to changes in hydrostatic pressure, *Science* **228**, 1101–1103.

Depix, M. S., Martinez, J., Santibanez, F., Rovirosa, J., San Martin, A., Maccioni, R. B. (1998), The compound 14-keto-stypodiol diacetate from the algae *Stypopodium flabelliforme* inhibits microtubules and cell proliferation in DU-145 human prostatic cells, *Mol. Cell. Biochem.* **187**, 191–199.

Desai, J. D., Banat, I. M. (1997), Microbial production of surfactants and their commercial potential, *Microbiol. Mol. Biol. Rev.* **61**, 47–64.

DeSilva, E. D., Scheuer, P. J. (1980), Manoalide, an antibiotic sesterterpenoid from the marine sponge *Luffariella variabilis*, *Tetrahedron Lett.* **21**, 1611–1614.

DeSwaaf, M. E., DeRijk, T. C., Eggink, G., Sijtsma, L. (1999), Optimisation of docosahexaenoic acid production in batch cultivations by *Crypthecodinium cohnii*, *J. Biotechnol.* **70**, 185–192.

DeVries, A. L. (1971), Glycoproteins as biological antifreeze agents in antarctic fishes, *Science* **172**, 1152–1155.

Distel, D. L., Lane, D. J., Olsen, G. J., Giovannoni, S. J., Pace, B. et al. (1988), Sulfur-oxidizing bacterial endosymbionts: analysis of phylogeny and specificity by 16S rRNA sequences, *J. Bacteriol.* **170**, 2506–2510.

Doshida, J., Hasegawa, H., Onuki, S., Shimidzu, N. (1996), Exophilin A, a new antibiotic from a marine microorganism *Exophiala pisciphila*, *J. Antibiot.* **49**, 1105–1109.

Dunlap, W. C., Masaki, K., Yamamoto, Y., Larsen, R. M., Karube, I. (1998), A novel antioxidant from seaweed, in: *New Developments in Marine Biotechnology* (LeGal, Y., Halvorson, H. O., Eds.), pp. 33–35. New York, London: Plenum Press.

Egorin, M. J., Sentz, D. L., Rosen, D. M., Ballesteros, M. F., Kearns, C. M. et al. (1996), Plasma pharmacokinetics, bioavailability, and tissue distribution in CD_2F_1 mice of halomon, an antitumor halogenated monoterpene isolated from the red algae *Portieria hornemannii*, *Cancer Chemother. Pharmacol.* **39**, 51–60.

EL SAYED, K. A., DUNBAR, D. C., PERRY, T. L., WILKINS, S. P., HAMANN, M. T. (1997), Marine natural products as prototype insecticidal agents, *J. Agric. Food Chem.* **45**, 2735–2739.

EL SAYED, K. A., DUNBAR, D. C., BARTYZEL, P., ZJAWIONY, J. K., DAY, W., HAMANN, M. T. (2000), Marine natural products as leads to develop new drugs and insecticides, in: *Biological Active Natural Products: Pharmaceuticals* (CUTLER, S. J., CUTLER, H. G., Eds.), pp. 233–252. Boca Raton, FL: CRC Press.

ELYAKOV, G. B., KUZNETSOVA, T. A., STONIK, V. A., MIKHAILOV, V. V. (1994), New trends of marine biotechnology development, *Pure Appl. Chem.* **66**, 811–818.

EVANS, D. A., CARTER, P. H., CARREIRA, E. M., CHARETTE, A. B., PRUNET, J. A., LAUTENS, M. (1999), Total synthesis of bryostatin 2, *J. Am. Chem. Soc.* **121**, 7540–7552.

FACCIOTTI, D., VALENTINE, R., METZ, J. et al. (1998), Cloning and characterisation of polyunsaturated fatty acids (PUFA) genes from marine bacteria, *Abstracts Int. Symp. Progress and Perspectives of Marine Biotechnology*, 5–10 October, Qingdao, China.

FAJARDO, M. A., ALVAREZ, F., PUCCI, O. H., MARTIN-DE-PORTELA, M. L. (1998), Contents of various nutrients, minerals and seasonal fluctuations in *Porphyra columbina*, an edible marine alga from the Argentinic Patagonian coast, *Arch. Latinam. Nutr.* **48**, 260–264.

FATMA, T., SULTAN, S. (1999), Significance of n-3 polyunsaturated fatty acids and algal potential as its source, in: *Cyanobacterial and Algal metabolism and environmental biotechnology* (FATMA, T., Ed.), New Delhi: Narosa Publishing House.

FATTORUSSO, E., MINALE, L., SODANO, G. (1970), Aeroplysinin-1; a bromo compound from the sponge *Aplysina aerophoba*, *J. Chem. Soc.; Chem. Comm.* 751–753.

FAULKNER, D. J. (2000a), Highlights of marine natural products chemistry (1972–1999), *Nat. Prod. Rep.* **17**, 1– 6.

FAULKNER, D. J. (2000b), Marine natural products, *Nat. Prod. Rep.* **17**, 7–55.

FAULKNER, D. J. (2000c), Marine pharmacology, *Antonie van Leeuwenhoek* **77**, 135–145.

FEENEY, R. E., YEH, Y. (1998), Antifreeze proteins: Current status and possible food uses, *Trends Food Sci. Technol.* **9**, 102–106.

FELLER, G., NARINX, E., ARPIGNY, J. L., AITTALEB, M., BAISE, E. et al. (1996), Enzymes from psychrophilic organisms, *FEMS Microbiol. Rev.* **18**, 189–202.

FENICAL, W. (1993), Chemical studies of marine bacteria: Developing a new resource, *Chem. Rev.* **93**, 1673–1683.

FENICAL, W. (1997), New pharmaceuticals from marine organisms, *TIBTECH* **15**, 339–341.

FENICAL, W., JENSEN, P. R. (1993), Marine microorganisms: a new biomedical resource, in: *Marine Biotechnology* Vol. 1, *Pharmaceutical and Bioactive Natural Products* (ATTAWAY, D. H., ZABORSKY, R. R., Eds.), pp. 419–457. New York, London: Plenum Press.

FODSTAD, O., BREISTOL, K., PETTIT, G. R., SHOEMAKER, R. H., BOYD, M. R. (1996), Comparative antitumor activities of halichondrins and vinblastine against human tumor xenografts, *J. Exp. Ther. Oncol.* **1**, 119–125.

FRANCIS, C. A., TEBO, B. M. (1999), Marine *Bacillus* spores as catalysts for oxidative precipitation and sorption of metals, *J. Mol. Microbiol. Biotechnol.* **1**, 71–78.

FRASER, P. D., MIURA, Y., MISAWA, N. (1997), *In vitro* characterization of astaxanthin biosynthetic enzymes, *J. Biol. Chem.* **272**, 6128–6135.

FRIEDRICH, A. B., MERKERT, H., FENDERT, T., HACKER, J., PROOKSCH, P., HENTSCHEL, U. (1999), Microbial diversity in the marine sponge *Aplysina cavernicola* (formerly *Verongia cavernicola*) analyzed by fluorescence *in situ* hybridisation (FISH), *Mar. Biol.* **134**, 461–470.

FULLER, R. W., CARDELLINA, J. H., KATO, Y., BRINEN, L. S., CLARDY, J. et al. (1992), A pentahalogenated monoterpene from the red alga *Portieria hornemannii* produces a novel cytotoxicity profile against a diverse panel of human tumor cell lines, *J. Med. Chem.* **35**, 3007–3011.

FULLER, R. W., CARDELLINA, J. H., JUREK, J., SCHEUER, P. J., ALVARADO-LINDNER, B. et al. (1994), Isolation and structure/activity features of halomon-related antitumor monoterpenes from the red alga *Portieria hornemannii*, *J. Med. Chem.* **37**, 4407– 4411.

GERDAY, C., AITTALEB, M., BENTAHIR, M., CHESSA, J. P., CLAVERIE, P. et al. (2000), Cold-adapted enzymes: from fundamentals to biotechnology, *TIBTECH* **18**, 103–107.

GILL, I., VALIVETY, R. (1997), Polyunsaturated fatty acids, Part 1: Occurrence, biological activities and applications, *TIBTECH* **15**, 401–409.

GLASER, K. B., JACOBS, R. S. (1986), Molecular pharmacology of manoalide, *Biochem. Pharmacol.* **35**, 449–453.

GLOCKNER, F. O., FUCHS, B. M., AMANN, R. (1999), Bacterioplankton compositions of lakes and oceans: a first comparison based on fluorescence *in situ* hybridization, *Appl. Environ. Microbiol.* **65**, 3721–3726.

GRABLEY, S., THIERICKE, R. (Eds.) (1999): *Drug Discovery from Nature*. Berlin, Heidelberg, New York: Springer-Verlag.

GROSS, M., KOSMOWSKY, I. J., LORENZ, R., MOLITORIS, H. P., JAENICKE, R. (1994), Response of bacteria and fungi to high-pressure stress as investigated by two-dimensional polyacrylamide gel electrophoresis, *Electrophoresis* **15**, 1559–1565.

GUNASEKERA, S. P., GUNASEKERA, M., LONGLEY, R. E., SCHULTE, G. K. (1990), Discodermolide: a new bioactive polyhydroxylated lactone from the marine sponge *Discodermia dissoluta*, *J. Org. Chem.* **55**, 4912–4915.

HAIGH, W. G., YODER, T. F., ERICSON, L., PRATUM, T., WINGET, R. R. (1996), The characterisation and cyclic production of a highly unsaturated homoserine lipid in *Chlorella minutissima*, *Biochim. Biophys. Acta* **1299**, 183–190.

HANSEL, A., LINDBLAD, P. (1998), Towards optimisation of cyanobacteria as biotechnologically relevant producers of molecular hydrogen, a clean and renewable energy source, *Appl. Microbiol. Biotechnol.* **50**, 153–160.

HASTINGS, J. W., POTRIKUS, C. J., GUPTA, S. C., KURFÜRST, M., MAKEMSON, J. C. (1985), Biochemistry and physiology of bioluminescent bacteria, *Adv. Microbiol. Physiol.* **26**, 235–291.

HAYGOOD, M. G., DISTEL, D. L. (1993), Bioluminescent symbionts of flashlight fishes and deep-sea anglerfishes form unique lineages related to the genus *Vibrio*, *Nature* **363**, 154–156.

HAYGOOD, M. G., DAVIDSON, S. K. (1997), Small-subunit rRNA genes and *in situ* hybridization with oligonucleotides specific for the bacterial symbionts in the larvae of the bryozoan *Bugula neritina* and proposal of "*Candidatus Endobugula sertula*", *Appl. Env. Microbiol.* **63**, 4612– 4616.

HAYS, L. M., FEENEY, R. E., CROWE, L. M., CROWE, J. H., OLIVER, A. E. (1996), Antifreeze glycoproteins inhibit leakage from liposomes during thermotropic phase transitions, *Proc. Natl. Acad. Sci. USA* **93**, 6835–6840.

HAZEL, J. R., WILLIAMS, E. E. (1990), The role of alterations in membrane lipid composition in enabling physiological adaptation of organisms to their physical environment, *Prog. Lipid. Res.* **29**, 167–227.

HELMHOLZ, H., ETOUNDI, P., LINDEQUIST, U. (1999), Cultivation of the marine basidiomycete *Nia vibrissa* (Moore & Meyers), *J. Biotechnol.* 203–206.

HIQASHI-OKAJ, K., OTANI, S., OKAI, Y. (1999), Potent suppressive effect of a Japanese edible seaweed, *Enteromorpha prolifera* (Sujiao-nori) on initiation and promotion phases of chemically induced mouse skin tumorigenesis, *Cancer Lett.* **140**, 21–25.

HIRANO, S., NAKAHIRA, T., NAKAGAWA, M., KIM, S. K. (1999), The preparation and applications of functional fibres from crab shell chitin, *J. Biotechnol.* **70**, 373–377.

HIRATA, Y., UEMURA, D. (1986), Halichondrins – antitumor polyether macrolides from a marine sponge, *Pure Appl. Chem.* **58**, 701–710.

HOPPE, H. A., LEVRING, T., TANAK, Y. (1979), *Marine Algae in Pharmaceutical Science*. Berlin, New York: Walter de Gruyter.

HOSHINO, T., HAYASHI, T., HAYASHI, K., HAMADA, J., LEE, J. B., SANKAWA, U. (1998), An antivirally active sulfated polysaccharide from *Sargassum horneri* (TURNER) C. AGARDH., *Biol. Pharm. Bull.* **21**, 730–734.

HU, M.-K., HUANG, W.-S. (1999), Synthesis and cytostatic properties of structure-simplified analogs of dolastatin 15, *J. Pept. Res.* **54**, 460–467.

IMAMURA, N., NISHIJIMA, T., TAKADERA, T., ADACHI, K., SAKAI, M., SANO, H. (1997), New anticancer antibiotics pelagiomicins, produced by a new marine bacterium *Pelagiobacter variabilis*, *J. Antibiot.* **50**, 8–12.

IRELAND, C. M., COPP, B. R., FOSTER, M. P., MCDONALD, L. A., RADISKY, D. C., SWERSEY, J. C. (1993), Biomedical potential of marine natural products, in: *Marine Biotechnology* Vol. 1, *Pharmaceutical and Bioactive Natural Products* (ATTAWAY, D. H., ZABORSKY, R. R., Eds.), pp. 1–43. New York, London: Plenum Press.

ISHIHARA, K., MURATA, M., KANENIWA, M., SAITO, H., SHINOHARA, K., MAEDA-YAMAMOTO, M. (1998), Inhibition of icosanoid production in MC/9 mouse mast cells by n-3 polyunsaturated fatty acids isolated from edible marine algae, *Biosci. Biotechnol. Biochem.* **62**, 1412–1415.

IWASHIMA, M., OKAMOTO, K., MIYAI, Y., IGUCHI, K. (1999), 4-Epiclavulones, new marine prostanoids from the Okinawan soft coral, *Clavularia viridis*, *Chem. Pharm. Bull.* **47**, 884–886.

IZUMIDA, H., ADACHI, K., MIHARA, A., YASUZAWA, T., SANO, H. (1996), Hydroxyakalone, a novel xanthine oxidase inhibitor produced by a marine bacterium, *Agrobacterium aurantiacum*, *J. Antibiot.* **49**, 76–80.

JACOBS, D. I., HAEMERS, S., VAN DER LEEDEN, M. C., FRENS, G., DUINE, J. A. (1998), Mussel adhesive proteins, possibilities for an adhesive in surgery, in: *Book of Abstracts Int. Symp. Marine Bioprocess Engineering*, sp. 76. Noordwijkerhout, The Netherlands.

JASPARS, M. (1998), Pharmacy of the deep–marine organisms as sources of anticancer agents, in: *Advances in Drug Discovery Techniques* (HARVEY, A. L., Ed.), pp. 65–84. Chichester: John Wiley & Sons.

JENSEN, P. R., FENICAL, W. (1994), Strategies for the discovery of secondary metabolites from marine bacteria: Ecological perspectives, *Ann. Rev. Microbiol.* **48**, 559–584.

JONES, K., HIBBERT, F., KEENAN, M. (1999) Glowing jellyfish, luminescence and a molecule called coelenterazine, *TIBTECH* **17**, 477–481.

KAEFFER, B., BENARD, C., LAHAYE, M., BLOTTIERE, H. M., CHERBUT, C. (1999), Biological properties of ulvan, a new source of green seaweed sulfated polysaccharides, on cultured normal and cancerous epithelial cells, *Planta Med.* **65**, 527–531.

KAKEYA, H., TAKAHASHI, I., OKADA, G., ISONO, K., OSADA, H. (1995), Epolactaene, a novel neuritogenic compound in human neuroblastoma cells produced by a marine fungus, *J. Antibiot.* **48**, 733–735.

KAWANO, Y., ASADA, M., INOUE, M., NAKAGOMI, K., OKA, S., HIGASHIRANA, T. (1998), Biological activity of thiotropocin produced by marine bacterium, *Caulobacter* sp. PK654, *J. Mar. Biotechnol.* **6**, 49–52.

KENDALL, J. M., BADMINTON, M. N. (1998), *Aequorea victoria* bioluminescence moves into exciting new era, *TIBTECH* **16**, 216–223.

KENDRICK, A., RATLEDGE, C. (1992), Lipids of selected molds grown for production of n-3 and n-6 polyunsaturated fatty acids, *Lipids* **27**, 15–20.

KERNAN, M. R., MOLINSKI, T. F., FAULKNER, D. J. (1988), Macrocyclic antifungal metabolites from the Spanish dancer nudibranch *Hexabranchus sanguineus* and sponges of the genus *Halichondria*, *J. Org. Chem.* **53**, 5014–5020.

KERR, R. G., KERR, S. S. (1999), Marine natural products as therapeutic agents, *Exp. Opin. Ther. Patents* **9**, 1207–1222.

KITAGAWA, I., KOBAYASHI, M., KITANAKA, K., KIDO, M., KYOGOKU (1983), Marine natural products. XII. On the chemical constituents of the Okinawan marine sponge *Hymeniacidon aldis*, *Chem. Pharm. Bull.* **31**, 2321–2328.

KOBAYASHI, J., ISHIBASHI, M. (1993), Bioactive metabolites of symbiotic marine microorganisms, *Chem. Rev.* **93**, 1753–1769.

KOHLMEYER, J., KOHLMEYER, E. (1979), *Marine Mycology – The Higher Fungi*. New York, San Francisco, London: Academic Press.

KOLENDER, A. A., PUJOL, C. A., DAMONTE, E. B., MATULEWICZ, M. C., CEREZO, A. S. (1997), The system of sulfated alpha-(1-3)-linked D-mannans from the red seaweed *Nohogenia fastigiata:* structure, antiherpetic and anticoagulant properties, *Carbohydr. Res.* **304**, 53–60.

KÖNIG, G. M., WRIGHT, A. D. (1996), Marine natural products research: Current directions and future potential, *Planta med.* **62**, 193–211.

KÖNIG, G. M., WRIGHT, A. D. (1997), Sesquiterpene content of the antibacterial dichloromethane extract of the marine red alga *Laurencia obtusa*, *Planta med.* **63**, 186–187.

KÖNIG, G. M., WRIGHT, A. D. (1999), Trends in marine biotechnology, in: *Drug Discovery from Nature* (GRABLEY, S., THIERICKE, R., Eds.), pp. 180–187. Berlin, Heidelberg, New York: Springer-Verlag.

KÖNIG, G. M., WRIGHT, A. D., DE NYS, R. (1999), Halogenated monoterpenes from *Plocamium costatum* and their biological activity, *J. Nat. Prod.* **62**, 383–385.

KOUSHAFAR, H., RUBINSKY, B. (1997), Effect of antifreeze proteins on frozen primary prostatic adenocarcinoma cells, *Urology* **49**, 421–425.

KREUTER, M. H., LEAKE, R. E., RINALDI, F., MULLERKLIESER, W., MAIDHOF, A. et al. (1990), Inhibition of intrinsic protein tyrosin kinase-activity of EGF-receptor kinase complex from human breast-cancer cells by the marine sponge metabolite aeroplysinin-1, *Comp. Biochem. Physiol. B, Comp. Biochem.* **97**, 151–158.

KREUTER, M. H., ROBITZKI, A., CHANG, S., STEFFEN, R., MICHAELIS, M. et al. (1992), Production of the cytostatic agent, aeroplysinin by the sponge *Veronica aerophoba* in *in vitro* culture, *Comp. Biochem. Physiol.* **101C**, 183–187.

LAW, B. A., GOODENOUGH, P. W. (1995), Enzymes in milk and cheese production, in: *Enzymes in Food Processing* 2nd Edn. (TUCKER, G. A., WOODS, L. F. J., Eds.), pp. 114–143. Bishopbriggs: Blackie Academic & Professionals.

LEA, A. G. H. (1995), Enzymes in production of beverages and fruit juices, in: *Enzymes in Food Processing* 2nd Edn. (TUCKER, G. A., WOODS, L. F. J., Eds.). pp. 223–249. Bishopbriggs: Blackie Academic & Professionals.

LEE, J. B., HAYASHI, K., HAYASHI, T., SANKAWA, U., MAEDA, M. (1999), Antiviral activities against HSV-1, HCMV, and HIV-1 of rhamnan sulfate from *Monostroma latissimum*, *Planta med.* **65**, 439–441.

LEWIS, T. E., NICHOLS, P. D., MCMEEKIN, T. A. (1999), The biotechnological potential of thraustochytrids, *Mar. Biotechnol.* **1**, 580–587.

LI, Z. Y., WARD, O. P. (1994), Production of docosahexaenoic acid by *Thraustochytrium roseum*, *J. Ind. Microbiol.* **13**, 238–241.

LIBERRA, K., JANSEN, R., LINDEQUIST, U. (1998), Corollosporine, a new phthalide derivate from the marine fungus *Corollospora maritima* Werderm. 1069. *Pharmazie* **53**, 578–581.

LIBERRA, K., LINDEQUIST, U. (1995), Marine fungi – a prolific resource of biologically active natural products? *Pharmazie* **50**, 583–588.

LINDEL, T., JENSEN, P. R., FENICAL, W., LONG, B. H., CASAZZA, A. M. et al. (1997), Eleutherobin, a new cytotoxin that mimics paclitaxel (taxol) by stabilizing microtubules, *J. Am. Chem. Soc.* **119**, 8744–8745.

LINDEQUIST, U., KUSNICK, C., HELMHOLZ, H., LIBERRA, K. (1999), Neue Wirkstoffe aus Pilzen des Meeres, *Z. Phytother.* **20**, 29–36.

LINKO, Y. Y., HAYAKAWA, K. (1996), Docosahexaenoic acid: a valuable nutraceutical? *Trends Food Sci. Technol.* **7**, 59–63.

LIPPERT, K., GALINSKI, E. A. (1992), Enzyme stabilisation by ectoine-type compatible solutes: protection against heating, freezing and drying, *Appl. Microbiol. Biotechnol.* **37**, 61–65.

LITAUDON, M., HART, J. B., BLUNT, J. W., LAKE, R. J., MUNRO, M. H. G. (1994), Isohomohalichondrin B, a new antitumour polyether macrolide from the New Zealand deep-water sponge *Lissodendoryx* sp., *Tetrahedron Lett.* **35**, 9435–9438.

LOEWEN, M. C., LIU, X., DAVIES, P. L., DAUGULIS, A. J. (1997), Biosynthetic production of type II fish antifreeze protein: fermentation by *Pichia pastoris*, *Appl. Microbiol. Biotechnol.* **48**, 480–486.

LOGEART, D., PRIGENT-RICHARD, S., JOZEFONVICZ, J., LETOURNEUR, D. (1997), Fucans, sulfated polysaccharides extracted from brown seaweed, inhibit vascular smooth muscle cell proliferation. I. Comparison with heparin for antiproliferative activity, binding and internalization, *Eur. J. Cell Biol.* **74**, 376–384.

LOOK, S. A., FENICAL, W., JACOBS, R. S., CLARDY, J. (1986), The pseudopterosins: anti-inflammatory and analgesic natural products from the sea whip *Pseudopterogorgia elisabethae*, *Proc. Natl. Acad. Sci. USA* **83**, 6238–6240.

LOUIS, P., GALINSKI, E. A. (1997), Characterization of genes for the biosynthesis of the compatible solute ectoine from *Marinococcus halophilus* and osmoregulated expression in *Escherichia coli*, *Microbiology* **143**, 1141–1149.

LOYA, S., HIZI, A. (1990), The inhibition of human immunodeficiency virus type I reverse transcriptase by avarol and avarone derivative, *FEBS Lett.* **269**, 131–134.

LYNDON, A. R. (1999), Fish growth in marine culture systems: a challenge for biotechnology, *Mar. Biotechnol.* **1**, 376–379.

MARGESIN, R., SCHINNER, F. (1999), Biodegradation of organic pollutants at low temperatures, in: *Biotechnological Applications of Cold-Adapted Organisms* (MARGESIN, R., SCHINNER, F., Eds.), pp. 271–289. Berlin, Heidelberg: Springer-Verlag.

MARSHALL, C. J. (1997), Cold-adapted enzymes, *TIBTECH* **15**, 359–364.

MARTINEZ, E. J., OWA, T., SCHREIBER, S. L., COREY, E. J. (1999), Phthalascidin, a synthetic antitumor agent with potency and mode of action comparable to ecteinascidin 743, *Proc. Natl. Acad. Sci. USA* **96**, 3496–3501.

MARWICK, J. D., WRIGHT, P. C., BURGESS, J. G. (1999), Bioprocess intensification for production of novel marine bacterial antibiotics through bioreactor operation and design, *Mar. Biotechnol.* **1**, 495–507.

MATSUDA, M., SHIGETA, S., OKUTANI, K. (1999), Antiviral activities of marine *Pseudomonas* polysaccharides and their oversulfated derivatives, *Mar. Biotechnol.* **1**, 68–73.

MEIGHEN, E. A. (1993), Bacterial bioluminescence: organization, regulation, and application of the lux genes, *FASEB J.* **7**, 1016–1022.

MEYERS, S. P. (1994), Developments in world aquaculture, feed formulations, and role of carotenoids, *Pure Appl. Chem.* **66**, 1069–1076.

MIKHAILOV, V. V., KUZNETSOVA, T. A., ELIAKOV, G. B. (1995) Bioactive compounds from marine actinomycetes, *Bioorg. Khim.* **21**, 3–8.

MIKI, W., KON-YA, K., MIZOBUCHI, S. (1996), Biofouling and marine biotechnology: New antifoulants from marine invertebrates, *J. Mar. Biotechnol.* **4**, 117–120.

MIYAKE, J., MIYAKE, M., ASADA, Y. (1999), Biotechnological hydrogen production: research for efficient light energy conversion, *J. Biotechnol.* **70**, 89–101.

MOHAN, R. R., ROBBINS, M. L., LASKIN, A. I., NASLUND, L. A. (1979), *U.S. Patent* 4 146 470.

MOORE, B. S. (1999), Biosynthesis of marine natural products: microorganisms and macroalgae, *Nat. Prod. Rep.* **16**, 653–674.

MORIKAWA, M., DAIDO, H., TAKAO, T., MURATA, S., SHIMONISHI, Y., IMANAKA, T. (1993), A new lipopeptide biosurfactant produced by *Arthrobacter* sp. strain MIS38, *J. Bacteriol.* **175**, 6459–6466.

MÜLLER, W. E. G. (1998), Origin of metazoa: Sponges as living fossils, *Naturwiss.* **85**, 11–25.

MÜLLER, W. E. G., KRUSE, M., BLUMBACH, B., SKOROKHOD, A., MÜLLER, I. M. (1999a), Gene structure and function of tyrosine kinase in the marine sponge *Geodia cydonium:* Autapomorphic characters of metazoa, *Gene* **238**, 179–193.

MÜLLER, W. E. G., WIMMER, W., SCHATTON, W., BÖHM, M., BATEL, R., FILIC, Z. (1999b), Initiation of an aquaculture of sponges for the sustainable production of bioactive metabolites in open systems: Example, *Geodium cydonium*, *Mar. Biotechnol.* **1**, 569–579.

MUNRO, M. H. G., BLUNT, J. W., DUMDEI, E. J., HICKFORD, S. J. H., LILL, R. E. et al. (1999), The discovery and development of marine compounds with pharmaceutical potential, *J. Biotechnol.* **70**, 15–25.

NAKAMURA, Y., KANEKO, T., HIROSAWA, M., MIYAJIMA, N., TABATA, S. (1998), CyanoBase, a www database containing the complete nucleotide sequence of the genome of *Synechocystis* sp. strain PCC6803, *Nucleic Acids Res.* **26**, 63–67.

NANDI, R., SENGUPTA, S. (1998), Microbial production of hydrogen: an overview, *Crit. Rev. Microbiol.* **24**, 61–84.

NICHOLS, D. S., RUSSELL, N. J. (1996), Fatty acid adaptation in an Antarctic bacterium – changes in primer utilization, *Microbiology* **142**, 747–754.

NICHOLS, D. S., BROWN, J. L., NICHOLS, P. D., MCMEEKIN, T. A. (1996), Enrichment of the rotifer *Brachionus plicatilis* fed an Antarctic bacterium containing polyunsaturated acids, *Aquaculture* **147**, 115–125.

NICHOLS, D. S., BOWMAN, J., SANDERSON, K.,

NICHOLS, C. M., LEWIS, T. et al. (1999), Developments with antarctic microorganisms: culture collections, bioactivity screening, taxonomy, PUFA production and cold-adapted enzymes, *Curr. Opin. Biotechnol.* **10**, 240–246.

NICOLAU, K. C., XU, J. Y., KIM, S., OHSHIMA, T., HOSOKAWA, S., PFEFFERKORN, J. (1997), Synthesis of the tricyclic core of eleutherobin and sarcodictyins and total synthesis of sarcodictyin A, *J. Am. Chem. Soc.* **119**, 11353–11354.

NUMATA, A., TAKAHASHI, C., ITO, Y., MINOURA, K., YAMADA, T. et al. (1995), Penochalasins, a novel class of cytotoxic cytochalasans from a *Penicillium* species separated from a marine alga: structure determination and solution conformation, *J. Chem. Soc. Perkin Trans.* **1**, 239–245.

OHGIYA, S., HOSHINO, T., OKUYAMA, H., TANAKA, S., ISHIZAKI, K. (1999), Biotechnology of enzymes from cold-adapted microorganisms, in: *Biotechnological Applications of Cold-Adapted Organisms* (MARGESIN, R., SCHINNER, F., Eds.), pp. 17–34. Berlin, Heidelberg: Springer-Verlag.

OHTA, K., MIZUSHINA, Y., HIRATA, N., TAKEMURA, M., SUGAWARA, F. et al. (1998), Sulfoquinovosyldiacylglycerol, KMO43, a new potent inhibitor of eukaryotic DNA polymerases and HIV reverse transcriptase type I from a marine red alga, *Gigartina tenella, Chem. Pharm. Bull.* **46**, 684–686.

OKAICHI, T., HASHIMOTO, Y. (1962) The structure of nereistoxin, *Agr. Biol. Chem.* **26**, 224–227.

OLIVER, J. D., COLWELL, R. R. (1973), Extractable lipids of gram-negative marine bacteria: phospholipid composition, *J. Bacteriol.* **114**, 897–908.

ONSOYEN, E. (1996), Commercial applications of alginates, *Carbohydr. Eur.* **14,** 26–31.

OSINGA, R., TRAMPER, J., WIJFFELS, R. H. (1998), Cultivation of marine sponges for metabolite production: applications for biotechnology? *TIBTECH* **16**, 130–134.

OSINGA, R., BEUKELAER, P. B., MEIJER, E. M., TRAMPER, J., WIJFFELS, R. H. (1999), Growth of the sponge *Pseudosuberites* (aff.) *andrewsi* in a closed system, *J. Biotechnol.* **70**, 155–161.

OSTLING, J., MCDOUGALD, D., MAROUGA, R., KJELLEBERG, S. (1997), Global analysis of physiological responses in marine bacteria, *Electrophoresis.* **18**, 1441–1450.

PAK, N., ARAYA, H. (1996), Chilean edible sea macroalgae as sources of dietary fiber: effect on apparent digestibility of protein, fiber, and energy and fecal weight of rats, *Arch. Latinam. Nutr.* **46**, 42–46.

PAPOV, V. V., DIAMOND, T. V., BIEMANN, K., WAITES, J. H. (1995), Hydroxyarginine-containing polyphenolic proteins in the adhesive plaques of the marine mussel *Mytilus edulis*, *J. Biol. Chem.* **270**, 20183–20192.

PATHIRANA, C., STEIN, R. B., BERGER, T. S., FENICAL, W., IANIRO, T. et al. (1995), Nonsteroidal human progesterone receptor modulators from the marine alga *Cymopolia barbata*, *Mol. Pharmacol.* **47**, 630–635.

PATIL, A. D., FREYER, A. J., KILLMER, L., HOFMANN, G., JOHNSON, R. K. (1997), Z-axinohydantoin and debromo-Z-axinohydantoin from the sponge *Stylotella aurantium:* inhibitors of protein kinase C, *Nat. Prod. Lett.* **9**, 201–207.

PAYNE, S. R., YOUNG, O. A. (1995), Effects of preslaughter administration of antifreeze proteins on frozen meat quality, *Meat Sci.* **41**, 147–155.

PETTIT, G. R., HERALD, C. L., DOUBEK, D. L., HERALD, D. L., ARNOLD, E., CLARDY, J. (1982), Isolation and structure of bryostatin 1, *J. Am. Chem. Soc.* **104**, 6846–6848.

PETTIT, G. R., KAMANO, Y., HERALD, C. L., FUJII, Y., KIZU, H. et al. (1993), Isolation of dolastatins 10–15 from the marine mollusc *Dolabella auricularia*, *Tetrahedron* **49**, 9151–9170.

PETTIT, G. R., GAO, F., BLUMBERG, P. M., HERALD, C. L., COLL, J. C. et al. (1996), Antineoplastic agents. 340. Isolation and structural elucidation of bryostatins 16–18, *J. Nat. Prod.* **59**, 286–289.

PIETRA, F. (1997), Secondary metabolites from marine microorganisms: bacteria, protozoa, algae and fungi. Achievements and prospects, *Nat. Prod. Rep.* **14**, 453–464.

POMPONI, S. A., WILLOUGHBAY, R., WRIGHT, A. E., PECORELLA, C., SENNETT, S. H. et al. (1998), in: *New Developments in Marine Biotechnology* (LE GAL, Y., HALVORSON, H. O., Eds.), pp. 73–76. New York: Plenum Press.

POMPONI, S. A. (1999), The bioprocess-technological potential of the sea, *J. Biotechnol.* **70**, 5–13.

PRINCE, R. C. (1997), Bioremediation of marine oil spills, *TIBTECH* **15**, 158–160.

PROKSCH, P. (1999), Chemical defense in marine ecosystems, *Ann. Plant Rev.* **3** (Functions of Plant Secondary Metabolites and Their Exploitation in Biotechnology), 134–154.

RAO, K. K., HALL, D. O. (1996), Hydrogen production by cyanobacteria: potential, problems and prospects *J. Mar. Biotechnol.* **4**, 10–15.

RATLEDGE, C. (1993), Single cell oils – have they a biotechnological future? *TIBTECH* **11**, 278–284.

REDDY, M., VENKATA, R., RAO, M. R., RHODES, D., HANSEN, M. S. T. et al. (1999), Lamellarin alpha 20-sulfate, an inhibitor of HIV-1 integrase active against HIV-1 virus in cell culture, *J. Med. Chem.* **42**, 1901–1907.

REHM, B. H. A., VALLA, S. (1997), Bacterial alginates: biosynthesis and applications, *Appl. Microbiol. Biotechnol.* **48**, 281–288.

REMAUT, E., BLIKI, C., ITURRIZA-GOMARA, M., KEYMEULEN, K. (1999), Development of regulatable expression systems for cloned genes in cold-adapted bacteria, in: *Biotechnological Applica-*

tions of Cold-Adapted Organisms (MARGESIN, R., SCHINNER, F., Eds.), pp. 1–16. Berlin, Heidelberg: Springer-Verlag.

RENN, D. (1997), Biotechnology and the red seaweed polysaccharide industry: status, needs and prospects, *TIBTECH* **15**, 9–14.

RIGUERA, R. (1997), Isolating bioactive compounds from marine organisms, *J. Mar. Biotechnol.* **5**, 187–193.

RINEHART, K. L., GLOER, B., HUGHES, R. G., RENIS, H. E., MCGOVREN, J. P. et al. (1981), Didemnins: antiviral and antitumor depsipeptides from a Caribbean tunicate, *Science* **212**, 933–935.

RINEHART, K. L., HOLT, T. G., FREGEAU, N. L., STROH, J. G., KIEFER, P. A. et al. (1990), Ecteinascidins 729, 743, 745, 759A, 759B and 770: potent antitumor agents from the Caribbean tunicate *Ecteinascidia turbinata*, *J. Org. Chem.* **55**, 4512–4515.

RINKEVICH, B. (1999), Cell cultures from marine invertebrates: obstacles, new approaches and recent improvements, *J. Biotechnol.* **70**, 133–153.

RODRIGUEZ, A. D., RAMIREZ, C., RODRIGUEZ, I. I. (1999), Elisabatins A and B: New amphilectane-type diterpenes from the West Indian sea whip *Pseudopterogorgia elisabethae*, *J. Nat. Prod.* **62**, 997–999.

RORRER, G. L., YO, H. D., HUANG, B., HAYDEN, C., GERWICK, W. H. (1997), Production of hydroxy fatty acids by cell suspension cultures of the marine brown alga *Laminaria saccharina*, *Phytochemistry* **46**, 871–877.

RORRER, G. L., MULLIKIN, R. K. (1999), Modeling and simulation of a tubular recycle photobioreactor for macroalgal cell suspension cultures, *Chem. Eng. Sci.* **54**, 3153–3162.

ROUDIER, M., BOUCHON, C., ROUVILLAIN, J. L., AMEDEE, J., BAREILLE, R. et al. (1995), The resorption of bone-implanted corals varies with porosity but also with the host reaction, *J. Biomed. Mat. Res.* **29**, 909–915.

ROUHI, A. M. (1995), Supply issues complicate trek of chemicals from sea to market, *C & EN*, **20**, Nov. 1995.

ROUSSIS, R., ZHONGDE, W., FENICAL, W. (1990), New antiinflammatory pseudopterosins from the marine octocoral *Pseudopterogorgia elisabethae*, *J. Org. Chem.* **56**, 4304–4307.

RUSSELL, N. J. (1998), Molecular adaptations in psychrophilic bacteria: potential for biotechnological applications, *Adv. Biochem. Eng. Biotechnol.* **61**, 1–21.

RUSSELL, N. J., NICHOLS, D. S. (1999), Polyunsaturated fatty acids in marine bacteria – a dogma rewritten, *Microbiology* **145**, 767–779.

SAITO, S., WATABE, S., OZAKI, H., KOBAYASHI, M., SUZUKI, T. et al. (1998), Actin-depolymerizing effect of dimeric macrolides, bistheonellide A and swinholide A, *J. Biochem.* **123**, 571–578.

SAKAI, R., RINEHART, K. L., KISHORE, V., KUNDU, B., FAIRCLOTH, G. et al. (1996), Structure–activity relationships of the didemnins, *J. Med. Chem.* **39**, 2819–2934.

SAKAMOTO, T., WADA, H., NISHIDA, I., OHMORI, M., MURATA, N. (1994), Delta 9 acyl-lipid desaturases of cyanobacteria. Molecular cloning and substrate specificities in terms of fatty acids, *sn*-positions, and polar head groups, *J. Biol. Chem.* **269**, 25576– 25580.

SALEH, M., MOTAWE, H. M., MAHMOUD, F., MAHRAN, G. H., SOLIMAN, F. M. (1992), Chemical and antimicrobial study of *Dictyota dichotoma* var. *implexa*, *Fitotherapia* **63**, 369–371.

SARIN, P. S., SUN, D., THORNTON, A., MÜLLER, W. E. G. (1987), Inhibition of replication of the etiologic agent of acquired immune deficiency syndrome (human T-lymphotropic retrovirus/lymphaedenopathy-associated virus) by avarol and avarone, *J. Natl. Cancer Inst.* **78**, 663–666.

SARMA, A. S., DAUM, T., MÜLLER, W. E. G. (1993), *Secondary Metabolites from Marine Sponges*. Berlin, Erfurt: Ullstein-Mosby.

SASIKALA, K., RAMANA, C. H. V., ROA, P. R. (1992), Photoproduction of hydrogen from the waste water of a distillary by *Rhodobacter sphaeroides* OU 001, *Int. J. Hydrogen Energy* **17**, 23–27.

SATA, N. U., MATSUNAGA, S., FUSETANI, N., VAN SOEST, R. W. M. (1999), Aurantiosides D, E, and F: New antifungal tetramic acid glycosides from the marine sponge *Siliquariaspongia japonica*, *J. Nat. Prod.* **62**, 969–971.

SAUER, T., GALINSKI, E. A. (1998), Bacterial milking: A novel bioprocess for production of compatible solutes, *Biotechnol. Bioeng.* **57**, 306–313.

SCHAUMANN, C. (1993), Marine Pilze, in: *Mikrobiologie des Meeresbodens* (MEYER-REIL, L. A., KÖSTER, M., Eds.), Jena. Gustav Fischer Verlag.

SCHLINGMANN, G., MILNE, L., WILLIAMS, D. R., CARTER, G. T. (1998), Cell wall active antifungal compounds produced by the marine fungus *Hypoxylon oceanicum* LL – 15G256. II. Isolation and structure determination. *J. Antibiot.* **51**, 303–316.

SCHMITT, T. M., LINDQUIST, N., HAY, M. E. (1998), Seaweed secondary metabolites as antifoulants. Effects of *Dictyota* spp. diterpenes on survivorship, settlement, and development of marine invertebrate larvae, *Chemoecology* **8**, 125–131.

SCHMITZ, F. J., BOWDEN, B. F., TOTH, S. I. (1993), Antitumor and cytotoxic compounds from marine organisms, in: *Marine Biotechnology*, Vol. 1, *Pharmaceutical and Bioactive Natural Products* (ATTAWAY, D. H., ZABORSKY, R. R., Eds.), pp. 197–308. New York, London: Plenum Press.

SCHÜLER, D. (1999), Formation of magnetosomes in magnetotatic bacteria, *J. Molec. Microbiol. Biotechnol.* **1**, 79–86.

SCHÜLER, D., BAEUERLEIN, E. (1998), Dynamics of

iron uptake and Fe3O4 biomineralization during aerobic and microaerobic growth of *Magnetospirillum gryphiswaldense*, *J. Bacteriol.* **180**, 159–162.

SCHÜLER, D., FRANKEL, R. B. (1999), Bacterial magnetosomes: microbiology, biomineralization and biotechnological applications, *Appl. Microbiol. Biotechnol.* **52**, 464–473.

SCHWEDER, T., BANDOW, J., KUSNICK, C., JÜLICH, W.-D., HECKER, M., LINDEQUIST, U. (2000), Antibacterial substances from marine fungi and investigation of their targets by proteome-based methods, *Proc. IMBC 2000*, Townsville, Australia.

SCHWEDER, T., KAAN, T. (1999), *German Patent* 19951765.7.

SEVERIN, J., WOHLFARTH, A., GALINSKI, E. A. (1992), The predominant role of recently discovered tetrahydropyrimidines for the osmoadaptation of halophilic eubacteria, *J. Gen. Microbiol.* **138**, 1629–1638.

SHARMA, G. W., BUYER, J. S., POMERANTZ (1980), Characterization of a yellow compound isolated from the marine sponge *Phakellia flabellata*, *J. C. S. Chem. Commun.* 435–456.

SIVAKUMAR, M., KUMAR, T. S., SHANTHA, K. L., RAO, K. P. (1996), Development of hydroxyapatite derived from Indian coral, *Biomaterials* **17**, 1709–1714.

SMITH, D. R., DOUCETTE-STAMM, L. A., DELOUGHERY, C., LEE, H., DUBOIS, J. et al. (1997), Complete genome sequence of *Methanobacterium thermoautotrophicum* deltaH: functional analysis and comparative genomics, *J. Bacteriol.* **179**, 7135–7155.

SOBEK, H., SCHMIDT, M., FREY, B., KALUZA, K. (1996), Heat-labile uracil-DNA glycosylase: purification and characterization, *FEBS Lett.* **388**, 1–4.

SODE, K., BURGESS, J. G., MATSUNAGA, T. (1996), Marine biotechnology based on marine microorganisms, *Adv. Mol. Cell Biol.* **15A**, 95–102.

SOLOMON, R. G., APPELS, R. (1999), Stable, high-level expression of a type I antifreeze protein in *Escherichia coli*, *Protein Expr. Purif.* **16**, 53–62.

SOOST, F., REISSHAUER, B., HERRMANN, A., NEUMANN, H. J. (1998), Natural coral calcium carbonate as alternative substitute in bone defects of the skull, *Mund-Kiefer-Gesichtschir.* **2**, 96–100.

STEIN, J. L., MARSH, T. L., WU, K. Y., SHIZUYA, H., DELONG, E. F. (1996), Characterization of uncultivated prokaryotes: isolation and analysis of a 40-kilobase-pair genome fragment from a planktonic marine archaeon, *J. Bacteriol.* **178**, 591–599.

SUETSUNA, K. (1998), Purification and identification of angiotensin I-converting enzyme inhibitors from the red alga *Porphyra yezoensis*, *J. Mar. Biotechnol.* **6**, 163–167.

SUGANA, M., SATO, A., IIJIMA, Y., OSHIMA, T., FURUYA, K. et al. (1991), Phomactin A: A novel PAF antagonist from a marine fungus *Phoma* sp., *J. Am. Chem. Soc.* **113**, 5463–5464.

SULLIVAN, E. R. (1998), Molecular genetics of biosurfactant production, *Curr. Opin. Biotechnol.* **9**, 263–269.

TAKAHASHI, C., NUMATA, A., ITO, Y., MATSUMURA, E., ARAKI, H. et al. (1994), Leptosins, antitumor metabolites of a fungus isolated from a marine alga, *J. Chem. Soc. Perkin. Trans.* **1**, 1859–1864.

TAKAHASHI, C., NUMATA, A., YAMADA, T., MINOURA, K., ENOMOTO, S. et al. (1996): Penostasins, novel cytotoxic metabolites from a *Penicillium* species separated from green alga, *Tetrahedron Lett.* **37**, 655–658.

TAKEBAYASHI, Y., POURQUIER, P., YOSHIDA, A., KOHLHAGEN, G., POMMIER, Y. (1999), Poisoning of human DNA topoisomerase I by ecteinascidin 743, an anticancer drug that selectively alkylates DNA in the minor groove, *Proc. Natl. Acad. Sci. USA* **96**, 7196–7201.

TAKEYAMA, H., TAKEDA, D., YAZAWA, K., YAMADA, A., MATSUNAGA, T. (1997), Expression of the eicosapentaenoic acid synthesis gene cluster from *Shewanella* sp. in a transgenic marine cyanobacterium, *Synechococcus* sp., *Microbiology* **143**, 2725–2731.

TAKIKAWA, M., UNO, K., OOI, T., KUSUMI, T., AKERA, S. et al. (1995), Study on the marine natural products and synthetic model compounds having neridae-cidal activity, *Tennen Yuki Kagobutsu Toronkai Koen Yoshishu* **37**, 660–665 (*Chem. Abstr.* **124**, 170134).

TAMAGNINI, P., TROSHINA, O., OXELFELT, F., SALEMA, R., LINDBLAD, P. (1997), Hydrogenases in *Nostoc* sp. strain PCC 73102, a strain lacking a bidirectional enzyme, *Appl. Environ. Microbiol.* **63**, 1801–1807.

TAN, L. T., WILLIAMSON, R. T., GERWICK, W. H., WATTS, K. S., MCGOUGH, K., JACOBS, R. (2000), *cis,cis*- and *trans,trans*-ceratospongamide, new bioactive cyclic heptapeptides from the Indonesian red alga *Ceratodictyon spongiosum* and symbiotic sponge *Sigmadocia symbiotica*, *J. Org. Chem.* **65**, 419–425.

TER HAAR, E., KOWALSKI, R. J., HAMEL, E., LIN, C. M., LONGLEY, R. E. et al. (1996), Discodermolide, a cytotoxic marine agent that stabilizes microtubules more potently than taxol, *Biochemistry* **35**, 243–250.

THIEL, T., LYONS, E. M., ERKER, J. C. (1997), Characterisation of the genes for a second Mo-dependent nitrogenase in the cyanobacterium *Anabaena variabilis*, *J. Bacteriol.* **179**, 5222–5225.

TRISCHMAN, J. A., TAPIOLAS, D. M., JENSEN, P. R., DWIGHT, R., FENICAL, W. et al. (1994), Salinamides A and B: Anti-inflammatory depsipeptides from a marine streptomycete, *J. Am. Chem. Soc.* **116**, 757–758.

TSUCHIYA, N., SATO, A., HARUYAMA, H., WATANABE, T., IIJIMA, Y. (1998), Nahocols and isonahocols, endothelin antagonists from the brown alga *Sargassum autumnale*, *Phytochemistry* **48**, 1001–1011.

TSYGANKOV, A. A., HIRATA, Y., MIYAKE, M., ASADA, Y., MIYAKE, J. (1993), Hydrogen evolution photosynthetic bacterium *Rhodobacter sphaeroides* RV immobilized on porous glass, in: *New Energy Systems and Conversion*, pp. 229–233. Universal Academy Press.

VALOTI, G., NICOLETTI, M. I., PELLEGRINO, A., JIMENO, J., HENDRIKS, H. et al. (1998), Ecteinascidin-743, a new marine natural product with potent antitumor activity on human ovarian carcinoma xenografts, *Clin. Cancer Res.* **4**, 1977–1983.

VAN DER WIELEN, L. A. M., CABATINGAN, L. K. (1999), Fishing products from the sea-rational downstream processing of marine bioproducts, *J. Biotechnol.* **70**, 363–371.

VLACHOS, V., CRITCHLEY, A. T., VON HOLY, A. (1996), Establishment of a protocol for testing antimicrobial activity in southern African macroalgae, *Microbios* **88**, 115–123.

VOLESKY, B. (1994), Advances in biosorption of metals: selection of biomass types, *FEMS Microbiol. Rev.* **14**, 291–302.

VOURLOUMIS, D., HOSOKAWA, S. (1998), Total synthesis of sarcodictyins A and B, *J. Am. Chem. Soc.* **120**, 8661–8673.

WALKER, J. D., COLWELL, R. R., PETRAKIS, L. (1975), Evaluation of petroleum-degrading potential of bacteria from water and sediment, *Appl. Microbiol.* **30**, 1036–1039.

WARREN, C. J., MUELLER, G. M., MCKNOWN, R. L. (1992), Ice crystal growth suppression polypeptides and methods of preparation. *U.S. Patent* 5 118 792.

WATANABE, K., ISHIKAWA, C., INOUE, H., CENHUA, D., YAZAW, K., KONDO, K. (1994), Incorporation of exogenous docosahexaenoic acid into various bacterial lipids, *J. Am. Oil Chem. Soc.* **71**, 325–330.

WEETE, J. D., KIM, H., GANDHI, S. R., WANG, Y., DUTE, R. (1997), Lipids and ultrastructure of *Thraustochytrium* sp. ATCC 26185, *Lipids* **32**, 839–845.

WEINER, R. M. (1997), Biopolymers from marine prokaryotes, *TIBTECH* **15**, 390–394.

WEINER, R. M., COLWELL, R. R., JARMAN, R. N., STEIN, D. C., SOMERVILLE, C. C., BONAR, D. B. (1985), Applications of biotechnology to the production, recovery and use of marine polysaccharides, *Biotechnology* **3**, 899–902.

WEINER, R. M., MANYAK, D. M. (1995), *U.S. Patent* 5 587 313.

WEINHEIMER, A. J., SPRAGGINS, R. L. (1969), The occurrence of two new prostaglandin derivatives (15-epi-PGA_2 and its acetate, methyl ester) in the gorgonian *Plexaura homomalla*. Chemistry of Coelenterates, *Tetrahedron Lett.* **15**, 5185–5188.

WENDER, P. A., HINKLE, K. W., KOEHLER, M. F., LIPPA, B. (1999), The rational design of potential chemotherapeutic agents: synthesis of bryostatin analogues, *Med. Res. Rev.* **19**, 388–407.

WEST, L. M., NORTHCOTE, P. T., BATTERSHILL, C. N. (2000), Peloruside A: a potent cytotoxic macrolide isolated from the New Zealand marine sponge *Mycale* sp., *J. Org. Chem.* **65**, 445–449.

WILSON, T., HASTINGS, J. W. (1998), Bioluminescence, *Annu. Rev. Cell Dev. Biol.* **14**, 197–230.

WRIGHT, A. E. (1998), Isolation of marine natural products, *Methods Biotechnol.* **4**, 365–408.

WRIGHT, A. E., FORLEO, D. A., GUNAWARDANA, G. P., GUNASEKERA, S. P., KOEHN, F. E., MCCONNELL, O. J. (1990), Antitumor tetrahydroisoquinoline alkaloids from the colonial ascidian *Ecteinascidia turbinata*, *J. Org. Chem.* **55**, 4508–4512.

XU, Y., YAKUSHIJIN, K., HORNE, D. A. (1997), Synthesis of C11N5 marine sponge alkaloids: (+/−)-hymenin, stevensine, hymenialdisine and debromohymenialdisine, *J. Org. Chem.* **62**, 456–464.

XU, J. Y., PFEFFERKORN, J., KIM, S. (1998), Total synthesis of eleutherobin and eleuthosides A and B, *J. Am. Chem. Soc.* **120**, 8674–8680.

YAGUCHI, T., TANAKA, S., YOKOCHI, T., NAKAHARA, T., HIGASHIHARA, T. (1997), Production of high yields of docosahexaenoic acid by *Schizochytrium limacinum* SR21, *J. Am. Oil Chem. Soc.* **74**, 1431–1434.

YAKIMOV, M. M., GIULIANO, L., BRUNI, V., SCARFI, S., GOLYSHIN, P. N. (1999), Characterization of Antarctic hydrocarbon-degrading bacteria capable of producing bioemulsifiers, *New Microbiol.* **22**, 249–256.

YAYANOS, A. A. (1995), Microbiology to 10,500 meters in the deep sea, *Annu. Rev. Microbiol.* **49**, 777–805.

YAZAWA, K. (1996), Production of eicosapentaenoic acid from marine bacteria, *Lipids* **31**, 297–300.

YOKOCHI, T., HONDA, D., HIGASHIHARA, T., NAKAHARA, T. (1998), Optimisation of docosahexaenoic acid production by *Schizochytrium limacinum* SR21, *Appl. Microbiol. Biotechnol.* **49**, 72–76.

YOKOYAMA, A., IZUMIDA, H., MIKI, W. (1994), Production of astaxanthin and 4-ketozeaxanthin by the marine bacterium, *Agrobacter aurantiacum*, *Biosci. Biotech. Biochem.* **56**, 1842–1844.

YONGMANITCHAI, W., WARD, O. P. (1991), Growth of and omega-3 fatty acid production by *Phaeodactylum tricornutum* under different culture conditions, *Appl. Environ. Microbiol.* **57**, 419–425.

YOON, H. S., GOLDEN, J. W. (1998), Heterocyst pattern formation controlled by a diffusible peptide, *Science* **82**, 935–938.

YOSHIKAWA, K., NAKAYAMA, Y., HAYASHI, M., UNEMOTO, T., MOCHIDA, K. (1999), Korormicin, an antibiotic specific for Gram-negative marine bacteria, strongly inhibits the respiratory chain-linked Na^+-translocating NADH:quinone reductase from the marine *Vibrio alginolyticus*, *J. Antibiot.* **52**, 182–185.

YU, C. M., CURTIS, J. M., WALTER, J. A., WRIGHT, J. L. C., AYER, S. W. et al. (1996), Potent inhibitors of cysteine proteases from the marine fungus *Microascus longirostris*, *J. Antibiot.* **49**, 395–397.

ZABORSKY, O. R. (1999), Marine bioprocess engineering: the missing link to commercialization, *J. Biotechnol.* **70**, 403–408.

16 Biotechnology in Plant Protection

FRANK NIEPOLD

Braunschweig, Germany

KLAUS RUDOLPH

Göttingen, Germany

1 Introduction

Biotechnology is defined by the European Federation of Biotechnology as "the integrated use of biochemistry, microbiology, and engineering sciences in order to achieve technological (industrial) applications of the capabilities of microorganisms, cultured tissue cells, and parts thereof"; and: "the term biotechnology includes the microbiological, biochemical and gene technological application of methods to change existing traits of enzymes, cell cultures, and microorganisms to obtain the desired properties" (AUST et al., 1991).

Several years ago a typical example of biotechnology in plant pathology would have been the production and application of pheromones functioning as attractants or repellents of insects. Nowadays, biotechnology is greatly influenced by the application of recombinant DNA or genetic engineering. The emerging subject of "gene technology" has become the dominant part of modern biotechnology, since it allowed an increased output by unraveling the mechanisms involved and pinpointing the efforts to the actual needs. By DNA transfer, plants can be created which may express genes from bacteria, fungi, or even insects to protect them at least partially against the attack of plant pathogens. Thus, disease-resistant cultivars can now be developed not only by conventional breeding methods but also by genetic engineering.

Earlier, naturally occurring microorganisms were used to control plant pathogenic organisms, but increasingly genetically engineered microorganisms or transgenic plants are being generated that express specific defense reactions towards pathogens. Since recently many books have been published on the broad field of biotechnology, this chapter will concentrate on biotechnological approaches to improve plant resistance towards pathogens. Finally, possible new strategies will be discussed. To actualize this review, contributions from the ISPMB (International Society for Plant Molecular Biology, Université Laval, Québec) Meeting in Canada have been included to cover the latest approaches of plant pathology.

Plant pathology in the broad sense includes diseases and pests caused by an array of diverse viruses, bacteria, fungi, nematodes, and insects. During a successful infection process the plant pathogen multiplies considerably and the host plant is partially or totally destroyed. These severely damaged plants are called susceptible. Resistant plants express defense reactions against the attacking pathogens, manifested by specific biochemical, physiological or anatomical features.

A traditional measure of plant protection against diseases and pests is – besides others – the application of chemical plant protection products or pesticides. However, according to THOMAS (1999) 30 to 40% of the yearly harvest worldwide is destroyed by plant pathogens although plant protection products are used. This tremendous loss of crops cannot be tolerated in the future in view of an ever increasing world population.

Also, the growing concerns on the harmful side effects of chemical pesticides on environment and human health have drawn the attention to safer, environmentally friendlier disease control measures. In this respect, the fast developing field of biotechnology may offer effective alternatives.

In several cases, plant protection products can no longer be applied because plant pathogens developed resistance against these chemicals, or the application of plant protection products was no longer allowed by the authorities because of harm to the environment.

Therefore, it can be anticipated that biotechnology will gain a key position in plant protection for this century; also because the development of new chemical plant protection products is getting increasingly costly. Several efforts have already been made to reduce application of plant protection products by different approaches, either by transferring a new gene into a cultivar resulting in resistance to a certain plant pathogen or by generating new types of plant protection products against pathogens which have been deduced from compounds secreted by microorganisms.

In order to achieve the required biotechnological results, a major part of this new area in plant pathology is the application of gene technology. After the desired gene has been constructed and integrated into the plant, the proper insertion into the plant genome has to be confirmed.

Another field of biotechnology in plant pathology is the development of techniques which allow a faster and more precise detection of plant pathogens, i.e., plant pathogenic viruses, bacteria, fungi, and nematodes.

Biotechnology in plant protection also deals with the cultivation of beneficial organisms to combat plant pathogenic bacteria, fungi, nematodes, or insects. Since this review article will address non-specialists in the area of plant pathology the key biological features of the main areas in plant pathology are introduced and explained by examples.

The term "biopesticide" is a general term including biochemical, microbial, and plant pesticides. Microbial pesticides are living organisms used as pesticides, i.e., viruses, bacteria, or fungi. Biochemical pesticides are naturally occurring pesticidal substances or their analogs, either with a toxic or with a non-toxic mode of action against the target insect or pest (e.g., pheromones and other semiochemicals used for mating disruption).

The application of biotechnology in plant pathology will be very helpful for the demands of an ever growing world population, which will increase from now 6 billion to 8 billion people within the next 30 years. Since nearly all tillable land is under cultivation, at present, food supply can only be improved by higher yields and more effective plant protection. Application of biotechnology may also improve food quality, preserve and extend existing natural resorts, which otherwise would have to be cultivated for farming (PINSTRUP-ANDERSEN and PANDYA-LORCH, 1999).

This chapter can only be an introduction into the topic of biotechnology in plant pathology and discuss the latest achievements in this field by selected examples. Since the main tools of biotechnology are recombinant DNA techniques, these techniques are not described here in detail. However, several literature quotations will help the interested reader to find more information on recent DNA techniques in gene technology.

2 Biocontrol of Plant Pathogens or Weeds by Living Organisms

Commercial applications of microorganisms in biocontrol will only be successful if (1) they are similar in efficacy to chemical pesticides, or (2) in specific situations, e.g., where chemicals are not allowed to be used (like in organic farming), in cases of fungicide resistance, or if no pesticides are authorized for a particular use. Generally, these organisms should have high virulence and a multiplication rate high enough to survive and spread under the conditions of use (KOCH, personal communication).

2.1 Bacteria Used for Disease Control

A major principle of biocontrol of plant diseases by bacteria is **occupation of ecological niches** in a manner of first comes first serve. This mechanism of occupying sites in or on plants, to prevent the settlement of plant pathogenic bacteria (competition), has early been tested, e.g., with *Pseudomonas putida* on potatoes, preventing an attack of *Erwinia carotovora*, the causal agent of black leg and soft rot of potatoes (XU and GROSS, 1986).

Some bacteria produce compounds which kill other microorganisms or inhibit their growth. Generally, these compounds (i.e., antibiotics) are products of secondary metabolism. One of the most successfully applied growth inhibitors in biological control is the **antibiotic Agrocin** secreted by the *Agrobacterium radiobacter* strain K84. Next to the ability of *Agrobacterium radiobacter* strain K84 to compete for the invasion sites of a host plant, Agrocin inhibits the growth of the plant pathogenic bacterium *Agrobacterium tumefaciens*, causing crown gall disease on woody plants (RAYDER and JONES, 1990). Therefore, since many years crown gall on stone fruit, rose, walnut, pecan, and *Euonymus* is controlled routinely by dipping planting material in a cell suspension of *Agrobacterium radiobacter* strain K84. However, the same disease on grapevine, apple, chrysanthemum, and blackberry is not controlled by this method (KERR et al., 1990).

A group of growth inhibitors produced by different bacteria are the **siderophores**. Siderophores are best studied in pseudomonads. The strategy is that the siderophores chelate iron which is essential for all microorganisms. The chelated iron can be taken up by siderophore producing bacteria but not by other microorganisms including plant pathogenic fungi (LOPER and ISHIMARU, 1991). This protection system is active in many **rhizobacteria** protecting roots from pathogens.

Several other anti-fungal compounds have been suggested to be involved in suppression of phytopathogens by rhizobacteria. Examples are *Pseudomonas fluorescens* suppressing *Gäumannomyces graminis* in wheat, and *Bacillus subtilis* controlling common scab of potatoes caused by the actinomycete *Streptomyces scabies* (CHET and INBAR, 1997). But also, several non-plant pathogenic *Streptomyces* species produce antibiotics which may potentially be exploited in biocontrol of fungal and bacterial diseases and even insects (microbial acaricides). The reader is referred to COPPING (1998) for further information.

2.2 Fungi Used for Disease Control

Many economically important plant diseases are caused by fungi. In agriculture, plant pathogenic fungi are usually controlled by **fungicides**. However, compounds extracted from naturally occurring fungi can also be used to protect plants against fungal attacks.

Fungal organisms exerting inhibitory effects on plant pathogens are for instance *Coniothyrium minitans*, *Gliocladium* sp., and *Trichoderma* sp. (CHET and INBAR, 1997). The mechanisms involved in biocontrol of plant pathogenic fungi are manifold. In the case of **antibiosis**, diffusible compounds are produced which inhibit the growth of the plant pathogen. Fungi with a strong competing ability may cause other microorganisms to starve by limiting their nutrition or other substantial growth factors.

Mycoparasitism is a direct attack on fungal organs followed by the utilization of their nutrients. This mechanism is the active principle of control of the plant pathogen *Sclerotinia sclerotiorum* with *C. minitans*. After its application to the soil, *C. minitans* parasitizes the resting structures (sclerotia) of its fungal host. Mycoparasitism is best documented in *Trichoderma* species. It could be shown that mycoparasitism is a complex process and involves a directed growth of *Trichoderma* toward the plant pathogenic fungus. Here, apparently a chemotropism takes place and the excretion of fungal cell wall lysing enzymes (chitinase, glucanase, protease) occurs, followed by the penetration of the *Trichoderma* hyphae (CHET and INBAR, 1997).

Several *Trichoderma* species have the ability to protect plants against soil-borne fungi like *Pythium ultimum*. In addition, specific *Trichoderma* strains have a plant growth-promoting effect and also protect the plant significantly from infection by other pathogens (NASEBY et al., 2000). Since these *Trichoderma* strains are naturally occurring strains, changes within the genome of the fungus might be of interest to increase their effectiveness. There are a lot of other fungi which have been used for disease control and the reader is referred to COPPING (1998) for more details.

Mycorrhiza fungi comprise a wide spectrum of species associated with plant roots. They have also been shown to control plant pathogens attacking roots (MUKERJI, 1999). One effect of mycorrhiza is a general growth stimulation resulting in improved plant resistance against pathogens. Apparently, colonization of roots by mycorrhiza fungi sites triggers mechanisms of plant defense such as stimulation of the secondary (phenolic) pathway of the plants (formation of phytoalexins) and activation of defense related genes coding for callose, peroxidase, chitinase, or other pathogenesis related proteins the function of which is not yet fully understood. These effects are part of the so-called **Systemic Acquired Resistance** which will be dealt with later. A challenging strategy is the biotechnological production of vesicular arbuscular mycorrhiza fungi to reduce plant diseases (MUKERJI, 1999), and further results for application in practical agriculture can be expected from this approach in the future.

2.3 Biocontrol of Plant Pathogenic Nematodes

Plant pathogenic nematodes are microscopically small roundworms which feed on their specific hosts. Nearly all cultivated plants can be infected by nematodes. Nematodes usually live in the soil and attack plants via their roots. Tremendous damage is caused worldwide by nematodes (GRUNDLER, personal communication).

A well known method to protect plants against attacks by nematodes is the use of **nematophagous fungi**, which infect nematodes. Several species with different infection mechanisms have been described: *Arthrobotrys oligospora* and *Dactyliaria candida* trap nematodes mechanically by hyphal loops, *Pleurotus pulmonaris* lives endoparasitically, destroying the intestinal tract of nematodes, or toxin producing fungi and fungi infecting only the eggs of the nematodes (*Verticillium* ssp., *Drechmeria coniospora*; BELLOWS, 1999). These fungi have been biotechnologically used for a long time. Another promising strategy to control nematodes in future might be the application of **toxins** and **antibiotics** (nematicidal fatty acids), produced by the nematophagous fungi. These compounds can be biotechnologically synthesized and can be spread onto nematode-contaminated fields (BELLOWS, 1999). With a more even spreading of these compounds the nematodes will be better controlled.

The effect of the nematophagous fungi can be enhanced by increasing the amount of compounds which cause the death of the nematodes (cutinases, proteases, and hydrolytic enzymes).

Controversially discussed is the controlling ability of the **vascular arbuscular mycorrhiza**, which seem to reduce the attractiveness of host plant roots for nematodes (MUKERJI, 1999).

Also, nematodes can be diseased by **bacteria**. For instance, the genus *Pasteuria* (BELLOWS, 1999) infects a large number of nematode species but does not attack other soil organisms, representing a **true obligate parasite** for nematodes. The bacterium produces spores (survival organs) which could be spread onto plants artificially in order to destroy plant pathogenic nematodes. *Pasteuria* spores have the advantage to withstand a wide range of temperatures and to remain viable in soil for more than 6 months. The drawback, however, is that this bacterium is an obligate parasite, so that the only way to generate spores is to multiply the bacterium in nematodes. This complication reduces the feasibility of this system for biotechnological purposes.

Another possibility for biological control of nematodes might be the use of **hyper-parasitic nematodes**. Four taxonomic groups of nematodes are known as **predators** to the phytopathogenic nematodes: Monochidae, Dorylaimidae, Aphelenchidae and Diplogasteridae. Some of these nematodes are omnivores, feeding on all kinds of nematodes but also on protozoa, bacteria, and other prey. However, it has been difficult so far to use this approach as a plant protection measure (COPPING, 1998; BELLOWS, 1999; ATKINSON et al., 1998).

Recently, experiments were carried out to utilize the naturally occurring resistance or the expression of genes which lead to the starvation of nematodes. During evolution several **genes with resistance against nematodes** (*Hs1pro1*, *Mi*) emerged, which are usually found in wild plants having no commercial value. The nematode-resistance genes cause the infected areas of the plant to die off much faster than the nematode is able to infect (JONES et al., 2000). The gene responsible for this reaction has been identified (ARMSTRONG et al., 2000) and similar genes are in the process of identification in potatoes (COPPOOLSE et al., 2000) and soybeans (DI MAURO et al., 2000). Recently, DNA was transferred into sweet potato coding for a cysteine containing protease inhibitor (CIPRIANI et al., 2000), acting as an **anti-feedant** killing nematodes by starvation. Efforts are ongoing to improve the starving effect by changing the coding DNA of this cysteine protease inhibitor (HARRISON and MAC PHERSON, 2000).

2.4 Control of Weeds by Bioherbicides

The artificial increase (mass production) of high inoculum levels of a **specific pathogen** to a target weed is often referred to as bioherbicide (controlling weeds). Until now, the success of bioherbicides has been limited, compared to chemical herbicides. However, the increasing occurrence of weed genotypes with resistance towards chemical herbicides makes the application of bioherbicides more attractive in the future in the frame of an integrated weed management.

Most of the bioherbicides damage leaves and are represented by the fungal or "**myco**"-**herbicides**. The mycoherbicides can be divided into two groups: (1) obligate parasites, colonizing their hosts by penetrating the plant cells and forming haustoria (organs for the nutritional uptake) within them (the rust, powdery and downy mildew fungi); (2) the non-obligate fungi, invading the intercellular space of the leaves by excreting toxins and hydrolytic enzymes. The most abundantly tested fungi on weeds are *Colletotrichum* sp., *Fusarium* sp., *Alternaria* sp., *Cercospora* sp., *Phoma* sp., and *Phomopsis* sp. (BOYETCHKO, 1999).

Weeds can also be damaged by **leaf spot causing bacteria**. Thus, a certain pathovar (*poae*) of *Xanthomonas campestris* is used to control a grass weed (annual blue grass) in cultivated Bermuda grass. The infection of the weed through wounds is improved by mowing during application of the bioherbicide. Since the host spectrum of this bacterium is very narrow, no simultaneous infection of the bermuda grass occurs (BOYETCHKO, 1999).

Weeds can also be controlled by infecting the roots with **soil-borne fungi** and thus reducing the competitiveness of the weeds. One of the most common applications is the fungus *Phytophthora palmivora* which is used to control stranglervine in citrus. But also other fungi, such as *Fusarium* sp., *Sclerotinia* sp., and *Gliocladium* sp., are used in controlling weeds of different crops (BOYETCHKO, 1999).

Deleterious rhizobacteria (bacteria, associated with roots) are also applied for biological weed control (TILAK et al., 1999). Bacterial species belonging to pseudomonads, flavobacteria, erwinias, and *Alcaligenes* comprise this group. These bacteria can easily be applied to soils as a granulate or as a seed treatment. The mode of action is a decrease of the weed emergence and a delay in the growth of the weeds. Usually, secondary products produced by the rhizobacteria are controlling the weeds, and there is an almost abundant source of these bacteria found in all types of soils everywhere (BOYETCHKO, 1999).

A very **crucial point** of applying bioherbicides is the pathogenic stability of the microorganism to the target weed. It would be detrimental when the host range of the bioherbicide would enlarge by mutation or selection and also include the cultivated crop.

Natural bioherbicidal substances are synthesized in the secondary metabolic pathway of certain microorganisms and include compounds such as curvulins, eremophilanes, or bialaphos (bilanafos). In the future, such compounds can be regarded as a potential to generate new herbicides because weeds are developing more and more resistance against the classical chemical herbicides (ROSSKOPF et al., 1999).

In order to further enhance the bioherbicidal activity it might be helpful to identify genes from microorganisms responsible for virulence, coding for synthesis of certain enzymes, phytotoxins or the competitiveness of the bioherbicidally active microorganisms against the indigenous soil population. For instance, an insertion of cutinase genes into these microorganisms might better destroy the cutin layer of the weed leaves, increasing the active process of weed destruction. However, care must be taken that the bioherbicidally active microorganisms do not widen their host range so that also the cultivated crop is infected.

Another problem is the **preservation of bioherbicides** when spraying onto the weeds because the viability of the microorganisms has to be ensured by using preserving chemicals, such as alginate. Alginate is now used for the encapsulation processes when pelleting plant pathogenic fungi to prevent the dry-out of the microorganisms (ROSSKOPF et al., 1999). As a kind of by-product during pathogenicity studies of plant pathogenic pseudomonads, alginate was found (RUDOLPH, 1995). To supply the increasing demand in several industrial branches, this polysaccharide is extracted from sea-algae, but it can also be produced biotechnologically by bacterial cultures.

A new group of bioherbicides might arise from the use of weed infective microorganisms: some of them produce compounds which are specifically toxic to certain weeds. Therefore, biotechnology can help to develop synthetic herbicides using the naturally occurring weed toxins as templates.

3 Plants Transformed with Bacterial, Fungal or Non-Indigenous Genes Reducing Diseases Caused by Fungi or Bacteria

In several cases, transgenic plants have been produced which synthesize **specific compounds** directed against phytopathogens. One example are transgenic pear trees producing lactoferrin which inhibits the growth of the fire blight pathogen, *Erwinia amylovora*, by competing with the iron chelating bacterial siderophores (MALNOY et al., 2000).

Another approach to protect plants from bacterial attacks is the utilization of **bacterial pathogenicity factors** of the bacteria, like enzymes which are involved in the destruction of plant cell walls. A gene coding for the pectate lyase of *Erwinia carotovora* has been integrated into potato plants which apparently reduces the outbreak of the soft rot disease by preactivating a certain plant defense mechanism (LEE et al., 2000b; WEGENER, 1999).

On the other hand, the strategy of using **non-plant-indigenous**, **synthetic compounds** derived from insects (melettines, cercropins) or amphibians (magainins; DESTÉFANO-BELTRÁN et al., 1993) is increasingly followed up. These compounds are surface active substances destroying the cell walls of plant pathogens. They have been integrated into tobacco to exhibit resistance to *Erwinia* ssp. (REYNOIRD et al., 2000), and after integration of these compounds into potato plants a broad-spectrum resistance against bacteria and fungi was reported (MISRA et al., 2000).

Cysteine-rich, synthetic antimicrobial peptides derived from naturally occurring precursors in plants have been shown to be effective against the attack of bacteria and fungi, when transformed into cotton (RAJASEKARAN et al., 2000).

The **enzyme lysozyme** can degrade the bacterial cell wall. The gene coding for lysozyme has been cloned into several host plants, including oil seed rape (from the bacteriophage T4 DNA; STAHL et al., 2000), wheat (from chicken DNA; BOURDON et al., 2000), tobacco (from human DNA; NAKAJIMA et al., 1997), or potato (from the bacteriophage T4; DÜRING et al., 1993) to prevent diseases caused by phytopathogenic bacteria. In the latter case, it has been reported that the transgenic potato plants possessed resistance against *Erwinia carotovora* (DÜRING et al., 1993). However, according to results of MAVRIDIS and RUDOLPH (1996), only a few (7 strains from 82 tested) proved to be weakly sensitive to 50 ppm lysozyme. The gram-positive phytopathogenic bacterium *Clavibacter michiganensis* ssp. *sepedonicus*, on the other hand, was very sensitive towards the lysozyme *in vitro*. But even in this case, the transgenic potato plants did not reveal a significant resistance against the disease. This example may indicate that before announcing transgenic plants with asserted resistance, a broad spectrum of different strains of the pathogen should be tested under natural conditions of disease incidence.

Another **enzyme** is the **glucose oxidase**, where the coding gene from the fungus *Aspergillus niger* was used to produce hydrogen peroxide which can play a role in controlling phytopathogenic bacteria. This gene has been cloned into cabbage to prevent infections by *Xanthomonas campestris* (LEE et al., 2000a). The gene coding for the enzyme chitinase of *Trichoderma* has been inserted into a number of plants, and in the case of grape it was found to give protection against mildew infections (KIKKERT et al., 2000).

In **conclusion**, the utilization of bacteria and fungi offers promising alternatives to control plant diseases. However, until now not enough information is available on a large-scale application and on the effectiveness of these measures under natural conditions. More extended studies in the future have to reveal how effective the use of bacteria and fungi and their products can be.

4 Virus Resistant Plants

The most abundant research on biotechnological approaches using recombinant DNA technology has been made for generating virus resistance in different plant species (KAWCHUK and PRÜFER, 1999). Most of the viruses contain RNA which is packed into a protein coat. Early observations showed that susceptible plants could be protected by preinoculating a similar, weakly plant pathogenic virus, or even by another related virus, which causes a delay of infection by the severe virus (**cross protection**; MACKINNEY, 1929). In a similar way HAMILTON (1980) suggested that this virus induced resistance may also be obtained with parts of a virus, such as the non-infective coat protein.

Therefore, many workers integrated parts of the virus into easily transformable plants (several dicotyledons). The complementary (c) DNA coding for the **coat proteins** of several plant pathogenic viruses was integrated into a **plasmid** of a plant pathogenic bacterium, *Agrobacterium tumefaciens*, the causative agent of crown galls. This bacterium can integrate its DNA and the added virus cDNA by transferring its plasmid via a naturally occurring process into the genomic DNA of the plant. The resulting plant is called a transgenic plant.

The second widely used method is the **particle bombardment** in which the cDNA is coated onto small gold particles which are then shot into plant cells by a burst of helium gas. Eventually, cDNA gets washed off the particles inside the cell and is integrated into the plant genome. This method is well suited for cereal crops which are not easily transformable by other methods. Transformed plants obtained by this method were protected against virus or showed at least a delay in symptom development. Currently this approach is applied to a number of plants, including monocotyledons like maize (MCGIVERN et al., 2000).

A problem can arise from the so-called **heterologous encapsidation** when another virus infects the transgenic plant. Its RNA may be packed into the empty coat proteins which are generated by the transgenic plant. However, this process also occurs in nature when two different viruses infect a plant simultaneously (BEACHY and BENDAHMANE, 1998). These cases are very rare and are subject to biosafety research, because the ecological concerns regarding these recombinations are very high. A summary of all the coat proteins from different viruses expressed in transgenic plants can be found in KAWCHUK and PRÜFER (1999). Examples are the papaya ring spot potyvirus, potato leaf roll virus, potato Y virus, and the squash virus (COPPING, 1998).

Meanwhile several other mechanisms were found which protect plants from virus attacks. Plants were constructed genetically with implanted genes coding for **defective proteins**, interfering with the assembly, movement or replication of a virus. One such target is the DNA replicase which is needed for the amplification of the viral nucleic acid in plant cells. For instance, the construction and expression of the faulty virus replicases in plants (by excising parts of the coding region for replicase) is interfering with the replication of the virus, and, therefore, the plant is protected against the attacking virus (KAWCHUK and PRÜFER, 1999).

The virus has to be transported from one plant cell to the other ("cell to cell movement"). This transport can be interrupted by a **defective movement protein** located at the connecting pathways of each plant cell (plasmodesmata). The generation of a defective movement protein in the plants can prevent the spread of the virus from one cell to the other, resulting in an even broader resistance against several viruses (KAWCHUK and PRÜFER, 1999).

Another strategy is to generate **antisense** or **untranslatable RNA** sequences which interfere with the virus RNA. Different regions of the viral genome have been used for generating these antisense RNA strands which include coat protein, movement protein or replicase genes. Untranslatable sense and antisense RNA strands seem to disrupt the translation at the ribosomes (KAWCHUK and PRÜFER, 1999).

Heterologous genes have been chosen from a distantly related plant like pokeweed, which code for a ribosome-inhibiting antiviral protein, or a double-stranded RNA was used coding for a yeast ribonuclease (cleaving the virus

RNA). Both strategies resulted in gaining resistance to a number of plant viruses (KAWCHUK and PRÜFER, 1999).

Another approach is the use of **ribozymes**. Ribozymes are small RNA sequences which function as an enzyme and can cleave or restrict single-stranded RNA. These sequence specific ribozymes have been integrated into plants, but the same efficiency was obtained by using antisense RNA (KAWCHUK and PRÜFER, 1999).

One of the most successful applications of transgenic virus resistant crops is papaya on the islands of Hawaii. Since 1992 the transgenic papaya cultivar "SunUp" synthesizing coat protein is grown without showing a decline in resistance against the papaya ring spot potyvirus. However, this resistance is unique to the local strain from Hawaii and the plants show less resistance to other virus strains from other countries, such as Japan, Thailand, or Malaysia (MAOKA and NODA, 2000).

5 Antibodies Produced in Plants as Plant Protection Products

Usually, antibodies can be generated in vertebrates against all antigens (in our case all plant pathogenic microorganisms). For the protection of crops the advantage of antibodies was utilized by integration into the plant genome so that the plants are able to produce the so-called "**plantibodies**".

The active parts of an antibody are large molecules, the variable domains of the light and heavy chains. However, smaller fractions were found to be sufficient. These are artificial molecules in which the domains are fused by a linker sequence, the so-called single chain variable fragments. By producing plantibodies against a virus the plants may become resistant against this specific virus (FECKER and KOENIG, 1999). However, the problem of virus targeting still remains, which means that the plantibodies are produced in one location of the plant cell but are not always delivered to the place where the viruses are located. Plantibodies are also used for generating resistance against plant pathogenic fungi (HARPER and ZIEGLER, 1999) and nematodes (SMANT et al., 1999). However, reports have not been published on successful, extensive applications of plantibodies as controlling agents. Plantibodies can also be produced in transformed bacteria and plants for diagnotic purposes (HARPER and ZIEGLER, 1999).

6 Herbicide Resistant Crops

Most of the presently applied herbicides for weed control are selective by killing weeds and saving cultivated plants. The selectivity is limited to a single crop or crop group, which possesses natural tolerance to the herbicide. However, not all weeds are controlled sufficiently. Therefore, transgenic plants resistant to herbicides with a broader spectrum of weed eradication have been generated. Several new cultivars have been released. These belong to the earliest biotechnology products which were successfully grown in fields. So far, several genes were cloned into the plants: the non-selective class II *EPSP* synthetase gene (**tolerance to glyphosate**) and the phosphinothricin acetyl transferase *pat* gene (tolerance to glucosinate-ammonium) are the most widely used genes to generate herbicide resistant crops, followed by the *als* 1 gene (**tolerance to sulfonylurea**), the bromoxynil *bxn* gene, and the *IMI* gene giving **tolerance to imidazolinone** (COPPING, 1998). These herbicide resistant plants offer the advantages that almost all weeds can be controlled, including those weeds which may be already resistant to selective herbicides now being used. However, the possibilities of a transfer of herbicide resistance genes by pollination into crop-related weeds or the upgrowth of herbicide resistant plants (volunteers) during the next growing season should be considered and are closely monitored by the agencies.

So far, the wide use of herbicide resistant plants is by far the most successful application of biotechnology, compared to other fields of biotechnology in phytopathology. Globally, herbicide-tolerant soybeans occupied 21.6 million ha representing 54% of the global area of 39.9 million ha for all transgenic crops in 1999 (JAMES, 2000).

7 Mechanisms of Plants for Protection against Pathogen Attack by "Systemic Acquired Resistance"

During evolution, plants have developed an array of different resistance reactions against the attacks of pathogens. Several terminas have been coined in this regard, for instance, the pair "horizontal" and "vertical" resistance. Horizontal resistance is a broad spectrum resistance, but does not inhibit the pathogen completely. **Vertical resistance** affects a very small range of pathogens. Often, only a few biotypes (races) of a pathogen are inhibited by vertical resistance, although very effectively. When the pathogen is able to change the spectrum of biotypes or races, vertical resistance can be overcome and does no longer protect the cultivar against the pathogen.

Therefore, an induction of the broad spectrum **horizontal resistance** has been the goal of many researchers all over the world. During horizontal resistance several biochemical pathways from the secondary metabolism are activated. When this reaction ends with the death of the infected plant tissue it is called **hypersensitive reaction** (HR), and the dead plant tissue does not allow the plant pathogens to further multiply by excluding the entrapped microorganisms from the rest of the living plant.

In aiming to improve the plants' disease resistance by biotechnological means the elicitation of HR does not appear suitable because it may lead to yield losses by reducing the assimilative plant tissue. However, a general stimulation of the horizontal resistance without leading to HR may be a successful strategy to improve the plants' resistance by biotechnology.

When the defense mechanism is activated throughout the entire plant, it is called **systemic acquired resistance** (SAR) or induced SAR. This effect was first observed by RAY and BEAUVERIE in 1901 (CHESTER, 1933). With this process many different plant pathogens can be unspecifically inhibited (KUĆ, 1987), giving the plant resistance against the attack of viruses, bacteria, fungi, and even insects (CHEN et al., 2000).

The effect is accomplished by the release of biostatic or biocidal compounds which are metabolized from non-toxic precursors to toxic metabolites (e.g., phenolics, isoprenes, polyacetylenes). The so-called **phytoalexins**, substances which inhibit the growth of certain microorganisms and which are produced in higher plants in response to biological and even chemical and physical stimuli, are associated with the general and specific resistance of plants (KEEN and BRUGGER, 1977).

Other biochemical reactions during development of resistance may be the phenomenon of "oxidative burst", i.e., the release of free radicals (including hydrogen peroxide), or the synthesis of new proteins, the so-called "pathogenesis related proteins", some of which show cutinase activity and may destroy plant pathogenic microorganisms (MITTLER et al., 1999) or release elicitors of disease resistance.

An **artificial induction** of systemic resistance against various plant pathogens has been achieved by previous treatments with a variety of chemical compounds or by inoculating non-pathogenic or even pathogenic microorganisms (RYALS et al., 1991). The inducing compounds are usually referred to as **elicitors**, that differ widely in their chemical nature. Elicitors can be proteins, oligosaccharides, glycoproteins, or lipids (GUPTA and MUKERJI, 1999). Even with compost or compost-water resistance against bacterial diseases was induced in cucumber (ZHANG et al., 1998).

SAR requires the activity of **signaling compounds**, such as salicylic acid (SA) or other chemically similar compounds (PIETERSE and VAN LOON, 1999). Only nanograms of SA produced artificially by *Pseudomonas aeruginosa* can activate the SAR in beans (DE MEYER et al., 1999), even though plant pathogenic bacteria are known to induce the synthesis of jasmonate, a plant hormone, which is involved in the activation of wound-responsive genes. The SAR was found to be initiated by aphids in cotton (LI et al., 2000). When an aphid was feeding on one cotton leaf it was found that molecular signals were emitted from there and that the defense mechanism was switched on, preventing further aphids from feeding.

The induction of the SAR is an alternative approach to conventional breeding to create resistance of plants against bacteria and fungi (DATTAROY et al., 2000). Two examples will be described here: the application of harpins and Bion.

Harpins belong to a group of proteins which are produced by several phytopathogenic bacteria (*Erwinia amylovora*, *Pseudomonas syringae* pathovars) and elicit the hypersensitive reaction (HR) in plants (WEI et al., 1992). Harpins are responsible for eliciting a complex natural defense mechanism in plants, when applied onto the leaf surface. After penetration harpins are bound to receptors in the plant, where a resistance is induced giving the plant a long lasting protection (*http://www.news.cornell.edu/releases/April00/FireBlight.bpf.html*). The U.S. Environmental Protection Agency has registered these proteins for commercial agricultural use and the commercial name is "messenger tm" (*http://www.epa.gov/pesticides*). Harpins do not directly kill plant pathogens but activate natural resistance without disturbing the balance of the plants' normal physiological mechanisms. Trials have been performed in the US to control bacterial spot disease on pepper, and the treatment with harpins was more effective than the application of copper. Also in fruit orchards application of harpins successfully controlled the fire blight causing bacterium *Erwinia amylovora* because plant resistance reactions were activated before an infection with the bacterium took place. In addition, harpins reduce infection by several other plant pathogens, such as fungi, viruses, and even insects. As a pleasant side effect, it is reported that harpins can increase the seed germination, resulting in earlier flowering and ripening by also obtaining higher yields and better qualities. Only two grams of harpins are necessary for the treatment of one acre, and because of its protein character harpins are degraded in soils very quickly. In trials in the US, the overall application of fungicides, bactericides, and insecticides could be reduced by 71%, while the overall marketable yield of the tested crops was improved by about 10–20% when harpins were applied, being an advancement for integrative plant protection (*http://www.epa.gov/oppbppd1/PESP/publications/vol3no1.htm*).

An extensive search for synthetic compounds which act as resistance inducers has led to the release of "**Bion**" by Novartis Co. This so-called plant resistance activator, Acibenzolar-S-methyl or benzothiadiazole (BTH), has shown effects in several host/parasite-relationships. Thus, Bion can be applied to prevent mildew in winter wheat (STADNIK and BUCHENAUER, 1999). However, due to the general effect on plant metabolism, treatment with Bion may lead to lowered yields when the inoculum pressure by pathogenic fungi is not very high. So far, SAR is mostly induced by Bion in minor crops controlling minor diseases where no chemical protectives are available. It is thought that a larger potential of a widespread use exists as a complementation of the conventional chemical plant protection reducing the amount of the applied pesticides (CUPPELS et al., 2000). Since the induction of SAR is temperature dependent, there is only a narrow window of optimal weather conditions when the application of Bion is successful. Another restriction is that the resistance inducer has to be applied before attack of a pathogen.

8 Use of Beneficial Arthropods and Sterile Insect Release

Beneficial arthropods are either parasitoids or predators of pest insects. In most cases, the application of beneficial arthropods implies also biotechnological methods, since the arthropods have to be produced in large quantities to be released in the field. One of the most commonly used techniques is the mass propagation of the insects in laboratories by feeding them on their natural host or prey (e.g., ***Trichogramma* subspecies**, ELZEN and KING, 1999). This is rather laborious since, first of all, the pest insect has to be mass produced. Therefore, several laboratories now work on the development of synthetic diets which may, in future, facilitate the mass production (HUBER, personal communication). Biological control by beneficial insects is increasing yearly, since the ever growing application of insecticides enhances the risk that pesticide resistant insect populations develop, especially in intensively used agricultural areas, like, e.g., in greenhouses. ***Phytoseiulus persimilis*** is used to control spider mites (PARRELLA et al., 1999). In the case of resistance in the pest, the only way of insect control is the application of beneficial arthropods that do not distinguish between resistant and non-resistant pest insects.

Another way of controlling pest insects biotechnologically is the release of sterile pest insects which have been irradiated by X-rays (**sterile insect release**, SIR). Also for this technique, first, a mass production of the pest insect has to be established. SIR was successful in Canada to control the codling moth (*Cydia pomonella*); even a total eradication of the codling moth was possible (THRISTLEWOOD and JUDD, 2000).

9 Insect Pathogenic Microorganisms

Insect pathogens have been successfully used over the last decades, and many data are available on this measure of biotechnological control. Effective insect pathogenic microorganisms can reduce insect populations down to a threshold below the economic damage of the crop. In addition, the ability of some insect pathogens to multiply in their host insects may reduce the costs of the farmer, since these microorganisms will also control insect pests emerging during the same or the next vegetation period.

Pathogens causing insect diseases may be viruses, bacteria, fungi, and protozoa, but also nematodes (SMITS, 2000). One of the most frequently applied and probably the best characterized insect pathogen is the bacterium *Bacillus thuringiensis* (*B. t.*) which is effective against many different insect species. But also a virus, the *Cydia pomonella* granulo-virus, controlling the codling moth, and the fungus *Metarhizium anisopliae*, controlling beetle larvae, have been successfully applied.

A special potential is seen in those **insect pathogenic fungi** which infect the insect by actively penetrating its chitin skeleton. This implies that the fungal spores can be applied in the field similar to insecticides.

Protozoa are not yet produced commercially on a large scale. However, these organisms may become important in the future since they represent alternatives in controlling insects which are resistant to the presently applied fungi, bacteria, or viruses.

Viruses pathogenic for insects occur in different virus groups. The Baculoviridae which are the most widely used insect pathogenic viruses only attack arthropods. Within this group the viruses pathogenic to Lepidoptera, the nucleopolyhedroviruses (NPVs), consist of more than 600 known varieties and are obligate intracellular parasites of their host insects. Cell culture media have been developed to cultivate the baculoviruses biotechnologically. However, in their hosts (target insects) the propagation of the baculoviruses results in a higher output and, therefore, is usually preferred. This virus group contains double-stranded RNA, is multiplying in several tissues of the insects, and causes a disease which leads to the death of the infected insect within several weeks. This is a disadvantage since during the incubation time of the baculovirus, the insect can still feed on the crop. Most of the genetic work was done with the NPV group of the baculoviruses with the aim to increase their pathogenicity and, thus, reduce the time needed for lethality (FEDERICI, 1999a).

Although Baculoviridae occur in different insect groups the host range is limited to closely related insects and closely related varieties. Therefore, the silk moth *Bombyx mori* can only be attacked by its NPV, and no other insect species is attacked by this specific virus. The application of the baculoviruses has the advantage that a directed control of the pest insects is possible without harming other useful insects, or destroying intact ecosystems.

However, by genetically manipulating the genome of the NPVs, host range or aggressiveness may be affected. A change of only one base pair in the genome of the *Bombyx mori* NPV resulted in an extended host range (KAMITA and MAEDA, 1997).

In order to register these genetically modified baculoviruses, regulatory agencies in almost all countries are testing the harmlessness of the newly generated organism. Readers interested in this issue are referred to the respective state agencies. A thorough review of the risk assessment and the registration of a genetically engineered organism in the USA is given by ROSSKOPF et al. (1999). For Germany, information on the registration of genetically engineered organisms is provided on the homepage of the Federal Biological Research Centre for Agriculture and Forestry under *www.BBA.de*.

Bacillus thuringiensis (*B. t.*) is a spore-producing bacterium. In contrast to the Baculoviridae, it is possible to generate a high biomass by cultivating the bacterium in fermenters. For generating spores, one of the major nutrients has to be reduced in the medium, and the bacterium switches from vegetative growth (simple cell division) to cells with spores, the survival organ of the bacilli. These spores can survive in the soil for several years. In combination with the spores the bacteria produce a parasporal crystal composed of proteins which are known as the *B. t.* toxin harmful to special groups of insects after activation.

The advantage of using *B. t.* as an insecticide is that it kills the target insect very fast (within 48 h). However, the host range is not as narrow as that of the viruses. Concerns have been raised about the **safety** of the *B. t.* toxins either in bacteria or in transgenic plants regarding side effects on non-target insects. It is known that *B. t.* toxins destroying certain harmful Lepidoptera, Coleoptera, or Diptera may also affect a few other non-harmful insects. However, the negative effects on non-target insects appear to be small in comparison to the side effects of chemical insecticides most of which have a much broader mode of action than *B. t.* In addition, the *B. t.* toxins also affect predator insects feeding on pest insects. Here again, the application of *B. t.* toxins has to be compared with non-selective chemical insecticides which would kill even more predator insects.

Several inactive preforms of the *B. t.* toxins have been characterized. Depending on the *B. t.* strain, one to four major **endotoxin proteins** occur in a crystalline form. More details on the structure and mode of action are described by FEDERICI (1999b). In the gut of the target insects these pretoxins are cleaved and, thus, activated by the insect digestion enzymes. The toxins are safe for vertebrates, because activation only occurs in the insect guts. The biotechnological production of the *B. t.* toxins can be performed in batches of up to 50,000 L with cheap and readily available growth media.

The choice of the applied *B. t.* depends on the insects to be controlled. The subspecies *B. t.* "*kurstaki*" is used for controlling Lepidoptera (butterflies) without the noctuids, the subspecies "*aizawai*" is active against Lepidoptera inclusive noctuids, the subspecies "*tenebrionis*" is used for controlling several Coleoptera (beetle) pests, especially larvae of the Colorado potato beetle, and the subspecies "*israelensis*" is applied against mosquitoes and several gnat species (e.g., Sciaridae). Until now about 60 subspecies of *B. t.* have been characterized and several strains are applied in agricultural crops, vegetables, orchards, and in forestry (FEDERICI, 1999b).

The finding that the genes coding for δ-endotoxin synthesis are located on a plasmid, a small extrachromosomal ring, allowed a characterization of the genes by DNA sequencing.

One way of enhancing the effectiveness of *B. t.* towards insects is the transfer of toxin coding plasmids to an already existing *B. t.* strain. This has been achieved with several Cry genes to obtain recombinant strains which have a higher toxin yield per cell.

Other insecticidal compounds are biotechnologically generated from **actinomycetes** (e.g., *Streptomyces* species), and the reader is referred to COPPING (1998) for further information.

9.1 Pest Resistance of Plants by Integrating *Bacillus thuringiensis* Toxin Genes into the Plant Genome

Another approach is the integration of the *B. t.* toxin into plants. This strategy started in the early eighties, and today mostly Cry proteins are synthesized by several commercially important cultivated plants (FEDERICI, 1999b).

However, there has been a concern about the widespread use of the *B. t.* toxins in the bactericidal form as well as in transgenic plants, because the target insects can develop **resistance towards the *B. t.* toxins**. If insects are controlled by only one mechanism this selective pressure can result in a fast build-up of resistance. This effect has been observed, e.g., when the Cry toxin was applied too often (BAUER, 1990). As most transgenic plants possess only one or two Cry genes in their genome, the appearance of resistance is only a question of time. A way of preventing upcoming resistance is a rotation of crops expressing different Cry proteins. But since the proteins are structurally related a cross-resistance can occur, causing a resistance to all known Cry proteins. This problem may be circumvented by integrating several Cry genes into the target plant. The desired tissue-specific secretion of the toxins (i.e., a synthesis of Cry toxins not all over the plant but only in the tissue on which the harmful insect is feeding) is yet far from field application.

In order to cope with the arising resistance problems a so-called crop management was established. Usually, *B. t.* plants are grown in the fields together with a certain percentage of non-*B. t.* plants which serve as a refugium for the insects. The idea is that toxin-susceptible insects survive on the non-*B. t.* plants. These *B. t.* susceptible insects casually mate with *B. t.* resistant insects resulting in offspring with "diluted" resistance genes. This measure keeps the acquired *B. t.* resistance genes of the insect population at a lower level, and has been applied successfully in cotton crops in the US (GOULD, 1998). In Germany it may be possible to prevent the spread of the corn borer from the South to the North by planting a small borderline of transgenic corn. Also small borderlines around the last year's corn fields shall keep the number of pest individuals in the corn field of the next year below the damage threshold (LANGENBRUCH, personal communication).

Globally, *B. t.*-maize occupied 7.5 million ha equivalent to 19% of the global area in 1999, planted as transgenic crops in USA, Canada, Argentina, South Africa, Spain, France, and Portugal (JAMES, 2000).

Also, several **other compounds** have been used to control pests or already *B. t.*-toxin resistant pest insects. For instance, lectins synthesized from the snowdrop lectin gene or enzymes generated by the cow pea trypsin inhibitor gene have been transferred into cultivated plants. However, these compounds do not exert the same selectivity as the Cry proteins (COPPING, 1998).

10 Pheromones

The application of pheromones to control pest insects has a long tradition. Insects communicate via chemical odors (pheromones) and each insect species has its own pheromone. The behavior of insects is mediated by pheromones, and, therefore, insect pests can be selectively controlled either by using the naturally occurring pheromones or by using chemically modified derivatives, the so-called "semiochemicals". Since a diverse range of pheromones are produced biotechnologically, the reader is referred to COPPING (1998).

Two main groups are of special interest for plant protection in agriculture and forestry: (1) sexual pheromones (mostly combinations of several compounds produced by female arthropods to attract males) and (2) aggregation pheromones (a mix of compounds produced by females as well as males to aggregate a species at a resource very well suited for development). Both kinds of pheromones are mainly used for monitoring of pest species due to their high species specificity (BATHON, personal communication).

The **Mating Disruption Technique** (SCHNEIDER, 1999) enables direct control of pests species by distribution of high quantities of pheromone dispensers, e.g., in vineyards where males of grape vine moths are no longer able to find the females for copulation. **Mass Trapping** has been experienced by use of aggregation pheromones, e.g., in forests to attract bark beetles or in small orchards for attraction of various leafroller moths (e.g., codling moth). In the **Attract and Kill** method pheromones (and sometimes also kairomones) act in concert with insecticides: the pest individuals are attracted by pheromones and after coming into contact with the dispensers are killed by an insecticide (DICKLER et al., 2000).

11 Serology and DNA Procedures as Detection Tools

The use of **antibodies** as detection tool is very common in plant pathology (MILLER and JOAQUIM, 1994). Antibodies are generated usually in mammalians when an antigen (like a virus or bacterium) is offered. Polyclonal antibodies are less specific (reacting with several closely related microorganisms) than highly specific monoclonal antibodies. When the DNA coding for a monoclonal antibody is known, these antibodies can be synthesized also in other organisms, such as bacteria or plants (see plantibodies, Sect. 4). Either intact antibodies or the single chain variable fragments of the antibodies are suitable for detection. Antibodies can be used to detect plant pathogens microscopically by **immunofluorescence**, where cells are recognized by the antibodies marked with fluorescent chemicals. This technique is time-consuming but very sensitive because single cells of the target plant pathogen can be detected.

Less time-consuming and sensitive is the enzyme-linked immunosorbent assay (**ELISA**), where the anchored antibodies first adsorb the plant pathogens out of a suspension. In the following step these microorganisms are visualized by a second batch of antibodies marked with an enzyme which initiates a color reaction (sandwich ELISA).

Microorganisms can also be detected directly in the plant tissue by the use of an adsorbing matrix (nitrocellulose, nylon). When incubating this tissue print with the antibodies the area of infection within the plant can be localized.

The serological procedures are applied routinely and are useful when many samples have to be tested. In addition, modern biotechnological methods have been introduced. The most prominent example of **biochemical procedures** is electrophoresis on polyacrylamide, cellulose acetate or starch gel to distinguish closely related organisms. The target organisms are differentiated by comparing protein or enzyme patterns, and electrofocusing can help to distinguish proteins via their indigenous electrical charge (UNRUH and WOOLLEY, 1999).

The most prominent gene technological method is the **polymerase chain reaction (PCR)** which is now replacing older techniques like DNA–DNA hybridizations of dot blot tests and tissue printings (MILLER and JOAQUIM, 1994).

The PCR technique amplifies and visualizes a small but distinct section of the DNA from a target organism. This requires that the DNA sequence (the order of the nucleotides) to be amplified, is known by DNA sequencing. The chosen DNA fragment should be unique to the target organism so that no cross-reactions occur with related organisms. A pair of short DNA fragments of about 20 nucleotides in length, the so-called primers, are generated which are complementary to the DNA sequence of each single DNA strand (sense- and antisense DNA strand). The primers anneal to one complementary DNA strand at a certain distance from each other at a primer-specific temperature. This double-stranded section is the starting point for the enzyme Taq-DNA-polymerase, which synthesizes a new DNA strand, using the four nucleotide triphosphates and the DNA strand as a matrix. This polymerase is heat-stable and, therefore, the steps necessary for amplifying the DNA in a tube can be performed, i.e., heating for denaturation of the double-stranded DNA, reducing the heat to about 55 °C to allow the annealing of the primers, and the synthesis step at 72 °C to generate new DNA. These three temperature shifts are a so-called cycle, and such a cycle can be repeated up to 40 times resulting in an amplification of the DNA up to 2^{40} DNA strand copies. A computerized thermocycler is performing these cycles automatically, and the automatization of the detection of any plant pathogenic microorganism is now possible. In general, the detection threshold depends on the extraction procedure of the DNA and the absence of enzyme inhibitors. Because of its easiness to amplify DNA, this technique can be used as a diagnostic tool for detecting a target microorganism out of a mixture of other microorganisms, since the amount of DNA generated by PCR is large enough to visualize the fragment on an agarose gel after electrophoretic separation. This technique is now routinely used for detecting all kinds of plant pathogenic microorganisms. For organisms which have RNA instead of DNA as their genetic information (RNA viruses) a pre-PCR step has to be inserted where first the RNA is transcribed via the enzyme reverse transcriptase into DNA, followed by the PCR amplification (RT-PCR).

When closely related microorganisms generate PCR fragments of the same size, an additional differentiation criterion can be used by restricting the DNA fragment with an enzyme, the so-called restriction endonuclease. This enzyme recognizes always the same nucleotide sequence within the DNA double strand and cleaves the DNA after this site. If the DNA sequences inside of equally long DNA fragments differ within the endonuclease recognition site, a distinction into digested or undigested DNA fragments can be made on an agarose gel (RFLP).

An additional method for discriminating target organisms by PCR is the **random amplified polymorphic DNA (RAPD)** technique. This method reveals variations in the DNA sequence of related organisms. Only a single 10 nucleotide primer of an arbitrary DNA sequence is used. Amplification of DNA occurs in the PCR when the sequence of this primer is complementary to both DNA strands of the organism and the distance between both annealing sites is not longer than 5,000 base pairs. This randomness of the complementary primer sites within the chromosome of the organism decides on the amplified DNA fragments allowing a differentiation. An increase of resolution of the synthesized PCR fragments is possible by using polyacrylamide gel electrophoresis (PAGE), such as denaturing gradient gel electrophoresis (DGGE), or temperature gradient gel electrophoresis (TGGE). The RAPDs have been applied for mapping fungal and insect biotypes and are now used for the taxonomy of plant pathogenic microorganisms (UNRUH and WOOLLEY, 1999).

The **amplified fragment length polymorphism** (AFLP) is another PCR-based differentiation method. When restricting the purified DNA from the target organism with a restriction enzyme, a large number of differently sized fragments are produced which all have the same nucleotide ends overhanging at both sides ("sticky" ends). By adding a small DNA fragment (ligation) of a known DNA sequence to these sticky ends, a complementary PCR primer can be generated containing the DNA sequences from the added fragment and the DNA sequence of the restriction cut site. For discrimination purposes, extra nucleotides are added to the primer which anneal beyond the restriction site into the unknown DNA sequence of the target organism. Successful amplification is only achieved when a primer reaches a 100% match of the DNA fragment which is to be amplified. Therefore, only those restricted DNA fragments of the target organism are amplified which have the correct complementary DNA sequence of the primer offered. This process is arbitrary and can be repeated with a choice of other randomly added nucleotides to reduce the number of amplificates produced. The obtained fragments are electrophoretically separated on a high-resolution gel (a thin polyacrylamide gel), and differences between strains become visible by the appearance of certain amplified DNA fragments differing in size. Additional methods used in biotechnology are described by O'DONNELL (1999) and BULL et al. (2000).

12 Biotechnologically Produced Plant Protection Products Authorized in Germany

Worldwide about 35 plant protection products of microbiological origin are authorized. Only a few of them have up to now been authorized by the German authorities as true plant protection products. These are 13 plant protection products (Bode, 2000) on the basis of *Bacillus thuringiensis*, 3 on the basis of granulose virus, and 2 of fungal origin (*Coniothyrium minitans* and *Metarhizium anisopliae*). In addition, the following microorganisms are now authorized EU-wide according to the Council directive of 15 July 1991 concerning the placing on the market of plant protection products (91/414/EEC, *Official Journal of the European Communities* L230, Vol. 34, 19. August 1991, 1–32): *Aschersonia aleyrodis*, *Beauveria bassiana*, *Phlebiopsis gigantea*, *Streptomyces griseoviridis*, 3 *Trichoderma* subspecies and 2 *Verticillium* subspecies.

Other biotechnologically produced plant protection products have already been applied in other countries. Besides extracts from microorganisms, **plant extracts** are also used. Two of the most frequently used plant extracts act as insecticides and are generated from chrysanthemum and from the tropical neem tree (Copping, 1998).

When releasing microbial plant protection products it is very important to know the environmental conditions for optimal effects, such as plant, soil, temperature, humidity, and other factors.

Even though it is envisaged that there is a market for biotechnologically produced plant protection products it is difficult to overview the present application of these compounds and to obtain an estimation on their financial output. Therefore, a compound's authorization as a plant protection product does not mean that it will sell well, and exact data are not easily available.

13 Discussion and Outlook

Our understanding of biological processes in phytopathology has grown rapidly due to the recent developments in gene technology. The ever increasing possibilities to manipulate DNA allows plant breeders to faster reach the goal of creating plants resistant to numerous pathogens.

Biotechnology in plant pathology comprises a wide diversity of techniques, from fermenting biomaterial to the latest advances in gene technology to protect crops from pathogen attacks. Hopefully, these transgenic plants will reduce application of chemical plant protection products, minimizing the input of chemicals in soil and water.

The chances are good that this strategy will be successful since conventional plant protection is faced with several problems, such as faster build-up of resistance in pathogens to new plant protection products, i.e., these products became ineffective after a few applications. In addition, impacts on the environment by chemical plant protection have become a serious public issue and, therefore, the biotechnologically produced living organisms or compounds are an attractive alternative, since they act more directly by saving the beneficial organisms or by controlling the target organisms with resistance against chemical plant protection products. The application of these organisms might also be useful for areas where no chemical plant protection products are allowed (e.g., water reserves) or simply are not available because the chemicals are too expensive to develop.

In the field of diagnosis the use of the so-called biochips or micro-arrays offers interesting applications. Since these miniaturized sensors give place to a number of DNA microsamples this technique seems to be very efficient. The DNA hybridizes with the corresponding DNA, finding its counterpart, which is then measurable by a light emitting device (Kroner and Schwerdtle, 2000). In future, this technique might be useful for a routine detection of plant pathogens since the biochips allow to screen for a number of traits at the same time.

Another interesting development for the PCR-procedure are the light-cyclers which can perform 32 cycles within 20 min and which allow to detect the target organism much faster (ZIEBOLZ, 2000).

An often heard critical argument is the risk that biotechnologically generated plant protection measures can easily be overcome by the pathogens by developing resistant mutants. Therefore, the strategy of integrated pest control, a combination of biological, biotechnological and chemical measures (AUST et al., 1991) should be followed to reduce the selective pressure on the pathogens, resulting in a longer period of developing resistance. This concept implies that the farmer has to accept a small number of pathogens in the field destroying parts of the harvest. Thus, the application of biopesticides is a problem of acceptance by the farmers who were accustomed to fields without visible disease symptoms in the past.

The transition from a biopesticide to a chemical produced by the industry has been shown in the case of pheromones, which are now produced as so-called semiochemicals (COPPING, 1998). In the future, methods should be available to cultivate the obligate parasites in fermenters as easy as *Bacillus thuringiensis* to increase the output of these parasites rather than the cultivation in their target insects.

On the other hand, several drawbacks of biotechnology in phytopathology should not be overlooked: Biotechnologically produced plant protection products need a proper handling, since their effects depend on environmental factors such as temperature, light, and moisture. Therefore, their effects cannot be guaranteed in every case, resulting in a possible reduction of the farmers income.

Another critical aspect is that the transgenic plants and their products are strongly rejected by an increasing portion of the population in Europe and the USA on the basis of emotional arguments and sometimes pseudo-scientific data. However, only a broad acceptance can promote the use of biotechnology in plant protection (LAPPÉ and BAILEY, 2000).

Biotechnologically produced plant protection products are just emerging from basic science and, therefore, are more expensive than their chemical competitors. Since the tools (genetic engineering) are expensive and will only be manageable by a few big companies, this fact will make farmers dependent from seed- and chemical companies selling the transgenic microorganisms or seeds. Another concern is that biodiversity is declining, when predominantly transgenic plants are cultivated (AVERY, 1999).

In future, we need deeper knowledge on the plant/pathogen-interaction to ensure that the released transgenic organisms are safe enough for practical application. A knowledge of the regulatory systems in the plant will become necessary to switch on the synthesis of resistance factors only under pathogen attack, since some of these compounds may affect human health (PADIDAM, 2000).

Another approach might be to integrate genes which detoxify compounds in resistant plants soon after synthesis to minimize the risk that pathogens develop a resistance against the transgenic plants, because selection of drug resistant microorganisms needs a certain time period of adaptation (SUPARNA et al., 2000). In order to control diverse diseases and pests by transgenic plants it may be necessary to incorporate several non-indigenous genes into the cultivated plants which may disturb the plant's regulatory equilibrium. Therefore, a well organized integrative plant protection will be necessary to maintain the resistance of the new cultivars for such a period that the arising costs can be paid off.

The overall possibilities of biotechnology in plant pathology are promising, but the limitations of these new techniques should be kept in mind, especially when living systems are used for plant protection measures.

Acknowledgements

We are grateful to our colleagues at the Federal Biological Research Centre for Agriculture and Forestry and at several universities for very helpful advice during the preparation of this review chapter.

14 References

ARMSTRONG, M. R., BLOK, V. C., BIRCH, P. R. J., PHILLIPS, M. S. (2000), Analysis of gene expression in incompatible plant nematode interactions, in: *Int. Soc. Plant Mol. Biol. and Université Laval, Québec*, Supplement to Reporter 18:2, Abstract S22–1. Congress of Plant Molecular Biology, June 18–24, 2000.

ATKINSON, H. J., LILLEY, C. J. , URWIN, P. E., MCPHERSON, M. J. (1998), Engineering resistance to plant parasitic nematodes, in: *The Physiology and Biochemistry of Free-Living and Plant-Parasitic Nematodes* (PERRY, R. N., WRIGHT, D. J., Eds.), pp. 381–413. Wallingford: CAB International Press.

AUST, H. J., BUCHENAUER, H., KLINGAUF, F., NIEMANN, P., PÖHLING, H. M., SCHÖNBECK, F. (1991), *Glossar Phytomedizinischer Begriffe*, German Phytomedical Society Series. Stuttgart: Ulmer.

AVERY, D. T. (1999), Why biotechnology might not represent the future in world agriculture, in: *National Agricultural Biotechnology Council Report* 11, pp. 49–53. Ithaca, NY: NABC Reports.

BAUER, L. S. (1990), Resistance: A threat to the insecticidal crystal proteins of *Bacillus thuringiensis*, *Florida Entomologist* **78**, 414–443.

BEACHY, R. N., BENDAHMANE, M. (1998), Pathogen derived resistance and reducing the potential to select viruses with increased virulence, in: *National Agricultural Biotechnology Council Report* 10, pp. 87–94. Ithaca, NY: NABC Reports.

BELLOWS, T. S. (1999), Controlling soil-borne pathogens, in: *Handbook of Biological Control* (BELLOWS, T. S., FISCHER, T. W., Eds.), pp. 669–712. San Diego, CA: Academic Press.

BODE, E. (2000), Development of microbial antagonists of plant diseases and their access to the market subject to the authorisation procedure, *Nachrichtenblatt des Deutschen Pflanzenschutzdienstes* **52**, 213–219.

BOURDON, V., LADBROOKE, Z., HARVEY, A., LONSDALE, D. (2000), Transgene expression in wheat: investigation and improvement, in: *Int. Soc. Plant Mol. Biol. and Université Laval, Québec*, Supplement to Reporter 18:2, Abstract S03–12. Congress of Plant Molecular Biology, June 18–24, 2000.

BOYETCHKO, S. M. (1999), Innovative applications of microbial agents for biological weed control, in: *Biotechnology Approaches in Biocontrol of Plant Pathogens* (MUKERJI, K. G., CHAMOLA, B. P., UPADHYAY, R. K., Eds.), pp. 73–97. Dordrecht: Kluwer Academic/Plenum Publisher.

BULL, A. T., WARD, A. C., GOODFELLOW, M. (2000), Search and discovery strategies for biotechnology: the paradigm shift, *Microbiol. Mol. Biol. Rev.* **64**, 573–605.

CHEN, X. Y., LUO, P., ZHOU, X.-J., JIA, J.-W., LU, S. et al. (2000), Regulation of biosynthesis of sesquiterpene aldehydes in cotton, in: *Int. Soc. Plant Mol. Biol. and Université Laval, Québec*, Supplement to Reporter 18:2, Abstract S06–10. Congress of Plant Molecular Biology, June 18–24, 2000.

CHESTER, K. S. (1933), The problem of acquired physiological immunity in plants, *Quart. Rev. Biol.* **8**, 275–324.

CHET, I., INBAR, J. (1997), Fungi, in: *Fungal Biotechnology* (ANKE, T., CHAPMAN, H., Eds.), pp. 65–80. Weinheim: Chapman & Hall.

CIPRIANI, P. G., BELLO, V., MICHAUD, D., ZHANG, D. (2000), Genetic transformation of sweet potato with oryzacystatin inhibitor I for nematode resistance, in: *Int. Soc. Plant Mol. Biol. and Université Laval, Québec*, Supplement to Reporter 18:2, Abstract S03–25. Congress of Plant Molecular Biology, June 18–24, 2000.

COPPING, L. G. (1998), *The Pesticide Manual*, Farnham, Surrey: British Crop Protection Council.

COPPOOLSE, E. R., VAN DER VOSSEN, E. A. G., KLEIN-LANKHORST, R., BAKKER, J., STIEKEMA, W. J. (2000), Map based cloning of the h1 resistance gene of potato against the nematode *Globodera rostochiensis*, in: *Int. Soc. Plant Mol. Biol. and Université Laval, Québec*, Supplement to Reporter 18:2, Abstract S22–18. Congress of Plant Molecular Biology, June 18–24, 2000.

CUPPELS, D. A., AINSWORTH, T., LU, H., HIGGINGS, V. (2000), Development of an integrated approach to the control of bacterial diseases of tomato and pepper, in: *10th Int. Conf. Plant Pathogenic Bacteria*, Charlottetown, PEI, Canada, Book of Abstracts, P 36.

DATTAROY, T., LI, Q., LAWRENCE, C., HUNT, A. (2000), Genetically engineering resistance to a broad range of pathogens. in: *Int. Soc. Plant Mol. Biol. and Université Laval, Québec*, Supplement to Reporter 18:2, Abstract S01–19. Congress of Plant Molecular Biology, June 18–24, 2000.

DE MEYER, G., CAPIEAU, K., AUDENAERT, K., BUCHALA, A., METRAUX, J. P., HOFTE, M. (1999), Nanogram amounts of salicylic acid produced by the rhizobacterium *Pseudomonas aeruginosa* 7NSK2 activate the systemic acquired resistance pathway in bean, *Mol. Plant–Microbe-Interact.* **12**, 450–458.

DESTÉFANO-BELTRÁN, L., NAGPALA, P. G., CETINER, S. M., DENNY, T., JAYNES, J. M. (1993), Using genes encoding novel peptides and proteins to enhance disease resistance in plants, in: *Biotechnology in Plant Disease Control* (CHET, I., Ed.), pp. 175–190. New York: Wiley-Liss.

DI MAURO, A. O., OLIVEIRA, R. C., DI MAURO, S. M. Z. (2000), Identification of RAPD molecular markers linked to genes conferring resistance to soybean cyst nematode (race 3), in: *Int. Soc. Plant

Mol. Biol. and Université Laval, Québec, Supplement to Reporter 18:2, Abstract S22–27. Congress of Plant Molecular Biology, June 18–24, 2000.

DICKLER, E., LÖSEL, P. M., VOGT, H., EBERT, A., EBBINGHAUS, D. (2000), Attract and Kill – an environmental method for codling moth control, *Mitt. Biol. Bundesanst. Land- Forstwirtsch.* **376**, 281–282.

DÜRING, K., PORSCH, P., FLADUNG, M., LÖRZ, H. (1993), Transgenic potato plants resistant to the phytopathogenic bacterium *Erwinia carotovora*, *Plant J.* **3**, 587–589.

ELZEN, G. W., KING, E. G. (1999), Periodic release and manipulation of natural enemies, in: *Handbook of Biological Control* (BELLOWS, T. S., FISHER, T. W., Eds.), pp. 253–270. San Diego, CA: Academic Press.

FECKER, L. F., KOENIG, R. (1999), Engineering of *Beet necrotic yellow vein virus* (BNYVV) resistance in *Nicotiana benthamiana*, in: *Recombinant Antibodies – Application in Plant Science and Plant Pathology* (HARPER, K., ZIEGLER, A., Eds.), pp. 157–170. London, Philadelphia, MD: Taylor and Francis.

FEDERICI, B. A. (1999a), A perspective on pathogens as biological control agent for insect pests, in: *Handbook of Biological Control* (BELLOWS, T. S., FISHER, T. W., Eds.), pp. 517–548. San Diego, CA: Academic Press.

FEDERICI, B. A. (1999b), *Bacillus thuringiensis* in biological control, in: *Handbook of Biological Control* (BELLOWS, T. S., FISHER, T. W., Eds.), pp. 575–593. San Diego, CA: Academic Press.

GOULD, F. (1998), Sustaining the efficacy of Bt toxins, in: *Agricultural Biotechnology and Environmental Quality: Gene Escape and Pest Resistance*, National Agricultural Biotechnology Council Report 10, pp. 77–86. Ithaca, NY: NABC Reports.

GUPTA, R., MUKERJI, K. G. (1999), Host parasite specification and pathogenesis, in: *Biotechnological Approaches in Biocontrol of Plant Pathogens* (MUKERJI, K. G., CHAMOLA, B. P., UPADHYAY, R. K., Eds.), pp. 1–29. Dordrecht: Kluwer Academic/Plenum Publishers.

HAMILTON, R. I. (1980), Defences triggered by previous invaders: Viruses, *Plant Dis.* 1977–1980, **5**, 279–303.

HARPER, K., ZIEGLER, A. (Eds.) (1999), Recombinant Antibodies–Application in Plant Science and Plant Pathology. London, Philadelphia, MD: Taylor and Francis.

HARRISON, D. J., MCPHERSON, M. J. (2000), Phytocystatins for enhanced transgenic resistance to plant parasitic nematodes, in: *Int. Soc. Plant Mol. Biol. and Université Laval, Québec*, Supplement to Reporter 18:2, Abstract S16–14. Congress of Plant Molecular Biology, June 18–24, 2000.

JAMES, C. (2000), *Global Status of Commercialised Transgenic Crops: 1999*, International Service for the Acquisition of Agri-Biotech Applications (ISAAA), Briefs 17. Ithaca, NY: ISAAA.

JONES, J. D. G., THOMAS, C., HAMMOND-KOSACK, K., VAN DER BIEZEN, E., DURRANT, W. et al. (2000), Plant disease resistance genes: Structure, function and evolution, in: Biotechnology 2000, 11th Int. Biotechnol. Symp. and Exhibition, September 3–8, 2000, Berlin, *Abstracts* Vol. 1, p. 290. Frankfurt/M.: DECHEMA e.V.

KAMITA, S. G., MAEDA, S. (1997), Sequencing of the putative DNA helicase-encoding gene of *Bombyx mori* nuclear polyhedrosis virus and fine-mapping of a region involved in host range expansion, *Gene* **190**, 173–179.

KAWCHUK, L. M., PRÜFER, D. (1999), Molecular strategies for engineering resistance to potato viruses, *Can. J. Plant Pathol.* **21**, 231–247.

KEEN, N. T., BRUGGER, B. (1977), Phytoalexins and chemicals that elicit their production in plants, in: *Host Plant Resistance to Pest* (HEDIN, P., Ed.), *Am. Chem. Soc. Symposium Ser.* **62**, 1–26.

KERR, A., BEER, S. V., SCHROTH, M. N., BAHME, J. B. (1990), Biological control, in: *Methods in Phytobacteriology* (KLEMENT, Z., RUDOLPH, K., SANDS, D. C., Eds.), pp. 369–379. Budapest, Hungary: Akadémiai Kiadó.

KIKKERT, J. R., VIDAL, J. R., REUSTLE, G. M., ALI, G. S., YU, L. X. et al. (2000), Genetic transformation of *Vitis vinifera* L. grapevines with genes encoding chitinases and antimicrobial peptides, in: *Int. Soc. Plant Mol. Biol. and Université Laval, Québec*, Supplement to Reporter 18:2, Abstract S03–61. Congress of Plant Molecular Biology, June 18–24, 2000.

KRONER, W. G., SCHWERDTLE, R. (2000), Biochips für die Arzneimittelentwicklung, *BioWorld* **2**, 4–9.

KUĆ, J. (1987): Plant immunisation and its applicability for disease control, in: *Innovative Approaches to Plant Disease Control* (CHET, I., Ed.), pp. 255–274. New York: Wiley-Liss.

LAPPÉ, M., BAILEY, B. (2000), *Machtkampf Biotechnologie*. München: Gerling Akademie Verlag.

LEE, Y.-H., YOON, I. S., KIM, H. I. (2000a), Transformation and expression of glucose oxidase gene in *Brassica oleracea* ssp. *capitata*, in: *Int. Soc. Plant Mol. Biol. and Université Laval, Québec*, Supplement to Reporter 18:2, Abstract S03–68. Congress of Plant Molecular Biology, June 18–24, 2000.

LEE, J., BLUME, B., KLESSIG, D., SCHEEL, D., NUERNBERGER, T. (2000b), Harpin perception and signal transduction mediating defense gene(s) expression in tobacco, in: *Int. Soc. Plant Mol. Biol. and Université Laval, Québec*, Supplement to Reporter 18:2, Abstract S09–47. Congress of Plant Molecular Biology, June 18–24, 2000.

LI, R., MAO, X., LI, C., LIU, Z. (2000), Analysis of the differential expression of genes associated with the induced resistance against aphids in cotton plants, in: *Int. Soc. Plant Mol. Biol. and Université Laval, Québec*, Supplement to Reporter 18:2, Abstract S09–49. Congress of Plant Molecular Biology, June 18–24, 2000.

LOPER, J. E., ISHIMARU, C. A. (1991), Factors influencing siderophore-mediated biocontrol activity of rhizosphere *Pseudomonas* ssp., in: *The Rhizosphere and Plant Growth* (KLEISTER, D. L., CREGAN, P. B., Eds.), pp. 253–261. Dordrecht: Kluwer Academic Publisher.

MACKINNEY, H. H. (1929), Mosaic disease in the Canary Islands, West Africa, and Gibraltar, *J. Agricult. Res.* **39**, 557–578.

MALNOY, M., REYNOIRD, J.-P., CHEVREAU, E. (2000), Strategies for improving pear resistance to fire blight through genetic engineering, in: *Int. Soc. Plant Mol. Biol. and Université Laval, Québec*, Supplement to Reporter 18:2, Abstract S26–14. Congress of Plant Molecular Biology, June 18–24, 2000.

MAOKA, T., NODA, C. (2000), Reactions of PRSV resistant papaya cultivar "SunUup" to Japanese isolates of PRSV and PLDMV, in: *Int. Soc. Plant Mol. Biol. and Université Laval, Québec*, Supplement to Reporter 18:2, Abstract S30–32. Congress of Plant Molecular Biology, June 18–24, 2000.

MAVRIDIS, A., RUDOLPH, K. (1996), Untersuchungen zur Reaktion kartoffelpathogener Bakterien und Pilze gegenüber dem T4-Lysozym *in vitro* und *in planta* bei transgenen Kartoffeln, *Mitt. Biol. Bundesanst. Land.- Forstw.* **321**, 139.

MCGIVERN, D., MAZITHULELA, G., PIROUX, N., HECKEL, T., BOULTON, M., DAVIES, J. (2000), Maize streak virus gene expression as a model for regulation of transcription in cereals, in: *Int. Soc. Plant Mol. Biol. and Université Laval, Québec*, Supplement to Reporter 18:2, Abstract S30–34. Congress of Plant Molecular Biology, June 18–24, 2000.

MILLER, S. A., JOAQUIM, T. R. (1994), Diagnostic techniques for plant pathogens, in: *Biotechnology in Plant Disease Control* (CHET, I., Ed.), pp. 322–340. New York: Wiley-Liss.

MISRA, S., OSUSKY, M., OSUSKA, L., KAY, W. W. (2000), Cationic peptide expression in transgenic potato confers broad-spectrum resistance to phytopathogens, in: *Int. Soc. Plant Mol. Biol. and Université Laval, Québec*, Supplement to Reporter 18:2, Abstract S22–71. Congress of Plant Molecular Biology, June 18–24, 2000.

MITTLER, R., HERR, E. H., ORVAR, B. L., VAN CAMP, W., WILLEKENS, H. et al. (1999), Transgenic tobacco plants with reduced capability to detoxify reactive oxygen intermediates are hyper-responsive to pathogen infection, *Proc. Natl. Acad. Sci. USA* **96**, 14165–14170.

MUKERJI, K. G. (1999), Mycorrhiza in control of plant pathogens: molecular approaches, in: *Biotechnology Approaches in Biocontrol of Plant Pathogens* (MUKERJI, K. G., CHAMOLA, B. P., UPADHYAY, R. K., Eds.), pp. 135–155. Dordrecht: Kluwer Academic/Plenum Publishers.

NAKAJIMA, H., MURANAKA, T., ISHIGE, F., AKUTSU, K. (1997), Fungal and bacterial disease resistance in transgenic plants expressing human lysozyme, *Plant Cell Reporter* **16**, 674–679.

NASEBY, D. C., PASQUAL, J. A., LYNCH, J. M. (2000), Effect of bio-control strains of *Trichoderma* on plant growth, *Pythium ultimum* populations soil microbial communities and soil enzyme activities, *J. Appl. Microbiology* **88**, 161–169.

O'DONNELL, K. J. (1999), Plant pathogen diagnosis: present status and future developments, *Potato Res.* **42**, 437–447.

PADIDAM, M. (2000), Chemical inducible system to activate or inactivate plant gene expression, in: *Int. Soc. Plant Mol. Biol. and Université Laval, Québec*, Supplement to Reporter 18:2, Abstract S03–94. Congress of Plant Molecular Biology, June 18–24, 2000.

PARRELLA, M. P., HANSEN, L. S., VAN LENTEREN, J. (1999), Glasshouse environments, in: *Handbook of Biological Control* (BELLOWS, T. S., FISHER, T. W., Eds.), pp. 819–839. San Diego, CA: Academic Press.

PIETERSE, C. M. J., VAN LOON, L. C. (1999), Salicylic acid independent plant defence pathways, *Trends Plant Sci.* **4**, 52–58.

PINSTRUP-ANDERSEN, P., PANDYA-LORCH, R. (1999), Securing and sustaining adequate world food production for the third millennium, in: *World Food Security and Sustainability: The Impacts of Biotechnology and Industrial Consolidation*, National Agricultural Biotechology Council Report 11, pp. 27–48. Ithaca, NY: NABC Reports.

RAJASEKARAN, K., CARY, J. W., CLEVELAND, T. E. (2000), Expression of a gene encoding a synthetic antimicrobial peptide confers fungal resistance *in vitro* and *in planta* in transgenic plants, in: *Int. Soc. Plant Mol. Biol. and Université Laval, Québec*, Supplement to Reporter 18:2, Abstract S03–96. Congress of Plant Molecular Biology, June 18–24, 2000.

RAYDER, M. H., JONES, D. A. (1990), Biological control of crown gall, in: *Biological Control of Soil Borne Plant Pathogens* (HORNBY, D., Ed.), pp. 45–63. Oxon: CAB International.

REYNOIRD, J. P., ABDUL-KADER, A. M., BAUER, D. W., BOREJSZA-WYSOCKA, E., NORELLI, J., ALDWINCKLE, H. S. (2000), Activation of the potato gst1 and tobacco hsr203j promoters by *Erwinia amylovora* in transgenic royal gala apple and m26 apple root-stock, in: *Int. Soc. Plant Mol. Biol. and Université Laval, Québec*, Supplement to Reporter 18:2, Abstract S22–81. Congress of Plant Molecular Biology, June 18–24, 2000.

ROSSKOPF, E. N., CHARUDATTAN, R., KADIR, J. B. (1999), Use of plant pathogens in weed control, in: *Handbook of Biological Control* (BELLOWS, T. S., FISHER, T. W., Eds.), pp. 891–918. San Diego, CA: Academic Press.

RUDOLPH, K. (1995), *Pseudomonas syringae* pathovars, in: *Pathogenesis and Host Specificity in Plant Diseases* Vol. I: Prokaryotes (SING, U. S., SING, R. P., KOHMOTO, K., Eds.), pp. 47–138. Oxford, New York, Tokyo: Pergamon/Elsevier Science Ltd.

RYALS, J., WARD, E., METRAUX, J.-P. (1991), Systemic acquired resistance: An inducible defence mechanism in plants, in: *The Biochemistry and Molecular Biology of Inducible Enzymes and Proteins in Higher Plants* (WRAY, J. L., Ed.). Cambridge: Cambridge University Press.

SCHNEIDER, D. (1999), Insect pheromone research: Some history and 45 years of personal recollections, in: *Scents in Orchards – Plant and Insect Semiochemicals from Orchard Environments* (WITZGALL, P., EL-SAYED, A., Eds.), Vol. 22 (9), pp. 1–8. IOBC wprs Bulletin, Bulletin OILB srop.

SMANT, G., BAKKER, J., GOMMERS, F. J., SCHOTS, A. (1999), Engineering antibody-mediated resistance to plant root nematodes – potential and prospects, in: *Recombinant Antibodies – Application in Plant Science and Plant Pathology* (HARPER, K., ZIEGLER, A., Eds.), pp. 171–181. London, Philadelphia, MD: Taylor and Francis.

SMITS, P. H. (2000), Insect pathogens and insect parasitic nematodes; *Capturing the Potential of Biological Control* Vol. 23 (2), IOBC wprs Bulletin, Bulletin OILB srop.

STADNIK, M. J., BUCHENAUER, H. (1999), Effects of benzothiadiazole, kinetin and urea on the severity of powdery mildew and yield of winter wheat, *Z. Pflanzenkrankheiten und Pflanzenschutz* **106**, 476–489.

STAHL, D. J., MAESER, A., DETTENDORFER, J., HOLTSCHULTE, B., BORCHARDT, D. et al. (2000), Increased fungal resistance of transgenic plants by heterologous expression of bacteriophage T4 lysozyme gene, in: *Int. Soc. Plant Mol. Biol. and Université Laval, Québec*, Supplement to Reporter 18:2, Abstract S03–109. Congress of Plant Molecular Biology, June 18–24, 2000.

SUPARNA, R., MUNDODI, N., PAIVA, L. (2000), Expression of a phytoalexin detoxification gene in *Alfalfa*, in: *Int. Soc. Plant Mol. Biol. and Université Laval, Québec*, Supplement to Reporter 18:2, Abstract S06–35. Congress of Plant Molecular Biology, June 18–24, 2000.

THOMAS, M. B. (1999), Ecological approaches and the development of "truly integrated" pest management, *Proc. Natl. Acad. Sci. USA* **96**, 5944–5951.

THRISTLEWOOD, H. M., JUDD, C. J. (2000), Sterile insect release in Canada: From eradication to area-wide management, *XXI Int. Congr. Entomol.*, Brazil, August 20–26, 2000, Abstract 2553, 645.

TILAK, K. V. B. R., SINGH, G., MUKERJI, K. G. (1999), Bio-control plant growth promoting rhizobacteria: Mechanism of action, in: *Biotechnology Approaches in Biocontrol of Plant Pathogens* (MUKERJI, K. G., CHAMOLA, B. P., UPADHYAY, R. K., Eds.), pp. 115–133. Dordrecht: Kluwer Academic/Plenum Publisher.

UNRUH, T. R., WOOLLEY, J. B. (1999), Molecular methods in classical biological control, in: *Handbook of Biological Control* (BELLOWS, T. S., FISHER, T. W., Eds.), pp. 57–85. San Diego, CA: Academic Press.

WEGENER, C. (1999), Field performance of transgenic potatoes bearing an *Erwinia* pectate lyase gene, in: *14th Triennial Conf. Eur. Ass. Potato Res.*, Sorrento, Italy, Assessorato Agricoltura Regione Campania, pp. 402–403.

WEI, Z.-M., LABY, R. J., ZUMOFF, C. H., BAUER, D. W., HE, S. Y. et al. (1992), Harpin, elicitor of the hypersensitive response, produced by the plant pathogen *Erwinia amylovora*, *Science* **257**, 85–88.

XU, G. W., GROSS D. C. (1986), Selection of fluorescent pseudomonads antagonistic to *Erwinia carotovora* and suppressive of potato seed piece decay, *Phytopathology* **76**, 414–422.

ZHANG, W., HAN, D. Y., DICK, W. A., HOITINK, H. A. J. (1998), Compost and compostwater extract induced systemic acquired resistance in cucumber and *Arabidopsis*, *Phytopathology* **88**, 450–455.

ZIEBOLZ, B. (2000), Automation enhancements boost efficiency of PCR, *Biotech J.* **5**, 10–11.

17 Biotechnology in Space

HEIDE SCHNABL

Bonn, Germany

1 Introduction

Gravity is known to act as a permanently and constantly effective abiotic factor which is related to plant growth, cell structure, reproduction, differentiation, biochemical and physiological metabolism, and to the adaptation strategies of organisms. A study of the role of gravity on living organisms has been under investigation for a long period. The biological effects of gravity are easier to understand and to interpret when this abiotic factor can be varied by means of simulated and real microgravity. As a research tool for the variation of the g vector a series of terms is being used such as "simulated microgravity" (clinostating: fast rotating, 50–120 rpm range), "altered gravity" (centrifugation), and "real microgravity" (spaceflight; see review of KORDYUM, 1997). In the fast rotating clinostat, very small cells are turned around their own axis. Thus the *g* vector changes direction in relation to the cell, creating simulated microgravity. The agreement with real microgravity is discussed in different ways. Restrictions of the employment of clinostats are described and discussed (VOLKMANN and SIEVERS, 1979; BRIEGLEB, 1984; SACK, 1991; KESSLER, 1992; SIEVERS and HEJNOWICZ, 1992; HEMMERSBACH-KRAUSE and BRIEGLEB, 1994). However, despite of these limits, the fast rotating clinostat proved to be a valuable tool to study the effects of microgravity on living organisms on earth with a higher frequency of experiments and a greater number of analytical methods than given in real flight opportunities, which are far too rare for a decent pace of scientific progress. Fortunately, in the last decade, a series of spaceflights – shuttle missions, orbital stations, biosatellite programs, spaceship missions, rocket and balloon programs – in low earth orbit under weightlessness environment made it possible to compare data obtained under real and simulated microgravity and to study the nature of changes occurring under altered g vectors. In future, the still open question has to be checked intensively, whether the two kinds of conditions, those in simulated and real microgravity, can be substituted.

As living systems which were often used in conditions of reduced gravity plants offer a lot of advantages. Unlike animals they have an open type of ontogenesis, they are immotile autotrophic organisms with a spatial orientation, which gives them the possibility to distinguish between up and down and to use the *g* vector as a means of orientation. Plants in different levels of specialization and organization were used in order to study the biological effects of microgravity in stages of differentiation, development, growth, and propagation. Lower plants (algae) and higher plants such as mosses, aquatic ferns, gymnosperms, angiosperms, as well as cells, especially tissue and protoplast cultures were used.

In this chapter biological changes in living organisms – microorganisms, cell-, tissue- and protoplast cultures – induced by microgravity are described as far as they are related to biotechnological processes; meaning that properties of organisms affected by microgravity are characterized using methods and processes based on molecular biological techniques and genetic engineering. Microgravity-induced phenomena such as variations in cellular functions, differentiation and regeneration processes, somatic hybridization events induced by electrical pulses, adaptation strategies, DNA damaging effectiveness, and mutation rates are brought together. Beside microgravity the absence of sedimentation processes, buoyancy, hydrostatic pressure, and capillarity as well as the existence of cosmic radiation, free convection, disturbances of magnetic and electrical fields are characteristic physical, abiotic features of space biotechnology (ZIMMERMANN et al., 1988).

In this context, it must be pointed out that the numerous experiments describing the mechanisms of graviperception in specialized gravisensitive cell systems are not presented here because a detailed discussion of these data is not within the scope of this chapter. I recommend to the reader to refer to related literature. Moreover, the connection of electrophoretic separation of cells and bioreactor systems in space with biotechnological aspects is not cited here because it has been discussed in detail elsewhere (ZIMMERMANN et al., 1988). Similarly, data concerning electrofusion processes prior to 1990 are not mentioned, again since they have already been documented.

2 Variations of Cellular Functions in Space

2.1 Effects of Microgravity on Cellular Functions of Active Prokaryotes, Unicellular Eukaryotes, and Cell Cultures

Data from previous spaceflights have indicated that unicellular organisms show differences in cellular functions when grown under ground or μg conditions. Interpreting these data it has to be kept in mind that the relevance of μg conditions has to be evaluated with respect to the existence of additional physical parameters (see above). Therefore, the μg data have to be compared not only to those obtained on the ground but also against a 1 g reference centrifuge aboard the spacecraft in order to exclude those abiotic effects in space.

Cultures of *Bacillus subtilis* and *Escherichia coli* were cultivated in a 1 g and μg-environment, on the D2-Mission without indicating any μg-induced difference in genetic stability (genomic mutation, loss of plasmid; MENNIGMANN and HEISE, 1995). This "proves" the genetic stability and "disproves" the importance of radiation. In the same mission, survival rate and mutagenic effects of *Saccharomyces cerevisiae* were studied (DONHAUSER et al., 1995). Phenotypically altered or respiration deficient clones were detected in orbit and ground samples; auxotrophic mutants were not found. The karyotopes of total yeast populations remained stable, however, the karyotopes of a few isolated clones from orbit were altered.

THEIMER et al. (1986) reported of an increase in cell differentiation in cell cultures of *Pimpinella anisum* under g and μg conditions during a 7 d spaceflight; in *Arabidopsis* tissue culture the number of proliferative activity and the cell sizes were smaller after a 63 d spaceflight than in the ground control (MERKYS et al., 1988). Whereas KLIMSHUK et al. (1992) and KORDYUM et al. (1995) could not find essential differences in the ultrastructural organization, the ploidy level, and the differentiation process in *Haplopappus* tissue cultures after spaceflights of 9–28 d varied. JUNG-HEILIGER et al. (1995) investigated the μg effects on secondary metabolite production in pharmaceutically relevant cell suspension cultures (*Aesculus hippocastanum*), however, without finding any change in ploidy level or in the content of secondary metabolites. SEITZER et al. (1995) cultured human fibroblasts under different gravity conditions in order to demonstrate a change in the synthetic capacity of collagen and DNA. Although in space bone mass is known to be reduced, a relevant difference was not identified.

Chromosomal anomalies – breakages, deletions, and translocations – and damage in embryogenic somatic cells of wild carrot, oats and sunflower were identified under different flight conditions (Soviet Biosatellite, 3 weeks, and U.S. shuttle as "mid-deck" experiment for 8 and 14 d) by the group of KRIKORIAN (KRIKORIAN, 1989, 1991, 1996). Using the model system of the liliaceous monocotyledon daylily (*Hemerocallis*) somatic embryogenic cells and those of the dicotyledon *Haplopappus gracilis*, karyological disturbances were found in root tips and basal meristems of the species grown in the NASA's plant growth unit equipped with a special air exchange system (LEVINE and KRIKORIAN, 1992; KRIKORIAN et al., 1995; KRIKORIAN, 1996). Both systems offered advantages because they could be managed on different cell and organ levels and could be cloned and grown in the low light environment characteristic of the growth unit. Attempts to repeat and extend the results with daylily embryogenic cells (8 d in NASA mid-deck and 14 d in IML-2, International Microgravity Laboratory 2 Mission) resulted in binucleate cells (more than 10%), in contrast to ground samples with uniformly uninucleate cells. The conditions of in-flight fixation emphasize that chromosomal damages were not due to reentry effects (KRIKORIAN, 1996). Further chromosomal aberrations with breaks and laggards were analyzed using division figures from wheat root cells stopped by prefixation with colchicine in space (TRIPATHY et al., 1996). Also in this experiment, no damage was detected in the ground controls. The reason for the chromosomal anomalies found especially in roots of space grown plants is still unclear. Numerous indirect effects are hypothesized to be responsible for the phenomenon of "space stress". Some perturbation factors, such as radiation, vibrations, changes in physical properties of fluids, of surface tension, geomagnetic and electrical disturbances in cellular processes have been discussed in this context (KRIKORIAN, 1996).

All of these partially contradictory data give an impression of the vast body of published information which has to be reviewed with respect to different hardware systems that often failed to provide rigorously controlled or defined environmental conditions. A longer flight duration (several weeks) is required because only short-term adaptation responses of cells have so far been measured instead of pure microgravity effects (HALSTEAD et al., 1991). The possibility of carrying out long-term experiments that should be offered by the existence of a space station has to be discussed in this context (KRIKORIAN, 1996). Moreover, quantitative approaches to distinguish between the effects of launch, flight, and re-entry of samples into the atmosphere are required. Experiments should be designed to separate these effects.

2.2 Effects of Microgravity on Cellular Functions of Plant Protoplasts – Adaptation Strategies

The presence of a cell wall protects the plant cell, minimizing the vulnerability of the "naked" protoplasts to environmental effects. Therefore, cell-wall free protoplasts present a system with a high sensitivity to a spectrum of effectors, such as spaceflight specific conditions in a non-optimal environment. They show short-term responses as symptoms of an adaptation mechanism. Under optimized conditions, however, they offer a well adapted, extremely sensitive system without showing any adaptation responses.

Since plants have evolved under the constant stimulus of gravity its presence is the prerequisite for growth and spatial orientation of plants. The sensitivity to gravitational control and the mechanisms of graviperception in intact plants as well as the detailed gravity effect on development and differentiation of single plant cells are mostly unknown. Without the permanent g vector, the framework of plant body is disrupted. Therefore, the removal of gravity, i.e., the condition under microgravity, is indicated as a factor for plants, disturbing the physiological balance. In this section the adaptation mechanisms in different cell systems to the conditions of microgravity are described.

On the basis of clinostat treatments many data using single cells show contradictory results (IVERSEN, 1985; IVERSEN et al., 1992). A lot of flight opportunities for plant cells were offered during the last 10 years using sounding rockets (7 and 15 min of weightlessness) and space shuttle (up to 10 and 15 d of weightlessness). Plant cell protoplasts are suitable objects for studying microgravity effects because they can be isolated from almost all higher plants in order to investigate processes of cell wall biosynthesis, cell division, callus formation, and plant regeneration. The protoplasts offer an elegant model system to apply biochemical and molecular techniques in order to analyze signal perception and transduction mechanisms for monitoring signals of abiotic and biotic stimuli and to indicate adaptation responses.

RASMUSSEN et al. (1992) showed a μg-induced delay in new cell wall synthesis using *Brassica napus* and *Daucus carota* protoplasts on board of the Soviet satellite "Biokosmos 9" in comparison with ground controls. There was, however, no 1 g control in orbit, so a further experiment with identical protoplast cultures was performed on board of the space shuttle with an 1 g centrifuge (IML-1 mission, NASA space shuttle "Discovery", 8 d; RASMUSSEN et al., 1994a). In orbit cells were swollen, showed larger vacuoles, formed incomplete cell walls, revealed a reduced specific activity of peroxidase, an increased concentration of soluble proteins per cell, and a retardation of regeneration processes. Calli derived from those protoplasts showed reduced growth without any signs of shoot formation in contrast to 1 g controls which developed highly regenerated shoots (RASMUSSEN et al., 1994b). An inhibition of cell wall synthesis in potato protoplasts treated under microgravity was also confirmed during clinostating by NEDUKHA et al. (1994). They showed only 10% of protoplasts being regenerated after clinostating. The authors assume that a retardation in the regeneration rate under microgravity is induced by an effect in the cytoskeleton system or by stress-induced *μg* factors, or both.

Using protoplasts from *Nicotiana tabacum* and *N. rustica*, these data could not be confirmed by HOFFMANN et al. (1996) and HAMPP et al. (1997). Their results on protoplast culture and regeneration ability gave no indication of any *μg*-induced modifications. They did not find any differences in protein contents and in the pattern of SDS electrophoresis. Furthermore, they investigated increased ratios in ATP/ADP and NADH/NAD and pool sizes of fructose-2,6-bisphosphate which they interpreted to be indicative of metabolic stress conditions given by short-term sounding rockets (microgravity 6 min). However, when protoplasts were exposed to microgravity for a long-term period (spacelab, 10 d) a reduction of energy and redox ratios as well as pool sizes of fructose-2,6-bisphosphate were observed. This decreased metabolic turnover was speculated to reflect a metabolic relaxation. As rates of protoplast regeneration were identical under 1 *g* and *μg*, the authors concluded that the demand for metabolic energy necessary in the *μg*-induced regeneration process is lower than that on the ground (HAMPP et al., 1997).

Several authors have proposed a close relationship between the treatment of protoplasts with microgravity and unspecific adaptation mechanisms (stress response), such as reduced growth rate of protoplast cultures grown under microgravity, elevated level of soluble proteins, and modified energy ratios. Cells under stress are known to produce a higher level of soluble proteins (SACHS and HO, 1986), the functional role of which is not clearly understood. In order to exclude adaptive strategies of protoplasts to environmental factors the experimental conditions have to be differentiated between short-term (sounding rockets) and long-term (spaceflight missions) conditions. This hypothesis is documented by experimental long-term conditions: When protoplasts cultivated in 10 d flight under microgravity were analyzed with respect to their protein levels and patterns (SDS), no gravity-dependent protein modification was found (HOFFMANN et al., 1996; IVERSEN et al., 1992).

In contrast, isolated protoplasts of *Vicia faba* which were exposed to short-term experimental conditions of a sounding rocket (TEXUS 30, 32, 6 min under microgravity), showed a dramatic modification in protein levels after some seconds (SCHNABL et al., 1996, 1997; SCHNABL, 1998). Western immunoblotting with ubiquitin antibodies revealed distinct ubiquitin conjugates (18 kDa, 19 kDa, 40 kDa) as well as the free ubiquitin level shown in Fig. 1A–D. The decrease and increase of free ubiquitin indicate an activation of the ubiquitin system. Free ubiquitin is linked to substrate proteins and bound ubiquitin is finally released by the proteasome complex (FINLEY and CHAU, 1991). As shown in Fig. 1, the 18, 19

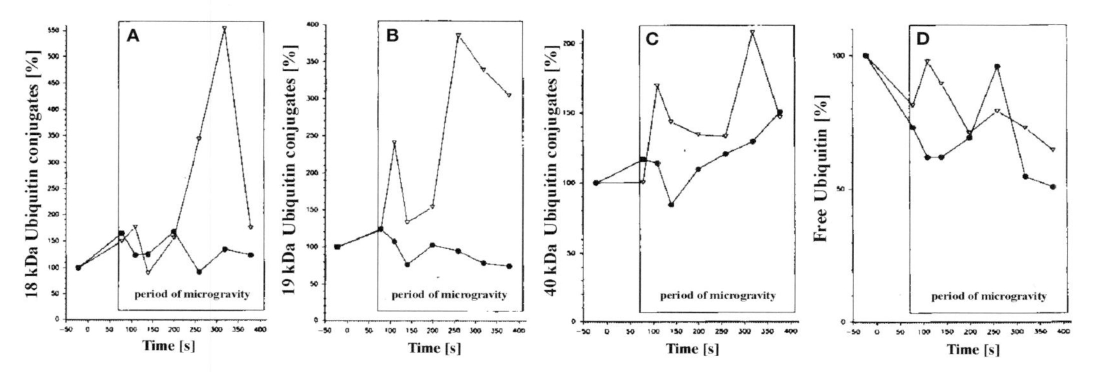

Fig. 1. Relative changes (%) of 18 **A**, 19 **B**, 40 kDa **C** ubiquitin conjugates, and free ubiquitin **D** in mesophyll protoplasts of *Vicia faba* during the TEXUS 32 flight (△) and control (●), box: *μg* period.

and 40 kDa proteins oscillating in real and in simulated microgravity conditions (data not shown) deliver hints for reversible ubiquitinylation, which represents an indicator for ubiquitin-dependent activities in protoplasts under stress. These effects are also shown for other stress parameters, such as temperature, dryness, heavy metals, and ozone (FERGUSON et al., 1990; SCHULZ et al., 1994). The experiments were confirmed by clinostating (SCHULZ et al., 1992; WOLF et al., 1993a, b; HUNTE et al., 1993, 1994).

Since the cytoskeleton system is suggested to be affected by gravity, modifications of actin isoforms have to be investigated under real (Fig. 2A–D) and simulated microgravity in comparison to *g* control. In both microgravitiy conditions (clinostating, as well as TEXUS 30 and 32) a decrease of the amount of isoforms was measured in contrast to the *g* control. The physiological meaning of the decrease in actin isoforms is not clearly understood. A part of the actin isoforms may be thought to loose function under μg and, as a consequence, is degraded. Another theory is that the actin isoform decline could be an adaptation effect induced by proteolytic activities (SCHNABL et al., 1996, 1997; SCHNABL, 1998). The four actin isoforms known to be constitutive polypeptides of different tissues and organs show a specific distribution in cellular compartments of protoplasts: 1 and 3 in soluble fractions, 2 in the tonoplast fraction, and 4 in the plasma membrane (JANßEN et al., 1996). On the basis of

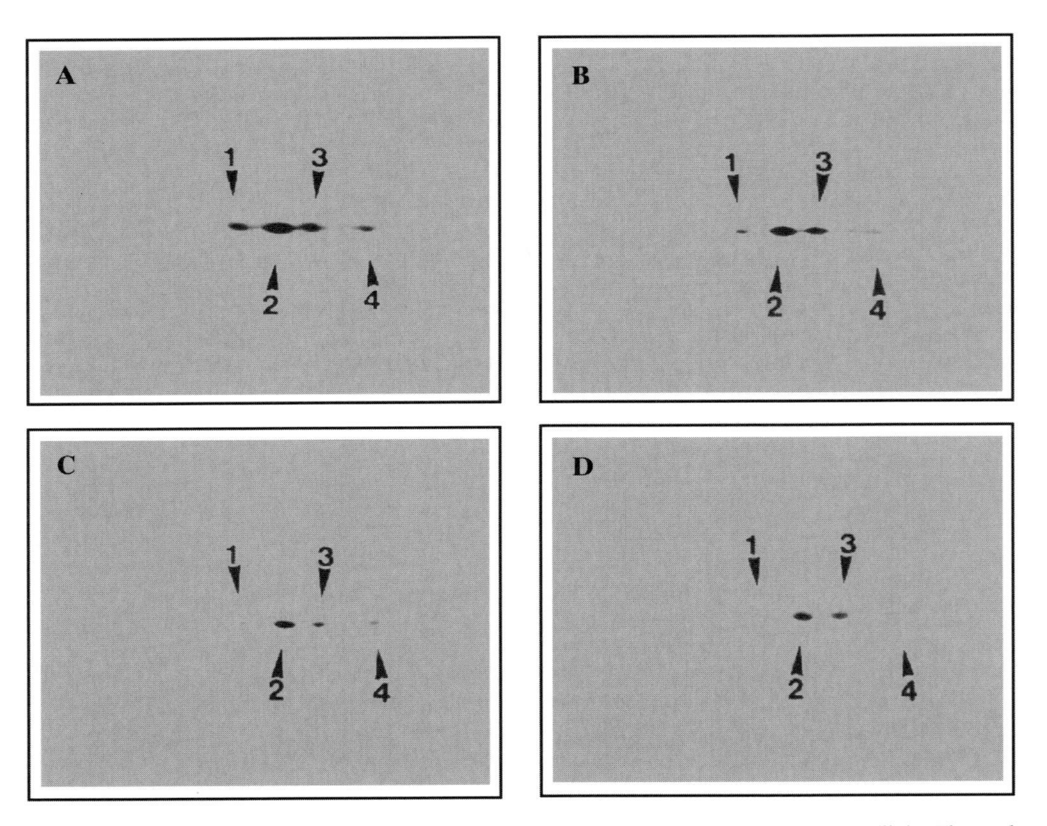

Fig. 2. Actin immunoblots of mesophyll protoplasts of *Vicia faba* during the TEXUS 32 flight. The molecular mass of the immunostained spots is 43 kDa. Arrows mark different actin isoforms with p*I* values of 6.05 (1), 6.0 (2), 5.95 (3), and 5.9 (4). Flight phases: before the launch **A**, after the boost phase **B**, during the μg phase **C**, at the end of the μg phase **D**.

these data actin isoforms are suggested to function as gravirelevant proteins of the microfilaments.

3 The Signal Perception and Transduction of Graviresponse by Cells

There is strong evidence that in higher plants the perception of gravity is coupled to the sedimentation of statoliths (VOLKMANN and SIEVERS, 1979). The coupling of the pressure-induced process to signal transduction responses following the early event of statolith sedimentation is still unclear. Although the bending of an organ (shoot/root) is the final event during gravitropism, analyzing the visible part of this response will provide more information about the time-scale of the biochemical reactions involved. Even the visible part of a typical bending reaction is rather complex (ISHIKAWA et al., 1990). Therefore, the current model of the gravisresponse pathway with statoliths and auxin as elements of a signal transduction chain is not absolutely adequate to explain the overall bending reaction of a hypocotyl in detail and needs further refinement. The displacement of starch statoliths in the endodermis of a hypocotyl is suggested to be responsible for the beginning and the direction of the graviresponse. Surprisingly, starchless mutants are also able to respond to gravity (KISS and SACK, 1989, 1990) supporting the hypothesis that there exists at least one alternative/additional starch-independent pathway. An elastic deformity of the hypocotyl on the basis of increased tissue pressure/tension on the lower/upper sides might result in an opening of stretch-activated channels during gravi-stimulation and the bending reaction (MILLET and PICKARD, 1988). Recent results obtained with flagellates and ciliates in rocket missions indicate that the pressure of sedimenting cytoplasm stimulates stretch-activated, mechanosensitive ion channels in the cytoplasmic membrane penetrated probably by calcium ions (HÄDER, 1999).

Thus, reactions coupled with sedimentation of statoliths are hypothesized to be associated with the signal transduction network equipped with a dual- or multisensory system. In this case, cells of an organ might use different or the same input signal(s) such as gravity, light, moisture, touch, pressure, or tension which are translated and integrated into a complex biochemical response. According to the gravitational pressure model the protoplast as the smallest functional unit may also perceive change in orientation of the gravity vector. By linking the extracellular matrix to the cytoskeleton or to the enzymes of a signal transduction pathway (protein kinases, phospholipid kinases, phospholipase C), integrins may transmit the modulation of gravitational forces across the plasma membrane. Integrins as putative elements of an alternative signal transduction chain have been discussed by RANJEVA et al. (1999) and identified in *Arabidopsis* (SWATZELL et al., 1999).

As far as the input signal of the integrin- and auxin-dependent signal transduction pathways is concerned, models for the two pathways differ significantly (CLARK and BRUGGE, 1995). However, both systems are mediated by inositol 1,4,5 trisphosphate (Ins) (1,4,5) P_3 as a second messenger. In both cases Ins (1,4,5) P_3 is produced by phospholipase C (PLC) which cleaves phosphatidyl inositol 4,5 diphosphate into 1,2 diacylglycerol and Ins (1,4,5) P_3. The latter is able to release Ca^{2+} from Ins (1,4,5) P_3-sensitive Ca^{2+} stores, also acting as a second messenger. Many experiments have led to the conclusion that Ca^{2+} functions in gravitropic signal transduction pathways, although the exact mechanism has not been elucidated up to now. Changes in Ca^{2+} levels seem to be necessary to create auxin gradients across gravistimulated organs (HASENSTEIN and EVANS, 1986; EVANS et al., 1990; YOUNG and EVANS, 1994). The group around TREWAWAS found strong similarities between growing rhizoid and orienting pollen tube growth. In a series of publications (MALHO et al., 1993, 1995; MALHO and TREWAWAS, 1996; FRANKLIN-TONG et al., 1996) they described a transient increase in Ca^{2+} which inhibited growth and induced pronounced changes in orientation some minutes later. The pollen tube reorientated its growth towards the side of the apical dome which contains the higher Ca^{2+} concentration. Release of Ca^{2+} induces a clear Ca^{2+} wave which is in part generated and maintained by Ins (1,4,5) P_3. There is evidence that Ca^{2+} ions affect gravitropic signal transduction by interacting with calcium binding proteins (SINCLAIR et al., 1996). LU et al. (1996) found a calcium/calmodulin-dependent protein kinase (see also WATILLON et al., 1995) homologously expressed in root caps, the genomic sequence of which was given. They hypothesize that in light-dependent gravitropism the kinase may be associated with the establishment of auxin gradients perhaps by activating a proton pump leading to a redistribution of auxin. RASHOTTE et al. (2000) showed that *Arabidopsis* mutants with agravitropic roots exhibited reduced basipetal auxin transport in contrast to wild types. Basipetally transported auxin is suggested to control root gravitropism in *Arabidopsis*.

First evidence for an auxin-dependent synthesis of Ins (1,4,5) P_3 was published by ETTLINGER and LEHLE (1988) showing a 3-fold increase within 60 s after auxin application. In another more direct approach PERERA et al. (1999) were able to detect a 5-fold transient increase of Ins (1,4,5) P_3 in the lower halves of maize pulvini within 10 s after gravistimulation. In parallel, they found a higher phosphatidylinositol-4-phosphate 5-kinase activity (30%) after 10 min. Furthermore, they measured a gradual increase in Ins (1,4,5) P_3 over a period of 2–7 h to 5 to 6-fold above the control in lower halves of pulvini. The data indicate that changes in the metabolism of phosphoinositides are involved in both the initial signaling events (10 s) as well as in the following process of commitment to differential growth with cell elongation (2–7 h). They suggest a short- and long-term involvement of Ins (1,4,5) P_3 signaling in gravistimulation and assuming a role for calcium in the gravity signal transduction cascade also in starch-deficient cells due to the concerted cell elongation.

Using gravistimulated hypocotyl segments (*Helianthus annuus*) as well as isolated protoplasts as a suitable model system, the phosphoinositid system was shown to be one element of the signal transduction chain (SCHNABL, 1998; MÜLLER et al., 1998). Sunflower hypocotyl protoplasts (HCP) stimulated by auxin (Fig. 3A), treated under simulated (clinostating; Fig. 3B) and real microgravity (TEXUS 35 flight; 6 min weightlessness; 1 *g*, 8 *g*, μg, 13 *g*; Fig. 3C) showed an increase of Ins (1,4,5) P_3 after 15 s (auxin), 180 s (clinostat), and about 6 min (sounding rocket) as detected with a highly isomer-specific mdd-HPLC method (metal dye detection) compared to the 1 *g* control. To test whether or not this increase was due to microgravity or hypergravity, protoplasts were exposed to hypergravity conditions (8 *g*, 13 *g* at different times), which however revealed no changes (not shown). Therefore, there is strong evidence that the late increase of Ins (1,4,5) P_3 during the TEXUS flight (Fig. 3C) was due to the μg period and not due to the short hypergravity stimulus at the end of the flight. These data provide strong

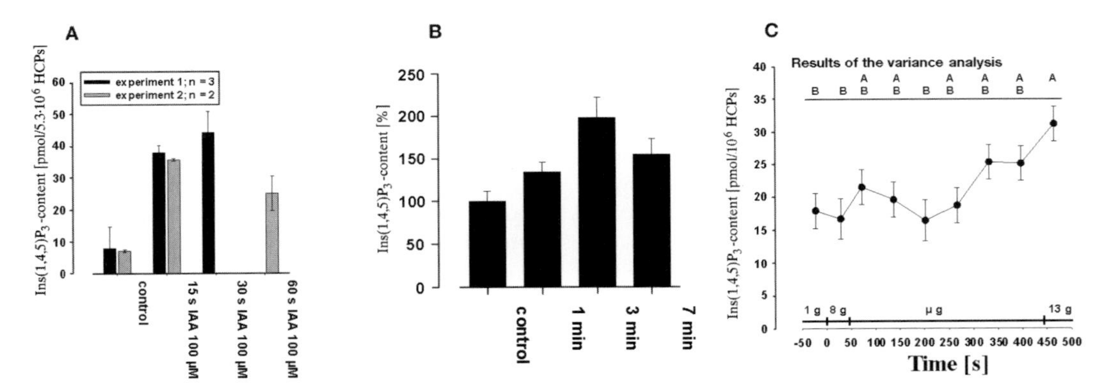

Fig. 3. Contents of Ins (1,4,5) P_3 in sunflower hypocotyl protoplasts stimulated by 100 µM auxin **A**, simulated μg conditions (clinostating) **B**, and real microgravity (TEXUS 35 flight; 6 min µg conditions **C**).

evidence of a direct involvement of a signal transduction chain, linking graviperception to graviresponse in starchless protoplasts, the smallest functioning cell unit of plants which are able to sense a g-modulation.

In this cascade the Ins (1,4,5) P_3 signal was shown to be transduced via Ca^{2+} modulating gene expression. The regulation of touch (TCH) genes from *Arabidopsis* demonstrated to be a rapid response to diverse stimuli such as touch, darkness, heat, and cold shock (BRAAM and DAVIS, 1990; BRAAM, 1992; BRAAM et al., 1996), as well as to the phytohormones auxin and brassinosteroids (ANTOSIEWICZ et al., 1995; XU et al., 1995), The induction of the TCH genes was revealed to occur very rapidly (within 10 min), to be transient with high turnover rates, and to have an increase in mRNA of high magnitude. There were good correlations between increases in Ca^{2+} levels and TCH gene upregulation of expression.

Evidence for alternative/additional gravisensing systems are provided by genetic studies. Six independent loci for shoot gravitropism were identified (FUKAKI et al., 1996; YAMAUCHI et al., 1997). A specifically expressed gene which was up/down-regulated during graviresponse of sunflower hypocotyls was found using differential display RT-PCR (THEISEN and SCHNABL, unpublished data; SCHNABL, 1998; MÜLLER et al., 1998). The identification of auxin-induced genes as opposed to genes which are expressed in response to an altered gravity vector might supply additional information about the mechanism involved.

4 Electrofusion of Plant and Animal Cells

Electro cell fusion is known as a powerful tool to generate somatic hybrids of different animal and plant cells (ZIMMERMANN, 1986; ZIMMERMANN et al., 1988). But there exist some inherent physical and technical parameters which reduce the yield of vital hybrids obtained under terrestrial conditions:

(1) separation of particles of different specific densities in suspension and the problems with alignment of heterospecific protoplasts involved,
(2) disruption of already fused protoplasts due to thermal convectional forces.

By means of increasing the field strength, the duration of the alignment phase, and the number of electrical pulses the improvement of the yield of hybrids could be obtained but at the expense of their vitality (ZIMMERMANN, 1986, 1987). Under microgravity conditions it would be possible to apply lower field strengths of the alternating field during the alignment phase and to minimize the number of electrical pulses because gravitational and thermal convection forces are eliminated (ZIMMERMANN et al., 1988).

Higher plant protoplasts released from cell walls are attractive objects for manipulating their genetic properties using somatic electro cell fusion between sexually incompatible parental species. Due to their potential for regeneration into intact plants the hybrids offer a convenient model system with agronomically essential traits such as pest resistance or cytoplasmic male sterility. Since major crops in the world are not protected sufficiently against pathogen attacks without applying pesticides, it is of great interest to introduce genetic properties of its wild-type species. Wild types are often equipped with a spectrum of interesting properties such as resistance against abiotic factors (salt, heavy metal, and temperature tolerance) as well as biotic parameters (insect, viral and fungal resistance). However, sexual incompatibilities often present an insurmountable barrier for traditional breeding technologies. The interspecific hybridization achieved by somatic cell fusion represents a potential biotechnological method for transferring desirable properties, thus avoiding crossing effects.

The sunflower is one of those interesting crops with a high sensitivity to many fungal attacks which lower annual yields substantially. A wide spectrum of wild-type species containing resistance against pathogens is offered as a fusion partner with cultivar sunflower. Suitable resistant perennial wild types of *Helianthus* were selected and generated by *in vitro* propagation. On the basis of a protocol for the isolation and regeneration of *Helianthus* wild type (HENN et al., 1998 a, b) and cultivar species (WINGENDER et al., 1996), both types of protoplasts served as a suitable source for fusion using chemicals (PEG) or electric pulses. Finally, two types of protoplasts from different source tissues with varying densities – the green mesophyll cell protoplasts of wild sunflower species with a diameter of about 33 μm and fairly brightly colored hypocotyl cell protoplasts of the sunflower cultivar with a diameter up to 70 μm – were electrically fused. As mentioned previously, rapid separation of particles with different sizes and densities results in a low yield of somatic hybrids under 1 *g* conditions. During the Second German Spacelab Mission D-2 the *μg* conditions of electrofusion processes were applied using sunflower protoplasts which were electrofused and subsequently cultured for 10 d under weightlessness (VON KELLER et al., 1995). A higher fusion rate under *μg* (about 26% and 15% compared with that under *g*) was achieved when fusion parameters such as alignment time (15 s at 1 *g* and 7 s at *μg*) and number of pulses (3 at 1 *g* and 2 at *μg*) were reduced. 50% of the fusion products generated under microgravity were heterospecific, representing a final hybrid yield of about 13%, whereas the rate of hybrids obtained under 1 *g* conditions was less than 4.5%. The absence of gravitational and thermal convection forces permitted a more heterogeneous pearl chain formation of protoplasts with different densities. Subsequent fusion, therefore, resulted in a higher rate and vitality of hybrids. Homo- and heterospecific fusion products were investigated for organelle structures in the home laboratory using electron microscopy. No significant differences between the fusion products derived from *μg* and *g* samples were detected (VON KELLER et al., 1995).

Similar data were found using plant cell protoplasts derived from leaf tissues of two different tobacco species (HOFFMANN et al., 1995; HAMPP et al., 1997). The time needed to establish close membrane contact between protoplasts (alignment time) was reduced from 15 s under 1 *g* to 5 s under μg. The number of fusion products increased by a factor of about 10. In addition the vitality of hybrids increased from 60% to more than 90%. The wide range of intermediate products was specified using isoenzyme analysis and RAPD-PCR. Also, EISENBEISS et al. (1995) could confirm similar results using protoplasts from *Digitalis lanata* and *D. purpurea* by finding an interspecific fusion product with the expression of a double hydroxylation ability of both clones fused. Under microgravity conditions 6.7% of the fusion pairs were found to be heterospecific compared with 0.7% under ground conditions. The cells regenerated well with no significant differences between *μg* and *g* conditions. However, no hybrid cell line was found among the regenerated *Digitalis* clones.

5 Exobiology and Radiation Biology in Space

To determine the radiation risk to humans in space as well as on the ground experiments are necessary which evaluate the impact of the extraterrestrial spectrum of solar UV radiation on chromosomal aberrations. QUINTERN et al. (1992) developed a biological monitoring system consisting of a biofilm of dried spores of the bacterium *Bacillus subtilis* immobilized on a plastic sheet. The UV radiation was monitored by biological UV dosimetry using these biofilms which determined the solar radiation according to its DNA damage and thereby carcinogenic potential (HORNECK, 1992, 1995; HORNECK et al., 1996). The biofilm treated under defined conditions (i.e., at the body of cosmonauts, inside or outside the space stations, or under a diminishing ozone layer) is developed in a nutrient medium to germinate the bacterial spores. The biomass produced was quantified using an image analysis system. The biofilm technique was applied during a long-term measurement campaign in Antarctica to study the harmful proportion of solar UV-radiation on the ground as a consequence of reduced ozone concentrations (QUINTERN et al., 1994). Experiments on different missions (MIR 97; IML-2; Spacelab D 2) presented information about the degree of UV-damage which is extremely low in the stations, but highly DNA damaging when penetrating. Experiments investigating the effects of microgravity on the efficiency and kinetics of repair mechanisms in radiated samples as well as the breakage of DNA strands did not elucidate the reasons of the combined effects of radiation and microgravity. Therefore, further studies are necessary to explain this phenomenon.

6 Conclusions

Studies at the biochemical and genetic level have demonstrated that microgravity has significant effects on fundamental biological processes such as stress metabolism, DNA stability, signal transduction, and gene expression. It is becoming more evident that organization of genes and signaling proteins within the cell nucleus, cytoplasm, cytoskeleton and the regulatory macromolecules within the extracellular matrix contribute to the physiological responses of gene expression. The functional interrelationships between cell structure and gene expression within the cell organization have to be studied more intensively under simulated and real microgravity and regular earth gravity conditions. On the basis of the data collected within the last 10 years potential areas of future investigations were recently developed by NASA, which should contribute to increase understanding of cell biology of terrestrial life in space and the effects of gravity on cell biology on earth. In this NASA program, microgravity effects are planned in detail to provide insight into fundamental aspects of biological regulation such as signal transduction and gene expression that will be important in terrestrial as well as extraterrestrial surroundings. In order to understand the gravity-induced perception mechanism and its physiological-biochemical responses in roots and shoots of plants, it is a prerequisite to treat plants or organelles under g and μg conditions. Since we are only able to elucidate effects of a biotic or abiotic stimulus by comparing its responses in plants with and without the corresponding signals, we have to establish experiments under both kinds of conditions. In the past the g-vector was frequently varied using "simulated microgravity" (clinostating). The data presented in Sect. 2.2 have shown that clinostating is not suitable to replace spaceflights, rocket programs, shuttle and spaceship missions because stress effects as adaptation mechanism can not be excluded.

Therefore, the author is convinced that more definitely equipped experiments under real microgravity should be available in the future for delivering significant results under true μg conditions. Only on the basis of those results the fundamental questions for gravity-regulated growth of plants on the ground can be answered. These experiments under "real microgravity" should offer flights of longer duration in order to minimize effects which are detected at the cellular level, since these might reflect short-term adaptation responses rather than true microgravity effects. Thus, only in a space shuttle which offers a longer duration of μg conditions, the real μg-induced effects on plant growth can be detected and evaluated fundamentally.

Well carried-out experiments on earth and in orbit in the field of plant biotechnology are the result of a long history of experience. Only when the environmental growth parameters can be reliably controlled in space will it be possible to pursue long-term work in a plant growth facility.

Acknowledgements.
The financial support and the flight opportunities in D2 mission and a series of rocket programs (TEXUS) is gratefully acknowledged to DLR (Deutsches Luft- und Raumfahrtzentrum, Köln-Porz, Germany) and to MWF (Ministerium für Wissenschaft und Forschung, NRW, Düsseldorf, Germany), Düsseldorf, NRW, Germany. I thank Dr. G. Müller for discussion and careful reading of the manuscript.

7 References

ANTOSIEWICZ, D. M., POLISENSKY, D. H., BRAAM, J. (1995), Cellular localization of the Ca^{2+}-binding TCH3 protein of *Arabidopsis,* Plant J. **8**, 623–636.

BRAAM, J. (1992), Regulated expression of the calmodulin-related TCH genes in cultured *Arabidopsis* cells: induction by calcium and heat shock, *Proc. Natl. Acad. Sci. USA* **89**, 3213–3216.

BRAAM, J., DAVIES, R. W. (1990), Rain-, wind- and touch-induced expression of calmodulin and calmodulin-related in *Arabidopsis, Cell* **60**, 357–364.

BRAAM, J., SISTRUNK, M. L., POLISENSKY, D. H., XU, W., PURUGGANAN, M. M. et al. (1996), Life in a changing world: TCH gene regulation of expression and responses to environmental signals, *Physiol. Plant* **98**, 909–916.

BRIEGLEB, W. (1984), Acceleration reactions of cells and tissues – Their genetic phylogenetic implications, *Adv. Space Res.* **4** (12), 5–7.

BRIEGLEB, W. (1992), Some qualitative and quantitative aspects of the fast-rotating clinostat as a research tool, *ASGSB Bull.* **5** (2), 23–32.

CLARK, A. C, BRUGGE J. S. (1995), Integrins and signal transduction pathways: The road taken, *Science* **268**, 232–239.

DONHAUSER, S., WAGNER, D., SPRINGER, R., BELLMER, H. G., GROMUS, J. (1995), D-2 Mission: results of the yeast experiment, in: *Scientific Results of the German Spacelab Mission D-2* (SAHM, P. R., KELLER, M. H., SCHIEWE, B., Eds.), pp. 551–557. Köln: DLR.

EISENBEISS, M., BAUMANN, T., REINHARD, E. (1995), Electrofusion of protoplasts from *Digitalis lanata* and *D. purpurea* suspension cultures under microgravity conditions and enzymatic characterization of the fusion products, in: *Scientific Results of the German Spacelab Mission D-2* (SAHM, P. R., KELLER, M. H., SCHIEWE, B., Eds.), pp. 657–663. Köln: DLR.

ETTLINGER, C., LEHLE, L. (1988), Auxin induces rapid changes in phosphatidylinositol metabolites, *Nature* **331**, 176–178.

EVANS, M. L., STEINMETZ, C. L., YOUNG, L. M., FONDREN, W. M. (1990), The role of calcium in the response of roots to auxin and gravity, in: *Plant Growth Substances 1988* (PHARIS, R. P., ROOD, S. B., Eds.), pp. 209–215. Heidelberg: Springer-Verlag.

FERGUSON, D. L., GUIKEMA, J. A., PAULSEN, G. M. (1990), Ubiquitin pool modulation and protein degradation in wheat roots during temperature stress, *Plant Physiol.* **92**, 740–749.

FINLEY, D., CHAU, V. (1991), Ubiquitination, *Annu. Rev. Cell. Biol.* **7**, 25–30.

FRANKLIN-TONG, V. E., DROBAK, B. K., ALLAN, A. C., WATKINS, P. A. C., TREWAWAS, A. J. (1996), Growth of pollen tubes of *Papaver rhoeas* is regulated by a slow moving calcium wave propagated by inositol-1,4,5-triphosphate, *Plant Cell* **8**, 1305–1321.

FUKAKI, H., FUJISAWA, H., TASAKA, M. (1996), SG1, SG2 and SG3: novel genetic loci involved in shoot gravitropism in *Arabidopsis thaliana*, *Plant Physiol.* **110**, 933–943.

HÄDER, D. (1999), Graviperception and graviorientation in flagellates and ciliates, in: *14th ESA Symp. Eur. Rockets and Balloon Programmes and Related Research.* Potsdam, Germany (ESA SP-437, September, 1999).

HALSTEAD, T. W., TODD, P., POWERS, J. (1991), Gravity and the Cell, *Am. Soc. Grav. Biol. Bull.* **4** (2), 1–260.

HAMPP, R., HOFFMANN, E., SCHÖNHERR, K., JOHANN, P., DE FILLIPIS, L. (1997), Fusion and metabolism of plant cells as affected by microgravity, *Planta* **203**, 42–53.

HASENSTEIN, K. H., EVANS, M. L. (1986), Calcium dependence of rapid auxin action in maize roots, *Plant Physiol.* **81**, 439–443.

HEMMERSBACH-KRAUSE, R., BRIEGLEB, W. (1994), Behaviour of free-swimming cells under various accelerations, *J. Gravit. Physiol.* **1**, 85–87.

HENN, H. J., WINGENDER, R., SCHNABL, H. (1998a), Regeneration of fertile plants from mesophyll protoplasts from *Helianthus giganteus* and *H. nutallii, Plant Cell Rep.* **18**, 288–291.

HENN, H. J., WINGENDER, R., SCHNABL, H. (1998b), Regeneration of fertile interspecific hybrids from cell fusions between *Helianthus annuus* L. and wild *Helianthus* species, *Plant Cell Rep.* **18**, 220–224.

HOFFMANN, E., SCHÖNHERR, K., JOHANN, P., HAMPP, R. (1995), Culture and electrofusion of plant cell protoplasts under microgravity: morphological and biochemical characterization, in: *Scientific Results of the German Spacelab Mission D-2* (SAHM, P. R., KELLER, M. H., SCHIEWE, B., Eds.), pp. 641–656. Köln: DLR.

HOFFMANN, E., SCHÖNHERR, K., HAMPP, R. (1996), Regeneration of plant protoplasts under microgravity: investigation of protein patterns by SDS page and immunoblotting, *Plant Cell Rep.* **15**, 914–919.

HORNECK, G. (1992), Radiobiological experiments in space: A Review, *Nucl. Tracks Radiat. Mears.* **20**, 185–205.

HORNECK, G. (1995), Quantification of biological effectiveness of environmental UV-radiation, *J. Photochem. Photobiol. B. Biol.* **31**, 43–49.

HORNECK, G., RETTBERG, P., RABBOW, E., STRAUCH, W., SECKMEYER, G. et al. (1996), Biological dosimetry of solar radiation for different simulated ozone column thicknesses, *J. Photochem. Photobiol. B. Biol.* **32**, 189–196.

HUNTE, C., SCHULZ, M., SCHNABL, H. (1993), Influence of clinostat rotation on plant proteins: Effect on membrane bound enzyme activities and ubiquitin conjugates of leaves of *Vicia faba, J. Plant Physiol.* **142**, 31–36.

HUNTE, C., WOLF, D., GHIENA-RAHLENBECK, C., SCHULZ, M., SCHNABL, H. (1994), Influence of clinostat rotation on soluble and membrane bound proteins of *Vicia faba* leaves, *Proc. 5th Eur. Symp. Life Sciences in Space*, Arachon, France (ESA SP-366).

ISHIKAWA, H., HASENSTEIN, K. H., EVANS, M. L. (1990), Computer-based video digitizer analysis of surface extension on maize roots, *Planta* **183**, 381–390.

IVERSEN, T. H. (1985), Protoplasts and gravireactivity, in: *The Physiological Properties of Plant Protoplasts* (PILET, P., Ed.), pp. 236–249. Berlin: Springer-Verlag.

IVERSEN, T. H., RASMUSSEN, O., GMUNDER, F., BAGGERUD, C., KORDYUM, E. L. et al. (1992), The effect of microgravity on the development of plant protoplasts flown on Biokosmos 9, *Adv. Space Res.* **12** (1), 123–131.

JANßEN, M., HUNTE, C., SCHULZ, M., SCHNABL, H. (1996), Tissue specification and intracellular distribution of actin isoforms in *Vicia faba, Protoplasma* **191**, 158–164.

JUNG-HEILIGER, H., MEYER, U., GALENSA, R., SCHRÖDER, M. B., LEHMANN H. et al. (1995), Influence of microgravity and calcium on the synthesis of secondary metabolites in cell suspension cultures of *Aesculus hippocastanum* L., in: *Scientific Results of the German Spacelab Mission D-2* (SAHM, P. R., KELLER, M. H., SCHIEWE, B., Eds.), pp. 558–569. Köln: DLR.

KESSLER, J. O. (1992), The internal dynamics of slowly rotating biological systems, *ASGSB Bull.* **5** (2), 11–22.

KISS, J. Z., SACK, F. D. (1989), Reduced gravitropic sensitivity in roots of a starch-deficient mutant of *Nicotiana sylvestris, Planta* **180**, 123–130.

KISS, J. Z., SACK, F. D. (1990), Severely reduced gravitropism in dark-grown hypocotyls of a starch-deficient mutant of *Nicotiana sylvestris, Plant Physiol.* **94**, 1867–1873.

KLIMSHUK, D. M., KORDYUM, E. L., DANEVICH, I. A., TAIRBEKOV, M. G., IVERSEN, T. H. et al. (1992), Structural and functional organisation of regenerated plant protoplasts exposed to microgravity on Biokosmos 9, *Adv. Space Res.* **12** (1), 133–140.

KORDYUM, E. L. (1997), Biology of plant cells in microgravity and under clinostating, *Int. Rev. Cyt.* **171**, 1–78.

KORDYUM, E. L., BARANENKO, V. V., NEDUKHA, E. M., SAMOILOV, V. M. (1995), Formation of *Solanum tuberosum* minitubers in microgravity, *Bot. Zh.* **80** (69), 74–80.

KRIKORIAN, A. D. (1989), Polarity establishment, morphogenesis and cultured plant cells in space, in: *Cells in Space* (SIBONGA, J. D., MAINS, R. C., FAST, T. N., CALLAHAN, P. X., WINGET, C. M., Eds.), pp. 87–95. NASA Conference Pub. 10034, Ames Res. Center Moffelt Field CA.

KRIKORIAN, A. D. (1991), Embryogenic plant cells in microgravity, *Am. Soc. Grav. Space Biol. Bull.* **5** (2), 65–72.

KRIKORIAN, A. D. (1996), Space and genome shock in developing plant cells, *Physiol. Plant* **98**, 901–908.

KRIKORIAN, A. D., KANN, R. P., SMITH, D. L. (1995), Somatic embryogenesis in daylily, in: *Biotechnology in Agriculture and Forestry* Vol. 31 (BAJAJ, Y. P. S., Ed.), pp. 285–293. Berlin: Springer-Verlag.

LEVINE, H. G., KRIKORIAN, A. D. (1992), Shoot growth on aseptically cultivated daylily and *Haplopappus* plantlets after a 5 day spaceflight, *Physiol. Plant.* **86**, 349–359.

LU, Y.-T., FELDMAN, L. J., HIDAKA, H. (1996), Characterization of a calcium/calmodulin-dependent protein kinase homolog from maize roots showing light-regulated gravitropism, *Planta* **199**, 18–24.

MALHO, R., TREWAWAS, A. J. (1996), Localised apical increases of cytosolic free calcium control pollen tube orientation, *Plant Cell* **8**, 1935–1949.

MALHO, R., READ, N. D., PAIS, S., TREWAWAS, A. J. (1993), Role of cytosolic free calcium in the reorientation of pollen tube growth, *Plant J.* **5**, 331–341.

MALHO, R., READ, N. D., TREWAWAS, A. J., PAIS, S. (1995), Calcium channel activity during pollen tube growth and reorientation, *Plant Cell* **7**, 1173–1184.

MENNIGMANN, H. D., HEISE, M. (1995), Influence of conditions in low earth orbit on expression and stability of genetic information in bacteria, in: *Scientific Results of the German Spacelab Mission D-2* (SAHM, P. R., KELLER, M. H., SCHIEWE, B., Eds.), pp. 547–550. Köln: DLR.

MERKYS, A. I., LAURINAVICHIUS, R. S., KENTAVICIENE, P. F., NECHITAILO, G. S. (1988), Formation and growth of *Arabidopsis* callus tissue under changed gravity, 27th COSPAR Plenary Meet, Helsinki, p. 58. ESPOO, Finland.

MILLET, B., PICKARD, B. G. (1988), Gadolinium ion is an inhibitor suitable for testing the putative role of stretch-activated ion channels in geomorphism and thigmomorphism, *Biophys. J.* **53**, 155–160.

MÜLLER, G., HÜBEL, F., SCHNABL H. (1998), Signal transduction during graviresponse, in: *14th ESA Symp. Eur. Rockets and Balloon Programmes and Related Research.* Potsdam, Germany, 487–488. ESA SP-437, September 1999.

NEDUKHA, E. M., SIDOROV, V. A., SAMOYKOV, V. M. (1994), Clinostat influence on regeneration of a cell wall on *Solanum tuberosum* protoplasts, *Adv. Space Res.* **14** (8), 97–101.

PERERA, I. Y., HEIMANN, I., BOSS, W. F. (1999), Transient and sustained increase in inositol-1,4,5-triphosphate precede the differential growth response in gravistimulated maize pulvini. *Proc. Natl. Acad. Sci. USA* **96**, 5838–5843.

QUINTERN, L. E., HORNECK, G., ESCHWEILER, U., BÜCKER, H. (1992), A biofilm used as ultraviolet dosimeter, *Photochem. Photobiol.* **55**, 389–395.

QUINTERN, L. E., PUSKEPPELEIT, M., RAINER, P., WEBER, S., EL NAGGAR, S. et al. (1994), Continuous dosimetry of the biologically harmful UV-radiation in Antarctica with the biofilm technique. *J. Photochem. Photobiol. B. Biol.* **22**, 59–66.

RANJEVA, R., GRAZIANA, A., MAZARS, C. (1999), Plant graviperception and gravitropism: a newcomer's view, *FASEB J.* **13**, 135–141.

RASHOTTE, A. M., BRADY, S. R., REED, R. C., ANTE, S. J., MUDAY, G. K. (2000), Basipetal auxin transport is required for gravitropism in roots of *Arabidopsis, Plant Physiol.* **122**, 481–490.

RASMUSSEN, O., KLIMCHUK, D. A., KORDYUM, E. L., DANEVICH, L. A., TARNAVSKAJA, E. B. et al. (1992), The effect of exposure to microgravity on the development and structural organisation of plant protoplasts flown on Biokosmos 9, *Physiol. Plant* **84**, 162–170.

RASMUSSEN, O., BAGGERUD, C. A., LARSSEN, H. C., EVJEN, K., IVERSEN, T. H. (1994a), The effect of 8 days of microgravity on regeneration of intact plants from protoplasts, *Physiol. Plant.* **92**, 404–411.

RASMUSSEN, O., BONDAR, R. L., BAGGERUD, C., IVERSEN, T. H. (1994b), Development of plant protoplasts during the IML 1-mission, *Adv. Space Res.* **14**, 189–196.

SACHS, M. M., HO, T. H. D. (1986), Alteration of gene expression during environmental stress in plants, *Annu. Rev. Plant. Physiol.* **37**, 363–376.

SACK, F. D. (1991), Plant gravity sensing, *Int. Rev. Cytol.* **127**, 193–251.

SCHNABL, H. (1998), Protein metabolism and phosphoinositid-system under microgravity in plants: Recent results from TEXUS-rockets (30, 32, 35), *Symp. Res. Under Space Conditions* (KELLER, M. H., Ed.), pp. 67–70. Bonn: DLR.

SCHNABL, H., HUNTE, C., SCHULZ, M., WOLF, D., GHIENA-RAHLENBECK, C. et al. (1996), Effects of fast clinostat treatment and microgravity on *Vicia faba* mesophyll cell protoplast ubiquitin pools and actin isoforms, *Microgravity Sci. Technol.* **IX/4**, 275–280.

SCHNABL, H., HUNTE, C., SCHULZ, M., WOLF, D., GHIENA-RAHLENBECK, C. et al. (1997), Microgravity affects the ubiquitin pools and actin isoforms of mesophyll protoplasts, in: *Proc. Eur. Symp. Life Sciences in Space,* Potsdam, Germany. ESA publication SP-1206, 105–108.

SCHULZ, M., SOLSCHEID, B., SCHNABL, H. (1992), Changes in the soluble protein pattern and evidence for stress reactions in the leaf tissue of *Vicia faba* after clinostat rotation. *J. Plant Physiol.* **140**, 502–505.

SCHULZ, M., JANßEN, M., KNOP, M., SCHNABL, H. (1994), Stress and age relates spots with immunoreactivity to ubiquitin-antibody at protoplast surfaces, *Plant Cell Physiol.* **35**, 551–556.

SEITZER, U., BODO, M., MÜLLER, P. K. (1995), Connective tissue synthesis in space: Gravity effects on collagen synthesis of cultured mesenchymal cells, in: *Scientific Results of the German Spacelab Mission D-2* (SAHM, P. R., KELLER, M. H., SCHIEWE, B., Eds.), pp. 570–574. Köln: DLR.

SIEVERS, A., HEJNOWICZ, Z. (1992), How well does the clinostat mimic the effect of microgravity on plant cells and organs? *ASGSB Bull.* **5** (2), 69–76.

SINCLAIR, W., OLIVER, I., MAHER, P., TREWAWAS, A. J. (1996), The role of calmodulin in the gravitropic response of *Arabidopsis thaliana* agr-3 mutant, *Planta* **199**, 343–351.

SWATZELL, L. J., EDELMANN, R. E., MAKAROFF, C. A., KISS, J. Z. (1999), Integrin-like proteins are localized to plasmamembrane fractions, not plastids, in *Arabidopsis, Plant Cell Physiol.* **40**, 173–183.

THEIMER, R. R., KUDIELKA, R. A., RÖSCH, I. (1986), Induction of somatic embryogenesis in anise in microgravity, *Naturwissenschaften* **73**, 442–443.

TRIPATHY, B. C., BROWN, C. S., LEVINE, H. G., KRIKORIAN, A. D. (1996), Growth and photosynthesis responses of wheat plants grown in space, *Plant Physiol.* **110**, 801–806.

VOLKMANN, D., SIEVERS, A. (1979), Graviperception in multicellular organs, in: *Encyclopedia of Plant Physiol.* New Series (HAUPT, W., FEINLEIB, M. E., Eds.) Vol. 7, pp. 573–600. Berlin: Springer-Verlag.

VON KELLER, A., VOESTE, D., SCHNABL, H. (1995), Electrofusion and regeneration of sunflower protoplasts under microgravity considering the ultrastructure, in: *Scientific Results of the German Spacelab Mission D-2* (SAHM, P. R., KELLER, M. H., SCHIEWE, B., Eds.), pp. 664–670. Köln: DLR.

WATILLON, B., KETTMANN, R., BOXUS, P., BURNY, A. (1995), Structure of a calmodulin-binding protein kinase gene from apple, *Plant Physiol.* **108**, 847–848.

WINGENDER, R., HENN, H. J., BARTH, S., VOESTE, D., MACHLAB, H., SCHNABL, H. (1996), A regeneration protocol of sunflower protoplasts, *Plant Cell Rep.* **15**, 742–745.

WOLF, D., SCHULZ, M., SCHNABL, H. (1993a), Influence of clinostat rotation on plant proteins. Effects on ubiquitinated polypeptides in the stroma and in thylakoid membranes of *Vicia faba* chloroplasts, *J. Plant Physiol.* **141**, 304–309.

WOLF, D., SCHULZ, M., SCHNABL, H. (1993b), Influence of clinostat rotation on root tip proteins of *Vicia faba, Plant Biochem. Physiol.* **31**, 717–724.

XU, W., PURUGGANAN, M. M., POLISENSKY, D. H., ANTOSIEWICZ, D. M., FRY, S. C., BRAAM, J. (1995), *Arabidopsis* TCH regulated by hormones and the environment, encodes a xyloglucan endotransglycosylase, *Plant Cell* **7**, 1555–1567.

YAMAUCHI, Y., FUKAKI, H., FUJISAWA, H., TASAKA, M. (1997), Mutations in the SGR4, SGR5 and SGR6 loci of *Arabidopsis thaliana* alter the shoot gravitropism, *Plant Cell Phys.* **38**, 530–535.

YOUNG, L. M., EVANS, M. L. (1994), Calcium-dependent asymmetric movement of 3-indole-3-acetic acid across gravistimulated isolated root caps of maize, *Plant Growth Regul.* **14**, 235–242.

ZIMMERMANN, U. (1986), Electrical breakdown, electropermeabilization and electrofusion, *Rev. Physiol. Biochem. Pharmacol.* **105**, 176–256.

ZIMMERMANN, U. (1987), Electrofusion of cells, in: *Methods of Hybridoma Formation* (BARTAL, A. H., HIRSAUT, Y., Eds.), pp. 97–150. Clifton, NJ: Humana Press.

ZIMMERMANN, U., SCHNETTLER, R., HANNIG, K. (1988), Biotechnology in space: Potentials and perspectives, in: *Biotechnology* 1st Edn., Vol. 6b (REHM, H. J., REED, G., Eds.), pp. 637–672. Weinheim: VCH.

18 Biotechnology in the Pulp and Paper Industry

LIISA VIIKARI
MANA TENKANEN
ANNA SUURNÄKKI
VTT, Finland

1 Introduction

Biocatalysts have significant potential for improving traditional pulp and paper manufacturing processes and in introducing environmental benefits and unique properties to the fiber raw materials. In spite of the promise of biocatalysts and their commercial success in other fields, such as food and textile industries, the use of biocatalysts in the forest sector, however, has so far increased only slowly. The general goals of the pulp and paper industry today are to increase the cost efficiency, to develop environmentally benign processes, and to improve the product quality. Biotechnical methods can help to reach these goals, but seldom solve problems alone. The modifications in fiber material, which can be achieved by enzymes or microorganisms, have so far been always combined with chemical or mechanical treatments. The real challenge for new commercial successes is to identify superior enzymes and their applications. Advances in molecular genetics have enabled the production and testing of the identified biocatalysts on a laboratory scale. In many applications, however, the identified enzymes represent products of the first generation, and may not act optimally under the harsh industrial conditions. The catalytic activity and stability of potential enzymes can further be improved by new powerful methods, such as directed evolution.

During the last years the number of new applications of enzymes in pulp and paper manufacture has grown steadily, and several have reached or are approaching commercial use. These include enzyme-aided bleaching with xylanases, direct delignification with oxidative enzymes, energy saving refining with cellulases, pitch reduction with lipases, freeness enhancement with cellulases and hemicellulases as well as enzymatic slime control of the paper machine (Fig. 1). The unique feature of enzymatic applications is their specificity. Enzymes are classified according to the reaction they catalyze, and the theoretical number of enzymes acting on wood derived components is quite high. Although enzymes acting on the major components, such as cellulose or hemicellulose are well known, the enzymology of lignin and extractives is still poorly known and exploited. The presently commercialized enzymatic methods are based on the action of hydrolytic enzymes, which act on different polysaccharides or extractives. These enzymatic applications have also been shown to be economically competitive leading to savings in chemical costs. First oxidative enzymes are being commercialized.

In addition to enzymes, microbial treatments are currently available or under development for increasing pulping efficiency, reduction of pitch problems, and enhancing process water reuse. The application of microbial treatments in wood chip treatment prior to pulping has traditionally been limited by the unspecificity of natural microbial strains, long treatment times needed, and negative effects of microbial growth on the optical properties of pulps. The search and genetic design for new, more specific, fast growing and efficient microbial strains is, however, expected to lead to further improvements and breakthroughs of new applications.

Today, molecular biology offers tools that allow to select and engineer superior trees. The pathway of lignin biosynthesis is one of the most intensively studied in higher plants. Fast growing, low lignin trees could provide significant practical benefits. Removal of lignin from the wood cell walls is the most energy intensive and environmentally damaging step in wood processing for pulp and paper, and thus reducing the lignin content in trees could provide both economic and environmental benefits. Recently, it was shown that an antisense construct of the gene coding for 4-coumarate:CoA ligase in the lignin pathway can greatly reduce lignin, increase cellulose, and dramatically stimulate the growth of transgenic aspen (Hu et al., 1999). Previous attempts to modify lignin by mutation or suppression of enzymes in the pathway for monolignol biosynthesis have resulted in striking changes in lignin composition, but usually with no effect on lignin content. Substantial opportunities exist for genetic improvement of forest tree species through molecular biology, with the key goal of accessing QTLs (Quantitative Trait Loci) governing complex traits such as yield, quality, and stress tolerance. Tree breeding is clearly the vehicle for introducing

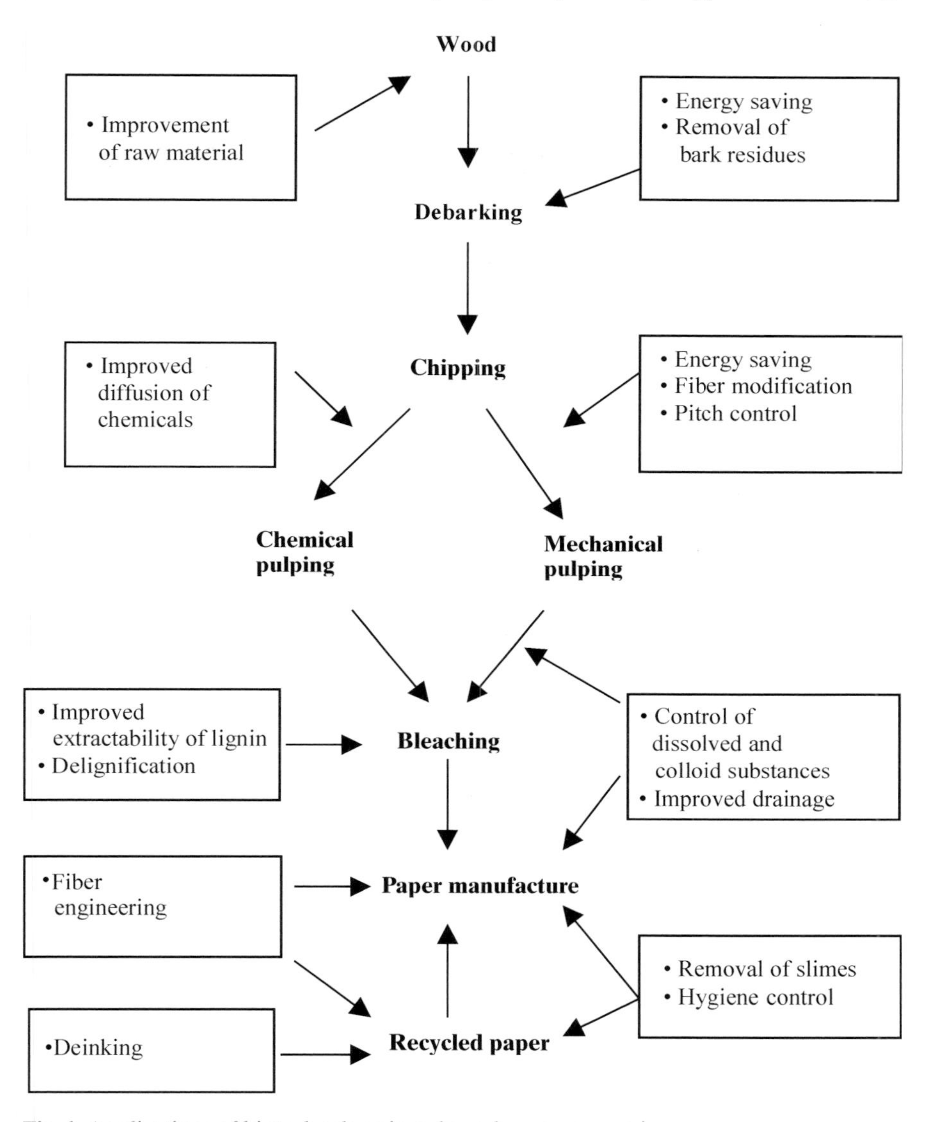

Fig. 1. Applications of biotechnology in pulp and paper processing.

molecular biology advances into forestry. The future attitudes of consumers is a concern for the industry, and will affect the realization of these possibilities. The uniformity and predictability of wood feedstock characteristics which is already achievable through clonal propagation, is presently considered as important as the introduction of new traits (ROBINSON, 1999).

2 Enzymes for Pulp and Paper Applications

Wood fibers are mainly composed of cellulose, hemicellulose, i.e., xylan, and glucomannan, lignin and extractives. In mechanical pulping no major chemical changes in the fiber components occur. During alkaline chemical pulping, i.e., kraft cooking, about 90% of lignin is removed from the fibers and the hemicellulose components are extensively modified due to their dissolution, partial degradation, and redeposition (SJÖSTRÖM, 1993). Consequently, the chemical composition and the structure of mechanical and chemical fibers is different (Tab. 1). The more open structure of chemical fibers as compared with mechanical fibers also renders them more susceptible to the action of macromolecular enzymes. The enzymatic action even in chemical pulps is, however, limited to accessible surfaces, i.e., to fines and to the outermost surface and accessible pores of long fibers (SUURNÄKKI et al., 1996a). Therefore, enzymatic treatments can be considered as surface specific modification methods of fibers.

Enzymes that have a number of potentially interesting applications in both mechanical and chemical pulps include native and modified monocomponent cellulases, hemicellulases, lignin modifying oxidative enzymes, and enzymes degrading extractives. First commercial enzymes designed for pulp and paper applications were introduced in the early 1990s. These were lipases, xylanases, and cellulases. Presently, first oxidative enzymes for delignification and designed monocomponent cellulases for energy saving are entering the markets. The successful implementation of biotechnical methods requires understanding of their reaction mechanisms and action in the fiber matrix and optimized combinations of biological and mechanical or chemical treatments.

2.1 Cellulases

Cellulose is the main carbohydrate in wood fibers. It is a chemically simple polymer, a homopolymer consisting of up to 1,000 β-1,4-linked anhydroglucopyranoside units. However, its physical state makes it a challenging substrate for enzymes. Single glucose polymers are packed onto each other to form highly crystalline microfibrils in which the individual cellulose chains are held together by hydrogen bonds. Cellulose microfibrils also contain some amorphous regions. In wood fibers the winding direction of cellulose microfibrils varies in different cell wall layers giving the fiber its unique strength and flexibility.

Efficient degradation of cellulose requires a mixture of different cellulases acting sequentially or in concert. The cellulolytic system of *Trichoderma reesei* has been studied in most detail (Tab. 2). Cellulases have traditionally been categorized into two different classes by the International Union of Biochemistry (IUB). Endoglucanases (EG; EC 3.2.1.4) cleave the cellulose chains internally. Exoglucanases (cellobiohydrolases, CBH; EC 3.2.1.91) degrade cellulose starting from free chain ends, producing cellobiose. The third enzyme needed for total cellulose hydrolysis, β-glucosidase (EC 3.2.1.21), hydrolyzes short cello-oligosaccharides to glucose. Endoglucanases act mainly on the amorphous parts of the cellulose fiber, whereas cellobiohydrolases are also able to degrade crystalline cellulose. However, this classification is not very satisfactory, as some exoglucanases have also been claimed to possess endoactivity (STÅHLBERG et al., 1993; TOMME et al., 1995). The classification of cellulases becomes even more complicated as some enzymes also have activity on other polysaccharides, such as xylan (BIELY et

Tab. 1. Chemical Composition of Fibers of Different Origins (SJÖSTRÖM, 1993)

Fiber Source	Average Composition [% of d.wt.]				
	Cellulose	Hemicellulose	Lignin	Extractives	Others
Mechanical pulp (softwood)	39	25	27	4	5
Chemical pulp (softwood)	74	19	6	1	0
Chemical pulp (hardwood)	63	32	4	1	0

Tab. 2. The Cellulolytic Enzymes of *T. reesei*

Enzyme[a]	New Name	Size [kDa]	pI	Remarks[b]	Optimal pH	Reference
EG I	Cel7B	50–55	4.6		4.5–5.5	PENTTILÄ et al. (1986)
EG II	Cel5A	48	5.5		4.5–5.5	SALOHEIMO et al. (1988)
EG III	Cel12A	25	7.4	no CBD		WARD et al. (1993)
EG IV	Cel61A	37[c]	nd		4.5–5.5	SALOHEIMO et al. (1997)
EG V	Cel45A	23[c]	2.8–3.0		4.5–5.5	SALOHEIMO et al. (1994)
CBH I	Cel7A	59–68	3.5–4.2	acts from reducing end	4.5–5.5	SHOEMAKER et al. (1983)
CBH II	Cel6A	50–58	5.1–6.3	acts from non-reducing end	4.5–5.5	CHEN et al. (1987)
BGL I		71	8.7	family 3	4.6	CHEN et al. (1992)
BGL II		114	4.8	family 1	4.0	CHEN et al. (1992), TAKASHIMA et al. (1999)

[a] EG, endoglucanase; CBH, cellobiohydrolase; BGL, β-glucosidase
[b] CBD, Cellulose Binding Domain
[c] Calculated according to the amino acid sequence

al., 1991). Therefore, a new classification of these enzymes, as well as of other glycosyl hydrolases has been initiated. Presently, the different glycosyl hydrolases are grouped according to the structures of their catalytic domains into more than 70 families (http//*afmb.cnrs-mrs.fr/~pedro/CAZY/db.html*). This, however, requires information on the amino acid sequence which is not always available. Currently, there are cellulases in 11 glycosyl hydrolase families.

Most cellulases have a multidomain structure consisting of a core domain separated from a cellulose binding domain (CBD) by a linker peptide. The core domain contains the active site whereas the CBD interacts with cellulose by binding the enzyme to it (VAN TILBEURGH et al., 1986; TOMME et al., 1988). The CBDs are particularly important in the hydrolysis of crystalline cellulose. It has been shown that the ability of cellobiohydrolases to degrade crystalline cellulose clearly decreases when the CBD is absent (LINDER and TEERI, 1997). However, the exact role and action mechanism of CBDs is still a matter of speculation. It has been suggested that the CBD enhances the enzymatic activity merely by increasing the effective enzyme concentration at the surface of cellulose (STÅHLBERG et al., 1991), and/or by loosening single cellulose chains from the cellulose surface (KNOWLES et al., 1987; TEERI et al., 1987; REINIKAINEN et al., 1995; TORMO et al., 1996). Most studies concerning the effects of cellulase domains on different substrates have been carried out with core proteins of cellobiohydrolases, as their core proteins can easily be produced by limited proteolysis with papain (TOMME et al., 1988).

2.2 Hemicellulases

In wood, the two most common hemicelluloses are xylans and glucomannans. The hardwoods contain mainly xylan, whereas the amount of glucomannan in softwoods is approximately twice the amount of xylan. Hardwood xylan is composed of β-D-xylopyranosyl units which may contain 4-O-methyl-α-D-glucuronic acid and acetyl side groups. 4-O-methylglucuronic acid is linked to the xylan backbone by O-(1→2) glycosidic bonds and the acetic acid is esterified at the carbon 2 and/or 3 hydroxyl group. The softwood xylan is arabino-4-O-methylglucuronoxylan in which the xylan backbone is substituted at carbon 2 and 3 with 4-O-methyl-α-D-glucuronic acid and α-L-arabinofuranosyl residues, respectively. Softwood galactoglucomannan has a backbone of β-(1→4)-linked β-D-glucopyranosyl and β-D-mannopyranosyl units which are partially substituted by α-D-galactopyranosyl and acetyl groups. The distribution of carbohydrates in the wood fibers varies depending on the wood species and growing conditions (SJÖSTRÖM, 1993). In softwoods, the xylan content of the innermost cell wall layer is generally very high but otherwise, the xylan is relatively uniformly distributed throughout the fiber walls in native wood. In hardwoods, the outermost layers of fiber walls are rich in xylan. In softwoods, the glucomannan content increases steadily from the outer parts to the inner parts of cell walls. The amount and composition of hemicellulose components is drastically altered in alkaline chemical pulping, i.e., kraft cooking. All acetyl groups are degraded and most of the glucuronic acid groups in xylan are converted to hexenuronic acid groups (TELEMAN et al., 1995). Softwood glucomannan is deacetylated and a considerable part is solubilized in the highly alkaline and hot cooking liquor.

2.2.1 Xylanases

The major part of the published work on hemicellulases deals with the production, properties, mode of action, and applications of xylanases (reviewed by BIELY, 1985; WONG and SADDLER, 1992; COUGHLAN and HAZLEWOOD, 1993; VIIKARI et al., 1994, 2000; SUURNÄKKI et al., 1997). Endoxylanases (1,4-β-D-xylan xylanohydrolases; EC 3.2.1.8) catalyze the random hydrolysis of 1,4-β-D-xylosidic linkages in xylans. Most xylanases belong to the two structurally different glycosyl hydrolase groups (families 10 and 11). The three-dimensional structures of xylanases in both of these families are available showing clear differences. *Trichoderma reesei* xylanase from family 11 is an ellipsoidal small enzyme with a diameter between 32 and 42 Å. It does not have any separate substrate binding domain (TÖRRÖNEN et al., 1994). Xylanases belonging to family 10 are generally much larger than the family 11 xylanases. This may reflect the substrate specificities of enzymes on fibers with different structures and pore sizes. The catalytic domain of the *Streptomyces lividans* family 10 xylanase is more loosely packed and about 1.5 times longer than the *T. reesei* xylanase (DEREWENDA, 1994). In addition, it contains a separate xylan binding domain (SHARECK et al., 1991). Some other xylanases have also been reported to contain either a xylan binding domain (IRWIN et al., 1994) or a cellulose binding domain (GILBERT and HAZLEWOOD, 1993; SAKKA et al., 1993). Some of the binding domains have been found to increase the degree of hydrolysis of fiber bound xylan whereas others have not had any effect on that. Neither xylan nor the cellulose binding domain had, however, any significant role on the biobleaching effect of xylanases (RIXON et al., 1996).

Most of the xylanases characterized are able to hydrolyze different types of xylans showing only differences in the spectrum of end products. The main products formed from the hydrolysis of xylans are xylobiose, xylotriose, and substituted oligomers of three to five xylosyl residues. The chain length and the structure of the substituted products depend on the mode of action of the individual xylanases. The family 10 and 11 xylanases differ also in their catalytic properties. Family 10 enzymes exhibit greater catalytic versatility than the family 11 xylanases and thus, they are, e.g., able to hydrolyze more efficiently highly substituted xylans (BIELY et al., 1997). However, no systematic studies on the substrate specificity of xylanases belonging to different families on fiber bound substrates have been carried out. From a practical point of view, the most important characteristics of xylanases are their high pH and temperature stability and activity. A number of enzymes produced by extremophilic organisms have been characterized (DA SILVA et al., 1994; LAPIDOT et al., 1996; GARG et al., 1996; SHAH et al., 2000), but few of these have reached commercial use (BARAZNENOK et al., 1999).

2.2.2 Mannanases

Endomannanases (1,4-β-D-mannan mannanohydrolase; EC 3.2.1.78) catalyze the random hydrolysis of β-D-1,4-mannopyranosyl linkages within the main chain of mannans and various polysaccharides consisting mainly of mannose, such as glucomannans, galactomannans, and galactoglucomannans. Mannanases seem also to be a more heterogeneous group of enzymes than xylanases. The mannanase of *Trichoderma reesei* has been found to have a similar multidomain structure as several cellulolytic enzymes, i.e., the protein contains a catalytic core domain which is separated by a linker from a cellulose binding domain (STÅLBRAND et al., 1995; TENKANEN et al., 1995a). The CBD has been found to increase the action of *T. reesei* mannanase on fiber bound glucomannan (Röhm Enzyme Finland, unpublished data) even though the catalytic domain is able to efficiently degrade crystalline mannan (SABINI et al., 2000). The mannanase of *Caldocellum saccharolyticum* has also been reported to be part of a multidomain protein that contains two catalytic domains (one with mannanase and another with endoglucanase activity) and two substrate binding domains (MORRIS et al., 1995). The three-dimensional structures of *Thermomonospora fusca* (HILGE et al., 1998) and *T. reesei* (SABINI et al., 2000) mannanases have recently been determined. The catalytic domains of these enzymes have diameters between 37 and 55 Å.

The main hydrolysis products from galactomannans and glucomannans are mannobiose, mannotriose, and various mixed oligosaccharides. The hydrolysis yield is dependent on the degree of substitution as well as on the distribution of the substituents (MCCLEARY, 1991). The hydrolysis of glucomannans is also affected by the glucose/mannose ratio. Some mannanases are able to hydrolyze not only the β-1,4-bond between two mannose units but also the bond between the mannose and glucose units (KUSAKABE et al., 1988; TENKANEN et al., 1997).

2.2.3 Other Hemicellulases

Enzymes needed for further hydrolysis of the short oligomeric compounds produced by endoenzymes from pulp hemicellulose are β-xylosidase (1,4-β-D-xyloside xylohydrolase; EC 3.2.1.37), β-mannosidase (1,4-β-D-mannoside mannohydrolase; EC 3.1.1.25), and β-glucosidase (EC 3.2.1.21). β-Xylosidases (EC 3.2.1.37) catalyze the hydrolysis of xylooligosaccharides by removing successive xylose residues from the non-reducing termini. β-Mannosidase catalyzes the hydrolysis of terminal, non-reducing mannose residues in mannans, respectively. Exoglycanases are generally larger proteins than endoglycanases and are often built up by two or more subunits.

The side groups connected to xylan and glucomannan main chains are removed by α-glucuronidase (EC 3.2.1.131), α-arabinosidase (α-L-arabinofuranoside arabinofuranohydrolase, EC 3.2.1.55) and α-D-galactosidase (α-D-galactoside galactohydrolase; EC 3.2.1.22). Acetyl substituents bound to hemicellulose are removed by esterases (EC 3.1.1.72) (COUGHLAND and HAZLEWOOD, 1993). There are clearly different types of side group cleaving enzymes. Some of them are able to hydrolyze only substituted short-chain oligomers which first must be produced by the backbone depolymerizing endoenzymes (xylanases and mannanases). Others are capable of also attacking intact polymeric substrates. Even most accessory enzymes of the latter type, however, prefer oligomeric substrates. The synergism between different hemicellulolytic enzymes is also observed by the accelerated action of endoglycanases in the presence of accessory enzymes. Some accessory enzymes, such as an α-arabinosidase and an esterase of *Pseudomonas fluorescens* ssp. *cellulosa* and an acetyl xylan esterase of *Trichoderma reesei* have been found to contain a multidomain structure with a separate cellulose binding domain (HAZLEWOOD and GILBERT, 1992; MARGOLLES-CLARK et al., 1996). Side group cleaving enzymes have been found to be most useful in the modification of solubilized carbohydrates, especially in the treatment of dissolved and colloid substances present in paper machines.

2.3 Lignin Modifying Enzymes

The enzymology of lignin has been the focus of enzymologists for more than 30 years, and has resulted in the identification of the major enzyme systems anticipated to participate in delignification: oxidases, peroxidases, dehydrogenases, and hydrogen peroxide generating enzymes. Due to the promising results in bleaching, laccases and manganese-dependent peroxidases have recently been the most extensively studied groups of enzymes in this area.

2.3.1 Peroxidases

Lignin biodegradation is initiated by several extracellular oxidative enzymes, including lignin peroxidase (LiP; EC 1.11.1.14) (TIEN and KIRK, 1984), manganese-dependent peroxidase (MnP; EC 1.11.1.13) (GLENN and GOULD, 1985) manganese-independent peroxidase (MIP) (DE JONG et al., 1992) and H_2O_2 generating oxidases, as well as laccase (EC 1.10.3.2). Purified ligninolytic enzymes have been shown to cause limited delignification provided that additives are supplemented: veratryl alcohol and H_2O_2 for LiP (ARBELOA et al., 1992) and manganese, H_2O_2, organic acids, and surfactants for MnP (PAICE et al., 1993; KONDO et al., 1994).

Since its discovery in *Phanerochaete chrysosporium* MnP has been found to be secreted by many white-rot fungi. Initially, MnP was thought to play only a limited role in lignin depolymerization, oxidizing only phenolic subunits, while LiP was regarded as the main enzyme (KIRK and FARRELL, 1987). MnP was found to depolymerize ^{14}C-labeled dehydrogenative polymerizate (DHP) lignin (WARIISHI et al., 1991) and to oxidize non-phenolic model compounds (BAO et al., 1994). Factors involved in the catalytic cycle of MnP are well understood; Mn(II) is oxidized to Mn(III), which is further complexed with an organic acid present in the medium and diffused into the matrix polymer depolymerizing lignin. The future success of MnP in practical delignification will obviously depend on the commercial availability of the enzyme.

2.3.2 Oxidases

Phenoloxidases, especially laccases, have been most intensively studied during recent years. Laccases catalyze the oxidation of a range of substances with the concomitant reduction of O_2 to water. Laccases have been isolated and characterized mainly from fungal sources. More than 40 different fungi have been reported to produce laccases. In addition, laccases have been detected in various plants and in several insects, while only one bacterium is known to produce laccase (THURSTON, 1994). Laccases belong to the copper metalloenzymes and to the blue oxidase subgroup, containing four copper atoms per molecule. The current knowledge of the structure of laccases is limited to the Cu-depleted native structure of *Coprinus cinereus* laccase (DUCROS et al., 1998). Laccases can couple four one-electron oxidations of a variety of substrates, such as di- and polyphenols, aromatic amines, and a range of other compounds to the reduction of O_2. In laccase-catalyzed oxidation, the substrate loses a single electron and forms a free radical which may undergo further laccase-catalyzed oxidation or non-enzymatic reactions, leading also to polymerization. Laccase can catalyze the alkyl-phenol and C_α–C_β cleavage of phenolic dimers which are used as model lignin substructures, and it can catalyze the demethoxylation of several lignin model compounds (ISHIHARA, 1980), but is not able to oxidize O-etherified phenols. Interestingly, it was found out that even this structure could be oxidized by laccase to the corresponding benzaldehyde in the presence of specific mediator molecules, such as 2,2′-azino-bis-(3-ethylbenzothiazoline-6-sulfonic acid) diammonium salt (ABTS) or 1-hydroxybenzotriazole (HBT) (BOURBONNAIS and PAICE, 1992; CALL, 1995). The concept of mediated oxidation (Fig. 2) has been exploited in practical delignification approaches.

2.4 Other Enzymes

Wood extractives are classified into terpenes and triterpenoids (including resin acids and steroids), esters of fatty acids (fats and waxes), fatty acids and alcohols and alkanes (SJÖSTRÖM, 1993). During both mechanical and chemical processing part of the wood extractives are liberated from wood to process water. In chemical pulping and bleaching wood extractives are extensively modified and degraded. The best known enzyme group acting on extractives is lipases. Lipases (triacylglycerol acylhydrolases; EC 3.1.1.3) catalyze the hydrolysis of triacylglycerols to fatty acids, mono- and diacylglycerols, and glycerol. Some lipases will also degrade a rather broad range of other compounds containing an ester linkage (DEREWENDA, 1994).

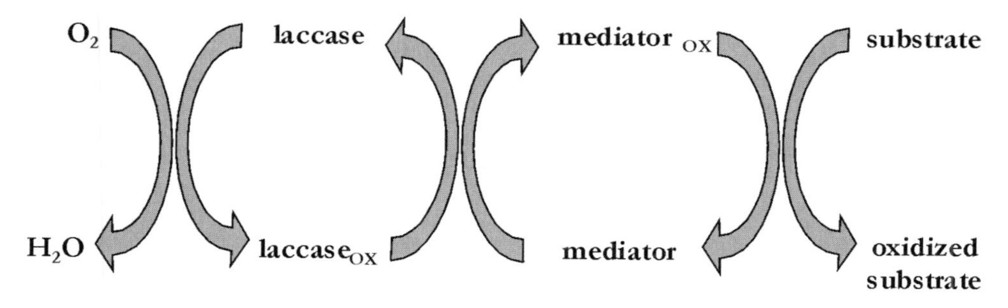

Fig. 2. Diagram of the mediated oxidation of substrate (lignin) by laccase.

Wood contains small amounts of other miscellaneous polysaccharides, not classified as hemicelluloses, such as arabinogalactans and pectins. Various enzymes acting on these polysaccharides can be used for the degradation of wood derived compounds. These polymers are solubilized during the mechanical refining processes and may be enzymatically modified. In addition, well known enzymes, such as amylases, can be used for degrading starch used in paper coating. Harmful compounds, including slimy polysaccharides and hydrogen peroxide, are produced by contaminating microorganisms in the paper machine, which can be degraded by a variety of still poorly known polysaccharases and by catalase, respectively.

3 Enzymes in Studying the Structure–Function of Fibers

The specificity of purified enzymes and the mild hydrolysis conditions can be exploited in the characterization of chemical pulp fibers. By combining enzymatic treatments with analysis methods, such as HPLC, ESCA (Electron Spectroscopy for Chemical Analysis), or NMR, new information about the structure and composition of fiber components, mainly hemicelluloses, has been obtained (Tenkanen et al., 1995b; Teleman et al., 1995; Laine et al., 1996; Buchert et al., 1996). Furthermore, the role of hemicellulose components in pulp properties has been studied using specific enzymes (Buchert et al., 1996, 1997; Oksanen et al., 1997a, b, 2000).

Enzymes can be used in the characterization of the composition of accessible carbohydrates in chemical pulps. The non-destructive action of enzymes was exploited in identification of the previously undetected, acid labile hexenuronic acid (Teleman et al., 1995). By using gradual enzymatic peeling and HPLC analysis it has been shown that the composition of xylan and glucomannan on the accessible surfaces of fibers differs from the composition of these polysaccharides in the overall pulp (Suurnäkki et al., 1996b). The content of galactose in the pine kraft pulp glucomannan and the content of hexenuronic acid in the birch kraft pulp xylan was found to be relatively high on the surfaces most accessible to enzymes. However, in pine kraft pulp, the degree of substitution of surface xylan was similar to that of the overall pulp xylan.

The organization of cellulose, hemicelluloses, and lignin on the outermost surface of kraft fibers has been studied by combining enzymatic peeling and ESCA analysis (Buchert et al., 1996). Xylan was found to partly cover lignin in pine but not in birch kraft fibers. Xylan was also found to overlap glucomannan and cellulose chains on the fiber surfaces in softwood kraft pulps.

The role of hemicelluloses in the properties of kraft pulps has been investigated by removing hemicellulose from pulp by specific enzymes. The brightness stability of kraft pulps has been found to be improved after enzymatic removal of xylan, probably mainly due to the concomitant reduction of the acidic content in pulps (BUCHERT et al., 1997). The enzymatic removal of both xylan and glucomannan has also been observed to increase the unwanted hornification of pulp fibers indicating that the hemicellulose on the pulp surfaces apparently affects the pulp fibers during drying (OKSANEN et al., 1997, 2000).

4 Use of Enzymes and Microorganisms in Processes

The main processes in the pulp and paper industry are pulping, bleaching, and papermaking. In addition, the paper recycling process with screening, cleaning, and deinking of recovered paper is an important part of the paper manufacturing process (Fig. 1). In pulping, the wood fibers are liberated from each other in mechanical or chemical processes and the wood chips are converted to pulp containing fibers and fines. In chemical pulping, or cooking, most of the fiber lignin and part of the hemicelluloses is additionally solubilized and removed. In bleaching the colored component originating mainly from lignin are either removed (chemical pulps) or converted to non-colored components (mechanical pulps). In papermaking mechanical, chemical or recycled pulp is converted to paper with the aid of papermaking chemicals, fillers, and additives. Due to the complex chemistry, a number of different enzymes can be used to improve the processes. Benefits obtained by enzymes are summarized in Tab. 3. Enzymes can be used at any process phase, whereas microbial treatments are most suited for pretreatments of wood chips.

Tab. 3. Benefits of Enzymes Acting on Different Wood Components

Component	Enzyme	Physicochemical Modification	Technical Benefit
Cellulose	cellobiohydrolase	microfibrillation	energy saving in refining flexibility of fibers
	endoglucanase	depolymerization	paper machine runnability speciality products
	mixed	depolymerization	release of ink particles fiber properties
Xylan	endoxylanase	depolymerization	extractability of lignin
Glucomannan	endomannanase	depolymerization	extractability of lignin
	endomannanase	decreased colloidal stability	runnability of paper machine
	acetyl glucomannan esterase	decreased solubility	increased pulp yield strength properties
Pectins	polygalacturonase	depolymerization	energy saving in debarking decreased cationic demand in papermaking
Lignin	laccases Mn-peroxidase	depolymerization, polymerization	increased brightness polymerization of lignin
Extractives	lipase	increased hydrophilicity	strength properties runnability of paper machine

4.1 Mechanical Pulping

Mechanical pulping involves the use of mechanical force as such or combined with chemicals, pressure, and temperature to separate the wood fibers. Mechanical processes, such as pressure grounding (PG) or thermomechanical pulping (TMP) have a high yield (up to 95%) and produce paper with high bulk, good opacity, and excellent printability. The drawbacks of these processes are the high energy intensity and the resulting fiber and paper quality with lower strength, higher pitch content, and higher color reversion rate as compared to chemical pulps.

4.1.1 Microbial Treatments

Research in biomechanical pulping, i.e., the treatment of lignocellulosic materials with lignin-degrading fungi prior to mechanical pulping started already in the 1970s. The first report that fungal pretreatment could result in significant energy savings and strength improvements for mechanical pulping was published by ANDER and ERIKSSON (1977). Although this research met limited success, primarily due to difficulties in scaling up the process, it provided valuable insights, as reviewed by AKHTAR et al. (1998). Biomechanical pulping of both hardwood and softwood species has been widely studied. Treatment of aspen, red alder or birch wood chips with fungi such as *Phlebia brevispora*, *Phlebia subserialis*, *Phanerochaete chrysosporium*, *Pholiota mutabilis*, and *Phlebia tramellosa* has been reported to be effective in reducing energy consumption (up to 47%), especially in primary refining (BAR-LEV et al., 1982; LEATHAM et al., 1990a, b; SETLIFF et al., 1990). Fungal treatment also enhances the strength properties of paper made of hardwood mechanical pulp (LEATHAM et al., 1990a). All white-rot fungi do not, however, affect the mechanical pulping efficiency in hardwoods even though they are aggressive wood degraders (LEATHAM et al., 1990a,b). In addition, the beneficial effect of fungal treatment cannot be obtained in all wood species (SETLIFF et al., 1990). Softwood lignin is more abundant and chemically more resistant for fungal degradation than hardwood lignin. Therefore, most of the reported fungal treatments have resulted in less pronounced energy savings in mechanical pulping of softwoods as compared to hardwoods (AKHTAR et al., 1998).

The mechanism of biopulping is based on the capability of fungal strains to produce a set of enzymes needed to degrade wood lignin. Removal or modification of lignin keeping the wood fibers together in wood chips reduces the energy consumption in pulping. Biopulped mechanical fibers have been shown to be more fibrillated and less stiff than untreated fibers, which leads to better binding properties (SACHS et al., 1990). The drawbacks of fungal treatments are the reduced brightness of wood chips that can, however, be recovered by pulp bleaching. Another drawback is the wood yield loss generally related to fungal treatments. In addition to lignin, white-rot fungi are capable to degrade cellulose and hemicellulose in wood. By using a suitable fungus and treatment time the yield loss can be limited to under 5% which is comparable to that in other mechanical pulping methods (SACHS et al., 1990).

Research aiming at industrially and economically feasible biomechanical pulping technology has been carried out at the Forest Products Laboratory, Madison, Wisconsin since the late 1980s. Currently the two-week microbial pretreatment with *Ceriporiopsis subvermispora* on spruce wood chips saves at least 30% electrical energy in mechanical pulping, and also improves paper strength properties (SCOTT et al., 1998). The economics of biomechanical pulping look attractive, as savings of about $ 5 million each year have been estimated based on the economical evaluation carried out for a 250 t d^{-1} TMP mill (SCOTT and SWANEY, 1998). The biopulping technology includes decontamination of wood chips from naturally occurring microorganisms by steaming, inoculation of pretreated chips with a selective lignin degrading fungus together with a commercially available nutrient source, and thereafter treatment of chips for two weeks with aeration. Investigations of biopulping technology on a laboratory scale have focused on the effect of species and strains of fungi, inoculum form and amount, species of wood, wood chip size, environmental factors, effect of added nutrients, and need to sterilize the chips (AKHTAR et al., 1998).

Traditional aging of wood chips, i.e., growth of naturally occurring microorganisms in chip piles can reduce the amount of lipophilic components already in the chip pile (FUJITA et al., 1992). Chip treatments with selected colorless fungi acting only on lipophilic components in wood have successfully been developed (FARRELL et al., 1997). In both chemical and mechanical pulping the decreased amount of triglycerides and fatty acids was reported along with increased pulping efficiency. The commercial, non-pathogenic fungal product, Cartapip™, has been used in several mills. In mechanical pulps, improvement of strength properties has also been observed.

4.1.2 Enzymatic Treatments

Because of the low accessibility of wood chips for enzymatic modification, incorporation of an enzymatic step into the mechanical pulping process can be expected to be successful only after the primary refining. A new process concept based on the treatment of coarse mechanical pulp fibers using monocomponent cellulases has been developed (PERE et al., 1996, 2000). The effects of enzymatic modification of coarse mechanical pulp with pure cellulases and hemicellulases were compared on reject refining. Experiments with different enzymes demonstrated that a slight modification of cellulose by CBH I gave rise to energy saving of 20% in a laboratory-scale disk refiner. Interestingly, no positive effect on energy consumption was detected with a cellulase mixture. Other reports have confirmed that unoptimized mixtures of cellulases do not lead to the same result (RICHARDSON et al., 1998). On a pilot scale two-stage secondary refining with a low-intensity refiner (wing defibrator) at a CSF (Canadian Standard Freeness, describing the extent of refining) level of 100 mL energy saving of 10–15% with CBH I was obtained (PERE et al., 1996). Lately, the method has been tested on an industrial scale, and is under further development.

Neither the cellulase mixture nor CBH I induced evident morphological modifications in the coarse and rigid TMP fibers during a short incubation (PERE et al., 2000). Based on the fibrillation index analyses it has been speculated that the action of CBH I possibly induces decreased interfibrillar cohesion inside the fiber wall and thus results in loosening and unraveling of the fiber structure. The cellulase treatment with CBH I did not have any detrimental effects on pulp quality. In fact, the tensile index was even higher for the CBH I treated pulp than for the reference. The increase in tensile index could be explained by the intensive fibrillation induced by the CBH I treatment. The good optical properties were also maintained after the CBH I treatment.

Modification of other components of chips may also lead to energy savings and improved fiber properties. Structural modifications of fibers have been studied by using laccases alone or in combinations with a mediator (CALL and STRITTMATTER, 1992), as well as with protease (MANSFIELD et al., 1999). Laccases appear to activate lignin, while proteases are thought to attack the primary cell walls enriched in proteins.

4.2 Chemical Pulping

Several methods have been studied to increase the diffusion and porosity in wood for improving the efficiency of chemical pulping processes. These methods also include biotechnical methods: use of microbial or enzymatic pretreatments. A long-term goal would be to totally replace sulfur chemicals. Before this, savings in chemical consumption could be gained by improving the impregnation and diffusion properties of chips. Impregnation of chemicals into wood and removal of dissolved lignin are governed by diffusion and sorption phenomena, by the porosity and structure of the cell wall matrix, as well as by the molecular size of extractable molecules. In general, the median size of pores in wood fibers accessible to external macromolecules is about 1 nm, thus quite limited to the access of enzymes. The penetration and capability of different enzymes to act may vary according to their size and surface properties, as well as the site of the specific substrate, i.e., cellulose, hemicellulose, or lignin. It is obvious that the penetration of enzymes is limited, and that softwood is much less accessible than hardwood for enzymes.

4.2.1 Microbial Treatments

Microbial treatments of wood chips prior to chemical pulping, i.e., sulfite, sulfate, or organosolv cooking have not been as beneficial as prior to mechanical pulping (MESSNER et al., 1992; SCOTT et al., 1996; WALL et al., 1996; CHRISTOV and AKHTAR, 1998; FERRAZ et al., 1998; MOSAI et al., 1999). This is due to the fact that the microbial treatment primarily loosens the structure of the wood cell wall which enhances fiber liberation and only secondarily increases the porosity of the cell structure, often at expense of fiber strength. The benefit of fungal pretreatment in chemical pulping seems to be dependent not only on the fungal strain but also on the pulping method used. Increases in delignification rates have been noted even at low pulp weight losses in organosolv pulping when selective lignin-degrading fungi had been used (FERRAZ et al., 1998). However, only few reports of increased pulp yield and strength and reduced *kappa* number and cooking time in kraft pulping have been published (MESSNER et al., 1998).

4.2.2 Enzymatic Treatments

Interestingly, it has been shown that even enzymes, including hemicellulases, pectinases and cellulases increase the diffusivity of sodium hydroxide in Southern pine sapwood (JACOBS et al., 1998). The mean diffusion rate was increased by 60% in the tangential direction, and by 90% in the longitudinal direction. This result was attributed to the dissolution of pit membranes, which are the main resistance to flow of liquids in wood. SEM pictures revealed the disruption of the pit membranes after the enzymatic treatment.

It was also demonstrated that enzyme pretreatment of sycamore chips resulted in enhanced delignification (JACOBS-YOUNG et al., 1998). The most efficient enzyme preparations for sycamore chips contained cellulases, hemicellulases, and pectinases. The Southern pine chips were delignified more efficiently only after acetone extraction, but were reported to produce pulps with lower *kappa* numbers and rejects. After the acetone extraction and enzyme treatment the pulps were reported to be more uniform, to have higher viscosity and yields and lower rejects (JACOBS-YOUNG et al., 2000). The enhanced pulp uniformity was attributed to more uniform delignification, caused by improved diffusion. The role of the degree of hydrolysis, i.e., loss of carbohydrates during hydrolysis, as well as the need of acetone extraction on softwood remains to be clarified.

4.3 Bleaching

Enzymes can be used to improve the bleaching process indirectly or directly. In the indirect method the bleachability of pulps is improved through the action of xylanases or mannanases, affecting the extractability of lignin, and the effect is limited. The most promising direct enzymatic bleaching system is the laccase-mediator system, which degrades lignin (Tab. 4). Xylanases have been used on an industrial scale for enhancing the bleachability of kraft pulps for about 10 years, whereas the laccase-mediator concept is still under development.

4.3.1 Xylanase-Aided Bleaching

The effect of xylanase in bleaching is based on the modification of pulp xylan enhancing the extractability of lignin in subsequent bleaching stages (VIIKARI et al., 1994). It has been proposed that the action of xylanases is due to the partial hydrolysis of reprecipitated xylan or to removal of xylan from the lignin–carbohydrate complexes (LCC). Both mechanisms would lead to enhanced diffusion of entrapped lignin from the fiber wall. In softwood kraft fibers, removal of xylan by xylanases was found to uncover lignin (BUCHERT et al., 1996). Thus, it can be expected that removal of xylan improves the extractability of lignin by exposing lignin surfaces. The partial hydrolysis of xylan or glucomannan may also degrade and improve the extractability of lignin–carbohydrate complexes. The action of xylanases on

Tab. 4. Enzymatic Systems Studied for Bleaching of Chemical Pulps

Enzyme	Mechanism	Benefits/problems	Status
Xylanase	degradation of redeposited xylan and lignin–carbohydrate complexes	10–20% saving in chemical consumption increased brightness	commercialized
Mannanase	degradation of glucomannan	benefit depends on pulp type	commercialized
Laccase (with mediators)	degradation of lignin in the presence of mediators	total replacement of chlorine chemicals specificity recyclability and price of mediator?	close to commercialization
Mn peroxidase (with additives)	degradation of lignin in the presence of additives	good performance costs of enzymes and additives?	not commercialized
Lipase and peroxidase (with additives)	degradation of lignin by dioxorane mimicking system	good performance costs?	not commercialized

both reprecipitated and LC-xylan in enhancing bleachability suggests that it is probably not only the type but also the location of the xylan that is important in the mechanism of xylanase-aided bleaching. Xylanases seem to be efficient on all types of fibers, whereas the effects by mannanases depend on the type of fibers used (SUURNÄKKI et al., 1996c). Hexenuronic acid has been shown to contribute to the *kappa* number (VUORINEN et al., 1996). The partial removal of xylan by xylanases also results in a small, about 15% decrease in the hexenuronic acid content, corresponding to the removal of substituted xylan from the pulp (VIIKARI et al., 1999). The effect of xylanase on bleachability has in most cases been independent of the origin of the enzyme and both fungal and bacterial xylanases have been reported to act on pulp xylan and result in enhanced bleachability (PATEL et al., 1993). The efficiency of xylanases from family-10 and family-11 in bleaching has been compared and it was proposed that some xylanases from family-11 were more effective in bleach boosting than family-10 xylanases (CLARKE et al., 1997).

The use of xylanases in different bleaching sequences uniformly leads to a reduction in chemical consumption. The benefits obtained by enzymes are dependent on the type of bleaching sequence used as well as on the residual lignin content of the pulp. Originally, xylanases were studied in order to reduce the consumption of elemental chlorine. Later, xylanases have been combined with various ECF (elementary chlorine free) and TCF (totally chlorine free) bleaching sequences to improve the otherwise lower brightness of pulp or to decrease the bleaching costs. In chlorine bleaching an average reduction of 25% in active chlorine consumption in prebleaching or a reduction of about 15% in total chlorine consumption has been reported both on a laboratory scale and in mill trials. As a result, the concentration of chlorinated compounds, measured as AOX (adsorbable organic halogens), in the bleaching effluent during mill trials was reduced by 15–20%. Today, xylanases are industrially used both in ECF and TCF sequences. In ECF sequences, the enzymatic step is often implemented due to the limiting chlorine dioxide production capacity. The use of enzymes allows bleaching to higher brightness values when chlorine gas is not used. In TCF sequences, the advantage of the enzymatic step is due to improved brightness, maintenance of fiber strength, and savings in bleaching costs (VIIKARI et al., 1994). Presently, about 20 mills in Northern America and Scandinavia use enzymes. Thermostable xylanase products for pulp bleaching have been on the market since 1995 (PALOHEIMO et al., 1998). The thermostabilities of new enzyme products are continuously improving and research on xylanases acting at both high pH and temperature (pH 10 and 90 °C) is ongoing. The approximate price of xylanase in 1999 was less than 2 US$ per ton of pulp. Calculations of the economic benefits in an ECF sequence reveal that reduction in the chlorine dioxide consumption leads to savings of at least 2 US$ per ton of pulp. The costs of oxygen based chemicals (ozone, peroxide) are even higher and the respective savings even more pronounced. Thus, the potential economic benefits of enzyme bleaching are significant to the user.

4.3.2 Laccase-Mediator Concept in Bleaching

In the laccase-mediator concept, the enzyme oxidized mediator acts directly on lignin and results in efficient delignification (Fig. 2). In the initial study, the common substrate of laccases, ABTS, was used as the first mediator (BOURBONNAIS and PAICE, 1992). The search for a more suitable mediator resulted in the use of 1-hydroxybenzotriazole (HBT) on a laboratory and pilot scale (CALL, 1995). This delignification procedure is commonly referred to as the LMS (laccase-mediator system) or Lignozyme process. Although the mediator HBT is very efficient, it suffers from some drawbacks, such as high production costs and limited biodegradability. A number of other new mediators with great structural variety have been discovered, indicating the unspecificity of laccases. The most effective mediators in delignification usually contain N–OH functional groups (AMANN, 1997; FREUDENREICH et al., 1998), such as the most promising present mediators, violuric acid (VIO) and N-hydroxy-N-phenylacetamide (NHA). Especially the latter mediator results in extremely fast delignification with no significant impact on cellulose structure. The delignification degree after an alkaline extraction is reported to be high, up to 40% (POPPIUS-LEVLIN et al., 1997). Several studies on the mechanisms of laccase-mediated delignification of pulps have been published (e.g., BOURBONNAIS et al., 1995; SEALY and RAGAUSKAS, 1997; BALAKSHIN et al., 1999; POPPIUS-LEVLIN et al., 1999; CHAKAR and RAGAUSKAS, 1999). The LMS system has been shown to be able to replace either the oxygen delignification or ozone stage (POPPIUS-LEVLIN et al., 1998; CHAKAR and RAGAUSKAS, 1999). The enzyme system resulted in the same brightness levels as obtained with oxygen or two alternative ozone treatments, but gave superior viscosities and yields (Tab. 5).

The beneficial effects of xylanase and laccase-mediator systems have shown to be additive when the treatments were used successively. The application of the LMS system employing HBT as mediator with xylanase treatment in one single stage was, however, found

Tab. 5. Comparison of the Effectiveness of Laccase-Mediator Treatment and Oxygen or Ozone Stage in Bleaching of Kraft Pine Pulp (*kappa* number 24.7)

Sequence[a]	Brightness [%]	*Kappa* no.	Viscosity [mg g^{-1}]	Yield [%]
OQPZ/QP	84.1	2.5	720	42.0
LQPZ/QP	83.2	2.4	730	42.6
OOQZP	85.6	2.3	670	42.1
OOLQP	87.6	3.0	790	42.9

[a] O: oxygen, Q: chelation, P: peroxide, Z: ozone, L: laccase-mediator.

to be ineffective, apparently due to the inactivation of xylanase by the HBT. This inactivating effect of HBT has also been observed towards laccases (FREUDENREICH et al., 1998). Due to the high reactivity of HBT radicals, they also undergo chemical reactions with the aromatic amino acid side chains of many laccases. Studies on new mediators have revealed that NHA caused less damage to enzymes (PFALLER et al., 1998; VIIKARI et al., 1999). In practice, it would be beneficial to combine these two treatments.

In addition to the delignification, the effect of LMS on the physical properties of pulps has been determined. In high *kappa* number chemical pulps both laccase-HBT treatment and HBT treatment alone enhanced the handsheet densification during PFI refining (WONG et al., 1999). The use of laccase with NHA and violuric acid resulted in similar bonding strength as compared with oxygen delignification, but without reduction in viscosity (HAYNES and RAGAUSKAS, 1998).

4.4 Paper Manufacture

Minimized water usage and effluent discharge is a growing challenge in paper manufacture. Currently, the amount of fresh water used in modern paper machines is 10–15 m^3 t^{-1}. Due to the legislation and environmental pressure there is, however, a constant aim towards closed water circuits in papermaking. One major problem in water closure is the accumulation of dissolved and colloid substances (DCS) in the process waters. These substances consist mainly of hemicelluloses, pectins, dispersed wood resin, and lignin derived aromatic components. They have a direct impact on the paper machine runnability and paper quality. The exact composition, however, depends on the type of pulp and process (HOLMBOM, 1998). The wood resin released from TMP is dipersed as colloid droplets with an anionic charge. Most of the lipophilic extractives released from unbleached TMP are dispersed as colloidal droplets, which in turn are partially stabilized by polysaccharides, i.e., glucomannan (SUNDBERG et al., 1994).

The effects of different enzymes on the chemistry and physicochemical behavior of extractives, glucomannan, pectin, and lignans are interesting. The small chemical changes achieved by enzymes may result in significant modifications in the behavior of DCS in white waters, and more significantly, may lead to potential improvements of technical parameters, such as yield, strength, or brightness. The composition and structure of dissolved and colloid substances can be specifically modified with esterases, mannanases, pectinases, and oxidative enzymes (TENKANEN et al., 1995c; FUJITA et al., 1992; KANTELINEN et al., 1995; THORNTON, 1994; ZHANG, 2000).

The glucomannans dissolved from mechanical pulp fibers have been shown to stabilize colloidal resin and therefore prevent the aggregation (SIHVONEN et al., 1998). This glucomannan can be modified by mannanase, which cleaves the polymer randomly (KANTELINEN et al., 1995). As a result of the enzymatic treatment, the turbidity of TMP filtrates was decreased, causing fixation of pitch onto fibers as single particles. The chemical structure of galactoglucomannan present in TMP water can be further modified with acetyl glucomannan esterase being able to cleave acetyl groups from polymeric glucomannan (TENKANEN et al., 1995c). The acetyl glucomannan esterase from *Aspergillus oryzae* could remove nearly 70% of the acetyl groups from the acetyl galactoglucomannan isolated from TMP water. By comparison, an alkaline treatment removed all acetyl groups. The deacetylation of soluble glucomannan has been found to result in decreased solubility and subsequent adsorption of glucomannan onto the fibers (THORNTON et al., 1994). In recycled fibers the action of endoglucanases has been found to enhance drainage and thus to affect the paper machine speed (STORK and PULS, 1996).

Alkaline peroxide bleaching of mechanical pulp efficiently solubilizes pectins. The presence of anionic polygalacturonic acids affects the need for cationic retention aids used in the paper machine. The ability of polygalacturonic acids to complex cationic polymers depends strongly on their degree of polymerization (DP); hexamers and longer polysaccharide chains have a high cationic demand. Pectinases depolymerize polygalacturonic acids, and consequently decrease the cationic demand in the filtrates from peroxide bleaching of TMP

(THORNTON et al., 1994). Experiments have also shown that the enzyme treatment improves the effectiveness of several cationic polymers to increase retention of fines and filler particles (REID and RICARD, 2000). No negative impact on strength properties has been observed.

Japanese scientists at Jujo Paper reported already in 1990 that lipases could reduce pitch problems by hydrolyzing the triglycerides to glycerol and free fatty acids in mechanical pulps. A commercial lipase product (Resinase) was developed to reduce pitch deposits from groundwood pine pulp. Using the enzymatic treatment, it is possible to produce mechanical pulp from fresh pinewood without problems. The enzymatic pitch control technology has been industrially used for several years (FUJITA et al., 1992). The triglycerides from other types of pulps were also efficiently hydrolyzed by lipases, reducing the stickiness and pitch problems (FISHER and MESSNER, 1992; MUSTRANTA et al., 1995). The lipase treatment allows savings in the consumption of white carbon, surface-active chemicals, and results in improved fiber property (FUJITA et al., 1992). The composition and structure of lipophilic and hydrophilic extractives have also been modified with oxidative enzymes, i.e., laccases (ZHANG, 2000; KARLSSON et al., in press). Laccases treatment resulted in the degradation of most of the extractives, while the lipase specifically hydrolyzed the ester-bonded extractives present in the colloidal fraction.

The effects of the individual cellulases on the properties of unbleached or bleached kraft pulp have been studied in detail (PERE et al., 1995). *T. reesei* cellobiohydrolases (CBH) have been found to have only a modest effect on pulp viscosity, whereas the endoglucanases (EG), and especially EG II dramatically decrease pulp viscosity and thus the strength properties after refining. Treatment of ECF bleached pine kraft pulp with *T. reesei* endoglucanases EG I and EG II has been reported to enhance the beatability considerably (OKSANEN et al., 1997a; KANTELINEN et al., 2000). The strength properties of the pulp were simultaneously impaired, presumably due to the endoglucanase attack on the amorphous cellulose present especially in the defects and irregular zones of the fibers, as proposed by PERE et al. (1995). However, in chemical pulp cellulases and cellulase–hemicellulase mixtures have been reported to enhance the beatability of the coarse fibers and thus to improve the paper properties (MANSFIELD et al., 1996).

The presence of cellulose binding domains has been found to enhance the hydrolysis of isolated cellulose and chemical pulps. As could be expected, the effect of CBD was more pronounced in cellobiohydrolases. The pulp properties, i.e., viscosity and strength after PFI refining, were equally affected by the treatment with intact enzymes or corresponding core proteins, suggesting that the presence of CBD in intact cellulases affects mainly the cellulose hydrolysis level, and less the mode of action of *T. reesei* cellulases in pulp (SUURNÄKKI et al., 2000).

4.5 De-Inking

Recycled pulp is increasingly used in newsprint, tissue paper, and in higher grades of graphic papers. Fibers in recovered paper must be de-inked, i.e., repulped and cleaned from dirt and ink before they can be used again in papermaking. In de-inking ink particles are detached and removed from fibers by combined mechanical and chemical action. When aiming at more efficient and environmentally friendly de-inking processes, enzyme-aided de-inking is a potential alternative. The application of cellulase and hemicellulase mixtures in de-inking has been studied on the laboratory, pilot, and mill scale (PRASAD, 1993; PRASAD et al., 1993; JEFFRIES et al., 1994; WELT and DINUS, 1995; BAJPAJ, 1997; MORKBAK and ZIMMERMANN, 1998; SUURNÄKKI et al., 1998; KNUDSEN et al., 1998; ZOLLNER and SCHROEDER, 1998). The use of enzymes in de-inking of recovered paper is one of the most promising biotechnical applications in the pulp and paper industry already tested on a mill scale.

There are two principal approaches for the use of enzymes in de-inking, i.e., the enzymatic liberation of ink particles from the fiber surface by carbohydrate hydrolyzing enzymes, such as cellulases, hemicellulases, or pectinases, or the hydrolysis of the ink carrier or coating layer (Fig. 3). The suitability of enzymes acting on soya oil based ink carriers,

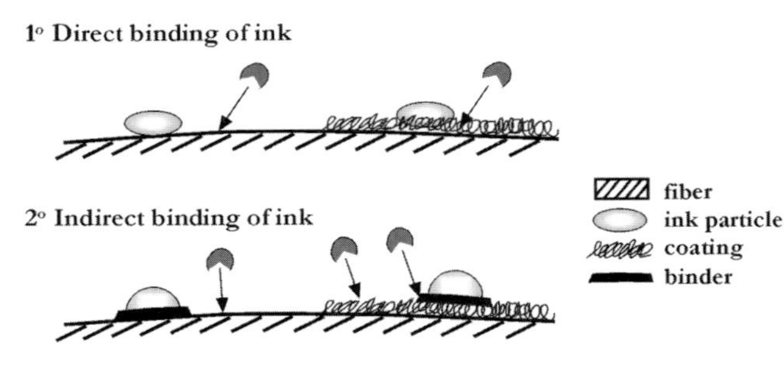

Fig. 3. Hypothetical mechanisms of enzyme-aided de-inking.

lignin and starch has been tested. The hypothesis on the enzyme-aided de-inking is that the enzymatic hydrolysis of the ink carrier, starch coating, or fiber surface liberates ink particles which are large enough to be removed by flotation de-inking. Enzyme mixtures designed for target paper grade are, however, needed in order to increase ink detachment and flotation of ink particles.

5 Future Prospects

The forest industry has great visions for the future. Among those are new fibers, intelligent packages, and tailor-made products. Enzymes are among the tools that will help the forest industry to reach these goals. Enzymes, due to their specificity, allow carrying out reactions that cannot be achieved by other means. Therefore, they should have great potential in the design of new products. Enzymes can be used to study the structure–function properties of fibers, useful when predicting the needs and consequences of possible modifications in the synthetic pathways of tree metabolism. Currently, the unique specificity, environmental safety, and cost savings are the driving forces behind the development of biotechnical applications for the pulp and paper industry.

In spite of these promises, the commercialization has been slow. Reasons for this are many. Few biotechnical applications have shown to be superior, as compared with traditional technologies. Rather, biotechnology has shown promise in small step improvements. As in the case of xylanases, factors, such as environmental aspects or a substantial increase in the energy price may, however, suddenly change the situation.

6 References

AKHTAR, M., BLANCHETTE, R., MYERS, G., KIRK, T. (1998), An overview of biomechanical pulping research, in: *Environmentally Friendly Technologies for the Pulp and Paper Industry* (YOUNG, R. A., AKTHAR, M., Eds.), pp. 309–340. New York. NY: John Wiley & Sons.

AMANN, M. (1997), The Lignozyme Process Coming Closer to the Mill, *Proc. ISWPC*, 1–5, F4, June 9–12, Montreal, Canada.

ANDER, P., ERIKSSON, K.-E. (1977), Selective degradation of wood components by white-rot fungi, *Physiol. Plant.* **41**, 239–248.

ARBEOLA, M., DE LESELEUC, J., GOMA, G., POMMIER, J. (1992), An evaluation of the potential of lignin peroxidases to improve pulps, *Tappi J.* **75**, 215–221.

BAJPAJ, P. (1997), Enzymatic deinking, *Adv. Appl. Microbiol.* **45**, 241–269.

BALAKSHIN, M., CAPANEMA, E., CHEN, C.-L., GRATZL, J., KIRKMAN, A., GRACZ, H. (1999), Lignin reactions in pulp biobleaching with laccase mediator system, *Proc. 10th Int. Symp. Wood and Pulp. Chem.* (ISWPC), Vol. 1, 572–577, Yokohama, Japan.

BAO, W., FUKUSHIMA, Y., JENSEN, K., MOEN, M. (1994), Oxidative degradation of non-phenolic lignin during lipid peroxidation by fungal manganese peroxidases, *FEBS Lett.* **354**, 297–300.

BARAZNENOK, V. A., BECKER, E. G., ANKUDIMOVA, N. V., OKUNEV, N. N. (1999), Characterization of neutral xylanases from *Chaetomium cellulolyticum* and their biobleaching effect on eucalyptus pulp, *Enzyme Microb. Technol.* **25**, 651–659.

BAR-LEV, S., KIRK, T., CHANG, H. (1982), Fungal treatment can reduce energy requirements for secondary refining of TMP, *Tappi J.* **65**, 111–113.

BIELY, P. (1985), Microbial xylanolytic systems, *Trends Biotechnol.* **3**, 286.

BIELY, P., VRSANSKA, M., CLAEYSSENS, M. (1991), The endo-1,4-β-D-glucanase from *Trichoderma reesei*: Action on β-1,4-oligomers derived from D-glucose and D-xylose, *Eur. J. Biochem.* **200**, 157–163.

BIELY, P., VRSANSKA, M., TENKANEN, M., KLUEPFEL, D. (1997), Endo-β-xylanase families: differences in catalytic properties, *J. Biotechnol.* **57**, 151–166.

BOURBONNAIS, R., PAICE, M. (1992), Demethylation and delignification of kraft pulp by *Trametes versicolor* laccase in the presence of 2,2′-azino-bis-(3-ethylbenzthiazoline-6-sulphonate), *Appl. Microbiol. Biotechnol.* **36**, 823–827.

BOURBONNAIS, R., PAICE, M., REID, I., LANTHIER, P., YAGUCHI, M. (1995), Lignin oxidation by laccase isozymes from *Trametes versicolor* and role of the mediator 2,2′-azino-bis-(3-ethylbenzthiazoline-6-sulfonate) in kraft lignin depolymerization, *Appl. Environ. Microbiol.* **61**, 1876–1880.

BUCHERT, J., CARLSSON, G., VIIKARI, L., STRÖM, G. (1996), Surface characterization of unbleached kraft pulps by enzymatic peeling and ESCA, *Holzforsch.* **50**, 69–74.

BUCHERT, J., BERGNOR, E., LINDBLAD, G., VIIKARI, L., EK, M. (1997), Significance of xylan and glucomannan in the brightness reversion of kraft pulps, *Tappi J.* **80**, 165–171.

CALL, H. P. (1995), Further improvements to the laccase-mediator system for enzymatic delignification and results from large-scale trials, *Proc. Int. Non-Chlorine Bleaching Conf.*, Paper 4-3, 16 p, March 5–9, Florida, USA.

CALL, H. P., STRITTMATTER, G. (1992), Application of ligninolytic enzymes in the paper and pulp industry – recent results, *Papier* **46**, 32.

CHAKAR, F., RAGAUSKAS, A. (1999), Fundamental investigations of laccase mediator delignification on high lignin content kraft pulps, *Proc. 10th Int. Symp. Wood and Pulp Chem.* (ISWPC), Vol. 1, 566–570, Yokohama, Japan.

CHEN, M., GRITZALI, M., STAFFORD, D. (1987), Nucleotide sequence and deduced primary structure of cellobiohydrolase II of *Trichoderma reesei*, *Bio/Technology* **5**, 274–278.

CHEN, H., HAYN, M., ESTERBAUER, H. (1992), Purification and characterization of two extracellular β-glucosidases from *Trichoderma reesei*, *Biochim. Biophys. Acta* **1121**, 54–60.

CHRISTOV, L., AKHTAR, M. (1998), Evaluation of selected *Cerioporiopsis subvermispora* strains for biosulfite pulping and bleaching to prepare dissolving pulp, *Proc. 7th Int. Conf. Biotechnology and Pulp and Paper Industry: Publication Clerk*, B15–B18, Montreal, Canada.

CLARKE, J. H., RIXON, J. E., CIRUELA, A., GILBERT, H. J. (1997), Family-10 and family-11 xylanases differ in their capacity to enhance the bleachability of hardwood and softwood paper pulps, *Appl. Microbiol. Biotechnol.* **48**, 177–183.

COUGHLAN, M., HAZLEWOOD, G. (1993), β-1,4-D-xylan-degrading enzyme systems: biochemistry, molecular biology and applications, *Biotechnol. Appl. Biochem.* **17**, 259.

DA SILVA, R., YIM, D. K., PARK, Y. K. (1994), Application of thermostable xylanase from *Humicola* sp. for pulp improvement, *J. Ferment. Bioeng.* **77**, 109–111.

DE JONG, E., DE VRIES, E., FIELD, J., VAN DER ZWAN, R., DE BONT, J. (1992), Isolation and screening of basidiomycetes with high peroxidative activity, *Mycol. Res.* **96**, 1098–1104.

DEREWENDA, Z. S. (1994), Structure and function of lipases, in: *Advances in Protein Chemistry* Vol. 45 (ANFINSEN, C. B., EDSALL, J. T., RICHARDS, F. M., EISENBERG, D.S., Eds.), pp. 1–52. New York, NY: Academic Press.

DUCROS, V., BRZOZOWSKI, A. M., WILSON, K. S., BROWN, S. H., OSTERGAARD, P. et al. (1998), Crystal structure of the type-2 Cu depleted laccase from *Coprinus cinereus* at 2.2 Å resolution, *Nature Struct. Biol.* **5**, 310–316.

FARRELL, R. L., HATA, K., WALL, M. B. (1997), Solving pitch problems in pulp and paper processes by the use of enzyme or fungi, in: *Advances in Biochemical Engineering/Biotechnology* Vol. 57 (SCHEPER, T., Ed.), pp. 197–212. Berlin, Heidelberg: Springer-Verlag.

FERRAZ, A., CHRISTOV, L. P., AKHTAR, M. (1998), Fungal pretreatment for organosolv pulping and dissolving pulp production, in: *Environmentally Friendly Technologies for the Pulp and Paper Industry* (YOUNG, R. A., AKHTAR, M., Eds.), pp. 421–447. New York, NY: John Wiley & Sons.

FISCHER, K., MESSNER, K. (1992), Reducing troublesome pitch in pulp mills by lipolytic enzymes, *Tappi J.* **75**, 130–134.

FREUDENREICH, I., AMANN, M., FRITZ-LANGHALS, E., STOHRER, J. (1998), Understanding the Lignozym-process, *Proc. Int. Pulp Bleaching Conf.*, 71–76, June 1–5, Helsinki, Finland.

FUJITA, Y., AWAJI, H., TANEDA, H., MATSUKURA, M., HATA, K. et al. (1992), Recent advances in enzymatic pitch control, *Tappi J.* **75**, 117–112.

GARG, A. P., MC CARTHY, A. J., ROBERTS, J. C. (1996), Biobleaching effect of *Streptomyces thermoviolaceus* xylanase preparations on birchwood kraft pulp, *Enzyme Microb. Technol.* **18**, 261–267.

GILBERT, H. J., HAZLEWOOD, G. P. (1993), Bacterial cellulases and xylanases, *J. Gen. Microbiol.* **139**, 187–194.

GLENN, J. K., GOULD, M. H. (1985), Purification and characterization of an extracellular Mn(II)-dependent peroxidase from the lignin-degrading basidiomycete *Phanerochaete chrysosporium*, *Arch. Biochem. Biophys.* **242**, 329–341.

HAYNES, K., RAGAUSKAS, A. (1998), Effects of laccase mediator delignification on fiber properties, *Proc. Int. Pulp Bleaching Conf.*, Book 2, 355–359, Helsinki, Finland, 1–5 June.

HAZLEWOOD, G. P., GILBERT, H. J. (1992), The molecular architecture of xylanases from *Pseudomonas fluorescens* ssp. *cellulosa*, in: *Xylan and Xylanases* (VISSER, J., BELDMAN, G., KUSTERS-VAN SOMEREN, N. A., VORAGEN, A. G. J., Eds.), p. 259. Amsterdam: Elsevier Science Publishers.

HILGE, M., GLOOR, S. M., RYPNIEWSKI, W., SAUER, O., HEIGHTMAN, T. D. et al. (1998), High resolution native and complex structures of thermostable β-mannanase from *Thermomonospora fusca* – substrate specificity in glycosyl hydrolase family 5, *Structure* **6**, 1433–1444.

HOLMBOM, B. (1998), Analysis of papermaking process waters and effluents, in: *Analytical Methods in Wood Chemistry, Pulping and Papermaking* (SJÖSTRÖM, E., ALEN, R., Eds.), pp. 269–285. Berlin: Springer-Verlag.

HU, W. J., HARDING, S., LUNG, J., POPKO, J., RALPH, J. et al. (1999), Repression of lignin biosynthesis promotes cellulose accumulation and growth in transgenic trees, *Nature Biotechnol.* **17**, 808.

IRWIN, D., JUNG, E. D., WILSON, D. B. (1994), Characterization and sequence of a *Thermomonospora fusca* xylanase, *Appl. Environ. Microbiol.* **60**, 763–770.

ISHIHARA, T. (1980), The role of laccase in lignin biodegradation, in: *Lignin Biodegradation: Microbiology, Chemistry and Potential Applications*, Vol. 2 (KIRK, T., HIGUCHI, T., CHAN, H., Eds.), pp. 17–31. Boca Raton, FL: CRC Press.

JACOBS, C. J., VENDITTI, R. A., JOYCE, T. W. (1998), Effect of enzyme 13. pretreatments on conventional kraft pulping, *Tappi J.* **81**, 260–266.

JACOBS-YOUNG, C. J., HEITMAN, J. A, VENDITTI, R. A. (1998), Conventional kraft pulping using enzyme pretreatment technology: role of chip thickness, specie and enzyme combinations, *Proc. AICHE Symp. Ser.* No. 319, Vol. 94, pp. 1–15.

JACOBS-YOUNG, C. J., GUSTAFSSON, R. R., HEITMAN, J. A. (2000), Conventional kraft pulping using enzyme pretreatment technology: role of diffusivity in enhancing pulp uniformity, *Paper Timber* **82**, 114–119.

JEFFRIES, T. W., KLUNGNESS, J. H., STYKES, M. S., RUTLEDGE-CROPSEY, K. (1994), Comparison of enzyme-enhanced deinking with conventional deinking of xerographic and laser-printed paper, *Tappi J.* **77**, 173–179.

KANTELINEN, A., JOKINEN, O., SARKKI, M.-L., PETTERSSON, C., SUNDBERG, K. et al. (1995), Effects of enzymes on the stability of colloidal pitch, *Proc. 8th Int. Symp. Wood Pulp Chem.* **1**, 605–612.

KANTELINEN, A., SARKAR, J., OKSANEN, T. (2000), Application of enzymes in production of release and high density paper, *Proc. Tappi Pulping Conf.*, Nov. 5–9, Boston, USA.

KARLSSON, S., HOLMBOM, S., SPETZ, P., MUSTRANTA, A., BUCHERT, J. (in press), Reactivity of *Trametes* laccases with fatty and resin acids, *Appl. Microbiol. Biotechnol.*

KIRK, T., FARRELL, R. (1987), Enzymatic "combustion": The microbial degradation of lignin, *Annu. Rev. Microbiol.* **41**, 465–505.

KNOWLES, J., LEHTOVAARA, P., TEERI, T. T. (1987), Cellulase families and their genes, *Trends Biotechnol.* **5**, 255–261.

KNUDSEN, O., YOUNG, J. D., YANG, J. L. (1998), Long term use of enzymatic deinking at Stora Dalum plant, *Proc. 7th Int. Conf. Biotechnology Pulp and Paper Industry*, A17–A20, Vancouver, Canada.

KONDO, R., KURASHIKI, K., SAKAI, K. (1994), *In vitro* bleaching of hardwood kraft pulp by extracellular enzymes excreted from white-rot fungi in a cultivation system using a membrane filter, *Appl. Environ. Microbiol.* **60**, 921–926.

KUSAKABE, I., PARK, G., KUMITA, N., YASUI, T., MURAKAMI, K. (1988), Specificity of β-mannanase from *Penicillium purpurogenum* for konjak glucomannan, *Agr. Biol. Chem.* **52**, 519.

LAINE, J., BUCHERT, J., VIIKARI, L., STENIUS, P. (1996), Characterisation of unbleached kraft pulps by enzymatic treatment, potentiometric titration and polyelectrolyte adsorption, *Holzforsch.* **50**, 208–214.

LAPIDOT, A., MECHALY, A., SHOHAM, Y. (1996), Overexpression and single step purification of a thermostable xylanase from *Bacillus stearothermophilus* T-6, *J. Biotechnol.* **51**, 259–264.

LEATHAM, G., MYERS, G., WEGNER, T., BLANCHETTE, R. (1990a), Biomechanical pulping of aspen chips: paper strength and optical properties resulting from different fungal treatments, *Tappi J.* **73**, 249–255.

LEATHAM, G., MYERS, G., WEGNER, T. (1990b), Biomechanical pulping of aspen chips: energy savings resulting from different fungal treatments, *Tappi J.* **73**, 197–200.

LINDER, M., TEERI, T. (1997), The roles and function of cellulose-binding domains, *J. Biotechnol.* **57**, 15–28.

Mansfield, S. D., Wong, K. K. Y., de Jong, E., Saddler, J. N. (1996), Modification of Douglas-fir mechanical and kraft pulps by enzyme treatment, *Tappi J.* **79**, 125–172.

Mansfield, S. D., Wong, K. K. Y., Richardson, J. D. (1999), Improvements in mechanical pulp processing with proteinase treatments, *Appita J.* **52**, 436–440.

Margolles-Clark, E., Tenkanen, M., Söderlund, H., Penttilä, M. (1996), Acetyl xylan esterase from *Trichoderma reesei* contains an active-site serine residue and a cellulose-binding domain, *Eur. J. Biochem.* **237,** 553–560.

McCleary, B. (1991), Comparison of endolytic hydrolases that polymerize 1,4-β-D-mannan, 1,5-α-L-arabinan, and 1,4-β-D-galactan, in: *Enzymes in Biomass Conversion* (Leatham, G. F., Himmel, M. E., Eds.), pp. 437–449. Washington, DC: American Chemical Society, ACS Symp. Ser. 460.

Messner, K., Masek, S., Srebotnik, E., Techt, G. (1992), Fungal pretreatment of wood chips for chemical pulping, *Proc. 5th Int. Conf. on Biotechnology in the Pulp and Paper Industry* (Kuwahara, M., Shimada, M., Eds.), pp. 9–13. Kyoto, Japan: Uni Publishers.

Messner, K., Koller, K., Wall, M. B., Akhtar, M., Scott, G. M. (1998), Fungal pretreatment of wood chips for chemical pulping, in: *Environmentally Friendly Technologies for the Pulp and Paper Industry* (Young, R. A., Akthar, M., Eds.), pp. 385–419. New York, NY: John Wiley & Sons.

Morbak, A., Zimmermann, W. (1998), Deinking of mixed office paper, old newspaper and vegetable oil based ink printed paper using cellulases, xylanases and lipases, *Prog. Ppa. Recycl.* Feb, 14–21.

Morris, D. D., Reeves, R. A., Gibbs, M. D., Saul, D. J., Bergqvist, P. L. (1995), Correction of the β-mannanase domain of the calC pseudogene from *Caldocellosiruptor saccharolyticus* and activity of the gene product on kraft pulp, *Appl. Environ. Microbiol.* **61**, 2262–2269.

Mosai, S., Wolfaardt, J., Prior, B., Christov, L. (1999), Evaluation of selected white-rot fungi for biosulfite pulping, *Bioresour. Technol.* **68**, 89–93.

Mustranta, A., Fagernäs, L., Viikari, L. (1995), Effects of lipases on birch extractives, *Tappi J.* **78**, 140–146.

Oksanen, T., Pere, J., Buchert, J., Viikari, L. (1997a), The effect of *Trichoderma reesei* cellulases and hemicellulases on the paper technical properties of never-dried bleached kraft pulp, *Cellulose* **4**, 329–339.

Oksanen, T., Buchert, J., Viikari, L. (1997b), The role of hemicelluloses in the hornification of bleached kraft pulps, *Holzforsch.* **51**, 355–360.

Oksanen, T., Pere, J., Paavilainen, L., Buchert, J., Viikari, L. (2000), Treatment of recycled kraft pulps with *Trichoderma reesei* hemicellulases and cellulases, *J. Biotechnol.* **78**, 39–48.

Paice, M., Reid, I., Bourbonnais, R., Archibald, F., Jurasek, L. (1993), Manganese peroxidase, produced by *Trametes versicolor* during pulp bleaching, demethylates and delignifies kraft pulp, *Appl. Environ. Microbiol.* **59**, 260–265.

Paloheimo, M., Mäntylä, A., Vehmaanperä, J., Hakola, S., Lantto, R. et al. (1998), Thermostable xylanases produced by recombinant *Trichoderma reesei* for pulp bleaching, in: *Carbohydratases from Trichoderma reesei and other micro-organisms. Structures, Biochemistry, Genetics and Applications* (Claeyssens, M., Nerinckx, W., Piens, K., Eds.), pp. 255–264. Cambridge: The Royal Society of Chemistry.

Patel, R. N., Grabski, A. C., Jeffries, T. W. (1993), Chromophore release from kraft pulp by purified *Streptomycetes roseiscleroticus* xylanase, *Appl. Microb. Biotechnol.* **39**, 405–412.

Penttilä, M., Lehtovaara, P., Nevalainen, H., Bhikhabhai, R., Knowles, J. (1986), Homology between cellulase genes of *Trichoderma reesei*: complete nucleotide sequence of the endoglucanase I gene, *Gene* **45**, 253–263.

Pere, J., Siika-aho, M., Buchert, J., Viikari, L. (1995), Effects of purified *Trichoderma reesei* cellulases on the fiber properties of kraft pulp, *Tappi J.* **78**, 71–78.

Pere, J., Liukkonen, S., Siika-aho, M., Gullichsen, J., Viikari, L. (1996), Use of purified enzymes in mechanical pulping. *Proc. 1996 Tappi Pulping Conf.*, Nashville, TN, 27–31 Oct 1996. pp. 693–696. Atlanta, GA: Tappi Press.

Pere, J., Siika-aho, M., Viikari, L. (2000), Biomechanical pulping with enzymes; susceptibility of coarse mechanical pulp to enzymatic modification and secondary refining, *Tappi J.* **83**, 1–8.

Pfaller, R., Amann, M., Freudenreich, J. (1998), Analysis of laccase and mediator interactions in the LMS®, *Proc. 7th Int. Conf. Biotechnology in the Pulp and Paper Industry*, A99–A102, 16–19 June, Vancouver, Canada.

Poppius-Levlin, K., Wang, W., Ranua, M., Niku-Paavola, M.-L., Viikari, L. (1997), Biobleaching of chemical pulps by laccase/mediator systems, *Proc. Biol. Sci. Symp.*, 327–333, Tappi Proc. Atlanta, GA: Tappi Press.

Poppius-Levlin, K., Wang, W., Ranua, M. (1998), TCF bleaching of laccase/mediator-treated kraft pulps, *Proc. Int. Pulp Bleaching Conference*, 77–85, June 1–5, Helsinki, Finland.

Poppius-Levlin, K., Wang, W., Tamminen, T., Hortling, B., Viikari, L., Niku-Paavola, M.-L. (1999), Effects of laccase/HBT treatment on pulp and lignin structures, *J. Pulp Paper Sci.* **25**, 90–94.

Prasad, D. Y. (1993), Enzymatic deinking of laser and zerographic office wastes, *Appita* **46**, 289–292.

PRASAD, D. Y., HEITMAN, J. A., JOYCE, T. W. (1993), Enzymatic deinking of coloured offset newsprint, *Nordic Pulp Paper Res. J.* **2**, 284–286.

RÄTTÖ, M., MATHRANI, I., AHRING, B., VIIKARI, L. (1996), Application of thermostable xylanase of *Dictyoglomus* sp. in enzymatic treatment of kraft pulps, *Appl. Microbiol. Biotechnol.* **41**, 130–133.

REID, I., RICARD, M. (2000), Pectinase in papermaking: solving retention problems in mechanical pulps bleached with hydrogen peroxide, *Enzyme Microb. Technol.* **26**, 115–123.

REINIKAINEN, T., TELEMAN, O., TEERI, T. T. (1995), Effects of pH and ionic strength on the adsorption and activity of native and mutated cellobiohydrolase I from *Trichoderma reesei*, *Protein* **22**, 392–403.

RICHARDSON, J. D., WONG, K. K. Y., CLARK, T. (1998), Modification of mechanical pulp using carbohydrate-degrading enzymes, *J. Pulp Paper Sci.* **24**, 125.

RIXON, J. E., CLARKE, J. H., HAZLEWOOD, G. P., HOYLAMD, R. W., MCCARTHY, A. J., GILBERT, H. J. (1996), Do the non-catalytic polysaccharide-binding domains and linker regions enhance the biobleaching properties of modular xylanases, *Appl. Microbiol. Biotechnol.* **46**, 514–520.

ROBINSON, C. (1999), Making forest biotechnology a commercial reality, *Nature Biotechnol.* **17**, 27.

SABINI, E., SCHUBERT, H., MURSHUDOV, G., WILSON, K. S., SIIKA-AHO, M., PENTTILÄ, M. (2000), The three-dimensional structure of a *Trichoderma reesei* β-mannanase from glycoside hydrolase family 5, *Acta Cryst.* **D56**, 3–13.

SACHS, I., LEATHAM, G., MYERS, G., WEGNER, T. (1990), Distinguishing characteristics of biomechanical pulp, *Tappi J.* **73**, 249–254.

SAKKA, K., KOJIMA, Y., KONDO, T., KARITA, S., OHMIYA, K., SHIMADA, K. (1993), Purification and characterization of xylanase A from *Clostridium stercorarium* F-9 and a recombinant *Escherichia coli*, *Biosci. Biotechnol. Biochem.* **57**, 273–277.

SALOHEIMO, M., LEHTOVAARA, P., PENTTILÄ, M., TEERI, T., STÅHLBERG, J. et al. (1988), EGIII, a new endoglucanase from *Trichoderma reesei*; the characterization of both gene and enzyme, *Gene* **63**, 11–21.

SALOHEIMO, A., HENRISSANT, B., HOFFRÉN, A. M., TELEMAN, O., PENTTILÄ, M. (1994), A novel, small endoglucanase gene, *egl5*, from *Trichoderma reesei* isolated by expression in yeast, *Mol. Microbiol.* **13**, 219–228.

SALOHEIMO, M., NAKARI, T., TENKANEN, M., PENTTILÄ, M. (1997), A new *Trichoderma reesei* cellulase: cDNA-cloning and demonstration of endoglucanase activity by yeast expression, *Proc. 4th Eur. Conf. Fungal Genetics*, 166, Leon, Spain, 4–8 April.

SCOTT, G., SWANEY, R. (1998), New technology for papermaking: biopulping economics, *Tappi J.* **81**, 153–157.

SCOTT, G., AKHTAR, M., LENTZ, M., SYKES, M., ABUBAKR, S. (1996), Biosulfite pulping using *Ceriporiopsis subvermispora*, *Proc. 6th Int. Conf. Biotechnology and Pulp and Paper Industry:* Advances in Applied and Fundamental Research (SREBOTNIK, E., MESSNER, K., Eds.), pp. 217–220. Vienna, Austria: Facultas-Universitätsverlag.

SCOTT, G. M., AKHTAR, M., LENTZ, M. J., SWANEY, R. E. (1998), Engineering, scale-up and economic aspects of fungal pretreatments of wood chips, in: *Environmentally Friendly Technologies for the Pulp and Paper Industry* (YOUNG, R. A., AKHTAR, M., Eds.), pp. 341–384. New York, NY: John Wiley & Sons.

SEALEY, J., RAGAUSKAR, A. (1997), Fundamental investigations into the chemical mechanisms involved in laccase-mediator biobleaching, *Proc. 9th Int. Symp. Wood and Pulp. Chem.* (ISWPC), Montreal, Canada.

SETLIFF, E., MARTON, R., GRANZOW, S., ERIKSON, K. (1990), Biomechanical biopulping with white-rot fungi, *Tappi J.* **73**, 141–147.

SHAH, A. K., COOPER, D., ADOLPHSON, R., ERIKSSON, K.-E. (2000), Xylanase treatment of oxygen-bleached hardwood kraft pulp at high temperature and high pH levels gives substantial savings in bleaching chemicals, *J. Pulp Paper Sci.* **26**, 8–11.

SHARECK, F., ROY, C., YAGUCHI, M., MOROSOLI, R., KLUEPFEL, D. (1991), Sequences of three genes specifying xylanases in *Streptomyces lividans*, *Gene* **107**, 75.

SHOEMAKER, S., SCHWEICKART, V., LADNER, M., GELFAND, D., KWOK, S. et al. (1983), Molecular cloning of exo-cellobiohydrolase I derived from *Trichoderma reesei* strain L27, *Bio/Technology* **1**, 691–696.

SIHVONEN, A.-L., SUNDBERG, K., SUNDBERG, A., HOLMBOM, B. (1998), Stability and deposition tendency of colloidal wood resin, *Nordic Pulp Paper Res. J.* **13**, 64–67.

SJÖSTRÖM, E. (1993), *Wood Chemistry, Fundamentals and Applications*. San Diego, CA: Academic Press.

STÅHLBERG, J., JOHANSSON, G., PETTERSSON, G. (1991), A new model for enzymatic hydrolysis of cellulose based on the two-domain structure of cellobiohydrolase, *Bio/Technology* **9**, 286–290.

STÅHLBERG, J., JOHANSSON, G., PETTERSSON, G. (1993), *Trichoderma reesei* has no true exocellulase: all intact and truncated cellulases produce new reducing end groups on cellulose, *Biochim. Biophys. Acta* **1157**, 107–113.

STÅLBRAND, H., SALOHEIMO, A., VEHMAANPERÄ, J., HENRISSAT, B., PENTTILÄ, M. (1995), *Appl. Environ. Microbiol.* **61**, 1090–1097.

STORK, G., PULS, J. (1996), Change in properties of

different recycled pulps by endoglucanase treatment, in: *Biotechnology in Pulp and Paper Industry* (SREBOTNIK, E., MESSNER, K., Eds.), pp. 145–150. Vienna: Facultas Universitätsverlag.

SUNDBERG, K., PETTERSSON, C., ECKERMAN, C., HOLMBOM, B. (1994), Preparation and properties of a model dispersion of colloidal wood resin from Norway spruce, *J. Pulp Paper Sci.* **22**, 226–230.

SUURNÄKKI, A., HEIJNESSON, A., BUCHERT, J., TENKANEN, M., VIIKARI, L., WESTERMARK, U. (1996a), Location of xylanase and mannanase action in kraft fibres, *J. Pulp Paper Sci.* **22**, J78–J83.

SUURNÄKKI, A., HEIJNESSON, A., BUCHERT, J., VIIKARI, L., WESTERMARK, U. (1996b), Chemical characterization of the surface layers of unbleached pine and birch kraft pulp fibers, *J. Pulp Paper Sci.* **22**, J43–J47.

SUURNÄKKI, A., CLARK, T. A., ALLISON, R. W., VIIKARI, L., BUCHERT, J. (1996c), Xylanase- and mannanase-aided ECF and TCF bleaching, *Tappi J.* **79**, 111–117.

SUURNÄKKI, A., TENKANEN, M., BUCHERT, J., VIIKARI, L. (1997), Hemicellulases in the bleaching of chemical pulps, in: *Adv. Biochem. Eng. Biotechnol.* Vol. 57 (SCHEPPER, T., Ed.), pp. 261–287. Berlin Heidelberg: Springer-Verlag.

SUURNÄKKI, A., PUTZ, H.-J., RENNER, K., GALLAND, G., BUCHERT, J. (1998), Enzyme-aided deinking of recovered papers, *Proc. EUCEPA 1998 Symp. Chemistry in Papermaking*, 215–224, Florence, 12–14 Oct. Florence: ATICELA-EUCEPA.

SUURNÄKKI, A., TENKANEN, M., SIIKA-AHO, M., NIKU-PAAVOLA, M.-L., VIIKARI, L., BUCHERT, J. (2000) *Trichoderma reesei* cellulases and their core domains in the hydrolysis and modification of chemical pulp, *Cellulose* **7**, 189–209.

TAKASHIMA, S., NAKAMURA, A., HIDAKA, M., MASAKI, H., UOZUMI, T. (1999), Molecular cloning and expression of the novel fungal *beta*-glucosidase genes from *Humicola grisea* and *Trichoderma reesei*, *J. Biochem.* **125**, 728–736.

TEERI, T., LEHTOVAARA, P., KAUPPINEN, S., SALOVUORI, I., KNOWLES, J. (1987), Homologous domains in *Trichoderma reesei* cellulolytic enzymes: gene sequence and expression of cellobiohydrolase II, *Gene* **51**, 42–52.

TELEMAN, A., HARJUNPÄÄ, V., TENKANEN, M., BUCHERT, J., HAUSALO, T. et al. (1995), Characterization of 4-deoxy-β-L-threo-hex-4-enopyranosyluronic acid attached to xylan in pine kraft pulp and pulping liquor by ^{1}H and ^{13}C NMR spectroscopy, *Carbohydr. Res.* **272**, 55–71.

TENKANEN, M., BUCHERT, J., VIIKARI, L. (1995a), Binding of hemicellulases on isolated polysaccharide substrates, *Enzyme Microb. Technol.* **17**, 499–505.

TENKANEN, M., THORNTON, J., VIIKARI, L. (1995b), An acetylglucomannan esterase of *Aspergillus oryzae*; purification, characterization and role in the hydrolysis of O-acetyl-galactoglucomannan, *J. Biotechnol.* **42**, 197–206.

TENKANEN, M., HAUSALO, T., SIIKA-AHO, M., BUCHERT, J., VIIKARI, L. (1995c), Use of enzymes in combination with anion exchange chromatography in the analysis of carbohydrate composition of kraft pulps, *Proc. 8th Int. Symp. Wood and Pulping Chem.*, Vol II, 189–194, Helsinki, Finland.

TENKANEN, M., MAKKONEN, M., PERTTULA, M. VIIKARI, L., TELEMAN, A. (1997), Action of *Trichoderma reesei* mannanase on galactoglucomannan in pine kraft pulp, *J. Biotechnol.* **57**, 191–204.

THORNTON, J. (1994), Enzymatic degradation of polygalacturonic acids released from mechanical pulp during peroxide bleaching, *Tappi J.* **77**, 161–167.

THORNTON, J., EKMAN, R., HOLMBOM, B., PETTERSSON, C. (1994), Effects of alkaline treatment on dissolved carbohydrates in suspensions of Norway spruce thermomechanical pulp, *J. Wood Chem. Technol.* **14**, 176–194.

THURSTON, C. F. (1994), The structure and function of fungal laccases, *Microbiology* **140**, 19–26.

TIEN, M., KIRK, T. (1984), Lignin-degrading enzyme from *Phanerochaete chrysosporium*: purification, characterization and catalytic properties of a unique H_2O_2-requiring oxygenase, *Proc. Natl. Acad. Sci. USA* **81**, 2280–2284.

TOMME, P., VAN TILBEURGH, H., PETTERSSON, G., VAN DAMME, J., VANDEKERCKHOVE, J. et al. (1988), Studies of the cellulolytic system of *Trichoderma reesei* QM9414. Analysis of domain function in two cellobiohydrolases by limited proteolysis, *Eur. J. Biochem.* **170**, 575–581.

TOMME, P., WARREN, R., GILKES, N. (1995), Cellulose hydrolysis by bacteria and fungi, *Adv. Microb. Physiol.* **37**, 1–81.

TORMO, J., LAMED, R., CHIRINO, A. J., MORAG, E., BAYER, E. A. et al. (1996), Chrystal structure of a bacterial family III cellulose binding domain: a general mechanism for attachment to cellulose, *EMBO J.* **15**, 5739–5751.

TÖRRÖNEN, A., HARKKI, A., ROUVINEN, J. (1994), Three dimensional structure of endo-1,4-β-xylanase II from *Trichoderma reesei*: two conformational states in the active site, *EMBO J.* **13**, 2493.

VAN TILBEURGH, H., CLAYSSENS, M., BHIKHABNAI, R., PETTERSSON, G. (1986), Limited proteolysis of the cellobiohydrolase I from *Trichoderma reesei*, *FEBS Lett.* **204**, 223–227.

VIIKARI, L., KANTELINEN, A., SUNDQUIST, J., LINKO, M. (1994), Xylanases in bleaching: From an idea to the industry, *FEMS Microbiol. Rev.* **13**, 335–350.

VIIKARI., L., OKSANEN, T., BUCHERT, J., AMANN, M.,

CANDUSSIO, A. (1999), Combined action of hemicellulases and oxidases in bleaching, *Proc. from the ISWPC*, Vol I, 504–507, June 7–10, Yokohama, Japan.

VIIKARI, L., TENKANEN, M., POUTANEN, K. (2000), Hemicellulases, in: *Encyclopedia of Bioprocess Technology: Fermentation, Biocatalysis, and Bioseparation* (FLICKINGER, M. C., DREW, S. W., Eds.), pp. 1383–1391. New York, NY: John Wiley & Sons.

VUORINEN, T., TELEMAN, A., FAGERSTRÖM, P., BUCHERT, J., TENKANEN, M. (1996), Selective hydrolyses of hexenuronic acid groups and its application in ECF and TCF bleaching of kraft pulps, *Proc. Int. Pulp Bleaching Conf.*, 43–51, Washington, DC., April 14–18.

WALL, M., STAFFORD, G., NOEL, Y., FRITZ, A., IVERSON, S., FARRELL, R. (1996), Treatment with *Ophiostuma piliferum* improves chemical pulping efficiency, *Proc. 6th Int. Conf. Biotechnology and Pulp and Paper Industry: Advances in Applied and Fundamental Research* (SREBOTNIK, E., MESSNER, K., Eds.), pp. 205–210. Vienna; Facultas-Universitätsverlag.

WARD, M., WU, S., DAUBERMAN, J., WEISS, G., LARENAS, E. et al. (1993), Cloning, sequence and preliminary analysis of a small, high pI endoglucanase (EG III) from *Trichoderma reesei*, in: *Trichoderma reesei Cellulases and Other Hydrolases* (SUORNINEN, P., REINIKAINEN, T., Eds.), Foundation for Biotechnical and Industrial Fermentation Research, **8**, 153–158.

WARIISHI, H., VALLI, K., GOLD, M. (1991), *In vitro* depolymerization of lignin by manganese peroxidase of *Phanerochaete chrysosporium*, *Biochem. Biophys. Res. Commun.* **176**, 269–275.

WELT, T., DINUS, R. (1995), Enzymatic de-inking – A review, *Prog. Pap. Recycl.* **4**, 36–47.

WONG, K. K. Y., SADDLER, J. N. (1992), *Trichoderma* xylanases, their properties and application, *Crit. Rev. Biotechnol.* **12**, 413.

WONG, K. K. Y., ANDERSON, K. B., KIBBLEWHITE, R. P. (1999), Effects of the laccase-mediator system on the handsheet properties of two high *kappa* pulps, *Enzyme Microbiol. Technol.* **25**, 125–131.

ZHANG, X. (2000), The effects of white water dissolved and colloid fractions on paper properties and effects of various enzyme treatments on the removal of organic compounds, *Pulp Paper Can.* **101**, 59–62.

ZOLLNER, H. K., SCHROEDER, L. R. (1998), Enzymatic de-inking of nonimpact printed white office waste paper with α-amylase, *Tappi J.* **81**, 166–170.

19 Microencapsulation-Based Cell Therapy

ULRICH ZIMMERMANN
HUBERT CRAMER
ANETTE JORK
FRANK THÜRMER

Würzburg, Germany

HEIKO ZIMMERMANN
GÜNTER FUHR

Berlin, Germany

CHRISTIAN HASSE
MATTHIAS ROTHMUND

Marburg, Germany

1 Introduction

Many diseases are closely tied with deficient or subnormal metabolic and secretory cell functions. Diabetes mellitus, Parkinson's disease, hemophilia, hypoparathyroidism, chronic pain, and hepatic failure are only a few examples for this kind of degenerative and disabling disorders. Milder forms of these diseases can be managed by a variety of treatments. However, very frequently it is extremely difficult or even impossible to imitate the "moment-to-moment" fine regulation and the complex roles of the hormone, factor, or enzyme that is not produced by the body (KÜHTREIBER et al., 1999). For example, patients who suffer from the insulin-dependent diabetes mellitus (IDDM) must take daily insulin injections. While such treatment can restore the average blood glucose level, true glucose homeostasis is not achieved. This failure leads to serious secondary side effects (such as micro- and macroangiopathy, diabetic neuropathy, nephropathy, and retinopathy), associated with a great reduction in life quality and expectancy. Also, the health care costs are staggering. Similarly, patients with chronic hypoparathyroidism show increased neuromuscular excitability (tetanic convulsions) resulting from a deficiency of the parathyroid hormone (parathormone; PTH) that regulates serum calcium. Patients with hypoparathyroidism are usually treated with oral calcium and vitamin D (calcitriol) when the symptoms of this disorder do not disappear and normocalcemia in the serum is not achieved. However, calcitriol lacks the complete renal calcium-retaining ability of parathormone. Accordingly, such patients have an increased risk of nephrolithiasis, nephrocalcinosis, and subsequent impairment of renal function (HASSE et al., 1999 and literature quoted there).

These two examples demonstrate the long-term failure and the high (partly unrealistic) costs of current therapies and the urgent need for alternative therapeutic strategies. Immunoisolated transplantation (i.e., encapsulated-cell therapy) is one of the most promising approaches to overcoming the limitations of the current treatment protocols (LIM and SUN, 1980; GEISEN et al., 1990; LANZA et al., 1996). Instead of drug administration or of engineering the patient's own cells (somatic-gene therapy), non-autologous standard laboratory cell lines, allogeneic (intraspecific), and xenogeneic (interspecific) cells/tissues are used that release the therapeutic substances that the body of the patient cannot itself produce – the only causal therapy. To avoid a life-time of immunosuppression therapy while excluding an immune response in the host, the transplants must be enclosed in immunoprotective capsules or devices (COLTON and AVGOUSTINIATOS, 1991).

Studies with macrocapsules (e.g., hollow fibers, diffusion chambers) made up of different materials have shown a number of drawbacks that stand in the way of their clinical use (LANZA et al., 1996; KÜHTREIBER et al., 1999). Aside from surgery and retrieval problems, non-specific fibrotic overgrowth, necrosis of the encapsulated cells due to unfavorable (disk and tube) geometries, and thus diffusion limitations, breakage and other problems resulted in the early failure of the grafts. In contrast, microcapsules that are produced from hydrogels offer potential solutions to the problems of macrocapsules.

First, because of their spherical configuration and their small size, microcapsules have much better surface-to-volume ratios than macrocapsules. Second, microcapsules allow precise tailoring of their permeability to allow diffusion of anabolic compounds (oxygen, glucose, etc.) and of cell-derived products (carbon dioxide, lactate, hormones, etc.) while, simultaneously excluding immunoglobulins. Third, microcapsules minimize the overall risk of immunoprotection failure by using thousands of them instead of a single large macrocapsule. Fourth, they can be injected directly or transplanted with minimal-invasive surgery into the muscle, peritoneal cavity, liver, or elsewhere.

Over the past two decades, a number of microcapsules made up of different hydrogels (e.g., alginate, agar, agarose, gellan gum, chitosan, synthetic polymers) have been developed and tested (DULIEU et al., 1999). This research has shown the feasibility of alginate-based microcapsules for transplantation of laboratory cell lines as well as of allo- and xenogeneic tissue. Numerous technical accomplishments of this immunoisolation method have recently made possible the successful therapeutic transplantation of allogeneic parathyroid tissue into patients with hypoparathyroidism (HASSE et al., 1997a). Other clinical applications (e.g., treatment of diabetes mellitus with encapsulated porcine islets) are currently under development as well (KÜHTREIBER et al., 1999).

Progress towards a routine clinical application of encapsulated-cell therapy requires continued rigorous investigation and elucidation of the processes underlying the interactions (and therefore the long-term risks and efficacy) of the tissue transplant and its immunoprotecting barrier with the immune system of the host. Thus, this review article does not pretend to give a complete overview of the enormous bulk of literature published to date on alginate-based microcapsules and related fields (see, e.g., KÜHTREIBER et al., 1999; ZIMMERMANN et al., 1999; HILLGÄRTNER et al., 1999). Rather, this article focuses on the exploitation of factors that have received limited attention in the past, but are important for the formulation of an alginate-based immunoisolation system that can gain medical approval.

2 Bioencapsulation Techniques

Although the concept of encapsulated-cell therapy is rather appealing, in practice a great deal of technology and know-how is needed for the production of long-term functional alginate-based transplants. Alginate microcapsules can be formulated by many different methods (MATTIASSON, 1982; KÜHTREIBER et al., 1999). The features of the capsules depend – among other things – on the medical application. Therefore, when discussing the optimum characteristics of alginate beads for encapsulation, we will focus mainly on immunoisolation of pancreatic islets, parathyroid tissue, and dopamine-secreting continuous cell lines for treatment of diabetes mellitus, hypoparathyroidism, and Parkinson's disease, respectively.

Alginate is one of the most abundant naturally occurring polymers; thus availability of the raw material imposes no problem for clinical application. Alginate constitutes a family of unbranched anionic polysaccharides, mainly extracted from brown algae. It is composed of 1,4-linked α-L-guluronic acid (G) and β-D-mannuronic acid (M). The monomers are sequenced in homopolymeric G–G and M–M blocks. These are interspersed with alternating or nearly random sequences containing M–G blocks (MCHUGH, 1987; KING, 1994). Both homopolymeric sequences are found together, although to a different extent, in all alginates independent of their origin.

Gelation is induced by cross-linking the alginate by oppositely charged, di- or trivalent counterions (HARTMEIER, 1986; BRODELIUS and VANDAMME, 1987). For immunoisolated transplantation, Ba^{2+} and Ca^{2+} are usually used for cross-linking of the polymeric chains (LIM and SUN, 1980; GEISEN et al., 1990; KÜHTREIBER et al., 1999). Fe^{3+} can also be employed, but the gels are not translucent. Very stable capsules can also be obtained (see Sect. 6.2) when high-viscosity alginate is cross-linked with FCSIII (HyClone Laboratories, Utah, USA) or when putrescine or other oligoamines are added to the alginate before bead formation (for further details about the features of oligoamines, see SULTZBAUGH and SPEAKER, 1996; MESSIAEN et al., 1997).

Ca^{2+} cross-linked beads have the disadvantage that they are sensitive to chelators such as citrate, phosphate, and lactate (PILWAT et al., 1980; SCHNABL and ZIMMERMANN, 1989). Thus, long-term survival of solid Ca^{2+} cross-linked alginate microcapsules is limited, but this can be advantageous when autologous (i.e., the patient's own) cells are transplanted, e.g., chondrocytes and osteoblasts for the restoration of cartilage and bone (JORK et al., 2000). Ca^{2+} in combination with chelators is also often used to produce liquid-core capsules (DULIEU et al., 1999). This encapsulation procedure involves forming alginate beads (usually in the range of 0.3–2 mm; GENTILE et al., 1995), coating them with poly-L-lysine or other polyelectrolytes (e.g., polyethylene imine, polybrene, polyethylene glycol) and then performing core reliquefaction (LIM and SUN, 1980; GOOSEN, 1993; KLÖCK et al., 1993; see also KÜHTREIBER et al., 1999). Frequently, a second alginate coat is applied by suspending the coated capsules in alginate solution (O'SHEA et al., 1984). The major shortcomings of alginate-poly-L-lysine capsules are their vulnerability (including the cells) to the reliquefaction process; their relatively weak stability (i.e., their high colloid osmotic pressure due to the dramatic reduction in water activity by the liquid polymers; see also Sect. 6), and the potential for an inflammatory response to capsule fragments (i.e., polyelectrolyte-alginate residues) upon breakage during transplantation (PLUMB and BRIDGMAN, 1972; FRITSCHY et al., 1991; ZIMMERMANN et al., 1994; DE VOS et al., 1996, 1997a; VAN SCHILFGAARDE and DE VOS, 1999; GÅSERØD et al., 1999). Moreover, poly-L-lysine is stringently cytotoxic and is used as an antineoplastic agent (ARNOLD et al., 1983).

The *in vitro* and *in vivo* long-term integrity of the capsules is greatly improved when Ba^{2+} is used (TANAKA and IRIE, 1988; SCHNABL and ZIMMERMANN, 1989), but this divalent cation is an inhibitor of the K^+ channels present in cell membranes. Thus, careful control of the bead manufacturing process and the subsequent removal of excessive Ba^{2+} after gelation (see Sect. 6) are important to maintain high cell viability.

For bead formation, cells (or tissue pieces) are first suspended in an iso-osmolar, saline-sodium alginate solution. The cell–alginate suspension is then forced (by using a syringe or a motor-driven piston) through a nozzle, which forms droplets of cell-containing alginate. These droplets fall into an iso-osmolar NaCl solution containing 20 mM Ba^{2+} (or Ca^{2+}) that complex with the alginate, resulting in formation of spherical beads. The pH of the dropping solutions must be adjusted to pH 7 by using histidine or other buffering biomolecules of very low molecular weight (JORK et al., 2000). Organic buffers (such as HEPES and MOPS) used by many authors in the past (e.g., KLÖCK et al., 1997; DE VOS et al., 1999; ZEKORN and BRETZEL, 1999) should be avoided because these buffers can be cytotoxic when released from the capsule during transplantation.

Drop formation is greatly improved by application of a coaxial air jet (GRÖHN et al., 1994), i.e., by using a two-channel bead generator (Fig. 1A). Studies have shown (JORK et al., 2000) that viscosity and concentration of the alginate, air flow rate, and the geometric properties of the channels and the nozzle are crucial for obtaining microcapsules that are (nearly) spherical, small in diameter, and with a uniform size distribution (see Sect. 6).

Optimum drop formation is also obtained by application of a high electrostatic potential between the nozzle and a stainless-steel ring placed between the nozzle and the bath solution (Fig. 1C; see also KRESTOW et al., 1991; COCHRUM et al., 1995; GOOSEN, 1999). When an axissymmetric and sinusoidal disturbance of a frequency of about 500 to 7,000 Hz is additionally imposed on the laminar jet flow, small droplets, with a very narrow size distribution, are formed (PLÜSS et al., 1997; BRANDENBERGER and WIDMER, 1998; HEINZEN, 1999).

There are several other (commercial) dropping techniques or modifications of the above devices (for an excellent overview the reader is referred to a recent review article of DULIEU et al., 1999). For transplantation, the most important one is the three-channel, air-jet bead generator (Fig. 1B; see also JORK et al., 2000). This device allows the one-step formation of microcapsules of homogeneous as well as of spatially heterogeneous composition. These include solid beads with a liquid core (e.g., oil or other hydrophobic fluids; see Sect. 6) or solid beads composed of a core of low alginate concentration (containing the cells) that is surrounded by a layer of higher alginate concentration ("layered solid microcapsules").

Drop formation under coaxial air flow and under electrostatic potential have been widely used in bioencapsulation (KÜHTREIBER et al., 1999). For medical applications the "coaxial-air-flow" technique is the method of choice because it allows – in contrast to the "electrostatic-potential" method – the use of highly viscous alginate, i.e., alginate of high molecular mass. Compared to low-viscosity alginate beads, microcapsules made up of high-viscosity alginate provide features that are advantageous for long-term transplantation (see Sect. 6). A disadvantage of the technique may be that tiny air bubbles can be included in the beads during the gelation process (if the procedure is not performed carefully). Such bubbles may lead to diffusion limitations and/or to long-term adverse side effects on the bead integrity.

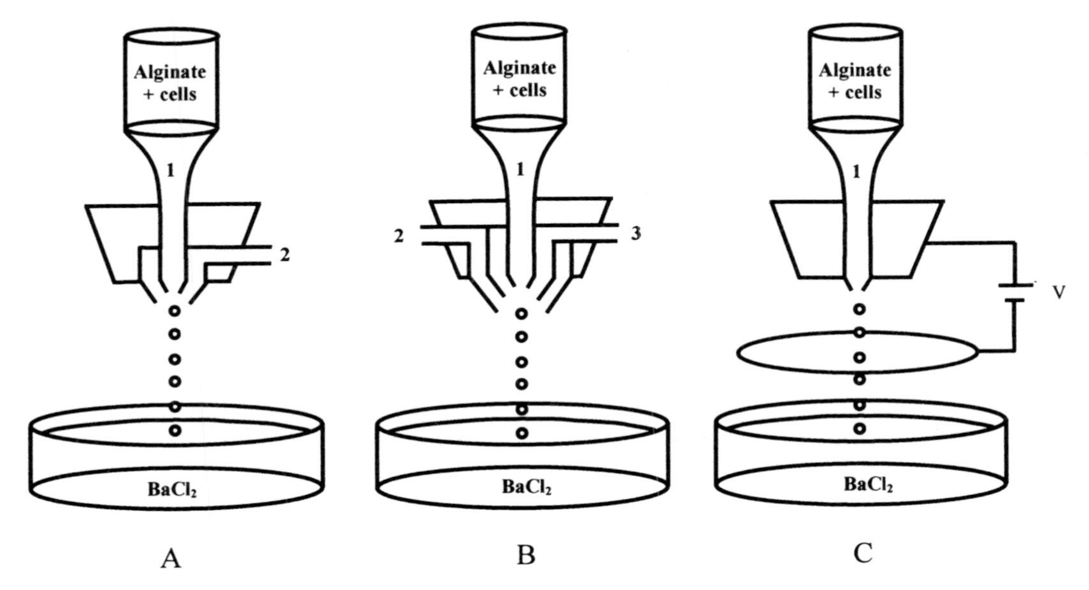

Fig. 1. Schematic diagrams of alginate capsule generators; (**A**) two-channel-, (**B**) three-channel coaxial air-jet bead generator, and (**C**) capsule formulation under high electrostatic potential, V. Channel 1 is fed with the alginate/cell suspension, channel 2 serves for air flow supply, and channel 3 is fed with an alginate solution, usually of higher concentration than that of channel 1 (for the formation of "layered beads"). The size of the beads is controlled by the speed of the air and alginate flow (**A**, **B**), or by the electrostatic potential (**C**). For further details, see Sect. 2.

3 Production of Transplantation-Grade Alginates

The demands on the alginate material to be used for transplantation are stringent. It should be produced with reproducible characteristics according to medical approval standards, and it should not elicit any inflammatory or fibrotic response from the host, i.e., it should not engender any cytotoxicity, and should be biocompatible for both the host and the cells it encloses.

Due to the harvesting and extraction process commercial alginates contain a fairly high number of impurities (ZIMMERMANN et al., 1992). Common contaminants are proteins, complex carbohydrates, fatty acids, phospholipids, lipopolysaccharides, toxins, and polyphenols (SKJÅK-BRÆK et al., 1989; DE VOS et al., 1993; SUN et al., 1996). These mitogenic and inflammation-provoking impurities engender ultimately fibrotic overgrowth (see, e.g., OTTERLEI et al., 1991; MAZAHERI et al., 1991; WIJSMAN et al., 1992; COLE et al., 1992; CLAYTON et al., 1993; DE VOS et al., 1993; KLÖCK et al., 1994) with the result that transport of nutrients and oxygen to the encapsulated cells is greatly impeded leading ultimately to cell necrosis. Removal of the impurities from the commercial alginate by free-flow electrophoresis or by chemical means (ZIMMERMANN et al., 1992; KLÖCK et al., 1994, 1997; DE VOS et al., 1997a; VAN SCHILFGAARDE and DE VOS, 1999), and subsequent implantation of the empty alginate gels into rodents did not evoke any significant foreign body reaction, even when alginate was implanted in diabetes-prone BB rats that exhibit elevated macrophage activity (ROTHE et al., 1990; GOTFREDSEN et al., 1990; WIJSMAN et al., 1992). Extensive research with purified high-M and high-G alginates gave further clear-cut evidence (see ZIMMERMANN et al., 1999) that neither the M–M nor the G–G blocks of alginate polymers of high molecular mass initiate an immunostimulatory response (cytokine production) as discussed very controversially in the literature (SOON-SHIONG et al., 1991; CLAYTON et al., 1991; OTTERLEI et al., 1991, 1993; ESPEVIK et al., 1993; JAHR et al., 1997; DE VOS et al., 1997a; KULSENG et al., 1999).

Purification of crude commercial alginate has the decisive disadvantage that many impurities have also to be removed which are not natural constituents of the brown algae, but rather present contaminants from the harvesting process (pollution by animal proteins, bacteria products, etc.). Treatment of the raw algal material by formaldehyde imposes further complications. Purification of alginate is not easy since high concentrations of alginates are difficult to work with because of the high viscosity of the solutions. Removal of mitogenic and inflammation-provoking contaminants requires, therefore, multiple-step and very time-consuming procedures. Because of the large number of operations the risk of further contamination is increased. As a final result, only small quantities of alginate of quite variable purity are obtained (KLÖCK et al., 1994).

Techniques of purification and monitoring have been recently improved sufficiently to allow the reproducible reduction of mitogenic and cytotoxic impurities to a negligible level. Progress was achieved by using clearly defined algal material for the production of highly purified alginate that fulfills the standards for medical application. Research in this direction has shown (HILLGÄRTNER et al., 1999; JORK et al., 2000) that fresh stipes of brown algae harvested directly from the sea or sporophytes of brown algae grown in bioreactors are ideal input sources. When using such material, the manufacturing process can be simplified considerably. Extraction and purification steps comprise (see flow chart in Fig. 2; for further information see JORK et al., 2000): extraction with 50 mM EDTA, removal of all visible aggregates by filtration in the presence of diatomaceous material, adjustment to 0.13 M KCl, precipitation with ethanol (37.5% v/v) under injection of air or nitrogen, manual sampling of the alginate layer accumulated at the surface of the liquid phase, redissolution in 0.5 M KCl under agitation, repetition of the precipitation (with an ethanol concentration of 45% v/v) and redissolution of the alginate in distilled water followed by dialysis, adjustment of the solution to 0.13 M KCl, precipitation (50% v/v ethanol), ethanol sterilization and drying of the snow-white alginate. All steps must be performed at room temperature because alginate solutions, particularly highly viscous ones, depolymerize when the temperature is raised (MCHUGH, 1987).

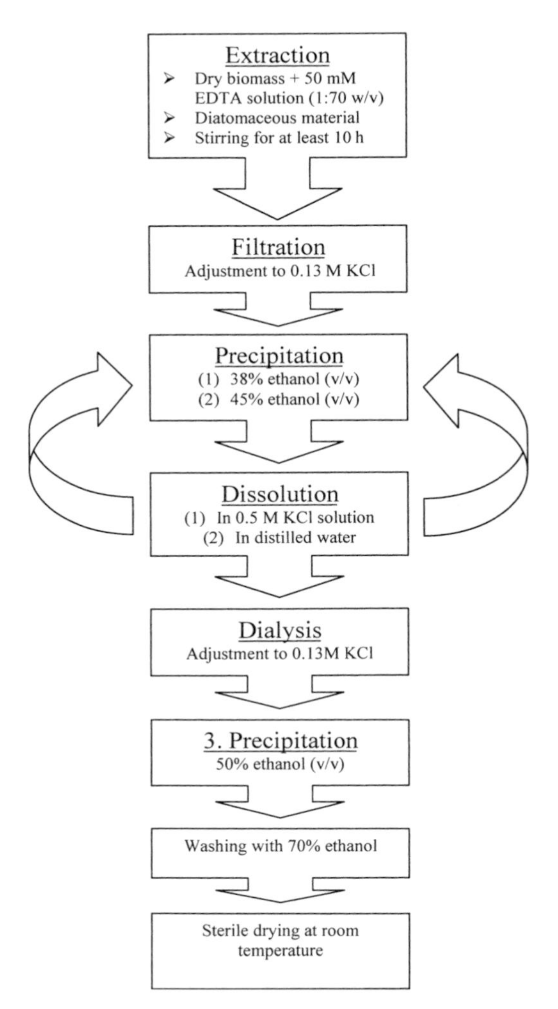

Fig. 2. Flow chart of the alginate purification process.

If the parameters of the process are carefully controlled, the above extraction and purification protocol leads to a final product with reproducible characteristics. Depending on the origin of the alginate, alginates with a high viscosity and thus with a high molecular mass much larger than about 300 kDa can be obtained. Such alginates have optimum transplantation properties (see Sect. 6). The above protocol can be applied to alginates of different uronic acid composition, and is, additionally, amenable to large-scale production. Thus, medically approved alginate can be produced in sufficient quantities for routine clinical applications and also at reasonable costs.

4 Biocompatibility Assays for Medically Approved Alginate Gels

Clinical application of alginate microcapsules must be based on adequate preclinical data including assays for the continuous control of the purification process. For practical, but also ethical reasons, *in vitro* evaluation of the alginates is required before implantation into animals and, ultimately human trials can be contemplated. Current routine analytical tests of the purity of the alginate comprise measurements of endotoxin, protein, phenolic-like compounds, and other contaminants by using the limulus-lysate assay, the Bradford test, fluorescence- and NMR-spectroscopy (SKJÅK-BRÆK et al., 1989; JORK et al., 2000; SCHILLER, ARNOLD, CRAMER, THÜRMER, ZIMMERMANN, unpublished data). Measurements with partly purified alginates in combination with implantation studies showed (ZIMMERMANN et al., 1999) that the above analytical tests are necessary, but not sufficient to exclude immunological reactions to the alginate under implantation and transplantation conditions. The reason for this is that fucoidan and other related mitogenic compounds (ARFORS and LEY, 1993) are difficult to detect by fluorescence- and NMR-spectroscopy.

Our laboratory has recently developed a "cell culture" assay for highly sensitive screening of mitogenic impurities in alginates. The assay is based on the activation and proliferation of murine splenocytes by mitogenic alginates ("mixed-lymphocyte test"; ZIMMERMANN et al., 1992; KLÖCK et al., 1994, 1997). The proliferation is considerably increased when the splenocytes are simultaneously costimulated by lipopolysaccharides (LPS; ZIMMERMANN et al., 1999). Growth and viability of the activated lymphocytes can be measured colorimetrically by the so-called XTT assay (SCUDIERO et al., 1988). However, long-term experience of our laboratory has shown that the rate of formazan production is not always proportional to the number of activated cells and thus very variable. Cell clumping, debris, and the viscosity of the alginate (lowering the water activity) do apparently interfere with the dye reaction. The reproducibility and sensitivity of the "mixed lymphocyte test" can be greatly improved (KÜRSCHNER, JORK, ZIMMERMANN, unpublished data) if the number and the size of the proliferating cells are determined electronically 3 d after co-stimulation and culture. By use of an electronic cell analyzer it is possible to distinguish clearly between dead cells, debris, and clumped cells on the one hand and viable, activated lymphocytes on the other because of the large differences in the signal amplitude of these populations (FRIEDRICH et al., 1998). Typical electronic size distribution measurements of murine splenocytes co-stimulated by various crude and purified alginate samples together with LPS are shown in Fig. 3. It is obvious from this figure that the purification process of Fig. 2 almost completely removed the mitogenic impurities. Most interestingly, alginate extracted from stipes collected at the beach of southern Africa as well as from fresh algal material showed no significant mitogenic activity according to this assay. When these alginate samples were implanted beneath the kidney capsule of rats, a (reduced) fibrotic overgrowth was, however, provoked by the "beach alginate". This clearly shows that assurance of biocompatibility still requires the implantation of empty microcapsules in rodents and that further efforts have to be made to develop efficient and simultaneously fast *in vitro* assays in order to replace the (sometimes very variable) animal models.

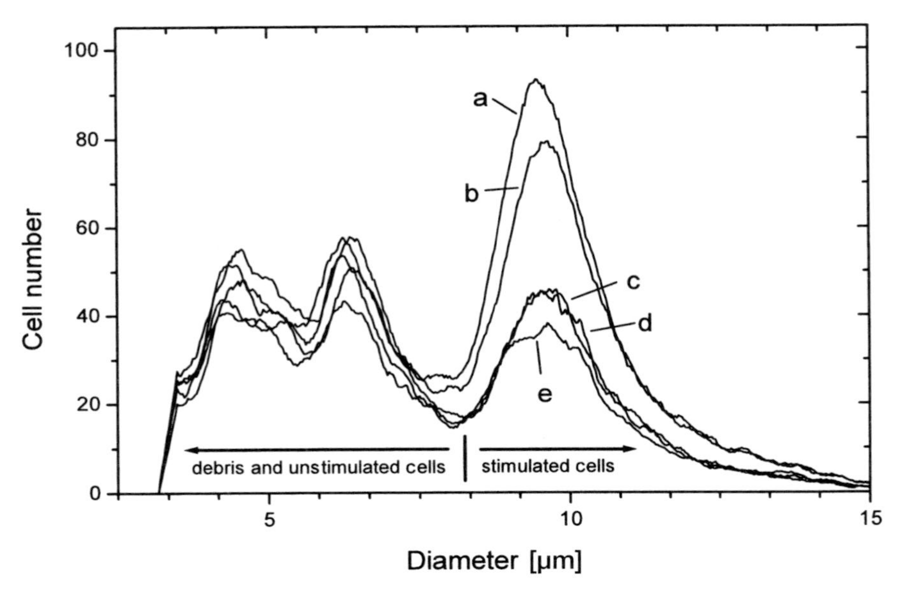

Fig. 3. *In vitro* bioassay for the detection of mitogenic impurities in alginates based on the activation and proliferation of murine splenocytes. Mixed lymphocytes prepared from 6- to 8-week-old male C3H/HEJ mice were co-stimulated with 10 µg ml^{-1} lipopolysaccharides and commercial, unpurified alginates (a, b) and with alginates purified from algae (c, d), respectively. After 3 d culture the ratio of dead cells/debris (diameter <8 µm) to viable proliferating cells (diameter >9 µm) was determined by using an electronic size analyzer. The starting algal material for the purification of the alginate according to the regime in Fig. 2 was: fresh algae (c) and algae collected at the beach of southern Africa (d). Curve e represents the control, i.e.,the stimulation of the mixed lymphocytes with lipopolysaccharide in the absence of alginate.
Note that the purified alginate of curve c, but not that of curve d exhibits a foreign body reaction after implantation under the kidney capsule of BB rats (Fig. 4) indicating the limitations of the *in vitro* bioassay (Kürschner, Jork, and Zimmermann, unpublished data).

The animal model has a major impact on the histological results of *in vivo* biocompatibility tests. Spontaneously diabetic BB rats are the most appropriate small animal models (Mazaheri et al., 1991; Pfeffermann et al., 1996; Zimmermann et al., 1999) even though the breeding of this strain is very man-power-intensive, time-consuming, and expensive. Many authors have used Lewis and other rats as well as mice (BALB/C) for biocompatibility tests (e.g., Weber et al., 1993; Cochrum et al., 1995). However, these animal models cannot be recommended because extensive studies in the laboratory of the authors have shown (Zimmermann et al., 1999; Kürschner, Jork, Zimmermann, unpublished data) that these rodents even tolerate alginates that evoked not only a strong non-specific foreign body reaction in BB rats, but also exhibited mitogenic activities in the "electronic activation/proliferation" assay. Typical results of Ba^{2+} cross-linked alginate beads (Fig. 4A) retrieved 3 weeks after implantation beneath the kidney capsules of BB rats are depicted in Fig. 4B. In contrast to commercial, unpurified alginate (Fig. 4D) no significant fibrotic overgrowth can apparently be detected when alginate purified according to the flow chart in Fig. 2 is used.

Similar results were obtained, when corresponding alginate gels were implanted for four weeks in the muscle of baboons (Fig. 4C;

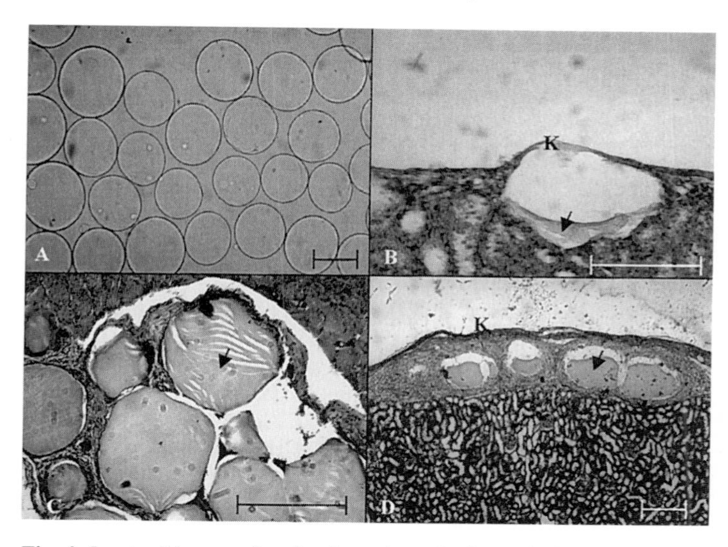

Fig. 4. *In vivo* bioassay for the detection of mitogenic impurities in alginates based on the induction of fibrotic overgrowth due to the immune response of the host. (**A**) Empty Ba^{2+} alginate beads (diameter 300–400 μm) were made from purified and highly viscous alginate (curve c in Fig. 3) by using the dropping device in Fig. 1A. (**B**) Capsules (arrow) retrieved 3 weeks after implantation beneath the kidney capsule (K) of a spontaneously diabetic BB rat and (**C**) capsules (arrow) retrieved 4 weeks after implantation in the muscle of a baboon. (**D**) Control capsules (arrow) made up of commercial, unpurified alginate (curve b in Fig. 3) and retrieved 3 weeks after implantation beneath the kidney capsule (K) of a BB rat (for fixation and staining of the tissue/beads see KLÖCK et al., 1997 and HILLGÄRTNER et al., 1999). Note that due to the fixation process the alginate beads may collapse (**B**) or may become deformed (**C**) (for further explanations, see text); Bars = 350 μm.

GEßNER, G. ZIMMERMANN, JORK, BOHRER, MELCHER, HASSE, ROTHMUND, ZIMMERMANN, unpublished data). In this case, a slight fibrotic reaction was observed. However, such a slight fibrotic overgrowth does not prevent nutrient and oxygen exchange between encapsulated cells and their environment. Rather, as will be demonstrated in the following section, such a reaction is advantageous to reduce capsule breakage and movement from the transplantation site.

5 Animal and Clinical Trials with Encapsulated Tissue

Immunoisolated-islet transplantation is an attractive therapy for insulin-dependent diabetes mellitus (IDDM) patients. Therefore, it is not surprising that extensive animal studies have been made with encapsulated rat, porcine, or human islets in the last two decades. In most of these experiments islets were entrapped in alginate-poly-L-lysine made up of commercial (non-purified), low-viscosity alginate. Intraperitoneal allo- and xenografts could normalize blood glucose of (diabetic) mice and rats for about 100–150 d, occasionally some grafts functioned up to 1 year (O'SHEA and SUN, 1986; WEBER et al., 1999; WANG, 1999; VAN SCHILFGAARDE and DE VOS, 1999). Normalization of hyperglycemia by xenotransplantation of microencapsulated (porcine) islets has also been reported for spontaneously diabetic dogs and cynomolgus monkeys (WARNOCK and RAJOTTE, 1988; SOON-SHIONG et al., 1992; ZHOU et al., 1994; SUN et al., 1996; LANZA et al., 1999). However, as a rule, the results were very variable and the success was always of limited duration as expected in the light of the above considerations (see also below). Graft failure could occur even 2 weeks after transplantation due mainly to foreign body reactions (MAZAHERI et al., 1991; WIJSMAN et al., 1992) but sometimes in their absence (DE VOS et al., 1997b).

Ba^{2+} microcapsules made up of alginate purified from commercial alginates (KLÖCK et al., 1994) have also been successfully used for the encapsulation of rat and porcine islets of Langerhans (Fig. 5; ZEKORN et al., 1992a, b). In glucose perifusion challenges, evaluation of insulin secretion by encapsulated rat islets showed the typical biphasic insulin release pattern of non-encapsulated islets. During static glucose challenge, the insulin release ranged from 40% to 70% as compared to the controls.

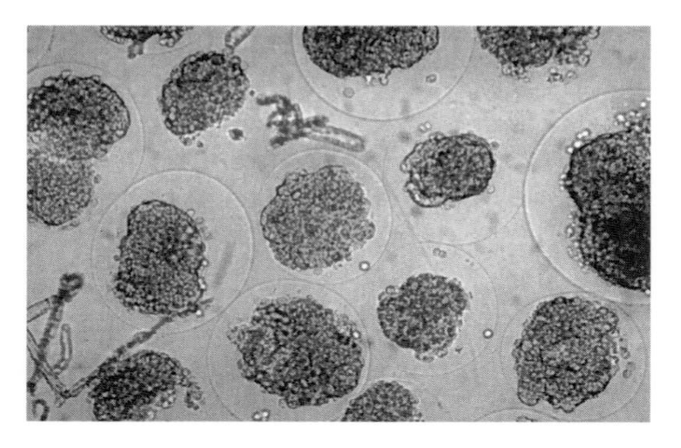

Fig. 5. Uniform preparation of porcine pancreatic islets immunoisolated by Ba^{2+} alginate capsules of small diameter. Encapsulation of the islets was performed by using the dropping device in Fig. 1A. Separation of empty beads (not shown) was achieved by discontinuous density gradient centrifugation according to the protocol of GRÖHN et al., 1994.

Accordingly, xenotransplantation of encapsulated rat and porcine islets in chemically induced diabetic mice demonstrated long-lasting graft function (up to 1 year) even though failure of some grafts was also observed (about 30%; ZEKORN et al., 1992a, b; SIEBERS et al., 1992, 1993). Histological examinations of long-term-functioning microcapsules demonstrated well preserved islets.

Despite the promising results of animal studies, there has been little success with the clinical allotransplantation of pancreatic islet cells into IDDM patients. This failure has generally been attributed to the inability to obtain large numbers of viable human pancreatic islets for grafting. About 1 million islets must be transplanted in order to cure diabetes (see below). However, the source of human organs is limited, thus only a small number of patients could benefit from the encapsulation method. Immunoisolation of porcine islets has the potential to fill the gap, but concerns remain about possible cross-species transmission of porcine endogenous retrovirus (PATIENCE et al., 1997; but see PARADIS et al., 1999; HUNKELER et al., 1999).

Animal and clinical data are available for allo- and xenotransplantation of encapsulated parathyroid glands. Parathyroid tissue excised from Lewis rats, encapsulated in Ba^{2+} alginate matrices and transplanted in parathyroidectomized Dark-Auita rats exhibited long-term function (HASSE et al., 1996, 1998). More than 6 months after allotransplantation (without systemic immunosuppression) nearly all animals that had received microcapsules made up of amitogenic alginate were normocalcemic. These results were independent of whether the alginate used was purified from commercial alginates according to the protocol of KLÖCK et al. (1994) or from fresh algal material (HASSE et al., unpublished data). Throughout the studies PTH and calcium concentrations were always concordant. Accordingly, histology of retrieved transplants revealed vital parathyroid tissue and intact microcapsules. The transplants were partly covered by a very thin fibrotic layer that apparently did not affect the function of the encapsulated tissue. Similar results were obtained for xenotransplantation of encapsulated human parathyroid tissue in rats with experimental hypoparathyroidism (HAS-

SE et al., 1997b). Evaluation of the Ca^{2+} level in the serum showed that about 100 d after transplantation 75% of the animals that received xenotransplanted human parathyroid tissue were still normocalcemic.

These and other results established the basis for pilot randomized clinical trials performed recently (HASSE et al., 1997a). The number of patients with clinical hypoparathyroidism is much lower than that of IDDM. Thus, there should be no shortage of donor human parathyroid tissue in the future. Cultured allogeneic parathyroid tissue immunoprotected by Ba^{2+} alginate was transplanted into the muscle of the right upper arm of two patients suffering from symptomatic persistent postoperative hypoparathyroidism (HASSE et al., 1997a). Shortly after allotransplantation, the two patients were normocalcemic and revealed normal levels of PTH without immunosuppression (Fig. 6). With ongoing transplantation both patients reported an impressive improvement of symptoms and sequelae of hypoparathyroidism. After about 90 d graft failure occurred. Residues of the alginate capsules and of the allogeneic tissue could not be found when – with permission of one of the patients – the transplantation site was re-examined 10 months after surgery (HASSE et al., unpublished data). Studies in animal models gave evidence (BOHRER et al., unpublished data) that activated macrophages degrade encapsulated parathyroid tissue when capsule breaks occur. However, this cannot be the only explanation because one year after transplantation, re-examination of one of the patients showed nearly normocalcemia and absence of symptoms, requiring much less Ca^{2+} and vitamin D than prior to surgery. Appropriate tests (CASANOVA et al., 1991) revealed that the PTH was released from the transplantation arm (HASSE et al., unpublished data). The authors of this re-

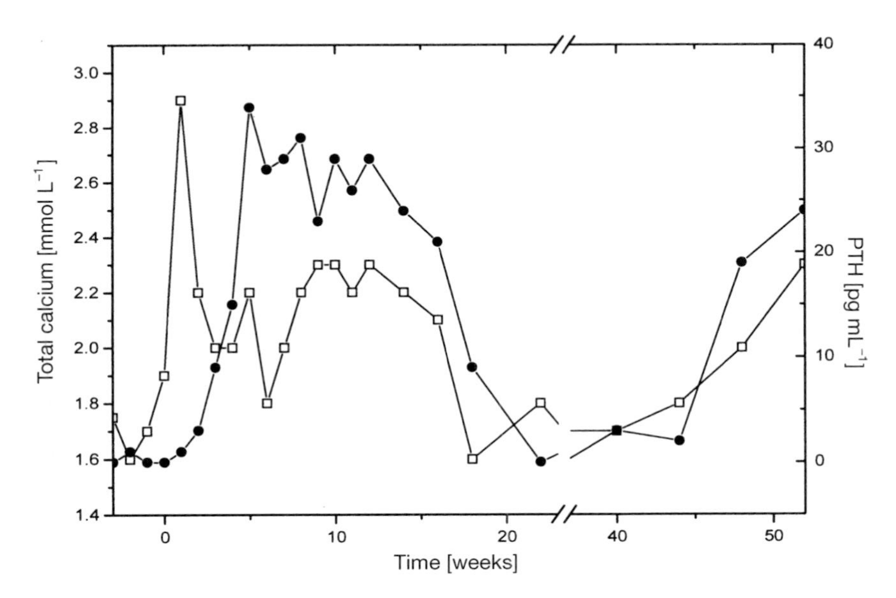

Fig. 6. Ca^{2+} (open squares) and parathormone (PTH; filled circles) levels in a patient with hypoparathyroidism after allotransplantation of parathyroid tissue into the muscle of the non-dominant forearm. The allogeneic tissue pieces were encapsulated by gelation of purified alginate (according to the protocol of KLÖCK et al., 1994) with Ba^{2+}.
Note that graft failure occurred after about 3 months, but that functional activity of the transplant was recorded again about 1 year after transplantation (for further details, see Sect. 5).

view believe that migration of the microcapsules from the transplantation site took place and/or that over a longer period of transplantation the alginate matrix was replaced by a fibrotic layer.

6 Conceptual Configuration of Microcapsules for Long-Term Transplantation

The above animal and, in particular, the first small-scale clinical studies are encouraging, but they have also clearly demonstrated that the duration of immunoprotection and/or function of the transplants were subjected to a large variability. There are strong indications (see also DE VOS et al., 1997b; ZIMMERMANN et al., 1999; ZEKORN and BRETZEL, 1999) that the biocompatibility of the beads is not only influenced by the chemical composition and purity of the material, but also by (1) physical and physicochemical imperfections of the capsule, (2) adverse properties of the precultured donor tissue, and even by (3) recipient- and operation-related factors.

The difficulty of designing a microcapsule system with optimum clinical properties is to adjust the numerous capsule parameters (such as size, permeability, mechanical strength, surface topography, swelling properties, etc.) independently. Therefore, seemingly minor modifications of one of the capsule parameters may have an important impact on other parameters of the cell-containing capsule, and thus ultimately on the outcome of the graft. Progress can only be expected when the underlying biophysical and immunological principles are thoroughly understood and documented. It is much more difficult and discouraging to troubleshoot problems that arise in cell therapy applications if one cannot fall back upon a solid foundation of first principles. This important issue was unfortunately ignored by many authors working with alginate-poly-L-lysine capsules (for notable exceptions, see, e.g. VAN SCHILFGAARDE and DE VOS, 1999) with the result that there are numerous different encapsulation procedures that have led to the formulation of capsules with a broad, but rarely documented spectrum of properties.

In contrast to alginate-poly-L-lysine capsules, systematic, well documented improvements of the solid Ba^{2+} alginate microcapsules have been reported in the last couple of years (DE VOS et al., 1997a; ZIMMERMANN et al. 1999, 2000; NÖTH et al., 1999; HILLGÄRTNER et al., 1999). Because of recent promising animal studies with this new generation of Ba^{2+} alginate microcapsules, we will focus mainly on the optimum design of this type of capsules in the following. However, it should be noted that many solutions found for the Ba^{2+} alginate barrier (summarized in Fig. 7) can also be assigned to the alginate-poly-L-lysine capsules.

6.1 Size and Diffusion

Immunoisolated cells lack intimate vascular access, and must be supplied with oxygen and nutrients by diffusion from the nearest blood vessels into the capsule interior, i.e., over distances greater than those normally encountered. Therefore, if the ratio of the capsule material to encapsulated tissue (cells) is too unfavorable, graft failure can occur (ZEKORN et al., 1992b). This is not a problem when encapsulated parathyroid tissue sections are transplanted because they are fairly large (about 1 mm^3), but functional survival of islet transplants may be affected. Islets are about 100–150 μm in diameter and are commonly entrapped in capsules of 500–800 μm in diameter for animal studies. In such cases, the large stagnant fluid environment between the tissue and the microcapsule wall can create pronounced diffusion resistances (by analogy to macrocapsules, see above). This was demonstrated both theoretically (GOOSEN, 1999) and experimentally in model cellular systems (BEUNINK et al., 1989; BÜCHNER and ZIMMERMANN, 1982; HARTMEIER, 1986; SCHUBERT, SCHNEIDER, ZIMMERMANN, unpublished data; see also below). Diffusion restrictions generally result in a dramatic reduction of nutrient and, particularly, of oxygen transfer, leading ultimately to necrosis of the encapsulated tissue (FAN et al., 1990; DE

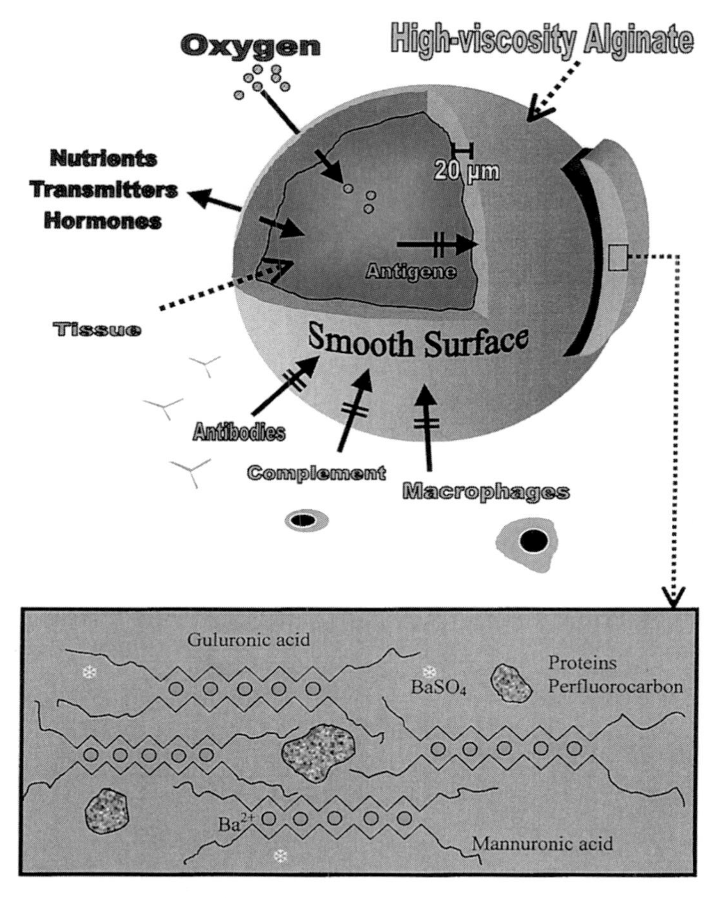

Fig. 7. Design of a Ba^{2+} alginate microcapsule that should fulfill the requirements for long-term, functional transplantation of allogeneic, xenogeneic, and transformed cells/tissues (for further details, see Sect. 6).

VOS et al., 1993; CLAYTON et al., 1993). Correspondingly, they can also engender a time-delayed secretion of regulating factors (such as insulin and parathormone) with the consequence that a "moment-to-moment" regulation of blood glucose and calcium, respectively, fails (see, e.g., CHICHEPORTICHE and REACH, 1988).

A further important drawback of large-sized (homogeneous or layered) microcapsules is that cross-linkage of the alginate polymers and, in turn, diffusion within the beads may vary considerably locally due to the inwardly directed cross-linking process in the gelation bath. This leads to the network narrower on the capsule surface than in the gel core (MARTINSEN et al., 1992). Unpredictable diffusion problems can additionally occur when the islets or other donor cells are not centered in the microcapsules or when an individual microcapsule contains more than one islet.

Another problem of large-sized capsules is that they can lead to huge transplant volumes, thus considerably reducing the number of possible transplantation sites. This is, in particular, very relevant for human diabetic-cell therapy that requires 1 million islets to be transplanted as mentioned above. This amount corresponds to a volume of 300 mL capsules or more, depending on the number of empty beads in the population (CLAYTON et al., 1993; DE VOS et al., 1993).

In contrast to these compromising effects of large-sized capsules, their mechanical strength and immunoprotection features are usually, but not always (see below), better than those of small-sized beads.

Several strategies have been developed recently to overcome the above problems of large-sized microcapsules – at least partially. One approach is to reduce the thickness of the immunoprotective alginate layer. This can be done by centrifugation of islets (or other donor cells) through three layers consisting of alginate, Ficol, and $BaCl_2$ (ZEKORN et al., 1992b). This leads to islets coated by a 10-μm-thick layer, but the yield is limited. The output number of coated islets can be increased by several orders of magnitude when microencapsulated islets are made by the two-channel, air jet droplet generator (Fig. 1A) and then subjected to discontinuous density gradient centrifugation (GRÖHN et al., 1994). By this procedure the various bead populations, particularly the abundant empty capsules, are separated because of their different buoyant density. A pure population of islets is obtained that are coated by a conformal, mechanically stable and immunoprotective thin alginate layer (provided that highly viscous alginate and Ba^{2+} is used). Because of the reduction of the alginate envelope together with the removal of empty capsules the islet transplant volume is only increased correspondingly by a factor of about 1.5 (GRÖHN et al., 1994). A large number of medium-sized microcapsules (about 300–400 μm thick), which envelop each individual islet by a thin alginate cast, can also be prepared quite satisfactorily by emulsification as shown by CALAFIORE et al. (1999) very recently.

The second approach to improve the crosslinking process and thus the homogeneity and diffusion properties of solid capsules is "internal gelation". This can be achieved by dropping an alginate/Ba^{2+} (or Ca^{2+}) carbonate/cell mixture into a buffered, iso-osmolar NaCl solution at a pH of about 5–6. Under these conditions, the divalent cations are liberated within the bead thus leading to a homogeneous cross-linkage of the alginate polymers throughout the entire bead volume. A milder procedure is to add caged Ba^{2+} or Ca^{2+} to the alginate solution before droplet formation (HILLGÄRTNER et al., 1999). The caged ions are non-reactive, but the free, reactive ions can be released by short-term UV irradiation at physiological pH (KAPLAN and ELLIS-DAVIES, 1988).

A third and very promising way to solve at least the oxygen diffusion problems of large-sized capsules is based on the incorporation of hydrophobic, non-toxic and medically approved perfluorocarbons (see Fig. 7; NÖTH et al., 1999; ZIMMERMANN et al., 2000). These compounds are known for their high oxygen capacity (relative to the surrounding medium) and their oxygen-transfer enhancement in the presence of cells (JUNKER et al., 1990; JU and ARMIGER, 1992). Accordingly, islet function and survival was improved by supplement of perfluorocarbons to the culture medium (ZEKORN et al., 1991). Implantation studies in rats showed (NÖTH et al., 1999; ZIMMERMANN et al., 2000) that microcapsules containing a liquid perfluorocarbon core disintegrated after about 50–100 d. In contrast, capsules in which the perfluorocarbon was emulsified homogeneously were stable over more than 2 years (and the animal studies are still going on; JORK, NÖTH, HAASE, ZIMMERMANN, unpublished data; see also below).

6.2 Swelling and Stability

A major concern in immunoisolated-cell therapy is the stability of the alginate microcapsules. The stability of the capsules is closely tied to the swelling properties of hydrogels. Freshly prepared Ca^{2+} and Ba^{2+} alginate capsules show a high tendency to take up large amounts of water, which causes them to swell in saline solutions. Swelling is less for Ba^{2+} than for Ca^{2+} beads, but even in the case of empty Ba^{2+} capsules swelling in saline solutions can lead to disintegration of most of the capsules after about 10 d (HILLGÄRTNER et al., 1999). Retrieved capsules from transplanted rodents also generally show an increase in size. Swelling is most probably one of the main causes for formation of cracks in the beads and capsule breakage leading ultimately to a failure of the graft (if the alginate barrier is not replaced by a permeable fibrotic layer; see Sect. 5).

The swelling properties of alginate gels were traced back by several authors (McHugh, 1987; Martinsen et al., 1989; Smidsrød and Skjåk-Bræk, 1990) to the M/G composition, sequential structure, and the molecular mass of the polymers as well as to the divalent cation used for gelation (Tanaka and Irie, 1988) and to the colloid osmotic pressure (Gåserød et al., 1999). Different strategies – but with limited success – were developed to stabilize the capsules and simultaneously to regulate their permeability, e.g., by using chitosan as complexing cation and alginates with high guluronic acid residues (Gåserød et al., 1999).

Recent work in our laboratory has shown (Fig. 7; Hillgärtner et al., 1999) that – independently of the molecular mass and composition of the alginate – *in vitro* swelling of Ba^{2+} beads can greatly be reduced if freshly formed microcapsules are incubated in a 6 mM Na_2SO_4 saline solution for 30 min. This treatment removes free, excessive divalent cations entrapped within the capsules under formation of very tiny $BaSO_4$ crystals. Swelling can be prevented (nearly) completely if proteins (e.g., 10% fetal calf serum, 10% human serum, etc.) are additionally added to the alginate solution before bead formation. Perfluorocarbons (see above) have a similar tendency as proteins to stabilize the microcapsules against swelling. Encapsulation of parathyroid tissue by using this novel capsule formulation procedure and transplantation of the encapsulated tissue into rats revealed superior properties compared to conventional capsules (unpublished data).

The mechanism of these additives is not completely understood, but they presumably interfere with the peculiar water structure in the microcapsules. Electrorotation studies of Ca^{2+} and Ba^{2+} beads (which consist of 98% water) have shown (Esch et al., 1999) that most of the water is highly structured, i.e., bound to the alginate polymers. This engenders a low electrical conductivity within the beads (relative to outside) and also a reduction in water activity (but less than in the case of liquid polymers; see Sect. 2), leading to an imbalance between the (electro-) chemical potential of the water and ions in the gel matrices and their surroundings. Thermodynamically, this is equivalent to the development of a colloid osmotic pressure (Zimmermann and Steudle, 1978), resulting in water uptake and swelling (similar to cells upon membrane electropermeabilization; see Zimmermann and Neil, 1996). Co-entrapment of proteins or hydrophobic compounds together with Ca^{2+} and Ba^{2+} reduces significantly, but far from completely, the amount of bound water molecules (Kürschner, Mussauer, Sukhorukov, Zimmermann, unpublished data).

The binding of water to the alginate polymers also provides a plausible explanation for the frequent finding that cell function and viability is greatly improved after immobilization in alginate matrices when compared with freely suspended cells (e.g., Schnabl et al., 1983). Because of the unphysiological aqueous environment (proteolytic), enzymes released from dead cells cannot develop full activity or are completely inactive, thus protecting other cells against enzymatic degradation.

6.3 Permeability

The prevention of swelling has not only an enormous impact on the mechanical strength, but also on the permeability of the capsules. There is no longer a need for permeation control by polyelectrolyte coating as called for by other authors (Kühtreiber et al., 1999; see also above). Solid (homogeneous and layered) microcapsules ensure a total immunoprotection against antibodies and complement components provided that excessive divalent cations are removed and proteins or perfluorocarbons are present (Fig. 7).

The viscosity of the alginate and thus the molecular mass of its polymers equally dictate the permeation properties of the capsules. When 1% w/v solutions of highly viscous alginate (0.1% w/v corresponds to 10–20 mPa·s) are used for encapsulation, capsules can even be formulated that are so tight that necrosis of the encapsulated cells (e.g., of parathyroid tissue) occurs very rapidly. However, if the concentration of the alginate is lowered to about 0.7% w/v stable capsules of optimum permeability for oxygen and nutrient transfer are obtained while the immunoprotection is maintained (HASSE, BOHRER, CRAMER, THÜRMER, ZIMMERMANN, ROTHMUND, unpublished data).

Highly viscous alginates became available only recently by the extraction and purification process described in Fig. 2 and their handling requires some expertise in order to prevent degradation. Therefore, most of the work in the field was made by using low-viscosity alginates of high G/M ratios (0.1% w/v corresponds to 1–5 mPa·s; see, e.g., VAN SCHILFGAARDE and DE VOS, 1999; KÜHTREIBER et al., 1999). Correspondingly, 2–3% w/v alginate solutions were employed in encapsulation. The molecular mass and thus the degree of cross-linking increase with the increase in viscosity (MCHUGH, 1987). Simultaneously, with increasing molecular mass toxicity of compounds generally decreases to negligible levels (MATTIASSON, 1982; SCHNEIDER et al., unpublished data). Thus, highly viscous alginate is advantageous because it allows optimum adaptation of permeation and immunoprotection properties to the desired needs while being biocompatible at the same time.

Retrospectively, the use of low-viscosity alginate, together with the chemical impurities of the commercial alginates, is most likely one of the main reasons for early graft failure in many animal studies reported in the literature.

6.4 Topography

Surface roughness of the alginate microcapsules has only recently been recognized as an important factor for the induction of a non-specific foreign body reaction (XU et al., 1998; HILLGÄRTNER et al., 1999). The surface topography is closely linked to the encapsulation procedure, the purity and viscosity of the alginate and the swelling properties of the capsules. Atomic force microscopy (AFM) provides the possibility for three-dimensional scanning of an alginate capsule surface by monitoring the forces of interaction experienced between the sample and a sharp probe. The forces of interaction may be repulsive or attractive and this gives rise to different modes of operation of the AFM (contact and non-contact mode, respectively; for a review article, see VANSTEENKISTE et al., 1998). The AFM can acquire images within an aqueous medium, and thus can be used for the characterization of the surface of alginate microcapsules without the need of dehydration. However, application of the AFM technique to alginate capsules is difficult because the beads are soft, elastic, and extremely adhesive, requiring slow scanning rates, careful force control, and a time-consuming elimination of artifacts that may be encountered during the imaging of such material (HILLGÄRTNER et al., 1999; see also FUHR et al., 1998; H. ZIMMERMANN et al., 1999). Typical AFM images of empty solid Ba^{2+} alginate capsules made up of purified alginate (see Fig. 4A) and produced by an air-jet droplet generator in the conventional way are given in Fig. 8. The images were recorded in the non-contact mode. Fig. 8A represents a capsule incubated for about 6 weeks and Fig. 8B a bead after 4-week implantation in a rat. There is apparently no significant difference in the appearance of the surfaces between the control and the implanted bead. The surfaces are quite homogeneous, but reveal some imperfections, such as wrinkles. These may originate from the spatial orientation of the gelation process (see above). In contrast, AFM images of alginate-perfluorocarbon beads retrieved from rats after about 15 months of implantation show deposition of material (presumably collagen fibrils) on the surface (Fig. 8C).

AFM studies have also most probably revealed the first steps towards fibrotic overgrowth of capsules. Purified alginate does not support the attachment, spreading, and growth of anchorage-dependent cells. However, crevices and other mechanical imperfections can present preferred sites of cellular attachment (CURTIS and WILKINSON, 1999; ITO, 1999). Similarly, adsorption of proteins on the surfaces of the alginate implants can lead to conformational changes being recognized as foreign by the immune system of the host (for further details, see BESSIS, 1973). When attachment and spreading of an individual macrophage, fibroblast, or another adhering cell is initiated, fibrotic overgrowth will proceed over time because the cells regularly release traces of material, while crawling on the surface (FUHR et al., 1998; H. ZIMMERMANN et al., 1999; ZIMMERMANN and FUHR, 1999), thus facilitating the attachment of other cells (see GRÖHN et al.,

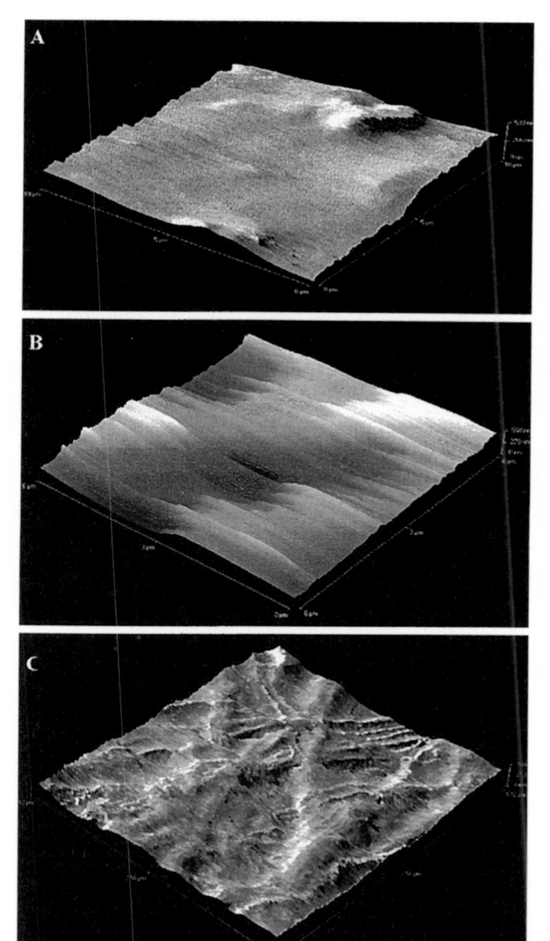

Fig. 8. Atomic force microscopy images of empty solid Ba^{2+} alginate capsules (made from purified alginate) before (**A**) and after transplantation into the peritoneum of Wistar rats (**B**, **C**). All images were acquired in an aqueous phase by using the non-contact mode.
A: Three-dimensional look onto the surface of an alginate bead stored in iso-osmolar NaCl solution for 47 d. The scan range is $10 \cdot 10\ \mu m^2$, the height scale is 0 to 533 nm. The surface is relatively smooth, but reveals some imperfections. In order to highlight small surface features the picture is left-shaded.
B: Image of the surface of an alginate capsule retrieved 4 weeks after implantation. The scan range is $6 \cdot 6\ \mu m^2$ and the height scale is 0 to 558 nm. As in (**A**), the surface is quite homogeneous.
C: Left-shaded picture of the surface of an alginate-perflurorocarbon capsule retrieved from a rat 15 months after implantation. Before bead formation perfluorocarbon had been emulsified homogeneously with the alginate solution (ratio 1:1). The scan range is $5.12 \cdot 5.12\ \mu m^2$ and the height scale is 0 to 506 nm. In contrast to (**A**) and (**B**), the topography is very rough and the surface is apparently covered by fibrils (for further explanations, see Sects. 6.2 and 6.4, for technical details, see H. Zimmermann et al., 1999).

1997; Jork et al., 2000). An AFM image of the migration of a primary anchorage-dependent cell on a glass surface is shown in Fig. 9. Obviously, there are basically two main classes of material traces. One class consists of characteristic high-order, dendritic structures that are well organized and enveloped by membranes (for more details, see Fuhr et al., 1998; H. Zimmermann et al., 1999). The other class is composed of smaller, randomly deposited material traces (arrows in Fig. 9) that cover the whole migration area of the respective cell. The origin and the chemical composition of the traces are not known in detail. However, there are some hints that the dendritic structures contain actin and are released from the pseudopods whereas the random traces are presumably made up of collagen and are released from the attachment sites of the crawling cell.

In principle, the occurrence of small imperfections during transplantation can never be excluded, even when microcapsules are used that are designed according to the suggestions made in this section. Therefore, it seems promising to bind agents that are toxic to macrophages and other adherent cells to the capsule surface.

In any case, deposition of material on the surfaces of microcapsules and, as a result, fi-

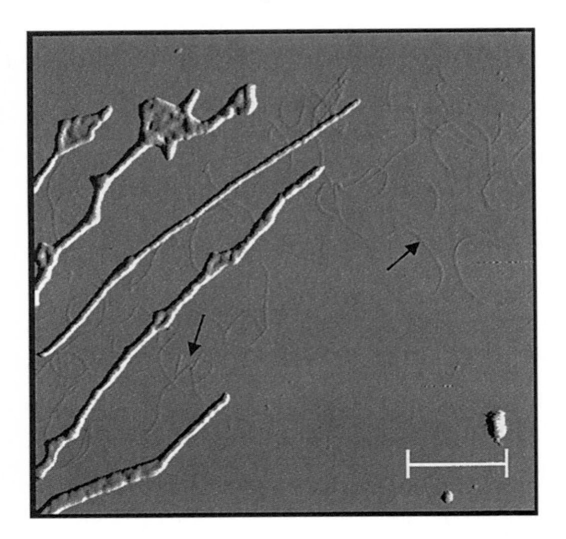

Fig. 9. Release of cellular material of an anchorage-dependent cell (not shown) during crawling over a glass surface visualized with atomic force microscopy. The measurements were performed in an aqueous environment by using the non-contact mode. The image is presented in the so-called "shadow mode" where height information is lost, but small corrugations are emphasized. There are two classes of material traces: relatively large dendritic, organized structures and background depositions (arrows) that are randomly oriented beneath the migration area of the crawling cell.
Bar = 3 μm.

brotic overgrowth is certainly a very important issue that deserves closer observation in future, in particular after availability of microcapsules with improved characteristics.

6.5 Transplantation Site and Monitoring of the Functional State of Transplants

Areas that also need to be explored in the light of the clinical trials with microencapsulated parathyroid tissue (see above) are the selection of the transplantation site and non-invasive methods to follow up continuously the fate of the microencapsulated transplants in the patients. ^{19}F-MRI (magnetic resonance imaging) offers a solution for both problems. Currently, most MRI studies are based on ^{1}H-MRI. However, because of the high water content of alginate microcapsules, the ^{1}H-MRI signal of the alginate cannot be distinguished from that of the surrounding tissue (Fig. 10A). In contrast, fluorine is only present in bones and teeth. Thus, perfluorocarbons are ideal MRI contrasting agents for microencapsulated transplants as shown in Fig. 10B.

^{19}F-MRI simultaneously allows to measure (after calibration) the oxygen partial pressure in the microcapsules via changes in the fluor-

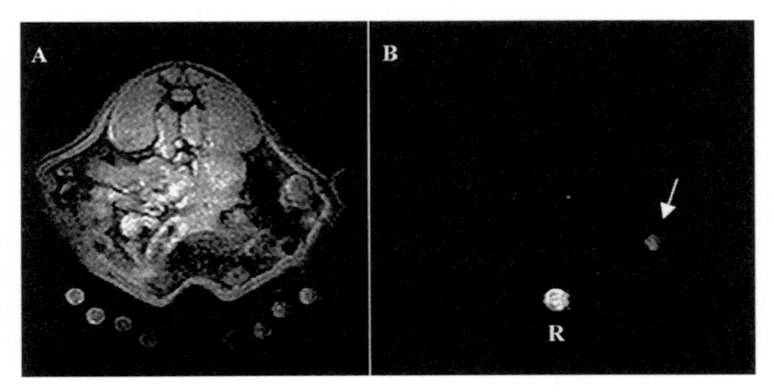

Fig. 10. ^{1}H-MR (**A**) and ^{19}F-MR (**B**) images of a 2-mm-thick transaxial slice through the abdominal cavity of a female Wistar rat. The images were acquired about 2 years after implantation of a single perfluorocarbon-loaded Ba^{2+} alginate capsule (4 mm in diameter) into the peritoneal cavity. The in-plane resolution was $0.5 \cdot 0.5$ mm^2. The arrow indicates the location of the implanted ^{19}F-labeled bead. The letter R denotes a glass capillary (5 mm in diameter) filled with pure perfluorocarbon placed under the rat during the experiment as an external reference (for further technical details, see Nöth et al., 1999 and Zimmermann et al., 2000).

ine spin-lattice relaxation time (NÖTH et al., 1999; ZIMMERMANN et al., 2000). This and the beneficial effect of perfluorocarbons on swelling and mechanical strength (see above) suggest that these compounds have a great potential for regular monitoring of oxygen partial pressure at a given transplantation site. Studies in this direction with rodents have shown (ZIMMERMANN et al., 2000 and unpublished data) that the peritoneum (where most of the islet transplantations had been conducted in the past), and also the kidney capsule are not recommendable transplantation sites because of low oxygen partial pressures. According to ^{19}F-MRI, transplantation in muscles seems to be more appropriate for long-term survival (despite the large mechanical stress exerted on the beads at these locations).

Regular MRI visualization of empty alginate ^{19}F-beads in rats for up to two years did not give any hints of migration of the capsules from the implantation site or for disintegration of the implants. This contrasts with the outcome of the clinical trials of patients with hypoparathyroidism (see above) and demonstrates the limited information that is sometimes received from animal studies. However, these studies have brought about convincing evidence that the fluorine/oxygen signal can be used as an early indicator for failure and thus for the necessity to retrieve the graft.

6.6 Xenogeneic Donor Cells

While there has apparently been a considerable progress in the design of novel, highly capable potential microcapsules (summarized in Fig. 7), there is another important issue that must be addressed. This is the supply of viable xenogeneic donor cells when human material is limited. For example, this will remain the case for human islets unless it becomes possible to propagate islets or β-cells *in vitro* while maintaining normal function. Therefore, many groups recommend discordant islet xenotransplantation.

Porcine islets

Porcine islets are considered to be a particularly promising donor source because there is only a difference of one amino acid between the structure of human and porcine insulin (VAN HAEFTEN, 1989), and because of the relatively low levels of preformed human xenoantibodies to pig tissue (KIRKMAN, 1989). New approaches include genetic manipulations of donor pigs (WEBER et al., 1999) (1) to completely eliminate natural human anti-pig antibody toxicity by producing transgeneic pigs that inhibit complement-mediated lysis by antibodies (ROSENGARD et al., 1995), (2) to inhibit glycosylation, which is a major target of natural human anti-pig antibodies (PARKER et al., 1996), and (3) to "knock out" pig MHC antigens (SEEBACH et al., 1996) that differ between human beings and pigs. However, even with the advent of manufacturing genetically engineered cells, it is agreed among the workers in the field that encapsulation will remain important to safely exclude an immune response of the host.

Islet viability is another factor which affects long-term viability of transplants. Isolation of viable porcine islets in high yields is still subjected to a large variability. They are usually isolated by the methods of RICORDI et al. (1990) and of WARNOCK and RAJOTTE (1988). After 30–60 min pancreas digestion exocrinic tissue is usually removed by discontinuous Ficoll gradient centrifugation. The freshly isolated islets initially look ragged, but regain a more compact appearance after 1–2 d culture. The success of commonly used isolation procedures depends critically on the collagenase used and on the donor pigs (ULRICHS et al., 1995; KÜHTREIBER et al., 1999). Preliminary studies in our laboratory have shown (CRAMER, LUDWIG, THÜRMER, ZIMMERMANN, unpublished data) that the isolation procedure of porcine islets can be improved by performing part of the digestion treatment under high pressure (up to 4 MPa). Subsequent rapid decompression results in cavitation with the formation of numerous tiny gas bubbles in the blood capillaries. These gas bubbles apparently facilitate the release of the islets from the pancreas tissue.

Permanent cell lines

The ultimate goal in cell therapy (including the treatment of IDDM patients) is the microencapsulation of continuous (genetically engineered) cell lines that deliver the factor that the body does not produce (see, e.g., NEWGARD, 1994). A key advantage of such cell systems over animal tissue is that they can be cultured under controlled conditions, that the encapsulated products can be characterized safely (e.g., for the presence of virus; see Sect. 5) and stored before clinical application. Therefore, it can be expected that medical approval of encapsulated, well defined cell lines is less difficult and expensive to obtain than that of cells isolated from animal tissue.

There are numerous animal studies with encapsulated cell lines for therapy of pain, Parkinson's disease, uremia, and other disorders (HAMA and SAGEN, 1993; AEBISCHER et al., 1991; CHANG, 1997; for more information, see KÜHTREIBER et al., 1999). Even though this work has yielded very promising results, success is limited – as in islet xenotransplantation – so far due to the use of improper capsule material and encapsulation techniques.

7 Concluding Remarks

The concept of alginate-based cell therapy has been shown to be benign, highly versatile, clinically efficient, potentially economical, and amenable to large-scale production and quality control. Clinical allotransplantations are planned to begin shortly in a group of patients with hypoparathyroidism to realize the potential of the novel microcapsule configurations described here and tested in animal studies in the last two years. Over the next years, as our knowledge of encapsulated cell systems increases, we can expect to see not only clinical xenotransplantation, but also many new areas of applications of this technology. A successful development of well documented, medically approved systems necessitate, however, close interdisciplinary collaborations among scientists with different areas of expertise such as biophysics, biotechnology, molecular biology, and medicine.

Acknowledgments

We are very grateful to P. AMERSBACH, P. GEßNER, and G. HOFMANN-PFANNES for skilful technical assistance. We would like to thank M. HILLGÄRTNER, H. SCHNEIDER, and S. MIMIETZ for performing some of the experiments and St. G. SHIRLEY for critical reading of the manuscript. Furthermore, we would like to acknowledge B. KUTTLER for performance of the implantation experiments in the BB rats. This work was supported by grants of the BMBF (0311716) and of the Deutsche Bundesstiftung Umwelt (13019) to U. Z., by grants of the Deutsche Forschungsgemeinschaft (HA 2611/1-1, GE 791-1/1) to C. H. and to the Lehrstuhl für Membranphysiologie, respectively.

8 References

AEBISCHER, P., WINN, S. R., TRESCO, P. A., JAEGER, C. B., GREENE, L. A. (1991), Transplantation of polymer encapsulated neurotransmitter secreting cells: effect of the encapsulation technique, *J. Biomech. Eng.* **113**, 178–183.

ARFORS, K. E., LEY, K. (1993), Sulfated polysaccharides in inflammation, *J. Lab. Clin. Med.* **121**, 201–202.

ARNOLD, L. J., DAGAN, A., KAPLAN, N. O. (1983), Poly (L-lysine) as an antineoplastic agent and tumor-specific drug carrier, in: *Targeted Drugs* Vol. 2 (GOLDBERG, E. P., Ed.), pp. 89–112. New York: John Wiley & Sons.

BESSIS, M. (1973), *Living Blood Cells and their Ultrastructure*. Berlin, Heidelberg, New York: Springer-Verlag.

BEUNINK, J., BAUMGÄRTL, H., ZIMELKA, W., REHM, H.-J. (1989), Determination of oxygen gradients in single Ca-alginate beads by means of oxygen-microelectrodes, *Experientia* **45**, 1041–1047.

BRANDENBERGER, H., WIDMER, F. (1998), A new multinozzle encapsulation/immobilization system to produce uniform beads of alginate, *J. Biotechnol.* **63**, 73–80.

BRODELIUS, P., VANDAMME, E. J. (1987), Immobilized cell systems, in: *Biotechnogy* Vol. 7a (REHM, H.-J., REED, G., Eds.), pp. 405–464. Weinheim: Verlag Chemie.

BÜCHNER, K. H., ZIMMERMANN, U. (1982), Water relations of immobilized giant algal cells, *Planta* **154**, 318–325.

CALAFIORE, R., BASTA, G., LUCA, G., BOSELLI, C., BUFALARI, A. et al. (1999), Transplantation of pancreatic islets contained in minimal volume microcapsules in diabetic high mammalians, *Ann. N.Y. Acad. Sci.* **875**, 219–232.

CASANOVA, D., SARFATI, E., DE FRANCISCO, A., AMADO, J. A., ARIAS, M., DUBOST, C. (1991), Secondary hyperparathyroidism: diagnosis of site of recurrence, *World J. Surg.* **15**, 546–550.

CHANG, T. M. S. (1997), Artificial cells, in: *Encyclopedia of Human Biology* 2nd Edn. (DULBECCO, R., Ed.), pp. 457–463. San Diego, CA: Academic Press.

CHICHEPORTICHE, D., REACH, G. (1988), *In vitro* kinetics of insulin release by microencapsulated rat islets: effect of the size of the microcapsules, *Diabetologia* **31**, 54–57.

CLAYTON, H. A., LONDON, N. J. M., COLLOBY, P. S., BELL, P. R. F., JAMES, R. F. L. (1991), The effect of capsule composition on the biocompatibility of alginate-poly-L-lysine capsules, *J. Microencapsul.* **8**, 221–233.

CLAYTON, H. A., JAMES, R. F. L., LONDON, N. J. M. (1993), Islet microencapsulation: a review, *Acta Diabetologica* **30**, 181–189.

COCHRUM, K., JEMTRUD, S., DORIAN, R. (1995), Successful xenografts in mice with microencapsulated rat and dog islets, *Transplant. Proc.* **27**, 3297–3301.

COLE, D. R., WATERFALL, M., MCINTYRE, M., BAIRD, J. D. (1992), Microencapsulated islet grafts in the BB/E rat: a possible role for cytokines in graft failure, *Diabetologia* **35**, 231–237.

COLTON, C. K., AVGOUSTINIATOS, E. S. (1991), Bioengineering in development of the hybrid artificial pancreas, *J. Biomech. Eng.* **113**, 152–170.

CURTIS, A, WILKINSON, C. (1999), New depths in cell behavior: reactions of cells to nanotopography, *Biochem Soc. Symp.* **65**, 15–26.

DE VOS, P., WOLTERS, G. H., FRITSCHY, W. M., VAN SCHILFGAARDE, R. (1993), Obstacles in the application of microencapsulation in islet transplantation, *Int. J. Artif. Organs* **16**, 205–212.

DE VOS, P., DE HAAN, B. J., WOLTERS, G. H. J., VAN SCHILFGAARDE, R. (1996), Factors influencing the adequacy of microencapsulation of rat pancreatic islets, *Transplantation* **62**, 888–893.

DE VOS, P., DE HAAN, B. J., VAN SCHILFGAARDE, R. (1997a), Effect of the alginate composition on the biocompatibility of alginate-polylysine microcapsules, *Biomaterials* **18**, 273–278.

DE VOS, P., DE HAAN, B. J., WOLTERS, G. H. J., STRUBBE, J. H., VAN SCHILFGAARDE, R. (1997b), Im proved biocompatibility but limited graft survival after purification of alginate for microencapsulation of pancreatic islets, *Diabetologia* **40**, 262–270.

DE VOS, P., VAN STRAATEN, J. F. M., NIEUWENHUIZEN, A. G., DE GROOT, M., PLOEG, R. J. et al. (1999), Why do microencapsulated islet grafts fail in the absence of fibrotic overgrowth? *Diabetes* **48**, 1381–1388.

DULIEU, C., PONCELET, D., NEUFELD, R. J. (1999), Encapsulation and immobilization techniques, in: *Cell Encapsulation Technoloy and Therapeutics* (KÜHTREIBER, W. M., LANZA, R. P., CHICK, W. L., Eds.), pp. 3–17. Boston, MA: Birkhäuser.

ESCH, M., SUKHORUKOV, V. L., KÜRSCHNER, M., ZIMMERMANN, U. (1999), Dielectric properties of alginate beads and bound water relaxation studied by electrorotation, *Biopolymers* **50**, 227–237.

ESPEVIK, T., OTTERLEI, M., SKJÅK-BRÆK, G., RYAN, L., WRIGHT, S. D., SUNDAN, A. (1993), The involvement of CD14 in stimulation of cytokine production by uronic acid polymers, *Eur. J. Immunol.* **23**, 255–261.

FAN, M. Y., LUM, Z. P., FU, X.-W., LEVESQUE, L., TAI, I. T., SUN, A. M. (1990), Reversal of diabetes in BB rats by transplantation of encapsulated pancreatic islets, *Diabetes* **39**, 519–522.

FRIEDRICH, U., STACHOWICZ, N., SIMM, A., FUHR, G., LUCAS, K., ZIMMERMANN, U. (1998), High efficiency electrotransfection with aluminum electrodes using microsecond controlled pulses, *Bioelectrochem. Bioenerg.* **47**, 103–111.

FRITSCHY, W. M., WOLTERS, G. H. J., VAN SCHILFGAARDE, R. (1991), Effect of alginate–polylysine–alginate microencapsulation on *in vitro* insulin release from rat pancreatic islets, *Diabetes* **40**, 37–43.

FUHR, G., RICHTER, E., ZIMMERMANN, H., HITZLER, H., NIEHUS, H., HAGEDORN, R. (1998), Cell traces – footprints of individual cells during locomotion and adhesion, *Biol. Chem.* **379**, 1161–1173.

GÅSERØD, O., SANNES, A., SKJÅK-BRÆK, G. (1999), Microcapsules of alginate-chitosan. II. A study of capsule stability and permeability, *Biomaterials* **20**, 773–783.

GEISEN, K., DEUTSCHLÄNDER, H., GORBACH, S., KLENKE, C., ZIMMERMANN, U. (1990), Function of barium alginate-microencapsulated xenogenic islets in different diabetic mouse models, in: *Frontiers in Diabetes Research. Lessons from Animal Diabetes III* (SHAFRIR, E., Ed.), II.8., pp. 142–148.

GENTILE, F. T., DOHERTY, E. J., REIN, D. H., SHOICHET, M. S., WINN, S. H. (1995), Polymer science for macroencapsulation of cells for central nervous system transplantation, *Reactive Polymers* **25**, 207–227.

GOOSEN, M. F. A. (1993), *Fundamentals of Animal Cell Encapsulation and Immobilization.* Boca Raton, FL: CRC Press.

GOOSEN, M. F. A. (1999), Mass transfer in immobilized cell systems, in: *Cell Encapsulation Technology and Therapeutics* (KÜHTREIBER, W. M., LANZA, R. P., CHICK, W. L., Eds.), pp. 18–28. Boston, MA: Birkhäuser.

GOTFREDSEN, C. F., STEWART, M. G., O'SHEA, G. M., VOSE, J. R., HORN, T., MOODY, A. J. (1990), The fate of transplanted encapsulated islets in spontaneously diabetic BB/Wor rats, *Diabetes Res.* **15**, 157–163.

GRÖHN, P., KLÖCK, G., SCHMITT, J., ZIMMERMANN, U., HORCHER, A. et al. (1994), Large-scale production of Ba^{2+} alginate-coated islets of Langerhans for immunoisolation, *Exp. Clin. Endocrinol.* **102**, 380–387.

GRÖHN, P., KLÖCK, G., ZIMMERMANN, U. (1997), Collagen-coated Ba^{2+} alginate microcarriers for the culture of anchorage-dependent mammalian cells, *BioTechniques* **22**, 970–975.

HAMA, A. T., SAGEN, J. (1993), Reduced pain-related behavior by adrenal medullary transplants in rats with experimental painful peripheral neuropathy, *Pain* **52**, 223–231.

HARTMEIER, W. (1986), *Immobilisierte Biokatalysatoren.* Berlin, Heidelberg, New York, Tokyo: Springer-Verlag.

HASSE, C., KLÖCK, G., ZIELKE, A., SCHLOSSER, A., BARTH, P. et al. (1996), Transplantation of parathyroid tissue in experimental hypoparathyroidism: *in vitro* and *in vivo* function of parathyroid tissue microencapsulated with a novel amitogenic alginate, *Int. J. Artif. Organs* **19**, 735–741.

HASSE, C., KLÖCK, G., SCHLOSSER, A., ZIMMERMANN, U., ROTHMUND, M. (1997a), Parathyroid allotransplantation without immunosuppression, *The Lancet* **350**, 1296–1297.

HASSE, C., ZIELKE, A., KLÖCK, G., BARTH, P., SCHLOSSER, A. et al. (1997b), First successful xenotransplantation of microencapsulated human parathyroid tissue in experimental hypoparathyroidism: long-term function without immunosuppression, *J. Microencapsulation* **14**, 617–626.

HASSE, C., ZIELKE, A., KLÖCK, G., SCHLOSSER, A., BARTH, P. et al. (1998), Amitogenic alginates: key to first clinical application of microencapsulation technology, *World J. Surg.* **22**, 659–665.

HASSE, C., ZIELKE, A., ZIMMERMANN, U., ROTHMUND, M. (1999), Transplantation of microencapsulated parathyroid tissue: clinical background, methods, and current status of research, in: *Cell Encapsulation Technology and Therapeutics* (KÜHTREIBER, W. M., LANZA, R. P., CHICK, W. L., Eds.), pp. 240–251. Boston, MA: Birkhäuser.

HEINZEN, C. (1999), Zellverkapselung bei 1000 Volt, *BioTec* **4**, 50–51.

HILLGÄRTNER, M., ZIMMERMANN, H., MIMIETZ, S., JORK, A., THÜRMER, F. et al. (1999), Immunoisolation of transplants by entrapment in ^{19}F-labeled alginate gels: production, biocompatibility, stability, and long-term monitoring of functional integrity, *Mat.-wiss. Werkstofftech.* **30**, 783–792.

HUNKELER, D., SUN, A. M., KORBUTT, S., RAJOTTE, R. V., GILL, R. G. et al. (1999), Bioartificial organs and acceptable risk, *Nature Biotechnol.* **17**, 1045.

ITO, Y. (1999), Surface micropatterning to regulate cell functions, *Biomaterials* **20**, 2333–2342.

JAHR, T. G., RYAN, L., SUNDAN, A., LICHENSTEIN, H. S., SKJÅK-BRÆK, G. et al. (1997), Induction of tumor necrosis factor production from monocytes stimulated with mannuronic acid polymers and involvement of lipopolysaccharide-binding protein, CD14, and bactericidal/permeability-increasing factor, *Infect. Immun.* **65**, 89–94.

JORK, A., THÜRMER, F., CRAMER, H., ZIMMERMANN, G., GESSNER, P. et al. (2000), Biocompatible alginate from freshly collected *Laminaria pallida* for implantation, *Appl. Microbiol. Biotechnol.* **53**, 224–229.

JU, L. K., ARMIGER, W. B. (1992), Use of perfluorocarbon emulsions in cell culture, *BioTechniques* **12**, 258–263.

JUNKER, B. H., WANG, D. I. C., HATTON, T. A. (1990), Oxygen transfer enhancement in aqueous/perfluorocarbon fermentation systems: II. Theoretical analysis, *Biotechnol. Bioeng.* **35**, 586–597.

KAPLAN, J. H., ELLIS-DAVIES, G. C. R. (1988), Photolabile chelators for the rapid photorelease of divalent cations, *Proc. Natl. Acad. Sci. USA* **85**, 6571–6575.

KING, K. (1994), Changes in the functional properties and molecular weight of sodium alginate following γ irradiation, *Food Hydrocolloids* **8**, 83–96.

KIRKMAN, R. L. (1989), Of swine and men: organ physiology in different species, in: *Xenograft* **25** (HARDY, M. A., Ed.), pp. 125–132. Amsterdam: Elsevier.

KLÖCK, G., SIEBERS, U., PFEFFERMANN, A., SCHMITT, J., HOUBEN, R. et al. (1993), Immunisolation transplantierter Langerhansscher Inseln: Mikrokapseln mit definierter molekularer Ausschlußgrenze, *Immun. Infekt.* **21**, 183–184.

KLÖCK, G., FRANK, H., HOUBEN, R., ZEKORN, T., HORCHER, A. et al. (1994), Production of purified alginates suitable for use in immunoisolated transplantation, *Appl. Microbiol. Biotechnol.* **40**, 638–643.

KLÖCK, G., PFEFFERMANN, A., RYSER, C., GRÖHN, P., KUTTLER, B. et al. (1997), Biocompatibility of mannuronic acid-rich alginates, *Biomaterials* **18**, 707–713.

KRESTOW, M., LUM, Z. P., TAI, I. T., SUN, A. (1991), Xenotransplantation of microencapsulated fetal rat islets, *Transplantation* **51**, 651–655.

KÜHTREIBER, W. M., LANZA, R. P., CHICK, W. L. (1999), *Cell Encapsulation Technology and Therapeutics*. Boston, MA: Birkhäuser.

KULSENG, B., SKJÅK-BRÆK, G., RYAN, L., ANDERSSON, A., KING, A. et al. (1999), Transplantation of alginate microcapsules, *Transplantation* **67**, 978–984.

LANZA, R. P., HAYES, J. L., CHICK, W. L. (1996), Encapsulated cell technology, *Nature Biotechnol.* **14**, 1107–1111.

LANZA, R. P., ECKER, D. M., KÜHTREIBER, W. M., MARSH, J. P., RINGELING, J., CHICK, W. L. (1999), Transplantation of islets using microencapsulation: studies in diabetic rodents and dogs, *J. Mol. Med.* **77**, 206–210.

LIM, F., SUN, A. M. (1980), Microencapsulated islets as bioartificial endocrine pancreas, *Science* **210**, 908–910.

MARTINSEN, A., SKJÅK-BRÆK, G., SMIDSRØD, O. (1989), Alginate as immobilization material: I. Correlation between chemical and physical properties of alginate gel beads, *Biotechnol. Bioeng.* **33**, 79–89.

MARTINSEN, A., STORRØ, I., SKJÅK-BRÆK, G. (1992), Alginate as immobilization material: III. Diffusional properties, *Biotechnol. Bioeng.* **39**, 186–194.

MATTIASSON, B. (1982), Immobilization methods, in: *Immobilized Cells and Organelles* Vol. I (MATTIASSON, B., Ed.), pp. 3–26. Boca Raton, FL: CRC Press.

MAZAHERI, R., ATKISON, P., STILLER, C., DUPRÉ, J., VOSE, J. et al. (1991), Transplantation of encapsulated allogeneic islets into diabetic BB/W rats. Effects of immunosuppression, *Transplantation* **51**, 750–754.

MCHUGH, D. J. (1987), Production, properties and uses of alginates, in: *Production and Utilization of Products from Commercial Seaweeds* (MCHUGH, D. J., Ed.), pp. 58–115. Rome: Food and Agriculture Organization of the United Nations.

MESSIAEN, J., CAMBIER, P., VAN CUTSEM, P. (1997), Polyamines and pectins, *Plant Physiol.* **113**, 387–395.

NEWGARD, C. B. (1994), Cellular engineering and gene therapy strategies for insulin replacement in diabetes, *Diabetes* **43**, 341–350.

NÖTH, U., GRÖHN, P., JORK, A., ZIMMERMANN, U., HAASE, A., LUTZ, J. (1999), ^{19}F-MRI *in vivo* determination of the partial oxygen pressure in perfluorocarbon-loaded alginate capsules implanted into the peritoneal cavity and different tissues, *Magn. Reson. Med.* **42**, 1039–1047.

O'SHEA, G. M., SUN, A. M. (1986), Encapsulation of rat islets of Langerhans prolongs xenograft survival in diabetic mice, *Diabetes* **35**, 943–946.

O'SHEA, G. M., GOOSEN, M. F. A., SUN, A. M. (1984), Prolonged survival of transplanted islets of Langerhans encapsulated in a biocompatible membrane, *Biochim. Biophys. Acta* **804**, 133–136.

OTTERLEI, M., ØSTGAARD, K., SKJÅK-BRÆK, G., SMIDSRØD, O., SOON-SHIONG, P., ESPEVIK, T. (1991), Induction of cytokine production from human monocytes stimulated with alginate, *J. Immunother.* **10**, 286–291.

OTTERLEI, M., SUNDAN, A., SKJÅK-BRÆK, G., RYAN, L., SMIDSRØD, O., ESPEVIK, T. (1993), Similar mechanisms of action of defined polysaccharides and lipopolysaccharides: characterization of binding and tumor necrosis factor alpha induction, *Infect Immun.* **61**, 1917–1925.

PARADIS, K., LANGFORD, G., LONG, Z., HENEINE, W., SANDSTROM, P. et al. (1999), Search for cross-species transmission of porcine endogenous retrovirus in patients treated with living pig tissue, *Science* **285**, 1236–1241.

PARKER, W., SAADI, S., LIN, S. S., HOLZKNECHT, Z. E., BUSTOS, M. (1996), Transplantation of discordant xenografts: a challenge revisited, *Immunology Today* **17**, 373–378.

PATIENCE, C., TAKEUCHI, Y., WEISS, R. A. (1997), Infection of human cells by an endogenous retrovirus of pigs, *Nature Med.* **3**, 282–286.

PFEFFERMANN, A., KLÖCK, G., GRÖHN, P., KUTTLER, B., HAHN, H. J., ZIMMERMANN, U. (1996), Evaluation of assay procedures for the assessment of the biocompatibility of alginate implants, *Cell. Eng.* **4**, 167–173.

PILWAT, G., WASHAUSEN, P., KLEIN, J., ZIMMERMANN, U. (1980), Immobilization of human red blood cells, *Z. Naturforsch.* **35c**, 352–356.

PLUMB, R. C., BRIDGMAN, W. B. (1972), Ascent of sap in trees, *Science* **176**, 1129–1131.

PLÜSS, R., HEINZEN, C., KOCH, M., WIDMER, F. (1997), Immobilisierte Biokatalysatoren, *BioWorld* **2**, 16–19.

RICORDI, C., SOCCI, C., DAVALLI, A. M., STAUDACHER, C., BARO, P. et al. (1990), Isolation of the elusive pig islet, *Surgery* **107**, 688–694.

ROSENGARD, A. M., CARY, N., HORSLEY, J., BELCHER, C., LANGFORD, G. et al. (1995), Endothelial expression of human decay accelerating factor in transgenic pig tissue: a potential approach for human complement inactivation in discordant xenografts, *Transplant. Proc.* **27**, 326–327.

ROTHE, H., FEHSEL, K., KOLB, H. (1990), Tumour necrosis factor alpha production is upregulated in diabetes prone BB rats, *Diabetologia* **33**, 573–575.

SCHNABL, H., YOUNGMAN, R. J., ZIMMERMANN, U. (1983), Maintenance of plant cell membrane integrity and function by the immobilization of protoplasts in alginate matrices, *Planta* **158**, 392–397.

SCHNABL, H., ZIMMERMANN, U. (1989), Immobilization of plant protoplasts, in: *Biotechnology in Agriculture and Forestry*, Vol. 8, *Plant Protoplasts and Genetic Engineering I.* (BAJAJ, Y. P. S., Ed.), pp. 63–96. Berlin, Heidelberg: Springer-Verlag.

SCUDIERO, D. A., SHOEMAKER, R. H., PAULL, K. D., MONKS, A., TIERNEY, S. et al. (1988), Evaluation of a soluble tetrazolium/formazan assay for cell growth and drug sensitivity in culture using human and other tumor cell lines, *Cancer Res.* **48**, 4827–4833.

SEEBACH, J. D., YAMADA, K., MCMORROW, I. S., SACHS, D. H., DERSIMONIAN, H. (1996), Xenogeneic human anti-pig cytotoxicity mediated by activated natural killer cells, *Xenotransplantation* **3**, 188–197.

SIEBERS, U., ZEKORN, T., HORCHER, A., HERING, B., BRETZEL, R. G. et al. (1992), *In vitro* testing of rat and porcine islets microencapsulated in barium alginate beads, *Transplant. Proc.* **24**, 950–951.

SIEBERS, U., HORCHER, A., BRETZEL, R. G., KLÖCK, G., ZIMMERMANN, U. (1993), Transplantation of free and microencapsulated islets in rats: evidence for the requirement of an increased islet mass for transplantation into the peritoneal site, *Int. J. Artif. Organs* **16**, 35–38.

SKJÅK-BRÆK, G., MURANO, E., PAOLETTI, S. (1989), Alginate as immobilization material. II: Determination of polyphenol contaminants by fluorescence spectroscopy, and evaluation of methods for their removal, *Biotechnol. Bioeng.* **33**, 90–94.

SMIDSRØD, O., SKJÅK-BRÆK, G. (1990), Alginate as immobilization matrix for cells, *Trends Biotechnol.* **8**, 71–78.

SOON-SHIONG, P., OTTERLEI, M., SKJÅK-BRÆK, G., SMIDSRØD, O., HEINTZ, R. et al. (1991), An immunologic basis for the fibrotic reaction to implanted microcapsules, *Transplant. Proc.* **23**, 758–759.

SOON-SHIONG, P., FELDMAN, E., NELSON, R., KOMTEBEDDE, J., SMIDSRØD, O. et al. (1992), Successful reversal of spontaneous diabetes in dogs by intraperitoneal microencapsulated islets, *Transplantation* **54**, 769–774.

SULTZBAUGH, K. J., SPEAKER, T. J. (1996), A method to attach lectins to the surface of spermine alginate microcapsules based on the avidin biotin interaction, *J. Microencapsul.* **13**, 363–375.

SUN, Y., MA, X., ZHOU, D., VACEK, I., SUN, A. M. (1996), Normalization of diabetes in spontaneously diabetic cynomolgus monkeys by xenografts of microencapsulated porcine islets without immunosuppression, *J. Clin. Invest.* **98**, 1417–1422.

TANAKA, H., IRIE, S. (1988), Preparation of stable alginate gel beads in electrolyte solutions using Ba^{2+} and Sr^{2+}, *Biotechnol. Techniques* **2**, 115–120.

ULRICHS, K., BOSSE, M., HEISER, A., ECKSTEIN, V., WACKER, H. H. et al. (1995), Histomorphological characteristics of the porcine pancreas as a basis for the isolation of islets of Langerhans, *Xenotransplantation* **2**, 176–187.

VAN HAEFTEN, T. W. (1989), Clinical significance of insulin antibodies in insulin-treated diabetic patients, *Diabetes Care* **12**, 641–648.

VAN SCHILFGAARDE, R., DE VOS, P. (1999), Factors influencing the properties and performance of microcapsules for immunoprotection of pancreatic islets, *J. Mol. Med.* **77**, 199–205.

VANSTEENKISTE, S. O., DAVIES, M. C., ROBERTS, C. J., TENDLER, S. J. B., WILLIAMS, P. M. (1998), Scanning probe miscroscopy of biomedical interfaces, *Prog. Surf. Sci.* **57**, 95–136.

WANG, T. G. (1999), Polymer membranes for cell encapsulation, in: *Cell Encapsulation Technology and Therapeutics* (KÜHTREIBER, W. M., LANZA, R. P., CHICK, W. L., Eds.), pp. 29–39. Boston, MA: Birkhäuser.

WARNOCK, G. L., RAJOTTE, R. V. (1988), Critical mass of purified islets that induce normoglycemia after implantation into dogs, *Diabetes* **37**, 467–470.

WEBER, C., COSTANZO, M., KREKUN, S., D'AGATI, V. (1993), Causes of destruction of microencapsulated islet grafts: Characteristics of a "double-wall" poly-L-lysine-alginate microcapsule, *Diabetes, Nutrition, Metabolism* **1**, 167–171.

WEBER, C. J., KAPP, J. A., HAGLER, M. K., SAFLEY, S., CHRYSSOCHOOS, J. T. et al. (1999), Long-term survival of poly-L-lysine-alginate microencapsulated islet xenografts in spontaneously diabetic NOD mice, in: *Cell Encapsulation Technology and Therapeutics* (KÜHTREIBER, W. M., LANZA, R. P., CHICK, W. L., Eds.), pp. 117–137. Boston, MA: Birkhäuser.

WIJSMAN, J., ATKISON, P., MAZAHERI, R., GARCIA, B., PAUL, T. et al. (1992), Histological and immunopathological analysis of recovered encapsulated allogeneic islets from transplanted diabetic BB/W rats, *Transplantation* **54**, 588–592.

XU, K., HERCULES, D. M., LACIK, I., WANG, T. G. (1998), Atomic force microscopy used for the surface characterization of microcapsule immunoisolation devices, *J. Biomed. Mat. Res.* **41**, 461–467.

ZEKORN, T., SIEBERS, U., BRETZEL, R. G., HELLER, S., MEDER, U. et al. (1991), Impact of the perfluorochemical FC43 on function of isolated islets, *Horm. metab. Res.* **23**, 302–303.

ZEKORN, T. D. C., HORCHER, A., SIEBERS, U., SCHNETTLER, R., KLÖCK, G. et al. (1992a), Barium-cross-linked alginate beads: A simple, one-step method for successful immunoisolated transplantation of islets of Langerhans, *Acta Diabetol.* **29**, 99–106.

ZEKORN, T., SIEBERS, U., HORCHER, A., SCHNETTLER, R., ZIMMERMANN, U. et al. (1992b), Alginate coating of islets of Langerhans: *in vitro* studies on a new method for microencapsulation for immunoisolated transplantation, *Acta Diabetol.* **29**, 41–45.

ZEKORN, T. D. C., BRETZEL, R. G. (1999), Immunoprotection of islets of Langerhans by microencapsulation in barium alginate beads, in: *Cell Encapsulation Technology and Therapeutics* (KÜHTREIBER, W. M., LANZA, R. P., CHICK, W. L., Eds.) pp. 90–96. Boston, MA: Birkhäuser.

ZHOU, D., SUN, Y. L., VACEK, I., MA, P., SUN, A. M. (1994), Normalization of diabetes in cynomolgus monkeys by xenotransplantation of microencapsulated porcine islets, *Transplant. Proc.* **26**, 1091.

ZIMMERMANN, H., FUHR, G. (1999), Zellspuren – neue Ansätze zur Diagnostik und Oberflächenanalyse, *BIOforum* **22**, 708–710.

ZIMMERMANN, H., HAGEDORN, R., RICHTER, E., FUHR, G. (1999), Topography of cell traces studied by atomic force microscopy, *Eur. Biophys. J.* **28**, 516–525.

ZIMMERMANN, U., NEIL, G. A. (1996), *Electromanipulation of Cells.* Boca Raton, FL: CRC-Press.

ZIMMERMANN, U., STEUDLE, E. (1978), Physical aspects of water relations, *Adv. Bot. Res.* **6**, 45–117.

ZIMMERMANN, U., KLÖCK, G., FEDERLIN, K., HANNIG, K., KOWALSKI, M. et al. (1992), Production of mitogen-contamination free alginates with variable ratios of mannuronic acid to guluronic acid by free flow electrophoresis, *Electrophoresis* **13**, 269–274.

ZIMMERMANN, U., ZHU, J. J., MEINZER, F. C., GOLDSTEIN, G., SCHNEIDER, H. et al. (1994), High molecular weight organic compounds in the xylem sap of mangroves: implications for long-distance water transport, *Bot. Acta* **107**, 218–229.

ZIMMERMANN, U., HASSE, C., ROTHMUND, M., KÜHTREIBER, W. (1999), Biocompatible encapsulation materials: fundamentals and application, in: *Cell Encapsulation Technology and Therapeutics* (KÜHTREIBER, W. M., LANZA, R. P., CHICK, W. L., Eds.), pp. 40–52. Boston, MA: Birkhäuser.

ZIMMERMANN, U., NÖTH, U., GRÖHN, P., JORK, A., ULRICHS, K. et al. (2000), Non-invasive evaluation of the location, the functional integrity and the oxygen supply of implants: ^{19}F nuclear magnetic resonance imaging of perfluorocarbon-loaded Ba^{2+}-alginate beads, *Artificial Cells, Blood Substitutes, and Immobilization Biotechnology* **28**, 129–146.

20 Comments on Other Fields of Biotechnology

HANS-JÜRGEN REHM

Münster, Germany

1 Introduction

In 16 volumes of this "Multi-Volume Comprehensive Treatise on Biotechnology" many fields of biotechnology have been described. The enormous advances of biotechnology in recent years made it impossible to treat all fields of biotechnology in separate chapters. Many other areas of biotechnology – sometimes of limited importance, but often also of industrial interest or with the potential for a great development – exist. In this chapter some further interesting fields of biotechnology will be described in very brief reviews with literature enabling the reader to get in depth information on the respective field.

In some cases the contents of some chapters which have been printed in the earlier volumes will be updated, provided that this information is important.

2 Influence of Stress on Cell Growth and Product Formation in Biotechnological Processes

In the past 20 years many research groups have investigated the influence of stress on microorganisms, plant and animal cells.

MERCHUK (1991) published a review in which shear effects were described in detail. Recent investigations indicate that stress of cells, caused by fluid dynamic stress/shear and turbulence in the bioreactor, extreme values of process parameters (e.g., temperature, pH, dissolved oxygen, carbon sources, and other substrate parameters) further metabolically harmful substances (e.g., organic acids, alcohols, ammonia, etc.) cause stress reactions in the fermenting cells. In a reviewing volume SCHÜGERL and KRETZMER (2000) have edited selected papers about various aspects of stress in biotechnological processes.

A sudden increase of temperature induces a heat shock response in cells while a decrease of temperature results in a cold shock response. The sudden increase of the growth temperature results in unfolding of proteins and hydrophobic amino acid residues. The cold shock response evokes two major threats to the cells, namely a drastic reduction in membrane fluidity and a transient complete stop of translation at least in *Escherichia coli* (for details, see SCHUMANN, 2000).

Particle stress in bioreactors as a low shear stress is of crucial importance for the optimal course of processes and was investigated by HENZLER (2000).

Stress has further an influence on adherent cells, especially on animal cells. The chemical environment of the cells in a bioreactor has to be considered very carefully. Also the osmotic pressure and the influence of mechanical forces on cell viability is of great importance for cells growing in an agitated system (KRETZMER, 2000).

The influence of stress producing parameters on plant cell suspension systems has been described by KIERAN et al. (2000).

3 High-Cell-Density Cultivation of Microorganisms

Since some years the aspect of high-cell-density cultivation of cells has been investigated by many research groups.

High-cell-density cultivation is required to improve microbial biomass and product formation substantially. An overview on the field is given for microorganisms including Bacteria, Archaea, and Eukaryota (yeasts) by RIESENBERG and GUTHKE (1999). Tab. 1 shows some results with literature from this excellent review.

The aspects of high-cell-density cultivation of microorganisms should be extensively considered in future fermentation technology.

Tab. 1. High-Cell Density Cultivation of Various Microorganisms Grown to Cell Densities Higher than 100 g dry Biomass per L Culture Volume. CDW: maximum cellular dry weight, *t*: cultivation time, *Pr*: overall biomass productivity, MM: minimal salt medium, CM: complex medium, DO: dissolved oxygen

Microorganism	Characteristics with Respect to HCDC	Medium for Batch Phase	Bioreactor Type	Culture Method	CDW [g L^{-1}]	*t* [h]	*Pr* [g L^{-1} d^{-1}]	Reference
Bacteria								
Methylobacterium extorquens	mesophile, high concentrations of methanol inhibit growth	methanol/MM	stirred-tank reactor	fed-batch with controls for methanol, DO, and C/N ratio	233	170	32.9	SUZUKI et al. (1986)
Escherichia coli	mesophile, formation of acetate during aerobic growth with excess of glucose	glucose/MM	dialysis reactor	DO-stat fed batch and dialysis	190	30	25.2	NAKANO et al. (1997)
		glycerol/MM	dialysis reactor	DO-stat fed batch and dialysis	180	27	26.6	NAKANO et al. (1997)
		glycerol/MM	stirred-tank reactor	glycerol-limited exponential feed	148	44	80.6	KORZ et al. (1995)
		glucose/MM	stirred-tank reactor	glucose-unlimited exponential feed	145	32	108.7	HORN et al. (1996)
Bacillus subtilis	mesophile	CM with glucose	stirred-tank reactor	fed-batch, pH-stat for glucose control	184	28	157.7	PARK et al. (1992)
Alcaligenes eutrophus NCIMB 11599	mesophile, glucose-utilizing mutant	glucose/MM	stirred-tank reactor	fed-batch, no ammonia limitation	184	50	88.3	LEE et al. (1994a)
Streptomyces laurentii	mesophile	CM with glucose	stirred-tank reactor	fed-batch, pH-stat with multiple substrate feed	157	220	17.0	SUZUKI et al. (1987)
Lactococcus lactis	mesophile, cell damage by high lactate level	CM with glucose	stirred ceramic-membrane reactor	intermittent feeding and filtering at high average dilution rate	141	238	0.2	SUZUKI (1996)
Pseudomonas putida BM01	mesophile, no inhibitory effect of glucose <40 g L^{-1}	glucose/MM	stirred-tank reactor	fed-batch, pH-stat with control of ammonia and glucose	100	30	79.9	KIM et al. (1996)
Archaea								
Marinococcus M52	halophile 10% NaCl 35 °C, pH 7.5	CM with glucose	dialysis reactor	dialysis fed-batch with nutrient split feeding	132	72	not calculatable	KRAHE et al. (1996)
Sulfolobus shibatae	thermoacidophile 75 °C, pH 3.5	CM with glucose	dialysis reactor	dialysis fed-batch with nutrient split feeding	114	358	0.9	KRAHE et al. (1996)
Eukarya								
Candida brassicae	mesophile	ethanol/MM	stirred-tank reactor	fed-batch, DO-stat, ethanol feed	268	28	229.7	YANO et al. (1985)
Saccharomyces cerevisiae	mesophile, formation of ethanol during aerobic growth with excess glucose	CM with glucose	stirred-tank reactor with internal filter	continuous culture with feed of glucose, total cell retention	208	77	2.4	LEE et al. (1994b)
		CM with glucose	shake flask with ceramic filter	shaking on ordinary shaker with fed/filtering	235	220	0.36	SUZUKI et al. (1997)
Pichia pastoris	mesophile, glycerol inhibits methanol consumption	glycerol/MM	stirred-tank reactor	fed-batch with glycerol and/or methanol feeding	~100	46–130	18.5–52.1	STRATTON et al. (1998)

4 Apoptosis

Programmed cell death or apoptosis have now been recognized as biological phenomena which are of fundamental importance to the integrity of organisms. LOCKSHIN and WILLIAMS (1964) introduced the term "programmed cell death". The morphological characteristics of apoptosis are accompanied by biochemical and genetic mediators of programmed cell death. The genetically regulated process has an inherent potential for manipulations and offers new possibilities for improvements in the biotechnology industries and also for other fields.

See MCKENNA et al. (1998) about molecular mechanisms of programmed cell death. Methods for measurement of apoptosis have been reviewed by DARZYNKIEWICZ and TRAGANOS (1998) and a summary of papers about apoptosis and bioprocess technology is presented by SINGH and RUBEAI (1998). Most research has been done about apoptosis with mammalian cell lines, but the view on apoptosis is increasingly challenged by reports of altruistic cell death in microbial cells, e.g., in *Schizosaccharomyces pombe* (see INK et al., 1997).

Research on programmed death in bacteria, yeasts, and other microorganisms may be important for biotechnological applications in the future. For overviews of apoptosis, see AL-RUBEAI (1998a, b).

5 Bacterial Magnetosomes and Biotechnological Applications

An example of biologically controlled mineralization is the formation of nanocrystals in magnetosomes within magnetotactic bacteria. It is a well documented example of the apparently widespread occurrence of magnetic minerals in the living world. Biomineralization of ferromagnetic materials, mainly magnetite, has also been reported for diverse organisms including algae, insects, mollusks, fishes, birds, and even humans. Literature has been reviewed by SCHÜLER and FRANKEL (1999), who wrote an overview of this field. The heterogeneous group of magnetotactic bacteria has been reviewed by BAZYLINSKI et al. (1994) and SPRING and SCHLEIFER (1995).

Magnetotactic bacteria can be easily enriched from natural samples, but their cultivation in the laboratory has proved difficult and only some species could be isolated in pure culture. One example of isolates that can be cultivated under laboratory conditions is *Magnetospirillum* with several strains (SCHLEIFER et al., 1991; SCHÜLER and KÖHLER, 1992).

Magnetic particles are formed within biological membranes. Small magnetic particles can also be formed synthetically, but these particles are often non-uniform, often not fully crystalline, and in an agglomerated state, which causes problems in application (SARIKAYA, 1994). Therefore, the application of biologically produced magnetic particles will be preferred. Fig. 1 shows purified magnetosomes from *Magnetospirillum gryphiswaldense* (from the review of SCHÜLER and FRANKEL, 1999). Magnetic cells were considered for the removal of heavy metals and radionuclides from wastewater (BAHAJ et al., 1994, 1998).

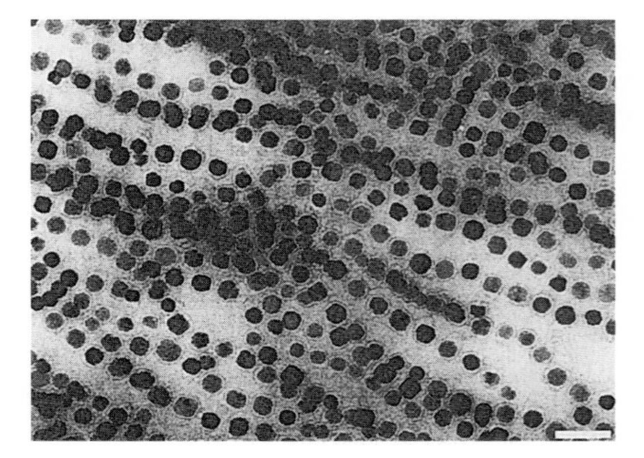

Fig. 1. Purified magnetosomes from *Magnetospirillum gryphiswaldense*. Owing to the presence of the enveloping membrane, isolated magnetosome particles form stable, well-dispersed suspensions (bar equivalent to 100 nm) (from SCHÜLER and FRANKEL, 1999).

Isolated magnetosome particles are useful as carriers for the immobilization of relatively large quantities of bioactive substances, which can then be separated by magnetic fields (MATSUNAGA, 1991). The glucose oxidase bound to biogenic magnetite particles had a 40-fold higher activity than that bound to artificial magnetite (MATSUNAGA and KAMIYA, 1987). Also antibodies and other substances could be immobilized with biogenic magnetite particles (NAKAMURA et al., 1993).

Bacterial magnetosomes have an application potential as a contrast agent for magnetic resonance imaging and tumor-specific drug carriers (BULTE and BROOKS, 1997). Perhaps magnetic extraction of cells may be possible (RAGHAVARAO et al., 2000).

Many other applications have been investigated. Problems are related to mass cultivation of the bacteria, which is still too expensive for many practical applications. Much research about the understanding of the biochemical and genetical principles is necessary.

6 Production of Sweet-Tasting Proteins by Biotechnology

In the past years sweet-tasting proteins have been investigated to replace sucrose and other low-molecular-weight sweetening agents. An informal short review about recent developments and the characterization and biological production of sweet-tasting proteins has been published recently by FAUS (2000). The carbohydrate-based sweeteners have been described by HEBEDA (1995).

Six known sweet-tasting proteins are thaumatin, monellin, mabinlin, pentadin, brazzein, and curculin. These compounds need only to be used in minute amounts in order to supply the same effect as sucrose. These effects result in an almost negligible addition to the calorie count, and in a product that does not contribute to tooth decay. Moreover, the alternative and intense sweetening additives can also be used in diabetic food, as they do not trigger a demand for insulin in these patients.

All of the sweet-tasting proteins have been isolated from fruits of tropical plants. In the past 30 years efforts have been made to exploit these sweet-tasting proteins commercially. In recent years many new sweet-tasting proteins have been discovered and characterized. The genes of some of these substances have been cloned and sequenced and some could be expressed in foreign hosts. The properties of the six important proteins are shown in Tab. 2.

Tabs. 3 and 4 show results of the expression of recombinant thaumatin and monellin (adapted from FAUS, 2000).

Many details and much literature should be consulted in the above review of FAUS. Several companies have groups mentioned working on research and development of sweet-tasting proteins, and it can be expected that in the near future such sweeteners will be on the market.

7 Biotechnological Capacities of Mushrooms

In the 1st Edition of *Biotechnology* the edible mushrooms have been described in detail by ZADRAZIL and GRABBE (1983). Advances in genetic analysis and biotechnology of *Agaricus bisporus* have been reviewed by STOOP and MOOIBROEK (1999). The recent availability of DNA technologies has resulted in a substantial increase in the number of *Agaricus bisporus* genes that have been identified and characterized. Some potential targets for strain improvement are discussed, such as the genes involved in brown discoloration, substrate utilization, carbon and nitrogen metabolism, and development of fruit bodies. An overview of economical data and the importance of the cultivated mushrooms *Agaricus bisporus*, *A. bitorquis*, *Lentinus edodes*, *Pleurotus* spp., *Auricularia* spp., *Flammulina velutipes*, and *Tremella fuciformis* was published by CHANG (1996).

The biodegradative and biosynthetic capacities of mushrooms have been reviewed by RAJARATHNAM et al. (1998). In this review much information about the nutritional value of many different mushroom species can be

Tab. 2. Comparison of Thaumatin, Monellin, Mabinlin, Pentadin, Brazzein, Curculin, and Miraculin (adapted from KURIHARA, 1994)

	Thaumatin	Monellin	Mabinlin	Pentadin	Brazzein	Curculin	Miraculin
Source	*Thaumatococcus danielli* Benth	*Dioscoreophyllum cumminsii* Diels	*Capparis masakai* Levl	*Pentadiplandra brazzeana* Baillon	*Pentadiplandra brazzeana* Baillon	*Curculingo latifolia*	*Richadella dulcifica*
Geographic distribution	West Africa	West Africa	China	West Africa	West Africa	Malaysia	West Africa
Variants	I, II, a, b, c[a]	–	I, II-a, III, IV[a]	–	–	–	–
Sweetness factor (weight basis)	3,000	3,000	100	500	2,000	550	–
Molecular mass (active form) [kDa]	22.2	10.7	12.4	12.0[b]	6.5	24.9	98.4
Amino acids	207	45 (A chain) 50 (B chain)	33 (A chain) 72 (B chain)	?	54	114	191
Active form	monomer	dimer (A + B)	dimer (A + B)	?	monomer	dimer (A + A)	tetramer (A + A + A + A)

[a] At least five different forms of thaumatin (LEE et al., 1988) and four different forms of mabinlin (NIRASAWA et al., 1994) have been identified.
[b] A chromatographic fraction containing a 12-kDa protein was sweet. This same fraction, when subjected to electrophoresis under non-reducing condition showed bands in the region between 22 and 41 kDa, suggesting the presence of subunits.

found, among others the non-protein amino acids of mushroom species. The polysaccharides of mushroom have many biological functions which have been described by RAJARATHNAM et al. (1998) with much literature. Many of these polysaccharides have been identified as antitumor active substances, especially against sarcoma.

Tab. 5 shows medical and biological properties of mushroom species.

Many other properties have recently been investigated. Patents have been published by many companies, among others for protoplast fusions, transformations, and interspecific hybrids of mushrooms (KÜES and LIU, 2000).

The hypocholesterolemic, hypolipidemic, antitumorous, antibacterial, antiviral activities including references on anti-AIDS viral properties show the potential possibilities for biotechnological applications.

Many mushrooms can be cultivated in mass culture and therefore many of the cited substances could be available in larger amounts. For many details and much literature see, RAJARATHNAM et al. (1998) and KÜES and LIU (2000).

8 Cultivation of Mycorrhizal Fungi

Mycorrhizae are symbiontic associations between plant roots and fungi. Most of the higher plants are infected with mycorrhizal fungi. In many cases they play a major role in many plant functions, e.g., in mineral nutrition, stress resistance, resistance against many plant pathogens, etc. Genetic manipulation of such mycorrhizal fungi would have a potential impact on forest production, as ESSER and MEINHARDT (1986) have stated. The biotechnological application of mycorrhizae is further expected to promote the production of food while maintaining ecologically and economically sustainable production systems.

More than 90% of all known plants have the potential to form mycorrhizal associations. For structure, function, molecular biology, and biotechnology see VARMA and HOCK (1999) and also VARMA (1998).

Tab. 3. Most Relevant Published Results on the Expression of Recombinant Thaumatin (adapted from FAUS, 2000)

Host	Promoter	Secretion	Yield	Sweet Phenotype	Reference
Escherichia coli	Trp/lac	no	very low	no	EDENS et al. (1982)
Saccharomyces cerevisiae	PgK	no	low	no	LEE et al. (1988)
Kluyveromyces lactis	Gapdh	yes	low	no	EDENS and VAN DER WEL (1985)
Bacillus subtilis	α-amy	yes	1 mg L^{-1}	yes	ILLINGWORTH et al. (1988)
Streptomyces lividans	β-gal	yes	0.2 mg L^{-1}	?	ILLINGWORTH et al. (1989)
Penicillium roquefortii	Gla	yes	1–2 mg L^{-1}	yes	FAUS et al. (1997)
Aspergillus niger var. *awamori*	Gla	yes	5–7 mg L^{-1}	yes	FAUS et al. (1998)
Solanum tuberosum	CaMV	no	low	yes	WITTY (1990)

Trp/lac, *E. coli* tryptophan and lactose promoters; PgK, *S. cerevisiae* 3-phosphoglycerate promoter; Gapdh, *K. lactis* glyceraldehyde-3-phosphate-dehydrogenase promoter, α-amy, *B. subtilis* α-amylase promoter; β-gal, *S. lividans* β-galactosidase promoter; Gla, *A. niger* glucoamylase promoter; CaMV, Cauliflower Mosaic Virus promoter for the 35S RNA.

Tab. 4. Monellin Gene Expression in Biotechnology (adapted from FAUS, 2000)

Host	Promoter	Yield	Reference
Escherichia coli	Trp	low	KIM et al. (1989)
Tomato	E8	23.9 µg monellin g^{-1} wet weight	PEÑARRUBIA et al. (1992)
Lettuce	CaMV	low	PEÑARRUBIA et al. (1992)
Candida utilis	gapdh	10 mg monellin g^{-1} wet weight	KONDO et al. (1997)
Saccharomyces cerevisiae	gapdh	low	KONDO et al. (1997)

Trp *E. coli* tryptophan promoter; E8 tomato fruit-ripening specific promoter; CaMV Cauliflower Mosaic Virus promoter for the 35S RNA; gapdh *C. utilis* or *S. cerevisiae* glyceraldehyde-3-phosphate dehydrogenase promoter. All the recombinant monellin proteins have been expressed as single-chain analogs.

The current knowledge of 15 individual genera of ectomycorrhizal fungi has been reviewed by CAIRNEY and CHAMBERS (1999).

Ectomycorrhizal fungi form structures on the roots of many economically important trees. Large-scale exploitation of the potential benefits of these fungi in improving plantation yields means that fermentation techniques for these fungi will be required. KUEK (1992) has produced inocula of *Laccaria laccata* by submerged aerobic culture of mycelia within hydrogel beads of high efficiency.

In later investigations with shake-flask cultures of *L. laccata* they got biomass yields of 12 g L^{-1}.

Further investigations of mass culture and inoculation techniques of mycorrhizal fungi are necessary to use this great potential for an improvement of agricultural crops by this method.

9 Further Fields of Biotechnology

Many other fields of biotechnology are important or of great interest. Some of them will be described here very briefly.

The biotechnological production of labeled compounds has been described in the 1st Edition of *Biotechnology* (SIMON and GÜNTHER,

Tab. 5. Medical and Biological Properties of Mushroom species (adapted from RAJARATHNAM et al., 1998)

Species	Component	Properties/ Chemical Nature	Biological Function	Reference
Agaricus blazei	lectin	60,000–70,000 mol wt.; stable at 65 °C; carbohydrates	agglutinated equally all types of human erythrocytes (11%) tested	KAWAGISHI et al. (1988)
Agaricus blazei	glycoprotein	sugar (50.2%), protein (43.3%)	antitumor against Sarcoma-180	MIZUNO et al. (1990)
Agaricus sp.	an extract	ash (5.54%), protein (43.19%), lipid (3.73%), crude fiber (6.61%) sugars (41.56%), ergosterol (6.14%)	improves liver functioning	ITO et al. (1990)
Coriolus versicolor	protein-polysaccharide		inhibits abnormal proliferation of blood vessels	KANON et al. (1990)
Coriolus versicolor	protein bound polysaccharide		immune enhancement 50.2%, in mice	LI et al. (1990)
Grifola frondosa	lectin from mushrooms	glycoprotein containing 50.2%, 3.3% total sugars	agglutinated all types of erythrocytes equally; cytotoxic against Hela cells	KAWAGISHI et al. (1990)
Grifola frondosa	copper binding peptide	mol wt. 2,240; acidic aspartate, glutamate, serine, and glycine occupied 84% of total residues	with properties to bind Cu or to maintain Cu in the soluble state at physiological pH; increased Cu absorption	SHIMAOKA et al. (1993, 1994)
Lentinus adhaerens	antibiotic	2-methoxy-5-methyl-1,4-benzoquinone	a thromboxane A_2 receptor antagonist	LAUER et al. (1991)
Lentinus edodes	extract of the culture medium	water-soluble lignins 50.2% + minor amounts of protein (3.2%) and sugars (12.2%)	immunostimulating activity	SUZUKI et al. (1989)
Lentinus edodes	extract of the culture medium	protein 2.5%, sugar 12–20%, lignin 70–85%	antiviral activity (against HIV)	IIZUKA et al. (1990)
Lentinus edodes	culture extract	lignin 65–75%, polysaccharide 15–30%, protein 10–20%	inhibited proliferation of herpes virus *in vitro* and *in vivo*	KOGA et al. (1991)
Lentinus edodes	lignin derivatives of culture medium		antiviral	SORIMACHI et al. (1990)
Pleurotus ostreatus	lectin	bind with galactose or galactosamine	useful as blood coagulation factor	HASHIMOTO et al. (1990)
Polyporus confluens	culture extract	xyloglucan–protein mixture	antitumor	ITO et al. (1991)
Tirmania pinoyl			mutagens and antimutagens by using different solvents	HANRAN et al. (1989)

1988). Methods for the production of labeled compounds are until now very common and a further review of this field is not intended. Much literature can be found in the excellent review of SIMON and GÜNTHER (1988).

The microbial production of hydrocarbons will also not be reviewed here. The review of BIRCH and BACHOFEN (1988) contains much literature.

Recently, MOOIBROEK and CORNISH (2000) reviewed alternative sources of natural rubber. Many genes involved in (poly)-isoprenoid biosynthesis in plants and yeasts have been identified. Squalene synthase (SQS) inhibits the ergosterol synthetic pathway and may cause an accumulation of isoprenoids and an increase of HMGR mRNA and was observed in rats (NESS et al., 1994). FEGUEUR et al. (1991) found squalene synthase in yeasts (ERG9). Tab. 6 shows observations on the ERG9-mutation in *Saccharomyces cerevisiae* (from MOOIBROEK and CORNISH, 2000).

Oleaginous yeasts, such as *Yarrowia lipolytica* or *Cryptococcus curvatus* have the capacity to accumulate up to 50% of storage carbohydrates in oil bodies. Rubber particles are analogous to oil bodies. Attempts are now underway to redirect the metabolic flux to the formation of (poly)-isoprenes, using multiple transgenesis (for details, see MOOIBROEK and CORNISH, 2000).

In Volume 6 of the 2nd Edition of *Biotechnology* (ROEHR, 1996) many primary products have been described, among others the production of succinic acid (ROEHR and KUBICEK, 1996). In the meantime some progress in fermentation of succinic acid could be observed, and the introduction of biotechnological processes in the industry instead of manufacturing this substance by chemical means will be discussed.

Yields of succinates as high as 110 g L^{-1} have been achieved from glucose by the newly discovered rumen organism *Actinobacillus succinogenes* (ZEIKUS et al., 1999). In a novel process the greenhouse gas carbon dioxide is fixed into succinate during glucose fermentation. New developments in end-product re-

Tab. 6. Observations on the ERG9 Mutation in *Saccharomyces cerevisiae* (the Key Regulatory Enzyme Squalene Synthase (SQS) is Denoted in *S. cerevisiae* (ERG9) (adapted from MOOIBROEK and CORNISH, 2000)

Year	Observations	Reference
1991	SQS (farnesyl-diphosphate : farnesyl-diphosphate farnesyltransferase, EC. 2.5.1.21) regulates flux of isoprene intermediates through sterol pathway; *SQS* (*ERG9*) gene complements the *S. cerevisiae erg9* mutation; *ERG9* (Genbank M63979) is a single copy gene, essential for cell growth; PEST (proline, glutamic acid, serine, threonine-rich) consensus motif; one or two membrane-spanning domains	JENNINGS et al. (1991)
1991	*ERG9* gene localized on 2.5-kb DNA fragment; functional expression in *E. coli*, protein 444 AA (51,600 Da), 1–4 transmembrane domains, localization in membrane of ER	FEGUEUR et al. (1991)
1993	structural and functional conservation between human, budding and fission yeast SQS, especially in regions involved in interaction with prenyl substrates = 2 × FPP); complementation of *erg9* mutation; inhibition of sterol synthesis by HMGR-specific lovastatin increases levels of *SQS* mRNA	ROBINSON et al. (1993)
1993	C-terminus-truncated protein expressed in *E. coli*; soluble enzyme is monomeric and catalyzes two-step conversion of FPP > presqualene diphosphate > squalene using Mg^{2+} and NADPH	ZHANG et al. (1993)
1995	nine of 11 genes involved in ergosterol pathway cloned, the first three genes *ERG9* (squalene synthase), *ERG1* (squalene epoxidase), *ERG7* (lanosterol synthase), being essential for aerobic growth	LEES et al. (1995)
1996	functional complementation of *erg9* mutation by *Nicotiana benthaminia* and *N. tabacum* cDNAs (1600 nt)	HANLEY et al. (1996)

covery technology have lowered the cost of succinate production by biotechnology to US$ 0.55 kg^{-1} at the 75,000 t a^{-1} level. However, the price per kg in large volumes must drop below $ 0.45 to open the market for a biotechnological process (for details, see ZEIKUS et al., 1999). Fig. 2 shows routes to succinic acid-based products.

The biotechnological production of 1,3-propanediol may also be of interest in the future, as it could compete with 1,3-propanediol made by petrochemistry. Glycerol conversion to 1,3-propanediol can be carried out by clostridia as well as enterobacteriaceae. The yield of 1,3-propanediol is determined by the availability of $NADH_2$. With *Klebsiella pneumoniae* yields of 60–70 g L^{-1} in batch and 50–60 g L^{-1} in continuous culture were observed (BIEBL et al., 1999). The yields with recombinant microorganisms are low until now, less than 0.1 g L^{-1} with recombinant *Saccharomyces cerevisiae* and 6.5 g L^{-1} with recombinant *Escherichia coli* AG 1 (SKRALY et al., 1998).

The combination of appropriate genes from yeast and bacteria offers the possibility to produce 1,3-propanediol directly from a cheap substrate and not from the expensive glycerol, if the yields could be increased.

Further information about the biotechnological production of 1,3-propanediol is published in the review of BIEBL et al. (1999) in which also much new literature has been cited.

10 Future Aspects

Many new developments in biotechnology can be expected, e.g., new substances of the secondary metabolism from East Asian or other exotic plants, which can be produced with microorganisms after genetic treatment, further bulk ware which can be produced perhaps much cheaper with transgenic plants than with microorganisms. Molecular plant ge-

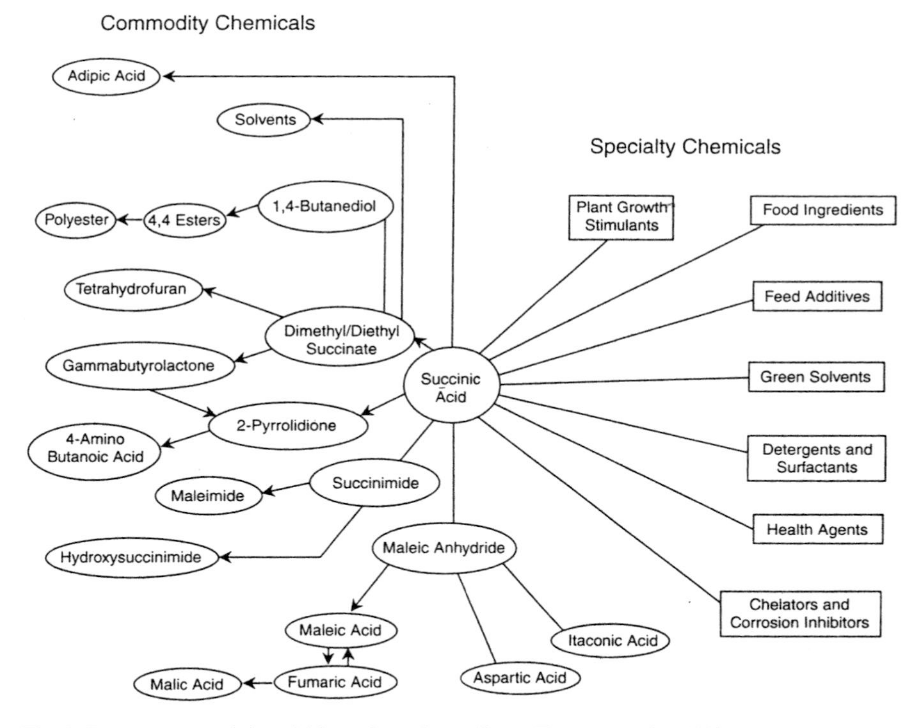

Fig. 2. Routes to succinic acid-based products (from ZEIKUS et al., 1999).

netics will challenge the chemical and the fermentation industry. Transgenic animals will be used to produce some fine chemicals, especially proteins, etc.

It can be expected that the future development of biotechnology will bring more health and welfare for mankind!

11 References

AL-RUBEAI, M. (1998a), Apoptosis and cell culture technology, *Adv. Biochem. Eng. Biotechnol.* **59**, 225–249.

AL-RUBEAI, M. (Ed.) (1998b), Apoptosis, *Adv. Biochem. Eng. Biotechnol.* **62**, 1–193.

BAHAJ, A. S., CROUDACE, I. W., JAMES, P. A. B. (1994), Treatment of heavy metal contaminants using magnetotactic bacteria, *IEEE Trans. Magnet.* **30**, 4707–4709.

BAHAJ, A. S., CROUDACE, I. W., JAMES, P. A. B., MOESCHLER, F. D., WARWICK, P. E. (1998), Continuous radionuclide recovery from wastewater using magnetotactic bacteria, *J. Magnet. Magnet. Mater.* **184**, 241–244.

BAZYLINSKI, D. A., GARRATT-REED, A., FRANKEL, R. B. (1994), Electron-microscopic studies of magnetosomes in magnetotactic bacteria, *Microsc. Res. Techn.* **27**, 389–401.

BIEBL, H., MENZEL, K., ZENG, A.-P., DECKWER, W.-D. (1999), Microbial production of 1,3-propanediol, *Appl. Microbiol. Biotechnol.* **52**, 289–297.

BIRCH, L. D., BACHOFEN, R. (1988), Microbial production of hydrocarbons, in: *Biotechnology* 1st Edn., Vol. 6b (REHM, H. J., REED, G., Eds.), pp. 71–99. VCH: Weinheim.

BULTE, J. W. M., BROOKS, R. A. (1997), Magnetic nanoparticles as contrast agents, for imaging, in: *Scientific and Clinical Applications of Magnetic Carriers* (HÄFELI, U., SCHÜTT, W., TELLER, J., ZBOROWSKI, M., Eds.), pp. 527–543. New York, London: Plenum Press.

CAIRNEY, J. W. G., CHAMBERS, S. M. (1999), *Ectomycorrhizal Fungi.* Berlin, Heidelberg, New York: Springer-Verlag.

CHANG, S. (1996), Mushroom research and development – equality and mutual benefit, in: *Proc. 2nd Int. Conf. Mushroom Biology and Mushroom Products* (ROYSE, D. J., Ed.), pp. 1–10. University Park, PA: Penn State University.

DARZYNKIEWICZ, Z., TRAGANOS, F. (1998), Measurement of apoptosis, *Adv. Biochem. Eng. Biotechnol.* **62**, 33–73.

EDENS, L., VAN DER WEL, H. (1985), Microbial synthesis of the sweet-tasting protein thaumatin, *Trends Biotechnol.* **3**, 61–64.

EDENS, L., HESLINGA, L., KLOK, R., LEDEBOER, A. M., MAAT, J. et al. (1982), Cloning of cDNA encoding the sweet tasting plant protein thaumatin and its expression in *Escherichia coli*, *Gene* **18**, 1–12.

ESSER, K., MEINHARDT, F. (1986), in: *Biomolecular Engineering in the European Community* (MAGNIEN, E., Ed.), pp. 831–838. Dordrecht: Martinus Nijhoff.

FAUS, J. (2000), Recent developments in the characterization and biotechnological production of sweet-tasting proteins, *Appl. Microbiol. Biotechnol.* **53**, 145–151.

FAUS, J., PATINO, C., DELRIO, J. L., DELMORAL, C., BARROSO, H. S. et al. (1997), Expression of a synthetic gene encoding the sweet-tasting protein thaumatin in the filamentous fungus *Penicillium roquefortii*, *Biotechnol. Lett.* **19**, 1185–1191.

FAUS, J., DELMORAL, C., ADROER, N., DELRIO, J. L., PATINO, C. et al. (1998), Secretion of sweet-tasting protein by recombinant strains of *Aspergillus niger*, var. *awamori*, *Appl. Microbiol. Biotechnol.* **49**, 393–398.

FEGUEUR, M., RICHARD, L., CHARLES, A. D., KARST, F. (1991), Isolation and primary structure of the ERG9 gene of *Saccharomyces cerevisiae* encoding squalene synthase, *Curr. Genet.* **20**, 365–372.

HANLEY, K. M., NICOLAS, O., DONALDSON, T. B., SMITH-MONROY, C., ROBINSON, G. W., HELLMANN, G. M. (1996), Molecular cloning, *in vitro* expression and characterization of a plant squalene synthase cDNA, *Plant Mol. Biol.* **30**, 1139–1151.

HANRAN, M. A., AL-DAKAN, A. A., ABOUT-ENEIN, H. Y., AL-OTHAIMEEN, A. A. (1989), Mutagenic and antimutagenic factor(s) extracted from a desert mushroom using different solvents, *Mutagenesis* **4** (2), 111–114.

HASHIMOTO, Y., SHIBATA, A., TANAKA, S. (1990), Extraction of lectins from mushrooms, *Jpn. Kokai. Tokkyo Koho* Jp 0227089 (90270890).

HEBEDA, R. E. (1995), Carbohydrate-based sweeteners, in: *Biotechnology*, 2nd Edn., Vol. 9 (REHM, H. J., REED, G., PÜHLER, A., STADLER, P., Eds.), pp. 737–765. Weinheim: VCH.

HENZLER, H. J. (2000), Particle stress in bioreactors, *Adv. Biochem. Eng. Biotechnol.* **67**, 35–82.

HORN, U., STRITTMATTER, W., KREBBER, A., KNÜPFER, U., KUJAU, M. et al. (1966), High volumetric yields of functional dimeric miniantibodies in *Escherichia coli*, using an optimized expression and high-cell-density fermentation under non-limited growth conditions, *Appl. Microbiol. Biotechnol.* **46**, 524–532.

IIZUKA, C., IIZUKA, H., ONASHI, Y. (1990), Extract of basidiomycetes especially *Lentinus edodes* for treatment of human immunodeficiency virus

(HIV) infection, *Eur. Patent Appl.* Ep 370673.

ILLINGWORTH, C., LARSON, G., HELLENKANT, G. (1988), Secretion of the sweet-tasting protein thaumatin by *Bacillus subtilis*, *Biotechnol. Lett.* **10**, 587–592.

ILLINGWORTH, C., LARSON, G., HELLENKANT, G. (1989), Secretion of the sweet-tasting protein thaumatin by *Streptomyces lividans*, *J. Ind. Microbiol.* **4**, 37–42.

INK, B., ZORNIG, M., BAUM, B., HAJIBAGHERI, N., JAMES, C. et al. (1997), *Mol. Cell. Biol.* **17**, 2468–2474.

ITO, H., SHIMURA, K., SUMITANI, T. (1990), Mushroom extracts for improving liver functions, *Jpn. Kokai Tokkyo Koho* Jp 02124829 (90124829).

ITO, H., SHIMURA, K., SUMITANI, T., MIRUNO, T. (1991), Peptic ulcer inhibiting glycoprotein manufacture with *Polyporus confluens*, *Jpn. Kokai Tokkyo Koho* Jp 03115224 (91115224).

JENNINGS, S. M., TSAY, Y. H., FISCHT, D. M., ROBINSON, G. W. (1991), Molecular cloning and characterization of the yeast gene for squalene synthase, *Proc. Natl. Acad. Sci. USA* **88**, 6038–6042.

KANON, T., MATSUNAGA, K., SAITO, K., FUJII, T. (1990), Angiogenesis inhibitor, *Eur. Patent Appl.* EP 353861.

KAWAGISHI, H., NOMURA, A., YUMEN, T., MIZUNO, T., HAGIWARA, T., NAKAMURA, T. (1988), Isolation and properties of a lectin from fruiting bodies of *Agaricus blazei*, *Carbohydr. Res.* **183**, 150–154.

KAWAGISHI, H., NOMURA, A., MIZUNO, T., KIMURA, A., CHIBA, S. (1990), Isolation and characterization of a lectin from *Grifola frondosa* fruiting bodies, *Biochem. Biophys. Acta* **1034** (3), 247–252.

KIERAN, P. M., MALONE, D. M., MACLOUGHLIN, P. E. (2000), Effects of hydrodynamic and interfacial forces on plant cell systems, *Adv. Biochem. Eng. Biotechnol.* **67**, 139–177.

KIM, S. H., KANG, C. H., KIM, R., CHO, J. M., LEE, Y. B., LEE, T. K. (1989), Redesigning a sweet protein: increased stability and renaturability, *Protein Eng.* **2**, 571–575.

KIM, G. J., LEE, I. Y., CHOI, D. K., YOON, S. C., PARK, Y. H. (1996), High cell density cultivation of *Pseudomonas putida* BM01 using glucose, *J. Microbiol. Biotechnol.* **6**, 221–224.

KOGA, J., OHASHI, Y., HIRATANI, H. (1991), Anti-viral fraction of aqueous *Lentinus edodes* extract, *Eur. Patent Appl.* Ep 437346.

KONDO, K., MIURA, Y., SONE, H., KOBAYASHI, K., IIJIMA, H. (1997), High-level expression of a sweet protein, monellin, in the food yeast *Candida utilis*, *Nature Biotechnol.* **15**, 453–457.

KORZ, D. J., RINAS, U., HELMUTH, K., SANDERS, E. A., DECKWER, W.-D. (1995), Simple fed-batch technique for a high cell density cultivation of *Escherichia coli*, *J. Biotechnol.* **39**, 59–65.

KRAHE, M., ANTRANIKIAN, G., MÄRKL, H. (1996), Fermentation of extremophilic microorganisms, *FEMS Microbiol. Rev.* **18**, 271–285.

KRETZMER, G. (2000), Influence of stress on adherent cells, *Adv. Biochem. Eng. Biotechnol.* **67**, 123–137.

KUEK, C. (1996), Shake-flask culture of *Laccaria laccata*, an ectomycorrhizal basidiomycete, *Appl. Microbiol. Biotechnol.* **45**, 319–326.

KÜES, U., LIU, Y. (2000), Fruiting body production in basidiomycetes, *Appl. Microbiol. Biotechnol.* **54**, 141–152.

KURIHARA, Y. (1994), Sweet proteins in general, in: *Thaumatin* (WITTY, M., HIGGINBOTHAM, J. D., Eds.), pp. 1–18. Boca Raton, FL: CRC Press.

LAUER, U., ANKE, T., HANSSKE, F. (1991), Antibiotics from basidiomycetes 38: 2-methoxy-5-methyl-1,4-benzoquinone, a thromboxane A2 receptor antagonist from *Lentinus adhaerens*, *J. Antibiot.* **44** (1), 59–65.

LEE, J. H., WEICKMANN, L. J., KODURI, R. K., GHOSH-DASTIDAR, P., SAITO, K. et al. (1988), *Biochemistry* **27**, 5101–5107.

LEE, I. Y., CHOI, E. S., KIM, G. J., NAM, S. W., SHIN, Y. C. et al. (1994a), Optimization of fed-batch fermentation for production of poly-β-hydroxybutyrate in *Alcaligenes eutrophus*, *J. Microbiol. Biotechnol.* **4**, 146–150.

LEE, W. G., LEE, Y.-S., CHANG, H. N., CHANG, Y. K. (1994b), A cell retention internal filter reactor for ethanol production using tapioca hydrolysates, *Biotechnol. Tech.* **8**, 817–820.

LEES, N. D., SKAGGS, B., KIRSCH, D. R., BARD, M. (1995), Cloning of the late genes of the ergosterol biosynthetic pathway of *Saccharomyces cerevisiae* – a review, *Lipids* **30**, 221–226.

LI, X., WANG, J., ZHU, P., LIU, L., GE, J., YANG, S. (1990), Immune enhancement of polysaccharide peptides isolated from *Coriolus versicolor*, *Zhongguo Yaoli Xuebao* **11** (6), 542–545.

LOCKSHIN, R. A., WILLIAMS, C. M. (1964), *J. Insect. Physiol.* **10**, 643.

MATSUNAGA, T. (1991), Applications of bacterial magnets, *Trends Biotechnol.* **9**, 91–95.

MATSUNAGA, T., KAMIYA, S. (1987), Use of magnetic particles isolated from magnetotactic bacteria for enzyme immobilization, *Appl. Microbiol. Biotechnol.* **26**, 328–332.

MCKENNA, S. L., MCGOWAN, A. J., COTTER, T. G. (1998), Molecular mechanisms of programmed cell death, *Adv. Biochem. Eng. Biotechnol.* **62**, 1–31.

MERCHUK, J. C. (1991), Shear effects of suspended cells, *Adv. Biochem. Eng. Biochnol.* **44**, 65–96.

MIZUNO, T., ITO, H., SHIMURA, K., SUMITANI, T., KAWAGASHI, H. et al. (1990), Antitumor glyco-proteins from mushroom, *Jpn. Kokai Tokkyo Koho* JP 0278630 (1978630).

MOOIBROEK, H., CORNISH, K. (2000), Alternative

sources for natural rubber, *Appl. Microbiol. Biotechnol.* **53**, 355–365.

NAKAMURA, N., BURGESS, J. G., YAGIUDA, K., KIUDO, S., SAGAGUCHI, T., MATSUNAGA, T. (1993), Detection and removal of *Escherichia coli* using fluorescein isothiocyanate conjugated monoclonal antibody immobilized on bacterial magnetic particles, *Ann. Chem.* **65**, 2036–2039.

NAKANO, R., RISCHKE, M., SATO, S., MÄRKL, H. (1997), Influence of acetic acid on the growth of *Escherichia coli* K12 during high-density cell cultivation in a dialysis reactor, *Appl. Microbiol. Biotechnol.* **48**, 597–601.

NESS, G. C., ZHAO, Z., KELLER, R. K. (1994), Effect of squalene synthase inhibition on the expression of the hepatic cholesterol biosynthetic enzymes, LDL receptor, and cholesterol 7-*alpha* hydroxylase, *Arch. Biochem. Biophys.* **311**, 277–285.

NIRASAWA, S., NISHINO, T., KATAHIRA, M., UESUGI, S., HU, Z., KURIHARA, Y. (1994), Structures of heat-stable and unstable homologues of the sweet protein mabinlin, *Eur. J. Biochem.* **223**, 989–995.

PARK, Y. S., KAI, K., IIJIMA, S., KOBAYASHI, T. (1992), Enhanced β-galactosidase production by high cell-density culture of recombinant *Bacillus subtilis* with glucose concentration control, *Biotechnol. Bioeng.* **40**, 686–696.

PEÑARRUBIA, L., KIM, R., GIOVANNONI, J., KIM, S. H., FISCHER, R. (1992), Production of the sweet protein monellin in transgenic plants, *Biotechnology* **10**, 561–564.

RAGHAVARAO, K. S. M. S., DUESER, M., TODD, P. (2000), Multistage magnetic and electrophoretic extraction of cells, particles and macromolecules, *Adv. Biochem. Eng. Biotechnol.* **68**, 139–190.

RAJARATHNAM, S., SHASHIREKHA, M. N., BANO, Z. (1998), Biodegradation and biosynthetic capacities of mushrooms: present and future strategies, *Crit. Rev. Biotechnol.* **18**, 91–236.

RIESENBERG, D., GUTHKE, R. (1999), High-cell-density cultivation of microorganisms, *Appl. Microbiol. Biotechnol.* **51**, 422–430.

ROBINSON, G. W., TSAY, Y. H., KIENZLE, B. K., SMITH-MONROY, C. A., BISHOP, R. W. (1993), Conservation between human and fungal squalene synthases: similarities in structure, function and regulation, *Mol. Cell. Biol.* **13**, 2706–2717.

ROEHR, M. (Ed.) (1996), Products of Primary Metabolism, *Biotechnology*, 2nd Edn., Vol. 6 (REHM, H.-J., REED, G., PÜHLER, A., STADLER, P., Eds.). Weinheim: VCH.

ROEHR, M., KUBICEK, C. P. (1996), Further organic acids, in: *Biotechnology*, 2nd Edn., Vol. 6 (REHM, H.-J., REED, G., PÜHLER, A., STADLER, P., Eds.), pp. 363–379. Weinheim: VCH.

SARIKAYA, M. (1994), An introduction to biomimetrics: a structural viewpoint, *Microsc. Res. Techn.* **27**, 360–375.

SCHLEIFER, K. H., SCHÜLER, D., SPRING, S., WEIZENEGGER, M., AMANN, R. et al. (1991), The genus *Magnetospirillum* gen. nov., description of *Magnetospirillum gryphiswaldense* sp. nov. and transfer of *Aquaspirillum magnetotacticum* to *Magnetospirillum magnetotacticum* comb. nov., *Syst. Appl. Microbiol.* **14**, 379–385.

SCHÜGERL, K., KRETZMER, G. (Eds.) (2000), Influence of stress on cell growth and product formation, *Adv. Biochem. Eng. Biotechnol.* **67**, 1–190.

SCHÜLER, D., FRANKEL, R. B. (1999), Bacterial magnetosomes and biotechnological applications, *Appl. Microbiol. Biotechnol.* **52**, 464–473.

SCHÜLER, D., KÖHLER, M. (1992), The isolation of a new magnetic spirillum, *Zbl. Microbiol.* **147**, 150–151.

SCHUMANN, W. (2000), Function and regulation of temperature-inducible bacterial proteins of the cellular metabolism, *Adv. Biochem. Eng. Biotechnol.* **67**, 1–34.

SHIMAOKA, I., KODAMA, J., NISHINO, K., ITOKAWA, Y. (1993), Purification of a copper-binding peptide from the mushroom *Grifola frondosa* and its effect on copper absorption, *J. Nutr. Biochem.* **4** (1), 33–38.

SHIMAOKA, I., ISHIYAMA, S., NISHINO, K., ITOKAWA, Y. (1994), Effects of copper-binding peptide from the mushroom *Grifola frondosa* on intestinal copper absorption, *Nutr. Res.* (N.Y.) **14** (9), 1331–1337.

SIMON, H., GÜNTHER, H. (1988), Microbial and enzymatic production of labelled compounds, in: *Biotechnology*, 1st Edn., Vol. 6b (REHM, H.-J., REED, G., Eds.), pp. 177–192. Weinheim: VCH.

SINGH, R. P., AL-RUBEAI, M. (1998), Apoptosis and bioprocess technology, *Adv. Biochem. Eng. Biotechnol.* **62**, 167–184.

SKRALY, F. A., LYTLE, B. L., CAMERON, D. C. (1998), Construction and characterization of a 1,3-propanediol operon, *Appl. Environ. Microbiol.* **64**, 98– 105.

SORIMACHI, K., NIWA, A., YAMAZAKI, S., TODA, S., YASUMURA, Y. (1990), Antiviral activity of water-solubilized lignin derivates *in vitro*, *Agric. Biol. Chem.* **54** (5), 1337–1339.

SPRING, S., SCHLEIFER, K. H. (1995), Diversity of magnetotactic bacteria, *Syst. Appl. Microbiol.* **18**, 147–153.

STOOP, J. M. H., MOOIBROEK, H. (1999), Advances in genetic analysis and biotechnology of the cultivated mushroom, *Agaricus bisporus*, *Appl. Microbiol. Biotechnol.* **52**, 474–483.

STRATTON, J., CHIRUVOLU, V., MEAGHER, M. (1998), High cell-density fermentation, in: *Pichia* protocol *Methods in Molecular Biology* Vol. 103 (HIGGINS, D. R., CREGG, J. M., Eds.), pp. 107–120. Totowa, NJ: Humana Press.

SUZUKI, T. (1996), A dense cell culture of microorganisms using a stirred ceramic membrane reactor incorporating asymmetric porous ceramic filters, *J. Ferment. Bioeng.* **82**, 264–271.

SUZUKI, T., YAMANE, T., SHIMIZU, S. (1986), Mass production of poly-β-hydroxy-butyric acid by fed-batch culture with controlled carbon/nitrogen feeding, *Appl. Microbiol. Biotechnol.* **24**, 370–374.

SUZUKI, T., YAMANE, T., SHIMIZU, S. (1987), Mass production of thiostrepton by fed-batch culture of *Streptomyces laurentii* with pH-stat modal feeding of multi-substrate, *Appl. Microbiol. Biotechnol.* **25**, 526–531.

SUZUKI, H., OKOBA, A., YAMAZAKI, S., SUZUKI, K., MITSUYA, H., TODA, S. (1989), Inhibition of the ineffectivity and cytopathic effect of human immunodeficiency virus bý water soluble lignin in an extract of the culture medium of *Lentinus edodes* mycelia, *Biochem. Biophys. Res. Commun.* **160** (1), 367–373.

SUZUKI, T., KAMOSHITA, Y., OCHASHI, R. (1997), A dense cell culture system for microorganisms using a shake flask incorporating a porous ceramic filter, *J. Ferment. Bioeng.* **84**, 133–137.

VARMA, A. (1998), *Mycorrhizal Manual*. Berlin, Heidelberg, New York: Springer-Verlag.

VARMA, A., HOCK, B. (1999), *Mycorrhiza*, 2nd Edn. Berlin, Heidelberg, New York: Springer-Verlag.

WITTY, M. (1990), Preprothaumatin II is processed to biological activity in *Solanum tuberosum*, *Biotechnol. Lett.* **12**, 131–136.

YANO, T., KOBAYASHI, T., SHIMIZU, S. (1985), High concentration cultivation of *Candida brassicae* in a fed-batch system, *J. Ferment. Technol.* **63**, 415–418.

ZADRAZIL, F., GRABBE, K. (1983), Edible Mushrooms, in: *Biotechnology* 1st Edn., Vol. 3 (REHM, H.-J., REED, G., Eds.), pp. 145–187. Weinheim, VCH.

ZHANG, D., JENNINGS, S. M., ROBINSON, G. W., POULTER, C. D. (1993), Yeast squalene synthase: Expression, purification and characterization of soluble recombinant enzyme, *Arch. Biochem. Biophys.* **304**, 133–143.

ZEIKUS, J. G., JAIN, M. K., ELANKOVAU, P. (1999), Biotechnology of succinic acid production and markets for derived industrial products, *Appl. Microbiol. Biotechnol.* **51**, 545–552.

Index

B

F

G

H

I

K

L

N

O

P

T

U

V